32. $\int \sqrt{u^2 \pm a^2}\, du = \dfrac{u}{2}\sqrt{u^2 \pm a^2} \pm \dfrac{a^2}{2}\ln|u + \sqrt{u^2 \pm a^2}|$

33. $\int \dfrac{du}{\sqrt{u^2 \pm a^2}} = \ln|u + \sqrt{u^2 \pm a^2}| + C$

34. $\int u\sqrt{u^2 \pm a^2}\, du = \dfrac{1}{3}(u^2 \pm a^2)^{3/2} + C$

35. $\int u^2 \sqrt{u^2 \pm a^2}\, du = \dfrac{u}{8}(2u^2 \pm a^2)\sqrt{u^2 \pm a^2} - \dfrac{a^4}{8}\ln|u + \sqrt{u^2 \pm a^2}| + C$

36. $\int \dfrac{u^2\, du}{\sqrt{u^2 \pm a^2}} = \dfrac{u}{2}\sqrt{u^2 \pm a^2} \mp \dfrac{a^2}{2}\ln|u + \sqrt{u^2 \pm a^2}| + C$

37. $\int \dfrac{u\, du}{\sqrt{u^2 \pm a^2}} = \sqrt{u^2 \pm a^2} + C$

38. $\int \dfrac{\sqrt{u^2 + a^2}}{u}\, du = \sqrt{u^2 + a^2} - a\ln\left|\dfrac{a + \sqrt{u^2 + a^2}}{u}\right| + C$

39. $\int \dfrac{\sqrt{u^2 - a^2}}{u}\, du = \sqrt{u^2 - a^2} - a\sec^{-1}\left|\dfrac{u}{a}\right| + C$

40. $\int \dfrac{\sqrt{u^2 \pm a^2}}{u^2}\, du = -\dfrac{\sqrt{u^2 \pm a^2}}{u} + \ln|u + \sqrt{u^2 \pm a^2}| + C$

41. $\int \dfrac{du}{u\sqrt{u^2 + a^2}} = -\dfrac{1}{a}\ln\left|\dfrac{a + \sqrt{u^2 + a^2}}{u}\right| + C$

42. $\int \dfrac{du}{u\sqrt{u^2 - a^2}} = \dfrac{1}{a}\sec^{-1}\left|\dfrac{u}{a}\right| + C$

43. $\int \dfrac{du}{u^2\sqrt{u^2 \pm a^2}} = \mp\dfrac{\sqrt{u^2 \pm a^2}}{a^2 u} + C$

44. $\int \dfrac{du}{(u^2 \pm a^2)^{3/2}} = \pm\dfrac{u}{a^2\sqrt{u^2 \pm a^2}} + C$

45. $\int (u^2 \pm a^2)^{3/2}\, du = \dfrac{u}{8}(2u^2 \pm 5a^2)\sqrt{u^2 \pm a^2} + \dfrac{3a^4}{8}\ln|u + \sqrt{u^2 \pm a^2}| + C$

FORMS CONTAINING $\sqrt{a^2 - u^2}$

46. $\int \sqrt{a^2 - u^2}\, du = \dfrac{u}{2}\sqrt{a^2 - u^2} + \dfrac{a^2}{2}\sin^{-1}\dfrac{u}{a} + C$

47. $\int \dfrac{du}{\sqrt{a^2 - u^2}} = \sin^{-1}\dfrac{u}{a} + C$

48. $\int u\sqrt{a^2 - u^2}\, du = -\dfrac{1}{3}(a^2 - u^2)^{3/2} + C$

49. $\int u^2 \sqrt{a^2 - u^2}\, du = \dfrac{u}{8}(2u^2 - a^2)\sqrt{a^2 - u^2} + \dfrac{a^4}{8}\sin^{-1}\left(\dfrac{u}{a}\right) + C$

50. $\int \dfrac{du}{u\sqrt{a^2 - u^2}} = -\dfrac{1}{a}\ln\left|\dfrac{a + \sqrt{a^2 - u^2}}{u}\right| + C$

51. $\int \dfrac{du}{u^2\sqrt{a^2 - u^2}} = -\dfrac{\sqrt{a^2 - u^2}}{a^2 u} + C$

52. $\int \dfrac{u\, du}{\sqrt{a^2 - u^2}} = -\sqrt{a^2 - u^2} + C$

53. $\int \dfrac{u^2}{\sqrt{a^2 - u^2}}\, du = -\dfrac{u}{2}\sqrt{a^2 - u^2} + \dfrac{a^2}{2}\sin^{-1}\dfrac{u}{a} + C$

54. $\int \dfrac{\sqrt{a^2 - u^2}}{u}\, du = \sqrt{a^2 - u^2} - a\ln\left|\dfrac{a + \sqrt{a^2 - u^2}}{u}\right| + C$

55. $\int \dfrac{\sqrt{a^2 - u^2}}{u^2}\, du = -\dfrac{\sqrt{a^2 - u^2}}{u} - \sin^{-1}\dfrac{u}{a} + C$

56. $\int (a^2 - u^2)^{3/2}\, du = \dfrac{u}{4}(a^2 - u^2)^{3/2} + \dfrac{3a^2 u}{8}\sqrt{a^2 - u^2} + \dfrac{3a^4}{8}\sin^{-1}\dfrac{u}{a} + C$

57. $\int \dfrac{du}{(a^2 - u^2)^{3/2}} = \dfrac{u}{a^2\sqrt{a^2 - u^2}} + C$

Mc

If Newton's mixture of
applesauce & the
moon's orbit is
giving you pain,
some rising sun
reagent can help you
pass through
the integration
symbolic Calculus
can make sense.

Call Mike
744-9547 ask for
mike in #22

or
943-1423
& leave a
message

Calculus with Analytic Geometry

Calculus with Analytic Geometry

MARVIN J. FORRAY

Professor of Mathematics, C. W. Post College of Long Island University

Macmillan Publishing Co., Inc.

New York

Collier Macmillan Publishers

London

Macmillan Publishing Co., Inc.
866 Third Avenue, New York, New York 10022

Collier Macmillan Canada, Ltd.

Library of Congress Cataloging in Publication Data

Forray, Marvin J
   Calculus with analytic geometry

   Includes index.
   1. Geometry, Analytic. 2. Calculus. I. Title.
QA551.F694   516'.3   77-1688
ISBN 0-02-338800-5

Printing: 1 2 3 4 5 6 7 8   Year: 8 9 0 1 2 3 4

Preface

This is a textbook on calculus and analytic geometry that is designed for (a) the prospective mathematics major, (b) a student interested in the physical sciences, (c) engineering students, or (d) anyone who simply wishes to develop an understanding of the concepts and applications of calculus. The author was guided by the following objectives:

(i) The student should be able to read this textbook and learn from it. A textbook must be more than a collection of theorems, examples, and exercises.

(ii) Definitions, theorems, and their proofs must be given in a straightforward manner with rigor that is appropriate for a beginner in this field. At the same time, this text is not written in a cookbook style because this would be an obvious disservice to the prospective reader.

(iii) The illustrative examples should be ample in number, properly ordered, easy to comprehend, and a definite asset in the solution of many of the exercises. A strong effort has been made to see that this correlation does exist.

(iv) The exercises should be sufficient in number, properly ordered, and generally graded as to order of difficulty. The author has solved each of the exercises so that he hopes there will be a minimum of numerical difficulties.

Chapter 1 starts with some basic facts about the real number system. Rational and irrational numbers and their decimal representations are discussed briefly. Particular attention is then given to the rules necessary for the manipulation of inequalities. Next, the solution of equalities and inequalities involving the absolute value is developed. The major portion of this chapter is devoted to the elements of plane analytic geometry involving studies of the straight line, the circle, and the parabola. The analysis of intercepts, symmetry, extent, and vertical and horizontal asymptotes, with applications to curve sketching, is presented in detail. In the last section of this chapter, there is a brief treatment of mathematical induction because of its application later in the text.

Chapter 2 is devoted to the concepts of (i) function, (ii) limit of a function, and (iii) continuity. The concept of function is defined in terms of rule of correspondence and then ordered pairs. The main thrust of this chapter is the development of limit of a function, given in Sections 2.4 and 2.5. The former section treats this material intuitively, avoiding the ϵ-δ concept. This was purposely done to make it simple for an instructor to bypass the ϵ-δ method (if desired) by skipping Section 2.5. Those instructors who want to present the ϵ-δ method can proceed with the intuitive discussion (or omit it altogether) and emphasize Section 2.5.

The concept of the derivative and the rules for differentiation are the primary subjects of Chapter 3. Introductory material on the tangent to a curve and velocity are given to motivate the definition of the derivative. An intuitive (heuristic) development of the chain rule is followed by a rigorous treatment. A brief historical introduction to calculus is given in the first section of this chapter. The author hopes that this material will be read and reread as maturity and technical know-how develop. Chapter 4 is devoted to applications of the derivative to problems of related rates, maxima and minima, curve sketching, and problems in economics.

In Chapter 5 the concepts of differentials and antiderivatives are treated where the standard notation for antiderivatives, $\int f(x)\, dx$, is used. The techniques of substitution and the generalized power rule are developed and the chapter closes with applications to elementary initial value and boundary value problems.

Summation notation is treated in the first section of Chapter 6 and is then applied to the concept of area under a curve as a limit of sums. This motivates the definition of the Riemann integral as a limit of a summation. However, as we are careful to point out, area is just one of the properties of the definite integral. The properties of the definite integral are then developed, leading to the fundamental theorems of calculus. These theorems bind together the two apparently (and historically) diverse ideas of differentiation and the definite integral.

Some of the numerous applications of the definite integral are investigated in Chapter 7. The objective is to explain each of these concepts in a lucid, straightforward, and interesting manner. The concept of element is the unifying idea that threads its way through this material. Hence we have an element of area, an element of volume, an element of work, and so on. These then yield the definite integral that pertains to the given application.

Chapter 8 begins with the modern definition of the natural logarithmic function by means of a definite integral. From this the properties of the logarithmic function are systematically deduced and application of the process known as logarithmic differentiation is given. Integrals yielding logarithmic functions are then considered. At this juncture, the Euler constant e is defined and interpreted geometrically, and relations that yield this number are developed briefly. The role of e in the case of continuous compound interest follows. Then the concept of functions and their inverses leads to the introduction of the exponential function and its useful properties. Applications of the exponential function are made to problems involving variations in atmospheric pressure, growth, and decay.

The first section of Chapter 9 constitutes a rather detailed review of trigonometry, necessary because of the wide variation of student preparation in this discipline. The other seven sections develop the differentiation and integration of the trigonometric, inverse trigonometric, and hyperbolic functions. Applications of these functions to rate problems, periodic motion, the analysis of a particle moving through a resistive medium, and the determination of deflection and tension in a hanging cable are given here. The usefulness of these functions in the evaluation of integrals is demonstrated in a number of instances.

The primary objective of Chapter 10 is to develop the important methods of formal integration. We start with the integration of trigonometric and algebraic functions by applying certain substitutions and the generalized power formula. The integration of rational functions is motivated and then treated in detail. It is followed by the very useful procedure of integration by parts. In Section 10.10 the practical use of integral tables is demonstrated. The last two sections are concerned with the derivation of the trapezoidal and Simpson's rules for the numerical integration of definite integrals.

Vectors in the plane is the subject of Chapter 11. After a brief historical introduction, vector algebra is developed and applied to problems in physics and geometry. Next, we investigate vector functions and the parametric representation of plane curves. This enables us to examine some of the more interesting plane curves such as cycloids. Vector calculus is introduced and applications are given to motion in two dimensions, where concepts such as velocity, acceleration, arc length, and curvature are treated.

In Chapter 12 the reader is introduced to plane polar coordinates. These coordinates are related to the Cartesian coordinates, and equations of simple plane curves are expressed in both coordinate systems. Various tests for symmetry of curves expressed in polar coordinates are derived. After this we consider the intersection of plane curves given in polar coordinates. This is a more intricate problem in polar coordinates than in rectangular coordinates because there is no longer a one-to-one correspondence between points and their polar coordinate representation. Then the formula for the length of a curve in polar coordinates is easily derived from the expression in rectangular coordinates. The chapter closes with problems of motion in the plane using vector calculus in the derivation of the velocity and acceleration components in polar coordinates.

Chapter 13 first presents a brief historical introduction illustrating the reason for the name "conic sections." In the next two sections the ellipse and hyperbola are treated in a traditional manner. Then the three conic sections—the ellipse, the hyperbola, and the parabola—are unified by using the concepts of focus, directrix, and eccentricity. Rotation of axes and the utility of both rotation and translation of axes in the simplification of conic sections in nonstandard form are discussed. In the last section, the polar coordinate representations of conic sections are developed.

The topics of the evaluation of indeterminate forms by L'Hospital's rules, improper integrals, the extended law of the mean (Taylor's formula), and the Newton-Raphson method for root determination constitute the subject matter of Chapter 14.

Chapter 15 gives a fairly comprehensive treatment of infinite sequences and infinite series. The ratio, root, integral, comparison, and Leibnitz tests are described. Absolute and conditional convergence and the question of rearrangement of terms of an infinite series are discussed. Also there is a development of interval of convergence for power series, and term by term differentiation and integration of power series are treated. Applications are given for the use of Taylor and Maclaurin series in the representation of functions, the evaluation of integrals, and the solution of elementary initial value problems.

The application of vectors in three-dimensional space to problems in solid analytic geometry is given in Chapter 16. The dot and cross products are defined and applications to the line in space and the plane are developed. Cylinders, surfaces of revolution, and quadratic surfaces are also discussed. Vector functions and their derivatives are applied to curves in space. Additional applications of vectors in space are given to velocity, acceleration, curvature, and the rudiments of differential geometry.

Chapter 17 starts with the concept of functions of two or more variables. Then we proceed to partial derivatives of first and higher orders, limits, continuity, differentiability, differentials, and the chain rule. The concepts of the directional derivative, gradient, and the tangent plane and the normal line to a surface are developed. The chapter closes with a discussion of the absolute maximum and minimum problem for a function of two variables, the method of least squares, and an introduction to the Lagrange multiplier method.

Double and triple integration and their applications to problems such as the determination of volumes, moments, center of mass, surface area, and moments of inertia are developed in Chapter 18.

Appendix A includes review formulas from algebra, plane geometry, trigonometry, and solid geometry.

Appendix B gives nine tables of numerical data associated with particular functions such as squares, cubes, the natural logarithmic function, the exponential function, trigonometric functions (radians and degrees), hyperbolic functions, and common logarithms.

Inside the front and back covers there is a table of derivatives and integrals. Of course, the table of integrals should be used *after* the reader is familiar with the formal integration process as discussed in Chapter 10.

A number of my colleagues aided me in achieving the final form for the first edition. Professors Joel Stemple, Robert McGuigan, James E. Anderson, James M. Stakkestad, and Richard Crownover served as reviewers of the manuscript. I wish to thank them for their effective and generous assistance in this project. Furthermore, Joel Stemple helped me revise my original manuscript in accordance with the comments of *all* the reviewers and, through our mutual interaction, a better book resulted. In addition, my thanks go to Professor Nathaniel R. Riesenberg and Stephanie M. White for their critical reading of the manuscript and their valuable suggestions for its improvement.

I am indebted to Professors Joel Stemple and Stephanie M. White for helping me review the manuscript in both the galley and page-proof stages.

I am also pleased to acknowledge the skill, interest, and cooperation of Mr. Leo Malek, production supervisor, and Mr. Everett W. Smethurst, mathematics editor, during the production of this book. My thanks also go to the other personnel at Macmillan for their interest and competent professional assistance.

Mr. Edward Lanaro and his staff also deserve acknowledgment for their most careful and beautifully executed art work.

To Mrs. Stasia Polster goes my gratitude for her most diligent and skillful typing of a considerable portion of the manuscript.

Last, but not least, I would like to thank my wife, Muriel, and children, Stephanie, Wendy, Elissa, and Anthony, for helping me with a number of tasks and being most cooperative during the development of this text. In particular, my thanks are due Wendy for her help in preparing the index listing—a most time-consuming endeavor.

MARVIN J. FORRAY

Contents

11   *Vectors in the Plane and Parametric Representation*   *603*

12   *Plane Polar Coordinates*   *660*

13   *Conic Sections*   *702*

14

Indeterminate Forms, Taylor's Formula, and Improper Integrals 754

15

Sequences and Series 802

16

Solid Analytic Geometry—Vectors in Space 880

Calculus with Analytic Geometry

1

Real Numbers and Elementary Analytic Geometry

1.1   REAL NUMBERS, INEQUALTIES, AND INTERVALS

Most aspects of calculus and analytic geometry involve real numbers. We have considerable experience with the operations of addition, subtraction, multiplication, and division and, therefore, we shall not list the axioms that explain these manipulations of real numbers. We also shall not state definitions that are by now familiar to all of us. Instead, our attention will be restricted to a few particularly important definitions and rules that will be used in our development of calculus.

For convenience, we use the symbol R_1 to denote the set of all real numbers. Unless stated otherwise, we shall assume that all numbers we work with are members of R_1.

The real numbers can be broken into two parts—"rational numbers" and "irrational numbers." We define these as follows:

Definition   A **rational number** is any number that can be expressed as a quotient of integers (that is, a rational number can be put in the form $\frac{a}{d}$ where a and d are integers and $d \neq 0$).

Examples of rational numbers are

$$\frac{3}{5} \qquad -\frac{1}{2} \qquad \frac{25}{8} \qquad \frac{14}{7} \qquad -21 \qquad \frac{9\sqrt{3}}{2\sqrt{3}} \qquad \frac{\sqrt{24}}{\sqrt{6}}$$

To see that $\dfrac{\sqrt{24}}{\sqrt{6}}$ is rational we note that

$$\frac{\sqrt{24}}{\sqrt{6}} = \frac{\sqrt{(4)(6)}}{\sqrt{6}} = \frac{\sqrt{4}\sqrt{6}}{\sqrt{6}} = \sqrt{4} = 2 = \frac{2}{1}$$

Definition   An *irrational number* is any real number that is not rational.

Examples of irrational numbers are

$$\sqrt{2} \qquad -\sqrt{7} \qquad \pi \qquad \frac{3\sqrt{5}}{2} \qquad \frac{\sqrt{6}}{\sqrt{2}}$$

$$\left(\text{Note that } \frac{\sqrt{6}}{\sqrt{2}} = \frac{\sqrt{(3)(2)}}{\sqrt{2}} = \frac{\sqrt{3}\sqrt{2}}{\sqrt{2}} = \sqrt{3}\right)$$

It can be shown that a number is rational if and only if it can be expressed as either a "terminating decimal" or a "repeating decimal." For example

$$\frac{5}{4} = 1.25 \qquad\qquad \left.\begin{array}{c} \\ \\ \end{array}\right\}$$

$$\frac{-127}{10000} = -0.0127$$

terminating decimals

$$\frac{13}{6} = 2.166666\cdots$$

$$\frac{1}{99} = 0.010101\cdots$$

repeating decimals

(the symbolism $\cdots$ means that the indicated pattern continues).

In a repeating decimal, the terms do not have to start repeating immediately—in $\frac{13}{6} = 2.16666\cdots$, only after we reach the first 6 do the numbers repeat.

The decimal expansion for irrational numbers is neither terminating nor repeating. There is no pattern to the expansion and the expansion never ends.

Definition   We say that *a is less than b* and *b is greater than a* if and only if $b - a$ is a positive number.

$$a < b \text{ is the notation for "a is less than b"}$$
$$b > a \text{ is the notation for "b is greater than a"}$$

We call $a < b$ and $b > a$ **strict inequalities.** If we wish to allow for the possibility that a and b might also be equal, we have the **nonstrict inequalities** defined by

$$a \leq b \text{ if and only if either } a < b \text{ or } a = b$$
$$b \geq a \text{ if and only if either } b > a \text{ or } b = a$$

Some numerical examples of inequalities are $3 < 6$, $8 > 2$, $4 \leq 5$, $-1 > -2$, $7 \geq 7$, $6 \geq -5$.

The following are rules that may be used when working with inequalities:

1. (A) if $a < b$ and $b < c$ then $a < c$ $\left.\begin{array}{c} \\ \end{array}\right\}$ the *transitive law*
    (B) if $a > b$ and $b > c$ then $a > c$
2. (A) if $a < b$ then $a + c < b + c$
    (B) if $a > b$ then $a + c > b + c$
3. (A) if $a < b$ then $a - c < b - c$
    (B) if $a > b$ then $a - c > b - c$

4. (A) if $a < b$ and $c > 0$ then $ac < bc$

   (B) if $a > b$ and $c > 0$ then $ac > bc$

5. (A) if $a < b$ and $c < 0$ then $ac > bc$

   (B) if $a > b$ and $c < 0$ then $ac < bc$

6. (A) if $0 < a < b$ and $0 < c < d$ then $0 < ac < bd$

   (B) if $a > b > 0$ and $c > d > 0$ then $ac > bd > 0$

Rules (2) and (3) say that you can add a number to or subtract a number from both sides of an inequality without changing the inequality sign; (4) and (5) say that, if both sides of an inequality are multiplied by the same number (not zero), then the inequality sign is the same if the number is positive and the inequality sign reverses if the number is negative. These rules can all be proven without much difficulty. However, the reader might find it more instructive first to construct some numerical examples to illustrate each of the rules.

We have already used the phrase "if and only if" (sometimes abbreviated "iff") several times. This phrase says that the two expressions it connects are equivalent; that is, a statement and its converse are simultaneously both true or both false. Thus, for example, the statement "$a < b$ if and only if $b - a$ is positive" means that "if $a < b$ then $b - a$ is positive" and also that "if $b - a$ is positive then $a < b$."

In the usual manner, we represent the real numbers R_1 geometrically by points on a horizontal line, which may be extended indefinitely (see Figure 1.1.1). An arbitrary point on the horizontal line (also called the axis) is chosen to represent the number 0. This point is taken to be the *origin* of coordinates. An arbitrary unit of distance is chosen. Then each positive number x is represented by a point x units to the right of the origin. Similarly, each negative number x is represented by a point at a distance $-x$ units to the left of the origin. Furthermore, it is postulated that there is a one-to-one correspondence between R_1 and the number line. This means that to each point on the number line there corresponds precisely one real number and, conversely, to each real number there is one and only one point on the line. Then two *distinct* points P and Q on the number line cannot represent the same number. Therefore the points on the number line may be identified with the numbers represented by them.

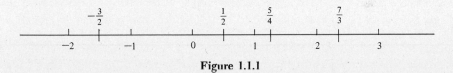

Figure 1.1.1

By our convention, $a < b$ if and only if the point representing the number a is to the left of the point representing the number b. Similarly, $c > d$ if and only if the point representing the number c is to the right of the point representing the number d. Thus, for example, the point representing $-\frac{3}{2}$ is to the left of the point representing -1. The point representing 2 is to the right of the point representing $\frac{1}{2}$.

A number x is said to be between the numbers a and b if and only if $a < x$ and $x < b$. We write this as a continued inequality as follows

$$a < x < b \tag{1}$$

The continued inequality (1) is called an **open interval** and is denoted (a, b). The set of all points x such that $a \leq x \leq b$ is called a **closed interval** and is denoted

[a, b]. When talking about intervals, "closed" means the endpoints are included whereas "open" means they are excluded. The open and closed intervals are shown in Figure 1.1.2.

(a) Open interval   (b) Closed interval

Figure 1.1.2

The interval $a < x \leq b$ is said to be **half-open on the left** and is denoted by $(a, b]$. Also, the interval $a \leq x < b$ is **half-open on the right** and is denoted by $[a, b)$ (Figure 1.1.3). In all cases $b - a$ is called the **length** of the interval and all such intervals are finite.

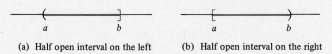

(a) Half open interval on the left   (b) Half open interval on the right

Figure 1.1.3

To denote the nonfinite or infinite intervals, symbols such as $-\infty$ and ∞ will be used. The reader must be cautioned not to confuse these symbols with real numbers because they do not obey the properties of real numbers. The set of all x such that $x > a$ is called the infinite interval (a, ∞). Similarly, $[a, \infty)$ is the set of all x with $x \geq a$, $(-\infty, b)$ the set of all $x < b$ and $(-\infty, b]$ the set of all x with $x \leq b$ (Figure 1.1.4). Finally $(-\infty, \infty)$ is the set of all real x and is represented by the whole number line.

$a < x < \infty$ $-\infty < x \leq b$

Figure 1.1.4

Let us now turn to the solution of inequalities.

EXAMPLE 1   Find all real numbers satisfying the inequality $-3x > 6$.

Solution   Division of both sides of the given inequality by -3 and reversing the sense of the inequality yields

$$x < -2$$

Therefore $-3x > 6$ if and only if $x < -2$. The solution interval is $(-\infty, -2)$ and is shown in Figure 1.1.5.   ●

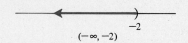

$(-\infty, -2)$

Figure 1.1.5

EXAMPLE 2   Find all real numbers satisfying the inequality $5x + 3 > x - 1$.

Solution   Add -3 to both sides of the inequality. Therefore $5x > x - 4$. Next, add $-x$ to both sides to obtain

$$4x > -4$$

Division of both sides by 4 yields

$$x > -1$$

and the interval solution $(-1, \infty)$ is shown in Figure 1.1.6.   ●

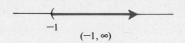

Figure 1.1.6

EXAMPLE 3   Find all real numbers satisfying the continued inequality $-3 < 5 - 2x \leq 7$.

Solution   We shall work with both inequalities (all three members) simultaneously. Subtract 5 from each term so that

$$-8 < -2x \leq 2$$

Divide each term by -2 (that is, multiply by $-\frac{1}{2}$) and reverse each inequality, obtaining

$$4 > x \geq -1$$

This is the same as $-1 \leq x < 4$. The interval solution is $[-1, 4)$ as shown in Figure 1.1.7.   ●

Figure 1.1.7

EXAMPLE 4   Find all real numbers satisfying $\dfrac{4}{x} > 3$, $x \neq 0$.

Solution   Our immediate inclination is to multiply both sides by x. However, the sense of the resulting inequality will depend upon whether x is positive or negative. Therefore, we must consider two cases:

Case 1: Suppose $x > 0$. Then multiplication of both sides by x results in the sense of the inequality being preserved.
Therefore

$$4 > 3x$$

and division by 3 yields

$$\tfrac{4}{3} > x$$

Thus the solution of Case 1 is the set of all numbers x such that both $x > 0$ and $\frac{4}{3} > x$; that is, $0 < x < \frac{4}{3}$ or, equivalently, the open interval $(0, \frac{4}{3})$.

Case 2: Suppose $x < 0$. Then multiplication by x reverses the sense of the inequality yielding

$$4 < 3x$$

and division by 3 results in

$$\tfrac{4}{3} < x$$

Therefore, for Case 2, $x < 0$ and $x > \frac{4}{3}$ must hold simultaneously; this is clearly impossible. Therefore, we have the null set.

From Cases 1 and 2 we conclude that the solution of the given inequality is the open interval $(0, \frac{4}{3})$ shown in Figure 1.1.8. ●

Figure 1.1.8

EXAMPLE 5 Find all real numbers satisfying the inequality

$$\frac{x + 3}{x - 2} < 5 \qquad x \neq 2$$

Solution Again we treat two cases.

Case 1: $x > 2$.

Multiply both sides by $x - 2 > 0$. Therefore

$$x + 3 < 5(x - 2) \qquad \text{or} \qquad x + 3 < 5x - 10$$

Combining terms yields

$$13 < 4x$$

Divide both sides by 4 and we have

$$\tfrac{13}{4} < x$$

Thus $x > 2$ and $x > \frac{13}{4}$. Their intersection is $x > \frac{13}{4}$ which gives the interval $(\frac{13}{4}, \infty)$

Case 2: $x < 2$

Multiply both sides by $x - 2 < 0$ and therefore reverse the sense of the inequality. Consequently,

$$x + 3 > 5(x - 2)$$

or multiplying and simplifying, we obtain

$$13 > 4x$$

Dividing both sides by 4, we have

$$\tfrac{13}{4} > x$$

Therefore $x < 2$ and $x < \tfrac{13}{4}$ which implies that $x < 2$, giving the interval $(-\infty, 2)$.

From Cases 1 and 2 we conclude that the inequality is satisfied if and only if $x > \tfrac{13}{4}$ *or* $x < 2$; that is, x is a number in either $(-\infty, 2)$ or $(\tfrac{13}{4}, \infty)$. Equivalently, we can say that the inequality is satisfied for all x *not* in $[2, \tfrac{13}{4}]$. This is shown in Figure 1.1.9. ●

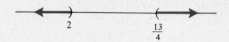

Figure 1.1.9

We could also have solved Example 5 by rewriting the given inequality as $\dfrac{x+3}{x-2} - 5 < 0$ which simplifies to $\dfrac{-4x+13}{x-2} < 0$ and then working from this new inequality.

EXAMPLE 6   Find all real numbers satisfying the inequality $(x-2)(x+7) < 0$.

Solution   The inequality will be satisfied when the two factors are of opposite signs. Therefore, once again there are two cases to be considered.

Case 1: $x - 2 > 0$ and $x + 7 < 0$

This implies that $x > 2$ *and* $x < -7$. Since it is impossible for x to satisfy both of these inequalities (intersection is the null set) we have no values of x for Case 1.

Case 2: $x - 2 < 0$ and $x + 7 > 0$

This implies that $x < 2$ *and* $x > -7$. Thus we have $-7 < x < 2$ or x is in the open interval $(-7, 2)$. The result is shown in Figure 1.1.10. ●

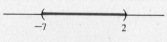

Figure 1.1.10

EXAMPLE 7   Find all real numbers satisfying the inequality $2x^2 - 3x \geq 5$

Solution   We add -5 to both sides and obtain

$$2x^2 - 3x - 5 \geq 0$$

Furthermore, we can factor the left side. Therefore

$$(2x - 5)(x + 1) \geq 0$$

The inequality is satisfied if both factors are nonnegative or if both factors are nonpositive. Therefore again we have two cases:

Case 1: $2x - 5 \geq 0$ and $x + 1 \geq 0$, which implies that

$$x \geq \tfrac{5}{2} \quad \text{and} \quad x \geq -1$$

The intersection of the two sets is $x \geq \tfrac{5}{2}$

Case 2: $2x - 5 \leq 0$ and $x + 1 \leq 0$, which implies that

$$x \leq \tfrac{5}{2} \quad \text{and} \quad x \leq -1$$

The intersection of the two sets is $x \leq -1$.

Therefore, the values of x for which the inequality is satisfied are the numbers *not* in the set $(-1, \tfrac{5}{2})$. This result is shown in Figure 1.1.11. ●

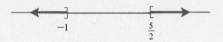

$$-1 \qquad \tfrac{5}{2}$$

Figure 1.1.11

Completion of the squares can be used to solve inequalities.

EXAMPLE 8 Find all real numbers satisfying $x^2 + 4x + 7 > 0$

Solution Adding -7 to each side yields

$$x^2 + 4x > -7$$

We add 4 to each side of this inequality, so that the left side is a square

$$x^2 + 4x + 4 > -3 \quad \text{or} \quad (x + 2)^2 > -3$$

Since the left side is nonnegative for all real numbers x, and the right side is negative, the given inequality is satisfied for *all* real numbers x. ●

EXAMPLE 9 Find all real numbers satisfying

$$x^2 + 8x + 17 < 0$$

Solution Adding -17 to each side yields

$$x^2 + 8x < -17$$

We add 16 to each side of this inequality, so that the left side is a square

$$x^2 + 8x + 16 < -1 \quad \text{or} \quad (x + 4)^2 < -1$$

Since the left side is nonnegative for all real numbers x, we conclude that this inequality, and hence the given inequality, is satisfied for *no* real numbers x. ●

In the next section, completion of the squares will be used to solve quadratic inequalities that cannot be factored.

Exercises 1.1

In Exercises 1 through 21, solve the following inequalities and show the solution on the real line.

1. $3x - 2 > 7$

2. $\frac{x}{2} - 3 \leq 9$

3. $-3x + 2 > -4$

4. $-x + 1 > -3$

5. $4x - 3 > -2x - 12$

6. $\frac{4}{x} < 0, \quad x \neq 0$

7. $\frac{3}{x - 1} > 0, \quad x \neq 1$

8. $x^2 - 3x + 2 < 0$

9. $x^2 - 7x + 6 \geq 0$

10. $x^2 - 4x + 4 \geq 0$

11. $4x^2 - 4x + 1 < 0$

12. $3(2x - 1) - 3x + 5 < 4 - 2x$

13. $\frac{3}{1 - x} \leq 1, \quad x \neq 1$

14. $\frac{5}{x} - 2 < \frac{2}{x} + 4, \quad x \neq 0$

15. $-x^2 + 25 \geq 0$

16. $x^3 - 4x \geq 0$

17. $x^4 \geq 13x^2 - 36$ (*Hint:* Factor the biquadratic)

18. $\frac{1}{(x - 3)^4} \geq 81, \quad x \neq 3$

19. $x + \frac{1}{x} \geq 2, \quad x \neq 0$

20. $\frac{x + 3}{4 - x} > 2, \quad x \neq 4$

21. $\frac{x + 4}{x + 2} > \frac{x + 3}{x - 1}$

22. Prove that if $a < b$ then $a < \dfrac{a + b}{2} < b$. Interpret the result.

23. Generalize the previous exercise by proving that if $a_1 < a_2 < a_3 < \cdots < a_n$, then

$$a_1 < \frac{a_1 + a_2 + \cdots + a_n}{n} < a_n$$

Interpret the result.

24. Prove that if $0 < a < b$, then $\dfrac{1}{a} > \dfrac{1}{b}$.

25. Prove that if $b > a \geq 0$, then $b^2 > a^2$. State and prove a converse of this.

26. Prove that for arbitrary real a and b,

$$ab \leq \frac{a^2 + b^2}{2}$$

In Exercises 27 through 30, find all real numbers satisfying the given inequality by applying the method of completing the squares.

27. $x^2 - 6x + 12 \geq 0$

28. $x^2 - 2x + 2 \leq 0$

29. $4x^2 - 20x + 29 < 0$

30. $6x^2 - 42x + 100 > 0$

31. Show that the following numbers are rational (a) $0.121212\cdots$; (b) $3.1313\cdots$; (c) $0.231231\cdots$. (*Hint:* Let $x = 0.121212\cdots$ and therefore $100x = 12.1212\cdots$ and subtract equals from equals, and so on.)

32. Show that the set of rational numbers is closed under (a) addition, (b) subtraction, (c) multiplication, and (d) division (except when the divisor is zero).

33. Prove that $\sqrt{7}$ is not rational. (*Hint:* Assume that $\sqrt{7}$ is rational by writing $\sqrt{7} = \dfrac{p}{q}$ where p and q are integers ($q \neq 0$) and $\dfrac{p}{q}$ is reduced to lowest terms. Square both sides and clear fractions. Then show that p and q are divisible by 7, thereby contradicting the hypothesis.)

34. Prove that $3 - \sqrt{7}$ is not rational.

The **absolute value** of a number a is either a or $-a$, whichever is nonnegative. The symbol for the absolute value of a is $|a|$. Therefore,

$$\begin{aligned} |a| &= a \text{ if } a > 0 \\ |a| &= -a \text{ if } a < 0 \\ |0| &= 0 \end{aligned} \qquad (1)$$

For example, $|5| = 5$, $|-4| = 4$, $|3 - 11| = |-8| = 8$, $|b - b| = |0| = 0$. This rather simple notation has very important consequences, which we shall be investigating. First however we shall develop some facility in handling equations and inequalities involving absolute values.

EXAMPLE 1   Solve for x: $|x + 4| = 7$.

Solution   From the definition of the absolute value, we know that $x + 4$ must be 7 or -7, since $|7| = |-7| = 7$.
Therefore,

$$x + 4 = 7 \qquad \text{or} \qquad x + 4 = -7$$

which implies that $x = 3$ or $x = -11$.   ●

EXAMPLE 2   Solve $|3x - 6| = 9$.

Solution   Since $|-9| = |9| = 9$, we have

$$3x - 6 = 9 \qquad \text{or} \qquad 3x - 6 = -9$$

Therefore $x = 5$ or $x = -1$.   ●

EXAMPLE 3   Simplify $\dfrac{x}{|x|}$ if x is a nonzero real number.

Solution

$$\frac{x}{|x|} = \frac{x}{x} = 1 \qquad \text{if } x > 0$$

$$\frac{x}{|x|} = \frac{x}{-x} = -1 \qquad \text{if } x < 0$$

The expression $\dfrac{x}{|x|}$ is not defined for $x = 0$ because $|0| = 0$ and division by zero is not permitted.   ●

EXAMPLE 4   Solve $|x + 3| = |2x - 4|$.

Solution   If two numbers have the same absolute value then the two numbers must be equal or one is the negative of the other. Thus

$$\pm(x + 3) = 2x - 4$$

If $x + 3 = 2x - 4$, then $x = 7$; whereas, if $-(x + 3) = 2x - 4$ or $-x - 3 = 2x - 4$, then $3x = 1$ and $x = \frac{1}{3}$. ●

EXAMPLE 5 Solve $|3 - 2x| = -1$.

Solution Since the absolute value of a number is nonnegative and the right side of the equation is negative, there is *no* value of x that satisfies the equation. ●

1.2.1 *Geometric Interpretation of the Absolute Value*

The following definition gives a geometric interpretation of the absolute value:

Definition Given any two real numbers x and y, the (undirected) **distance** between x and y is $|x - y|$.

To see that this definition is reasonable, we note the following:
1. If $x = y$ then $|x - y| = |0| = 0$ (which is clearly the distance from x to y).
2. If $x \neq y$ then the distance between x and y should be the length of the closed interval having x and y as endpoints. We have:
   if $x > y$, then the length of $[y, x]$ is $x - y = |x - y|$ (see Figure 1.2.1);
   if $x < y$, then the length of $[x, y]$ is $y - x = -(x - y) = |x - y|$ (see Figure 1.2.2).

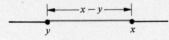

Figure 1.2.1   Figure 1.2.2

Theorem 1 $|-x| = |x|$ for any real number x.

Proof If $x > 0$ then $|x| = x$ and $|-x| = x$, also.
If $x < 0$ then $|x| = -x$ and $|-x| = -x$, also, since $-x > 0$.
If $x = 0$ then $|-0| = |0| = 0$,
and the proof is complete. ∎

The geometric interpretation of this result is that the distance between 0 and x is the same as the distance between 0 and $-x$. (See Figure 1.2.3 for the case in which $x > 0$).

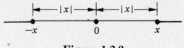

Figure 1.2.3

Corollary $|y - x| = |x - y|$ for any real numbers x and y.

Since $|x|$ represents the distance of x from the origin then a condition such as $|x| < 6$ is equivalent to the statement that x be any number in the open interval extending from -6 to 6 (Figure 1.2.4); that is, x must lie in $(-6, 6)$.

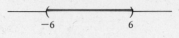

Figure 1.2.4

In a similar manner, $|x| < a$ where a is an arbitrary positive number is equivalent to the double inequality $-a < x < a$. Furthermore, an inequality such as $|x - 3| < 4$ is equivalent to $-4 < x - 3 < 4$. If we add 3 to each of the three members, there results

$$-1 < x < 7$$

Therefore x must lie in the interval $(-1, 7)$ as shown in Figure 1.2.5.

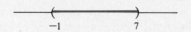

Figure 1.2.5

EXAMPLE 6   Solve for x: $|4x - 3| \leq 9$

Solution   We write the inequality in the equivalent form

$$-9 \leq 4x - 3 \leq 9$$

Now add 3 to each member, obtaining

$$-6 \leq 4x \leq 12$$

and division of each member by 4 yields

$$-\tfrac{3}{2} \leq x \leq 3$$

The solution set consists of all numbers in the closed interval $[-\tfrac{3}{2}, 3]$ as shown in Figure 1.2.6. ●

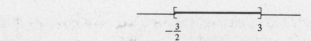

Figure 1.2.6

EXAMPLE 7   Solve for x: $\left| \dfrac{3x + 2}{2x + 5} \right| < 2, \; x \neq -\dfrac{5}{2}$

Solution   The given inequality is equivalent to

$$-2 < \frac{3x + 2}{2x + 5} < 2$$

Case 1: $2x + 5 > 0$

Multiply the three members of the double inequality by $2x + 5$. Thus

$$-2(2x + 5) < 3x + 2 < 2(2x + 5)$$

or

$$-4x - 10 < 3x + 2 < 4x + 10$$

From the left inequality,

$$-12 < 7x \quad \text{or} \quad -\tfrac{12}{7} < x$$

From the right inequality,

$$-8 < x$$

But,

$$2x + 5 > 0 \quad \text{or} \quad x > -\tfrac{5}{2}$$

Thus the intersection of these three inequalities $x > -\tfrac{12}{7}, x > -8,$ and $x > -\tfrac{5}{2}$ is $x > -\tfrac{12}{7}$.

Case 2: $2x + 5 < 0$

Multiply the three members of the double inequality by $2x + 5$ reversing the sense of the inequality because $2x + 5 < 0$. Thus

$$-2(2x + 5) > 3x + 2 > 2(2x + 5)$$

or

$$-4x - 10 > 3x + 2 > 4x + 10$$

From the left inequality,

$$-12 > 7x \quad \text{or} \quad -\tfrac{12}{7} > x$$

From the right inequality,

$$-8 > x$$

But from $2x + 5 < 0$ we have $x < -\tfrac{5}{2}$, and the intersection of $x < -\tfrac{5}{2}, x < -8,$ and $x < -\tfrac{12}{7}$ is $x < -8$.

Therefore the solution is

$$x > -\tfrac{12}{7} \quad \text{or} \quad x < -8 \qquad \bullet$$

If $|x| > 3$ then either $x > 3$ or $x < -3$. More generally, if $|x| > a$ and a is positive then either $x > a$ or $x < -a$. An inequality such as $|x + 2| > 5$ is then equivalent to the statement

$$\text{either } x + 2 > 5 \text{ or } x + 2 < -5$$

This implies that $x > 3$ or $x < -7$; that is, x is either in the interval $(-\infty, -7)$ or in $(3, \infty)$ as shown in Figure 1.2.7.

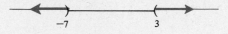

Figure 1.2.7

1.2.2 Absolute Value and Inequalities

From elementary algebra, we may recall that for $a \geq 0$, $\sqrt{a}$ is the nonnegative number x such that $x^2 = a$. This is also referred to as the "principal square root of a." For example, $\sqrt{9} = 3$, $\sqrt{\frac{1}{4}} = \frac{1}{2}$ and $\sqrt{0} = 0$. Note that $\sqrt{9} \neq -3$, even though $(-3)^2 = 9$. From our definition, $\sqrt{9} = \sqrt{3^2} = \sqrt{(-3)^2} = 3$. More generally, $\sqrt{x^2} = |x|$ for *all* x.

We now utilize this simple result to establish two important properties.

PROPERTY 1 If a and b are any two numbers, then $|ab| = |a|\,|b|$. In other words, *the absolute value of a product equals the product of the absolute values.*

Proof
$$|ab| = \sqrt{a^2 b^2} = \sqrt{a^2}\,\sqrt{b^2} = |a|\,|b| \qquad\blacksquare$$

PROPERTY 2 If a is any number and b is any nonzero number then $\left|\dfrac{a}{b}\right| = \dfrac{|a|}{|b|}$. In other words, *the absolute value of a quotient equals the quotient of the absolute values* (with division by zero excluded as usual).

Proof
$$\left|\frac{a}{b}\right| = \sqrt{\frac{a^2}{b^2}} = \frac{\sqrt{a^2}}{\sqrt{b^2}} = \frac{|a|}{|b|} \qquad \text{if } b \neq 0 \qquad\blacksquare$$

The proof may also be done by considering the various possibilities (signs of a and b) and verifying that the two sides are equal. This investigation is left to the reader.

Next, we have the very interesting and significant result.

Theorem 2 **Triangle Inequality** If a and b are any numbers, then
$$|a + b| \leq |a| + |b|$$

In words, the absolute value of the sum is less than or equal to the sum of the absolute values.

We given two simple illustrations of this theorem:

1. If $a = 4$ and $b = 2$ then $|a + b| = |4 + 2| = 6$ and $|a| + |b| = |4| + |2| = 6$ so $|a + b| = |a| + |b|$.
2. If $a = 5$ and $b = -3$ then $|a + b| = |5 - 3| = 2$ and $|a| + |b| = |5| + |-3| = 5 + 3 = 8$ so $|a + b| < |a| + |b|$.

Proof of
Theorem 2
Since $|a| = a$ or $-a$ it follows that $-|a| \leq a \leq |a|$. Similarly, $-|b| \leq b \leq |b|$. Addition of these inequalities yields $-(|a| + |b|) \leq a + b \leq |a| + |b|$. Therefore $|a + b| \leq |a| + |b|$. $\qquad\blacksquare$

Note that a more "obvious" proof is to treat three cases: (i) $a \geq 0$ and $b \geq 0$, (ii) $a < 0$ and $b < 0$, (iii) one number say $a \geq 0$ and $b < 0$. This is left to the reader as an exercise.

From the triangle inequality we next establish two important corollaries.

Corollary 1   If a and b are any numbers then

$$|a - b| \leq |a| + |b|$$

Proof   We write $a - b$ as $a + (-b)$ and apply Theorem 2 to obtain

$$|a - b| = |a + (-b)| \leq |a| + |-b| = |a| + |b|$$ ∎

Corollary 2   If a and b are any numbers, then

$$|a| - |b| \leq |a - b|$$

Proof $$|a| = |a - b + b| \leq |a - b| + |b|$$

and by subtracting $|b|$ from both sides of the inequality, there results

$$|a| - |b| \leq |a - b|$$ ∎

EXAMPLE 8   Estimate how large the expression $|x^3 - 8|$ can become if x is restricted to the interval $[-2, 2]$.

From Corollary 1,

$$|x^3 - 8| \leq |x^3| + |8|$$

Solution   But $|x^3| = |x^2 \cdot x| = |x^2||x| = |x \cdot x||x| = |x||x||x|$ by applying Property 1 twice. Therefore

$$|x^3 - 8| \leq (|x|)^3 + 8$$

where $|x| \leq 2$. Thus

$$|x^3 - 8| \leq 2^3 + 8 = 16$$

if x is any number in the interval $[-2, 2]$.

In this particular case the estimate is exact; that is, it can be shown that $|x^3 - 8|$ possesses a largest value, and this occurs when $x = -2$. Furthermore,

$$|x^3 - 8|_{x=-2} = |-8 - 8| = |-16| = 16†$$

However, in general, this method will yield an upper bound and not necessarily the least upper bound.   ●

EXAMPLE 9   If $|x - 2| < \delta$ and $\delta < 1$, show that $|x^3 - 8| < 19\delta$.

Solution $x^3 - 8 = (x - 2)(x^2 + 2x + 4)$ is an identity. [This follows from $x^3 - y^3 = (x - y)(x^2 + xy + y^2)$ with $y = 2$]. Therefore

$$|x^3 - 8| = |(x - 2)(x^2 + 2x + 4)| = |x - 2||x^2 + 2x + 4|$$

and it follows that

$$|x^3 - 8| \leq |x - 2| \cdot (|x^2| + |2x| + |4|)$$
$$= |x - 2| \cdot (|x|^2 + 2|x| + 4)$$

---

† $(\quad)_{x=a}$ or $| \quad |_{x=a}$ means evaluate the expression at $x = a$.

Now $|x - 2| < \delta < 1$ or $1 < x < 3$. Therefore

$$|x^3 - 8| < \delta(3^2 + 2(3) + 4) = 19\delta$$ ●

We now turn our attention to solving quadratic inequalities that cannot be factored.

EXAMPLE 10   Find all x satisfying $x^2 + 6x + 4 < 0$.

Solution   The given inequality is equivalent to

$$x^2 + 6x < -4$$

We add 9 to each side of this inequality, so that the left side is a perfect square

$$x^2 + 6x + 9 < 5 \qquad \text{or} \qquad (x + 3)^2 < 5$$

Taking the positive square root of each side yields

$$|x + 3| < \sqrt{5}$$

This yields

$$-\sqrt{5} < x + 3 < \sqrt{5} \qquad \text{or} \qquad -3 - \sqrt{5} < x < -3 + \sqrt{5}$$

as the desired solution. ●

EXAMPLE 11   Find all x satisfying $-2x^2 + 8x + 3 < 0$.

Solution   Dividing by -2 reverses the sense of the inequality, thus we obtain

$$x^2 - 4x - \tfrac{3}{2} > 0$$

Adding $\tfrac{3}{2}$ to both sides yields

$$x^2 - 4x > \tfrac{3}{2}$$

To make the left side a perfect square we add 4 to each side of the inequality

$$x^2 - 4x + 4 > \tfrac{3}{2} + 4 = \tfrac{11}{2}$$

Thus

$$(x - 2)^2 > \tfrac{11}{2}$$

Taking the positive square root of each side yields

$$|x - 2| > \sqrt{\tfrac{11}{2}}$$

From the last inequality we obtain

$$x - 2 > \sqrt{\tfrac{11}{2}} \qquad \text{or} \qquad x - 2 < -\sqrt{\tfrac{11}{2}}$$

or, equivalently,

$$x > 2 + \sqrt{\tfrac{11}{2}} \qquad \text{or} \qquad x < 2 - \sqrt{\tfrac{11}{2}}$$

This is the solution we require. ●

Exercises 1.2

In Exercises 1 through 8, solve for x.

1. $|3x| = 8$

3. $|3x - 1| = 2$

5. $|2x + 7| = |-x - 1|$

7. $|3x - 1| = 5x + 7$

2. $|2x + 5| = 3$

4. $|3x + 2| = |2x - 1|$

6. $\left|\dfrac{x + 2}{4x - 1}\right| = 3$

8. $\left|x - \dfrac{1}{3}\right| = |4x|$

In Exercises 9 through 24, find the values of x, if any, satisfying the inequalities. Give the interval solution and illustrate the solution on the real line.

9. $|x - 5| < 3$

11. $|2x - 3| < 7$

13. $|x + 9| \geq 3$

15. $|3x + 7| \leq 0$

17. $\sqrt{8 - 3x} > 0$

19. $|3 - 2x| \leq |x + 4|$

21. $|x - c| < a, \quad a > 0$

22. $|x - 3| < |2x + 10|$ (*Hint:* $\sqrt{(x - 3)^2} < \sqrt{2^2(x + 5)^2}$; square both sides; combine and factor.)

23. $|x| + |x - 2| \geq 3$ (*Hint:* Consider separately, $x \leq 0$, $0 < x < 2$, $x \geq 2$.)

24. $|x| + |x - 1| + |x + 1| < 3$ (*Hint:* Consider separately, $x \leq -1$, $-1 < x < 0$, $0 \leq x < 1$, $x \geq 1$.)

10. $|x + 4| < 1$

12. $|3 + 2x| \leq 1$

14. $|2x - 7| \geq -1$

16. $\sqrt{49 - x^2} \geq 4\sqrt{3}$

18. $\sqrt{8 - 3x^2} \geq 3$

20. $|5x - 7| \leq 0.01$

25. Solve for x: $\dfrac{x - 3}{x + 4} < 0$

26. Solve for x: $\dfrac{x + a}{x - a} < 0, \quad$ if $a \neq 0$

27. Solve for x: $\dfrac{3b + x}{3b - x} \geq 0, \quad$ if $b \neq 0$

28. Prove that $|a + b + c| \leq |a| + |b| + |c|$

29. Generalize the previous exercise; that is, prove the inequality

$$|a_1 + a_2 + \cdots + a_n| \leq |a_1| + |a_2| + \cdots + |a_n|$$

30. Prove that $\left|a + \dfrac{1}{a}\right| \geq 2, \quad$ if $a \neq 0$

In Exercises 31 through 34, find a positive number M, if it exists, such that the absolute value of the expression is $\leq M$ if x is in the given interval.

31. $x^2 - 2x + 6, \quad x$ in $[-4, 4]$

32. $x^3 + 2x^2 - x + 3, \quad x$ in $[-1, 2)$

33. $\dfrac{x + 3}{x - 2}, \quad x$ in $[4, 6]$

34. $\dfrac{x^2 + 3x + 1}{x^2 + 4x}, \quad x$ in $(0, 2)$

In Exercises 35 through 38 solve the given inequality by the method of completing the squares.

35. $4x^2 - 4x - 35 \leq 0$

36. $x^2 - 10x + 23 \geq 0$

37. $9x^2 - 6x - 8 > 0$

38. $49x^2 + 28x - 143 > 0$

We observed that the real line provided us with a geometric representation of real numbers as points on a line. This geometric representation was used to enable us to obtain a pictorial idea of the order of real numbers and to introduce the idea of distance between two points on the line in terms of absolute values of differences. In generalizing our discussion so that we may represent points in the plane, it is necessary to use ordered pairs (x, y) of real numbers.†

We use the symbol R_2 to designate the number plane; that is, the points in two dimensional space. The description of this geometric plane, which we shall briefly review, was developed independently (but simultaneously) by René Descartes (1596–1650), a philosopher and mathematician, and Pierre de Fermat (1601–1665), a judge and great mathematician. These two French mathematicians are the primary claimants to the title of father of analytic geometry in its current form. Descartes is credited with the important concept of locating a point relative to two perpendicular axes, to which we now turn.

Let us draw two perpendicular lines in the plane and call one the ***horizontal axis*** and the other the ***vertical axis.*** The intersection O of these lines "of infinite extent" is called the ***origin*** of the axes. The horizontal line is chosen as the x-axis or axis of ***abscissas,*** and the vertical axis becomes the y-axis or axis of ***ordinates.*** A unit of length is chosen on both axes and, although the unit lengths do not have to be equal, we generally will take them to be the same. The positive direction on the x-axis is taken to the right, whereas the upward direction is the positive direction on the y-axis. This of course is merely our arbitrary choice. The two axes with the unit of length are called a ***rectangular cartesian coordinate system*** in the plane (Figure 1.3.1). Cartesius is the Latinized form of the name Descartes.

To each point P in the geometric plane, we associate an ordered pair of real numbers (x, y) by drawing two lines from P, one perpendicular to the x-axis and

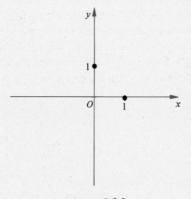

Figure 1.3.1

---

† In representing a set consisting of the elements $\{a, b\}$ it makes no difference whether we list the set as $\{a, b\}$ or $\{b, a\}$. In other words, order is not significant. In contrast, we say (a, b) is an ***ordered pair*** consisting of *first the element a and second the element b*. Thus we say that the ordered pair $(2, 5)$ is different from $(5, 2)$. Therefore, by definition, $(a, b) = (c, d)$ if and only if $a = c$ and $b = d$.

one perpendicular to the y-axis. The values of x and y where these perpendiculars meet the horizontal and vertical axes respectively give us the ordered pair (x, y). We call these values the *x and y coordinates* of P. The x coordinate is also called the *abscissa* and the y coordinate is called the *ordinate* of the point. Each point in the plane corresponds to exactly one ordered pair (x, y) and vice versa. Figure 1.3.2 shows a rectangular cartesian coordinate system with some representative points plotted.

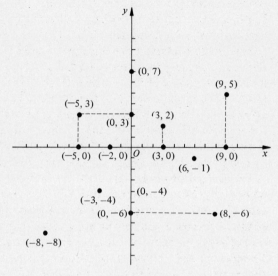

Figure 1.3.2

The x- and y-axes are called the *coordinate axes,* the points on the x-axis have zero ordinates and the points on the y-axis have zero abscissas. All points to the right of the y-axis have positive first coordinates and this set, defined by $x > 0$, is called the *right half-plane.* Similarly, the set $x < 0$ is the *left half-plane;* the set $y > 0$ is the *upper half-plane,* and $y < 0$ is the *lower half-plane.*

The intersection of the upper and right half-planes is the set of positive first and positive second coordinates. This is called the first *quadrant.* The second, third, and fourth quadrants, respectively, are the set $x < 0$ and $y > 0$, $x < 0$ and $y < 0$, and $x > 0$ and $y < 0$. The four quadrants are shown in Figure 1.3.3. Note that the coordinate axes are not in any of the quadrants.

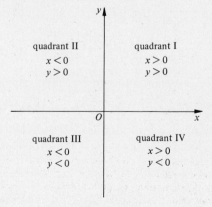

Figure 1.3.3

EXAMPLE 1 Locate all points with abscissas greater than 2 and ordinates less than or equal to -1.

Solution We seek the region of the plane containing all points (x, y) for which $x > 2$ and $y \leq -1$. The region is shown shaded in Figure 1.3.4. Note that the unbroken line (or solid line) is part of the region, whereas the broken line is not part of the region. ●

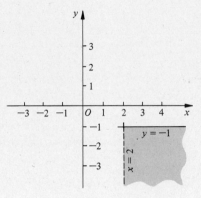

Figure 1.3.4

EXAMPLE 2 Locate all points with abscissas equal to 3 and ordinates greater than -2 and less than 1.

Solution We seek the region of the plane containing all points (x, y) for which $x = 3$ and $-2 < y < 1$. The region is shown in Figure 1.3.5. It is a line segment with endpoints not included (shown by small circles). ●

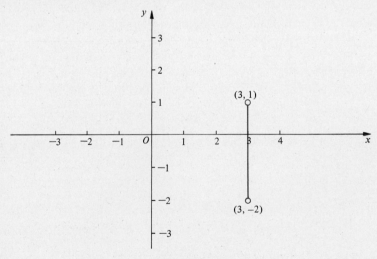

Figure 1.3.5

Consider the equation

$$y = x^3 - 3x \tag{1}$$

where (x, y) is a point in R_2. This is called an equation in R_2. By a solution of such an equation we mean ordered pairs of numbers that satisfy the equation. For example $(0, 0)$, $(1, -2)$, $(2, 2)$, and so on, are such solutions. In fact, there are an unlimited number of solutions. We can simply choose a value for x and find the corresponding value of y. Table 1.3.1 gives several such solutions.

Table 1.3.1

x	0	1	2	3	-1	-2	-3	$\sqrt{3}$	$-\sqrt{3}$
$y = x^3 - 3x$	0	-2	2	18	2	-2	-18	0	0

If we plot the points having as coordinates the number pairs (x, y) that satisfy (1), then, by connecting these points with a smooth curve, we have a sketch of the graph of (1). This is shown in Figure 1.3.6. Note that two different scales were chosen in the x- and y-directions. Furthermore, it should also be noted that ordered pairs of R_2 that do not satisfy (1) are *not* on the curve defined by (1). In general, we define the **graph** of an equation to be the set of all points in R_2 (and only those points) that satisfy the given equation. A graph of an equation is also referred to as the **locus** of the equation.

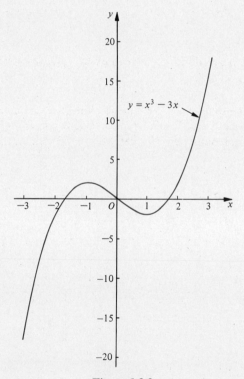

Figure 1.3.6

EXAMPLE 3   Draw a sketch of the graph of the equation

$$x^2 - y^2 = 0 \qquad (2)$$

Solution   Factor the left side of (2) to obtain

$$(x - y)(x + y) = 0 \qquad (3)$$

However, a product of two real numbers is zero if and only if at least one of the factors is zero. Therefore (3) is satisfied if and only if

$$x - y = 0 \quad \text{or} \quad x + y = 0$$

that is,

$$y = x \quad \text{or} \quad y = -x \tag{4}$$

Equations (4) are the equations of two straight lines inclined 45° and 135° with the positive x-direction and intersecting at the origin. The lines are shown in Figure 1.3.7. ●

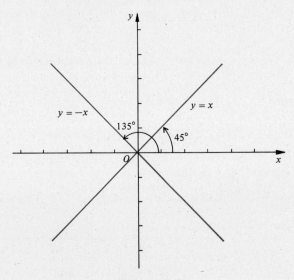

Figure 1.3.7

EXAMPLE 4   Draw a sketch of the graph of the equation

$$y = |x - 1| \tag{5}$$

Solution   From the definition of the absolute value of a number, it follows that

$$\begin{cases} y = x - 1 & \text{if } x - 1 \geq 0 \\ y = -(x - 1) = 1 - x & \text{if } x - 1 < 0 \end{cases} \tag{6}$$

From this information, we set up Table 1.3.2 giving us some values of x and y satisfying (6) and, equivalently, (5). A sketch of (5) is shown in Figure 1.3.8. ●

Table 1.3.2

x	1	2	3	4	0	-1	-2
y	0	1	2	3	1	2	3

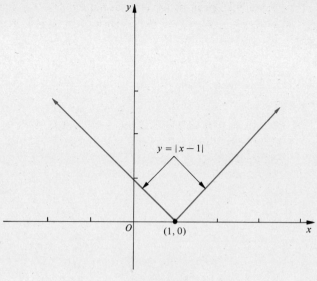

$$y = |x - 1|$$

$(1, 0)$

Figure 1.3.8

EXAMPLE 5   Draw a sketch of the graph of the equation

$$y = 2^{-x} \quad Liz \tag{7}$$

Solution   We set up Table 1.3.3, showing some values of x and y satisfying (7):

Table 1.3.3

x	0	1	2	3	-1	-2	-3
y	1	$\frac{1}{2}$	$\frac{1}{4}$	$\frac{1}{8}$	2	4	8

A sketch of (7) is shown in Figure 1.3.9. Note that $y > 0$ for all values of x and that y approaches zero as x becomes larger positively. ●

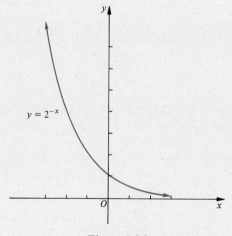

$$y = 2^{-x}$$

Figure 1.3.9

EXAMPLE 6 Sketch the graph of $|x| + |y| = 4$.

Solution From $|y| = 4 - |x|$, the values of x for which y has a value is restricted to $-4 \leq x \leq 4$. If $0 \leq x \leq 4$, then $|x| = x$ and $|y| = 4 - x$ or $y = \pm(4 - x)$. Also if $-4 \leq x < 0$, we have $|x| = -x$ and $|y| = 4 + x$. Hence $y = \pm(4 + x)$ in $-4 \leq x < 0$. Table 1.3.4 shows some of the values of x and y which satisfy the given relation and its graph is the square displayed in Figure 1.3.10. ●

Table 1.3.4

x	0	1	2	3	4	-1	-2	-3	-4
y	±4	±3	±2	±1	0	±3	±2	±1	0

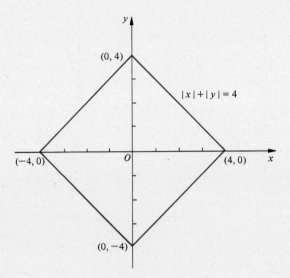

Figure 1.3.10

Exercises 1.3

1. Plot the following points: $(0, 1)$, $(-5, 0)$, $(-3, 4)$, $(6, -2)$.
2. Plot the points: $(4, 3)$, $(4, -3)$, $(-4, 3)$, and $(-4, -3)$.
3. Plot (approximately) the points: $(1, \sqrt{3})$, $(\sqrt{5}, \sqrt{2})$, $(-\pi, \sqrt{6})$, and $\left(\frac{1}{2}, \sqrt[3]{4}\right)$.
4. Describe the location in the xy-plane of the points $P(x, y)$ for which
   (a) $x < -2$
   (b) $x > 0$ and $y \leq 0$
   (c) x and y have opposite signs
   (d) $|y| \leq 3$
   (e) $x = 5$ and $y = -2$
   (f) $|x| + |y| = 0$
   (g) $y = \sqrt{y^2}$
5. Describe the location in the xy-plane of the points $P(x, y)$ for which
   (a) $x = 3$, $y \leq 0$
   (b) $|x| \geq 2$

(c) $xy > 0$
(d) $xy = 0$
(e) $x \leq -2$ and $y \geq 1$
(f) $y < |x|$
(g) $-|x| \leq y \leq |x|$

6. Where do all the points lie if the abscissa plus the ordinate equals -1? Sketch the locus.
7. A line segment with one end at the point $(-5, 3)$ is bisected at the origin. What are the coordinates of the other end of the segment?
8. An isosceles triangle with altitude 10 has its base vertices at $(-3, -4)$ and $(7, -4)$. Find the coordinates of the third vertex if it is in the fourth quadrant.
9. A straight line is drawn through $(0, 0)$ and the point $(-3, -3)$. Find the acute angle between that line and the x-axis. Sketch.
10. A straight line is drawn through the point $A(2, 5)$ and the point $B(4, 10)$. Find $\tan \beta$ where β is the acute angle that this line makes with the horizontal line through A. Sketch.

In Exercises 11 through 30, draw a sketch of the graph of the equation.

11. $y = 5x - 3$
12. $y = 4 + 3x$
13. $x = -1$
14. $y = 1/2$
15. $y^2 = x$
16. $y = 2x^2 - 7$
17. $y = |3x|$
18. $3x^2 + 4y^2 = 0$

19. $16x^2 + 9y^2 = 144$
20. $y = \left|\dfrac{x}{2}\right| + 3$

21. $y = |x - 2|$
22. $y^2 + |x| + 3 = 0$
23. $4x^2 - y^2 = 0$
24. $4x^2 - y^2 = 1$
25. $y = 2x^3$
26. $x^2 = y^3$
27. $y^2 - 3xy + 2x^2 = 0$
28. $6y^2 - xy - 2x^2 = 0$
29. $(x + 2y - 3)(y - x^2) = 0$
30. $y = x^3 - 12x$

1.4

INTERCEPTS, SYMMETRY, EXTENT AND ASYMPTOTES

When graphing equations, it is useful to examine the equation of the curve with regard to

(i) intercepts
(ii) symmetry
(iii) extent
(iv) horizontal and vertical asymptotes

1.4.1 Intercepts

It is helpful to determine where the curve or locus crosses the x- and y-axes. A point at which the curve crosses the x-axis is called an *x-intercept.* Similarly, a point at which the curve crosses the y-axis is called a *y-intercept.* Of course it may happen that the curve does not cross a given axis (or either axis). To find the x-intercepts, let $y = 0$ in the equation of the curve and solve for x; to find the y-intercepts, let $x = 0$ and solve for y.

EXAMPLE 1   Find the intercepts of the graph of $y = x^2 - 4$ (Figure 1.4.1).

Solution   When $x = 0$, we find the y-intercept is $(0, -4)$. To find where the curve crosses the x-axis, set $y = 0$ and find that $x^2 - 4 = 0$ which yields $x = \pm 2$. Thus $(\pm 2, 0)$ are the x-intercepts.   ●

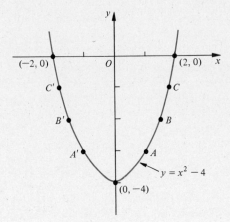

Figure 1.4.1

1.4.2 *Symmetry*

By definition, two points A and A' are **symmetric** *with respect to a given line L* if the line is the perpendicular bisector of the line segment connecting the two points. In Figure 1.4.2 points A and A' are symmetric with respect to the line L. Also, a curve is said to be **symmetric** *with respect to a given line* if, corresponding to each point of the curve, there is a symmetric point with respect to the line, which also lies on the curve.

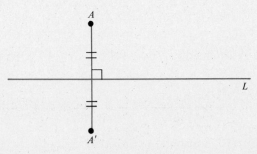

Figure 1.4.2

In Figure 1.4.1 the curve $y = x^2 - 4$ is symmetric with respect to the y-axis. This means that if (a, b) lies on the curve so does $(-a, b)$. Thus a test for symmetry with respect to the y-axis is to replace x by $-x$ and, if the same equation results, we have symmetry with respect to the y-axis. Similarly, if $-y$ is substituted for y without altering the equation, the curve is symmetric with respect to the x-axis.

Thus $y = x^2 - 4$ is symmetric with respect to the y-axis since $(-x)^2 - 4 = x^2 - 4 = y$. In fact points A, A'; B, B' and C, C' are symmetric points with respect to the y-axis in Figure 1.4.1.

EXAMPLE 2   Test the locus of the equation $x = 3y^2 - 5$ for symmetry with respect to the coordinate axes.

Solution   Substitution of $-x$ for x yields $-x = 3y^2 - 5$, which is not the same as the original equation. Therefore, the curve is not symmetric with respect to the y-axis. If we substitute $-y$ for y, there results $x = 3(-y)^2 - 5$ or $x = 3y^2 - 5$. Since the equation is unchanged, the curve is symmetric with respect to the x-axis.   ●

Two points are said to be symmetric with respect to a point if the latter point is the midpoint of the line segment joining the two given points. In Figure 1.4.3 points A and A' are symmetric with respect to P if P is the midpoint of segment AA'. In particular, a locus is symmetric with respect to the origin if, whenever (a, b) lies on the locus, so does $(-a, -b)$. Thus, to test for symmetry with respect to the origin, replace x by $-x$ and y by $-y$. If the equation is unaltered by this replacement, then the resulting curve is symmetric with respect to the origin.

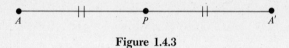

Figure 1.4.3

EXAMPLE 3   Test the locus of the equation $xy = 4$ for symmetry with respect to the origin.

Solution   Substitution of $-x$ for x and $-y$ for y yields $(-x)(-y) = 4$ or $xy = 4$. Therefore the curve governed by $xy = 4$ is symmetric with respect to the origin (Figure 1.4.4).   ●

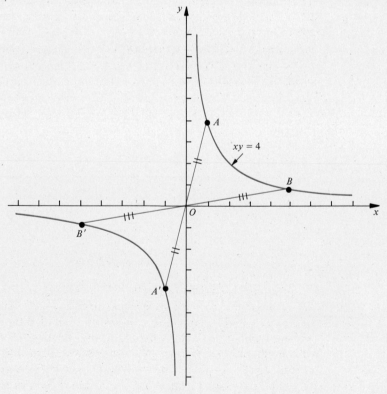

Figure 1.4.4

A little elementary plane geometry yields the fact that the two points (a, b) and (b, a) are located symmetrically with respect to the line $y = x$ for arbitrary a and b. This means that, if x and y are interchanged (that is, x is replaced by y and y is replaced by x) and if the equation relating x and y is unchanged, then we have symmetry with respect to the line $y = x$. Thus the curve in Example 3 is symmetric with respect to the line $y = x$ since $xy = 4$ becomes $yx = 4$ as a result of interchanging x and y. However, the curve described by $y^2 = x^3$ would not have this symmetry because $x^2 = y^3$ (which results by interchanging x and y) is not the same equation as $y^2 = x^3$.

If a curve possesses symmetry about an axis, say the x-axis, then only the portion of the curve above it must be plotted in detail. The remainder of the graph is obtained as the mirror image with respect to the x-axis of the part above the x-axis.

1.4.3 *Extent*

Only real values of the variables x and y are considered in determining the points (x, y) whose coordinates satisfy a given equation. Thus we are interested in the values of x that determine one or more values of y, and, similarly, the values of y that yield one or more values of x.

EXAMPLE 4   Find the extents in the x- and y-directions for the graph of $x^2 + y^2 = 25$.

Solution   When this equation is solved for y, we obtain

$$y = \pm \sqrt{25 - x^2}$$

The quantity under the square root sign is nonnegative if and only if $25 - x^2 \geq 0$ or, equivalently, $|x| \leq 5$. Thus the extent of the curve in the x-direction is limited to $-5 \leq x \leq 5$. By symmetry about the line $y = x$, the curve is also limited in the y-direction to $-5 \leq y \leq 5$ (see Figure 1.4.5).   ●

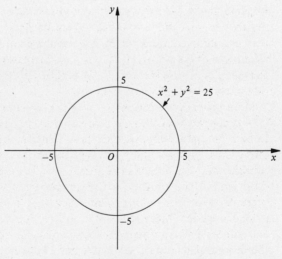

Figure 1.4.5

EXAMPLE 5   Find the extents in the x- and y-directions for the graph of $y^2 = x^3$.

Solution $y^2 \geq 0$ so that $x^3 \geq 0$ or $x \geq 0$. Also $x = y^{2/3}$ and there is no restriction on y. The curve is in the right half-plane and also includes the origin which is on the boundary of that half-plane (see Figure 1.4.6).   ●

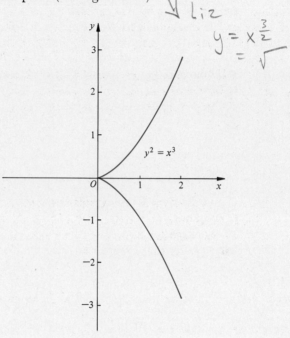

Figure 1.4.6

1.4.4   *Horizontal and Vertical Asymptotes*

The graph of the equation of Example 3, $xy = 4$ (Figure 1.4.4), is a very simple illustration of asymptotes. Note that y is not defined when $x = 0$. However, as x approaches zero through positive values, y becomes unbounded positively. Similarly, as x approaches zero through negative values, y becomes unbounded negatively. Also, the distance of the points $P(x, y)$ from the y-axis is approaching zero as y becomes numerically greater and greater. We call the y-axis a vertical asymptote of the curve. Similarly, as $|x|$ increases indefinitely, the distance from the curve $xy = 4$ to the x-axis, that is, $|y|$ is approaching zero. The x-axis is called a horizontal asymptote of the curve.

More generally, a line $x = c$ is said to be a ***vertical asymptote*** of a curve if $|x - c|$ approaches zero as $|y|$ increases indefinitely. Similarly, a line $y = b$ is said to be a ***horizontal asymptote*** if $|y - b|$ approaches zero as $|x|$ increases indefinitely. In Chapter 2, we shall explain more carefully what we mean by "approaches zero" and "increases indefinitely." In the meantime, we shall deal with these expressions intuitively. Curves can also have oblique asymptotes, but at present we shall not concern ourselves with these.

One procedure that may be used to find the horizontal and vertical asymptotes is as follows. Solve for y in terms of x (if possible) and find the values of x for which the denominator is zero while the numerator is not zero (because, when x gets close to one of these values, $|y|$ increases indefinitely). If c is such a value, then

$x = c$ is a vertical asymptote. Similarly, to locate the horizontal asymptotes, solve for x in terms of y (if possible) and find those values of y that make the denominator zero while the numerator is not zero. Then if $y = b$ is such a value, the line $y = b$ is a horizontal asymptote. Applying this method to Example 3, we have $y = \dfrac{4}{x}$ and $x = \dfrac{4}{y}$, from which we see that $x = 0$ is a vertical asymptote and $y = 0$ is a horizontal asymptote.

EXAMPLE 6   Find the asymptotes and sketch the graph of

$$y = \frac{2x + 1}{x - 1}$$

Solution   When $x = 1$, the denominator (but not the numerator) is zero. Then $x = 1$ is a vertical asymptote.

Solving for x, we have $x = \dfrac{y + 1}{y - 2}$ from which we see that $y = 2$ is a horizontal asymptote.

The intercepts are $(0, -1)$ and $(-\frac{1}{2}, 0)$. We plot these and a few other points, such as $(2, 5)$ and $(3, \frac{7}{2})$, and use the information about the asymptotes. This gives the curve in Figure 1.4.7.   ●

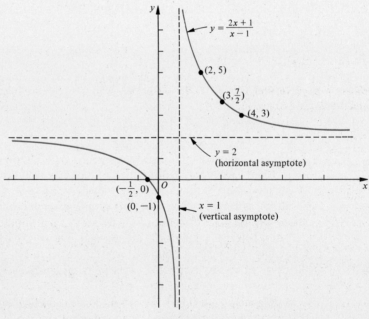

Figure 1.4.7

EXAMPLE 7   Discuss $x^2 y + 4y = 8$ as to intercepts, symmetry, extent, and asymptotes. Plot the curve.

Solution   Solving for y and x, respectively, there results

$$y = \frac{8}{x^2 + 4} \qquad \text{and} \qquad x = \pm 2 \sqrt{\frac{2 - y}{y}}$$

INTERCEPTS: No x-intercept because, when we set $y = 0$, we obt
from the given equation. The y-intercept is 2.

SYMMETRY: With respect to the y-axis.

EXTENT: The extent in x is clearly $-\infty < x < \infty$, whereas, for the ex
y, we must have $\dfrac{2 - y}{y} \geq 0$ which yields $0 < y \leq 2$.

ASYMPTOTES: There is no vertical asymptote since $x^2 + 4 = 0$ has no real
roots. There is a horizontal asymptote, namely, $y = 0$ or the x-axis.

This curve is bell shaped and is a special case of a more general curve governed
by $x^2 y + 4a^2 y = 8a^3$ known as the **witch of Agnesi** after Maria Gaetana Agnesi
(1718–1799). The curve for our special case $a = 1$ is shown in Figure 1.4.8.

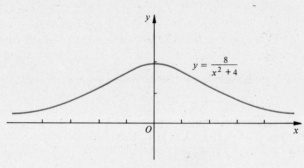

Figure 1.4.8

Exercises 1.4

In Exercises 1 through 16, find in each case the intercepts, symmetries, extent in x and y, and the vertical and
horizontal asymptotes. Sketch the graph.

1. $y = 3x + 1$
2. $y = 2x^2 - 1$
3. $y^2 + 2x = 0$
4. $y = 4 - x^2$
5. $x^2 y = 5$
6. $xy = 2$
7. $x^2 + y^2 = 9$
8. $y^2 = x + 3$
9. $y^2 = x^2 + 3$
10. $2x^2 + 3y^2 = 6$
11. $xy = x + 3$
12. $x^2 y + 2y = 1$
13. $x^2 y = x - 5$ (*Hint:* To find extent in y-direction, solve the quadratic equation for x in terms of y.)
14. $y(x - 1)(x - 3) = 2$ (*Hint:* To find extent in y-direction, solve $x^2 - 4x + 3 = \dfrac{2}{y}$ $(y \neq 0)$ by complet-
    ing the squares.)
15. $x^4 + y^4 = 1$
16. $x^2 = \dfrac{y^2 + 1}{y^2 - 1}$
17. Show that if the graph of a curve has any two of the three types of symmetry, namely, about the x-axis,
    about the y-axis and about the origin, then it must have the third. (*Hint:* Suppose for example that the
    curve is symmetric about both axes. Then consider any point (a, b) on the curve. It follows from
    symmetry with respect to the x-axis that $(a, -b)$ also lies on the curve, and so on.)
18. Show that if two perpendicular lines are lines of symmetry of a given curve, then their point of
    intersection is a point of symmetry.
19. Propose an equation for a curve C that has the following properties:
    (a) has symmetry with respect to the y-axis, intercepts at $(\pm 2, 0)$, $(\pm 3, 0)$, $(0, 4)$ with no vertical or
        horizontal asymptotes.
    (b) has symmetry with respect to the origin, intercepts at $(0, 0)$, $(\pm 3, 0)$ with no vertical or horizontal
        asymptotes.

20. Propose an equation for a curve C that has the following properties:

(a) has symmetry with respect to the x-axis, intercepts at $(0, 0)$, $(5, 0)$, does not exist for $0 < x < 5$, and possesses no vertical or horizontal asymptotes.

(b) has intercepts at $(-2, 0)$ and at $(3, 0)$ and lines $x = 5$ and $y = -1$ as asymptotes.

1.5   DISTANCE AND MIDPOINT FORMULAS

Let P_1 be the point $P_1(x_1, b)$ and P_2 the point $P_2(x_2, b)$ then the **directed distance** from P_1 to P_2, denoted by $\overline{P_1P_2}$, is defined as $x_2 - x_1$. Here the ordinates of P_1 and P_2 are the same while the abscissas are generally different. Examples are shown in Figure 1.5.1. Note that $\overline{P_1P_2}$ is positive if P_2 is to the right of P_1, whereas $\overline{P_1P_2}$ is negative if P_2 is to the left of P_1.

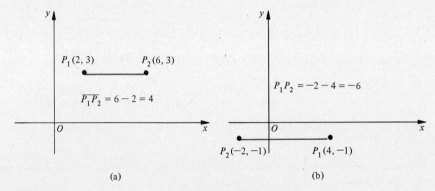

Figure 1.5.1

Similarly, if P_1 is the point $P_1(a, y_1)$ and P_2 is the point $P_2(a, y_2)$, the **directed distance** from P_1 to P_2 is denoted by $\overline{P_1P_2} = y_2 - y_1$.

For example, if $P_1(2, 1)$ and $P_2(2, 4)$ then $\overline{P_1P_2} = 4 - 1 = 3$, whereas, if $P_1(0, 3)$ and $P_2(0, -2)$ then $\overline{P_1P_2} = -2 - 3 = -5$. The number $\overline{P_1P_2}$ is positive if P_2 is above P_1 (with the same abscissa), whereas $\overline{P_1P_2}$ is negative if P_2 is below P_1 (with the same abscissa). The examples in this paragraph are illustrated in Figure 1.5.2.

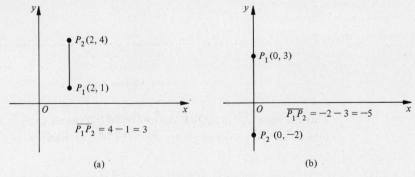

Figure 1.5.2

Consider a directed line segment $\overline{P_1 P_2}$ along the x-axis or parallel to it, say $P_1(x_1, b)$, $P_2(x_2, b)$. Then $\overline{P_1 P_2}$ may be thought of as the path traversed by a moving particle that starts at P_1 and travels in one direction to P_2. In such a motion, the abscissa of the particle changes from x_1 to x_2. We denote this change by the symbol

$$\Delta x = x_2 - x_1$$

read "delta x." It is important to note that Δx denotes the "change in x" or "increment in x." It does *not* mean a number multiplied by x. Thus if $P_1(3, 6)$ and $P_2(-2, 6)$ then

$$\overline{P_1 P_2} = \Delta x = -2 - 3 = -5$$

Analogously, if $P_1(a, y_1)$ and $P_2(a, y_2)$, we define the change in y by

$$\overline{P_1 P_2} = \Delta y = y_2 - y_1$$

Now let $P_1(x_1, y_1)$ and $P_2(x_2, y_2)$ be any two points in the xy-plane. We seek a formula for finding the nonnegative distance between these two points. If the same unit of measurement is used on both axes, all distances in the plane may be expressed in terms of this fundamental unit. Application of the theorem of Pythagoras to the right triangle $P_1 Q P_2$ in Figure 1.5.3 yields

$$|\overline{P_1 P_2}| = \sqrt{(\Delta x)^2 + (\Delta y)^2}$$

or

$$|\overline{P_1 P_2}| = \sqrt{(x_2 - x_1)^2 + (y_2 - y_1)^2} \qquad (1)$$

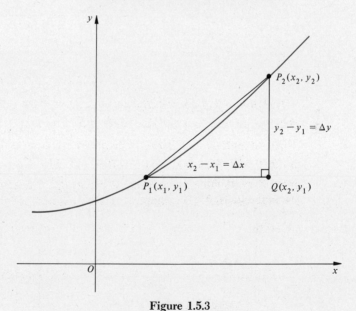

Figure 1.5.3

This distance $|\overline{P_1 P_2}|$ is called the **undirected distance** from P_1 to P_2. The examples that follow illustrate some applications of this distance formula.

EXAMPLE 1   Find the distance between the points $P_1(4, 3)$ and $P_2(-6, 5)$.

Solution Substitution in the distance formula between two points yields

$$|\overline{P_1P_2}| = \sqrt{(-6-4)^2 + (5-3)^2} = \sqrt{100+4} = \sqrt{104} \qquad \bullet$$

EXAMPLE 2 The point $P_1(-1, 1)$ is 5 units away from a second point whose y-coordinate is 4. Locate the point P_2.

Solution The point P_2 will have coordinates $(x_2, 4)$. Application of the distance formula yields

$$5 = \sqrt{(x_2 + 1)^2 + (4-1)^2}$$

To solve for x_2, we eliminate the square root by squaring both sides. Therefore

$$25 = (x_2 + 1)^2 + 3^2 \qquad \text{or} \qquad 16 = (x_2 + 1)^2$$

Thus $x_2 + 1 = \pm 4$, which yields

$$x_2 = 3 \qquad \text{and} \qquad x_2 = -5$$

It is easily verified that both solutions $P_2(3, 4)$ and $P_2(-5, 4)$ are valid. $\qquad \bullet$

EXAMPLE 3 If a point $P(x, y)$ is such that it is the same distance from $(-1, 2)$ and $(-3, 4)$, find an equation that the coordinates of P must satisfy. Interpret the result geometrically.

Solution We equate the two distances:

$$\sqrt{(x+1)^2 + (y-2)^2} = \sqrt{(x+3)^2 + (y-4)^2}$$

Square both sides and expand the indicated squares to obtain

$$x^2 + 2x + 1 + y^2 - 4y + 4 = x^2 + 6x + 9 + y^2 - 8y + 16$$

Simplification yields

$$4x - 4y + 20 = 0$$

or

$$x - y + 5 = 0$$

is the required equation. This is the equation of the line that is the perpendicular bisector of the segment joining the two given points. This result may also be obtained from the fact that the locus of points equidistant from two points must lie on the perpendicular bisector of the line segment connecting the points. $\qquad \bullet$

EXAMPLE 4 Show that the points $P_1(1, -1)$, $P_2(4, 3)$ and $P_3(9, -7)$ are the vertices of a right triangle.

Solution The theorem of Pythagoras states that, if triangle ABC is a right triangle with right angle at C, the square of the hypotenuse equals the sum of the squares of its legs; that is, $c^2 = a^2 + b^2$. The converse states that if $c^2 = a^2 + b^2$ then triangle ABC is a right triangle (angle $C = 90°$). We shall employ the converse for this solution.

$$|\overline{P_1P_2}|^2 = (4-1)^2 + (3-(-1))^2 = 9 + 16 = 25$$
$$|\overline{P_2P_3}|^2 = (9-4)^2 + (-7-3)^2 = 25 + 100 = 125$$
$$|\overline{P_1P_3}|^2 = (9-1)^2 + (-7+1)^2 = 64 + 36 = 100$$

Thus $|\overline{P_2P_3}|^2 = |\overline{P_1P_2}|^2 + |\overline{P_1P_3}|^2$ and the result follows. ●

EXAMPLE 5   Prove analytically that the lengths of the diagonals of a rectangle are equal.

Solution   Draw a general rectangle (Figure 1.5.4). For convenience (in order to reduce the analytic work) the coordinate axes are chosen so that $O(0, 0)$ is at one vertex and the x- and y-axes are along two adjacent sides. This may be done because the sides of a rectangle meet at right angles. Let a and b denote the lengths of the adjacent sides. The coordinates of the four vertices $OABC$ are shown in the figure. From the distance formula

$$|\overline{OB}| = \sqrt{(a-0)^2 + (b-0)^2} = \sqrt{a^2 + b^2}$$
$$|\overline{AC}| = \sqrt{(0-a)^2 + (b-0)^2} = \sqrt{a^2 + b^2}$$

Therefore,

$$|\overline{OB}| = |\overline{AC}|$$

That is, the diagonals of a rectangle are equal. ●

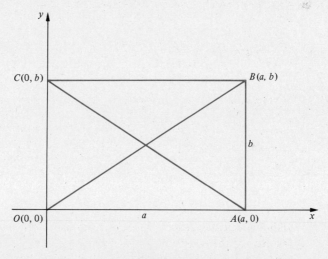

Figure 1.5.4

Let $P_1(x_1, y_1)$ and $P_2(x_2, y_2)$ be the endpoints of a line segment (Figure 1.5.5). Let $M(\overline{x}, \overline{y})$ denote the midpoint of the line segment P_1P_2. Projecting parallels to the coordinate axes as shown in the figure, we observe that the triangle P_1QM and MRP_2 are congruent. Therefore the corresponding sides are equal, that is, $|\overline{P_1Q}| = |\overline{MR}|$. This implies that $\overline{x} - x_1 = x_2 - \overline{x}$ or

$$\overline{x} = \frac{x_1 + x_2}{2} \qquad (2)$$

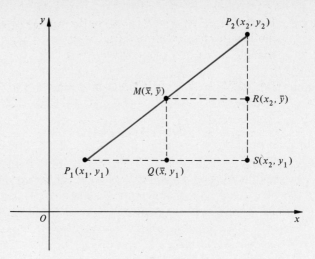

Figure 1.5.5

Similarly, $|\overline{MQ}| = |\overline{P_2R}|$, which yields

$$\bar{y} = \frac{y_1 + y_2}{2} \tag{3}$$

Therefore the coordinates of the midpoint M are, respectively, the arithmetic means of the abscissas and ordinates of the endpoints.

Although the above derivation was for the instance when $x_2 > x_1$ and $y_2 > y_1$, a similar argument may be employed for other orderings to yield (2) and (3). It is recommended that the reader verify this assertion.

EXAMPLE 6 Find the coordinates of the midpoint of the line segment joining $P_1(3, -1)$ and $P_2(-2, 5)$.

Solution From (2) and (3),

$$\bar{x} = \frac{3 + (-2)}{2} = \frac{1}{2} \qquad \bar{y} = \frac{-1 + 5}{2} = \frac{4}{2} = 2$$

Thus the coordinates of M are $(\frac{1}{2}, 2)$. ●

EXAMPLE 7 Find the length of the line segment joining the point $P_1(3, -5)$ to the midpoint of the line segment between the points $P_2(4, 2)$ and $P_3(0, 6)$.

Solution The midpoint of the segment between P_2 and P_3 is at

$$\bar{x} = \frac{4 + 0}{2} = 2 \qquad \bar{y} = \frac{2 + 6}{2} = 4$$

and the distance between P_1 and this point is, from (1),

$$d = \sqrt{(2 - 3)^2 + (4 - (-5))^2} = \sqrt{(-1)^2 + 9^2} = \sqrt{82}$$ ●

EXAMPLE 8 Let $P_1(x_1, y_1)$, $P_2(x_2, y_2)$, $P_3(x_3, y_3)$ be the vertices of an arbitrary triangle. Show that the length of the line segment joining the midpoints of any two sides of this triangle is one half the length of the third side.

Solution Let $M(\bar{x}, \bar{y})$, $N(x^\star, y^\star)$ be the midpoints of the line segments P_1P_2 and P_2P_3, respectively. Therefore from (2) and (3), we have

$$\bar{x} = \frac{x_1 + x_2}{2} \qquad \bar{y} = \frac{y_1 + y_2}{2} \qquad x^\star = \frac{x_2 + x_3}{2} \qquad y^\star = \frac{y_2 + y_3}{2}$$

Thus

$$|\overline{MN}| = \sqrt{\left(\frac{x_2 + x_3}{2} - \frac{x_1 + x_2}{2}\right)^2 + \left(\frac{y_2 + y_3}{2} - \frac{y_1 + y_2}{2}\right)^2}$$

$$= \sqrt{\left(\frac{x_3 - x_1}{2}\right)^2 + \left(\frac{y_3 - y_1}{2}\right)^2}$$

$$= \frac{1}{2}\sqrt{(x_3 - x_1)^2 + (y_3 - y_1)^2} = \frac{1}{2}|\overline{P_1P_3}|$$

The results for the other pairs of midpoints follows analogously. ●

EXAMPLE 9 Prove that the diagonals of a parallelogram bisect each other.

Solution Draw a general parallelogram. For convenience we take the origin at one vertex, the x-axis along one side. Furthermore, when we label the coordinates, we make use of the fact that the opposite sides of a parallelogram are equal and parallel.
Thus let $OP_1P_2P_3$ be the general parallelogram (Figure 1.5.6). From (2) and (3) the midpoint of diagonal OP_2 has the coordinates $\left(\frac{a + b}{2}, \frac{c}{2}\right)$ and the midpoint of diagonal P_1P_3 has the coordinates $\left(\frac{a + b}{2}, \frac{c}{2}\right)$. Thus the midpoints coincide and we conclude that the diagonals of a parallelogram bisect each other. ●

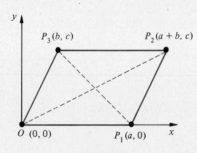

Figure 1.5.6

Exercises 1.5

In Exercises 1 through 6, find the distance d between each pair of points.

1. $P_1(2, -1)$, $P_2(4, 3)$
3. $P_1(-2, -3)$, $P_2(-4, -5)$
5. $P_1(x_1, 0)$, $P_2(x_1 + b, 0)$

2. $P_1(0, 3)$, $P_2(4, 0)$
4. $P_1(a, b)$, $P_2(m, n)$
6. $P_1(a, y_1)$, $P_2(a, y_1 + k)$

7. Prove that the triangle with vertices $P_1(1, 2)$, $P_2(5, 4)$ and $P_3(4, 1)$ is an isosceles triangle.

8. Prove that the triangle with vertices $P_1(1, 2)$, $P_2(5, -2)$ and $P_3(11, 4)$ is a right triangle and find its area.

9. Prove that the points $A(-2, -7)$, $B(3, 3)$ and $C(7, 11)$ lie on a straight line. [*Hint:* Apply the distance formula (1) three times.]

10. Find the lengths of the medians of the triangle with vertices at $P_1(-3, -2)$, $P_2(-1, 2)$, and $P_3(5, 4)$.

11. Show that the quadrilateral with vertices at $D(2, 3)$, $A(-1, 0)$, $C(5, 0)$, and $B(2, -3)$ is a square.

12. The four points $A(2, 2)$, $B(3, 1)$, $C(1, 0)$, $D(0, 2)$ form the vertices of a quadrilateral. Show that the midpoints of the sides are the vertices of a parallelogram.

13. Generalize the previous exercise by proving analytically that the line segments joining the midpoints of the sides of an arbitrary quadrilateral form a parallelogram.

14. Find the coordinates of the two points that divide the line segment from $P_1(-4, 2)$ to $P_2(6, 8)$ into three equal parts.

15. If r and s are positive integers, prove that the coordinates of the point $P(x^\star, y^\star)$, which divides the line segment P_1P_2 in the ratio $\dfrac{r}{s}$ (that is, $|\overline{P_1P}|/|\overline{PP_2}| = r/s$), are given by

$$x^\star = \frac{rx_2 + sx_1}{r + s} \qquad y^\star = \frac{ry_2 + sy_1}{r + s}$$

where $P_1(x_1, y_1)$ and $P_2(x_2, y_2)$.

16. If a point $P(x, y)$ is such that its distance from $P_1(2, 1)$ is three times its distance from $P_2(0, 4)$, find an equation that the coordinates of P must satisfy.

17. Given the points $P_1(0, 0)$ and $P_2(a, 0)$, find a third point $P_3(b, c)$ such that $|\overline{P_1P_2}| = |\overline{P_2P_3}| = |\overline{P_3P_1}|$. In other words, find the coordinates of P_3 so that the resulting triangle with vertices P_1, P_2, and P_3 is equilateral.

18. Verify the following result first developed by Euclid. Show that the three points $P_1(x_1, y_1)$, $P_2(x_2, y_2)$, and $P_3(x_3, y_3)$, where

$$x_3 = \frac{1}{2}(x_1 + x_2) - \frac{\sqrt{3}}{2}(y_2 - y_1)$$

$$y_3 = \frac{1}{2}(y_1 + y_2) + \frac{\sqrt{3}}{2}(x_2 - x_1)$$

are the vertices of an equilateral triangle.

19. Prove analytically that the diagonals of a parallelogram are equal if and only if the parallelogram is a rectangle.

20. Prove analytically that the midpoint of the hypotenuse of any right triangle is equidistant from each of the three vertices.

21. Prove analytically that the sum of the squares of the four sides of a parallelogram is equal to the sum of the squares of the two diagonals.

22. Let $A(0, 0)$, $B(c, 0)$, and $C(b \cos A, b \sin A)$ denote the vertices of triangle ABC. Utilize the distance formula to derive the *law of cosines:* $a^2 = b^2 + c^2 - 2bc(\cos A)$, where a, b, c are the lengths of the sides of the triangle and side a is opposite angle A.

1.6 THE STRAIGHT LINE

1.6.1 *Inclination and Slope of a Line*

Consider a line that intersects the x-axis; that is, the line is not parallel to the x-axis. Choose the "upward" direction along this line to be the positive direction. The **inclination** of a line is defined to be the (directed) angle measured counterclockwise from the positive direction along the x-axis (which is taken to be the initial side of the angle) to the positive direction of the given line (which is the terminal side of the angle). Figure 1.6.1 shows two angles of inclination—one acute and the other obtuse.

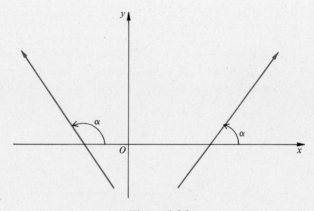

Figure 1.6.1

If the given line is parallel to the x-axis, the inclination is defined to be zero. With these conventions we observe that the permissible values of α are $0° \leq \alpha < 180°$.

The *slope* of a line is defined to be the tangent of the inclination, providing that the units of measurement are the same on both axes. Denoting the slope by the letter m we have

$$m = \tan \alpha$$

If α is an acute angle, the slope is positive; and if α is obtuse, the slope is negative. If $\alpha = 0°$, that is, the line is parallel to the x-axis and the slope $m = 0$. If $\alpha = 90°$, that is, the line is perpendicular to the x-axis (or parallel to the y-axis), then the slope is undefined because $\tan 90°$ is not defined.

Consider the line passing through the two points $P_1(x_1, y_1)$ and $P_2(x_2, y_2)$ in Figure 1.6.2. The slope of the line is given by

$$m = \tan \alpha = \frac{y_2 - y_1}{x_2 - x_1} = \frac{\Delta y}{\Delta x} \tag{1}$$

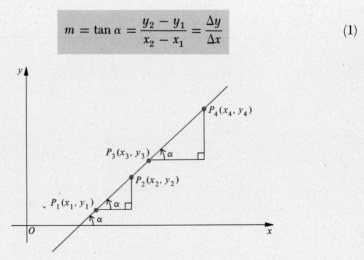

Figure 1.6.2

The formula $m = \dfrac{y_2 - y_1}{x_2 - x_1}$ is valid for any choice of the two distinct points on the line. This follows from similar triangles as illustrated in Figure 1.6.2 since, in that figure

$$m = \frac{y_2 - y_1}{x_2 - x_1} = \frac{y_4 - y_3}{x_4 - x_3}$$

EXAMPLE 1   Find the slope and inclination of the line through $P_1(4, -3)$ and $P_2(2, -1)$ as shown in Figure 1.6.3.

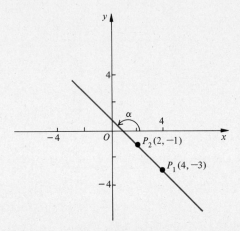

Figure 1.6.3

Solution   From (1),

$$m = \tan \alpha = \frac{-1 - (-3)}{2 - 4} = \frac{-1 + 3}{2 - 4} = \frac{2}{-2} = -1$$

and $\alpha = 135°$ $\left(\text{or } \dfrac{3\pi}{4} \text{ radians}\right)$.   ●

1.6.2 Parallel and Perpendicular Lines

If L_1 and L_2 are two lines that intersect the x-axis and if L_1 and L_2 are parallel, then the angles of inclination and slopes must be equal, that is

$$\alpha_1 = \alpha_2 \quad \text{and} \quad m_1 = m_2 \tag{2}$$

This follows from "corresponding angles of parallel lines are equal." Conversely, if $m_1 = m_2$, then $\tan \alpha_1 = \tan \alpha_2$, which implies that $\alpha_1 = \alpha_2$ because the inclination angles are between $0°$ and $180°$.

Next, we wish to determine a relationship between m_1 and m_2 if the lines are perpendicular: If L_1 and L_2 are perpendicular to each other, either

$$\alpha_2 = \alpha_1 + 90° \quad \text{or} \quad \alpha_1 = \alpha_2 + 90°$$

depending upon which is larger. In Figure 1.6.4, $\alpha_2 > \alpha_1 > 0$ so that

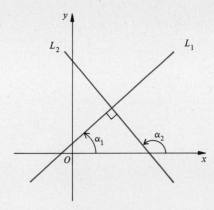

Figure 1.6.4

$\alpha_2 = \alpha_1 + 90°$. Furthermore, we recall from trigonometry† that if two angles α_1 and α_2 differ by 90° then

$$\tan \alpha_2 = -\frac{1}{\tan \alpha_1}$$

(providing $\tan \alpha_1 \neq 0$).

The last formula, in terms of slopes, states that

$$m_2 = -\frac{1}{m_1} \tag{3}$$

In words: *if two lines are perpendicular, their slopes are negative reciprocals of each other; and, conversely (because the steps are reversible), if their slopes are negative reciprocals of each other, the two lines are perpendicular.* It should be noted that this result does not apply if the two perpendicular lines are parallel to the coordinate axes.

EXAMPLE 2   Show that the line through $P_1(2, 5)$ and $Q_1(4, 8)$ is parallel to the line through $P_2(-1, -3)$ and $Q_2(1, 0)$.

Solution   The slope of the line through P_1 and Q_1 is from (1)

$$m_1 = \frac{8 - 5}{4 - 2} = \frac{3}{2}$$

Similarly, the line through P_2 and Q_2 has slope

$$m_2 = \frac{0 - (-3)}{1 - (-1)} = \frac{3}{2}$$

Since the slopes are the same, the lines are parallel.   ●

EXAMPLE 3   Determine whether or not the three points $P(1, -1)$, $Q(3, -5)$, and $R(5, -10)$ lie on the same line.

---

† A detailed review of trigonometry is given in Section 9.1.

Solution The line through P and Q has slope

$$m_1 = \frac{-5 - (-1)}{3 - 1} = \frac{-4}{2} = -2$$

If point R is on the line, the line joining Q and R must have the same slope. The slope of the line joining Q and R is

$$m_2 = \frac{-10 - (-5)}{5 - 3} = \frac{-5}{2}$$

and this is different from m_1. Therefore P, Q, and R do *not* lie on the same line.

        ●

EXAMPLE 4 Is the line through the points $P_1(1, 4)$ and $Q_1(8, 6)$ perpendicular to the line through the points $P_2(-3, 2)$ and $Q_2(-1, -5)$?

Solution The line through P_1 and Q_1 has slope

$$m_1 = \frac{6 - 4}{8 - 1} = \frac{2}{7}$$

The line through P_2 and Q_2 has slope

$$m_2 = \frac{-5 - 2}{-1 - (-3)} = \frac{-7}{2}$$

The slopes are negative reciprocals of each other and the lines are perpendicular.

        ●

1.6.3 Equations of a Line

Let us first note that, if the slope of a line is undefined, the line is vertical and all x-values are the same. Therefore, it has an equation of the form

$$x = c \tag{4}$$

We now seek the equation of a line passing through a given point $P_1(x_1, y_1)$ with a given slope m. Let $Q(x, y)$ be any point on the line (Figure 1.6.5). The slope of the line connecting P_1 and Q is $\dfrac{y - y_1}{x - x_1}$. However, this must be equal to m, giving us the **point-slope formula** for a straight line:

$$\frac{y - y_1}{x - x_1} = m \tag{5}$$

or

$$y - y_1 = m(x - x_1) \tag{6}$$

EXAMPLE 5 Find the equation of the line passing through the point $(-3, 4)$ and having the slope $-\frac{2}{5}$.

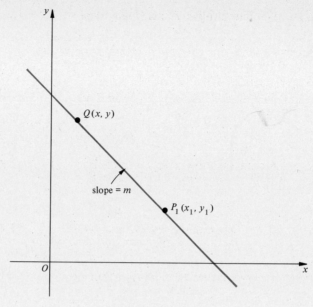

Figure 1.6.5

Solution   Substitution of the known data for x_1, y_1, and m into Equation (6) yields

$$y - 4 = -\tfrac{2}{5}(x - (-3))$$

or

$$5y - 20 = -2x - 6$$

and finally

$$2x + 5y - 14 = 0$$ ●

The result (6) yields another result almost immediately. Suppose that we have a straight line passing through two points, say, $P_1(x_1, y_1)$ and $P_2(x_2, y_2)$. Then $m = \dfrac{y_2 - y_1}{x_2 - x_1}$ and (6) yield the *two-point form* for the equation of a line

$$y - y_1 = \frac{y_2 - y_1}{x_2 - x_1}(x - x_1) \qquad (7)$$

Although we may use (7) directly, we should recognize that (7) is really (6) with the expression for the slope substituted into it.

A useful variation of the point-slope formula is obtained by the introduction of the y-intercept. Every line that is not parallel to the y-axis must intersect it at a point, say, $(0, b)$. By definition the number b is the y-intercept. Suppose then that a line with slope m has y-intercept b. In order to find its equation, we substitute into the point-slope formula (6) to obtain

$$y - b = m(x - 0)$$

or

$$y = mx + b \qquad (8)$$

Equation (8) is the *slope-intercept form* for the equation of a straight line.

EXAMPLE 6   A line has slope -2 and y-intercept of 5. Find its equation.

Solution   Substitution of $m = -2$ and $b = 5$ into the slope-intercept form (8) yields

$$y = -2x + 5 \qquad \bullet$$

By this time, the reader may have observed that the equations of lines could be expressed in the more general form

$$Ax + By + C = 0 \qquad (9)$$

where A, B, and C are real constants. Equation (9) is the most general equation of the first degree in x and y. In fact, we shall establish the theorem:

Theorem 1   Every equation of the form

$$Ax + By + C = 0$$

is the equation of a straight line, provided that A and B are not both zero.

Proof   **Case (1)**: $B = 0$, $A \neq 0$ and thus

$$x = -\frac{C}{A}$$

which is the equation of a line parallel to the y-axis.

Case (2): $B \neq 0$

Division by B and solving for y yields

$$y = -\frac{A}{B}x - \frac{C}{B}$$

which from the slope-intercept form (8) is a line with slope $-\dfrac{A}{B}$ and y-intercept

$-\dfrac{C}{B}$.   ∎

EXAMPLE 7   Given the linear equation

$$4x - 3y - 5 = 0$$

find the slope and y-intercept.

Solution   Solving for y, there results,

$$y = \tfrac{4}{3}x - \tfrac{5}{3}$$

so that $m = \tfrac{4}{3}$ is the slope and $b = -\tfrac{5}{3}$ is the y-intercept. $\bullet$

Quite appropriately an equation of the first degree in x and y, that is, an equation of the form (9) is called a *linear equation.* If (9) is solved for y in terms of x ($B \neq 0$) we obtain a function called a *linear function*† of x. Analogously, any line not parallel to the x-axis yields x as a linear function of y.

---

† The concept of function will be dealt with systematically in Chapter 2.

EXAMPLE 8   Find the equation of the line which is the perpendicular bisector of the line segment joining $P(-1, 4)$ and $Q(5, -6)$. See Figure 1.6.6.

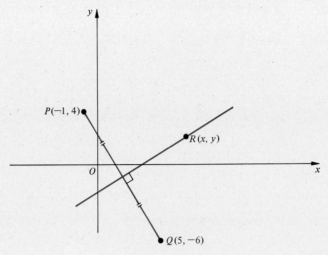

Figure 1.6.6

Solution   We give two methods.

Method 1:   First we obtain the slope of the line segment joining P and Q. It is

$$m_1 = \frac{-6 - 4}{5 - (-1)} = \frac{-10}{6} = \frac{-5}{3}$$

The slope of the perpendicular bisector is the negative reciprocal of m_1 and therefore must be $m_2 = \frac{3}{5}$. Furthermore, the coordinates of the midpoint of the line segment PQ are

$$\bar{x} = \frac{-1 + 5}{2} = 2 \quad \text{and} \quad \bar{y} = \frac{4 + (-6)}{2} = -1$$

The equation of the line through $(2, -1)$ with slope $\frac{3}{5}$ is

$$\frac{y - (-1)}{x - 2} = \frac{3}{5} \qquad \text{Slope + point}$$

or

$$3x - 5y - 11 = 0$$

Method 2:   We recall the fact that any point on the perpendicular bisector of a line segment must be equidistant from the ends of a line segment. Thus, if $R(x, y)$ is any such point then application of the distance formula yields

$$\sqrt{[x - (-1)]^2 + (y - 4)^2} = \sqrt{(x - 5)^2 + [y - (-6)]^2}$$

If we square both sides and multiply out, there results

$$x^2 + y^2 + 2x - 8y + 17 = x^2 + y^2 - 10x + 12y + 61$$

The second degree terms cancel, and combination of like terms gives

$$12x - 20y - 44 = 0$$

Then division by 4 yields

$$3x - 5y - 11 = 0$$

in agreement with the previous result.

EXAMPLE 9   By way of our last illustration, the intercept form of an equation of a straight line will be derived.

Solution   By definition, the x-intercept of a line is defined as the abscissa of the point at which the line intersects the x-axis. Let a be the x-intercept; that is, the line passes through $(a, 0)$. If, furthermore, the y-intercept is b, then the line passes through $(0, b)$. Thus from (7), the two-point form, we have

$$y - 0 = \frac{b - 0}{0 - a}(x - a)$$

and simplification yields

$$bx + ay = ab$$

Dividing by ab, assuming that both a and b are non-zero, yields

$$\frac{x}{a} + \frac{y}{b} = 1 \qquad \qquad (10)$$

Equation (10) is called the **intercept form** of the line. It says

$$\frac{x}{x\text{-intercept}} + \frac{y}{y\text{-intercept}} = 1$$

It does not apply if the line passes through the origin ($a = b = 0$) or if the line is parallel to a coordinate axis.

Exercises 1.6

In Exercises 1 through 12, find the equation of the line which satisfies the given conditions.

1. Slope 3 and passes through $(1, -1)$.
2. Slope $-\frac{1}{2}$ and passing through $(10, -2)$.
3. Through the two points $(1, -1)$ and $(4, 8)$.
4. Through the point $(-1, -4)$ and parallel to the x-axis.
5. Through the point $(5, -6)$ and parallel to the y-axis.
6. Through the point $(8, 3)$ and inclined at $45°$ to the x-axis (in the positive direction).
7. Through the origin and bisecting the angle between the axes in the second and fourth quadrants.
8. Slope -10 and y-intercept 9.
9. The intercepts are equal and passing through $(-4, 3)$.
10. The x-intercept is twice the y-intercept and passing through $(1, 3)$.
11. The sum of the x- and y-intercepts is 10 and passing through $(6, -3)$.
12. The slope is -3 and the x-intercept is 4.

In each of the exercises 13 through 16 find the slope of the given line.

13. $3x + 2y = 10$ (*Hint:* write the equation in slope intercept form by solving for y.)

14. $-x = 3y - 7$

15. $x_1x + y_1y - 2 = 0$

16. $\dfrac{x}{b} + \dfrac{y}{a} = 1, \quad ab \neq 0$

17. If A, B, C, and D are constants, for which $AB \neq 0$, show that
   (a) The lines $Ax + By + C = 0$ and $Ax + By + D = 0$ are parallel.
   (b) The lines $Ax + By + C = 0$ and $Bx - Ay + D = 0$ are perpendicular.

18. (a) Let C and F denote, respectively, corresponding centigrade (also called Celsius) and Fahrenheit temperature readings. Given that the graph of F versus C is a straight line, determine the linear relationship between F and C from the following information: $F = 32$ when $C = 0$ and $F = 212$ when $C = 100$.
   (b) Find C when $F = 140$.

19. The perpendicular distance ON from the origin to line L is p. The line ON makes an angle α with the positive x-axis. Show that L has an equation

$$x(\cos \alpha) + y(\sin \alpha) = p$$

This form is called the **normal form** for an equation of a straight line.

20. Find the distance from the origin to the straight line $x + 3y = 12$. (*Hint:* use the result of the previous exercise. Put the straight line in normal form by dividing both sides by an appropriate number. Recall that $\cos^2 \alpha + \sin^2 \alpha = 1$.)

21. For a falling body, the velocity v at any time t is a linear function of t. Given that $v = 36$ ft/sec, when $t = 0$ sec, and $v = 100$ ft/sec, when $t = 2$ sec, find the linear function $v(t)$.

22. Find the point of intersection of the straight lines whose equations are $2x + 3y + 8 = 0$ and $10x - 8y = 29$.

23. Draw the graph of $9x^2 - y^2 = 0$. (*Hint:* factor the expression on the left.)

24. Draw the graph of $3x^2 - 2xy - y^2 = 0$.

25. Find the equation of a straight line through $(-2, 4)$ that forms an isosceles triangle with the coordinate axes.

26. Prove that the perpendicular bisectors of the sides of an arbitrary triangle meet in a point. (*Hint:* without any loss of generality, it is permissible to choose the coordinate axes so that the vertices are located at the points $(a, 0)$, $(-a, 0)$ and (b, c), where a, b, and c are arbitrary positive numbers.)

27. Let L_1 be the line having the equation $A_1x + B_1y + C_1 = 0$. Similarly, let L_2 be the line having the equation $A_2x + B_2y + C_2 = 0$. If the two lines L_1 and L_2 are not parallel and if k is any constant, verify that the equation

$$A_1x + B_1y + C_1 + k(A_2x + B_2y + C_2) = 0$$

represents an unlimited number of lines passing through the intersection of lines L_1 and L_2.

28. Given an equation of L_1 is $x + y - 3 = 0$ and an equation of L_2 is $2x + 3y - 7 = 0$, by application of the previous exercise and without finding the coordinates of the point of intersection of L_1 and L_2, find an equation of the line through this point of intersection **(a)** having slope -3, **(b)** passing through the point $(4, 5)$, **(c)** parallel to the y-axis, **(d)** perpendicular to the line $x + 3y - 5 = 0$.

29. Find the tangent of the angle between the lines $3x - y - 11 = 0$ and $x + 2y + 1 = 0$. $\left[\textit{Hint: If}\right.$
$\theta = \alpha_1 - \alpha_2, \ \tan \theta = \tan(\alpha_1 - \alpha_2) = \dfrac{\tan \alpha_1 - \tan \alpha_2}{1 + \tan \alpha_1 \tan \alpha_2}. \left.\right]$

30. Prove analytically that the diagonals of a rhombus are perpendicular.

31. Prove that the diagonals of a rectangle are perpendicular if and only if the rectangle is a square.

32. Prove that the three medians of a triangle meet at a common point.

1.7   THE CIRCLE

Definition   A **circle** is the locus of points in a plane at a given distance from a fixed point.
The fixed point is called the **center** of the circle and the given distance is the **radius.**

EQUATION OF A CIRCLE: Let $C(h, k)$ be the center of the circle and r the radius (Figure 1.7.1). Let $P(x, y)$ be an arbitrary point on the circumference of the circle. Then $|\overline{CP}| = r$ so that

$$\sqrt{(x - h)^2 + (y - k)^2} = r$$

or, squaring both sides

$$(x - h)^2 + (y - k)^2 = r^2 \qquad\qquad (1)$$

Equation (1) is satisfied by the coordinates of those and only those points that lie on the given circle. Therefore (1) is an equation of the circle.

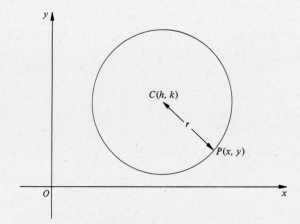

Figure 1.7.1

EXAMPLE 1   Find the equation of the circle with center at the origin and radius r.

Solution   In this instance, $h = k = 0$ and (1) becomes

$$x^2 + y^2 = r^2 \qquad\qquad \bullet \quad (2)$$

EXAMPLE 2   Find the equation of the circle through the origin with center at $C(-3, -2)$.

Solution   Substitute $h = -3$ and $k = -2$ into (1). Thus

$$(x + 3)^2 + (y + 2)^2 = r^2$$

Also, $x = y = 0$ must satisfy the equation because the circle passes through $(0, 0)$. Thus

$$3^2 + 2^2 = r^2 \qquad \text{or} \qquad 13 = r^2$$

and the required equation is

$$(x + 3)^2 + (y + 2)^2 = 13$$

This is a circle with center at $(-3, -2)$ and radius $= \sqrt{13}$. ●

EXAMPLE 3   Find an equation of a circle with center at $(4, 3)$ and passing through $(-1, 2)$.

Solution   Since the coordinates of the center are at $(4, 3)$, an equation of the circle is

$$(x - 4)^2 + (y - 3)^2 = r^2$$

And because $(-1, 2)$ is on the circumference of the circle, we have $r^2 = (-1 - 4)^2 + (2 - 3)^2 = 26$. The required equation is

$$(x - 4)^2 + (y - 3)^2 = 26$$ ●

EXAMPLE 4   Find an equation of a circle with points $(-4, 5)$ and $(0, 1)$ at the ends of a diameter.

Solution   The center of the circle is at the midpoint of the diameter connecting the points $(-4, 5)$ and $(0, 1)$. Its coordinates from the midpoint formula are $(-2, 3)$. The square of the distance from $(-2, 3)$ to $(0, 1)$ is 8. Therefore an equation of the circle is

$$(x + 2)^2 + (y - 3)^2 = 8$$ ●

EXAMPLE 5   What is the locus of points $P(x, y)$ whose coordinates satisfy the inequality

$$(x - h)^2 + (y - k)^2 < r^2 \tag{3}$$

Solution   The left side of (3) is the square of the distance from the center of the circle $C(h, k)$ to $P(x, y)$. Thus (3) is satisfied if and only if

$$|\overline{CP}|^2 < r^2$$

or since $\overline{CP}$ and r are nonnegative

$$|\overline{CP}| < r$$

This implies that point P lies inside the circle of radius r with center at $C(h, k)$. (See Figure 1.7.2.) ●

Now, suppose that we have a circle with center at $(7, -1)$ and radius 2. Its equation is

$$(x - 7)^2 + (y + 1)^2 = 4 \tag{4}$$

If we multiply out and combine terms there results

$$x^2 + y^2 - 14x + 2y + 46 = 0 \tag{5}$$

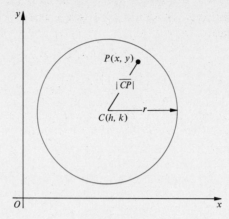

Figure 1.7.2

Now we start the other way around. Suppose that we are given (5). The question is then posed: Does (5) represent a circle and, if so, what is its center and radius? We can readily answer this question by using the method of **completing the squares.** To this end, the x and y terms are combined separately in (5) so that we have

$$(x^2 - 14x \qquad) + (y^2 + 2y \qquad) = -46$$

in which the appropriate spaces are left as shown. Then we add the square of half the coefficient of x and the square of half the coefficient of y to *both sides* and so obtain

$$(x^2 - 14x + 49) + (y^2 + 2y + 1) = -46 + 49 + 1 = 4$$

Equation (4) is now recovered, that is

$$(x - 7)^2 + (y + 1)^2 = 4$$

EXAMPLE 6 Analyze the equation $x^2 + y^2 - 6x + 8y + 26 = 0$.

Solution Collecting x and y terms we have

$$(x^2 - 6x \qquad) + (y^2 + 8y \qquad) = -26$$

and completion of squares yields

$$(x^2 - 6x + 9) + (y^2 + 8y + 16) = -26 + 9 + 16$$

or

$$(x - 3)^2 + (y + 4)^2 = -1$$

However, the left side is nonnegative, being the sum of squares, and the right side is -1. Therefore there are no values of x and y that will satisfy the relation; that is, no real locus exists. ●

EXAMPLE 7 Analyze the equation $x^2 + y^2 + 4x - 2y + 5 = 0$.

Solution   Writing $(x^2 + 4x \quad) + (y^2 - 2y \quad) = -5$ and completing the squares gives

$$(x^2 + 4x + 4) + (y^2 - 2y + 1) = -5 + 4 + 1$$

or

$$(x + 2)^2 + (y - 1)^2 = 0$$

The only way in which the sum of two squares can equal zero is if each of them is zero. Thus

$$x + 2 = 0 \quad \text{and} \quad y - 1 = 0$$

or

$$x = -2 \quad \text{and} \quad y = 1$$

Therefore, the point $(-2, 1)$ is the only point that satisfies the original equation. We sometimes refer to this as a *point circle*.   ●

Examples 6 and 7 show that an equation of the form

$$Ax^2 + Ay^2 + Dx + Ey + F = 0 \quad A \neq 0 \tag{6}$$

can often be reduced to the form (1) by dividing by A and completion of the squares. However, while the equations of all circles are expressible in the form $x^2 + y^2 + Dx + Ey + F = 0$, the converse is *not* true. Instead, the quantities D, E, and F must satisfy an inequality, which we will now derive. First, both sides of (6) are divided by A. Then the terms are rearranged to yield

$$\left(x^2 + \frac{D}{A}x \quad \right) + \left(y^2 + \frac{E}{A}y \quad \right) = -\frac{F}{A}$$

To complete the squares, we add $\left(\dfrac{D}{2A} \right)^2 = \dfrac{D^2}{4A^2}$ and $\left(\dfrac{E}{2A} \right)^2 = \dfrac{E^2}{4A^2}$ to both sides, thus obtaining

$$\left(x + \frac{D}{2A} \right)^2 + \left(y + \frac{E}{2A} \right)^2 = -\frac{F}{A} + \frac{D^2 + E^2}{4A^2}$$

$$= \frac{D^2 + E^2 - 4AF}{4A^2} \tag{7}$$

Comparison of (7) with (1) implies that we have a circle of radius r, where

$$r = \frac{\sqrt{D^2 + E^2 - 4AF}}{2|A|} \tag{8}$$

provided that $D^2 + E^2 - 4AF$ is positive. Also, the center of the circle has coordinates

$$h = -\frac{D}{2A} \quad \text{and} \quad k = -\frac{E}{2A} \tag{9}$$

If $D^2 + E^2 - 4AF = 0$, then the locus is a single point; if that quantity is less than zero, there are no points in the xy-plane that satisfy the relation (6). It is recommended that the reader use the method of completing the squares to determine the locus rather than memorizing Equation (8). In summary, *an equation of the form (6) represents a circle, a single point, or no locus.*

If Equation (6) is divided by A there results an equation of the form

$$x^2 + y^2 + ax + by + e = 0 \qquad (10)$$

where a, b, and e are constants. The three coefficients a, b, and e may be determined from three prescribed conditions, such as the circle must pass through three given noncollinear points, or be tangent to three nonparallel lines, and so on.

EXAMPLE 8 Find the equation of the circle that passes through the three points $(0, 0)$, $(3, 0)$ and $(-1, 2)$.

Solution Substitution of the three sets of coordinates into Equation (10) yields

$$0^2 + 0^2 + a(0) + b(0) + e = 0$$
$$3^2 + 0^2 + 3a + b(0) + e = 0$$
$$(-1)^2 + 2^2 - a + 2b + e = 0$$

from which

$$e = 0 \qquad a = -3 \qquad b = -4$$

Therefore (10) becomes

$$x^2 + y^2 - 3x - 4y = 0$$

Alternative Solution Consider the segment joining $(0, 0)$ and $(3, 0)$. Any point on the perpendicular bisector of this segment must be equidistant from the ends of the line segment and therefore must pass through the center of the circle. The equation of the perpendicular bisector is $x = \frac{3}{2}$ so that $h = \frac{3}{2}$. We similarly consider the segment joining $(0, 0)$ and $(-1, 2)$. Any point on the perpendicular bisector including (h, k), the center of the circle, must be equidistant from the ends of the segment, thus

$$h^2 + k^2 = (h + 1)^2 + (k - 2)^2$$

or

$$2h - 4k + 5 = 0$$

But $h = \frac{3}{2}$ so that $k = 2$. Further, $h^2 + k^2 = r^2$ or $r^2 = \frac{25}{4}$ and the equation of the circle is

$$\left(x - \frac{3}{2}\right)^2 + (y - 2)^2 = \frac{25}{4}$$

or, expanding,

$$x^2 + y^2 - 3x - 4y = 0$$

in agreement with the previous result. Finally, we observe that the relation between h and k could have been determined by finding the slope of the line joining $(0, 0)$ and $(-1, 2)$, using the midpoint formula, and then the perpendicular bisector has a slope that is the negative reciprocal of the slope of the given segment. ●

EXAMPLE 9 A point moves so that its distance from $(3, 4)$ and $(-2, -3)$ is always in the ratio 1 to 2. Find the equation of the locus of this point.

Solution   Let $P(x, y)$ be such a required point. Then

$$\frac{\sqrt{(x-3)^2 + (y-4)^2}}{\sqrt{(x+2)^2 + (y+3)^2}} = \frac{1}{2}$$

or squaring and cross multiplying

$$4[(x-3)^2 + (y-4)^2] = [(x+2)^2 + (y+3)^2]$$

Thus

$$4x^2 + 4y^2 - 24x - 32y + 100 = x^2 + y^2 + 4x + 6y + 13$$

or

$$3x^2 + 3y^2 - 28x - 38y + 87 = 0$$

which is the equation of a circle.   ●

EXAMPLE 10   Prove analytically that an angle inscribed in a semicircle is a right angle.

Solution   For convenience, the circle is placed with its center at the origin. We take its radius to be a (Figure 1.7.3). Let $P(x, y)$ be any point on the circle and join it to the ends of the diameter $A(-a, 0)$ and $B(a, 0)$. Then the result will be proved if we can show that PA is perpendicular to PB. The equation of the circle with center at the origin and radius a is

$$x^2 + y^2 = a^2$$

The slope of PA is $m_1 = \dfrac{y}{x+a}$ and the slope of PB is $m_2 = \dfrac{y}{x-a}$. Then

$$m_1 m_2 = \frac{y^2}{(x+a)(x-a)} = \frac{y^2}{x^2 - a^2} = -1$$

since $x^2 - a^2 = -y^2$ from the equation of the circle. Therefore, PA is perpendicular to PB because the product of the slopes is -1.   ●

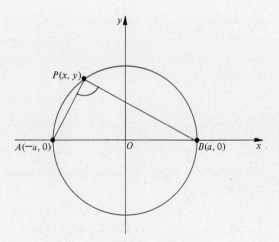

Figure 1.7.3

Exercises 1.7

1. Write an equation of each of the following circles
   (a) center $(0, 0)$, radius 7
   (b) center $(2, 3)$, radius 5
   (c) center $(-5, 1)$, radius 4
   (d) center $(-8, -2)$, radius $\sqrt{11}$

2. Find an equation of the circle whose center is at the origin and which passes through $(-6, 5)$.

3. Find an equation of the circle whose center is at $(-3, 2)$ and which passes through the origin.

4. Find an equation of the circle with center at $(4, 0)$ and tangent to the y-axis.

5. Find an equation of the circle with center at $(-4, 5)$ and tangent to the x-axis.

6. Find an equation of the circle having the line segment from $(-3, 1)$ to $(5, 3)$ as ends of a diameter.

7. The points $(-4, -2)$ and $(6, 8)$ are ends of a diameter of a circle, find an equation of the circle.

8. Find an equation of the circle with center at $(5, 5)$ and tangent to the line $y = 3x$.

9. Find an equation of the circle with center at $(3, 5)$ and tangent to the line $y = \frac{x}{2} + 1$.

10. Find an equation of the circle that is in the second quadrant, has radius 5, and is tangent to both the x- and y-axes.

In Exercises 11 through 16, determine the locus of the given equation by completing the squares.

11. $x^2 + y^2 - 2x - 4y - 11 = 0$
12. $x^2 + y^2 + 6x - 10y - 15 = 0$
13. $4x^2 + 4y^2 - 4x - 3 = 0$
14. $4x^2 + 4y^2 - 24x + 4y + 21 = 0$
15. $x^2 + y^2 + 8x - 6y + 32 = 0$
16. $9x^2 + 9y^2 - 6x + 18y = 8$
17. Find an equation of the circle tangent to the x-axis at the point $(3, 0)$ with radius 4.

In Exercises 18 through 21, find an equation of the circle which passes through the given points.

18. $(3, -1)$, $(0, 8)$, and $(6, 0)$
19. $(-1, 0)$, $(0, 4)$, and $(-6, 0)$
20. $(1, 1)$, $(2, -1)$, and $(-2, -3)$
21. $(3, -1)$, $(1, 3)$, and $(7, 1)$
22. Find the coordinates of the points of intersection (if any) of the line $y - x + 1 = 0$ and the circle $x^2 + y^2 - x - 3y = 0$.
23. Find the points of intersection and the equation of the common chord of the circles $x^2 + y^2 + 4x - 25 = 0$ and $x^2 + y^2 - 8y + 3 = 0$. *Note:* The equation of the common chord may be found by taking the difference of the two equations. Why does this method apply in general?
24. Find an equation of the line that is tangent to the circle $x^2 + y^2 - 6x - 2y - 15 = 0$ at the point $(6, 5)$.
25. Find an equation of the circle that passes through the points $(6, 4)$ and $(-4, -2)$ and has its center on the line $2x - 3y = 20$.
26. Prove analytically, *without* the use of slopes, that any angle inscribed in a semicircle is a right angle. (*Hint:* Use the formula for the distance between two points.)
27. Prove analytically that a radius of a circle perpendicular to a chord bisects the chord.
28. Prove analytically that a straight line cannot cut a circle in more than two points.
29. Prove that the common chord of two circles is perpendicular to their line of centers. (*Hint:* Let (h, k) and $(h^\star, k^\star)$ be the coordinates of the centers of the circles. Find the slope of the line joining (h, k) and $(h^\star, k^\star)$ and then compare this slope with that of the common chord.)
30. The distance between two fixed points is $2c$. Find the locus of a point that moves so that the sum of the squares of its distances from the fixed points is a constant, k. (*Hint:* Choose the coordinates of the fixed points at $(c, 0)$ and $(-c, 0)$ and let (x, y) be the variable point. Then use the formula for the square of the distances between two points.)

31. Prove analytically that the locus of points whose distance from one fixed point is k times the distance from a second fixed point, where k is positive and unequal to 1, is a circle. What is the locus if $k = 1$?

32. Chords are drawn from one end of a diameter of a given radius a. Find the locus of the midpoints of the chords.

1.8   THE PARABOLA

Consider a fixed point F and a fixed line L that does not pass through the point.

Definition   A *parabola* is the locus of points equidistant from the fixed point F and the fixed line L. The fixed point is called the *focus* and the fixed line is called the *directrix* (Figure 1.8.1).

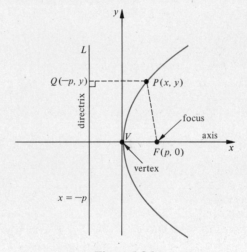

Figure 1.8.1

Clearly, there are many different parabolas depending upon the choice of focus and directrix. Any parabola, however, can be shown to be symmetric about the line that passes through its focus and is perpendicular to its directrix. This line of symmetry, called the *axis* of the parabola must cross the parabola in one point called the *vertex* (Figure 1.8.1). These observations will now be derived mathematically by choosing a convenient coordinate system from which a particularly simple Cartesian equation of the parabola is obtained.

Let the focus of a parabola be the point $(p, 0)$ on the x-axis and let the equation of the directrix be the line $x = -p$, which is parallel to the y-axis. Then any point $P(x, y)$ is on the parabola if and only if

$$|\overline{FP}| = |\overline{QP}|$$

where $Q(-p, y)$ is the projection of point $P(x, y)$ on the directrix $x = -p$. But, by the distance formula between two points $|\overline{FP}| = \sqrt{(x - p)^2 + y^2}$ and

$|\overline{QP}| = |x - (-p)|$. Therefore P is on the parabola if and only if

$$\sqrt{(x - p)^2 + y^2} = |x + p|$$

If we square both sides there results

$$(x - p)^2 + y^2 = (x + p)^2$$

and simplification yields

$$y^2 = 4px \tag{1}$$

This result may be stated as a theorem:

Theorem 1 An equation of the parabola having its focus at $(p, 0)$ and the line $x = -p$ as its directrix is given by Equation (1).

Figure 1.8.2 shows the parabola when $p > 0$ while Figure 1.8.3 shows the parabola when $p < 0$.

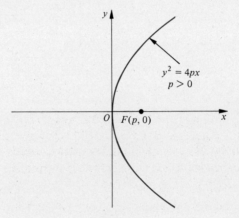

Figure 1.8.2

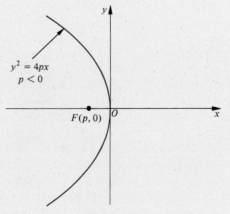

Figure 1.8.3

A parabola is said to be in **standard position** when its vertex is at the origin and its axis is one of the coordinate axes. The equation of a parabola in standard position is called a ***standard equation.*** Thus, for example, Equation (1) is a standard equation of a parabola.

EXAMPLE 1   Find the coordinates of the focus and the equation of the directrix of the parabola $y^2 = 12x$. Sketch the parabola, its focus, and its directrix.

Solution   Comparing the given equation with (1), we have $4p = 12$ or $p = 3$. Thus the focus is at $(3, 0)$ and its directrix is $x = -3$. Its vertex is at the origin. The parabola opens to the right as shown in Figure 1.8.4 because p is positive. ●

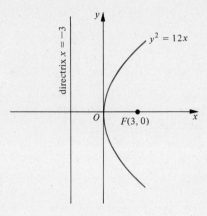

Figure 1.8.4

An equally simple standard equation of a parabola is obtained by interchanging the x- and y-axes in the preceding derivation. In other words, we have the focus on the y-axis, the directrix $y =$ constant and the vertex at the origin. Thus the y-axis coincides with the axis of the parabola. In short, we have

Theorem 2   The standard equation of the parabola with focus at $(0, p)$ and directrix $y = -p$ is

$$x^2 = 4py \qquad\qquad (2)$$

where if $p > 0$, the parabola opens upward, as in Figure 1.8.5, and if $p < 0$ the parabola opens downward (Figure 1.8.6).

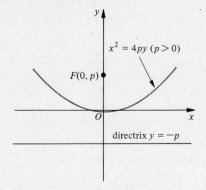

Figure 1.8.5

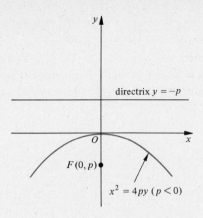

directrix $y = -p$

$F(0, p)$

$x^2 = 4py \ (p < 0)$

Figure 1.8.6

EXAMPLE 2   Find the coordinates of the focus and the equation of the directrix of the parabola $x^2 = -6y$. Sketch the parabola, its focus and its directrix.

Solution   Comparing $x^2 = -6y$ with (2) yields $4p = -6$ or $p = -\frac{3}{2}$. The focus is at $(0, -\frac{3}{2})$ and the directrix is $y = \frac{3}{2}$. The vertex is at the origin. The parabola opens downward because $p < 0$ (Figure 1.8.6 with $p = -\frac{3}{2}$).   ●

When the parabola is *not* in one of the standard positions, the equation is more complicated. If the directrix is horizontal or vertical, that is, parallel to the x- or y-axis, the equation is only slightly different; however, if the directrix is oblique to the coordinate axes, the form of the equation is changed extensively. Let us next illustrate how the equation of a parabola may be derived directly from its definition.

EXAMPLE 3   Find an equation of the parabola with focus at $F(3, -4)$ and with the line $y = 2$ as directrix.

Solution   From the definition, $P(x, y)$ is a point on the parabola if and only if $|\overline{PF}| = \sqrt{(x - 3)^2 + (y + 4)^2}$ is equal to the distance from point P to the directrix, which is $|2 - y|$. Therefore

$$\sqrt{(x - 3)^2 + (y + 4)^2} = |2 - y|$$

If we square both sides and note that $|2 - y|^2 = (2 - y)^2$, there results

$$(x - 3)^2 + (y + 4)^2 = (2 - y)^2$$

or, expanding,

$$x^2 + y^2 - 6x + 8y + 25 = 4 - 4y + y^2$$

so that

$$x^2 - 6x + 12y = -21$$

This relation may be written in a more suggestive form if we complete the squares on x:

$$x^2 - 6x + 9 = -12y - 21 + 9$$

or
$$(x - 3)^2 = -12(y + 1)$$

A sketch of the equation is shown in Figure 1.8.7. The axis of the parabola is the line $x = 3$, the vertex is at $(3, -1)$, and the curve opens downward. Furthermore, we note that $p = 3$ is the distance from the vertex to the focus or, equivalently, from the vertex to the directrix. ●

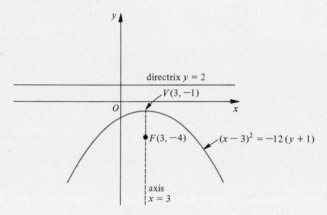

Figure 1.8.7

By analogy with the equation of a parabola in standard form, we will establish (see Section 13.3, Translation of Axes) that

$$(y - k)^2 = 4p(x - h) \qquad (3)$$

represents a parabola with vertex $V(h, k)$, with the line $y = k$ as axis. If $p > 0$ the parabola opens to the right and if $p < 0$ it opens to the left.

Similarly,

$$(x - h)^2 = 4p(y - k) \qquad (4)$$

has the line $x = h$ as axis. It opens upward or downward if p is positive or negative, respectively. It should be noted that (1) and (2) are special cases of (3) and (4), respectively, when $h = k = 0$.

In general, a parabola with axis parallel to one of the coordinate axes is quadratic in one variable and linear in the other. The axis of the parabola is always parallel to the axis of the linear variable.

It is helpful when we sketch the graph of a parabola to draw the chord through the focus, perpendicular to the axis of the parabola. This chord is called the **latus rectum** or **right focal chord** of the parabola. Its length is easily obtained as we show next.

EXAMPLE 4   Find the length of the latus rectum of a parabola.

Solution   For our purposes, let us use form (1). The focus is at $(p, 0)$ so that if $x = p$ we find from (1) that $y^2 = 4p(p) = 4p^2$ or $y = \pm 2p$. The magnitude of the difference of the ordinates is $|2p - (-2p)| = 4|p|$. This result is clearly the same for forms (2), (3), and (4) for the parabola. ●

EXAMPLE 5 Find the equation of the parabola with vertex $V(0,0)$ and focus $F(1,0)$. What is the axis, directrix, and length of the latus rectum. Sketch the curve.

Solution The axis is the x-axis and the directrix is the line $x = -1$. Since $p = 1$, the equation of the parabola is

$$y^2 = 4x$$

and the length of the latus rectum is $4|p| = 4$. The graph of the parabola is sketched in Figure 1.8.8. ●

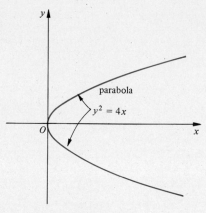

Figure 1.8.8

EXAMPLE 6 Find the equation of the parabola with vertex $V(-2, 1)$ and focus $F(-2, -3)$.

Solution The axis is $x = -2$, and $p = -4$, that is, the curve opens downward. The equation of the parabola is

$$y - 1 = -16(x + 2)^2$$

and the length of the latus rectum is $4|p| = 16$. ●

EXAMPLE 7 Find the lowest point on the parabola $y = x^2 - 10x - 7$.

Solution The question is easily answered by using the method of completing the squares. To this end, we write

$$y = (x^2 - 10x \qquad) - 7$$

and to complete the square we must add 25 and neutralize this by taking away 25. Thus

$$y = (x^2 - 10x + 25) - 7 - 25$$
$$y = (x - 5)^2 - 32$$

Now, the least value of y occurs when $(x - 5)^2$ is least; that is, when $x = 5$, since any square of a real number is ≥ 0. Therefore the least value of y is $y|_{x=5} = -32$. ●

EXAMPLE 8 Geometrical Construction of a Parabola.

Solution   A parabola may be constructed point by point as indicated in Figure 1.8.9. Point F is the focus and line L is the directrix. The axis of the parabola is constructed by dropping a perpendicular from F to L. Now draw a line K perpendicular to the axis (and thus parallel to L). With F as center and radius equal to the distance between K and L draw arcs intersecting K in points A and B. Then A and B are points of the parabola because $AF = AM$ and $BF = BN$ and this agrees with the definition of the parabola.

Additional points may be obtained in a similar fashion by drawing other lines parallel to the directrix L.   ●

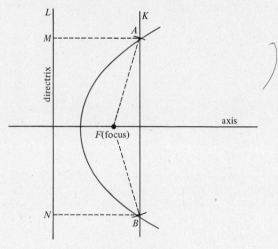

Figure 1.8.9

Exercises 1.8

For each of the parabolas in Exercises 1 through 6, find the coordinates of the focus, an equation of the directrix, and the length of the latus rectum. Draw a sketch of the curve.

1. $y^2 = 8x$
3. $y = x^2$
5. $-6x = y^2$

2. $y^2 + 4x = 0$
4. $x^2 + 12y = 0$
6. $2x^2 + 5y = 0$

In Exercises 7 through 16, find an equation of the parabola having the given properties.

7. Focus $(6, 0)$; directrix: $x = -6$
8. Focus $(0, 3)$; directrix: $y = -3$
9. Focus $(-5, 0)$; directrix: $x = 5$
10. Focus $(0, -4)$; directrix: $y = 4$
11. Focus $(\frac{5}{2}, 0)$; directrix: $x + \frac{5}{2} = 0$
12. Focus $(0, \frac{5}{3})$; directrix: $3y + 5 = 0$
13. Vertex $(0, 0)$, opens upward; length of latus rectum $= 7$
14. Vertex $(0, 0)$, opens to the left; length of latus rectum $= \frac{1}{2}$
15. Vertex $(0, 0)$, axis along the x-axis and passing through $(-1, 4)$
16. Vertex $(0, 0)$, axis along the y-axis and passing through $(4, 2)$
17. Find the points of intersection of the parabolas $x^2 = 4y$ and $y^2 = 4x$.
18. Find the points of intersection of the straight line $4x + y = 12$ and the parabola $y = x^2$.
19. Find the lowest point on the parabola $y = x^2 - 14x + 52$.

20. Find the lowest point on the parabola $y = 16x - x^2$.

21. Find the lowest point of the parabola $y = ax^2 + bx + c$, $(a > 0)$.

22. Using the definition of a parabola, find an equation of a parabola having as its directrix the line $y = -2$ and having $F(3, 4)$ as its focus.

23. From the definition of a parabola, find an equation of a parabola having line $x = 3$ as its directrix and point $(-5, 2)$ as its focus.

24. A ship is steered so that its path will keep it equidistant from a battery of guns located at point F (Figure 1.8.10) and the shoreline 50 miles away. Write an equation for the ship's path.

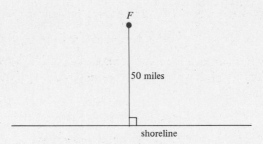

F

50 miles

shoreline

Figure 1.8.10

25. A high voltage cable is supported by two towers 3200 ft apart and 400 ft high. The cable hangs in the shape of a parabola, and the lowest point of the cable is 240 ft above the ground. Find the equation of the parabola. (*Hint:* Choose the origin of the coordinates as the lowest point of the parabola.)

26. A parabola with an axis of symmetry parallel to the y-axis passes through the points $(0, 7)$, $(1, 4)$ and $(-3, 40)$. Find an equation of the parabola and an equation of its axis of symmetry. (*Hint:* The parabola must have an equation of the form $y = ax^2 + bx + c$. Why? Find a, b, and c by using the given data.)

27. What is the locus of the equation $(3x - 2y + 7)(x^2 + 6y) = 0$? Explain your answer.

1.9   MATHEMATICAL INDUCTION

Mathematical induction is a useful method for proving propositions for all natural numbers n. The idea behind this procedure is as follows: in order to prove a formula or theorem true for all natural numbers n, we first verify by actual substitution that the formula is true for $n = 1$. Next, we establish that, if the formula is true for $n = k$, where k is a *particular* (but arbitrary) natural number, then it must be true for the next natural number $n = k + 1$. Having these two steps, we conclude that the formula is true for all natural numbers n because it was verified true for 1 and thus must be true for $n = 2$. This, in turn, implies that it is true for $n = 3$, $n = 4, \ldots$.

Symbolically, let $P(n)$ be such a proposition. We first show that $P(1)$ is true, meaning that the proposition is true for $n = 1$. Next, we demonstrate that the proposition is true for $n = k + 1$ whenever it is true for $n = k$; that is, $P(k)$ implies $P(k + 1)$. [We indicate this implication by the notation $P(k) \to P(k + 1)$.] Upon completion of these two steps it follows that $P(n)$ is true for all natural numbers n.

EXAMPLE 1 Prove by mathematical induction that

$$2 + 4 + 6 + \cdots + 2n = n(n + 1) \qquad (1)$$

for all natural numbers n.

Solution We first verify that the proposition is true for $n = 1$,

$$2 = 1(2)$$

Next, we assume that the propositon is true for $n = k$,

$$2 + 4 + 6 + \cdots + 2k = k(k + 1) \qquad (2)$$

where k is an arbitrary natural number. It must be shown that, if (1) is true for $n = k$ [that is, (2) holds], then (1) is true also for $n = k + 1$. In other words, we must prove that

$$2 + 4 + 6 + \cdots + 2k + 2(k + 1) = (k + 1)(k + 2) \qquad (3)$$

In order to establish (3) from (2), it is suggested that we add $2(k + 1)$ to both sides of (2), hence

$$2 + 4 + \cdots + 2k + 2(k + 1) = k(k + 1) + 2(k + 1)$$
$$= (k + 2)(k + 1)$$

which is (3). Thus, in the preceding notation, we have established that $P(k) \rightarrow P(k + 1)$ and also verified that $P(1)$ is true. Therefore, (1) is true for all natural numbers n. ●

EXAMPLE 2 Prove by mathematical induction that

$$1 + 3 + 5 + \cdots + (2n - 1) = n^2 \qquad (4)$$

for all natural numbers n.

Solution First, the formula is verified for $n = 1$,

$$1 = 1^2$$

Second, it is assumed that the formula is true for $n = k$, that is

$$1 + 3 + 5 + \cdots + (2k - 1) = k^2 \qquad (5)$$

Then, in order to verify that it holds for $n = k + 1$, we add $2k + 1$ to both sides of (5).

$$1 + 3 + 5 + \cdots + (2k - 1) + (2k + 1) = k^2 + (2k + 1) = (k + 1)^2 \quad (6)$$

and Equation (6) is (5) with k replaced by $k + 1$. This completes the proof. ●

EXAMPLE 3 Prove by mathematical induction that

$$(1)(2) + (2)(3) + \cdots + n(n + 1) = \tfrac{1}{3}n(n + 1)(n + 2) \qquad (7)$$

Solution We first verify (7) for $n = 1$,

$$(1)(2) = \tfrac{1}{3}(1)(2)(3)$$

Next, we assume that the formula is true for $n = k$,

$$(1)(2) + (2)(3) + \cdots + k(k + 1) = \tfrac{1}{3}(k)(k + 1)(k + 2) \tag{8}$$

To show that it must also be true for $n = k + 1$, add $(k + 1)(k + 2)$ to both sides of (8). Thus

$$(1)(2) + (2)(3) + \cdots + (k)(k + 1) + (k + 1)(k + 2)$$
$$= \tfrac{1}{3}k(k + 1)(k + 2) + (k + 1)(k + 2)$$
$$= (k + 1)(k + 2)\left(\frac{k}{3} + 1\right)$$
$$= \tfrac{1}{3}(k + 1)(k + 2)(k + 3) \tag{9}$$

and Equation (9) is the same as (8) with k replaced by $k + 1$. Therefore (7) is true for all natural numbers n. ●

EXAMPLE 4   A useful inequality is

$$(1 + p)^n \geq 1 + np \tag{10}$$

where p is an arbitrary real number > -1 and n is a natural number. We shall prove this for arbitrary natural numbers n by mathematical induction.

Solution   If $n = 1, (1 + p)^1 \geq 1 + p$, which is certainly true. If (10) is true for $n = k$, then $(1 + p)^k \geq 1 + kp$. Multiply both sides of (10) by $1 + p > 0$ to obtain

$$(1 + p)^{k+1} \geq (1 + kp)(1 + p) = 1 + (k + 1)p + kp^2$$

or, since kp^2 is positive, we have

$$(1 + p)^{k+1} \geq 1 + (k + 1)p \tag{11}$$

Therefore if (10) is true for $n = k$, it must of necessity, be true for $n = k + 1$; hence (10) is true for all natural numbers n. ●

Exercises 1.9

Prove Exercises 1 through 14 by mathematical induction for all natural numbers n.

1. $5 + 10 + 15 + \cdots + 5n = \tfrac{5}{2}n(n + 1)$
2. $1 + 4 + 7 + \cdots + (3n - 2) = \tfrac{1}{2}n(3n - 1)$
3. $2^1 + 2^2 + 2^3 + \cdots + 2^n = 2^{n+1} - 2$
4. $n^2 + n$ is divisible by 2
5. $n^3 + 2n$ is divisible by 3
6. $5^n - 1$ is divisible by 4
7. For all real numbers x and y, $(xy)^n = x^n \cdot y^n$
8. $3^{2n} - 1$ is divisible by 8
9. $1^2 + 3^2 + 5^2 + \cdots + (2n - 1)^2 = \dfrac{n(4n^2 - 1)}{3}$
10. $x^n - 1$ is divisible by $x - 1$ if $x \neq 1$
11. $2^n \geq 1 + n$
12. $1 \cdot 3 + 2 \cdot 4 + 3 \cdot 5 + \cdots + n(n + 2) = \tfrac{1}{6}n(n + 1)(2n + 7)$

13. $1^2 + 2^2 + 3^2 + \cdots + n^2 = \dfrac{n(n+1)(2n+1)}{6}$

14. $1^3 + 2^3 + 3^3 + \cdots + n^3 = \dfrac{n^2(n+1)^2}{4}$

Review and Miscellaneous Exercises

In Exercises 1 through 12, find the real numbers x such that the following equality or inequality holds. Express your answer in terms of finite, semiinfinite, or infinite intervals when applicable.

1. $|x| = 100$

2. $|x - 4| = 10$

3. $|x + 3| = 6$

4. $|x - 7| \leq 3$

5. $|x + 1| > 5$

6. $\dfrac{1}{x} < \dfrac{1}{7}$

7. $x^2 + a^2 \geq 2ax$

8. $(x + 1)(x + 2) \leq 12$

9. $3x^2 \geq 2 - 5x$

10. $|x - a| \leq b, \quad b > 0$

11. $|x + 4| + |x| \geq 6$

12. $\dfrac{|x - 1|}{|x - 3|} > 1$

(*Hint:* consider 3 cases)

13. If $0 < a < b$, show that $a < \sqrt{ab} < b$ where $\sqrt{ab}$ is called the **geometric mean** of a and b.

14. Show that if $0 < a \leq b$ then $\sqrt{ab} \leq \dfrac{a + b}{2}$, thus the geometric mean of two positive numbers is less than or equal to their arithmetic mean. Show that the equality holds if and only if $a = b$. (*Hint:* Use Exercise 13.)

15. Show that $|a - b| \leq |a| + |b|$.

16. Show that $|a + b + c| \leq |a| + |b| + |c|$.

17. Show that $|abc| = |a| \, |b| \, |c|$.

18. Prove that $\big||a| - |b|\big| \leq |a - b|$.

19. Show that of all rectangles of a given perimeter P, the square has the largest area. $\Big($*Hint:* use the result of Exercise 14. Let x be the length of one side and $\dfrac{P}{2} - x$ represent the adjacent side.$\Big)$

20. Prove the dual problem of Exercise 19; namely, that of all rectangles of given area A, the square has the minimum perimeter. $\Big($*Hint:* Use the result of Exercise 14. Let x be the length of one side and $\dfrac{A}{x}$ represent the adjacent side.$\Big)$

21. Find all the roots of the equation $3x^3 + x^2 - 5x + 2 = 0$.

22. Prove that $0.232323 \cdots$ is a rational number. (*Hint:* Let $x = 0.232323 \cdots$ then $100x = 23.2323 \cdots$ and form the difference.)

23. Prove that the sum of a rational number and an irrational number is irrational.

24. Let β be a positive irrational number. Prove that $\sqrt{\beta}$ is also irrational. (*Hint:* Suppose that $\sqrt{\beta}$ is rational, then write $\beta = \sqrt{\beta}\sqrt{\beta}$.)

25. Prove the Pythagorean theorem by Euclidean geometry, that in a right triangle, the square of the hypotenuse equals the sum of the squares of the legs. (*Hint:* A proof often given in Euclidean geometry texts uses similar triangles.) It has been stated that this is the most proved theorem in Euclidean geometry—a recent book presented 370 proofs of the theorem.

26. State and prove the converse of the theorem of Pythagoras. (*Hint:* Use the theorem of Pythagoras and two triangles, which must be proved congruent.)

27. Give two proofs that the three given points $(1, -8)$, $(5, 4)$ and $(10, 19)$ lie on a straight line.

28. Find an equation of a line whose x-intercept is twice its y-intercept and which passes through $(4, \tfrac{3}{2})$.

29. Find equations of the three medians of the triangle with vertices $A(-6, -1)$, $B(4, 5)$ and $C(2, -3)$ and show that the medians intersect in one point.

30. Find an equation that must be satisfied by the coordinates of any point equidistant from the two points $(0, 0)$ and (a, b). Interpret your result geometrically.

31. Find the coordinates of the two points that divide the segment joining $A(-3, 2)$ and $B(6, 9)$ into three equal parts.

32. Find the coordinates of the $(n - 1)$ points that divide the segment joining $A(a, b)$ and $B(c, d)$ into n equal parts.

33. If a point $P(x, y)$ is such that its distance from $P_1(3, 2)$ is twice its distance from $P_2(1, -5)$, find an equation that the coordinates of P must satisfy. Identify the curve.

34. Find the distance between the parallel lines $3x + 4y = 48$ and $3x + 4y + 30 = 0$.

35. Find an equation of the circle that passes through the three points $(-1, 1)$, $(3, 2)$ and $(-2, 4)$.

36. Describe the graph of $x^2 + y^2 + 4x + 6y + 28 = 0$.

37. It is conjectured that the equation $x^2 + y^2 + Dx + Ey + F = 0$ is a circle if and only if $F < 0$. Prove or disprove this.

38. Show that the points $A(-1, 9)$, $B(4, 4)$, and $C(1, 3)$ are the vertices of a right triangle. What is its area?

39. Find the point (or points) of intersection of the two curves $x^2 + y^2 = 52$ and $2y - x - 2 = 0$. Also sketch the figure.

40. Find an equation of a circle with center at (h, k) and which passes through the origin.

41. Find the point (or points) of intersection of the two curves $x^2 + y^2 - 2y - 3 = 0$ and $x^2 + y^2 - 20x + 6y + 93 = 0$.

42. Given the points $A(-1, 0)$ and $B(1, 0)$ on the x-axis. Find an equation that must be satisfied by points $P(x, y)$ such that $|\overline{AP}| + |\overline{PB}| = 4$.

43. Find an equation of a parabola having $y = -6$ as its directrix and $(0, 2)$ as its focus by starting with the definition of a parabola. Where is its vertex and what is its axis of symmetry?

44. Find the locus of all points P such that the product of the slopes of the lines through P and $A(-a, 0)$, and $B(a, 0)$ is -1. Identify the locus.

45. Find the intersection(s) of the graphs for $x^2 = ay$ and $y^2 = ax$, where $a \neq 0$ is a constant.

46. Sketch the graph of $|x| + |2y| = 1$.

47. Determine the coordinates of a point P on the segment connecting $P_1(x_1, y_1)$ and $P_2(x_2, y_2)$ such that

$$\frac{|\overline{P_1P}|}{|\overline{PP_2}|} = \frac{2}{5}$$

48. Prove by using slopes that the diagonals of a rectangle are perpendicular to each other if and only if the rectangle is a square.

49. Find the distance from the point $P(2, 3)$ to the line $x + 2y + 1 = 0$ by first finding the point of intersection of the perpendicular from P to the given line.

50. Sketch the curve $|xy| = 5$.

51. Sketch the curve $y = \dfrac{x + 2}{x - 2}$.

52. Find the values of x that satisfy the inequality $3x - 2 > \dfrac{2}{x + 1}$ by

(i) an analytic method

(ii) a graphic method $\left(\text{that is, sketch } y_1 = 3x - 2 \text{ and } y_2 = \dfrac{2}{x + 1} \text{ and read off when } y_1 > y_2.\right)$

53. Show that the curve $y = \dfrac{x^2}{x - 3}$ does not have a horizontal asymptote. Does it have a vertical asymptote?

Prove Exercises 54 through 58 by mathematical induction for all natural numbers n.

54. $1 + \dfrac{1}{2} + \dfrac{1}{2^2} + \cdots + \dfrac{1}{2^{n-1}} = 2 - 2^{1-n}$

55. $1 + r + r^2 + \cdots + r^{n-1} = \dfrac{r^n - 1}{r - 1}, \quad r \neq 1$

56. $\dfrac{1}{1 \cdot 3} + \dfrac{1}{3 \cdot 5} + \cdots + \dfrac{1}{(2n - 1)(2n + 1)} = \dfrac{n}{2n + 1}$

57. $5^n - 2^n$ is divisible by 3 where n is an arbitrary natural number.

58. $2^n > n^2$ for $n \geq 5$.

2

Functions, Limits, and Continuity

THE CONCEPT OF FUNCTION

The notion of *function* was introduced into mathematical language by the mathematician Leibnitz, one of the discoverers of calculus. This was done in the seventeenth century. The term referred originally to certain kinds of mathematical formulas where one quantity depended upon another quantity. Let us consider some examples.

EXAMPLE 1   The area of a square is a function of its edge length. The area K is given by x^2, where x is the length of a side. Note that x and K must be positive. ●

EXAMPLE 2   The force F necessary to stretch a steel spring a distance x beyond its natural length is proportional to x. Thus $F = kx$, where k, the spring constant, is positive and is independent of x. This formula was discovered by Robert Hooke in the midseventeenth century and is therefore known as *Hooke's law*. It expresses the spring force as a function of the spring displacement. ●

EXAMPLE 3   The equation $y = \sqrt{x^2 + 4}$ assigns a positive value of y to each value of x that is chosen. Thus if $x = 0$, $y = \sqrt{4} = 2$; whereas if $x = -1$, $y = \sqrt{(-1)^2 + 4} = \sqrt{5}$; and so on. Since $x^2 \geq 0$ for any real value of x, y will be smallest when $x = 0$; that is, the minimum value of y is $y|_{\min} = 2$. ●

EXAMPLE 4   The preceding examples of functions all involve formulas. However, the concept of function (to be defined shortly) transcends the notion that a function must be associated with a particular formula. For example, consider the cost C in cents of mailing a first class letter which weighs x ounces. The cost for each ounce or fraction thereof we shall label as b cents.† The following table indicates how the cost C as a function of weight x is determined.

Table 2.1.1

weight x(oz)	$0 < x \leq 1$	$1 < x \leq 2$	$2 < x \leq 3$	$3 < x \leq 4$
cost C(cents)	b	$2b$	$3b$	$4b$

This table may be continued in an obvious manner until, say, $x = 320$ oz or 20 lb (we arbitrarily chose 320 oz—in practice this is dictated by postal regulations!). It is important to note that for each weight $0 < x \leq 320$ there corresponds a unique cost C. Thus, for example, if the letter weighs 2.5 oz, the cost is $3b$ cents. ●

From the foregoing discussion, the following definition of the term function may be abstracted.

Definition   We say that y is a ***function*** of x if there is a rule or procedure by which each value of x determines a unique value of y.

The set of all permissible values of x is called the ***domain*** of the function, and the set of corresponding values of y is called the ***range*** of the function.

We call x the ***independent variable*** and allow it to have any value in the domain; y is called the ***dependent variable*** since its value depends upon the chosen x value.

Our definition is stated in terms of x and y because these are the variables that we use most often. However, different variables may be used, or the roles of x and y could be altered as is the case in the following example.

EXAMPLE 5   The equations $x = t, y = t^2$ assign values to x and y corresponding to each t. Thus we have two different functions. In both of them, t is the independent variable; x is the dependent variable in the first function whereas y is the dependent variable in the second function.

If x and y are interpreted as the rectangular coordinates of a point P in the plane and t is the time, then the equations describe the location of P at the time t. In fact, if $t \geq 0$ then a portion of the parabola $y = x^2$ is traced out by the movement of P as t increases. ●

In Example 1, the domain is the set of positive numbers and the range is also the set of positive numbers. Notice that the correspondence is one-to-one in that to each element x of the domain there corresponds one element K of the range and conversely, to each $K > 0$ there is one and only one $x > 0$.

Example 2 is also one-to-one in that to each x there is associated one value of F. In fact, for many real materials (steel, aluminum, glass, and so on) a Hooke's law holds for x positive (caused by tensile or positive force) as well as x negative

---

† Because the postal service has been changing the cost of mailing a first class letter so often in recent years, we have used the letter b rather than a specific numerical value for this cost per ounce.

(compression or negative force). However, the precise domain depends on the particular material involved.

In Example 3, the formula $y = \sqrt{x^2 + 4}$ defines a function whose domain is unrestricted x (that is, all real numbers) and whose range is $y \geq 2$. To each real number x there corresponds one real number $y \geq 2$; however, the correspondence is not one-to-one because, for example, $y = \sqrt{5}$ is obtained by substitution of $x = +1$ or $x = -1$.

In Example 4, we have a table that gives us a rule of correspondence between the set $A = \{x \mid 0 < x \leq 320\}$ and the set $B = \{b, 2b, 3b, \ldots, 320b\}$ such that to each x in A there corresponds one and only one number C. Thus the correspondence is a function whose domain is the set A and whose range is the set B. The correspondence is not one-to-one. In fact, an infinite number of x's such that $0 < x \leq 1$ map into $C = b$. The function C is a **step** function derived from its graphical representation shown in Figure 2.1.1.

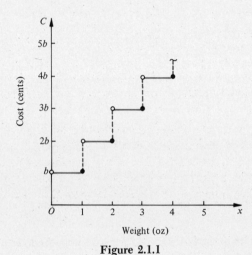

Figure 2.1.1

Example 5 yields a one-to-one correspondence for the two functions of t if t is nonnegative. Also, in this instance, the domain and ranges are $t \geq 0, x \geq 0$, and $y \geq 0$. But if the domain of t is unrestricted (that is, t can assume any real value) then the ranges are all real x, and y nonnegative. Furthermore, the correspondence between y and t is not one-to-one. For example, if $y = 9$ then $t = \pm 3$ since $9 = (\pm 3)^2$.

In the preceding examples, both the independent and dependent variables were real numbers. This does not have to be the case. An alternative, and perhaps less restrictive way of defining function is the following.

Definition A **function** is a set of ordered pairs of elements such that no two ordered pairs of the set have the same first element. The set of all first members is the **domain** of the function, and the **range** of the function is the set of all second members.

If we represent the ordered pairs as (x, y), the statement "no two ordered pairs of the set have the same first element" is seen to be equivalent to the requirement in our original definition that "each value of x determines a unique value of y."

Note that the term function is defined as a set of ordered pairs of "elements" rather than as an ordered pair of real numbers. The latter would be more restrictive, although we will primarily (but not exclusively) be interested in functions that relate real numbers. A function is said to be a **real-valued function of a real variable** if and only if both its domain and range are subsets of the set R of all real numbers. For example, the set of ordered pairs

$$\{(-3, 1), (0, 2), (3, -1), (5, 4)\}$$

is a function with domain $\{-3, 0, 3, 5\}$ and range $\{1, 2, -1, 4\}$. However, $\{(4, 3), (-1, 2), (5, 0), (4, 6)\}$ is not a function because the ordered pairs $(4, 3)$ and $(4, 6)$ have the same first element, 4.

EXAMPLE 6   Let X be the set of colleges in the United States on a particular date. Let Y be the set of positive integers and $y \in Y$ be the enrollment at x on that day (x is an element of X, that is, x is any one of the colleges). Then for each college x there is associated a positive integer $y \in Y$. This again defines a function: However, the domain would be the set of colleges and the range would be a finite subset of positive integers, so this is not a real-valued function of a real variable.   ●

We often find it convenient to use a letter to represent the function itself. Although it does not matter which letter is chosen, the most common choices are $f, g, h, F, G,$ and H. In our first definition of function, the letter can be thought of as representing the rule for determining y, given a value of x. If our second definition is used, this letter represents the set of all ordered pairs in the function.

Definition   The **graph** of a function f, in the particular case of a real-valued function of a real variable, is the set of all points in the plane associated with ordered pairs of f. For each number x in the domain of f there is a unique point (x, y) in the graph of f. Also, for every number y in the range of f there is associated *at least* one point (x, y) in its graph, where x is an element in the domain of f.

We shall find it convenient to refer to a typical element of the range as a function of the typical element of the domain. Thus we will say that the area of a square is a function of the length of its side. Also, the enrollment in a college on a particular day is a function of the college to which we refer.

EXAMPLE 7   The equation $y = |x|$, $-2 \le x \le 3$ defines a function of x whose domain is the set of real numbers $\{x \mid x \in [-2, 3]\}$ and whose range is the set $\{y \mid y \in [0, 3]\}$. In this instance, x is the independent variable and y is the dependent variable. Thus, for example, $|-1| = 1$, $|-\frac{3}{4}| = \frac{3}{4}$, $|0| = 0$, $|2.3| = 2.3$, and so on. The graph of the function is given in Figure 2.1.2.   ●

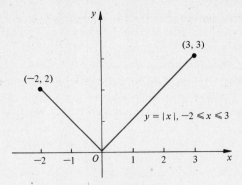

Figure 2.1.2

EXAMPLE 8   Let g be the function which is the set of ordered pairs (x, y) defined by

$$g = \{(x, y) \mid y = \sqrt{100 - x^2}\}$$

the domain of g is the set of numbers x, which yield real values for y. This occurs if and only if the quantity under the square root sign is nonnegative. Hence the condition $100 - x^2 \geq 0$ must hold. From this, we obtain the domain $-10 \leq x \leq 10$ or, equivalently, $|x| \leq 10$. The corresponding range is $0 \leq y \leq 10$ and its graph is a semicircle in the upper half-plane $(y \geq 0)$ with radius 10 and center at the origin of coordinates. The graph of g is shown in Figure 2.1.3. ●

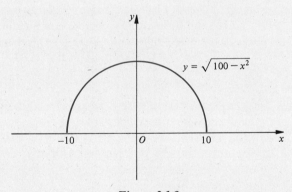

Figure 2.1.3

EXAMPLE 9   Let h be the function which is the set of ordered pairs (x, y) defined by

$$y = \begin{cases} \dfrac{2x^2 - 9x + 4}{x - 4} & \text{if } x \neq 4 \\ 5 & \text{if } x = 4 \end{cases}$$

We observe that the value of the numerator $2x^2 - 9x + 4$ is zero when $x = 4$. This implies that $x - 4$ is a factor of the numerator. In fact, our function h may be replaced by

$$y = \begin{cases} \dfrac{(x - 4)(2x - 1)}{x - 4} & \text{if } x \neq 4 \\ 5 & \text{if } x = 4 \end{cases}$$

and therefore, since $x - 4$ cancels $(x \neq 4)$, we obtain

$$y = \begin{cases} 2x - 1 & \text{if } x \neq 4 \\ 5 & \text{if } x = 4 \end{cases}$$

Thus the graph consists of the point $(4, 5)$ and all the points on the line $y = 2x - 1$ with the exception of the point $(4, 7)$ (see Figure 2.1.4). The domain of h is all real numbers and the range is all real numbers except for $y = 7$.   ●

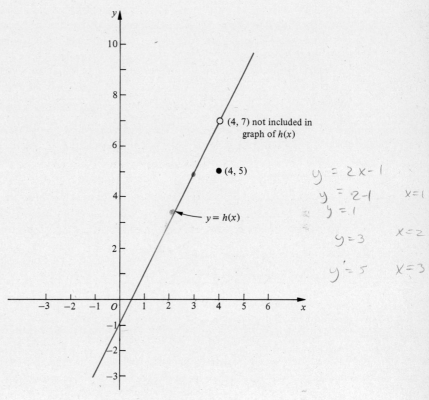

Figure 2.1.4

Exercises 2.1

In each of Exercises 1 through 22, the function is the set of ordered pairs (x, y) that satisfies the given equation. Find the domain and range for each function and sketch its graph.

1. $y = 2x - 3$

2. $y = x^2 + 3$

3. $y = 2x^2 - 5$

4. $y = \sqrt{x + 1}$

5. $y = \sqrt{36 - x^2}$

6. $y = \sqrt{x^2 - 36}$

7. $y = \sqrt{2x + 3}$

8. $y = \dfrac{x^2 - 5x + 4}{x - 4}, \quad x \neq 4$

9. $y = \begin{cases} \dfrac{x^2 - 36}{x + 6}, & x \neq -6 \\ 0 & x = -6 \end{cases}$

10. $y = \dfrac{3x^2 + x}{x}, \quad x \neq 0$

11. $y = \dfrac{x^2 - a^2}{x - a}, \quad x \neq a$

12. $y = \begin{cases} \dfrac{x^3 - 2x^2 - 4x + 8}{x - 2}, & x \neq 2 \\ 5 & x = 2 \end{cases}$

13. $y = |x - 4|$

14. $y = |2x - 3|$

15. $y = \dfrac{|x|}{x}, \quad x \neq 0$

16. $y = -\dfrac{1}{x}, \quad x \neq 0$

17. $y = \begin{cases} -3, & \text{if } x < -4 \\ 0, & \text{if } -4 \leq x \leq 4 \\ 2, & \text{if } x > 4 \end{cases}$

18. $y = \begin{cases} 3x + 1, & \text{if } x \neq 4 \\ -1 & \text{if } x = 4 \end{cases}$

19. $y = \begin{cases} x^2 & \text{if } x < 1 \\ 3x - 1, & \text{if } x \geq 1 \end{cases}$

20. $y = \begin{cases} 4x - 1 & \text{if } x \leq -1 \\ x + 3 & \text{if } x > -1 \end{cases}$

21. $y = \sqrt{x(x - 3)}$

22. $y = \sqrt{(x - 3)(x^2 - 1)}$

23. For all positive integers n, we define a function that is the set of all ordered pairs $(n, n!)$, where $n! = n(n - 1)(n - 2) \cdots 3 \cdot 2 \cdot 1$. This function is called the **factorial function.** What is its domain and range? List it for $n = 1, 2, \ldots, 8$.

24. An electric light company has the following rate schedule for n kilowatt hours:

> 10 cents per kilowatt hour if $0 < n \leq 20$
> 6 cents per kilowatt hour if $20 < n \leq 60$
> 3 cents per kilowatt hour if $n > 60$

There is a minimum charge of \$3. Express the total cost C in dollars as a function of the number of kilowatt hours and sketch the graph of the function.

25. A *relation* is simply a set of ordered pairs of elements; thus all functions are relations. But the converse is not true. In each case, indicate whether the relation below is a function. What is the domain and range? Graph the relation.

    (a) $\{(1, 3), (2, 3), (3, 4), (4, 6)\}$
    (b) $\{(x, y) \mid y > 3x + 1\}$
    (c) $y = x^2 - 1, x \in \{-2, -1, 0, 3\}$
    (d) $x = 2$

26. The area A of a square is a function of the perimeter p. Find a formula for this function. Is the perimeter p also a function of the area A? If your answer is yes, find the function.

27. The surface area S of a cube is a function of the edge x. Find a formula for this function. Express the edge as a function of S.

28. The volume V of a cube is a function of the edge x. Find a formula for this function. Express the edge x as a function of V.

29. The volume V of a cube is a function of the surface area S. Find a formula for this function. Is the surface area S also a function of the volume V? If your answer is yes, find the function.

2.2 FUNCTION NOTATION AND COMPOSITION OF FUNCTIONS

Suppose now that we are given a particular function f. If x is an element of the domain of f, the corresponding element in the range is denoted by the symbol

$$y = f(x)$$

(read "y equals f of x" or "y equals f at x"). Note that f represents the rule or procedure by which y is determined when x is given. Thus, once the domain and the rule is specified, the set of all (x, y) can, at least in theory, be computed by a machine into which each of the elements x are fed. The machine (or person) computes the value of y (the output) corresponding to the given input x.

Let us reconsider Example 3 of Section 2.1.

$$y = f(x) = \sqrt{x^2 + 4}$$

In particular, suppose that $x = -3$ is fed into the machine: It must square -3, add 4 and then take the square root of the resulting number. We have $f(-3) = \sqrt{(-3)^2 + 4} = \sqrt{13} \doteq 3.605551$ to six decimal places. Also, if $x = 0$ is inserted into the machine, there results $f(0) = \sqrt{0^2 + 4} = 2$, and so forth.

To illustrate further, if g is the function defined by

$$g(x) = x^3 - x + 3$$

then, unless otherwise stated, the domain is all real numbers. If we wish to find $g(-2)$ we simply take the expression for $g(x)$ and replace each x by -2 (that is, the substitution $x = -2$ is made throughout the expression). Thus

$$g(-2) = (-2)^3 - (-2) + 3 = -3 \qquad (-8) + 2 + 3 = -3$$

Similarly

$$g(\sqrt[3]{3}) = (\sqrt[3]{3})^3 - (\sqrt[3]{3}) + 3 = 6 - \sqrt[3]{3} \qquad \begin{aligned} &= (\sqrt[3]{3})^3 - \sqrt[3]{3} + 3 = \\ &= 3 - \sqrt[3]{3} + 3 = 6 - \sqrt[3]{3} \end{aligned}$$

In Example 3 of Section 2.1 and in the function $g(x)$ just considered, the domain was the set of all real numbers. This is because there were no restrictions, either implied or stated, placed on the values of the independent variable.

On the other hand, if Example 3 of Section 2.1 is modified so that we are given the function: $F(x) = \sqrt{x^2 + 4}$, defined on $|x| \leq 1$ then the values of x are restricted to the closed interval $-1 \leq x \leq 1$, and this would be our domain.

To exemplify further, the function h defined by

$$h(x) = \sqrt{7 - x} \quad \geq 0 \qquad \begin{aligned} &\sqrt{7 - x} \geq 0 \\ &\phantom{\sqrt{}} 7 - x \geq 0 \quad 7 - x \geq 0 \\ &\phantom{\sqrt{7-x}} 7 \geq x \end{aligned}$$

requires that $7 - x \geq 0$ since we cannot take square roots of negative numbers. Therefore the domain is all $x \leq 7$.

Finally, if u is a function defined by

$$u(z) = \frac{3z^3 + 2}{z^2 - 49} \qquad \text{Domain } u \quad z \neq \pm 7$$

then the domain of u is all $z \neq \pm 7$, because the quotient is defined if and only if the denominator is not zero.

EXAMPLE 1   Suppose that f is the function defined by the equation

$$f(x) = x^2 + 3x - 2$$

Find $f(-2)$, $f(0)$, $f(3)$, $f(m)$, $f(m^2)$, $f(f(x))$.

Solution   We have

$$\begin{aligned} f(-2) &= (-2)^2 + 3(-2) - 2 = -4 \\ f(0) &= 0^2 + 3(0) - 2 = -2 \\ f(3) &= 3^2 + 3(3) - 2 = 16 \\ f(m) &= m^2 + 3m - 2 \\ f(m^2) &= (m^2)^2 + 3(m^2) - 2 = m^4 + 3m^2 - 2 \end{aligned}$$

To find $f(f(x))$, we note that

$$f(\quad) = (\quad)^2 + 3(\quad) - 2$$

so that

$$f(f(x)) = (f(x))^2 + 3(f(x)) - 2$$

However, $f(x) = x^2 + 3x - 2$ and consequently,

$$f(f(x)) = (x^2 + 3x - 2)^2 + 3(x^2 + 3x - 2) - 2$$
$$= x^4 + 6x^3 + 5x^2 - 12x + 4 + 3x^2 + 9x - 6 - 2$$
$$= x^4 + 6x^3 + 8x^2 - 3x - 4$$

●

EXAMPLE 2   Given $F(x) = x^3 - 3x + 1$, find

$$\frac{F(x + h) - F(x)}{h} \qquad h \neq 0$$

Solution $F(x + h) = (x + h)^3 - 3(x + h) + 1$. Then

$$F(x + h) - F(x) = [(x + h)^3 - 3(x + h) + 1] - [x^3 - 3x + 1]$$
$$= x^3 + 3x^2h + 3xh^2 + h^3 - 3x - 3h + 1 - x^3 + 3x - 1$$
$$= 3x^2h + 3xh^2 + h^3 - 3h$$
$$= 3h(x^2 - 1) + h^2(3x + h)$$

Thus

$$\frac{F(x + h) - F(x)}{h} = 3(x^2 - 1) + h(3x + h)$$

where the common factor, $h \neq 0$, was divided out from numerator and denominator.   ●

Next, some basic operations or combinations of functions will be treated. First, we define the operations of sum, difference, product, and quotient of two given functions.

Definition   Given two functions $f(x)$ and $g(x)$ we define
(a) their **sum,** denoted $f + g$, by

$$(f + g)(x) = f(x) + g(x)$$

(b) their **difference,** denoted $f - g$, by

$$(f - g)(x) = f(x) - g(x)$$

(c) their **product,** denoted $f \cdot g$, by

$$(f \cdot g)(x) = f(x) \cdot g(x)$$

(d) their **quotient,** denoted $\dfrac{f}{g}$, by

$$\frac{f}{g}(x) = \frac{f(x)}{g(x)}$$

In cases (a), (b) and (c) the domain of the resulting function consists of the set of values x common to the domains of f and g. The same is true for case (d) except that we must also exclude the values of x that cause $g(x)$ to be zero.

EXAMPLE 3   If $f(x) = x^3 + 1$ and $g(x) = \dfrac{x - 1}{x + 2}$, $x \ne -2$, find each of the following functions and state their domains:

(i) $f + g$, (ii) $f - g$, (iii) $f \cdot g$, (iv) $\dfrac{f}{g}$.

Solution

(i) $(f + g)(x) = f(x) + g(x) = x^3 + 1 + \dfrac{x - 1}{x + 2}$ $x \ne -2$

(ii) $(f - g)(x) = f(x) - g(x) = x^3 + 1 - \dfrac{x - 1}{x + 2}$ $x \ne -2$

(iii) $(f \cdot g)(x) = f(x) \cdot g(x) = (x^3 + 1)\left(\dfrac{x - 1}{x + 2}\right)$ $x \ne -2$

(iv) $\left(\dfrac{f}{g}\right)(x) = \dfrac{f(x)}{g(x)} = \dfrac{x^3 + 1}{\dfrac{x - 1}{x + 2}} = \dfrac{(x^3 + 1)(x + 2)}{x - 1}$ $x \ne -2, 1$

In determining the domain of $\dfrac{f}{g}$, $x \ne 1$ because $g(1) = 0$ and division by zero must be excluded. We do not include -2 in the domain because $g(-2)$ is not defined. Note that -2 is excluded even though our final expression for $\left(\dfrac{f}{g}\right)(x)$ would make sense if $x = -2$ were substituted.   ●

EXAMPLE 4   Given that $f(x) = \sqrt{x + 4}$ and $g(x) = |x| + x$, find (i) $(f + g)(x)$, (ii) $(f - g)(x)$, (iii) $(f \cdot g)(x)$, (iv) $\left(\dfrac{f}{g}\right)(x)$. State the domain of each of these functions.

Solution   First of all we note that the domain of $f(x)$ is $[-4, \infty)$ and the domain of $g(x)$ is $(-\infty, \infty)$. Also $g(x) = 0$ whenever $x \le 0$. Therefore

(i) $(f + g)(x) = \sqrt{x + 4} + |x| + x$ $[-4, \infty)$
(ii) $(f - g)(x) = \sqrt{x + 4} - |x| - x$ $[-4, \infty)$
(iii) $(f \cdot g)(x) = \sqrt{x + 4}\,(|x| + x)$ $[-4, \infty)$
(iv) $\left(\dfrac{f}{g}\right)(x) = \dfrac{\sqrt{x + 4}}{|x| + x}$ $(0, \infty)$ why not $(-4, \infty)$?   ●

why?   ∂parc   why?   Dr.Call

The four methods of combining functions that we have so far defined are quite straightforward. There is a fifth method of forming a new function from two given ones that is quite useful, but it involves a new type of operation.

Definition   If f and g are two functions then the **composite** or **compound function,** denoted by $f \circ g$, is defined by

$$(f \circ g)(x) = f(g(x))$$

where the domain of $f \circ g$ is the set of all numbers x in the domain of g such that $g(x)$ is in the domain of f.

EXAMPLE 5 Given that f is defined by $f(x) = \sqrt{x}$ and $g(x) = x^4 + 1$, find $h(x)$ if $h = f \circ g$ and find the domain of $h(x)$.

Solution $h(x) = (f \circ g)(x) = f(g(x)) = f(x^4 + 1)$. Thus,

$$h(x) = \sqrt{x^4 + 1}$$

The domain of g is $(-\infty, \infty)$ and the domain of f is $[0, \infty)$. The domain of h is the set of all real numbers for which $x^4 + 1 \geq 0$ or, equivalently, $(-\infty, \infty)$. ●

EXAMPLE 6 Given that f is defined by $f(x) = \sqrt{x - 2}$ and $g(x) = 4x - 3$, find $h(x)$ if $h = f \circ g$ and determine the domain of $h(x)$. What is $F(x)$ if $F = g \circ f$?

Solution $h(x) = (f \circ g)(x) = f(g(x)) = f(4x - 3)$, so

$$h(x) = \sqrt{4x - 5}$$

The domain of g is $(-\infty, \infty)$ and the domain of f is $[2, \infty)$. The domain of h is the set of all x such that $4x - 3 \geq 2$, or equivalently $[\frac{5}{4}, \infty)$.

$F(x) = (g \circ f)(x) = g(\sqrt{x - 2}) = 4\sqrt{x - 2} - 3$, and the domain of F is $x \geq 2$. ●

EXAMPLE 7 Given that f is defined by $f(x) = \sqrt{4 - x}$ and $g(x) = 4 - x$, find (a) $f \circ f$, (b) $f \circ g$, and (c) $g \circ f$ giving the domain of each of the composite functions.

Solution (a) $$(f \circ f)(x) = f(f(x)) = f(\sqrt{4 - x}) = \sqrt{4 - \sqrt{4 - x}}$$

Domain: $-12 \leq x \leq 4$ since $f(x)$ is in the domain of f if $\sqrt{4 - x} \leq 4$.

(b) $$(f \circ g)(x) = f(g(x)) = f(4 - x) = \sqrt{4 - (4 - x)} = \sqrt{x}$$

Domain of $f \circ g$ is $x \geq 0$ since $g(x)$ is in the domain of f if $4 - x \leq 4$ or $x \geq 0$.

(c) $$(g \circ f)(x) = g(f(x)) = 4 - f(x) = 4 - \sqrt{4 - x}$$

Domain of $g \circ f$ is $x \leq 4$ since $f(x)$ is in the domain of g if $4 - x \geq 0$ or, equivalently, $x \leq 4$. ●

Definition A function F is said to be **even** if $F(-x) = F(x)$ for all x in the domain of F and is said to be **odd** if $F(-x) = -F(x)$.

EXAMPLE 8 For each of the functions given state whether it is even, odd, or neither, and discuss the symmetry of the graphs.

(a) $g(x) = 3|x|$
(b) $h(x) = 2x - x^3$
(c) $G(x) = 4x + 3x^2$

Solution (a) $$|x| = \sqrt{x^2} = \begin{cases} x & \text{if } x \geq 0 \\ -x & \text{if } x < 0 \end{cases}$$

and therefore

$$g(-x) = 3|-x| = 3\sqrt{(-x)^2} = 3\sqrt{x^2} = 3|x| = g(x)$$

Thus $g(x)$ is even and the graph of $g(x)$ is symmetric with respect to the y-axis.

(b) $$h(-x) = -2x - (-x)^3 = -2x + x^3 = -(2x - x^3) = -h(x)$$

Thus $h(x)$ is odd and the graph of $h(x)$ is symmetrical with respect to the origin.

(c) $$G(-x) = 4(-x) + 3(-x)^2 = -4x + 3x^2$$

Thus, $G(x) \neq G(-x)$ and also $G(-x) \neq -G(x)$. Therefore $G(x)$ is neither even nor odd and the graph of $G(x)$ is not symmetrical with respect to the y-axis or the origin. ●

Exercises 2.2

1. Given $f(x) = 3x + 8$, find
   (a) $f(-3)$
   (b) $f(0)$
   (c) $f(4)$
   (d) $f(\frac{1}{2})$
   (e) $f(x^2)$
   (f) $f(x + 3)$
   (g) $f(x + h)$
   (h) $f(x) + f(h)$
   (i) $\dfrac{f(x + h) - f(x)}{h}$, $h \neq 0$

2. Given $F(x) = 2x^2 - 3$, find
   (a) $F(-3)$
   (b) $F(\frac{1}{3})$
   (c) $F(-x)$
   (d) $F(x^2)$
   (e) $F(3x + 2)$
   (f) $F(x - h)$
   (g) $F(x) - F(h)$
   (h) $\dfrac{F(x + h) - F(x)}{h}$, $h \neq 0$

3. Given $g(x) = \sqrt{3x + 2}$, find
   (a) $g(2)$
   (b) $g(\frac{1}{3})$
   (c) $g(-2)$
   (d) $g(2x + 5)$
   (e) $\dfrac{g(x + h) - g(x)}{h}$, $h \neq 0$

4. Given $f(x) = x|x|$, find
   (a) $f(-2)$
   (b) $f(3)$
   (c) $f(-2x)$
   (d) $f(x^2)$
   (e) $f(-x^3)$

5. Given $f(x) = |x + 1|$, find:
   (a) $f(-10)$
   (b) $f(-\frac{3}{2})$
   (c) $f(\pi)$
   (d) $f(x - 1)$
   (e) the solutions of $f(2x - 1) = 5$

6. Given $f(x) = \dfrac{|x - 2| - |x + 2|}{x}$, $x \neq 0$, express $f(x)$ without absolute value bars if
   (a) $x \leq -2$
   (b) $-2 < x < 2$
   (c) $x \geq 2$

7. Given $f(u) = \dfrac{|u - 5| - |u| + 5}{u}$, $u \neq 0$, express $f(u)$ without absolute bars if
   (a) $u < 0$
   (b) $0 < u < 5$
   (c) $u \geq 5$

In Exercises 8 through 12, the functions $f(x)$ and $g(x)$ are defined. In each exercise find each of the following functions and determine their domains.

   (a) $f(x) + g(x)$
   (b) $f(x) - g(x)$
   (c) $f(x) \cdot g(x)$
   (d) $\dfrac{f(x)}{g(x)}$
   (e) $\dfrac{g(x)}{f(x)}$
   (f) $(f \circ g)(x)$
   (g) $(g \circ f)(x)$

8. $f(x) = x^2$, $g(x) = x + 1$
9. $f(x) = \sqrt{x}$, $g(x) = x - 1$
10. $f(x) = |x|$, $g(x) = 2x - 1$

11. $f(x) = |x + 1|$, $g(x) = \dfrac{1}{x}$

12. $f(x) = -\dfrac{1}{x^2}$, $g(x) = x + 3$

13. For each of the following functions, determine whether the function is even, odd, or neither.

(a) $f(x) = x^3 - 3x$

(b) $g(x) = \dfrac{1}{x^2 + 2}$

(c) $f(x) = x|x|$

(d) $F(z) = \dfrac{z + 1}{z}$

(e) $g(y) = y^{100} - y^2 - 2$

(f) $h(x) = x - |x|$

14. (a) Prove that if $f(x)$ and $g(x)$ are even functions, then $f(x) + g(x)$ and $f(x) - g(x)$ are also even; that is, the sum and differences of even functions are even.

(b) Formulate and prove a similar theorem for odd functions.

15. Prove that for any function $f(x)$ that $\frac{1}{2}[f(x) + f(-x)]$ is an even function, whereas $\frac{1}{2}[f(x) - f(-x)]$ is odd. [*Hint:* Let $g(x) = \frac{1}{2}[f(x) + f(-x)]$ and find $g(-x)$. Compare $g(-x)$ with $g(x)$, and so on.]

16. (a) Prove that *any* function $f(x)$ is expressible as the sum of an even and odd function by observing the identity

$$f(x) = \frac{f(x) + f(-x)}{2} + \frac{f(x) - f(-x)}{2}$$

(b) Write 3^x as the sum of an even function and an odd function.

17. Show that a function is expressible as the sum of an even and an odd function in one and only one way by showing that the only function that is both even and odd is the zero function. (*Note:* the zero function is the function h such that $h(x) \equiv 0$.)

18. (a) If $f(x) = kx$ where k is a constant, prove that $f(x) + f(y) = f(x + y)$ for all x and y.

(b) Find a function $g(x)$ such that $g(x) \cdot g(y) = g(x + y)$ for all x and y.

19. If $g(x) = \sqrt{x}$, prove that $\dfrac{g(x + h) - g(x)}{h} = \dfrac{1}{\sqrt{x + h} + \sqrt{x}}$ for $x > 0$, $x + h \geq 0$ and $h \neq 0$. (*Hint:* multiply by the conjugate of the numerator.)

20. If $g(x) = \sqrt{x}$, prove that $\dfrac{g(x) - g(a)}{x - a} = \dfrac{1}{\sqrt{x} + \sqrt{a}}$ for $x > 0$, $a > 0$, and $x \neq a$.

21. Let $g(x) = ax^2 + bx + c$, where a, b, and c are constants. Prove that $g(x + 2) - 2g(x + 1) + g(x) = 2a$ for all x.

22. Let $F(x) = Ax + B$ where A and B are constants. Prove that $F(x + 2h) - 2F(x + h) + F(x) = 0$ for all x and h.

23. A right circular cylinder is inscribed in a sphere of radius a. Find the volume V and total surface area S in terms of the altitude h of the cylinder.

24. If $f(x) = x^2 - 2x + 4$, find the two functions $g(x)$ for which $(f \circ g)(x) = x^2 - 4x + 7$.

25. Solve the inequality

$$|F(x) + G(x)| < |F(x)| + |G(x)|$$

if $F(x) = 3x - 5$ and $G(x) = 7 - x$.

26. Find the zeros and the domain of the function f defined by

$$f(t) = \frac{t^3 - 6t^2 + 11t - 6}{t^2 - 7t + 10}$$

provided that the denominator $\neq 0$.

27. A function f is defined by

$$f(x) = \frac{3x - 2}{x + 1} \qquad x \neq -1$$

Find a function g such that $(f \circ g)(x) = x$. For this function g, determine $(g \circ f)(x)$.

2.3

2.3.1 Linear Functions

A **linear function** is a function of the form $f(x) = mx + b$ where m and b are constants. The name is used because the graph of this function is a straight line. As was seen in Chapter 1, every nonvertical line is the graph of a linear function.

If $m = 0$, the function becomes $f(x) = b$. A function of this type is called a **constant function,** due to the fact that the value of the function is unchanged for all choices of x. The graph is a horizontal line.

If $m = 1$ and $b = 0$, we have another special function, the **identity function,** $f(x) = x$.

2.3.2 Quadratic Functions

A function of the form

$$f(x) = ax^2 + bx + c$$

where a, b, c are constants and $a \neq 0$ is called a **quadratic function.** The graph of a quadratic function is a parabola with the line $x = -\dfrac{b}{2a}$ as the vertical axis of symmetry. This can be verified by completing the square, and we can draw the graph by using the methods learned in Section 1.7.

There are times when we are specifically interested in determining for what values of x a quadratic function is equal to zero. This means we wish to solve the equation

$$ax^2 + bx + c = 0$$

When possible, this is done by factoring the equation. Also, we may use the method of completing the square, which leads to the quadratic formula

$$x = \frac{-b \pm \sqrt{b^2 - 4ac}}{2a}$$

The term $\Delta = b^2 - 4ac$, which appears under the square root sign in the quadratic formula, is called the **discriminant.** The quadratic equation has two roots, one root, or no roots if $\Delta > 0, \Delta = 0$, or $\Delta < 0$ respectively (Figure 2.3.1 exhibits these cases for $a > 0$). If $\Delta < 0$, we say that the quadratic function is irreducible.

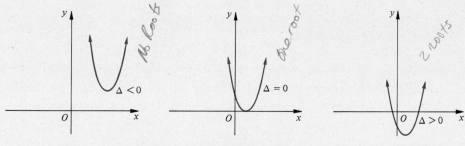

Figure 2.3.1

It is useful to note that, if a quadratic function is never zero, then it is always positive if $a > 0$ and it is always negative if $a < 0$.

EXAMPLE 1   For which values of k does $f(x) = x^2 - 4kx + 9$ have two distinct real roots?

Solution $\Delta = (-4k)^2 - 4(1)(9) = 16k^2 - 36 > 0$ must   hold.   Thus $k^2 > \frac{36}{16} = \frac{9}{4}$ and therefore $|k| > \frac{3}{2}$.   ●

2.3.3 *Polynomials*

An expression of the form

$$a_n x^n + a_{n-1} x^{n-1} + \cdots + a_1 x + a_0 \tag{1}$$

where $a_n, a_{n-1}, \ldots, a_1, a_0$ are real numbers and n is a nonnegative integer, is called a *polynomial* in x. Examples of polynomials are

$$3x^3 - x^2 - \frac{x}{2} + 4, \qquad \sqrt{3}\,x + \sqrt{2}, \qquad -3 \tag{2}$$

The domain of all polynomials is all real values of x.

The number a_k in the term $a_k x^k$ in (1) is called the *coefficient* of x^k. The integer n in (1) is called the *degree* of the polynomial provided that $a_n \neq 0$. Thus the expressions in (2) are polynomials of degree 3, 1 and 0, respectively. The polynomial 0 is *not* considered to have a degree.

A detailed study of polynomials is not part of our course work. However, some of the results describing their behavior will be useful in our future discussions. For convenience, we shall state (without proof) certain important theorems concerning polynomials. These are probably familiar results and can be easily understood by working out some examples.

Theorem 1   If f and g are polynomials, then so are $f \pm g$, $f \cdot g$ and $f \circ g$.

Theorem 2   Let $f(x)$ and $g(x)$ be polynomials of degree n and m, respectively, where $n \geq m > 0$. Then there exists a polynomial $Q(x)$, called the *quotient,* of degree $n - m$ and also a polynomial $R(x)$, the remainder, of degree less than m, such that

$$f(x) = g(x)\,Q(x) + R(x) \tag{3}$$

In the special case $R \equiv 0$, we say that $f(x)$ is *divisible* by $g(x)$ or that $g(x)$ is a *divisor* of $f(x)$.

EXAMPLE 2   If we divide $f(x) = x^4 + 3x^3 - 2x^2 - 2x + 10$ by $g(x) = x^2 + x - 1$, the long division process yields $Q(x) = x^2 + 2x - 3$ and $R(x) = 3x + 7$. This should be verified.   ●

An important special case occurs if $n > 1$ and $g(x) = x - r$, where r is some number. This we cite as

Theorem 3   If a polynomial $f(x)$ of degree $n > 1$ is divided by the linear function $x - r$, the remainder is the number $f(r)$.

Theorem 4   A polynomial of degree n has at most n real roots.

This theorem is true even if each root is counted as many times as it occurs. (We refer to this as the **multiplicity** of the root.)

EXAMPLE 3   Solve the polynomial equation $x^4 - 16 = 0$.

Solution
$$x^4 - 16 = (x^2 - 4)(x^2 + 4) = (x - 2)(x + 2)(x^2 + 4) = 0$$
The real roots are $x = 2$ and $x = -2$. (There is no real root of $x^2 + 4 = 0$.)   ●

EXAMPLE 4   Find a polynomial that has a double root at $x = -8$ and simple roots $\pm\sqrt{2}$.

Solution
$$f(x) = (x + 8)^2(x - \sqrt{2})(x + \sqrt{2})$$
$$= (x + 8)^2(x^2 - 2)$$ ●

EXAMPLE 5   Let $f(x)$ and $g(x)$ be polynomials in x of degrees less than or equal to n. Prove that if $f(x) = g(x)$ for more than n different values of x that the polynomials must be identical.

Solution   Let $h(x) = f(x) - g(x)$, then $h(x)$ is a polynomial of degree at most n, and therefore it has at most n zeros. But we are given that $h(x)$ has more than n zeros. Thus $h(x) \equiv 0$ or $f(x) \equiv g(x)$. Conversely, if
$$a_n x^n + a_{n-1} x^{n-1} + \cdots + a_1 x + a_0 \equiv b_n x^n + b_{n-1} x^{n-1} + \cdots + b_1 x + b_0$$
then $a_k = b_k$ ($k = 0, 1, \ldots, n$); that is, two polynomials are identically equal if and only if the coefficients of like powers are equal.   ∎

2.3.4   Other Special Functions

A **rational function** is a function that can be expressed as the quotient of two polynomial functions with no variable common factor. It can be defined by
$$f(x) = \frac{a_n x^n + a_{n-1} x^{n-1} + \cdots + a_0}{b_m x^m + b_{m-1} x^{m-1} + \cdots + b_0} \tag{4}$$
Its domain is all x for which the denominator is not zero. Thus
$$f(x) = \frac{1}{x^2}, x \neq 0 \qquad \text{and} \qquad g(x) = \frac{1}{x} - \frac{x^2 - 1}{(x + 3)^2}, \qquad x \neq 0, -3$$
are examples of rational functions. The following is easily established.

Theorem 5   Let $f(x)$ and $g(x)$ be rational functions, then $f \pm g, f \cdot g$ and $f \circ g$ are rational functions. Furthermore, if $g \not\equiv 0$ then $\dfrac{f}{g}$ is a rational function.

Algebraic operations include addition, subtraction, multiplication, division, raising to a power, and taking roots. Any function that can be formed by a finite number of algebraic operations on polynomials is called an **algebraic function.** Examples of algebraic functions are:

$$f(x) = \sqrt{3x + 2}$$

$$g(x) = \frac{(5x^2 + 3x + 1)\sqrt{2x + 3}}{3x - (7x^4 + x + 1)^{1/3}}$$

Not all functions are algebraic. The simplest nonalgebraic or **transcendental functions** are exponential, logarithmic, and trigonometric functions, to be discussed in Chapters 8 and 9.

2.3.5   *The Greatest Integer Function*

The notation $[\![x]\!]$ means the greatest integer in x or, equivalently, the greatest integer $\leq x$. Thus, if x is any number and n is an integer such that $n \leq x < n + 1$, then $[\![x]\!] = n$. For example

$$[\![\pi]\!] = 3, \; [\![5/4]\!] = 1, \; [\![-1/2]\!] = -1, \text{ and } [\![4]\!] = 4.$$

EXAMPLE 6   Sketch the graph of $F(x) = [\![x]\!]$.

Solution   If $-1 \leq x < 0$ then $F(x) = -1$,
if $0 \leq x < 1$ then $F(x) = 0$,
if $1 \leq x < 2$ then $F(x) = 1$.

Noting that this pattern continues, we find that our graph is the step function shown in Figure 2.3.2.   ●

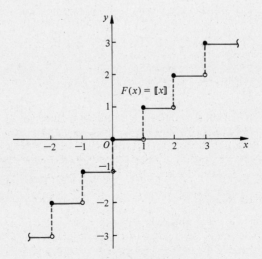

Figure 2.3.2

Exercises 2.3

1. For which values of b does $x^2 - 2bx + 8 = 0$ have two distinct roots? (*Hint:* For what values of b are the two roots equal?)
2. For which values of k does $f(z) = z^2 + z + k = 0$ have no real roots?
3. Write $g(m) = 2m^2 - 2m - \frac{3}{2}$ as a product of two linear functions.
4. For what values of α does $z^2 - 2\alpha z + \alpha = 0$ have real roots?
5. Divide $m^5 - 2m^4 + 3m^2 - m - 6$ by $m^2 + 1$.
6. Divide $z^2 + z - 2$ by $z^2 - z + 1$.
7. Find a polynomial that has a double root at -3, single roots at -4 and 0, and no others.
8. Find a polynomial whose roots are 3 with multiplicity 2, and $\pm\sqrt{2}$, each with multiplicity 1.
9. The polynomial $g(x) = 2x^4 + 9x^3 + 11x^2 - 4$ has a double root at $x = -2$. Find the other roots.
10. The polynomial $v(x) = x^4 + x^3 + 3x^2 + 9x - 54$ has roots 2 and -3. Are there any other real roots?
11. Find *all* the real zeros of $f(x) = (x - 1)^2(x^3 + 27)(x^2 - 16)^3$.
12. Find all the *real* roots of $g(x) = (x^2 - x + 1)^4(x - 1)(3x - 2)$.
13. Find all the real zeros of $y^3 - 8y^2 + 25y - 26 = 0$ by finding a root that is an integer.
14. Solve $3z^3 - z^2 - 6z + 2 = 0$ by first finding a rational root.
15. The graph of a polynomial of degree 2 in x can be made to pass through the three points $(0, 5)$, $(1, 7)$ and $(-1, 9)$. Find the polynomial. (*Hint:* Form $y = ax^2 + bx + c$ and determine the coefficients a, b, and c by substituting the given data.)
16. The graph of a polynomial of degree 3 in x (a cubic) can be made to pass through the four points $(0, -5)$, $(1, -3)$, $(-1, -11)$ and $(2, 1)$. Find the polynomial.
17. Show that the equation $x^3 - a = 0$ has only one real root if a is real. (*Hint:* Write $a = b^3$ where b is a real number, and then factor $x^3 - b^3$ and show that the quadratic equation formed by setting the quadratic factor equal to 0 has no real roots.)
18. Find all polynomials of degree ≤ 2 which satisfy for all x the condition that $f(x) = f(1 - x)$.
19. For which positive integers n is $x^n + a^n$ divisible by $x + a$? Prove your assertion. (*Hint:* Consider the equation $x^n + a^n = 0$.)
20. The equation $4x^3 - 4x^2 - 31x + 10 = 0$ has a rational root. Solve the equation.
21. Factor $y^4 + 3y^3 - 8y - 24$.
22. Form the cubic equation in x that has roots a, b, and c.
23. Prove that $\sqrt{1 + 2x}$ cannot be a polynomial. (*Hint:* Compare domains.)

In Exercises 24 through 29, find the domain and range of the given function and sketch its graph.

24. $g(x) = [\![x - 1]\!]$
25. $h(x) = \dfrac{x}{2} - [\![x]\!]$
26. $F(x) = \left[\!\!\left[\dfrac{1}{1 + x^2}\right]\!\!\right]$
27. $H(x) = \begin{cases} 1 & \text{for } x > 0 \\ \dfrac{1}{2} & \text{for } x = 0 \\ 0 & \text{for } x < 0 \end{cases}$

   $H(x)$ is called the *Heaviside unit step function* and is useful in the analysis of vibrating sytems.
28. $H(x) - H(x - 1)$, where $H(x)$ is the function defined in Exercise 27.
29. $3H(x) + 2H(x - 5)$, where H is defined in Exercise 27.

30. The function $g(x) = \dfrac{x^n - a^n}{x - a}$, $x \neq a$, where n is a positive integer, is a rational function. Is it also a polynomial? Prove your assertion.
31. Prove that a polynomial $P(x)$ (a) is an even function if and only if it consists of even powers of x, (b) is an odd function if and only if odd powers of x are present, and (c) is neither even nor odd if both even and odd powers are in the expression.

2.4

THE LIMIT OF A FUNCTION— INTUITIVE APPROACH

The fundamental concept upon which calculus is based is that of the limit of a function. It is our objective to investigate this topic at first intuitively and then to take a more precise view of this question.

EXAMPLE 1   Let $f(x) = 2x + 3$. We pose two questions about this function:

(i) What is the value of f when $x = 1$?
(ii) How does $f(x)$ behave when x is close to 1?

Solution   The answer to (i) we know, namely, $f(1) = 5$. To determine the answer to (ii), let us first make a table of values (Table 2.4.1) showing $f(x)$ versus x for several values of x near 1 but not equal to 1. Table 2.4.1 certainly suggests that, when x is near 1, whether a little less than 1 or a little greater than 1, the function $f(x)$ is close to 5. In fact, as x gets close to 1, $2x$ is near 2 and $2x + 3$ is near 5. Intuitively, it appears that $f(x)$ can be made to come as close to 5 as we like by simply choosing x sufficiently close to 1. Symbolically, we express this by writing $\lim_{x \to 1} f(x) = 5$, or "the limit of f as x approaches 1 is 5." Alternatively, we write $\lim_{x \to 1} (2x + 3) = 5$ as a concise way of asserting that $\lim_{x \to 1} f(x) = 5$ where f is the function defined by $f(x) = 2x + 3$.   ●

Table 2.4.1

x	$f(x)$
0.9	4.8
0.95	4.9
0.99	4.98
0.999	4.998
1.1	5.2
1.01	5.02
1.001	5.002

EXAMPLE 2   Let g be the function defined by $g(x) = x^2 - 1$. We ask for the limit of g as x approaches 2. Again, we set up a table of values (Table 2.4.2). It seems plausible that, as $x \to 2$ from values of x slightly less than or a little greater than 2, $\lim_{x \to 2} g(x) = 3$. We also observe that $g(2) = 3$. However, we have not proved that $\lim_{x \to 2} g(x) = 3$—all that we did was to try a few values on either side of $x = 2$. There is no guarantee that if we continued our computations by choosing x

Table 2.4.2

x	1	1.5	1.9	1.99	1.999	2.001	2.01	2.1	2.5
$g(x)$	0	1.25	2.61	2.96	2.996	3.004	3.04	3.41	5.25

"sufficiently close" to 2 that $|g(x) - 3| = |(x^2 - 1) - 3|$ can be made still closer and closer to zero. ●

EXAMPLE 3   Let h be the function defined by

$$h(x) = \frac{x^2 - 4x + 3}{x - 1} \qquad \text{if } x \neq 1$$

$$h(1) = 3$$

We seek the limit of $h(x)$ as x approaches 1. Furthermore is $\lim\limits_{x \to 1} h(x) = h(1)$?

To answer this, we observe that $h(1) = 3$, and for $x \neq 1$,

$$h(x) = \frac{(x - 1)(x - 3)}{x - 1} = x - 3$$

Thus, we can set up a table of values (Table 2.4.3)

Table 2.4.3

x	0.9	0.95	0.99	0.999	1.001	1.01	1.05	1.1
$h(x)$	-2.1	-2.05	-2.01	-2.001	-1.999	-1.99	-1.95	-1.9

It is therefore suggested that, as $x \to 1$ from values of x slightly less than or a little greater than 1, $\lim\limits_{x \to 1} h(x) = -2$ and $\lim\limits_{x \to 1} h(x) \neq h(1)$ since $h(1) = 3$. A graph of $h(x)$ is shown in Figure 2.4.1. ●

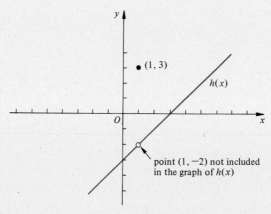

point $(1, -2)$ not included
in the graph of $h(x)$

Figure 2.4.1

EXAMPLE 4   Consider the greatest integer or bracket function $f(x) = [\![x]\!]$. Its value was defined to be the greatest integer less than or equal to x. Does $f(x)$ have a limit at $x = 3$? To answer this, consider values of x very close to 3 but less than 3. For these values, $[\![x]\!] = 2$. However, for values of x greater than 3 and very close to 3, $[\![x]\!] = 3$. Thus, no matter what number is proposed for the limit of $[\![x]\!]$ as $x \to 3$, $[\![x]\!]$ will always differ from the proposed limit by at least $\frac{1}{2}$ for all numbers x on (at least) one side of 3, no matter how close x comes to 3. Thus the limit does *not* exist. ●

The reader should observe that $\lim_{x \to a} [\![x]\!]$ exists if and only if a is not an integer.

In the next section, a formal definition of the statement $\lim_{x \to a} f(x) = L$ will be given. For the present, we have at least some intuitive idea of the concept. Keeping in mind that *$\lim_{x \to a} f(x)$ refers to the behavior of $f(x)$ when x is close to but not equal to a*, we are going to state (without proof) some rules that will enable us to find most limits (if they exist).

Rule 1: If $f(x)$ is a polynomial then $\lim_{x \to a} f(x) = f(a)$.

Both Examples 1 and 2 are illustrations of this rule. This says that, for a polynomial, we can find the limit by "direct substitution" (that is, by substituting $x = a$ into the function). This method did not work in Examples 3 or 4.

Rule 2: Let c be a constant and assume that $\lim_{x \to a} f(x) = L$ and $\lim_{x \to a} g(x) = M$,

then   (A) $\lim_{x \to a} [f(x) + g(x)] = L + M$

       (B) $\lim_{x \to a} [f(x) - g(x)] = L - M$

       (C) $\lim_{x \to a} [cf(x)] = cL$

       (D) $\lim_{x \to a} [f(x) \cdot g(x)] = L \cdot M$

       (E) $\lim_{x \to a} \left[\dfrac{f(x)}{g(x)}\right] = \dfrac{L}{M}$, providing $M \neq 0$.

These rules assert that, when combining functions, we take the limit of each part and then combine in the indicated manner.

Rule 3: Let $\lim_{x \to a} f(x) = L$ and let n be a positive integer.

(A)   If $L > 0$ then $\lim_{x \to a} \sqrt[n]{f(x)} = \sqrt[n]{L}$,

(B)   If $L \leq 0$ and n is odd then $\lim_{x \to a} \sqrt[n]{f(x)} = \sqrt[n]{L}$,

(C)   If $L \leq 0$ and n is even then $\lim_{x \to a} \sqrt[n]{f(x)}$ does not exist.

This rule says that, when working with radicals, if you are not taking an even root of a negative number, the limit exists and is the value to be expected.

EXAMPLE 5   If k is a constant, find $\lim_{x \to a} k$.

Solution $\lim_{x \to a} k = k$ because k is a constant and thus does not change when x changes. Also, k is a polynomial so this result follows from Rule 1.   ●

EXAMPLE 6   Find $\lim\limits_{x\to 2} (x^3 - 2x + 3)$

Solution $\lim\limits_{x\to 2} (x^3 - 2x + 3) = (2)^3 - 2(2) + 3 = 7$ ●

EXAMPLE 7   Find $\lim\limits_{x\to 6} \sqrt{2x - 8}$ if it exists.

Solution $\lim\limits_{x\to 6} \sqrt{2x - 8} = \sqrt{\lim\limits_{x\to 6} (2x - 8)} = \sqrt{4} = 2$ ●

EXAMPLE 8   Find $\lim\limits_{x\to 3} \left[5(x^2 + 1) - \dfrac{8}{x + 1} \right]$ if it exists.

Solution $\lim\limits_{x\to 3} \left[5(x^2 + 1) - \dfrac{8}{x + 1} \right] = \lim\limits_{x\to 3} [5(x^2 + 1)] - \lim\limits_{x\to 3} \left[\dfrac{8}{x + 1} \right]$

$$= [\lim\limits_{x\to 3} 5][\lim\limits_{x\to 3} (x^2 + 1)] - \frac{\lim\limits_{x\to 3} 8}{\lim\limits_{x\to 3} (x + 1)}$$

$$= [5] \cdot [10] - \frac{8}{4} = 48$$

This example was worked out step by step, showing the use of Rules 1, 2(B), 2(C) and 2(E). In general, it is not necessary to be so detailed.   ●

EXAMPLE 9   Find $\lim\limits_{u\to 2} \dfrac{u^2 - 4}{u - 2}$ if it exists.

Solution   If we try to let

$$\lim\limits_{u\to 2} \frac{u^2 - 4}{u - 2} = \frac{\lim\limits_{u\to 2} (u^2 - 4)}{\lim\limits_{u\to 2} (u - 2)}$$

we are stuck because the limit of the denominator is zero. However, recall that $\lim\limits_{u\to 2} f(u)$ means that u is never equal to 2. Then

$$\frac{u^2 - 4}{u - 2} = \frac{(u - 2)(u + 2)}{u - 2} = u + 2$$

Hence

$$\lim\limits_{u\to 2} \frac{u^2 - 4}{u - 2} = \lim\limits_{u\to 2} (u + 2) = 4$$ ●

It turns out that the limits of many types of functions (not just polynomials) can be evaluated by "direct substitution." In general, we must be careful about such things as zero denominators, an even root of a negative number, and a function that is defined in an unusual manner (as in Examples 3 and 4). Note that "direct substitution" works in Example 8. However, due to the zero denominator in Example 9, some simplification was first necessary.

Example 10 Find $\lim\limits_{y\to 1}\dfrac{y^2+3}{y-1}$ if it exists.

Solution The limit of this quotient does not exist. One can observe this by constructing a table of values for the function as y gets close to 1. Note that in this example the limit of the denominator is zero, but the limit of the numerator is not zero. In all such cases, the limit does not exist. (We shall study this situation in more detail in Section 2.8.) In Example 9, both numerator and denominator had zero limits, and there we were able to simplify the fraction and try again. ●

Example 11 Find $\lim\limits_{h\to 0}\dfrac{3(x+h)^2-3x^2}{h}$

Solution
$$\lim_{h\to 0}\frac{3(x+h)^2-3x^2}{h}=\lim_{h\to 0}\frac{3x^2+6xh+3h^2-3x^2}{h}$$
$$=\lim_{h\to 0}\frac{6xh+3h^2}{h}=\lim_{h\to 0}\frac{h(6x+3h)}{h}$$
$$=\lim_{h\to 0}(6x+3h)=6x+0=6x$$

since x is independent of h. ●

Example 12 If $f(x)=x^3-3x+20$, find $\lim\limits_{x\to a}\dfrac{f(x)-f(a)}{x-a}$ if it exists.

Solution
$$\frac{f(x)-f(a)}{x-a}=\frac{(x^3-3x+20)-(a^3-3a+20)}{x-a}$$
$$=\frac{x^3-a^3-3(x-a)}{x-a}$$
$$=\frac{x^3-a^3}{x-a}-3\qquad\text{if }x\neq a$$

Consequently,
$$\frac{f(x)-f(a)}{x-a}=\frac{(x^2+ax+a^2)(x-a)}{x-a}-3$$
$$=x^2+ax+a^2-3$$
$$\lim_{x\to a}\frac{f(x)-f(a)}{x-a}=\lim_{x\to a}(x^2+ax+a^2-3)$$
$$=a^2+a^2+a^2-3=3a^2-3\qquad●$$

Exercises 2.4

In each of Exercises 1 through 15, a function f is defined and a number a is given. Compute the values of the function for values of the independent variable approaching a (but ≠ a) and try to determine whether or not f has a limit L at a. If so, what is the value of that limit? Sketch the function in the vicinity of a.

1. $f(x) = x^2, \quad a = 5$

2. $f(x) = \dfrac{x}{x + 1}, \quad a = 0$

3. $f(x) = \sqrt{x}, \quad a = 0$

4. $f(x) = x[\![x]\!], \quad a = 1$

5. $f(x) = \dfrac{x^2 - 1}{x - 1}, \quad a = 1$

6. $f(x) = \dfrac{x^2 - 3x + 2}{x - 2}, \quad a = 2$

7. $f(x) = \dfrac{x^2 - x - 6}{x + 2}, \quad a = -2$

8. $f(x) = \dfrac{x - 4}{\sqrt{x} - 2}, \quad a = 4$

9. $f(u) = \dfrac{2u - 1}{u - \dfrac{1}{2}}, \quad a = \dfrac{1}{2}$

10. $f(x) = \dfrac{3x^2 + 13x + 4}{x + 4}, \quad a = -4$

11. $f(x) = \dfrac{6x^2 + x - 2}{3x + 2}, \quad a = -\dfrac{2}{3}$

12. $f(x) = \dfrac{x^3 - 27}{x - 3}, \quad a = 3$

13. $f(x) = \dfrac{(2 - x)^3 - 8}{x}, \quad a = 0$

14. $f(t) = \sqrt{t - 1}, \quad a = 1$

15. $f(t) = \dfrac{1 - \cos t}{t}; \quad t$ is measured in radians, $\quad a = 0$

In Exercises 16 through 37, find the limit (if it exists) using the rules given in this section.

16. $\lim\limits_{x \to 4} (x^2 - 7x + 5)$

17. $\lim\limits_{x \to 0} (x^4 + 3x^2 - x + 6)$

18. $\lim\limits_{x \to 5} \dfrac{x + 3}{2x - 1}$

19. $\lim\limits_{x \to 2} \dfrac{x^2 - x + 1}{x^2 + 2x + 3}$

20. $\lim\limits_{x \to -1} \dfrac{x^3 + x^2 - x + 4}{x^2 - 5x + \sqrt{3}}$

21. $\lim\limits_{x \to -2} \sqrt{2x + 5}$

22. $\lim\limits_{u \to 4} \sqrt{7u - 25}$

23. $\lim\limits_{x \to -2} \sqrt{3x + 5}$

24. $\lim\limits_{u \to -2} \sqrt[3]{3u + 5}$

25. $\lim\limits_{x \to 4} \dfrac{x^2 - 16}{x - 4}$

26. $\lim\limits_{t \to -4} \dfrac{t^2 - 16}{t - 4}$

27. $\lim\limits_{x \to 4} \sqrt{3x - 12}$

28. $\lim\limits_{x \to 1} \dfrac{x^2 + x}{x - 1}$

29. $\lim\limits_{x \to 1} \dfrac{x^3 - x}{x - 1}$

30. $\lim\limits_{x \to 3} \dfrac{x^2 + 2x - 15}{x - 3}$

31. $\lim\limits_{z \to 0} \dfrac{\sqrt{z^2 + z^4}}{|z|}$

32. $\lim\limits_{h \to 0} \dfrac{5(t + h)^2 - 5t^2}{h}$

33. $\lim\limits_{h \to 0} \dfrac{\dfrac{1}{u + h} - \dfrac{1}{u}}{h}, \quad u \neq 0$

34. If $f(t) = t^2 - 4t - 9$, find $\lim\limits_{t \to t_0} \dfrac{f(t) - f(t_0)}{t - t_0}$

35. If $f(t) = t^2 - 4t - 9$, find $\lim\limits_{h \to 0} \dfrac{f(t_0 + h) - f(t_0)}{h}$

36. If $f(x) = x^3 + 7x^2 - x - 4$, find $\lim\limits_{x \to a} \dfrac{g(x) - g(a)}{x - a}$

37. If $F(x) = x^2 + 7x + 11$, find $\lim\limits_{h \to 0} \dfrac{F(a + h) - F(a - h)}{2h}$

2.5   THE LIMIT OF A FUNCTION— FORMAL APPROACH

In our previous development, we have discussed in an intuitive way what is meant by the statement

$$\lim_{x \to a} f(x) = L \qquad (1)$$

Equation (1) is taken to mean that the values of $f(x)$ will be as close to L as desired, provided that x is sufficiently close to (but not equal to) a. In other words (1) means that we can make the magnitude of $f(x) - L$ as small as desired by choosing $|x - a|$ sufficiently small or, equivalently, by restricting x to a suitably small interval about $x = a$ (excluding $x = a$ itself). To make this concept precise, we introduce the following notation. Consider the Greek letter ε (read "epsilon"). This stands for a positive number (chosen as small as desired). Consider then the inequality $L - \varepsilon < f(x) < L + \varepsilon$, which geometrically is shown in Figure 2.5.1.

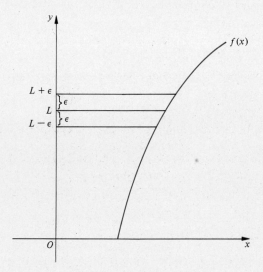

Figure 2.5.1

The smaller the positive value of ε, the smaller the interval $(L - \varepsilon, L + \varepsilon)$. Now, having specified what is meant by "$f(x)$ is close to L" by a choice of ε, we must next make precise what is meant by "all values of x sufficiently close to a (but $x \neq a$)." This is done by introducing an interval about $x = a$ on the x-axis. If a small positive number δ (read "delta") is chosen, then $|x - a| < \delta$ implies that $a - \delta < x < a + \delta$. This interval is shown in Figure 2.5.2. The length of this interval can be made smaller by choosing δ smaller.

The idea of limit is made precise by the following definition.

Definition

$$\lim_{x \to a} f(x) = L$$

if for *any* $\varepsilon > 0$ there can be found a $\delta > 0$ such that for all $x \neq a$ satisfying $|x - a| < \delta$, we have $|f(x) - L| < \varepsilon$.

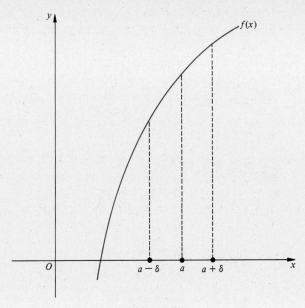

Figure 2.5.2

This may be construed as an "ε–δ game." One player specifies the number $\varepsilon > 0$ as small as desired. With this choice of ε, the second player must be able to determine a $\delta > 0$, so that for all $x \neq a$ satisfying $|x - a| < \delta$, we have $|f(x) - L| < \varepsilon$. This definition is exhibited in Figure 2.5.3 where $f(h) = L - \varepsilon$, $f(k) = L + \varepsilon$ and, in general, $a - h \neq k - a$. Now, if a positive number δ is chosen so that $\delta < a - h$ and $\delta < k - a$, then, when $x \neq a$ is in the interval $a - \delta < x < a + \delta$, we also must have $|f(x) - L| < \varepsilon$.

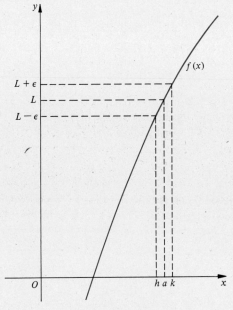

Figure 2.5.3

We sometimes use the phrase **deleted neighborhood** of a to refer to an open interval centered at a and excluding the point a itself. This is what our definition requires when we ask for all $x \neq a$ satisfying $|x - a| < \delta$. This can also be written $0 < |x - a| < \delta$. The inequality $0 < |x - a|$ guarantees that $x \neq a$.

Let us return to Example 1 of Section 2.4 where $f(x) = 2x + 3$ and we are concerned with whether or not $\lim_{x \to 1} f(x)$ exists. Our numerical plausibility approach suggested that $\lim_{x \to 1} (2x + 3) = 5$; that is, $L = 5$. Now form

$$|f(x) - 5| = |2x + 3 - 5| = |2x - 2| = 2|x - 1|$$

so that a challenge number $\varepsilon = 0.01$ can be met by taking $\delta = 0.005$ since, if $0 < |x - 1| < 0.005$, then $|f(x) - 5| = 2|x - 1| < 2(0.005) = 0.01$. Similarly a challenge number $\varepsilon = 0.001$ can be met by taking $\delta = 0.0005$ because, if $0 < |x - 1| < 0.0005$, then $|f(x) - 5| < 0.001$. In general, for any **arbitrary** positive ε, no matter how small, if we choose $0 < |x - 1| < \delta = \dfrac{\varepsilon}{2}$ then $|f(x) - 5| < \varepsilon$. Thus we have **proved** that $\lim_{x \to 1} (2x + 3) = 5$. Note that by writing $0 < |x - 1|$, we are insisting that $x \neq 1$, even though in this case $x = 1$ causes no problem (in fact $|f(1) - 5| = 0$).

It follows from our geometric interpretation of the definition of limit that, once one value of δ is found to be correct, any smaller positive value may also be used. Therefore in this example we could have chosen $\delta = \dfrac{\varepsilon}{3}$ or $\delta = \dfrac{\varepsilon}{10}$ or any $\delta \leq \dfrac{\varepsilon}{2}$.

EXAMPLE 1   Prove that $\lim_{x \to 3} (5x - 2) = 13$.

Solution   We are going to require $|x - 3| < \delta$. Then $|f(x) - L| = |5x - 2 - 13| = |5x - 15| = 5|x - 3|$. Thus, for any positive challenge number ε (no matter how small), if we choose $\delta = \dfrac{\varepsilon}{5}$, then, for $0 < |x - 3| < \delta$, $|f(x) - L| < \varepsilon$.

Any positive $\delta < \dfrac{\varepsilon}{5}$ can be used in place of $\dfrac{\varepsilon}{5}$ as the required δ in this example.

●

EXAMPLE 2   Does $\lim_{x \to 3} \dfrac{x^2 - 9}{x - 3}$ exist? Prove your assertion.

Solution   The function $f(x) = \dfrac{x^2 - 9}{x - 3}$ is defined for all x, except $x = 3$. In fact, $f(3)$ is *not* defined, and it does not have to be in order for the limit to exist. We factor the numerator

$$f(x) = \frac{x^2 - 9}{x - 3} = \frac{(x - 3)(x + 3)}{x - 3} = x + 3 \qquad x \neq 3$$

Thus it is suggested that $\lim_{x \to 3} f(x) = 6$. Form

$$|f(x) - 6| = |x + 3 - 6| = |x - 3| \qquad x \neq 3$$

Therefore, if $0 < |x - 3| < \varepsilon$, then $|f(x) - 6| < \varepsilon$; that is, in this special case we take $\delta = \varepsilon$. Therefore, $\lim\limits_{x \to 3} \dfrac{x^2 - 9}{x - 3} = 6$. The graph of the function $f(x) = \dfrac{x^2 - 9}{x - 3}$ is shown in Figure 2.5.4 and is a straight line with the point $(3, 6)$ deleted.

●

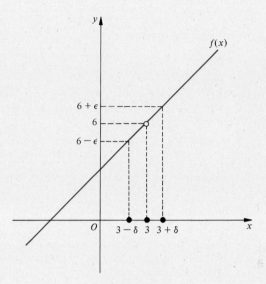

Figure 2.5.4

EXAMPLE 3   Prove that $\lim\limits_{x \to 2} (x^2 + 1) = 5$.

Solution   Form the difference $|x^2 + 1 - 5| = |x^2 - 4|$. We must show that for any $\varepsilon > 0$ there exists a $\delta > 0$ such that

$$|x^2 - 4| < \varepsilon \quad \text{whenever} \quad 0 < |x - 2| < \delta$$

Now, $|x^2 - 4| = |(x - 2)(x + 2)| = |x - 2||x + 2|$. We must show that $|x^2 - 4|$ can be made as small as desired provided that x is sufficiently close to 2. Now $|x - 2|$ can certainly be made as small as we like, but what about $|x + 2|$? If x is close to 2 then $|x + 2|$ is close to 4. Because values close to 2 are under consideration, we can concern ourselves with those values of x for which $|x - 2| < 1$; thus, we are imposing a restriction on δ, namely, that $\delta \leq 1$. The inequality $|x - 2| < 1$ is equivalent to $-1 < x - 2 < 1$ or, by addition of 2 to each of the three members, $1 < x < 3$. Also, $3 < x + 2 < 5$.
Thus

$$|x - 2| < 1 \quad \text{implies that} \quad 3 < |x + 2| < 5$$

Therefore

$$5 < x < 3$$

$$|x^2 - 4| = |x - 2||x + 2| < 5|x - 2| \quad \text{provided that } |x - 2| < 1.$$

Furthermore, we must also have

$$|x - 2| \cdot 5 < \varepsilon$$

or, equivalently,

$$|x - 2| < \frac{\varepsilon}{5}$$

Thus if δ is chosen as the smaller of the two numbers 1 and $\frac{\varepsilon}{5}$, then whenever $|x - 2| < \delta$, it follows that $|x - 2| < \frac{\varepsilon}{5}$ and $|x + 2| < 5$ so that

$$|x^2 - 4| < \frac{\varepsilon}{5} \cdot 5 = \varepsilon$$

Therefore we conclude that if $\delta = \min\left(1, \frac{\varepsilon}{5}\right)$ that

$$|x^2 - 4| < \varepsilon \qquad \text{when } 0 < |x - 2| < \delta \qquad \bullet$$

EXAMPLE 4   Show that $\lim_{x \to 2} (x^3 - 5) = 3$.

Solution
$$|x^3 - 5 - 3| = |x^3 - 8| = |(x - 2)(x^2 + 2x + 4)|$$
$$= |x - 2||x^2 + 2x + 4|$$

where $x^3 - 8$ was factored as the difference of two cubes $(x)^3 - (2)^3$.

Next, we must deal with $|x^2 + 2x + 4|$ when x is in the vicinity of 2. Suppose we confine our attention to the values of x in $[1.5, 2.5]$. Therefore

$$|x^2 + 2x + 4| \le |x|^2 + 2|x| + 4 \le (2.5)^2 + 2(2.5) + 4$$

so that

$$|x^2 + 2x + 4| \le 15.25$$

and also

$$|x^3 - 5 - 3| \le 15.25 \, |x - 2|$$

if $1.5 \le x \le 2.5$. Thus

$$|x^3 - 5 - 3| < \varepsilon$$

provided that

$$|x - 2| < \frac{\varepsilon}{15.25}$$

Therefore we choose δ to be the smaller of the two numbers $\frac{\varepsilon}{15.25}$ and $\frac{1}{2}$, and we require that $0 < |x - 2| < \delta$ thereby completing the task.

Note that in this case it was *not* necessary to insist that $x \ne 2$ because $f(2) = 3$ and $f(x) \to 3$ as $x \to 2$. $\qquad \bullet$

EXAMPLE 5   Prove that $\lim_{u \to 5} \dfrac{4}{u - 3} = 2$.

Solution   Form the absolute value of the difference

$$\left| \frac{4}{u - 3} - 2 \right| = \left| \frac{4 - 2u + 6}{u - 3} \right| = 2 \left| \frac{5 - u}{u - 3} \right| = \frac{2}{|u - 3|} |5 - u|$$

We must establish that $\left| \dfrac{4}{u - 3} - 2 \right|$ can be made as small as desired when u is close to 5. To this end, we seek an upper bound for $\dfrac{2}{|u-3|}$. Let us, for convenience sake, require that $\delta \leq 1$. This means that u is at first restricted to the interval $4 < u < 6$. Thus $1 < u - 3 < 3$ and $1 < |u - 3| < 3$ since $u - 3 > 0$ in this interval. If we take reciprocals and multiply by 2 then $2 > \dfrac{2}{|u-3|} > \dfrac{2}{3}$ so that 2 is a required upper bound for the quantity $\dfrac{2}{|u-3|}$. Therefore

$$\left| \frac{4}{u-3} - 2 \right| < 2|5 - u|$$

We seek to make $2|5 - u| < \varepsilon$ or, equivalently, $|5 - u| < \dfrac{\varepsilon}{2}$. Therefore if we choose δ to be the smaller of 1 and $\dfrac{\varepsilon}{2}$, we are assured that $\left| \dfrac{4}{u-3} - 2 \right| < \varepsilon$. Since this is true for *any* $\varepsilon > 0$, it follows that $\lim\limits_{u \to 5} \dfrac{4}{u-3} = 2$. ●

In Section 2.4, we stated a number of rules that are useful in finding limits. Having used the definition of limit to verify specific limit examples, we are now capable of proving most of these rules along with some other results. We shall state each rule as a theorem and prove some of these results.

Theorem 1   If $\lim\limits_{x \to a} f(x) = L$ and $\lim\limits_{x \to a} g(x) = M$ (2a)

then

$$\lim_{x \to a} [f(x) \pm g(x)] = L \pm M \qquad (2b)$$

Proof   We must prove that

$$\lim_{x \to a} [f(x) + g(x)] = L + M \qquad (3)$$

By the definition of a limit we are to prove that for any $\varepsilon > 0$ there is a $\delta > 0$ such that if $0 < |x - a| < \delta$ then

$$|f(x) + g(x) - (L + M)| < \varepsilon$$

This is the meaning of (3). From the hypotheses (2a), there exists a $\delta_1 > 0$ such that if $0 < |x - a| < \delta_1$, then

$$|f(x) - L| < \frac{\varepsilon}{2}$$

Similarly, from the fact that $\lim\limits_{x \to a} g(x) = M$ it follows that a $\delta_2 > 0$ exists such that if $0 < |x - a| < \delta_2$, then

$$|g(x) - M| < \frac{\varepsilon}{2}$$

Now, let δ be the minimum of the two numbers δ_1 and δ_2. Thus if $0 < |x - a| < \delta$, it follows that

$$|f(x) + g(x) - (L + M)| = |f(x) - L + g(x) - M|$$

$$\leq |f(x) - L| + |g(x) - M| < \frac{\varepsilon}{2} + \frac{\varepsilon}{2} = \varepsilon$$

The proof that $\lim_{x \to a} [f(x) - g(x)] = L - M$ follows in the same manner and it is suggested that the reader establish it as an exercise.

In words, we have proved that the limit of the sum of two functions equals the sum of the limits of the functions (assuming the existence of these limits). ∎

Theorem 2 If $\lim_{x \to a} f(x) = L$ and c is *any* constant, then

$$\lim_{x \to a} c\, f(x) = cL \tag{4}$$

This is an easy theorem to prove (try it). However, it easily follows as a special case of the next theorem.

Theorem 3 If $\lim_{x \to a} f(x) = L$ and $\lim_{x \to a} g(x) = M$, then

$$\lim_{x \to a} f(x)\, g(x) = LM$$

Proof In order to utilize the differences of the respective functions and their limits, we use the algebraic identity

$$f(x)\, g(x) - LM = [f(x) - L]\, g(x) + [g(x) - M]L \tag{5}$$

On taking the absolute value of both sides and using the fact that $|A + B| \leq |A| + |B|$, and $|A \cdot B| = |A| \cdot |B|$ for arbitrary real A and B, there results

$$|f(x)\, g(x) - LM| \leq |f(x) - L|\,|g(x)| + |g(x) - M|\,|L| \tag{6}$$

If $\varepsilon > 0$ is prescribed (no matter how small), it is our objective to make each of the two terms on the right side of (6) less than $\frac{\varepsilon}{2}$ when x is suitably restricted.

First, suppose that both L and M are *not* zero. Therefore there is a $\delta_1 > 0$ such that, if $0 < |x - a| < \delta_1$, then

$$|g(x)| < 2|M| \tag{7}$$

Since $\lim_{x \to a} f(x) = L$, there exists a $\delta_2 > 0$ such that, if $0 < |x - a| < \delta_2$, then

$$|f(x) - L| < \frac{\varepsilon}{4|M|} \tag{8}$$

Also, from $\lim_{x \to a} g(x) = M$, there is a $\delta_3 > 0$ such that, if $0 < |x - a| < \delta_3$, we have

$$|g(x) - M| < \frac{\varepsilon}{2|L|} \tag{9}$$

Now, let $\delta = \min(\delta_1, \delta_2, \delta_3)$. If $0 < |x - a| < \delta$, then all three of the inequalities (7) through (9) hold. Thus

$$|f(x) - L||g(x)| + |g(x) - M||L| < \frac{\varepsilon}{4|M|} 2|M| + \frac{\varepsilon}{2|L|} |L|$$

$$= \frac{\varepsilon}{2} + \frac{\varepsilon}{2} = \varepsilon \qquad (10)$$

Therefore, from (6) and (10) we obtain

$$|f(x)\,g(x) - LM| < \varepsilon \qquad (11)$$

which implies that $\lim_{x \to a} f(x)\,g(x) = LM$ if $L \neq 0$ and $M \neq 0$.

The proofs of the other cases where $LM = 0$ are left to the reader. ∎

Theorems 1 and 3 can be extended to any finite number of functions by use of mathematical induction to yield

Theorem 4 If $\lim_{x \to a} f_1(x) = L_1$, $\lim_{x \to a} f_2(x) = L_2, \ldots$, and $\lim_{x \to a} f_n(x) = L_n$, then

$$\lim_{x \to a} [f_1(x) \pm f_2(x) \pm \cdots \pm f_n(x)] = L_1 \pm L_2 \pm \cdots \pm L_n$$

Theorem 5 If $\lim_{x \to a} f_1(x) = L_1$, $\lim_{x \to a} f_2(x) = L_2, \ldots$, and $\lim_{x \to a} f_n(x) = L_n$, then

$$\lim_{x \to a} [f_1(x) \cdot f_2(x) \cdot \cdots \cdot f_n(x)] = L_1 \cdot L_2 \cdot \cdots \cdot L_n$$

The following special cases are easily established

Theorem 6 If $f(x) \equiv k$, the constant function, then

$$\lim_{x \to a} f(x) = k$$

Proof For any positive ε and any value of x,

$$|f(x) - k| = |k - k| = |0| = 0 < \varepsilon \qquad ∎$$

Theorem 7 $\lim_{x \to a} x = a$

Proof $$|x - a| < \varepsilon \qquad \text{if } 0 < |x - a| < \delta = \varepsilon \qquad ∎$$

Theorem 8 If $P(x)$ is any polynomial; that is, if

$$P(x) = a_n x^n + a_{n-1} x^{n-1} + \cdots + a_1 x + a_0$$

where $a_n, a_{n-1}, \ldots, a_0$ are arbitrary real constants, then

$$\lim_{x \to a} P(x) = P(a)$$

The proof follows from limit Theorems 2, 4, 5, 6, and 7.

Theorem 9   If

$$\lim_{x \to a} f(x) = L \quad \text{and} \quad \lim_{x \to a} g(x) = M \tag{12}$$

and if $M \neq 0$, then

$$\lim_{x \to a} \frac{f(x)}{g(x)} = \frac{L}{M} \tag{13}$$

The proof involves a manipulation similar to that used to prove Theorem 3 and is left as an exercise (see Exercise 36).

Theorem 10 $\lim\limits_{x \to a} \sqrt{x} = \sqrt{a}$ if $a > 0$.

Proof   Restrict x to $\dfrac{a}{2} \leq x \leq \dfrac{3a}{2}$; that is $\delta \leq \dfrac{a}{2}$.

$$\sqrt{x} - \sqrt{a} = \frac{(\sqrt{x} - \sqrt{a})(\sqrt{x} + \sqrt{a})}{\sqrt{x} + \sqrt{a}} = \frac{x - a}{\sqrt{x} + \sqrt{a}}$$

Thus

$$|\sqrt{x} - \sqrt{a}| = \frac{|x - a|}{\sqrt{x} + \sqrt{a}}$$

since the denominator is greater than 0. Therefore, for any $\varepsilon > 0$,

$$|\sqrt{x} - \sqrt{a}| \leq \frac{|x - a|}{\sqrt{\dfrac{a}{2}} + \sqrt{a}} < \varepsilon$$

provided that $0 < |x - a| < \left(\sqrt{\dfrac{a}{2}} + \sqrt{a}\right)\varepsilon$. Consequently,

$$|\sqrt{x} - \sqrt{a}| < \varepsilon \quad \text{if } \delta = \min\left(\frac{a}{2}, \left(\sqrt{\frac{a}{2}} + \sqrt{a}\right)\varepsilon\right) \qquad \blacksquare$$

EXAMPLE 6   Prove that $\lim\limits_{x \to 5} \sqrt[3]{x} = \sqrt[3]{5}$.

Solution   If $4 < x < 6$

$$|\sqrt[3]{x} - \sqrt[3]{5}| = \left| \frac{\sqrt[3]{x} - \sqrt[3]{5}}{1} \cdot \frac{\sqrt[3]{x^2} + \sqrt[3]{5x} + \sqrt[3]{25}}{\sqrt[3]{x^2} + \sqrt[3]{5x} + \sqrt[3]{25}} \right| < |x - 5|$$

since the product of the numerator terms equals $|x - 5|$ and the denominator is greater than 1. Thus for any $\varepsilon > 0$,

$$|\sqrt[3]{x} - \sqrt[3]{5}| < \varepsilon \quad \text{if } 0 < |x - 5| < \delta = \min(1, \varepsilon) \qquad \bullet$$

More generally, by the same approach we can show that $\lim\limits_{x \to a} \sqrt[3]{x} = \sqrt[3]{a}$ if $a > 0$ and for any positive integer n, $\lim\limits_{x \to a} \sqrt[n]{x} = \sqrt[n]{a}$ where $a > 0$. The proofs of these facts are left as exercises. These are special cases of Rule 3, Section 2.4. For convenience, we now restate this rule as a theorem.

Theorem 11 If $\lim\limits_{x \to a} f(x) = L$ and if n is a positive integer, then

(A) $\lim\limits_{x \to a} \sqrt[n]{f(x)} = \sqrt[n]{L}$ if $L > 0$

(B) $\lim\limits_{x \to a} \sqrt[n]{f(x)} = \sqrt[n]{L}$ if $L \leq 0$ and n is odd

(C) $\lim\limits_{x \to a} \sqrt[n]{f(x)}$ does not exist if $L \leq 0$ and n is even

Two other useful results are:

Theorem 12 If $\lim\limits_{x \to a} f(x) = L$ and $L > 0$, then there is a deleted neighborhood of a where $f(x)$ is positive.

Theorem 13 **(Squeeze Theorem):** If in some deleted neighborhood of $x = a$
$$h(x) \leq f(x) \leq g(x)$$
and if $\lim\limits_{x \to a} h(x) = \lim\limits_{x \to a} g(x) = L$, then $\lim\limits_{x \to a} f(x)$ exists and $\lim\limits_{x \to a} f(x) = L$.

Proofs of Theorems 12 and 13 as well as other related results are left as exercises for the reader (see Exercises 30 through 35).

Exercises 2.5

In Exercises 1 through 7, we are given $f(x)$, a, and L and that $\lim\limits_{x \to a} f(x) = L$. For the given specific $\varepsilon > 0$, find a δ such that $|f(x) - L| < \varepsilon$ when $0 < |x - a| < \delta$.

1. $\lim\limits_{x \to 2} (4x + 3) = 11$, $\varepsilon = 0.02$

2. $\lim\limits_{x \to 4} (3 - 2x) = -5$, $\varepsilon = 0.01$

3. $\lim\limits_{x \to 3} (2 - 5x) = -13$, $\varepsilon = 0.005$

4. $\lim\limits_{x \to -1} (3 + 7x) = -4$, $\varepsilon = 0.035$

5. $\lim\limits_{x \to 5} \dfrac{x^2 - 25}{x - 5} = 10$, $\varepsilon = 0.01$

6. $\lim\limits_{x \to 9} \sqrt{x} = 3$, $\varepsilon = 0.001$

7. $\lim\limits_{x \to 1/2} \dfrac{4x^2 - 1}{2x - 1} = 2$, $\varepsilon = 0.02$

In Exercises 8 through 20, establish the following limits directly from the definition. This means that for any $\varepsilon > 0$, find a $\delta > 0$ such that $|f(x) - L| < \varepsilon$ whenever $0 < |x - a| < \delta$.

8. $\lim\limits_{x \to 2} 6x = 12$

9. $\lim\limits_{x \to 4} (2x - 3) = 5$

10. $\lim_{x \to -1} (3 - 4x) = 7$

11. $\lim_{x \to 4} \dfrac{x^2 - 16}{x - 4} = 8$

12. $\lim_{x \to 6} x^2 = 36$

13. $\lim_{x \to \frac{1}{2}} x^2 = \frac{1}{4}$

14. $\lim_{x \to 3} \dfrac{2}{x - 1} = 1$

15. $\lim_{x \to 1} \dfrac{x}{x + 1} = \dfrac{1}{2}$

16. $\lim_{x \to 2} (x^2 - 2x) = 0$

17. $\lim_{x \to 16} \sqrt{x} = 4$

18. $\lim_{x \to 2} (x^2 - x - 1) = 1$

19. $\lim_{x \to 1} x^3 = 1$

20. $\lim_{x \to 5} \sqrt{x + 4} = 3$

21. If $g(x) = \dfrac{|x|}{3x}$, does $\lim_{x \to 0} g(x)$ exist? If so, what is it? Prove your answer.

22. If $h(x) = \dfrac{x - |x|}{x}$, does $\lim_{x \to 0} h(x)$ exist? Prove your answer.

23. A function g is defined by

$$g(x) = \begin{cases} \dfrac{x^3}{a^2} & \text{if } 0 < x < a \\ a & \text{if } x = a \\ 3a - \dfrac{2a^3}{x^2} & \text{if } x > a \end{cases}$$

Prove that $\lim_{x \to a} g(x) = g(a)$.

24. Prove that if $\lim_{x \to a} f(x) = L$, then $\lim_{x \to a} |f(x)| = |L|$. (*Hint:* Establish that $\big||a| - |b|\big| \leq |a - b|$.)

25. Prove that a function cannot possess two different limits. This is an example of a **uniqueness theorem**. It guarantees that if a function has a limit it must be unique. (*Hint:* Assume that $\lim_{x \to a} f(x) = L_1$ and $\lim_{x \to a} f(x) = L_2$, where $L_1 \neq L_2$ and establish a contradiction.)

In Exercises 26 through 29, determine the indicated limit if it exists.

26. $\lim_{x \to 1} \dfrac{x - 1}{x^3 - 1}$

27. $\lim_{x \to 1} \dfrac{(x + 2)(\sqrt{x} - 1)}{x^2 + x - 2}$

28. $\lim_{x \to 0} \dfrac{\sqrt{1 + x} - \sqrt{1 - x}}{x}$

29. $\lim_{x \to b} \dfrac{x(x^2 + 2bx - 3b^2)}{x^2 + bx - 2b^2}$

Exercises 30 through 35 are interrelated.

30. If $\lim_{x \to a} f(x) = L$ and $L > 0$, then there is a deleted neighborhood of a $(x \neq a)$ where $f(x) > 0$. In fact, if we choose $\varepsilon = \dfrac{L}{2}$ show that $\dfrac{L}{2} < f(x) < \dfrac{3L}{2}$ when $0 < |x - a| < \delta$.

31. Show that a limit of a function whose values are nonnegative is nonnegative. (*Hint:* Assume that the limit is negative and establish a contradiction.)

32. If $f(x) \leq g(x)$ in some deleted neighborhood of $x = a$ and $\lim_{x \to a} f(x) = L$ and $\lim_{x \to a} g(x) = M$, then $L \leq M$.

   (*Hint:* Use Exercise 23.)

33. Prove the **sandwich theorem:** If in some deleted neighborhood of $x = a$,

$$h(x) \leq f(x) \leq g(x)$$

   and if $\lim_{x \to a} h(x) = A$ and $\lim_{x \to a} g(x) = B$, then if $\lim_{x \to a} f(x)$ exists, we have $A \leq \lim_{x \to a} f(x) \leq B$.

34. Prove the *squeeze theorem:* If, in some deleted neighborhood of $x = a$,

$$h(x) \leq f(x) \leq g(x)$$

and if $\lim\limits_{x \to a} h(x) = \lim\limits_{x \to a} g(x) = L$, then $\lim\limits_{x \to a} f(x)$ exists and $\lim\limits_{x \to a} f(x) = L$. (Note that we do not have to know beforehand that $\lim\limits_{x \to a} f(x)$ exists, which is required in the more general sandwich theorem.)

35. Prove that if $\lim\limits_{x \to a} f(x) = 0$ and $g(x)$ is bounded in a neighborhood of $x = a$, then $\lim\limits_{x \to a} f(x) \, g(x) = 0$. (*Hint:* Use the squeeze theorem.)

36. Prove Theorem 9. *Hint:* Start with the following

$$\frac{f(x)}{g(x)} - \frac{L}{M} = \frac{f(x)\,M - L\,g(x)}{M\,g(x)} = \frac{(f(x) - L)M + L(M - g(x))}{M\,g(x)}$$

2.6   ONE-SIDED LIMITS

The function f defined by

$$f(x) = \sqrt{x}$$

has as its domain $[0, \infty)$. Therefore

$$\lim_{x \to 0} f(x)$$

does not exist because no deleted neighborhood of $x = 0$ exists in the domain of f. However, for $\delta = \varepsilon^2$ we have for $0 < x < \delta$ that $0 < f(x) < \varepsilon$. Thus $f(x)$ can be made as close to 0 as desired by restricting x to be positive and sufficiently close to 0. In many applications we need consider only one-sided limits. In this case the *one-sided limit was from the right* or the *right-hand limit,* which we now define.

Definition   Let f be a function that is defined at every number in some open interval (a, c), where $a < c$. Then the *limit of $f(x)$ as x approaches a from the right* is L, written

$$\lim_{x \to a^+} f(x) = L \tag{1}$$

if for any $\varepsilon > 0$ there exists a $\delta > 0$ such that

$$|f(x) - L| < \varepsilon \quad \text{whenever} \quad 0 < x - a < \delta$$

Thus we say

$$\lim_{x \to 0^+} \sqrt{x} = 0$$

Similarly, when $x \to a^-$, that is, the independent variable is restricted to values less than a number a, we say that x approaches a from the left. The limit (if it exists) is said to be a *one-sided limit from the left* or the *left-hand limit.*

Definition   Let f be a function that is defined in the open interval (c, a) where $c < a$. Then the *limit of $f(x)$ as x approaches a from the left* is L, written

$$\lim_{x \to a^-} f(x) = L \tag{2}$$

if for any positive ε there exists a $\delta > 0$ such that

$$|f(x) - L| < \varepsilon \quad \text{whenever} \quad -\delta < x - a < 0$$

These one-sided limits are less restrictive than the limit as defined in Section 2.5. In fact, the limit as defined there is a ***two-sided limit*** to distinguish it from the one-sided limit.

It should be noted that the rules stated in Section 2.4 and the limit theorems stated and proved in Section 2.5 are also valid for one-sided limits where $x \to a$ is replaced by $x \to a^+$ or $x \to a^-$.

EXAMPLE 1   For a function

$$F(x) = \begin{cases} 2x & 0 \leq x < 3 \\ 9 - x & x \geq 3 \end{cases}$$

find (a) $\lim\limits_{x \to 3^-} F(x)$ and (b) $\lim\limits_{x \to 3^+} F(x)$; (c) does $\lim\limits_{x \to 3} F(x)$ exist? (See Figure 2.6.1.)

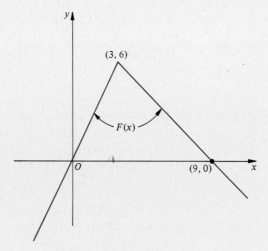

Figure 2.6.1

Solution   (a) For $x < 3$, $F(x) = 2x$ so that

$$\lim_{x \to 3^-} F(x) = \lim_{x \to 3^-} 2x = 2 \left(\lim_{x \to 3^-} x \right) = 2(3) = 6$$

(b) For $x > 3$, $F(x) = 9 - x$ and we have

$$\lim_{x \to 3^+} F(x) = \lim_{x \to 3^+} (9 - x) = 9 - 3 = 6$$

(c) $\lim\limits_{x \to 3} F(x)$ exists since

$$\lim_{x \to 3^-} F(x) = \lim_{x \to 3^+} F(x)$$

(see Theorem 1 of this section and Exercise 18). In fact,

$$\lim_{x \to 3} F(x) = F(3) = 6$$

EXAMPLE 2   Find $\displaystyle\lim_{x\to 3^+} \frac{x-3}{\sqrt{x^2-9}}$

Solution   The domain of $f(x) = \dfrac{x-3}{\sqrt{x^2-9}}$ is $|x| > 3$ and the graph for $x > 3$ is shown in Figure 2.6.2. Now for $x > 3$,

$$\frac{x-3}{\sqrt{x^2-9}} = \frac{x-3}{\sqrt{x-3}\cdot\sqrt{x+3}} = \frac{\sqrt{x-3}}{\sqrt{x+3}}$$

$$\lim_{x\to 3^+}\frac{\sqrt{x-3}}{\sqrt{x+3}} = \frac{\displaystyle\lim_{x\to 3^+}\sqrt{x-3}}{\displaystyle\lim_{x\to 3^+}\sqrt{x+3}} = \frac{\sqrt{\displaystyle\lim_{x\to 3^+}(x-3)}}{\sqrt{\displaystyle\lim_{x\to 3^+}(x+3)}} = \frac{\sqrt{0}}{\sqrt{6}} = 0$$

Note that $\displaystyle\lim_{x\to 3^-} \frac{x-3}{\sqrt{x^2-9}}$ does not exist because $0 < x < 3$ are not in the domain of the function.   ●

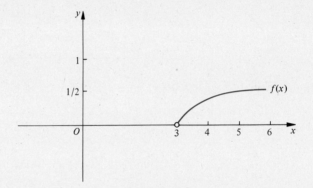

Figure 2.6.2

An important result relating one-sided and two-sided limits is the following.

Theorem 1 $\displaystyle\lim_{x\to a} f(x) = L$ if and only if $\displaystyle\lim_{x\to a^+} f(x) = L$ and $\displaystyle\lim_{x\to a^-} f(x) = L$.

EXAMPLE 3   For $g(x) = |x-3|$, find (a) $\displaystyle\lim_{x\to 3^+} g(x)$, (b) $\displaystyle\lim_{x\to 3^-} g(x)$, and (c) $\displaystyle\lim_{x\to 3} g(x)$ if the limits in question exist.

(See Figure 2.6.3)

Solution
$$g(x) = \begin{cases} x-3 & \text{if } x \geq 3 \\ 3-x & \text{if } x < 3 \end{cases}$$

(a) $\displaystyle\lim_{x\to 3^+} g(x) = \lim_{x\to 3^+}(x-3) = \lim_{x\to 3^+}x - \lim_{x\to 3^+}3 = 3 - 3 = 0$

(b) $\displaystyle\lim_{x\to 3^-} g(x) = \lim_{x\to 3^-}(3-x) = \lim_{x\to 3^-}3 - \lim_{x\to 3^-}x = 3 - 3 = 0$

(c) Since $\displaystyle\lim_{x\to 3^+} g(x) = \lim_{x\to 3^-} g(x) = 0$, this implies that $\displaystyle\lim_{x\to 3} g(x)$ exists and also equals 0.   ●

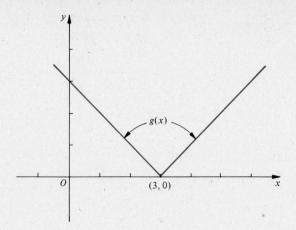

Figure 2.6.3

EXAMPLE 4

$$f(x) = \begin{cases} x^2 + x - 3 & \text{if } x < 2 \\ 3x - 1 & \text{if } x > 2 \end{cases}$$

Find (a) $\lim\limits_{x \to 2^-} f(x)$, (b) $\lim\limits_{x \to 2^+} f(x)$, and (c) $\lim\limits_{x \to 2} f(x)$ if the limits in question exist.

Solution   Refer to Figure 2.6.4

(a)
$$\lim_{x \to 2^-} f(x) = \lim_{x \to 2^-} (x^2 + x - 3)$$

$$= (\lim_{x \to 2^-} x)(\lim_{x \to 2^-} x) + \lim_{x \to 2^-} x - \lim_{x \to 2^-} 3$$

$$= 2^2 + 2 - 3 = 3$$

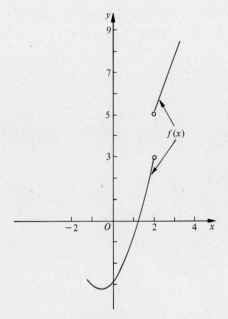

Figure 2.6.4

(b)
$$\lim_{x \to 2^+} f(x) = \lim_{x \to 2^+} (3x - 1)$$

$$= \lim_{x \to 2^+} 3x - \lim_{x \to 2^+} 1$$

$$= 3(\lim_{x \to 2^+} x) - \lim_{x \to 2^+} 1$$

$$= 3(2) - 1 = 6 - 1 = 5$$

(c) Since $\lim_{x \to 2^-} f(x) \neq \lim_{x \to 2^+} f(x)$, it follows that $\lim_{x \to 2} f(x)$ cannot exist.

●

EXAMPLE 5   Find $\lim\limits_{x \to 0^+} \dfrac{\sqrt{x}}{\sqrt{4 + \sqrt{x}} - 2}$ if it exists.

Solution $\lim\limits_{x \to 0^+} \dfrac{\sqrt{x}}{\sqrt{4 + \sqrt{x}} - 2} \cdot \dfrac{\sqrt{4 + \sqrt{x}} + 2}{\sqrt{4 + \sqrt{x}} + 2} = \lim\limits_{x \to 0^+} \dfrac{(\sqrt{x})(\sqrt{4 + \sqrt{x}} + 2)}{\sqrt{x}}$

$$= \lim_{x \to 0^+} (\sqrt{4 + \sqrt{x}} + 2) = \lim_{x \to 0^+} \sqrt{4 + \sqrt{x}} + \lim_{x \to 0^+} 2$$

$$= \sqrt{\lim_{x \to 0^+}(4 + \sqrt{x})} + 2 = \sqrt{\lim_{x \to 0^+} 4 + \sqrt{\lim_{x \to 0^+} x}} + 2$$

$$= \sqrt{4 + 0} + 2 = 4$$

Note that the given function is not defined at $x = 0$.

●

Exercises 2.6

Find each of the following limits (if it exists).

1. $f(x) = \begin{cases} 3 & \text{if } x > 5 \\ 2 & \text{if } x = 5 \\ 0 & \text{if } x < 5 \end{cases}$

   (a) $\lim\limits_{x \to 5^+} f(x)$ (b) $\lim\limits_{x \to 5^-} f(x)$ (c) $\lim\limits_{x \to 5} f(x)$

2. $g(x) = \begin{cases} 4 & \text{if } x > -2 \\ -3 & \text{if } x = -2 \\ 4 & \text{if } x < -2 \end{cases}$

   (a) $\lim\limits_{x \to -2^+} g(x)$ (b) $\lim\limits_{x \to -2^-} g(x)$ (c) $\lim\limits_{x \to -2} g(x)$

3. $h(x) = \begin{cases} x^2 & \text{if } x > 0 \\ 1 & \text{if } x = 0 \\ 0 & \text{if } x < 0 \end{cases}$

   (a) $\lim\limits_{x \to 0^+} h(x)$ (b) $\lim\limits_{x \to 0^-} h(x)$ (c) $\lim\limits_{x \to 0} h(x)$

4. $F(z) = \begin{cases} z & \text{if } z > 2 \\ 2 & \text{if } z = 2 \\ 5 - z & \text{if } z < 2 \end{cases}$

   (a) $\lim\limits_{z \to 2^+} F(z)$ (b) $\lim\limits_{z \to 2^-} F(z)$ (c) $\lim\limits_{z \to 2} F(z)$

5. $f(s) = \begin{cases} 4 + s & \text{if } s \geq 3 \\ 10 - s & \text{if } s < 3 \end{cases}$

   (a) $\lim\limits_{s \to 3^+} f(s)$ (b) $\lim\limits_{s \to 3^-} f(s)$ (c) $\lim\limits_{s \to 3} f(s)$

6. $g(u) = \begin{cases} 2 + u^2 & \text{if } u \le 3 \\ 8 + [\![u]\!] & \text{if } u > 3 \end{cases}$

    (a) $\lim\limits_{u \to 3^+} g(u)$ (b) $\lim\limits_{u \to 3^-} g(u)$ (c) $\lim\limits_{u \to 3} g(u)$

7. $f(x) = |x + 4|$

    (a) $\lim\limits_{x \to -4^+} f(x)$ (b) $\lim\limits_{x \to -4^-} f(x)$ (c) $\lim\limits_{x \to -4} f(x)$

8. $h(x) = \dfrac{|2x|}{x}$

    (a) $\lim\limits_{x \to 0^+} h(x)$ (b) $\lim\limits_{x \to 0^-} h(x)$ (c) $\lim\limits_{x \to 0} h(x)$

9. $f(x) = [\![x]\!] + 2x$

    (a) $\lim\limits_{x \to 3^+} f(x)$ (b) $\lim\limits_{x \to 3^-} f(x)$ (c) $\lim\limits_{x \to 3} f(x)$

10. $g(x) = \dfrac{x - [\![x]\!]}{x - 4}$

    (a) $\lim\limits_{x \to 4^+} g(x)$ (b) $\lim\limits_{x \to 4^-} g(x)$ (c) $\lim\limits_{x \to 4} g(x)$

11. $h(x) = \sqrt[3]{x} \cdot [\![x]\!]$

    (a) $\lim\limits_{x \to 0^+} h(x)$ (b) $\lim\limits_{x \to 0^-} h(x)$ (c) $\lim\limits_{x \to 0} h(x)$

12. $F(x) = \sqrt[4]{x} \cdot [\![x]\!]$

    (a) $\lim\limits_{x \to 0^+} F(x)$ (b) $\lim\limits_{x \to 0^-} F(x)$ (c) $\lim\limits_{x \to 0} F(x)$

13. $g(z) = \dfrac{|z|}{\sqrt{z^2}}$

    (a) $\lim\limits_{z \to 0^+} g(z)$ (b) $\lim\limits_{z \to 0^-} g(z)$ (c) $\lim\limits_{z \to 0} g(z)$

14. $h(u) = \dfrac{[\![u^2]\!] - [\![u]\!]^2}{u^2 - 9}$

    (a) $\lim\limits_{u \to 3^+} h(u)$ (b) $\lim\limits_{u \to 3^-} h(u)$ (c) $\lim\limits_{u \to 3} h(u)$

15. $G(y) = \sqrt[3]{y} \text{ sgn } y$, where $\text{sgn } y = \begin{cases} -1 & \text{if } y < 0 \\ 0 & \text{if } y = 0 \\ 1 & \text{if } y > 0 \end{cases}$

    (a) $\lim\limits_{y \to 0^+} G(y)$ (b) $\lim\limits_{y \to 0^-} G(y)$ (c) $\lim\limits_{y \to 0} G(y)$

16. $u(x) = \dfrac{x}{x^2 + |x|}$

    (a) $\lim\limits_{x \to 0^+} u(x)$ (b) $\lim\limits_{x \to 0^-} u(x)$ (c) $\lim\limits_{x \to 0} u(x)$

17. $g(h) = \dfrac{1}{h}\left(\dfrac{1}{x + h} - \dfrac{1}{x}\right)$ for given $x \ne 0$

    (a) $\lim\limits_{h \to 0^+} g(h)$ (b) $\lim\limits_{h \to 0^-} g(h)$ (c) $\lim\limits_{h \to 0} g(h)$

18. Prove that $\lim\limits_{x \to a} f(x) = L$ if and only if $\lim\limits_{x \to a^+} f(x) = L$ and $\lim\limits_{x \to a^-} f(x) = L$.

2.7 LIMITS AT INFINITY

Suppose that we are given a function f defined by the equation

$$f(x) = \frac{x}{2x + 1} \tag{1}$$

Let x take on the values 1, 3, 5, 10, 100, 1000, and so on. The corresponding function values are shown in Table 2.7.1.

Table 2.7.1

x	1	3	5	10	100	1000
$f(x)$	$\frac{1}{3}$	$\frac{3}{7}$	$\frac{5}{11}$	$\frac{10}{21}$	$\frac{100}{201}$	$\frac{1000}{2001}$

From Table 2.7.1 we may surmise that, as x increases through positive values, the value of the function gets closer and closer to $\frac{1}{2}$. In fact, if we form the difference

$$\frac{1}{2} - \frac{x}{2x + 1} = \frac{2x + 1 - 2x}{2(2x + 1)} = \frac{1}{2(2x + 1)} \tag{2}$$

then at $x = 10$ we obtain $\frac{1}{42}$, and at $x = 1000$ we have $\frac{1}{4002}$. From (2), the expression for the difference can be made as small as desired by choosing x sufficiently large. Symbolically, for any positive ε we can find a positive number A such that $|\frac{1}{2} - f(x)| < \varepsilon$ provided $x > A$.

When an independent variable x is increasing without bound through positive values we write "$x \to \infty$." Thus for the function (1) we have

$$\lim_{x \to \infty} \frac{x}{2x + 1} = \frac{1}{2} \tag{3}$$

In general, we have the following definition:

Definition Let f be a function that is defined for all $x \geq a$. The **limit of $f(x)$ as x increases beyond all bounds** is L, written

$$\lim_{x \to \infty} f(x) = L \tag{4}$$

provided that for an arbitrary $\varepsilon > 0$ (no matter how small) there exists an $A > 0$, such that

$$|f(x) - L| < \varepsilon \qquad \text{when } x > A \tag{5}$$

Note that we are *not* saying that infinity is a real number. For example, we do *not* imply that infinity is a point on the number line. Thus when we say that x is approaching 10 or $x \to 10$ it has quite a different meaning from $x \to \infty$.

Next, let us consider the same function as x assumes negative values, -1, -5, -10, -100, -1000, and so on. Table 2.7.2 gives these values of x and the corresponding values of $f(x)$

Table 2.7.2

x	-1	-5	-10	-100	-1000
$f(x)$	1	$\dfrac{5}{9}$	$\dfrac{10}{19}$	$\dfrac{100}{199}$	$\dfrac{1000}{1999}$

Intuitively we see from the table that the values of $f(x) \to \frac{1}{2}$ from above as x assumes negative values without bound. Thus we say for any $\varepsilon > 0$, there exists an $A < 0$ such that $|f(x) - \frac{1}{2}| < \varepsilon$ when $x < A$. With the introduction of the symbolism "$x \to -\infty$," to signify that the variable x is decreasing beyond all bound, we may write

$$\lim_{x \to -\infty} \frac{x}{2x+1} = \frac{1}{2} \tag{6}$$

Note that $y = \frac{1}{2}$ is a horizontal asymptote of the graph of $f(x)$.

More generally we have

Definition   Let f be a function which is defined in some interval $(-\infty, a)$. The **limit of** $f(x)$ **as** x **decreases without bound** is L, written

$$\lim_{x \to -\infty} f(x) = L \tag{7}$$

if for any $\varepsilon > 0$ there exists a number $A < 0$ such that

$$|f(x) - L| < \varepsilon \text{ when } x < A \tag{8}$$

Equation (7) is read "the limit of $f(x)$ as x becomes negatively infinite is L."

Both (4) and (7) tell us that the line $y = L$ is a horizontal asymptote of the graph of $y = f(x)$.

The theorems on limit of a constant, sum, product, quotient, and so on (of Section 2.5) remain valid with simple modifications in notation. Thus by way of an example we shall prove the following theorem:

EXAMPLE 1   If $\lim_{x \to \infty} f(x) = L$ and $\lim_{x \to \infty} g(x) = M$, then $\lim_{x \to \infty} [f(x) + g(x)] = L + M$.

Proof   From the hypothesis, for any $\varepsilon > 0$ there exist positive real numbers N_1 and N_2 such that

$$|f(x) - L| < \frac{\varepsilon}{2} \qquad \text{when } x > N_1$$

and

$$|g(x) - M| < \frac{\varepsilon}{2} \qquad \text{when } x > N_2$$

Now, when $N = \max(N_1, N_2)$, then, if $x > N$

$$|f(x) + g(x) - (L+M)| = |f(x) - L + g(x) - M|$$

$$\leq |f(x) - L| + |g(x) - M| < \frac{\varepsilon}{2} + \frac{\varepsilon}{2} = \varepsilon$$

which proves the theorem.   ∎

Theorems 7 and 8 involving $\lim\limits_{x \to a} x$ have no direct counterpart. Instead we have

Theorem 14†   If k is any positive integer, then

$$\text{(a)}\ \lim_{x \to \infty} \frac{1}{x^k} = 0 \qquad \text{(b)}\ \lim_{x \to -\infty} \frac{1}{x^k} = 0$$

Proof of (a)   The proof is very simple. We just use the definition of $\lim\limits_{x \to \infty} \frac{1}{x^k}$. We must show that for any $\varepsilon > 0$ a number A can be found so that

$$\left| \frac{1}{x^k} - 0 \right| < \varepsilon \qquad \text{whenever } x > A$$

or equivalently (by taking reciprocals)

$$|x^k| > \frac{1}{\varepsilon} \qquad \text{when } x > A$$

or since $k > 0$

$$|x| > \left(\frac{1}{\varepsilon} \right)^{1/k} \qquad \text{when } x > A$$

The above will hold if we choose $A = \left(\frac{1}{\varepsilon} \right)^{1/k}$, for then

$$\left| \frac{1}{x^k} - 0 \right| < \varepsilon \qquad \text{when } x > A$$

The proof of (b) is similar and is left to the reader as an exercise.   ∎

EXAMPLE 2   Find $\lim\limits_{x \to \infty} \dfrac{5x - 7}{3x + 2}$, citing the limit theorem used at each step.

Solution   In order to use limit Theorem 14, we divide each term in the numerator and denominator by x. Thus

$$\frac{5x - 7}{3x + 2} = \frac{5 - \dfrac{7}{x}}{3 + \dfrac{2}{x}} \qquad \text{for } x \neq 0, -\frac{2}{3}$$

then

$$\lim_{x \to \infty} \frac{5x - 7}{3x + 2} = \lim_{x \to \infty} \frac{5 - \dfrac{7}{x}}{3 + \dfrac{2}{x}}$$

$$= \frac{\lim\limits_{x \to \infty} \left(5 - \dfrac{7}{x} \right)}{\lim\limits_{x \to \infty} \left(3 + \dfrac{2}{x} \right)} \qquad \text{(Theorem 9)}$$

---

† The number of this theorem follows the sequence started in Section 2.5.

$$= \frac{\lim\limits_{x\to\infty} 5 - \lim\limits_{x\to\infty} \dfrac{7}{x}}{\lim\limits_{x\to\infty} 3 + \lim\limits_{x\to\infty} \dfrac{2}{x}} \qquad \text{(Theorem 1)}$$

$$= \frac{\lim\limits_{x\to\infty} 5 - \lim\limits_{x\to\infty} 7 \cdot \lim\limits_{x\to\infty} \dfrac{1}{x}}{\lim\limits_{x\to\infty} 3 + \lim\limits_{x\to\infty} 2 \cdot \lim\limits_{x\to\infty} \dfrac{1}{x}} \qquad \text{(Theorem 3)}$$

$$= \frac{5 - 7(0)}{3 + 2(0)} = \frac{5}{3} \qquad \text{(Theorems 6 and 14)} \qquad \bullet$$

EXAMPLE 3 Evaluate $\lim\limits_{x\to\infty} \dfrac{\sqrt{x^2+1}}{4x+3}$ citing the limit theorems used.

Solution First we observe that $\sqrt{x^2} = x$ if $x > 0$. Thus we divide the numerator and denominator by x,

$$\lim_{x\to\infty} \frac{\sqrt{x^2+1}}{4x+3} = \lim_{x\to\infty} \frac{\dfrac{\sqrt{x^2+1}}{x}}{\dfrac{4x+3}{x}} = \lim_{x\to\infty} \frac{\sqrt{\dfrac{x^2+1}{x^2}}}{4+\dfrac{3}{x}}$$

$$= \frac{\lim\limits_{x\to\infty} \sqrt{1+\dfrac{1}{x^2}}}{\lim\limits_{x\to\infty} \left(4+\dfrac{3}{x}\right)} \qquad \text{(Theorem 9)}$$

$$= \frac{\sqrt{\lim\limits_{x\to\infty} \left(1+\dfrac{1}{x^2}\right)}}{\lim\limits_{x\to\infty} \left(4+\dfrac{3}{x}\right)} \qquad \text{(Theorem 10)}$$

$$= \frac{\sqrt{\lim\limits_{x\to\infty} 1 + \lim\limits_{x\to\infty} \dfrac{1}{x^2}}}{\lim\limits_{x\to\infty} 4 + \lim\limits_{x\to\infty} \dfrac{3}{x}} \qquad \text{(Theorem 1)}$$

$$= \frac{\sqrt{1+0}}{4 + \lim\limits_{x\to\infty} 3 \cdot \lim\limits_{x\to\infty} \dfrac{1}{x}} \qquad \text{(Theorems 2 and 14)}$$

$$= \frac{1}{4 + 3(0)} = \frac{1}{4} \qquad \text{(Theorem 14)} \qquad \bullet$$

EXAMPLE 4 Evaluate $\lim\limits_{x\to-\infty} \dfrac{\sqrt{x^2+1}}{4x+3}$

Solution The function is the same as in Example 3. However, since we are interested only in negative values of x, then $\sqrt{x^2} = -x$. Divide numerator and denominator by $-x$. Thus

$$\lim_{x \to -\infty} \frac{\sqrt{x^2+1}}{4x+3} = \lim_{x \to -\infty} \frac{\sqrt{1+\dfrac{1}{x^2}}}{-4-\dfrac{3}{x}}$$

$$= \frac{\lim\limits_{x \to -\infty} \sqrt{1+\dfrac{1}{x^2}}}{\lim\limits_{x \to -\infty} \left(-4-\dfrac{3}{x}\right)} \qquad \text{(Theorem 9)}$$

$$= \frac{\sqrt{\lim\limits_{x \to -\infty} \left(1+\dfrac{1}{x^2}\right)}}{\lim\limits_{x \to -\infty} \left(-4-\dfrac{3}{x}\right)} \qquad \text{(Theorem 10)}$$

$$= \frac{\sqrt{\lim\limits_{x \to -\infty} 1 + \lim\limits_{x \to -\infty} \dfrac{1}{x^2}}}{\lim\limits_{x \to -\infty}(-4) - \lim\limits_{x \to -\infty} 3 \cdot \lim\limits_{x \to -\infty} \dfrac{1}{x}} \qquad \text{(Theorems 2 and 14)}$$

$$= \frac{1+0}{-4-0} = -\frac{1}{4} \qquad \bullet$$

When taking limits, the terminology "indeterminate form" is used to indicate an expression whose form (upon direct substitution) is ambiguous—it could have many possible values. One such indeterminate form is $\dfrac{0}{0}$. In Section 2.4, Examples 9, 11, and 12 were each of this form; yet all three had different limits.

The next examples illustrate that $\infty - \infty$ is also indeterminate and is not necessarily zero.

EXAMPLE 5   Find $\lim\limits_{x \to \infty} [(x+1) - x]$.

Solution   Note that if we try using Theorem 1 we get

$$\lim_{x \to \infty} [(x+1) - x] = \lim_{x \to \infty} (x+1) - \lim_{x \to \infty} x = \infty - \infty$$

However,

$$\lim_{x \to \infty} [(x+1) - x] = \lim_{x \to \infty} 1 = 1 \qquad \bullet$$

EXAMPLE 6   Evaluate $\lim\limits_{x \to \infty} (\sqrt{x^2+a^2} - x)$

Solution

$$\sqrt{x^2+a^2} - x = \frac{\sqrt{x^2+a^2} - x}{1} \cdot \frac{\sqrt{x^2+a^2} + x}{\sqrt{x^2+a^2} + x}$$

$$= \frac{a^2}{\sqrt{x^2+a^2} + x}$$

Divide numerator and denominator by x noting that $\sqrt{x^2} = x$ when $x > 0$.

$$\sqrt{x^2 + a^2} - x = \frac{\dfrac{a^2}{x}}{\sqrt{1 + \dfrac{a^2}{x^2}} + 1}$$

$$\lim_{x \to \infty} \sqrt{x^2 + a^2} - x = \frac{\displaystyle\lim_{x \to \infty} \frac{a^2}{x}}{\displaystyle\lim_{x \to \infty}\left[\sqrt{1 + \dfrac{a^2}{x^2}} + 1\right]} \qquad \text{(Theorem 9)}$$

$$= \frac{a^2 \displaystyle\lim_{x \to \infty} \frac{1}{x}}{1 + \displaystyle\lim_{x \to \infty} \sqrt{1 + \dfrac{a^2}{x^2}}} \qquad \text{(Theorems 2 and 1)}$$

$$= \frac{a^2 \displaystyle\lim_{x \to \infty} \frac{1}{x}}{1 + \sqrt{\displaystyle\lim_{x \to \infty}\left(1 + \dfrac{a^2}{x^2}\right)}} \qquad \text{(Theorem 10)}$$

$$= \frac{a^2(0)}{1 + \sqrt{\displaystyle\lim_{x \to \infty} 1 + a^2 \displaystyle\lim_{x \to \infty} \frac{1}{x^2}}} \qquad \text{(Theorems 1 and 14)}$$

$$= \frac{0}{1 + \sqrt{1 + a^2(0)}} \qquad \text{(Theorem 14)}$$

$$= \frac{0}{2} = 0$$

Exercises 2.7

In Exercises 1 through 16, find the limits in each instance stating the limit theorems used.

1. $\displaystyle\lim_{x \to \infty} \frac{3x + 9}{4x - 7}$

2. $\displaystyle\lim_{x \to \infty} \frac{2 - 5x}{4x + 11}$

3. $\displaystyle\lim_{x \to -\infty} \frac{6x + 5}{2x - 1}$

4. $\displaystyle\lim_{x \to -\infty} \frac{x - 3}{x^2 + 9}$

5. $\displaystyle\lim_{x \to \infty} \frac{3x^2 - x - 4}{8x^2 + 3}$

6. $\displaystyle\lim_{u \to -\infty} \frac{5u^2 - u + 1}{u^3 + 6u}$

7. $\displaystyle\lim_{t \to \infty} \frac{3t - 4}{\sqrt{t^2 + 5}}$

8. $\displaystyle\lim_{w \to -\infty} \frac{3w^2 - 7w - 1}{4 - 8w - 2w^2}$

9. $\displaystyle\lim_{x \to \infty} \left(\frac{1}{x} - \frac{1}{x + 3}\right)$

10. $\displaystyle\lim_{x \to \infty} \frac{[\![x]\!] - \sqrt{7}}{x}$

11. $\displaystyle\lim_{z \to -\infty} \frac{\sqrt{z^2 + 7}}{2z - 3}$

12. $\displaystyle\lim_{x \to \infty} \frac{\sqrt{x^4 + 3}}{x^3 + x + 1}$

13. $\displaystyle\lim_{x \to \infty} \frac{8x^3 - x^2 - 6}{5x(3x^2 - 4)}$

14. $\displaystyle\lim_{x \to \infty} \frac{1}{\sqrt[3]{x}}$

15. $\displaystyle\lim_{x \to \infty} \frac{x\sqrt{x} + \sqrt[3]{x}}{\sqrt{x^3 - 2}}$

16. $\lim\limits_{x\to\infty} \dfrac{x^{-1/2} + x^{-1}}{(2x + 3)^{-1/2} + 2x^{-1}}$ (*Hint:* Multiply numerator and denominator by $x^{1/2}$)

17. Let f be the power function defined by $f(x) = x^n$ where n is any natural number. Prove that f does *not* have a limit at ∞.

18. Prove that for a polynomial function $P = P(x)$ that P is either a constant function or has no limit at ∞. (*Hint:* Use the conclusion of Exercise 17.)

19. Let R be a rational function; that is, R is in the form $R = \dfrac{P}{Q}$, where P and Q are polynomial functions.

   (a) Prove that R has a limit at plus infinity if and only if the degree of $P \leq$ the degree of Q, assuming that P is not the zero function.

   (b) Prove that R has a nonzero limit at plus infinity if and only if the degree of P equals the degree of Q.

2.8   INFINITE LIMITS

Consider the function f defined by

$$f(x) = 1/x^2 \tag{1}$$

To examine the behavior of f as x approaches 0, let us set up Table 2.8.1, where the approach to 0 is made from both sides.

Table 2.8.1

x	1	$\frac{1}{2}$	$\frac{1}{10}$	$\frac{1}{100}$	-1	$-\frac{1}{2}$	$-\frac{1}{10}$	$-\frac{1}{100}$
$f(x)$	1	4	100	10000	1	4	100	10000

First, as $x \to 0^+$ (that is, as x approaches 0 from positive values), the value of $f(x)$ is increasing beyond all bounds. This behavior is indicated symbolically by

$$\lim_{x\to 0^+} f(x) = \lim_{x\to 0^+} 1/x^2 = \infty \tag{2}$$

Again, we remind the reader that the symbol ∞ is not considered as a real number; rather, Equation (2) expresses the fact that $f(x)$ increases without bound as $x \to 0^+$.

Similarly, as $x \to 0^-$, that is, x approaches zero through negative values, the value of $f(x)$ increases without limit so that

$$\lim_{x\to 0^-} 1/x^2 = \infty \tag{3}$$

In this case the statements (2) and (3) can be combined into one, namely

$$\lim_{x\to 0} 1/x^2 = \infty \tag{4}$$

A sketch of $f(x) = 1/x^2$ is shown in Figure 2.8.1. More formally, we have:

Definition   Let f be a function defined in an open interval containing $x = a$, but not necessarily at $x = a$ itself. Then if, when $x \to a$, $f(x)$ *increases without bound,* we write

$$\lim_{x \to a} f(x) = \infty \qquad (5)$$

This means that for any $A > 0$ there exists a $\delta > 0$ such that $f(x) > A$ whenever $0 < |x - a| < \delta$.

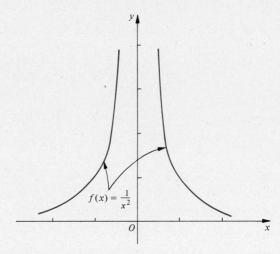

$$f(x) = \frac{1}{x^2}$$

Figure 2.8.1

It should be noted that this is not the limit in the usual sense. Also $x = a$ is a vertical asymptote of the graph of $y = f(x)$.

Similarly, consider the function g defined by

$$g(x) = -\frac{1}{(x - 1)^2} \qquad (6)$$

We observe that as $x \to 1$ from either the right or the left the value of $g(x)$ is decreasing without bound. This is expressed symbolically by

$$\lim_{x \to 1} -\frac{1}{(x - 1)^2} = -\infty \qquad (7)$$

A sketch of the graph of this function is given in Figure 2.8.2. In general, we have:

Definition   Let f be a function defined in an open interval containing $x = a$ but not necessarily at $x = a$ itself. Then if, when $x \to a$, $f(x)$ *decreases without bound,* we write

$$\lim_{x \to a} f(x) = -\infty \qquad (8)$$

This means that for any number $A < 0$ there exists a $\delta > 0$ such that $f(x) < A$ whenever $0 < |x - a| < \delta$.

The line $x = a$ is a vertical asymptote of the graph of $f(x)$ when (8) is satisfied.

Next, suppose that h is a function given by

$$h(x) = \frac{x}{x + 1} \qquad (9)$$

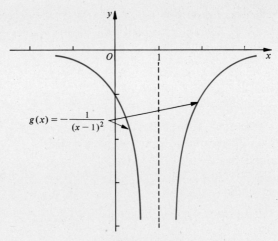

$$g(x) = -\frac{1}{(x-1)^2}$$

Figure 2.8.2

This function is sketched in Figure 2.8.3. Note that the values taken on by $h(x)$ in a neighborhood of $x = -1$ are different from the previous behaviors of f and g. As $x \to -1^+$, $h(x) \to -\infty$, whereas as $x \to -1^-$, we have $h(x) \to \infty$. To distinguish this situation from its predecessors, we write

$$\lim_{x \to -1^+} h(x) = -\infty \qquad \text{and} \qquad \lim_{x \to -1^-} h(x) = \infty$$

Also, $x = -1$ is a vertical asymptote of the graph. Since $\lim\limits_{x \to \pm\infty} \dfrac{x}{x+1} = 1$, by the results of Section 2.7, $y = 1$ is a horizontal asymptote of the graph of $y = h(x)$.

Some functions possess infinite limits from one side only. For example, the function g defined by

$$g(x) = \frac{1}{\sqrt{x-5}}$$

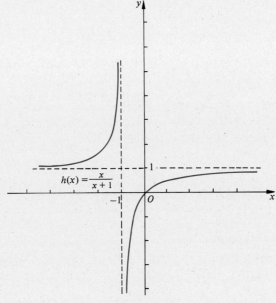

$$h(x) = \frac{x}{x+1}$$

Figure 2.8.3

becomes positively infinite as $x \to 5^+$; however, the function is not defined for $x < 5$. Therefore the limit from the left cannot exist and, consequently, there can be no two-sided limit as $x \to 5$.

Many of the infinite limits that we shall encounter are of the following type:

$$\lim_{x \to a^+} \frac{1}{(x-a)^p} = \infty \qquad \text{if } p > 0 \qquad (10)$$

Proof of (10) It must be shown that for any positive number A, no matter how large, there exists an open interval $(a, a + \delta)$ such that

$$\frac{1}{(x-a)^p} > A \qquad \text{for all } x \text{ in } (a, a + \delta)$$

If $x - a > 0$, then the inequality $\dfrac{1}{(x-a)^p} > A$ is equivalent to

$$0 < (x-a)^p < \frac{1}{A}$$

which in turn is equivalent to

$$0 < x - a < \left(\frac{1}{A}\right)^{1/p}$$

Thus we choose $\delta = \left(\dfrac{1}{A}\right)^{1/p}$ and (10) is proved. ∎

The limit (10) is also valid if x approaches a *from the left* and if $p = \dfrac{m}{n} > 0$ where m and n are, respectively, even and odd integers.

If p is either an odd positive integer or the ratio of two odd positive integers, then $(x-a)^p < 0$ if $x < a$ and

$$\lim_{x \to a^-} \frac{1}{(x-a)^p} = -\infty \qquad (11)$$

may be established in a manner similar to (10).

Care should be exercised when operating with limits when one of the quantities becomes infinite. We may recall that infinity is not a number and, consequently, cannot be treated as such. However, there are a number of rules that can be stated and proved. For example, suppose that

$$\lim_{x \to a} f(x) = \infty \qquad \text{and} \qquad \lim_{x \to a} g(x) = L$$

where L is any real number, then

$$\lim_{x \to a} [f(x) + g(x)] = \infty$$

If $L \neq 0$ we can establish that

$$\lim_{x \to a} [f(x)\, g(x)] = \begin{cases} \infty & \text{if } L > 0 \\ -\infty & \text{if } L < 0 \end{cases}$$

However, if $L = 0$ further investigation is necessary. In fact, this is when the problem becomes interesting and is an example of an "indeterminate form."

To see why further investigation is required, consider $f(x) = \dfrac{1}{(x-1)^2}$ and

$g(x) = x - 1$ then

$$\lim_{x \to 1} f(x)\, g(x) = \lim_{x \to 1} \frac{1}{x - 1}$$

and the limit does not exist. Now if instead, $g(x) = k(x - 1)^2$ where k is *any* constant, then

$$\lim_{x \to 1} f(x)\, g(x) = \lim_{x \to 1} \frac{1}{(x - 1)^2} \cdot k(x - 1)^2 = k$$

and the result varies as k varies. It is left to the reader to construct examples in which $f \cdot g \to 0$ as $x \to 1$.

In cases where $f(x) \to \infty$ and $g(x) \to -\infty$ as $x \to a$, nothing can be said about the behavior of $f(x) + g(x)$ without scrutinizing the functions more closely.

The principal conclusions about combining functions with infinite limits are summarized in Table 2.8.2. The first entry is the rule for $\lim (f + g)$ when $\lim f = \infty$ and $\lim g = L$. Then, if $h(x) = f(x) + g(x)$, it can be established that $\lim h(x) = \infty$. Symbolically we have "$\infty + L = \infty$." The fourth entry is the rule for $\lim (f - g)$ when $\lim f = \infty$ and $\lim g = \infty$, and we observe that, without any further information, the result is in question. This conclusion follows quite readily from simple examples such as

$$\lim_{x \to 3} \left(\frac{1}{x - 3} - \frac{1}{x - 3} \right) = 0$$

since the quantity in parentheses is 0 for all $x \neq 3$, whereas $\lim_{x \to 3} \left(\dfrac{2}{x - 3} - \dfrac{1}{x - 3} \right)$ does not exist.

Note that $\lim g = 0^+$ means that g tends towards zero through positive values; that is, $g > 0$ for x sufficiently close to the value it is approaching. Furthermore it is noted that the first four entries in Table 2.8.2 involve addition or subtraction,

Table 2.8.2   Limits Involving Infinity

Entry	lim f	lim g	function $h(x)$	lim $h(x)$	Symbolic Relation
1	∞	L	$f + g$	∞	$\infty + L = \infty$
2	$-\infty$	L	$f + g$	$-\infty$	$-\infty + L = -\infty$
3	∞	∞	$f + g$	∞	$\infty + \infty = \infty$
4	∞	∞	$f - g$	?	$\infty - \infty = ?$
5	∞	$L > 0$	$f \cdot g$	∞	$\infty \cdot L = \infty \quad (L > 0)$
6	∞	$L < 0$	$f \cdot g$	$-\infty$	$\infty \cdot L = -\infty \quad (L < 0)$
7	∞	∞	$f \cdot g$	∞	$\infty \cdot \infty = \infty$
8	∞	$-\infty$	$f \cdot g$	$-\infty$	$\infty \cdot (-\infty) = -\infty$
9	∞	0	$f \cdot g$	?	$\infty \cdot 0 = ?$
10	L	∞	f/g	0	$L \div \infty = 0$
11	∞	∞	f/g	?	$\infty \div \infty = ?$
12	∞	0^+	f/g	∞	$\infty \div (0^+) = \infty$
13	∞	0^-	f/g	$-\infty$	$\infty \div (0^-) = -\infty$
14	$L > 0$	0^+	f/g	∞	$L \div (0^+) = \infty \quad (L > 0)$
15	$L > 0$	0^-	f/g	$-\infty$	$L \div (0^-) = -\infty \quad (L > 0)$
16	$L < 0$	0^+	f/g	$-\infty$	$L \div (0^+) = -\infty \quad (L < 0)$
17	$L < 0$	0^-	f/g	∞	$L \div (0^-) = \infty \quad (L < 0)$
18	0	0	f/g	?	$0 \div 0 = ?$

the next five involve multiplication, and the remaining, division. In five of the cases, the outcome cannot be decided without further investigation. This is usually determined by methods of differential calculus under the heading "indeterminate forms," and these methods will be developed later.

Finally, we remark that we purposely wrote $\lim f(x)$ rather than $\lim\limits_{x \to a} f(x)$ because the results are also valid if a is replaced by a^+ or a^- or ∞ or $-\infty$. Thus, for example, entry 12 in the table becomes, for the case $x \to \infty$:

$$\text{if } \lim_{x \to \infty} f(x) = \infty \text{ and } \lim_{x \to \infty} g(x) = 0^+ \text{ then } \lim_{x \to \infty} \frac{f(x)}{g(x)} = \infty$$

Most problems where the limit is infinite can be worked out quite easily by referring to entries 14 through 17 in the table.

EXAMPLE 1 Find $\lim\limits_{x \to \infty} \dfrac{x^2}{2x + 1}$

Solution Dividing the numerator and denominator by x^2, there results

$$\lim_{x \to \infty} \frac{x^2}{2x + 1} = \lim_{x \to \infty} \frac{1}{\dfrac{2}{x} + \dfrac{1}{x^2}}$$

Now, the limit of the numerator is 1 because $\lim\limits_{x \to \infty} 1 = 1$.

Evaluating the limit of the denominator, we have

$$\lim_{x \to \infty} \left(\frac{2}{x} + \frac{1}{x^2} \right) = 2 \lim_{x \to \infty} \frac{1}{x} + \lim_{x \to \infty} \frac{1}{x^2} = 2(0) + 0 = 0$$

and, furthermore, the denominator is approaching zero through positive values. Therefore from entry 14 of Table 2.8.2 it follows that

$$\lim_{x \to \infty} \frac{x^2}{2x + 1} = \infty$$

Alternative Solution Divide the numerator and denominator by x so that

$$\lim_{x \to \infty} \frac{x^2}{2x + 1} = \lim_{x \to \infty} \frac{x}{2 + \dfrac{1}{x}} = \lim_{x \to \infty} x \cdot \left(\frac{1}{2 + \dfrac{1}{x}} \right)$$

From entry 5 in Table 2.8.2 $\left(\text{since } \lim\limits_{x \to \infty} x = \infty \text{ and } \lim\limits_{x \to \infty} \dfrac{1}{2 + \dfrac{1}{x}} = \dfrac{1}{2} \right)$ we have

$$\lim_{x \to \infty} \frac{x^2}{2x + 1} = \infty \qquad \bullet$$

EXAMPLE 2 Find $\lim\limits_{x \to \infty} \dfrac{4 - 2x - 7x^2}{10x + \sqrt{3}}$

Solution Divide the numerator and denominator by x^2, yielding

$$\lim_{x \to \infty} \frac{4 - 2x - 7x^2}{10x + \sqrt{3}} = \lim_{x \to \infty} \frac{\dfrac{4}{x^2} - \dfrac{2}{x} - 7}{\dfrac{10}{x} + \dfrac{\sqrt{3}}{x^2}}$$

Consider then the limits of the numerator and denominator separately.

$$\lim_{x \to \infty} \left(\frac{4}{x^2} - \frac{2}{x} - 7 \right) = \lim_{x \to \infty} \frac{4}{x^2} - \lim_{x \to \infty} \frac{2}{x} - \lim_{x \to \infty} 7 = 0 - 0 - 7 = -7$$

$$\lim_{x \to \infty} \left(\frac{10}{x} + \frac{\sqrt{3}}{x^2} \right) = 10 \lim_{x \to \infty} \frac{1}{x} + \sqrt{3} \lim_{x \to \infty} \left(\frac{1}{x} \right)^2 = 10(0) + \sqrt{3}(0)^2 = 0$$

Also the denominator is approaching zero through positive values and entry 16 of Table 2.8.2 applies. Therefore

$$\lim_{x \to \infty} \frac{4 - 2x - 7x^2}{10x + \sqrt{3}} = -\infty$$

$\bullet$

EXAMPLE 3   Find $\displaystyle \lim_{x \to \infty} \frac{x^3 - 100x^2}{x + 1}$

Solution   One way to approach this problem is to divide $x + 1$ into each of the terms so that

$$\lim_{x \to \infty} \frac{x^3 - 100x^2}{x + 1} = \lim_{x \to \infty} \left(\frac{x^3}{x + 1} - \frac{100x^2}{x + 1} \right)$$

However, a brief analysis shows that this is entry 4 of Table 2.8.2, namely, $\infty - \infty$, and it is indeterminate as it stands.

Instead, we factor the numerator

$$\lim_{x \to \infty} \frac{x^3 - 100x^2}{x + 1} = \lim_{x \to \infty} \left[x(x - 100) \frac{x}{x + 1} \right]$$

and identify $x(x - 100)$ with f and $\dfrac{x}{x + 1}$ with g. Now, $\displaystyle\lim_{x \to \infty} f = \infty$ and $\displaystyle\lim_{x \to \infty} g = 1$ so that this is entry 5. Therefore

$$\lim_{x \to \infty} \frac{x^3 - 100x^2}{x + 1} = \infty$$

$\bullet$

EXAMPLE 4   Find $\displaystyle \lim_{x \to 2^+} \frac{x^2 + x + 1}{x^2 - 6x + 8}$

Solution   We factor the denominator to obtain

$$\lim_{x \to 2^+} \frac{1}{x - 2} \cdot \frac{x^2 + x + 1}{x - 4}$$

As $x \to 2^+$, the first factor is approaching ∞ while the second factor is approaching $-\frac{7}{2}$. Therefore entry 6 of Table 2.8.2 applies and we have

$$\lim_{x \to 2^+} \frac{x^2 + x + 1}{x^2 - 6x + 8} = -\infty$$

$\bullet$

LE 5   Prove that, if $\lim\limits_{x \to a^+} f(x) = \infty$ and $\lim\limits_{x \to a^+} g(x) = \infty$, $\lim\limits_{x \to a^+} (f(x) + g(x)) = \infty$.

ution   For any $A > 0$, we choose $\delta_1 > 0$ so that $f(x) > \dfrac{A}{2}$ for x in $(a, a + \delta_1)$. Also, we can choose a $\delta_2 > 0$ so that $g(x) > \dfrac{A}{2}$ for x in $(a, a + \delta_2)$. Then if $\delta = \min(\delta_1, \delta_2)$, we have

$$f(x) + g(x) > A \qquad \text{if } x \text{ is in } (a, a + \delta)$$

Therefore $\lim\limits_{x \to a^+} (f(x) + g(x)) = \infty$. ∎

Exercises 2.8

In Exercises 1 through 18, evaluate the limit giving reasons for each step.

1. $\lim\limits_{x \to -3} \dfrac{1}{(x + 3)^2}$

2. $\lim\limits_{x \to 1} \dfrac{-x}{(x - 1)^4}$

3. $\lim\limits_{x \to \infty} \dfrac{4x^2 - x + 1}{x^3 + 1}$

4. $\lim\limits_{x \to -\infty} \dfrac{3x + 1}{4x + 5}$

5. $\lim\limits_{x \to \infty} \dfrac{x^2 + 100}{x}$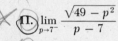

6. $\lim\limits_{x \to 4^+} \dfrac{x - 3}{\sqrt{x - 4}}$

7. $\lim\limits_{u \to \infty} \dfrac{u^2 + 100u + 17}{u^3 - 2}$

8. $\lim\limits_{y \to 0^-} \dfrac{\sqrt{9 + y^2}}{y}$

9. $\lim\limits_{m \to -\infty} \dfrac{4 - 7m + 5m^2}{-3m^2 + 2}$

10. $\lim\limits_{r \to 7^-} \dfrac{49 - r^2}{r - 7}$

11. $\lim\limits_{p \to 7^-} \dfrac{\sqrt{49 - p^2}}{p - 7}$

12. $\lim\limits_{x \to 5^+} \dfrac{[\![x]\!] - x}{x - 5}$

13. $\lim\limits_{x \to 5^-} \dfrac{[\![x]\!] - x}{x - 5}$

14. $\lim\limits_{x \to 3^+} \dfrac{[\![x^2]\!] - 9}{x - 3}$

15. $\lim\limits_{x \to 3^-} \dfrac{[\![x^2]\!] - 9}{x - 3}$

16. $\lim\limits_{x \to 0} \left(\dfrac{2}{x} - \dfrac{1}{x^2} \right)$

17. $\lim\limits_{s \to 1} \left(\dfrac{1}{s - 1} - \dfrac{2}{(s - 1)^2} \right)$

18. $\lim\limits_{x \to 1^-} \dfrac{x^2 - 3x + 2}{\sqrt{x - 1}}$

19. With regard to the indeterminate form $\infty - \infty$, choose functions f and g so that f has limit ∞ and g has limit ∞ as $x \to 0$ and for which a prescribed one of the following holds true:
    (a) $\lim\limits_{x \to 0} [f(x) - g(x)] = L$

    (b) $\lim\limits_{x \to 0} [f(x) - g(x)] = \infty$

    (c) $f(x) - g(x)$ has no limit finite or infinite at $x = 0$.

20. Choose functions $f(x)$ and $g(x)$ so that both have limit ∞ as x approaches 6 and for which a prescribed one of the following holds true:

    (a) $\lim\limits_{x \to 6} \dfrac{f(x)}{g(x)} = L$

    (b) $\lim\limits_{x \to 6} \dfrac{f(x)}{g(x)} = \infty$

21. Prove entry 1 in Table 2.8.2 for $x \to a^+$.
22. Prove entry 5 in Table 2.8.2 for $x \to \infty$.
23. Prove entry 6 in Table 2.8.2 for $x \to a$.
24. Prove entry 14 in Table 2.8.2 for $x \to a^-$.

Let us consider the function f defined by

$$f(x) = \frac{x^2 - 2x}{2(x - 2)}, \quad x \neq 2 \tag{1}$$

This function is defined for all x except at $x = 2$. In fact, $f(2) = \dfrac{0}{0}$ which is meaningless. The function is easily graphed. We may factor the numerator to obtain

$$f(x) = \frac{x(x - 2)}{2(x - 2)} = \frac{x}{2} \quad \text{if } x \neq 2 \tag{2}$$

Thus the graph of (1) (Figure 2.9.1) is just the straight line $y = \dfrac{x}{2}$ minus the single point $(2, 1)$. This gap may be filled by defining $f(2) = 1$. However, if instead we defined $f(2) = 3$, then the function would be discontinuous at $x = 2$, and this is manifested by a break or jump at $x = 2$. In fact, in this latter instance $\lim_{x \to 2} f(x) = 1 \neq f(2)$.

This illustrates the following definition of a continuous function at a point $x = a$.

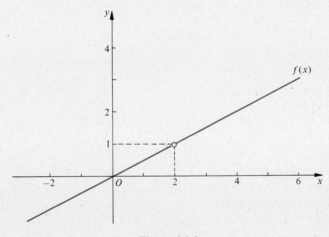

Figure 2.9.1

Definition   A function $f(x)$ is said to be ***continuous*** at $x = a$ if and only if the following three conditions are satisfied

    (i) $f(a)$ exists
    (ii) $\lim_{x \to a} f(x)$ exists
    (iii) $\lim_{x \to a} f(x) = f(a)$

If any one of the three conditions is not met, $f(x)$ is said to be **discontinuous** or **not continuous** at $x = a$.

It is customary in writing the definition of a continuous function to write only (iii), because (iii) makes sense only if (i) and (ii) are satisfied.

We can therefore state the following, which follows from the definition of limit.

Definition   A function $f(x)$ is said to be **continuous** at $x = a$ if and only if for each $\varepsilon > 0$ there exists a $\delta > 0$ such that $|x - a| < \delta$ implies that $|f(x) - f(a)| < \varepsilon$.

EXAMPLE 1   The function f defined by $f(x) = x^2$ is continuous at $x = 4$ because $\lim\limits_{x \to 4} x^2 = (\lim\limits_{x \to 4} x)(\lim\limits_{x \to 4} x) = 4^2 = 16 = f(4)$. More generally, x^2 is continuous at any number $x = a$ because $\lim\limits_{x \to a} x^2 = a^2 = f(a)$.   ●

EXAMPLE 2   The function f defined by $f(x) = \dfrac{1}{x}$ is continuous for any $x = a \neq 0$ since

$$\lim_{x \to a} \frac{1}{x} = \frac{\lim\limits_{x \to a} 1}{\lim\limits_{x \to a} x} = \frac{1}{a} = f(a) \qquad \text{if } a \neq 0$$

However, $\dfrac{1}{x}$ is not continuous at $x = 0$ because $\dfrac{1}{0}$ is *not* defined.   ●

EXAMPLE 3   In this example, we investigate the question of continuity for $g(x) = |x - 2|$ at $x = 2$.

Solution   The value of $g(2) = |2 - 2| = |0| = 0$. By definition,

$$g(x) = \begin{cases} x - 2 & \text{if } x \geq 2 \\ 2 - x & \text{if } x < 2 \end{cases}$$

and thus

$$\lim_{x \to 2^+} g(x) = \lim_{x \to 2^+} (x - 2) = \lim_{x \to 2^+} x - \lim_{x \to 2^+} 2 = 2 - 2 = 0$$

$$\lim_{x \to 2^-} g(x) = \lim_{x \to 2^-} (2 - x) = \lim_{x \to 2^-} 2 - \lim_{x \to 2^-} x = 2 - 2 = 0$$

Therefore $\lim\limits_{x \to 2} g(x) = 0 = g(2)$ and the function (see Figure 2.9.2) is continuous at $x = 2$. Although the function has a corner at $x = 2$, it has no break there.   ●

EXAMPLE 4   Determine whether or not the function h given by

$$h(x) = \begin{cases} x & \text{if } x < 3 \\ 4 & \text{if } x = 3 \\ 2x - 1 & \text{if } x > 3 \end{cases}$$

is continuous at $x = 3$.

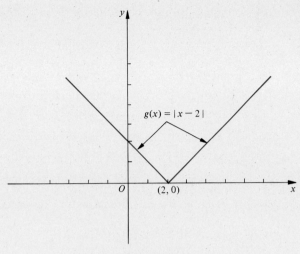

Figure 2.9.2

Solution $\lim\limits_{x \to 3^-} h(x) = \lim\limits_{x \to 3^-} x = 3 \neq h(3)$ so that $h(x)$ is not continuous at $x = 3$ (Figure 2.9.3).

●

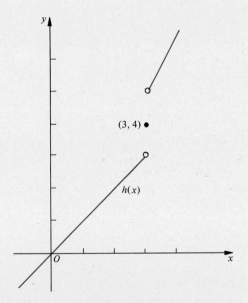

Figure 2.9.3

Suppose now that we return to the function f defined by

$$f(x) = \frac{x^2 - 2x}{2(x - 2)} \qquad x \neq 2$$

but we also stipulate that

$$f(2) = 3$$

Then our function (Figure 2.9.4) is defined for all x; however, it is not continuous at $x = 2$ because

$$\lim_{x \to 2} f(x) = \lim_{x \to 2} \frac{x}{2} = 1 \neq f(2)$$

Now, if instead, $f(2)$ were defined to be 1, the discontinuity would be removed. This illustrates the concept of a ***removable discontinuity.*** In general, suppose that $f(x)$ is not continuous at $x = a$; however, the $\lim_{x \to a} f(x)$ exists and $\neq f(a)$ (possibly $f(a)$ will not exist). Then this discontinuity can be removed by defining or redefining $f(a) = \lim_{x \to a} f(x)$. Such a discontinuity is called a removable discontinuity. If the discontinuity is not removable because $\lim_{x \to a} f(x)$ fails to exist as a real number, then we say the discontinuity is ***essential.***

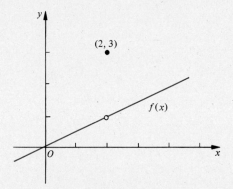

Figure 2.9.4

EXAMPLE 5   Let us return to Example 4, exhibited in Figure 2.9.3. For this case, $\lim_{x \to 3^-} h(x) = 3$ while $\lim_{x \to 3^+} h(x) = \lim_{x \to 3^+} (2x - 1) = 5$. Therefore $\lim_{x \to 3} h(x)$ does not exist, and it is impossible to define the function at $x = 3$ in order to make it continuous there. Therefore we have an essential discontinuity at $x = 3$.   ●

EXAMPLE 6   Let f be defined by

$$f(x) = \frac{3}{(x - 4)^2} \quad \text{if } x \neq 4 \quad \text{and} \quad f(4) = 0$$

Determine whether or not there is an essential discontinuity at $x = 4$.

Solution

$$\lim_{x \to 4} f(x) = \lim_{x \to 4} \frac{3}{(x - 4)^2} = \infty$$

But the only way we can remove a discontinuity is when the limit is finite. Thus there is an essential discontinuity at $x = 4$ (∞ is *not* a number).   ●

EXAMPLE 7   Discuss the continuity of f where

$$f(x) = \frac{x^3 + 1}{x + 1} \quad \text{if } x \neq -1$$

Solution

$$f(x) = \frac{x^3 + 1}{x + 1} = x^2 - x + 1 \qquad \text{if } x \neq -1$$

so that f is continuous for all values of x except $x = -1$. Also,

$$\lim_{x \to -1} f(x) = \lim_{x \to -1} (x^2 - x + 1) = (-1)^2 - (-1) + 1 = 3$$

Therefore we have a removable discontinuity that may be removed by defining $f(-1)$ to be 3. The graph of the original function is shown in Figure 2.9.5.

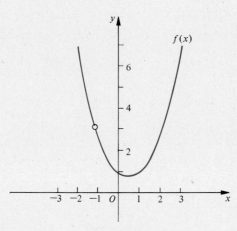

Figure 2.9.5

Exercises 2.9

*In each of the Exercises 1 through 20, a function is defined (sometimes in a prescribed domain). Determine the values of the independent variable at which the function is **not** continuous, and show in what sense the definition of continuity at a point is not satisfied at each discontinuity. Sketch the graph.*

1. $f(x) = \dfrac{3}{x - 2}$ if $x \neq 2$; $f(2) = 3$

2. $g(x) = \dfrac{1}{x^2 + 1}$ if $-2 < x < 2$

3. $F(u) = \dfrac{u^2 - 4}{u - 2}$ if $u \neq 2$; $F(2) = 3$

4. $h(z) = \dfrac{z^2 - 7z + 6}{z - 1}$ if $z \neq 1$; $h(1) = -5$

5. $G(x) = \dfrac{x^3 - 27}{x - 3}$ if $x \neq 3$; $G(3) = 9$

6. $f(r) = |3r - 1|$

7. $f(x) = \begin{cases} 0 & \text{if } x \leq 0 \\ 3x & \text{if } x > 0 \end{cases}$

8. $g(x) = \begin{cases} \dfrac{|2x|}{x} & \text{if } x > 0 \\ |-x| & \text{if } x \leq 0 \end{cases}$

9. $h(x) = \begin{cases} x - 3 & -4 < x \leq 2 \\ x^2 + x - 7 & 2 < x < 5 \end{cases}$

10. $f(x) = \begin{cases} \dfrac{x + 1}{|x + 1|} & \text{if } x \neq -1 \\ 0 & \text{if } x = -1 \end{cases}$

11. $F(x) = [\![x]\!]$ if $\dfrac{1}{2} < x < \dfrac{9}{10}$

12. $g(x) = x^2[\![x]\!]$ if $\dfrac{1}{2} < x < \dfrac{3}{2}$

13. $h(x) = \sqrt{x}$ if $x \geq 0$

14. $F(r) = \dfrac{r - 1}{\sqrt{r} - 1}$ if $r \geq 0$

15. $g(t) = \begin{cases} \sqrt{t - 6} & \text{if } t \geq 6 \\ 0 & \text{if } t < 6 \end{cases}$

16. $f(x) = [\![x]\!] + [\![-x]\!]$

17. $g(x) = \begin{cases} x + 1 & \text{if } x < 2 \\ 3 & \text{if } x = 2 \\ x^2 - 1 & \text{if } x > 2 \end{cases}$ **18.** $F(t) = \begin{cases} t & \text{if } t < 3 \\ 3 & \text{if } t = 3 \\ 10 + t - t^2 & \text{if } t > 3 \end{cases}$

19. $G(t) = \begin{cases} t(1 + t) & \text{if } t < -4 \\ -3t & \text{if } -4 \le t < 5 \\ t^2 - 3t - 20 & \text{if } t \ge 5 \end{cases}$ **20.** $h(u) = \begin{cases} u^2 - u + 2 & \text{if } u < -1 \\ 6 + 2u & \text{if } -1 \le u < 2 \\ u(u + 3) & \text{if } u \ge 2 \end{cases}$

21. Prove that $f(x) = 2x - [\![x]\!]$ is continuous for all nonintegral x and discontinuous for x equal to an integer.

In each of the Exercises 22 through 26, define a new function that agrees with the given one for $x \ne a$ and is continuous at $x = a$.

22. $f(x) = \dfrac{x^4 - 81}{x - 3}$ $a = 3$ **23.** $f(x) = \dfrac{x^3 - 8}{x^2 - 4}$ $a = 2$

24. $g(x) = \dfrac{2 - \sqrt{x}}{4 - x}$ $a = 4$ **25.** $h(x) = \dfrac{x^n - 1}{x - 1}$ n a positive integer; $a = 1$

26. $F(x) = \dfrac{(x - 3)^2}{x^2 - 2x - 3}$ $a = 3$

In each of the Exercises 27 through 29, if possible define a new function that agrees with the given function for $x \ne 0$ and is continuous at $x = 0$. If this is not possible, state why.

27. $f(x) = \dfrac{\sqrt{x^2}}{|x|}$

28. $f(x) = \dfrac{|x|}{x}$

29. $g(x) = x^2 - [\![x]\!]$

30. If f is a function whose domain is the set of all real numbers and if f is *not* continuous at $x = a$, what can be said about $\lim_{x \to a} |f(x) - f(a)|$? Illustrate your argument.

31. Each of the functions f, g, and h is defined for all real x. Which of the functions is continuous at $x = 0$?

(a) $f(x) = \begin{cases} 0 & \text{for } x \text{ a rational number} \\ 1 & \text{for } x \text{ irrational} \end{cases}$

(b) $g(x) = \begin{cases} 3 & \text{for } x \text{ irrational} \\ -3 & \text{for } x \text{ rational} \end{cases}$

(c) $h(x) = \begin{cases} 0 & \text{for } x \text{ rational} \\ x & \text{for } x \text{ irrational} \end{cases}$

32. A function $f(x)$ is defined by

$$f(x) = \lim_{n \to \infty} \frac{x^{2n}}{1 + x^{2n}}$$

provided the limit exists.
(a) Find $f(-\frac{1}{3})$, $f(\frac{1}{3})$, $f(0)$, $f(1)$, and $f(3)$.
(b) For which values of x is $f(x)$ defined? Sketch the graph of $y = f(x)$.
(c) For which values of x is $f(x)$ continuous?

33. A function $f(x)$ is defined by

$$f(x) = \lim_{n \to \infty} \frac{n^2 x^2}{1 + n^2 x^2} \qquad -1 \le x \le 1$$

provided the limit exists.
(a) Find $f(-\frac{1}{2})$, $f(\frac{1}{2})$, $f(0)$, and $f(1)$.
(b) For which values of x is $f(x)$ defined? Sketch the graph of $y = f(x)$.
(c) For which values of x is $f(x)$ continuous?

2.10   CONTINUITY THEOREMS

From the definition of continuity of a function at a number and the algebra of limits, that is, the limit of a sum equals the sum of the limits, and so on, it is a simple exercise to establish the following theorem.

Theorem 1   If f and g are two functions that are continuous at $x = a$, then

    (i) $f \pm g$ are continuous at a

    (ii) $f \cdot g$ is continuous at a

    (iii) f/g is continuous at a, provided that $g(a) \neq 0$

To illustrate the nature of the proof required, we shall establish (ii).

Proof   Because f and g are continuous at $x = a$, it follows that $\lim\limits_{x \to a} f(x) = f(a)$ and $\lim\limits_{x \to a} g(x) = g(a)$.

Therefore multiplication yields

$$\lim_{x \to a} f(x) \cdot \lim_{x \to a} g(x) = f(a)\, g(a)$$

But, $\lim\limits_{x \to a} \left(f(x)\, g(x) \right) = \lim\limits_{x \to a} f(x) \cdot \lim\limits_{x \to a} g(x)$ provided the latter limits exist, which we know is true. Therefore

$$\lim_{x \to a} \left(f(x)\, g(x) \right) = f(a)\, g(a)$$

and it follows (by definition) that $f(x) \cdot g(x)$ is continuous at $x = a$.

The proofs of (i) and (iii) are left to the reader.   ∎

Next, we establish a second theorem.

Theorem 2   A polynomial function is continuous for all values of the independent variable.

Proof   Consider the polynomial function

$$P(x) = b_n x^n + b_{n-1} x^{n-1} + \cdots + b_1 x + b_0 \qquad (b_n \neq 0) \qquad (1)$$

where n is a nonnegative integer and the coefficients b_k, $(k = 0, 1, 2, \ldots, n)$ are real numbers.

Now,

$$
\begin{aligned}
\lim_{x \to a} P(x) &= \lim_{x \to a} [b_n x^n + b_{n-1} x^{n-1} + \cdots + b_1 x + b_0] \\
&= \lim_{x \to a} b_n x^n + \lim_{x \to a} b_{n-1} x^{n-1} + \cdots + \lim_{x \to a} b_1 x + \lim_{x \to a} b_0 \\
&= b_n (\lim_{x \to a} x^n) + b_{n-1} (\lim_{x \to a} x^{n-1}) + \cdots + b_1 (\lim_{x \to a} x) + \lim_{x \to a} b_0 \\
&= b_n a^n + b_{n-1} a^{n-1} + \cdots + b_1 a + b_0 = P(a)
\end{aligned}
$$

and the theorem is proved.   ∎

Theorem 3   A rational function is continuous at every number of its domain.

Proof   If R is a rational function, it can be expressed as the quotient of two polynomial functions P and Q. Therefore we write

$$R(x) = \frac{P(x)}{Q(x)} \qquad (2)$$

and the domain of R consists of all x for which $Q(x) \neq 0$. Thus if a is any number in the domain of R we have $Q(a) \neq 0$ so that

$$\lim_{x \to a} R(x) = \frac{\lim\limits_{x \to a} P(x)}{\lim\limits_{x \to a} Q(x)} \qquad (3)$$

or

$$\lim_{x \to a} R(x) = \frac{P(a)}{Q(a)} = R(a)$$

since $P(x)$ and $Q(x)$ are continuous at $x = a$ from Theorem 2. Therefore, by definition, $R(x)$ is continuous at $x = a$ where a is any number in its domain (it should be noted that if $Q(x_0) = 0$ then $R(x_0)$ is not defined and therefore $R(x)$ is not continuous at $x = x_0$.) ∎

EXAMPLE 1 $f(x) = x^2(x + 1)^3$ is a polynomial in x. Therefore $f(x)$ is continuous for all values of x.   ●

EXAMPLE 2 $g(x) = \dfrac{x^2 - 3x - 1}{x^2 - 4x + 3}$ is continuous at all x except for the real roots of the denominator, $x^2 - 4x + 3 = 0$, which are at $x = 1$ and $x = 3$.   ●

EXAMPLE 3 $h(x) = \dfrac{1}{|x|}$ is not a rational function over its whole domain. However,

$$h(x) = \begin{cases} \dfrac{1}{x} & \text{if } x > 0 \\[2mm] -\dfrac{1}{x} & \text{if } x < 0 \end{cases}$$

and $h(0)$ is not defined. Thus $h(x)$ is a rational function with a nonzero denominator for $x > 0$ and $x < 0$. Consequently, $h(x)$ is continuous for all $x \neq 0$.   ●

Theorem 4   If $\lim\limits_{x \to a} g(x) = b$ and if the function f is continuous at b, then

$$\lim_{x \to a} (f \circ g)(x) = f(b)$$

or, equivalently

$$\lim_{x \to a} f(g(x)) = f(\lim_{x \to a} g(x)) \qquad (4)$$

Proof   Since f is continuous at b, it follows that for any $\varepsilon > 0$, there exists a $\delta_1 > 0$, such that

$$|f(z) - f(b)| < \varepsilon \qquad \text{if } |z - b| < \delta_1$$

Also, $\lim\limits_{x \to a} g(x) = b$ so that

$$|g(x) - b| < \delta_1 \qquad \text{if } 0 < |x - a| < \delta_2$$

Thus if $0 < |x - a| < \delta_2$, and we identify $z \equiv g(x)$, then $|z - b| < \delta_1$ and $|f(z) - f(b)| < \varepsilon$. This implies that

$$|f(g(x)) - f(b)| < \varepsilon \text{ when } 0 < |x - a| < \delta_2$$

so that

$$\lim_{x \to a} f(g(x)) = f(b) = f(\lim_{x \to a} g(x)) \qquad \blacksquare$$

EXAMPLE 4   By application of Theorem 4, find $\lim\limits_{x \to 2} \sqrt{3x - 2}$.

Solution   Let functions g and f be defined by

$$g(x) = 3x - 2 \qquad \text{and} \qquad f(x) = \sqrt{x}$$

The composite function $f \circ g$ is defined by

$$f(g(x)) = \sqrt{3x - 2} \qquad \text{domain } x \geq 2/3.$$

Now, $\lim\limits_{x \to 2} g(x) = 4$ and $f(x)$ is continuous at $x = 4$. Thus

$$\lim_{x \to 2} \sqrt{3x - 2} = \lim_{x \to 2} f(g(x)) = f(\lim_{x \to 2} g(x)) = f(4) = \sqrt{4} = 2 \qquad \bullet$$

Another application of Theorem 4 is the following important theorem.

Theorem 5   If the function g is continuous at a and the function f is continuous at $g(a)$, then the composite function $f \circ g$ is continuous at a.

Proof   Since $g(x)$ is continuous at $x = a$, we must have

$$\lim_{x \to a} g(x) = g(a) \tag{5}$$

Now f is continuous at $g(a)$ so that from Theorem 4 applied to this composite function $f \circ g$, we obtain

$$\lim_{x \to a} (f \circ g)(x) = \lim_{x \to a} f(g(x)) \qquad \text{by definition of a composite function}$$

$$= f(\lim_{x \to a} g(x)) \qquad \text{Theorem 4}$$

$$= f(g(a)) \qquad \text{from (5)}$$

$$= (f \circ g)(a) \qquad \text{by definition of a composite function}$$

which proves that $f \circ g$ is continuous at $x = a$. $\qquad \blacksquare$

EXAMPLE 5 Given the function $F(x) = \sqrt{3 + x}$, determine the values of x for which F is continuous.

Solution Let $g(x) = 3 + x$ and let $f(u) = \sqrt{u}$. Now $g(x)$ is continuous for all x and $f(u)$ is continuous for $u > 0$. Thus $f(g(x))$ is continuous for $3 + x > 0$ or $x > -3$. Therefore $F(x) = f(g(x))$ is continuous in the semiinfinite open interval $x > -3$.

Exercises 2.10

In Exercises 1 through 4, apply Theorem 4 to establish the following limits.

1. $\lim\limits_{x \to 5} \sqrt{x + 4}$

2. $\lim\limits_{x \to 2} \sqrt{x^2 + 12}$

3. $\lim\limits_{z \to 2} \sqrt[3]{5z - 2}$

4. $\lim\limits_{y \to -3} \sqrt{\dfrac{y + 8}{y + 5}}$

In Exercises 5 through 16, determine all values of the independent variable for which the given function is continuous. Justify each step by citing the appropriate theorem used.

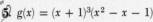

5. $g(x) = (x + 1)^3(x^2 - x - 1)$

6. $h(x) = \sqrt{1 + x^2}$

7. $F(u) = \dfrac{u^2 - 2u + 4}{u^2 - 3}$

8. $G(z) = \sqrt[3]{z + 1}$

9. $g(y) = \sqrt{9 - y^2}$

10. $h(y) = \sqrt{y^2 - 49}$

11. $F(x) = \dfrac{3x + 2}{x^2 + 4x + 5}$

12. $G(x) = \sqrt{\dfrac{2 - x}{2 + x}}$

13. $f(x) = |x^2 - 9|$

14. $g(x) = |10 - 3x + x^3|$

15. $h(x) = \begin{cases} 4x - 3 & \text{if } |x| < 1 \\ x^2 & \text{if } |x| \geq 1 \end{cases}$

16. $f(x) = \max\{2x, x^2\}$; sketch $f(x)$

17. Find a function that is defined for all x and is discontinuous only at $x = 2$.

18. Find a function that is defined for all x and is discontinuous only at $x = 0$, $x = 2$, and $x = 5$.

19. Find a function that is continuous in $-10 < x < 10$ only and is not defined for $|x| > 10$.

20. Find a function that is defined only in $[0, 1]$ and is continuous in $[0, 1]$.

21. Find a function that is defined everywhere and discontinuous from the left only at $x = 0$; that is, $\lim\limits_{x \to 0^-} f(x) \neq f(0)$.

22. Prove that if functions f_k $(k = 1, 2, \ldots, n)$ are continuous at $x = a$ then also the sum
    (i) $f_1 + f_2 + \cdots + f_n$ is continuous at $x = a$
    (ii) $c_1 f_1 + c_2 f_2 + \cdots + c_n f_n$ is continuous at $x = a$, where $c_1, c_2, \ldots, c_n$ are arbitrary constants.

23. Prove that if functions f_k $(k = 1, 2, \ldots, n)$ are continuous at $x = a$, then also the product $f_1 \cdot f_2 \cdots \cdot f_n$ is continuous at $x = a$.

24. Suppose that f and g are functions of x that are defined at all points of a closed interval $[a, b]$. If f is continuous at all points in (a, b) and if g is not continuous at all points in (a, b), then
    (i)   Can $f + g$ be continuous in (a, b)? [*Hint:* $g = (f + g) - f$.]
    (ii) Can $f - g$ be continuous in (a, b)?
    (iii) Can $f \cdot g$ be continuous in (a, b)? (*Hint:* Choose $f \equiv 0$.)
    (iv) Can $\dfrac{f}{g}$ be continuous in (a, b)?

25. Suppose that f and g are two functions that are defined in (a, b) and are both not continuous in (a, b), then
    (i)   Can $f + g$ be continuous in (a, b)?
    (ii) Can $f - g$ be continuous in (a, b)?
    (iii) Can $f \cdot g$ be continuous in (a, b)?
    (iv) Can $\dfrac{f}{g}$ be continuous in (a, b)?

2.11

Consider the function g defined by

$$g(x) = \sqrt{25 - x^2} \qquad \text{where} \quad -5 \le x \le 5 \qquad (1)$$

We know that $g(x)$ is continuous at each x in the interval $-5 < x < 5$. Since $g(x)$ is continuous at all points in $(-5, 5)$, we say that $g(x)$ is continuous in the open interval $-5 < x < 5$. More generally, we define continuity in an open interval as follows:

Definition   A function f is said to be ***continuous in (a, b)*** if it is continuous at all points of (a, b).

closed — end point included
[ ] included

open — not including endpoints
( )

EXAMPLE 1   The function g defined by

$$g(x) = \frac{1}{x(x - 3)}$$

is continuous in three open intervals $(-\infty, 0)$, $(0, 3)$ and $(3, \infty)$. It is also continuous in any open interval that does not contain the numbers $x = 0$ and $x = 3$. ●

If we return to (1), we do not have "two-sided continuity" at $x = 5$ or $x = -5$ because the function is not defined if $|x| > 5$. Therefore to discuss the question of continuity at the end points of the interval, the concept of continuity must be extended. This is accomplished by defining one-sided continuity, that is, right-hand and left-hand continuity. Thus we say that $g(x)$ is continuous from the right at $x = -5$ if $\lim\limits_{x \to -5^+} g(x) = g(-5)$. Similarly, $g(x)$ is said to be continuous from the left at $x = 5$ if $\lim\limits_{x \to 5^-} g(x) = g(5)$. In the case of (1),

$$R H \qquad \lim_{x \to -5^+} g(x) = \lim_{x \to -5^+} \sqrt{25 - x^2} = g(-5)$$

and

$$L H \qquad \lim_{x \to 5^-} g(x) = \lim_{x \to 5^-} \sqrt{25 - x^2} = g(5)$$

We say that $g(x)$ defined by (1) is continuous in the closed interval $[-5, 5]$ because $g(x)$ is continuous in $(-5, 5)$ and, furthermore, $\lim\limits_{x \to -5^+} g(x) = g(-5)$ and $\lim\limits_{x \to 5^-} g(x) = g(5)$.

More generally, we have

know for test

Definition $f(x)$ is continuous in $[a, b]$ where $a < b$ if

(i) $f(x)$ is continuous in (a, b)
(ii) $\lim\limits_{x \to a^+} f(x) = f(a)$ (right hand continuity at a)

(iii) $\lim\limits_{x \to b^-} f(x) = f(b)$ (left hand continuity at b)

EXAMPLE 2   Determine the intervals on which the function $g(x) = \sqrt{-x^2 + 5x - 4}$ is continuous.

Solution $g(x) = \sqrt{(4 - x)(x - 1)}$ is defined if and only if the expression under the square root sign is nonnegative. Now $(4 - x)(x - 1)$ is nonnegative if

$$4 - x \geq 0 \qquad \text{and} \qquad x - 1 \geq 0 \qquad \text{(i)}$$

or if

$$4 - x \leq 0 \qquad \text{and} \qquad x - 1 \leq 0 \qquad \text{(ii)}$$

This yields $1 \leq x \leq 4$ from (i) and from (ii), $x \geq 4$ and $x \leq 1$. This latter compound requirement is contradictory.

Now $g(x)$ is continuous in $1 < x < 4$, and furthermore,

$$\lim_{x \to 4^-} g(x) = g(4) = 0 \qquad \text{and} \qquad \lim_{x \to 1^+} g(x) = g(1) = 0$$

Therefore, $g(x)$ is continuous in $[1, 4]$ and, of course, in any subinterval of $[1, 4]$.

●

Next, suppose that we are given a function whose domain includes the half open interval $[a, b)$, that is $a \leq x < b$. Such a function is said to be continuous in $[a, b)$ if it is continuous in (a, b) and continuous from the right at a. Similarly, a function whose domain includes $(a, b]$ is said to be continuous in $(a, b]$ if it is continuous in (a, b) and continuous from the left at b.

EXAMPLE 3   Given the function h defined by

$$h(x) = \sqrt{\frac{4 - x}{3 + x}}$$

determine whether or not h is continuous or discontinuous on each of the following intervals:

(i) $(-3, 4)$ (ii) $[-3, 4)$ (iii) $(-3, 4]$
(iv) $[-3, 4]$ (v) $[-2, 3]$

Solution $h(x)$ is defined if and only if $\dfrac{4 - x}{3 + x}$ is nonnegative. This occurs if $4 - x \geq 0$ and $3 + x > 0$ or $4 - x \leq 0$ and $3 + x < 0$. This yields $-3 < x \leq 4$ or $(-3, 4]$. The function is continuous on $(-3, 4]$ because it is continuous on $(-3, 4)$ and $\lim_{x \to 4^-} h(x) = 0 = h(4)$. Thus $h(x)$ is continuous on $(-3, 4)$, $(-3, 4]$, and $[-2, 3]$, but discontinuous on $[-3, 4)$ and $[-3, 4]$.

●

We shall need two very important theorems involving continuous functions on closed intervals.

Theorem 1   (*Extreme Value Theorem*).   If f is a continuous function defined on the closed interval $[a, b]$, there is (at least) one point c_1 in $[a, b]$ for which f has a largest value or absolute maximum. (This means that $f(x) \leq f(c_1)$ for all x in $[a, b]$.) Similarly, there is (at least) one point c_2 in $[a, b]$ for which f has a smallest value or absolute minimum. (This implies that $f(x) \geq f(c_2)$ for all x in $[a, b]$.)

Know
for
Test

This theorem is fairly clear intuitively. If we think of a continuous function as one with no breaks or gaps, then, as the curve is traversed from the point $(a, f(a))$ to the point $(b, f(b))$ there must be a value of x at which $f(x)$ is largest and a value of x at which $f(x)$ is least. Despite the "obviousness" of the theorem, its proof is not simple and would delay our progress too long. Instead, we will confine ourselves to a critical evaluation of the theorem by examining special situations.

EXAMPLE 4   Find the absolute maximum and absolute minimum for $f(x) = |x|$ in $[-2, 3]$.

Solution $f(x) = |x|$ is continuous in $[-2, 3]$ and thus the hypothesis of Theorem 1 is satisfied. Also, $|x| \geq 0$ so that the absolute minimum occurs at $x = 0$ where $f(0) = 0$. The absolute maximum of $f(x)$ in $[-2, 3]$ is $|3| = 3$.   ●

EXAMPLE 5   Find the absolute maximum and absolute minimum for $f(x) = x$ in $(0, 4)$.

Solution $f(x) = x$ is certainly continuous in $(0, 4)$. However, the interval is open. Therefore the hypothesis of Theorem 1 is not satisfied, and there is no guarantee that we do have an absolute maximum and minimum (although we may). Actually for this case we do not, since $f(x)$ comes as close to 4 as desired but never attains the value 4. Similarly, $f(x) > 0$ and can be made as close to 0 as desired, but never attains the value 0. Hence there is no absolute maximum or absolute minimum.   ●

EXAMPLE 6   Find the absolute maximum and minimum for

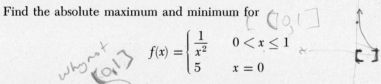

$$f(x) = \begin{cases} \dfrac{1}{x^2} & 0 < x \leq 1 \\ 5 & x = 0 \end{cases}$$

Solution $f(x)$ is defined in $[0, 1]$; however, it is not continuous on the closed interval since $\lim_{x \to 0^+} f(x)$ does not exist. In fact, as $x \to 0^+$, $f(x) \to \infty$, and there is no absolute maximum in $[0, 1]$; however, $f(1) = 1$ is the absolute minimum in this interval.   ●

Corollary to   (*Boundedness Theorem for Continuous Functions*).   Let f be continu-
Theorem 1   ous on a closed interval $[a, b]$. Then f is bounded on $[a, b]$; that is, there is a number $K \geq 0$ such that $|f(x)| \leq K$ for all x in $[a, b]$.

For example, $$f(x) = \begin{cases} \dfrac{1}{x} & \text{if } 0 < x \leq 1 \\ 1 & \text{if } x = 0 \end{cases}$$

is not bounded in $[0, 1]$ since $\lim_{x \to 0^+} \dfrac{1}{x} = \infty$ and therefore $f(x)$ is not continuous in $[0, 1]$.

The next theorem also is concerned with a function that is continuous in a closed interval $[a, b]$.

Theorem 2   (*Intermediate Value Theorem*) Assume the function $f(x)$ is continuous in $[a, b]$ and $f(a) \neq f(b)$. Then for any number L between $f(a)$ and $f(b)$ there exists at least one number c such that $a < c < b$ for which $f(c) = L$.

The intermediate value theorem states that a continuous function defined in $[a, b]$ takes on all values between $f(a)$ and $f(b)$ at least once. Geometrically the line $y = L$ intersects $f(x)$ for at least one value of x in (a, b) (in Figure 2.11.1 there are three numbers that may serve as c.)

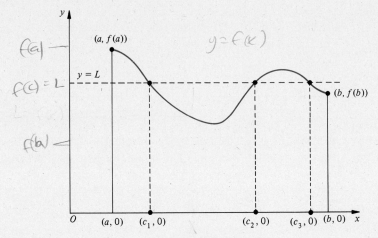

Figure 2.11.1

EXAMPLE 7   Consider the function f given by

$$f(x) = x^3 - 3x + 1$$

Certainly $f(x)$ is continuous for all x because it is a polynomial. Also, we observe that $f(1) = -1$ while $f(2) = 3$. Therefore the function $f(x)$ must take on all real values between -1 and 3 at least once. In particular, $f(x)$ has a zero in $(1, 2)$. Now $f(1.5) > 0$, which means that at least one real root exists in $(1, 1.5)$ because $f(1) < 0$ and $f(1.5) > 0$. The process may be continued by finding $f(1.1)$, $f(1.2)$, $f(1.3)$, $f(1.4)$, and so on. This process of root searching is easily automated and the root can be found to any order of approximation desired.   ●

Exercises 2.11

In Exercises 1 through 12, determine whether or not the function is continuous on each of the indicated intervals.

1. $f(x) = \dfrac{3}{x}$; $[0, 1]$, $[-1, 0)$, $(-1, 2)$, $(0, \infty)$

2. $g(x) = \dfrac{1}{x^2 - 25}$; $[0, 5)$, $[-5, 0]$, $(-5, 5)$, $(-\infty, -5)$

3. $h(u) = \dfrac{u + 3}{u^2 + 2u - 8}$; $(-\infty, 2)$, $[0, 1]$, $[1, 2]$, $(2, \infty)$

4. $g(y) = \sqrt{16 + y}$; $(-\infty, 0)$, $[-20, -16]$, $[-16, 16]$, $[-16, 100]$

5. $f(z) = \sqrt{z^2 - 100}$; $(-\infty, -10], [-10, 0), (0, 15), [-10, \infty)$

6. $g(x) = \dfrac{x^2 - 9}{x^2 - 5x + 6}$; $(-\infty, 2), (2, 3), (0, 3), [3, \infty)$

7. $h(x) = [\![x]\!]$; $(-\frac{1}{4}, \frac{1}{4}), (\frac{1}{4}, \frac{3}{4}), (0, 1), (0, 1]$ *Class*

8. $f(u) = \sqrt{-u^2 + 5u - 6}$; $(-\infty, 3), (2, 3], [2, 3], (3, 4)$

9. $g(y) = \sqrt{\dfrac{y - 3}{y + 5}}$; $(-\infty, -5), (-5, 3), (0, 5), [3, \infty)$

10. $h(z) = \sqrt{\dfrac{4 - z}{4 + z}}$; $(4, \infty), [-1, 2], (-4, 4], [-4, 4)$

11. $f(x) = \sqrt{\dfrac{x + 3}{x^2 - 5x + 4}}$; $(4, \infty), (1, 4), [-3, 0], (-\infty, -3]$

12. $g(x) = \dfrac{1}{\sqrt{4x^2 + x - 3}}$; $(-\infty, -1), [-3, -1), (0, 1), [1, \infty)$

In Exercises 13 through 18, determine the intervals on which each of the functions is continuous.

13. $g(x) = \dfrac{1}{2x^2 - 7x + 3}$ *Dr Cole* *Why include the interval $(\frac{1}{2}, 3)$ when the f is not cont. in this interval?*

14. $h(x) = \dfrac{x^2 + 1}{2x^2 - x - 2}$

15. $f(x) = \sqrt{(x - a)(b - x)}$ where $a < b$

16. $g(x) = \sqrt{\dfrac{x - a}{b - x}}$ where $a < b$

17. $h(x) = \sqrt[3]{(x - a)(b - x)}$

18. $F(x) = \sqrt[3]{\dfrac{x - a}{x - b}}$

In Exercises 19 through 23, determine the absolute maximum and minimum (if possible) for each of the functions defined on the given interval. A sketch of the given function should be helpful.

19. $f(x) = |2 - x|$ in $[0, 3]$

20. $g(x) = \sqrt{17 - x^2}$ in $[-2, 3]$

21. $f(x) = \begin{cases} 6 - x & \text{in } [0, 3] \\ \dfrac{x^2}{3} & \text{in } (3, 5] \end{cases}$

How can we draw graph with little experience without taking time to plot points?

22. $h(x) = \begin{cases} \dfrac{[\![x]\!]}{x} & \text{in } (0, 1] \\ \dfrac{1}{2} & \text{at } x = 0 \end{cases}$

23. $g(u) = u^2 - 2u + 5$ in $[-3, 2]$ (*Hint:* Complete the square.)

24. The equation $x^3 - 4x - 7 = 0$ has only one real root. Find an interval of length 1 that contains the root.

25. The equation $x^3 - 6x^2 + 24 = 0$ has three real roots. Isolate each root of the given equation by exhibiting an interval containing this root and no others.

26. Suppose that we are given a thin circular ring for which the temperature is a continuous function of angle θ (which identifies the position on the ring). Show that there exists at least one diameter for which the end points must be at the same temperature. (*Hint:* Work with the difference of the temperatures at the ends of the diameter, that is, form $g(\theta) = f(\theta + \pi) - f(\theta)$ where $f(\theta)$ is the value of the temperature as a function of θ.)

Review and Miscellaneous Exercises

In Exercises 1 through 6, determine the domain of each of the functions whose formulas are defined for real values of the variable and which yield real numbers.

1. $f(x) = \sqrt{5x - 4}$

2. $g(x) = x^{-2}(x - 1)$

3. $h(x) = \dfrac{x^2 - 2x - 1}{x^2 - 5x}$

4. $F(x) = |2x - 7|$

5. $G(x) = (x^2 - 4x + 5)^{1/4}$

6. $H(x) = \dfrac{\sqrt{x^2 - 4}}{\sqrt{x^2 - 9}}$

In Exercises 7 and 8, for each of the following pairs of functions, find $f + g$, $f \cdot g$, $\dfrac{f}{g}$, $\dfrac{g}{f}$, $f \circ g$ and $g \circ f$. State the domain in each case.

7. $f(x) = 2x$, $g(x) = 4 - 3x$

8. $f(x) = 3x + 5$, $g(x) = x^2$

In Exercises 9 and 10, find $f(g(x))$ and $g(f(x))$ for each of the following pairs of functions.

9. $f(x) = \sqrt{x}$ and $g(x) = x^2$

10. $f(x) = x^{1/3} + 1$ and $g(x) = (x - 1)^3$

11. Prove that the product of two odd functions must be even.

12. Find $\lim\limits_{t \to 1} \dfrac{t^n - 1}{t - 1}$ where n is an arbitrary positive integer. What is the limit when $n = 0$? (*Hint:* Divide out.)

In Exercises 13 through 18, determine whether the indicated limit exists. If it does exist, find its value.

13. $\lim\limits_{x \to 1} \dfrac{(x - 3)(x^2 + 8x - 9)}{\sqrt{x} - 1}$

14. $\lim\limits_{x \to 0} \dfrac{\sqrt{16 + x} - \sqrt{16 - x}}{x}$

15. $\lim\limits_{u \to 3^-} \dfrac{[\![u]\!]^2 - 9}{u - 3}$

16. $\lim\limits_{y \to 4^-} (y - 5 + [\![3 - y]\!] + [\![y]\!])$ $[\![\ ]\!]$ = greatest integer function

17. $\lim\limits_{h \to 0} \dfrac{(u + h)^3 - u^3}{h}$

18. $\lim\limits_{t \to 0} \dfrac{|t^3|}{t^3}$

19. If $f(x) = \dfrac{1}{x}$, $x \neq 0$, find

    (a) $\lim\limits_{x \to 2} \dfrac{f(x) - f(2)}{x - 2}$ **(b)** $\lim\limits_{h \to 0} \dfrac{f(2 + h) - f(2)}{h}$

20. If $g(t) = t^{-2}$, $t \neq 0$, find

    (a) $\lim\limits_{t \to a} \dfrac{g(t) - g(a)}{t - a}$ **(b)** $\lim\limits_{h \to 0} \dfrac{g(a + h) - g(a)}{h}$, $a \neq 0$

21. If $h(u) = \dfrac{u}{u + 1}$, $u \neq -1$, find

$$\lim\limits_{u \to 0} \dfrac{h(u) - h(0)}{u}$$

In Exercises 22 through 30, at what numbers (if any) is each of the functions discontinuous.† Indicate which theorems are being applied. Sketch the function.

22. $F(t) = \dfrac{t^2 - 3t - 1}{t^2 - 4t + 6}$

23. $g(z) = \dfrac{z^3 - 7z + 5}{z^2 - 2z - 6}$

24. $h(u) = \begin{cases} \dfrac{u^3 - 1}{u - 1} & u \neq 1 \\ 2 & u = 1 \end{cases}$

25. $G(t) = |3t + 1|$

26. $f(x) = |2x - 5| + |x^2 - 3|$

27. $g(x) = \left| \dfrac{x + 1}{3x - 1} \right|$

28. $u(t) = \begin{cases} t^2 - t & t < 3 \\ 6 & t = 3 \\ t + 3 & t > 3 \end{cases}$

29. $h(z) = \sqrt{-10 + 7z - z^2}$ (*Hint:* Factor the quadratic.)

30. $F(x) = 3x - [\![x]\!]$

In Exercises 31 through 36, evaluate the limit by use of the limit theorems.

31. $\lim\limits_{x \to \infty} \dfrac{10x + 37}{3x + 5}$

32. $\lim\limits_{x \to -\infty} \dfrac{8x + 13}{5x + 29}$

33. $\lim\limits_{u \to \infty} \dfrac{29\sqrt{u} + 100}{u - 4}$

34. $\lim\limits_{t \to -\infty} \dfrac{2t^4 + 12t^3 + 7}{11t^4 - t^2 + 19}$

35. $\lim\limits_{v \to -\infty} \dfrac{\sqrt{v^2 + 10}}{9v + 8}$

36. $\lim\limits_{v \to \infty} \dfrac{\sqrt{7v^4 + 100v^3}}{3v^2 + v + 10}$

In Exercises 37 through 40, find the horizontal and vertical asymptotes of the graph of the given equation.

37. $y = \dfrac{3x + 1}{2x - 7}$

38. $9xy - 4y + 2x - 5 = 0$ (*Hint:* Solve for y in terms of x.)

39. $x^2y - 7xy - x + 6y - 3 = 0$

40. $x^2(y - 3) + xy - x^3 - 2 = 0$

41. The function $3x^2 + x + 6$ is continuous in the interval $(0, 4)$. However, it does not have a largest and a smallest value there. Explain why this occurs.

42. Two functions $f(x)$ and $g(x)$ are continuous in $[a, b]$. Furthermore, $f(a) > g(a)$ while $f(b) < g(b)$ where $a < b$. What is implied about the equation $f(x) = g(x)$?

43. The function

$$f(x) = \begin{cases} x^2 & -\infty < x \leq 0 \\ -(1 + x) & x > 0 \end{cases}$$

is defined for all values of x. **(a)** For what value of x will $f(x)$ be $-\tfrac{1}{2}$? **(b)** Determine the range of the function.

44. Show that all cubic equations of the form $x^3 + ax^2 + bx + c = 0$ where a, b, and c are real constants, possess at least one real root. State a generalization to polynomials of higher degree.

---

† Two sided continuity.

3

The Derivative

BRIEF HISTORICAL BACKGROUND OF CALCULUS

The fusion of the two types of calculus, known as differential and integral calculus, was developed quite independently of each other by two great mathematicians, Isaac Newton (1642–1727) and Gottfried Leibnitz (1646–1716), as a needed tool for problems in geometry and mechanics. However, the origin of integral calculus goes back to Archimedes (287–212 BC) when the Greeks attempted to determine areas of circles and other special figures by what they called the *method of exhaustion.* Unfortunately, the development of the method of exhaustion beyond the point to which Archimedes carried it did *not* occur for nearly eighteen centuries, until the effective application of algebraic symbolism became standard equipment for the mathematicians of the sixteenth and seventeenth centuries AD. Then the method of exhaustion was gradually transformed into what is currently known as integral calculus (to be discussed later).

We are going to begin the study of differential calculus in this chapter. The central idea of differential calculus is the concept of the *derivative.* Analogous to integral calculus, the notion of a derivative stemmed from a geometrical problem, namely that of determining the tangent line to a "smooth curve" at a point on the curve. However, the derivative was introduced into mathematics very late in the history of the development of calculus. The French mathematician Pierre de Fermat (1601–1665) was concerned early in the seventeenth century with the problem of maximizing and minimizing certain special functions. He recognized that the slope of the tangent line played a significant role in such problems. In particular, Fermat observed that, at certain points where the curve has a maximum or minimum (Figure 3.1.1), the tangent line must be horizontal. This would occur at points such as P_1 and P_2 with abscissas c_1 and c_2 as shown in Figure 3.1.1. Therefore, the problem of determining such extreme values seems to depend on the problem of determining those points on the curve at which the tangents are horizontal.

The question then occurs: How does one define a tangent line to a given curve

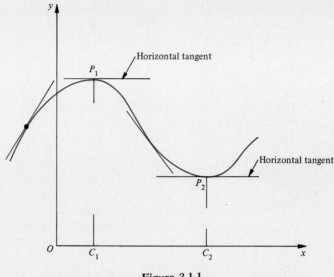

Figure 3.1.1

at any point of the curve? This was resolved by Fermat for special curves and led to the idea of a derivative.

But what has the slope of the tangent line got to do with the determination of the area of plane figures? That these two apparently remote ideas are in fact intimately related was first seen by Isaac Barrow (1630–1677), Newton's teacher.

From the positive accomplishments of Leibnitz notationally and the exciting creativity of Newton towards the solution of many physical problems in mechanics, optics, and planetary motion, calculus developed extensively in the next 125 years into the early nineteenth century. Two of the greatest mathematicians of that period were Leonhard Euler (1707–1783) and Joseph Louis Lagrange (1736–1813). Euler wrote popular books in trigonometry and algebra. He also wrote the first widely read texts on calculus. Furthermore, Euler made advances in all phases of mathematics, including, in particular, significant work in optics, the three-body problem of mechanics, that is, the effect of mutual attraction of the sun, moon, and the Earth, and important developments in the subject now known as the calculus of variations. He also did very creative work in the field of fluid mechanics. Lagrange also worked in these fields, achieving even greater generality and unity. His most magnificent accomplishment was his monumental *Mechanique Analytique* (1788), which brought the science of mechanics close to its present form.

In the first half of the nineteenth century, the great mathematician Carl Friedrich Gauss (1777–1855) made significant contributions to number theory, theory of surfaces, complex numbers and complex functions, numerical methods, electricity and magnetism, and the development of the electric telegraph. This genius also did work on line and surface integrals of calculus and on the underlying logic of calculus. The latter was done to overcome valid objections to the work of the founders of calculus. This penetrating analysis was continued further by another great mathematician and contemporary Augustin-Louis Cauchy (1789–1857). In 1821 Cauchy wrote and published his *Cours d'analyse*. This was a book on analysis based upon the course of lectures given at the Polytechnique. In this book Cauchy gave the definitions of limit, continuity, convergence of sequences and series, and so on, which are the definitions given in most current textbooks. Further developments of the foundations of calculus were due to Karl Weierstrass (1815–1897), who pointed out logical oversights at strategic points in the development of calculus. He also did extensive work in the development of infinite sequences and series and used these tools effectively in the analysis of functions of a real and complex variable. Finally, Richard Dedekind (1831–1916) contributed a penetrating analysis of the nature of the real number system. Of course, further refinements and extensions of the theory and application of calculus to the social as well as the physical sciences are being developed in contemporary mathematics.

3.2   TANGENT TO A CURVE

The problem of finding an equation of the *tangent* line to a curve at a point on the curve is significant to the development of differential calculus. We recall from plane geometry that a straight line intersects a circle in two points, or is tangent to the circle (that is, touches the circle at one point) or does not intersect the circle at all. This may suggest that we should utilize this definition for more general curves. However this definition of touching at one point is not satisfactory, as seen in Figure 3.2.1, where T, the tangent line to the curve at P intersects the curve at three points.

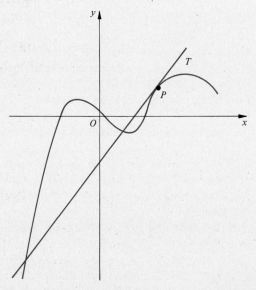

Figure 3.2.1

Let us investigate a procedure for obtaining an equation of the tangent line to a curve at an arbitrary point on the curve. Let f be the given continuous function and let $P(x_1, f(x_1))$ be any point on the graph of f (Figure 3.2.2). Any nonvertical line passing through $(x_1, f(x_1))$ has an equation of the form (point-slope formula)

$$y = f(x_1) = m(x - x_1) \tag{1}$$

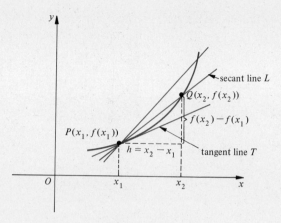

Figure 3.2.2

where m is the slope of the required line. This number m is what must be determined, for the other constants in the equation x_1 and $f(x_1)$ are already known.

Fermat's procedure for the evaluation of the slope m is illustrated in Figure 3.2.2. Let the line L pass through $P(x_1, f(x_1))$ and some other point $Q(x_2, f(x_2))$ on the graph. Any line through two points on a curve is called a **secant line** so that L is a secant line. The slope of the line through P and Q is given by

$$m_L = \frac{f(x_2) - f(x_1)}{x_2 - x_1} \qquad (x_2 \neq x_1) \tag{2}$$

If the difference between the abscissas at Q and P is denoted by h then

$$h = x_2 - x_1 \tag{3}$$

where h in the figure is positive; however, h may be negative (if $x_2 < x_1$). From $x_2 = x_1 + h$ we may rewrite (2) in the form

$$m_L = \frac{f(x_1 + h) - f(x_1)}{h} \qquad h \neq 0 \tag{4}$$

Next consider P as fixed and move Q along the curve toward P; that is, let the quantity h approach zero. As $h \to 0$ the secant line pivots about point P. Now, if the secant line has a limiting position, it is this limiting position that we postulate is the tangent line. In other words the slope of the tangent line is *defined* by the relation

$$m(x_1) = \lim_{h \to 0} \frac{f(x_1 + h) - f(x_1)}{h} \tag{5}$$

if the limit exists. If $m(x_1)$ does not exist in the ordinary sense, but

$$\lim_{h \to 0} \left| \frac{f(x_1 + h) - f(x_1)}{h} \right| = \infty \tag{6}$$

then the secant line L is getting steeper and steeper as $h \to 0$. In this case the tangent line to the graph of f at the point $P(x_1, f(x_1))$ appears to be the vertical line $x = x_1$. These remarks lead us to state the following definition.

Definition The ***tangent line*** to the graph of f at the point $P(x_1, f(x_1))$ is
    (i)   The line passing through P with slope

$$m(x_1) = \lim_{h \to 0} \frac{f(x_1 + h) - f(x_1)}{h} \tag{5}$$

        provided that the limit exists or
    (ii) The line $x = x_1$ if (6) holds.
    (iii) If neither (5) nor (6) holds then the graph of f does not have a tangent line at the point $P(x_1, f(x_1))$.

In case $m(x_1)$ defined by (5) exists, then

$$y - f(x_1) = m(x_1)(x - x_1) \tag{7}$$

is an equation of the tangent line T to the graph of f at the point $P(x_1, f(x_1))$.

Definition The ***normal line*** N to the graph of a function f at the point $P(x_1, f(x_1))$ is defined to be the line through P perpendicular to the tangent line.

It follows that, if the slope of the tangent line $m(x_1)$ given by (5) is not equal to zero, then the slope of N is its negative reciprocal $-\dfrac{1}{m(x_1)}$ and an equation of N is

$$y - f(x_1) = -\frac{1}{m(x_1)} (x - x_1). \tag{8}$$

If $m(x_1) = 0$, then N is the vertical line $x = x_1$; and if the tangent line is vertical (that is, $x = x_1$), N is the horizontal line $y = f(x_1)$.

EXAMPLE 1 Find the slope of the tangent line to the curve $y = f(x) = x^2 - 2x + 3$ at $(2, 3)$.

Solution A sketch of the curve corresponding to $f(x) = x^2 - 2x + 3$ is shown in Figure 3.2.3.
    The slope at $x = 2$ is defined by

$$\begin{aligned}
m(2) &= \lim_{h \to 0} \frac{f(2 + h) - f(2)}{h} \quad (h \neq 0) \\[2mm]
&= \lim_{h \to 0} \frac{(2 + h)^2 - 2(2 + h) + 3 - 3}{h} \\[2mm]
&= \lim_{h \to 0} \frac{4 + 4h + h^2 - 4 - 2h}{h} \\[2mm]
&= \lim_{h \to 0} \frac{h(h + 2)}{h} = \lim_{h \to 0} (h + 2) \\[2mm]
&= 2
\end{aligned}$$

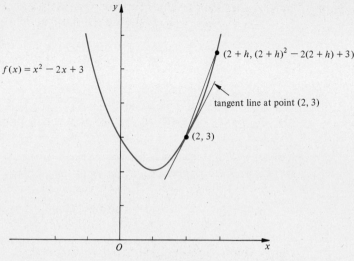

$f(x) = x^2 - 2x + 3$

$(2 + h, (2 + h)^2 - 2(2 + h) + 3)$

tangent line at point $(2, 3)$

$(2, 3)$

O

Figure 3.2.3

EXAMPLE 2 (a) Find the slope of the tangent line to an arbitrary point (x_1, y_1) on the curve $y = f(x) = x^2 - 2x + 3$. (b) What is the slope at $(2, 3)$? (c) For what value of x_1 is the slope equal to 3?

Solution (a) $$m(x_1) = \lim_{h \to 0} \frac{f(x_1 + h) - f(x_1)}{h} \qquad h \neq 0$$

$$= \lim_{h \to 0} \frac{[x_1^2 + 2x_1 h + h^2 - 2x_1 - 2h + 3] - [x_1^2 - 2x_1 + 3]}{h}$$

$$= \lim_{h \to 0} \frac{2x_1 h + h^2 - 2h}{h} = \lim_{h \to 0} \frac{h(2x_1 - 2 + h)}{h}$$

$$= \lim_{h \to 0} (2x_1 - 2 + h) = 2x_1 - 2$$

(b) In particular, $m(2) = 2(2) - 2 = 2$ in agreement with the previous example.
(c) Finally $m(x_1) = 2x_1 - 2 = 3$ is required so that $x_1 = \frac{5}{2}$. Therefore, the slope of the tangent line to the curve is 3 at $x_1 = 5/2$. ●

EXAMPLE 3 If $f(x) = \sqrt{x}$, find the slope of the tangent line at any point with abscissa $x = a > 0$. Find the equations of the tangent and normal lines at $(9, 3)$.

Solution The function is shown in Figure 3.2.4.

$$\frac{f(a + h) - f(a)}{h} = \frac{\sqrt{a + h} - \sqrt{a}}{h} \qquad h \neq 0$$

for any nonzero h for which $a + h \geq 0$.
The slope of the tangent line at $x = a$ is

$$m(a) = \lim_{h \to 0} \frac{\sqrt{a + h} - \sqrt{a}}{h}$$

If we let $h \to 0$ both numerator and denominator are approaching zero and it is virtually impossible to see what the limit becomes as $h \to 0$ (as the expression

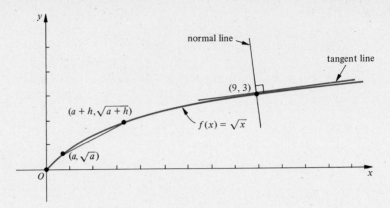

Figure 3.2.4

stands). However, the following device, called *rationalizing the numerator,* enables us to resolve this problem. We multiply both numerator and denominator by $\sqrt{a+h} + \sqrt{a}$ so that

$$m(a) = \lim_{h \to 0} \frac{\sqrt{a+h} - \sqrt{a}}{h}\left(\frac{\sqrt{a+h} + \sqrt{a}}{\sqrt{a+h} + \sqrt{a}}\right)$$

$$= \lim_{h \to 0} \frac{a + h - a}{h(\sqrt{a+h} + \sqrt{a})} = \lim_{h \to 0} \frac{1}{\sqrt{a+h} + \sqrt{a}}$$

$$= \frac{1}{2\sqrt{a}} \qquad a > 0$$

In particular, $m(9) = \frac{1}{6} = $ slope of tangent line and the point-slope equation of the tangent line is

$$\frac{y - 3}{x - 9} = \frac{1}{6} \qquad \text{or} \qquad x - 6y + 9 = 0$$

The slope of the normal line is the negative reciprocal of the slope of the tangent line—that is, $-1/m(a)$, where $m(a) \neq 0$, and its equation is

$$\frac{y - 3}{x - 9} = -6 \qquad \text{or} \qquad 6x + y - 57 = 0 \qquad\qquad \bullet$$

EXAMPLE 4 Find equations of the tangent line T and the normal line N to the graph of g defined by $g(x) = \dfrac{1}{x}$ at $(2, \frac{1}{2})$.

Solution The function g is shown in Figure 3.2.5 together with the lines T and N determined below. We form

$$\frac{g(2+h) - g(2)}{h} = \frac{\dfrac{1}{2+h} - \dfrac{1}{2}}{h} \qquad h \neq 0$$

$$= \frac{-h}{2h(2+h)}$$

$$= \frac{-1}{2(2+h)}$$

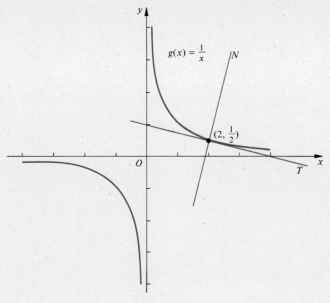

Figure 3.2.5

The slope of the tangent line at $(2, \frac{1}{2})$ is

$$m(2) = \lim_{h \to 0} \frac{g(2 + h) - g(2)}{h} = -\frac{1}{4}$$

and an equation of the tangent line is, from (7)

$$y - \tfrac{1}{2} = -\tfrac{1}{4}(x - 2) \qquad \text{or} \qquad x + 4y - 4 = 0$$

The slope of the normal line is 4 and an equation of it is

$$y - \tfrac{1}{2} = 4(x - 2)$$

or (clearing the fraction)

$$8x - 2y - 15 = 0 \qquad\qquad\qquad\qquad \bullet$$

EXAMPLE 5 Find equations of the tangent line T and the normal line N to the graph of the function f defined by $f(x) = \sqrt[3]{x}$ at the origin.

Solution We form

$$\frac{f(h) - f(0)}{h} = \frac{\sqrt[3]{h} - 0}{h} = \frac{1}{h^{2/3}} \qquad h \neq 0$$

Therefore

$$\lim_{h \to 0} \frac{f(h) - f(0)}{h} = \infty$$

It follows that T is the vertical line $x = 0$ (y-axis) and N is the horizontal line $y = 0$ (x-axis). The graph of f is shown in Figure 3.2.6. $\qquad \bullet$

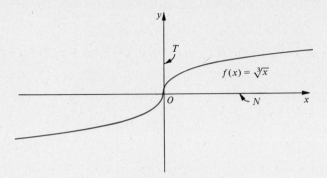

Figure 3.2.6

EXAMPLE 6 Is there a tangent line to $f(x) = |x|$ at $x = 0$?

Solution Recall that

$$f(x) = |x| = \begin{cases} -x & \text{if } x < 0 \\ x & \text{if } x \geq 0 \end{cases}$$

$$m(0) = \lim_{h \to 0} \frac{f(0 + h) - f(0)}{h} = \lim_{h \to 0} \frac{f(h) - f(0)}{h} = \lim_{h \to 0} \frac{f(h)}{h}$$

But

$$\lim_{h \to 0^-} \frac{f(h)}{h} = \lim_{h \to 0^-} \frac{-h}{h} = -1$$

and

$$\lim_{h \to 0^+} \frac{f(h)}{h} = \lim_{h \to 0^+} \frac{h}{h} = 1$$

So the limit does not exist either as a real number or as an infinite limit. Hence there is no tangent line at $x = 0$. The graph is drawn in Figure 3.2.7. ●

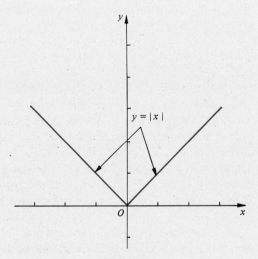

Figure 3.2.7

Exercises 3.2

Finding the Derivative
Finding slope of for line

In Exercises 1 through 10, find the slope of the curve at the point indicated.

1. $f(x) = x^2$ at $(-2, 4)$

2. $f(x) = x^2 - x$ at $(2, 2)$

3. $f(x) = 2x^2 - 3x + 1$ at $(1, 0)$

4. $f(x) = (x - 3)^2$ at $(1, 4)$

5. $f(x) = (x + 2)(x - 1)$ at $(3, 10)$

6. $f(x) = x^3 - 3x$ at $(-1, 2)$

7. $f(x) = \dfrac{x + 2}{x - 2}$ at $(0, -1)$

CLASS ✗ 8. $f(x) = \dfrac{3 - x}{2 + 3x}$ at $\left(2, \dfrac{1}{8}\right)$ $-\dfrac{11}{64}$

9. $f(x) = \sqrt{1 + x}$ at $(2, \sqrt{3})$

CLASS ✗ 10. $f(x) = \dfrac{1}{\sqrt{x}}$ at $\left(4, \dfrac{1}{2}\right)$

In Exercises 11 through 14, find equations of the tangent and normal lines to the given curve at the indicated point. Sketch the curve, the tangent, and normal lines at the indicated point.

11. $y = x^2 - 3x$ at $(-1, 4)$

12. $y = x^2 + 1$ at $(0, 1)$

13. $y = x^3$ at $(-1, -1)$

14. $y = x\,|x|$ at $(0, 0)$

15. Let $f(x) = 3x^2 + 4x + 2$. At which points (if any) is the tangent line to the graph of f (a) parallel to the x-axis; (b) parallel to the line $y = 2x + \frac{1}{2}$; (c) perpendicular to the line $y = -2x$.

16. Find the slope of the tangent to the curve $f(x) = \sqrt{x + 1}$ at $(1, \sqrt{2})$.

3.3   VELOCITY IN RECTILINEAR MOTION

Suppose that we drive from one town to another which is located 200 miles away. If the total time for the trip is 4 hours then we say that the average speed (that is, the total distance traversed divided by the total elapsed time) is $\frac{200}{4} = 50$ miles per hour (mi/hr). Of course, this does not necessarily imply that the automobile's speed was uniform. In fact, the speed was initially zero because the car started from a rest position. When the car was cruising, its speed may have varied from 50 to 55 mi/hr. The automobile's speed was again zero whenever it stopped for a light or at a toll booth.

In general if a distance D is traveled in a time T, the **average speed** for this time interval is defined as D/T. Thus the problem of determining the average speed is resolved very easily. However, this does not give any information about the speed at any instant in time. Note also that the average speed is always positive, since D and T are both positive.

A body moving in a straight line is said to undergo a **rectilinear motion.** If the dimensions of the body are small relative to the distances traveled by it, the body is usually referred to as a **particle.** This means that it is assumed that at each instant of time the entire body is taken to be located at a point, and at times we endow such a particle with attributes such as mass, kinetic energy, and so on. We call the straight line along which the particle moves the s-axis and an origin O is arbitrarily selected where $s = 0$. A positive s-axis is selected (where $s > 0$) and the points on the s-axis are taken to be in a one-to-one correspondence with the real numbers.

An appropriate mathematical model describing this idealized physical problem consists of the introduction of the **displacement** s as a function of the time t; that is, to each t in $t_0 \leq t \leq t_1$ say there exists one and only one value of s. We write

Sec. 3.3 / Velocity in Rectilinear Motion   149

this position function as $s = f(t)$ where $[t_0, t_1]$ is the domain of f. The function f is called the position function of the particle, and the displacement s may be positive, zero, or negative depending on the location of the point relative to the origin (Figure 3.3.1). We also refer to $s = f(t)$ as the *equation of motion*. Suppose then that a particle moves from position $s_0 = f(t_0)$ to $s_1 = f(t_0 + h)$. The *average velocity* during the time interval $[t_0, t_0 + h]$ is, by definition,

$$\text{average velocity} = \frac{s_1 - s_0}{h} = \frac{f(t_0 + h) - f(t_0)}{h} \tag{1}$$

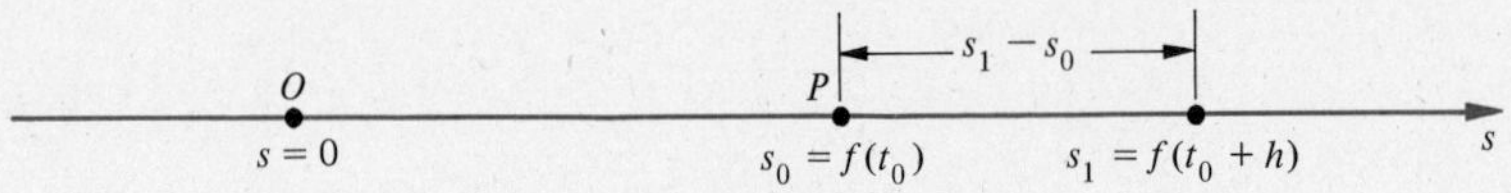

Figure 3.3.1

Formula (1) yields the average velocity over a time interval of duration $|h|$ no matter how small $|h|$ may be, except that $h = 0$ must be excluded. We define the *velocity* at P (often called the instantaneous velocity at P) to be

$$v\big|_{t=t_0} = \lim_{h \to 0} \frac{s_1 - s_0}{h} = \lim_{h \to 0} \frac{f(t_0 + h) - f(t_0)}{h} \tag{2}$$

Another form for the velocity at $t = t_0$ may be obtained as follows

$$v\big|_{t=t_0} = \lim_{t \to t_0} \frac{s(t) - s(t_0)}{t - t_0} = \lim_{t \to t_0} \frac{f(t) - f(t_0)}{t - t_0} \tag{3}$$

The equivalence of (2) and (3) follows if we replace $t_0 + h$ by t so that $h = t - t_0 \neq 0$. Thus as $h \to 0$, we also must have $t - t_0 \to 0$ or, equivalently, $t \to t_0$.

If we return to the example of a car and assume that it is moving in the positive direction, the velocity at a point P tells how fast the car is moving at that instant—this is what registers on a car's speedometer.

Note that assuming the velocity $v\big|_{t=t_0}$ exists implies that the right members in (2) and (3) must exist. Of course this means that $f(t)$ is not arbitrary and must be suitably restricted. This question will be investigated as we proceed further into this chapter.

The velocity may be positive, negative, or zero if $v > 0$, $v < 0$, or $v = 0$, respectively. When $v = 0$ the particle is at rest at that particular time. If v changes sign in passing through zero, the direction of motion reverses.

The *speed* of a particle is defined as the absolute value of the velocity. Hence the speed is a nonnegative number. Thus if $v = -6$, the speed $= |v| = |-6| = 6$. The term speed indicates how fast a particle is moving but it does *not* give the direction of motion. The velocity, however, yields both the speed and the direction of motion.

Since both s and v are functions of t, we may write these as $s(t)$ and $v(t)$.

EXAMPLE 1   Let $s = t^2 + 2t - 1$ where s is the displacement in feet and t is the time in seconds. Find $s(0)$, $s(1)$, $s(2)$, and $s(5)$. Also find $v(0)$, $v(1)$, $v(2)$, and $v(5)$. Describe the direction of motion if $t \geq 0$.

Solution By direct substitution, $s(0) = -1$, $s(1) = 2$, $s(2) = 7$, and $s(5) = 34$, where s is in feet. Also

$$v(t_0) = \lim_{h \to 0} \frac{s(t_0 + h) - s(t_0)}{h}$$

$$= \lim_{h \to 0} \frac{[(t_0 + h)^2 + 2(t_0 + h) - 1] - [t_0^2 + 2t_0 - 1]}{h}$$

$$= \lim_{h \to 0} \frac{t_0^2 + 2t_0 h + h^2 + 2t_0 + 2h - 1 - t_0^2 - 2t_0 + 1}{h}$$

$$= \lim_{h \to 0} \frac{(2t_0 + 2)h + h^2}{h}$$

$$= 2t_0 + 2$$

Therefore $v(0) = 2$ ft/sec, $v(1) = 4$ ft/sec, $v(2) = 6$ ft/sec, and $v(5) = 12$ ft/sec. Since $v(t_0) > 0$ if $t_0 \geq 0$, the velocity is positive and the motion is to the right (in the direction of positive s.) Also the velocity increases linearly with increasing t_0.

●

EXAMPLE 2 A particle is moving along a straight line according to the equation of motion

$$s = t^2 - 3t + 2$$

(a) For what value(s) of t is $s = 0$? (b) Find $v(t)$ and determine the value(s) of t that yield $v = 0$ (that is, the particle stops). (c) When is the particle moving to the right and when does the particle move to the left? Describe the motion.

Solution (a) We set $s = 0$ and therefore $t^2 - 3t + 2 = 0$. Factoring the left side, we obtain $(t - 1)(t - 2) = 0$ or $t = 1$ and $t = 2$.

(b)
$$v(t) = \lim_{h \to 0} \frac{s(t + h) - s(t)}{h}$$

$$= \lim_{h \to 0} \frac{[(t + h)^2 - 3(t + h) + 2] - [t^2 - 3t + 2]}{h}$$

$$= \lim_{h \to 0} \frac{t^2 + 2th + h^2 - 3t - 3h + 2 - t^2 + 3t - 2}{h}$$

$$= \lim_{h \to 0} \frac{2th + h^2 - 3h}{h}$$

or, finally,

$$v(t) = \lim_{h \to 0} (2t + h - 3) = 2t - 3$$

$$v = 0 \quad \text{if and only if} \quad 2t - 3 = 0 \text{ or } t = \tfrac{3}{2}$$

(c) $v > 0$ if $t > \tfrac{3}{2}$, whereas $v < 0$ if $t < \tfrac{3}{2}$. Therefore, the particle moves to the left if $t < \tfrac{3}{2}$, stops at $t = \tfrac{3}{2}$, and moves to the right for $t > \tfrac{3}{2}$. To describe the motion we construct the Table 3.3.1, giving s and v for specific values of t in $[-2, 4]$. The motion of the particle indicated in Figure 3.3.2 is along the horizontal line; however, the behavior of the motion is indicated (schematically) above the line. Note the reversal in the direction of motion that occurs when $t = \tfrac{3}{2}$ and $s = -\tfrac{1}{4}$.

●

Table 3.3.1

t	s	v
-2	12	-7
-1	6	-5
0	2	-3
1	0	-1
$\frac{3}{2}$	$\frac{-1}{4}$	0
2	0	1
3	2	3
4	6	5

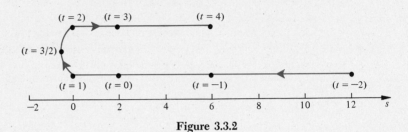

Figure 3.3.2

EXAMPLE 3   A ball is thrown vertically upward from the ground with an initial velocity of 96 ft/sec. If the displacement s is measured from the ground level positively upward, it may be shown that the equation of motion is

$$s = 96t - 16t^2$$

(where air resistance is considered to be negligible). If t is the time elapsed since the ball was thrown and s is the number of feet of the ball from ground level at t sec, find (a) the velocity of the ball at the end of 2 sec; (b) the velocity of the ball at the end of 5 sec; (c) how many seconds it will take the ball to reach its highest point; (d) how high the ball will go; (e) the velocity and speed at the end of 4 sec; (f) the time to reach the ground; and (g) the velocity when the ball reaches the ground. What is the speed of impact?

Solution   Application of formula (3) yields

$$v(t_0) = \lim_{t \to t_0} \frac{s(t) - s(t_0)}{t - t_0}$$

$$= \lim_{t \to t_0} \frac{96t - 16t^2 - (96t_0 - 16t_0{}^2)}{t - t_0}$$

$$= \lim_{t \to t_0} \frac{96(t - t_0) - 16(t^2 - t_0{}^2)}{t - t_0}$$

$$= \lim_{t \to t_0} [96 - 16(t + t_0)]$$

$$= 96 - 32t_0$$

(a) $v(2) = 96 - 32(2) = 32$; so, at the end of 2 sec the ball is rising with a velocity of 32 ft/sec.

(b) $v(5) = 96 - 32(5) = -64$; so, at the end of 5 sec the ball is moving downward with a speed of 64 ft/sec.

(c) The ball reaches its highest point when the direction of motion reverses from upward to downward, that is, when $v(t_0) = 0$. Setting $v\vert_{t_0} = 0$, there results $96 - 32t_0 = 0$ or $t_0 = 3$. Thus, it takes 3 sec for the ball to reach its highest point.

(d) When $t_0 = 3$, $s(3) = 96(3) - 16(3^2) = 144$; therefore, the ball reaches a maximum height of 144 ft above the ground.

(e) $v(4) = 96 - 32(4) = -32$ ft/sec
$\vert v(4)\vert = 32$ ft/sec is the speed at the end of 4 sec.

(f) The ball will hit the ground when $s = 0$. Setting $s = 0$, there results $96t - 16t^2 = 0$ or $16t(6 - t) = 0$ this yields $t = 0$ and $t = 6$. Therefore, the ball will reach the ground in 6 sec.

(g) $v(6) = 96 - 32(6) = -96$ ft/sec. When the ball reaches the ground its velocity is -96 ft/sec and its speed of impact with the ground is $\vert v(6)\vert = 96$ ft/sec.

Table 3.3.2 gives the displacement s and the velocity v for integral values of t in $[0, 6]$. The motion of the ball is given in Figure 3.3.3, where the actual motion is assumed to be along a straight line. ●

Table 3.3.2

t (sec)	s (ft)	v (ft/sec)
0	0	96
1	80	64
2	128	32
3	144	0
4	128	-32
5	80	-64
6	0	-96

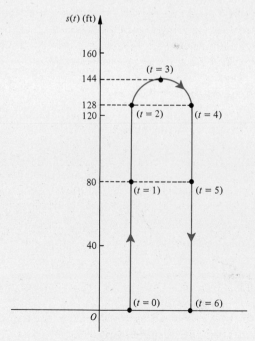

Figure 3.3.3

Exercises 3.3

In Exercises 1 through 4 assume that an object starting from rest falls s ft in t sec where $s = 16t^2$. Find the velocity of the falling object at the end of t_0 sec, for the given value of t_0.

1. $t_0 = 4$ **2.** $t_0 = 1.5$ **3.** $t_0 = 0.75$ **4.** $t_0 = 7.3$
5. How many seconds does it take the falling object of the preceding exercises to attain a velocity of
(a) 160 ft/sec, (b) 112 ft/sec and (c) 70 ft/sec?

In Exercises 6 through 10, a particle moves along a coordinate line. The quantity s is the directed distance in feet from the origin at the end of t sec. Find the velocity of the particle at the end of b sec.

6. $s = -10t + 7$, $b = 4$ **7.** $s = 3t^2 - 4t + 50$, $b = 2$
8. $s = 400 - 16t^2$, $b = 3$ **9.** $s = \sqrt{3t + 2}$, $b = \frac{14}{3}$
10. $s = (t^2 + 1)/3t$, $t > 0, b = 1$

In Exercises 11 through 14 a particle is moving along a straight line according to the given equation of motion, where s ft is the directed distance of the particle from the origin O at t sec. Find the velocity $v(t_0)$ ft/sec at t_0 sec, and then find $v(t_0)$ for the particular value of t specified using the alternative (but equivalent) definition to (2) or (3) in the text:

$$v(t_0) = \lim_{\Delta t \to 0} \frac{s(t_0 + \Delta t) - s(t_0)}{\Delta t} \tag{4}$$

11. $s = 10t^2 - 3$; $t_0 = 2$ **12.** $s = 600 - 16t^2$; $t_0 = 3$
13. $s = \sqrt{t + 5}$; $t_0 = 4$ **14.** $s = t^3 - 4t^2 + 2t + 2$; $t_0 = 4$
15. The height of a ball dropped from a building 640 ft high is given by the equation $s = 640 - 16t^2$, $0 \le t \le \sqrt{40}$ where t is time (sec) and s (ft) is measured from ground level positively to points above ground. Find (a) the velocity for any t in $[0, \sqrt{40}]$; (b) the impact velocity with the ground; (c) the time to reach the halfway point between the top of the building and the ground.
16. A ball is thrown upward from the edge of a roof 256 ft high with an initial velocity = 96 ft/sec. Its height above ground t sec later is given by the formula $s(t) = -16t^2 + 96t + 256$. Find (a) the time to reach the ground; (b) the velocity at any time until impact; (c) the maximum height above the ground; (d) the velocity at impact with the ground.
17. When a ball rolls down an inclined plane its distance from the initial rest point is given by $y = 10t^2$. Find (a) its velocity at any time t; (b) at what time will its velocity be 50 ft/sec; (c) the velocity attained after the ball has rolled 10 ft, 90 ft, y ft.
18. A particle moves along a horizontal line according to the formula $s = t^3 - 3t^2 - 9t + 10$ where s is positive to the right. During which time intervals is the particle moving to the right, and during which is it moving to the left? Show the motion schematically as in Example 2 of the text.
19. A rocket is fired vertically upward and it is s ft above the ground t sec after being fired where $s = 960t - 16t^2$ and the positive direction is upward. Find (a) the average velocity of the rocket during the first 10 sec of travel; (b) the velocity at any time t during flight and at $t = 10$; (c) the time to reach maximum height; (d) the maximum height; (e) the total time of travel.
20. An object at the surface of the earth is shot upward with an initial velocity v_0 ft/sec. Its displacement measured from ground level positively upward is $s = v_0 t - 16t^2$. Show that its maximum height above the earth's surface is given by $\dfrac{v_0^2}{64}$ ft.

3.4   THE DERIVATIVE OF A FUNCTION

The slope of the tangent line to the graph of $y = f(x)$ at the point $(x_1, f(x_1))$ was *defined* in Section 3.2 to be

$$m(x_1) = \lim_{h \to 0} \frac{f(x_1 + h) - f(x_1)}{h} \tag{1}$$

In Section 3.3 we discussed motion along a straight line so that, if $s = f(t)$ represents directed distance from a given point on this line, the velocity at $t = t_0$ is defined by

$$v(t_0) = \lim_{h \to 0} \frac{f(t_0 + h) - f(t_0)}{h} \tag{2}$$

The limits in formulas (1) and (2) are of the same form. Furthermore this type of limit has many other applications. Because of its great significance and wide applicability we have the following definition.

Definition   The **derivative** of a function f is a new function f' defined by

$$f'(x) = \lim_{h \to 0} \frac{f(x + h) - f(x)}{h} \tag{3}$$

provided that the limit exists.

In many uses, the symbol h is replaced by Δx (to denote a change in x) and we have

$$f'(x) = \lim_{\Delta x \to 0} \frac{f(x + \Delta x) - f(x)}{\Delta x} \tag{4}$$

A function is said to be **differentiable** at each value of x for which this limit exists. The set of all such values of x forms the domain of the function f'.

There are many other notations for the derivative of $y = f(x)$. Some of the most common ones are

$$f'(x) \qquad y' \qquad D_x f \qquad D_x f(x) \qquad D_x y \qquad \frac{dy}{dx} \qquad \frac{df}{dx}$$

When we are working with y as a function of x, we shall sometimes call the derivative "the derivative of y with respect to x" or "the derivative of f with respect to x."

Another useful formula for the derivative is

$$f'(x) = \lim_{z \to x} \frac{f(z) - f(x)}{z - x} \tag{5}$$

That this is equivalent to (3) follows from letting $h = z - x$ so that $z = x + h$. Then as $z \to x$, $h \to 0$, and conversely.

If our function is $y = f(x)$, we can observe that the numerator of the limit fraction in (4) is just the difference between two y values. This leads us to write

$$\Delta y = f(x + \Delta x) - f(x)$$

and formula (4) can then be written:

$$D_x y = \lim_{\Delta x \to 0} \frac{\Delta y}{\Delta x} \qquad (6)$$

If we wish to find the derivative for a specific value $x = a$, we may either find an expression for $f'(x)$ and then substitute $x = a$ or we can use one of the following formulas, which are restatements of (3), (4), and (5).

$$f'(a) = \lim_{h \to 0} \frac{f(a + h) - f(a)}{h} \qquad (7)$$

$$f'(a) = \lim_{\Delta x \to 0} \frac{f(a + \Delta x) - f(a)}{\Delta x} \qquad (8)$$

$$f'(a) = \lim_{x \to a} \frac{f(x) - f(a)}{x - a} \qquad (9)$$

In formula (9), we have replaced x in formula (5) by a, and then replaced z by x. Relating this new terminology to our introductory work in this chapter, we see

(i)   If there is a nonvertical tangent line to the curve $y = f(x)$ at the point where $x = x_1$, then the slope of this tangent line is $m(x_1) = f'(x_1)$.

(ii) If $s = f(t)$ is an equation of motion of a particle, then the velocity at the time $t = t_0$ is $v(t_0) = f'(t_0)$.

The following examples show how to find the derivatives of some particular functions. Note that we are employing the same type of manipulations used in Sections 3.2 and 3.3.

EXAMPLE 1   If $f(x) = 3x + 7$, find (i) $f'(x)$, (ii) $f'(2)$.

Solution   (i) By definition (3),

$$f'(x) = \lim_{h \to 0} \frac{f(x + h) - f(x)}{h} = \lim_{h \to 0} \frac{[3(x + h) + 7] - [3x + 7]}{h}$$

$$= \lim_{h \to 0} \frac{3h}{h} = 3$$

(ii) $f'(x) = 3$ for all values of x. So, $f'(2) = 3$. ●

EXAMPLE 2   If $f(x) = 3x^2 - 2x + 4$, find (i) $f'(x)$, (ii) $f'(-5)$.

Solution   (i) Now we shall use formula (4), although (3), (5), or (6) would apply just as well.

$$f'(x) = \lim_{\Delta x \to 0} \frac{f(x + \Delta x) - f(x)}{\Delta x} =$$

$$= \lim_{\Delta x \to 0} \frac{[3(x + \Delta x)^2 - 2(x + \Delta x) + 4] - [3x^2 - 2x + 4]}{\Delta x}$$

$$= \lim_{\Delta x \to 0} \frac{[3x^2 + 6x\,\Delta x + 3(\Delta x)^2 - 2x - 2\,\Delta x + 4] - [3x^2 - 2x + 4]}{\Delta x}$$

$$= \lim_{\Delta x \to 0} \frac{6x\,\Delta x + 3(\Delta x)^2 - 2\,\Delta x}{\Delta x} = \lim_{\Delta x \to 0} (6x + 3\,\Delta x - 2)$$

$$= 6x - 2$$

(ii) Since $f'(x) = 6x - 2$, we let $x = -5$ to obtain

$$f'(-5) = 6(-5) - 2 = -32 \qquad \bullet$$

EXAMPLE 3   If $f(x) = x^3$ find $f'(x)$ by using formula (5).

Solution
$$f'(x) = \lim_{z \to x} \frac{f(z) - f(x)}{z - x} = \lim_{z \to x} \frac{z^3 - x^3}{z - x}$$

$$= \lim_{z \to x} \frac{(z - x)(z^2 + zx + x^2)}{z - x} = \lim_{z \to x} (z^2 + zx + x^2)$$

$$= x^2 + x^2 + x^2 = 3x^2 \qquad \bullet$$

EXAMPLE 4   If $g(x) = \dfrac{1}{x}$, $x \neq 0$, find $g'(x)$.

Solution
$$g'(x) = \lim_{h \to 0} \frac{g(x + h) - g(x)}{h} = \lim_{h \to 0} \frac{\dfrac{1}{x + h} - \dfrac{1}{x}}{h}$$

$$= \lim_{h \to 0} \frac{x - (x + h)}{h(x + h)x} = \lim_{h \to 0} \frac{-h}{h(x + h)x}$$

$$= \lim_{h \to 0} \frac{-1}{(x + h)x} = -\frac{1}{x^2} \qquad x \neq 0 \qquad \bullet$$

EXAMPLE 5   If $f(x) = \sqrt{x}$, $x > 0$, then $f'(x) = \dfrac{1}{2\sqrt{x}}$

Solution   This is the same as Example 3, Section 3.2, except that a is replaced by x. In that example the slope of the tangent line was determined to be $\dfrac{1}{2\sqrt{a}}$ and we now know that this is a geometric interpretation of the derivative. $\qquad \bullet$

EXAMPLE 6   Given $g(x) = \dfrac{(2x + 3)}{(x + 1)}$ find $g'(x)$ if $x \neq -1$.

Solution

$$g'(x) = \lim_{h \to 0} \frac{g(x + h) - g(x)}{h}$$

$$= \lim_{h \to 0} \frac{\dfrac{2(x + h) + 3}{x + h + 1} - \dfrac{2x + 3}{x + 1}}{h}$$

$$= \lim_{h \to 0} \frac{(x + 1)[2(x + h) + 3] - (2x + 3)(x + h + 1)}{h(x + h + 1)(x + 1)}$$

$$= \lim_{h \to 0} \frac{(x + 1)(2x + 3 + 2h) - (2x + 3)(x + 1 + h)}{h(x + h + 1)(x + 1)}$$

$$= \lim_{h \to 0} \frac{(x + 1)(2x + 3) + 2h(x + 1) - (x + 1)(2x + 3) - h(2x + 3)}{h(x + h + 1)(x + 1)}$$

Simplification and division by h yields

$$g'(x) = \lim_{h \to 0} \frac{2(x + 1) - (2x + 3)}{(x + h + 1)(x + 1)}$$

$$= \frac{-1}{(x + 1)^2} \qquad (x \neq -1) \qquad \bullet$$

EXAMPLE 7 $f(x) = |x|$ is not differentiable at $x = 0$.

Solution   This is the same as Example 6, Section 3.2, where we showed that there is no tangent line to the graph of $y = |x|$ at $x = 0$. $\qquad \bullet$

The result of Example 7 is strongly "suggested" by the graph of $f(x) = |x|$ in Figure 3.2.7, which has a corner at $x = 0$. Note also that $|x|$ is continuous at $x = 0$. Thus, continuity in this instance does not imply differentiability. Furthermore, we will establish next that in all cases differentiability *does* imply continuity (continuity is necessary for differentiability). Therefore, the requirement of differentiability is more stringent than that of continuity.

Theorem 1   If $f(x)$ is differentiable at $x = a$ then f is also continuous at $x = a$.

Proof

$$f(x) - f(a) = \frac{f(x) - f(a)}{x - a}(x - a)$$

is true for all $x \neq a$. Therefore

$$\lim_{x \to a} [f(x) - f(a)] = \lim_{x \to a} \left[\frac{f(x) - f(a)}{x - a}(x - a) \right]$$

provided the limit on the right exists. But

$$\lim_{x \to a} \left[\frac{f(x) - f(a)}{x - a}(x - a) \right] = \lim_{x \to a} \frac{f(x) - f(a)}{x - a} \cdot \lim_{x \to a} (x - a)$$

$$= f'(a) \cdot 0 = 0$$

and it follows that

$$\lim_{x \to a} [f(x) - f(a)] = 0 \qquad \text{or} \qquad \lim_{x \to a} f(x) = f(a)$$

This last result is just the definition of continuity of $f(x)$ at $x = a$. Thus differentiability at a point implies continuity at that point. ∎

Since the converse of this theorem does not hold in general, it is useful to introduce formally the concept of a **one-sided derivative.** If the function $f(x)$ is defined at $x = a$, then the **derivative from the right** $f'_+(a)$ is defined by

$$f'_+(a) = \lim_{\Delta x \to 0^+} \frac{f(a + \Delta x) - f(a)}{\Delta x} \tag{10}$$

or, equivalently,

$$f'_+(a) = \lim_{x \to a^+} \frac{f(x) - f(a)}{x - a} \tag{11}$$

Similarly we define the **derivative from the left** $f'_-(a)$ by

$$f'_-(a) = \lim_{\Delta x \to 0^-} \frac{f(a + \Delta x) - f(a)}{\Delta x} \tag{12}$$

or, equivalently,

$$f'_-(a) = \lim_{x \to a^-} \frac{f(x) - f(a)}{x - a} \tag{13}$$

Of course, if $f'(a)$ exists, then $f'(a) = f'_+(a) = f'_-(a)$; and, conversely, if $f'_+(a) = f'_-(a)$, then $f'(a)$ exists and $f'(a) = f'_+(a) = f'_-(a)$.

EXAMPLE 8   Given

$$g(x) = \begin{cases} -x & \text{if } x < 0 \\ x^2 & \text{if } x \geq 0 \end{cases}$$

discuss the continuity and differentiability at $x = 0$.

Solution   If $x < 0$ then $\lim_{x \to 0^-} g(x) = \lim_{x \to 0^-} -x = 0$

and

if $x > 0$ then $\lim_{x \to 0^+} g(x) = \lim_{x \to 0^+} x^2 = 0$

Since $\lim_{x \to 0^-} g(x) = \lim_{x \to 0^+} g(x) = 0 = g(0)$ then $g(x)$ is continuous at $x = 0$. Now

$$g'_-(0) = \lim_{x \to 0^-} \frac{g(x) - g(0)}{x} = \lim_{x \to 0^-} \frac{-x - 0}{x} = -1$$

that is the left hand derivative $= -1$. Also

$$g'_+(0) = \lim_{x \to 0^+} \frac{g(x) - g(0)}{x} = \lim_{x \to 0^+} \frac{x^2 - 0}{x}$$

$$= \lim_{x \to 0^+} x = 0$$

that is the right hand derivative $= 0$. Since $g'_+(0) \neq g'_-(0)$ it follows that $g'(0)$ does not exist. The graph of $g(x)$ is shown in Figure 3.4.1. ●

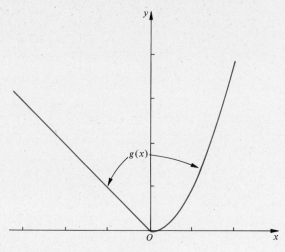

Figure 3.4.1

The function $g(x) = \dfrac{2x + 3}{x + 1}$ is said to be ***differentiable*** because for each value of x in its domain of definition its derivative exists. The function $f(x) = \sqrt{x}, x \geq 0$ is *not* differentiable because $f'(0)$ fails to exist. In order that $f'(0)$ exist, $f(x)$ must be defined for $x < 0$ as well as $x \geq 0$. However $f(x) = \sqrt{x}, x > 0$ is differentiable in this modified domain.

In case $f'(a)$ exists then

$$y - f(a) = f'(a)(x - a) \tag{14}$$

is an equation of the tangent line to the graph of f at the point $P(a, f(a))$.

If $f'(a) \neq 0$, then the slope of the normal line is $-1/f'(a)$ and

$$y - f(a) = -\frac{1}{f'(a)}(x - a) \tag{15}$$

is an equation of the normal line. If $f'(a) = 0$, then the normal line is the vertical line $x = a$; whereas, if tangent line is vertical, then the horizontal line $y = f(a)$ is the normal line.

EXAMPLE 9   Find an equation of the tangent line to the graph of $f(x) = x^2 - 3x - 4$ at $(3, -4)$.

Solution

$$
\begin{aligned}
f'(x) &= \lim_{h \to 0} \frac{f(x + h) - f(x)}{h} \\
&= \lim_{h \to 0} \frac{[(x + h)^2 - 3(x + h) - 4] - [x^2 - 3x - 4]}{h} \\
&= \lim_{h \to 0} \frac{x^2 + 2xh + h^2 - 3x - 3h - 4 - x^2 + 3x + 4}{h} \\
&= \lim_{h \to 0} (2x + h - 3) = 2x - 3
\end{aligned}
$$

Therefore, $f'(3) = 3$ is the slope m of the tangent line at the given point $(3, -4)$. An equation of the tangent line is

$$y - (-4) = 3(x - 3) \qquad \text{or} \qquad 3x - y - 13 = 0$$

Exercises 3.4

In Exercises 1 through 14, find $f'(x)$ for the given function $f(x)$ by applying the definition for the derivative

$$f'(x) = \lim_{h \to 0} \frac{f(x + h) - f(x)}{h}$$

1. $f(x) = 2x - 6$
2. $f(x) = \frac{1}{2}x + 9$
3. $f(x) = ax + b$
4. $f(x) = -3x^2 + 4$
5. $f(x) = 2x^2 - 3x - 5$
6. $f(x) = (3x + 5)^2$
7. $f(x) = ax^2 + bx + c$
8. $f(x) = (ax + b)^2$
9. $f(x) = x^3 + 80$
10. $f(x) = 3x^3 - 7x$
11. $f(x) = \dfrac{1}{x}$
12. $f(x) = \dfrac{3}{4\sqrt{x}}$
13. $f(x) = \dfrac{3}{2x} - \sqrt{2}$
14. $f(x) = \dfrac{1}{3x - 1} + \sqrt{3x}$

In Exercises 15 through 20, find $D_x F$

15. $F(x) = \dfrac{x}{x + 1}$
16. $F(x) = \dfrac{2}{(x + 1)^2}$
17. $F(x) = \dfrac{2x^2 - 4x + 9}{x}$ (Hint: For $x \neq 0$ divide denominator into numerator)
18. $F(x) = \dfrac{x}{x^2 + 1}$
19. $F(x) = \dfrac{5x - 8}{x^2}$
20. $F(x) = \sqrt{5x - 6}$

In Exercises 21 through 24 find $f'(a)$ for the given a by applying formula (7) of this section.

21. $f(x) = -3x^2 + 14, \quad a = 2$
22. $f(x) = 7 - x^3, \quad a = -3$
23. $f(x) = 2x^2 - 3x + \sqrt{5}, \quad a = 0$
24. $f(x) = \sqrt{6 + 5x}, \quad a = 2$
25. Is $f(x) = \sqrt{x - 1}$ differentiable at $x = 1$? If it is, find $f'(1)$. Sketch the graph of the function.
26. Is $f(x) = x\,|x|$ differentiable at $x = 0$? If it is, find $f'(0)$. Draw a sketch of the graph of the function.
27. Is $g(x) = |2x - 3|$ differentiable for all x? Find $g'(x)$.
28. Find $h'(x)$ if $h(x) = |x^2 - 5|$.
29. Examine each of the following functions to determine if it is differentiable at the number $x = 0$

   (a) $f(x) = \begin{cases} 0 & \text{for } x \leq 0 \\ 3x & \text{for } x > 0 \end{cases}$ (b) $g(x) = \begin{cases} 0 & \text{for } x \leq 0 \\ x^2 & \text{for } x > 0 \end{cases}$ (c) $F(x) = \begin{cases} 0 & \text{for } x \text{ rational} \\ x^2 & \text{for } x \text{ irrational} \end{cases}$

30. Let f be a function whose domain is the set of all real numbers. Also f is assumed to satisfy the relation $f(h + k) = f(h) \cdot f(k)$ for arbitrary h and k. Furthermore suppose that $f(0) = 1$ and $f'(0)$ exists. Prove that $f'(x)$ exists for all x and that $f'(x) = f'(0) \cdot f(x)$.

The operation of finding the derivative of a function is called **differentiation.** The reader may have been impressed with the fact that the performance of differentiations by going back to the definition of a derivative can lead to algebraically lengthy details. Much of this can be circumvented by developing differentiation formulas. It is the purpose of this section to systematically develop these basic rules for differentiation.

Theorem 1   If c is a constant and if $f(x) = c$ for all values of x, then $f'(x) = 0$

Proof

$$f'(x) = \lim_{h \to 0} \frac{f(x + h) - f(x)}{h}$$

$$= \lim_{h \to 0} \frac{c - c}{h}$$

$$= \lim_{h \to 0} 0 = 0$$

In other words

$$D_x c = 0 \tag{1}$$

The derivative of any constant is zero.   ∎

EXAMPLE 1   If $y = 2\pi$ for all x, find $D_x y$.

Solution

$$D_x(2\pi) = 0$$

Theorem 2   If n is a positive integer and if $f(x) = x^n$, then

$$f'(x) = nx^{n-1}$$

Proof

$$f'(x) = \lim_{h \to 0} \frac{(x + h)^n - x^n}{h}$$

Applying the binomial theorem to $(x + h)^n$ there results

$$f'(x) = \lim_{h \to 0} \frac{x^n + nx^{n-1}h + \dfrac{n(n - 1)}{1 \cdot 2}x^{n-2}h^2 + \cdots + nxh^{n-1} + h^n - x^n}{h}$$

$$= \lim_{h \to 0} \left[nx^{n-1} + \frac{n(n - 1)}{1 \cdot 2}x^{n-2}h + \cdots + nxh^{n-2} + h^{n-1} \right]$$

$$= nx^{n-1}$$

since each term after the first tends to zero with h and there are actually $n - 1$ such terms (that is, a finite number of terms). Thus

$$D_x(x^n) = nx^{n-1} \qquad ∎ \tag{2}$$

Equation (2) is the formula for the differentiation of a power function x^n. Note that n thus far must be restricted to positive integers.

EXAMPLE 2   If $y = x^{15}$ find $D_x y$.

Solution   In this instance, $n = 15$ so that (2) yields

$$D_x y = 15x^{14}$$ ●

EXAMPLE 3   If $y = x$ find $D_x y$.

Solution   Here $x = x^1$ so that $n = 1$. Thus

$$D_x y = 1x^0 = 1 \cdot 1 = 1$$ ●

Theorem 3   If f is a function, c is a constant, and g is a function defined by

$$g(x) = c f(x)$$

then, if $f'(x)$ exists, we have

$$g'(x) = c f'(x) \qquad (3)$$

Proof
$$
\begin{aligned}
g'(x) &= \lim_{h \to 0} \frac{g(x+h) - g(x)}{h} \\
&= \lim_{h \to 0} \frac{c f(x+h) - c f(x)}{h} \\
&= c \lim_{h \to 0} \frac{f(x+h) - f(x)}{h} = c f'(x)
\end{aligned}
$$

Thus

$$D_x(c f(x)) = c D_x f(x) \qquad (4)$$

This says that *the derivative of a constant times a function is the constant times the derivative of the function.*   ■

By combining (2) and (3), we see that if $g(x) = cx^n$ then $g'(x) = D_x(cx^n) = D_x c(x^n) = cnx^{n-1}$ for any positive integer n.

EXAMPLE 4   Given $F(x) = -3x^9$ find $F'(x)$.

Solution
$$F'(x) = -3(9x^8) = -27x^8$$ ●

Theorem 4   If f and g are differentiable functions and $F(x) = f(x) + g(x)$ then in their common domain of differentiability we must have

$$F'(x) = f'(x) + g'(x). \qquad (5)$$

Proof

$$F'(x) = \lim_{h \to 0} \frac{F(x + h) - F(x)}{h}$$

$$= \lim_{h \to 0} \frac{[f(x + h) + g(x + h)] - [f(x) + g(x)]}{h}$$

$$= \lim_{h \to 0} \frac{f(x + h) - f(x)}{h} + \lim_{h \to 0} \frac{g(x + h) - g(x)}{h}$$

$$= f'(x) + g'(x)$$

or

$$D_x F(x) = D_x[f(x) + g(x)] = D_x f(x) + D_x g(x) \tag{6}$$

The derivative of the sum of two functions is the sum of their derivatives provided that each of the derivatives exists. ∎

Equations (3) and (6) can be combined to yield the **linearity property of the derivative operator,** namely, that for arbitrary constants c_1 and c_2,

$$D_x[c_1 f(x) + c_2 g(x)] = c_1 D_x f(x) + c_2 D_x g(x) \tag{7}$$

Letting $c_1 = 1$ and $c_2 = -1$ in (7), we see that Theorem 4 is also true if we take the difference between the two functions. That is

$$D_x[f(x) - g(x)] = D_x f(x) - D_x g(x)$$

The result of Theorem 4 can be extended to any *finite* number of functions by mathematical induction and is stated as another theorem.

Theorem 5   The derivative of the sum of a finite number of functions is equal to the sum of their derivatives, provided that each of the derivatives exists.

From Theorems 2, 3, and 5 we can now easily differentiate any polynomial.

EXAMPLE 5   Given $f(x) = -3x^4 + 7x^3 - 9x + 6$ find $f'(x)$.

Solution

$$f'(x) = D_x(-3x^4 + 7x^3 - 9x + 6)$$

$$= D_x(-3x^4) + D_x(7x^3) + D_x(-9x) + D_x(6)$$

$$= -12x^3 + 21x^2 - 9 \qquad \bullet$$

Our next problem is to develop a formula for the derivative of a product of two functions. One might be tempted to say that the derivative of a product must be equal to the product of their derivatives. That this is *wrong* is easily seen by examining a particular example. We know that $D_x(x^2) = 2x$. However, $x^2 = x \cdot x$ and the application of the incorrect rule would yield $D_x(x \cdot x) = D_x x \cdot D_x x = 1 \cdot 1 = 1$. Thus it is verified that the conjecture is indeed not valid. Instead we will establish the following theorem.

Theorem 6   If f and g are differentiable functions and if $F(x)$ is defined by

$$F(x) = f(x)\,g(x)$$

then

$$F'(x) = f(x)\,g'(x) + g(x)\,f'(x) \tag{8}$$

Proof By definition,

$$F'(x) = \lim_{h \to 0} \frac{F(x + h) - F(x)}{h}$$

and, consequently

$$F'(x) = \lim_{h \to 0} \frac{f(x + h)\,g(x + h) - f(x)\,g(x)}{h}$$

The trick is to subtract and add $f(x + h)\,g(x)$ in the numerator. Thus

$$F'(x) = \lim_{h \to 0} \frac{f(x + h)\,g\,(x + h) - f(x + h)\,g(x) + f(x + h)\,g(x) - f(x)\,g(x)}{h}$$

$$= \lim_{h \to 0} \left\{ f(x + h)\frac{g(x + h) - g(x)}{h} + g(x)\frac{f(x + h) - f(x)}{h} \right\}$$

$$= \lim_{h \to 0} \left[f(x + h)\frac{g(x + h) - g(x)}{h} \right] + g(x) \lim_{h \to 0} \frac{f(x + h) - f(x)}{h}$$

$$= \lim_{h \to 0} f(x + h) \lim_{h \to 0} \frac{g(x + h) - g(x)}{h} + g(x) \lim_{h \to 0} \frac{f(x + h) - f(x)}{h}$$

but $f'(x)$ exists so that f is continuous at x or $\lim_{h \to 0} f(x + h) = f(x)$. Also

$$\lim_{h \to 0} \frac{g(x + h) - g(x)}{h} = g'(x) \quad \text{and} \quad \lim_{h \to 0} \frac{f(x + h) - f(x)}{h} = f'(x)$$

by definition of a derivative, and therefore

$$F'(x) = f(x)\,g'(x) + g(x)\,f'(x)$$

The derivative of a product of two differentiable functions is the first function times the derivative of the second plus the second function times the derivative of the first function. ∎

EXAMPLE 6 Find $D_x(x^2)$ by using the formula for the differentiation of a product.

Solution
$$D_x(x^2) = D_x(x \cdot x) = x\,D_x x + x\,D_x x$$
$$= x(1) + x(1) = 2x$$

EXAMPLE 7 Find $D_x[(x^2 + 4)(x^3 - x + 1)]$.

Solution
$$D_x[(x^2 + 4)(x^3 - x + 1)] = (x^2 + 4)(3x^2 - 1) + (x^3 - x + 1)(2x)$$
$$= 3x^4 + 11x^2 - 4 + 2x^4 - 2x^2 + 2x$$
$$= 5x^4 + 9x^2 + 2x - 4$$

The reader should verify that the result is correct by multiplying out first and then differentiating the resulting polynomial. It is illuminating to do problems by more than one method and then to compare the results. It also serves as an excellent check when the answers agree.

Next we establish a formula for the derivative of a quotient of two functions.

Theorem 7 If f and g are differentiable functions and if $F(x)$ is defined by

$$F(x) = \frac{f(x)}{g(x)} \qquad \text{where } g(x) \neq 0$$

then

$$F'(x) = \frac{g(x) f'(x) - f(x) g'(x)}{[g(x)]^2} \qquad (9)$$

Proof

$$F'(x) = \lim_{h \to 0} \frac{F(x + h) - F(x)}{h}$$

$$= \lim_{h \to 0} \frac{\dfrac{f(x + h)}{g(x + h)} - \dfrac{f(x)}{g(x)}}{h}$$

$$= \lim_{h \to 0} \frac{f(x + h) g(x) - f(x) g(x + h)}{h \, g(x + h) \, g(x)}$$

Subtracting and adding $f(x) g(x)$ results in

$$F'(x) = \lim_{h \to 0} \frac{f(x + h) g(x) - f(x) g(x) + f(x) g(x) - f(x) g(x + h)}{h \, g(x + h) \, g(x)}$$

$$= \lim_{h \to 0} \frac{g(x) \left[\dfrac{f(x + h) - f(x)}{h} \right] - f(x) \left[\dfrac{g(x + h) - g(x)}{h} \right]}{g(x + h) \, g(x)}$$

$$= \frac{g(x) \lim\limits_{h \to 0} \left[\dfrac{f(x + h) - f(x)}{h} \right] - f(x) \lim\limits_{h \to 0} \left[\dfrac{g(x + h) - g(x)}{h} \right]}{\lim\limits_{h \to 0} g(x + h) \, g(x)}$$

$$F'(x) = \frac{g(x) f'(x) - f(x) g'(x)}{g^2(x)}$$

thus we have established that

$$D_x \left[\frac{f(x)}{g(x)} \right] = \frac{g(x) D_x f(x) - f(x) D_x g(x)}{[g(x)]^2}$$

The derivative of a quotient of two differentiable functions is the fraction whose denominator is the square of the original denominator, and whose numerator is the denominator times the derivative of the numerator minus the numerator times the derivative of the denominator. It is assumed that the denominator is not zero for the value of x in question. ∎

EXAMPLE 8 If $F(x) = \dfrac{x^2 + 3}{x + 1}$ where $x \neq -1$ find $F'(x)$.

Solution
$$F'(x) = \frac{(x+1)(2x) - (x^2+3)(1)}{(x+1)^2}$$

$$= \frac{2x^2 + 2x - x^2 - 3}{(x+1)^2}$$

$$= \frac{x^2 + 2x - 3}{(x+1)^2}$$

$$= \frac{(x+3)(x-1)}{(x+1)^2} \qquad (x \neq -1) \qquad \bullet$$

EXAMPLE 9   If $F(x) = \dfrac{6}{x+3}$, $(x \neq -3)$, find $F'(x)$.

Solution
$$F'(x) = \frac{(x+3)(0) - 6(1)}{(x+3)^2} = \frac{-6}{(x+3)^2} \qquad (x \neq -3) \qquad \bullet$$

From the method of this last example, it is a simple matter to establish the following theorem.

Theorem 8   If $f(x) = x^{-n}$ where n is a positive integer, then for $x \neq 0$,

$$f'(x) = -nx^{-n-1} \qquad (10)$$

(In other words the power formula holds true for negative as well as positive integers.)

Proof
$$f(x) = \frac{1}{x^n} \qquad (x \neq 0)$$

$$f'(x) = \frac{x^n \cdot 0 - 1(nx^{n-1})}{x^{2n}} = \frac{-nx^{n-1}}{x^{2n}}$$

$$= -nx^{-n-1} \qquad (x \neq 0) \qquad \blacksquare$$

EXAMPLE 10   If $f(x) = \dfrac{5}{x^7}$, $x \neq 0$, find $f'(x)$.

Solution
$$f'(x) = 5(x^{-7}) = 5(-7)x^{-8} = \frac{-35}{x^8} \qquad (x \neq 0) \qquad \bullet$$

Thus if m is any positive or negative integer and c is any constant, we have

$$D_x(cx^m) = cmx^{m-1} \qquad (11)$$

In fact the formula holds true even when $m = 0$ since we then have

$$D_x(c) = 0$$

Exercises 3.5

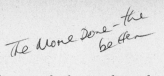
The More Done – the better

In Exercises 1 through 27 differentiate the function by applying the theorems of this section.

1. $f(x) = x^3 + 6x^2 - 8x - 9$

2. $f(x) = x^4 + 9x - 1$

3. $f(x) = x^8 - 6x^5 + x^4 - 3x$

4. $f(x) = \dfrac{x^6}{6} - x^3 - x + 4$

5. $u(s) = 3s^3 - 10s^2 - 6s + \sqrt{3}$

6. $v(r) = 4\pi r^2 + \sqrt{2}r$

7. $g(y) = \dfrac{y^3 - 6y + 9}{3}$

8. $F(x) = \dfrac{3}{4}x^4 - \dfrac{2}{3}x^2 + \dfrac{\pi}{2} - 1$

9. $g(x) = x^3 - 10x + \dfrac{2}{x}$

10. $u(x) = (x^2 + 1)(3x^2 - 2)$

11. $g(x) = (x^2 - 2)^2$

12. $h(x) = \dfrac{10}{x^2} - \dfrac{7}{x^3}$ $\dfrac{10}{x^2} = \dfrac{10}{x^2} = 10x^{-2}$

13. $f(y) = \dfrac{y^2}{y + 1}$

14. $F(t) = 4t^2 - \sqrt{3}t$

15. $g(x) = \dfrac{4 - 3x}{2x + 5}$

16. $f(z) = \dfrac{3z^2}{z - 2}$

17. $u(x) = \dfrac{x^3 + 5}{x^3 - 5}$

18. $v(x) = \dfrac{x^3 - 3x^2 + x - 2}{x}$

19. $w(x) = \dfrac{x^2 - 4x}{x^2}$

20. $f(x) = (x^2 - 3)(x^3 + 2)$

21. $v(x) = \dfrac{x^2 - 1}{3x + 2}$

22. $G(x) = (3x - 2)^2(2x + 3)$

23. $F(t) = \dfrac{at^2 + b}{-t^2 + c}$

24. $G(x) = (x^m + 1)(x^{2m} - 3)$, $(m = \text{integer})$

25. $g(t) = \dfrac{t^2 - 1}{t + 2}(t^2 + 3)$

26. $h(y) = \dfrac{y^2(3y^{-1} + 2y^{-2})}{y^2 + 1}$

27. $f(x) = (2x - 1)^3 = (2x - 1)^2(2x - 1)$

28. Assume for the moment that we are given $D_x(x^3) = 3x^2$ but that we do not know $D_x(x^4)$. Show how $D_x(x^4)$ can be found from the given information.

29. Generalize the previous problem by showing that if $D_x(x^{n-1}) = (n - 1)x^{n-2}$ then $D_x(x^n) = nx^{n-1}$ where n is an arbitrary positive integer. (*Hint:* $x^n = x^{n-1} \cdot x$)

30. Prove that if $f(x) = u(x)\,v(x)\,w(x)$ then

$$f'(x) = u(x)\,v(x)\,w'(x) + u(x)\,v'(x)\,w(x) + u'(x)\,v(x)\,w(x)$$

assuming that $u(x)$, $v(x)$, and $w(x)$ are differentiable functions of x. (*Hint:* $f(x) = [u(x)\,v(x)]\,w(x)$.) If $f(x)$ is a product of four differentiable functions of x, what is $f'(x)$?

In Exercises 31 through 34, use the result of problem 30 to find the derivative.

31. $f(x) = x^2(x - 3)(x + 2)$

32. $g(x) = (x^2 + 1)(x + 3)(x + 2)$

33. $h(x) = (x^2 + 2)(2x + 1)(x^2 - 2)$

34. $F(x) = (x - 1)(x - 2)(x - 3)(x - 5)$

35. If $f(x) = x^n$ where n is a positive integer show that $f'(a) = na^{n-1}$ directly from the alternative definition

$$f'(a) = \lim_{x \to a} \frac{f(x) - f(a)}{x - a}$$

[*Hint:* $x^n - a^n = (x - a)(x^{n-1} + x^{n-2}a + \cdots + xa^{n-2} + a^{n-1})$.]

36. For a given differentiable function f, what are $D(f^2)$ and $D(f^3)$? Generalize your result to $D(f^n)$ where n is any positive integer.

3.6

CHAIN RULE

By far the most important of the general rules for calculating derivatives is known as the ***chain rule.*** This rule has to do with the differentiation of composite functions that is, differentiation of function of functions.

For example, suppose that we are faced with the task of determining $D_x(x^2 + 7)^{80}$. One way to proceed is to expand the binomial so that

$$(x^2 + 7)^{80} = x^{160} + ax^{158} + bx^{156} + \cdots + kx^2 + l$$

and then differentiate term by term. However, the explicit determination of the 80 coefficients is a most tedious task and the probability of coming up with the correct answer is empirically very close to zero. Actually, a special case of the chain rule can be used here, namely, that for $n =$ positive integer,

$$D_x(f(x))^n = n(f(x))^{n-1} f'(x) \tag{1}$$

A derivation of (1) was requested in Exercise 36 of Section 3.5. The development of (1) as a special case of the chain rule will be established later.

From (1) with $f(x) = x^2 + 7$ we have

$$D_x(x^2 + 7)^{80} = 80(x^2 + 7)^{79} \cdot 2x = 160x(x^2 + 7)^{79}$$

which is a very compact solution indeed.

Suppose next that we are given the two linear relations $y = 4t + 5$ and $t = 7x - 12$. We seek $D_x y$, that is, the rate of change of y with respect to x. This may be solved directly by writing

$$y = 4t + 5 = 4(7x - 12) + 5 = 28x - 43$$

so that

$$D_x y = 28$$

Now this same problem could have been approached in a different way. Note that $D_t y = 4$, that is, the rate of change of y with respect to t, is 4. Also $D_x t = 7$ that is, the rate of change of t with respect to x is 7. But then the rate of change of y with respect to x is the product of the two rates, namely

$$D_x y = D_t y \cdot D_x t \tag{2}$$

and in this case

$$D_x y = (4)(7) = 28$$

Equation (2) is actually a statement of the chain rule, however, this intuitive discussion should *not* be construed as a proof.

By way of a little more complicated expressions involving nonlinear relations, suppose that $y = t^2 + 1$ and $t = x^3 - x$. We seek $D_x y$. Again this may be obtained by eliminating t between the two relations. Thus

$$y = (x^3 - x)^2 + 1$$

and expansion of the binomial yields

$$y = x^6 - 2x^4 + x^2 + 1$$

so that

$$D_x y = 6x^5 - 8x^3 + 2x$$
$$= 2x(x^2 - 1)(3x^2 - 1)$$
$$= 2(x^3 - x)(3x^2 - 1)$$

But $D_t y = 2t = 2(x^3 - x)$ and $D_x t = 3x^2 - 1$ so that equation (2) is verified once again.

With these preliminary examples in mind let us state the basic chain rule theorem.

Theorem 1
(Chain Rule) Let $y = f(t)$ and $t = g(x)$, and assume that f and g are differentiable. If $y = f(g(x))$ then $D_x y$ exists and is given by

$$D_x y = D_t y \cdot D_x t \qquad (3)$$

Note that the way we expressed y as a function of x is the way a composite function is defined; that is, $y = (f \circ g)(x)$. Thus, the chain rule tells us how to differentiate composite functions.

One can attempt to justify the chain rule as follows. Let $y = f(g(x)) = F(x)$. Then

$$D_x y = \lim_{\Delta x \to 0} \frac{F(x + \Delta x) - F(x)}{\Delta x} = \lim_{\Delta x \to 0} \frac{\Delta y}{\Delta x} \qquad (4)$$

Also

$$D_t y = \lim_{\Delta t \to 0} \frac{f(t + \Delta t) - f(t)}{\Delta t} = \lim_{\Delta t \to 0} \frac{\Delta y}{\Delta t} \qquad (5)$$

and

$$D_x t = \lim_{\Delta x \to 0} \frac{g(x + \Delta x) - g(x)}{\Delta x} = \lim_{\Delta x \to 0} \frac{\Delta t}{\Delta x} \qquad (6)$$

In this last expression $\Delta t = g(x + \Delta x) - g(x)$ and g is continuous (since it is differentiable). Thus $\Delta t \to 0$ if $\Delta x \to 0$.

Algebraically, we can write

$$\frac{\Delta y}{\Delta x} = \frac{\Delta y}{\Delta t} \frac{\Delta t}{\Delta x} \qquad (7)$$

If we take the limit of this expression as $\Delta x \to 0$ we get

$$\lim_{\Delta x \to 0} \frac{\Delta y}{\Delta x} = \left(\lim_{\Delta x \to 0} \frac{\Delta y}{\Delta t} \right) \left(\lim_{\Delta x \to 0} \frac{\Delta t}{\Delta x} \right) = \left(\lim_{\Delta t \to 0} \frac{\Delta y}{\Delta t} \right) \left(\lim_{\Delta x \to 0} \frac{\Delta t}{\Delta x} \right)$$

But, from formulas (4), (5), and (6), this becomes $D_x y = D_t y \cdot D_x t$, the desired result.

Unfortunately, the preceding is not a formal proof. A problem arises in that there are functions $t = g(x)$ for which $\Delta t = 0$ infinitely often. In such cases, we cannot write formula (7) because it is invalid when $\Delta t = 0$. The preceding "proof" is valid for most functions occurring in practice.

We shall now give another form of the statement of the chain rule (equivalent to our first statement) and follow it with a correct proof of the result.

Theorem 1'
(Chain Rule)
If $g(x)$ and $f(g)$ are differentiable functions of their respective arguments, the function $F(x) = f[g(x)]$ is differentiable also and its derivative is given by

$$F'(x) = f'(g) \cdot g'(x) \qquad (8)$$

Proof   We know that

$$\lim_{\Delta g \to 0} \frac{\Delta f}{\Delta g} = f'(g)$$

by definition of the derivative of f with respect to g. Therefore $\varepsilon = \dfrac{\Delta f}{\Delta g} - f'(g)$ is a function of Δg for a given g which must approach zero as $\Delta g \to 0$. If we *define* $\varepsilon = 0$ for $\Delta g = 0$ then we have, without restriction on Δg

$$\Delta f = [f'(g) + \varepsilon] \Delta g$$

Similarly for any x in the domain of differentiability for $g(x)$,

$$\Delta g = g(x + \Delta x) - g(x) = [g'(x) + \eta] \Delta x$$

where $\eta \to 0$ as $\Delta x \to 0$. Then for any $\Delta x \neq 0$,

$$\frac{\Delta f}{\Delta x} = [f'(g) + \varepsilon] \frac{\Delta g}{\Delta x} = [f'(g) + \varepsilon][g'(x) + \eta]$$

Now let $\Delta x \to 0$, and note that (i) $\lim\limits_{\Delta x \to 0} \dfrac{\Delta f}{\Delta x} = F'(x)$ and (ii) $\lim\limits_{\Delta x \to 0} \varepsilon = 0$ and $\lim\limits_{\Delta x \to 0} \eta = 0$; therefore $F'(x) = f'(g) \cdot g'(x)$ and the chain rule is thereby established. ∎

By successive application of our rule, the chain rule can be extended to more than two differentiable functions, say

$$y = g(u) \qquad u = \phi(t) \qquad t = \psi(x)$$

Then

$$D_x y = g'(u) \cdot \phi'(t) \cdot \psi'(x) \qquad (9)$$

where each differentiation is with respect to the corresponding argument.

EXAMPLE 1   If $y = (x^2 - 3x^3)^{14}$, find $D_x y$.

Solution   Let $t = x^2 - 3x^3$ and $y = t^{14}$ then

$$D_t y = 14 t^{13} \qquad \text{and} \qquad D_x t = 2x - 9x^2$$

and from (2),

$$D_x y = 14 t^{13}(2x - 9x^2) = 14(x^2 - 3x^3)^{13}(2x - 9x^2) \qquad \bullet$$

EXAMPLE 2   Prove (1).

Solution   Write $t = f(x)$ then, if $y = t^n$, we have

$$D_x y = D_t y \cdot D_x t = n t^{n-1} \cdot f'(x) = n(f(x))^{n-1} f'(x) \qquad \bullet$$

EXAMPLE 3   Find $D_x(x^2 + 1)^{-3}$

Solution   Let $y = u^{-3}$ and $u = x^2 + 1$.

$$D_x y = D_u y \, D_x u$$
$$= -3u^{-4} \, 2x$$
$$= \frac{-6x}{(x^2 + 1)^4}$$

EXAMPLE 4   If $y = \dfrac{1 - u}{1 + u^2}$, $u = t^2 + 3$, and $t = 3x + 5$, find $D_x y \Big|_{x=-1}$.

Solution
$$D_x y = D_u y \cdot D_t u \cdot D_x t$$

By the quotient rule

$$D_u y = \frac{(1 + u^2)(-1) - (1 - u)(2u)}{(1 + u^2)^2}$$

$$= \frac{-1 - u^2 - 2u + 2u^2}{(1 + u^2)^2} = \frac{u^2 - 2u - 1}{(1 + u^2)^2}$$

$$D_t u = 2t \quad \text{and} \quad D_x t = 3$$

Thus

$$D_x y = \frac{u^2 - 2u - 1}{(1 + u^2)^2} (2t)(3) = 6t \frac{(u^2 - 2u - 1)}{(1 + u^2)^2}$$

When $x = -1$, $t = 2$, and $u = 7$; therefore

$$D_x y \Big|_{x=-1} = \frac{(6)(2)(49 - 14 - 1)}{(1 + 49)^2} = \frac{408}{2500} = \frac{102}{625}$$

EXAMPLE 5   If $y = [4x + (x^3 - 2)^5]^8$, find $D_x y$.

Solution   Let $t = 4x + (x^3 - 2)^5$. Then $y = t^8$. So
$$D_x y = D_t y \, D_x t = 8t^7 \, D_x t$$

Letting $u = x^3 - 2$ gives $t = 4x + u^5$. Then

$$D_x t = 4 + 5u^4 \, D_x u = 4 + 5u^4 \, 3x^2 = 4 + 15x^2(x^3 - 2)^4$$

Thus

$$D_x y = 8t^7 \, D_x t = 8[4x + (x^3 - 2)^5]^7[4 + 15x^2(x^3 - 2)^4]$$

Exercises 3.6

In each of the following Exercises 1 through 14 find the derivative with respect to the indicated independent variable.

1. $f(x) = (3x + 2)^7$
2. $f(x) = (3 - 7x)^5$
3. $g(x) = (4 + x)^{-3}$
4. $F(y) = (3y + 1)^{-1}$

5. $h(u) = (3u^2 - u + \pi)^8$

6. $f(x) = (3x^2 - 7x - \sqrt{2})^{20}$

7. $G(x) = (3x - 2)^4(2x + 3)^3$

8. $g(x) = \left(\dfrac{3x - 4}{2x + 3}\right)^9$

9. $F(x) = (x^2 + 2x - 3)^9(3x - 7)^{12}$

10. $G(x) = \dfrac{(x^2 + 2)^{-2}}{(x^2 + 1)^{-3}}$

11. $f(u) = \left(u - \dfrac{1}{u^2}\right)^{5/2}$

12. $F(y) = \sqrt{2y} - \sqrt{\dfrac{y}{2}}$

13. $F(x) = \sqrt{x + \sqrt{x}}$

14. $f(y) = \dfrac{1 + \sqrt{1 - 2y}}{\sqrt{1 - 2y}}$

15. Find an equation of the tangent line to $y = 3x\sqrt{x^2 + 1}$ at $(0, 0)$.

16. Find an equation of the normal line to $y = \sqrt{1 + \sqrt{x}}$ at $(9, 2)$.

17. If $w = z^2 - 3z + 1$ and $z = \dfrac{1}{x^2}$ find $D_x w$ in two ways: (a) by the chain rule; (b) by determining the composite function and then differentiating.

18. Prove as a corollary to the chain rule that

$$[f(g(h))]' = f'(g(h))\, g'(h)\, h'(x)$$

where $h = h(x)$, g and f are differentiable functions of their respective arguments.

19. Let $w = 2z + 3$, $z = \sqrt{y^2 + 5}$, and $y = 1 - x^3$. Find $D_x w$ when $x = -1$.

20. Let $w = v^2 + 1$, $v = \sqrt{u^2 + 40}$, and $u = t^3 + 1$. Find $D_t w$ when $t = 2$.

21. If $f(x) = \dfrac{x + 1}{x}$ and $g(x) = x^3$ find (a) $D\,(f \circ g)\,(x)$; (b) $D\,(g \circ f)\,(x)$; (c) $D\,(f \circ f)(x)$; (d) $D\,(g \circ g)(x)$.

22. If $f = \dfrac{x^2}{1 + x}$ find (a) $D\,f(x)$ and (b) $D\,(f \circ f)\,(x)$.

23. Starting with identity $x = \sqrt{x} \cdot \sqrt{x}$ where $x > 0$ derive a formula for $D_x \sqrt{x}$.

24. Derive a formula for $D_x x^{3/2}$, $x > 0$ by starting with the identity $x^3 = x^{3/2} \cdot x^{3/2}$ $(x > 0)$.

25. Two functions f and g are such that $f(x) = f'(x)$ and $(f \circ g)\,(x) = x$. Assume that $g'(x)$ exists and determine it.

26. Prove that for differentiable functions the derivative of an even function is odd and the derivative of an odd function is even. (*Hint:* Use the definition of even and odd functions in conjunction with the chain rule.)

27. Find $f'(x)$ if $f(x) = 2\sqrt[3]{x} + |x|$.

28. Show that it is impossible to find polynomials $p(x)$ such that (a) $p'(x) = \dfrac{1}{x}$, and (b) $\left[\dfrac{1}{p(x)}\right]' = \dfrac{1}{x}$.

3.7

DIFFERENTIATION OF
ALGEBRAIC FUNCTIONS

With the aid of the chain rule our ability to easily differentiate functions such as $\sqrt{x^2 + 1}$ and $\sqrt[5]{3x + 10}$ depends upon the differentiation of $\sqrt[n]{x}$ where n is a positive integer > 1. We showed previously that

$$D_x(\sqrt{x}) = \frac{1}{2\sqrt{x}} = \frac{1}{2}x^{-1/2} \qquad (x > 0)$$

This was done by finding the limit of the difference quotient, that is, by going back to the definition of the derivative.

In a similar manner we will now show that the nth root function

$$f(x) = \sqrt[n]{x} \qquad (n \text{ an integer} > 1)$$

may be differentiated in accordance with the power formula. By definition

$$f'(a) = \lim_{x \to a} \frac{\sqrt[n]{x} - \sqrt[n]{a}}{x - a}$$

Let $u = \sqrt[n]{x}$, $v = \sqrt[n]{a}$, then $x = u^n$ and $a = v^n$. Furthermore

$$\frac{u - v}{u^n - v^n} = \frac{1}{u^{n-1} + u^{n-2}v + \cdots + uv^{n-2} + v^{n-1}}$$

so that

$$f'(a) = \lim_{x \to a} \frac{1}{(\sqrt[n]{x})^{n-1} + (\sqrt[n]{x})^{n-2}\sqrt[n]{a} + \cdots + \sqrt[n]{x}(\sqrt[n]{a})^{n-2} + (\sqrt[n]{a})^{n-1}}$$

Since $\lim_{x \to a} \sqrt[n]{x} = \sqrt[n]{a}$ it follows by application of the limit theorems that

$$f'(a) = \frac{1}{n(\sqrt[n]{a})^{n-1}} = \frac{1}{n}a^{\frac{1}{n}-1}$$

Thus if $r = 1/n$ we have

$$D_x(x^r) = rx^{r-1} \tag{1}$$

where $r = \frac{1}{2}, \frac{1}{3}, \frac{1}{4}, \ldots$; that is (1) holds for reciprocals of positive integers as well as for positive and negative integers.

The result (1) may be used to enable us to find the derivative of the more general function

$$g(x) = x^{m/n} = (\sqrt[n]{x})^m$$

where m and n are integers and $n > 0$. All we need do is apply the chain rule since $g(x) = h^m$ and $h = \sqrt[n]{x}$. Thus

$$D_x g = D_h g \, D_x h = mh^{m-1} \cdot \frac{1}{n}x^{\frac{1}{n}-1}$$

$$= m(\sqrt[n]{x})^{m-1}\frac{1}{n}x^{\frac{1}{n}-1}$$

$$= \frac{m}{n}x^{\frac{m-1}{n}+\frac{1}{n}-1}$$

so that

$$D_x(x)^{m/n} = \frac{m}{n}x^{\frac{m}{n}-1} \tag{2}$$

Thus $D_x(x^r) = rx^{r-1}$ holds for *any rational number*. More generally by application of the chain rule

$$D_x[f(x)]^r = D_f[f(x)]^r \, D_x f(x) = r[f(x)]^{r-1}D_x f(x) \tag{3}$$

where again r is any rational number.

EXAMPLE 1   Find $D_x\left(\dfrac{1}{\sqrt{x}}\right)$, $\qquad (x > 0)$.

Solution Application of (2) with $r = \dfrac{m}{n} = -\dfrac{1}{2}$ yields

$$D_x\left(\frac{1}{\sqrt{x}}\right) = D_x(x^{-1/2}) = -\frac{1}{2}x^{-3/2} = \frac{-1}{2x\sqrt{x}} \qquad (x > 0) \qquad \bullet$$

EXAMPLE 2 Find $D_x(\sqrt{x^2 + 9})$.

Solution Application of (3) with $f(x) = x^2 + 9$ and $r = \frac{1}{2}$ yields

$$D_x\sqrt{x^2 + 9} = \frac{1}{2}(x^2 + 9)^{-1/2}D_x(x^2 + 9)$$

$$= \frac{1}{2}(x^2 + 9)^{-1/2}2x$$

$$= \frac{x}{\sqrt{x^2 + 9}} \qquad \bullet$$

It only remains to discuss the domains of validity of (2). For example, we know that the domain of $\sqrt{x}$ is $x \geq 0$ and the domain of its derivative $\dfrac{1}{\sqrt{x}}$ is $x > 0$. More generally if $g(x) = x^r = x^{m/n}$ then the domains A of g and B of Dg depend upon the quantities m and n.

Case 1. n odd (that is, $n = 1, 3, 5, \ldots$).

If $m \geq 0$, then A is the set of all real numbers; whereas A is the set of all real numbers with the exception of $x = 0$ if $m < 0$. Also B (the domain of Dg) is the set of all real numbers if $m \geq n$; the set of all real numbers $\neq 0$ if $m < n$.

Case 2. n even.

A is the set $[0, \infty)$ if $m > 0$ and A is the set $(0, \infty)$ if $m < 0$. Also B is the set $[0, \infty)$ if $m > n$, whereas $B = (0, \infty)$ if $m < n$.

It is easily established for example that $g'(0)$ fails to exist if $g = x^r$ and r is in $(0, 1)$. This follows from the fact that $g(0) = 0$ and

$$g'(0) = \lim_{x \to 0}\frac{x^r - 0}{x - 0} = \lim_{x \to 0}x^{r-1}$$

which fails to exist if $0 < r < 1$.

EXAMPLE 3 Find $D_x\sqrt[3]{x^3 + x + 2}$

Solution We use (3) with $f = x^3 + x + 2$, $r = \frac{1}{3}$, and $D_x f = 3x^2 + 1$. Therefore

$$D_x\sqrt[3]{x^3 + x + 2} = \frac{1}{3}(x^3 + x + 2)^{-2/3}(3x^2 + 1) = \frac{3x^2 + 1}{3(x^3 + x + 2)^{2/3}} \qquad \bullet$$

EXAMPLE 4 Find an equation of the tangent line to the graph of $y = \dfrac{x^2}{\sqrt{x^2+1}}$ at the point $\left(2, \dfrac{4}{\sqrt{5}}\right)$.

Solution The slope of the curve at any point on it is given by

$$y' = \frac{\sqrt{x^2+1}\,(2x) - x^2\,D_x\sqrt{x^2+1}}{x^2+1} = \frac{2x\sqrt{x^2+1} - \dfrac{x^3}{\sqrt{x^2+1}}}{(x^2+1)}$$

$$= \frac{2x(x^2+1) - x^3}{(x^2+1)^{3/2}} = \frac{x^3 + 2x}{(x^2+1)^{3/2}} = \frac{x(x^2+2)}{(x^2+1)^{3/2}}$$

$$y'\Big|_{(2,\,4/\sqrt5)} = \frac{2(2^2+2)}{(2^2+1)^{3/2}} = \frac{12}{5^{3/2}}$$

Therefore an equation of the tangent line at the given point is

$$y - \frac{4}{\sqrt5} = \frac{12}{5\sqrt5}\,(x-2)$$

or

$$12x - 5\sqrt5\,y - 4 = 0 \qquad \bullet$$

Exercises 3.7

In each of Exercises 1 through 20, find the derivative of the given function with respect to the independent variable.

1. $f(x) = x^{3/2} + 2x^{7/2} + 5$

2. $g(x) = x^{1/2} + 2x^{3/2} - \frac{1}{2}$

3. $F(x) = \dfrac{x^{5/2} - 3x^2}{x}$

4. $h(x) = \dfrac{x^{7/2} - 2x}{x^{1/2}}$

5. $G(y) = \sqrt[3]{y^2 + 3y - 2}$

6. $f(t) = \sqrt[3]{t+1}\sqrt{t+1}$

7. $f(u) = \dfrac{(u^2 - u - 2)\sqrt{u^2 - u - 2}}{2}$

8. $f(t) = \dfrac{2}{t}\sqrt{\dfrac{1}{t}}$

9. $f(y) = y^2\sqrt{y^2 - a^2}$

10. $F(x) = x^2\sqrt{x^4 + 1}$

11. $G(x) = \dfrac{x}{\sqrt{x^2 + 1}}$

12. $f(x) = \sqrt{4 + \sqrt{4 + x}}$

13. $h(t) = \dfrac{b}{t}\sqrt{b^2 - t^2}$

14. $f(z) = (z^2 + 1)(z^3 + 3)^{-2/3}$

15. $F(x) = \dfrac{x^2 + 10}{\sqrt{x+4}}$

16. $u(x) = 2x^{1/2} - x^{2/3} + 3x^{-1/4}$

17. $g(t) = \sqrt[3]{\dfrac{t^2+1}{t^2-1}}$

18. $h(s) = \sqrt{\dfrac{3s-4}{2s+5}}$

19. $f(x) = x(\sqrt{x} + 1)^3$

20. $g(x) = (1 + \sqrt{x} + x)^{10}$

21. Find an equation of the tangent line to the graph of $y = \sqrt{x} + 2$ at the point $(2, 2)$.
22. Find an equation of the tangent line to $y = x\sqrt{x^2 + 7}$ at $(3, 12)$.
23. Find equations of the tangent and normal lines to $y = 1 + \sqrt[3]{x}$ at $(8, 3)$.
24. Find equations of the tangent and normal lines to $g(x) = x - \sqrt[3]{x}$ at $(1, 0)$. At what points of the graph if any is the tangent horizontal?
25. Find $g'(x)$ for $g(x) = |x^2 - 1|$. (*Hint*: $|x^2 - 1| = \sqrt{(x^2 - 1)^2}$.)

26. Find $F'(x)$ for $F(x) = \dfrac{3|x|}{x}$.

27. If $G(x) = 5x|x|$, find $G'(x)$.

28. Find $\dfrac{d}{dx}\,|g(x)|$ where $g(x)$ is a differentiable function of x.

29. Find $\dfrac{d}{dx}\,\sqrt{h(x^2 + 5)}$ where h is a differentiable function of its argument.

30. Find $\dfrac{d}{dx}\,h(\sqrt{x^2 + 5})$ where h is a differentiable function of its argument.

3.8   IMPLICIT DIFFERENTIATION

Functions such as

$$y = x^3 + x + 3$$
$$y = \sqrt{x^2 + 16}$$
$$y = ax + b$$

are called *explicit* functions, where y is expressed explicitly in terms of x. A function can also be defined indirectly or *implicitly* as follows. Consider the relation

$$x^2 + y^2 = 25 \tag{1}$$

Corresponding to each x in $[-5, 5]$ there is at least one value of y that satisfies the relation. In this case the relation yields two explicit functions

$$y_1(x) = \sqrt{25 - x^2} \quad \text{and} \quad y_2(x) = -\sqrt{25 - x^2} \tag{2}$$

The equation $F(x, y) = 0$† defines y implicitly as a function of x if a function $f(x)$ exists such that $F(x, f(x)) = 0$ is satisfied for all x in some interval. For example, the equation

$$x^2 + y^2 + 25 = 0 \tag{3}$$

is satisfied by *no pair* of real numbers [since the left side of (3) is positive for all (x, y)]. Thus, in this case the equation does *not* define y as a function of x. In advanced calculus, methods are developed to test functions in order to determine whether or not implicit relations of the form $F(x, y) = 0$ do or do not imply the existence of a function $y = f(x)$. These tests for this and more general situations are known as *implicit function theorems*. We will not pursue this subject any further here.

Let us instead return to (1). We seek the derivative of y with respect to x without attempting to solve for y in terms of x. Assuming that (1) defines at least one function of x, $y = y(x)$ then we have the *identity*

$$x^2 + [y(x)]^2 = 25$$

---

† $F(x, y)$ is a notation used to denote a function of two variables. Equation (1) may be expressed in the form $F(x, y) = 0$ where $F(x, y) = x^2 + y^2 - 25$. This type of function will be studied in detail in Chapter 17 of this text.

Therefore we may differentiate both sides with respect to x:

$$D_x(x^2 + [y(x)]^2) = D_x(25) = 0$$

Now $D_x(x^2) = 2x$ and to find $D_x(y^2)$ we use the chain rule

$$D_x(y^2) = 2yy'$$

so that

$$2x + 2yy' = 0$$

This may be solved for y' to yield (for $y \neq 0$)

$$y' = \frac{-x}{y} \tag{4}$$

We now ask ourselves which function of x has been differentiated. The answer is *both*. If y_1 is substituted for y on the right side of the expression, we have the derivative of y_1 namely $y_1' = \dfrac{-x}{y_1}$ and similarly for y_2; $y_2' = \dfrac{-x}{y_2}$.

To see this explicitly:

$$y_1' = D_x(\sqrt{25 - x^2}) = \frac{-x}{\sqrt{25 - x^2}} = \frac{-x}{y_1}$$

$$y_2' = D_x(-\sqrt{25 - x^2}) = \frac{x}{\sqrt{25 - x^2}} = \frac{-x}{y_2} \tag{5}$$

In the future, we shall find it convenient to let $[y(x)]^n$ be denoted by $y^n(x)$ or just y^n, where it will be understood that y is being treated as a function of x.

EXAMPLE 1 $y^2 + xy + x^2 = 3$, find y' at $(1, 1)$ by implicit differentiation.

Solution   Note that $(1, 1)$ satisfies the relation; that is $(1, 1)$ lies on its graph. Form

$$D_x(y^2(x) + xy(x) + x^2) = D_x(3) = 0$$

so that

$$2yy' + xy' + y + 2x = 0$$

where the derivative of the product xy was obtained by the product formula. Collecting terms,

$$y'(2y + x) = -(y + 2x)$$

and then solving for y',

$$y' = -\frac{y + 2x}{2y + x}$$

so that

$$y' \bigg|_{(1, 1)} = -1$$

EXAMPLE 2   Find an equation of the tangent line to the curve $x^3 + y^2 = 3$ at $(-1, 2)$.

Solution $D_x(x^3 + y^2) = D_x(3)$ so that

$$3x^2 + 2yy' = 0 \qquad \text{or} \qquad y' = \frac{-3x^2}{2y}$$

and

$$y'\Big|_{(-1,\,2)} = \frac{-3}{4}$$

An equation of the tangent line at $(-1, 2)$ is

$$y - 2 = -\tfrac{3}{4}(x + 1)$$

or

$$3x + 4y = 5 \qquad\qquad\qquad \bullet$$

EXAMPLE 3 Find y' if $x^3 + 6xy + 5y^3 = 10$.

Solution To find y in terms of x is extremely difficult (in some cases it will be impossible), so we proceed immediately to implicit differentiation.

$$D_x(x^3 + 6xy + 5y^3) = D_x 10 = 0$$

Therefore

$$3x^2 + 6xy' + 6y + 15y^2y' = 0$$

Collecting terms and solving for y' results in

$$y'(6x + 15y^2) = -(3x^2 + 6y)$$

or

$$y' = -\frac{3x^2 + 6y}{6x + 15y^2} = -\frac{x^2 + 2y}{2x + 5y^2} \qquad\qquad \bullet$$

EXAMPLE 4 Suppose that $y = x^r = x^{m/n}$ where r is a rational number $= m/n$ (m and n are integers for which $n \neq 0$). We seek $D_x(x^r)$ by using implicit differentiation.

Solution From $y = x^{m/n}$, if we raise both sides to the nth power, there results

$$y^n = x^m$$

Differentiation of both sides with respect to x yields

$$ny^{n-1} D_x y = mx^{m-1}$$

or

$$D_x y = \frac{m}{n} \frac{x^{m-1}}{y^{n-1}} = \frac{m}{n} \frac{x^{m-1}}{(x^{m/n})^{n-1}}$$

$$= \frac{m}{n} \frac{x^{m-1}}{x^{m(n-1)/n}} = \frac{m}{n} \frac{x^{m-1}}{x^{m-\frac{m}{n}}}$$

$$= \frac{m}{n} x^{\frac{m}{n}-1} = rx^{r-1}$$

which again establishes the power formula for rational exponents. $\bullet$

Exercises 3.8

In Exercises 1 through 18 find y' by implicit differentiation.

1. $x^2 + y^2 = 100$
2. $x^2 - y^2 = 16$
3. $x^3 + y^3 = 1$
4. $3x^2 - y^3 = 2$
5. $\sqrt{x} + \sqrt{y} = 3$
6. $x^{1/3} - y^{1/3} = 1$
7. $xy = 7$
8. $2x^2y + 3xy^2 = 4$
9. $x^2 + y^3 - xy = 1$
10. $x^2 + xy + y^2 - 6 = 0$
11. $\dfrac{3}{x} - \dfrac{2}{y} = 1$
12. $3x^2 - 2xy + y^2 - 3x + 5 = 0$
13. $x^{2/3} + y^{2/3} = a^{2/3}$
14. $x^{1/3} + y^{1/3} = a^{1/3}$
15. $(x + y)^2 - (x - y)^2 - 3x - y = 1$
16. $(x - 2)^3 - 2x^2y + xy^2 = 0$
17. $y + \sqrt{xy} - x^2 - 3 = 0$
18. $\sqrt{xy} - 3x = \sqrt{y} + 1$
19. Find an equation of the line tangent to $x^2 + y^2 = 100$ at $(8, -6)$.
20. Find an equation of the line normal to $x^2 + y^2 - 8x - 2y - 8 = 0$ at $(1, 5)$.
21. Find equations of the tangent and normal lines to $3x^2 + xy + 2y^2 = 9$ at $(-1, 2)$.
22. Find an equation of the normal line to $x^3 - y^3 + 3xy = 3$ at $(2, -1)$.
23. Show that an equation of the line tangent to the ellipse $\dfrac{x^2}{a^2} + \dfrac{y^2}{b^2} = 1$ at any point (x_0, y_0) on the curve is

$$\frac{xx_0}{a^2} + \frac{yy_0}{b^2} = 1$$

24. Find an equation of a line tangent to the hyperbola $\dfrac{x^2}{a^2} - \dfrac{y^2}{b^2} = 1$ at any point (x_0, y_0) on it.

25. Prove that the sum of the x and y intercepts of *any* tangent line to the curve $x^{1/2} + y^{1/2} = a^{1/2}$ is constant and equal to a.

26. Find y' by implicit differentiation:

$$Ax^2 + Bxy + Cy^2 + Dx + Ey + F = 0$$

27. Suppose that we are given a polynomial of degree n of the form $P_n(x) = 1 + x + \dfrac{x^2}{2!} + \cdots + \dfrac{x^n}{n!}$ for each positive integer n. Assume that $x = P_n(y)$ determines y as a differentiable function of x, say, $y = f_n(x)$. Show that

$$f_n'(x) = \frac{n!}{n!x - [f(x)]^n}$$

28. In elementary geometry it is proved that a tangent line to a circle is perpendicular to the radial line at the point of contact. Give an independent proof of this theorem using implicit differentiation on the function $x^2 + y^2 - r^2 = 0$.

29. Find y' by implicit differentiation if

$$x^2 + y^2 - 2x - 4y + 6 = 0 \tag{6}$$

However show that this result is meaningless because the "curve" defined by (6) is nonexistent. (*Hint:* use the method of completing the squares.)

3.9 DERIVATIVES OF HIGHER ORDER

If a function f is differentiable, then a new function f' can be formed. If f' itself is a differentiable function, then we can form its derivative, which is called the **second derivative** of f, and denoted by f''. So long as there is differentiability, this process may be continued, forming f''', and so on. The prime notation is not used beyond the third order, rather we write $f^{(4)}, f^{(5)}, \ldots$, and more generally, for the nth derivative, $f^{(n)}$. Thus, for example, if $f(x) = x^3 - 3x^2 + 7x + 100$, there results

$$f'(x) = 3x^2 - 6x + 7 \qquad f''(x) = 6x - 6 \qquad f'''(x) = 6 \qquad f^{(4)}(x) = 0$$

and, in general,

$$f^{(n)}(x) = 0 \qquad \text{if } n \geq 4.$$

In fact, a polynomial has derivatives of all orders, and the domain of the polynomial in x and all its derivatives is $(-\infty, \infty)$. Furthermore the derivatives of order greater than the degree of the polynomial are all identically zero. For example if $P(x)$ is an arbitrary polynomial of degree 5 then $P^{(n)}(x) = 0$ if $n \geq 6$.

Some common notations for the nth derivative of $y = f(x)$ with respect to x are $f^{(n)}, y^{(n)}$, and $D_x^n y$. Thus, the second derivative of $y = f(x)$ with respect to x may be written in any one of the forms $f^{(2)}, f'', y^{(2)}$, or $D_x^2 y$, and the higher order derivatives may be expressed similarly.

EXAMPLE 1   If

$$f(x) = x^{-1}$$

then $f'(x) = -x^{-2}, f''(x) = 2x^{-3}, f'''(x) = -6x^{-4}, f^{(4)} = 24x^{-5}$, and by induction it can be shown that

$$f^{(n)}(x) = (-1)^n (n!) x^{-(n+1)}$$

This last claim is left as an exercise for the reader. ●

EXAMPLE 2   If $y = (1 + 3x)^3$ find y', y'', y''' and $y^{(n)}$ for $n > 3$.

Solution $y' = 3(1 + 3x)^2(3)$ by the chain rule

$$y' = 9(1 + 3x)^2$$

Continuing,

$$y'' = 18(1 + 3x)(3) = 54(1 + 3x)$$
$$y''' = 162 \qquad \text{and} \qquad y^{(n)} = 0 \qquad \text{if } n > 3$$

Of course we could alternatively expand the cubic first and then differentiate term by term. ●

EXAMPLE 3   If $x^2 + y^2 = 25$ find y' and y'' at $(3, 4)$.

Solution By implicit differentiation, we have $2x + 2yy' = 0$ or $y' = -\dfrac{x}{y}$ so that $y'\Big|_{(3,4)} = -\dfrac{3}{4}.$

The second derivative y'' may be found by differentiating the first derivative by the quotient rule

$$y'' = -\frac{((y)(1) - xy')}{y^2} = \frac{xy' - y}{y^2}$$

but $y' = -\dfrac{x}{y}$ so that

$$y'' = \frac{x(-x/y) - y}{y^2} = -\frac{x^2 + y^2}{y^3}$$

This may be further simplified, since $x^2 + y^2 = 25$, and we have

$$y'' = -\frac{25}{y^3}$$

Therefore

$$y''\Big|_{(3,4)} = -\frac{25}{64}$$

We could have obtained y'' in a different manner by writing

$$yy' + x = 0$$

and taking $D_x(yy' + x) = D_x 0 = 0$. Thus, by the product rule,

$$yy'' + (y')^2 + 1 = 0$$

which implies that

$$y'' = -\frac{1 + (y')^2}{y} = -\frac{1 + (x/y)^2}{y} = -\frac{1 + x^2/y^2}{y}$$

$$= -\frac{x^2 + y^2}{y^3}$$

or

$$y'' = -\frac{25}{y^3}$$

in agreement with the previous results. ●

We recall that if the displacement of a particle moving along a straight line is given by the directed distance $s(t)$ then the velocity of the particle $v(t)$ is given by

$$v(t) = s'(t) = \lim_{h \to 0} \frac{s(t + h) - s(t)}{h}$$

Similarly we define the **acceleration** of the particle at any time t to be

$$a(t) = v'(t) = \lim_{h \to 0} \frac{v(t + h) - v(t)}{h} \qquad (1)$$

It follows that $a(t) = v'(t) = s''(t)$.

EXAMPLE 4 A ball thrown vertically upward with an initial velocity of 128 ft/sec will have $s(t) = 128t - 16t^2$ as its equation of motion. Find $v(t)$ and $a(t)$.

Solution
$$v(t) = s'(t) = 128 - 32t$$
$$a(t) = v'(t) = -32$$

Note that the acceleration is a negative constant, which means that the velocity is decreasing algebraically as t increases. ●

The *second law of Newtonian mechanics* states that if $F(t)$ is the resultant force acting on the mass m then

$$F(t) = D_t(mv) \tag{2}$$

and if the mass is constant (assumed here) then

$$F(t) = mD_t v = ma(t) \tag{3}$$

that is, the force acting on the mass is proportional to its acceleration. Alternatively,

$$F(t) = ms''(t) \tag{4}$$

Note that, if the acceleration is constant, the force acting on the particle must also be constant. Thus, in the Example 4 where $a(t) = -32$, we must have $F = -32m$, which the reader may recognize as the negative of the weight of the particle ($g = $ acceleration in a gravitational field $= 32$ ft/sec^2). The reason for the minus sign is that the force is due to the Earth's attraction on the particle, and this force acts in the opposite direction to the positive direction of s (see Figure 3.9.1).

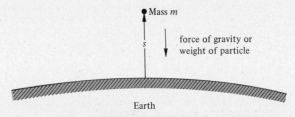

Figure 3.9.1

EXAMPLE 5 If $f(x) = u(x)\,v(x)$ find $f'(x)$, $f''(x)$ and $f'''(x)$ assuming u and v are differentiable functions to all the orders required to make the formalism correct.

Solution By the product rule,

$$f'(x) = u(x)\,v'(x) + u'(x)\,v(x)$$

so that application of the product rule again yields

$$f''(x) = u(x)\,v''(x) + u'(x)\,v'(x) + u'(x)\,v'(x) + u''(x)\,v(x).$$
$$f''(x) = u(x)\,v''(x) + 2u'(x)\,v'(x) + u''(x)\,v(x).$$

Similarly,

$$f'''(x) = u(x)\,v'''(x) + u'(x)\,v''(x) + 2u'(x)\,v''(x) + 2u''(x)\,v'(x)$$
$$+ u''(x)\,v'(x) + u'''(x)\,v(x)$$
$$= u(x)\,v'''(x) + 3u'(x)\,v''(x) + 3u''(x)\,v'(x) + u'''(x)\,v(x)$$

where u and v are assumed to possess derivatives of order 3.   ●

It is instructive to note that the coefficients in f', f'', f''' are

$$1 \quad 1$$
$$1 \quad 2 \quad 1$$
$$1 \quad 3 \quad 3 \quad 1$$

that is, the coefficients in $a + b$, $(a + b)^2$, $(a + b)^3$ forming the beginning of an array known as Pascal's triangle. Observe that each coefficient is the sum of two coefficients in the row above immediately to the left and to the right of the required number (add in zero if the number is missing). Thus if this pattern continues

$$f^{(4)}(x) = u(x)\,v^{(4)}(x) + 4u'(x)\,v'''(x) + 6u''(x)\,v''(x) + 4u'''(x)\,v'(x) + u^{(4)}(x)\,v(x)$$

the coefficients are 1, 4, 6, 4, 1, the next row in Pascal's triangle. If we denote the numbers in the nth row of Pascal's triangle by

$$C_0^n = 1,\ C_1^n,\ C_2^n,\ C_3^n,\ \ldots,\ C_{n-1}^n,\ C_n^n = 1$$

then we have

$$(a + b)^n = a^n + C_1^n a^{n-1}b + C_2^n a^{n-2}b^2 + \cdots + C_{n-1}^n ab^{n-1} + b^n$$

Furthermore by mathematical induction it can be established that

$$C_r^n = \frac{n(n-1)(n-2)\cdots(n-r+1)}{r!} = \frac{n!}{r!(n-r)!}$$

This enables us to establish Leibnitz' formula for the nth derivative of a product namely

$$D^n[u(x)\,v(x)] = \sum_{r=0}^{n} C_r^n\, u^{(r)}(x)\, v^{(n-r)}(x) \tag{5}\dagger$$

The proof of this formula can be established by mathematical induction. A more interesting exercise is that (when time permits) the reader review or investigate the significance of Pascal's triangle in the elements of permutations, combinations, and probability.

---

† The notation $\displaystyle\sum_{r=0}^{n}$ is called the "sigma notation" or "summation notation" and is a shorthand way of indicating that terms of a certain form are being added together. This notation will be studied in some detail in Section 6.1. Briefly, we mean that $\displaystyle\sum_{r=0}^{n} A_r = A_0 + A_1 + A_2 + \cdots + A_{n-1} + A_n.$ In formula (5), $A_r = C_r^n u^{(r)}(x)\, v^{(n-r)}(x).$ Students who have not seen the sigma notation before may find it instructive to return to this example after studying Section 6.1.

EXAMPLE 6   This example will show us how higher derivatives may be used in the study of polynomials. In elementary algebra we learn to simplify expressions such as $(x - 2)^2 + 3(x - 2) + 5$. In fact, if the expression is expanded and terms combined we have the identity

$$(x - 2)^2 + 3(x - 2) + 5 = x^2 - x + 3 \tag{6}$$

Suppose on the other hand we are given $x^2 - x + 3$ and would like to express it as a polynomial in $x - 2$; that is, we equate

$$x^2 - x + 3 = a_2(x - 2)^2 + a_1(x - 2) + a_0 \tag{7}$$

How do we find the coefficients a_0, a_1, and a_2?

One way to proceed is to expand the right side of (7) and equate coefficients of like powers. Thus

$$x^2 - x + 3 = a_2 x^2 + (a_1 - 4a_2)x + 4a_2 - 2a_1 + a_0$$

or

$$a_2 = 1 \qquad a_1 - 4a_2 = -1 \qquad 4a_2 - 2a_1 + a_0 = 3$$

Therefore $a_2 = 1$, $a_1 = 3$, and $a_0 = 5$ in agreement with (6). However this technique may be quite cumbersome, and we shall now show how higher derivatives may be used to find the coefficients. Assume that there are numbers a_2, a_1, and a_0 such that

$$f(x) = x^2 - x + 3 = a_2(x - 2)^2 + a_1(x - 2) + a_0 \tag{8}$$

To find a_0 we set $x = 2$, thus

$$a_0 = f(2) = 5$$

To find a_1 differentiate both sides of the identity. Thus

$$f'(x) = 2x - 1 = 2a_2(x - 2) + a_1$$

and set $x = 2$, so that

$$a_1 = f'(2) = 3$$

The quantity a_2 may be found by differentiating both sides with respect to x once more

$$f''(x) = 2a_2 \qquad \text{or} \qquad a_2 = \frac{f''(x)}{2} = \frac{f''(2)}{2} = 1$$

This technique, of finding coefficients one at a time by successive differentiation rather than by solving simultaneous linear equations, is the substance of the next two theorems.   ●

Theorem 1   Any polynomial in x can be expressed as a polynomial in powers of $(x - a)$ where a is any fixed number.

Proof   Let $f(x) = a_0 + a_1 x + a_2 x^2 + \cdots + a_n x^n$ be a polynomial in x. Then

$$f(x + a) = a_0 + a_1(x + a) + a_2(x + a)^2 + \cdots + a_n(x + a)^n$$

is also a polynomial in x which becomes, upon expansion and simplification

$$f(x + a) = b_0 + b_1 x + b_2 x^2 + \cdots + b_n x^n \tag{9}$$

Now replace x by $x - a$ in both sides of (9) then (9) becomes

$$f(x) = b_0 + b_1(x - a) + b_2(x - a)^2 + \cdots + b_n(x - a)^n$$

which is the desired polynomial in $x - a$. ∎

This theorem assures us that any polynomial in x may also be written as a polynomial in $x - a$. The next theorem shows us how to determine the coefficients efficiently by application of higher derivatives at $x = a$.

Theorem 2 Let f be a polynomial in $(x - a)$,

$$f(x) = a_0 + a_1(x - a) + a_2(x - a)^2 + \cdots + a_n(x - a)^n \qquad (10)$$

Then the successive coefficients a_k, $(k = 0, 1, \ldots, n)$ may be found by means of the formula

$$a_k = \frac{f^{(k)}(a)}{k!} \qquad (k = 0, 1, \ldots, n) \qquad (11)$$

Proof To find a_0 replace x with a on both sides of (10). Thus

$$f(a) = a_0$$

Then differentiate both sides of (10) to obtain

$$f'(x) = a_1 + 2a_2(x - a) + 3a_3(x - a)^2 + \cdots + na_n(x - a)^{n-1} \qquad (12)$$

and set $x = a$. There results

$$f'(a) = a_1$$

To find a_2 we differentiate (12) to obtain

$$f''(x) = 2a_2 + 3 \cdot 2a_3(x - a) + \cdots + n(n - 1)a_n(x - a)^{n-2} \qquad (13)$$

and then replace x by a in (13). Therefore

$$a_2 = \frac{f''(a)}{2} = \frac{f''(a)}{2!}$$

Another differentiation yields

$$f'''(x) = 3 \cdot 2 \cdot 1a_3 + \cdots + n(n - 1)(n - 2)a_n(x - a)^{n-3} \qquad (14)$$

and if x is replaced by a, we obtain

$$a_3 = \frac{f'''(a)}{3!}$$

Continuing this way it follows that (11) holds. ∎

Thus a polynomial is completely determined by its value and the values of its derivatives at a single number $x = a$ where a may be arbitrarily chosen. Furthermore, we shall later utilize the significant formula $a_k = f^{(k)}(a)/k!$ where f may be a function such as $\sin x$ or $\log (1 + x)$ in our study of "infinite series" representations of such functions.

Exercises 3.9   odd 1–17

In Exercises 1 through 14, find the indicated derivative.

1. $f(x) = 10x^2 - 7x + 100$, $f''(x)$

2. $g(x) = (5x - 2)^2$, $g'''(x)$

3. $F(v) = v^3 - 3v^2 + 7v - 1$, $F''(v)$

4. $G(x) = \dfrac{x^2 - 5}{\sqrt{x}}$, $G''(x)$

5. $f(y) = \sqrt{2y + 1}$, $f'''(y)$

6. $F(x) = \dfrac{x - 1}{x + 1}$, $F''(x)$

7. $G(x) = \dfrac{ax + b}{cx + d}$, $G''(x)$

What happens if $bc - ad = 0$? Why?

8. $H(x) = (2x + 1)^4$, $H^{(4)}(x)$

9. $g(u) = (u^3 - \sqrt{2}u^2 + u - 7)^2$, $g^{(7)}(u)$

10. $h(x) = x^{-1}$, $h^{(n)}(x)$

11. $s(t) = \dfrac{a}{2}t^2 + bt + c$, $s''(t)$

12. $f(t) = (at + b)^{-1}$, $f^{(n)}(t)$

13. If $D_x F(x) = G(x)$ and $D_x G(x) = F(x)$, $D^{(43)}F(x)$

14. $F(t) = \dfrac{1}{t^2 - a^2}$, $F^{(n)}(t)$ $\left[\text{Hint: } \dfrac{1}{t^2 - a^2} = \dfrac{1}{2a}\left(\dfrac{1}{t - a} - \dfrac{1}{t + a}\right).\right]$

In Exercises 15 through 20 find y' and y'' by implicit differentiation.

15. $x^2 + y^2 = a^2$

16. $x^2 - y^2 = a^2$

17. $xy + y^2 = 2$

18. $b^2x^2 + a^2y^2 = a^2b^2$

19. $x^{1/2} + y^{1/2} = a^{1/2}$ question

20†. $x^n + y^n = a^n$

21. If $xy = a^2$ find y', y'', y''' at (a, a).

22. Given the following functions S and C that satisfy the following derivative equations (later we will use the term differential equations): $DS = C$ and $DC = -S$. Show that
   (a) $D^2S = -S$ and $D^2C = -C$.
   (b) $D^2(S^2 + C^2) = 0$.
   (c) $D^{(2n)}S = \begin{cases} -S & (n \text{ odd}) \\ S & (n \text{ even}) \end{cases}$

23. Given a function f such that $Df = af$, where a is a constant. Show that $D^{(n)}f$ exists and find an expression for it.

24. Find an expression for $D^n(xf)$ in terms of derivatives of f.

25. A certain function f has $f(4) = 3$, $f'(4) = -1$, $f''(4) = 4$ and $f'''(4) = 6$ and $f^{(k)}(4) = 0$ for $k > 3$. Find a formula for the function.

26. A certain function f has $f(1) = -7$, $f'(1) = 10$, $f''(1) = 6$, and $f^{(n)}(1) = 0$ if $n > 2$. Find the polynomial expression in powers of $(x - 1)$. Find $f(1.01)$ approximately by using the linear approximation $f(1) + a_1(x - 1)$. What is $f(1.01)$ exactly?

27. For a function

$$f(x) = \begin{cases} x^2 & \text{if } x > 0 \\ 0 & \text{if } x \le 0 \end{cases}$$

(a) Find $f'(x)$ for $x \ne 0$. (b) Does $f'(0)$ exist? (c) Does $f''(x)$ exist for $x \ne 0$? (d) Does $f''(0)$ exist?

---

† Note that this includes Exercises 15 and 19 as special cases and serves as a nice check on the answers.

In each of Exercises 28 through 30 the position function of a particle moving on a straight line is given. Discuss the motion of the particle.

28. $s(t) = 160t - 16t^2, \quad 0 \le t \le 10$

29. $s(t) = t^3 - 12t$

30. $s(t) = 3t^2 + \dfrac{48}{t}, \quad (t \ge 1)$

31. Let s, v, and a be the position, velocity, and acceleration of a particle moving on a straight line. Show that

$$a = v\frac{dv}{ds} = s''(t)$$

Find an expression for $s'''(t)$ involving just v and its first and second derivatives with respect to s.

32. A function $E(ax)$ is such that $D\,E(ax) = a\,E(ax)$. Use Leibnitz's formula (5) to find $D^n[x^n\,E(ax)]$. What is $D^3[x^3\,E(ax)]$?

Review and Miscellaneous Exercises

In Exercises 1 through 22 find $D_x y$:

1. $y = (x + \sqrt{x})^2$

2. $y = \dfrac{x^2}{x + 1}$

3. $xy + y^3 = 3$

4. $y = (x^3 + 4x - 6)^5$

5. $xy = -40$

6. $y = \dfrac{3x^2 + 2}{4x^2 - 1}$

7. $y = \dfrac{x^2 - 10x + 1}{3x}$

8. $y = \dfrac{x^2 + a^2}{x + a}, \quad (a \text{ constant})$

9. $y = \sqrt{10x^2 + 7x - 2}$

10. $y = \sqrt{\dfrac{x^2 + b}{2x + b}}, \quad (b \text{ constant})$

11. $x^{1/2} + y^{1/2} = b^{1/2}, \quad (b \text{ constant})$

12. $(x^2 + y)^3 = x + 8$

13. $y = \sqrt{x + 3\sqrt{x}}$

14. $x^3 - x^2 y + y^3 = 3$

15. $y = |x| - |x - 2|$
Where (if anywhere) does $D_x y$ fail to exist?

16. $y = x^2|x|$
Where (if anywhere) does $D_x y$ fail to exist?

17. $y = (x^3 + x^2 + 5)^3 (x^2 - x + 6)^4$

18. $y = (ax^3 - bx^2 + cx - e)^9$
$(a, b, c, \text{ and } e \text{ are constants})$

19. $y = \dfrac{3x^2 + 3x + 8}{x^2 + x + 1}$

20. $y = \dfrac{x^4 - b^4}{x - b}, \quad (b \text{ constant})$

21. $y = |x^2 - a^2|, \quad (a \text{ constant})$

22. $y = x\sqrt[3]{2x + b}, \quad (b \text{ constant})$

23. Find an equation of the tangent line to the curve $y = \dfrac{x}{x + 3}$ at the point $(0, 0)$.

24. Find equations of the tangent and normal lines to the curve $y = \sqrt{7 + \sqrt{x}}$ at the point $(4, 3)$.

25. Find equations of the tangent and normal lines to the curve $x^3 + y^3 - 5x^2 y + 25 = 0$ at the point $(2, 3)$.

26. A ball is thrown upward from the top of a building 512 ft high with an initial velocity of 64 ft/sec. Its height above the ground t sec later is given by the formula $s(t) = 512 + 64t - 16t^2$. Find
(a) the time to reach the maximum height
(b) the maximum height above the ground
(c) the time to reach the ground
(d) the velocity of impact with the ground

27. From the definition of a derivative as a limit of a quotient find $f'(x)$ if $f(x) = \dfrac{(2x+3)}{(x+1)}$ if $x \neq -1$.

28. From the definition of a derivative as a limit of a quotient find $f'(x)$ if $f(x) = \sqrt{3x}$ if $x > 0$.

29. From the definition of the derivative as a limit of a quotient find $f'(x)$ if $f(x) = x^3 - 4x + 3$.

30. A function is defined as follows:

$$f(x) = \begin{cases} 3x^2 & \text{if } x \leq x_0 \\ mx + b & \text{if } x > x_0 \end{cases}, \quad (x_0, m, b \text{ constants})$$

Find m and b in terms of x_0 so that $f'(x_0)$ exists.

31. A polynomial of the third degree in x, $f(x)$, has the following properties at $x = 2$: $f(2) = 3$, $f'(2) = 7$, $f''(2) = -8$, $f'''(2) = 6$. Find it.

32. A polynomial of the third degree in x, $f(x)$, has the following properties: $f(0) = 2$, $f'(0) = 4$, $f(1) = 6$, and $f'(1) = 5$. Find it.

33. A polynomial of the third degree in x, $f(x)$, has the following properties: $f(0) = -10$, $f(1) = -11$, $f(2) = 0$, and $f(3) = 29$. Find it.

34. (a) If $f(x) = (x - x_1)(x - x_2)(x - x_3)$ prove that (if $x \neq x_1, x_2,$ or x_3)

$$\frac{f'(x)}{f(x)} = \frac{1}{x - x_1} + \frac{1}{x - x_2} + \frac{1}{x - x_3}$$

(b) Generalize to the case $f(x) = (x - x_1)(x - x_2) \cdots (x - x_n)$.

35. If $g(x) = (x^2 + 3x + 4)^7$, find (a) $D_x g(-x)$, (b) $D_x g(x^2)$, (c) $D_x g(1/x)$.

36. If $y = u^3 + 2u - 1$ and $u = x^2 + 3$ find $D_x y$ in terms of x by two methods and check the results.

37. If $y = \dfrac{1 + 3u}{2 + u}$, $u = 3 + x^2$, and $x = 2t + 7$ find $D_t y$.

38. Find $D_x \sqrt[3]{f(x^2 + 4)}$ where f is an arbitrary differentiable function of $x^2 + 4$.

39. If

$$f(x) = \frac{x^3 + x^2 + 10x + 1}{x} \qquad x \neq 0$$

find (a) f'', (b) $f^{(10)}$, (c) $f^{(n)}$, $(n > 2)$.

40. Let $P(x) = (x - a)^2 Q(x)$ where $Q(x)$ is a polynomial in x for which $Q(a) \neq 0$. This means that the polynomial $P(x)$ is divisible by $(x - a)^2$ [but not by $(x - a)^3$] and $P(x)$ is said to have a double root at $x = a$. Show that $P'(x)$ is divisible by $x - a$ and interpret your results geometrically.

41. For what value(s) of k does the equation $x^3 - x^2 - 8x + k = 0$ have a double root? Write the polynomial(s) with this special property in factored form.

42. Given the relation $x^2 + 2xy - 5y^2 = 1$ find $D_x y$ and $D_x^2 y$ by implicit differentiation. (*Hint:* Use the given relation to simplify the formula for $D_x^2 y$.)

43. It is given that the derivative $f'(a)$ exists. Let

$$g(t) = \frac{f(a + 3t) - f(a)}{t} \qquad t \neq 0$$

for any given a. Find a formula for $\lim\limits_{t \to 0} g(t)$ in terms of $f'(a)$.

44. Same question as Exercise 43 applied to

$$g(t) = \frac{f(a + 3t) - f(a + t)}{2t} \qquad t \neq 0$$

4

Applications of the Derivative

THE DERIVATIVE AS A RATE OF CHANGE

If $y = f(x)$ is a differentiable function of x then the graph is a curve whose slope

$$y' = \lim_{\Delta x \to 0} \frac{f(x + \Delta x) - f(x)}{\Delta x} = \lim_{\Delta x \to 0} \frac{\Delta y}{\Delta x} \qquad (1)$$

gives the rate of change of y with respect to x. In general this rate of change will vary from point to point. Thus at $x = x_1$, the rate of change is $f'(x_1)$, and at $x = x_2$ the rate of change is $f'(x_2)$, and so on. Of course by definition the rate of change is the slope of the tangent at the given point (Figure 4.1.1).

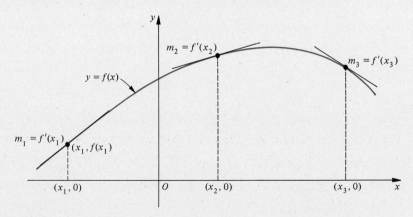

Figure 4.1.1

It is important to understand that this rate of change concept is fundamental to calculus and the reader should keep this notion in mind whenever a derivative is encountered. The rate of change of y with respect to x, namely y', is also called the *instantaneous rate of change* of y with respect to x. The ratio $\dfrac{\Delta y}{\Delta x}$ is called the

average rate of change of y with respect to x as x changes by an amount Δx. Thus if $y = f(x)$, then the average change of y per unit change in x as x changes from $x = x_1$ to $x = x_1 + \Delta x$ is given by

$$\left.\frac{\Delta y}{\Delta x}\right|_{x=x_1} = \frac{f(x_1 + \Delta x) - f(x_1)}{\Delta x} \tag{2}$$

Also

$$\left.y'\right|_{x_1} = \lim_{\Delta x \to 0} \left.\frac{\Delta y}{\Delta x}\right|_{x=x_1}$$

is the instantaneous rate of change of y with respect to x at $x = x_1$.

Velocity is another example of an instantaneous rate of change that we express as a derivative. If $s = f(t)$ is an equation of motion, then the average velocity is

$$\frac{\Delta s}{\Delta t} = \frac{f(t + \Delta t) - f(t)}{\Delta t}.$$

The (instantaneous) velocity is defined by

$$v(t) = \lim_{\Delta t \to 0} \frac{\Delta s}{\Delta t} = D_t s = f'(t) \tag{3}$$

EXAMPLE 1 Find the instantaneous rate of change of the area A of a square with respect to the length s of its side. Evaluate this rate of change when $s = 3$ in.

Solution The area of a square is

$$A = s^2$$

and the rate of change of A with respect to s is given by

$$A'(s) = 2s$$

In particular

$$\left.A'(s)\right|_{s=3} = 6 \text{ in.}^2/\text{in.}$$

EXAMPLE 2 Find the rate of change of the area of an equilateral triangle with respect to the length of a side. Evaluate this rate of change when the side is $4\sqrt{3}$ in.

Solution If s denotes the length of a side then

$$A = \frac{1}{2}(\text{base})(\text{height}) = \frac{s}{2}\left(\frac{s\sqrt{3}}{2}\right) = \frac{\sqrt{3}}{4}s^2$$

The rate of change of the area with respect to the length of a side is given by

$$A'(s) = \frac{\sqrt{3}}{4}\,2s = \frac{\sqrt{3}s}{2}$$

so that

$$A'(4\sqrt{3}) = \frac{\sqrt{3}}{2}\,4\sqrt{3} = 6 \text{ in.}^2/\text{in}$$

This means that, when the side has length $4\sqrt{3}$, the area is changing six times as fast as the length of side. ●

EXAMPLE 3   There are many other applications of the notion of average rate and instantaneous rate in physics and chemistry. For example, suppose that the quantity or volume of water Q (gal) in a reservoir at time t (min) is a function of t. Water may flow into or be drained out of the reservoir. In either case, suppose that Q changes by an amount ΔQ as time t changes to time $t + \Delta t$. Then the average rate of change of Q with respect to t is

$$\frac{\Delta Q}{\Delta t} \text{ [gal/min]}$$

and the instantaneous rate of change of Q with respect to t is

$$Q'(t) = \lim_{\Delta t \to 0} \frac{\Delta Q}{\Delta t} \text{ [gal/min]} ●$$

EXAMPLE 4   The radius r of a right circular cone is always three times the altitude h. Find the rate of change of the volume V (a) with respect to the altitude h and (b) with respect to the radius r.

Solution   (a) The volume of a right circular cone is given by

$$V = \frac{\pi}{3} r^2 h$$

In this case $r = 3h$ so that V may be expressed in terms of h only,

$$V = \frac{\pi}{3}(3h)^2 h = 3\pi h^3$$

Therefore,

$$D_h V = \lim_{\Delta h \to 0} \frac{\Delta V}{\Delta h} = 3\pi 3h^2 = 9\pi h^2 \tag{i}$$

(b) Next we express the volume in terms of the radius alone,

$$V = \frac{\pi}{3} r^2 h = \frac{\pi}{3} r^2 \frac{r}{3} = \frac{\pi r^3}{9}$$

Therefore

$$D_r V = \lim_{\Delta r \to 0} \frac{\Delta V}{\Delta r} = \frac{3\pi r^2}{9} = \frac{\pi r^2}{3}$$

or in terms of h, $D_r V$ becomes

$$D_r V = \pi \frac{(3h)^2}{3} = 3\pi h^2 \tag{ii}$$

Note that comparison of (i) with (ii) reveals that

$$D_h V = 3 D_r V \qquad \text{(iii)}$$

This is not surprising because $D_h V = D_r V \, D_h r$ by the chain rule. However $r = 3h$, which yields $D_h r = 3$ so that (iii) follows. This analysis serves as a partial check on our calculations. ●

Exercises 4.1

1. Find the rate of change of the area of a circle with respect to its radius r. Evaluate this rate of change when $r = 5$ in.

2. Find the rate of change of the area of a circle with respect to its diameter b; that is, $b = 2r$. Evaluate this rate of change when $r = 5$ in.

3. Use the chain rule to verify that the results of Exercises 1 and 2 should be in the ratio of 2 to 1.

4. Find the rate of change of the volume of a cube with respect to the length s of one of its sides. Evaluate this rate of change when $s = 10$ in.

5. Find the rate of change of the area of a square with respect to the length u of one of its diagonals. Evaluate this rate when $u = 8$ in.

6. Find the rate of change of $y = \dfrac{1}{x^3}$ with respect to x if $x = -2$.

7. The radius r and height h of a cylinder vary in such a way that the volume $V = \pi r^2 h$ remains fixed. Find the rate of change of r with respect to h when $r = 2h$.

8. Two like electric charges repel each other in accordance with Coulomb's law $F = \dfrac{k}{x^2}$ where $k > 0$ is a constant. Find the average rate of change of F with respect to x when x changes from $x = x_1$ to $x = 2x_1$. Also find the instantaneous rate of change of F with respect to x at $x = x_1$.

9. What is the rate of change of the reciprocal of the fourth power of a number with respect to the number when the number is 2?

10. What is the rate of change of the reciprocal of the fourth power of a number with respect to the square of the number when the number is 2?

11. The volume $V(\text{ft}^3)$ of a sphere of radius $r(\text{ft})$ is given by $V = \frac{4}{3}\pi r^3$. Find (a) the rate of change of V with respect to the radius; (b) the rate of change of V with respect to the diameter.

12. The radius of a right circular cylinder is always one half the altitude. Find the rate of change of the volume $V = \pi r^2 h$ (a) with respect to the altitude h; (b) with respect to the radius r.

13. The cost, C, in dollars to produce x TV units is given by $C = 100x + \dfrac{96}{x}$. What is the average cost per unit produced (also known as the average marginal cost) for the units from $x = 16$ to $x = 24$?

14. The average cost of producing an item is defined to be the total cost divided by the number of items produced. The cost function for producing x refrigerators is $C(x) = 100x + \dfrac{12000}{x}$, where C is in dollars. Find the average cost for producing (a) the first 10; (b) the first 100.

15. Water is pouring into a cylindrical tank whose radius is 4 ft at the rate of 32 ft³/min. How fast is the water level rising? (*Hint:* Express the depth of the water h as a function of time t and then find $D_t h$.)

16. Boyle's law for a confined gas kept at constant temperature is $PV = C$ where the pressure P is the number of pounds per square unit of area, V is the number of cubic units in the volume of the gas, and C is a constant. Find the instantaneous rate of change of the pressure with respect to the volume that is, $D_V P$ when $P = 3$ and $V = 5$.

17. Determine the average speed of a car for a round trip if it averages 60 mi/hr going and 30 mi/hr returning. (*Note:* the answer is *not* 45 mi/hr.)

18. An oil tank is being emptied. If there are Q gal of oil in the tank at time t, where $Q(t) = 60{,}000 - 6000t + 150t^2$ and t is measured in minutes, what is the average rate of outflow (a) during the first 10 min? (b) in the last 10 min?

4.2   RELATED RATES

In problems of physics, geometry, or economics it frequently occurs that two quantities are implicitly related so that each may be considered as a function of the other. Furthermore, each quantity is also a differentiable function of the time t. In such a situation, if the time rate of change of one of these quantities is known, we can find the rate of change of the other quantity with respect to time without needing to know explicitly how either quantity depends on the time.

More generally, when two or more quantities are related implicitly by an equation and each is a differentiable function of time, their derivatives are also related by an equation obtained from the original equation by implicit differentiation with respect to the time. This relation between the derivatives of the quantities is called **related rates.**

EXAMPLE 1   A conical tank, with vertical axis of symmetry open at the top and with vertex at the bottom, has base radius 4 ft and altitude 10 ft. If water is flowing into the tank at 9 ft³/min, how fast is the level rising when the water is 5 ft deep?

Solution   Referring to Figure 4.2.1, we let r and h be, respectively, the radius of the circle on the water surface and the height of the water at time t. If V is the volume of water in the tank, then, by the formula for the volume of a cone,

$$V = \tfrac{1}{3}\pi r^2 h$$

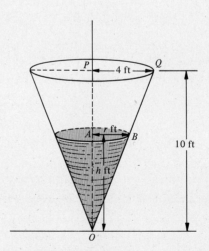

Figure 4.2.1

The fact that water is flowing into the tank at the rate of 9 ft³/min means that $D_t V = 9$ ft³/min. We are asked to find $D_t h$ when $h = 5$ ft. To do this, we first express V in terms of h. Since triangles OAB and OPQ are similar, we have

$$\frac{r}{h} = \frac{4}{10} \qquad \text{or} \qquad r = \frac{2}{5}h$$

Then

$$V = \frac{\pi}{3}\left(\frac{2}{5}h\right)^2 h = \frac{4\pi}{75}h^3$$

We differentiate implicitly with respect to t, getting

$$D_t V = \frac{4\pi}{75}\, 3h^2\, D_t h = \frac{4\pi h^2}{25}\, D_t h$$

Solving for $D_t h$, we have

$$D_t h = \frac{25\, D_t V}{4\pi h^2} = \frac{(25)(9)}{4\pi h^2} = \frac{225}{4\pi h^2}$$

In particular

$$D_t h\,\bigg|_{h=5} = \frac{225}{(4\pi)(25)} = \frac{9}{4\pi}\ \text{ft/min} \qquad \bullet$$

EXAMPLE 2   A ladder 26 ft long is leaning against a vertical wall. If the bottom of the ladder is pulled horizontally away from the wall at a uniform rate of 2 ft/sec, what is the velocity of the top of the ladder when the lower end is 24 ft from the wall?

Solution   We refer to Figure 4.2.2 where we let t = number of seconds elapsed since the ladder began to slide, x = number of feet from the bottom of the ladder to the wall at t sec, and y = number of feet from the top of the ladder to the ground at t sec.

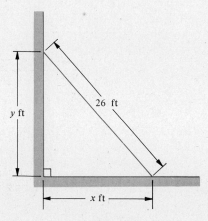

26 ft

y ft

x ft

Figure 4.2.2

Since the ladder is being pulled horizontally away from the wall at the constant rate of 2 ft/sec, we have $D_t x = 2$. We seek $D_t y$ when $x = 24$ ft. Now, from the theorem of Pythagoras, $x^2 + y^2 = (26)^2 = 676$ so that when $x = 24$, $y = 10$.
    Next we differentiate implicitly, with respect to t, the relationship between x and y to obtain

$$2x\, D_t x + 2y\, D_t y = 0 \qquad \text{or} \qquad D_t y = \frac{-x}{y}\, D_t x$$

Substitution of $D_t x = 2$, $x = 24$, and $y = 10$ yields

$$D_t y = \frac{-24}{10}(2) = -4.8 \text{ ft/sec}$$

Thus, the top of the ladder is moving downward with a speed of 4.8 ft/sec. The significance of the minus sign is that y is decreasing as t increases so that $D_t y$ must be less than zero. ●

EXAMPLE 3　A man 5 ft tall walks at the rate of 4 ft/sec directly away from a lamp 15 ft above the ground. Find the rate at which the end of his shadow is moving when he is 30 ft away from the lamp post.

Solution　We refer to Figure 4.2.3 where the man and the lamp post are shown. Let x be the distance from the man to the base and y be the length of his shadow. We are given $D_t x = 4$ and we seek $D_t(x + y)$, since the tip of his shadow is at distance $x + y$ from the base of the lamp post. Thus, all we need obtain is $D_t y$ because $D_t(x + y) = D_t x + D_t y$. In general, from similar triangles: $\dfrac{y}{5} = \dfrac{y + x}{15}$ or $y = \dfrac{x}{2}$.

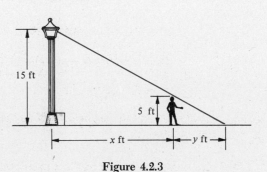

Figure 4.2.3

Therefore $D_t y = \frac{1}{2} D_t x$, which implies that $D_t y = 2$.

It follows that the end of his shadow is moving at the rate of 6 ft/sec and this rate is independent of the distance the man is from the base of the lamp post. ●

EXAMPLE 4　An automobile travelling at a rate of 60 ft/sec is approaching an intersection. When the automobile is 1200 ft from the intersection, a second automobile moving at a rate of 40 ft/sec, crosses the intersection travelling in a direction at right angles to the path of the first vehicle. How fast are the two automobiles separating 10 sec after the second vehicle crosses the intersection?

Solution　Let A and B be the two automobiles t seconds after the second vehicle, B, crosses the intersection O (Figure 4.2.4). Then $|\overline{OA}| = 1200 - 60t$ and $|\overline{OB}| = 40t$ (both measured in feet) so that the distance s between them is given by

$$s^2 = (40t)^2 + (1200 - 60t)^2$$

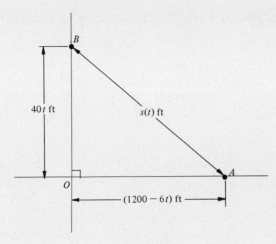

Figure 4.2.4

We seek $s' = D_t s$ when $t = 10$.

Differentiation of both sides with respect to t yields

$$2ss' = 2(40t)40 + 2(1200 - 60t)(-60)$$

or

$$ss' = 1600t - 72000 + 3600t = 5200t - 72000$$

$$s' \Big|_{t=10} = \frac{52000 - 72000}{s|_{t=10}} = \frac{-20000}{\sqrt{(400)^2 + (600)^2}} = \frac{-200}{\sqrt{52}} = \frac{-200}{7.211}$$

$$\doteq -27.7 \text{ ft/sec}$$

that is, the two automobiles are approaching each other at the rate of 27.7 ft/sec at that particular instant. ●

There are occasionally related rates problems that can be done more easily if one first finds a rate of change that is not actually requested but can easily be expressed in terms of given quantities. This method is indicated in the following example.

EXAMPLE 5   A right circular cylinder has a constant height of 5 in. If the volume is increasing at the rate of 8 in.³/min, how fast is the lateral surface area increasing when the radius of the base is 4 in.?

Solution   If r is the base radius and h is the height of the cylinder, then volume $= V = \pi r^2 h$ and lateral surface area $= A = 2\pi rh$.

We are given that $D_t V = 8$ in.³/min and are asked to find $D_t A$ when $r = 4$. Since $h = 5$ throughout the problem, we have

$$V = 5\pi r^2 \qquad \text{and} \qquad A = 10\pi r.$$

Differentiating the first equation implicitly with respect to t gives $D_t V = 10\pi r\, D_t r$ or, solving for $D_t r$,

$$D_t r = \frac{D_t V}{10\pi r} = \frac{8}{10\pi r}$$

Then

$$D_t r \bigg|_{r=4} = \frac{8}{40\pi} = \frac{1}{5\pi} \text{ in./min}$$

From $A = 10\pi r$, we have $D_t A = 10\pi D_t r$. Therefore

$$D_t A \bigg|_{r=4} = 10\pi \frac{1}{5\pi} = 2 \text{ in.}^2/\text{min}$$

This problem could also have been done by first expressing V in terms of A (eliminating r) and then differentiating implicitly with respect to t. ●

Exercises 4.2

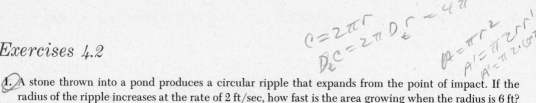

1. A stone thrown into a pond produces a circular ripple that expands from the point of impact. If the radius of the ripple increases at the rate of 2 ft/sec, how fast is the area growing when the radius is 6 ft?

2. A man 6 ft tall walks away from a 24-ft lamp post at the rate of 5 ft/sec. How fast is his shadow lengthening when he is 30 ft from the lamp post?

3. The volume of a spherical balloon is increasing at the constant rate of 5 ft³/min. At what rate is the radius increasing when the radius is 10 ft?

4. A cylinder is expanding in such a way that the height and radius increase at the rates of 1 % and 2 %/day, respectively. At what rate is the volume increasing?

5. The surface area of a cube is changing at the rate of K in.²/sec. How fast is the volume changing when the surface is A in.²?

6. A ship steams along a course parallel to a straight beach, always maintaining a distance of 9 mi from the shore line. The ship is travelling with uniform speed of 10 mi/hr and is approaching a radar station located on the beach. At what rate is it approaching the radar station at the instant its distance from the station is 15 mi?

7. A baseball player runs from homeplate to first base at a rate of 30 ft/sec. At what rate is his distance from third base changing when he is 60 ft from first base? (A baseball diamond is a square whose sides are 90 ft.)

8. One ship is 24 mi north of a certain point O and is steaming north at 20 mi/hr. Another ship is at that moment 7 mi east of point O and is travelling west at 15 mi/hr. Find the rate at which the distance between the ships is changing at that instant.

9. A light source 9 ft above the ground is moving with a velocity of 4 ft/min toward a 4-ft wall (Figure 4.2.5). How fast is the shadow receding when (a) the light is 8 ft from the top of the wall? (b) the light is 6 ft from the top of the wall?

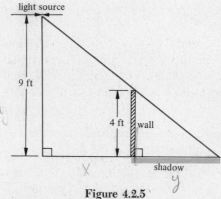

light source

9 ft

4 ft   wall

shadow

Figure 4.2.5

10. The force of repulsion between two like charges is inversely proportional to the square of the distance s between them. It is found that the force is 6 dynes when the distance between the charges is 3 cm. One charge is held stationary while the other is moved away at the rate of 0.2 cm/sec. Find the rate at which the force is changing when $s = 3$ cm.

11. A 25-ft length of metal pipe is leaning against a vertical wall. If the bottom of the pipe is pulled along the level pavement, directly away from the wall at 3 ft/sec, how fast is the height of the midpoint of the pipe decreasing when the foot of the pipe is 20 ft from the wall?

12. Water is poured at 3 ft³/min into a hemispherical container that is 16 ft in diameter. At what rate is the liquid surface rising when the depth is 4 ft? [*Note:* The volume of a spherical segment of altitude h is $\pi h^2 (3r - h)/3$ for a sphere of radius r. This will be proved later.]

13. A particle is moving along the curve $4x^2 + y^2 = 36$. Determine how fast the y coordinate is changing when the particle is at $(2, 2\sqrt{5})$ and it is given that $D_t x = 3$ in./sec.

14. A spherical snowball is melting in such a way that the rate of change of its volume is proportional to its surface area. Show that the rate of change of its radius is constant.

15. The pressure and volume of a gas are related by the formula $PV^n = k$ where n and k are constants. Find a formula for the ratio of $D_t P/P$ to $D_t V/V$.

16. For any thin convex lens, the focal length f, the distance u of the object from the lens, and the image distance v are related by the equation $\dfrac{1}{u} + \dfrac{1}{v} = \dfrac{1}{f}$. Find the rate of change of v with respect to time for a lens of focal length 15 in. when $u = 5$ ft and u is changing at the rate of 20 in./sec.

17. In order to produce w units of a certain commodity, z workers are needed. Furthermore the quantities w and z are related by $w = 16z^2$. If the current production is 160,000 units/year and is increasing at the rate of 32,000 units/year, what is the current rate at which the labor force should be increased?

18. The cost in dollars of selling x items is given by $C = 300 + 3x + \dfrac{1000}{x}$. When the 50th item is sold, it is noted that the rate of sales is 10 items/hr. What is the rate of change of cost per hour at that moment?

19. A trough 20 ft long has cross sections that are equilateral triangles with vertical axes of symmetry and vertices at the bottom. Water is flowing into the trough at the rate of K ft³/min. Find K if the water level is rising at the rate of 3 in./min when the depth is 36 in.

20. At a certain instant of time the length, width, and height of a rectangular parallelepiped are a, b, and c in. Find the relative rate of increase of the volume V, that is $\dfrac{D_t V}{V}$ in terms of the relative rate of increase of the dimensions of the edges.

21. If $z = x^2 + 3xy^2$ find $D_t z$ when $x = 2$, $y = 3$, $D_t x = -1$, and $D_t y = 2$.

22. A dock stands 12 ft above the deck of a boat. The boat is being pulled towards the dock by means of a rope attached to the deck at the front of the boat. If 3 ft of rope are drawn in each minute, at what rate is the boat moving towards the dock when the boat is 16 ft away?

4.3   MAXIMUM AND MINIMUM VALUES OF A FUNCTION

This and the next section will be concerned with the tremendously important topic of determining the largest and smallest values of a function defined on a prescribed domain. We will start by stating and discussing a theorem that is intuitively plausible but whose proof is generally part of an advanced calculus course.

Theorem 1   (*Extreme Value Theorem*). If f is a continuous function throughout the closed interval $[a, b]$, then f takes on a largest value (absolute or global maximum) and a smallest value (absolute or global minimum).

This means that if f is continuous in $[a, b]$ there exists at least one number X in $[a, b]$ for which $f(x)$ takes on a maximum value, that is

$$f(X) \geq f(x) \text{ for all } x \text{ in } [a, b]$$

Similarly, f takes on a minimum value for at least one value of x in $[a, b]$.

EXAMPLE 1   Given

$$f(x) = \frac{3x}{2} \quad \text{in } [2, 6]$$

closed including end pts

Find the absolute extrema of f if any.

Solution   The function is continuous in the given closed interval. Thus the extreme value theorem applies and f must possess an absolute maximum and an absolute minimum in $[2, 6]$. A sketch of the function is shown in Figure 4.3.1. Since it is linear and has a positive slope, the extrema occur at the ends of the interval with $f(2) = 3$ and $f(6) = 9$ the absolute minimum and maximum, respectively.   ●

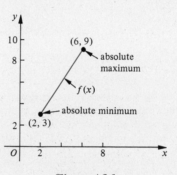

Figure 4.3.1

EXAMPLE 2   Given

$$f(x) = \frac{3x}{2} \quad \text{in } (2, 6]$$

Find the absolute extrema of f if any.

Solution   First we note that the interval is open at one end and closed at the other. Therefore the extreme value theorem cannot be applied. Actually the function does have an absolute maximum at $x = 6$, $f(6) = 9$; however, there is no absolute minimum. The $\lim_{x \to 2^+} f(x) = 3$ and yet the function does *not* take on the value 3.

●

EXAMPLE 3   Given

$$f(x) = |x| \quad \text{in } [-3, 2]$$

Find the absolute extrema if any.

Solution   The function is shown in Figure 4.3.2. Since $|x|$ is continuous (for all x) it follows that the given function is continuous on the closed interval. Therefore the extreme

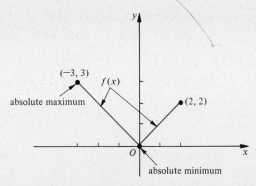

Figure 4.3.2

value theorem applies. Since the $|x| \geq 0$, the absolute minimum is zero and occurs at $x = 0$ (an interior point in the given interval). Also the absolute maximum value occurs at $x = -3$ and is $|-3| = 3$ (at the left end of the interval). ●

EXAMPLE 4   Given

$$f(x) = \begin{cases} \dfrac{1}{x-2} & x \neq 2 \\ 0 & x = 2 \end{cases}$$

Find the absolute extrema of f on the interval $[0, 4]$, if any.

Solution   A sketch of the graph of f is shown in Figure 4.3.3. The function f fails to be continuous at $x = 2$. In fact $\lim\limits_{x \to 2^+} f(x) = \infty$ whereas $\lim\limits_{x \to 2^-} f(x) = -\infty$. Since $f(x)$ can be made as large positively and negatively as desired by choosing any neighborhood about $x = 2$, it follows that there is neither an absolute maximum nor an absolute minimum in $[0, 4]$. ●

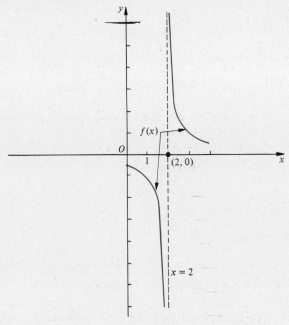

Figure 4.3.3

Example 5 Given

$$f(x) = \begin{cases} x & \text{if } 0 \le x < 1 \\ x + 1 & \text{if } 1 \le x \le 2 \end{cases}$$

Find the absolute extrema of f on $[0, 2]$, if any.

Solution A sketch of the graph of f is shown in Figure 4.3.4. The function f is not

Def. of — continuous at $x = 1$ since $\lim\limits_{x \to 1^-} f(x) = 1 \ne f(1)$ and therefore the extreme value

Continuity theorem does not apply. However, there is an absolute maximum and an absolute minimum of f on $[0, 2]$, namely, $f(2) = 3$ and $f(0) = 0$, respectively. Therefore although the extreme value theorem, when applicable, does imply that f attains a largest and a smallest value in the interval, it is not necessary to meet the continuity requirements in order for an extremum to exist. ●

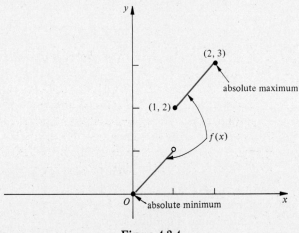

Figure 4.3.4

Example 6 Given

$$f(x) = x^3 - 6x^2 + 9x + 2 \qquad \text{on } [\tfrac{1}{2}, \tfrac{9}{2}]$$

Find the absolute maximum and minimum of f, if any.

Partial Since f is a polynomial, it must be continuous in any interval. Therefore the
Solution extreme value theorem applies and we are guaranteed that f must have absolute extrema in the interval $\tfrac{1}{2} \le x \le \tfrac{9}{2}$. But where? The theorem certainly gives no indication as to how to determine the values of x which extremize the function.

Theorem 2 will enable us to answer this equation, and we will then return to Example 6. ●

Definition 1 A function f is said to have a **_relative maximum_** at $x = c$ if there exists an open interval containing $x = c$ on which $f(c) \ge f(x)$ for all x in this interval.

Figures 4.3.5 and 4.3.6 show sketches of a portion of $f(x)$ having a relative maximum at $x = c$.

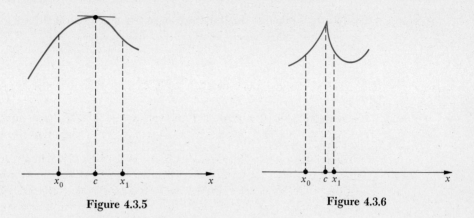

Figure 4.3.5

Figure 4.3.6

Definition 2 A function f is said to have a **_relative minimum_** at $x = c$ if there exists an open interval containing $x = c$ on which $f(c) \leq f(x)$ for all x in this interval.

Figures 4.3.7 and 4.3.8 show sketches of a portion of $f(x)$ having a relative minimum at $x = c$.

If the function has either a relative maximum or a relative minimum at $x = c$ it is said to have a **_relative extremum_** at $x = c$.

Now let us assume that the function f is differentiable at $x = c$ and that f takes on a relative maximum or minimum at $x = c$. If we look at Figure 4.3.5 and 4.3.7 it appears that at such a point the tangent must be horizontal; that is, $f'(c) = 0$. We now state and prove this analytically.

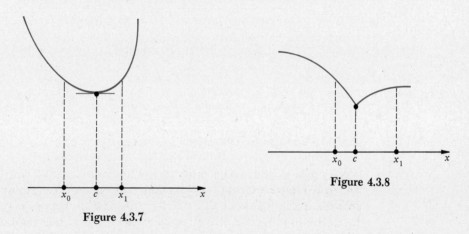

Figure 4.3.7

Figure 4.3.8

Theorem 2 Let f be a function defined on an interval. If f takes on a relative maximum (or minimum) at an interior point $x = c$ and if $f'(c)$ exists, then $f'(c) = 0$.

Proof The plan is to show that $f'(c) \leq 0$ and also that $f'(c) \geq 0$. This then implies that $f'(c) = 0$ must hold.

To this end, consider the difference quotient

$$\frac{f(c + h) - f(c)}{h}$$

where $|h|$ is taken so small that we are inside a neighborhood of c for which $f(c)$ is an absolute maximum. Therefore it follows that

$$f(c + h) \leq f(c)$$

or

$$f(c + h) - f(c) \leq 0$$

Therefore when h is positive,

$$\frac{f(c + h) - f(c)}{h} \leq 0$$

Thus, if we take

$$\lim_{h \to 0^+} \frac{f(c + h) - f(c)}{h}$$

the result must be ≤ 0 because the limit of nonpositive numbers cannot be positive. (Prove this last assertion.) Thus

$$f'(c) \leq 0 \qquad (i)$$

If, on the contrary, h is negative, the denominator of $[f(c + h) - f(c)]/h$ is negative while the numerator is still ≤ 0. Therefore for $h < 0$,

$$\frac{f(c + h) - f(c)}{h} \geq 0$$

Since the limit of nonnegative numbers is nonnegative let $h \to 0^-$ and we conclude that

$$f'(c) \geq 0 \qquad (ii)$$

Thus $0 \leq f'(c) \leq 0$ or $f'(c) = 0$.

It is left as an exercise for the reader to establish that at a relative minimum point of a differentiable function the derivative must be zero. ∎

The geometric interpretation of Theorem 2 is that, if f has a relative extremum at c and if $f'(c)$ exists, the graph of $y = f(x)$ must have a horizontal tangent at the point where $x = c$.

If f is a differentiable function, the only possible values of x for which f can have a relative extrema are those (and only those) for which $f' = 0$. However, the converse is not true, namely, that if $f'(c) = 0$ where $a < c < b$ then $f(x)$ must have a relative extremum at $x = c$. For example, consider the function f defined by

$$f(x) = x^3 \qquad -1 \leq x \leq 2.$$

This function is graphed in Figure 4.3.9. If we form f', then $f'(x) = 3x^2$ so that if we set $f' = 0$ the only solution of $3x^2 = 0$ is $x = 0$. But $f(0) = 0$ while $f(x) > 0$ if $x > 0$ and $f(x) < 0$ if $x < 0$. Therefore f does not have relative extremum at $x = 0$. Furthermore, f may have a relative extremum at a number and f' may fail to exist there. For example, let us go back to $f(x) = |x|$ in $[-3, 2]$. This function has a relative (and absolute) minimum at $x = 0$ even though $f'(0)$ fails to exist.

The situation may be summarized as follows: If a function f is defined at a number c, a *necessary condition that f possesses an extremum at c is that* (i)

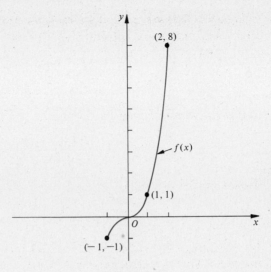

(2, 8)

f(x)

(1, 1)

O

x

(−1, −1)

Figure 4.3.9

$f'(c) = 0$ or (ii) $f'(c)$ *fails to exist.* Such numbers c are called ***critical numbers*** of f. Of course, if we know that f is differentiable, the only critical numbers are the roots of the equation $f'(c) = 0$.

Let us now return to the question of determining the absolute extrema (that is, absolute maximum and minimum) values of a continuous function defined on a closed interval. The absolute extrema must be either a relative extremum of a functional value or at an end point of the interval. But the necessary condition for a relative extremum at a point c is that $x = c$ be a critical point. Thus we have the following procedure for locating absolute extrema:

1. Find the critical numbers of f on $[a, b]$.
2. Find the functional values at these critical numbers.
3. Find $f(a)$ and $f(b)$.
4. The largest of the values from steps (2) and (3) yields the absolute maximum whereas the smallest of the values gives the absolute minimum.

We return to Example 6 for which we had only a partial solution.

EXAMPLE 6   Given

$$f(x) = x^3 - 6x^2 + 9x + 2 \qquad \text{on } [\tfrac{1}{2}, \tfrac{9}{2}]$$

find the absolute maximum and minimum of f.

Solution   Since $f(x)$ is differentiable, the only critical values are the roots of $f' = 0$. Now

$$f' = 3x^2 - 12x + 9 = 3(x - 3)(x - 1)$$

1. Set $f' = 0$ and find critical numbers $c_1 = 1$ and $c_2 = 3$, both interior points of the given integral.
2. $f(1) = 1 - 6 + 9 + 2 = 6$
   $f(3) = 27 - 54 + 27 + 2 = 2$

3. $f\left(\dfrac{1}{2}\right) = \dfrac{1}{8} - \dfrac{3}{2} + \dfrac{9}{2} + 2 = 5\tfrac{1}{8}$

$f\left(\dfrac{9}{2}\right) = \dfrac{729}{8} - \dfrac{243}{2} + \dfrac{81}{2} + 2 = \dfrac{97}{8} = 12\tfrac{1}{8}$

4. The absolute maximum is $12\tfrac{1}{8}$ at $x = 4\tfrac{1}{2}$ while the absolute minimum is 2 at $x = 3$. A sketch of the function is shown in Figure 4.3.10. ●

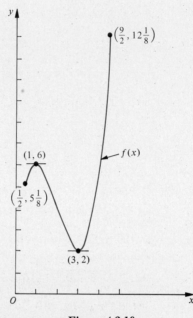

Figure 4.3.10

In all of the examples looked at so far, the interval was bounded. If we are working on an unbounded interval, we must then check what happens to the function as $x \to -\infty$ or $x \to \infty$ (or both). The following example is such an illustration.

EXAMPLE 7 Find the absolute extrema of $f(x) = 6x - x^2$ on $[1, \infty)$ if they exist.

Solution $f'(x) = 6 - 2x$

Setting $f'(x) = 0$ gives $x = 3$ as the only critical number.

$$f(3) = (6)(3) - (3)^2 = 9,$$

$$f(1) = (6)(1) - (1)^2 = 5,$$

$$\lim_{x \to \infty} (6x - x^2) = \lim_{x \to \infty} x(6 - x) = -\infty$$

Thus, $f(3) = 9$ is the absolute maximum value. There is no absolute minimum value since $f(x) \to -\infty$ as $x \to \infty$. A graph of this function is drawn in Figure 4.3.11. ●

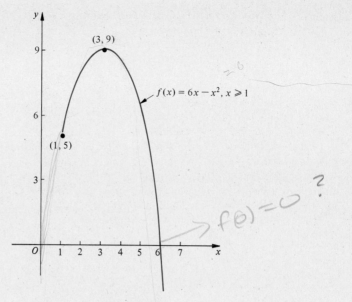

$f(x) = 6x - x^2,\ x \geqslant 1$

Figure 4.3.11

Exercises 4.3

In Exercises 1 through 24 find the absolute maximum and absolute minimum of the given function over the given interval if there are any.

1. $f(x) = 3x + 2,\ \ [-3, 5]$

2. $g(x) = 4 - 5x,\ \ [0, 6]$

3. $h(x) = 4x,\ \ [0, 3)$

4. $F(x) = 7 - x,\ \ (-2, 8)$

5. $G(x) = 3x - x^2,\ \ [0, 4]$

6. $H(x) = x^2 - 2x + 3,\ \ [-2, 3]$

7. $f(x) = 2x^2 - 4x - 6,\ \ (0, 5)$

8. $h(x) = x^3 - 12x + 2,\ \ (-2, 3]$

9. $F(x) = x^{-1},\ \ [-3, -1]$

10. $G(x) = x^2 - 6x + 10,\ \ (-\infty, \infty)$

11. $H(x) = \sqrt{x - 5},\ \ [5, \infty)$

12. $f(x) = \sqrt{36 - x^2},\ \ [-6, 6]$

13. $g(x) = \dfrac{x}{x^2 + 2},\ \ [-1, 4]$ class

14. $H(x) = \dfrac{x}{4} + \dfrac{1}{x},\ \ [1, 5]$

15. $F(y) = y\sqrt{1 - y},\ \ [0, 1]$

16. $f(x) = \dfrac{x}{1 + x^2},\ \ (-\infty, \infty)$

17. $g(x) = 1 + 3(x - 2)^{2/3},\ \ [0, 4]$

18. $h(y) = y[\![y]\!],\ \ [0, 2)$

19. $F(x) = \begin{cases} 4 - 2x & 0 \leq x \leq 2 \\ \dfrac{x}{2} & 2 < x \leq 3 \end{cases}$

20. $G(x) = x\sqrt{a^2 - x^2},\ \ [-a, a]$

21. $H(u) = \dfrac{u}{u^2 - u + 1},\ \ (-\infty, \infty)$

22. $H(x) = |x^2 - 25|,\ \ (-\infty, \infty)$

23. $F(x) = \begin{cases} \dfrac{x}{x - 1} & x \neq 1 \\ 0 & x = 1 \end{cases}\ \ [0, 6]$

24. $f(x) = x^{2/3}(x - 2)^2,\ \ [-1, 3]$

25. Find constants a, b, and c so that the curve $y = ax^2 + bx + c$ passes through $(0, 1)$ and has a relative extremum at $(1, -2)$.

26. Show that the condition that a triangle be isosceles is a necessary condition but not a sufficient condition for the triangle to be equilateral.

27. In order that two triangles be similar, it is sufficient but not necessary that the triangles be congruent. Is this valid? Prove your assertion.

28. Compare or contrast the following:
    (a) If A is true, then B is true.
    (b) B is a necessary condition for A.
    (c) A is true only if B is true.
    (d) A is a sufficient condition for B.

29. Comment on the assertion "if a continuous function defined on $[a, b]$ possesses an absolute maximum at x_0 such that $a < x_0 < b$ then $f'(x_0) = 0$."

30. Discuss the character of the point $(0, 0)$ on the curve $y = x^n$, $(-\infty, \infty)$, for each positive integer n.

31. Without using calculus find the minimum of $g(x) = (x + 6)^4 + 19$.

32. Without using calculus find the minimum of $h(x) = x^4 - 4x^3 + 6x^2 - 4x - 7$. (*Hint:* Observe the coefficients and put $h(x)$ into more compact form.)

4.4 APPLICATIONS OF MAXIMA AND MINIMA

The results of the last section will now be applied to a number of interesting and practical situations. This is best illustrated by working out examples.

EXAMPLE 1   Find two *positive* numbers whose sum is 50 if their product is to be maximized.

Solution   If x denotes the first then $50 - x$ is the second where $x > 0$ and $50 - x > 0$ or, equivalently, $0 < x < 50$. Let $P(x)$ denote the product so that

$$P = x(50 - x) = 50x - x^2$$

We seek the value of x in $(0, 50)$ that maximizes the differentiable function P. To this end we form

$$P' = 50 - 2x$$

and for critical values set $P' = 0$. Thus $x = 25$ is the only critical value. But $P(25) = (25)^2 = 625$ while $P \to 0$ as $x \to 0^+$ and $x \to 50^-$ so that $P(25) = 625$ is the absolute maximum of P in $(0, 50)$.   ●

EXAMPLE 2   What is the largest possible area of a rectangular rug whose perimeter is 100 ft?

Solution   In Figure 4.4.1 let x and y denote the sides of the rectangle. The perimeter of the rectangle is

$$2x + 2y = 100 \qquad \text{or} \qquad x + y = 50$$

and the area is

$$A = xy$$

But $y = 50 - x$ so that

$$A(x) = x(50 - x) = 50x - x^2$$

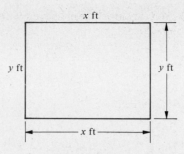

Figure 4.4.1

where $0 < x < 50$. Thus we see that this problem is mathematically identical to that of the previous example. Therefore $x = 25$ maximizes A, and if $x = 25$ then $y = 25$ also. Consequently the largest rectangle would be a square in this case with area of 625 ft.^{2} ●

EXAMPLE 3   The stiffness of a beam is a measure of its resistance to being bent or deflected by applied forces transverse to its longitudinal axis. If the strength of a beam of rectangular cross section is proportional to the breadth and to the cube of the depth, find the shape of the stiffest rectangular beam that can be cut from a circular cylindrical log of radius a.

Solution   In Figure 4.4.2 we have the rectangular cross section inscribed in the circle of diameter $2a$. Let x be the breadth and y the depth (the y direction is parallel to the applied loads). The stiffness S of the beam is given by

$$S = kxy^3$$

where k is a constant of proportion. Now from the theorem of Pythagoras, $y^2 = 4a^2 - x^2$ so that

$$S(x) = kx(4a^2 - x^2)^{3/2} \qquad \text{where } 0 < x < 2a \text{ and } 0 < y < 2a$$

Thus

$$S'(x) = k[(-3x^2)(4a^2 - x^2)^{1/2} + (4a^2 - x^2)^{3/2}]$$

and if we set $S' = 0$ and factor out $(4a^2 - x^2)^{1/2} \neq 0$, then we obtain $x = a$ and $y = a\sqrt{3}$. The ratio of the depth, to the width is $y/x = \sqrt{3}$.

Since $S \to 0$ as $x \to 0$ or as $x \to 2a$, it follows that the preceding dimensions will indeed maximize the strength of the beam.   ●

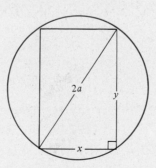

Figure 4.4.2

EXAMPLE 4   An open box is constructed by removing a small square from each corner of a tin sheet and then folding up the sides (see Figure 4.4.3). If the sheet is L in. on each side, what is the largest possible volume of the box?

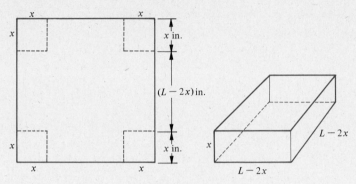

Figure 4.4.3

Solution   Let each cut out square have side x. We must express the volume of the box as a function of x:

$$V(x) = (L - 2x)^2 x = L^2 x - 4Lx^2 + 4x^3 \qquad 0 < x < \frac{L}{2}$$

$$V'(x) = L^2 - 8Lx + 12x^2 = (L - 2x)(L - 6x)$$

But $L - 2x > 0$ so that when we set $V' = 0$, the only critical value is $x = \dfrac{L}{6}$. Also

as $x \to 0^+$ and as $x \to \dfrac{L^-}{2}$, $V \to 0$. Thus $V(x)$ has its maximum value for $x = \dfrac{L}{6}$:

$$V\left(\frac{L}{6}\right) = \left(L - \frac{2L}{6}\right)^2 \frac{L}{6} = \left(\frac{2L}{3}\right)^2 \left(\frac{L}{6}\right) = \frac{2L^3}{27} \text{ in.}^3 \qquad \bullet$$

EXAMPLE 5   Determine the largest right circular cylinder that can be inscribed in a sphere of radius a.

Solution   Figure 4.4.4(a) shows a right circular cylinder in a sphere of radius a. Figure 4.4.4(b) illustrates a longitudinal section of the cylinder of radius r and altitude h. The volume V of the inscribed cylinder is to be maximized where

$$V = \pi r^2 h$$

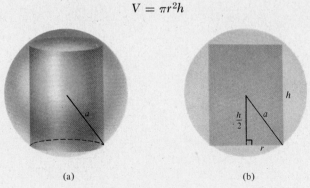

(a)   (b)

Figure 4.4.4

But r and h are related since $r^2 + \left(\dfrac{h}{2}\right)^2 = a^2$ so that r^2 may be eliminated easily and we obtain

$$V(h) = \pi\left(a^2 - \frac{h^2}{4}\right)h = \pi\left(a^2 h - \frac{h^3}{4}\right)$$

Differentiation with respect to h yields

$$V'(h) = \pi\left(a^2 - \frac{3h^2}{4}\right) \quad \text{what happens to } \pi$$

so that if we set $V' = 0$ we obtain

$$h = \frac{2}{\sqrt{3}}a \doteq 1.15a \qquad \text{and} \qquad \boxed{r \doteq 0.82a} \quad \text{how ?}$$

Since $V \to 0$ as $h \to 0^+$ and as $h \to 2a^-$ and V is positive in $0 < h < 2a$, then we must have a maximum volume when $h = \dfrac{2a}{\sqrt{3}}$ given by $\quad 0 < x < 2a$

$$V = \frac{\pi 2a}{\sqrt{3}}\left(a^2 - \frac{a^2}{3}\right) = \frac{4\pi\sqrt{3}}{9}a^3 \qquad \bullet$$

$h \quad r^2$

EXAMPLE 6   Given two points A and B on one side of a straight line find a point P on the line such that the sum of the distances $|\overline{AP}| + |\overline{PB}|$ is a minimum.

Solution   Refer to Figure 4.4.5. The given line is taken as the x-axis and the coordinates of points A and B are $(0, h)$ and (a, h_1), respectively. Then the sum of the two distances in question is given by

$$g(x) = \sqrt{x^2 + h^2} + \sqrt{(x - a)^2 + h_1^2}$$

Differentiation with respect to x yields

$$g'(x) = \frac{x}{\sqrt{x^2 + h^2}} + \frac{x - a}{\sqrt{(x - a)^2 + h_1^2}}$$

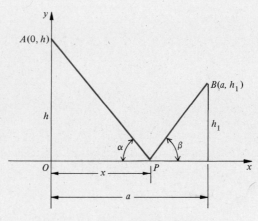

Figure 4.4.5

For critical values find c such that $g'(c) = 0$ or

$$\frac{c}{\sqrt{c^2 + h^2}} = \frac{a - c}{\sqrt{(c - a)^2 + h_1^2}}$$

But the left side is precisely $\cos \alpha$ and the right side is $\cos \beta$ so that at the critical value,

$$\cos \alpha = \cos \beta$$

which implies that $\alpha = \beta$. ●

The solution of this problem is closely connected with the **optical law of reflection.** It is known experimentally that a light ray emanating from A and striking a mirror at P will be reflected so that the angle of incidence is equal to the angle of reflection. Therefore, the time for the light ray to go from A to P to B will be least. This is an instance of Fermat's **principle of least time,** which has very wide application.

Actually we did not rigorously establish that an absolute minimum has been obtained. This can be done by calculus, but at this point we will go no further with this argument. Instead we will give an ingenious geometric proof due to Heron of Alexandria.

Reflect point B with respect to the given line L; that is, drop the perpendicular from B to the L-axis and extend it to B' so that $|\overline{BN}| = |\overline{NB'}|$ (Figure 4.4.6). Draw AB' intersecting line L in P_0. It is claimed that P_0 is the required point that minimizes $|\overline{AP}| + |\overline{PB}|$. From the figure, $|\overline{AP}| + |\overline{PB}| = |\overline{AP}| + |\overline{PB'}|$. Thus we wish to minimize $|\overline{AP}| + |\overline{PB'}|$. But $|\overline{AP}| + |\overline{PB'}| \geq |\overline{AP_0}| + |\overline{P_0B'}|$ since the shortest distance between the two points A and B' is the straight line connecting them. The equality holds if and only if P coincides with P_0. The fact that the incidence angle α equals the reflection angle β is easily established and is left to the reader.

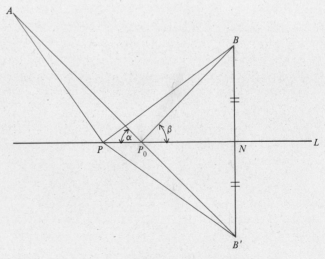

Figure 4.4.6

EXAMPLE 7 (SNELL'S LAW OF REFRACTION). A significant application of maxima and minima is to the derivation of *Snell's law of refraction.* It is again based upon the minimum principle of Fermat, which states that the time elapsed in the passage of light between two fixed points is an absolute minimum with respect to possible paths connecting the two points. To illustrate this concept, suppose that light travels with speed c_1 in air and with speed c_2 in water. In what follows consider only those paths that lie in the xy plane containing the two fixed points A and B (Figure 4.4.7).

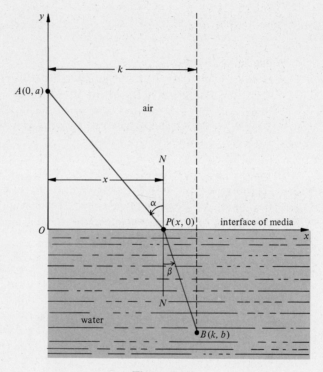

Figure 4.4.7

From the principle that the shortest distance between two points in a plane is a straight line connecting the points and that the two media are homogeneous, this shortest path must consist of two straight line segments intersecting at some point P on the line of separation of the two media. For convenience, we place the x-axis along this line and the y-axis through point A as shown in the figure. The lengths $|\overline{AP}|$ and $|\overline{PB}|$ are $\sqrt{a^2 + x^2}$ and $\sqrt{(k - x)^2 + b^2}$, respectively, and the time of passage is determined by dividing these distances by the corresponding speeds and adding. This yields for the total time T:

$$T(x) = \frac{\sqrt{a^2 + x^2}}{c_1} + \frac{\sqrt{(k - x)^2 + b^2}}{c_2} \qquad 0 \leq x \leq k$$

Differentiation with respect to x yields

$$T'(x) = \frac{x}{c_1 \sqrt{a^2 + x^2}} - \frac{k - x}{c_2 \sqrt{(k - x)^2 + b^2}}$$

For critical values we set $T'(x) = 0$, which implies that

$$\frac{x}{c_1\sqrt{a^2 + x^2}} = \frac{k - x}{c_2\sqrt{(k - x)^2 + b^2}}$$

or if α and β denote the respective angles between the path segments AP and PB and the normal NN to the interface $y = 0$:

$$\frac{\sin \alpha}{c_1} = \frac{\sin \beta}{c_2}$$

or

$$\frac{\sin \alpha}{\sin \beta} = \frac{c_1}{c_2}$$

which is Snell's law.

It will be established rigorously later that this yields an absolute minimum. Meanwhile, it is plausible physically that a minimum will not occur at the end points $x = 0$ and $x = k$ so that it must be at an interior point.

Finally the reader should note that this problem (unlike the previous example) is not easily disposed of without calculus. ●

EXAMPLE 8   A manufacturer of refrigerator–freezer units will, on the average, sell 800 units/month at \$400 per unit. If it has been determined that the company can sell an additional 100 units for each reduction of \$20 in price, what price should be set in order to bring the largest income?

Solution   Let x = number of \$20 reductions in price so that the price/unit = $400 - 20x$. Also it is anticipated that the number of units sold will be $800 + 100x$. If R stands for the total revenue then

Why multiply? $R(x) = (400 - 20x)(800 + 100x)$

and $R(x)$ is to be maximized. Multiplication yields

$$R(x) = 320{,}000 + 24{,}000x - 2000x^2$$

Differentiation yields

$$R'(x) = 24000 - 4000x$$

and by setting $R'(x) = 0$ we find that $x = 6$. This means that the price per unit that maximizes income is $400 - 20(6) = \$280$. It is easily verified that this yields an absolute maximum. ●

EXAMPLE 9   A covered box with a square base is to hold 270 yd^3. The material for the base will cost \$12 per yd^2 and the sides and the top will cost \$8 per yd^2. Find the dimensions of the box that minimize the cost.

Solution   Let x yd be the side of the square base and y yd be the height. (Figure 4.4.8) The cost of each rectangular surface is the cost per square yard times the area of that surface. Thus the cost C is (for bottom, sides, and top surfaces).

$$C = 12x^2 + 8(4xy + x^2)$$
$$= 20x^2 + 32xy$$

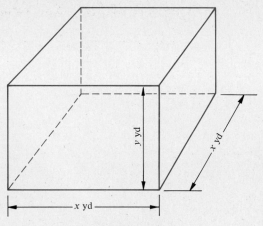

Figure 4.4.8

However $x^2y = 270$, so that $y = \dfrac{270}{x^2}$ and therefore

$$C(x) = 20x^2 + \frac{8640}{x}$$

Differentiation yields

$$C'(x) = 40x - \frac{8640}{x^2}$$

and if we set $C' = 0$ then

$$x^3 = 216 \qquad \text{or} \qquad x = 6 \text{ yd} \qquad \text{and} \qquad y = 7\tfrac{1}{2} \text{ yd}$$

It is left to the reader to verify that the cost is actually minimized.   ●

In most application problems, the absolute extreme value we are looking for occurs at a critical number (it did in all of our examples so far). It is, however, possible for such a value to occur at an end point of the interval under consideration, and we must take this into account in our work.

EXAMPLE 10   Find two nonnegative numbers whose sum is 50 if their product is to be minimized.

Solution   This is a variation of Example 1. Letting x and $50 - x$ be the numbers, we have $0 \leq x \leq 50$.

$$P = x(50 - x) = 50x - x^2.$$
$$P' = 50 - 2x.$$
$$P' = 0 \text{ if } x = 25 \qquad \text{and} \qquad P|_{x=25} = 625.$$

However, this is the absolute maximum value of P (as was seen in Example 1).

If $x = 0$, $50 - x = 50$ and $P = 0$,
If $x = 50$, $50 - x = 0$ and again $P = 0$.

So $P = 0$ is the absolute minimum value and it occurs at the end points of the interval.   ●

Exercises 4.4

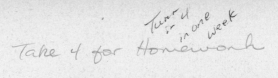
Take 4 for Homework *Turn'd in one week*

1. A farmer has 800 yd of fencing and wishes to enclose a rectangular field on land adjoining a straight river. If no fencing is required on the river side, what should be the dimensions of the field if the area is to be maximized?

2. Find two numbers x and y whose sum is S such that the sum of the squares is minimum.

3. A farmer wants to fence in 15,000 ft^2 of land in a rectangular plot and then divide it in half with a fence parallel to one of the sides. Find the dimensions to be chosen in order to minimize the amount of fencing.

4. If a rectangle is inscribed in a circle of radius R, express its dimensions in terms of R if
   (a) the area of the rectangle is to be maximized
   (b) the perimeter of the rectangle is to be maximized

5. A rectangular box, open at the top, with a square base is to have a volume of 32,000 in.^{3} What must be its dimensions if the box is to require the least amount of material?

6. A horizontal gutter is to be made from a long piece of sheet iron 12 in. wide by turning up equal widths along the edges into vertical position. How many inches should be turned up at each side to yield maximum carrying capacity?

7. A tin can is to be made with a prescribed capacity of V in.^{3} What dimensions for the can will require the least amount of tin?

8. Find the area of the largest rectangle which can be inscribed in the ellipse

$$\frac{x^2}{a^2} + \frac{y^2}{b^2} = 1$$

   (*Hint:* the dimensions that maximize the area also maximize the square of the area—using this idea will save work.)

9. Assume that the strength of a rectangular beam (that is, resistance to breaking and not bending as in illustrated Example 3) is directly proportional to the width and the square of the depth. What rectangular beam cut from a circular log of radius R will have the maximum strength?

10. The post office does not accept a package if the sum of the length and the perimeter of cross section exceeds 60 in. Find the circular cylindrical box of largest volume that meets this restriction.

11. What should be the dimensions of a circular sector of perimeter 24 in. if the area is to be a maximum?

12. The area of the printed text on a page is 96 in^2. The left-hand and the right-hand margins are each $1\frac{1}{2}$ in. wide while the upper and lower margins are each 1 in. Find the most economical dimensions of the page if only the amount of paper matters.

13. Find the volume of the largest right circular cylinder that can be inscribed in a right circular cone of radius R and altitude H. Show that the ratio of the volume of the largest cylinder to that of the volume of the cone is $\frac{4}{9}$.

14. Find the point on the line $2x - 3y = 3$ which is closest to the point $(1, 4)$.

15. A Norman window consists of a rectangle surmounted by a semicircle. If the perimeter is to be a fixed value P, what are the dimensions of the window admitting the most light?

16. A uniformly loaded beam of length L, for which $0 \leq x \leq L$, is simply supported at its ends; that is, displacements are prevented and there is no restraint to rotation at each end. For such a beam the bending moment M at a distance x in. from one end is given by $M = \frac{q}{2} x(L - x)$ where q lb/in. is the given transverse load. Prove that the bending moment is a maximum at the center of the beam.

17. A beam of uniform cross section of length L, for which $0 \leq x \leq L$, is clamped or built in at each end preventing deflections and rotations at the ends. It supports a uniform load of q lb/in. Under this load, the beam deflects in accordance with the equation

$$EIy = \frac{qx^4}{24} - \frac{qLx^3}{12} + \frac{qL^2x^2}{24}$$

where E and I are assumed to be constants (E is the modulus of elasticity and depends upon the material of which the beam is made. I is the moment of inertia with respect to the neutral axis of the cross section. I is a constant for a beam of constant cross section—that is, a prismatic beam, and its value depends upon the shape of the cross section.) At what section is the deflection the largest? What is the deflection at that section?

18. A club maintains a membership of from 300 to 500 members. When the membership is 300, each member must pay $400 per year dues. However, for each additional member above 300 the dues is reduced by $1. What membership will yield a maximum yearly income to the club from members' dues?

19. When travelling at u mi/hr, the cost in dollars per hour of operating a truck is $10 + \dfrac{u^2}{200}$, which includes the driver's wages, cost of fuel, maintenance, depreciation, and so on. At what speed is the cost least for a 500-mi trip?

20. The cost of fuel in running a helicopter is proportional to the square of the speed and is $160 per hr when the speed is 80 mi/hr. Other costs amount to $90 per hr independently of the speed. Find the speed that will make the cost per mile a minimum.

21. Given the cost function $C(x) = 5x + 1000$ and the revenue function $R(x) = 15x - 0.001x^2$ in dollars where $x \geq 1$ is the number of items, find the value of x that will maximize the profit and determine the maximum profit.

22. A man is in a boat at a point P which is 8 mi from the nearest point A on the shore (Figure 4.4.9). He wishes to go to a point B, which is further along the shore line 8 mi from A. If he can row at 3 mi/hr and walk at the rate of 5 mi/hr, toward what point C between points A and B should he row in order to reach point B in least time? Find the minimum time of travel.

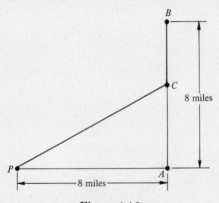

Figure 4.4.9

23. The conditions are the same as in Exercise 22 except that the man can row at the rate of 4 mi/hr.

24. Find the maximum possible area of an isosceles triangle with two sides equal to b.

25. A conical drinking cup has a fixed volume. What is the ratio of the altitude h to the radius r in order to use a minimum amount of material? The lateral surface of a cone is $\pi r L$, where L is the slant height.

26. When measurements $L_1, L_2, \ldots, L_n$ of a certain length or other quantity are made, a common practice is to take $\dfrac{L_1 + L_2 + \cdots + L_n}{n}$ as the "true length." The quantities $L_1 - L, L_2 - L, \ldots, L_n - L$ are the errors in the individual measurements. Show that one justification for using the arithmetic mean is that it makes the sum of the squares of the errors a minimum.

27. Find the point on $y^2 = 2x$ closest to $(3, 0)$.

28. Find the point in the first quadrant on $xy = 1$ closest to the origin.

29. Find the points on the circle $x^2 + y^2 = 9$ that are nearest and furthest from $(2, 0)$.

30. A wire of length L is to be cut into two pieces. One piece is bent to form an equilateral triangle whereas the second piece is to be shaped into a circle. Determine where to cut the wire in order that the sum of the areas of the triangle and the circle be a minimum.

31. A metal sphere of radius a rests on level ground. To protect the sphere a conical tent is to be erected over it. Find the minimum volume of this tent. (*Hint:* Use two similar right triangles to eliminate the radius of the base of the cone.)

32. Two heat sources are 30 ft apart, one giving out four times as much heat as the other. At a point on the

line between the sources of heat and at a distance x from the larger, the amount of heat received from the stoves is given by

$$H(x) = \frac{40}{x^2} + \frac{10}{(30 - x)^2}$$

Find the point on the line between the stoves that receives the least amount of heat.

33. Show that of all triangles with a given base and a given perimeter, the isosceles triangle has the greatest area. (*Hint:* Use Heron's formula for the area of a triangle $A = \sqrt{s(s-a)(s-b)(s-c)}$ where a, b, and c are the sides and s is half the perimeter.)

34. Show that, of all triangles with given base and given area, the isosceles triangle has the least perimeter. (*Hint:* Refer to Figure 4.4.10. Let AB be the given base and choose the x-axis along AB and the y-axis bisecting AB. Because the area of the triangle is prescribed, then altitude h is fixed for all competing triangles. Finally show that for the triangle of least perimeter $|\overline{CA}| = |\overline{CB}|$.)

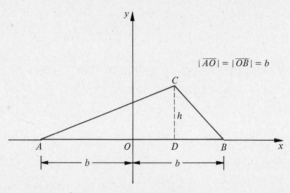

Figure 4.4.10

4.5 MAXIMUM AND MINIMUM VALUES BY IMPLICIT DIFFERENTIATION

When the function that is to be extremized is expressed as a function of two variables which in turn are related by a given equation, then, instead of eliminating one of the variables by use of the given relation (which may be difficult, if not impossible), it is frequently expedient to carry along both functional relations and use implicit differentiation on them. This is best explained by means of illustrative examples.

EXAMPLE 1 Find the largest and smallest value of $u(x, y) = 2x + y$ on the circle $x^2 + y^2 = 1$ (Figure 4.5.1).

Solution Since the u we seek is to be greatest or least, then $D_x u = 0$ so that $D_x(2x + y) = 0$ at the extrema. Therefore

$$2 + y' = 0 \tag{1}$$

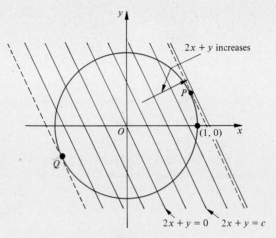

Figure 4.5.1

at the maximum and minimum points. Also $D_x(x^2 + y^2) = D_x(1) = 0$ so that at all points on the circle (as well as at the extrema)

$$2x + 2yy' = 0 \qquad \qquad (2)$$

Thus at the extrema (eliminating y' between (1) and (2))

$$x + y(-2) = 0$$

From $x = 2y$ and $x^2 + y^2 = 1$ we find the common solutions

$$\left(\frac{2}{\sqrt{5}}, \frac{1}{\sqrt{5}}\right) \qquad \text{and} \qquad \left(\frac{-2}{\sqrt{5}}, \frac{-1}{\sqrt{5}}\right)$$

and

$$u\bigg|_{\max} = \sqrt{5} \qquad \text{and} \qquad u\bigg|_{\min} = -\sqrt{5}$$

These points are points P and Q in Figure 4.5.1.   ●

EXAMPLE 2   An open cylindrical can is to contain 200π in^3. What dimensions will require the least material?

Solution   If r and h are the base radius and height, respectively, then the volume V is

$$V = \pi r^2 h = 200\pi$$

while the surface area

$$S = \pi r^2 + 2\pi r h$$

Now $r^2 h = 200$ so that $D_r(r^2 h) = 0$, where h is considered as a differentiable function of r, which results in

$$2rh + r^2 D_r h = 0 \qquad \qquad (3)$$

Also S is to be a minimum and therefore $D_r S = 0$ must hold. Thus

$$D_r S = \pi(2r + 2h + 2r D_r h) = 0 \qquad \qquad (4)$$

From (3), $D_r h = \dfrac{-2h}{r}$, and substitution into (4) yields

$$r + h - 2h = 0 \quad \text{or} \quad r = h$$

If $r = h$, then

$$r^3 = 200 \quad r = h = \sqrt[3]{200} \doteq 5.85 \text{ in.}$$

and it can be shown that an absolute minimum for the surface area has been obtained.

It should be noted that we could have formed $D_h(r^2 h) = 0$ and considered r as a differentiable function of h. ●

EXAMPLE 3   Find the rectangle of maximum area that can be inscribed in the ellipse

$$\frac{x^2}{a^2} + \frac{y^2}{b^2} = 1 \tag{5}$$

Solution   Let (x, y) be a point in the first quadrant on the ellipse and form the rectangle with vertices $(x, \pm y)$ and $(-x, \pm y)$. The area A of the rectangle is $4xy$. Thus $4xy$ is to be maximized subject to the constraint (5).

$D_x(xy) = 0$ must hold at the extremum so that *interpret*

$$xy' + y = 0 \tag{6}$$

Also

$$D_x \left(\frac{x^2}{a^2} + \frac{y^2}{b^2} \right) = 0$$

or

$$2 \left(\frac{x}{a^2} + \frac{yy'}{b^2} \right) = 0 \tag{7}$$

Now $y' = \dfrac{-y}{x}$ from (6) so that $\dfrac{x}{a^2} + \dfrac{y}{b^2} \left(\dfrac{-y}{x} \right) = 0$ or, at the extremum

$$y^2 = \frac{b^2 x^2}{a^2} \tag{8}$$

Substitution of (8) into (5) yields

$$\frac{x^2}{a^2} + \frac{x^2}{a^2} = 1 \quad \text{or} \quad x = \frac{a}{\sqrt{2}} \quad \text{and} \quad y = \frac{b}{\sqrt{2}}$$

so that $A_{\max} = 4 \dfrac{a}{\sqrt{2}} \dfrac{b}{\sqrt{2}} = 2ab$ ●

EXAMPLE 4   Find the point on the hyperbola $x^2 - y^2 = 1$ which is closest to the point $(0, 2)$.

Solution   Form

$$u(x, y) = x^2 + (y - 2)^2$$

which is the square of the distance from $(0, 2)$ to the point (x, y) on the hyperbola.

At extrema

$$D_x u = 2x + 2(y - 2)y' = 0 \tag{9}$$

Also from $x^2 - y^2 = 1$ we have $D_x(x^2 - y^2) = 0$ or

$$2x - 2yy' = 0 \tag{10}$$

from which $y' = \dfrac{x}{y}$ so that

$$2x + 2(y - 2)\frac{x}{y} = 0$$

Therefore $x(y - 1) = 0$ and we obtain

$$x = 0 \quad \text{or} \quad y = 1$$

Now $x = 0$ is discarded since, from $x^2 - y^2 = 1$ we have $x^2 = 1 + y^2$ or $|x| \geq 1$. If $y = 1$, then $x = \pm\sqrt{2}$, and the distance $= \sqrt{3}$. The function u has an absolute minimum at $(\pm\sqrt{2}, 1)$ and $\sqrt{3}$ is the shortest distance between the given point and the hyperbola. ●

EXAMPLE 5   An open box of given volume is to have a square base. The material used for the bottom costs twice as much per unit area as that of the sides. Find the proportions that minimize the cost.

Solution   Let k be the cost per unit of area for the sides. Then if x and y are the lengths of an edge of the base and height, respectively, the total cost C is

$$C = 2kx^2 + 4kxy$$

or

$$C = 2k(x^2 + 2xy)$$

Also $V = x^2 y$ is the fixed volume of the box. Now for C to be a maximum or minimum with x considered as the independent variable

$$D_x C = 0 = 2k(2x + 2xy' + 2y)$$

from which

$$y' = -\frac{(x + y)}{x} = -1 - \frac{y}{x}$$

Also V is fixed so that $D_x(x^2 y) = 0$ which yields

$$x^2 y' + 2xy = 0$$

or solving for y'

$$y' = -\frac{2y}{x}$$

Therefore

$$-1 - \frac{y}{x} = -\frac{2y}{x} \quad \text{or} \quad y = x$$

and, in order to minimize the cost, the shape is to be a cube. ●

Exercises 4.5

1. Find the value of x that maximizes $u = xy$ if $3x + y = 6$.
2. Find the value of x that maximizes $u = xy$ if $x^3 + y = 32$.
3. Find the point on $x + y = \frac{1}{2}$ that is closest to the origin.
4. Find the maximum and minimum of xy on the circle $x^2 + y^2 = 4$.
5. Find the maximum and minimum of x^2y on $x^2 + y^2 = 1$.
6. A cylindrical tin can is to be made to contain 100 in.³ Find the relative dimensions so that the least amount of tin is used.
7. Show that the rectangle of given perimeter with shortest diagonal is a square.
8. Find the relative dimensions of the largest rectangle that can be cut from a semicircle whose radius is a.
9. An open metal water trough with semicircular end plates is to be constructed with a given volume V (Figure 4.5.2). Find the ratio of the radius r to the length h of the trough if a minimum total surface area S is required.

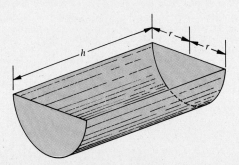

Figure 4.5.2

10. Find the extrema of $x^2 + y^2$ if x and y are subject to the constraint $x^2 + y^2 - 4x - 2y + 4 = 0$. Give a geometric interpretation of your result.
11. Let the general equation of a straight line L be given in the form $ax + by = c$ and let $P(x_0, y_0)$ be a point not on L. Determine the point Q on L that is closest to P. Show also that the line joining P to Q is perpendicular to L.
12. A package to be sent parcel post must have a combined girth (distance around) and length of at most 60 in. What are the dimensions of the largest rectangular parallelepiped (or box) that can be sent if its base is a square?
13. A storage tank with a given volume is to be made in the form of a right circular cylinder surmounted by a hemisphere (Figure 4.5.3). Determine the relative dimensions so that the total surface area is least.

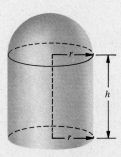

Figure 4.5.3

14. A tin can encloses a volume of 200 in.³ The top and bottom of the can are made of material twice as expensive per square inch as the material in the side. Find the relative dimensions which minimizes the cost.

15. Find the largest and smallest value of $x^2 + 24xy + 8y^2$ when $x^2 + y^2 = 100$.

16. Find the shortest distance from $(\frac{3}{2}, 0)$ to the curve $y^2 = x$.

17. Show that the largest rectangle that can be inscribed in the segment of the parabola $y^2 = 2px$ bounded by the vertical line $x = a$ has a horizontal dimension $\dfrac{2a}{3}$ (where $a > 0$ and $p > 0$ are assumed).

18. A right circular cone of maximum volume is inscribed in a sphere of radius a. Show that the volume of the cone is

$$\tfrac{32}{81}\pi a^3 = \tfrac{8}{27} \cdot \text{volume of the sphere}$$

4.6   ROLLE'S THEOREM AND THE MEAN-VALUE THEOREM

This section is concerned with the statement and proofs of two significant theorems: (1) Rolle's theorem and (2) the basic mean value theorem of differential calculus. These theorems are used to establish many important results in both differential and integral calculus and furthermore will lead directly to the establishment of sufficient conditions for relative extrema (Sections 4.7 and 4.8).

Theorem 1
Rolle's Theorem

Let $f(x)$ be a function such that

(i) $f(x)$ is continuous on $a \leq x \leq b$,
(ii) $f(x)$ is differentiable on $a < x < b$,
(iii) $f(a) = f(b)$.

Then there exists at least one number c in (a, b) for which $f'(c) = 0$.

The following definition helps us obtain a geometric interpretation of Rolle's theorem.

Definition

If f is a function, then the line segment joining two points of the graph of f is called a **chord of f**.

Thus in Rolle's theorem, since $f(a) = f(b)$, the line joining $(a, f(a))$ and $(b, f(b))$ is a horizontal chord. Rolle's theorem says that there is at least one value of x in (a, b), say, $x = c$, at which the tangent line to the graph is also horizontal, that is, parallel to the horizontal chord (Figure 4.6.1).

A proof of Rolle's theorem is obtained by formalizing our geometric observations as follows.

Proof

Since $f(x)$ is continuous in the closed interval $[a, b]$, by the extreme value theorem f must possess an absolute maximum M and an absolute minimum m where clearly $m \leq M$.

(i) If $m = M$ then f is constant and $f' = 0$ for all x in (a, b) so that any number x in (a, b) will serve as the desired number c.

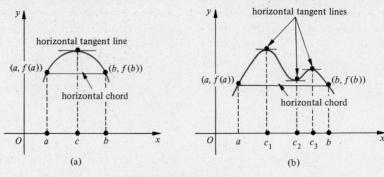

Figure 4.6.1

(ii) If $m < M$ then the minimum and maximum cannot both occur at the ends of the interval since $f(a) = f(b)$. One of them at least occurs at a point c such that $a < c < b$. But at this interior point, since f' exists, it must be zero; that is, $f'(c) = 0$. This proves Rolle's theorem, which was first established by the French mathematician Michel Rolle (1652–1719). ∎

In some other books, parts (iii) of the hypothesis of Rolle's theorem is stated: $f(a) = 0$ and $f(b) = 0$. This statement of Rolle's theorem is implied by ours (since ours is true as long as $f(a)$ and $f(b)$ are the same). The statement given in this book is a slightly more general result.

EXAMPLE 1 Verify Rolle's theorem for $f(x) = x^2 - 3x + 2$ where $a = 0$ and $b = 3$.

Solution First we observe that $f(0) = f(3) = 2$. Also f is continuous in $[0, 3]$ and f' exists in $(0, 3)$. (In fact $f, f', f'', \ldots$, exist for all x since f is a polynomial.) In accordance with Rolle's theorem, there exists at least one value of c in $(0, 3)$ such that $f'(c) = 0$. Let us find c in this case. The function $f' = 2x - 3$ and, if we set $f' = 0$, there results the unique value $c = \frac{3}{2}$. ●

EXAMPLE 2 Verify Rolle's theorem for

$$g(x) = x^3 + 5x^2 - x - 3 \qquad a = -5 \text{ and } b = 1$$

Solution $g(-5) = g(1) = 2$ and, since g is a polynomial, the continuity and differentiability requirements are met.

$$g'(x) = 3x^2 + 10x - 1$$

$$g'(c) = 0 \rightarrow c = \frac{-10 \pm \sqrt{100 + 12}}{6} = \frac{-5 \pm 2\sqrt{7}}{3}$$

Now

$$\frac{-5 + 2\sqrt{7}}{3} \doteq \frac{-5 + 2(2.65)}{3} = \frac{0.30}{3} = 0.10$$

$$\frac{-5 - 2\sqrt{7}}{3} \doteq \frac{-5 - 2(2.65)}{3} = \frac{-10.30}{3} \doteq -3.43$$

and we have two values of c in $(-5, 1)$ where the tangent to the graph of $g(x)$ is horizontal.

If in this example we had been given $a = -5$ and $b = -1$, then again $g(-5) = g(-1) = 2$. The calculations would be the same as before, but this time only $c = \dfrac{-5 - 2\sqrt{7}}{3}$ would be correct, since, of the two roots of the quadratic equation, only this one is in the interval $(-5, -1)$. ●

EXAMPLE 3 Attempt to verify Rolle's theorem for $F(x) = |x|$ where $a = -2$ and $b = 2$.

Solution Since $F'(0)$ does not exist and $x = 0$ is in $(-2, 2)$ Rolle's theorem does not apply. In fact, in this case $F'(x)$ is never zero.

It is also possible that in some cases there may be one or more values of c in (a, b) for which $f'(c) = 0$ even though the function does not satisfy one or more of the conditions of Rolle's theorem. ●

EXAMPLE 4 Verify Rolle's theorem for $f(x) = \sqrt{25 - x^2}$ where $a = -5$ and $b = 5$.

Solution Now $f(-5) = f(5) = 0$ and we observe that $f(x)$ is continuous in $[-5, 5]$. Also $f'(x) = \dfrac{-x}{\sqrt{25 - x^2}}$ in $(-5, 5)$. Therefore Rolle's theorem applies and, if we set $f'(c) = 0$, then $\dfrac{-c}{\sqrt{25 - c^2}} = 0$ which yields $c = 0$. ●

Rolle's theorem asserts that if the graph of a function (satisfying the appropriate continuity and differentiability requirements) has a horizontal chord, it has a tangent line parallel to that chord. The law of the mean (also called the mean value theorem) is a generalization of Rolle's theorem in that it is concerned with arbitrary chords of f rather than just horizontal chords.

In geometric terms the mean value theorem states that, if a chord is drawn for the graph of a function $f(x)$ (satisfying continuity and differentiability requirements), in the open interval defined by the abscissas of the chordal end points there exists at least one point c in (a, b) such that the tangent line is parallel to the chord (Figure 4.6.2). Let us translate this geometric statement into analytical language.

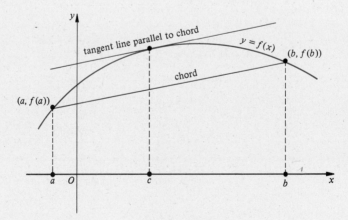

Figure 4.6.2

Referring to Figure 4.6.2 the slope of a typical chord is

$$\frac{f(b) - f(a)}{b - a}$$

while the slope of the tangent at any point $(x, f(x))$ on the graph is $f'(x)$. The law of the mean then asserts that there exists at least one number c in (a, b) for which

$$f'(c) = \frac{f(b) - f(a)}{b - a}$$

Now let us state the law of the mean precisely and prove it with the aid of Rolle's theorem.

Theorem 2
Mean Value
Theorem

Let $f(x)$ be a function such that

(i) $f(x)$ is continuous on the closed interval $[a, b]$,
(ii) $f(x)$ is differentiable on the open interval (a, b).

Then there is at least one number c in the open interval (a, b) such that

$$f'(c) = \frac{f(b) - f(a)}{b - a} \tag{1}$$

Proof   To prove this a function must be introduced to which Rolle's theorem may be applied. In Figure 4.6.3 we have two functions $y = f(x)$ and $y = g(x)$ where $g(x)$ is the function whose graph is the straight line L containing the chord connecting $(a, f(a))$ and $(b, f(b))$. Let $h(x) = f(x) - g(x)$, which represents the difference between the ordinates of the graphs of $f(x)$ and $g(x)$. Certainly $h(a) = h(b) = 0$, since $f(a) = g(a)$ and $f(b) = g(b)$. Also $g(x)$ is continuous and differentiable for all x. Thus Rolle's theorem applies to the difference function $h(x)$ and we know that $h'(x) = 0$ for at least one value of x, say $x = c$ in (a, b).

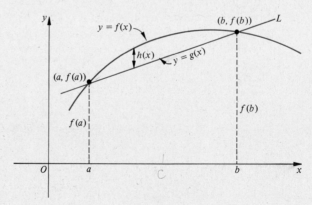

Figure 4.6.3

Thus

$$h'(c) = 0$$

But

$$h'(c) = f'(c) - g'(c)$$

Also

$$g'(c) = g'(x) = \frac{f(b) - f(a)}{b - a} = \text{slope of } L$$

Therefore

$$f'(c) = \frac{f(b) - f(a)}{b - a} \qquad a < c < b$$

and the law of the mean is proved. ∎

EXAMPLE 5   Verify the law of the mean for

$$f(x) = x^2 - 3x + 4 \qquad a = -1 \text{ and } b = 2$$

Solution

$$f(b) = f(2) = 2^2 - 3(2) + 4 = 2$$

and

$$f(a) = f(-1) = (-1)^2 - 3(-1) + 4 = 8$$

According to the law of the mean there exists at least one value c in $(-1, 2)$ such that

$$f'(c) = \frac{2 - 8}{2 - (-1)} = \frac{-6}{3} = -2 \quad \text{slope}$$

Now $f'(x) = 2x - 3$ so that we must solve the equation

$$2c - 3 = -2$$

which yields

$$c = \frac{1}{2}$$

Indeed $-1 < \frac{1}{2} < 2$ and the law of the mean is verified. ●

EXAMPLE 6   Verify the mean value theorem for

$$g(x) = \frac{1}{x} \qquad a = 1 \text{ and } b = 4$$

Solution $g(x)$ is continuous and differentiable for all $x \neq 0$ so that the mean value theorem applies. We have

$$g(4) = \tfrac{1}{4} \qquad \text{and} \qquad g(1) = 1$$

and therefore

$$\frac{g(4) - g(1)}{4 - 1} = \frac{-3/4}{3} = -\frac{1}{4}$$

Now $g'(x) = -\dfrac{1}{x^2}$ so that

$$-\frac{1}{c^2} = -\frac{1}{4}$$

In this case $c = 2$ is the only value of x in $(1, 4)$ that satisfies the requirements of the mean value theorem. ●

EXAMPLE 7 Let

$$g(x) = \begin{cases} -x & \text{if } x < 0 \\ x^2 & \text{if } x \geq 0 \end{cases}$$

Does the mean value theorem apply if $a = -1$ and $b = 2$? Is the conclusion of the mean value theorem satisfied on the interval $[-1, 2]$?

Solution In Example 8, Section 3.4, we saw that this function is not differentiable at $x = 0$. Therefore the mean value theorem does not apply.

$$\frac{g(b) - g(a)}{b - a} = \frac{g(2) - g(-1)}{2 - (-1)} = \frac{4 - 1}{2 + 1} = \frac{3}{3} = 1$$

If $x > 0$, $g'(x) = 2x$ and, in particular, $g'(\frac{1}{2}) = 1$.
Then, letting $c = \frac{1}{2}$ gives

$$g'(c) = \frac{g(b) - g(a)}{b - a}$$

Thus, the conclusion of the mean value theorem is satisfied even though the hypotheses are not. ●

EXAMPLE 8 Verify the mean value theorem for

$$f(x) = x|x| \qquad \text{if } a = -3 \text{ and } b = 3$$

Solution

$$f(x) = \begin{cases} x^2 & x \geq 0 \\ -x^2 & x < 0 \end{cases}$$

and $f(x)$ is continuous in $[-3, 3]$ and differentiable in $(-3, 3)$ so that the mean value theorem applies

$$\frac{f(b) - f(a)}{b - a} = \frac{3^2 - (-3^2)}{3 - (-3)} = \frac{9 - (-9)}{6} = 3$$

$$f'(x) = \begin{cases} 2x & \text{if } x \geq 0 \\ -2x & \text{if } x < 0 \end{cases}$$

$2c = 3$ if $c \geq 0$, and $-2c = 3$ if $c < 0$. Then

$$c = 3/2 \qquad \text{and} \qquad c = -3/2$$

Therefore we have obtained two values for c in $(-3, 3)$ for which the mean value theorem holds. The reader is invited to sketch the function $f(x)$ and verify the result geometrically. ●

A useful alternative way of writing the law of the mean is easily obtained. From

$$f'(c) = \frac{f(b) - f(a)}{b - a}$$

we obtain the equivalent equation

$$f(b) - f(a) = (b - a)f'(c)$$

and by adding $f(a)$ to both sides,

$$f(b) = f(a) + (b-a)f'(c) \qquad (2)$$

The following corollary utilizes form (2).

Corollary 1 If the derivative of a function is 0 throughout an interval, then the function is constant throughout that interval.

Proof Let x_0 and x_1 be any two points in the given interval. To establish the corollary it suffices to show that $f(x_0) = f(x_1)$.

By the law of the mean, there is a number c in (x_0, x_1) such that

$$f(x_1) = f(x_0) + (x_1 - x_0)f'(c)$$

But $f'(c) = 0$ by hypothesis, so that $f(x_1) = f(x_0)$. ∎

Next we establish Corollary 2.

Corollary 2 If two functions have the same derivatives throughout an interval, they must differ by a constant. Symbolically if $f(x)$ and $g(x)$ are two functions and if $f'(x) = g'(x)$ for all x in an interval, then $f(x) = g(x) + C$ where C is a constant.

Proof Let $h(x) = f(x) - g(x)$. Then

$$h'(x) = f'(x) - g'(x) = 0$$

for all x in the interval. From Corollary 1 it follows that

$$h(x) = C$$

where C is a constant. This implies that

$$f(x) - g(x) = C \qquad \text{or} \qquad f(x) = g(x) + C$$

and the corollary is proved. ∎

EXAMPLE 9 Find all functions such that their derivatives equal x^2 everywhere.

Solution One such function is $\dfrac{x^3}{3}$. Another is $\dfrac{x^3}{3} + 100$. For any constant C,

$$D_x\left(\frac{x^3}{3} + C\right) = x^2$$

Are there any other possibilities? Corollary 2 says that there are no other possibilities. For if $g(x)$ is such that $D_x g(x) = x^2$ then $g(x) - \dfrac{x^3}{3} = C$, say. Thus

$$g(x) = \frac{x^3}{3} + C$$

is the totality of functions (one for each C) that satisfy $g'(x) = x^2$. ●

Exercises 4.6

In Exercises 1 through 8, verify that all the conditions of Rolle's theorem are satisfied by the given function on the indicated interval. Then find the value(s) for c that satisfies the conclusion of Rolle's theorem.

1. $f(x) = x^2 - 6x + 7$, $[1, 5]$

2. $g(x) = x^3 - 4x + 3$, $[0, 2]$

3. $h(x) = x^3 - 7x + 6$, $[-3, 2]$

4. $F(x) = x^2 - 2bx$, $[0, 2b]$

5. $G(x) = x^3 - a^2x + b$, $a > 0$ $[0, a]$

6. $H(x) = x^3 + 3x^2 - x - 2$, $[-3, 1]$

7. $f(x) = |x^2 - 16|$, $[-4, 4]$

8. $g(x) = 3 + [\![x]\!]$, $[1, \frac{3}{2}]$

9. (a) Does the function $F(x) = |1 + 2x| + 5$ satisfy the hypotheses of Rolle's theorem on $[-2, 1]$?
   (b) Is there a number c in $(-2, 1)$ such that $F'(c) = 0$?   (c) Does this contradict Rolle's theorem? Explain.

10. Are the hypotheses of Rolle's theorem necessary in order that there exist a number c in (a, b) such that $f'(c) = 0$? Are they sufficient?

11. A displacement function $s = f(t)$ is continuous for all t in $[t_0, t_1]$ and differentiable in (t_0, t_1). Furthermore $f(t_0) = f(t_1)$. What does this indicate about the velocity function $v = f'(t)$?

In Exercises 12 through 18, verify that the hypotheses of this mean value theorem are satisfied for the given function on the given interval. Then find a value for c that satisfies the conclusion of the mean value theorem.

12. $f(x) = x^2 - 5x + 4$, $[1, 3]$

13. $g(x) = x^3 - 7x^2 + 5x + 3$, $[1, 5]$

14. $h(x) = \dfrac{1}{x^2}$, $[2, 4]$

15. $F(x) = x^{2/3}$, $[0, 8]$

16. $G(x) = \sqrt{x}$, $[0, 9]$

17. $H(x) = \dfrac{x + 3}{x}$, $[1, 6]$

18. $f(x) = \dfrac{x^2 - 4x + 3}{x + 2}$, $[1, 3]$

In Exercises 19 through 23 determine whether the hypotheses of the mean value theorem are satisfied by the given function in the given interval. If they are not satisfied, does the conclusion nevertheless hold?

19. $F(x) = x^{1/3}$, $[-8, 1]$

20. $G(x) = \begin{cases} x^2 & x \geq 2 \\ 2x & x < 2 \end{cases}$, $[1, 4]$

21. $H(x) = \begin{cases} x^2 & x \geq 1 \\ 2x & x < 1 \end{cases}$, $[0, 4]$

22. $f(x) = \begin{cases} x^2 & x \geq 0 \\ -x^2 & x < 0 \end{cases}$, $[-2, 2]$

23. $f(x) = x + \dfrac{1}{x}$, $[a, b]$, $0 < a < b$

24. The equation $x^3 - 5x + 1 = 0$ has a root in $[0, 1]$ because $f(x) = x^3 - 5x + 1$ is continuous in $[0, 1]$ and $f(0) > 0$ while $f(1) < 0$. Prove that the equation cannot have two distinct real roots in $[0, 1]$.

25. The equation $x^5 + x + a = 0$ must have at least one real root where a is any constant. Why? Prove that the equation cannot have distinct real roots.

26. Let f be a function that is continuous in the closed interval $[a, b]$. Furthermore it is known that $f(a) = f(b)$. Show that f has at least one critical number in the open interval (a, b).

27. In each of the following cases show that $f'(x) = g'(x)$. Find the relationship between $f(x)$ and $g(x)$.
   (a) $f(x) = x(x - 6)$, $g(x) = (x - 3)^2$
   (b) $f(x) = x^2/1 + x^2$, $g(x) = (3x^2 + 2)/1 + x^2$

28. State the mean value theorem that would imply that, under suitable hypotheses, the equation

$$f'(x_1) - f'(x_0) = (x_1 - x_0)f''(c) \qquad x_0 < c < x_1.$$

   Is a "new" proof necessary?

29. If $b^2 - 4ae > 0$ and $a \neq 0$, then the quadratic function $f(x) = ax^2 + bx + e$ has two distinct real zeros. Verify Rolle's theorem in this instance and relate the zero of f' with the zeros of f.

30. Use Rolle's theorem to prove that, if $a > 0$, the quintic equation $x^5 + ax + b = 0$ cannot have more than one real root no matter what value is assigned to b.

31. (a) Assume that every polynomial of degree 3 has at most three real roots. Use Rolle's theorem to prove that any polynomial of degree 4 has at most four real roots.
    (b) Generalize the technique of part (a) to prove that every polynomial of degree n has at most n real roots.

32. Prove that the equation $f(x) = x^n + ax + b = 0$ cannot have more than two real solutions for an even integer n nor more than three real solutions for n odd.

33. Given that $|f'(x)| \le 1$ for all numbers x, show that $|f(x_2) - f(x_1)| \le |x_2 - x_1|$ for all numbers x_1, x_2.

34. Find a function $f(x)$ for which
    (a) $f'(x) = x - 1$, $f(0) = 2$
    (b) $f''(x) = 3$, $f(0) = -1$, $f'(0) = 1$

35. A vehicle is moving on the x-axis and its position at time t is given by $x = f(t)$. At time $t = t_0$ its position is $f(t_0)$, whereas at a later time t_1 its position is $f(t_1)$. What important fact can be deduced about the motion by application of the mean value theorem?

4.7 SUFFICIENCY TESTS FOR RELATIVE MAXIMA AND MINIMA

In this section we return to the maximum and minimum problem. With the law of the mean at our disposal, we shall develop two very useful sufficiency tests for relative extrema involving, respectively, first and second derivatives.

To this end we start with two definitions.

Definition 1   A function f defined on an interval is said to be ***increasing*** on that interval if and only if

$$f(x_1) < f(x_2) \qquad \text{whenever } x_1 < x_2$$

where x_1 and x_2 are any numbers in that interval (see Figure 4.7.1(a)).

Definition 2   A function f defined on an interval is said to be ***decreasing*** on that interval if and only if

$$f(x_1) > f(x_2) \qquad \text{whenever } x_1 < x_2$$

where x_1 and x_2 are any numbers in that interval (see Figure 4.7.1(b)).

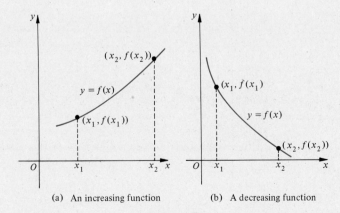

(a) An increasing function   (b) A decreasing function

Figure 4.7.1

Definition 3   If a function f is either increasing or decreasing on an interval, then f is said to be **monotonic** on f.

With these definitions and the mean value theorem it is easy to establish the following theorem.

Theorem 1   If f is continuous on $[a, b]$ and has a positive derivative on the open interval (a, b), then f is increasing on the interval $[a, b]$.

Proof   First we note that the hypotheses of the mean value theorem are satisfied. Take two numbers x_1 and x_2 such that

$$a \leq x_1 < x_2 \leq b$$

By the law of the mean,

$$f(x_2) = f(x_1) + (x_2 - x_1)f'(c) \qquad \text{where } x_1 < c < x_2$$

Since $x_2 - x_1$ is positive and since $f'(c)$ is also assumed to be positive, it follows that

$$(x_2 - x_1)f'(c) > 0$$

or

$$f(x_2) > f(x_1) \qquad\qquad ∎$$

Theorem 2   If f is continuous on $[a, b]$ and has a negative derivative on the open interval (a, b), then f is decreasing on the interval $[a, b]$.

The proof of Theorem 2 is similar to that of Theorem 1 and is left as an exercise for the reader.

Theorem 1 states that an increasing function has a positive slope, whereas Theorem 2 asserts that a decreasing function has negative slope. We can observe these properties in Figure 4.7.1 (a) and (b).

EXAMPLE 1   Prove that $f(x) = x^3$ is an increasing function of x.

Solution $$f'(x) = 3x^2 > 0 \qquad \text{if } x \neq 0$$

$f(x)$ is continuous in $[0, \infty)$ and $f' > 0$ if $x > 0$ so that by Theorem 1, $f(x)$ is increasing in $[0, \infty)$.

Similarly $f(x)$ is continuous in $(-\infty, 0]$ and $f' > 0$ if $x < 0$, which implies that $f(x)$ is increasing in $(-\infty, 0]$. Consequently $f(x)$ is increasing in $(-\infty, \infty)$. Note that this is true despite the fact that $f' = 0$ at $x = 0$.   ●

EXAMPLE 2   Graph the function $f(x) = 1 + \frac{1}{2}x^2$ showing where the function is increasing and decreasing.

Solution $f'(x) = x$, which implies that $f' > 0$ if $x > 0$ and $f' < 0$ if $x < 0$. Also $f'(x)$ is zero only at $x = 0$. Therefore the function is increasing for $x > 0$ and decreasing for $x < 0$. At $x = 0$ the graph has a horizontal tangent line. This occurs at the point $(0, 1)$. With this information, a sketch of f can be made (Figure 4.7.2). Furthermore, $(0, 1)$ is a relative minimum and also an absolute minimum of $f(x)$.   ●

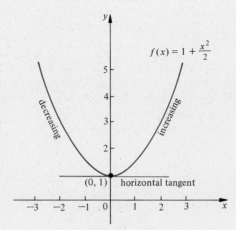

Figure 4.7.2

EXAMPLE 3   Discuss and sketch the graph of $f(x) = \frac{1}{3}(x^3 - 9x)$ showing the relative extrema.

Solution   If we write $y = f(x) = \dfrac{x(x - 3)(x + 3)}{3}$ then $y = 0$ when $x = 0$ and $x = \pm 3$. Also $f'(x) = \frac{1}{3}(3x^2 - 9) = x^2 - 3$, which means that $f' > 0$ if $|x| > \sqrt{3}$ and $f' < 0$ when $|x| < \sqrt{3}$. Therefore the function is decreasing when $-\sqrt{3} < x < \sqrt{3}$ and increasing for $x > \sqrt{3}$ or $x < -\sqrt{3}$. Also $f' = 0$ when $x = \pm\sqrt{3}$. Since f is decreasing for $x < \sqrt{3}$ (in a sufficiently small one-sided neighborhood of $\sqrt{3}$) and f is increasing for $x > \sqrt{3}$, it follows that $(\sqrt{3}, -2\sqrt{3})$ is a relative minimum (but not an absolute minimum) for f. Similarly, f is increasing for $x < -\sqrt{3}$ and decreasing for $x > -\sqrt{3}$ (in a sufficiently small neighborhood of $-\sqrt{3}$) so that $(-\sqrt{3}, 2\sqrt{3})$ is a relative maximum (but not an absolute maximum). The function is sketched in Figure 4.7.3.   ●

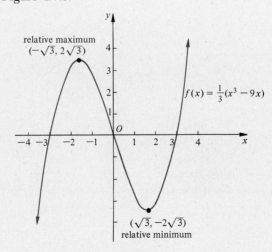

Figure 4.7.3

Examples 1 to 3 illustrate Theorem 3, which follows directly from Theorems 1 and 2.

Theorem 3   ***First Derivative Test for Relative Extrema.***   Let the function f be continuous at all points of the open interval (a, b) containing the number c, and furthermore suppose that f' exists at all points in (a, b) except perhaps at c itself then

    (i) If $f'(x) > 0$ for all x in some open interval having c as a right-hand end point, and if $f'(x) < 0$ for all x in some open interval having c as a left-hand end point, then f has a relative maximum value at c.

    (ii) If $f'(x) < 0$ for all x in some interval having c as a right-hand end point, and if $f'(x) > 0$ for all x in some open interval having c as a left-hand end point, then f has a relative minimum value at c.

Proof of (i)   Let (d, c) be an open interval having c as its right end point, for which $f'(x) > 0$ for every value of x in the interval. Thus from Theorem 1 it follows that f is increasing on $[d, c]$. Similarly, let (c, e) be the interval having c as a left-hand end point, for which $f'(x) < 0$ for all x in the interval. By Theorem 2, f is decreasing on $[c, e]$. This means that if x_1 is any point in $[d, c]$ and $x_1 \neq c$, then $f(x_1) < f(c)$. Also if x_2 is any point in $[c, e]$ and $x_2 \neq c$, then, since $f(x)$ is decreasing in $[c, e]$, $f(c) > f(x_2)$. Thus, by definition, $f(x)$ has a relative maximum value at c.

    The proof of (ii) is similar to that of (i) and is left as an exercise for the reader.   ■

    The first derivative test states that, if f is continuous at c and $f'(x)$ changes algebraic sign from positive to negative as x increases through the number c, then f has a relative maximum at c. Similarly if $f'(x)$ changes sign from negative to positive in increasing through c, then f has a relative minimum at c.

    Figures 4.7.4 and 4.7.5 illustrate parts (i) and (ii) of Theorem 3 when $f'(c)$ exists. Figure 4.7.6 shows a sketch of the graph of a function f that has a relative maximum at c, but $f'(c)$ does not exist. However, in a sufficiently small neighborhood about $x = c$, $f'(x) > 0$ if $x < c$ and $f'(x) < 0$ if $x > c$.

    Finally Figure 4.7.7 is a graph of $y = f(x)$ that has a critical number at $x = c$, since $f'(c) = 0$. But $f' > 0$ if $x > c$ and also when $x < c$ so that f does *not* have a relative extremum at $x = c$. The sign of f' did not change as x increased through $x = c$—in fact f is an increasing function of x.

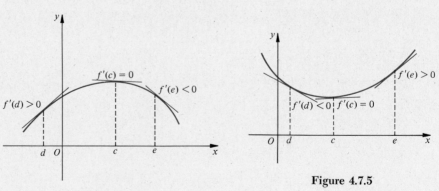

Figure 4.7.4

Figure 4.7.5

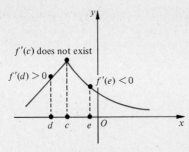

Figure 4.7.6

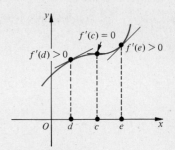

Figure 4.7.7

EXAMPLE 4   Find the extrema of the function

$$f(x) = 3x^4 - 4x^3$$

using the first derivative test.

Solution $f'(x) = 12x^3 - 12x^2 = 12x^2(x - 1)$

Thus

$$f'(x) < 0 \quad \text{if } x < 1 \text{ and } x \neq 0$$
$$f'(x) > 0 \quad \text{if } x > 1$$
$$f'(x) = 0 \quad \text{if } x = 0 \text{ and } x = 1$$

and the function is decreasing for $x < 1$ and increasing for $x > 1$.

At $x = 1$, therefore, we have a relative and also an absolute minimum. Also $f'(x)$ does not change sign as x increases through $x = 0$; that is, $f'(x) < 0$ if $x < 0$ and when $0 < x < 1$. Thus $f(x)$ does not have a maximum or minimum at $x = 0$. A sketch of the function is shown in Figure 4.7.8.   ●

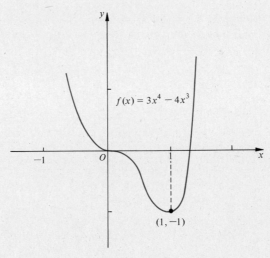

$$f(x) = 3x^4 - 4x^3$$

(1, −1)

Figure 4.7.8

EXAMPLE 5   Find the extrema of the function $f(x) = x^{2/3}$ using the first derivative test.

Solution $f'(x) = \dfrac{2}{3}x^{-1/3} = \dfrac{2}{3x^{1/3}}$

The derivative does not exist if $x = 0$, and this is the only critical number. If $x < 0$, then $f'(x) < 0$; if $x > 0$, then $f'(x) > 0$. Thus $x = 0$ yields a relative and absolute minimum point. A sketch of the graph is shown in Figure 4.7.9.

●

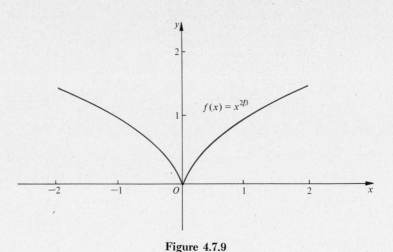

$f(x) = x^{2/3}$

Figure 4.7.9

If $f'(x)$ is continuous then, when using the first derivative test, it is not actually necessary to look at the sign of $f'(x)$ in entire intervals to the left and to the right of a critical number c. Instead, we can evaluate f' at just one x-value to the left of c and just one x-value to the right of c, as long as these two values are not past any other critical numbers. The two specific values of $f'(x)$ found in this manner tell us the sign of the derivative immediately to the left and to the right of c. This method is illustrated in the following example.

EXAMPLE 6 Find the extrema of the function

$$f(x) = \tfrac{1}{6}(2x^3 - 9x^2 + 12x)$$

using the first derivative test.

Solution

$$f'(x) = x^2 - 3x + 2 = (x - 2)(x - 1)$$

$f' = 0$ if $x = 1$ and $x = 2$ so that these values of x are the only critical numbers. Now $f'(0) = 2 > 0$ and $f'(\tfrac{3}{2}) = -\tfrac{1}{4} < 0$. Also $f'(3) = 1(2) = 2 > 0$. Therefore $x = 1$ yields a relative maximum $f(1) = \tfrac{5}{6}$ and $x = 2$ yields a relative minimum $f(2) = \tfrac{2}{3}$. These are not absolute extrema since $f(x) \to \infty$ as $x \to \infty$ and $f(x) \to -\infty$ as $x \to -\infty$.

●

We turn now to the statement and proof of another test for relative maxima and minima that is often the easiest to apply. Just as the first derivative is the rate of change of the function, the second derivative measures the rate of change of the first derivative. When the second derivative is positive at a number $x = c$, it means that the first derivative is increasing as x increases through $x = c$.

Therefore if $f'(c) = 0$ while $f''(c) > 0$ then $f'(x)$ is increasing from negative values to positive values as x increases through c. This means that we have a situation similar to that of Figure 4.7.5; that is $f(c)$ is a relative minimum of f.

Similarly if $f'(c) = 0$ and $f''(c) < 0$ then $f'(x)$ is decreasing from positive values to negative values as x increases through c, indicating that $f(c)$ is a relative maximum of f (Figure 4.7.4). We now back up these observations by stating and proving the following theorem.

Theorem 4   ***Second Derivative Test for Relative Extrema.***   Let c be a critical number at which $f'(c) = 0$ and let f' exist at all points in some open interval I containing c as an interior point. Then if $f''(c)$ exists and

(i)   if $f''(c) < 0$, $f(c)$ is a *relative maximum* of f
(ii)   if $f''(c) > 0$, $f(c)$ is a *relative minimum* of f

Proof of (i)   Since by definition and hypothesis,

$$f''(c) = \lim_{x \to c} \frac{f'(x) - f'(c)}{x - c} < 0$$

there exists an open interval (d, e) in I and containing c for which

$$\frac{f'(x) - f'(c)}{x - c} < 0 \tag{1}$$

for all $x \neq c$ in (d, e).

When $d < x < c$, it follows that $x - c < 0$ and thus $f'(x) - f'(c) > 0$ in (1). Since $f'(c) = 0$, we have $f'(x) > 0$ for x in (d, c).

Similarly, if $c < x < e$, we have $x - c > 0$ and therefore $f'(x) - f'(c) < 0$ in (1). Since $f'(c) = 0$, $f'(x) < 0$ for all x in (c, e).

Therefore by the first derivative test (Theorem 3), $f(c)$ is a relative maximum of f.

The proof of (ii) is similar and is left as an exercise for the reader.   ∎

EXAMPLE 7   Find the relative extrema of the function f defined by

$$f(x) = x^3 - 3x + 10$$

Solution $f'(x) = 3x^2 - 3$ and, if we set $f' = 0$ there results $x = \pm 1$ as the critical numbers.

$f''(x) = 6x$, $f''(1) = 6$ so that $x = 1$ yields a relative minimum. Also, $f''(-1) = -6$, which implies that we have a relative maximum at $x = -1$.   ●

EXAMPLE 8   Find the relative maxima and minima of the function f defined by

$$f(x) = Ax^2 + Bx + E \qquad A \neq 0$$

Solution $f'(x) = 2Ax + B$ and if we set $f'(c) = 0$ there results $c = \dfrac{-B}{2A}$ as the critical number.

Furthermore $f''(x) = 2A$ and therefore we conclude that $f\left(\dfrac{-B}{2A}\right)$ is a relative maximum of f if $A < 0$ and a relative minimum of f if $A > 0$.

Therefore, for example, $f(x) = x^2 - 4x + 7$ has a relative minimum at $x = 2$.   ●

EXAMPLE 9   Find the relative extrema of the function f defined by

$$f(x) = x^4$$

Solution $f'(x) = 4x^3$ so that $c = 0$ is the only critical number. However $f''(0) = 0$ and the second derivative test cannot be applied. So let us go back to the first derivative test. We find that $f' < 0$ if $x < 0$ and $f' > 0$ when $x > 0$. Therefore, the function decreases in $(-\infty, 0)$ and increases in $(0, \infty)$ which means that $f(0) = 0$ is both a relative and an absolute minimum for $f(x) = x^4$.   ●

Note that the conditions $f'(c) = 0$ and $f''(c) \neq 0$ provide *sufficient but not necessary conditions* for extrema.

Let us observe that in every example we looked at in this section where there was *just one* relative extreme value, this value also turned out to be an absolute extreme value. It seems reasonable for this to be the case. If $f(c)$ is a relative maximum value, the only way it could fail to be an absolute maximum value is if the graph of the function turns someplace and increases above $f(c)$. However, wherever the graph of the function turns would result in a second relative extreme value. (A similar argument applies if $f(c)$ is a relative minimum value.) We are then led to the following result.

Theorem 5   Let $f(x)$ be continuous on an interval I containing the number c, and assume that $f(c)$ is the only relative extreme value of $f(x)$ on I. Then $f(c)$ is an absolute extreme value. In particular, if $f(c)$ is a relative maximum value, then it is an absolute maximum value; if $f(c)$ is a relative minimum value, then it is an absolute minimum value.

The principal use of Theorem 5 is in application exercises of the type covered in Sections 4.4 and 4.5. In such problems, if there is only one critical number we can apply the first or second derivative test to see what type of relative extreme value (if any) occurs. Theorem 5 then yields that this is an absolute extreme value. Assuming it is the type of absolute extreme value required, we would no longer have to check the ends of the interval.

In Example 1, Section 4.4, we could have noted (after finding $x = 25$ to be the only critical number) that, since $P'' = -2$, $P(25)$ is both a relative and an absolute maximum value.

EXAMPLE 10   For our last example we return to Example 6 of Section 4.4 in which the optical law of reflection was derived. In trying to minimize the sum of these two distances $|\overline{AP}| + |\overline{PB}|$ we set up the function $g(x)$, found g' and set $g' = 0$ obtaining our critical number which implied that $\angle\alpha = \angle\beta$. But we never established that a minimum had been actually secured.

Now we had obtained

$$g'(x) = \frac{x}{\sqrt{x^2 + h^2}} + \frac{x - a}{\sqrt{(x - a)^2 + h_1^2}}$$

so that

$$g''(x) = \frac{\sqrt{x^2 + h^2} - \dfrac{x^2}{\sqrt{x^2 + h^2}}}{x^2 + h^2} + \frac{\sqrt{(x - a)^2 + h_1^2} - \dfrac{\sqrt{(x - }}{}}{(x - a)^2 + h_1^2}$$

$$g''(x) = \frac{h^2}{[x^2 + h^2]^{3/2}} + \frac{h_1^2}{[(x - a)^2 + h_1^2]^{3/2}} > 0 \qquad \text{for all } x$$

This guarantees that we have obtained a relative minimum at the critical point. Furthermore because $g'' > 0$ for all x, both a relative and an absolute minimum has been found. ●

Exercises 4.7

In Exercises 1 through 20, find the relative extrema of f by applying the first derivative test and the absolute extrema in the given interval.

1. $f(x) = x^2 - 6x + 1$, $[0, 7]$

2. $g(x) = 1 - 3x - x^2$, $[-3, 4]$

3. $h(x) = x^3 - 6x^2 + 12x - 7$, $[-1, 5]$

4. $F(x) = x^3 - 3x + 5$, $[-4, 2]$

5. $G(x) = x + \dfrac{1}{x}$, $[\frac{1}{2}, 4]$

6. $H(x) = \dfrac{x + 4}{x - 4}$, $[-1, 3]$

7. $f(x) = x^3 - 6x^2 + 9x - 5$, $[0, 4]$ *hope*

8. $g(x) = (x + 2)^2(x - 3)^2$, $[-3, 4]$

9. $h(x) = x\sqrt{4 - x^2}$, $[-2, 2]$

10. $F(x) = 3x^4 - 8x^3 - 18x^2 + 10$, $[-2, 4]$

11. $G(x) = \dfrac{x}{x^2 + 1}$, $[-3, 10]$

12. $H(x) = (x^2 - 3)(x - 1)^2$, $[-2, 2]$

13. $f(x) = 3 + 2(x - 1)^{2/3}$, $[0, 9]$

14. $f(x) = x^{1/3}(x - 2)^{1/3}$, $[-1, 4]$

15. $f(x) = \begin{cases} x + 3 & -2 \le x \le 1 \\ 5 - x & 1 < x \le 3 \end{cases}$

16. $g(x) = \begin{cases} 3 - x & 0 \le x \le 2 \\ x^2 - 4x + 5 & 2 < x < \infty \end{cases}$

17. $h(x) = \begin{cases} x & -1 \le x < 0 \\ 4x - x^2 & 0 \le x \le 3 \end{cases}$

18. $G(x) = x\sqrt[3]{3x - 4}$, $[0, 4]$

19. $H(x) = \sqrt[3]{x}(x - 7)^2$, $[0, 6]$

20. $F(x) = |x^2 - 2|$, $[-3, 2]$

In Exercises 21 through 36, find the relative extrema of the given function by using the second derivative test. If the second derivative test cannot be applied then utilize the first derivative test.

21. $f(x) = 2x^2 - 4x + 7$

22. $g(x) = x^3 - 3x + 2$

23. $h(x) = (x - 3)^2$

24. $F(x) = (x + 5)^3$

25. $H(x) = (2x - 1)^4$

26. $G(x) = x^3 - 5x^2 - 8x + 18$

27. $f(x) = \dfrac{bx}{b^2 + x^2}$ $(b \ne 0)$

28. $g(x) = \sqrt{1 - x^2} + \dfrac{1 + x}{2}$

29. $h(x) = \sqrt[3]{x - 1} - 1$

30. $f(x) = \begin{cases} -3x & x < 1 \\ -1 - 2x & 1 \le x \le 2 \\ x - 7 & 2 < x \end{cases}$

31. $g(x) = x^2 + \dfrac{1}{x^2}$

32. $h(x) = \dfrac{3x + 1}{x + 1}$

33. $F(x) = (1 + 3x)^4$

34. $G(x) = (2 - x)^5$

35. $H(x) = \begin{cases} 3x - 6 & x > 2 \\ 4 - x^2 & |x| \le 2 \\ -(x + 2)^2 & x < -2 \end{cases}$

36. $f(x) = \dfrac{x}{3}\sqrt[3]{4 - x}$

w that the function F defined by

$$F(x) = \frac{ax + b}{cx + d} \qquad ad - bc \neq 0$$

has no extrema regardless of the values of a, b, c, d.

38. If $g(x) = ax^2 + 2bx + c$, $a > 0$, show that $g \geq 0$ for all x if and only if $b^2 - ac \leq 0$. (*Hint:* Complete squares.)

39. Prove Schwarz's inequality

$$(a_1b_1 + a_2b_2 + \cdots + a_nb_n)^2 \leq (a_1^2 + a_2^2 + \cdots + a_n^2)(b_1^2 + b_2^2 + \cdots + b_n^2)$$

for all real $a_1, a_2, \ldots, a_n$ and $b_1, b_2, \ldots, b_n$. This inequality is significant in more advanced analysis. (*Hint:* Set $g(x) = (a_1x + b_1)^2 + (a_2x + b_2)^2 + \cdots + (a_nx + b_n)^2$ and use Exercise 38.)

40. If possible, determine a line through the point $(3, 4)$ such that the area of the triangle formed in the first quadrant with the positive coordinate axes is a positive number k. For what values of k is it impossible to construct such a triangle?

4.8 CONCAVE CURVES, INFLECTION POINTS, AND CURVE SKETCHING

We know that, if a function f is such that $f' > 0$ in an interval I, the function *increases* with increasing x. If the second derivative f'' is positive for all x in some interval, we would also like to know what can be said about this graph of f.

If f'' is positive in (a, b) then, since $f'' = (f')'$, this means that f' is an increasing function of x in (a, b). In geometric terms, the slope of the tangent line increases as x increases. The slope may be positive and increasing in (a, b) as shown in Figure 4.8.1, or the slope may increase from negative to positive (Figure 4.8.2). Finally, the slope may be negative in (a, b) but increasing algebraically (Figure 4.8.3).

The curve in all cases where $f'' > 0$ is shaped like the inside of a cup (holding fluid) and is said to be concave upward. More precisely the graph of $y = f(x)$ is said to be **concave upward** in (a, b) if the tangent line to this portion of the graph lies below the curve. It can be proved that (if we start with this latter definition of concave upward), if $f'' > 0$ in (a, b), the graph is concave upward. It can also be shown that when a curve is concave upward all the chords in that interval lie above the graph.

If, on the contrary, $f'' < 0$ throughout (a, b), then f' is a decreasing function and the graph of f appears as shown in Figure 4.8.4. The curve lies below its

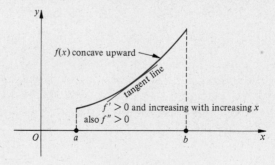

Figure 4.8.1

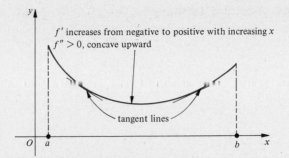

Figure 4.8.2

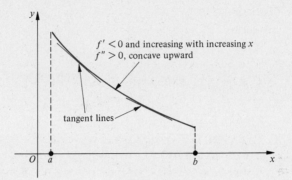

Figure 4.8.3

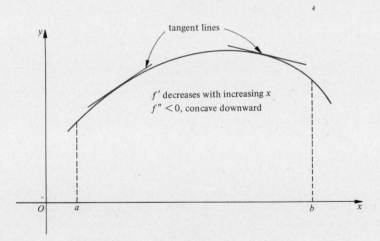

Figure 4.8.4

tangent lines and is called *concave downward* there. It can be shown that if $f'' < 0$ the curve must lie below the tangent lines and be concave downward.

Before we formalize our treatment of the subject of concavity let us consider two examples.

EXAMPLE 1   Examine the graph of $f(x) = x^2 - 2x - 2$ for concavity.

Solution $f' = 2x - 2, f'' = 2$, and the curve is concave upward for all x. In this case f' increases from negative to positive as x increases through $x = 1$ (Figure 4.8.5).

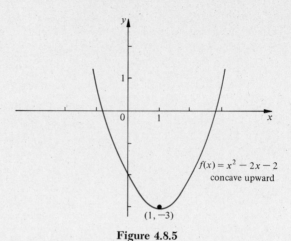

Figure 4.8.5

EXAMPLE 2 Examine the concavity of the graph of $f(x) = x^3 - 3x$.

Solution $f' = 3x^2 - 3$ and $f'' = 6x$ so that $f'' > 0$ if $x > 0$, whereas $f'' < 0$ if $x < 0$. Therefore, the graph is concave upward if $x > 0$ and concave downward if $x < 0$ (Figure 4.8.6).

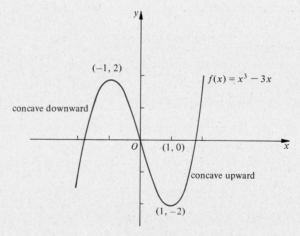

Figure 4.8.6

Now let us firm up our concepts by defining concavity and establishing certain results.

Definition 1 The graph of a function f is said to be **concave upward** at the point $(c, f(c))$ if $f'(c)$ exists and if there exists an open interval I containing c such that for all values of $x \neq c$ in I the point $(x, f(x))$ on the graph is above the tangent line to the graph at $(c, f(c))$. **Downward concavity** is defined analogously.

The equation of the tangent line T to the graph of f at the point $(c, f(c))$ is

$$y = f(c) + f'(c)(x - c) \tag{1}$$

so that the directed distance $g(x)$ from T to the graph of f at x measured parallel to the y axis is given by (Figure 4.8.7)

$$g(x) = f(x) - [f(c) + f'(c)(x - c)] \tag{2}$$

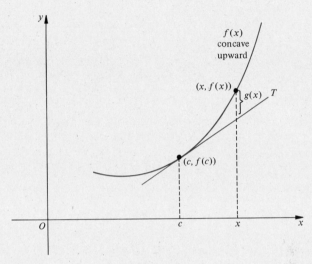

Figure 4.8.7

Furthermore the point $(x, f(x))$ is above T if $g(x) > 0$ and below T if $g(x) < 0$. Also $g(c) = 0$. Therefore the graph of f at the point $(c, f(c))$ is concave upward if $g(x) > 0$ for all x in some deleted neighborhood of c. Similarly, the graph of f is concave downward if, in some deleted neighborhood of c, $g(x) < 0$.

A very useful test for concavity is the following.

Theorem 1   If a function f is such that f' exists in some neighborhood of c, then

    (i)   the graph of f is concave upward at $(c, f(c))$ if $f''(c) > 0$

   (ii) the graph of f is concave downward at $(c, f(c))$ if $f''(c) < 0$.

Proof of (i)   If $f''(c) > 0$ then

$$f''(c) = \lim_{x \to c} \frac{f'(x) - f'(c)}{x - c} > 0$$

Thus

$$\frac{f'(x) - f'(c)}{x - c} > 0 \text{ for all } x \text{ in some deleted neighborhood}$$
$$I \text{ of } c$$

This means that in this neighborhood $f'(x) > f'(c)$ if $x > c$ and $f'(x) < f'(c)$ if $x < c$. Now, starting with (2) apply the mean value theorem of Section 4.6 to obtain

$$\begin{aligned}
g(x) &= [f(x) - f(c)] - f'(c)(x - c) \\
&= f'(d)(x - c) - f'(c)(x - c) \\
&= [f'(d) - f'(c)](x - c)
\end{aligned}$$

where d is some number between c and x. If x is in I and $x < c$, then $x < d < c$, $f'(d) < f'(c)$, and $g(x) > 0$. If x is in I and $x > c$, then $c < d < x$, $f'(c) < f'(d)$, and, once again, $g(x) > 0$. Therefore $g(x) > 0$ for every x in I, $(x \neq c)$, and the graph of f is concave upward at $(c, f(c))$ by definition.

The proof of (ii) is similar and is therefore left as an exercise. ∎

The converse of Theorem 1 is not true. For example, if f is the function defined by $f(x) = x^4$, the graph of f is concave upward everywhere and in particular at $(0, 0)$ where the tangent line is horizontal. However, $f''(0) = 0$. Accordingly a sufficient condition for the graph of a function to be concave upward at a point $(c, f(c))$ is that $f''(c) > 0$; however, this is not necessary. Analogously, a sufficient but not necessary condition that a function be concave downward at a point $(c, f(c))$ is that $f''(c) < 0$.

If there is a point on the graph of f for which the sense of the concavity changes, then, since $g(x)$ must change sign, the graph crosses the tangent line at such a point (Figure 4.8.8 and Figure 4.8.9). Such a point is called a *point of inflection.*

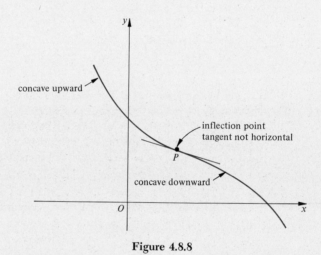

concave upward

inflection point
tangent not horizontal

P

concave downward

O

Figure 4.8.8

concave upward

inflection point
tangent horizontal

concave downward

O

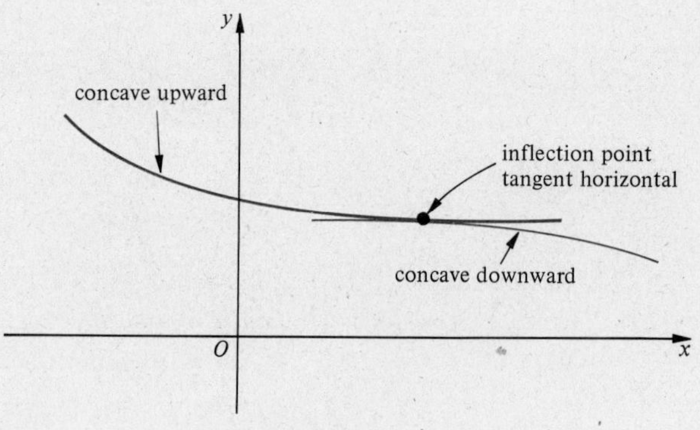

Figure 4.8.9

Definition 2　The point $(c, f(c))$ is a **point of inflection** of the graph of the function f if the graph has a tangent line there and if there exists an open interval I containing c such that for x in I either

(i)　$f''(x) < 0$ if $x < c$ and $f''(x) > 0$ for $x > c$ or
(ii)　$f''(x) > 0$ if $x < c$ and $f''(x) < 0$ for $x > c$.

Figures 4.8.8 and 4.8.9 illustrate condition (ii).

Note that Definition 2 indicates nothing about the value of f'' at the point of inflection. The following theorem establishes that, if the second derivative exists at the point of inflection, it must be zero there.

Theorem 2　If the function f is differentiable on some open interval containing the point of inflection $(c, f(c))$ of the graph of f, then if $f''(c)$ exists $f''(c) = 0$.

Proof　Let g be a function such that $g = f'$. Then differentiation yields $g' = f''$. Since $(c, f(c))$ is a point of inflection of f then f'' changes sign at $x = c$ which implies that g' also changes sign at c and $g'(c)$ exists. From the first derivative test (Theorem 3 of Section 4.7), g has a relative extremum at c. This implies that $g'(c) = 0$ (Theorem 2 of Section 4.3). But this means that $f''(c) = 0$. ∎

The converse of Theorem 2 does *not* hold. That is, if c is a number at which $f''(c) = 0$, it does not necessarily follow that c is an inflection point. This fact is easily established by considering the function $f(x) = x^4$ which has an absolute minimum at $x = 0$. However $f''(x) = 12x^2$ and $f''(0) = 0$. But $f'' \geq 0$ and does not change sign as x increases through $x = 0$. Thus $(0, 0)$ is not an inflection point.

Also the graph of a function may have a point of inflection and not have a second derivative at that point. For example, consider $f(x) = x^{1/3}$. Then for $x \neq 0, f'(x) = \frac{1}{3}x^{-2/3}$ and $f''(x) = -\frac{2}{9}x^{-5/3}$. Now $f'(0)$ and of course $f''(0)$ fail to exist while $f'' < 0$ if $x > 0$ and $f'' > 0$ when $x < 0$. Hence, f has a point of inflection at $(0, 0)$, and it has a vertical tangent at that point (Figure 4.8.10).

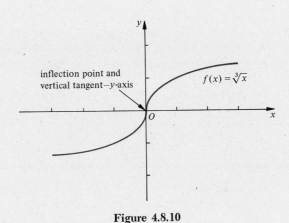

inflection point and
vertical tangent—y-axis

$f(x) = \sqrt[3]{x}$

Figure 4.8.10

EXAMPLE 3 Find the points of inflection of the graph of the function f given by

$$f(x) = x^3 - 3x^2 + 4$$

and determine where the graph is concave upward and where it is concave downward.

Solution

$$f(x) = x^3 - 3x^2 + 4$$
$$f'(x) = 3x^2 - 6x \qquad (f' = 0 \text{ implies that } x = 0 \text{ and } x = 2)$$
$$f''(x) = 6x - 6$$
$$f'' > 0 \text{ if } x > 1 \qquad \text{(graph is concave upward)}$$
$$f'' < 0 \text{ if } x < 1 \qquad \text{(graph is concave downward)}$$
$$f'' = 0 \text{ if } x = 1 \qquad \text{and } (1, 2) \text{ is the only point of inflection.}$$

Furthermore at this point the graph of f has a tangent line with slope $f'(1) = -3$. Figure 4.8.11 shows a sketch of the graph of f where the preceding information has been utilized. ●

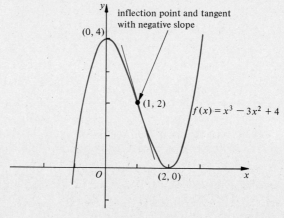

Figure 4.8.11

EXAMPLE 4 Determine the graph of the function g defined by

$$g(x) = \frac{3x^2}{2} + x|x|$$

Where is $g(x)$ concave upward and where is it concave downward? Where are the points of inflection?

Solution (See Figure 4.8.12)

$$g(x) = \begin{cases} \dfrac{5}{2}x^2 & \text{if } x \geq 0 \\[2mm] \dfrac{x^2}{2} & \text{if } x < 0 \end{cases}$$

and therefore

$$g'(x) = \begin{cases} 5x & \text{if } x \geq 0 \\ x & \text{if } x < 0 \end{cases}$$

and, in particular, the fact that $g'(0) = 0$ is included in the formulas for g'. Also

$$g''(x) = \begin{cases} 5 & \text{if } x > 0 \\ 1 & \text{if } x < 0 \end{cases}$$

and $g''(0)$ does not exist. Thus g is concave upward for all x and $g(0)$ is both a relative and an absolute minimum. There are no points of inflection. ●

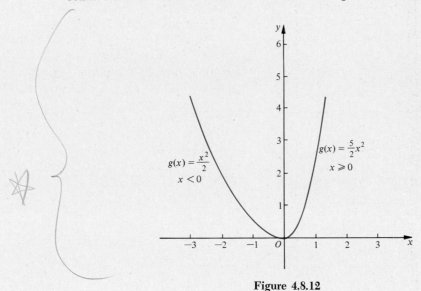

Figure 4.8.12

We have learned many techniques in this chapter and in earlier ones that enable us to draw a fairly accurate sketch of the graphs for many functions. We shall now summarize the methods that can be used and illustrate with two additional examples.

To graph the function $y = f(x)$ we proceed as follows:

1. Use $f(x)$ to determine the intercepts, the domain, and in some cases the range, to check for symmetry, and to find any vertical or horizontal asymptotes.
2. Use $f'(x)$ to find the critical numbers and to determine where the graph is increasing and where it is decreasing.
3. Use $f''(x)$ to locate possible points of inflection and to determine where the graph is concave upward and where it is concave downward.

EXAMPLE 5 Graph the function $f(x) = 3x^5 - 10x^3$.

Solution We follow the steps outlined previously.

1. Intercepts: If $x = 0$ then $y = 0$. Conversely, if $y = 0$ then $x^3(3x^2 - 10) = 0$ which yields $x = 0$ (three times) and $x = \pm\sqrt{\dfrac{10}{3}} \doteq \pm 1.83$.

Domain: $-\infty < x < \infty$.

There are no vertical or horizontal asymptotes.

Symmetry: Writing the equation as $y = 3x^5 - 10x^3$ and then replacing x by

$-x$ and y by $-y$, an equivalent equation is obtained. The graph is thus symmetric with respect to the origin. It would therefore be sufficient to investigate just the graph for $x \geq 0$ and then use the symmetry. However, for the sake of better illustrating steps (2) and (3), we shall look at all values of x.

2. $f'(x) = 15x^4 - 30x^2 = 15x^2(x^2 - 2)$
   $f'(x) = 0$ if $x = 0, \sqrt{2}, -\sqrt{2}$ so these are our critical numbers. Substituting into $f(x)$ gives $f(0) = 0$, $f(\sqrt{2}) = -8\sqrt{2} \doteq -11.3$, and $f(-\sqrt{2}) = 8\sqrt{2} \doteq 11.3$.
   If $x < -\sqrt{2}$, $f'(x) > 0$ (the function is increasing); if $-\sqrt{2} < x < 0$ or $0 < x < \sqrt{2}$, $f'(x) < 0$ (the function is decreasing), and if $x > \sqrt{2}$, $f'(x) > 0$ (the function is increasing).

3. $f''(x) = 60x^3 - 60x = 60x(x^2 - 1)$
   $f''(x) = 0$ if $x = 0, 1, -1$ and $f(0) = 0$, $f(1) = -7$, $f(-1) = 7$.
   If $x < -1$, $f''(x) < 0$ (the graph is concave downward); if $-1 < x < 0$, $f''(x) > 0$ (the graph is concave upward); if $0 < x < 1$, $f''(x) < 0$ (the graph is concave downward); and if $x > 1$, $f''(x) > 0$ (the graph is concave upward).

By either the first or second derivative test, $(-\sqrt{2}, 8\sqrt{2})$ is a relative maximum point and $(\sqrt{2}, -8\sqrt{2})$ is a relative minimum point.

The points $(-1, 7), (0, 0)$, and $(1, -7)$ are points of inflection since the concavity reverses at each of these points.

We sketch the graph (Figure 4.8.13) by first plotting these five points and the intercepts, and then connecting them according to the information we have found. We observe that the graph is symmetric with respect to the origin as was indicated earlier. ●

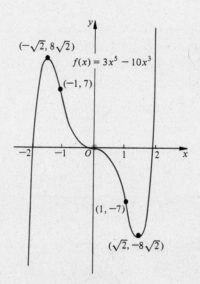

Figure 4.8.13

EXAMPLE 6   Graph the function $f(x) = x^2 + 1/x$.

Solution   1. The only intercept is $(-1, 0)$.
   Domain: All real numbers x where $x \neq 0$.

There are none of the standard symmetries.

Asymptotes: $x = 0$ is a vertical asymptote because $\lim\limits_{x \to 0^+} f(x) = \infty$ and $\lim\limits_{x \to 0} f(x) = -\infty$.

There are no horizontal asymptotes because $\lim\limits_{x \to \pm\infty} f(x) = \infty$

2. $f'(x) = 2x - \dfrac{1}{x^2}$

$f'(x)$ does not exist if $x = 0$, but this is not a critical number since $x = 0$ is not in the domain of $f(x)$.

$f'(x) = 0$ if $0 = 2x - \dfrac{1}{x^2} = \dfrac{2x^3 - 1}{x^2}$. Then $2x^3 - 1 = 0$, yielding

$x = \sqrt[3]{\dfrac{1}{2}} \doteq 0.79$. Also $f\left(\sqrt[3]{\dfrac{1}{2}}\right) = 3\sqrt[3]{\dfrac{1}{4}} \doteq 1.89$ (it takes several arithmetic steps to get this answer).

If $x < 0$ or $0 < x < \sqrt[3]{\dfrac{1}{2}}$, $f'(x) = \dfrac{2x^3 - 1}{x^2} < 0$ (the function is decreasing),

if $x > \sqrt[3]{\dfrac{1}{2}}$, $f'(x) > 0$ (the function is increasing).

3. $f''(x) = 2 + \dfrac{2}{x^3} = \dfrac{2x^3 + 2}{x^3}$

Setting $f''(x) = 0$ gives $x = -1$. Also $f(-1) = 0$.

If $x < -1$, $f''(x) > 0$ (the graph is concave upward); if $-1 < x < 0$, $f''(x) < 0$ (the graph is concave downward); and if $x > 0$, $f''(x) > 0$ (the graph is concave upward).

By either the first or second derivative test, $\left(\sqrt[3]{\dfrac{1}{2}}, 3\sqrt[3]{\dfrac{1}{4}}\right)$ is a relative minimum point.

$(-1, 0)$ is a point of inflection since the concavity reverses at that point. The graph is drawn in Figure 4.8.14. ●

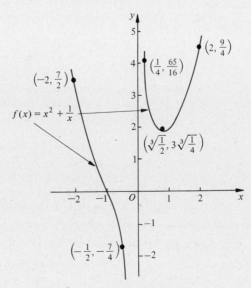

Figure 4.8.14

Note that, in any problem requiring the graph of $y = f(x)$ to determine where $f(x)$ is increasing or decreasing, we check the sign of $f'(x)$ between those values of x at which either $f'(x) = 0$ or $f'(x)$ does not exist. Similarly, for the concavity we determine the sign of $f''(x)$ between those values of x for which $f''(x) = 0$ or where $f''(x)$ does not exist.

Exercises 4.8

In Exercises 1 through 12 determine where the graph of the function is concave upward, concave downward, and locate all of the points of inflection if any.

1. $y = (x - 1)^3$

2. $y = x^6$

3. $f(x) = 2x^3 - 7x^2 + 8x - 10$

4. $g(x) = x + \dfrac{1}{x}$

5. $h(x) = (x + 2)^{3/2} - 12$

6. $F(x) = (x - 3)^{1/5}$

7. $G(x) = x^4 - 4x^3 + 8x + 2$

8. $H(x) = \dfrac{3x + 2}{x^2}$

9. $f(x) = 3 - \dfrac{4}{x} + \dfrac{4}{x^2}$

10. $g(x) = 3\sqrt{x} - 4\sqrt[4]{x} + 6$

11. $h(x) = -3x|x|$

12. $F(x) = x^{4/3} - 8x^{1/3}$

13. Show that $(1, 4)$ is not a point of inflection on the curve $y = g(x) = x^4 - 4x^3 + 6x^2 + 4x - 3$ even though $g''(1) = 0$. Is $(1, 4)$ a relative extrema?

14. Show that the graph of a cubic polynomial $f(x) = ax^3 + bx^2 + c + d$ $(a \neq 0)$ has one and only one point of inflection. Find the point of inflection.

15. Find a and b given that the graph of $f(x) = ax^3 + bx^2 + x - 1$ passes through $(-1, 1)$ and has an inflection point when $x = 2$.

16. Show that, if $f(x) = ax^3 + bx^2 + cx + d$ has a point of inflection at the origin, its graph is symmetric with respect to the origin—that is, for every x, $f(-x) = -f(x)$.

17. Show that the graph of the general cubic $y = ax^3 + bx^2 + cx + d$ is centrosymmetric about its inflection point. (A graph is said to be *centrosymmetric* with respect to a point O if for every point P of the graph there is a corresponding point P' on the graph and such that O is the midpoint of the line segment PP'.) (*Hint:* If (x_1, y_1) is the point of inflection for the general cubic write $x = u + x_1$ and $y = v + y_1$ then the relationship between v and u assumes the form $v = au^3 + \left(c - \dfrac{b^2}{3a}\right)u$ and the result follows from Exercise 16.)

18. An inflection point of a graph is called a *horizontal inflection* point if the slope at that point is zero. Find the horizontal inflection points of (a) $(x + 5)^3$ and (b) $(x - a)^3 + b$.

19. Show that a cubic polynomial has a horizontal inflection point if and only if it can be expressed in the form $A(x - a)^3 + b$.

20. Let a be a root of the polynomial $P(x)$. Prove that a is a double root of $P(x)$ if and only if it is a root of the derivative $P'(x)$. (*Hint:* Use the factor theorem $P(x) = (x - a) G(x)$, and so on.)

21. Prove that the graph of the polynomial $P(x)$ is tangent to the x-axis at $(a, 0)$ if and only if a is a root of multiplicity greater than 1. (*Hint:* Use Exercise 20.)

22. Prove that if a is a root of multiplicity 3 of a polynomial $P(x)$ then $(a, 0)$ is a horizontal point of inflection for the graph of $P(x)$.

23. Prove that given $g(x) = a + bx + cx^n$, $(c \neq 0)$, where n is an integer > 1, then the graph of g has an inflection point if and only if n is odd.

24. Show that the function $f(x) = x^3 + ax^2 + bx + c$ has no local extrema if $a^2 \leq 3b$ but has one local maximum and one local minimum if $a^2 > 3b$.

4.9 APPLICATION OF THE DERIVATIVE TO ECONOMICS

Calculus has been applied to problems in economics with considerable success. In this section, there will be no new mathematical ideas; however, the specialized terminology must be defined and exemplified.

The variation of one quantity with respect to another may be described by using the *average* or the *marginal* concept. In the case of the motion of a particle, we used the term average velocity. If a particle travelled a total distance of s ft in t sec then $\frac{s}{t}$ ft/sec is the average velocity and $s'(t)$ is the velocity (or instantaneous velocity). It is the velocity that corresponds to the marginal concept.

Let $C(x)$ be the total cost in dollars of producing x units of a commodity. The function C is called the *total cost function*. Normally $C \geq 0$ and $x \geq 0$. Also $C(0)$ is called the *overhead cost or fixed cost of production*. In order to apply calculus, x must vary continuously; that is, we will assume that x is a nonnegative and otherwise arbitrary real number even though x actually assumes integral values only. (In many instances our solutions to problems will yield nonintegral x. In that event we will round off our answers to the nearest integer.)

The *average cost* of producing x items of a commodity is given by

$$F(x) = \frac{C(x)}{x} = \frac{\text{total cost}}{\text{number of items produced}} \tag{1}$$

and $F(x)$ is called the *average cost function.*
The *marginal cost* is defined by

$$\lim_{\Delta x \to 0} \frac{C(x + \Delta x) - C(x)}{\Delta x} = C'(x) \tag{2}$$

It is the rate of increase in cost per unit of output. In particular, $C'(x_1)$ is the marginal cost when $x = x_1$ is the number of units produced. $C'(x)$ is called the *marginal cost function.* Note that

$$C'(x_1) \doteq \frac{C(x_1 + \Delta x) - C(x_1)}{\Delta x}$$

for small increments Δx; in other words

$$C'(x_1) \doteq \frac{\text{cost of increasing production from } x_1 \text{ to } x_1 + \Delta x}{\text{increment } \Delta x \text{ in production}} \tag{3}$$

In many practical situations we deal with x very large relative to $\Delta x = 1$. In such cases $C'(x_1) \doteq C(x_1 + 1) - C(x_1) =$ cost of increasing production by one unit when x_1 units are produced. Many economists interpret marginal cost in this way.

EXAMPLE 1　Suppose the cost in dollars for a weekly production of steel is given by the formula

$$C(x) = 8000 + 3x + \frac{x^2}{10}$$

when x is in tons. Find (a) the total cost in producing 1000 tons; (b) the average cost in producing 1000 tons; (c) the marginal cost when $x = 1000$ and the cost of producing the 1001th ton of steel.

Solution　(a)
$$C(1000) = 8000 + 3(1000) + \frac{(1000)^2}{10} = 111{,}000 \text{ dollars}$$

(b)
$$\frac{C(1000)}{1000} = \frac{111000}{1000} = 111 \text{ dollars/ton}$$

(c)
$$C'(x) = 3 + \frac{x}{5}$$

$$C'(1000) = 203 \text{ dollars/ton}$$

The cost of producing the 1001th ton of steel is given by

$$C(1001) - C(1000) = \left[8000 + 3(1001) + \frac{(1001)^2}{10} \right]$$

$$- \left[8000 + 3(1000) + \frac{(1000)^2}{10} \right]$$

$$= 3(1) + \frac{1}{10}[(1001)^2 - (1000)^2]$$

$$= 3 + \frac{1}{10}(1001 + 1000)(1001 - 1000)$$

$$= 3 + \frac{2001}{10} = 3 + 200.1 = 203.1 \text{ dollars/ton}$$

Thus $C'(1000)$ is an excellent approximation to the increment

$$\frac{C(1001) - C(1000)}{1001 - 1000} = C(1001) - C(1000) \qquad \bullet$$

By the **marginal average cost** we mean the derivative of the average cost or $F'(x)$ in our notation. Geometrically $F'(x_1)$ gives the slope of the tangent line to the average cost curve at the point where $x = x_1$.

EXAMPLE 2　Suppose that the cost function C is of the form $C(x) = mx + b$ (where m and b are assumed to be constants), then the marginal cost is $C' = m$. The average cost function is

$$F(x) = \frac{C(x)}{x} = m + \frac{b}{x}$$

and the marginal average cost function is given by

$$F'(x) = \frac{-b}{x^2} \qquad \bullet$$

EXAMPLE 3 Suppose that $C(x)$ is the total cost in dollars of producing x units of a commodity and that

$$C(x) = x^2 - 2x + 9 \qquad \text{if } x \geq 1$$

Find (a) the average cost; (b) the marginal cost; (c) the marginal average cost; (d) the absolute minimum for the average cost function. Sketch the graph of the cost function, the average cost function, and the marginal cost function on the same set of coordinate axes.

Solution (a) The average cost function

$$F(x) = \frac{C(x)}{x} = x - 2 + \frac{9}{x}$$

(b) The marginal cost function is C' and is given by

$$C'(x) = 2x - 2$$

(c) The marginal average cost function F' is

$$F'(x) = 1 - \frac{9}{x^2}$$

(d) In order to minimize the average cost function F we set $F' = 0$ which yields $x = \sqrt{9} = 3$. Furthermore, $F''(x) = \dfrac{18}{x^3} > 0$ since $x > 0$. Since $x = 3$ is the only critical value, then, from Theorem 5 of Section 4.7, it follows that F has an absolute minimum at $x = 3$. This means that the average cost is least when 3 units are produced and is given by \$4.

The graphs of the three functions are shown in Figure 4.9.1. Note that the lowest point on the graph of the average cost function occurs when the curves for

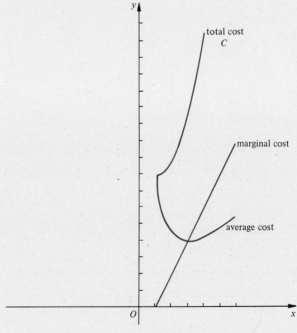

Figure 4.9.1

the marginal and average cost functions intersect. This is not a coincidence but rather is a general result. To see this we observe that from $F = \dfrac{C}{x}$ we have by differentiation

$$F' = \frac{xC' - C}{x^2}$$

and thus $F' = 0$ if and only if $xC' - C = 0$ or

$$C' = \frac{C}{x}$$

This says that the average cost is least when the average and marginal costs are equal. ●

If x units of a commodity are sold at a unit price p then $p(x)$ is called the **demand function.** Generally p is a decreasing function of x; that is, $p(x_2) < p(x_1)$ if $x_2 > x_1$. Thus if we treat x as a continuous variable, then $p'(x) < 0$ generally. The **total revenue** or **income** function from the sale of x units of a commodity $R(x)$ is given by

$$R(x) = x \cdot p(x) \tag{4}$$

and its derivative R' is the **marginal revenue function.** It measures approximately the change in revenue for unit change in the number of items sold. Also from (4), $p(x) = \dfrac{R(x)}{x}$ so that the demand function is the same as the average revenue.

EXAMPLE 4   A manufacturer of movie projectors can produce x projectors per day at a total cost $C = 100 + 50x$. The demand equation for this commodity is $x + \dfrac{p}{4} = 90$. How many projectors should be produced per diem in order to maximize the profit?

Solution
$$p = 360 - 4x \qquad \text{and} \qquad R = 360x - 4x^2 \qquad x \geq 0$$

Also $C = 100 + 50x$ so that the daily profit $u(x)$ is

$$u = R - C = (360x - 4x^2) - (100 + 50x) = -4x^2 + 310x - 100$$
$$u' = 310 - 8x \qquad u'' = -8$$

Set $u' = 0$ and find $x = \frac{310}{8} = 38\frac{3}{4} \doteq 39$ so that, in order to maximize his daily profit, the manufacturer should produce and sell 39 projectors per day at a price of 204 dollars each. ●

Another application of extrema theory is obtained by examining the classical inventory model. This model is used to determine the **economic order quantity,** which is defined as the order quantity that minimizes the inventory costs.

There are two basic factors that make up inventory costs. The first is the holding cost per item. If x is the order size and assuming uniform selling of the items, the average inventory level will be $\frac{x}{2}$. If a is the holding cost per item, then the holding cost is $\frac{ax}{2}$. If it is assumed that orders are received exactly when the inventory level reduces to zero, we have the situation exhibited in Figure 4.9.2.

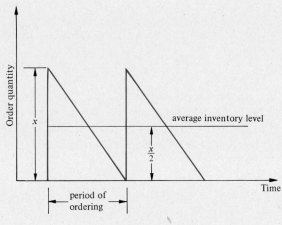

Figure 4.9.2

The other factor affecting inventory cost is the ordering cost. If, say, the total annual demand is q, then $\frac{q}{x}$ is the number of orders per annum. If b is the cost of placing a single order, the ordering cost is $\frac{bq}{x}$.

In summary, the cost function C is given by

$$C = \frac{ax}{2} + \frac{bq}{x}$$

where a = holding cost per item
b = cost of placing an order
q = periodic demand (often the annual demand)
x = order size
We seek the value of x that minimizes the cost C.

$$C' = \frac{a}{2} - \frac{bq}{x^2} \qquad C'' = \frac{2bq}{x^3} > 0$$

$C' = 0$ implies that $x = \sqrt{\frac{2bq}{a}}$. This is the order size that minimizes the cost. It is interesting to note that in this instance C is least when the holding cost and the ordering cost are equal; that is $\frac{ax}{2} = \frac{bq}{x}$. Graphically these costs appear as shown in Figure 4.9.3.

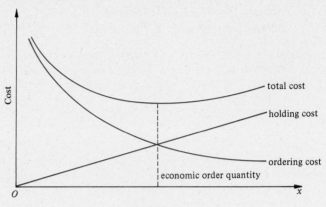

Figure 4.9.3

Exercises 4.9

1. The number of dollars in the total cost of manufacturing x radios is given by $C(x) = 600 + 40x + \dfrac{90}{x}$, where $x \geq 1$. Find (a) the average cost, (b) the marginal cost when $x = 10$, (c) the cost of manufacturing the 11th radio.

2. For the cost function $C(x) = 600 + 0.40x + \dfrac{90}{x}$, $x \geq 1$, find the value of x that minimizes the total cost. What is the least cost?

3. In the production of a certain commodity the number of dollars in the total cost of producing x units is $C(x) = 36\sqrt{x} + 10x + 80$. Find (a) the marginal cost when 100 units are produced and (b) the number of units produced when the marginal cost is $12.25.

4. The number of hundreds of dollars in the total cost of producing x units of a certain commodity is $C(x) = x^2 + x + 4$, $x \geq 1$. Find the function giving (a) the average cost, (b) the marginal cost, (c) find the absolute minimum average cost, (d) sketch the total cost, average cost, and the marginal cost on the same set of axes. Verify that the average cost is least when it equals the marginal cost.

5. If $C(x)$ dollars is the total cost of producing x units of a commodity and $C(x) = \dfrac{x^3}{3} - x^2 - 8x + 4$, $x \geq 1$. Find (a) the range of C, (b) the average cost function, (c) the marginal cost function, (d) sketch the total cost function. Find the points of inflection.

6. A manufacturer of transistor radios finds that his fixed cost per week is $500 and that it costs $10 for each radio produced. If x denotes the number of radios produced in a week, find (a) the total cost function, (b) the average cost function, (c) the marginal cost function, (d) show that the average cost function has no absolute minimum, (e) what is the smallest number of radios that must be produced so that the average cost per radio be less than $15?

7. Given the cost function $C = x^2 - 20x + 200$ and the revenue function $R = 80x - 10x^2$, where x, C, and R are in hundreds, find the value of x where the marginal revenue equals the marginal cost.

8. Given the cost function $C = 400 + 1.60x$ and the revenue function $R = 5.5x - 0.002x^2$, both in dollars. Find the maximum profit.

9. A manufacturer estimates that he can sell 2000 toys per month if he sets the unit price at $5.00. Furthermore he estimates that for each $0.20 decrease in price his sales will go up by 200. Find (a) the demand and revenue functions, (b) the number of units he should sell each month in order to maximize the revenue per month, (c) the maximum monthly revenue.

10. A steel mill can produce x tons of steel at a cost of C dollars where $C = 800 + 92x + 0.05x^2$. The steel can be sold at $132 per ton. How many tons should be produced so as to maximize profits? What is the maximum profit?

11. A manufacturer of ladies tailored suits estimates that to produce a certain style his costs will consist of $2200 fixed overhead plus $100 per suit. The price he can sell them for depends upon their "exclusive-

ness," and it is estimated that he can get in dollars $350 - 4\sqrt{x}$ for each one if x suits are placed on the market. How many suits should the manufacturer produce for maximum profit? What is the maximum profit? What price should be set by the manufacturer?

12. The cost of holding one item in inventory is $1.00, and the cost of placing an order $2.00. Assume a yearly demand for the item of 400 units. Find the value of x for which costs are minimized. What is the minimum cost?

13. An umbrella manufacturer finds that he can sell x umbrellas a day at p dollars each. His production costs are $100 + 2x + 0.05x^2$. If the demand function is $x = 500 - 25p$, find the production level that produces the maximum profit. What price should be set? What is the maximum daily profit?

14. A woman has a total of $3000 to invest in two enterprises. An investment of u dollars in the first enterprise will give a return of $2\sqrt{u}$ dollars, whereas an investment of v dollars in the second enterprise will give a return of $\sqrt{v}$ dollars. Determine how much she should invest in each enterprise in order to maximize her return.

15. Let x denote a quantity of a product demanded. Also let $p(x) = \alpha - \beta x$ and $C(x) = ax^2 + bx + k$ be the demand and cost functions, respectively. Find the output level that maximizes the profit assuming that all constants α, β, a, b, and k are positive.

16. A fabric manufacturer must supply a customer with 24,000 yd of a certain cloth during the coming year. There is a uniform demand for this material and the manufacturer plans to produce the same number of yards at each production run. Furthermore the following information is known

   (i)   cost per set up = $300
   (ii) inventory costs per yard = $0.40
   (iii) production costs per yard = $1.00

Determine the number of yards to be made in each production run so that the total cost to the manufacturer be minimized. What is the minimum cost?

17. The demand equation for a certain commodity is $p = 6.8 - 0.0003x$, where x is the number of units produced each month and p dollars is the unit price. The cost to the monopolist is 800 dollars overhead and 3 dollars per unit. Furthermore the government taxes the monopolist 20 cents for each item produced. If the monthly profit is to be as large as possible, find (a) the number of units that should be produced each month, (b) the unit price, (c) the monthly profit.

18. A monopolist determines that if $C(x)$ dollars is the total cost of producing x units of a certain commodity, then $C(x) = 30x + 2000 + tx$ where t dollars/unit is the tax levied by the government. The demand equation is $p + 0.02x = 100$, where p is the unit price in dollars. Find the value of t that should be imposed by the government to maximize its tax revenue. What is the maximum revenue?

Review and Miscellaneous Exercises

1. The volume of a sphere is given by $V = \frac{4}{3}\pi r^3$ where r is its radius. Find the rate of change of the volume V (a) with respect to its radius r and (b) with respect to its diameter.

2. The height h of a cylinder is always $\frac{2}{3}$ the radius r. Find the rate of change of the volume V with respect to the height h.

3. A right triangle has hypotenuse 50 in. and variable legs x and y respectively. If y increases at the rate of 4 in./sec, at what rate is x changing when $y = 40$ in.?

4. The cost in dollars of selling x items is given by $C = 300 + 20x + \frac{100}{x}$. When the 20th item is sold, it is observed that the rate of sales is 5 items/hr. What is the rate of change of the cost at that moment?

5. The cost in dollars of selling x items is given by $C = 0.0016x^3 - 0.20x^2 + 40x + 10,000$. Find the rate of change of the cost when the 500th item is sold if the rate of sales is 50 per day.

6. A man stands near the top of a 17-ft ladder that is leaning against a vertical wall. The foot of the ladder is slipping away from the wall at the rate of $\frac{3}{4}$ ft/sec. At what rate is the top of the ladder coming down when it is 8 ft off the ground?

7. The base radius of a conical filter is 4 in. and the altitude is 9 in. Coffee is dripping out at the rate of $\frac{1}{2}$ in.^{3}/min. At what rate is the coffee level changing when the coffee level is 3 in. deep?

8. Find two numbers x and y such that $x + 2y = 10$ and xy is as large as possible.
9. Find the absolute maximum and absolute minimum for $f(x) = x^3 + 2x + 10$ where x is any real number.
10. Find the absolute maximum and absolute minimum for $f(x) = 3|x - 2|$ if $0 < x \le 6$.
11. Find the absolute maximum and absolute minimum for

$$f(x) = \begin{cases} \dfrac{1}{3}(22 - 4x) & 1 \le x \le 4 \\ \dfrac{1}{4}(5x - 12) & 4 \le x \le 8 \end{cases}$$

Are the absolute extrema also relative extrema?

12. A spherical weather balloon is expanding due to heating of its gas by solar radiation. Its volume is increasing at 5 ft^3/min. At what rate is its surface area changing when the radius is 16 ft? (*Given data:* $V = \frac{4}{3}\pi r^3$, $S = 4\pi r^2$. *Hint:* Use the chain rule.)
13. A man 6 ft tall walks at 4 ft/sec directly away from a lamp post supporting a lamp 22 ft above the ground. How fast is his shadow lengthening, and with what speed does the tip of his shadow move?
14. The average speed of a car for a round trip is $37\frac{1}{2}$ mph. If the car averaged 30 mi/hr going, what was the average rate for the return trip?
15. Verify Rolle's theorem for each function on the indicated interval:
    (a) $f(x) = x^2 - 8x - 20$ on $[-2, 10]$
    (b) $f(x) = x^3(x - 4)$ on $[0, 4]$
16. Does Rolle's theorem apply to the function $f(x) = \sqrt{x(3 - x)}$ in $[0, 3]$? Explain your conclusion.
17. Does Rolle's theorem apply to the function f defined by $f(x) = |4 - x^2|$ on $[-3, 3]$? Explain your conclusion.
18. Two ships leave from the same point at different times. The first leaves at noon and sails due west at a speed of 20 mi/hr. The second leaves at 1 P.M. and sails $30°$ west of south at a speed of 30 mi/hr. At what rate are the ships separating at 3 P.M.?
19. Suppose that f is continuous on $[a, b]$ where $a < b$ and $f' < 0$ on (a, b). Prove that f is strictly decreasing on $[a, b]$. (*Hint:* Use the basic mean value theorem of differential calculus.)
20. Discuss the graph of

$$f(x) = \frac{x^3}{3} - x^2 + \frac{4}{3} \qquad \text{where } -\infty < x < \infty$$

(*Hint:* Utilize the signs of f', relative extrema, and so on.)
21. Find a and b given that $y = x^3 + ax^2 + 2x + b$ has an inflection point at $(1, 2)$.
22. Show that the equation $x^5 - 80x - 19 = 0$ has no more than three real roots.
23. A particular response u of the body to a dosage x of a drug is given by the formula

$$u = ax^2 - \frac{x^4}{4} \qquad 0 \le x \le 2\sqrt{a}, \ (a > 0 \text{ and constant})$$

  (i) What is the dosage that maximizes u?
  (ii) What is the dosage that maximizes the rate of change of u?
24. A firm manufactures solid state radios and wholesales its product at $30 per radio. The total cost C in dollars of producing x radios is given by

$$C(x) = 200 + 5x + 0.01x^2$$

Find the number of radios to be produced to yield the maximum profit. What is the maximum profit?
25. Consider the cubic polynomial

$$g(x) = x^3 + ax^2 + bx + c \qquad -\infty < x < \infty$$

(a) For what values of a, b, and c will g have critical numbers at $x = 1$ and $x = 4$?
(b) Describe the nature of the critical values.
(c) Are the values absolute extrema?
26. For the function

$$f(x) = x - 3x^{1/3} \qquad -\infty < x < \infty$$

find the critical values, nature of extrema, and inflection points.

27. Find the area of the largest isosceles triangle that can be inscribed in a circle of radius a.
28. It has been established (both empirically and analytically using a model of idealized fluid flow in circular tubes) that the velocity v of the flow of air through the respiratory system during coughing is given by the formula

$$v(r) = \frac{k}{\pi} r^2 (r_0 - r)$$

where k is a positive constant. Also r_0 is the radius of the windpipe in a nonexcited state, that is, when there is no differential pressure (no excess above atmospheric pressure). Thus r_0 is also constant. The quantity r is the variable windpipe radius, which is pressure dependent. Find the value of r that maximizes the velocity v and most effectively cleans the lungs.
29. The function

$$f(x) = x^9 - 6x^5 + 3x^3 + 400$$

is such that $f'(0) = 0$. Decide upon whether $f(0)$ is a relative maximum, a relative minimum, or neither for $f(x)$. (*Hint:* Investigate the sign of $f'(x)$ in a neighborhood of $x = 0$.)
30. An oil tank is to have a prescribed volume V and is to be made in the shape of a right circular cylinder with hemispherical ends. The material for the ends costs three times as much per square foot as that for the cylindrical part. Find the most economical dimensions. (*Hint:* If h is the altitude of the cylinder and r is the radius of the base as well as the radius of hemisphere then $V = \pi r^2 h + \frac{4}{3}\pi r^3$ is the given volume of the tank. Next write the expressions for the surface areas of the cylinder and hemispheres, then for the cost function, and so on.)
31. A man can rent all of his 100 apartments if he rents them for $400 per month. He estimates that for each $10 increase in rent he will rent one less apartment. What rent should he charge to maximize his return? What would this maximum monthly return be?
32. Find the right circular cylinder of largest volume V that can be inscribed in a sphere of radius R. (*Hint:* Maximize V^2 and simplify the algebra.)
33. (a) Find the minimum value of

$$\frac{\frac{a_1 + x}{2}}{\sqrt{a_1 x}} \qquad \text{where } a_1 > 0 \text{ and } x > 0$$

(b) Establish as a corollary to (a) that

$$\frac{a_1 + a_2}{2} \geq \sqrt{a_1 a_2}$$

(c) Interpret the result in (b).
34. This is a generalization of Exercise 33.
(a) Find the minimum value of

$$\frac{\frac{a_1 + a_2 + x}{3}}{\sqrt[3]{a_1 a_2 x}} \qquad \text{where } a_1 > 0,\ a_2 > 0 \text{ and } x > 0$$

(b) Establish as a corollary to (a) that

$$\frac{a_1 + a_2 + a_3}{3} \geq \sqrt[3]{a_1 a_2 a_3}$$

(c) Interpret the result in (b).

5

Differentials Antiderivatives, and the Substitution Method

THE DIFFERENTIAL

Suppose that we are given

$$y = f(x) = x^2 \tag{1}$$

so that

$$\Delta y = f(x + \Delta x) - f(x) = (x + \Delta x)^2 - x^2 \tag{2}$$

or expanding and simplifying

$$\Delta y = 2x \, \Delta x + (\Delta x)^2$$

If $|\Delta x|$ is much less than $|x|$, then $(\Delta x)^2$ is insignificant in comparison to $2x \, \Delta x$. It thus appears that $2x \, \Delta x$ would be a "good approximation" to Δy. We say that

$$\Delta y \approx 2x \, \Delta x = f'(x) \, \Delta x \tag{3}$$

which simply means that the right side is approximately equal to the left side under certain conditions to be made precise a little later.

Two new symbols are introduced at this point. First we have the *differential of the independent variable dx*, which is defined by

$$dx = \Delta x \tag{4}$$

Since Δx is arbitrary, dx must also of course be an arbitrary real number. However, in applications $|dx| = |\Delta x|$ is usually chosen to be much less than one. Secondly, the *differential dy of the dependent variable y* is defined by

$$dy = f'(x) \, dx = f'(x) \, \Delta x \tag{5}$$

Note that dy, the differential of f, depends not only upon x but also upon dx. Furthermore, we may assign to x any real number in the domain of f' and again dx may be any real number whatsoever. The differential of f is often denoted by df,

Differentials, Antiderivatives, and Substitution Method / Ch. 5

that is

$$df = f'dx \qquad (6)$$

by definition. This differential notation is due to Leibnitz.

EXAMPLE 1 Find the differential of the function $f(x) = x^2 - 2x - 6$ when $x = 3$ and $dx = -4$.

Solution
$$df = f'dx = (2x - 2)\,dx$$

and therefore

$$df\Big|_{\substack{x=3 \\ dx=-4}} = 4(-4) = -16 \qquad \bullet$$

EXAMPLE 2 Find the differential of the function given by $y = g(x) = \sqrt{x}$ when $x = 4$ and $dx = 0.1$.

Solution
$$dy = g'dx = \frac{1}{2\sqrt{x}}\,dx$$

and therefore

$$dy\Big|_{\substack{x=4 \\ dx=0.1}} = \frac{1}{2(2)}(0.1) = \frac{0.1}{4} = 0.025 \qquad \bullet$$

Figures 5.1.1 and 5.1.2 illustrate the relationship among the quantities defined. The reason that df, or equivalently dy, is equal to the vertical side of the triangle PRT in the figure whose horizontal base is $\Delta x = dx > 0$ is that the slope of the tangent line is equal simultaneously to $f'(x)$ and to the quotient of the lengths of the vertical side to the horizontal. Therefore

$$f'(x) = dy \div dx = \frac{dy}{dx} \qquad (7)$$

provided that $dx \neq 0$.

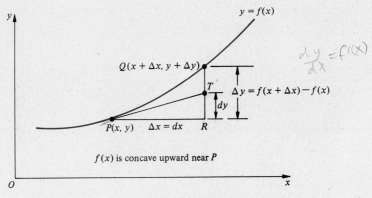

$f(x)$ is concave upward near P

Figure 5.1.1

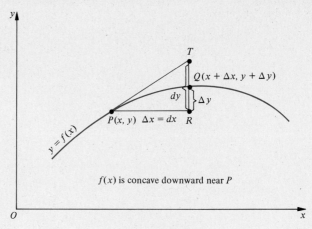

$f(x)$ is concave downward near P

Figure 5.1.2

In fact, the symbol $\dfrac{dy}{dx}$ for the derivative was due to Leibnitz at the end of the seventeenth century and was used in most texts until recently. Since then its popularity has waned because of the correct criticism that students tend to think of $\dfrac{dy}{dx}$ as a quotient, $dy \div dx$, before the symbols dx and dy are given meanings individually. The derivative is the limit of a quotient and it is only because we defined dy by the formula $dy = f'(x)\,dx$ that (7) is justified. It is the author's viewpoint that Leibnitz' notation, although very convenient, at times tends to mask what limits are being taken, and it is best to delay using it as we have done.

In the Leibnitz notation the derivatives of higher order are written

$$\frac{d^2y}{dx^2} = \frac{d}{dx}\left(\frac{dy}{dx}\right),\ \frac{d^3y}{dx^3} = \frac{d}{dx}\left(\frac{d^2y}{dx^2}\right),\ \dots,\ \frac{d^ny}{dx^n} = \frac{d}{dx}\left(\frac{d^{n-1}y}{dx^{n-1}}\right)$$

or if $y = f(x)$,

$$\frac{d^2f}{dx^2} = \frac{d}{dx}\left(\frac{df}{dx}\right),\ \frac{d^3f}{dx^3} = \frac{d}{dx}\left(\frac{d^2f}{dx^2}\right),\ \dots,\ \frac{d^nf}{dx^n} = \frac{d}{dx}\left(\frac{d^{n-1}f}{dx^{n-1}}\right)$$

The following relationships can be easily derived

$$dc = 0 \qquad\qquad (c \text{ constant}) \qquad (8)$$

$$d(x^n) = nx^{n-1}\,dx \qquad\qquad (n \text{ constant}) \qquad (9)$$

$$d(f + g) = df + dg \qquad\qquad (10)$$

$$d(cf) = c\,df \qquad\qquad (c \text{ constant}) \qquad (11)$$

$$d(f - g) = df - dg \qquad\qquad (12)$$

$$d(fg) = f\,dg + g\,df \qquad\qquad (13)$$

$$d(f/g) = \frac{g\,df - f\,dg}{g^2} \qquad\qquad (14)$$

$$d(g^n) = ng^{n-1}\,dg \qquad\qquad (n \text{ constant}) \qquad (15)$$

We will prove two of these and the rest are left as exercises.

Proof of (10)   By definition

$$d(f + g) = (f + g)' \, dx$$
$$= (f' + g') \, dx$$
$$= f' \, dx + g' \, dx$$
$$= df + dg \qquad\qquad \blacksquare$$

Proof of (13)

$$d(fg) = (fg)' \, dx$$
$$= (fg' + gf') \, dx$$
$$= fg' \, dx + gf' \, dx$$
$$= f \, dg + g \, df \qquad\qquad \blacksquare$$

The reader is encouraged to establish the remaining formulas.

How close is df to Δf? Certainly if $\Delta x = dx$ are small in magnitude so are df and Δf. To claim that Δf and df differ by a little is therefore meaningless. What is significant is that their ratio is near 1 as Δx tends to zero. In fact we establish the following as a theorem.

Theorem 1   Let f be differentiable at a number x. Also assume that $f'(x) \neq 0$. Then

$$\lim_{\Delta x \to 0} \frac{\Delta f}{df} = 1 \qquad\qquad (16)$$

Proof   Write $\dfrac{\Delta f}{df} = \dfrac{\Delta f / \Delta x}{df / \Delta x} = \dfrac{\Delta f / \Delta x}{df / dx}$

since $dx = \Delta x$. Therefore

$$\lim_{\Delta x \to 0} \frac{\Delta f}{df} = \frac{\lim_{\Delta x \to 0} \Delta f / \Delta x}{\lim_{\Delta x \to 0} df / dx} = \frac{df / dx}{df / dx} = 1$$

since $\lim_{\Delta x \to 0} \Delta f / \Delta x = f' = df / dx$ (by definition of f'). $\qquad \blacksquare$

Although it is not significant that $\Delta f - df$ tends to zero as $\Delta x \to 0$, the following important theorem shows that $\Delta f - df$ approaches zero *more rapidly* than Δx as $\Delta x \to 0$.

Theorem 2   Let f be differentiable at a number x then

$$\lim_{\Delta x \to 0} \frac{\Delta f - df}{\Delta x} = 0 \qquad\qquad (17)$$

Proof

$$\frac{\Delta f - df}{\Delta x} = \frac{\Delta f}{\Delta x} - \frac{df}{\Delta x} \qquad \text{if } \Delta x \neq 0$$
$$= \frac{\Delta f}{\Delta x} - \frac{df}{dx}$$

Take limits of both sides as $\Delta x \to 0$,

$$\lim_{\Delta x \to 0} \frac{\Delta f - df}{\Delta x} = \lim_{\Delta x \to 0} \frac{\Delta f}{\Delta x} - \lim_{\Delta x \to 0} \frac{df}{dx}$$
$$f' \quad - \quad f' \; = \; 0 \qquad\qquad \blacksquare$$

If we write $\dfrac{\Delta f - df}{\Delta x} = \varepsilon$ then $\varepsilon \to 0$ as $\Delta x \to 0$. Also if the relation is solved for Δf,

$$\Delta f = df + \varepsilon \, \Delta x \tag{18}$$

Differentials are frequently useful in approximations. Suppose that we wish to find Δf and, as a first approximation, df is calculated. It is natural to call

$$\Delta f - df \tag{19}$$

the *error.* If the error is divided by $\Delta x \neq 0$ then we obtain

$$\frac{\Delta f - df}{\Delta x} \tag{20}$$

which is called the **relative error.** Of course if we compare (18) with (20), ε is the relative error; thus, the relative error approaches zero as Δx approaches zero. The *percentage relative error* is 100 times the relative error.

EXAMPLE 3   For the function $y = f(x) = x^2 + 5x$ find Δf and df
(a) in general, and (b) at $x = 2$. (c) At $x = 2$ compute Δf and df for $\Delta x = 1$, $\Delta x = 0.1$, and $\Delta x = 0.01$. Find the error and the relative error when Δf is replaced by df.

Solution   (a) $\Delta f = f(x + \Delta x) - f(x) = [(x + \Delta x)^2 + 5(x + \Delta x)] - [x^2 + 5x]$
$ = x^2 + 2x \, \Delta x + (\Delta x)^2 + 5x + 5 \, \Delta x - x^2 - 5x$
$ = (2x + 5) \, \Delta x + (\Delta x)^2$

Also $df = f' \, dx = f' \, \Delta x = (2x + 5) \, \Delta x$

Therefore

$$\Delta f - df = (\Delta x)^2$$

Hence, in this case, $\varepsilon \, \Delta x = (\Delta x)^2$ and for $\Delta x \neq 0$, $\varepsilon = \Delta x$. Clearly as $\Delta x \to 0$ so does $\varepsilon \to 0$.
(b) In particular, at $x = 2$, $\Delta f = 9 \, \Delta x + (\Delta x)^2$ and $df = 9 \, \Delta x$ so that $\Delta f - df = (\Delta x)^2$
(c) The results for specific values of Δx are presented in Table 5.1.1. Note that the relative error is going to zero as Δx approaches zero. In this special case $\varepsilon = $ relative error $= \Delta x$.   ●

Table 5.1.1

Δx	$\Delta f = 9 \, \Delta x + (\Delta x)^2$	$df = 9 \, \Delta x$	error $\Delta f - df = \varepsilon \, \Delta x$	relative error
1.0	10	9	1	1
0.1	0.91	0.9	0.01	0.1
0.01	0.0901	0.09	0.0001	0.01

EXAMPLE 4   Find an approximate value for $\sqrt[3]{27.3}$ without using tables.

Solution   We note that 27.3 is close to 27 which is a perfect cube. Thus the application of differentials is suggested. This will tell us approximately how much $\sqrt[3]{x}$ changes as

x changes from 27 to 27.3. Thus $\Delta x = dx = 27.3 - 27 = 0.3$ in this instance. Set

$$y = f(x) = \sqrt[3]{x} = x^{1/3}$$

so that

$$y' = f' = \frac{1}{3}x^{-2/3} = \frac{1}{3(\sqrt[3]{x})^2}$$

Let $x = 27$ and $\Delta x = dx = 0.3$ to yield

$$dy = df = f' \, dx = \frac{1}{3(\sqrt[3]{27})^2}(0.3) = \frac{1}{3(9)}(0.3) = \frac{1}{90} \doteq 0.011$$

Since $\Delta y = f(x + \Delta x) - f(x) = \sqrt[3]{27 \cdot 3} - \sqrt[3]{27}$. We have

$$\sqrt[3]{27 \cdot 3} = \sqrt[3]{27} + \Delta y = 3 + \Delta y \doteq 3 + dy \doteq 3 + 0.0111 = 3 \cdot 0111$$

The correct answer to seven significant figures is 3.011070, so that the method of differentials gave the first five figures correctly, which is very satisfactory. The error here is approximately 0.00004. ●

EXAMPLE 5   If $y = t^2 - 3t + 1$ and $t = 3x + 7$ find dy in terms of x and dx by two methods.

Solution   **Method 1**

$$\begin{aligned} y &= t^2 - 3t + 1 = (3x + 7)^2 - 3(3x + 7) + 1 \\ &= 9x^2 + 42x + 49 - 9x - 21 + 1 \\ &= 9x^2 + 33x + 29 \\ dy &= y' \, dx = (18x + 33)\, dx = 3(6x + 11)\, dx \end{aligned}$$

Method 2

$$dy = D_t y \, dt \qquad \text{and} \qquad dt = D_x t \, dx$$

so that

$$\begin{aligned} dy &= D_t y \, D_x t \, dx \\ &= (2t - 3)\, 3 \, dx \\ &= 3[2(3x + 7) - 3]\, dx = 3(6x + 11)\, dx \end{aligned}$$ ●

Note that the chain rule in terms of Leibnitz' notation has the very suggestive form

$$\frac{dy}{dx} = \frac{dy}{dt}\frac{dt}{dx} \qquad (21)$$

and

$$dy = \frac{dy}{dt}\frac{dt}{dx}\, dx = D_t y \, D_x t \, dx \qquad (22)$$

EXAMPLE 6   If $x^2 - 3xy + y^2 = 4$ find $\dfrac{dy}{dx}$ using differentials.

Solution

$$d(x^2 - 3xy + y^2) = d(4) = 0$$
$$2x\,dx - 3x\,dy - 3y\,dx + 2y\,dy = 0$$
$$(2x - 3y)\,dx + (2y - 3x)\,dy = 0$$

or

$$\frac{dy}{dx} = \frac{3y - 2x}{2y - 3x}$$

●

EXAMPLE 7   Show that the relative error in calculating the volume of a sphere due to an error in measuring the radius is approximately three times the relative error in the measurement of the radius.

Solution $V = \frac{4}{3}\pi r^3$ is the volume of a sphere in terms of its radius r. $dV = 4\pi r^2\,dr$ so that

$$\frac{dV}{V} = \frac{4\pi r^2\,dr}{\frac{4}{3}\pi r^3} = 3\frac{dr}{r}$$

Now $\dfrac{dr}{r}$ is the relative error in the radius and $\dfrac{dV}{V}$ is approximately the relative error in the volume; therefore, the stated result follows.   ●

Exercises 5.1

In Exercises 1 through 12 find the differential of the function defined by each of the following equations:

1. $y = x + 3$

2. $y = x^2 - 7x + 9$

3. $y = x^4 - 2x^3 + x - 3$

4. $y = 3x + \sqrt{x}$

5. $y = \sqrt{4 + x^2}$

6. $s = 16t^2 - 100t + 5$

7. $V = e^3$

8. $S = 4\pi r^2$

9. $V = \frac{4}{3}\pi r^3$

10. $T = 2\pi\sqrt{\dfrac{L}{32}}$

11. $v = u\sqrt{u + 1}$

12. $w = \dfrac{v + 1}{v + 2}$

In Exercises 13 through 16 we are given $y = x^2 - 3x - 5$, find dy when

13. $x = 3$ and $dx = 0.02$

14. $x = 5$ and $dx = -0.01$

15. $x = -2$ and $dx = 0$

16. $x = 0$ and $dx = -3$

In Exercises 17 through 22 find (a) Δy, (b) dy, and (c) $\Delta y - dy$. Also find (d) $\varepsilon = \dfrac{\Delta y - dy}{\Delta x}$, $(\Delta x \neq 0)$

17. $y = 3x + 2$

18. $y = ax + b$

19. $y = x^2 - 5x - 8$

20. $y = ax^2 + bx + c$

21. $y = x^3$

22. $y = \dfrac{1}{x}$

23. Prove formula (11), namely that $d(cf) = c\,df$ where c is a constant.

24. Prove formula (14), namely that $d(f/g) = \dfrac{g\,df - f\,dg}{g^2}$

In Exercises 25 through 30 use differentials to find an approximate value for the given quantity. Give each answer to four significant figures.

25. $\sqrt{65}$

27. $\sqrt{97}$

29. $\sqrt{34.7}$

26. $\sqrt[3]{65}$

28. $\sqrt[5]{33}$

30. $(1.97)^4$

In Exercises 31 through 34 use differentials to prove the approximation formulas for small $|h|$; that is, $|h| \ll 1$

31. $\sqrt[3]{1 + h} \approx 1 + \dfrac{h}{3}$

32. $\dfrac{1}{1 + h} \approx 1 - h$

33. $\dfrac{1}{\sqrt{1 + h}} \approx 1 - \dfrac{h}{2}$

34. $(1 + h)^n \approx 1 + nh$

35. Prove that in Exercise 32 the relative error is $\dfrac{h}{1 + h}$

In Exercises 36 through 39, find dy in terms of t and dt by two different methods (see Example 5).

36. $y = x^2 - 7, \quad x = 4t + 3$

37. $y = x^3 + 2x - 1, \quad x = t^2$

38. $y = \sqrt{x + 2}, \quad x = 3 - 2t$

39. $y = ax^2 + bx + c, \quad x = \dfrac{k}{t}$

In Exercises 40 through 44, find $\dfrac{dy}{dt}$ using differentials:

40. $y = u^2 - u + 3, \quad u = \dfrac{t}{1 + t}$

41. $y = \sqrt{u}, \quad u = s^2 + 1, \quad s = t^3 - t$

42. $5y^2 - 3ty + t^2 = 20$

43. $y^3 + 2t^2y - 4t^3 = 1$

44. $y^2 - sy + 3s^2 = 6, \quad s^3 - st + 2t^2 = -4$

45. Show that the relative error in the volume of a cube, due to an error in measuring the edge, is approximately three times the relative error in the edge.

46. A square of side x is increased to a square of side $x + dx$. Show analytically and geometrically the difference between ΔA and dA where A is the area of the square.

47. If a body falls from rest a distance s ft in t sec, then $s = \dfrac{gt^2}{2}$. Approximately how much will a small error Δt in t affect the computation of the displacement s and the velocity v?

48. The force of attraction between two heavenly bodies of masses m and m' and distance r apart is given by $F = \dfrac{Gmm'}{r^2}$, where G is a universal constant called the constant of gravitation. This is called *Newton's law of gravitation.* Find the approximate change in the force F due to a change in the distance r and find the ratio of $\dfrac{dF}{F}$ to $\dfrac{dr}{r}$.

49. The period T of oscillation of a simple pendulum (that is, the time for one complete oscillation) is given by $T = 2\pi \sqrt{L/g}$ where L is the length of the pendulum in feet, g is the acceleration due to gravity in feet per second per second, and T is in seconds. If the percentage error in measuring L is at most 3%, find approximately the greatest possible percentage error in T.

50. If the three linear dimensions of a box are measured with percentage errors of 2%, $1\frac{1}{2}$%, and $\frac{1}{2}$%, respectively, what is the approximate maximum percentage error in the calculated volume?

51. A spherical balloon leaks gas and reduces in size from a radius of 8 ft to 7 ft 10 in. Use differentials to estimate the total volume of gas in cubic feet that has escaped.

52. From an assortment of steel balls, it is required to select all those with diameter of 2 cm. The permissible percentage deviation in the diameters is 2%, and the balls are to be selected by their mass. What is the approximate permissible deviation in the mass, if steel has mass density of 7.8 g/cm³?

The reader is already familiar with the concepts of *inverse operations.* For example, addition and subtraction are operations that are the inverse of each other. Also multiplication and division are operations inverse to each other. A further example of such operations is squaring of positive numbers and taking the square root, and more generally raising to the nth power and taking the nth root. Each operation in a pair of inverse operations undoes the effect of the other. The inverse operation to differentiation is called *antidifferentiation.*

Often in calculus we are given the derivative of an unknown function and required to determine the function.

In Section 4.6 we established, with the help of the law of the mean, two important corollaries that are restated as theorems for our convenience.

Theorem 1   If the derivative of a function is zero throughout an interval, the function is constant throughout that interval.

Theorem 2   If two functions have the same derivatives throughout an interval, the functions must differ by a constant.

Symbolically, if $f(x)$ and $g(x)$ are two functions and if $f'(x) = g'(x)$ for all x in an interval, then $f(x) = g(x) + C$, where C is a constant, in that interval.

Definition   A function F is called an *antiderivative* of a function f in an interval I if $F'(x) = f(x)$ for all x in I.

For example, if $F(x) = x^3 - 4x + 7$ then $F'(x) = 3x^2 - 4$. Thus, if f is the function defined by $f(x) = 3x^2 - 4$, $F(x)$ is an antiderivative of $f(x)$ in $(-\infty, \infty)$. In fact, there are an infinite number of antiderivatives of $f(x)$, namely, $x^3 - 4x + C$ where C is an arbitrary constant. Furthermore, from Theorem 2, there are no other antiderivatives; that is, $x^3 - 4x + C$ represents the totality of antiderivatives of $f(x)$. This follows from the fact that, if $G(x)$ is any other antiderivative of $f(x)$, then

$$G' - F' = f - f = 0$$

or

$$(G - F)' = 0$$

from which

$$G - F = C \quad \text{or} \quad G = F + C$$

Thus, any two antiderivatives of f in an interval I can differ by, at most, a constant. Therefore, we have the following theorem.

Theorem 3 If F is any particular antiderivative of f on an interval I, then the most general antiderivative of f on I is given by

$$F(x) + C \qquad (1)$$

where C is an arbitrary constant; all antiderivatives of f on I may be obtained by assigning particular values to C.

Therefore, for example, the most general antiderivative of x^4 is $\dfrac{x^5}{5} + C$, since $D_x \left(\dfrac{x^5}{5} + C \right) = x^4$.

We define **antidifferentiation** as the process of finding the most general antiderivative of a given function. Furthermore we introduce the symbol $\int$ to denote the operation of antidifferentiation, and write

$$\int f(x)\, dx = F(x) + C \qquad (2)$$

where

$$F'(x) = f(x) \qquad (3)$$

or equivalently

$$d(F(x)) = f(x)\, dx \qquad (4)$$

From (2) and (4) we have

$$\int d(F(x)) = F(x) + C \qquad (5)$$

The notation $\int f(x)\, dx$ is read "the antiderivative of $f(x)$ with respect to x."

To find $\int x^7\, dx$, we note that $D_x(x^8) = 8x^7$ and thus $D_x \left(\dfrac{x^8}{8} \right) = x^7$. This means that $\dfrac{x^8}{8}$ is an antiderivative of x^7 so that

$$\int x^7\, dx = \frac{x^8}{8} + C$$

If the previous examples are generalized we obtain

Theorem 4

$$\int x^n\, dx = \frac{x^{n+1}}{n+1} + C \qquad \text{if } n \neq -1 \qquad (6)$$

Proof
$$D_x \left(\frac{x^{n+1}}{n+1} + C \right) = \frac{(n+1)x^n}{n+1} = x^n \qquad \blacksquare$$

Note that n must not be equal to -1 since division by zero is excluded. The question of $\int x^{-1}\, dx$ will be discussed later.

EXAMPLE 1   Evaluate $\int x^{60}\,dx$.

Solution   From (6) with $n = 60$ we have

$$\int x^{60}\,dx = \frac{x^{61}}{61} + C$$ ●

EXAMPLE 2   Evaluate $\int \sqrt{x}\,dx$

Solution   From (6) with $n = \frac{1}{2}$ there results

$$\int \sqrt{x}\,dx = \int x^{1/2}\,dx = \tfrac{2}{3}x^{3/2} + C$$ ●

EXAMPLE 3   Evaluate $\int \dfrac{dx}{x^4}$

Solution   From (6) with $n = -4$,

$$\int \frac{dx}{x^4} = \int x^{-4}\,dx = \frac{x^{-3}}{-3} + C$$

$$= -\frac{1}{3x^3} + C$$ ●

EXAMPLE 4A   If $F'(x) = 1$, find $F(x)$.

Solution $$F(x) = \int 1\,dx = \int x^0\,dx = \frac{x^1}{1} + C = x + C$$ ●

EXAMPLE 4B   If $F'(t) = 1$, find $F(t)$.

Solution $$F(t) = \int 1\,dt = \int t^0\,dt = \frac{t^1}{1} + C = t + C$$ ●

Note that Examples 4A and 4B are similar in that both asked for the antiderivative of 1. However, x is the variable in 4A, whereas t is the variable in 4B. This illustrates a need for dx or dt in our antiderivative notation—it names the variable.

Because antidifferentiation is the inverse operation of differentiation, we can obtain antidifferentiation formulas from differentiation formulas.

Theorem 5 $$\int k\,f(x)\,dx = k\int f(x)\,dx, \qquad \text{where } k \text{ is a constant} \tag{7}$$

Proof   Let $\int f(x)\,dx = G(x)$, then $G'(x) = f(x)$

The right side of (7) is $k\,G(x)$. We must show that the left side is also $k\,G(x)$. Now

$$D_x(kG) = kG' = kf \text{ so that } \int k\,f(x)\,dx = kG \qquad \blacksquare$$

EXAMPLE 5 $\quad \int \sqrt{2x}\,dx$

Solution $\qquad \int \sqrt{2x}\,dx = \int \sqrt{2}\,\sqrt{x}\,dx = \sqrt{2}\int x^{1/2}\,dx$

$$= \sqrt{2}(\tfrac{2}{3}x^{3/2} + C_1) \qquad (C_1 = \text{arbitrary constant})$$
$$= \tfrac{2}{3}\sqrt{2}x^{3/2} + C \quad (C = \sqrt{2}C_1 \text{ is also an arbitrary constant}) \quad \bullet$$

Theorem 6 $\qquad \boxed{\int [f(x) + g(x)]\,dx = \int f(x)\,dx + \int g(x)\,dx} \qquad (8)$

provided that each of the antiderivatives on the right side of (8) exist.

Proof $\quad$ Let $\int f(x)\,dx = F(x)$ and $\int g(x)\,dx = G(x)$ so that the right side of (8) is $F(x) + G(x)$. But $D_x(F(x) + G(x)) = F'(x) + G'(x) = f(x) + g(x)$. However, the left side of (8) is $F(x) + G(x)$ also, which proves (8). $\qquad \blacksquare$

EXAMPLE 6 $\quad$ Evaluate $\int (3x + 2x^4)\,dx$

Solution $\qquad\qquad \int (3x + 2x^4)\,dx = \int 3x\,dx + \int 2x^4\,dx \qquad \text{from (8)}$

$$= 3\int x\,dx + 2\int x^4\,dx \qquad \text{from (7)}$$

$$= \frac{3x^2}{2} + \frac{2x^5}{5} + C \qquad \text{from (6)} \quad \bullet$$

Note that we might argue that we should have two arbitrary constants, one for each antidifferentiation. However the two constants can be combined into one. Thus since $\int x\,dx = \dfrac{x^2}{2} + C_1$ and $\int x^4\,dx = \dfrac{x^5}{5} + C_2$, where C_1 and C_2 are arbitrary constants, it follows that,

$$\int (3x + 2x^4)\,dx = 3\left(\frac{x^2}{2} + C_1\right) + 2\left(\frac{x^5}{5} + C_2\right)$$

$$= \frac{3x^2}{2} + \frac{2x^5}{5} + 3C_1 + 2C_2$$

Let $C = 3C_1 + 2C_2$ be a new arbitrary constant, and we have

$$\int (3x + 2x^4)\,dx = \frac{3x^2}{2} + \frac{2x^5}{5} + C$$

We can readily generalize the result of Theorems 5 and 6 to the case of n functions that have antiderivatives in a common interval. Therefore we have

Theorem 7

$$\int [c_1 f_1(x) + c_2 f_2(x) + \cdots + c_n f_n(x)] \, dx$$

$$= c_1 \int f_1(x) \, dx + c_2 \int f_2(x) \, dx + \cdots + c_n \int f_n(x) \, dx \qquad (9)$$

provided that each of the antiderivatives on the right side exist, where c_1, $c_2, \ldots, c_n$ are constants.

EXAMPLE 7   Find $\int (x^3 - 2x^2 + 7) \, dx$

Solution

$$\int (x^3 - 2x^2 + 7) \, dx = \int x^3 \, dx - 2 \int x^2 \, dx + 7 \int dx \qquad \text{from (9)}$$

$$= \frac{x^4}{4} - \frac{2x^3}{3} + 7x + C \qquad \text{from (6)}$$

Note that we can readily check our calculation by differentiating back. Thus, in this instance

$$D_x \left(\frac{x^4}{4} - \frac{2x^3}{3} + 7x + C \right) = \frac{4x^3}{4} - \frac{6x^2}{3} + 7 = x^3 - 2x^2 + 7$$

which is our original function. ●

EXAMPLE 8   Evaluate $\int \left(x^2 + \frac{1}{x} \right)^2 dx$

Solution   Expand the binomial first to obtain

$$\int \left(x^2 + \frac{1}{x} \right)^2 dx = \int \left(x^4 + 2x^2 \frac{1}{x} + \frac{1}{x^2} \right) dx$$

$$= \int \left(x^4 + 2x + \frac{1}{x^2} \right) dx$$

$$= \int x^4 \, dx + \int 2x \, dx + \int \frac{1}{x^2} \, dx$$

$$= \frac{x^5}{5} + x^2 - \frac{1}{x} + C$$ ●

EXAMPLE 9   Evaluate $\int \frac{2x^3 - x^2 - 3}{\sqrt{x}} \, dx$

Solution   Divide $\sqrt{x}$ into each term of the numerator to obtain

$$\int \frac{2x^3 - x^2 - 3}{\sqrt{x}} \, dx = \int (2x^{5/2} - x^{3/2} - 3x^{-1/2}) \, dx$$

$$= 2 \int x^{5/2} \, dx - \int x^{3/2} \, dx - 3 \int x^{-1/2} \, dx$$

$$= 2(\tfrac{2}{7})x^{7/2} - \tfrac{2}{5}x^{5/2} - 3(2)x^{1/2} + C$$

$$= \tfrac{4}{7}x^{7/2} - \tfrac{2}{5}x^{5/2} - 6x^{1/2} + C$$ ●

Exercises 5.2

In Exercises 1 through 16 find the most general antiderivative:

1. $\int x^4 \, dx$

2. $\int 3x^3 \, dx$

3. $\int x^{1/4} \, dx$

4. $\int \sqrt[3]{x} \, dx$

5. $\int x^3 \sqrt{x} \, dx$

6. $\int (4 - 6u + u^5) \, du$

7. $\int \sqrt{11w} \, dw$

8. $\int \dfrac{u^2 + 5}{u^4} \, du$

9. $\int (s^2 + 3)^2 \, ds$

10. $\int (ax^3 - bx^2 + ex - d) \, dx$

11. $\int \left(\sqrt{7x} + \dfrac{3}{\sqrt{5x}} \right) dx$

12. $\int \dfrac{4r^2 + 7}{\sqrt[3]{r}} \, dr$

13. $\int (x - 1)^2 x \, dx$

14. $\int (x + a)(x + b) \, dx$

15. $\int (x + 2)^3 \, dx$

16. $\int \dfrac{y^3 + 3y^2 - 7}{y^2} \, dy$

In each of Exercises 17 through 20 find the most general function f that satisfies the given equation:

17. $f'' = 3x - 5$

18. $f'' = 12x^2 - 30x + 6$

19. $f'' = \dfrac{3x + 4}{\sqrt{x}}$

20. $f'' = x - \dfrac{1}{x^3}$

21. Find $D_x \int \dfrac{x^4 + 3x}{\sqrt{x}} \, dx$

22. Find $\int D_x (ax^2 + bx + e) \, dx$

23. Find $\int f' \, dx$ where $f = f(x)$

24. Find $\int f f' \, dx$ where $f = f(x)$

25. Verify that $\int (3x - 5)^4 \, dx = \dfrac{(3x - 5)^5}{15} + C$ by differentiating the right side.

26. Find $\int (4x + 1)^6 \, dx$.

27. For all values of x, find $\int |x| \, dx$

28. For all values of x, find

$$\int (x + |x|)^2 \, dx$$

and give your answer as a single expression that is valid for all values of x.

5.3

THE SUBSTITUTION METHOD

In many cases, antiderivatives cannot be found directly by the formulas given in Section 5.2. However, sometimes a change in variable can be introduced such that the required antiderivative may be obtained by the previous formulas applied to the new variables. This change of variable or **substitution technique** will first be illustrated by means of examples. It will then be generalized and proved. This technique is the most important tool in determining antiderivatives.

EXAMPLE 1   Find $\int \sqrt{2x + 3}\, dx$

Solution   Let $u = 2x + 3$ then $du = 2\, dx$ or $dx = \dfrac{du}{2}$ so that

$$\int \sqrt{2x + 3}\, dx = \int u^{1/2} \frac{du}{2}$$

$$= \frac{1}{2} \frac{u^{3/2}}{\frac{3}{2}} + C$$

$$= \frac{1}{3} u^{3/2} + C$$

$$= \frac{1}{3}(2x + 3)^{3/2} + C \qquad\qquad \bullet$$

EXAMPLE 2   Evaluate $\int (x^2 - 7)^4 x\, dx$

Solution   This problem can be done by first expanding the expression $(x^2 - 7)^4 x$ and then finding the antiderivative of each term. We can also do this by substitution as follows.

Let $u = x^2 - 7$ then $du = 2x\, dx$ or $x\, dx = \dfrac{du}{2}$. Therefore

$$\int (x^2 - 7)^4 x\, dx = \int u^4 \frac{du}{2}$$

$$= \frac{1}{2} \frac{u^5}{5} + C$$

$$= \frac{u^5}{10} + C$$

$$= \frac{(x^2 - 7)^5}{10} + C$$

CHECK: Of course this result can be checked by differentiating the answer. Thus by the chain rule of differentiation,

$$D_x \left[\frac{(x^2 - 7)^5}{10} + C \right] = \frac{5(x^2 - 7)^4 2x}{10} + 0 = (x^2 - 7)^4 x$$

and the original function has been recovered.

EXAMPLE 3   Determine
$$\int \frac{x^2}{(1 + x^3)^3} \, dx$$

Solution   Let $u = 1 + x^3$, then $du = 3x^2 \, dx$ or $x^2 \, dx = \dfrac{du}{3}$

$$\int \frac{x^2 \, dx}{(1 + x^3)^3} = \int \frac{du/3}{u^3}$$

$$= \tfrac{1}{3} \int u^{-3} \, du$$

$$= -\tfrac{1}{6} u^{-2} + C$$

$$= -\frac{1}{6u^2} + C$$

$$= -\frac{1}{6(1 + x^3)^2} + C$$

CHECK

$$D_x \left[-\frac{1}{6(1 + x^3)^2} + C \right] = -\frac{1}{6} D_x (1 + x^3)^{-2} + 0$$

$$= -\tfrac{1}{6}(-2)(1 + x^3)^{-3}(3x^2)$$

$$= (1 + x^3)^{-3} x^2$$

$$= \frac{x^2}{(1 + x^3)^3}$$

EXAMPLE 4   Find $\int x^2 \sqrt{4 + x} \, dx$

Solution   Let $u = 4 + x$. Then $x = u - 4$ so that $x^2 = (u - 4)^2 = u^2 - 8u + 16$. Also $dx = du$. Therefore

$$\int x^2 \sqrt{4 + x} \, dx = \int (u^2 - 8u + 16) u^{1/2} \, du$$

$$= \int (u^{5/2} - 8u^{3/2} + 16u^{1/2}) \, du$$

$$= \frac{u^{7/2}}{\frac{7}{2}} - \frac{8u^{5/2}}{\frac{5}{2}} + \frac{16u^{3/2}}{\frac{3}{2}} + C$$

$$= \tfrac{2}{7}(4 + x)^{7/2} - \tfrac{16}{5}(4 + x)^{5/2} + \tfrac{32}{3}(4 + x)^{3/2} + C$$

$$= 2 \left[\frac{(4 + x)^{7/2}}{7} - \frac{8(4 + x)^{5/2}}{5} + \frac{16(4 + x)^{3/2}}{3} \right] + C$$

CHECK: Checking by differentiation yields

$$D_x\left\{2\left[\frac{(4+x)^{7/2}}{7} - \frac{8(4+x)^{5/2}}{5} + \frac{16(4+x)^{3/2}}{3}\right] + C\right\}$$

$$= 2[\tfrac{1}{2}(4+x)^{5/2} - 4(4+x)^{3/2} + 8(4+x)^{1/2}] + 0$$

$$= 2(4+x)^{1/2}\left[\frac{(4+x)^2}{2} - 4(4+x) + 8\right]$$

$$= 2\sqrt{4+x}\left[\frac{16+8x+x^2}{2} - (16+4x) + 8\right]$$

$$= 2\sqrt{4+x}\,\frac{16+8x+x^2-32-8x+16}{2}$$

$$= x^2\sqrt{4+x} \qquad\qquad\qquad \bullet$$

In a particular example, there is not necessarily just one substitution that works. Example 4 could also have been done by using the substitution $u = \sqrt{4+x}$. The next example can also be done by more than one method.

EXAMPLE 5   Evaluate $\int x^3\sqrt{x^2+3}\,dx$.

Solution   We could let $u = x^2 + 3$. However, for variety, we shall instead let $u = \sqrt{x^2+3}$ so that $u^2 = x^2 + 3$. Taking the differential of both sides we have

$$2u\,du = 2x\,dx$$

Now

$$x^3\,dx = x^2(x\,dx) = (u^2-3)u\,du$$

so that

$$\int x^3\sqrt{x^2+3}\,dx = \int (u^2-3)u\,u\,du$$

$$= \int (u^4 - 3u^2)\,du$$

$$= \frac{u^5}{5} - u^3 + C$$

$$= u^3\left(\frac{u^2}{5} - 1\right) + C$$

$$= (x^2+3)^{3/2}\left(\frac{x^2+3}{5} - 1\right) + C$$

$$= \tfrac{1}{5}(x^2+3)^{3/2}(x^2-2) + C \qquad\qquad \bullet$$

EXAMPLE 6   Find $\int 7x\sqrt[3]{x+1}\,dx$

Solution   Let $u = \sqrt[3]{x+1}$ so that $x = u^3 - 1$ and $dx = 3u^2\,du$. (This example could also be done by using the substitution $u = x + 1$). Therefore

$$\int 7x \sqrt[3]{x+1}\, dx = 7 \int (u^3 - 1)u\, 3u^2\, du$$

$$= 21 \int (u^6 - u^3)\, du$$

$$= 21 \left(\frac{u^7}{7} - \frac{u^4}{4} \right) + C$$

$$= \tfrac{21}{28}(4u^7 - 7u^4) + C$$

$$= \tfrac{3}{4}u^4(4u^3 - 7) + C$$

$$= \tfrac{3}{4}(x+1)^{4/3}[4(x+1) - 7] + C$$

$$= \tfrac{3}{4}(x+1)^{4/3}(4x - 3) + C \qquad \bullet$$

EXAMPLE 7   Evaluate $\int \sqrt{x^2 + 1}\, dx$

Attempted
Solution
   Let $u = \sqrt{x^2 + 1}, (x > 0)$; then $u^2 = x^2 + 1$, $2u\, du = 2x\, dx$, and
$$dx = \frac{u\, du}{x} = \frac{u\, du}{\sqrt{u^2 - 1}}, (x > 0).$$

$$\int \sqrt{x^2 + 1}\, dx = \int \frac{u^2\, du}{\sqrt{u^2 - 1}}$$

We observe that there is still a square root in the denominator. Undaunted, let us try again with

$$\sqrt{u^2 - 1} = v \qquad \text{or} \qquad u^2 = v^2 + 1 \qquad 2u\, du = 2v\, dv$$

Therefore

$$\int \frac{u^2\, du}{\sqrt{u^2 - 1}} = \int \frac{(v^2 + 1)}{v}\frac{v\, dv}{u} = \int \frac{v^2 + 1\, dv}{\sqrt{v^2 + 1}}$$

$$= \int \sqrt{v^2 + 1}\, dv$$

Thus we have run full circle and achieved nothing. $\bullet$

We might next try a different substitution in Example 7, perhaps $u = x^2 + 1$. However, we would still not be able to solve this problem. We can then conclude that the method of algebraic substitution *does not always work*. In other words we will not always end up with a problem that can be handled by algebraic substitution.

Note that $\int x\sqrt{x^2 + 1}\, dx$ is easily found by making the substitution $u = x^2 + 1$. This is a simple exercise because $d(x^2 + 1) = 2x\, dx$ and our problem is of the form

$$\tfrac{1}{2}\int (x^2 + 1)^{1/2}\, d(x^2 + 1) = \tfrac{1}{2}\tfrac{2}{3}(x^2 + 1)^{3/2} + C = \frac{(x^2 + 1)^{3/2}}{3} + C$$

In each preceding example given the method is basically the same. We are given $\int f(x)\, dx$ and seek a change in variable $u = h(x)$ such that for some function g,

$$f(x)\,dx = g(u)\,du = g(h(x))h'(x)\,dx$$

Then the problem of finding

$$\int f(x)\,dx$$

is replaced by the problem of determining

$$\int g(u)\,du$$

If the problem of finding $\int g(u)\,du$ is straightforward, our task is essentially accomplished. If, on the contrary, $\int g(u)\,du$ is not easier than $\int f(x)\,dx$, either try a different substitution or a different approach. Other techniques will be discussed later. Unfortunately there is no routine procedure for handling antiderivatives. This is in contrast with the routine that indeed does exist for finding derivatives of a function. If this is the case, how is the ability to find antiderivatives developed? It is the author's very strong conviction that this skill is acquired by doing many problems with effective supervision. In this way the reader will often recognize which method is most promising.

Examples 1 through 6 suggest one substitution that frequently works: If there is a term under a radical sign or a term raised to a power, try letting this term equal a new variable (we usually choose u).

Later on we will also learn how to use extensive tables of antiderivatives that are available to the scientist. It will be apparent that the ability to work efficiently with the tables is enhanced by being familiar with the techniques for finding antiderivatives and the functions involved.

In the previous examples, the substitution method was developed. Now let us justify it.

Theorem 1 **(Substitution Theorem).** If $f(x) = g(h(x))\,h'(x)$, where $u = h(x)$, and if $G(u) = \int g(u)\,du$, then

$$G(h(x)) = \int f(x)\,dx$$

Proof It is only necessary to establish that

$$\frac{d}{dx}[G(h(x))] = f(x)$$

By the chain rule of differentiation

$$\frac{d}{dx}[G(h(x))] = \frac{dG(u)}{du}\frac{du}{dx}$$
$$= g(u)\,u'(x)$$
$$= g(h(x))\,h'(x)$$
$$= f(x)$$

so that

$$G(h(x)) = \int f(x)\, dx$$

and the theorem is proved. ∎

A special case of the substitution theorem and a very useful result is the following.

Theorem 2 **(Generalized Power Rule).** If f is a differentiable function, then

$$\int [f(x)]^n f'(x)\, dx = \frac{[f(x)]^{n+1}}{n+1} + C, \text{ where } n \neq -1$$

Proof Let $u = f(x)$. Then $du = f'(x)\, dx$.

$$\int [f(x)]^n f'(x)\, dx = \int u^n\, du = \frac{u^{n+1}}{n+1} + C$$

$$= \frac{[f(x)]^{n+1}}{n+1} + C \qquad n \neq -1 \qquad ∎$$

Examples 1, 2, and 3 are all illustrations of the use of the generalized power rule, provided that the constant is adjusted. With experience, you will be able to do such problems without actually having to make a substitution. Example 2 could be done as follows:

$$\int (x^2 - 7)^4 x\, dx = \frac{1}{2} \int (x^2 - 7)^4 (2x\, dx) \qquad \text{Note: } f(x) = x^2 - 7$$

$$= \frac{1}{2} \frac{(x^2 - 7)^5}{5} + C$$

$$= \frac{(x^2 - 7)^5}{10} + C$$

The generalized power rule can be used to do Example 3 as follows:

$$\int \frac{x^2\, dx}{(1 + x^3)^3} = \frac{1}{3} \int (1 + x^3)^{-3} (3x^2\, dx) \qquad \text{Note: } f(x) = 1 + x^3$$

$$= \frac{1}{3} \frac{(1 + x^3)^{-2}}{-2} + C$$

$$= -\frac{1}{6(1 + x^3)^2} + C$$

Exercises 5.3

In Exercises 1 through 20 find the most general antiderivative. Check Exercises 6 through 10 by differentiating the obtained result.

1. $\int \sqrt{3x - 5}\, dx$

2. $\int \sqrt{10x + 6}\, dx$

3. $\int \sqrt{x^3 + 4}\, x^2\, dx$

4. $\int \sqrt{x^2 + 16}\, x\, dx$

5. $\displaystyle\int \frac{t}{\sqrt{5t^2+4}}\,dt$

6. $\displaystyle\int x\sqrt{x+5}\,dx$

7. $\displaystyle\int x\sqrt{x-3}\,dx$

8. $\displaystyle\int \frac{x}{\sqrt{x+7}}\,dx$

9. $\displaystyle\int \frac{x}{(x-4)^3}\,dx$ Book Answer

10. $\displaystyle\int \frac{x\,dx}{(x^2-9)^3}$

11. $\displaystyle\int \frac{x\,dx}{(x-6)^4}$

12. $\displaystyle\int (5x-3)^{1/4}\,dx$

13. $\displaystyle\int \frac{(x^3+4)^2}{x^2}\,dx$

14. $\displaystyle\int \frac{x^3}{\sqrt{2+x^2}}\,dx$

15. $\displaystyle\int \frac{\left(3+\dfrac{1}{x}\right)^5}{x^2}\,dx$

16. $\displaystyle\int t\sqrt[3]{8-t}\,dt$

17. $\displaystyle\int y\sqrt[3]{2y+1}\,dy$

18. $\displaystyle\int z(z-a)^{2/3}\,dz$

19. $\displaystyle\int \sqrt{w+a}(w+b)\,dw$

20. $\displaystyle\int \frac{t\,dt}{\sqrt[4]{t-a}}$

21. Evaluate $\displaystyle\int (3x-1)^2\,dx$ using two methods.

(a) By expanding the binomial and taking antiderivatives term by term.
(b) By making the substitution $u = 3x - 1$.
Reconcile the two apparently different results.

22. Evaluate $\displaystyle\int \sqrt{x-a}\,x\,dx$ by using two substitutions

(a) $u = \sqrt{x-a}$ and (b) $v = x - a$. In this case both substitutions lead to antiderivatives that are readily found.

5.4   APPLICATIONS OF ANTIDERIVATIVES

EXAMPLE 1   Find an equation of the curve that passes through $(2, 3)$ and whose slope at any point on the curve is equal to the abscissa at that point.

Solution   Since the slope of a curve at a point (x, y) on the curve is y' and the abscissa of that point is x, we require that

$$y' = x \tag{1}$$

Equation (1) is an equation that involves a derivative and is an example of a *differential equation.* To solve the differential equation is to find a function f such that $y = f(x)$ satisfies (1) identically.

If the antiderivative of both sides is taken, there results the totality of solutions

$$y = \frac{x^2}{2} + C \tag{2}$$

where C is an arbitrary constant.

The graph of (2) is a family of *parallel curves,* one curve for each particular value of C (Figure 5.4.1). By parallel curves we mean that, for any particular x,

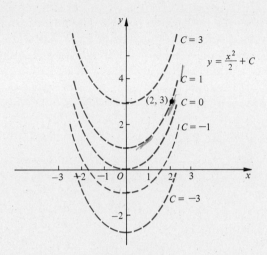

Figure 5.4.1

the slopes of all curves of the family are the same. To determine the particular curve that passes through the required point $(2, 3)$, we must determine C from the condition that $y = 3$ when $x = 2$. Thus

$$3 = \frac{2^2}{2} + C \quad \text{or} \quad C = 1$$

This means that the required curve is

$$y = \frac{x^2}{2} + 1$$

●

EXAMPLE 2   Solve the differential equation

$$y'' = 6x - 2 \tag{3}$$

subject to the initial conditions

$$y(0) = -3 \qquad y'(0) = 1 \tag{4}$$

Solution   Since $y'' = (y')'$, the antiderivative of each side of (3) gives

$$y' = 3x^2 - 2x + C_1 \tag{5}$$

where C_1 thus far is an arbitrary constant. However $y'(0) = 1$, which yields upon substitution of $y' = 1$ and $x = 0$ into (5), $C_1 = 1$ so that

$$y' = 3x^2 - 2x + 1 \tag{6}$$

Taking antiderivatives once more,

$$y = x^3 - x^2 + x + C_2 \tag{7}$$

where C_2 is found by substitution into $y(0) = -3$. Therefore, $C_2 = -3$, and the required solution of the *initial value problem* is

$$y = x^3 - x^2 + x - 3† \qquad\qquad ● \tag{8}$$

---

† In this instance, $C_1 = y'(0)$ and $C_2 = y(0)$. However, this will not always be the case.

EXAMPLE 3   A metal rod 4 meters long has one end maintained at a temperature of 30°C while the other end is maintained at 100°C. An x-axis is chosen so that the temperature u is 30° when $x = 0$ and 100° when $x = 4$. Furthermore, since the temperature u is independent of the time (so called steady state) then $u(x)$ must satisfy $u''(x) = 0$ at all interior points of the rod. Determine the temperature of the rod for all x in $[0, 4]$ and, in particular, what is the midsection temperature?

Solution   From

$$u'' = 0 \qquad 0 < x < 4 \tag{9}$$

we have by taking antiderivatives

$$u' = C_1 \tag{10}$$

and then

$$u = C_1 x + C_2 \tag{11}$$

Now $u(0) = 30$ and $u(4) = 100$ are the prescribed boundary conditions. This is generally called a **boundary value problem** because conditions are given at more than one value of the independent variable.

From $u(0) = 30$ we obtain, from (11)
$$30 = C_1(0) + C_2 \qquad \text{or} \qquad C_2 = 30$$

Also $u(4) = 100$ so that from (11) again,

$$100 = 4C_1 + 30 \qquad \text{or} \qquad C_1 = 17.5$$

and $$u(x) = 30 + 17.5x$$

is the required temperature distribution. In particular, the midsection temperature is

$$u(2) = 30 + 35 = 65°C$$

This result is not surprising because [from (11)] the temperature u varies linearly with the spatial coordinate x so that the midsection temperature must be the arithmetic average of the endsection temperatures.   ●

EXAMPLE 4   A ball is thrown straight up with an initial speed of 128 ft/sec from the edge of a cliff 160 ft above the ground. Find the equation of motion of the ball.

Solution   The acceleration due to gravity is -32 ft/sec² provided that displacement is measured positively upward from the ground. This example is a rectilinear motion problem in which the axis is vertical. We shall choose ground level (not the top of the cliff) to be the zero position. The notation of Section 3.3 will be used in our solution.

$$\frac{dv}{dt} = a = -32$$

Then $v = -32t + C_1$ and we write $v = v(t)$, since velocity is a function of time.

When $t = 0$, $v = 128$ (that is, $v(0) = 128$) which implies that $128 = C_1$.

$$S \, (e) \quad \frac{ds}{dt} = v = -32t + 128$$

$$s = -16t^2 + 128t + C_2$$

When $t = 0$, $s = 160$, so $160 = C_2$. Then

$$s = -16t^2 + 128t + 160$$

is the equation of motion.   ●

In many rectilinear motion problems the acceleration is determined as a given function of the time. Furthermore, there usually are two initial conditions involving velocity and displacement. A basic problem of this type is the **falling body problem** already illustrated in Example 4. Usually it is assumed that the object is moving vertically near the surface of the Earth. Furthermore in its simplest form air resistance is taken to be negligible. This means that the Earth's gravitational force is the only force acting on the particle and is directed downward toward the center of the Earth. Therefore if the displacement s is measured positively upward then the acceleration $a = -g \doteq -32 \, \text{ft/sec}^2$ if the units used are feet and seconds.† Therefore we have

$$a(t) = v'(t) = s''(t) = -g \tag{12}$$

where g is taken to be a constant. Taking antiderivatives of both sides of (12), there results

$$v(t) = -gt + C_1 \tag{13}$$

Now suppose that t is measured from a certain instant denoted as the initial time $t = 0$, and, if v is known at $t = 0$, we call $v|_{t=0} = v_0 =$ the **initial velocity.** Thus

$$v_0 = -g(0) + C_1 \qquad \text{or} \qquad C_1 = v_0$$

and substitution into (13) yields

$$v(t) = v_0 - gt \tag{14}$$

Taking the antiderivative again results in

$$s(t) = -\frac{gt^2}{2} + v_0 t + C_2$$

and if $s(0) = s_0$, a known **initial displacement,** then

$$s(t) = -\frac{gt^2}{2} + v_0 t + s_0 \tag{15}$$

More generally, if $a = $ constant is the constant acceleration in the positive direction, then, proceeding analogously, we obtain

$$s(t) = \frac{at^2}{2} + v_0 t + s_0 \tag{16}$$

---

† Actually, a better approximation for g is 32.17 ft/sec². Also g varies slightly with altitude and position on the Earth's surface.

In particular, if the displacement is taken positively toward the center of the Earth then $a = g$ and

$$s(t) = \frac{gt^2}{2} + v_0 t + s_0 \qquad (17)$$

EXAMPLE 5   A stone is dropped down a vertical mine shaft and hits the bottom in 12 sec. How deep is the shaft?

Solution   If the displacement is measured downward from the opening of the shaft and with the initial velocity v_0 and s_0 both zero, (17) becomes (with $g = 32$); $s(t) = 16t^2$. Therefore

$$s \bigg|_{t=12} = 16(12)^2 = 16(144) = 2304 \text{ ft}$$

and this is the required depth of the shaft.   ●

EXAMPLE 6   A ball is thrown straight up with an initial speed of 128 ft/sec from the edge of a cliff 160 ft above the ground. How long does it take for the ball to reach its maximum height? What is the maximum height? At what time and with what velocity does the ball strike the ground?

Solution   In Example 4 we found that $v = 128 - 32t$ and $s = -16t^2 + 128t + 160$. We could also now have obtained these equations from formulas (14) and (15), respectively.

The velocity is zero when the ball reaches its maximum height. Therefore

$$128 - 32t = 0 \qquad \text{or} \qquad t = 4 \sec$$

is the time to reach the maximum height.

The maximum height of the ball relative to the ground is

$$s \bigg|_{t=4} = -256 + 512 + 160 = 416 \text{ ft}$$

To find the time to reach the ground we set $s = 0$ so that

$$-16t^2 + 128t + 160 = 0$$

Division by -16 yields the simpler equation

$$t^2 - 8t - 10 = 0$$

which has two solutions

$$t = \frac{8 \pm \sqrt{64 + 40}}{2} = \frac{8 \pm \sqrt{104}}{2} = 4 \pm \sqrt{26}$$

Now $4 - \sqrt{26}$ is negative, and we reject this since the ball cannot land before it is thrown. The ball hits the ground $(4 + \sqrt{26})$ sec after it is thrown; that is, it is in the air for about 9.1 sec.

Finally

$$v\Big|_{9.1} = 128 - 32(9.1) = -163.2 \text{ ft/sec}$$

that is, 163.2 ft/sec is the appropriate impact speed.

The graphs of $s(t)$, $v(t)$, and the speed $= |v(t)|$ are shown in Figure 5.4.2. ●

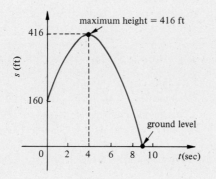
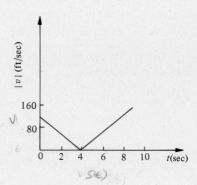

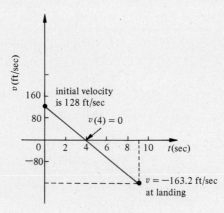

Figure 5.4.2

Exercises 5.4

In Exercises 1 through 10 find a solution of the differential equation satisfying the given initial or boundary conditions:

1. $f'(x) = -4x$; $f(2) = 3$

2. $F'(t) = 1 - 3t$; $F(0) = -1$

3. $s'(t) = \dfrac{1}{2\sqrt{1+t}}$; $s(3) = 5$

4. $y' = 9x^2 - 1$; $y(-1) = 0$

5. $y' = \dfrac{3x}{\sqrt{3x^2+1}}$; $y(1) = 2$

6. $y' = \dfrac{1}{2\sqrt{x}} - \dfrac{3}{x^2}$; $y(4) = \dfrac{7}{4}$

7. $g''(t) = -\dfrac{1}{4t\sqrt{t}}$; $g(1) = 3, \; g'(1) = \dfrac{5}{2}$

8. $h''(t) = bt$ (b constant); $h(0) = 1, \; h'(0) = 3$

9. $F''(u) = -\dfrac{2}{u^3}\left(1 + \dfrac{3}{u}\right)$; $F(2) = -\dfrac{3}{4}, \; F'(2) = \dfrac{1}{2}$

10. $G''(x) = -2k$ (k constant); $G(0) = 0$, $G(L) = 0$
11. Find an equation of a curve with slope x^2 at each point (x, y) on it and passing through $(-2, 0)$.
12. Find an equation of a curve with slope $\sqrt{x}$ at each point (x, y) on it and passing through $(4, 4)$.
13. Find an equation $y = g(x)$ of a curve such that $g'' = 6$ at each point (x, y) on it and passing through $(1, 4)$ with slope 5 at that point.
14. Find an equation $y = h(x)$ of a curve such that $h'' = \dfrac{3x}{2} - 2$ at each point (x, y) on it and passing through $(2, -1)$ with horizontal tangent at that point.
15. Find an equation $y = F(x)$ of a curve such that $F'' = \dfrac{2}{x^3} - \dfrac{12}{x^4}$ at each point (x, y) on it and passing through the points $(1, -1)$ and $(\frac{1}{2}, -5)$.
16. A stone is dropped from the top of a building 484 ft high. After how many seconds will the stone reach the ground? What is the velocity of the stone when it strikes the ground?
17. An arrow is shot straight up into the air from ground level with an initial velocity of 96 ft/sec. How high will it go and when will it strike the ground? What is the velocity when it hits the ground?
18. If a gun is capable of firing a bullet at the speed of 800 ft/sec and if the gun is fired straight up, how high will the bullet go?
19. Determine the minimum initial velocity with which a shell must be fired to strike a target 1600 ft high and directly overhead.
20. A stone is thrown down an abandoned well with an initial speed of 40 ft/sec, and hits bottom at 120 ft/sec. Determine the depth of the well.
21. A driver of a car is travelling along a straight road at a uniform speed of 60 mi/hr when he suddenly notices an obstruction 500 ft in front of him. If his reaction time (that is, the time to reach the brake pedal with his foot) is 1 sec and his brakes cause a uniform deceleration (negative acceleration) of 8 ft/sec², will the driver avoid hitting the obstruction? (*Hint:* 60 mi/hr = 88 ft/sec)
22. The kinetic energy K of a body of mass m moving with velocity v is defined to be $K = \dfrac{mv^2}{2}$. If a mass is thrown downward with an initial kinetic energy K_0 show that its kinetic energy K is given by $K = K_0 + mgs$, where s is the distance it has fallen. What is the interpretation of $K - K_0 = mgs$?
23. A ball is thrown from ground level vertically upward. Show that the ball takes as long to go up as it does to come down. Furthermore show that the initial and impact speeds are equal.
24. A block slides down a smooth inclined plane starting from rest. Suppose that its initial position is h ft above the ground level, which is assumed to be horizontal. If it moves L ft along the plane, determine the time for the block to reach the ground. What velocity has the block acquired upon reaching the ground? Note that the acceleration is $32 \sin \theta$, where θ is the angle the plane makes with the horizontal.
25. A block slides down an inclined plane 150 ft long with an acceleration of 10 ft/sec². If the block is started from rest at the top, how long does it take to reach the end of the plane? Determine the initial velocity required to reach the end of the plane in 5 sec.
26. A car is moving at 40 ft/sec. What must be its constant acceleration in order to travel 512 ft in 8 sec?
27. A vehicle is moving with constant acceleration. Suppose that its displacement at times t_1 and t_2 are, respectively, s_1 and s_2. Find the acceleration in terms of the given quantities.

Review and Miscellaneous Exercises

1. If $y = 3x + 5$, find Δy, dy, and the difference $\Delta y - dy$ when (a) $x = 2$ and $\Delta x = 0.3$; (b) $x = 3$ and $\Delta x = -0.01$.

2. If $y = \dfrac{1}{x}$ find y, dy, and the difference $\Delta y - dy$, when $x = 3$ and $\Delta x = 0.1$ to five decimal places.

3. Use differentials to obtain the linear approximation: $\sqrt[4]{1 + \Delta x} \doteq 1 + \dfrac{\Delta x}{4}$ for $|\Delta x| \ll 1$.

In Exercises 4 through 14 find the most general antiderivative

4. $\int \left(3x - \frac{1}{x}\right)^2 dx$

5. $\int 6x \sqrt[3]{x^2 + 2}\, dx$

6. $\int \frac{x\, dx}{\sqrt{1 + 2x}}$

7. $\int x(2 + x)^{1/3}\, dx$

8. $\int \frac{8x}{(5x + 2)^{3/2}}\, dx$

9. $\int \frac{t^2 + 3}{\sqrt[4]{t}}\, dt$

10. $\int t \left(t + \frac{1}{\sqrt{t}}\right)^3 dt$

11. $\int \frac{v\, dv}{(5 + v^2)^{4/3}}$

12. $\int \frac{\sqrt{5x^3}}{\sqrt[3]{6x}}\, dx$

13. $\int \frac{dx}{(3 - 4x)^3}$

14. $\int \frac{\sqrt{5 + \sqrt{t}}}{\sqrt{t}}\, dt$

15. Find an approximate value for $\sqrt[3]{26.1}$ without using tables.

16. Prove that the relative error in the surface area of a sphere is approximately two times the relative error in the measurement of the diameter.

17. If the three linear dimensions of a box are measured to errors within 0.5%, 1%, and 1.5%, respectively, find the approximate maximum per cent error in the calculated volume.

18. If $uv = C$ (a constant $\neq 0$) find a relationship between $\frac{du}{u}$ and $\frac{dv}{v}$.

19. Without attempting to calculate exactly, find a reasonable approximation to $(1.001)^4 + (1.003)^{-1}$.

20. Find a solution of the initial value problem:

$$\frac{dy}{dx} = x^2 - x \qquad y(3) = \frac{19}{2}$$

21. Find a solution of the initial value problem

$$x^2 y \frac{dy}{dx} = 4 \qquad y(4) = 2$$

22. Find a particular solution of $x^4 \frac{d^2y}{dx^2} + 12 = 0$ for which $y(2) = 2$ and $y'(2) = \frac{3}{2}$.

23. Find a particular solution of $\sqrt{2x + 3}\, y \frac{dy}{dx} = 1$ for which $y = -4$ when $x = -1$.

24. Find a particular solution of the boundary value problem

$$y'' = -\frac{1}{4(x + 4)^{3/2}}$$

subject to $y(0) = 2$ and $y(5) = -12$.

25. Find a particular solution of $\sqrt{(x + 2)(3y + 1)}\, dx - dy = 0$ for which $y = 0$ when $x = 2$.

26. Given two functions f and g such that $\frac{df}{dx} = \frac{dg}{dx}$ for all x and $f(3) = 4 + g(3)$. Find $g(0)$ if $f(0) = 5$.

27. Given two functions f and g such that $\frac{df}{dx} = x + 6\frac{dg}{dx}$ for all x, find the most general relationship between f and g.

28. At any point (x, y) on a curve

$$D_x^2 y = \frac{2}{3} - \frac{3}{4}x^{-3/2} \qquad (x > 0)$$

Find an equation of the curve if it passes through the point $(1, 3)$ with slope $\frac{13}{6}$.

29. A car is cruising at 60 mi/hr (or 88 ft/sec) when the brakes are suddenly applied. If the car has a

constant acceleration -16 ft/sec^2, how long does it take for the vehicle to stop? How far did the car travel once the brakes were applied?

30. A stone is thrown downward from the top of a building 160 ft high with an initial speed of 48 ft/sec. How long does it take to strike the ground and what is the velocity of impact?

31. From a balloon 32,000 ft above the ground, an object is shot directly upward and hits the Earth 100 sec later. Find the initial velocity of the object and its maximum height above the ground.

32. If $x = t^3 - 3t + 1$ and $y = x^2 + 2x - 4$ find dy in terms of dt when $t = 2$, (a) without eliminating x, (b) by eliminating x.

6

Area, Definite Integral, and the Fundamental Theorems of Calculus

6.1 SUMMATION NOTATION

The definition of the definite integral requires the use of sums involving many terms. A notation called the *sigma notation* is used to facilitate our writing and evaluating such sums. It is the objective of this section to explain and illustrate it.

The symbol $\sum$ is a capital sigma in the Greek alphabet and it corresponds to the letter S in our English alphabet. It stands for summation. Thus the symbol

$$\sum_{k=1}^{5} k$$

means the sum

$$1 + 2 + 3 + 4 + 5$$

Other examples of summation notation are

$$\sum_{k=1}^{4} k^2 = 1^2 + 2^2 + 3^2 + 4^2$$

$$\sum_{r=-1}^{3} (2r + 1) = [2(-1) + 1] + [2(0) + 1] + [2(1) + 1]$$
$$+ [2(2) + 1)] + [2(3) + 1]$$
$$= -1 + 1 + 3 + 5 + 7$$

$$\sum_{m=1}^{n} m^4 = 1^4 + 2^4 + 3^4 + \cdots + (n - 1)^4 + n^4$$

$$\sum_{t=1}^{100} \frac{1}{t^2} = \frac{1}{1^2} + \frac{1}{2^2} + \cdots + \frac{1}{(99)^2} + \frac{1}{(100)^2}$$

In general,

$$\sum_{k=m}^{n} f(k) = f(m) + f(m + 1) + f(m + 2) + \cdots + f(n) \qquad (1)$$

where m and n are integers for which $m \leq n$. The right-hand side of (1) consists of the sum of $n - m + 1$ terms obtained by replacing k by m, then by $m + 1$, and so forth, until the last term where k is replaced by n. The number m is called the **lower limit** of the sum, and n is called the **upper limit.** The symbol k is called the **index of summation.** It is a **dummy** symbol because any other letter could be used without affecting the expression. Thus

$$\sum_{k=1}^{n} k = \sum_{r=1}^{n} r = \sum_{j=1}^{n} j = 1 + 2 + \cdots + n$$

Another example is

$$\sum_{k=2}^{5} \frac{k}{k + 3} = \sum_{r=2}^{5} \frac{r}{r + 3} = \frac{2}{5} + \frac{3}{6} + \frac{4}{7} + \frac{5}{8}$$

A sum such as $\sum_{k=1}^{7} 5$ is to be interpreted as

$$\sum_{k=1}^{7} 5 = 5 + 5 + 5 + 5 + 5 + 5 + 5 = 7(5) = 35$$

and

$$\sum_{k=3}^{6} B_k = B_3 + B_4 + B_5 + B_6$$

and

$$\sum_{k=1}^{3} f(x_k) \, \Delta x_k = f(x_1) \, \Delta x_1 + f(x_2) \, \Delta x_2 + f(x_3) \, \Delta x_3$$

In many applications it is necessary to change the set of values that the summing index can assume, as illustrated in the following

$$\sum_{k=2}^{n+1} a_{k-1} = \sum_{r=1}^{n} a_r \qquad (i)$$

$$\sum_{k=3}^{n+2} (k - 2)^2 = \sum_{r=1}^{n} r^2 \qquad (ii)$$

$$\sum_{s=2}^{5} \frac{x^{s-1}}{s^2} = \sum_{u=1}^{4} \frac{x^u}{(u + 1)^2} \qquad (iii)$$

Although each of the preceding formulas may be verified by writing out the terms, we may directly prove the equality. For example in (i) let $k - 1 = r$, then, when $k = 2$, $r = 1$ and when $k = n + 1$, $r = (n + 1) - 1 = n$ so that the result follows. Similarly for (ii) let $k - 2 = r$. Finally for (iii) the correspondence is $s - 1 = u$.

The following properties are readily established by writing out the terms

PROPERTY 1
$$\sum_{k=1}^{n} c = nc \tag{2}$$

(Proof left as exercise.)

PROPERTY 2
$$\sum_{k=1}^{n} c\,f(k) = c \sum_{k=1}^{n} f(k) \qquad \text{for any constant } c \tag{3}$$

(Proof left as exercise.)

PROPERTY 3
$$\sum_{k=1}^{n} [f(k) + g(k)] = \sum_{k=1}^{n} f(k) + \sum_{k=1}^{n} g(k) \tag{4}$$

Proof
$$\sum_{k=1}^{n} [f(k) + g(k)] = f(1) + g(1) + f(2) + g(2) + \cdots + f(n) + g(n)$$
$$= [f(1) + f(2) + \cdots + f(n)] + [g(1) + g(2) + \cdots + g(n)]$$
$$= \sum_{k=1}^{n} f(k) + \sum_{k=1}^{n} g(k) \qquad\blacksquare$$

PROPERTY 4
$$\sum_{k=1}^{n} [f(k) - f(k - 1)] = f(n) - f(0) \tag{5}$$

Proof
$$\sum_{k=1}^{n} [f(k) - f(k - 1)] = [f(1) - f(0)] + [f(2) - f(1)] + [f(3) - f(2)]$$
$$+ \cdots + [f(n - 1) - f(n - 2)] + [f(n) - f(n - 1)]$$
$$= -f(0) + [f(1) - f(1)] + [f(2) - f(2)] + \cdots +$$
$$[f(n - 1) - f(n - 1)] + f(n)$$
$$= f(n) - f(0) \qquad\blacksquare$$

This is known as a *telescoping* series because the sums neutralize in pairs and we are left only with the contribution from $f(n)$ and $-f(0)$. Thus, for example, with $f(k) = k^3$ substituted in (5) there results

$$\sum_{k=1}^{n} [k^3 - (k - 1)^3] = n^3 - 0^3 = n^3$$

We now apply these properties to some further illustrations.

EXAMPLE 1 Find a formula for $\displaystyle\sum_{k=1}^{n} k$ where n is an arbitrary positive integer.

Solution Let $\displaystyle S_n = \sum_{k=1}^{n} k = 1 + 2 + 3 + \cdots + (n - 1) + n$. Write the terms that make up S_n in reverse order; that is, $S_n = n + (n - 1) + (n - 2) + \cdots + 2 + 1$. If the two equations are added term by term, we obtain

$$2S_n = (n + 1) + (n + 1) + (n + 1) + \cdots + (n + 1)$$
$$= n(n + 1)$$

Therefore

$$S_n = \frac{n(n + 1)}{2} \qquad (6)$$

For example,

$$S_6 = \frac{6(7)}{2} = 21 = 1 + 2 + 3 + 4 + 5 + 6.$$

More significantly,

$$S_{100} = \frac{100(101)}{2} = 5050 = 1 + 2 + 3 + \cdots + 99 + 100 \qquad \bullet$$

EXAMPLE 2 Find a formula for $\displaystyle\sum_{k=1}^{n} (3k - 1)$.

Solution Let $\displaystyle g_n = \sum_{k=1}^{n} (3k - 1)$. Then, from Properties 1 through 3 and Example 1,

$$g_n = 3 \sum_{k=1}^{n} k - \sum_{k=1}^{n} 1$$

$$= 3 \frac{n(n + 1)}{2} - n$$

$$= (n) \left(\frac{3n + 3 - 2}{2} \right) = \frac{n(3n + 1)}{2}$$

For example,

$$g_5 = \sum_{k=1}^{5} (3k - 1) = \frac{5(16)}{2} = 40 \qquad \bullet$$

EXAMPLE 3 We now give an alternative derivation of the formula for $\displaystyle S_n = \sum_{k=1}^{n} k$. This derivation is preferable to that of Example 1 since it can be generalized. Consider the difference of two consecutive squares

$$k^2 - (k - 1)^2 = 2k - 1$$

This last relation is an identity. Sum both sides over k from $k = 1$ to $k = n$,

$$\sum_{k=1}^{n} [k^2 - (k-1)^2] = 2 \sum_{k=1}^{n} k - n$$

If the summation on the left is written out, the terms neutralize in pairs (telescoping series) and we obtain

$$n^2 - 0^2 = 2 \sum_{k=1}^{n} k - n$$

Therefore

$$\sum_{k=1}^{n} k = \frac{n^2 + n}{2} = \frac{n(n+1)}{2} \qquad \bullet$$

Some formulas we will find useful follow. The first was derived in Example 1 (and again in Example 3). The others can be verified without too much difficulty.

$$\sum_{k=1}^{n} k = \frac{n(n+1)}{2} \tag{7}$$

$$\sum_{k=1}^{n} k^2 = \frac{n(n+1)(2n+1)}{6} \tag{8}$$

$$\sum_{k=1}^{n} k^3 = \frac{n^2(n+1)^2}{4} \tag{9}$$

Formula (8) can be verified by starting with the difference between two consecutive cubes $k^3 - (k-1)^3 = 3k^2 - 3k + 1$ and employing the method used in Example 3. A similar procedure can be used for formula (9).

EXAMPLE 4   Find the sum to n terms

$$(1)(3) + (2)(4) + (3)(5) + \cdots$$

Solution   By inspection we seek $\sum_{k=1}^{n} k(k+2)$. Now,

$$\sum_{k=1}^{n} k(k+2) = \sum_{k=1}^{n} k^2 + 2 \sum_{k=1}^{n} k$$

$$= \frac{n(n+1)(2n+1)}{6} + n(n+1)$$

$$= n(n+1)\left[\frac{2n+1}{6} + 1\right]$$

$$= \frac{n(n+1)(2n+7)}{6}$$

Therefore for all positive integers n,

$$(1)(3) + (2)(4) + (3)(5) + \cdots + n(n + 2) = \frac{n(n + 1)(2n + 7)}{6}$$ ●

EXAMPLE 5 Prove by mathematical induction that for arbitrary positive integer n

$$\sum_{i=1}^{n} \frac{1}{i(i + 1)} = \frac{n}{n + 1} \tag{10}$$

Proof The proposition is true for $n = 1$ because $\dfrac{1}{1(2)} = \dfrac{1}{1 + 1}$. Now assume that the proposition is true for some positive integer $n = k$, that is,

$$\sum_{i=1}^{k} \frac{1}{i(i + 1)} = \frac{k}{k + 1} \tag{i}$$

Then it must be shown that it is also true for $n = k + 1$. To this end, we add $\dfrac{1}{(k + 1)(k + 2)}$ to both sides of (i) and obtain

$$\sum_{i=1}^{k+1} \frac{1}{i(i + 1)} = \frac{k}{k + 1} + \frac{1}{(k + 1)(k + 2)} = \frac{k(k + 2) + 1}{(k + 1)(k + 2)} = \frac{k^2 + 2k + 1}{(k + 1)(k + 2)}$$

$$= \frac{(k + 1)^2}{(k + 1)(k + 2)} = \frac{k + 1}{k + 2} \tag{ii}$$

Now (ii) implies that, if the proposition (10) is true for any positive integer $n = k$, it must also be true for $n = k + 1$. But the proposition is true for $n = 1$ so that it must also be true for $n = 2, 3, 4, \ldots$. This establishes (10) for *all* positive integers. ∎

Exercises 6.1

In Exercises 1 through 8, write in expanded form

1. $\displaystyle\sum_{k=1}^{4} (3k + 1)$

2. $\displaystyle\sum_{m=1}^{5} \frac{1}{m^2}$

3. $\displaystyle\sum_{s=1}^{5} \frac{s}{s + 1}$

4. $\displaystyle\sum_{r=1}^{6} \frac{1}{r!}$

5. $\displaystyle\sum_{k=2}^{6} \frac{k(k - 1)}{k!}$

6. $\displaystyle\sum_{k=1}^{n} (k^3 + k)$

7. $\displaystyle\sum_{i=1}^{n^2} f(i)$

8. $\displaystyle\sum_{j=1}^{m} [f(i) - g(i)]$

In Exercises 9 through 13, write the formulas or expressions in summation notation:

9. $1 + 3 + 5 + 7 + \cdots + (2n - 1) = n^2$
10. $3 + 6 + 9 + \cdots + 3n = \frac{3}{2}n(n + 1)$

11. $3 + 3^2 + 3^3 + \cdots + 3^n = \frac{3}{2}(3^n - 1)$

12. $f(\xi_1)(\Delta x_1) + f(\xi_2)(\Delta x_2) + \cdots + f(\xi_n)(\Delta x_n)$

13. $y_1 \Delta x_1 + y_2 \Delta x_2 + \cdots + y_n \Delta x_n$

In Exercises 14 through 26, find the given sum.

14. $\displaystyle\sum_{k=1}^{100} \frac{k}{10}$

15. $\displaystyle\sum_{k=1}^{10} 3k^2$

16. $\displaystyle\sum_{j=1}^{n} (2j + 3)$

17. $\displaystyle\sum_{k=1}^{n} (3k - 2)$

18. $\displaystyle\sum_{k=1}^{n} (6k^2 - 2k)$

19. $\displaystyle\sum_{k=0}^{n} (5^{k+1} - 5^k)$

20. $\displaystyle\sum_{k=1}^{n} (3^{k-1} - 3^k)$

21. $\displaystyle\sum_{k=2}^{n} \left(\frac{1}{k+1} - \frac{1}{k} \right)$

22. $\displaystyle\sum_{k=3}^{100} \left(\frac{1}{k+2} - \frac{1}{k+1} \right)$

23. $\displaystyle\sum_{k=3}^{100} (\sqrt{3k+1} - \sqrt{3k-2})$

24. $\displaystyle\sum_{n=1}^{m} \left(\frac{1}{n+3} - \frac{1}{n+2} \right)$

25. $\displaystyle\sum_{r=2}^{m} [h(r) - h(r-1)]$

26. $\displaystyle\sum_{r=2}^{m} f(r - (r-1))$

In Exercises 27 through 30 simplify the given expression

27. $\displaystyle\sum_{k=1}^{n} \left(\frac{k}{n}\right)^2 \left(\frac{1}{n}\right)$

28. $\displaystyle\sum_{k=1}^{n} \left[\left(\frac{k}{n}\right)^2 + \frac{2k}{n} \right] \left(\frac{1}{n+1}\right)$

29. $\displaystyle\sum_{k=1}^{n} \left(\frac{3k}{n}\right)^2 \left(\frac{5}{n}\right)$

30. $\displaystyle\sum_{k=1}^{n} \left(\frac{k}{n}\right)^3 \left(\frac{1}{n+1}\right)$

In Exercises 31 through 36 establish the given formula by mathematical induction for all positive integers n.

31. $1 + 2 + 3 + \cdots + n = \dfrac{n(n+1)}{2}$

32. $1^2 + 2^2 + 3^2 + \cdots + n^2 = \dfrac{n(n+1)(2n+1)}{6}$

33. $1^3 + 2^3 + 3^3 + \cdots + n^3 = \dfrac{n^2(n+1)^2}{4}$

34. $a + (a + d) + (a + 2d) + \cdots + (a + (n-1)d) = \dfrac{n}{2}(2a + (n-1)d)$

(arithemtic progression)

35. $a + ar + ar^2 + \cdots + ar^{n-1} = \dfrac{a - ar^n}{1 - r}, \quad (r \neq 1)$

(geometric progression)

36. $1^2 + 3^2 + 5^2 + \cdots + (2n-1)^2 = \dfrac{n(4n^2 - 1)}{3}$

37. By starting with $k^4 - (k-1)^4 \equiv 4k^3 - 6k^2 + 4k - 1$ and using the method shown in Example 3 show that

$$\sum_{k=1}^{n} k^3 = \frac{n^2(n+1)^2}{4} = \left[\frac{n(n+1)}{2}\right]^2 = \left[\sum_{k=1}^{n} k\right]^2$$

38. Prove by mathematical induction $\displaystyle\sum_{i=1}^{n} \frac{1}{(2i-1)(2i+1)} = \frac{n}{2n+1}$

39. Find the sum to n terms: $(1)(4) + (2)(5) + (3)(6) + \cdots$

40. Find the sum to n terms: $(1)(2)(3) + (2)(3)(4) + (3)(4)(5) + \cdots$

41. If $A = \displaystyle\sum_{i=1}^{n} a_i^2, B = \sum_{i=1}^{n} a_i b_i,$ and $C = \sum_{i=1}^{n} b_i^2$ where a_i and b_i are real numbers $(i = 1, 2, 3, \ldots, n)$, prove

that $At^2 + 2Bt + C \geq 0$ for all real numbers t. $\left(Hint: At^2 + 2Bt + C = \displaystyle\sum_{i=1}^{n} (a_i t + b_i)^2\right)$

42. Prove the **Cauchy-Schwarz inequality**

$$\left(\sum_{i=1}^{n} a_i b_i\right)^2 \leq \sum_{i=1}^{n} a_i^2 \sum_{i=1}^{n} b_i^2$$

for arbitrary real numbers a_i and b_i $(i = 1, 2, 3, \ldots, n)$. (*Hint:* Use completing the square and the result of the previous exercise.)

6.2 INTRODUCTION TO AREA

Associated with each rectangle having length l and width w is the number lw called the ***area*** of the rectangle. The area of a triangle is shown to be $\dfrac{bh}{2}$ where b is a base (length of one side) and h is the altitude to that base (Figure 6.2.1). Since any polygon can be subdivided into triangles, the area of the polygon is defined to be the sum of the areas of the constituent triangles (Figure 6.2.2).

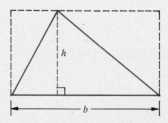

Figure 6.2.1

Over 2000 years ago the Greek mathematician Eudoxus investigated the area of a circle by a method known as the ***method of exhaustion.*** Imagine that a regular polygon of n sides is inscribed in a circle. The area of the regular polygon will be less than the area of the circle. Now suppose that the number of sides of the polygon increases. Then the area will increase and, as $n \to \infty$ it appears that the

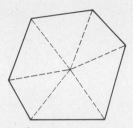

Figure 6.2.2

regular polygon's area will "exhaust" the area of the circle. Since it is easy to find a formula for the area of a regular polygon with n sides, the area of a circle may be obtained by taking the limit as $n \to \infty$. In particular, the following sequence of areas $a(n)$ is obtained for regular polygons with n sides of the form $n = 2^k (k = 2, 3, \ldots, 8)$ with the radius of the circle equal to r (Table 6.2.1 and Figure 6.2.3)

Table 6.2.1

n	$a(n)$
4	$2.000r^2$
8	$2.8284r^2$
16	$3.0614r^2$
32	$3.1214r^2$
64	$3.1363r^2$
128	$3.1405r^2$
256	$3.1411r^2$

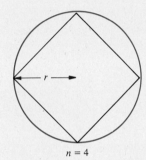
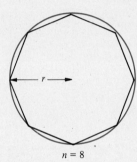

Figure 6.2.3

Now suppose we employ circumscribed regular polygons (Figure 6.2.4) rather than inscribed regular polygons. Then the area of any particular circumscribed regular polygon having n sides would always exceed the area of the circle. However as n increases, this excess will decrease and will tend to zero as n becomes infinite.

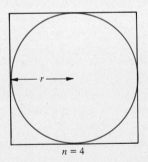
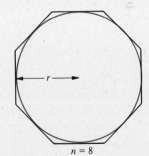

Figure 6.2.4

In fact if $A(n)$ is the area of the circumscribed regular polygon of n sides, Table 6.2.2 can be constructed.

Thus if B is the area of the circle,

$$3.1411r^2 < B < 3.1418r^2$$

and, in fact, $B = \pi r^2 \doteq 3.1416r^2$

Table 6.2.2

n	$A(n)$
4	$4.0000r^2$
8	$3.3137r^2$
16	$3.1826r^2$
32	$3.1517r^2$
64	$3.1441r^2$
128	$3.1422r^2$
256	$3.1418r^2$

How can we define the area of a plane region R with a curved boundary (Figure 6.2.5)? In fact it is not clear that the region has an area. We could attack the problem by placing the region on a piece of graph paper and (1) under-approximate this area A of R by counting the number of squares of the graph paper completely contained in R and (2) overapproximate its area by counting the number of squares containing some part of a square or a whole square belonging to R. The area of R should be between these two numbers. By using finer and finer grids, a closer and closer approximation to the area of R may be determined.

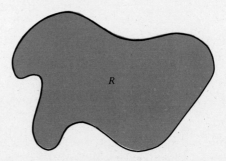

Figure 6.2.5

In order to make the problem of determining the area of a region with curved boundaries mathematically tractable, our considerations will be restricted to the case where the curved boundary is the graph of a continuous function. To this end, let f be a continuous and nonnegative function (that is, $f(x) \geq 0$) in the interval $[a, b]$. The region R of interest (Figure 6.2.6) is bounded by the graph of f, the lines $x = a$ and $x = b$, and the x-axis. It is termed *the region under the graph of f from a to b.*

Let us now consider the problem of finding the area of the region bounded by the parabola $f(x) = x^2$, the x-axis, and the vertical line $x = 3$ (Figure 6.2.9). In accordance with the method of exhaustion, we seek an upper estimate and a lower estimate so that the required area is squeezed between these two quantities. To this end we *partition* the interval $[0, 3]$ into n subintervals

$$[x_0, x_1], \ [x_1, x_2], \ [x_2, x_3], \ldots, [x_{n-1}, x_n]$$

where $0 = x_0 < x_1 < x_2 < x_3 < \cdots < x_{n-1} < x_n = 3$ and, for convenience of calculation only, take the n intervals of the partition to be of equal length, Δx (Figure 6.2.7). Therefore in this instance

$$\Delta x = \frac{3}{n} \qquad (1)$$

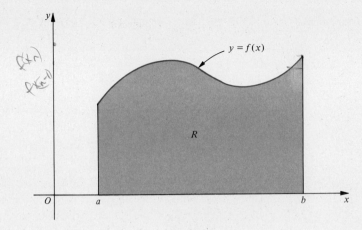

Figure 6.2.6

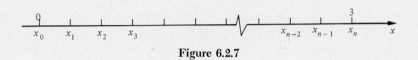

Figure 6.2.7

Now $f(x) = x^2$ is an increasing function for $x \geq 0$, and, in particular, on $[0, 3]$. Thus in the interval $[x_{k-1}, x_k]$, $f(x_{k-1})$ is the absolute minimum of f while $f(x_k)$ is the absolute maximum of f in that subinterval (Figure 6.2.8).

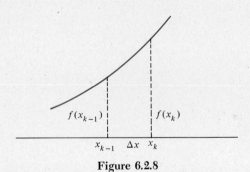

Figure 6.2.8

The sum of the areas of the rectangles having base $[x_{k-1}, x_k]$ and altitude $f(x_{k-1})$, where $k = 1, 2, \ldots, n$ is

$$s_n = f(x_0)\,\Delta x + f(x_1)\,\Delta x + f(x_2)\,\Delta x + \cdots + f(x_{n-1})\,\Delta x = \sum_{k=0}^{n-1} f(x_k)\,\Delta x \quad (2)$$

The summation on the right side of (2) gives the sum of the areas of n *inscribed* rectangles (Fig. 6.2.9(a)). It is called the ***lower sum*** of f on $[a, b]$ with respect to the given partitioning. No matter what the definition of A may be, we must have

$$s_n \leq A \quad (3)$$

Similarly if we form the sum of the areas of the n *circumscribed* rectangles with base $[x_{k-1}, x_k]$ and altitude $f(x_k)$ where $k = 1, 2, \ldots, n$ then we have the ***upper sum*** (Fig. 6.2.9(b)) of f, on $[a, b]$ with respect to the given partitioning

$$S_n = f(x_1)\,\Delta x + f(x_2)\,\Delta x + \cdots + f(x_n)\,\Delta x = \sum_{k=1}^{n} f(x_k)\,\Delta x \qquad (4)$$

in that

$$A \le S_n \qquad (5)$$

Thus for each partition of $[0, 3]$ we must have

$$s_n \le A \le S_n \qquad (6)$$

for all positive integers n. We now seek explicit formulas for s_n and S_n.

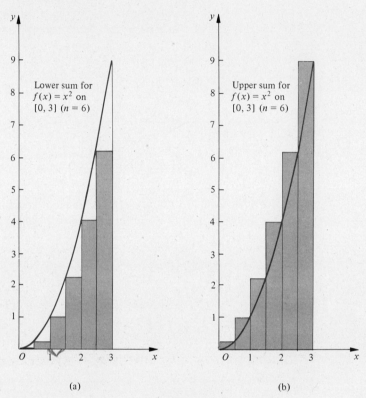

Lower sum for $f(x) = x^2$ on $[0, 3]$ $(n = 6)$

Upper sum for $f(x) = x^2$ on $[0, 3]$ $(n = 6)$

(a)

(b)

Figure 6.2.9

Because all the intervals $[x_{k-1}, x_k]$ have the same length Δx,

$$x_k = k\,\Delta x \qquad (7)$$

so that $x_0 = 0\,\Delta x = 0$, $x_1 = 1\,\Delta x$, $x_2 = 2\,\Delta x$, and so on. Now,

$$f(x_k)\,\Delta x = x_k^2\,\Delta x = (k\,\Delta x)^2\,\Delta x = k^2(\Delta x)^3 = \frac{3^3}{n^3}k^2 \qquad \text{from (7) and (1).}$$

Thus, substitution into (2) yields

$$s_n = \sum_{k=0}^{n-1} f(x_k)\,\Delta x = \sum_{k=0}^{n-1} \frac{3^3}{n^3}k^2$$

$$= \frac{27}{n^3}\sum_{k=0}^{n-1} k^2 = \frac{27}{n^3}\frac{(n-1)(n)(2n-1)}{6}$$

from (8) in Section 6.1 with n replaced by $n - 1$. Therefore

$$s_n = \frac{9}{2}\left(2 - \frac{3}{n} + \frac{1}{n^2}\right)$$

Again from (1) and (7),

$$S_n = \sum_{k=1}^{n} f(x_k)\,\Delta x = \sum_{k=1}^{n} \frac{3^3}{n^3} k^2$$

$$= \frac{27}{n^3}(1^2 + 2^2 + \cdots + n^2) = \frac{27}{n^3} \frac{n(n+1)(2n+1)}{6}$$

from (8) in Section 6.1. Thus

$$S_n = \frac{9}{2}\left(2 + \frac{3}{n} + \frac{1}{n^2}\right)$$

But then

$$\lim_{n\to\infty} s_n = \lim_{n\to\infty} \frac{9}{2}\left(2 - \frac{3}{n} + \frac{1}{n^2}\right) = 9 \leq A$$

and

$$\lim_{n\to\infty} S_n = \lim_{n\to\infty} \frac{9}{2}\left(2 + \frac{3}{n} + \frac{1}{n^2}\right) = 9 \geq A$$

which implies that

$$\lim_{n\to\infty} s_n = \lim_{n\to\infty} S_n = 9 = A$$

This suggests that we might *define* the area A of the region R to be

$$A = \lim_{n\to\infty} s_n \qquad \text{or} \qquad A = \lim_{n\to\infty} S_n$$

From either definition, the area of the region R is 9 square units.

EXAMPLE 1   Find the area A of the region bounded by the function $f(x) = 6 - x$, the line $x = 1$, the x-axis, and line $x = 5$, using circumscribed and inscribed rectangles (Figure 6.2.10).

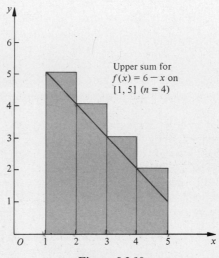

Figure 6.2.10

Solution   Divide the interval $[1, 5]$ into n equal subintervals so that $\Delta x = \dfrac{4}{n}$ and $x_k = 1 + \dfrac{4k}{n}$. Now $f(x_k) = 6 - x_k = 6 - \left(1 + \dfrac{4k}{n}\right) = 5 - \dfrac{4k}{n}$. In this instance, $f(x)$ is monotonically decreasing (since $f' < 0$) so that S_n is obtained by using the ordinates of the left end point of each subinterval. Therefore an upper estimate for A is

$$S_n = \sum_{k=0}^{n-1} \left(5 - \frac{4k}{n}\right) \Delta x = 5n\, \Delta x - \frac{4\, \Delta x}{n} \sum_{k=0}^{n-1} k$$

$$= 5(4) - \frac{4\, \Delta x}{n} \frac{(n-1)(n)}{2}$$

$$= 20 - 2(n\, \Delta x - \Delta x) = 20 - 8 + 2\, \Delta x$$

$$= 12 + 2\, \Delta x$$

Also, s_n is found by using the ordinates of the right end point of each subinterval. Thus

$$s_n = \sum_{k=1}^{n} \left(5 - \frac{4k}{n}\right) \Delta x = 5n\, \Delta x - \frac{4\, \Delta x}{n} \sum_{k=1}^{n} k$$

$$= 5(4) - \frac{4\, \Delta x}{n} \frac{n(n+1)}{2}$$

$$= 20 - 2(n\, \Delta x + \Delta x) = 20 - 8 - 2\, \Delta x$$

$$= 12 - 2\, \Delta x$$

so that

$$\lim_{n\to\infty} (12 - 2\, \Delta x) \le A \le \lim_{n\to\infty} (12 + 2\, \Delta x)$$

Now $\lim\limits_{n\to\infty} \Delta x = 0$, which implies that

$$A = 12 \text{ square units}$$

●

Exercises 6.2

In Exercises 1 throughout 8 use the method of this section to determine the area of the given region by using inscribed (lower sums) or circumscribed (upper sums) rectangles as indicated.

1. The region bounded by $y = x^2$, the x-axis, and the line $x = 1$, using circumscribed rectangles.
2. The region of Exercise 1, using inscribed rectangles.
3. The region bounded by $y = x$, $x = 4$, and the x-axis, using circumscribed rectangles.
4. The region of Exercise 3, using inscribed rectangles.
5. The region bounded by $y = 2x$, $x = 3$, and the x-axis, using inscribed rectangles.
6. The region of Exercise 5, using circumscribed rectangles.
7. The region bounded by $y = x$, the x-axis, and the lines $x = 1$ and $x = 3$, using circumscribed rectangles.
8. The region bounded by $y = \dfrac{x^2}{4} + 1$, the y-axis, the x-axis and $x = 4$, using circumscribed rectangles.
9. Starting with the expression for $(k + 1)^5 - k^5$, prove that

$$\sum_{k=1}^{n} k^4 = \frac{n}{30}(n+1)(6n^3 + 9n^2 + n - 1)$$

using the formulas for $\sum_{k=1}^{n} k$, $\sum_{k=1}^{n} k^2$, and $\sum_{k=1}^{n} k^3$.

10. Find the area of the region bounded by $y = x^4$, the x-axis, and the line $x = 2$, using circumscribed rectangles.

11. Find the area of the region bounded by $y = x$, the x-axis, and the lines $x = a$ and $x = b$, where $0 < a < b$, using circumscribed rectangles. What is the geometric interpretation of the result?

12. Find the area of the region bounded by $y = x^2$, the x-axis, and the lines $x = a$ and $x = b$, where $0 < a < b$, using circumscribed rectangles.

13. Find the area of the region bounded by $y = x^3$, the x-axis, and the lines $x = 1$ and $x = 3$, using circumscribed rectangles.

14. Find the area of the region bounded by $y = x^2 + 3x$, the x-axis, and the lines $x = 0$ and $x = 1$, using circumscribed rectangles.

15. Find the area of the region bounded by $y = x^2 - x$, the x-axis, and the lines $x = 1$ and $x = 3$, using circumscribed rectangles.

16. Find the area of the region bounded by $y = (x + 1)^2$, the x-axis, the y-axis, and the line $x = 2$ using circumscribed rectangles.

17. The region bounded by $y = h$ and the parabola $y = 4h\dfrac{x^2}{b^2}$ is a parabolic segment (where h and b are assumed to be positive). Show that its area is given by $A = \dfrac{2bh}{3}$ square units.

6.3

THE DEFINITE INTEGRAL

In the previous section, we emphasized the geometric concepts involved in determining area as a limit of sums. Now we shall generalize this and approach the problem in an analytic rather than geometric manner.

Our generalization involves modifying the type of sum used to find s_n or S_n in two respects:

1. When we partition the interval into a number of subintervals, we shall not assume that the subintervals have equal lengths.
2. In each subinterval, we shall evaluate the function at any point; not necessarily the point that determines the inscribed or the circumscribed rectangle.

We formalize this in the following discussion.

Let f be a function defined on $[a, b]$ where $a < b$. Divide this interval into n subintervals by choosing any $n - 1$ intermediate points between a and b. Let $x_0 = a$, $x_n = b$, and let $x_1, x_2, \ldots, x_{n-1}$ be the intermediate points so that

$$x_0 < x_1 < x_2 < \cdots < x_{n-1} < x_n$$

(the points $x_0, x_1, x_2, \ldots, x_{n-1}, x_n$ are not necessarily equidistant).

The set of closed intervals

$$P = \{[x_0, x_1], [x_1, x_2], [x_2, x_3], \ldots, [x_{n-1}, x_n]\}$$

determined by these points is called a **partition** or **subdivision** of $[a, b]$. The points $x_1, x_2, \ldots, x_{n-1}$ are called the **points of division** of the partition P (See

Figure 6.3.1). For convenience we sometimes also refer to the set of points $\{x_0, x_1, \ldots, x_n\}$ as a partition of $[a, b]$.

The lengths of these subintervals are given by

$$\Delta x_i = x_i - x_{i-1} \qquad (i = 1, 2, \ldots, n)$$

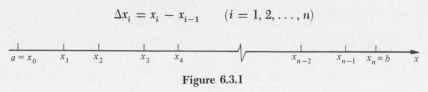

Figure 6.3.1

EXAMPLE 1   Consider the closed interval $[-4, 1.5]$. One partition of the interval $[-4, \frac{3}{2}]$ is

$$P = \{[-4, -3.1], [-3.1, -1.6], [-1.6, 0], [0, 1.5]\}$$

(Figure 6.3.2(a)) and another is (Figure 6.3.2(b))

$$P = \{[-4, -2.5], [-2.5, -1.3], [-1.3, 0.5], [0.5, 1.5]\}$$

There are of course infinitely many possible partitions of $[-4, \frac{3}{2}]$ including the interval itself when $n = 1$. ●

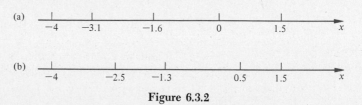

Figure 6.3.2

The length of the longest interval in a partition P is called the **norm** of the partition P and is symbolized by $|P|$.

EXAMPLE 2   The norm of the partition of $[-4, 1.5]$ shown in Figure 6.3.2(a) is $|P| = 0 - (-1.6) = 1.6$ and the norm of the partition in Figure 6.3.2(b) is $|P| = 0.5 - (-1.3) = 1.8$. ●

Let us return to an arbitrary partition of an interval $[a, b]$ over which a function f has been defined. In the first interval $[x_0, x_1]$ choose any point ξ_1, so that $x_0 \le \xi_1 \le x_1$; in the second interval select any point ξ_2, so that $x_1 \le \xi_2 \le x_2$; and continue this process of choosing one point in each of n intervals so that

$$x_{i-1} \le \xi_i \le x_i \qquad (i = 1, 2, \ldots, n)$$

Then form

$$\sum_{i=1}^{n} f(\xi_i)\, \Delta x_i = f(\xi_1)\, \Delta x_1 + f(\xi_2)\, \Delta x_2 + \cdots + f(\xi_n)\, \Delta x_n \qquad (1)$$

Such a sum is called a **Riemann sum**† for the function f on the interval $[a, b]$.

---

† After G. F. B. Riemann (1826–1866) who did fundamental work in analysis and is the founder of what is known today as Riemannian geometry.

EXAMPLE 3   Given $f(x) = 2 + x^2$ with $-1 \leq x \leq 2$, find the Riemann sum for the function f on $[-1, 2]$ for the partition P: $x_0 = -1$, $x_1 = 0$, $x_2 = 0.5$, $x_3 = 1.5$, $x_4 = 2$ and $\xi_1 = -\frac{1}{2}$, $\xi_2 = 0.3$, $\xi_3 = 1$, and $\xi_4 = 1.8$. What is the norm of the partition?

Solution

$$\sum_{i=1}^{4} f(\xi_i) \, \Delta x_i = f(\xi_1) \, \Delta x_1 + f(\xi_2) \, \Delta x_2 + f(\xi_3) \, \Delta x_3 + f(\xi_4) \, \Delta x_4$$
$$= f(-0.5)(0 - (-1)) + f(0.3)(0.5 - 0)$$
$$+ f(1)(1.5 - 0.5) + f(1.8)(2 - 1.5)$$
$$= (2.25)(1) + (2.09)(0.5) + 3(1) + (5.24)(0.5)$$
$$= 2.25 + 1.045 + 3 + 2.62$$
$$= 8.915$$

The norm of $|P|$ is the length of the longest subinterval. In this case $|P| = 1$ (shared by two subintervals.)

Since the function $f(x) = 2 + x^2$ is nonnegative, the Riemann sum can be interpreted as the sum of areas of rectangles that approximates the area bounded by the curve $f(x) = 2 + x^2$, the axis and the lines $x = -1$ and $x = 2$.   ●

In general we shall not restrict $f(x)$ to nonnegative values; that is, some of $f(\xi_i)$ may be negative. In such a case the geometric interpretation of a Riemann sum would be the sum of the areas of the rectangles lying above the x-axis plus the negatives of the areas lying below the x-axis. Thus, for example, in Figure 6.3.3 we would have for the Riemann sum,

$$\sum_{i=1}^{8} f(\xi_i) \, \Delta x_i = A_1 + A_2 - A_3 - A_4 - A_5 + A_6 + A_7 - A_8$$

because $f(\xi_3)(x_3 - x_2) = -A_3$ since $f(\xi_3) < 0$, and so on.

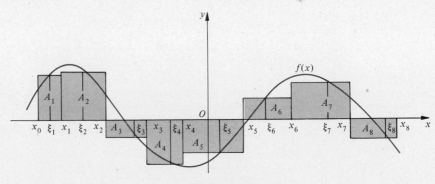

Figure 6.3.3

In Section 6.2, we found areas by forming a sum similar to the Riemann sum and then taking a limit as the lengths of the intervals approached zero. We now repeat this procedure, but use the Riemann sum, to obtain one of the basic definitions in calculus.

Definition Let f be defined on $[a, b]$. We say that f is **integrable** on $[a, b]$ where $a < b$ if there is a number L such that for every $\varepsilon > 0$ there exists a $\delta > 0$ which guarantees that

epsilon   *delta*

$$\left| \sum_{i=1}^{n} f(\xi_i)\, \Delta x_i - L \right| < \varepsilon$$

for each partition P with $|P| < \delta$ and for any choice of ξ_i in $[x_{i-1}, x_i]$, $i = 1, 2, \ldots, n$.

[handwritten margin notes: if the Limit of ξ as δx→0 is same as overestimate & underestimate the A is correct – this means area same as size of rectangles vary]

This definition states that the function is integrable on $[a, b]$ if the Riemann sums can be made as close to L as desired by making the norms $|P|$ of the partitions sufficiently small for all possible choices of the numbers ξ_i for which $x_{i-1} \leq \xi_i \leq x_i$; that is

$$\lim_{|P| \to 0} \sum_{i=1}^{n} f(\xi_i)\, \Delta x_i = L. \tag{2}$$

It should be noted that, for each $\delta > 0$, there exist infinitely many partitions P whose norm $|P| < \delta$. Also, in the formation of the Riemann sums, the choices of the ξ_i in $[x_{i-1}, x_i]$ is arbitrary (and there are infinitely many possible choices). It is not difficult to show that if the limit (2) exists, it is unique (see Exercise 28).

The limit L that we get in (2) for an integrable function is of such importance that we give it a special name as indicated in the following definition.

Definition If f is integrable on $[a, b]$, then the **definite integral of f from a to b,** denoted by $\int_a^b f(x)\, dx$, is given by

$$\int_a^b f(x)\, dx = \lim_{|P| \to 0} \sum_{i=1}^{n} f(\xi_i)\, \Delta x_i \tag{3}$$

The function f is called the **integrand,** a is called the **lower limit,** and b the **upper limit** for the definite integral $\int_a^b f(x)\, dx$.

The symbol $\int_a^b f(x)\, dx$ is read "the integral from a to b of f of x dx."

The symbols x and dx in (3) are "dummy variables" because any other letter could be substituted for x. The value of the integral depends only on the function f and the interval. It does not depend upon the symbol used to represent the independent variable. Thus

$$\int_a^b f(x)\, dx = \int_a^b f(u)\, du \tag{4}$$

Then why do we bother to write the x and dx? Why not simply write $\int_a^b f$? One reason is that the dx reminds us of the Δx_i and is significant in applications. Also the similar notation for the antiderivative $\int f(x)\,dx$ has already proved very useful.

In Section 6.5, we shall find an important relationship between antiderivatives and definite integrals.

EXAMPLE 4   Find $\int_a^b k\,dx$, where k is a constant.

Solution   For any partition of $[a, b]$ we have, since $f(\xi_i) = k$, that

$$\sum_{i=1}^{n} f(\xi_i)\,\Delta x_i = \sum_{i=1}^{n} k\,\Delta x_i = k\sum_{i=1}^{n}(x_i - x_{i-1}) = k(x_n - x_0)$$
$$= k(b - a)$$

Thus

$$\lim_{|P| \to 0} \sum_{i=1}^{n} f(\xi_i)\,\Delta x_i = k(b - a)$$

that is,

$$\int_a^b k\,dx = k(b - a) \qquad (a < b) \qquad \bullet \qquad (5)$$

By methods similar to those used in Section 6.2, we can establish the following formulas:

$$\int_a^b x\,dx = \frac{b^2}{2} - \frac{a^2}{2} \qquad (a < b) \tag{6}$$

$$\int_a^b x^2\,dx = \frac{b^3}{3} - \frac{a^3}{3} \qquad (a < b) \tag{7}$$

$$\int_a^b x^3\,dx = \frac{b^4}{4} - \frac{a^4}{4} \qquad (a < b) \tag{8}$$

and, as a generalization of these, for arbitrary positive integer m,

$$\int_a^b x^m\,dx = \frac{b^{m+1}}{m+1} - \frac{a^{m+1}}{m+1} \qquad (a < b) \tag{9}$$

We shall not prove any of these results now because the work is quite tedious. In Section 6.5, we shall discover an easy method of verifying these formulas.

Since the concept of definite integrals was motivated in Section 6.2 by the discussion of area under a curve, we are led to state the following definition.

Definition   If $f(x)$ is continuous and nonnegative on $[a, b]$, where as usual $a < b$, then the **area** A under the curve $f(x)$ from $x = a$ to $x = b$ is defined by

$$A = \int_a^b f(x)\,dx \tag{10}$$

$\left(\text{The proof that } \int_a^b f(x)\, dx \text{ exists for such functions is best given in an advanced}\right.$

$\left.\text{calculus course.}\right)$

In Section 6.5, using the concept of area as an intuitive guide, we shall develop a very useful formula for the determination of definite integrals for a wide class of continuous functions. This will then be applied to the practical computation of area.

Although our definition of the definite integral allows for subintervals of different sizes, it is usually sufficient to work just with equal subintervals as we did in Section 6.2. Assuming this to be the case, Example 1 of Section 6.2 tells us that

$$\int_1^5 (6 - x)\, dx = 12$$

The introductory example of Section 6.2, asking us to find the area of the region bounded by $f(x) = x^2$, the x-axis, and $x = 3$, can be written $\int_0^3 x^2\, dx = 9$. Note that this result also follows from formula (7).

Comment: The definite integral is also called the ***Riemann integral***, and a function that is integrable in the sense defined in this section is called ***Riemann integrable.*** In more advanced work in mathematics, other types of integrals are studied.

Exercises 6.3

1. Find the norm of each of these partitions of $[-2, 5]$
   (a) $x_0 = -2, x_1 = -0.5, x_2 = 0, x_3 = 3, x_4 = 5$
   (b) $x_0 = -2, x_1 = -1, x_2 = 0.5, x_3 = 2, x_4 = 4, x_5 = 5$
   (c) $x_0 = -2, x_1 = 0, x_2 = 1, x_3 = 3, x_4 = 4.5, x_5 = 5$

Using formulas (5) through (9), evaluate the definite integrals in Exercises 2 through 6.

2. $\int_1^6 4\, dx$

3. $\int_{-1}^3 -5\, dx$

4. $\int_2^6 0\, dx$

5. $\int_2^6 x\, dx$

6. $\int_3^5 x^2\, dx$

7. If $\int_a^b x^2\, dx = \dfrac{b^3}{3} - \dfrac{a^3}{3}$ where $b > a \geq 0$ show that $\int_a^c x^2\, dx = \int_a^b x^2\, dx + \int_b^c x^2\, dx$, where $c > b > a \geq 0$.

8. A space vehicle moves with variable speed given by $g(t)$ mi/sec at time t. Let $t_0, t_1, \ldots, t_n$ be a partition of $[a, b]$ and let $T_1, T_2, \ldots, T_n$ be sampling numbers such that $t_{i-1} \leq T_i \leq t_i$, $(i = 1, 2, \ldots, n)$. What is the physical interpretation of
   (a) $t_i - t_{i-1}$
   (b) $g(T_i)$
   (c) $g(T_i)(t_i - t_{i-1})$
   (d) $\sum_{i=1}^n g(T_i)(t_i - t_{i-1})$
   (e) $\int_a^b g(t)\, dt$

9. Find $\int_{-b}^{b} x\, dx$, $(b > 0)$. Interpret your result.

10. What is the relationship between $\int_{-b}^{-a} x\, dx$ and $\int_{a}^{b} x\, dx$ where $0 \leq a < b$?

11. Interpreting the integral in terms of area, find the relationship between $\int_{-b}^{-a} x^2\, dx$ and $\int_{a}^{b} x^2\, dx$, where $0 \leq a < b$. What is the formula for $\int_{-b}^{-a} x^2\, dx$?

12. Evaluate each of the following definite integrals

(a) $\int_{-5}^{-3} x\, dx$ (b) $\int_{-5}^{-3} x^2\, dx$ (c) $\int_{-2}^{2} x^2\, dx$

using the symmetries of the integrands.

13. Compute a Riemann sum for the function $f(x) = 2x - x^2$ over the interval $[0, 4]$ for the partition $[0, 1, 2.5, 3, 4]$, choosing $\xi_1 = 0.5$, $\xi_2 = 1.5$, $\xi_3 = 2.8$, and $\xi_4 = 3.2$.

For the functions in Exercises 14 through 19, find three Riemann sums for the given partition by choosing ξ_i so that (a) ξ_i is the left-hand end point of the subinterval $[x_{i-1}, x_i]$, (b) ξ_i is the right-hand end point of the subinterval $[x_{i-1}, x_i]$, and (c) ξ_i is the midpoint of the subinterval $[x_{i-1}, x_i]$.

14. $f(x) = 3 + 2x$ $[0, 1, 2, 3, 4]$
15. $f(x) = x^2$ $[1, 2, 3, 4]$
16. $f(x) = x^2$ $[1, 1.5, 2, 2.5, 3, 3.5, 4]$
17. $g(x) = [\![x]\!]$ $[1, 1.3, 1.8, 2.4, 3]$

18. $h(x) = \dfrac{1}{x}$ $[1, 2, 3, 4]$

19. $F(z) = \dfrac{1}{1 + z^2}$ $[0, 1, 2, 3, 4, 5, 6]$

20. Express as a definite integral: $\displaystyle\lim_{n\to\infty} \sum_{k=1}^{n} \frac{k}{n^2}$. $\left(\text{Hint: Consider the function } f(x) = x, \text{ with } \Delta x = \frac{1}{n}\right)$

21. Express as a definite integral: $\displaystyle\lim_{n\to\infty} \sum_{k=1}^{n} \frac{k^2}{n^3}$. (Hint: Consider the function $f(x) = x^2$.)

22. Express as a definite integral: $\displaystyle\lim_{n\to\infty} \sum_{k=1}^{n} \frac{1}{n + k}$.

23. Express as a definite integral:

$$\lim_{n\to\infty} \left[\frac{n}{n^2 + 1^2} + \frac{n}{n^2 + 2^2} + \cdots + \frac{n}{n^2 + 4n^2} \right]$$

24. Try to find $\int_{0}^{1} \sqrt{x}\, dx$ using the method of exhaustion by subdividing $[0, 1]$ into n equal parts. What is S_n, the sum of the areas of the circumscribed rectangles?

25. Let us try once again to find $\int_{0}^{1} \sqrt{x}\, dx$. Instead of subdividing the interval uniformly we subdivide the interval $[0, 1]$ into n subintervals where $x_k = \left(\dfrac{k}{n}\right)^2$, $(k = 0, 1, 2, \ldots, n - 1, n)$. Find S_n the sum of the areas and then determine $\int_{0}^{1} \sqrt{x}\, dx$.

26. Apply the method of exhaustion to determine the value of the integral $\int_{0}^{1} \sqrt[3]{x}\, dx$, by making a suitable subdivision of the interval $[0, 1]$ into subintervals of unequal size.

27. Find $\int_0^b \sqrt{x}\, dx$ where b is any number greater than zero.

28. Prove that if $\lim\limits_{|P|\to 0} \sum\limits_{i=1}^{n} f(x_i)\, \Delta x_i$ exists, the limit is unique. (*Hint:* Assume two distinct limits and show that this situation is untenable.)

29. Prove that if $f(x)$ is integrable on $[a, b]$ and if k is a constant, then $k\, f(x)$ is integrable on $[a, b]$ and

$$\int_a^b k\, f(x)\, dx = k \int_a^b f(x)\, dx$$

6.4   PROPERTIES OF THE DEFINITE INTEGRAL

In order to expedite the calculation of the definite integral, we shall develop some of its properties.

In the definition of the definite integral, the closed interval $[a, b]$ was prescribed where $a < b$. To consider the definite integral of a function f from a to b when $a > b$, or when $a = b$, let us have the following definitions

Definition 1   If $a > b$, then

$$\int_a^b f(x)\, dx = -\int_b^a f(x)\, dx \tag{1}$$

provided that $\int_b^a f(x)\, dx$ exists.

Definition 2   If f is defined at the number a then

$$\int_a^a f(x)\, dx = 0 \tag{2}$$

It can be shown that the limit as $|P| \to 0$ satisfies some of the limit theorems established in Section 2.5; in particular, Theorems 1 and 2 of that section. We shall assume this in proving some of the following results.

Theorem 1   If $f(x)$ is integrable on $[a, b]$ and if k is a constant, then $k\, f(x)$ is integrable on $[a, b]$ and

$$\int_a^b k\, f(x)\, dx = k \int_a^b f(x)\, dx \tag{3}$$

Proof   Since f is integrable on $[a, b]$, $\displaystyle\int_a^b f(x)\, dx = \lim_{|P| \to 0} \sum_{i=1}^n f(\xi_i)\, \Delta x_i$

Then
$$k \int_a^b f(x)\, dx = k \lim_{|P| \to 0} \sum_{i=1}^n f(\xi_i)\, \Delta x_i = \lim_{|P| \to 0} k \sum_{i=1}^n f(\xi_i)\, \Delta x_i$$

$$= \lim_{|P| \to 0} \sum_{i=1}^n k\, f(\xi_i)\, \Delta x_i = \int_a^b k\, f(x)\, dx \qquad \blacksquare$$

Theorem 2   If $f(x)$ and $g(x)$ are integrable on $[a, b]$, then the sum $f(x) + g(x)$ is integrable on $[a, b]$ and

$$\int_a^b [f(x) + g(x)]\, dx = \int_a^b f(x)\, dx + \int_a^b g(x)\, dx \qquad (4)$$

Proof
$$\int_a^b f(x)\, dx = \lim_{|P| \to 0} \sum_{i=1}^n f(\xi_i)\, \Delta x_i \qquad \text{and} \qquad \int_a^b g(x)\, dx = \lim_{|P| \to 0} \sum_{i=1}^n g(\xi_i)\, \Delta x_i$$

So
$$\int_a^b f(x)\, dx + \int_a^b g(x)\, dx = \lim_{|P| \to 0} \sum_{i=1}^n f(\xi_i)\, \Delta x_i + \lim_{|P| \to 0} \sum_{i=1}^n g(\xi_i)\, \Delta x_i$$

$$= \lim_{|P| \to 0} \left[\sum_{i=1}^n f(\xi_i)\, \Delta x_i + \sum_{i=1}^n g(\xi_i)\, \Delta x_i \right]$$

$$= \lim_{|P| \to 0} \sum_{i=1}^n [f(\xi_i) + g(\xi_i)]\, \Delta x_i$$

$$= \int_a^b [f(x) + g(x)]\, dx \qquad \blacksquare$$

EXAMPLE 1   Find $\displaystyle\int_1^3 (3x^2 + 7x)\, dx$.

Solution   From Theorems 2 and 1, and formulas (6) and (7) of Section 6.3,

$$\int_1^3 (3x^2 + 7x)\, dx = \int_1^3 3x^2\, dx + \int_1^3 7x\, dx$$

$$= 3 \int_1^3 x^2\, dx + 7 \int_1^3 x\, dx$$

$$= 3 \left(\frac{3^3}{3} - \frac{1}{3} \right) + 7 \left(\frac{3^2}{2} - \frac{1^2}{2} \right)$$

$$= 3 \left(\frac{26}{3} \right) + 7 \left(\frac{8}{2} \right) = 26 + 28 = 54 \qquad \bullet$$

Example 1 is an illustration of the following result, which is developed from Theorems 1 and 2. The proof is left to the reader (see Exercise 22).

Corollary 1 If $f(x)$ and $g(x)$ are integrable on $[a, b]$ and k_1 and k_2 are constants then $k_1 f(x) + k_2 g(x)$ is integrable on $[a, b]$ and

$$\int_a^b [k_1 f(x) + k_2 g(x)]\, dx = k_1 \int_a^b f(x)\, dx + k_2 \int_a^b g(x)\, dx \tag{5}$$

If in Corollary 1 we let $k_1 = 1$ and $k_2 = -1$, we have the following.

Corollary 2 If $f(x)$ and $g(x)$ are integrable on $[a, b]$, then $f(x) - g(x)$ is integrable on $[a, b]$ and

$$\int_a^b [f(x) - g(x)]\, dx = \int_a^b f(x)\, dx - \int_a^b g(x)\, dx \tag{6}$$

The definite integral of a nonnegative function represents the area under the curve defined by that function. Since the area is nonnegative, it follows that the definite integral should also be nonnegative. This is indicated in the following theorem.

Theorem 3 If $h(x)$ is integrable and $h(x) \geq 0$ on $[a, b]$, where $a < b$, then

$$\int_a^b h(x)\, dx \geq 0 \tag{7}$$

Proof For any partition P and any choice of ξ_i in $[x_{i-1}, x_i]$ we have $h(\xi_i) \geq 0$ and $\Delta x_i > 0$. Therefore $\sum_{i=1}^{n} h(\xi_i)\, \Delta x_i \geq 0$.

So

$$\int_a^b h(x)\, dx = \lim_{|P| \to 0} \sum_{i=1}^{n} h(\xi_i)\, \Delta x_i \geq 0 \qquad \blacksquare$$

Corollary 3 If $f(x)$ and $g(x)$ are integrable on $[a, b]$ where $a < b$, and if $f(x) \geq g(x)$ on this interval then

$$\int_a^b f(x)\, dx \geq \int_a^b g(x)\, dx \tag{8}$$

Proof Let $h(x) = f(x) - g(x)$. Then $h(x) \geq 0$ on $[a, b]$. Therefore

$$\int_a^b f(x)\, dx - \int_a^b g(x)\, dx = \int_a^b [f(x) - g(x)]\, dx = \int_a^b h(x)\, dx \geq 0$$

So

$$\int_a^b f(x)\, dx \geq \int_a^b g(x)\, dx \qquad \blacksquare$$

Figure 6.4.1 gives a geometric interpretation of Corollary 3 when $f(x) \geq g(x) \geq 0$ for all x in $[a, b]$. It says that the area bounded by the curve $y = f(x)$, the x-axis, and the lines $x = a$ and $x = b$ is greater than or equal to the area bounded by $y = g(x)$, the x-axis, and the lines $x = a$ and $x = b$.

EXAMPLE 2 Prove that $\int_0^1 \sqrt{x}\, dx \geq \int_0^1 x\, dx$ assuming the existence of both definite integrals.

Solution If $0 \leq x \leq 1$, $x = \sqrt{x} \cdot \sqrt{x} \leq \sqrt{x}$ since $\sqrt{x} \leq 1$. Therefore from $\sqrt{x} \geq x$ in $[0, 1]$ and Corollary 3 the required result follows. ●

EXAMPLE 3 Find upper and lower bounds on $\displaystyle\int_2^3 \frac{dx}{1 + x^2}$

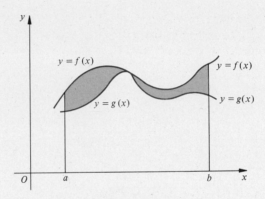

Figure 6.4.1

Solution $\dfrac{1}{1 + x^2}$ is monotonically decreasing as x increases from $x = 2$ to $x = 3$. Thus

$$\frac{1}{5} \geq \frac{1}{1 + x^2} \geq \frac{1}{10} \qquad \text{for } x \text{ in } [2, 3]$$

Therefore from Corollary 3,

$$\int_2^3 \frac{1}{5}\, dx \geq \int_2^3 \frac{dx}{1 + x^2} \geq \int_2^3 \frac{1}{10}\, dx$$

or

$$\frac{1}{5} \geq \int_2^3 \frac{dx}{1 + x^2} \geq \frac{1}{10}$$

which gives us the required upper and lower bounds. ●

Theorem 4   If $f(x)$ is integrable in an interval $[a, b]$ and m and M are numbers such that

$$m \leq f(x) \leq M \qquad \text{for } a \leq x \leq b$$

then

$$m(b - a) \leq \int_a^b f(x)\, dx \leq M(b - a) \qquad (9)$$

A geometric interpretation of Theorem 4 is given in Figure 6.4.2 where $f(x) \geq 0$ for all x in $[a, b]$. The integral $\int_a^b f(x)\, dx$ gives the area bounded by the curve $y = f(x)$, the x-axis, and the lines $x = a$ and $x = b$. This area is greater than that of the rectangle whose dimensions are m and $(b - a)$ and less than that of the rectangle whose dimensions are M and $(b - a)$.

Theorem 4 can be proved by using the method employed in Example 3. ∎

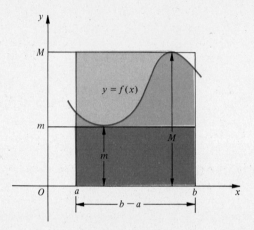

Figure 6.4.2

Theorem 5   If $a < b < c$ and if $f(x)$ is integrable on the intervals $[a, b]$ and $[b, c]$, then it is integrable on the interval $[a, c]$ and

$$\int_a^c f(x)\, dx = \int_a^b f(x)\, dx + \int_b^c f(x)\, dx \qquad (10)$$

We shall interpret Theorem 5 geometrically for the case where $f(x) \geq 0$ for all x in $[a, c]$. In this case, the theorem states that the area of the region bounded by the curve $y = f(x)$ and the x-axis from $x = a$ to $x = c$ is equal to the sum of the areas of the regions from a to b and from b to c (Figure 6.4.3).

While Theorem 5 is true when $f(x)$ is not always nonnegative on $[a, c]$, we shall not give a formal proof of this result. ∎

Theorem 5 was stated and illustrated for the particular order $a < b < c$. However, it can now be shown by virtue of Definition 1 that (10) holds true for any order of the numbers a, b, and c. For example, suppose that $a < c < b$, then from Theorem 5,

$$\int_a^b f(x)\, dx = \int_a^c f(x)\, dx + \int_c^b f(x)\, dx$$

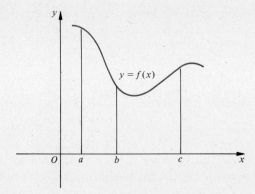

Figure 6.4.3

Thus

$$\int_a^c f(x)\,dx = \int_a^b f(x)\,dx - \int_c^b f(x)\,dx$$

$$= \int_a^b f(x)\,dx + \int_b^c f(x)\,dx$$

It is left to the reader to examine other order relations among a, b, and c including possible equalities of some of the letters.

By mathematical induction, relation (10) can be extended to an arbitrary (finite) number of terms

$$\int_{a_0}^{a_n} f(x)\,dx = \sum_{k=1}^{n} \int_{a_{k-1}}^{a_k} f(x)\,dx \qquad (11)$$

where $a_0, a_1, \dots, a_n$ are any $n+1$ real numbers and where each integral of (11) is assumed to exist. The proof is left as an exercise for the reader.

An important point is that the Riemann integral defined by

$$\lim_{|\Delta P| \to 0} \sum_{i=1}^{n} f(\xi_i)\,\Delta x_i \qquad (12)$$

never exists for an unbounded function. In other words, *every integrable function is bounded.*

Proof This is easily proved. Let $f(x)$ be unbounded in the kth subinterval $[x_{k-1}, x_k]$. Then whatever may be the choice of points ξ_i for $i \neq k$, the point ξ_k can be selected so that the sum (1) is numerically larger than any preassigned quantity. ∎

On the other hand, *not* all bounded functions are integrable. For example, consider the function f defined in $[0, 1]$ and given by

$$f(x) = \begin{cases} 1 & \text{if } x \text{ is a rational number} \\ 0 & \text{if } x \text{ is not rational} \end{cases}$$

Thus $f(\tfrac{1}{2}) = 1$, $f(0) = 1$, while $f\left(\dfrac{1}{\sqrt{2}}\right) = 0$, and so on. This is indeed a bizarre or "pathological function." Its range consists of two numbers namely 0 and 1. It is really impossible to adequately plot this function, which is composed of a dense set of points on the parallel lines $y = 0$ and $y = 1$. Furthermore this function is *not continuous* at any value of x in $[0, 1]$. This follows because any interval, no matter how small, contains an infinite number of rational and irrational points so that $\lim_{x \to x_0} f(x)$ does not exist for any x_0 in $[0, 1]$. Furthermore the $\int_0^1 f(x)\, dx$ does *not* exist. This is due to the fact that, no matter how small the norm $|P|$ of a partition P may be, every subinterval contains both rational and irrational points and the sums $\sum_{i=1}^n f(\xi_i)\, \Delta x_i$ can be made to have either the extreme value 1 (if every point ξ_i is chosen to be rational) or the extreme value 0 (if every point ξ_i is chosen to be irrational). Thus the limit of the Riemann sum as given by (12) does not exist. (Take any ε such that $0 < \varepsilon < 1$).

We are naturally led to ask: "what kinds of functions are integrable?" The preceding illustration might lead us to expect that continuous functions are integrable, and, indeed, this is true. But so are some discontinuous functions. We state without proof three theorems that tell us that certain classes of functions are integrable (proofs of these results can be found in most advanced calculus texts).

Theorem 6   If $f(x)$ is either a nondecreasing function or a nonincreasing function on $[a, b]$, then $\int_a^b f(x)\, dx$ exists.

Theorem 7   If $f(x)$ is continuous on $[a, b]$ then $\int_a^b f(x)\, dx$ exists.

Theorem 8   If $f(x)$ is bounded and has a finite number of discontinuities on $[a, b]$ then $\int_a^b f(x)\, dx$ exists.

Furthermore, if the discontinuities occur at c_1, $c_2, \ldots, c_n$ where $a \le c_1 < c_2 < c_3 < \cdots < c_n \le b$, then

$$\int_a^b f(x)\, dx = \int_a^{c_1} f(x)\, dx + \int_{c_1}^{c_2} f(x)\, dx + \int_{c_2}^{c_3} f(x)\, dx + \cdots + \int_{c_n}^b f(x)\, dx \quad (13)$$

To evaluate the integrals on the right in (13), we use the following theorem.

Theorem 9   Let $g(x)$ be continuous on the closed interval $[c, d]$ and let $f(x)$ be bounded on $[c, d]$ such that $f(x) = g(x)$ for all values of x on (c, d); that is, $f(x)$ and $g(x)$ differ at most at c and d. Then

$$\int_c^d f(x)\, dx = \int_c^d g(x)\, dx$$

Some functions with an infinite number of discontinuities are integrable. For example, if $f(x)$ is a nondecreasing function on $[a, b]$ with infinitely many discontinuities, then Theorem 6 still guarantees that it is integrable. There are also other types of functions that are integrable but do not satisfy the hypotheses of Theorems 6, 7, or 8. A discussion of this is usually reserved for an advanced calculus course.

EXAMPLE 4 Consider the function

$$f(x) = \frac{1}{3 - x}$$

in the interval $0 \leq x < 3$. This function is monotonically increasing in $0 < x < 3$ but is not defined at $x = 3$. In fact $f(x) \to \infty$ as $x \to 3^-$ and $\int_0^3 \frac{1}{3 - x} \, dx$ does not exist. However if b is any fixed number in $0 < x < 3$, then $\int_0^b \frac{1}{3 - x} \, dx$ exists in accordance with Theorem 6. ●

EXAMPLE 5 Does $\int_1^4 \frac{3x^2 - 1}{x + 2} \, dx$ exist?

Solution The function $\frac{3x^2 - 1}{x + 2}$ is a rational function that is continuous except at those values of x at which the denominator is zero. Now $x + 2 = 0$ if and only if $x = -2$, which is outside the interval of integration. Therefore $\int_1^4 \frac{3x^2 - 1}{x + 2} \, dx$ exists. Note that Theorem 7 does not tell us how to evaluate this integral—it only guarantees its existence. ●

EXAMPLE 6 $f(x) = [\![x]\!]$ is bounded on $[0, 4]$ and has discontinuities at $x = 1, 2, 3,$ and 4. By Theorem 8, $\int_0^4 [\![x]\!] \, dx$ exists. Furthermore, by (13) and Theorem 9,

$$\int_0^4 [\![x]\!] \, dx = \int_0^1 [\![x]\!] \, dx + \int_1^2 [\![x]\!] \, dx + \int_2^3 [\![x]\!] \, dx + \int_3^4 [\![x]\!] \, dx$$

$$= \int_0^1 0 \, dx + \int_1^2 1 \, dx + \int_2^3 2 \, dx + \int_3^4 3 \, dx$$

$$= 0 + 1 + 2 + 3 = 6$$ ●

EXAMPLE 7
$$f(x) = \begin{cases} \dfrac{1}{x} & x \neq 0 \\ 0 & x = 0 \end{cases}$$

is not sectionally continuous in $[0, 1]$. In fact $f(x) \to \infty$ as $x \to 0^+$ and $\int_0^1 \frac{1}{x} \, dx$ does not exist. ●

In Section 4.6, we stated and proved the mean value theorem for derivatives. This was an important result in that it enabled us to prove many of the theorems relating to curve sketching. This mean value theorem has many more uses which we shall encounter in our study of calculus.

A similar theorem, the mean value theorem for integrals, is stated next.

Theorem 10   **(Mean Value Theorem for Integrals).** If $f(x)$ is continuous on $[a, b]$, then there exists a number c in $[a, b]$ such that

$$\int_a^b f(x)\, dx = f(c)(b - a) \tag{14}$$

Proof   Since f is continuous in the closed interval, it follows from the extreme value theorem that $f(x)$ assumes a smallest value m and a largest value M in $[a, b]$. This means that there are two numbers x_1 and x_2 in $[a, b]$ such that $f(x_1) = m$ and $f(x_2) = M$. Therefore $m \leq f(x) \leq M$ for all x in $[a, b]$. Thus for $a < b$,

$$\int_a^b m\, dx \leq \int_a^b f(x)\, dx \leq \int_a^b M\, dx$$

or

$$m(b - a) \leq \int_a^b f(x)\, dx \leq M(b - a)$$

Division by $b - a$ yields

$$m \leq \frac{\int_a^b f(x)\, dx}{b - a} \leq M$$

Since $f(x)$ is continuous in $[a, b]$, by the intermediate value theorem (Section 2.11) it must assume all values between m and M at least once in $[a, b]$ and more specifically in $[x_1, x_2]$. Thus there is a number c in $[a, b]$ at which

$$f(c) = \frac{\int_a^b f(x)\, dx}{b - a}$$

or

$$\int_a^b f(x)\, dx = f(c)\,(b - a) \qquad\qquad \blacksquare$$

Note that this theorem does not offer any easy way to determine $\int_a^b f(x)\, dx$. In fact, the value c is often found by equating $f(c)$ to $\int_a^b f(x)\, dx/(b - a)$ and then solving for c. However, it is a significant result that has many applications.

To interpret (14) geometrically consider $f(x) \geq 0$ for all values of x in $[a, b]$. Then $\int_a^b f(x)\, dx$ yields the area bounded by the curve $y = f(x)$, the x-axis, and the lines $x = a$ and $x = b$ (Figure 6.4.4). Theorem 10 states that there is a number c

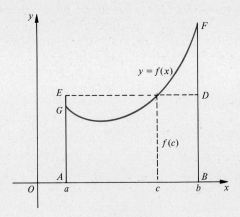

Figure 6.4.4

in $[a, b]$ such that the area of the rectangle $ABDE$ of height $f(c)$ and width $b - a$ is equal to the area of the region $ABFG$.

EXAMPLE 8   Find the value of c such that

$$\int_2^6 (3x + 2)\, dx = f(c)\,(6 - 2) = 4f(c)$$

if $f(x) = 3x + 2$.

Solution
$$\int_2^6 (3x + 2)\, dx = 3 \int_2^6 x\, dx + 2 \int_2^6 dx$$

Now
$$\int_a^b x\, dx = \frac{b^2}{2} - \frac{a^2}{2} \qquad \text{and} \qquad \int_a^b dx = b - a$$

so that
$$\int_2^6 (3x + 2)\, dx = 3\left(\frac{6^2}{2} - \frac{2^2}{2}\right) + 2(6 - 2)$$
$$= 3(18 - 2) + 2(4) = 48 + 8 = 56$$

Therefore
$$56 = 4f(c)$$

or
$$f(c) = 14 = 3c + 2$$

Finally, $c = 4$. This result is to be expected since the region involved is a trapezoid and the area of a trapezoid is equal to the height times the length of its midsection (the length of the midsection is one half the sum of its bases.)

The value $f(c)$ given in Theorem 10 is called the **average value** of f over $[a, b]$, that is, we have

Definition 3   The **average value** of f over $[a, b]$ is the quotient

$$\frac{\int_a^b f(x)\, dx}{b - a} \qquad (15)$$

To motivate this definition, divide the interval $[a, b]$ into n equal parts so that $x_i - x_{i-1} = (b - a)/n$, $(i = 1, 2, \ldots, n)$ where as usual $x_0 = a$ and $x_n = b$. Now

$$\frac{f(x_1) + f(x_2) + \cdots + f(x_n)}{n} \tag{16}$$

would be a reasonable estimate of the average value of f over $[a, b]$. To make (16) look more like (15), multiply the numerator and denominator of (16) by $b - a$ and replace $(b - a)/n$ by $x_i - x_{i-1}$ so that (15) becomes

$$\frac{1}{b - a}[f(x_1)(x_1 - x_0) + f(x_2)(x_2 - x_1) + \cdots + f(x_n)(x_n - x_{n-1})] \tag{17}$$

Now the bracketed expression in (17) tends to $\int_a^b f(x)\,dx$ as $n \to \infty$, thus suggesting Definition 3.

EXAMPLE 9   Find the average value of $f(x) = \sqrt{a^2 - x^2}$ for x in $[0, a]$.

Solution   Interpreted geometrically $\int_0^a \sqrt{a^2 - x^2}\,dx$ is one fourth the area of a circle of radius a. Thus

$$\int_0^a \sqrt{a^2 - x^2}\,dx = \frac{\pi a^2}{4}$$

and the average value $\overline{f}$ is

$$\overline{f} = \frac{\pi a^2/4}{a} = \frac{\pi a}{4} \qquad \bullet$$

Exercises 6.4

In Exercises 1 through 8 assume that all integrals involved exist and prove the following

1. $\int_3^{10} f(x)\,dx = \int_3^{14} f(x)\,dx - \int_{10}^{14} f(x)\,dx$

2. $-\int_7^{-5} g(x)\,dx + \int_7^{10} g(x)\,dx = \int_{-5}^{10} g(x)\,dx$

3. $\int_a^c h(x)\,dx = -\int_b^0 h(x)\,dx + \int_a^0 h(x)\,dx - \int_c^b h(x)\,dx$

4. $\int_d^a F(x)\,dx - \int_b^a F(x)\,dx + \int_b^c F(x)\,dx - \int_d^c F(x)\,dx = 0$

In Exercises 5 through 8, simplify

5. $\int_{-10}^0 f(x)\,dx + \int_2^6 f(x)\,dx - \int_2^0 f(x)\,dx$

6. $\int_4^1 g(x)\,dx + \int_3^3 F(x)\,dx + \int_{-1}^4 g(x)\,dx$

Next Page

7. $\int_b^{b+3} h(x)\,dx - \int_b^a h(x)\,dx + \int_{-a}^a h(x)\,dx$

8. $\int_a^{x+\Delta x} f(t)\,dt - \int_a^x f(t)\,dt$

9. Given that $\int_{-2}^0 x^2\,dx = \frac{8}{3}$, $\int_0^1 x^2\,dx = \frac{1}{3}$, and $\int_1^2 x^2\,dx = \frac{7}{3}$

   Find (using only these formulas)

   (a) $\int_0^2 5x^2\,dx$

   (b) $\int_{-2}^2 3x^2\,dx$

   (c) $\int_2^8 3x^2\,dx - \int_0^8 3x^2\,dx$

   (d) $\int_{-4}^2 x^3\,dx + \int_2^{-4} x^3\,dx$

   (e) $\int_1^{-2} \frac{x^2}{3}\,dx$

In Exercises 10 through 21, apply Theorem 4 to find a smallest and a largest possible value of the given integral

10. $\int_3^6 (4x - 1)\,dx$

11. $\int_{-2}^3 x^2\,dx$

12. $\int_1^3 \frac{dx}{x^2}$

13. $\int_{-1}^4 |x|\,dx$

14. $\int_{-4}^{-2} |2 - x|\,dx$

15. $\int_{-2}^2 |9 - x^2|\,dx$

16. $\int_{-2}^1 x|x|\,dx$

17. $\int_0^1 \frac{2x^2}{1 + x^2}\,dx$

18. $\int_{-1}^3 \sqrt{2 + x}\,dx$

19. $\int_0^4 \frac{\sqrt{x}}{1 + x}\,dx$

20. $\int_{-3}^1 \frac{x + 4}{x - 2}\,dx$

21. $\int_1^2 (x^4 - 6x^2 + 11)\,dx$

22. If k_1 and k_2 are arbitrary constants prove that

$$\int_a^b (k_1 f(x) + k_2 g(x))\,dx = k_1 \int_a^b f(x)\,dx + k_2 \int_a^b g(x)\,dx$$

provided each of the integrals on the right side exists.

23. Interpreting the definite integral in terms of area find

   (a) $\int_0^1 [\![x]\!]\,dx$, where $[\![x]\!]$ is the bracket function

   (b) $\int_0^2 [\![x]\!]\,dx$

   (c) $\int_0^n [\![x]\!]\,dx$, where n is an arbitrary positive integer

   Note that the integrand is discontinuous at $x = 1$ in (b) and at $x = 1, 2, \ldots, n - 1$ in (c).

24. Employ a geometric argument to find $\int_0^6 \sqrt{36 - x^2}\,dx$

25. From Theorem 5 and Definitions 1 and 2 prove that

$$\int_a^c f(x)\,dx = \int_a^b f(x)\,dx + \int_b^c f(x)\,dx$$

where $c \le b < a$ and the integrals on the right side are assumed to exist.

26. A function $f(x)$ is said to be *even* if and only if the equality $f(x) = f(-x)$ holds for all x in the domain of the definition. Examples of even functions are constants, x^{2n} ($n = $ nonnegative integer), $\sqrt{x^2 + 16}$, and $|x|$. Prove that if $f(x)$ is even on $[-a, a]$ and integrable on $[0, a]$ where $a > 0$, then $f(x)$ is integrable on $[-a, a]$ and $\int_{-a}^{a} f(x)\, dx = 2 \int_{0}^{a} f(x)\, dx$.

27. A function $f(x)$ is said to be *odd* if and only if the equality $f(-x) = -f(x)$ holds for all x in the domain of definition of the function. Examples of odd functions are $x, x - x^3, x^{2n-1}$ ($n = $ positive integer). Prove that if $f(x)$ is odd on $[-a, a]$ and integrable on $[0, a]$, then $f(x)$ is integrable on $[-a, a]$ and $\int_{-a}^{a} f(x)\, dx = 0$. (See the exercises in Section 2.3 for additional information on even and odd functions.)

28. Criticize the following argument: $f(x) = x^{-1}$ is an odd function since $f(-x) = (-x)^{-1} = -x^{-1}$ and therefore $\int_{-a}^{a} x^{-1}\, dx = 0$.

29. Find the value of $\int_{-3}^{3} \dfrac{x}{9 + x^4}\, dx$

30. If f is continuous on $[a, b]$, $(a < b)$, prove that

$$\left| \int_{a}^{b} f(x)\, dx \right| \leq \int_{a}^{b} |f(x)|\, dx$$

(*Hint:* $-|f(x)| \leq f(x) \leq |f(x)|$)

31. If the functions f and g are continuous on $[a, b]$ show that for any real constant t,

(a) $\displaystyle\int_{a}^{b} (t\,f(x) + g(x))^2\, dx$ exists; and

(b) $t^2 \displaystyle\int_{a}^{b} [f(x)]^2\, dx + 2t \int_{a}^{b} f(x)\, g(x)\, dx + \int_{a}^{b} [g(x)]^2\, dx \geq 0$ if $a < b$

[*Hint:* Note that the square of any real function is nonnegative and that the left member of (b) is identical with (a).]

32. Let f and g be functions that are continuous on $[a, b]$. Prove the **Cauchy-Schwarz inequality** for integrals,

$$\left(\int_{a}^{b} f(x)\, g(x)\, dx \right)^2 \leq \int_{a}^{b} [f(x)]^2\, dx \int_{a}^{b} [g(x)]^2\, dx$$

(*Hint:* See the previous exercise and the Cauchy-Schwarz inequality for sums.)

In Exercises 33 through 39 find the value of c satisfying the mean value theorem for integrals. Draw a figure that illustrates the application of the theorem.

33. $\displaystyle\int_{0}^{4} (2x + 3)\, dx$

34. $\displaystyle\int_{a}^{b} (\alpha x + \beta)\, dx$, ($\alpha$ and β are constants)

35. $\displaystyle\int_{0}^{2} (4 - x^2)\, dx$

36. $\displaystyle\int_{0}^{3} f(x)\, dx$, where $f(x) = \begin{cases} 2x & 0 \leq x < 1 \\ 3 - x & 1 \leq x \leq 3 \end{cases}$

37. $\displaystyle\int_{-1}^{1} (x^3 - 2x)\, dx$

38. $\displaystyle\int_{0}^{2} [\![x]\!]\, dx$

39. $\displaystyle\int_{1}^{3} (x^2 - 3x + 4)\, dx$

40. A chord is drawn in a circle of radius a units. Determine the average length $\bar{L}$ of this chord. (*Hint:* Consider an arbitrary chord at a distance x units from the center. Use the theorem of Pythagoras.)

41. If f and p are continuous functions on $[a, b]$ and $p(x) \geq 0$ on $[a, b]$ where $a < b$, show that there exists a number c on (a, b) such that

$$\int_a^b f(x)\, p(x)\, dx = f(c) \int_a^b p(x)\, dx$$

In what way does this generalize Theorem 10? This theorem is usually called the **second mean value theorem of integral calculus.**

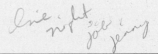

6.5

FUNDAMENTAL THEOREMS OF THE CALCULUS

It was noted in Section 6.3 that the symbols x and dx in $\int_a^b f(x)\, dx$ are "dummy variables." The value of the integral depends only on f, a, and b. Thus

$$\int_a^b f(x)\, dx = \int_a^b f(t)\, dt = \int_a^b f(u)\, du$$

If t is taken as the dummy variable (along with dt), then formula (6) of Section 6.3 becomes $\int_a^b t\, dt = \dfrac{b^2}{2} - \dfrac{a^2}{2}$ $(b > a \geq 0)$. Now suppose that we replace b by x, then, for each x, it is known that

$$\int_a^x t\, dt = \frac{x^2}{2} - \frac{a^2}{2} = A(x) \tag{1}$$

The notation $A(x)$ has been introduced because a trapezoidal region has been generated by $f(t) = t$ from $t = a$ to $t = x$ (Figure 6.5.1) and $A(x)$ is the area of the trapezoid. Furthermore, if we differentiate both sides of (1) with respect to x

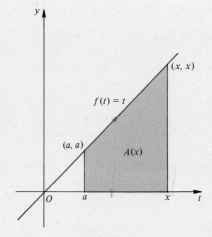

Figure 6.5.1

$$A'(x) = \frac{d}{dx}\left(\frac{x^2}{2} - \frac{a^2}{2}\right) = x \qquad (2)$$

But $f(t) = t$ so that

$$A'(x) = \frac{d}{dx}\int_a^x t\,dt = x$$

This indicates a remarkable connection between the area function and a nonnegative function $f(t)$, namely, that the derivative of the area function is the integrand function evaluated at the upper limit x. This was first discovered by the English mathematician Isaac Barrow (1630–1677), who was Newton's tutor. In fact, it was Barrow who resigned his professorship in favor of his remarkable student who was only 26 at that time.

EXAMPLE 1   Consider the curve $f(t) = t^2$ and its associated area predicting function $A(x)$ (Figure 6.5.2) given by

$$A(x) = \int_a^x t^2\,dt \qquad (x > a \geq 0)$$

Find $A'(x)$.

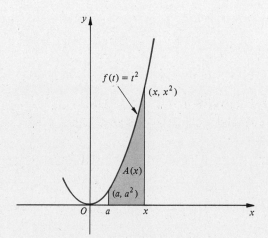

Figure 6.5.2

Solution   We know that [Formula (7), Section 6.3]

$$A(x) = \frac{x^3}{3} - \frac{a^3}{3}$$

so that

$$A'(x) = \frac{d}{dx}\int_a^x t^2\,dt = x^2$$

Again it is observed that the integrand function is the derivative of the area function. Alternatively, the area function $A(x)$ is an antiderivative for the given integrand function $f(x)$ ●

These examples lead us to expect the following result.

Theorem 1   **(First Fundamental Theorem of Calculus).** Let the function $f(t)$ be continuous on $[a, b]$ and let x be any number in (a, b). Let a_0 be any fixed number in $[a, b]$. Suppose that a function $F(x)$ is defined by

$$F(x) = \int_{a_0}^{x} f(t)\, dt \tag{3}$$

then

$$F'(x) = f(x) \tag{4}$$

Proof   Consider two numbers x and $x + h$ in (a, b). Then

$$F(x) = \int_{a_0}^{x} f(t)\, dt$$

and

$$F(x + h) = \int_{a_0}^{x+h} f(t)\, dt$$

so that

$$F(x + h) - F(x) = \int_{a_0}^{x+h} f(t)\, dt - \int_{a_0}^{x} f(t)\, dt$$

or

$$F(x + h) - F(x) = \int_{a_0}^{x} f(t)\, dt + \int_{x}^{x+h} f(t)\, dt - \int_{a_0}^{x} f(t)\, dt$$

$$= \int_{x}^{x+h} f(t)\, dt$$

But from the mean value theorem for integrals (which applies since $f(t)$ is continuous in $[x, x + h]$ by hypothesis) there exists a number c in $[x, x + h]$ such that

$$\int_{x}^{x+h} f(t)\, dt = f(c) \cdot h$$

or

$$F(x + h) - F(x) = f(c) \cdot h$$

Divide by $h \neq 0$ so that

$$\frac{F(x + h) - F(x)}{h} = f(c) \qquad c \text{ is in } [x, x + h]$$

Now take $\lim_{h \to 0}$ of both sides, thus

$$\lim_{h \to 0} \frac{F(x + h) - F(x)}{h} = \lim_{h \to 0} f(c)$$

But, by the definition of continuity of f at x, $\lim_{h \to 0} f(c) = f(x)$

Also by the definition of $F'(x)$

$$F'(x) = \lim_{h \to 0} \frac{F(x + h) - F(x)}{h}$$

and thus (4) follows.   ∎

Note that Theorem 1 says that $\int_{a_0}^{x} f(t)\, dt$ is an antiderivative of $f(t)$ (with t replaced by x). In fact it is that antiderivative which is zero at a_0 whatever the value of a_0 may be.

EXAMPLE 2   Find $F(x) = \int_{1}^{x} t^2\, dt$

Solution   Observe first that $f(t) = t^2$ is continuous for all t and thus $F(x)$ must be such that

$$F'(x) = x^2 \tag{i}$$

and

$$F(1) = 0 \tag{ii}$$

Now to satisfy (i)

$$F(x) = \frac{x^3}{3} + C \tag{5}$$

with arbitrary C represents the totality of functions for which $F'(x) = x^2$. But $F(1) = 0$ must also hold, and thus C may be found. Substitute $x = 1$ into (5),

$$0 = \frac{1^3}{3} + C \quad\text{or}\quad C = -\tfrac{1}{3}$$

and

$$F(x) = \frac{x^3}{3} - \frac{1}{3} \qquad \bullet \tag{6}$$

EXAMPLE 3   If $G(x) = \int_{0}^{x} \dfrac{dt}{\sqrt{1 + t^2}}$ find $G'(x)$.

Solution   It should be noted that it is not necessary to find $G(x)$ in order to determine $G'(x)$. From Theorem 1, $G'(x) = \dfrac{1}{\sqrt{1 + x^2}}$ because $f(t) = \dfrac{1}{\sqrt{1 + t^2}}$ is continuous for all t. This is of course very helpful in giving insight into the behavior of $G(x)$. Thus $G'(0) = 1$, $G'(\pm 1) = \dfrac{1}{\sqrt{2}}$, $G'(\pm 2) = \dfrac{1}{\sqrt{5}}$, and so on. Also $G(0) = 0$. In this way, knowing the slopes of the tangents to $G(x)$ in as much detail as desired and also the fact that $G(0) = 0$, we can piece together the function $G(x)$. It can be shown that $G(x)$ is an odd function, that is, the graph of $G(x)$ is symmetric with respect to the origin. Figure 6.5.3 is a graph of $G(x)$. $\qquad \bullet$

EXAMPLE 4   If $H(x) = \int_{x}^{4} \sqrt{1 + t^3}\, dt$ find $H'(x)$.

Solution   We first write $H(x) = -\int_{4}^{x} \sqrt{1 + t^3}\, dt = \int_{4}^{x} [-\sqrt{1 + t^3}]\, dt$ so that Theorem 1 is applicable. (This theorem requires the constant to be the lower limit and the variable to be the upper limit.)

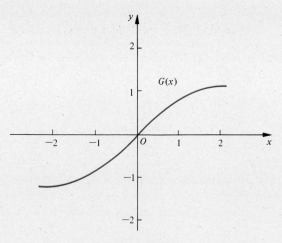

Figure 6.5.3

Then

$$H(x) = \int_4^x f(t)\, dt \qquad \text{where } f(t) = -\sqrt{1 + t^3}$$

So

$$H'(x) = f(x) = -\sqrt{1 + x^3} \qquad \bullet$$

Perhaps the most important theorem of the calculus is the following.

Theorem 2   (**Second Fundamental Theorem of Calculus**). Let the function f be continuous on the closed interval $[a, b]$ and G be a function such that $G'(x) = f(x)$ for all x in $[a, b]$. Then

$$\int_a^b f(t)\, dt = G(b) - G(a) \qquad (7)$$

Proof   By hypothesis, $f(t)$ is continuous at all numbers in $[a, b]$ and thus by Theorem 1, choosing $a_0 = a$, if we set

$$F(x) = \int_a^x f(t)\, dt$$

it follows that

$$F'(x) = f(x)$$

Therefore since $G'(x) = f(x)$ also,

$$G(x) = \int_a^x f(t)\, dt + C$$

for some constant C. But

$$G(a) = \int_a^a f(t)\, dt + C = C$$

because $\int_a^a f(t)\, dt = 0$. Thus,

$$G(x) - G(a) = \int_a^x f(t)\, dt$$

and if we replace x by b,

$$G(b) - G(a) = \int_a^b f(t)\, dt \quad \blacksquare$$

With Theorem 2 we can evaluate a definite integral provided that an antiderivative of the integrand can be found. In applying the theorem we denote $G(b) - G(a)$ by any of the following:

$$G(x)\,|_a^b \qquad G(x)]_a^b \qquad [G(x)]_a^b$$

EXAMPLE 5 Evaluate $\int_1^2 x^3\, dx$ by application of Theorem 3.

Solution An antiderivative of x^3 is $\dfrac{x^4}{4}$. From this we choose $G(x) = \dfrac{x^4}{4}$. Thus from Theorem 3,

$$\int_1^2 x^3\, dx = \frac{x^4}{4}\bigg|_1^2 = \frac{16}{4} - \frac{1}{4} = \frac{15}{4} = 3\frac{3}{4}$$

Now what would happen if we chose $G(x) = \dfrac{x^4}{4} + 10$ instead? Then

$$\int_1^2 x^3\, dx = \left[\frac{x^4}{4} + 10\right]_1^2 = \left(\frac{16}{4} + 10\right) - \left(\frac{1}{4} + 10\right)$$
$$= \frac{15}{4}$$

as before.

More generally if $G(x) = \dfrac{x^4}{4} + C$ where C is an arbitrary constant, the result will still be the same because

$$G(b) - G(a) = \left(\frac{b^4}{4} + C\right) - \left(\frac{a^4}{4} + C\right) = \frac{b^4}{4} - \frac{a^4}{4}$$

What occurred here happens all the time—the constant from the antiderivative drops out. Thus, in practice we usually use the simplest antiderivative of the integrand (if it can be found). $\bullet$

EXAMPLE 6 Find $\int_2^3 \dfrac{dx}{x^2}$

Solution $$\int_2^3 \frac{dx}{x^2} = \int_2^3 x^{-2}\, dx = \frac{x^{-1}}{-1}\bigg]_2^3 = -\frac{1}{x}\bigg]_2^3 = -\frac{1}{3} - \left(-\frac{1}{2}\right) = \frac{1}{6} \quad \bullet$$

EXAMPLE 7   Find $\int_{-2}^{3} \dfrac{dx}{x^2}$.

Solution   An antiderivative of $\dfrac{1}{x^2}$ is $-x^{-1}$ so that

$$\int_{-2}^{3} \frac{dx}{x^2} = -\frac{1}{x}\bigg]_{-2}^{3} = -\frac{1}{3} - \frac{1}{2} = -\frac{5}{6}$$

However, this solution is *wrong!* The integrand is *not* continuous at $x = 0$, which is in the interval $[-2, 3]$. Therefore we have no justification in using Theorem 2. In fact, the integral does *not* exist since $\dfrac{1}{x^2}$ is unbounded in any neighborhood of $x = 0$.   ●

EXAMPLE 8   Find $D_x \int_{0}^{x^2} (t^2 + 3)\, dt$ by two methods.

Solution   **Method 1:**   An antiderivative of $t^2 + 3$ is $\dfrac{t^3}{3} + 3t$ so that

$$\int_{0}^{x^2} (t^2 + 3)\, dt = \frac{t^3}{3} + 3t\bigg]_{0}^{x^2} = \frac{x^6}{3} + 3x^2 - (0 + 0)$$

$$= \frac{x^6}{3} + 3x^2$$

$$D_x\left(\frac{x^6}{3} + 3x^2\right) = 2x^5 + 6x = 2x(x^4 + 3)$$

Method 2:   Let $u = x^2$ so that from Theorem 1 and the chain rule of differentiation

$$D_x \int_{0}^{u} (t^2 + 3)\, dt = D_u \int_{0}^{u} (t^2 + 3)\, dt \cdot D_x u$$

$$= (u^2 + 3) \cdot 2x$$

$$= ((x^2)^2 + 3) \cdot 2x$$

$$= 2x(x^4 + 3)$$

in agreement with the solution by Method 1.   ●

EXAMPLE 9   Find $\int_{0}^{1} (x^3 + 1)^3 x^2\, dx$

Solution   **Method 1:**   To find an antiderivative of $(x^3 + 1)^3 x^2$, let $u = x^3 + 1$ then $du = 3\, x^2\, dx$ so that

$$\int (x^3 + 1)^3 x^2\, dx = \int u^3 \frac{du}{3} = \frac{u^4}{12} + C$$

$$= \frac{(x^3 + 1)^4}{12} + C$$

Therefore $\dfrac{(x^3 + 1)^4}{12}$ is an antiderivative of $(x^3 + 1)^3 x^2$. Consequently,

$$\int_0^1 (x^3 + 1)^3 x^2 \, dx = \dfrac{(x^3 + 1)^4}{12}\bigg|_0^1 = \dfrac{2^4}{12} - \dfrac{1^4}{12}$$

$$= \dfrac{15}{12} = \dfrac{5}{4}$$

Method 2: We again substitute $u = x^3 + 1$ and $du = 3x^2 \, dx$. This time we shall change the limits of integration to correspond to the new variable u. When $x = 0$, $u = 1$ and when $x = 1$, $u = 2$. Thus

$$\int_0^1 (x^3 + 1)^3 x^2 \, dx = \int_1^2 u^3 \cdot \dfrac{du}{3} = \dfrac{u^4}{12}\bigg]_1^2 = \dfrac{16}{12} - \dfrac{1}{12} = \dfrac{15}{12} = \dfrac{5}{4} \quad \bullet$$

The second fundamental theorem of calculus can be applied to the type of area problem introduced in Section 6.2. We shall now be able to do many problems of this type that we would not have wanted to attempt when the topic was first introduced. The tedious computations of Section 6.2 are now replaced by a few relatively simple steps.

EXAMPLE 10   Find the area of the region bounded by the curve $f(x) = x^4 + 1$, the x-axis, and the lines $x = -1$ and $x = 2$.

Solution   The region is drawn in Figure 6.5.4. We know that $A = \displaystyle\int_{-1}^2 f(x) \, dx$. So

$$A = \int_{-1}^2 (x^4 + 1) \, dx = \left[\dfrac{x^5}{5} + x\right]_{-1}^2 = \left[\dfrac{32}{5} + 2\right] - \left[\dfrac{-1}{5} - 1\right]$$

$$= \dfrac{33}{5} \div 3 = \dfrac{48}{5} \text{ square units} \qquad \bullet$$

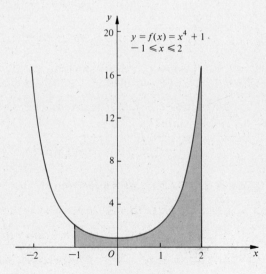

$$y = f(x) = x^4 + 1$$
$$-1 \leqslant x \leqslant 2$$

Figure 6.5.4

EXAMPLE 11   Find the area of the region bounded by the curve $y = 3x - x^2$ and the x-axis.

Solution   The region is drawn in Figure 6.5.5. Note that we are finding the area under the curve where $0 \leq x \leq 3$. Letting $f(x) = 3x - x^2$ we have

$$A = \int_0^3 f(x)\, dx = \int_0^3 (3x - x^2)\, dx = \left[\frac{3x^2}{2} - \frac{x^3}{3} \right]_0^3 = \frac{27}{2} - 9$$

$$= \frac{9}{2} \text{ square units} \qquad \bullet$$

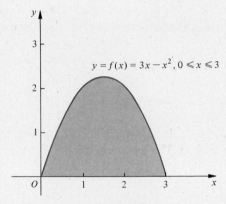

Figure 6.5.5

We shall start the next chapter by learning how to find areas of more general types of regions than the ones considered so far. After that, we shall discover many other applications for the definite integral.

In Chapter 5 the symbol $\int f(x)\, dx$ was introduced for the antiderivative. This similar notation to that for the definite integral $\int_a^b f(x)\, dx$ was chosen because of the close relationship between these two concepts as indicated by the fundamental theorems.

The antiderivative $\int f(x)\, dx$ is also called the **indefinite integral of f.** The process of finding either the definite or indefinite integral is called **integration.** One should, however, carefully distinguish these two types of integrals in that

1. the indefinite integral is a set of functions that differ from each other by a constant; that is,

$$\int f(x)\, dx = G(x) + C \qquad \text{where } G'(x) = f(x)$$

2. the definite integral of a function from a to b, where a and b are constants, is a number. Even though a function may be used to help evaluate $\int_a^b f(x)\, dx$, the final result is just a number. This is the way the definite integral was defined in Section 6.3. Furthermore, if $G'(x) = f(x)$ then $\int_a^b f(x)\, dx = G(b) - G(a)$ from the second fundamental theorem of calculus.

Exercises 6.5

In Exercises 1 through 4 find $\dfrac{dF}{dx}$

1. $F(x) = \displaystyle\int_1^x \dfrac{dt}{t}$ $(x > 0)$

2. $F(x) = \displaystyle\int_3^x \sqrt[3]{t}\, dt$

3. $F(x) = \displaystyle\int_{-3}^x (t^4 - 3t + 7)\, dt$

4. $F(x) = \displaystyle\int_{-100}^x \sqrt{t^2 + 4t + 5}\, dt$

In Exercises 5 through 9 find $\dfrac{dG}{dx}$ using the chain rule when needed

5. $G(x) = \displaystyle\int_x^2 \sqrt{t^2 + 16}\, dt$

6. $G(x) = \displaystyle\int_1^{x^2} \sqrt{3t^2 + 7}\, dt$

7. $G(x) = \displaystyle\int_{-x}^x \dfrac{dt}{t^4 + 9}$

8. $G(x) = \displaystyle\int_x^{x^3} (t^3 - 4t)\, dt$, by two methods

9. $G(x) = \displaystyle\int_3^{\sqrt{x}} g(t)\, dt$ $(x > 0)$

10. Find $\displaystyle\lim_{h \to 0} \left[\dfrac{\displaystyle\int_1^{2+h} t^2\, dt - \int_1^2 t^2\, dt}{h} \right]$

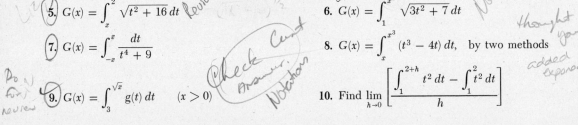

In Exercises 11 through 27 evaluate the following definite integrals using the second fundamental theorem of calculus

11. $\displaystyle\int_{-1}^{-2} \dfrac{dx}{x^2}$

12. $\displaystyle\int_{-2}^2 (x^3 + 10x)\, dx$

13. $\displaystyle\int_4^{16} \dfrac{dx}{\sqrt{x}}$

14. $\displaystyle\int_{-3}^0 (x^2 - 6x - 5)\, dx$

15. $\displaystyle\int_0^1 (\sqrt{x} + 2)^2\, dx$

16. $\displaystyle\int_2^3 \dfrac{x^2 - 2}{x^2}\, dx$

17. $\displaystyle\int_4^9 \left(\dfrac{3\sqrt{x}}{2} + \dfrac{1}{\sqrt{x}} \right) dx$

18. $\displaystyle\int_1^2 (x^2 - 1)^4 x\, dx$ $\left(\text{Hint: To find an antiderivative let } u = x^2 - 1, \dfrac{du}{dx} = 2x, \text{ and then use the appropriate power rule.}\right)$

19. $\displaystyle\int_0^2 \sqrt{1 + u^2}\, u\, du$

20. $\displaystyle\int_0^3 \dfrac{y}{\sqrt{1 + y}}\, dy$

21. $\displaystyle\int_0^b x\sqrt{b^2 - x^2}\, dx$

22. $\displaystyle\int_0^4 \dfrac{(t + 2)}{\sqrt{2t + 1}}\, dt$ $\left(\text{Hint: To find an antiderivative of } \dfrac{t + 2}{\sqrt{2t + 1}}, \text{ let } \sqrt{2t + 1} = u, \text{ and so on.}\right)$

23. $\displaystyle\int_0^3 z\sqrt{z + 1}\, dz$

24. $\int_0^1 (t-1)^2 \sqrt{t}\, dt$

25. $\int_{-1}^3 |x-1|\, dx$ $\quad\left(Hint: |x-1| = \begin{cases} x-1 & \text{if } x \geq 1 \\ 1-x & \text{if } x < 1 \end{cases} \text{ and write } \int_{-1}^3 |x-1|\, dx = \int_{-1}^1 x-1\, dx + \right.$
$\left. \int_1^3 x-1\, dx\right).$

26. $\int_{-2}^2 \sqrt{2+|x|}\, dx$

27. $\int_1^2 \dfrac{x^4 + 2x^3 - 2x + 2}{(x+1)^2}\, dx$ $\quad$ (Hint: Divide the numerator by the denominator.)

In Exercises 28 through 40, find the area bounded by the given curve, the x-axis, and the given vertical lines using the second fundamental theorem of calculus.

28. $y = x^2 + 3x; \quad x = 1, x = 3$

29. $y = \sqrt{x}; \quad x = 0, x = 4$

30. $y = x^2 + 2x + 4; \quad x = -2, x = 1$

31. $y = (2x-3)^2; \quad x = -2, x = 0$

32. $y = \dfrac{x^2 + 6}{x^2}; \quad x = \dfrac{1}{2}, x = 2$

33. $y = x(x+1)(x+2); \quad x = 0, x = 3$

34. $y = \sqrt{x}(x+2); \quad x = 1, x = 4$

35. $y = |3x|; \quad x = -4, x = 5$

36. $y = (x^3 + 3)^4 x^2; \quad x = -1, x = 1$

37. $y = \dfrac{x}{\sqrt{4x^2 - 1}}; \quad x = 1, x = 2$

38. $y = x\sqrt{x+2}; \quad x = 0, x = 2$

39. $y = \sqrt{5 + |x|}; \quad x = -4, x = 4$

40. $y = \dfrac{x+2}{\sqrt[3]{x+3}}; \quad x = -2, x = 5$

Review and Miscellaneous Exercises

In Exercises 1 through 3 find the given sum

1. $\sum_{k=0}^5 3^{-k}$

2. $\sum_{k=2}^{100} (\sqrt{k} - \sqrt{k-1})$

3. $\sum_{k=1}^{39} \dfrac{1}{k(k+1)}$ $\quad\left(Hint: \dfrac{1}{k(k+1)} = \dfrac{1}{k} - \dfrac{1}{k+1}\right)$

4. Find a formula for $(1)(5) + (2)(6) + (3)(7) + \cdots + n(n+4)$ using properties and formulas developed in this chapter.

5. Find a formula for $\sum_{k=1}^n (k-2)(k)(k+2)$.

6. Prove by mathematical induction that $1 + 3 + 5 + \cdots + (2n-1) = n^2$.

7. Prove by mathematical induction that $1^3 + 2^3 + 3^3 + \cdots + n^3 = n^2(n+1)^2/4$.

8. Determine whether each of the following statements is true or false. In each instance give a reason for your decision.

(a) $\sum_{i=1}^{60} 3 = 180$

(b) $\displaystyle\sum_{i=1}^{n} (3 - 2i) = 3 - 2 \sum_{i=1}^{n} i$ for all n

(c) $\displaystyle\sum_{k=0}^{98} (k + 2)^3 = \sum_{n=2}^{100} n^3$

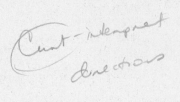

(d) $\displaystyle\sum_{k=1}^{n} k^2 = \left(\sum_{k=1}^{n} k\right)^2$ for all n

In Exercises 9 through 11 find the values of the definite integral directly from the definition of the definite integral as a limit of a sum. Do not use the concept of antiderivatives.

9. $\displaystyle\int_{-3}^{2} x \, dx$ **10.** $\displaystyle\int_{0}^{1} x(x + 3) \, dx$ **11.** $\displaystyle\int_{-1}^{1} x(x + 1)(x + 2) \, dx$

In Exercises 12 through 27 evaluate the definite integral by application of the Second Fundamental Theorem of the Calculus.

12. $\displaystyle\int_{1}^{2} \frac{t^3 - 3}{t^2} \, dt$ **13.** $\displaystyle\int_{0}^{1} \sqrt{2 + 5t} \, dt$

14. $\displaystyle\int_{0}^{2} (4 - t^2)^7 t \, dt$ **15.** $\displaystyle\int_{-1}^{2} \frac{x}{\sqrt{8 - x^2}} \, dx$

16. $\displaystyle\int_{1}^{4} \frac{x^5 + 3x^3 - 7}{x^2} \, dx$ **17.** $\displaystyle\int_{1}^{4} \sqrt{\sqrt{x} + 3} \, \frac{dx}{\sqrt{x}}$

18. $\displaystyle\int_{-2}^{1} \sqrt{3 + x} \, x \, dx$ **19.** $\displaystyle\int_{0}^{2} \frac{dt}{\sqrt[3]{2 + 3t}}$

20. $\displaystyle\int_{-1}^{1} (t^2 + 1)^3 t^2 \, dt$ **21.** $\displaystyle\int_{-1}^{1} \left(t^2 + \frac{1}{t^2}\right)^3 dt$

22. $\displaystyle\int_{1}^{3} 10t|t + 5| \, dt$ **23.** $\displaystyle\int_{0}^{2} \frac{x^2 - 6x - 16}{x^2 - 6x + 9} \, dx$

24. $\displaystyle\int_{-1}^{3} \frac{x^2}{\sqrt{2x + 3}} \, dx$ **25.** $\displaystyle\int_{a/2}^{a} (x - a)\left(x - \frac{a}{2}\right) dx$ where a is constant

26. $\displaystyle\int_{a}^{2a} (3x - a)^3 \, dx$ where a is constant   **27.** $\displaystyle\int_{a}^{2a} x\sqrt{x - a} \, dx$ where a is a positive constant

28. Find $\displaystyle\frac{d}{dx} \int_{2}^{x} (3t^2 + 10)^6 \, dt$

29. Find $\displaystyle\frac{d}{dx} \int_{0}^{x^2} \sqrt{3 + t^2} \, dt$ (*Hint:* Use the first fundamental theorem of calculus and the chain rule of differentiation.)

30. If a function $F(x)$ is defined by

$$F(x) = \int_{1}^{x} \frac{du}{u} \qquad x > 0$$

Find

(i) upper and lower bounds for $F(2)$ and (ii) $\dfrac{dF}{dx}$.

31. Define a function F by

$$F(x) = \int_{1}^{3x} \frac{du}{u} - \int_{1}^{x} \frac{du}{u} \qquad \text{where } x > 0$$

show that (i) $F'(x) = 0$, which implies that $F(x) = C$, and (ii) Find an expression for this constant.

32. Evaluate $\displaystyle\int_0^3 |x - 1|\, dx$.

33. If $F(x) = \displaystyle\int_2^{x^2} (x^3 t^2 + x)\, dt$, find (i) $F(x)$ explicitly, (ii) $F'(x)$, (iii) $\dfrac{d}{dx}(F(1))$, (iv) $\displaystyle\int_{-1}^1 F(x)\, dx$

34. Find $\dfrac{d}{dx} \displaystyle\int_3^x [u(t)\, v'(t) + v(t)\, u'(t)]\, dt$ by two methods.

35. Find the average value of the function f defined by $y = 16 - x^2$ from $(-3, 7)$ to $(2, 12)$.

36. The relation $x^2 + y = 9$, $x \geq 0$ defines x as a function of y say $x = g(y)$. Find the average value of g in $0 \leq y \leq 9$, that is, the average value of x with respect to y in the specified interval.

37. An airplane starting from rest accelerates in accordance with the formula $a = 3t$ ft/sec^2 during the first 20 sec of travel. Find the average velocity with respect to time in this interval.

38. The integral

$$Q(u) = \int_a^b (f(x) - u)^2\, dx \qquad (b > a)$$

where u is independent of x is a quadratic function of u. Show that the value of u that makes $Q(u)$ an absolute minimum is given by the mean value of $f(x)$ with respect to x on $[a, b]$. What is the minimum value of Q?

39. (a) What theorem guarantees that $f(x) = 2x^3 + 7x^2 + 14x + 12$ possesses at least one real zero?
(b) Locate the two consecutive integers that straddle one root of $f(x) = 0$.
(c) Prove that $f(x) = 0$ has only one real root.

7

Applications of the Definite Integral

7.1   AREA BOUNDED BY CURVES

The definite integral was introduced in Chapter 6 as a result of our attempts to find the area of the region under the graph of a function from a to b, where the function was nonnegative on $[a, b]$. In Section 6.5, we used the second fundamental theorem of calculus to find some of these areas. We now wish to extend this so that definite integrals can be used to find areas of other types of regions.

Consider two functions f and g, which are both continuous on an interval $[a, b]$. If f and g are both positive and if $f(x) \geq g(x)$ for all x in $[a, b]$ (see Figure 7.1.1), then the area between the graphs is equal to the area bounded by $f(x)$, the x-axis, and the vertical lines $x = a$ and $x = b$ minus the corresponding area bounded by

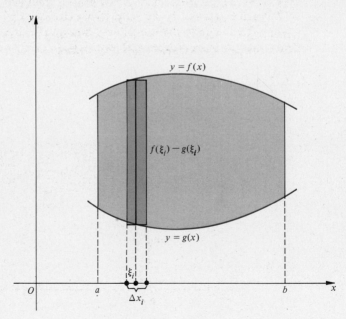

Figure 7.1.1

$g(x)$, the x-axis, and the lines $x = a$ and $x = b$. Thus this difference is given by

$$A = \int_a^b f(x)\,dx - \int_a^b g(x)\,dx \tag{1}$$

or, equivalently, by

$$A = \int_a^b [f(x) - g(x)]\,dx \tag{2}$$

Formulas (1) and (2) are also valid even if $g(x)$ or $f(x)$ is negative over part of the interval provided that $f(x) \geq g(x)$ in $[a, b]$. To see this, note that, if a partition P of the interval $[a, b]$ is introduced, the expression $[f(\xi_i) - g(\xi_i)]\,\Delta x_i$ represents the area of the shaded rectangle in Figure 7.7.1. Thus

$$\int_a^b [f(x) - g(x)]\,dx = \lim_{|P| \to 0} \sum_{i=1}^{n} [f(\xi_i) - g(\xi_i)]\,\Delta x_i \tag{3}$$

Since this integral does give the area between the curves by (2), it is useful to think of the area as the limit of the sum of the areas of elementary rectangles of the type shown in Figure 7.1.1. It is important to remember that in applying (2) the equation of the *lower* curve must be subtracted from the equation of the *upper* curve so that the computation for the area will be nonnegative.

Figure 7.1.2 shows a still more general situation where the curves $y = f(x)$ and $y = g(x)$ intersect several times. Then the required area between the graphs of $y = f(x)$ and $y = g(x)$ is given by

$$A = \int_a^c [f(x) - g(x)]\,dx + \int_c^d [g(x) - f(x)]\,dx + \int_d^b [f(x) - g(x)]\,dx \tag{4}$$

since $f(x) \geq g(x)$ in $[a, c]$, $g(x) \geq f(x)$ in $[c, d]$ and $f(x) \geq g(x)$ in $[d, b]$.

When finding areas, it is a good idea to include in the diagram of the region rectangular elements of the type used in approximating the area. The length of the rectangle then corresponds to the integrand $f(x) - g(x)$ and the width of the

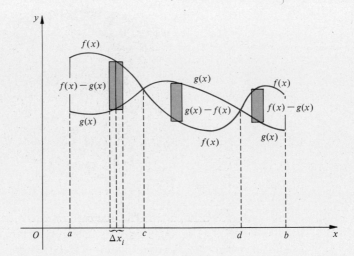

Figure 7.1.2

rectangle, Δx_i, corresponds to the dx in the integral. Furthermore, the rectangular elements show us which is the top curve and which is on the bottom. This is especially important if we have a situation such as the one illustrated in Figure 7.1.2 where the upper and lower curves change.

EXAMPLE 1   Find the area of the region R bounded by the lines $x = 1$, $x = 3$, $y = 3x + 4$, and the curve $y = x^2$.

Solution   A sketch makes it clear that $f(x) = 3x + 4$ is above $g(x) = x^2$ in $[1, 3]$ (see Figure 7.1.3). Thus the required area is given by

$$A = \int_1^3 (3x + 4 - x^2)\, dx = \left[\frac{3x^2}{2} + 4x - \frac{x^3}{3} \right]_1^3$$

$$= \left(\frac{3}{2}(9) + 12 - 9 \right) - \left(\frac{3}{2} + 4 - \frac{1}{3} \right)$$

$$= \frac{33}{2} - \frac{31}{6} = \frac{34}{3} \text{ square units} \qquad \bullet$$

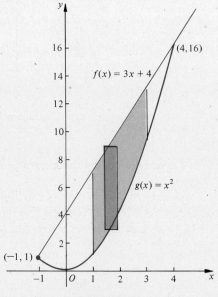

Figure 7.1.3

EXAMPLE 2   Find the area of the region bounded by the graphs of $f(x) = x^3$ and $g(x) = x$.

Solution   The graph of $f(x) = x^3$ and $g(x) = x$ is given in Figure 7.1.4. The points of intersection are obtained by equating x and x^3. Thus if $x = x^3$ then $x(1 - x^2) = 0$ or $x = 0$, ± 1. Consequently the curves intersect at $(0, 0)$, $(1, 1)$ and $(-1, -1)$. There are two regions, one in the first quadrant labeled R_1 and the other in the third called R_2. Let A_1 and A_2 denote their respective areas. For R_1, $x \geq x^3$, for example, when $x = \frac{1}{2}$ we obtain $\frac{1}{2} > \frac{1}{8}$, while for R_2, $x^3 \geq x$ since $(-\frac{1}{2})^3 = -\frac{1}{8} > -\frac{1}{2}$. Therefore

$$A_1 = \int_0^1 (x - x^3)\, dx = \left[\frac{x^2}{2} - \frac{x^4}{4}\right]_0^1 = \frac{1}{2} - \frac{1}{4} - 0 = \frac{1}{4}$$

and

$$A_2 = \int_{-1}^0 (x^3 - x)\, dx = \left[\frac{x^4}{4} - \frac{x^2}{2}\right]_{-1}^0 = 0 - \left(\frac{1}{4} - \frac{1}{2}\right) = \frac{1}{4}$$

The total area between the curves is $A_1 + A_2 = \frac{1}{2}$ square units

Let us also observe that if $\phi(x) = x - x^3$ then $\phi(-x) = -\phi(x)$. Thus $\phi(x)$ is an odd function and the graph is symmetric with respect to the origin. Hence A_1 and A_2 must be equal. Such information may be utilized whenever applicable. ●

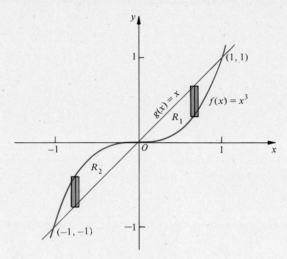

Figure 7.1.4

EXAMPLE 3   Find the area bounded by the curves $y + 2x = 2$, $x - y = 1$, and $x + 2y = 7$.

Solution   The curves are straight lines (shown in Figure 7.1.5) so that the area of the triangle bounded by these lines is required. The intersection points of the lines in pairs are $(-1, 4)$, $(1, 0)$ and $(3, 2)$. Although the line $x + 2y = 7$ is always on top, the

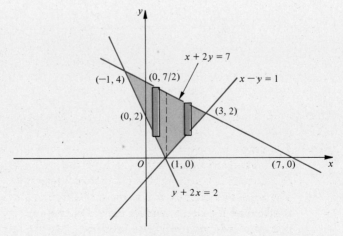

Figure 7.1.5

rectangular elements show us that the bottom curve changes. Therefore, in the interval $-1 \leq x \leq 1$ we take the difference in the y values due to the lines $x + 2y = 7$ and $y + 2x = 2$, whereas in $1 \leq x \leq 3$ we must take the difference in the y values due to the lines $x + 2y = 7$ and $x - y = 1$. Therefore the required area A is

$$A = \int_{-1}^{1} \left[\frac{7 - x}{2} - (2 - 2x) \right] dx + \int_{1}^{3} \left[\frac{7 - x}{2} - (x - 1) \right] dx$$

$$= \int_{-1}^{1} \left(\frac{3}{2} + \frac{3x}{2} \right) dx + \int_{1}^{3} \left(\frac{9}{2} - \frac{3x}{2} \right) dx$$

$$= \left. \left(\frac{3x}{2} + \frac{3x^2}{4} \right) \right|_{-1}^{1} + \left. \left(\frac{9x}{2} - \frac{3x^2}{4} \right) \right|_{1}^{3}$$

$$= \frac{3}{2} + \frac{3}{4} - \left(\frac{-3}{2} + \frac{3}{4} \right) + \left(\frac{9(3)}{2} - \frac{3(3)^2}{4} \right) - \left(\frac{9(1)}{2} - \frac{3(1)}{4} \right)$$

$$= \frac{3}{2} + \frac{3}{4} + \frac{3}{2} - \frac{3}{4} + \frac{27}{2} - \frac{27}{4} - \frac{9}{2} + \frac{3}{4}$$

$$= 6 \text{ square units}$$

●

EXAMPLE 4   Find the area of the region bounded by the x-axis, the curve $f(x) = x^2 - 6x$, and the lines $x = 1$ and $x = 4$.

Solution   In Figure 7.1.6, we see that the upper curve is the line $y = 0$ (the x-axis) and the lower curve is the parabola $y = x^2 - 6x$. Thus

$$A = \int_{1}^{4} [0 - (x^2 - 6x)] \, dx = -\int_{1}^{4} (x^2 - 6x) \, dx = -\left[\frac{x^3}{3} - 3x^2 \right]_{1}^{4}$$

$$= -\left[\frac{64}{3} - 48 \right] + \left[\frac{1}{3} - 3 \right] = -\frac{63}{3} + 45 = -21 + 45 = 24 \text{ square units}$$

●

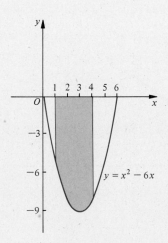

Figure 7.1.6

This example illustrates the following result:
If R is a region bounded by the x-axis, the curve $y = f(x)$ where $f(x) \leq 0$, and the lines $x = a$ and $x = b$, then the area of R is given by

$$A = -\int_a^b f(x)\,dx \qquad (5)$$

As in Example 4, this formula follows immediately from (2) and the fact that the upper curve is $y = 0$.

EXAMPLE 5 Find the area bounded by the curves $y^2 = 5 - x$ and $y^2 = 4x$.

Solution The curves are parabolas and are sketched in Figure 7.1.7(a). The equation $y^2 = 5 - x$ is a relation in x and y that may be expressed as two functions, namely, $y_1(x) = \sqrt{5 - x}$ and $y_2(x) = -\sqrt{5 - x}$. Similarly, $y^2 = 4x$ may be expressed as $y_3(x) = 2\sqrt{x}$ and $y_4(x) = -2\sqrt{x}$. The intersections of the two curves is easily found by forming the equation

$$5 - x = 4x \qquad \text{or} \qquad x = 1$$

so that

$$y = \pm 2$$

Thus $(1, 2)$ and $(1, -2)$ are the intersection points. The required area A is given by

$$A = \int_0^1 [2\sqrt{x} - (-2\sqrt{x})]\,dx + \int_1^5 [\sqrt{5 - x} - (-\sqrt{5 - x})]\,dx$$

$$= 4\int_0^1 \sqrt{x}\,dx + 2\int_1^5 \sqrt{5 - x}\,dx$$

$$= 4\left(\frac{2}{3}\right)x^{3/2}\,\Big|_0^1 + 2\left(-\frac{2}{3}\right)(5 - x)^{3/2}\,\Big|_1^5$$

$$= \frac{8}{3}(1 - 0) - \frac{4}{3}[0 - (4)^{3/2}] = \frac{8}{3} + \frac{32}{3}$$

$$= \frac{40}{3} \text{ square units} \qquad \bullet$$

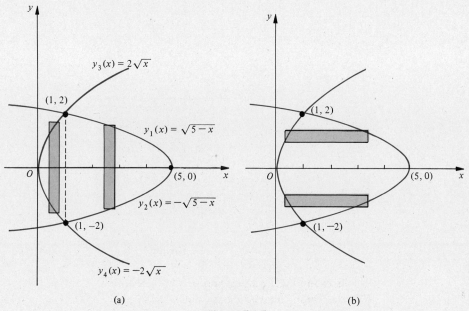

Figure 7.1.7

Example 5 is solved more easily if the rectangular elements are drawn horizontally (see Figure 7.1.7(b)). In this case, the length of the rectangle always extends from one parabola to the other. Figures 7.1.8 and 7.1.9 show similar types of regions in which the boundary curves are functions of y (rather than of x). For a given partition of $[c, d]$, the area of a rectangular element in Figure 7.1.8 is $[f(\eta_i) - g(\eta_i)]\,\Delta y_i$ since $f(y) \geq g(y)$ in $[c, d]$. Then, since

$$\lim_{|P| \to 0} \sum_{i=1}^{n} [f(\eta_i) - g(\eta_i)]\,\Delta y_i = \int_{c}^{d} [f(y) - g(y)]\,dy$$

the area is given by

$$A = \int_{c}^{d} [f(y) - g(y)]\,dy \qquad (6)$$

Note that in (6), as in (2), the integrand corresponds to the length of the rectangle and the dy corresponds to the width of the rectangle, Δy_i.

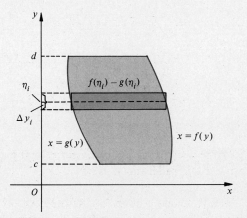

Figure 7.1.8

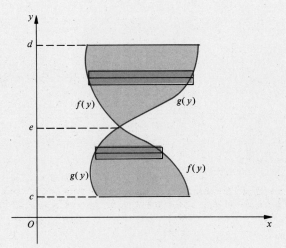

Figure 7.1.9

In Figure 7.1.9, the area is determined by

$$A = \int_c^e [f(y) - g(y)] \, dy + \int_e^d [g(y) - f(y)] \, dy \qquad (7)$$

where it is assumed that the functions $f(y)$ and $g(y)$ are continuous in each of the given intervals.

EXAMPLE 6   Find the area bounded by the curves $y^2 = 5 - x$ and $y^2 = 4x$ by using elements parallel to the x-axis.

Solution   This is the same problem as Example 5. Now, if we solve for x in each case

$$x = f(y) = 5 - y^2 \qquad \text{and} \qquad x = g(y) = y^2/4$$

and the required area A (see Figure 7.1.7(b)) is given by

$$A = \int_{-2}^2 [f(y) - g(y)] \, dy$$

$$= \int_{-2}^2 \left(5 - y^2 - \frac{y^2}{4}\right) dy$$

$$= \int_{-2}^2 \left(5 - \frac{5}{4} y^2\right) dy = \left(5y - \frac{5y^3}{12}\right)\Big|_{-2}^2$$

$$= \left(10 - \frac{40}{12}\right) - \left(-10 + \frac{40}{12}\right) = 20 - \frac{40}{6} = \frac{40}{3} \text{ square units} \qquad \bullet$$

EXAMPLE 7   Find the area of the region bounded by the graphs of $y = x^3$ and $y = x$ by using elements parallel to the x-axis.

Solution   This is the same problem as Example 2 (refer to Figure 7.1.4). If we solve $y = x^3$ for x then $x = y^{1/3}$. Therefore the required area is given by

$$A = \int_{-1}^0 (y - y^{1/3}) \, dy + \int_0^1 (y^{1/3} - y) \, dy$$

$$= \left(\frac{y^2}{2} - \frac{3}{4} y^{4/3}\right)\Big|_{-1}^0 + \left(\frac{3y^{4/3}}{4} - \frac{y^2}{2}\right)\Big|_0^1$$

$$= 0 - \left(\frac{1}{2} - \frac{3}{4}\right) + \left(\frac{3}{4} - \frac{1}{2}\right) - 0$$

$$= \frac{1}{4} + \frac{1}{4} = \frac{1}{2} \text{ square units} \qquad \bullet$$

Exercises 7.1

In Exercises 1 through 20 find the areas bounded by the following curves. In each case draw a graph showing the required area and the elementary rectangle or element of area utilized.

1. $y = x^2$, $y = 4$
2. $y = x^2$, $y = x$
3. $y = x^2$, $x + y - 2 = 0$

4. $y = x^3$, $y = x^4$
5. $y = 8 - x^2$, $y = 2x$
6. $y = 2x^2$ and $y = -2x^2$ and the lines $x = -1$ and $x = 2$
7. $x = 4 - y^2$, $x = 0$
8. $y = x^2 - 6x + 6$, $y = -2$
9. $x + y^2 = 4$, $x = 4$, $y = 2$, and $y = -2$

10. $y = \sqrt{x}$, $y = \dfrac{x}{2}$

11. $y^3 - x = 0$, $x^2 - y = 0$
12. $xy^2 = 4$, $x + 3y = 7$
13. $y = x$, $y = 4 - x$, and $y = -1$
14. $x - y = 0$, $x - 2y = 0$, and $x + 2y = 6$
15. $x + y = 3$, $2x - y = 9$, and $7x + y - 9 = 0$
16. $y = x^2 - 4x + 5$, $x - y = 3$, $x = 1$, and $x = 3$

17. $y = \dfrac{x^2}{4}$, y-axis, and the tangent to $y = \dfrac{x^2}{4}$ at $(2, 1)$.

18. $y = \dfrac{x}{4} \sqrt{x^2 + 12}$, the x-axis, $x = -2$, and $x = 2$

19. $y = \sqrt{x}$, $y = -2\sqrt{x}$, and $x = 4$
20. $x + y^2 = 12$, $x + y = 0$
21. A square is formed by the coordinate axes and parallel lines through the point (a, a) where $a > 0$. Calculate the ratio of the larger to the smaller of the two areas into which it is divided by each of the following curves
    (a) $ay = x^2$
    (b) $\sqrt{x} + \sqrt{y} = \sqrt{a}$
    (c) $y = a^{1-n}x^n$, $n > 0$
22. Find the area of the region bounded by one loop of the graph of the equation $y^2 = 9x^2 - x^4$.
23. Find the area of the lens-shaped region bounded by the two parabolas $y^2 = 4bx$ and $x^2 = 4by$, where $b > 0$.
24. Find the area of the curvilinear triangle formed by the curves $y = 1 - |x|$, $3x - y - 1 = 0$, and $y + 1 = x^2$ in $-1 \le x \le 1$.

7.2 VOLUME OF A SOLID OF REVOLUTION: DISK AND RING METHODS

The utility of the integral extends well beyond the determination of area. In the present section we shall show how the definite integral is used to find the volumes of solids of revolution. By definition a *solid of revolution* is obtained by revolving a region in a plane about a line in the plane called the *axis of revolution,* which either touches the boundary of the region or does not intersect the region at all. Thus, for example, if a rectangular region *EFGH* is rotated about side *EF* (Figure 7.2.1) then a right circular cylinder is generated. If the region inside a right triangle *ABC* is rotated about one of its legs, say *AB*, then a right circular cone (Figure 7.2.2) is developed. By way of another illustration, if the region bounded by a semicircle and its diameter is revolved about that diameter, it sweeps out a sphere (Figure 7.2.3).

We call a solid *C* a *right cylinder* if *C* is bounded by two congruent regions R_1 and R_2 lying in parallel planes and by a lateral surface *S*, which consists of line

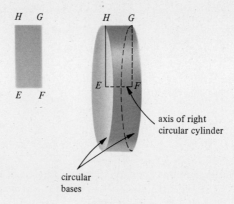

Figure 7.2.1

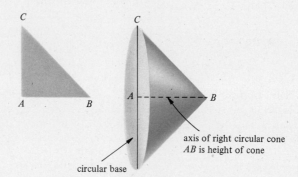

Figure 7.2.2

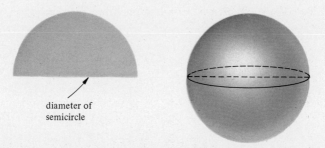

Figure 7.2.3

segments perpendicular to the planes of R_1 and R_2 and connecting corresponding points of the boundaries of R_1 and R_2 (see Figure 7.2.4). Each of the regions R_1 and R_2 is called a **base** of cylinder C and the distance between the planes of R_1 and R_2 is called the **height** h of C.

The **right circular cylinder** is a particular right cylinder having a circle as a base. Other examples of cylinders are the rectangular parallelepiped, which is a cylinder having a rectangle as a base. It is simply a box. Another example illustrating the right cylinder is the triangular prism (Figure 7.2.5).

If the area of the base of a right cylinder is A square units and the height is h units, then by definition the volume V is given by

$$V = Ah \qquad (1)$$

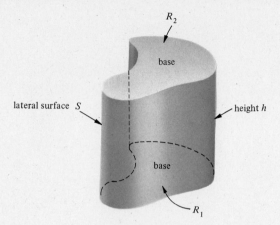

Figure 7.2.4

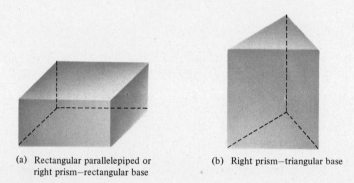

(a) Rectangular parallelepiped or
   right prism–rectangular base

(b) Right prism–triangular base

Right Cylinders

Figure 7.2.5

Thus in the particular case of a right circular cylinder of radius r and height h,

$$V = \pi r^2 h \tag{2}$$

and for the rectangular parallelepiped with edges of lengths a, b, and c we have

$$V = abc \tag{3}$$

as anticipated.

We seek a definition of the word "volume" for a solid of revolution. To this end we consider a function $f(x)$ which is assumed to be continuous and nonnegative in $a \le x \le b$. Let R be the region bounded by the curve $y = f(x)$, the x-axis, and the lines $x = a$ and $x = b$. Let S be the solid of revolution obtained by revolving the region R about the x-axis. The volume V of S must now be defined.

To this end, we partition $[a, b]$ as usual by

$$a = x_0 < x_1 < x_2 < \cdots < x_{n-1} < x_n = b$$

which yields n subintervals of length $\Delta x_i = x_i - x_{i-1}$. Choose a point ξ_i in each $[x_{i-1}, x_i]$ and draw the rectangles having altitudes $f(\xi_i)$ and widths Δx_i. In Figure 7.2.6 we have region R shown together with the ith rectangle.

When the ith rectangle is revolved about the x-axis, a cylindrical disk is obtained whose base radius is given by $f(\xi_i)$ and whose altitude is Δx_i, which

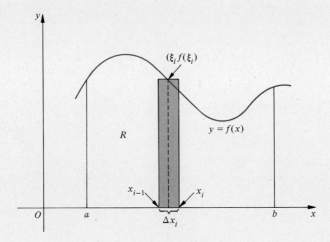

Figure 7.2.6

implies that the volume of this disk, ΔV_i, is

$$\Delta V_i = \pi [f(\xi_i)]^2 \, \Delta x_i \tag{4}$$

(see Figure 7.2.7)

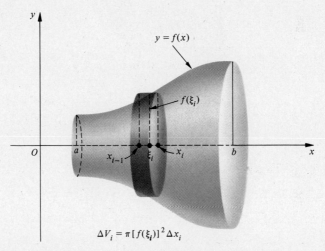

$$\Delta V_i = \pi [f(\xi_i)]^2 \, \Delta x_i$$

Figure 7.2.7

The sum of the volumes of the n cylinders is the Riemann sum

$$\sum_{i=1}^{n} \Delta V_i = \pi \sum_{i=1}^{n} [f(\xi_i)]^2 \, \Delta x_i \tag{5}$$

This Riemann sum approximates what we would intuitively think of as the volume of the solid of revolution. Furthermore, this approximation improves as the norm $|P|$ of the partition tends to zero. We are then led to define the volume of the solid of revolution under consideration to be

$$V = \lim_{|P| \to 0} \sum_{i=1}^{n} \Delta V_i = \lim_{|P| \to 0} \sum_{i=1}^{n} \pi [f(\xi_i)]^2 \, \Delta x_i \tag{6}$$

Since this function f is continuous on the closed interval then $f^2 = f \cdot f$ must also be continuous. Therefore the limit (6) must exist and is equal to $\pi \int_a^b [f(x)]^2 \, dx$. The situation may be summarized by the following definition.

Definition   If f is a continuous function on $[a, b]$ where, furthermore, $f(x) \geq 0$ for all $a \leq x \leq b$, then the ***volume of the solid*** generated by revolving the region bounded by the curve $y = f(x)$, the x-axis, and the lines $x = a$ and $x = b$ about the x-axis is

$$V = \lim_{|P| \to 0} \pi \sum_{i=1}^{n} [f(\xi_i)]^2 \, \Delta x_i = \pi \int_a^b [f(x)]^2 \, dx \qquad (7)$$

In working out examples, we again find it useful to include at least one elementary rectangle in the diagram. This is usually quite helpful in setting up the integral.

EXAMPLE 1   A sphere of radius a can be obtained by rotating about the x-axis the region bounded by

$$f(x) = \sqrt{a^2 - x^2} \qquad -a \leq x \leq a$$

and the x-axis (refer to Figure 7.2.8). Therefore from (7), the volume V of the sphere is obtained easily as follows

$$V = \pi \int_{-a}^{a} (a^2 - x^2) \, dx = \pi \left(a^2 x - \frac{x^3}{3} \right) \Big|_{-a}^{a}$$

$$= \pi \left[\left(a^3 - \frac{a^3}{3} \right) - \left(-a^3 + \frac{a^3}{3} \right) \right]$$

$$= \tfrac{4}{3} \pi a^3 \text{ cubic units} \qquad \bullet$$

EXAMPLE 2   Find the volume of a right circular cone of base radius a and height h.

Solution   The cone is a solid of revolution obtained by revolving the right triangle shown in Figure 7.2.9 about the x-axis. From similar triangles

$$\frac{f(x)}{x} = \frac{a}{h} \qquad \text{or} \qquad f(x) = \frac{a}{h} x$$

Thus the volume of the right circular cone is given by

$$V = \pi \int_0^h f^2(x) \, dx$$

$$= \pi \frac{a^2}{h^2} \int_0^h x^2 \, dx$$

$$= \pi \frac{a^2}{h^2} \frac{h^3}{3}$$

$$= \pi \frac{a^2 h}{3} \text{ cubic units} \qquad \bullet$$

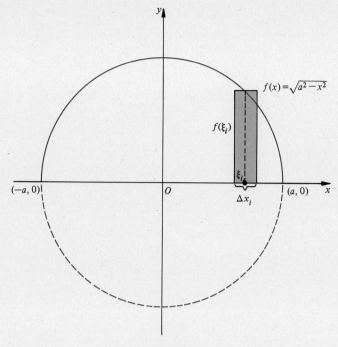

Figure 7.2.8

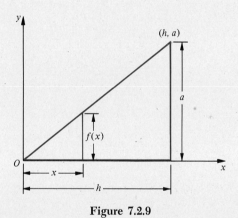

Figure 7.2.9

EXAMPLE 3   Find the volume of the solid of revolution obtained by rotating the region bounded by $y^2 = 4px$ ($y \geq 0$ and $p > 0$), $x = p$ and the x-axis about the x-axis (Figure 7.2.10)

Solution   The solid obtained by rotating a parabola about its axis is called a *paraboloid of revolution.* In fact, automobile headlights are built in this shape because of the "reflection property" of the parabola, namely, that rays emanating from a light source at the focus of the parabola will be reflected forward by the curve along lines parallel to the axis of the parabola. This provides a concentrated beam of illumination.

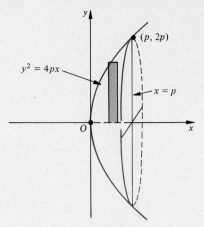

Figure 7.2.10

The required volume is easily found by

$$V = \pi \int_0^p y^2 \, dx = \pi \int_0^p 4px \, dx = 4\pi p \left. \frac{x^2}{2} \right|_0^p$$

$$= 2\pi p^3 \text{ cubic units}$$

Analogously, if g is continuous in $[c, d]$ and $g(y) \geq 0$ in $[c, d]$, the **volume of the solid of revolution** generated by rotating about the y-axis the region bounded by $x = g(y)$, the y-axis, $y = c$, and $y = d$ is defined by

$$V = \lim_{|P| \to 0} \sum_{i=1}^{n} \pi [g(\eta_i)]^2 \, \Delta y_i = \pi \int_c^d [g(y)]^2 \, dy \qquad (8)$$

where η_i is any point in the ith interval $[y_{i-1}, y_i]$ of a partition P of $[c, d]$ and $\Delta y_i = y_i - y_{i-1}$.

EXAMPLE 4   The region bounded by $y = x^2$, the y-axis and $y = 4$ is rotated about the y-axis. Find the volume of the resulting solid.

Solution   In this case the plane region may be subdivided into horizontal strips and these are rotated about the y-axis. Thus our element of volume is a disk with representative radius $x = g(y)$ and height Δy so that (8) applies (see Figure 7.2.11). Thus

$$V = \pi \int_0^4 [g(y)]^2 \, dy$$

$$= \pi \int_0^4 (\sqrt{y})^2 \, dy$$

$$= \pi \int_0^4 y \, dy$$

$$= \pi \left. \frac{y^2}{2} \right|_0^4$$

$$= 8\pi \text{ cubic units}$$

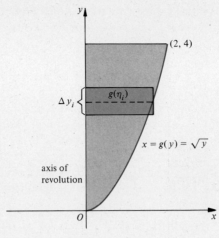

Figure 7.2.11

Up to this point the axis of revolution has been a boundary of the region being revolved. Now let us consider the case where it is not part of the boundary of the given region. Suppose that $f(x)$ and $g(x)$ are two continuous functions on $a \leq x \leq b$ for which it is further assumed that $f(x) \geq g(x) \geq 0$ for all x in $[a, b]$. Let R be the region bounded by the curves $y = f(x)$ and $y = g(x)$ and the lines $x = a$ and $x = b$. We seek the volume V of the solid of revolution obtained by rotating the region R about the x-axis (see Figure 7.2.12(a)).

We can think of this volume as the difference between the volumes of the solids obtained by rotating the region under f and the region under g about the x-axis. Then

$$V = \pi \int_a^b [f(x)]^2 \, dx - \pi \int_a^b [g(x)]^2 \, dx = \pi \int_a^b [(f(x))^2 - (g(x))^2] \, dx$$

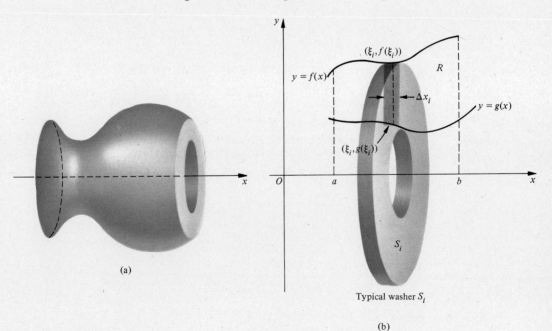

(a)

Typical washer S_i

(b)

Figure 7.2.12

We could also obtain this formula from the rectangular elements that would be used for this region. As before, we partition the interval $[a, b]$ into n subintervals $[x_{i-1}, x_i]$ and let ξ_i be any point in the ith subinterval. Thus, the region R is replaced by n rectangles and when each of these rectangles is rotated about the x-axis a circular ring or washer shaped solid is formed. Figure 7.2.12(b) shows a typical washer S_i.

The volume of S_i, which we denote ΔV_i, is the difference between the volumes of the two concentric cylindrical disks. Both cylinders have height Δx_i; the outer cylinder has radius $f(\xi_i)$ and the inner cylinder's radius is $g(\xi_i)$. Thus

$$\Delta V_i = \pi [f(\xi_i)]^2 \Delta x_i - \pi [g(\xi_i)]^2 \Delta x_i = \pi [[f(\xi_i)]^2 - [g(\xi_i)]^2] \Delta x_i$$

With this in mind, we have the following.

Definition Let the functions $f(x) \geq g(x) \geq 0$ be defined for $a \leq x \leq b$. The **volume V of the solid of revolution** generated by revolving the region R bounded by $y = f(x)$, $y = g(x)$, $x = a$, and $x = b$ about the x-axis is

$$V = \lim_{|P| \to 0} \sum_{i=1}^{n} \pi([f(\xi_i)]^2 - [g(\xi_i)]^2 \Delta x_i$$

which yields

$$V = \pi \int_a^b ([f(x)]^2 - [g(x)]^2) \, dx \tag{9}$$

Note that (9) reduces to (6) if $g(x) \equiv 0$ on $[a, b]$. As before, a similar definition applies when the axis of revolution is the y-axis or any line parallel to the x-axis or the y-axis. For example, assume f and g are continuous and $f(y) \geq g(y) \geq 0$ for all y in $[c, d]$. If R is the region bounded by $x = f(y)$, $x = g(y)$, $y = c$, and $y = d$, then the volume of the solid of revolution obtained by rotating R about the y-axis is

$$V = \pi \int_c^d [[f(y)]^2 - [g(y)]^2] \, dy \tag{10}$$

The reader is encouraged *not* to memorize these formulas. These problems can be handled more effectively by drawing a figure showing an elementary rectangle and the axis of rotation. The width of the rectangle, Δx_i or Δy_i, gives us the variable of integration (indicated by dx or dy). The functions that correspond to the length of the rectangle determine the integrand.

The formulas for the volume of a solid of revolution are sometimes written in the following single form:

$$V = \pi \int_a^b [r_2^2 - r_1^2] \, dh \tag{11}$$

Here h corresponds to the height of the rotated cylinder (that is, $dh = dx$ if Δx_i is the height and $dh = dy$ if Δy_i is the height); r_2 and r_1 represent the distances of the ends of the rectangle from the axis of revolution where $r_2 \geq r_1 \geq 0$ (the r is used to suggest the radius of a rotated cylinder). Formulas (7), (8), (9), and (10) are all special cases of (11). In addition, (11) can be used even if we rotate about a line other than one of the coordinate axes.

EXAMPLE 5   Find the volume generated by revolving about the x-axis the region bounded by $y^2 = 4x$ and $x^2 = 4y$.

Solution   Both curves are parabolas and intersect at the two points $(0, 0)$ and $(4, 4)$ as shown in Figure 7.2.13. Thus $[f(x)]^2 = 4x$ so that $f(x) = 2\sqrt{x}$ since $f \geq 0$. Also $g(x) = \dfrac{x^2}{4}$. Direct application of (9) yields

$$V = \pi \int_0^4 \left[4x - \left(\frac{x^2}{4} \right)^2 \right] dx$$

$$= \pi \int_0^4 \left[4x - \frac{x^4}{16} \right] dx$$

$$= \pi \left(2x^2 - \frac{x^5}{80} \right) \Big|_0^4$$

$$= \pi \left(32 - \frac{4^5}{80} \right) = \pi \left(32 - \frac{1024}{80} \right) = \pi \left(32 - \frac{64}{5} \right) = \frac{96\pi}{5} \text{ cubic units} \quad \bullet$$

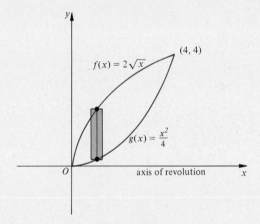

Figure 7.2.13

EXAMPLE 6   Find the volume generated by revolving about the y-axis the region bounded by $x = y^2$ and $y = x - 2$.

Solution   Solving $x = y^2$ and $y = x - 2$ simultaneously tells us that the curves intersect at $(1, -1)$ and $(4, 2)$. Figure 7.2.14 shows the region in question along with an elementary rectangle.
    Applying formula (10) with $f(y) = y + 2$ and $g(y) = y^2$ gives

$$V = \pi \int_{-1}^{2} [(y + 2)^2 - (y^2)^2] \, dy = \pi \int_{-1}^{2} [y^2 + 4y + 4 - y^4] \, dy$$

$$= \pi \left[\frac{y^3}{3} + 2y^2 + 4y - \frac{y^5}{5} \right]_{-1}^{2}$$

$$= \pi \left[\frac{8}{3} + 8 + 8 - \frac{32}{5} \right] - \pi \left[-\frac{1}{3} + 2 - 4 + \frac{1}{5} \right]$$

$$= \pi \left[\frac{9}{3} + 18 - \frac{33}{5} \right] = \pi \left[21 - \frac{33}{5} \right] = \frac{72\pi}{5} \text{ cubic units}$$

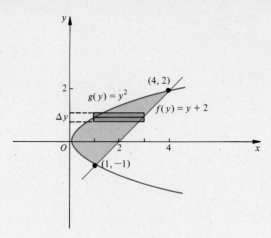

Figure 7.2.14

Note that in finding $f(y)$, we had to write the equation $y = x - 2$ in the form $x = y + 2$ since we specifically needed x as a function of y. ●

EXAMPLE 7   Find the volume obtained by revolving the region bounded by $y = x^2$ and $y = x^3$ about the line $y = 1$.

Solution   The area bounded by $F(x) = x^3$ and $G(x) = x^2$ is to be rotated about $y = 1$ (see Figure 7.2.15). Since the points on $F(x) = x^3$ are further away from the axis of

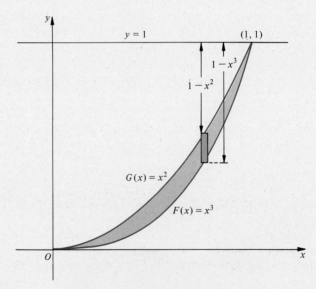

Figure 7.2.15

rotation than the corresponding points on $G(x) = x^2$ (for the same value of x) let $f(x) = 1 - x^3$ and $g(x) = 1 - x^2$. That is, $f(x)$ is the distance between a point on $y = 1$ and a point on $y = x^3$, whereas $g(x)$ is the distance between a point on $y = 1$ and a point on $y = x^2$. In formula (11), we have $r_2 = f(x)$ and $r_1 = g(x)$.

Thus we have

$$V = \pi \int_0^1 [(1-x^3)^2 - (1-x^2)^2]\, dx$$

$$= \pi \int_0^1 [(1 - 2x^3 + x^6) - (1 - 2x^2 + x^4)]\, dx$$

$$= \pi \int_0^1 [2x^2 - 2x^3 - x^4 + x^6]\, dx$$

$$= \pi \left(\frac{2x^3}{3} - \frac{x^4}{2} - \frac{x^5}{5} + \frac{x^7}{7} \right) \Big|_0^1$$

$$= \pi \left(\frac{2}{3} - \frac{1}{2} - \frac{1}{5} + \frac{1}{7} \right)$$

$$= \frac{23\pi}{210} \text{ cubic units} \qquad\qquad \bullet$$

Exercises 7.2

In each of the Exercises 1 through 10, find the volume generated by revolving about the x-axis the region bounded by the given curves. Make a sketch.

1. $y = \dfrac{x^2}{2}$, x-axis, $x = 2$

2. $y = x - x^2$, x-axis

3. $y = 3x$, x-axis, $x = 2$

4. $y = \dfrac{1}{x}$, x-axis, $x = 1$, $x = 5$

5. $y = \sqrt{9 - x^2}$, $x = 0$, x-axis

6. $y = 2x$, y-axis, $y = 2$

7. $y = x^2$, y-axis, $y = 4$

8. $x = y^2$, x-axis, $x = 1$

9. $y = 1 - |x|$, $x = 1$, $x = -1$, $y = 1$

10. First quadrant portion of $4x^2 + 9y^2 = 36$, x-axis, y-axis.

In each of Exercises 11 through 16, find the volume generated by revolving about the y-axis the region bounded by the given curves. Make a sketch.

11. $y = x^2$, y-axis, $y = 4$

12. $y = x^2$, x-axis, $x = 3$

13. $y = x^2 - 1$, x-axis, $y = 3$

14. $xy = 1$, $y = 1$, $y = b > 1$

15. $y = \sqrt{1 - x^2}$, x-axis, $y = x$

16. $\sqrt{x} + \sqrt{y} = \sqrt{b}$ $(b > 0)$, x-axis, y-axis

In Exercises 17 through 20, find the volume of the solid of revolution, when the given region of Figure 7.2.16 is revolved about the indicated line. An equation of the curve in the figure is $y = x^2$.

17. OAP about the x-axis.

18. OBP about $y = 4$

19. OBP about $x = 2$

20. OAP about $x = 2$

21. Find the volume generated by revolving about the x-axis the region bounded by the upper half of the ellipse

$$\frac{x^2}{a^2} + \frac{y^2}{b^2} = 1 \qquad a > b$$

and the x-axis, and therefore find the volume of a prolate spheroid.

22. Find the volume of a frustum of a right circular cone by the methods of this section. For uniformity of notation let r and R be the radii of the upper and lower bases respectively, and let h be the distance between the bases.

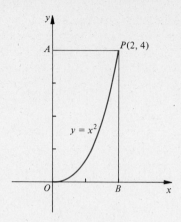

Figure 7.2.16

23. A sphere of radius r is cut by a plane so that a segment of a sphere of height h is formed. Prove that the volume of the segment is $\dfrac{\pi h^2}{3}(3r - h)$ cubic units.

24. A hole of radius $\dfrac{r}{2}$ is bored through the center of a sphere of radius r. Find the remaining volume.

7.3 VOLUME OF A SOLID OF REVOLUTION BY THE SHELL METHOD

Last section was devoted to the determination of the volume of a solid of revolution by taking the rectangular elements of area perpendicular to the axis of revolution. Upon rotating, the element becomes either a circular disk or ring. Unfortunately, there are solids of revolution whose volume is very difficult or impossible to compute by this method. In Figure 7.3.1, we have a region that,

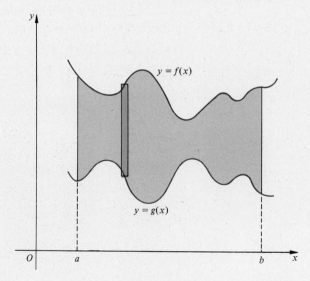

Figure 7.3.1

when rotated about the x-axis, gives a solid whose volume can be easily computed by the ring method. However, what would happen if we wished to rotate this region about the y-axis? It would be almost impossible to work with rectangular elements perpendicular to the y-axis. We are thus led to ask whether we could use the rectangular elements parallel to the y-axis (that is, perpendicular to the x-axis) to compute the volume for this case. This leads to what is called the **shell method** for finding volumes.

If we choose elements of area parallel to the axis of revolution, a cylindrical shell is obtained. This cylindrical shell can be pictured as a tin can with no top and no bottom. To get insight into this shell technique for computing volume consider first a rectangle of dimensions $2\pi r$ and h. This rectangle can be rolled to form a cylindrical shell of base radius r and height h (see Figure 7.3.2). Therefore we say that such a cylindrical shell has a lateral surface area given by

$$S = 2\pi rh \tag{1}$$

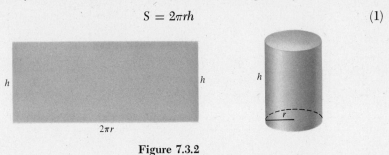

Figure 7.3.2

Next consider a solid cylinder with base radius b and height h. Its volume is the area of the base times its height or $\pi b^2 h$. If we remove a cylindrical core of radius a from this cylinder, the resulting solid (Figure 7.3.3) has volume

$$V = \pi b^2 h - \pi a^2 h \tag{2}$$

Now (2) may be rewritten in the form

$$V = \frac{2\pi bh + 2\pi ah}{2}(b - a) \tag{3}$$

The quantity $2\pi bh$ is the lateral area of the outer surface, and $2\pi ah$ is the lateral area of the inner surface. Therefore it follows that

$$V = \text{average lateral area} \times \text{wall thickness} \tag{3a}$$

This is the underlying idea upon which the method of shells is based.

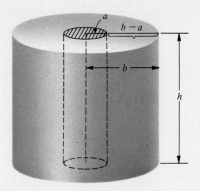

Figure 7.3.3

Let R be the region bounded by the curve $y = f(x)$, the x-axis, and the vertical lines $x = a$ and $x = b$, where $0 \le a < b$. Furthermore, it is assumed that $f(x)$ is continuous and nonnegative on $[a, b]$ (see Figure 7.3.4). Suppose that the region R is revolved about the y-axis so that a solid of revolution S is generated (Figure 7.3.5). It is our objective to find the volume V of S by using rectangular elements parallel to the y-axis.

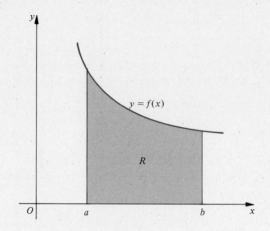

Figure 7.3.4

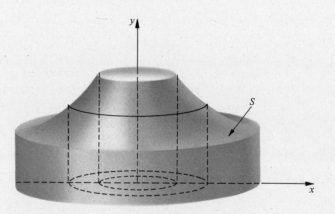

Figure 7.3.5

Let P be a partition of $[a, b]$ into n subintervals where

$$a = x_0 < x_1 < x_2 < \cdots < x_n = b$$

Let w_i be the midpoint of the ith subinterval $[x_{i-1}, x_i]$ so that $w_i = \frac{1}{2}(x_{i-1} + x_i)$. If the rectangle of height $f(w_i)$ and width $\Delta x_i = x_i - x_{i-1}$ is revolved about the y-axis, a cylindrical shell is generated. The average lateral area of the shell is

$$\frac{2\pi x_i f(w_i) + 2\pi x_{i-1} f(w_i)}{2} = 2\pi \frac{x_i + x_{i-1}}{2} f(w_i) = 2\pi w_i f(w_i)$$

and the thickness of the wall is Δx_i. If ΔV_i denotes the volume of the shell, application of (3a) yields

$$\Delta V_i = 2\pi w_i f(w_i) \Delta x_i$$

Thus for our partition, the sum of the volumes of the n cylindrical shells is the Riemann sum

$$\sum_{i=1}^{n} \Delta V_i = \sum_{i=1}^{n} 2\pi w_i f(w_i) \Delta x_i$$

It then follows that the volume V of S must be

$$V = \lim_{|P| \to 0} \sum_{i=1}^{n} 2\pi w_i f(w_i) \Delta x_i$$

Since $f(x)$ is continuous on $[a, b]$, so also is $2\pi x f(x)$ continuous there. Thus the definite integral $2\pi \int_{a}^{b} x f(x) \, dx$ exists. Our discussion leads us to the following.

Let $f(x)$ be continuous and nonnegative on $[a, b]$ where $0 \le a < b$. Let the region bounded by $y = f(x)$, the x-axis, and the vertical lines $x = a$ and $x = b$ be revolved about the y-axis. The *volume of the solid of revolution* thus generated is given by

$$V = 2\pi \int_{a}^{b} x f(x) \, dx \tag{4}$$

We cannot call (4) a definition because the volume of this type of solid of revolution was already defined in Section 7.2, formula (7). It can be shown that the two representations of volume are equivalent. Later in this book, we shall state other formulas for volume (using multiple integrals). In some advanced courses, it is proved that these different formulations of volume are consistent (that is, if two methods can be used to find the volume of a given solid, both methods will give the same answer).

An analogous formula may be written for the volume V of a solid of revolution obtained by rotating the region bounded by $x = g(y)$, the y-axis and the horizontal lines $y = c$ and $y = d$. Again we assume that $g(y)$ is continuous and nonnegative in $[c, d]$, where $0 \le c < d$. The volume V is given by

$$V = 2\pi \cdot \int_{c}^{d} y \, g(y) \, dy \tag{5}$$

We shall now treat some illustrative examples. It is hoped, as we progress, the reader will recognize that the formulas, such as (4) and (5), should not be used without careful interpretation. If the shell method is used, it is necessary to draw an element parallel to the axis of revolution. Then, if a formula such as (4) is to be applied, the "x" in the "formula" is the distance from the axis of rotation, $f(x)$ is the height of the element, and dx is the symbolic representation of the thickness of the element. A similar interpretation applies to (5). We sometimes find the following more suggestive notation convenient:

$$V = 2\pi \int_{a}^{b} rh \, dr \tag{6}$$

where h denotes the height or length of an element and r is the distance of the element from the axis of rotation (that is, r is the radius of the rotated cylindrical shell).

EXAMPLE 1   Find the volume of a sphere of radius a by the shell method.

Solution   Refer to Figure 7.3.6. The volume of a hemisphere is given by

$$V_1 = \int_0^a 2\pi x \sqrt{a^2 - x^2}\, dx$$

$$= 2\pi \int_0^a (a^2 - x^2)^{1/2} x\, dx = -\pi \int_0^a (a^2 - x^2)^{1/2}(-2x\, dx)$$

and by the generalized power formula,

$$V_1 = \frac{-\pi 2}{3} (a^2 - x^2)^{3/2} \Big|_0^a$$

$$= \frac{-2\pi}{3} [0 - a^3]$$

$$= \frac{2\pi a^3}{3}$$

Therefore the required volume V is

$$V = 2V_1 = \frac{4\pi a^3}{3} \text{ cubic units}$$

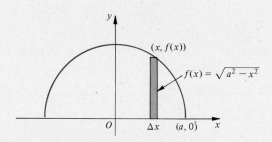

Figure 7.3.6

EXAMPLE 2   Find the volume of a right circular cone of base radius r and height h.

Solution   Application of the equation of a straight line passing through two given points yields (see Figure 7.3.7).

$$f(x) = h - \frac{h}{r} x \qquad 0 \le x \le r$$

The region bounded by $f(x)$, the y-axis, and the x-axis is to be rotated about the y-axis. Therefore the volume V by the shell method is

$$V = \int_0^r 2\pi x \left(h - \frac{h}{r} x \right) dx$$

$$= 2\pi \left(h\frac{x^2}{2} - \frac{h}{r}\frac{x^3}{3} \right)\Bigg|_0^r$$

$$= 2\pi \left(h\frac{r^2}{2} - \frac{h}{3r}r^3 - 0 \right)$$

$$= \frac{2\pi h r^2}{6}$$

$$= \frac{\pi r^2 h}{3} \text{ cubic units}$$

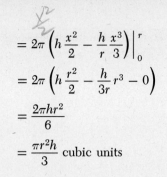

Figure 7.3.7

EXAMPLE 3 The region bounded by the parabola $y = 3x - x^2$ and the line $y = 2$ is rotated about the y-axis. Find the volume of the resulting solid.

Solution The line and the parabola intersect at $(1, 2)$ and $(2, 2)$. We shall take the elements parallel to the y-axis (see Figure 7.3.8) and use the shell method.

The length of a rectangular element is $(3x - x^2) - 2$ and the distance from the y-axis is x [that is, in (6), $r = x$ and $h = 3x - x^2 - 2$]. Thus

$$V = 2\pi \int_1^2 x(3x - x^2 - 2)\,dx = 2\pi \int_1^2 (3x^2 - x^3 - 2x)\,dx$$

$$= 2\pi \left[x^3 - \frac{x^4}{4} - x^2 \right]_1^2 = 2\pi(8 - 4 - 4) - 2\pi\left(1 - \frac{1}{4} - 1\right) = \frac{\pi}{2} \text{ cubic units}$$

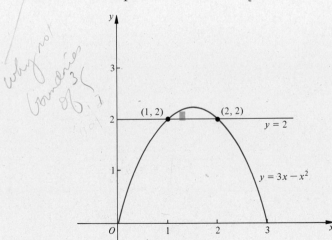

Figure 7.3.8

Although this example can also be done by the ring method, the shell method is much easier. ●

EXAMPLE 4   The region R bounded by the line $x + y = 1$, the x-axis, and the y-axis is revolved about the line $y = -1$. Find the volume of the solid generated by (a) the shell method of this section and (b) by the ring method of Section 7.2.

Solution   (a) The region R is shown in Figure 7.3.9. The equation of the line is $x = 1 - y$. The distance from an element (parallel to the x-axis) to the line $y = -1$ is $y + 1$. Then in (6) we have $r = y + 1$, $dr = dy$, and $h = x = 1 - y$. By the shell method,

$$V = 2\pi \int_0^1 (1 + y)(1 - y)\, dy$$

$$= 2\pi \int_0^1 (1 - y^2)\, dy$$

$$= 2\pi \left(y - \frac{y^3}{3}\right)\Big|_0^1$$

$$= \frac{4}{3}\pi \text{ cubic units}$$

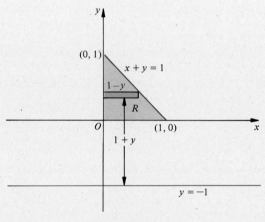

Figure 7.3.9

(b) Next we use the ring method of the previous section (see Figure 7.3.10)

$$V = \pi \int_0^1 (r_2^2 - r_1^2)\, dx$$

In this case,

$$r_2 = 1 + y = 1 + f(x)$$
$$= 1 + (1 - x)$$
$$= 2 - x$$

and $r_1 = 1$ so that

$$V = \pi \int_0^1 [(2 - x)^2 - 1]\, dx$$

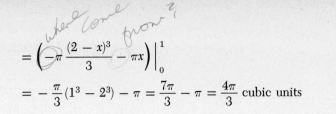

$$= \left(-\pi \frac{(2-x)^3}{3} - \pi x \right) \Bigg|_0^1$$

$$= -\frac{\pi}{3}(1^3 - 2^3) - \pi = \frac{7\pi}{3} - \pi = \frac{4\pi}{3} \text{ cubic units} \qquad \bullet$$

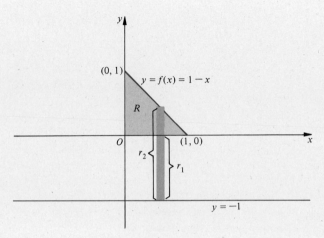

Figure 7.3.10

Exercises 7.3

In Exercises 1 through 8, find by the shell method the volume generated by rotating the area bounded by the given curves about the y-axis. Make a sketch.

1. $y = x^2$, $y = 0$, $x = 3$
3. $y = 4x - x^2$, $y = 0$
5. $y = x^3$, $y = x$, first quadrant
7. $y = \sqrt{x}$, $x = \sqrt{y}$

2. $y = x^2$, $y = 9$, $x = 0$
4. $y = x^2 - 4x + 2$, $y = x - 2$
6. $y = x^2$, $x = a$ where $a > 0$
8. $xy = 10$, $x = a$, $x = b$, where $b > a > 0$

In Exercises 9 through 16, find by the shell method the volumes generated by revolving the area bounded by the given curves about the x-axis. Make a sketch.

9. $y = x$, $y = 2$, $x = 0$
11. $x + y - 1 = 0$, $2y - x - 2 = 0$, x-axis
13. $x = y^2 - 6y + 8$, y-axis
15. $x^2 + y^2 = 9$, $8x = y^2$

10. $x = y^2 - 2y$, $x = 0$
12. $b^2x^2 + a^2y^2 = a^2b^2$, first quadrant
14. $y^2 - x^2 - a^2 = 0$, $y = 2a$, $(a > 0)$
16. $y = x^3$, x-axis, $x = 1$

In Exercises 17 through 20 use the method of shells to find the indicated volume.

17. Find the volume of the solid of revolution obtained by rotating about the x-axis the region bounded by $y^2 = 4px$, $p > 0$, $x = p$ and the x-axis.
18. Find the volume of a frustum of a right circular cone. Let r and R be the radii of the upper and lower bases respectively, and let h be the altitude of the frustum.
19. A segment of height h is cut by a plane from a sphere of radius r. Find the volume of the segment.
20. A hole of radius $\frac{r}{2}$ is bored through the center of a sphere of radius r. Find the remaining volume.

In Exercises 21 through 24 find by the shell method the volume generated by rotating the area bounded by the given curves about the indicated axis. Make a sketch.

21. $x = 2y^2$, $x = 2$; about $y = -2$

22. $y = x^2$, $y = x^4$, $x \geq 0$; about $y = 1$

23. $y = \sqrt{x}$, $32y = x^3$; about $x = 4$

24. $y^2 = x$, $y = x^2$; about $y = -2$

7.4 VOLUME OF SOLID WITH KNOWN CROSS-SECTIONAL AREA

In many cases we are interested in finding the volumes of solids that are *not* solids of revolution. These can be found by using double or triple integrals (see Chapter 18). However, certain important special cases can be evaluated by using the single integrals we have studied, provided that parallel cross sections all have the same simple shape (such as all are squares, isosceles triangles, and so forth.)

Consider then the problem of defining the volume of a solid S. A plane intersecting S cuts S is a plane region called a cross section of S. Furthermore, it is assumed that the areas of all cross sections of S perpendicular to some fixed line are known and change continuously. This implies that there exists a coordinate line L such that the solid S lies between the planes drawn perpendicular to L at some numbers $x = a$ and $x = b$ and that the cross section of S in the plane perpendicular to L at each number of x in $[a, b]$ has a known area $A(x)$ for which $A(x)$ is continuous in $a \leq x \leq b$ (Figure 7.4.1).

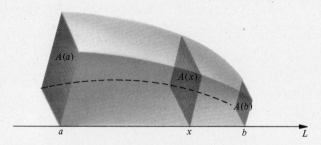

Figure 7.4.1

To write an expression for the volume V of the solid S, we form the partition P of $[a, b]$ into n intervals $[x_0, x_1]$, $[x_1, x_2]$, ..., $[x_{n-1}, x_n]$ with norm $|P|$. Select arbitrary numbers w_1, w_2, ..., w_n in $[x_0, x_1]$, $[x_1, x_2]$, ..., $[x_{n-1}, x_n]$, respectively. Construct n cylinders with cross-sectional areas $A(w_1)$, $A(w_2)$, ..., $A(w_n)$ and heights $\Delta x_1 = x_1 - x_0$, $\Delta x_2 = x_2 - x_1$, ..., $\Delta x_n = x_n - x_{n-1}$. Therefore the Riemann sum

$$V_n = \sum_{i=1}^{n} A(w_i)\, \Delta x_i$$

is the sum of the volume of n cylinders and approximates the volume of S. Now the smaller we take the norm $|P|$ of the partition P, the larger will be n and the closer this approximation will be to the number we wish to assign to the required volume V. Thus we define the volume V of S as follows.

Definition   Let S be a solid such that S lies between planes drawn perpendicular to the x-axis at a and b. For each x in $[a, b]$ let $A(x)$ be the area of the cross section of S drawn perpendicular to the x-axis at x. Furthermore it is assumed that $A(x)$ is a continuous function of x in $[a, b]$. Then the **volume** V of S is defined by

$$V = \int_a^b A(x)\, dx \qquad\qquad (1)$$

If the solid S is a solid of revolution obtained by rotating the region under the curve $y = f(x)$, $a \le x \le b$ (where $f(x) \ge 0$ on $[a, b]$) about the x-axis, then the cross sections are circles. Thus, $A(x) = \pi[f(x)]^2 = \pi[f(x)]^2$. Formula (1) then becomes $V = \pi \int_a^b [f(x)]^2\, dx$. This agrees with the definition of volume of a solid of revolution given in Section 7.2, formula (7) and thus, the definitions are consistent.

EXAMPLE 1   Find the volume of a right circular cone with base radius r and height h.

Solution   The cone is shown in Figure 7.4.2 with the axis coincident with the x-axis and with the vertex at the origin. The cross sections are circles and at any section x the radius $f(x)$ is found from similar triangles to be given by

$$\frac{f(x)}{x} = \frac{r}{h} \qquad \text{or} \qquad f(x) = \frac{r}{h}x$$

Then the area $A(x)$ is determined from

$$A(x) = \pi[f(x)]^2 = \pi \frac{r^2}{h^2}x^2$$

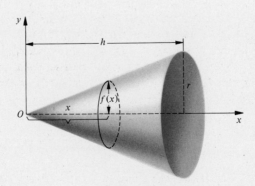

Figure 7.4.2

Therefore the volume V is found from

$$V = \int_0^h A(x)\,dx$$

$$= \frac{\pi r^2}{h^2} \int_0^h x^2\,dx$$

$$= \frac{\pi r^2}{h^2} \frac{h^3}{3}$$

$$= \frac{\pi r^2 h}{3} \text{ cubic units} \qquad \bullet$$

EXAMPLE 2   Find the volume V of a pyramid of height h with a square base of length a on each side.

Solution   The pyramid is shown in Figure 7.4.3(a). The axis is chosen to be vertical with origin at the apex. The x-axis cuts the center of the base at $x = h$.

We seek an expression for the cross sectional area $A(x)$. Now if a longitudinal cut is made containing BC as well as the x-axis, we have a triangular section as shown in Figure 7.4.3(b). Since corresponding parts of similar triangles are proportional, then

$$\frac{y}{x} = \frac{a}{h}$$

from which

$$A(x) = y^2 = \frac{a^2 x^2}{h^2}$$

Therefore

$$V = \int_0^h A(x)\,dx = \frac{a^2}{h^2} \int_0^h x^2\,dx$$

$$= \frac{a^2}{h^2} \frac{h^3}{3} = \frac{a^2 h}{3} \text{ cubic units}$$

This is the known result from solid geometry that the volume of a pyramid (not necessarily with square base) is equal to one third the product of the height by the area of the base. $\qquad \bullet$

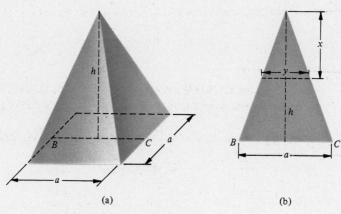

(a)   (b)

Figure 7.4.3

EXAMPLE 3 A solid has a circular base of radius a units. Find the volume of the solid if every plane section perpendicular to a fixed diameter is an equilateral triangle.

Solution Choose the coordinate system as shown in Figure 7.4.4 with the x-axis as the fixed diameter. The equation of the circle is $x^2 + y^2 = a^2$. As given, the cross section PQR of the solid is an equilateral triangle of side $2y$ and the corresponding area.

$$A(x) = \sqrt{3}y^2 = \sqrt{3}(a^2 - x^2)$$

Thus the volume V of the solid is

$$V = \int_{-a}^{a} \sqrt{3}(a^2 - x^2)\, dx$$

$$= \sqrt{3}\left(a^2 x - \frac{x^3}{3}\right)\Big|_{-a}^{a}$$

$$= \sqrt{3}\left[\left(a^3 - \frac{a^3}{3}\right) - \left(-a^3 + \frac{a^3}{3}\right)\right]$$

$$= \frac{4}{3}\sqrt{3}a^3 \text{ cubic units}$$

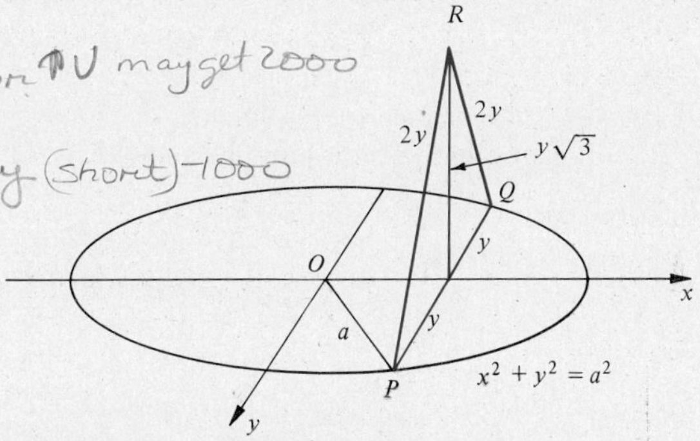

Figure 7.4.4

EXAMPLE 4 A *prismatoid* is defined as a solid for which the area of a cross section parallel to and at a distance u from a fixed plane can be expressed as a quadratic polynomial

$$A(u) = au^2 + bu + c$$

Prove that the volume V of a prismatoid is given by

$$V = (B_1 + B_2 + 4M)\frac{h}{6} \tag{2}$$

where B_1 and B_2 are the areas of the bases, M is the area of a cross section parallel to the bases and midway between them, and h is the distance between the bases.

Solution For convenience take $u = 0$ as the position of one base. Thus

$$V = \int_0^h A(u)\, du$$

$$= \int_0^h (au^2 + bu + c)\, du$$

$$= \frac{au^3}{3} + \frac{bu^2}{2} + cu \Big|_0^h$$

$$= \frac{ah^3}{3} + \frac{bh^2}{2} + ch - 0$$

$$= \frac{h}{6}(2ah^2 + 3bh + 6c) \text{ cubic units}$$

We write

$$V = \frac{h}{6}[c + (ah^2 + bh + c) + ah^2 + 2bh + 4c]$$

But $B_1 = A(0) = c$

$$B_2 = A(h) = ah^2 + bh + c$$

and

$$4M = 4A\left(\frac{h}{2}\right) = 4\left[a\frac{h^2}{4} + b\frac{h}{2} + c\right]$$

$$= ah^2 + 2bh + 4c$$

Thus

$$V = \frac{h}{6}[B_1 + B_2 + 4M]$$

and the verification is complete.

For example, the right circular cone of Example 1 is a special prismatoid with upper base area $B_1 = 0$, lower base area $B_2 = \pi r^2$ and middle section $M = \pi\left(\frac{r}{2}\right)^2$ so that $4M = \pi r^2$. Also h is the height of the cone. Thus

$$V = \frac{h}{6}[0 + \pi r^2 + \pi r^2]$$

or

$$V = \frac{\pi r^2 h}{3} \text{ cubic units} \qquad \bullet$$

Exercises 7.4 (Cont)

1. A solid S has a flat base which is the plane region bounded by the parabola $y = x^2$ and the line $y = 4$. Each cross section perpendicular to the y-axis is a square with one edge lying in the base. Find the volume of S.

2. A solid S is constructed so that its cross section is an isosceles triangle with constant height 6. The base of the triangle is a chord of the parabola $y^2 = 4x$ perpendicular to the x-axis for $0 \le x \le 4$. Find the volume of S.

3. A solid has a circular base of radius a units. Find the volume of the solid if every plane section perpendicular to a fixed diameter is a semicircle.

4. The base of a solid is the segment of a parabola $y^2 = 4px$ cut off by the latus rectum. A section of the solid is an isosceles right triangle with the hypotenuse in the plane of the base. Find the volume of the solid.

5. Find the volume of a wedge cut from a right circular cylinder of radius b by a plane that passes through a diameter of the base and makes an angle of $45°$ with the plane of the base.

6. A monument is 80 ft tall. A horizontal cross section x ft above the base is an equilateral triangle whose sides are $s = \frac{80 - x}{4}$ ft. Find the volume of the monument.

7. A tetrahedron is a solid with four vertices and four flat triangular faces. Let T be a tetrahedron that has three mutually perpendicular edges of length 5, 8, and 12 meeting at a vertex. Sketch T and find its volume.

8. Find the volume of a tetrahedron T having three mutually perpendicular edges of lengths a, b, and c meeting at a vertex. Sketch T.

9. Show that the following solids are prismatoids and use formula (2) of Example 4 to determine their volumes
   (a) a sphere of radius r.
   (b) the frustum of a right circular cone with base radii r and R and altitude h.

10. Show that the formula for volume in Example 4 also holds for more general solids called prismatoids whose cross-sectional area is given by

$$A(u) = eu^3 + au^2 + bu + c$$

11. Find the volume common to two right circular cylinders each having base radius r, if the axes of the cylinders intersect at right angles.

7.5

WORK

Suppose that a *constant* force F is applied to an object and moves it a distance s in the direction of the force. Then we say that the work done by such a force is given by

$$W = F \cdot s \tag{1}$$

That is, in its simplest form, work = force $\times$ distance. If the unit of force is pounds and the distance is measured in feet, the work done has units foot-pounds, written ft-lb. Thus if a constant force of 700 lb is applied to pushing an automobile 80 ft (with constant velocity) then the work done by the pushing force is 56000 ft-lb. Other possible units of work are inch-pounds, foot-tons, and so forth.

In lifting a body vertically with uniform velocity through a height h, the force F exerted upward must be equal to the weight of the body and the work done is

$$W = Fh \tag{2}$$

(Note that if the lifting force exceeds the weight, then, in accordance with Newton's second law of motion, the body must accelerate in the upward direction.)

Work problems are not all so simple. For example, consider the work done in stretching a spring. At first, relatively little force is needed to stretch the spring. However, as the spring is stretched it requires larger and larger forces to continue to stretch it. Thus, in order to compute the work done by this variable force we must know how the force varies with the stretching of the spring. The definite integral may then be used to determine the work done by the force as described in the following paragraphs.

Suppose that $F(x)$, where F is continuous on $[a, b]$, is the number of units of force acting in the direction of motion on an object as it moves along the x-axis from $x = a$ to $x = b$ where $a < b$. We select a partition

$$P = \{x_0, x_1, x_2, \ldots, x_n\}$$

of $[a, b]$. In each subinterval $[x_{i-1}, x_i]$ of P, let $F(u_i)$ be the minimum and $F(v_i)$ be the maximum value of the force F. Thus if W_i is the actual work done on the object as it moves from x_{i-1} to x_i, it is reasonable to assert that

$$F(u_i)\,\Delta x_i \le W_i \le F(v_i)\,\Delta x_i$$

Then since $F(x)$ is continuous on $[a, b]$ and in particular on $[x_{i-1}, x_i]$, by the intermediate value theorem there exists a ξ_i in this interval such that

$$W_i = F(\xi_i)\,\Delta x_i \qquad x_{i-1} \le \xi_i \le x_i$$

Now the total amount W of work done in moving the object from a to b is

$$W = \sum_{i=1}^{n} W_i$$

that is

$$W = \sum_{i=1}^{n} F(\xi_i)\,\Delta x_i \tag{3}$$

Thus W is a Riemann sum of the function F on $[a, b]$ so that, if we let the norm of the partition P tend to zero, we have

$$W = \int_a^b F(x)\,dx \tag{4}$$

This leads us to the formal definition.

Definition Let F be a continuous function on the closed interval $[a, b]$ and $F(x)$ be the number of units of force acting on an object at the point x on the x-axis. Then the **work** W done by the force F in moving the object from $x = a$ to $x = b$ is given by

$$W = \int_a^b F(x)\,dx$$

EXAMPLE 1 A spring can be stretched 4 in. by a force of 60 lb. How much work is done in stretching the spring 1 ft? Assume the validity of Hooke's law (see Figure 7.5.1).

Solution *Hooke's law* says that the force $F(x)$ required to extend a spring (within its elastic limit) x units beyond its natural length is given by

$$F(x) = kx$$

where k is a constant (independent of x), which is called the **spring constant.** The magnitude of k depends upon the material and the geometry of the system (that is, spring and support(s)). We are given that $F(4) = 60$, so that

$$60 = k \cdot 4 \qquad \text{or} \qquad k = 15\,\text{lb/in.}$$

The amount W of work done in stretching the spring 12 in. (that is, x varies from $x = 0$ to $x = 12$) is given by

$$W = \int_0^{12} 15x\,dx = \left.\frac{15x^2}{2}\right|_0^{12}$$

$$= 1080\ \text{in.-lb}$$

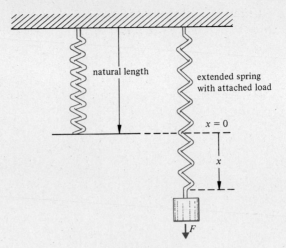

natural length

extended spring
with attached load

$x = 0$

x

$\downarrow F$

Figure 7.5.1

The reader is urged to do the same problem using feet as the unit of length (*Ans.*: $W = 90$ ft-lb).

EXAMPLE 2   A chain 60 ft long and weighing 5 lb/ft is hanging vertically. How much work is done in raising the chain to a horizontal position level with the top of the chain?

Solution   The force needed to move the chain (with constant velocity) is equal to the weight of the chain hanging down. Initially the force must be $(60)(5) = 300$ lb, and if x is the number of feet raised to the level of the top of the chain then the force $F(x)$ is given by

$$F(x) = (60 - x)5 = 300 - 5x \qquad 0 \le x \le 60$$

This means that the force function varies linearly from 300 lb down to zero as x increases from $x = 0$ to $x = 60$. Thus

$$W = \int_0^{60} (300 - 5x)\, dx$$

$$= \left(300x - \frac{5x^2}{2}\right)\Bigg|_0^{60}$$

$$= 18000 - 9000$$

$$= 9000 \text{ ft-lb}$$

EXAMPLE 3   A cylindrical tank 8 ft in diameter and 10 ft high is full of water, which weighs 62.5 lb/ft³. How much work is required to pump the water out over the top?

Solution   Refer to Figure 7.5.2. It is observed that different layers must be raised different distances to the top. Let x denote distance to the top of the tank and partition the interval $[0, 10]$ into n subintervals. If Δx_i is the thickness of the ith layer, then the weight of that ith layer is

$$F = (62.5)\pi 4^2\, \Delta x_i$$
$$= 1000\pi\, \Delta x_i$$

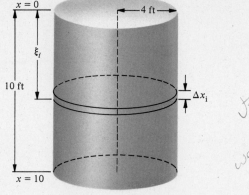

$J = \pi r^2 h$

$weight = d \cdot v$

$d = \frac{m}{v}$

Figure 7.5.2

(We have used the fact that weight = density × volume. The density of water is 62.5 lb/ft³ and the volume is $\pi r^2 h$ where $r = 4$ and $h = \Delta x_i$.)

The work done W_i in lifting this layer to the top of the cylinder is (for some ξ_i in $[x_{i-1}, x_i]$)

$$W_i = 1000\pi \, \Delta x_i \, \xi_i \qquad (x_{i-1} \le \xi_i \le x_i)$$

If we let the norm of the partitions tend to zero there results

$$W = 1000\pi \int_0^{10} x \, dx$$

$$= 1000\pi \left. \frac{x^2}{2} \right|_0^{10}$$

$$= 50{,}000\pi \ \text{ft-lb}$$

$$\doteq 157{,}080 \ \text{ft-lb}$$

Alternative Solution Figure 7.5.3 shows a piston that is pushing the water out of the tank. The force $F(x)$ on the piston after it has moved a distance of x ft is the weight of the water

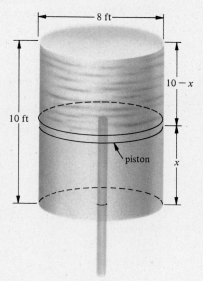

Figure 7.5.3

Applications of the Definite Integral / Ch. 7

remaining in the tank given by

$$F(x) = \pi(4^2)(62.5)(10 - x)$$

so that the work done is

$$W = 1000\pi \int_0^{10} (10 - x)\, dx$$
$$= 1000\pi \left(10x - \frac{x^2}{2} \right) \Big|_0^{10}$$
$$= 1000\pi \, (100 - 50)$$
$$= 50,000\pi$$
$$\doteq 157,080 \text{ ft-lb} \qquad\qquad \bullet$$

EXAMPLE 4   According to Newton's law of gravitation, two bodies of mass M and m are attracted to each other by a force of magnitude $\dfrac{GMm}{r^2}$, where r is the distance between the masses and G is a universal constant. If the Earth has mass M, find the work done in projecting a missile of mass m 2000 mi from the surface of the Earth.

Solution   It can be shown that, if the Earth's shape is spherical, the Earth attracts a particle as if its entire mass were concentrated at the center so that the distance between the missile and the Earth is the distance to the center of the Earth. Assuming this spherical shape, let the center of the Earth be the origin and let x be the radial coordinate measured positively outward as shown in Figure 7.5.4. By our sign convention, the gravitational force acting on the missile at any position x is given by

$$F(x) = -\frac{GMm}{x^2}$$

since it is actually directed toward the center of the Earth. The radius of the Earth $a = 4000$ mi, and the final position is $b = 4000 + 2000 = 6000$ mi. Thus the work done to project the missile from a to b (with uniform velocity) is given by

$$W = -\int_a^b F(x)\, dx$$

since the thrust force must be equal and opposite to the attracting force $F(x)$.

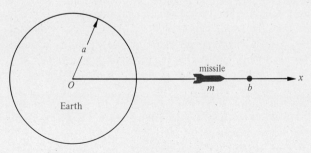

Figure 7.5.4

Therefore

$$W = GMm \int_a^b \frac{dx}{x^2}$$

$$= GMm \left(-\frac{1}{x} \right) \Big|_a^b$$

$$= GMm \left(\frac{1}{a} - \frac{1}{b} \right)$$

$$= GMm \left(\frac{1}{4000} - \frac{1}{6000} \right)$$

$$= \frac{GMm}{12,000} \text{ mi-lb}$$

How much work is required to project the missile from the surface of the Earth to points successively farther and farther away? In other words, how much work is necessary to send the missile to infinity? This is easily answered as follows:

$$\lim_{b \to \infty} W = \lim_{b \to \infty} GMm \left(\frac{1}{a} - \frac{1}{b} \right)$$

$$= \frac{GMm}{a} \text{ mi-lb}$$

This, then, is the work necessary to send the missile completely out of the Earth's gravitational field. In our analysis, we used the so-called two-body model, thereby assuming that the effect of the other bodies in the universe is negligible. ●

Exercises 7.5

1. A spiral spring stretches 1.8 in. when a force of 9 lb is applied. The natural length of the spring is 14 in. Find the work done in stretching the spring from 16 in. to 20 in.

2. Is the work necessary to stretch the spring in Exercise 1 from 16 to 20 in. the same as required to stretch it from 20 to 24 in.? If not, what is the required work done in the latter case?

3. A spring has a natural length of 10 in. A 6000-lb force compresses the spring to $8\frac{1}{2}$ in. Find the work done in compressing it from 10 in. to 8 in. It is found experimentally that Hooke's law holds for compression as well as for extension.

4. A bucket of water weighing 70 lb is lifted vertically at a uniform rate of 2 ft/sec. The weight of the bucket itself is negligible. Find the work done in lifting the bucket 80 ft.

5. A block weighing 50 lb is to be lifted from the ground level to the top of a building, which is 30 ft high, by a cable attached to a windlass on top of the building. If the weight of the cable is 4 lb/ft, find the work done.

6. A rectangular tank 8 ft high and with a rectangular base 4 ft by 5 ft is full of water (weight density 62.5 lb/ft³). Find the work done in pumping the water over the top of the tank.

7. Two electrons repel each other with a force inversely proportional to the square of the distance between them. If one electron is held fixed at the origin of the coordinate line, find the work done by the fixed electron in moving the other electron from $x = a$ to $x = b$, where a and b are arbitrary positive numbers.

8. Any two electrons repel each other with a force inversely proportional to the square of the distance between them. If two electrons are held stationary at the points $(\pm 8, 0)$ on the x-axis, find the work done in moving a third electron:
   (a) from $(0, 0)$ to $(6, 0)$ along the x-axis
   (b) from $(-6, 0)$ to $(6, 0)$ along the x-axis

9. If the weight of a space platform is 15 tons on the surface of the Earth, then how much work must be done against the gravitational force in propelling this platform to a height of 800 mi above the Earth's surface?

10. Calculate the work done in pumping out oil (weight density 60 lb/ft^3) in a full hemispherical reservoir of radius r ft.

11. A right circular conical tank, height 20 ft and radius 12 ft (at the top), is full of water. How much work is done in pumping all the water to the top of the tank?

12. A right circular conical tank, height h ft and radius r ft, is full of a liquid of weight density b lb/ft^3. Show that the work done in pumping all the liquid to the top is given by

$$W = \tfrac{1}{12}b\pi r^2 h^2 \text{ ft-lb}$$

This is a generalization of Exercise 11.

13. Find the work required to pump a fluid filling a hemispherical tank of radius r ft to a height h ft above the top of the tank. Let k lb/ft^3 be the constant weight density of the fluid.

14. A vat a ft long, b ft wide, and h ft deep has the shape shown in Figure 7.5.5. The vat is filled initially to the depth $\tfrac{2}{3}h$ (as shown) with liquid of weight density k lb/ft^3. Determine the work done in pumping all the liquid to the top of the vat.

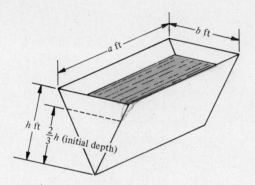

Figure 7.5.5

15. A bucket of sand weighing 120 lb is on the ground attached to one end of a cable of weight 5 lb/ft. The other end of the cable is attached to a windlass 16 ft above the top of the bucket. The cable is wound around the windlass to lift the sand. As the bucket is being hauled up, it is observed that the sand is pouring out of a hole in the bottom at a constant rate. When the bucket reached the top there was only 88 lb of sand left. What amount of work was done in lifting the bucket of sand 16 ft?

16. Air at pressure of 15 lb/in.^{2} is compressed from an initial volume of 200 ft^3 to 50 ft^3. Find the final pressure and the amount of work done if the pressure and volume of a gas are related by

$$pV^{1.4} = C$$

where C is a constant. [*Hint:* Show that the work (by an external source) in compressing a gas in a rigid cylindrical container from V_1 to V_2 is given by

$$W = \int_{V_2}^{V_1} p\, dV = \int_{V_2}^{V_1} f(V)\, dV$$

where $p = f(V)$ is the pressure (in pounds per square foot) as a function of volume. Then use the specific equation between pressure and volume given in Exercise 16.]

17. Find the work done in compressing a gas from volume V_1 to V_2 if the relationship between the pressure and volume is given by the formula

$$pV^\gamma = \text{constant} = C$$

where γ is a constant > 1.

18. A metal bar of length L and cross-sectional area A obeys Hooke's law when stretched. If it is stretched x units, the force required to hold it elongated is

$$F = \frac{EA}{L} x$$

where E is Young's modulus of elasticity and depends only upon the particular metal used. Find the work done in stretching a metal δ units. In particular, find the work done in stretching a steel bar 0.1 in. given the following data: $E = 30 \times 10^6$ lb/in.^{2}, $A = \frac{1}{2}$ in.^{2}, and $L = 60$ in.

19. A cube a in. on an edge and having weight density γ lb/in.^{3} is floating in water whose weight density is w lb/in.^{3} Archimedes' law states that a floating body is buoyed up by a force equal to the weight of the water displaced. How much work is required to force the cube down so that its top surface is in the plane of the surface of the water?

20. Determine the work needed to raise an object of mass m from the surface of the Earth (sphere of radius R) to a height h above the surface. Use the fact that the gravitational force is mg (where g is the acceleration due to gravity) when the mass m is at the surface of the Earth. Show that if $h \ll R$, then $W \doteq mgh$, whereas if $h \gg R$ then $W \doteq mgR$. The latter expression is the work necessary to send the payload to "infinity."

7.6   FLUID PRESSURE

Anybody who has tried to submerge a block of wood under water is aware of the fact that the pressure of the water increases with increase in depth. Consideration of the weight of a column of liquid with a free surface easily yields the fact that the pressure p due to this column is given by

$$p = wh \tag{1}$$

where w is the weight density of the fluid and h is the distance below the surface of the fluid in compatible units. Thus if the units of w are pounds per cubic foot and the distance h is measured in feet, then the units of the pressure p are pounds per square foot. In the derivation of (1) it is assumed that the density of the fluid is constant; that is, the fluid is taken to be incompressible. For water, the weight density w is approximately 62.5 lb/ft.^{3}

What is not at all obvious is the fact that the pressure at a point P does *not* depend on the direction through P. This implies that, if a flat plate is submerged in a liquid, no matter how that plate is orientated in the fluid formula (1) yields the pressure normal (or perpendicular) to the face of the plate at each point of the plate. In general the pressure will vary from point to point as h varies.

In the special case of a horizontally submerged plate of area A the total force F on the plate is given by

$$F = pA = whA \tag{2}$$

However, if the plate is not horizontal, h is variable. In particular, when the submerged plate is *vertical*, the total force on its face is more complicated because the pressure on the plate varies with the depth; that is, it is greater near the bottom of the plate than near the top.

In fluid pressure problems, we usually label the vertical axis so that the downward direction is positive. Often, we choose this vertical axis to be our x-axis. Let f and g be functions that are continuous on a closed interval $[a, b]$ where

$g(x) \le f(x)$ for $a \le x \le b$. Let R be the plane region $\{P(x, y) | g(x) \le y \le f(x), a \le x \le b\}$, bounded by $y = g(x)$, $y = f(x)$, $x = a$, and $x = b$. Assume that the surface of the liquid is along $x = k$, where $k \le a$ (Figure 7.6.1). We wish to define the force of liquid pressure on the vertically submerged region R.

We partition the interval $[a, b]$ on the x-axis into n subintervals $[x_{i-1}, x_i]$ and let ξ_i be an arbitrarily chosen point in $[x_{i-1}, x_i]$; that is, $x_{i-1} \le \xi_i \le x_i$, where $i = 1, 2, \ldots, n$. Form n approximating rectangles R_i ($i = 1, 2, \ldots, n$), a typical one R_i is shown in Figure 7.6.1. The area of the rectangle R_i is $\Delta A_i = [f(\xi_i) - g(\xi_i)] \Delta x_i$.

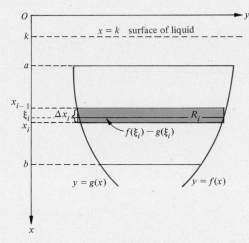

Figure 7.6.1

The pressure on the rectangle R_i at the top is $w(x_{i-1} - k)$ and $w(x_i - k)$ at the bottom. Thus it follows that $w(x_{i-1} - k) \le w(\xi_i - k) \le w(x_i - k)$ since $x_{i-1} \le \xi_i \le x_i$. Therefore $w(\xi_i - k)$ is an approximation to the pressure at any point of R_i and $w(\xi_i - k) \cdot \Delta A_i$ is approximately the total force on R_i. It follows that the Riemann sum

$$\sum_{i=1}^{n} w(\xi_i - k) \Delta A_i = \sum_{i=1}^{n} w(\xi_i - k)[f(\xi_i) - g(\xi_i)] \Delta x_i$$

is an approximation to the total force on the submerged region R. Furthermore the approximation improves as the norm of the partition decreases. Thus we have the following definition.

Definition   The **force of liquid pressure on the vertically submerged region R** described previously is given by

$$F = \lim_{|P| \to 0} \sum_{i=1}^{n} w(\xi_i - k)[f(\xi_i) - g(\xi_i)] \Delta x_i$$

and thus

$$F = w \int_a^b (x - k)[f(x) - g(x)] \, dx \qquad (3)$$

where w is the weight density of the liquid.

It is recommended that the reader set up each problem with an appropriate coordinate system rather than memorize (3). This can be done by using (2) to suggest that fluid pressure be found by

$$F = w \int_a^b h \, dA \qquad (4)$$

where h represents the variable depth and dA (the differential of area) corresponds to the area of a rectangular element of the cross-sectional region.

EXAMPLE 1   A dam contains a vertical gate 10 ft wide and 8 ft high (Figure 7.6.2). The top of the gate is horizontal and is 14 ft below the surface of the water. If water weighs 62.5 lb/ft³, find the total force on the gate.

Solution   Figure 7.6.2 shows the submerged rectangular gate. The element of force on the shaded rectangle is

$$F_i = 62.5\xi_i \, 10 \, \Delta x_i \qquad x_{i-1} \le \xi_i \le x_i$$

here $w = 62.5$, $h = \xi_i$, and ΔA_i = area of the ith rectangular element = $10 \, \Delta x_i$ (so $dA = 10 \, dx$) and the total force

$$F = 625 \int_{14}^{22} x \, dx$$

$$= 625 \left. \frac{x^2}{2} \right|_{14}^{22}$$

$$= \frac{625}{2} (484 - 196)$$

$$= 90,000 \text{ lb} \qquad \bullet$$

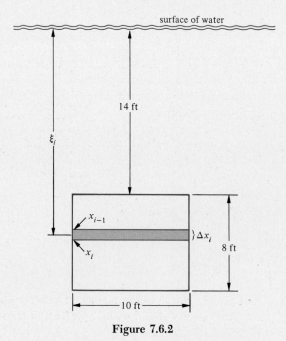

Figure 7.6.2

EXAMPLE 2   An oil tank is in the shape of a right circular cylinder 6 ft in diameter with its axis horizontal. If the tank is half full of oil weighing 50 lb/ft^3, find the total force on one end due to liquid pressure.

Solution   One end of the half full oil tank is shown in Figure 7.6.3. The element of force on the shaded rectangle is

$$\Delta F_i = 50\xi_i\, 2\, f(\xi_i)\, \Delta x_i$$

so that

$$F = 100 \int_0^3 x\, f(x)\, dx$$

But the equation of the circle is $x^2 + y^2 = 9$ so that $f(x) = \sqrt{9 - x^2}$ or

$$F = 100 \int_0^3 x\sqrt{9 - x^2}\, dx$$

$$= \frac{100}{-2} \int_0^3 (9 - x^2)^{1/2}(-2x\, dx)$$

$$= -50 \frac{2}{3}(9 - x^2)^{3/2}\Big|_0^3$$

$$= -\frac{100}{3}(0 - 27)$$

Therefore the total force $F = 900$ lb.

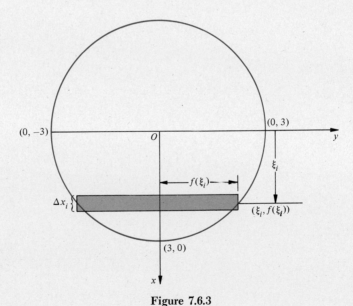

Figure 7.6.3

EXAMPLE 3   Find the total force on the surface of a dam in the shape of an inverted isosceles triangle with altitude 16 ft and base 24 ft, if the surface level is at the 10-ft mark (Figure 7.6.4).

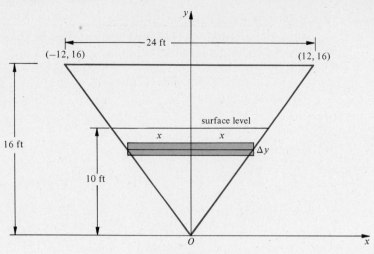

Figure 7.6.4

Solution   The equation of the line connecting the origin with $(16, 24)$ is $y = \dfrac{4x}{3}$ so that

$x = \dfrac{3}{4}y$. The element of area of a typical strip is

$$\Delta A = 2x\,\Delta y = \tfrac{3}{2}y\,\Delta y$$

and the corresponding element of force is

$$\Delta F = (10 - y)(62.5)\,\Delta A$$

or

$$\Delta F = (62.5)(10 - y)\tfrac{3}{2}y\,\Delta y$$

so that the total force F is given by

$$F = 93.75 \int_{0}^{10} (10y - y^2)\,dy$$

$$= 93.75 \left(5y^2 - \frac{y^3}{3}\right)\Bigg|_{0}^{10}$$

$$= 93.75 \left(\frac{500}{3}\right)$$

$$= 15{,}625 \text{ lb} \qquad\qquad \bullet$$

EXAMPLE 4   A swimming pool is 30 ft long and 20 ft wide. The sides are vertical and the bottom is flat but not horizontal. The depth of the water varies linearly from 3 ft on one end to 9 ft at the other. Find the total force of the water on one 30-ft side.

Solution   Figure 7.6.5 shows the rectangular coordinate axes chosen for this problem. The equation of the line connecting points $(30, 3)$ and $(0, 9)$ is $x + 5y = 45$. Also the cross section may be divided by a horizontal line into a rectangle and a right triangle. Thus we have two sets of horizontal strips and two integrals are required—that is $x = 30$ when $0 \le y \le 3$ and $x = 5(9 - y)$ when $3 \le y \le 9$. For $0 \le y \le 3$, $dA = 30\,dy$ and for $3 \le y \le 9$, $dA = 5(9 - y)\,dy$.

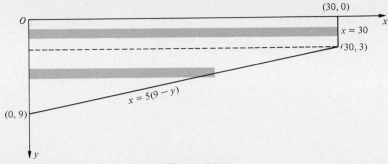

Figure 7.6.5

Therefore

$$F = \int_0^3 62.5y\,30\,dy + \int_3^9 62.5y5(9-y)\,dy$$

$$= 1875 \int_0^3 y\,dy + 312.5 \int_3^9 (9y - y^2)\,dy$$

$$= 1875\,\frac{y^2}{2}\bigg|_0^3 + 312.5\left(\frac{9y^2}{2} - \frac{y^3}{3}\right)\bigg|_3^9$$

$$= 8437.5 + 28{,}125$$

$$= 36{,}562.5\text{ lb} \qquad\qquad\bullet$$

Exercises 7.6

1. A dam contains a vertical gate 8 ft wide and 6 ft high. The top of the gate is horizontal and 12 ft below the surface of the water. Find the total force on the gate.
2. The surface of water is 6 ft from the top of a rectangular dam, which is 50 feet wide and 24 ft deep. Find the total force on the dam.
3. A circular water main 8 ft in diameter is half full of water. Find the total force on the gate that closes the main.
4. Find the total force on the end of a trough in the form of a semicircle of radius a ft when filled with a liquid weighing w lb/ft^3.
5. Find the total force on one end of a tank if the end is an inverted triangle 16 ft wide at the top and 12 ft high and the tank is full of water.
6. A dam is in the form of an inverted triangle b ft wide at the top and with an altitude (maximum depth) of h ft. If the weight density is w lb/ft^3, find the expression for the maximum force that the water behind the dam can exert on the dam.
7. A vertical dam has the form of a segment of a parabola 300 ft wide at the top and 50 ft high at the center. Find the maximum force that the water behind the dam can exert on the dam. (*Hint:* Put the parabola in standard position so that the equation has the form $y = cx^2$ and find c from the given data.)
8. A rectangular plate of dimensions 6 by 8 ft is immersed vertically with the longer side horizontal. Find the total force on one side of the plate if
   (a) the top is in the surface of the liquid.
   (b) the top is 4 ft below the surface of the liquid.
   Take w to be the weight density of the liquid.
9. A rectangular gate in a dam is 10 ft wide and 6 ft high, with the longer side horizontal. If 60,000 lb is the largest force the gate can safely withstand, how high above the top of the gate can the water be allowed to rise?
10. A vertical dam has the form of a parabola a ft wide at the top and h ft deep at the center. Show that the maximum force that the water behind the dam can exert on the dam is $4wah^2/15$ lb where w is the weight of a cubic foot of water.

11. Find the total force on one face of a vertical triangular plate shown in Figure 7.6.6 when the plate is submerged in water.

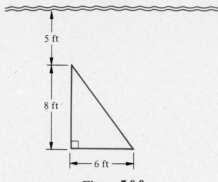

Figure 7.6.6

12. Find a general formula for the total force on one face of a vertical triangular plate submerged in water as shown in Figure 7.6.6 when the horizontal base is b ft, the altitude a ft, and the top is h ft below the surface of the water. Let w be the weight density of water.
13. A vertical gate in a dam is an isosceles trapezoid with the dimensions and position shown in Figure 7.6.7. Find the total force on the gate when the water level is 10 ft above the top of the gate.

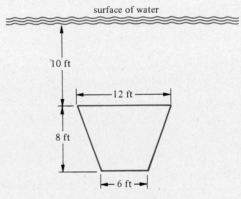

Figure 7.6.7

14. Generalize Exercise 13 by finding the total force on a vertical gate in a dam that is in the shape of an isosceles trapezoid, the upper base is B ft, the lower base is b ft, the altitude is h ft, and the water level is H ft above the top of the gate. Let w be the weight density of water.

7.7   MOMENTS, EQUILIBRIUM, AND MASS CENTER

7.7.1   Point or Lumped Mass Distribution on a Line

The terms "mass" and "weight" are often thought of as interchangeable quantities, but actually they are quite different. Mass is an inherent property of an object or particle and does not change with location. Weight, on the other hand, can vary tremendously. For example, an astronaut's weight on the Moon is approximately one-sixth his weight on the Earth and while in flight he experiences "weightlessness." During all this, the astronaut's mass remains constant.

Mass may be defined by considering a hypothetical experiment in which two particles are assumed to be moving free from all influences except their mutual interaction. Since force is a measure of this interaction, the third law of Newton requires that the two particles will be acted on by two equal and opposite collinear forces. (For every action there is an equal and opposite reaction.) In general, however, the accelerations of the two particles will differ in magnitude; that is, one particle will be more inert (more sluggish or more massive) than the other. Stated precisely, the ratio of the masses is defined as the reciprocal of the ratios of the magnitudes of the acceleration. Therefore, once a standard mass is selected, all other masses may be determined by comparison with it by means of the preceding experiment. In engineering calculations, mass is often measured in slugs. An idea of the magnitude of a slug is obtained as follows.

Consider a particle of constant mass m moving along a straight line and acted upon by a force F in the direction of this motion. In this instance Newton's second law of motion may be expressed as force $F = $ mass $m \times$ acceleration a when appropriate units are employed. In fact a force of 1 lb is defined as the force necessary to give a mass of 1 slug an acceleration of 1 ft/sec². Thus, for example, if a mass of 10 slugs is accelerated 8 ft/sec² then the force must be of magnitude 80 lb. We investigate now an elementary introductory problem.

EXAMPLE 1   Consider two boys of weight 80 lb and 60 lb, respectively, on opposite sides of a fulcrum of a seesaw of length 14 ft (Figure 7.7.1). If the seesaw is regarded as weightless, find the distance of each boy from the fulcrum if the seesaw is in equilibrium. What is the reaction at the fulcrum?

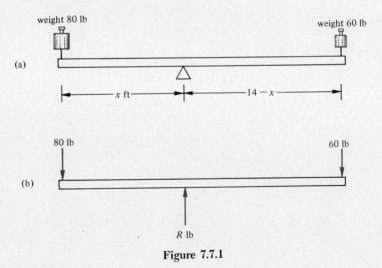

Figure 7.7.1

Solution   Experience has shown us that a light person can balance a heavier person if the latter is closer to the balance point (fulcrum). The rule of physics that gives balance (equilibrium) is that the weight of one person multiplied by his distance from the fulcrum must equal the other person's weight times his distance from the fulcrum.

If x and $14 - x$ are the distances of the 80 lb boy and the 60 lb boy from the fulcrum respectively then

$$80x = 60(14 - x)$$

So $80x = 840 - 60x$, giving $x = 6$ ft and $14 - x = 8$ ft   ●

The weight (force) times the distance to the fulcrum is called a ***moment about the fulcrum.*** In Figure 7.7.1, we see that the 80-lb weight causes a moment of $80x$ about the fulcrum in the counterclockwise direction; the moment of the 60-lb weight about the fulcrum is $60(14 - x)$ in the clockwise direction.

As an alternative solution to Example 1 (and one more convenient for generalization) we assert that for equilibrium the sum of the moments is zero in, say, the clockwise direction. Thus

$$-80x + 60(14 - x) = 0 \qquad \text{or} \qquad x = 6 \text{ ft}$$

The reaction of the fulcrum on the seesaw is R and for equilibrium the algebraic sum of all the applied and reactive forces (in the up or down direction) must be zero. (See 7.7.1(b)—free body diagram.)

$$R = 80 + 60 = 140 \text{ lb}$$

in the direction shown in Figure 7.7.1(b).

Note that we could also have obtained R by taking moments about either end of the seesaw. For example, if we take moments about the left end, then

$$60(14) - R(6) = 0$$

$$R = \frac{840}{6} = 140 \text{ lb}$$

What we have just done for two weights on a seesaw, we can also do for masses in a more general setting as follows. Suppose next that a mass m is located on a line (rod of negligible mass) at a coordinate x. We define its moment about the point having coordinate a to be the product $m(x - a)$ (see Figure 7.7.2). This can be generalized in an apparent manner. Suppose that we have n masses m_i located at points x_i $(i = 1, 2, \ldots, n)$. Then $\sum\limits_{i=1}^{n} m_i(x_i - a)$ is the total moment of all the masses about the point $x = a$. Now, if a fulcrum is placed at $x = a$, the reaction R must be vertical and balance the weights, thereby preventing up or down (that is,

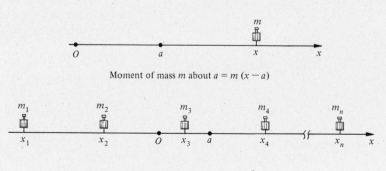

Moment of mass m about $a = m\,(x - a)$

Sum of moments of masses m_i at x_i about $a = \sum\limits_{i=1}^{n} m_i(x_i - a)$

Figure 7.7.2

translational) motion. However, a rotation about the fulcrum will take place in a clockwise direction if the total moment is greater than zero and in a counterclockwise direction if the total moment is less than zero. The system will be in equilibrium if and only if the total moment about a is zero.

EXAMPLE 2   Determine the position of the fulcrum so that the five masses shown in Figure 7.7.3 will be in equilibrium.

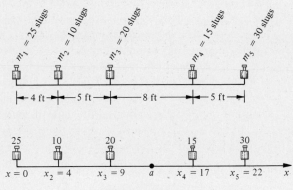

Figure 7.7.3

Solution   Introduce a fulcrum at point $x = a$ and let the origin of coordinates be placed at the position of the first mass m_1 (Figure 7.7.3). For equilibrium, it is necessary that the sum of the moments about a be zero. Therefore

$$25(0 - a) + 10(4 - a) + 20(9 - a) + 15(17 - a) + 30(22 - a) = 0$$

or

$$-25a + 40 - 10a + 180 - 20a + 255 - 15a + 660 - 30a = 0$$

$$100a = 1135$$

$$a = 11.35 \text{ ft to the right of the mass } m_1 \qquad \bullet$$

Note that in Example 2 for five masses m_i located at x_i the position of the center of mass at a is found from

$$\sum_{i=1}^{5} m_i(x_i - a) = 0$$

or

$$a \sum_{i=1}^{5} m_i = \sum_{i=1}^{5} m_i x_i$$

$$a = \frac{\displaystyle\sum_{i=1}^{5} m_i x_i}{\displaystyle\sum_{i=1}^{5} m_i}$$

More generally if there are n masses m_i at points x_i then the center of mass is located from

$$a = \frac{\sum\limits_{i=1}^{n} m_i x_i}{\sum\limits_{i=1}^{n} m_i} \tag{1}$$

In fact (1) is taken as definition of the center of mass for n mass points on the x-axis. Thus we have an alternative and very useful interpretation of the center of mass. It is that point on the x-axis such that, if we concentrated the entire mass there, it would have the same moment as the total moment of the distributed mass $m_1, m_2, \ldots, m_n$ about *any* origin on the x-axis. Note that the center of mass of a system of mass points $m_1, m_2, \ldots, m_n$ is independent of the choice of origin. Also in Exercise 9 we shall show that the center of mass is invariant or unchanged if each of the masses is multiplied by the same number k. Thus the position of the center of mass is independent of the units by which mass is measured. Also, since the weight w_i of a mass m_i is given by $w_i = m_i g$ where g is a constant at a particular location on the surface of the Earth, the position of the **center of weight or gravity** (defined analogously) is identical with the position of the center of mass.

7.7.2 *Continuous Mass Distribution on a Line*

We now extend our previous discussion to a rigid horizontal rod having a continuously distributed mass. Consider the mass Δm of a rod between the positions $x = x_0$ and $x = x_0 + \Delta x$ where $\Delta x > 0$. Then the ratio $\dfrac{\Delta m}{\Delta x}$ is called the average linear mass density of the rod in $[x_0, x_0 + \Delta x]$. We assume that $\lim\limits_{\Delta x \to 0} \dfrac{\Delta m}{\Delta x}$ exists and denote this limit by $\rho(x_0)$ called the **linear mass density** at the point $x = x_0$. Furthermore it is assumed that the density $\rho(x)$ is a continuous function of x. In the special case where ρ is constant, the rod is said to be **homogeneous;** otherwise it is said to be **nonhomogeneous.** If the unit of mass is slug and the unit of distance is feet, it follows that the unit of linear density is slugs per foot (slug/ft).

Consider then a rod of length L ft, which is placed along the x-axis extending from $x = 0$ to $x = L$ (Figure 7.7.4). The density of the rod is $\rho(x)$ where ρ is a continuous function of x in $[0, L]$. We seek the mass M of the rod. To this end, we introduce a partition P of the interval $[0, L]$ into n subintervals. The ith subinterval is $[x_{i-1}, x_i]$, where $\Delta x_i = x_i - x_{i-1}$ is its length. If ξ_i is *any* point in $[x_{i-1}, x_i]$, an approximation to the mass of the part of the rod contained in the ith subinterval is $\Delta m_i = \rho(\xi_i)\, \Delta x_i$ slugs. The mass M of the rod is approximated by

$$\sum_{i=1}^{n} \Delta m_i = \sum_{i=1}^{n} \rho(\xi_i)\, \Delta x_i$$

If we let the norm $|P|$ of the partition tend to zero, the Riemann sum tends to the Riemann integral. Accordingly, we have the following definition for the mass M of a rod.

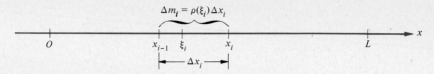

Figure 7.7.4

Definition   A rod of length L ft extends from $x = 0$ to $x = L$. If the linear density $\rho(x)$ (slugs/ft) is continuous in $[0, L]$, then the **mass** M (slugs) of the rod is given by

$$M = \lim_{|P| \to 0} \sum_{i=1}^{n} \rho(\xi_i)\,\Delta x_i = \int_0^L \rho(x)\,dx \qquad (2)$$

EXAMPLE 3   The length of a rod is 10 ft, and the linear density of the rod at a point x ft from the left end is $(6 + 2x)$ slugs/ft. Find the mass M of the rod.

Solution
$$\rho(x) = 6 + 2x \qquad 0 \le x \le 10$$

so that from (2),

$$M = \int_0^{10} (6 + 2x)\,dx = (6x + x^2)\Big|_0^{10}$$

$$= 160 \text{ slugs} \qquad\qquad \bullet$$

EXAMPLE 4   The length of a rod is L ft. The linear density $\rho(x)$ is $\rho = \rho_0 =$ constant over the left third and $\rho = \rho_1 =$ constant over the right third, and the density varies linearly with the spatial coordinate x in the middle third of the rod. Find the mass of the rod assuming that the density is a continuous function of x over the length of the beam.

Solution
$$\rho(x) = \begin{cases} \rho_0 & 0 \le x \le \dfrac{L}{3} \\[2mm] a + bx & \dfrac{L}{3} \le x \le \dfrac{2L}{3} \\[2mm] \rho_1 & \dfrac{2L}{3} \le x \le L \end{cases}$$

where a and b are readily found from the conditions that $\rho = \rho_0$ when $x = L/3$ and $\rho = \rho_1$ when $x = 2L/3$. Thus

$$\begin{cases} a + b\dfrac{L}{3} = \rho_0 \\[2mm] a + b\dfrac{2L}{3} = \rho_1 \end{cases}$$

from which

$$b = \frac{3}{L}(\rho_1 - \rho_0) \qquad a = 2\rho_0 - \rho_1$$

Thus the mass M is found from

$$M = \int_0^L \rho(x)\, dx = \int_0^{L/3} \rho_0\, dx + \int_{L/3}^{2L/3} (a + bx)\, dx + \int_{2L/3}^L \rho_1\, dx$$

$$= (\rho_0 + \rho_1)\frac{L}{3} + \left(ax + \frac{bx^2}{2}\right)\Bigg|_{L/3}^{2L/3}$$

$$= (\rho_0 + \rho_1)\frac{L}{3} + a\frac{L}{3} + \frac{b}{6}L^2$$

and from the formulas for a and b this reduces to

$$M = (\rho_0 + \rho_1)\frac{L}{2}$$

a result which perhaps is not very shocking (sketch $\rho(x)$). ●

In order to define the center of mass of a rod it is necessary to define the moment of mass of the rod with respect to the origin. To this end, we place the rod of length L along the x-axis in the interval $[0, L]$. Let P denote a partition of $[0, L]$ into n subintervals $[x_{i-1}, x_i]$ for $i = 1, 2, \ldots, n$, in the usual manner. If ξ_i is any point in $[x_{i-1}, x_i]$ an approximation to the moment of mass with respect to the origin of that piece of the rod in the ith subinterval is $\xi_i\, \Delta m_i$, where $\Delta m_i = \rho(\xi_i)\, \Delta x_i$. Thus the entire moment of mass is approximated by the Riemann sum

$$\sum_{i=1}^n \xi_i\, \Delta m_i = \sum_{i=1}^n \xi_i\, \rho(\xi_i)\, \Delta x_i$$

The smaller we take the norm of the partition, the better, in general, the approximation will be. Thus we have the following definition.

Definition   A rod of length L has its left end at the origin and the linear mass density is taken to be a continuous function $\rho(x)$ on $[0, L]$ (slugs/ft). The **moment of mass M_0 with respect to the origin** is

$$M_0 = \lim_{|P| \to 0} \sum_{i=1}^n \xi_i \rho(\xi_i)\, \Delta x_i = \int_0^L x\, \rho(x)\, dx \qquad (3)$$

The unit of M_0 is slug-feet.

The **center of mass** of the rod is by definition the point $\bar{x}$ such that $\bar{x}M = M_0$ or from (2) and (3),

$$\bar{x} = \frac{\displaystyle\int_0^L x\, \rho(x)\, dx}{\displaystyle\int_0^L \rho(x)\, dx} \qquad (4)$$

EXAMPLE 5   Find the center of mass for the rod of Example 3.

Solution   In Example 3 it was found that $M = 160$ slugs. With

$$\rho(x) = 6 + 2x,$$

$$M_0 = \int_0^{10} x(6 + 2x)\,dx = \int_0^{10} [6x + 2x^2]\,dx$$

$$= \left(3x^2 + \frac{2x^3}{3}\right)\Bigg|_0^{10}$$

$$= 300 + \frac{2000}{3} - 0$$

$$= \frac{2900}{3}$$

and

$$\bar{x} = \frac{2900}{480} = \frac{145}{24} \doteq 6.04 \text{ ft}$$ ●

Exercises 7.7

In Exercises 1 through 6 a system of particles is located on the x-axis. The number of slugs in the mass of each particle and the coordinates of its position are given, where distance is measured in feet. Determine the center of mass $\bar{x}$ of each system.

1. $m_1 = 6$ at 3; $m_2 = 2$ at 4; $m_3 = 7$ at 6
2. $m_1 = 2$ at -2; $m_2 = 4$ at 1; $m_3 = 4$ at 5
3. $m_1 = 3$ at -4; $m_2 = 6$ at -1; $m_3 = 12$ at 1; $m_4 = 9$ at 2
4. $m_1 = 2$ at -3; $m_2 = 10$ at -2; $m_3 = 5$ at 2; $m_4 = 1$ at 5
5. $m_1 = 6$ at -4; $m_2 = 4$ at -1; $m_3 = 6$ at 0; $m_4 = 3$ at 2; $m_5 = 5$ at 3; $m_6 = 1$ at 7
6. $m_1 = m$ at a, $m_2 = 2m$ at b; $m_3 = m$ at $-b$; $m_4 = 2m$ at $-a$
7. Where should the fulcrum be placed in Figure 7.7.5 so that the three masses shown will be in equilibrium?

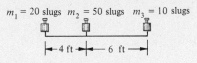

Figure 7.7.5

8. Where should the fulcrum be placed in Figure 7.7.6 so that the four masses shown will be in equilibrium?

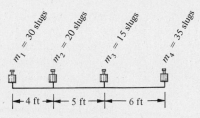

Figure 7.7.6

9. Given n masses m_i at position x_i ($i = 1, 2, \ldots, n$) relative to an arbitrary origin O, show that the center of mass is invariant (that is, unchanged) if each of the masses is multiplied by the same nonzero constant k.

10. A bar 12 ft long supports three weights of magnitude 20 lb, 50 lb, and 30 lb located as shown in Figure 7.7.7. Determine the magnitude and placement of a single force F in order to support this load.

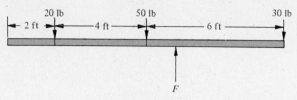

Figure 7.7.7

11. A professor gives four examinations to her mathematics class and decides to weight them chronologically as follows 1:2:2:3 (Note that the final is weighted most heavily.)
   (a) What is the weighted average for a student who obtains the following respective grades 56, 74, 60, and 76?
   (b) If a student gets a 46, 56, and 62 on the first three examinations, what must be the minimum final examination grade in order to pass the course, where a 60 weighted average is to be minimum achievement for passing?

In Exercises 12 through 16 find the mass of the given rod and the center of mass.

12. The length of the rod is 6 ft, and the linear density of the rod at a point x ft from one end is $(3 + 2x)$ slugs/ft.

13. The length of the rod is 8 ft, and the linear density of the rod at a point x ft from one end is $\left(20 - \dfrac{3x}{2}\right)$ slugs/ft.

14. A bar that is 4 ft long has a density that is linear with respect to the distance from one end. The density is 5 slugs/ft at one end and 17 slugs/ft at the other end.

15. A bar that is L ft long has a density that is linear with respect to the distance from one end. The density is ρ_0 slugs/ft at one end and ρ_1 slugs/ft at the other end.

16. A bar is 12 ft long. Its density is a linear function of the distance from the center of the bar. The density is 6 slugs/ft at the center and 2 slugs/ft at each end.

7.8   CENTER OF MASS ON A PLANE

7.8.1   *Center of Mass for Point or Lumped Mass Distribution on a Plane*

Suppose that we have n mass particles located at the points (x_1, y_1), (x_2, y_2), $(x_3, y_3), \ldots, (x_n, y_n)$ in the xy-plane. Furthermore let $m_1, m_2, \ldots, m_n$ be their respective masses. We seek an expression for the center of mass of this system. The masses may be imagined as being supported by a thin plate of negligible mass and thickness with each particle located at a single point in this plane. The center of mass of the system of particles is that point at which the system can be supported in balance. This point is denoted by $(\bar{x}, \bar{y})$ and we seek formulas for the determination of $\bar{x}$ and $\bar{y}$.

To this end, we define the moment of a system of particles with respect to an axis. If a particle with mass m is at (x, y), the **moment** M_y of the mass of the particle with respect to the y-axis is defined by

$$M_y = mx \qquad (1)$$

Similarly the **moment** M_x of the mass of the particle with respect to the x-axis is defined by

$$M_x = my \qquad (2)$$

Thus in the case of n particles, assuming that the moment of the system is the sum of the moments of the individual particles with respect to these axes, it follows that

$$M_y = \sum_{i=1}^{n} m_i x_i \qquad (3)$$

and

$$M_x = \sum_{i=1}^{n} m_i y_i \qquad (4)$$

If the masses are in slugs and the distances $|x_i|$ and $|y_i|$ are in feet, the units of M_y and M_x are both slug-feet.

The total mass of the system is M slugs where

$$M = \sum_{i=1}^{n} m_i \qquad (5)$$

The center of mass of the system is at the point $(\bar{x}, \bar{y})$ where

$$\bar{x} = \frac{M_y}{M} \qquad (6)$$

and

$$\bar{y} = \frac{M_x}{M} \qquad (7)$$

Thus the point $(\bar{x}, \bar{y})$ can be interpreted as the point such that, if the total mass is concentrated there, its moment of mass with respect to the y-axis and x-axis would be the same as M_y and M_x defined by (3) and (4), respectively.

EXAMPLE 1   Find the center of mass of the five particles having masses 3, 5, 4, 1, and 6 slugs located at the points $(-2, 3)$ $(1, 2)$ $(4, -1)$, $(6, 5)$ and $(-4, 0)$, respectively.

Solution

$$M_y = \sum_{i=1}^{5} m_i x_i = 3(-2) + 5(1) + 4(4) + 1(6) + 6(-4)$$

$$= -6 + 5 + 16 + 6 - 24 = -3$$

$$M_x = \sum_{i=1}^{5} m_i y_i = 3(3) + 5(2) + 4(-1) + 1(5) + 6(0)$$

$$= 9 + 10 - 4 + 5 + 0 = 20$$

$$M = \sum_{i=1}^{5} m_i = 3 + 5 + 4 + 1 + 6 = 19$$

Therefore,

$$\bar{x} = \frac{M_y}{M} = \frac{-3}{19}$$

$$\bar{y} = \frac{M_x}{M} = \frac{20}{19}$$

The center of mass is at $\left(\dfrac{-3}{19}, \dfrac{20}{19}\right)$.

EXAMPLE 2   Equal masses are placed on the coordinate plane at (x_1, y_1), (x_2, y_2), and (x_3, y_3). Find the center of mass of the system.

Solution   Let m be the mass placed at each of the given points.

$$M_y = \sum_{i=1}^{3} m x_i = m \sum_{i=1}^{3} x_i$$

$$M = 3m$$

so that

$$\bar{x} = \frac{M_y}{M} = \frac{\displaystyle\sum_{i=1}^{3} x_i}{3} = \frac{x_1 + x_2 + x_3}{3} = \begin{array}{l} \text{arithmetic mean of the} \\ \text{abscissas at the mass} \\ \text{points} \end{array}$$

Similarly,

$$\bar{y} = \frac{M_x}{M} = \frac{\displaystyle\sum_{i=1}^{3} y_i}{3} = \frac{y_1 + y_2 + y_3}{3} = \begin{array}{l} \text{arithmetic mean of the} \\ \text{ordinates at the mass} \\ \text{points} \end{array}$$

More generally, if equal masses m are placed at (x_i, y_i), where $i = 1, 2, \ldots, n$, then

$$\bar{x} = \frac{\displaystyle\sum_{i=1}^{n} x_i}{n} \quad \text{and} \quad \bar{y} = \frac{\displaystyle\sum_{i=1}^{n} y_i}{n}$$

EXAMPLE 3   Show that if two axes u and v are drawn parallel to the x and y axes through point $(\bar{x}, \bar{y})$ then $M_v = M_u = 0$ (Figure 7.8.1)

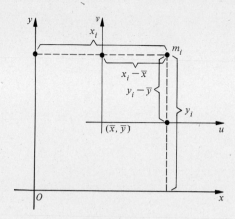

Figure 7.8.1

Solution

$$M_v = \sum_{i=1}^{n} m_i(x_i - \overline{x})$$

$$= \sum_{i=1}^{n} m_i x_i - \overline{x} \sum_{i=1}^{n} m_i$$

$$= M_y - \frac{M_y}{M}(M) = M_y - M_y = 0$$

Similarly,

$$M_u = 0 \qquad\qquad\qquad \bullet$$

7.8.2 *Distributed Mass on a Plane—Centroid*

Next, we extend our consideration to continuously distributed mass systems on a plane such as a flat strip of tin. This flat strip of tin or any other very thin flat plate is called a *lamina.* Suppose that we have a lamina of matter extending over some region R in the xy-plane Figure 7.8.2). We define the *mass density ρ* at a point $P(x, y)$ in R to be given by

$$\rho = \lim_{\Delta A \to 0} \frac{\Delta m}{\Delta A} \qquad\qquad (8)$$

where Δm is the mass of a subregion ΔR containing the point $P(x, y)$, ΔA is the area of that subregion, and the limit is taken as the diameter of the subregion (that is, the maximum distance between any two points in ΔR) tends to zero.

Generally the mass density ρ may change from point to point or $\rho = \rho(x, y)$; that is, ρ is a function of both x and y. However, in many of the applications, the density ρ is constant and the plate is said to be *homogeneous.* The problem of determining the mass center where $\rho = \rho(x, y)$ can be done more efficiently by techniques in the calculus of more than one variable. Therefore the balance of this discussion will be restricted to plates or areas of uniform density.

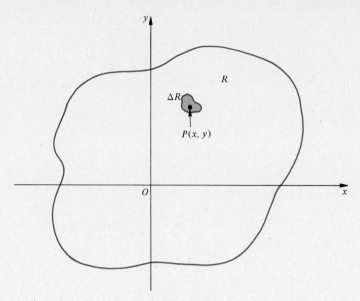

Figure 7.8.2

Our entire development will be based upon our knowing the center of mass of a homogeneous rectangular plate. From simple physical considerations, we assume that the mass center will be at the center of the rectangle, that is, halfway between the bases and halfway between the sides. This point is the **geometric center** of the rectangle—it is also the intersection of the diagonals. (Figure 7.8.3).

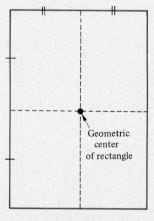

Geometric
center
of rectangle

Figure 7.8.3

Let us consider a homogeneous lamina with constant mass density ρ which is bounded by the curve $y = f(x) \geq 0$, the x-axis, and the lines $x = a$ and $x = b$. The function $f(x)$ is assumed to be continuous on the closed interval $a \leq x \leq b$. We seek the center of mass of this homogeneous lamina.

Let P denote a partition of the interval $[a, b]$ into n subintervals $[x_{i-1}, x_i]$ where $i = 1, 2, \ldots, n$, such that $\Delta x_i = x_i - x_{i-1}$. Denote the midpoint of $[x_{i-1}, x_i]$ by z_i. Thus, for each subinterval there is a rectangular lamina whose height, width, and area density are given by $f(z_i)$, Δx_i, and ρ, respectively. Furthermore, the mass center is at $\left(z_i, \dfrac{f(z_i)}{2}\right)$ (see Figure 7.8.4). Therefore the area of the rectangular

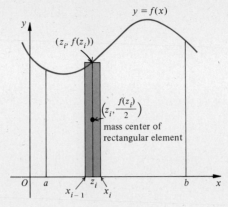

Figure 7.8.4

element is $f(z_i)\,\Delta x_i$, and its mass is given by $\rho\, f(z_i)\,\Delta x_i$ where ρ is a constant. Thus if $(\Delta M_y)_i$ is the moment of mass of the element with respect to the y-axis, we have (by formula (1) with $x = z_i$)

$$(\Delta M_y)_i = (\rho\, f(z_i)\,\Delta x_i)z_i$$

so that, by superposition, the total moments of such rectangular laminae is given by $\sum\limits_{i=1}^{n} \rho z_i\, f(z_i)\,\Delta x_i$. Therefore if M_y denotes the moment of mass of the lamina with respect to the y-axis, we define M_y as follows.

Definition

$$M_y = \lim_{|P|\to 0} \sum_{i=1}^{n} \rho z_i f(z_i)\,\Delta x_i = \rho \int_a^b x\, f(x)\,dx \qquad (9)$$

In a similar manner $\left(\text{since } \dfrac{f(z_i)}{2} \text{ is the moment arm from the center of the mass}\right.$ of the ith rectangle to the x-axis $\Big)$ if $(\Delta M_x)_i$ denotes the moment of mass of the ith rectangular element with respect to the x-axis,

$$(\Delta M_x)_i = \tfrac{1}{2} f(z_i)\, \rho\, f(z_i)\,\Delta x_i$$

and again, by superposition, the total moments of such rectangular laminae with respect to the x-axis are $\sum\limits_{i=1}^{n} \dfrac{\rho}{2}\,[\,f(z_i)]^2\,\Delta x_i$. Thus if M_x is the moment of mass of the lamina with respect to the x-axis, we define M_x as follows.

Definition

$$M_x = \lim_{|P|\to 0} \sum_{i=1}^{n} \frac{\rho}{2}\,[f(z_i)]^2\,\Delta x_i = \frac{\rho}{2}\int_a^b [f(x)]^2\,dx \qquad (10)$$

Furthermore if M denotes the total mass of the lamina,

$$M = \lim_{|P| \to 0} \sum_{i=1}^{n} \rho \, f(z_i) \, \Delta x_i = \rho \int_a^b f(x) \, dx \tag{11}$$

Therefore if $(\bar{x}, \bar{y})$ is the mass center of the lamina we define

$$\bar{x} = \frac{M_y}{M} \quad \text{and} \quad \bar{y} = \frac{M_x}{M}$$

which from (9), (10), and (11) yields (since the density ρ divides out)

$$\bar{x} = \frac{\displaystyle\int_a^b x \, f(x) \, dx}{\displaystyle\int_a^b f(x) \, dx} \tag{12}$$

and

$$\bar{y} = \frac{\displaystyle\frac{1}{2} \int_a^b [\, f(x)]^2 \, dx}{\displaystyle\int_a^b f(x) \, dx} \tag{13}$$

For a homogeneous plate, ρ does not appear in the formulas for $\bar{x}$ and $\bar{y}$; that is, the location of the center of mass depends only on the geometry not on the mass density. In such cases the point $(\bar{x}, \bar{y})$ is called the **centroid** of the plane region. Furthermore, if the notion of mass is ignored and we introduce the **first moments of area** M_y and M_x of the plane region defined by

$$M_y = \int_a^b x \, f(x) \, dx \tag{14}$$

and

$$M_x = \frac{1}{2} \int_a^b [\, f(x)]^2 \, dx \tag{15}$$

then the coordinates of the centroid $(\bar{x}, \bar{y})$ are by definition

$$\bar{x} = \frac{M_y}{A} \quad \text{and} \quad \bar{y} = \frac{M_x}{A}$$

where A is the area of the region. Thus (12) and (13) yield the centroid of an area as well as the center of gravity of a homogeneous lamina.

The reader is again advised *not* to memorize these formulas. There are different situations that can occur. It is important to understand and to utilize the method that yielded formulas such as (12) and (13) by constructing representative elements such as ΔM_y and ΔM_x and then taking the appropriate limit of a sum.

EXAMPLE 4   Find the centroid of the region bounded by $y = x^2$, $y = 0$ and $x = 2$.

Solution Refer to Figure 7.8.5. This case is exactly in the form illustrated previously so that (12) and (13) may be used. Thus, with $f(x) = x^2$, $a = 0$, and $b = 2$, we have from (12),

$$\bar{x} = \frac{\displaystyle\int_0^2 x(x^2)\,dx}{\displaystyle\int_0^2 x^2\,dx} = \frac{\displaystyle\int_0^2 x^3\,dx}{\displaystyle\int_0^2 x^2\,dx} = \frac{\left.\dfrac{x^4}{4}\right|_0^2}{\left.\dfrac{x^3}{3}\right|_0^2}$$

$$= \frac{4}{8/3} = \frac{3}{2}$$

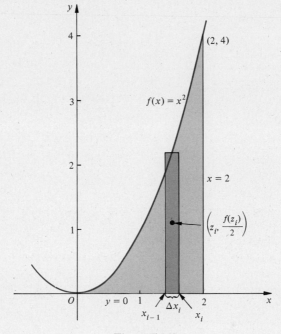

Figure 7.8.5

Then from (13),

$$\bar{y} = \frac{\dfrac{1}{2}\displaystyle\int_0^2 (x^2)^2\,dx}{\displaystyle\int_0^2 x^2\,dx} = \frac{\dfrac{1}{2}\displaystyle\int_0^2 (x^4)\,dx}{8/3} = \frac{\left.\dfrac{x^5}{10}\right|_0^2}{8/3}$$

$$= \left(\frac{32}{10}\right)\left(\frac{3}{8}\right)$$

$$= \frac{6}{5}$$

EXAMPLE 5 Find the centroid of the region bounded by $x = -y^2 + 2y + 8$ and the y-axis.

Solution The graph is a parabola with $y = 1$ as an axis of symmetry. The curve has the y intercepts at $y = -2$ and $y = 4$ (see Figure 7.8.6).

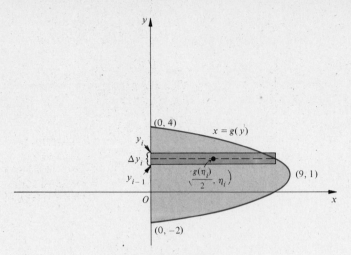

Figure 7.8.6

We first find the area bounded by the parabola and the y-axis

$$A = \lim_{|P| \to 0} \sum_{i=1}^{n} g(\eta_i)\,\Delta y_i$$

$$= \int_{-2}^{4} g(y)\,dy$$

$$= \int_{-2}^{4} (-y^2 + 2y + 8)\,dy$$

$$= \left(-\frac{y^3}{3} + y^2 + 8y\right)\Bigg|_{-2}^{4}$$

$$= \left(-\frac{64}{3} + 16 + 32\right) - \left(\frac{8}{3} + 4 - 16\right)$$

$$= 36$$

Next we compute M_y [note that (9) and (14) are not applicable here because the rectangular element is horizontal].

$$M_y = \lim_{|P| \to 0} \sum_{i=1}^{n} \frac{[g(\eta_i)]^2}{2}\,\Delta y_i$$

$$= \int_{-2}^{4} \frac{[g(y)]^2\,dy}{2}$$

$$= \frac{1}{2}\int_{-2}^{4} (-y^2 + 2y + 8)^2\,dy$$

$$= \frac{1}{2}\int_{-2}^{4} (y^4 - 4y^3 - 12y^2 + 32y + 64)\,dy$$

$$= 129.6$$

$$\bar{x} = \frac{M_y}{A} = \frac{129.6}{36} = 3.6$$

Similarly we calculate M_x [we cannot use (10) or (15)].

$$M_x = \lim_{|P| \to 0} \sum_{i=1}^{n} \eta_i \, g(\eta_i) \, \Delta y_i$$

$$= \int_{-2}^{4} y(-y^2 + 2y + 8) \, dy$$

$$= \int_{-2}^{4} (-y^3 + 2y^2 + 8y) \, dy$$

$$= \left(-\frac{y^4}{4} + \frac{2y^3}{3} + 4y^2 \right) \Big|_{-2}^{4}$$

$$= 36$$

$$\bar{y} = \frac{M_x}{A} = \frac{36}{36} = 1$$

This last result is anticipated since $y = 1$ is the axis of symmetry of the parabola and the centroid must lie on this axis (see Exercise 25). ●

Formulas (9) through (15) only apply when we are working with a region under a curve $y = f(x)$, $f(x) \geq 0$ on $[a, b]$. In Example 5, we needed different formulas because we had a different type of region. We shall learn more efficient means of finding centroids in Chapter 18. In the meantime, we can handle these problems better if we write (14) and (15) in the more general form:

$$M_y = \int_a^b X \, dA \tag{16}$$

$$M_x = \int_a^b Y \, dA \tag{17}$$

Here, dA corresponds to the area of a rectangular element, and X and Y correspond, respectively, to the x and y coordinates of the center of a rectangular element.

In Example 4, $\Delta A_i = f(z_i) \, \Delta x_i$, and the center of a rectangular element is $(z_i, \frac{1}{2} f(z_i))$. Into (16) and (17) we would then have substituted $dA = f(x) \, dx$, $X = x$, and $Y = \frac{1}{2} f(x)$.

In Example 5, $\Delta A_i = g(\eta_i) \, \Delta y_i$ and $(\frac{1}{2} g(\eta_i), \eta_i)$ is the center of a rectangular element. We then have $dA = g(y) \, dy$, $X = \frac{1}{2} g(y)$, and $Y = y$.

EXAMPLE 6   Find the centroid of the area in the first quadrant bounded by $y = 4x$ and $y = x^3$ (see Figure 7.8.7).

Solution   The curves $f(x) = 4x$ and $g(x) = x^3$ meet at $(0, 0)$ and $(2, 8)$ in the first quadrant. The area A is given by

$$A = \int_0^2 (f(x) - g(x)) \, dx = \int_0^2 (4x - x^3) \, dx$$

$$= \left[2x^2 - \frac{x^4}{4} \right]_0^2 = 4$$

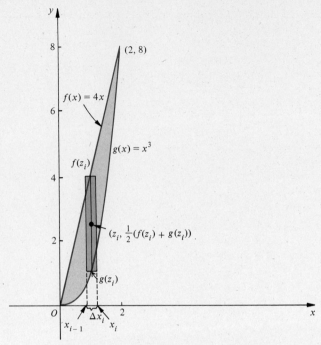

Figure 7.8.7

Our area formula uses the fact that $\Delta A_i = (f(z_i) - g(z_i))\,\Delta x_i$. Then $d\Lambda = (f(x) - g(x))\,dx$. Since the midpoint of a rectangular element is $\left(z_i, \dfrac{f(z_i) + g(z_i)}{2}\right)$, we have $X = x$ and $Y = \frac{1}{2}(f(x) + g(x))$. Therefore

$$M_y = \int_0^2 x(4x - x^3)\,dx = \int_0^2 (4x^2 - x^4)\,dx$$

$$= \left[\frac{4x^3}{3} - \frac{x^5}{5}\right]_0^2 = \frac{64}{15}$$

Thus

$$\bar{x} = \frac{M_y}{A} = \frac{16}{15}$$

Also

$$M_x = \int_0^2 \frac{1}{2}(4x + x^3)(4x - x^3)\,dx = \frac{1}{2}\int_0^2 (16x^2 - x^6)\,dx$$

$$= \frac{1}{2}\left[\frac{16x^3}{3} - \frac{x^7}{7}\right]_0^2 = \frac{1}{2}\left(\frac{128}{3} - \frac{128}{7}\right) = \frac{256}{21}$$

and

$$\bar{y} = \frac{M_x}{A} = \frac{64}{21}$$

●

Exercises 7.8

1. Find the center of mass of the two particles having masses of 2 and 5 slugs located at the points $(-2, 3)$ and $(4, 6)$, respectively. Does the mass center lie on the line through the given points?

2. Find the center of mass of the two particles having masses m and $2m$ slugs located at the points $(6, 2)$ and $(-3, -1)$, respectively.

3. Find the center of mass of the three particles having masses 3, 1, and 4 slugs at the points $(0, 1)$, $(3, 0)$ and $(2, 1)$, respectively.

4. Find the center of mass of the four particles having masses of 2, 4, 5, and 7 slugs at the points $(-3, -2)$, $(-1, 2)$, $(3, 4)$, and $(6, 5)$, respectively.

5. Four equal masses are placed at the vertices of a rectangle. Show that the center of mass is at the intersection of the diagonals.

6. Four equal masses are placed at the vertices of a parallelogram. Show that the center of mass is at the intersection of the diagonals.

7. Three equal masses of m slugs each are located at (x_1, y_1), (x_2, y_2), and (x_3, y_3). Find the center of mass. Prove that the center of mass coincides with the intersection of the medians of the triangle formed by these points.

8. There is a 5-slug mass at $(-1, -5)$, a 10-slug mass at $(2, 4)$, and a 15-slug mass at $(1, 5)$. Where should a 20-slug mass be placed in order to have the system balance at $(2, 4)$?

9. There is an m-slug mass at (a, b) and a $2m$-slug mass at $(-a, -b)$. Where should a $3m$-slug mass be placed in order to have the system balance at $\left(\dfrac{a}{2}, \dfrac{b}{4}\right)$?

10. Find the centroid of the lamina in Figure 7.8.8.

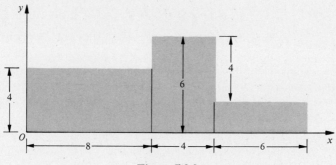

Figure 7.8.8

11. Find the centroid of the lamina in Figure 7.8.9.

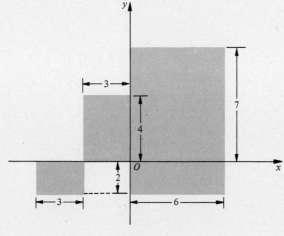

Figure 7.8.9

In Exercises 12 through 23 find the centroid of the region with the indicated boundaries.

12. $y = 2x$, $x = 4$, and the x-axis.
13. $y = 3x$, $y = 8 - x$, and the x-axis.

14. $y = 16 - x^2$ and the x-axis.
15. $y = x^2$, $y = 0$, and $x = a$, where $a > 0$.
16. $y = \sqrt{x}$, $y = 0$, and $x = a$, where $a > 0$.
17. $y = x^2 - x^3$ and the x-axis.
18. $y = x^4 + 1$, $x = 1$, the x-axis, and the y-axis.
19. $y = x^2$, $y = 9$, and the y-axis.
20. $y = x^3$ and $y^2 = x$.
21. $x^2 + y^2 = a^2$ (first quadrant). (*Hint:* Area of a circle of radius a is πa^2.)
22. $y^2 = 2x$ and $x - y - 4 = 0$.
23. Parabola $\sqrt{x} + \sqrt{y} = \sqrt{a}$ and the coordinate axes, where a is positive.
24. Find the centroid of the area bounded by the parabola $y^2 = 4px$ and the line $y = mx$, where m and p are both positive.
25. Prove the significant fact that if a region R has the line L as an axis of symmetry, then the centroid of R must lie on L. Assume for simplicity that the region is bounded by the curve $y = f(x)$ on $-b \leq x \leq b$ and $y = 0$ so that L is on the y-axis.

7.9 CENTER OF MASS OF A SOLID OF REVOLUTION

In general the problem of determining the center of mass of solids requires the application of multiple integration. This will be developed in Chapter 18. However, if the shape of the solid is of revolution and if the volume mass density (that is, mass per unit volume) is constant, then we can determine the center of mass by procedures analogous to that of finding the center of mass of a homogeneous lamina. By symmetry it is reasonable to assume that the mass center is on the axis of revolution. We shall postulate that this is true. It can be established by a method analogous to the procedure for proving that the centroid of an area must lie on an axis of symmetry.

In order to proceed, we set up a three-dimensional rectangular coordinate system. The x- and y-axes are taken as in the two-dimensional case. We now append a third axis, the z-axis, which is perpendicular to both the x- and y-axes at the origin. In Figure 7.9.1 we have drawn (in perspective) the positive x-, y-, and z-axes. A point in three dimensions is given by (x, y, z) where x is the directed distance from the yz-plane, which is the plane containing the y- and z-axes. Similarly y and z are directed distances from the zx- and xy-planes, respectively.

Let us choose the x-axis to coincide with the axis of revolution. Then, from the

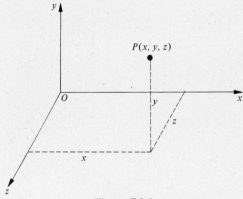

Figure 7.9.1

fact that the mass center must lie on the axis of revolution, it follows that the y and z coordinate of the mass center must be zero. Therefore it is only necessary to find the x coordinate of the mass center, which we call $\bar{x}$. In order to find $\bar{x}$ we must find the moment of the solid of revolution with respect to the yz-plane.

The solid of revolution is obtained by rotating a region R in the xy-plane about the x-axis. We assume that the region R is bounded by a continuous nonnegative function $f(x)$ on the interval $[a, b]$, the x-axis, and the lines $x = a$ and $x = b$ (Figure 7.9.2). Let S be the homogeneous solid of revolution with constant mass density ρ slugs/ft^3. We introduce a partition P of the closed interval $[a, b]$ and let $[x_{i-1}, x_i]$ be the ith subinterval ($i = 1, 2, 3, \ldots, n$). Let w_i be the midpoint of $[x_{i-1}, x_i]$. Form n rectangles with altitudes $f(w_i)$ and bases with width $\Delta x_i = x_i - x_{i-1}$. Figure 7.9.2 shows the ith such rectangle. If the n rectangles are rotated about the x-axis, then n right circular cylindrical disks are formed. The volume of the ith disk is $\pi[f(w_i)]^2 \Delta x_i$ ft^3 and its mass is $\pi\rho[f(w_i)]^2 \Delta x_i$ slugs. Figure 7.9.3 shows the solid of revolution and the ith circular disk. The center of mass of this disk must be at $(w_i, 0, 0)$ so that the moment of mass of the disk with respect to

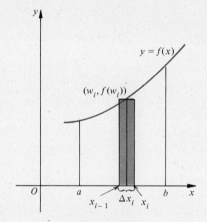

Figure 7.9.2

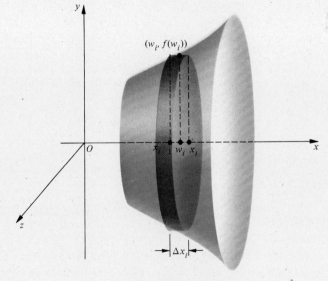

Figure 7.9.3

the yz-plane is then ΔM_{yz} where

$$\Delta M_{yz} = w_i \rho \pi [f(w_i)]^2 \Delta x_i$$

By addition over i from $i = 1$ to $i = n$ we obtain the moment of mass of the n right circular disks with respect to the yz-plane

$$\sum_{i=1}^{n} w_i \rho \pi [f(w_i)]^2 \Delta x_i$$

Thus we define the **math moment** with respect to the yz-plane by

$$M_{yz} = \lim_{|P| \to 0} \sum_{i=1}^{n} w_i \rho \pi [f(w_i)]^2 \Delta x_i = \rho \pi \int_a^b x [f(x)]^2 \, dx \qquad (1)$$

The mass M is the density ρ times the volume so that

$$M = \rho \pi \int_a^b [f(x)]^2 \, dx \qquad (2)$$

The **center of mass** of S is thus by definition given by $(\bar{x}, 0, 0)$ where from (1) and (2)

$$\bar{x} = \frac{M_{yz}}{M} = \frac{\rho \pi \displaystyle\int_a^b x [f(x)]^2 \, dx}{\rho \pi \displaystyle\int_a^b [f(x)]^2 \, dx}$$

or, equivalently,

$$\bar{x} = \frac{\displaystyle\int_a^b x [f(x)]^2 \, dx}{\displaystyle\int_a^b [f(x)]^2 \, dx} \qquad (3)$$

From (3) it is clear that $\bar{x}$ depends on the shape but not on the substance of the material in the case of a homogeneous solid of revolution. Thus, in our calculations we can use

$$\bar{x} = \frac{M_{yz}}{V}$$

where M_{yz} is the **moment of volume** defined now by

$$M_{yz} = \lim_{|P| \to 0} \sum_{i=1}^{n} w_i \pi [f(w_i)]^2 \Delta x_i = \pi \int_a^b x [f(x)]^2 \, dx \qquad (4)$$

so that (3) applies.

EXAMPLE 1   Find the centroid of the solid of revolution generated by revolving the region bounded by the curve $y = x^2$, the x-axis, and the line $x = 2$ about the x-axis.

Solution   The region involved and the rectangular element to be rotated about the x-axis are shown in Figure 7.9.4. The solid of revolution and a typical element of volume is shown in Figure 7.9.5. In this case $f(x) = x^2$, $0 \le x \le 2$.

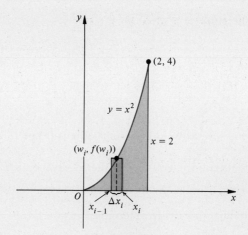

Figure 7.9.4

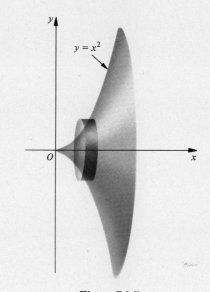

Figure 7.9.5

$$M_{yz} = \lim_{|P| \to 0} \sum_{i=1}^{n} w_i \pi [f(w_i)]^2 \, \Delta x_i$$

$$= \pi \int_0^2 x[f(x)]^2 \, dx = \pi \int_0^2 x(x^2)^2 \, dx$$

$$= \pi \int_0^2 x^5 \, dx = \frac{32\pi}{3}$$

$$V = \lim_{|P| \to 0} \sum_{i=1}^{n} \pi [f(w_i)]^2 \, \Delta x_i$$

$$= \pi \int_0^2 [f(x)]^2 \, dx = \pi \int_0^2 x^4 \, dx$$

$$= \frac{32\pi}{5}$$

Therefore

$$\bar{x} = \frac{M_{yz}}{V} = \frac{32\pi}{3} \left| \frac{32\pi}{5} = \frac{5}{3} \right.$$

so that the centroid is at the point $(\frac{5}{3}, 0, 0)$.

●

EXAMPLE 2   Find the centroid of a right circular cone of radius r and altitude h.

Solution   The right circular cone is the solid generated by revolving about the x-axis the region bounded by the x-axis and the lines $y = \dfrac{r}{h}x$ and $x = h$ (Figure 7.9.6). The volume V is easily determined from

$$V = \lim_{|P| \to 0} \sum_{i=1}^{n} \pi \left(\frac{r}{h} w_i \right)^2 \Delta x_i$$

$$= \pi \int_0^h \frac{r^2}{h^2} x^2 \, dx$$

$$= \pi \frac{r^2 h}{3}$$

Next we calculate M_{yz}, the moment of the right circular cone with respect to the yz-plane

$$M_{yz} = \lim_{|P| \to 0} \sum_{i=1}^{n} \pi w_i \left(\frac{r}{h} w_i \right)^2 \Delta x_i$$

$$= \pi \int_0^h \frac{r^2}{h^2} x^3 \, dx$$

$$= \pi \frac{r^2 h^2}{4}$$

The centroid is at the point $(\bar{x}, 0, 0)$ where

$$\bar{x} = \frac{M_{yz}}{V} = \frac{\frac{1}{4}\pi r^2 h^2}{\frac{1}{3}\pi r^2 h} = \frac{3}{4}h$$

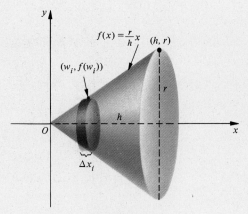

Figure 7.9.6

Therefore the centroid is on the axis of the cone at a distance of $\dfrac{3h}{4}$ units from the vertex. ●

The method of cylindrical shells can also be applied to the determination of the centroid of a solid of revolution. Thus let R be the region bounded by the graph of $y = f(x)$, where f is a continuous function and $f \geq 0$ on $[a, b]$, and the three lines $y = 0$, $x = a$ and $x = b$. When R is revolved about the y-axis a solid of revolution S is obtained. We seek the centroid of S, which must lie on the y-axis due to symmetry; that is, its coordinates must be of the form $(0, \overline{y}, 0)$. If the rectangular elements are chosen parallel to the y-axis (Figure 7.9.7), the element of volume is a cylindrical shell element of thickness $\Delta x_i = x_i - x_{i-1}$ with inner and outer radii x_{i-1} and x_i, respectively. Thus if w_i is the midpoint of the interval $[x_{i-1}, x_i]$, the centroid of the cylindrical shell of height $f(w_i)$ is at $\left(0, \dfrac{f(w_i)}{2}, 0\right)$, and the element of moment of the cylindrical shell with respect to the xz-plane is given by (Figure 7.9.8)

$$\Delta M_{xz} = \frac{f(w_i)}{2} 2\pi w_i \, f(w_i) \, \Delta x_i$$

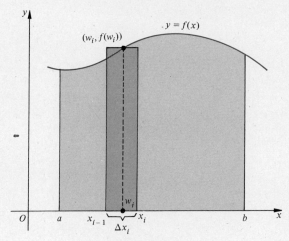

Figure 7.9.7

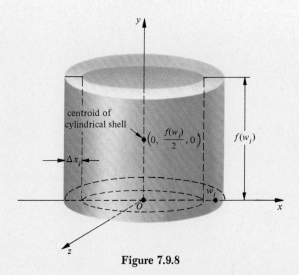

Figure 7.9.8

Thus by taking the limit of the sum in the usual manner we have

$$M_{xz} = \lim_{|P| \to 0} \sum_{i=1}^{n} \pi w_i [f(w_i)]^2 \, \Delta x_i = \pi \int_a^b x[f(x)]^2 \, dx \qquad (5)$$

Also if V is the volume of the solid of revolution S, then

$$V = \lim_{|P| \to 0} \sum_{i=1}^{n} 2\pi w_i \, f(w_i) \, \Delta x_i = 2\pi \int_a^b x \, f(x) \, dx \qquad (6)$$

so that

$$\bar{y} = \frac{M_{xz}}{V} \qquad (7)$$

EXAMPLE 3   Use the cylindrical shell method to find the centroid of the solid of revolution obtained by rotating the region bounded by $y = x^3$, $y = 0$, and the line $x = 1$ about the y-axis.

Solution   The region and a rectangular element of area is shown in Figure 7.9.9. Also Figure 7.9.10 shows the solid of revolution and a cylindrical shell element. From (5),

$$M_{xz} = \pi \int_0^1 x[f(x)]^2 \, dx$$

$$= \pi \int_0^1 x[x^3]^2 \, dx = \pi \int_0^1 x^7 \, dx$$

$$= \pi \frac{x^8}{8} \bigg|_0^1 = \frac{\pi}{8}$$

Also from (6)

$$V = 2\pi \int_0^1 x \, f(x) \, dx$$

$$= 2\pi \int_0^1 x(x^3) \, dx$$

$$= 2\pi \int_0^1 x^4 \, dx = 2\pi \frac{x^5}{5} \bigg|_0^1 = \frac{2\pi}{5}$$

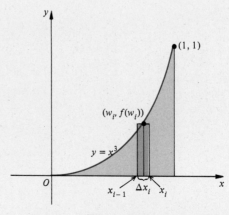

Figure 7.9.9

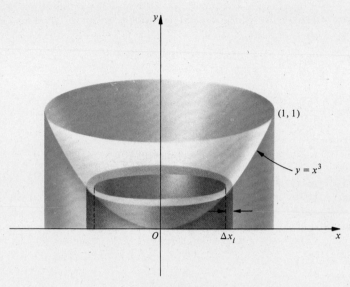

Figure 7.9.10

so that

$$\bar{y} = \frac{M_{xz}}{V} = \frac{\pi}{8} \left| \frac{2\pi}{5} = \frac{5}{16} \right.$$

Therefore the centroid is at the point $(0, \frac{5}{16}, 0)$.　　　　　　　●

We do not mean to imply that the center of gravity of a solid of revolution is always found by the disk method when rotating about the x-axis, nor are we saying that the shell method applies only for rotation about the y-axis. The ring, disk, or shell methods can each be used for both types of rotation—the rectangular elements suggest which is best. However, if a different method is used, we shall need new formulas for M_{yz} and M_{xz} instead of (4) and (5). The reader will find it instructive to obtain the formulas that apply in each of these cases.

Exercises 7.9

In each of Exercises 1 through 8, the region bounded by the graphs of the given equations is rotated about the x-axis. Find the centroid of the solid generated.

1. $y = \sqrt{x}$, $y = 0$, $x = 9$
2. $y = 3x$, $y = 0$, $x = 5$
3. $y = 3 - x$, $y = 0$, $x = 0$
4. $y = 1 - x^2$ (first quadrant), $y = 0$, $x = 0$
5. $y = \sqrt{x^2 + 1}$, $y = 0$, $x = 2$
6. $y = \dfrac{1}{x^2}$, $y = 0$, $x = 1$ and $x = 2$
7. $x^2 + y^2 = 9$ (first quadrant), $y = 0$, and $x = 0$
8. $y^2 - x^3 = 0$ (first quadrant), $y = 0$, and $x = a > 0$

In Exercises 9 through 16 find the centroid of the solid generated by rotating about the y-axis the regions bounded by the curves below.

9. $2y - x = 0$ (first quadrant), $x = 0$, and $y = 3$
10. $y^2 = 12x$, $x = 0$, $y = 5$
11. $y^2 = 4px$, $x = 0$, $y = k > 0$
12. $x^2 + y^2 = 4$ (first quadrant), $x = 0$, above the line $y = 1$.
13. $y = (x - 1)^2$ and the coordinate axes.
14. $25x^2 + y^2 = 25$ (first quadrant) and the coordinate axes.
15. $y = \sqrt{x}$, $x = 4$, and $y = 0$. Use cylindrical shells.
16. $y = ax - x^2$ $(a > 0)$, and $y = 0$. Use cylindrical shells.
17. Find the centroid of the solid generated by revolving the region bounded by the curve $x^2 + y^2 = a^2$ in the first quadrant and the coordinate axes about the x-axis. Identify the solid of revolution.
18. The area bounded by a parabola, its axis, and its latus rectum is revolved about the latus rectum. Find the centroid of the solid generated.
19. The radii of the bases of a frustum of a right circular cone are r and R. The altitude is h. Find the centroid.

7.10 ARC LENGTH OF A PLANE CURVE IN RECTANGULAR COORDINATES

Let f be a function that is continuous on a closed interval $[a, b]$. Consider the arc AB of the graph of $y = f(x)$ where A and B are the points on the graph with coordinates $(a, f(a))$ and $(b, f(b))$, respectively (Figure 7.10.1). It is our objective to assign a number to what we intuitively think of as the length of such an arc.

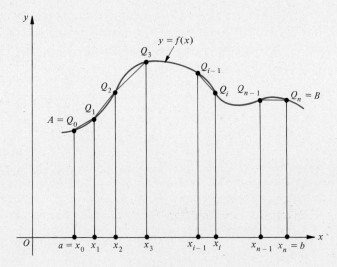

Figure 7.10.1

To this end, we partition $[a, b]$ in the usual fashion into n subintervals by a partition P and denote the length of the ith subinterval $[x_{i-1}, x_i]$ by $\Delta x_i = x_i - x_{i-1}$ where $i = 1, 2, \ldots, n$. Then if $|P|$ be the norm of the partition, by definition each $\Delta x_i \leq |P|$. Let Q_i be the point on the curve with coordinates $(x_i, f(x_i))$. Also denote the points A: $(a, f(a))$ by Q_0 and B: $(b, f(b))$ by Q_n. The length of the line segment joining points Q_{i-1} and Q_i is denoted by $|\overline{Q_{i-1}Q_i}|$ and is given by the distance formula

$$|\overline{Q_{i-1}Q_i}| = \sqrt{(x_i - x_{i-1})^2 + (y_i - y_{i-1})^2} \tag{1}$$

The sum of the lengths of the chords is

$$|\overline{Q_0Q_1}| + |\overline{Q_1Q_2}| + \cdots + |\overline{Q_{n-1}Q_n}| = \sum_{i=1}^{n} |\overline{Q_{i-1}Q_i}| \tag{2}$$

It is suggested that, as $|P| \to 0$ (which implies that $n \to \infty$), the sum in (2) will approach a limit L. More precisely, by the notation

$$\lim_{|P| \to 0} \sum_{i=1}^{n} |\overline{Q_{i-1}Q_i}| = L \tag{3}$$

we mean that corresponding to each positive number ε (as small as desired) there exists a $\delta > 0$ such that

$$\left| \sum_{i=1}^{n} |\overline{Q_{i-1}Q_i}| - L \right| < \varepsilon$$

for all partitions P of the interval $[a, b]$ with $|P| < \delta$. By definition, if there exists a number L, such that

$$\lim_{|P| \to 0} \sum_{i=1}^{n} |\overline{Q_{i-1}Q_i}| = L$$

then the arc of the curve $y = f(x)$ from the point $(a, f(a))$ to the point $(b, f(b))$ is said to be *rectifiable,* and its **length** is L.

Now we shall prove that, if the derivative of f is continuous on $[a, b]$, the curve is rectifiable. Also a formula for the length L will be derived simultaneously.

By the distance formula (1), the length of a typical chord $Q_{i-1}Q_i$ is

$$|\overline{Q_{i-1}Q_i}| = \sqrt{(\Delta x_i)^2 + (\Delta y_i)^2} \tag{4}$$

where $\Delta y_i = f(x_i) - f(x_{i-1})$; or, equivalently, since $\Delta x_i > 0$,

$$|\overline{Q_{i-1}Q_i}| = \sqrt{1 + \left(\frac{\Delta y_i}{\Delta x_i}\right)^2} \, \Delta x_i \tag{5}$$

Since f' is continuous (and therefore exists) on $[x_{i-1}, x_i]$ the mean value theorem applies. Therefore, there exists a number z_i in the open interval (x_{i-1}, x_i) such that

$$f(x_i) - f(x_{i-1}) = f'(z_i)(x_i - x_{i-1}) \tag{6}$$

or

$$\frac{\Delta y_i}{\Delta x_i} = f'(z_i) \tag{7}$$

Therefore substitution of (7) into (5) yields

$$|\overline{Q_{i-1}Q_i}| = \sqrt{1 + [f'(z_i)]^2} \, \Delta x_i \qquad x_{i-1} < z_i < x_i \tag{8}$$

and the length of arc AB is

$$\lim_{|P|\to 0} \sum_{i=1}^{n} \sqrt{1 + [f'(z_i)]^2}\, \Delta x_i \qquad (9)$$

provided that the limit exists.

Let h be the function defined by $h(x) = \sqrt{1 + [f'(x)]^2}$. Since f' is continuous on $[a, b]$, the composite function h is also continuous on $[a, b]$. Thus, the limit (9) must exist and is given by

$$\lim_{|P|\to 0} \sum_{i=1}^{n} \sqrt{1 + [f'(z_i)]^2}\, \Delta x_i = \int_a^b \sqrt{1 + [f'(x)]^2}\, dx = L$$

In summary we have the following theorem.

Theorem 1   If f is a continuous function and f' is also a continuous function on the closed interval $[a, b]$, then the length of the arc of the graph of $y = f(x)$ from the point $(a, f(a))$ to the point $(b, f(b))$ is

$$L = \int_a^b \sqrt{1 + [f'(x)]^2}\, dx \qquad (10)$$

We also have the analogous theorem that gives the length of arc of a plane curve when x is expressed as a function of y, say, $x = g(y)$.

Theorem 2   If g is a function such that g and its derivative g' are continuous on the closed interval $[c, d]$, then the length of arc of the curve $x = g(y)$ from the point $(g(c), c)$ to the point $(g(d), d)$ is given by

$$L = \int_c^d \sqrt{1 + [g'(y)]^2}\, dy \qquad (11)$$

If the reader will select a curve at random and try to compute its length by means of (10) or (11) it will be rapidly observed that the integrals are difficult to evaluate. Therefore at this stage of our development the function must be chosen carefully so that the integrals can be evaluated. This difficulty will be somewhat removed when we learn more about the techniques of integration in Chapter 10. Furthermore in that chapter we will develop some numerical methods for evaluating definite integrals.

EXAMPLE 1   Compute the length of the straight line segment joining the two points $(2, 4)$ and $(5, 13)$.

Solution 1   The simplest way to find this length is to use the distance formula

$$L = \sqrt{(5 - 2)^2 + (13 - 4)^2} = \sqrt{90}$$
$$= 3\sqrt{10}$$

Solution 2   An equation of the straight line is found from the point slope formula

$$\frac{y-4}{x-2} = \frac{13-4}{5-2} \qquad \text{or} \qquad \frac{y-4}{x-2} = 3$$

Cross multiplication and simplification yields

$$y = f(x) = 3x - 2 \qquad \text{so that} \qquad y' = f'(x) = 3$$

By the length formula (10)

$$L = \int_2^5 \sqrt{1 + (f')^2}\, dx = \int_2^5 \sqrt{1 + (3)^2}\, dx \qquad dx = 3$$
$$= 3\sqrt{10}$$

in agreement with the first solution.   ●

Example 2   Find the length of the portion of the upper half of the semicubical parabola $y^2 = x^3$ from the origin to the point $(5, 5\sqrt{5})$.

Solution   The curve under consideration is shown in Figure 7.10.2. Now for $y \geq 0$ we have

$$y = f(x) = x^{3/2}$$

so that $f'(x) = \frac{3}{2}x^{1/2}$. By the length formula (10)

$$\left(\frac{3}{2}x^{\frac{1}{2}}\right)^2 = \frac{9}{4}x$$

$$L = \int_0^5 (1 + \tfrac{9}{4}x)^{1/2}\, dx$$

$$(1 + \tfrac{9}{4}x)^{\frac{1}{2}} \quad \frac{4}{9}\frac{2}{3}\left(1 + \tfrac{9}{4}x\right)^{3/2}$$

$$\frac{2}{3}\frac{3}{2}\left(1 + \tfrac{9}{4}x\right)^{\frac{1}{2}}\frac{9}{4}$$

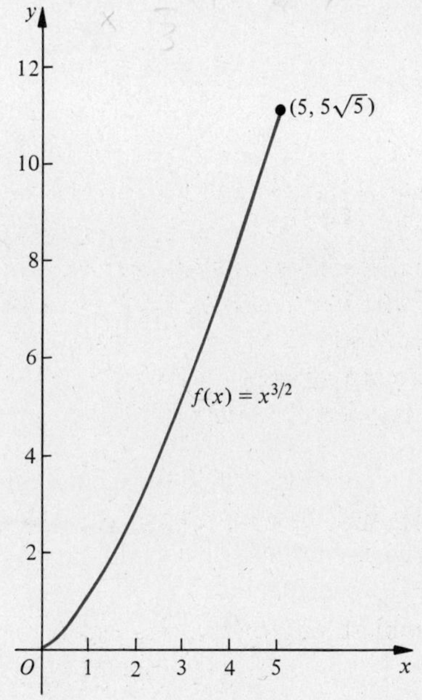

Figure 7.10.2

which is easily integrated. We obtain

$$L = \left(\frac{4}{9}\right)\left(\frac{2}{3}\right)\left(1 + \frac{9x}{4}\right)^{3/2}\Big|_0^5$$

$$= \frac{8}{27}\left[\left(\frac{49}{4}\right)^{3/2} - (1)^{3/2}\right]$$

$$= \frac{8}{27}\left[\left(\frac{7}{2}\right)^3 - 1\right]$$

$$= \frac{8}{27}\left(\frac{343}{8} - 1\right)$$

$$= \frac{335}{27}$$

EXAMPLE 3   Find the length of the arc defined by

$$x = \tfrac{1}{3}(2 + y^2)^{3/2}$$

between the points $(\sqrt{3}, 1)$ and $(2\sqrt{6}, 2)$.

Solution   From (11) with

$$x = g(y) = \tfrac{1}{3}(2 + y^2)^{3/2}$$

and therefore

$$g'(y) = y(2 + y^2)^{1/2}$$

we find that

$$L = \int_1^2 \sqrt{1 + y^2(2 + y^2)}\, dy$$

$$= \int_1^2 \sqrt{1 + 2y^2 + y^4}\, dy$$

$$= \int_1^2 \sqrt{(1 + y^2)^2}\, dy$$

$$= \int_1^2 (1 + y^2)\, dy$$

$$= \left(y + \frac{y^3}{3}\right)\Big|_1^2$$

$$= \left(2 + \frac{8}{3}\right) - \left(1 + \frac{1}{3}\right)$$

$$= \frac{10}{3}$$

Note: If a curve is defined by a function whose derivative is not continuous, we separate the curve into segments such that along each segment the function's derivative is continuous. We find the lengths of these segments by (10) or (11) and then add them together.

If the function f' is continuous on $[a, b]$, we know that the definite integral $\int_a^x \sqrt{1 + [f'(t)]^2}\, dt$ is a function of x that represents the length of the arc from point $(a, f(a))$ to the point $(x, f(x))$. If $s(x)$ denotes the function that represents the length of arc then

$$s(x) = \int_a^x \sqrt{1 + [f'(t)]^2}\, dt \tag{12}$$

From Theorem 2 of Section 6.5 (first fundamental theorem of calculus) it follows that

$$s'(x) = \sqrt{1 + [f'(x)]^2} \tag{13}$$

In Leibnitz' notation

$$\frac{ds}{dx} = \sqrt{1 + \left(\frac{dy}{dx}\right)^2}$$

since $\dfrac{ds}{dx} \equiv s'(x)$ and $\dfrac{dy}{dx} \equiv f'(x)$. Furthermore $\dfrac{ds}{dx} = ds \div dx$ so that

$$ds = \sqrt{1 + \left(\frac{dy}{dx}\right)^2}\, dx \tag{14}$$

Analogously, if we are interested in the length of arc of the curve $x = g(y)$ from the fixed point $(g(c), c)$ to the variable point $(g(y), y)$ on the arc, it follows that

$$ds = \sqrt{1 + \left(\frac{dx}{dy}\right)^2}\, dy \tag{15}$$

From (14) or (15) by squaring both sides we obtain

$$(ds)^2 = (dx)^2 + (dy)^2 \tag{16}$$

which is interpreted geometrically in Figure 7.10.3. In this figure, the line PT is the tangent to the graph of $y = f(x)$ at the point P. Also, since x is the independent variable, $|\overline{PR}| = \Delta x = dx$, $|\overline{RQ}| = \Delta y$, and $|\overline{RM}| = dy$. The length $|\overline{PM}| = ds$, the hypotenuse of the right triangle, and the length of arc $PQ = \Delta s$.

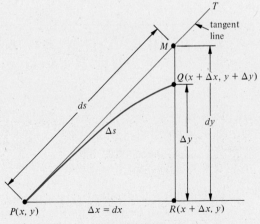

Figure 7.10.3

Exercises 7.10

In Exercises 1 through 10, find the arc length of the graph of each of the following equations between the indicated points.

1. $y = 3x - 5$, $(2, 1)$ to $(4, 7)$
2. $2y - x = 4$, $(0, 2)$ to $(6, 5)$
3. $y = mx + b$, $(c, mc + b)$ to $(d, md + b)$, where $d > c$
4. $y^2 = x^3$, $(0, 0)$ to $(2, 2\sqrt{2})$
5. $y = x^{2/3}$, $(0, 0)$ to $(8, 4)$ [Hint: Solve for x in terms of y and use formula (11).]
6. $y = \dfrac{x^3}{6} + \dfrac{1}{2x}$, $\left(1, \dfrac{2}{3}\right)$ to $\left(3, \dfrac{14}{3}\right)$
7. $y = \dfrac{x^3}{3} + \dfrac{1}{4x}$, $\left(1, \dfrac{7}{12}\right)$ to $\left(2, \dfrac{67}{24}\right)$
8. $3y = 2(1 + x^2)^{3/2}$, $\left(1, \dfrac{4\sqrt{2}}{3}\right)$ to $\left(3, \dfrac{20\sqrt{10}}{3}\right)$
9. $9y = (2 + 9x^2)^{3/2}$, $\left(0, \dfrac{2\sqrt{2}}{9}\right)$ to $\left(1, \dfrac{11\sqrt{11}}{9}\right)$
10. $x = \dfrac{y^4}{4} + \dfrac{1}{8y^2}$, (a, c) to (b, d), where $0 < c < d$
11. Sketch and find the length of the hypocycloid

$$x^{2/3} + y^{2/3} = a^{2/3} \qquad a > 0$$

(This is referred to as the hypocycloid of four cusps or astroid.)

12. Sketch and find the length of the loop of the graph of

$$3ay^2 = x(x - a)^2 \qquad a > 0$$

(Hint: Use implicit differentiation and the fact that the curve is symmetric with respect to the x-axis.)

In Exercises 13 through 16 set up the integrals for the following lengths, but do not try to evaluate them.

13. $f(x) = x^2$ between $(0, 0)$ and $(2, 4)$
14. $x = y^2 + 3$ between $(3, 0)$ and $(4, 1)$
15. The portion of $f(x) = -x^2 + 7x - 10$, lying above the x-axis.
16. $f(x) = \sqrt{25 - x^2}$, $-2 \leq x \leq 3$.

7.11   THE AREA OF A SURFACE OF REVOLUTION

If a plane curve is rotated about an axis in the plane it generates a surface called a **surface of revolution**. Our problem is the determination of the area of such a surface.

To this end we consider some important special cases.

Right circular cylinder: If a straight line of length h, parallel to the x-axis and r units from it is revolved about the x-axis, a right circular cylinder is generated (Figure 7.11.1). The area of its surface is

$$S = 2\pi rh \tag{1}$$

This may be verified intuitively by cutting the cylinder along a line parallel to the axis of symmetry. If the cut cylinder is then unrolled, a rectangle is formed with dimensions h by $2\pi r$ (since $2\pi r$ is the circumference of a base of the cylinder).

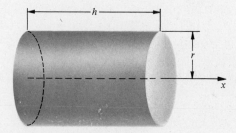

Figure 7.11.1

Right circular cone: If a right circular cone has radius of base r and slant height L, the area of its curved surface is (see Figure 7.11.2)

$$S = \pi rL \tag{2}$$

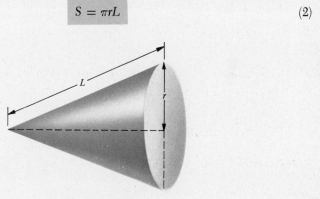

Figure 7.11.2

Formula (2) may be justified intuitively by cutting the cone along a straight line from the vertex to the base—unroll and spread out on a plane. The resulting area will be a portion of a circular disk of radius L (Figure 7.11.3). The perimeter of the base of the cone is $2\pi r$ and therefore the area S of the shaded portion of the disk is the fraction $\dfrac{2\pi r}{2\pi L}$ of the whole area of the disk of radius L. Thus

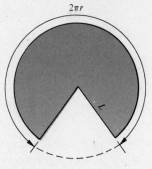

Figure 7.11.3

$$S = \frac{2\pi r}{2\pi L}\pi L^2 = \pi r L$$

and (2) is verified.

For a more rigorous demonstration of (1) and (2), the reader should consult books on solid geometry.

Next we cut off part of the vertex-end of the right circular cone by a plane that is perpendicular to the axis of symmetry. The part of the conical surface between the two circular sections is a ***frustum of a cone*** (Figure 7.11.4). If r and R are the base radii and L is the slant height, we shall prove that the surface area S of the conical frustum is

$$S = \pi(r + R)L \qquad (3)$$

Proof of (3)   From Figure 7.11.4 and (2)

$$S = \pi R L_2 - \pi r L_1$$

where L_2 and L_1 are the two slant heights shown. Now, $L_2 = L_1 + L$, so that

$$S = \pi R(L_1 + L) - \pi r L_1$$
$$= \pi(R - r)L_1 + \pi R L$$

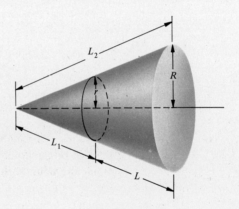

Figure 7.11.4

Also from similar triangles (Figure 7.11.5)

$$\frac{R - r}{L} = \frac{r}{L_1} \qquad \text{or} \qquad (R - r)L_1 = rL$$

Thus

$$S = \pi r L + \pi R L = \pi(r + R)L$$

and our proof is complete.   ∎

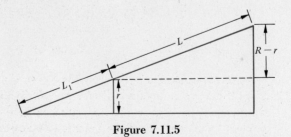

Figure 7.11.5

Note that we can rewrite (3) in the form

$$S = 2\pi \left(\frac{r + R}{2}\right) L \qquad (4)$$

so that the surface area of the frustum of the cone is the perimeter of the midsection times the slant height.

Now we seek to generalize our previous development using formula (3) for the surface area of the frustum of a right circular cone. Consider a function f that is continuous and nonnegative on $[a, b]$. Furthermore it is assumed that its first derivative f' is also continuous on $[a, b]$. If the arc from the point $(a, f(a))$ to the point $(b, f(b))$ is revolved about the x-axis, a surface of revolution is swept out (Figure 7.11.6). We seek an expression for the surface area S.

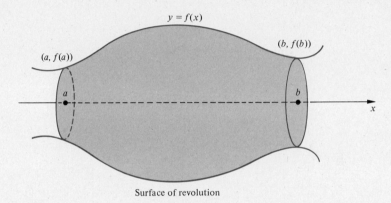

Surface of revolution

Figure 7.11.6

Analogous to Section 7.10, we partition $[a, b]$ into n subintervals $[x_{i-1}, x_i]$, where $i = 1, 2, 3, \ldots, n$. Let Q_i be the point on the curve whose coordinates are $(x_i, f(x_i))$ and let Q_0 represent the point with coordinates $(a, f(a))$ (Figure 7.11.7 shows a case with $n = 4$).

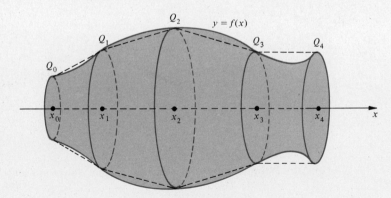

Figure 7.11.7

If the polygonal line formed by the n chords $Q_{i-1}Q_i$ of the curve is revolved about the x-axis, it sweeps out n frustums of right circular cones. The sum of the n surfaces approximates the surface of revolution and this approximation improves as the norm $|P|$ of the partition decreases.

From (3) we obtain the surface of the ith frustum of a cone obtained when the chord connecting Q_{i-1} and Q_i is rotated about the x-axis. Its area is

$$\pi[f(x_{i-1}) + f(x_i)]\,|\overline{Q_{i-1}Q_i}| \tag{5}$$

But

$$|\overline{Q_{i-1}Q_i}| = \sqrt{(\Delta x_i)^2 + (\Delta y_i)^2}$$

$$= \sqrt{1 + \left(\frac{\Delta y_i}{\Delta x_i}\right)^2}\,\Delta x_i$$

and by the mean value theorem

$$\frac{\Delta y_i}{\Delta x_i} = \frac{f(x_i) - f(x_{i-1})}{x_i - x_{i-1}} = f'(z_i) \tag{6}$$

where z_i is some value of x in (x_{i-1}, x_i). Thus, substitution of (6) into (5) yields

$$\pi[f(x_{i-1}) + f(x_i)]\sqrt{1 + [f'(z_i)]^2}\,\Delta x_i$$

and we have the expression

$$S_n = \pi \sum_{i=1}^{n} [f(x_{i-1}) + f(x_i)]\sqrt{1 + [f'(z_i)]^2}\,\Delta x_i \tag{7}$$

for the area of the approximating surface swept out by rotating the polygonal line formed by the n chords $Q_{i-1}Q_i$ $(i = 1, 2, \ldots, n)$ about the x axis. Now

$$S_n = \pi \sum_{i=1}^{n} f(x_{i-1})\sqrt{1 + [f'(z_i)]^2}\,\Delta x_i + \pi \sum_{i=1}^{n} f(x_i)\sqrt{1 + [f'(z_i)]^2}\,\Delta x_i \tag{8}$$

and it is suggested that the two sums tend to the same definite integral as the norm of the partition $|P| \to 0$. However the reader should be warned that the two sums are *not* Riemann sums, and this step requires formal justification. This is because, in general, $z_i \neq x_{i-1}$ in the first summation and $z_i \neq x_i$ in the second. That the limit of (8) is precisely

$$2\pi \int_{a}^{b} f(x)\sqrt{1 + [f'(x)]^2}\,dx$$

can be established by a theorem known as Bliss' theorem†. This depends on the concept of uniform continuity and will *not* be given here.

At this point we simply adopt the definition that the surface area S of the surface of revolution about the x-axis is given by

$$S = 2\pi \int_{a}^{b} f(x)\sqrt{1 + [f'(x)]^2}\,dx = 2\pi \int_{a}^{b} y\sqrt{1 + (y')^2}\,dx \tag{9}$$

and this formula will be applied in a number of examples and problems to follow.

---

† The Bliss or Duhamel theorem states that

$$\lim_{|P| \to 0} \sum_{i=1}^{n} f(\xi_i)\,g(\eta_i)\,\Delta x_i = \int_{a}^{b} f(x)\,g(x)\,dx$$

if f and g are continuous on $[a, b]$ where ξ_i and η_i are arbitrary values of x in $[x_{i-1}, x_i]$

EXAMPLE 1   Find the area of the surface of revolution obtained by rotating $y = \dfrac{x^3}{3}, 0 \leq x \leq 3$
about the x-axis.

Solution   From $y = \dfrac{x^3}{3}$ we obtain $y' = x^2$ and $(y')^2 = x^4$ so that

$$S = 2\pi \int_0^3 y \sqrt{1 + (y')^2} \, dx$$

$$= 2\pi \int_0^3 \frac{x^3}{3} \sqrt{1 + x^4} \, dx$$

Multiplication and division by 4 yields

$$S = \frac{2\pi}{3(4)} \int_0^3 (1 + x^4)^{1/2} 4x^3 \, dx$$

$$= \frac{\pi}{6} \frac{2}{3} (1 + x^4)^{3/2} \Big|_0^3$$

$$= \frac{\pi}{9} ((82)^{3/2} - 1) \text{ square units}$$

EXAMPLE 2   Find the area of the part of a spherical surface of radius a between parallel planes
h units apart. The region is called a *zone* of a sphere.

Solution   The sphere may be generated by revolving the semicircle $x^2 + y^2 = a^2$, $y \geq 0$
around the x-axis. Note that the two planes cutting the sphere must be arbitrarily
placed. Let the planes be perpendicular to the x-axis cutting it at $x = c$ and
$x = c + h$ (Figure 7.11.8). We must determine

$$S = 2\pi \int_c^{c+h} y \sqrt{1 + (y')^2} \, dx$$

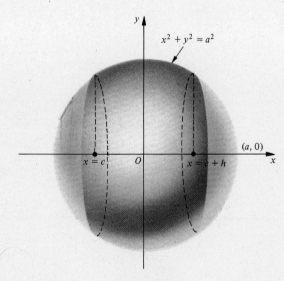

Figure 7.11.8

Now $x^2 + y^2 = a^2$ so that differentiation yields $2x + 2yy' = 0$ or $y' = -\dfrac{x}{y}$ for $y > 0$

Thus

$$S = 2\pi \int_c^{c+h} y\sqrt{1 + \frac{x^2}{y^2}}\, dx$$

$$= 2\pi \int_c^{c+h} y\, \frac{\sqrt{y^2 + x^2}}{y}\, dx \qquad \text{since } y > 0$$

$$= 2\pi a \int_c^{c+h} dx \qquad (\text{since } \sqrt{y^2 + x^2} = \sqrt{a^2} = a)$$

$$= 2\pi a h \text{ square units}$$

The result is independent of c. In particular, if we set $h = a$ we have the surface area $2\pi a^2$ of a hemisphere. Also if we consider the limiting case $h = 2a$ then we have the surface of a sphere of radius a, namely, $S = 4\pi a^2$.   ●

Next, assume that C is a curve defined by $x = g(y)$, $c \le y \le d$ where g is continuous and nonnegative on $[c, d]$ and g' is continuous on $[c, d]$. If C is rotated about the y-axis we again obtain a surface of revolution. We adopt a definition similar to (9) for the surface area of this surface, namely

$$S = 2\pi \int_c^d g(y)\sqrt{1 + [g'(y)]^2}\, dy = 2\pi \int_c^d x\sqrt{1 + (x')^2}\, dy \qquad (10)$$

EXAMPLE 3   Find the area of the surface of revolution obtained by rotating $x = \sqrt{y}$ from $y = 2$ to $y = 20$ about the y-axis.

Solution   From $x = g(y) = \sqrt{y}$, $g'(y) = \dfrac{1}{2\sqrt{y}}$. The required surface is given by

$$S = 2\pi \int_2^{20} x\sqrt{1 + (g'(y))^2}\, dy$$

$$= 2\pi \int_2^{20} \sqrt{y}\sqrt{1 + \frac{1}{4y}}\, dy$$

$$= \pi \int_2^{20} \sqrt{4y + 1}\, dy$$

$$= \frac{\pi}{4}\frac{2}{3}(4y + 1)^{3/2}\Big|_2^{20}$$

$$= \frac{\pi}{6}\left[(81)^{3/2} - (9)^{3/2}\right]$$

$$= \frac{\pi}{6}(729 - 27)$$

$$= 117\pi \text{ square units}$$ ●

Exercises 7.11

In each of Exercises 1 through 10 find the area of the surface of revolution obtained by rotating the given curve about the x-axis.

1. $f(x) = k = \text{constant}, \quad (k > 0), x \in [a, b]$

2. $f(x) = 3x, \quad x \in [0, 3]$

3. $f(x) = 4x, \quad x \in [2, 5]$

4. $f(x) = 3x + 2, \quad x \in [0, 4]$

5. $\dfrac{x}{a} + \dfrac{y}{b} = 1, \quad x \in [0, a]$ where $a > 0$ and $b > 0$

6. $x + y = 1, \quad x \in [-2, \frac{1}{2}]$

7. $y = 2\sqrt{2}\sqrt{x}, \quad x \in [0, 2]$

8. $y = \sqrt{16 - x^2}, \quad x \in [2, 3]$

9. $y = b\sqrt{\dfrac{x}{a}}, \quad (a > 0 \text{ and } b > 0), x \in [0, a]$

10. $y = \dfrac{x^3}{3} + \dfrac{1}{4x}, \quad x \in [1, 2]$

11. Verify the formula for the lateral surface area S of a right circular cone of base radius r and slant height L.

12. Verify the formula for the lateral surface area S of a frustum of a right circular cone of base radii r and R and slant height L.

13. The arc of $x^{2/3} + y^{2/3} = a^{2/3}$ from $(0, a)$ to $(a, 0)$ is rotated about the y-axis. Find the area of the surface generated.

14. The arc in the first quadrant of $9x^2 = y(3 - y)^2$ from $(0, 0)$ to $(0, 3)$ is rotated about the y-axis. Find the area of the surface generated.

15. The arc of $f(x) = |x - 4|, \ 2 \le x \le 7$ is rotated about the x-axis. Find the area of the resulting surface.

Review and Miscellaneous Exercises

1. Find the area bounded by the curves $x^2 = 2y + 1$ and $y - x = 1$. Sketch the curves.

2. Find the area bounded by the curves $x = y(y - 2)^2$ and $x = 0$. Sketch the curves.

3. Find the volume obtained by rotating about the x-axis the region in the first quadrant bounded by $y = x\sqrt{b^2 - x^2}$ and $y = 0$ where $b > 0$.

4. Find the volume obtained by rotating about the y-axis the region in the first quadrant bounded by $y = \sqrt{4px} \ (p > 0), x = 4p$, and $y = 0$. Use the shell method.

5. A force of 60 lb is required to compress a spring 26 in. long to 25 in. What is the work done in compressing the spring from a length of 24 in. to 22 in.? Assume that Hooke's law applies throughout the compression.

6. A tank in the shape of an inverted right circular cone 16 ft high and with base radius 8 ft contains water. The surface of the liquid is 4 ft from the top of the cone. Find the work done in pumping the water to the top of the tank if water weighs 62.5 lb/ft^3.

7. A cable 100 ft long and weighing 4 lb/ft is hanging from a windlass. A weight of 300 lb is attached to the bottom of the cable. Find the work done in lifting the cable and weight to the windlass.

8. An electron is attracted to a nucleus by a force F, which is inversely proportional to the square of the distance r between them; that is, $F = -\dfrac{k}{r^2} \ (k > 0)$. If the nucleus does not move, compute the work done by the nucleus (a) in moving the particle from $r = 3b$ to $r = 2b$ and (b) in moving the particle from $r = 3b$ to $r = b$.

9. A vertical plane plate in the shape of a triangle is immersed in a liquid of weight density γ such that an edge of length b lies in the surface of the liquid. If the altitude to this base is of length h, find the total force on one face of the plate. (*Hint:* Choose the origin of coordinates at the vertex opposite b and let the altitude h coincide with one of the coordinate axes.)

10. A vertical tank submerged in water contains a circular window in its wall with radius 2 ft and whose center is 6 ft below the surface of the liquid. Find the total force due to liquid pressure on the window, assuming that the density of water is 62.5 lb/ft^3.

11. Five particles having masses 3, 2, 5, 8, and 4 slugs are located on the x-axis respectively at $x = 1$, $x = 6$, $x = -2$, $x = 3$, and $x = 4$. Determine the center of mass of the system.

12. Find the center of mass of the five particles with masses 1, 3, 4, 5, and 2 slugs located respectively at the points $(0, 1)$, $(2, 1)$, $(1, 3)$, $(-1, 2)$, and $(2, 4)$.

13. A bar that is 8 ft long has a density that is linear with respect to the distance from one end. The density is 4 slugs/ft at one end and 20 slugs/ft at the other end. Find **(a)** the total mass of the bar and **(b)** the center of mass.

14. Two boys of weights w_1 lb and w_2 lb are at opposite ends of a fulcrum of a seesaw of length L ft. If the seesaw is regarded as weightless, find the formula for the distance of each boy from the fulcrum required for equilibrium. What is the reaction at the fulcrum?

15. Find the centroid of the region bounded by $x + y = 2$, $y = 0$, and $x = 0$.

16. Find the centroid of the region bounded by the curves $y^2 = 4x$ and $y^3 = 8x$.

17. Find the center of mass of the homogeneous solid formed when the area between the parabola $y^2 = 4x$ and the x-axis from $x = 0$ to $x = 4$ is revolved about the x-axis.

18. Find the center of mass of the homogeneous solid formed when the area between the parabola $y^2 = 4x$ and the y-axis from $y = 0$ to $y = 4$ is revolved about the x-axis.

19. Find the length of arc of the curve $y = ax + b$ connecting the two points $(0, b)$ and $(2, 2a + b)$.

20. Find the length of arc of the curve that satisfies the equation $3y = (x^2 + 2)^{3/2}$ and connecting the two points $(0, \frac{2}{3}\sqrt{2})$ and $(1, \sqrt{3})$.

21. Find the area of the surface of revolution by rotating the arc of the curve $y = \sqrt{R^2 - x^2}$ from $x = a$ to $x = b$ around the x-axis $(-R \leq a < b \leq R)$. Interpret your result.

22. The parabola $y^2 = 4px$, $p > 0$, is revolved about the x-axis and a paraboloid of revolution is produced. Find its surface area if $0 \leq x \leq p$.

23. Find the area bounded by $y = 4 - x^2$ and $x - y + 2 = 0$. Sketch the curves.

24. Find the area bounded by $y = x$, $x - 4y - 9 = 0$, and $y + x - 4 = 0$. Sketch the area involved.

25. A solid figure F has a base in the xy plane that is a circle of radius a, and each section of F formed by a plane perpendicular to the x-axis is an isosceles triangle with altitude equal to $\frac{3}{2}$ times the base. Find the volume of F.

26. Find the volume generated by revolving about the line $x = 3$ the area bounded by $y = x^2 + 2$, the x-axis, the y-axis, and the line $x = 3$.

27. Set up but do not evaluate integrals for the length of the following curves:
    (a) $y = \sqrt{x}$ from $(0, 0)$ to $(9, 3)$.
    (b) $xy = 1$ from $(1, 1)$ to $(5, \frac{1}{5})$.

28. A pail weighs 8 lb and contains 20 lb of sand at the start. The pail is lifted slowly a distance 6 ft but, as it is lifted, the sand leaks out of a hole at a uniform rate of $\frac{3}{2}$ lb per foot lifted. Find the work done in lifting the pail.

29. Find the natural length of a heavy spring, given that the work done in stretching it from 2 ft to $2\frac{1}{2}$ ft is 60% of the work required to stretch it from $2\frac{1}{2}$ ft to 3 ft. Assume that Hooke's law holds throughout.

30. Find the surface area obtained when the curve $y = |2x - 5|$, $0 \leq x \leq 6$ is revolved about the x-axis.

31. A solid of revolution is obtained by rotating the area bounded by $y = f(x)$ $(f(x) \geq 0)$, the x-axis, and the lines $x = a$ and $x = u$. Its volume $V(u)$ for all u in $[2, b]$ where $b \geq u > 2$ is $\dfrac{3u^2}{2} - 2u - 2$. Find $y = f(x)$. Make a sketch of the region being rotated.

32. Let $F(x)$ denote the resultant of all forces acting on a particle of mass m in the direction of x along which the particle is moving. In accordance with Newton's second law of motion: $F = m\dfrac{d^2x}{dt^2} = m\dfrac{dv}{dt}$, where x is the displacement, v the velocity, and t the time. Show that

$$\int_{x_0}^{x_1} F(x)\, dx = \frac{mv^2}{2}\Bigg|_{x=x_1} - \frac{mv^2}{2}\Bigg|_{x=x_0}$$

This states that the work done by the resultant force $F(x)$ in moving the particle from $(x_0, 0)$ to $(x_1, 0)$ on the x-axis is equal to the change in kinetic energy of the particle. (*Hint:* Recall that the kinetic energy of

a particle is given by $\dfrac{mv^2}{2}$ and that by the chain rule

$$\frac{d^2x}{dt^2} = \frac{dv}{dt} = \frac{dv}{dx}\frac{dx}{dt} = v\frac{dv}{dx}$$

Integrate both sides of $F = m\dfrac{d^2x}{dt^2}$ with respect to x. This is a special case of a very significant theorem in the study of mechanics.)

8

Logarithmic and Exponential Functions

THE DEFINITION AND PROPERTIES OF THE NATURAL LOGARITHM

The elements of calculus may be used to develop a more adequate theory of the logarithmic and exponential functions than is ordinarily done in the precalculus courses taught in high school or in college. We may recall that our previous instruction began with the integral powers of a positive number a, and then various rules were developed such as

$$a^m \cdot a^n = a^{m+n} \qquad \frac{a^m}{a^n} = a^{m-n} \qquad (a^m)^n = a^{mn}$$

where n and m are positive integers. Furthermore, we defined $a^{1/n} = \sqrt[n]{a}$ and this led to rules involving a^r where $r = \dfrac{m}{n}$ is a rational number. Then additional relations such as $a^r \cdot a^s = a^{r+s}$, $(a^r)^s = a^{rs}$, and so on, were proved where $a > 0$, and r and s are arbitrary rational numbers.

The extension to numbers of the form $a^{\sqrt{2}}$ is very difficult, and the generalization to a^x where x is an arbitrary real number is also delicate. Consequently, no justification of rules such as $a^{x_1} \cdot a^{x_2} = a^{x_1+x_2}$ is ordinarily given. At this point, the logarithm of x to the base a, written $y = \log_a x$, $(a > 0, a \neq 1)$, is usually introduced, by its definition, as the inverse function of $y = a^x$.

It is our objective to deveop the theory of these functions through the application of calculus. We will start with the logarithmic function, develop some of its important properties, and then obtain the exponential function. Thus, the order in which these functions will be considered is the reverse of our previous experience.

Let us start with the basic power rule of integral calculus, for $0 < a < b$,

$$\int_a^b x^n \, dx = \frac{x^{n+1}}{n+1} \Bigg|_a^b = \frac{b^{n+1} - a^{n+1}}{n+1} \qquad \text{if } n \neq -1$$

This still leaves the problem of integrating x^{-1}. We recognize that $f(x) = \dfrac{1}{x}$, being a rational function of x, is continuous for all x in its domain, that is, all

$x \neq 0$. Thus the integral $\int_a^b \dfrac{1}{x}\, dx$ certainly exists when a and b are both positive. Therefore we define a "new and interesting function"

$$\ln x = \int_1^x \frac{1}{t}\, dt \qquad x > 0 \tag{1}$$

called the **natural logarithm** (Figure 8.1.1). Thus, for example, $\ln 2$ is the area bounded by the curve $y = \dfrac{1}{t}$, the t-axis, and the vertical lines $t = 1$ and $t = 2$.

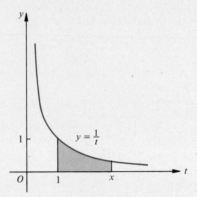

Figure 8.1.1

The properties of this function $\ln x$ may seem less apparent than is usually the case with functions given by algebraic formulas; however, this is due to some extent to our being unaccustomed to working with functions of an integral's upper limit. We start by making use of the first fundamental theorem of calculus, which yields

$$D_x(\ln x) = \frac{1}{x} \qquad x > 0 \tag{2}$$

This says first of all that $\ln x$ is differentiable for all positive x. Furthermore, from (2), the sign of the first derivative is always positive, which implies that $\ln x$ is an increasing function of x. This means that the graph of $\ln x$ rises as x increases. Also from (1),

$$\ln 1 = \int_1^1 \frac{dt}{t} = 0 \tag{3}$$

which states that $y = \ln x$ crosses the x-axis at $(1, 0)$. Since the value of y increases as x increases, $\ln x$ is negative in $0 < x < 1$, and positive when $x > 1$.

The sign of the second derivative indicates the direction in which the curve bends. Differentiation of both sides of (2) yields

$$D_x^2(\ln x) = -\frac{1}{x^2} < 0 \qquad x > 0$$

which implies that the graph of the curve $y = \ln x$ is concave downward over its entire domain. Also there are no inflection points.

Next, we establish the significant properties for the natural logarithm function.

Theorem 1   If x and a are any positive numbers, then

$$\ln ax = \ln a + \ln x \tag{4}$$

$$\ln \frac{x}{a} = \ln x - \ln a \tag{5}$$

and if r is any rational number, then

$$\ln x^r = r \ln x \tag{6}$$

Proof   Let u be any differentiable function of x, then, by the chain rule and (2)

$$D_x(\ln u) = \frac{1}{u} D_x u \qquad (u(x) > 0) \tag{7}$$

In particular, substitute ax for u in (7), where a is an arbitrary positive number. Then

$$D_x(\ln ax) = \frac{1}{ax} D_x(ax) = \frac{1}{ax} a = \frac{1}{x}$$

Therefore we conclude that the two functions $\ln ax$ and $\ln x$ possess the same derivative $\frac{1}{x}$, which implies that they must differ by a constant. Thus

$$\ln ax = \ln x + C, \quad a > 0, \, x > 0 \tag{8}$$

In particular, set $x = 1$ in (8) and obtain from (3),

$$\ln a = \ln 1 + C = 0 + C = C$$

Therefore

$$\ln ax = \ln x + \ln a = \ln a + \ln x \qquad a > 0, \, x > 0$$

and (4) is proved.

Equation (5) follows readily from (4) by setting $x = \frac{1}{a}$ in (4). We obtain

$$\ln \left(a \cdot \frac{1}{a} \right) = \ln 1 = 0 = \ln a + \ln \frac{1}{a}$$

or

$$\ln \frac{1}{a} = -\ln a$$

Thus

$$\ln \frac{x}{a} = \ln \left(x \cdot \frac{1}{a} \right) = \ln x + \ln \frac{1}{a} = \ln x - \ln a$$

and (5) is proved.

To obtain (6), we again apply the chain rule (7) with $u = x^r$. Thus

$$D_x(\ln x^r) = \frac{1}{x^r} D_x(x^r) = \frac{rx^{r-1}}{x^r} = \frac{r}{x} \qquad (r \text{ rational})$$

However, $D_x(r \ln x) = r D_x(\ln x) = \dfrac{r}{x}$ for any constant r. Therefore the two functions $\ln x^r$ and $r \ln x$ have the same derivative $\dfrac{r}{x}$ and, therefore, differ by a constant; that is,

$$\ln x^r = r \ln x + C \tag{9}$$

Set $x = 1$ in (9) and obtain (since $1^r = 1$)

$$\ln 1 = r \ln 1 + C \qquad \text{or} \qquad 0 = 0 + C$$

Thus $C = 0$ and equation (6) follows. ∎

We know that $\ln (x^n) = n \ln x, (x > 0)$ where n is an arbitrary positive integer. If x is chosen to be 2, for example, $\ln 2^n = n \ln 2$, where $\ln 2$ is a positive constant. Thus as $n \to \infty$, $\ln 2^n$ also tends to infinity. This means that $\ln x$ is a monotonically increasing function that tends to infinity as $x \to \infty$. Also $\ln x \to -\infty$ as $x \to 0^+$. This is easily proved by making the substitution $x = \dfrac{1}{u}$ and using the identity $\ln \dfrac{1}{u} = -\ln u$. In detail,

$$\ln x = \ln \frac{1}{u} = -\ln u$$

so that $\ln x$ becomes larger and larger negatively as $x \to 0^+$ and correspondingly $u \to \infty$. Therefore the y-axis is a vertical asymptote of the curve $y = \ln x$. From the accumulated information on the properties of $\ln x$ we can easily make a sketch of $y = \ln x$. This is shown in Figure 8.1.2.

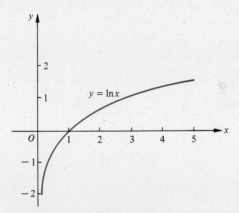

Figure 8.1.2

EXAMPLE 1   If $y = \ln (x^2 + 3x + 11)$, find $D_x y$.

Solution   This is a direct application of (7) with $u = x^2 + 3x + 11 > 0$. Also $D_x u = 2x + 3$ so that

$$D_x y = \frac{1}{u} D_x u = \frac{2x + 3}{x^2 + 3x + 11} \qquad \bullet$$

EXAMPLE 2   If $y = \ln \dfrac{x^2 + 3}{x^2 + x + 2}$, find $D_x y$.

Solution   It is easy to see that both the numerator and denominator are positive for all x so that

$$u = \frac{x^2 + 3}{x^2 + x + 2} > 0$$

Therefore (7) is directly applicable. However, it is more convenient to employ property (5) first, thus

$$y = \ln(x^2 + 3) - \ln(x^2 + x + 2)$$

Then applying property (7) to each term, we obtain

$$D_x y = \frac{2x}{x^2 + 3} - \frac{2x + 1}{x^2 + x + 2}$$

which simplifies to

$$D_x y = \frac{x^2 - 2x - 3}{(x^2 + 3)(x^2 + x + 2)}$$

●

EXAMPLE 3   If $y = \dfrac{x^4 \sqrt{x^2 + 9}}{(x^2 + 5)^3}$, find $D_x y$.

Solution   This is a somewhat bothersome problem because of the many rules that must be utilized in order to obtain the derivative in the conventional manner. Fortunately, the ln function can be employed to our advantage. We note that each of the expressions is positive.

We take the ln of both sides, and, using properties (4), (5), and (6), we obtain

$$\ln y = \ln x^4 + \ln \sqrt{x^2 + 9} - \ln(x^2 + 5)^3$$

$$= 4 \ln x + \frac{1}{2} \ln(x^2 + 9) - 3 \ln(x^2 + 5)$$

Form D_x of both sides using (7):

$$\frac{1}{y} D_x y = \frac{4}{x} + \frac{x}{x^2 + 9} - \frac{6x}{x^2 + 5}$$

After simplification

$$D_x y = \frac{y[-x^4 + 7x^2 + 180]}{x(x^2 + 9)(x^2 + 5)}$$

$$= \frac{x^4 \sqrt{x^2 + 9}}{(x^2 + 5)^3} \cdot \frac{[-x^4 + 7x^2 + 180]}{x(x^2 + 9)(x^2 + 5)} = \frac{x^3[-x^4 + 7x^2 + 180]}{(x^2 + 5)^4 \sqrt{x^2 + 9}}$$

●

The technique that has just been developed is known as **logarithmic differentiation.** It utilizes the fact that the ln function converts products, quotients, powers and roots into sums, differences, and constant multiples, thereby resulting in a considerable reduction of labor. Thus we have received a substantial side benefit from our new function $\ln x$.

In our development of the formula

$$D_x(\ln x) = \frac{1}{x}$$

it has been assumed that x is positive. We started with 1 as the lower limit and, in order for the Riemann integral to exist, x was of necessity restricted to positive values. Suppose now that we choose -1 as the lower limit, then x of necessity must be less than zero in order that

$$\int_{-1}^{x} \frac{dt}{t} \text{ exists.}$$

Thus if we start with the integral

$$\int_{-1}^{x} \frac{dt}{t} \qquad (x < 0)$$

and the substitution $t = -u$ is made, there results

$$\int_{-1}^{x} \frac{dt}{t} = \int_{1}^{-x} \frac{d(-u)}{-u} = \int_{1}^{-x} \frac{du}{u} = \ln(-x) = \ln|x|$$

Therefore we have

$$\int \frac{dx}{x} = \ln|x| + C \qquad (x \neq 0) \tag{10}$$

which implies that

$$D_x(\ln|x| + C) = \frac{1}{x} \qquad (x \neq 0) \tag{11}$$

More generally, if u is any differentiable function of x, we obtain from (10) and (11)

$$\int \frac{du}{u} = \ln|u| + C \qquad u \neq 0 \tag{12}$$

and

$$D_x(\ln|u| + C) = \frac{1}{u} D_x u \qquad u \neq 0 \tag{13}$$

EXAMPLE 4   If $f(x) = \dfrac{5x \sqrt[3]{x-2}}{x+4}$ find $f'(x)$.

Solution $f(x)$ is sometimes negative and is nonnegative for other values of x. To accommodate this, we take the absolute value of both sides

$$|f(x)| = \left| \frac{5x \sqrt[3]{x-2}}{x+4} \right|$$

and then apply the ln function to both sides

$$\ln |f(x)| = \ln \left| \frac{5x \sqrt[3]{x-2}}{x+4} \right| = \ln |5x| + \frac{1}{3} \ln |x-2| - \ln |x+4|$$

where some of the properties of the logarithmic function are utilized. Next, we differentiate both sides with respect to x, in accordance with (12) to obtain

$$\frac{1}{f(x)} f'(x) = \frac{1}{x} + \frac{1}{3(x-2)} - \frac{1}{x+4}$$

and we solve for $f'(x)$ by multiplying both sides by $f(x)$:

$$f'(x) = \frac{5x \sqrt[3]{x-2}}{x+4} \left(\frac{x^2 + 16x - 24}{3x(x-2)(x+4)} \right)$$

$$= \frac{5(x^2 + 16x - 24)}{3(x-2)^{2/3}(x+4)^2} \qquad \bullet$$

EXAMPLE 5   Evaluate $\displaystyle\int_{-4}^{-1} \frac{dx}{5x - 1} = I$

Solution   The function $\dfrac{1}{5x-1}$ is continuous for all $x \neq \frac{1}{5}$, and $\frac{1}{5}$ is not in the given interval. Thus, the definite integral exists and (12) applies

$$I = \frac{1}{5} \int_{-4}^{-1} \frac{d(5x-1)}{5x-1} = \frac{1}{5} \ln |5x - 1| \Big|_{-4}^{-1} \qquad \text{(here } u = 5x - 1\text{)}$$

$$= \frac{1}{5} (\ln|-6| - \ln|-21|) = \frac{1}{5} (\ln 6 - \ln 21)$$

$$= -\frac{1}{5} (\ln 21 - \ln 6) = -\frac{1}{5} \ln 3.5$$

[Note that the result is negative because the integrand is negative in the given interval and the integration is from a smaller number (-4) to a larger number (-1).] $\qquad \bullet$

EXAMPLE 6   Find $\displaystyle\int \frac{x^3}{x+5} dx$.

Solution   Since the rational fraction is an improper fraction, we must divide the denominator into the numerator and obtain

$$\frac{x^3}{x+5} = x^2 - 5x + 25 - \frac{125}{x+5}$$

Term by term integration then yields

$$\int \frac{x^3}{x+5} dx = \int \left(x^2 - 5x + 25 - \frac{125}{x+5} \right) dx$$

$$= \frac{x^3}{3} - \frac{5x^2}{2} + 25x - 125 \ln |x+5| + C \qquad \bullet$$

EXAMPLE 7 Find $\displaystyle\int_0^2 \frac{x}{x^2 - 3}\,dx = I$.

Solution The "obvious" approach is to note that $D_x(x^2 - 3) = 2x$ so that

$$I = \frac{1}{2}\int_0^2 \frac{d(x^2 - 3)}{x^2 - 3} = \frac{1}{2}\left(\ln|x^2 - 3|\right)\Big|_0^2$$

$$= \frac{1}{2}\left(\ln|1| - \ln|-3|\right) = -\frac{1}{2}\ln 3$$

However, this solution is *wrong* because the magnitude of the integrand is approaching infinity as x tends to $\sqrt{3}$ which is in $[0, 2]$. Therefore the Riemann integral does *not* exist. ●

EXAMPLE 8 Find $\displaystyle\int \frac{\ln x^5}{x}\,dx = I$

Solution $I = 5\displaystyle\int \ln x\,d(\ln x)$, because $d(\ln x) = \dfrac{dx}{x}$ and hence $I = \dfrac{5(\ln x)^2}{2} + C$ by application of the generalized power rule. ●

EXAMPLE 9 Find $D_x \displaystyle\int_5^{x^2} \frac{dt}{2t + 3}$.

Solution **Method 1.** $\dfrac{1}{2t + 3}$ is continuous in the given interval and the method of this section applies. Thus

$$\int_5^{x^2} \frac{dt}{2t + 3} = \frac{1}{2}\ln(2t + 3)\Big|_5^{x^2}$$

$$= \frac{1}{2}\left[\ln(2x^2 + 3) - \ln 13\right]$$

Thus by Equation (7),

$$D_x \int_5^{x^2} \frac{dt}{2t + 3} = \frac{2x}{2x^2 + 3}$$

Method 2. The first fundamental theorem of calculus together with the chain rule yields

$$D_x \int_5^{x^2} \frac{dt}{2t + 3} = D_u\left(\int_5^u \frac{dt}{2t + 3}\right)D_x u \qquad \text{where } u = x^2$$

$$= \frac{1}{2u + 3}\cdot 2x = \frac{2x}{2x^2 + 3}$$ ●

Exercises 8.1

1. Given that $\ln 2 \doteq 0.6931$, $\ln 3 \doteq 1.0986$ and $\ln 5 \doteq 1.6094$. To four decimal places find, approximately (without the use of tables) (a) $\ln 4$, (b) $\ln 18$, (c) $\ln 10$, (d) $\ln \sqrt{15}$ and (e) $\ln \frac{1}{12}$.
2. Solve the inequality $\ln(x^2 - 3x) \geq \ln x$.

In Exercises 3 through 14, differentiate each of the following functions.

3. $f(x) = \ln(x^2 + 16)$

4. $g(x) = \ln|5x - 7|$

5. $F(t) = \ln(t^2 + 3t)$

6. $f(x) = \ln\sqrt{3 + 4x^2}$

7. $G(t) = \ln(\ln t)$

8. $F(x) = x^3 \ln x$

9. $g(x) = \ln\left(\dfrac{x^2 - 8}{x^2 + 1}\right)$

10. $h(t) = \ln\dfrac{t^3}{\sqrt{t^2 + 9}}$

11. $F(t) = \ln^5(2t + 9)$

12. $G(x) = \ln(x + \sqrt{x^2 + 16})$

13. $f(t) = \ln(\ln^2 t)$

14. $g(x) = \ln^3(\ln 2x)$

In Exercises 15 through 20, use logarithmic differentiation to find the derivatives of the following functions.

15. $y = x^3 \sqrt[4]{x^2 + 9}$

16. $y = \dfrac{x\sqrt{x^2 + 4}}{\sqrt[3]{x^2 + 1}}$

17. $y = \dfrac{(x + 3)(x - 2)}{(x - 1)}$

18. $y = (x + a)(x + b)(x + c)$ $(a, b, c$ constants$)$

19. $y = \sqrt{\dfrac{ax^2 + b}{cx^2 + d}}$ $(a, b, c,$ and d are positive constants$)$

20. $y = (5x + 4)^3 (x^2 + 3)^2 \sqrt[3]{6x + 1}$

In Exercises 21 through 24, find the derivative at the indicated value of x.

21. $y = x^2 \ln x$, $x = 1$

22. $y = x \ln\sqrt{x}$, $x = 4$

23. $y = \dfrac{\ln x}{x^2}$, $x = 2$

24. $y = \ln\sqrt{\dfrac{2x + 1}{x + 2}}$, $x = 1$

In Exercises 25 through 38, evaluate the integrals.

25. $\displaystyle\int \dfrac{dt}{3t + 1}$

26. $\displaystyle\int \dfrac{t + 1}{t^2 + 16}\, dt$

27. $\displaystyle\int \dfrac{x^2}{x + 1}\, dx$

28. $\displaystyle\int \dfrac{\ln^3 x}{x}\, dx$

29. $\displaystyle\int \dfrac{t^3}{t - 2}\, dt$

30. $\displaystyle\int_{-2}^{0} \dfrac{dx}{(11 + 5x)^3}$

31. $\displaystyle\int_{0}^{4} \dfrac{x}{1 - x^2}\, dx$

32. $\displaystyle\int_{2}^{4} \dfrac{t}{1 - t^2}\, dt$

33. $\displaystyle\int_{3}^{x} \dfrac{dt}{t}$, $x > 0$

34. $\displaystyle\int_{1}^{x} \dfrac{(t + \sqrt{t})^2}{t^3}\, dt$

35. $\displaystyle\int_{1}^{3} \dfrac{t^2 - 5t + 2}{6 - t}\, dt$

36. $\displaystyle\int_{1}^{4} \dfrac{x}{3 - x}\, dx$

37. $\displaystyle\int \dfrac{5 + \ln^2 x}{x(3 + \ln x)}\, dx$

38. $\displaystyle\int \dfrac{x}{(3x + 5)^2}\, dx$

In Exercises 39 and 40, find the derivative y′ by implicit differentiation.

39. $\ln(x + 2y) = y + 2x$

40. $\ln(xy) + 2x - 3y = 1$

In Exercises 41 and 42, find an equation of the tangent line to the given curve at the indicated point.

41. $y = \dfrac{2 \ln x}{x^2}$, $(2, \frac{1}{2} \ln 2)$

42. $y = (x + \ln x)^{10}$, $(1, 1)$

43. Derive a formula for $D_x^n(\ln x)$, $x > 0$, for arbitrary positive integer n.

44. Derive the formula $D_x^{n+1}(x^n \ln x) = \dfrac{n!}{x}$, $x > 0$, where n is an arbitrary positive integer.

45. Find a formula for $\int \ln x\, dx$ by starting with $D_x(x \ln x)$. This formula will be obtained more directly in Chapter 10 using the method of integration by parts.

46. If $y = \ln x$, then we know that $y \to \infty$ as $x \to \infty$. However, it can also be proved that $\lim\limits_{x \to \infty} \dfrac{\ln x}{x} = 0$.

Show this by comparing $\displaystyle\int_1^x \dfrac{dt}{t}$ with $\displaystyle\int_1^x \dfrac{1}{\sqrt{t}}\, dt$ for $x > 1$.

47. Prove that $\lim\limits_{x \to 0^+} (x \ln x) = 0$ using the result of Exercise 46. $\left(Hint:\ \text{Let } x = \dfrac{1}{u} \text{ so that as } x \to 0^+, \right.$ $\left. u \to \infty. \right)$

48. Prove that for all positive values of x,

$$x - \frac{x^2}{2} < \ln(1 + x) < x$$

(*Hint:* Compare derivatives and use the basic mean value theorem of calculus.) This establishes that $\lim\limits_{x \to 0^+} \dfrac{\ln(1 + x)}{x} = 1$. Why?

8.2   THE NUMBER e

The function ln has been defined by the formula

$$\ln x = \int_1^x \frac{dt}{t}, \quad x > 0 \tag{1}$$

In particular, we may use numerical integration to approximate $\ln x$ for various values of x. For example, by the trapezoidal or Simpson's rule of Chapter 10, or by using the approximating rectangles of Chapter 6, we can easily estimate that to two decimal places

$$\ln 2 \doteq 0.69 \quad \text{and} \quad \ln 3 \doteq 1.10$$

Thus it follows that

$$\ln 2 < 1 < \ln 3$$

We know that $\ln x$ increases monotonically with x and, furthermore, $\ln x$ is differentiable and consequently continuous for all positive x. This means that the equation $\ln x = 1$ has one (and only one) solution in $(2, 3)$. We therefore introduce a number e in $(2, 3)$ by the following definition: The number e is *defined* to be the unique solution of the equation $\ln x = 1$; that is, $\ln e = 1$.

The number e is named for the eighteenth century Swiss mathematician Euler who introduced it in his publication *Introductio in Analysin Infinitorum*. It plays an important role in advanced mathematics and ranks with the number π in its universality.

The number e is easily interpreted geometrically. It may be determined from the requirement that the area under the curve $y = \dfrac{1}{x}$ between $x = 1$ and $x = e$ (Figure 8.2.1) will be exactly 1 square unit:

$$\ln e = \int_1^e \frac{dt}{t} = 1 \qquad (2)$$

Hence the number e could be approximated by a trial and error procedure of partitioning the interval $[1, 3]$ and utilizing rectangles or perhaps trapezoids starting from the left side until 1 square unit is exceeded. However, this is a rather tedious technique, especially if we are interested in determining e to, for example, four or five decimal places.

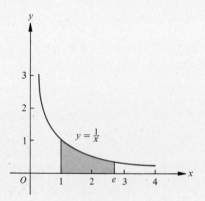

Figure 8.2.1

Therefore, let us try another method. We recall that the slope of the tangent line to the curve $y = \ln x$ is $\dfrac{1}{x}$ and, in particular, at $(1, 0)$ the slope of the tangent line is 1. Now instead of using the tangent line at $(1, 0)$ consider the slopes of the secant lines connecting the points $(1, \ln 1)$ and $\left[1 + \dfrac{1}{n}, \ln \left(1 + \dfrac{1}{n} \right) \right]$, $(n = 1, 2, 3, \ldots)$, on the curve $y = \ln x$ (Figure 8.2.2). By definition, the slope of the tangent line at $(1, 0)$ is the limit of the slopes of the secant lines as $n \to \infty$. Thus

$$1 = \lim_{n \to \infty} \frac{\ln \left(1 + \dfrac{1}{n} \right) - \ln 1}{1 + \dfrac{1}{n} - 1} = \lim_{n \to \infty} n \ln \left(1 + \frac{1}{n} \right)$$

or, by an application of one of the ln function's basic properties,

$$1 = \lim_{n \to \infty} \ln \left(1 + \frac{1}{n} \right)^n$$

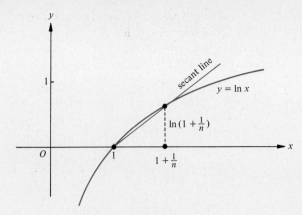

Figure 8.2.2

However, $1 = \ln e$ and hence

$$\ln e = \lim_{n \to \infty} \ln \left(1 + \frac{1}{n}\right)^n \overset{?}{=} \ln \lim_{n \to \infty} \left(1 + \frac{1}{n}\right)^n \tag{3}$$

The question mark is required because we don't know whether the order of the two operations $\lim_{n \to \infty}$ and $\ln$ can be interchanged. The formulation of (3) suggests the following valid relation

$$e = \lim_{n \to \infty} \left(1 + \frac{1}{n}\right)^n \tag{4}$$

In Table 8.2.1 we give the result of calculating $\left(1 + \frac{1}{n}\right)^n$ for a number of different values of n. Note that the values of $\left(1 + \frac{1}{n}\right)^n$ appear to be increasing with increase in n, however the increase is not rapid.

Table 8.2.1

n	1	2	3	4	5	10	100	1000	10,000
$\left(1 + \dfrac{1}{n}\right)^n$	2	2.25	2.3704	2.4414	2.4883	2.5938	2.7048	2.7169	2.7181

Fortunately it will be shown later that e can be calculated quickly and accurately by the infinite series

$$1 + 1 + \frac{1}{2!} + \frac{1}{3!} + \frac{1}{4!} + \cdots$$

and, for example, by adding the terms up to and including $\frac{1}{12!}$ we find $e \doteq 2.7182818$ correct to the seven decimal places shown.

EXAMPLE 1   Find $\lim_{n \to \infty} \left(1 + \frac{1}{n}\right)^{n+3}$.

Solution
$$\lim_{n\to\infty}\left(1 + \frac{1}{n}\right)^{n+3} = \lim_{n\to\infty}\left(1 + \frac{1}{n}\right)^{n} \cdot \lim_{n\to\infty}\left(1 + \frac{1}{n}\right)^{3}$$

provided that both limits exist on the right side. But,

$$\lim_{n\to\infty}\left(1 + \frac{1}{n}\right)^{n} = e \quad \text{and} \quad \lim_{n\to\infty}\left(1 + \frac{1}{n}\right)^{3} = 1$$

$$\left[\lim_{n\to\infty}\left(1 + \frac{1}{n}\right)^{3} = \lim_{n\to\infty}\left(1 + \frac{1}{n}\right) \cdot \lim_{n\to\infty}\left(1 + \frac{1}{n}\right) \cdot \lim_{n\to\infty}\left(1 + \frac{1}{n}\right) = (1)(1)(1) = 1\right]$$

Thus

$$\lim_{n\to\infty}\left(1 + \frac{1}{n}\right)^{n+3} = e$$

●

EXAMPLE 2   Find $\lim_{n\to\infty}\left(1 + \frac{4}{n}\right)^{n}$.

Solution   The exponent n is not the reciprocal of $\frac{4}{n}$. Therefore we rewrite the exponent n as

$$\left(1 + \frac{4}{n}\right)^{n} = \left[\left(1 + \frac{4}{n}\right)^{n/4}\right]^{4}$$

Letting $u = \frac{n}{4}$, we have

$$\lim_{n\to\infty}\left(1 + \frac{4}{n}\right)^{n/4} = \lim_{u\to\infty}\left(1 + \frac{1}{u}\right)^{u} = e$$

Using: limit of product = product of limits, we obtain

$$\lim_{n\to\infty}\left(1 + \frac{4}{n}\right)^{n} = \left[\lim_{n\to\infty}\left(1 + \frac{4}{n}\right)^{n/4}\right]^{4} = e^{4}$$

●

EXAMPLE 3   **Compound Interest Problem.** Suppose that P dollars is placed in a savings account that pays $r\%$ per year compounded quarterly. Write a formula for the amount of money in the account after n years (that is, $4n$ periods) have elapsed.

Solution   The interest for the first period (3 months) is the rate per period $\frac{r}{4}$ times the initial investment of P dollars. Thus the interest is $\frac{rP}{4}$ and the value of the account at the end of one period is $P + \frac{rP}{4} = P\left(1 + \frac{r}{4}\right)$ dollars. The interest for the second period is $\frac{r}{4}P\left(1 + \frac{r}{4}\right)$ and the value of the account at the end of two periods is $P\left(1 + \frac{r}{4}\right) + \frac{r}{4}P\left(1 + \frac{r}{4}\right) = P\left(1 + \frac{r}{4}\right)\left(1 + \frac{r}{4}\right) = P\left(1 + \frac{r}{4}\right)^{2}$ dollars.   At the end of the third period its value is

$$\frac{r}{4}P\left(1+\frac{r}{4}\right)^{2} + P\left(1+\frac{r}{4}\right)^{2}$$

$$\underbrace{\qquad\qquad}_{\text{interest}}\quad\underbrace{\qquad\qquad}_{\substack{\text{amount at end of}\\\text{second period}}}$$

or

$$P\left(1+\frac{r}{4}\right)^{2}\left(1+\frac{r}{4}\right) = P\left(1+\frac{r}{4}\right)^{3}\text{ dollars}$$

Continuing in this way, the value of the account at the end of n years, that is $4n$ periods, is

$$P\left(1+\frac{r}{4}\right)^{4n}\text{ dollars}$$

Extensive tables are available to the user of mathematics giving the value of $(1 + i)^{n}$ for values of i and n. This gives the amount after n periods on unit original principal $(P = 1)$ at the rate of interest i per period. ●

EXAMPLE 4 **Continuous Compound Interest.** Today different banks offer varying interest policies such as interest being compounded monthly, weekly, daily, and, in some cases, even *continuously*. Of course the more often a given interest rate is compounded the larger will be the accumulated total. Thus, for a given principal, 6% per year compounded continuously will be worth more than 6% per year compounded quarterly. Our objective now is to determine how much more.

PROBLEM: Suppose that $1000 is deposited in a savings bank at 6% per annum. Find the value of the account at the end of 1 year if (a) interest is compounded quarterly and (b) if interest is compounded continuously.

Solution   (a) From Example 3 we have

$$A = 1000\left(1+\frac{0.06}{4}\right)^{4} = 1000(1 + 0.015)^{4}$$

$$\doteq 1000(1.06136) = \$1061.36$$

from compound interest tables.

(b) If n is the number of periods per year, the amount at the end of the year is

$$A_{n} = 1000\left(1+\frac{0.06}{n}\right)^{n}$$

and we seek $\lim\limits_{n\to\infty} A_{n}$. But this is

$$\lim\limits_{n\to\infty} A_{n} = 1000\lim\limits_{n\to\infty}\left(1+\frac{0.06}{n}\right)^{n}$$

$$= 1000\lim\limits_{n\to\infty}\left[\left(1+\frac{0.06}{n}\right)^{n/0.06}\right]^{0.06}$$

Application of (4) and utilizing our tables yields

$$\lim\limits_{n\to\infty} A_{n} = 1000(e^{0.06}) \doteq 1000(1.0618) = \$1061.80$$

The difference is very small in this instance and will become more significant only if a much larger principal is placed in the bank or if the duration of the investment is much longer. ●

Exercises 8.2

In Exercises 1 through 5, find the limits if they exist.

1. $\lim\limits_{n \to \infty} \left(1 + \dfrac{1}{n}\right)^{100}$

2. $\lim\limits_{n \to \infty} \left(1 + \dfrac{1}{n}\right)^{n-2}$

3. $\lim\limits_{n \to \infty} \left(1 + \dfrac{1}{3n}\right)^{n}$

4. $\lim\limits_{n \to \infty} \left(\dfrac{3}{2} + \dfrac{1}{n}\right)^{n}$

5. $\lim\limits_{n \to \infty} \left(1 + \dfrac{2}{n}\right)^{n+1}$

6. Compute $\left(1 + \dfrac{1}{n}\right)^{n}$ to two decimal places for n equal to (a) -2; (b) -3; (c) -4; and (d) -10

7. (a) Show that $\left(1 - \dfrac{1}{n}\right)^{-n} = \left(1 + \dfrac{1}{n-1}\right)^{n-1}\left(1 + \dfrac{1}{n-1}\right)$

(b) Thus from the fact that $\lim\limits_{n \to \infty} \left(1 + \dfrac{1}{n}\right)^{n} = e$ what can you conclude about $\lim\limits_{n \to \infty} \left(1 + \dfrac{1}{n}\right)^{-n}$?

(c) What does this suggest about $\lim\limits_{x \to 0} (1 + x)^{1/x}$?

8. Calculate $\left(1 - \dfrac{1}{n}\right)^{n}$ to two decimal places for n equal to (a) 2; (b) 3; (c) 4; and (d) 10. We shall show in

Exercise 9 that $\lim\limits_{n \to \infty} \left(1 - \dfrac{1}{n}\right)^{n} = \dfrac{1}{e} \doteq 0.3679 \cdots$.

9. (a) Show that $\left(1 - \dfrac{1}{n}\right)^{n} = \left(\dfrac{n-1}{n}\right)^{n}$

(b) Let $n - 1 = k$, then verify that

$$\left(1 - \dfrac{1}{n}\right)^{n} = \left(\dfrac{k}{k+1}\right)^{k+1} = \left(\dfrac{1}{1 + \dfrac{1}{k}}\right)^{k+1}$$

(c) From (a) and (b) show that $\lim\limits_{n \to \infty} \left(1 - \dfrac{1}{n}\right)^{n} = \dfrac{1}{e}$

10. Suppose that \$15,000 is deposited in a savings account at a bank whose annual interest rate is 7%. What amount will be in the account at the end of 10 years if
(a) interest is compounded quarterly?
(b) interest is compounded continuously?

11. A parent decides to place \$$P$ in a bank today so that in 15 years he will have \$20,000 for educational expenses. Find P if the interest rate is 6% per year compounded continuously.

12. Show that the numbers of the form $\left(1 + \dfrac{1}{n}\right)^{n}$ are monotonically increasing as n increases for $n = 1, 2,$

$3, \ldots.$ $\Bigg($*Hint:* Consider the function F obtained by starting with $\ln\left(1 + \dfrac{1}{n}\right)^{n}$ and replacing n by a

continuous variable x; that is, $F(x) = x \ln\left(1 + \dfrac{1}{x}\right)$. Show that $F'(x) > 0$ when $x \geq 1.$$\Bigg)$

8.3   INVERSE FUNCTIONS

In Section 2.2 we discussed the composition of functions, that is, the formation of functions of functions. In particular, we have seen examples of two functions f and g such that $f(g(x)) \neq g(f(x))$. Now we will focus on a special case of major significance where not only does $g(f(x)) = f(g(x))$ but also both of these expressions are equal to x.

EXAMPLE 1   Suppose that $f(x) = 4x$ and we seek a second function $g(x)$ such that $g(f(x)) = x$.

Solution   The function $g(x)$ that undoes or reverses the effect of $f(x) = 4x$ is $g(x) = \dfrac{x}{4}$. It is readily verified that

$$g(f(x)) = \frac{f(x)}{4} = \frac{4x}{4} = x$$

and

$$f(g(x)) = 4\,g(x) = 4 \cdot \frac{x}{4} = x$$

The function g is said to be the ***inverse*** of the function f. The graphs of f and g are straight lines passing through the origin with slopes 4 and $\frac{1}{4}$, respectively (Figure 8.3.1). ●

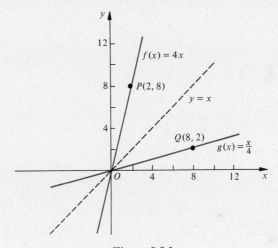

Figure 8.3.1

EXAMPLE 2   Let $f(x) = 3x + 2$ and a second function $g(x)$ is required such that $g(f(x)) = x$.

Solution   For any given "input" x, that is, any real number x, we obtain the "output" $f(x)$ by multiplying x by 3 and then adding 2 to the result. Thus it appears that to reverse the effect of f we should subtract 2 from the output $f(x)$ and then divide by 3, from which it is suggested that $g(x) = \dfrac{x-2}{3}$. To verify, we calculate

$$g(f(x)) = \frac{f(x) - 2}{3} = \frac{3x + 2 - 2}{3} = x$$

and

$$f(g(x)) = 3g(x) + 2 = 3\left(\frac{x-2}{3}\right) + 2 = x$$

The graphs of f and g are straight lines not passing through the origin with slopes 3 and $\frac{1}{3}$, respectively (Figure 8.3.2). ●

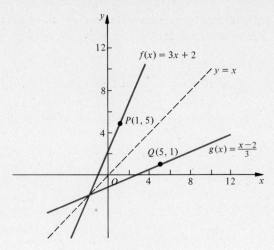

Figure 8.3.2

Suppose we next consider the function $f(x) = x^2$ (Figure 8.3.3). It seems natural to select $g(x) = \sqrt{x}$ to be the inverse of f. However, if we take a negative value of x (which is in the domain of f) we find that g is not real. Also $g(x) = \sqrt{-x}$ will not be real if x is positive. In fact,

$$f(x) = x^2 \qquad (-\infty, \infty)$$

does *not* possess an inverse function. For example,

$$g(f(5)) = g(5^2) = g(25) = 5 \quad \text{is required}$$

and

$$g(f(-5)) = g((-5)^2) = g(25) = -5 \quad \text{is also required.}$$

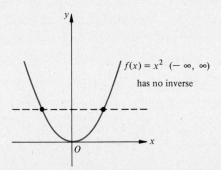

Figure 8.3.3

The difficulty is that $f(x) = x^2$ is not a one-to-one function. Interpreted geometrically, lines parallel to the x-axis and above it intersect the graph of $f(x)$ in two points. Thus we observe that a necessary condition for a function f to possess an inverse is that it be one-to-one. The preceding discussion leads to the following definition.

Definition   Let f be a one-to-one function. The function g such that

$$g(f(x)) = x \tag{1a}$$

is said to be the ***inverse*** of f.

442   *Logarithmic and Exponential Functions / Ch. 8*

Note that from (1a) the range of f must be the same as the domain of g and also from (1a) the range of g must be the domain of f (since g maps into x). Thus

$$f(g(x)) = x \qquad (1b)$$

must hold also; that is, functions f and g must be mutually inverse.

If $f(x)$ and $g(x)$ are mutually inverse functions, we may write f^{-1} instead of g to denote the inverse of the function f. The symbol f^{-1} is pronounced "f inverse." Note that the -1 is not to be construed as an exponent; that is, f^{-1} does not mean $\dfrac{1}{f}$ as it would if f were a number. Thus we have

$$f^{-1}(f(x)) = x \qquad (2a)$$

for all x in the domain of f and

$$f(f^{-1}(x)) = x \qquad (2b)$$

for all x in the domain of f^{-1}.

Analytically if $(x, y) \in f$ then $(y, x) \in f^{-1}$; that is, f^{-1} can be constructed from f by reversing the members in each ordered pair of f. This gives rise to two significant observations:

1. A necessary and sufficient condition that a function f possess an inverse is that it be one-to-one.
2. A simple geometric interpretation enables us to graph the inverse f^{-1} of a given one-to-one function $f(x)$. If point $P(a, b)$ lies on $f(x)$, then a corresponding point $Q(b, a)$ must lie on the graph of its inverse function. From elementary Euclidean geometry it is known that the line segment PQ is perpendicular to the line $y = x$ and is bisected by it (show this). Thus, point Q is a reflection of point P with respect to the line $y = x$, and conversely. This implies that, if we know the graph of a one-to-one function $y = f(x)$, the graph of its inverse is obtained by reflecting the graph of $y = f(x)$ with respect to the line $y = x$ (see Figures 8.3.1 and 8.3.2).

These observations suggest a simple procedure for finding f^{-1} if it exists:

1. Solve the equation $y = f(x)$ for x in terms of y, getting $x = g(y)$. This function is the inverse function f^{-1}; that is, $f^{-1}(y) = g(y)$.
2. If we wish to write f^{-1} as a function of x, simply interchange the x and y variables in $x = f^{-1}(y)$ to get $y = f^{-1}(x)$.

EXAMPLE 3 Given the function $f(x) = x^2$ where $x \geq 0$. Find the inverse function $g(x) = f^{-1}(x)$ and sketch the two functions.

Solution Since x is nonnegative, the given function is one-to-one. Solving $y = x^2$ for x gives $x = \sqrt{y}$ ($\sqrt{y}$ exists since $y = x^2 \geq 0$; we only use the positive square root because we were given $x \geq 0$).
Thus

$$g(y) = f^{-1}(y) = \sqrt{y}$$

Written in terms of x, we have

$$g(x) = f^{-1}(x) = \sqrt{x}$$

The graphs of functions f, g, and the line of symmetry are shown in Figure 8.3.4. ●

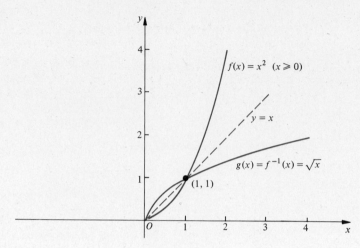

Figure 8.3.4

EXAMPLE 4   Show that the function $f(x) = \dfrac{5x - 2}{3x - 5}$ possesses an inverse function and determine it.

Solution   First let us show that the given function is one-to-one with domain all $x \neq \frac{5}{3}$. Let x_1 and x_2 be any two values of x with the same image; that is, $f(x_1) = f(x_2)$. We must show that $x_1 = x_2$. By hypothesis

$$\frac{5x_1 - 2}{3x_1 - 5} = \frac{5x_2 - 2}{3x_2 - 5}$$

or

$$15x_1x_2 - 25x_1 - 6x_2 + 10 = 15x_1x_2 - 25x_2 - 6x_1 + 10$$

from which

$$19x_1 = 19x_2 \qquad \text{or} \qquad x_1 = x_2$$

Hence, the inverse function $g(x) = f^{-1}(x)$ must exist.

Solving $y = \dfrac{5x - 2}{3x - 5}$ for x we find that $3xy - 5y = 5x - 2$, giving $(3y - 5)x = 5y - 2$ so that $x = \dfrac{5y - 2}{3y - 5} = f^{-1}(y)$

Thus

$$f^{-1}(x) = \frac{5x - 2}{3x - 5}$$

In this example, we have $f(x) = f^{-1}(x)$. Therefore $f(x)$ is its own inverse and its graph has $y = x$ as an axis of symmetry. We might have anticipated this result since in the equation $3xy - 5y = 5x - 2$ (obtained in the process of solving for x) if x and y are interchanged an identical expression results. ●

In Example 4, an alternative procedure to establish that the given function $f(x) = \dfrac{5x - 2}{3x - 5}$ is one-to-one is to examine the sign of the derivative. We have

$$f'(x) = \frac{(3x - 5)(5) - (5x - 2)(3)}{(3x - 5)^2} = \frac{-19}{(3x - 5)^2} < 0 \qquad \left(x \neq \frac{5}{3}\right)$$

Thus we conclude that $f(x)$ is a decreasing function of x in the semiinfinite open intervals $(-\infty, \frac{5}{3})$ and $(\frac{5}{3}, \infty)$. However, without any additional information there is still a possibility that the functional values may overlap. In this instance, we know that this is not the case. In fact the graph of the given function (Figure 8.3.5) verifies this—note that $y = \frac{5}{3}$ is a horizontal asymptote and $x = \frac{5}{3}$ is a vertical asymptote.

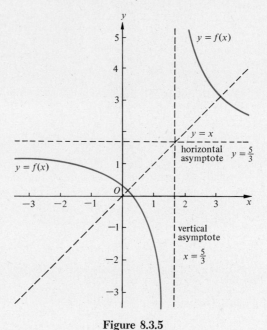

Figure 8.3.5

Next, we shall state without proof the following theorems. Their proofs are somewhat lengthy, but not difficult. They are perhaps best given in an advanced calculus text.

Theorem 1 Let f be a function of x which is continuous and increasing on the interval $[a, b]$. Then f has an inverse function f^{-1} which is continuous and increasing on $[f(a), f(b)]$.

A diagram will probably convince the reader that this result is "intuitively obvious" and also helps suggest the method of proof.

Theorem 2 Let f be a function of x that is continuous and decreasing on the interval $[a, b]$. Then f has an inverse function f^{-1} which is continuous and decreasing on $[f(a), f(b)]$.

An interesting relationship between the derivative of a function and the derivative of its inverse is indicated in the following theorem.

Theorem 3 Let f be continuous and also strictly monotone on the interval $[a, b]$ (that is, either monotonically increasing or decreasing on that interval) and let $y = f(x)$. If $D_x f(x)$ exists and is not zero for all x on $[a, b]$, the value of the derivative of the inverse function $x = f^{-1}(y)$ for any y in $[f(a), f(b)]$ is

$$D_y f^{-1}(y) = \frac{1}{D_x f(x)} \qquad (3)$$

or, more simply

$$D_y x = \frac{1}{D_x y} \qquad (4)$$

We shall justify this theorem by some simple algebraic manipulation instead of giving a formal proof. Writing

$$D_x y = \frac{dy}{dx} \qquad \text{and} \qquad D_y x = \frac{dx}{dy}$$

and recalling that a derivative can be treated as a quotient of differentials, we have

$$D_y x = \frac{dx}{dy} = \frac{1}{\dfrac{dy}{dx}} = \frac{1}{D_x y}$$

EXAMPLE 5 If $y = x^2 + 3x - 1$, find $D_y x$ at $(2, 9)$.

Solution $D_x y = 2x + 3$ so that from (4),

$$D_y x \Big|_{y=9} = \frac{1}{D_x y|_{x=2}} = \frac{1}{(2x + 3)|_{x=2}} = \frac{1}{7}$$

Alternative Solution Differentiate both sides of the given relation with respect to y. Thus by the chain rule,

$$1 = 2x \cdot D_y x + 3 D_y x = (2x + 3) D_y x$$

or

$$D_y x = \frac{1}{2x + 3} \qquad x \neq -\frac{3}{2}$$

Hence

$$D_y x \Big|_{(2, 9)} = \frac{1}{2x + 3} \Big|_{x=2} = \frac{1}{7} \qquad \bullet$$

EXAMPLE 6 Let $y = f(x)$ possess an inverse function $x = f^{-1}(y)$. Find a formula for $D_y^2 x$ in terms of $D_x y$ and $D_x^2 y$ assuming that all operations are valid.

Solution

$$D_y x = \frac{1}{D_x y} \qquad (i)$$

By the quotient and chain rules

$$D_{y^2}^2 x = \frac{(D_x y)(0) - (1)(D_x^2 y)}{(D_x y)^2} D_y x$$

Using (i),

$$D_{y^2}^2 x = -\frac{D_x^2 y}{(D_x y)^3} \qquad D_x y \neq 0 \qquad (5)$$

Since $y = f(x)$ and $x = f^{-1}(y)$ are mutually inverse functions, we have

$$D_x^2 y = -\frac{D_{y^2}^2 x}{(D_y x)^3} \qquad (6)$$

●

EXAMPLE 7   If $y = 5x^3 + x + 8$, find $D_{y^2}^2 x$.

Solution

$$y = 5x^3 + x + 8$$

and

$$D_x y = 15x^2 + 1 > 0$$

Thus the inverse function $x = f^{-1}(y)$ must exist. But $D_x^2 y = 30x$, and substitution of the expressions for $D_x y$ and $D_x^2 y$ into (5) yields

$$D_{y^2}^2 x = -\frac{30x}{(15x^2 + 1)^3}$$

●

Exercises 8.3

In each of Exercises 1 through 12, find the inverse of the given function, if it exists, and determine the domains of each. If the given function possesses an inverse, sketch the graphs of the functions on the same set of coordinate axes. If the function does not possess an inverse, justify your conclusion either analytically or graphically.

1. $f(x) = 4x$

2. $f(x) = x + 7$

3. $f(x) = -x + 3$

4. $f(x) = \dfrac{1}{5x + 3}$

5. $f(x) = \sqrt{x + 2}, \quad x > 0$

6. $g(x) = |1 - 2x|, \quad x > 0$

7. $h(x) = x^2 - 4$

8. $F(x) = x^3 - 10$

9. $G(x) = x^5 + 1$

10. $h(x) = (x + 4)^3$

11. $f(x) = \dfrac{3x + 5}{2x}$

12. $F(x) = \dfrac{2x + 1}{4x - 3}$

Exercises 13 through 16 refer to the previous exercises. Verify analytically that the given function and the determined inverse function are mutually inverse functions.

13. Exercise 3

14. Exercise 4

15. Exercise 8

16. Exercise 11

17. (a) Show that the function $F(x) = 5x^4 + 8$ is not a one-to-one function.
    (b) How may the domain of F be restricted so that the function be made one-to-one?
    (c) Find the inverse of F, restricted as in part (b). What is its domain?

18. Given that $F(t) = t(t^2 - 16)^{1/2}$, find a formula for $F^{-1}(t)$.

19. Given an even function $g(x)$; that is, $g(-x) = g(x)$. What can be said about its inverse function, $g^{-1}(x)$?

20. Find the inverse of the function $h(t) = 7 - (t - 2)^{1/3}$.

21. A function $f(x)$ is defined by the equation $x^5 + 3x + 2y = 0$.
    (a) Find the domain and range of $f(x)$.
    (b) Does $f(x)$ possess an inverse function? What is the domain and range (if it exists)?

22. A function $F(x)$ is defined by

$$F(x) = \int_1^x \frac{1}{1 + t^3}\, dt \qquad \text{where } x \text{ is in } [1, 2].$$

    (a) Find the domain and an expression for the range.
    (b) Does the inverse function exist?

In Exercises 23 through 25, find $D_x F^{-1}(2)$.

23. $F(x) = \int_1^x t\, dt, \quad x > 0$

24. $F(x) = \int_{-7}^x (t + 7)^{1/2}\, dt, \quad x > 0$

25. $F(x) = \int_x^0 \frac{1}{(t + 4)^{1/2}}\, dt, \quad x > -4$

26. If $F(x) = \int_1^x \frac{dt}{t}, \quad x > 0$ show that $D_x F^{-1}(x) = F^{-1}(x)$.

27. Suppose that a function $F(x)$ satisfies $F'(x) = F(x)$ for all x in $(-\infty, \infty)$. Suppose further that $F^{-1}(x)$ exists. Show that $D_x F^{-1}(x) = \dfrac{1}{x}$.

28. A formula for x as a function of y is

$$x = \int_0^y \frac{dt}{\sqrt{1 + 9t^2}}$$

    Show that $D_x^2 y$ is proportional to y and determine the constant of proportionality.

29. Given the function $y = \dfrac{ax + b}{cx + d}$ where $c \neq 0$, under what conditions is the given function identical with its inverse?

30. If $x = 1 + 3y + 2y^5$, find formulas for $D_x y$ and $D_x^2 y$.

31. If $y = x^3 + 3x - 2$, find formulas for $D_y x$, $D_y^2 x$, and $D_y^3 x$.

8.4

THE EXPONENTIAL FUNCTION
AND ITS PROPERTIES

From our definition of $y = \ln x$ we found that $\ln x$ is defined for all positive x and that $\ln x$ is an increasing function with range $(-\infty, \infty)$. Therefore $\ln x$ possesses an inverse function that we denote $y = \exp(x)$.

The function $y = \exp(x)$, called the **exponential function,** is by definition the inverse function of $y = \ln x$. This means that $y = \exp(x)$ if and only if $x = \ln y$.

The domain of $\exp(x)$ is the same as the range of $\ln x$ and thus consists of all real numbers. The range of $\exp(x)$ is all positive numbers. Since the functions $y = \ln x$ and $y = \exp(x)$ are inverses of each other, the graph of $y = \exp(x)$ may be found by reflection of $y = \ln x$ in the line $y = x$ (Figure 8.4.1).

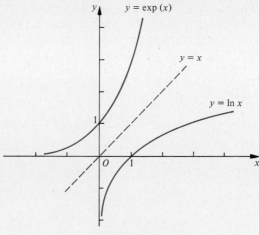

Figure 8.4.1

From the fact that $y = \exp(x)$ and $y = \ln x$ are mutually inverse functions, we have

$$\exp(\ln x) = x \qquad \text{and} \qquad \ln[\exp(x)] = x \tag{1}$$

Also since $\ln x$ is a continuous, monotonically increasing function of x, it follows that $\exp(x)$ is also an increasing function of x which is continuous for all x (Theorem 1 of Section 8.3). The following theorems will establish important properties of the exponential function.

Theorem 1 If a and b are any two real numbers, then

$$\exp(a) \cdot \exp(b) = \exp(a + b) \tag{2}$$

Proof Let $A = \exp(a)$ and $B = \exp(b)$, then the left side of (2) is AB. It must be established that the value of the right side is also AB. This, of course, will require using the properties of the ln function. From $A = \exp(a)$, we have $a = \ln A$ and similarly, $b = \ln B$ follows from $B = \exp(b)$. Therefore $a + b = \ln A + \ln B = \ln AB$ from which $AB = \exp(a + b)$. ∎

As a corollary to Theorem 1, we substitute $a = -b$ into (2) and thus $\exp(-b) \cdot \exp(b) = \exp(0) = 1$ (since $\ln 1 = 0$). Therefore $\exp(-b)$ and $\exp(b)$ are reciprocals of each other for arbitrary values of b,

$$\exp(-b) = \frac{1}{\exp(b)} \tag{3}$$

Theorem 2 If a and b are any two real numbers, then

$$\frac{\exp(a)}{\exp(b)} = \exp(a - b) \tag{4}$$

Proof   We use (3) and then (2) with b replaced by $-b$ to obtain (4):

$$\frac{\exp{(a)}}{\exp{(b)}} = \exp{(a)} \cdot \exp{(-b)} = \exp{(a + (-b))} = \exp{(a - b)} \qquad \blacksquare$$

We can also establish (4) by a procedure analogous to that of the proof of Theorem 1. This is left for the reader to verify as an exercise.

Theorem 3   If a is any real number and r is any rational number, then

$$\exp{(ra)} = [\exp{(a)}]^r \tag{5}$$

Proof   From the first equation of (1) with $x = [\exp{(a)}]^r$,

$$[\exp{(a)}]^r = \exp{\{\ln{[\exp{(a)}]^r}\}}$$

But $\ln{[\exp{(a)}]^r} = r\ln{[\exp{(a)}]}$ for any rational number r [Section 8.1, equation (6)]. Therefore

$$[\exp{(a)}]^r = \exp{[r\ln{(\exp{(a)})}]}$$

Also, from the second equation of (1), $\ln{[\exp{(a)}]} = a$, so that we have the required result

$$[\exp{(a)}]^r = \exp{(ra)} \tag{5} \qquad \blacksquare$$

We know that $\exp{(0)} = 1$ and now inquire as to the value of $\exp{(1)}$. To answer this, we let $y_1 = \exp{(1)}$; then, $1 = \ln{y_1}$. But e is the only number with a natural logarithm equal to 1. Thus $\exp{(1)} = e$.

We may now apply equation (5). Let x be any rational number, then

$$e^x = [\exp{(1)}]^x = \exp{(1 \cdot x)} = \exp{(x)} \tag{6}$$

The fact that $\exp{(x)} = e^x$ was anticipated in calling $\exp{(x)}$ the exponential function.

What about e^x when x is an arbitrary real number? Note that if a is any positive number our previous experience with a^x was generally restricted only to x being a rational number. Thus, for example, we probably have no idea as to what $2^{\sqrt{3}}$ means. For the number e, however, there is a very natural procedure for defining e^x for arbitrary real numbers x. It has been shown that if x is rational, then $e^x = \exp{(x)}$. Since the function exp is defined for all real x, we *define* e^x to be $\exp{(x)}$ for x irrational. Therefore,

$$e^x = \exp{(x)} \tag{6}$$

for arbitrary real x. Formulas (1) can now be written:

$$e^{\ln{x}} = x \qquad \text{and} \qquad \ln{(e^x)} = x \tag{7}$$

It is now a simple step to define a^x for any positive number a. This is done in Section 8.5 and then numbers of the form $2^{\sqrt{3}}$ will make sense.

The function $y = \ln{x}$ is increasing and differentiable and, furthermore,

$D_x y = \dfrac{1}{x}$ is never zero. Therefore its inverse function $\exp{(x)}$ or e^x is also increasing and differentiable for all x. We seek $D_x[\exp{(x)}] = D_x(e^x)$.

If $y = e^x$, then $x = \ln y$. Form D_x of both sides implicitly considering y as a function of x,

$$1 = \frac{1}{y} D_x y \qquad \text{or} \qquad D_x y = y$$

Therefore we have a most important formula

$$D_x(e^x) = e^x \tag{8}$$

This says that e^x is a function that is its own derivative. In fact, it will be shown in Example 4 that the only functions that possess this property are given by $f(x) = ce^x$ where c is an arbitrary constant. In particular, if $c = 0$ we have the constant function zero, which is its own derivative.

If $u(x)$ is a differentiable function of x, then from (8) and the chain rule

$$D_x(e^u) = D_u(e^u) \cdot D_x u$$

or

$$D_x(e^u) = e^u \cdot D_x u \tag{9}$$

EXAMPLE 1   Given $y = e^{3x+1}$ find $D_x y$.

Solution   We use (9) with $u = 3x + 1$ and $D_x u = 3$. Thus

$$D_x y = 3e^{3x+1} \qquad\qquad \bullet$$

EXAMPLE 2   Given $y = e^{\sqrt[3]{x}}$, find $D_x y$.

Solution   We use (9) with $u = \sqrt[3]{x} = x^{1/3}$ and $D_x u = \frac{1}{3}x^{-2/3}$, $x \neq 0$
Therefore

$$D_x y = \frac{1}{3} x^{-2/3} e^{\sqrt[3]{x}} \qquad x \neq 0 \qquad\qquad \bullet$$

EXAMPLE 3   Given $y = \dfrac{x}{e^{2x}}$, find $D_x y$.

Solution   We write $y = xe^{-2x}$ and then use the product rule and (9). Therefore

$$D_x y = xe^{-2x}(-2) + e^{-2x}(1) = e^{-2x}(1 - 2x) = \frac{1 - 2x}{e^{2x}} \qquad \bullet$$

EXAMPLE 4   Find the complete solution of $D_x y = ay$ where a is a constant.

Solution   We assume that the functions we seek will never be zero for all x. Thus we write $\dfrac{D_x y}{y} = a$, and integration of both sides yields $\ln|y| = ax + b$ where b is a second constant. This implies that

$$|y| = e^{ax+b} = e^b \cdot e^{ax}$$

Thus $y = Be^{ax}$, where $B = \pm e^b$, is an arbitrary nonzero constant.

The preceding procedure suffers from the deficiency that we do not know whether or not all solutions have been obtained. Indeed we did not obtain the solution $y = 0$ for all x. An interesting and rigorous procedure is to multiply the given equation by e^{-ax}. Thus

$$e^{-ax}(D_x y - ay) = 0$$

or

$$D_x(ye^{-ax}) = 0$$

(Check this by differentiating the product.)

Therefore $ye^{-ax} = c$ where c is an arbitrary constant or $y = ce^{ax}$, c arbitrary is the *totality* of solutions. In particular, the discussion following formula (8) refers to the case $a = 1$. ●

Since $D_x(e^x) = e^x$, it follows that

$$\int e^x \, dx = e^x + C \tag{10}$$

More generally,

$$\int e^u \, du = \int e^u D_x u \, dx = e^u + C \tag{11}$$

EXAMPLE 5 Find $\int e^{-4x} \, dx$.

Solution Let $u = -4x$, then $du = D_x u \, dx = -4 \, dx$ so that in order to utilize (11) we multiply and divide by -4:

$$\int e^{-4x} \, dx = -\tfrac{1}{4} \int e^{-4x}(-4 \, dx) = -\tfrac{1}{4} e^{-4x} + C$$ ●

EXAMPLE 6 Find $\displaystyle\int_0^1 xe^{3x^2+2} \, dx$.

Solution Let $u = 3x^2 + 2$, then $du = 6x \, dx$ and thus to use (11) we multiply and divide by 6:

$$\int_0^1 xe^{3x^2+2} \, dx = \tfrac{1}{6} \int_0^1 e^{3x^2+2}(6x \, dx) = \tfrac{1}{6} e^{3x^2+2} \bigg]_0^1 = \tfrac{1}{6}[e^5 - e^2]$$ ●

EXAMPLE 7 Find $\displaystyle\int \frac{e^x}{7e^x + 8} \, dx$.

Solution   This example involves both logarithms and exponentials. We observe that $D_x(7e^x + 8) = 7e^x$ so that except for the factor 7, we have the form

$$\int \frac{D_x u}{u}\, dx = \int \frac{du}{u} = \ln |u| + C.$$

Thus

$$\int \frac{e^x}{7e^x + 8}\, dx = \frac{1}{7} \int \frac{7e^x}{7e^x + 8}\, dx = \frac{1}{7} \ln (7e^x + 8) + C$$

where the absolute value sign is not needed since $7e^x + 8$ is positive for all x.

●

EXAMPLE 8   This example is devoted to a physical application of the exponential function. We will now establish that the atmospheric pressure and the height above the Earth's surface are related by a simple exponential rule. To derive this, let us imagine that we have a static vertical column of air above a surface of Earth of 1 square unit of area (Figure 8.4.2). We seek a formula for the pressure p as a function of the height h above the Earth's surface.

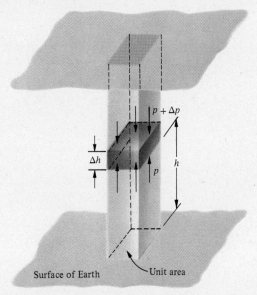

Figure 8.4.2

Consider an element of volume of thickness Δh and cross-sectional area 1 square unit. Then if p is the pressure (that is, force/unit area) acting on its bottom face and $p + \Delta p$ is the corresponding pressure on the top face, static equilibrium requires that

$$\Delta p = -\overline{\gamma}\, \Delta h \tag{i}$$

where $\overline{\gamma}$ is the average weight density of the air between the levels. If the temperature is assumed to be constant, then, in accordance with Boyle's law, the average weight density $\overline{\gamma}$ and the average pressure $\overline{p}$ for the interval h to $h + \Delta h$ are proportional. Thus we have

$$\overline{\gamma} = k\overline{p} \tag{ii}$$

where k is a positive constant of proportionality depending on the properties of the gas. Substitution of (ii) into (i) yields

$$\Delta p = -k\bar{p}\,\Delta h \tag{iii}$$

Divide both sides of (iii) by Δh and let $\Delta h \to 0$ so that

$$\frac{\Delta p}{\Delta h} \to D_h p \quad \text{and} \quad \bar{p} \to p.$$

This yields, for all nonnegative h,

$$D_h p = -kp \tag{iv}$$

Integration of (iv) yields

$$p = p_0 e^{-kh} \tag{12}$$

where p_0 is the pressure when $h = 0$. Equation (12) implies that the atmospheric pressure decreases with increase in height above the surface of the Earth. ●

Example 8 is a particular application of the case when the rate of change of a variable is proportional to itself. There is a large class of problems in physics, chemistry, biology, economics, and so on, where the equation

$$\frac{dy}{dt} = ky \tag{13}$$

occurs. In (13), k is a constant and y is a positive function of t which is being sought. As we already know from Example 4, the totality of solutions may be expressed in the form

$$y = y_0 e^{kt} \tag{14}$$

where y_0 is the value of y when $t = 0$. Thus the law is completely determined once the values of y_0 and k are known. If k is positive, we have the law of *exponential growth.* If k is negative the law is that of *exponential decay.* Application of (13) and (14) will be given in Examples 9 and 10.

EXAMPLE 9   A colony of bacteria grows at a rate proportional to the number present. If 500,000 bacteria are present at a certain time and 1,500,000 are present 2 hours later, find the number present 6 hours later.

Solution   Let y be the number of bacteria present at any time t (in hours). Then

$$D_t y = ky$$

holds, from which

$$y = y_0 e^{kt}$$

Let $t = 0$ correspond to the time when 500,000 bacteria are present. Therefore

$$500,000 = y_0 e^{k \cdot 0} = y_0$$

Furthermore, at time $t = 2$, $y = 1,500,000$, and so

$$1,500,000 = 500,000e^{k \cdot 2}$$

or

$$e^{2k} = 3.$$

Take the natural logarithm of both sides and find

$$k = \frac{1}{2}\ln 3$$

$$y = 500,000e^{\frac{1}{2}\ln 3} = 500,000e^{\ln 3^{t/2}}$$
$$= 500,000(3^{t/2}) \qquad (\text{since } e^{\ln f(t)} = f(t))$$

and

$$y|_{t=6} = 500,000(3^3) = 13,500,000$$

bacteria at the end of 6 hr. ●

EXAMPLE 10   A tank contains 100 gal of water. In error 150 lb of salt are poured into the tank instead of 75 lb. This condition is to be corrected by pumping fresh water into the tank at the rate of 2 gal/min and, simultaneously, the brine solution is allowed to run out of the tank at the same rate of 2 gal/min. If the mixture is kept uniform by constant stirring, how long will it take for the brine solution to contain the desired 75 lb of salt?

Solution   Let x be the number of pounds of salt at any time t(min). Then $\dfrac{x}{100}$ lb/gal represents the concentration of salt at any time t. The number of pounds per minute that is draining out is $\dfrac{2x}{100}$. Therefore $-\dfrac{2x}{100}$ is the rate of increase of salt. However, $D_t x$ is also the rate of increase of salt; that is,

$$D_t x = -\frac{2x}{100} = -\frac{x}{50}$$

Hence from (14),

$$x = x_0 e^{-t/50}$$

$x = 150$ at $t = 0$ yields $x_0 = 150$, or

$$x = 150e^{-t/50}$$

When $x = 75$ then

$$\tfrac{1}{2} = e^{-t/50}$$

or

$$t = 50 \ln 2 = 50\,(0.69315) \doteq 34.66 \text{ min}$$ ●

Exercises 8.4

In Exercises 1 through 8, simplify the given expression.

1. $\ln e^{3x}$

2. $e^{-\frac{2}{3}\ln x}$

3. $e^{-4 \ln \sqrt{x}}$

4. $\ln (e^{-\ln x})$

5. $(e^{2\ln 6})^2$

6. $\ln(x^5 e^{2x})$

7. $\ln\left(\dfrac{e^{3x}}{e^{x+5}}\right)^2$

8. $e^{6\ln\sqrt{x^2+1}}$

9. Sketch the graph of $y = e^{-2x}$.

10. Sketch the graph of $y = \frac{1}{4}(e^x + e^{-x})$.

In Exercises 11 through 22, differentiate the indicated functions.

11. $y = e^{3x-1}$

12. $y = e^{2-7x}$

13. $y = e^{\sqrt{x+5}}$

14. $y = e^{ax^2+bx}$

15. $y = xe^{10x}$

16. $y = e^{2/x}$

17. $y = e^{bx^2} - e^{-bx^2}$

18. $y = \dfrac{x^3}{e^{2x}}$

19. $y = e^{2x}\ln 5x$

20. $\displaystyle\int_4^x e^{-t^2}\,dt$

21. $y = \dfrac{e^x + e^{-x}}{e^x - e^{-x}}, \quad x \neq 0$

22. $y = (e^x + e^{-x})^3$

In Exercises 23 through 34, determine the following integrals.

23. $\displaystyle\int e^{3x}\,dx$

24. $\displaystyle\int e^{bx}\,dx$, (b is a nonzero constant)

25. $\displaystyle\int \dfrac{e^{6x}}{e^x}\,dx$

26. $\displaystyle\int \dfrac{e^{3x}+1}{e^x}\,dx$

27. $\displaystyle\int e^{-x^2}x\,dx$

28. $\displaystyle\int \dfrac{e^{1/x}}{x^2}\,dx$

29. $\displaystyle\int e^{\ln x}\,dx$

30. $\displaystyle\int (e^{-4x}+2)^2\,dx$

31. $\displaystyle\int (e^x+5)^3\,dx$

32. $\displaystyle\int (e^{2x}-3)^4 e^{2x}\,dx$

33. $\displaystyle\int g'(x)\,e^{g(x)+1}\,dx$

34. $\displaystyle\int \dfrac{e^{10x}}{100+e^{10x}}\,dx$

In Exercises 35 through 42, evaluate the definite integral.

35. $\displaystyle\int_{-1}^1 e^4\,dx$

36. $\displaystyle\int_1^2 xe^{x^2+1}\,dx$

37. $\displaystyle\int_0^1 (e^{3x}-e^{-3x})^2\,dx$

38. $\displaystyle\int_2^3 xe^{-4\ln x}\,dx$

39. $\displaystyle\int_0^a \frac{1}{2}(e^{x/a}+e^{-x/a})\,dx$

$(a$ is a nonzero constant)

40. $\displaystyle\int_{-2}^{-1} \dfrac{e^{-x}}{1+e^{-x}}\,dx$

41. $\displaystyle\int_0^1 \dfrac{dx}{e^x+1}$ (Hint: How is the integrand of this problem related to that of Exercise 40?)

42. $\displaystyle\int_0^a \left(\dfrac{x}{\pi}\right)^2 e^{-x^3}\,dx$ (a is constant)

In Exercises 43 through 46, find $D_x y$ by the method of implicit differentiation.

43. $e^x + e^y = 3$

44. $e^{x+y} + \ln x - 5 = 0$

45. $e^{3y} + \ln(2x+5) = 10x - 7y$

46. $x^2 e^{-y} + (y-x)e^{2x} - e^{-1} = 100$

47. Determine all the functions $F(x)$ whose derivative is three times the value of the function for any x.

48. Determine the totality of functions $g(x)$ whose derivative is the negative of the value of the function for any x.

49. Find the area bounded by the curve $y = \dfrac{1}{\pi} e^{-3x}$ and the lines $y = 0$, $x = 0$, and $x = 2$. Sketch the region.

50. Find the area bounded by $y = \dfrac{e^x - e^{-x}}{e^x + e^{-x}}$ and the lines $x = 1$ and $y = 0$. Sketch the region.

51. The demand law for a certain product is given by the formula $p = 100 e^{-0.001x}$ where x is the number of items sold and p is the price in dollars per item. Find the number of items that will yield maximum revenue and determine the maximum revenue.

52. Consider an electric circuit consisting of a resistance R, a self-inductance L and an electromotive force E. If $i(t)$ is the current in the circuit then Ohm's and Kirchhoff's laws require that

$$L \frac{di}{dt} + Ri = E \tag{i}$$

Verify that if the current at time $t = 0$ is zero and if L, R, and E are constants, then

$$i = \frac{E}{R} \left(1 - e^{-\frac{R}{L}t} \right)$$

satisfies (i) subject to the initial condition $i(0) = 0$. Interpret this result graphically.

53. A firm decides to invest capital u at time t, at a rate $\dfrac{du}{dt}$ such that

$$\frac{du}{dt} = k(u - u_0)$$

where k is a negative constant and u_0 is a constant known as the equilibrium investment. Find $u(t)$. Show that regardless of the magnitude of the initial capital investment u approaches u_0 as $t \to \infty$. Sketch some representative curves showing, in particular, $u(t)$ when $u(0) < u_0$ and when $u(0) > u_0$.

54. A radioactive substance decays according to the law

$$u'(t) = -\tfrac{1}{4} u(t)$$

where u is the mass of the substance at the end of t years. How long will it take the substance to reduce to 10% of its original mass?

8.5 MORE GENERAL LOGARITHMIC AND EXPONENTIAL FUNCTIONS

We want to give a definition for a^x where $a > 0$. Our results on the exponential function suggest a reasonable way of doing this. Recall that for any real number r,

$$r = \exp(\ln(r)) = e^{\ln r}$$

Letting $r = a^x$ gives $a^x = e^{\ln a^x} = e^{x \ln a}$. The term on the right in this equation has already been defined. This allows us to state the following.

Definition If a is a positive number and x is any real number then

$$a^x = e^{x \ln a} \tag{1}$$

From (1) we can show that the function a^x satisfies the rules already established in the special case $a = e$, namely:

$$a^x \cdot a^y = a^{x+y} \tag{2}$$

$$a^x \div a^y = a^{x-y} \tag{3}$$

$$(a^x)^y = a^{xy} \tag{4}$$

$$(ab)^x = a^x \cdot b^x \tag{5}$$

$$a^0 = 1 \tag{6}$$

To illustrate the procedure, we will prove (3) and leave the others as exercises. From (1) $a^x \div a^y = (e^{x \ln a}) \div (e^{y \ln a}) = e^{(x-y) \ln a}$. But, from definition (1), $a^{x-y} = e^{(x-y) \ln a}$. Therefore

$$a^x \div a^y = e^{(x-y) \ln a} = a^{x-y}$$

The derivative of a^x is readily obtained from (1) by application of the chain rule. We have

$$D_x(a^x) = D_x(e^{x \ln a}) = e^{x \ln a} D_x(x \ln a)$$
$$= e^{x \ln a}(\ln a) = a^x \ln a$$

More generally, if u is any differentiable function of x,

$$D_x(a^u) = a^u (D_x u) \ln a \tag{7}$$

EXAMPLE 1   Find $D_x[5^{x^3+2x}]$.

Solution   We apply (7) with $a = 5$, $u = x^3 + 2x$ and $D_x u = 3x^2 + 2$. Thus
$$D_x[5^{x^3+2x}] = 5^{x^3+2x}(3x^2 + 2) \ln 5 \qquad \bullet$$

EXAMPLE 2   Find $D_x[4 + \sqrt[3]{x}]^e$.

Solution   Formula (7) does *not* apply because this is an instance of a variable raised to a constant power rather than a constant raised to a variable power. Thus by the power and chain rules,
$$D_x[4 + \sqrt[3]{x}]^e = e(4 + \sqrt[3]{x})^{e-1} D_x[4 + \sqrt[3]{x}]$$
$$= e(4 + \sqrt[3]{x})^{e-1} \tfrac{1}{3}x^{-2/3}$$
$$= \frac{e(4 + x^{1/3})^{e-1}}{3x^{2/3}} \qquad \bullet$$

From (7) we obtain the integration formula

$$\int a^u \, du = \frac{a^u}{\ln a} + C, \, a \neq 1 \tag{8}$$

EXAMPLE 3   Find $\int 9^{x^2} x \, dx$.

Solution   We let $u = x^2$, $du = 2x \, dx$ and let I denote the integral so that from (8),

$$I = \frac{1}{2} \int 9^u \, du = \frac{9^u}{2 \ln 9} + C = \frac{9^{(x^2)}}{2 \ln 9} + C \qquad \bullet$$

Next, we derive the general power rule

$$\boxed{D_x(x^a) = ax^{a-1} \qquad x > 0} \tag{9}$$

where a is an arbitrary real number. From (1), (interchanging the roles of a and x), $x^a = e^{a \ln x}$, $x > 0$. Therefore

$$D_x(x^a) = e^{a \ln x} D_x(a \ln x) = e^{a \ln x} \left(\frac{a}{x}\right) = x^a \left(\frac{a}{x}\right) = ax^{a-1}, \qquad x > 0$$

This proves the power rule for irrational as well as rational exponents.

EXAMPLE 4   Find $D_x(x^e + e^x)^2$.

Solution   If we square first, then

$$D_x(x^e + e^x)^2 = D_x(x^{2e} + 2x^e e^x + e^{2x})$$
$$= 2ex^{2e-1} + 2(x^e e^x + ex^{e-1}e^x) + 2e^{2x}$$
$$= 2(ex^{2e-1} + x^e e^x + ex^{e-1}e^x + e^{2x})$$

Alternative   Better still, we use the form $D_x(u^2) = 2u \cdot D_x u$, thus
Solution

$$D_x(x^e + e^x)^2 = 2(x^e + e^x) \cdot D_x(x^e + e^x)$$
$$= 2(x^e + e^x)(ex^{e-1} + e^x)$$

and this is a factored form of our previous result. $\bullet$

From $a^x = e^{x \ln a}$, $(a > 0, \ a \neq 1)$, we know that the exponential function $f(x) = a^x$ is a one-to-one function with domain $(-\infty, \infty)$ and range $(0, \infty)$. Since the function f is a one-to-one function, it has an inverse which we call the logarithm function to the base a, written $\log_a x$. Thus

$$y = \log_a x \quad \text{if and only if} \quad x = a^y \tag{10}$$

If $x = a^y$, then $\ln x = \ln (a^y) = y \ln a$ and, thus

$$\log_a x = \frac{\ln x}{\ln a} \tag{11}$$

In particular, if $x = e$, we have (since $\ln e = 1$)

$$\log_a e = \frac{1}{\ln a} \tag{12}$$

From (11) and the known properties of the ln function we may readily establish the following basic rules for the logarithmic function to the base a, $(x > 0$ and $y > 0)$:

$$\log_a (xy) = \log_a x + \log_a y \tag{13}$$

$$\log_a \left(\frac{x}{y}\right) = \log_a x - \log_a y \tag{14}$$

$$\log_a(x^y) = y \log_a x \tag{15}$$

$$\log_a 1 = 0 \tag{16}$$

and the significant rule for changing bases

$$\log_a x = \log_b x \cdot \log_a b \qquad (b > 0,\ b \neq 1) \tag{17}$$

The proofs of (13) through (17) are left as exercises for the reader.

From (11) we deduce a formula for $D_x(\log_a x)$. If both sides of (11) are differentiated with respect to x,

$$D_x(\log_a x) = D_x \left(\frac{\ln x}{\ln a} \right) = \frac{1}{\ln a} D_x(\ln x) = \frac{1}{x \ln a}$$

$$D_x \log_a x = \frac{1}{x} (\log_a e) \tag{18}$$

where (12) was utilized in obtaining the last equation.

More generally, if $u(x)$ is any differentiable function of x, by the chain rule

$$D_x(\log_a u) = \frac{1}{u} (\log_a e)\, D_x u \tag{19}$$

and, in particular, if $a = e$,

$$D_x(\ln u) = \frac{1}{u} D_x u \tag{20}$$

EXAMPLE 5   If $y = \log_5 (x^2 + e^{3x} + 4)$ find $D_x y$.

Solution   Using (19) with $a = 5$ and $u = x^2 + e^{3x} + 4$ gives

$$D_x y = \frac{1}{x^2 + e^{3x} + 4} (\log_5 e)(2x + 3e^{3x}) = \frac{(\log_5 e)(2x + 3e^{3x})}{x^2 + e^{3x} + 4} \qquad \bullet$$

EXAMPLE 6   If $y = x^x$, $(x > 0)$, find $D_x y$.

Solution   If $y = x^x$, then, by taking the natural logarithm of each side,

$$\ln y = x \ln x$$

Differentiation of both sides with respect to x yields

$$\frac{1}{y} D_x y = x \left(\frac{1}{x} \right) + \ln x = 1 + \ln x$$

from which

$$D_x y = y(1 + \ln x) = x^x(1 + \ln x)$$

We have used here the method of *logarithmic differentiation* introduced in Section 8.1.

| Alternative | Let $y = x^x = e^{x \ln x}$, thus |
| Solution | |

$$D_x y = e^{x \ln x} D_x(x \ln x)$$
$$= e^{x \ln x}(1 + \ln x)$$
$$= x^x(1 + \ln x) \qquad \bullet$$

It is often convenient to express a law of growth or decay in terms of an exponential function to some other base a. If $q(t)$ denotes the amount of the substance, number of bacteria in a culture, and so on, then

$$q(t) = q(0)\, a^{kt} \tag{21}$$

where $a > 0$ and k are constants to be determined from the data of the specific problem.

EXAMPLE 7   The half-life of radium is approximately 1600 years, that is, given any initial mass of radium, one half of it will be left in approximately 1600 years. Find a formula for the amount $q(t)$ remaining from 80 milligrams (mg) of radium after t years. When will there be 30 mg left?

Solution   From the given information and the notation of (21), $q(0) = 80$ and $q(1600) = 40$. Also since the given data involves the factor $\frac{1}{2}$, it is convenient to select $a = \frac{1}{2}$ in (21). Thus

$$q(t) = 80(\tfrac{1}{2})^{kt}$$

and

$$q(1600) = 40$$

that is, when $t = 1600$, $q = 40$

$$40 = 80(\tfrac{1}{2})^{1600k}$$

from which $k = \frac{1}{1600}$. Therefore

$$q(t) = 80(\tfrac{1}{2})^{t/1600}$$

We must find $t = t_1$ when $q = 30$. This yields the equation

$$30 = 80(\tfrac{1}{2})^{t_1/1600} = 80(2)^{-t_1/1600}$$

or

$$\tfrac{8}{3} = 2^{t_1/1600}$$

Taking the natural logarithm of both sides and solving for t_1,

$$t_1 = \frac{1600}{\ln 2} \ln \frac{8}{3} = \frac{1600}{\ln 2}(\ln 8 - \ln 3)$$

$$\doteq \frac{1600(2.07944 - 1.09861)}{0.69315} \doteq 1600(1.41503)$$

$$\doteq 2264 \text{ years} \qquad \bullet$$

EXAMPLE 8   Under ideal circumstances bacteria grow at a rate proportional to their number. Suppose that the number of bacteria in a specific culture at a certain instant is 10,000 and the number present 4 hours later was 30,000. Find the law of growth for this culture and determine the number present 6 hours after the first count.

Solution   With the notation of (21), we have

$$q(t) = q(0) a^{kt}$$

We are given that $q(0) = 10,000$. Furthermore the number of bacteria is tripled in 4 hours. Thus we choose $a = 3$, yielding

$$q(t) = 10,000(3^{kt})$$

Now, $q(4) = 30,000$ from which we obtain $4k = 1$ or $k = \frac{1}{4}$. Thus the required law of growth is

$$q(t) = 10,000(3^{t/4})$$

We want $q(6)$. Therefore

$$q(6) = 10,000(3^{6/4}) = 10,000[3\sqrt{3}]$$
$$\doteq 51,960$$

We might wonder at this point why e can't be used rather than a. Actually it can. In fact $a^{kt} = (a^k)^t$ and we can write $a^k = e^r$, say. Thus

$$q(t) = q(0) e^{rt} \qquad (22)$$

and proceed with the use of the tables for the exponential function. Let's give a second solution of this example using (22).

Alternative Solution   Using the initial data,

$$q(t) = q(0) e^{rt} = 10,000 e^{rt}$$

Now, $q = 30,000$ when $t = 4$, which yields $e^{4r} = 3$ or $r = \frac{1}{4}\ln 3$. Thus the growth law is

$$q(t) = 10,000\, e^{\frac{1}{4}(\ln 3)t}$$

$$q(6) = 10,000\, e^{\frac{3}{2}(\ln 3)}$$
$$\doteq 10,000(e^{1.648}) \doteq 51,960$$

in agreement with the first solution. (Note that in the absence of accurate tables a small discrepancy in the results obtained by the two procedures may sometimes occur.)   ●

Exercises 8.5

In Exercises 1 through 5, solve for x. Check all answers.

1. $\log_2 4x + \log_2 x = 1$

2. $2^{x+1} + 4^x = 24$

3. $6^x - 2^{x+1} = 0$

4. $15^{1-2x} - 3^{-x} = 0$

5. $\log_6 (2x - 3) + \log_6 (4x) = 1 + \log_6 (x + 3)$

In Exercises 6 through 23, differentiate the given function.

6. $F(x) = 7^x$

7. $G(t) = 3^{2t}$

8. $H(y) = 5^{-y}$

9. $f(x) = 4^{-\sqrt{x}}$

10. $g(x) = x \log_{10} x$

11. $h(x) = \log_{10} |4x + 3|$

12. $F(t) = (2 + 3e)^{t^2}$

13. $G(z) = \dfrac{\log_{10} z}{z}$

14. $H(v) = \log_{10} \sqrt[3]{2v + 1}$

15. $f(t) = \dfrac{\log_{10}(t^2)}{\ln t}$

16. $g(x) = \log_{10} [\ln(x^3 + x - 1)]$

17. $h(y) = \log_b [\log_b (y + 1)]$

18. $F(v) = (3v + 2)^{\log_{10} v}$

19. $f(x) = (x^4 + 9)^{\log_{10} x}$

20. $g(t) = \dfrac{\log_{10} \sqrt{t + 2}}{\log_{10}(3t - 1)}$

21. $h(z) = z^{(2^z)}$

22. $F(x) = x^{(e^{\sqrt{z}})}$

23. $G(x) = (\log_b x)^{\log_b x}$

In Exercises 24 through 36, evaluate the integral.

24. $\displaystyle\int 6^x \, dx$

25. $\displaystyle\int \pi^x \, dx$

26. $\displaystyle\int 4^{7x} \, dx$

27. $\displaystyle\int 2^{x/5} \, dx$

28. $\displaystyle\int_0^1 (e - 1)^x \, dx$

29. $\displaystyle\int x4^{x^2} \, dx$

30. $\displaystyle\int 10^x e^x \, dx$

31. $\displaystyle\int \sqrt{8^{5x}} \, dx$

32. $\displaystyle\int \sqrt[3]{9^{2x}} \, dx$

33. $\displaystyle\int \dfrac{1}{x \log_b x} \, dx$

34. $\displaystyle\int_1^2 \dfrac{dx}{(x \log_{10} 5)^3}$

35. $\displaystyle\int \dfrac{2^x}{5 + 2^x} \, dx$

36. $\displaystyle\int \dfrac{1}{1 + 3^x} \, dx$

In Exercises 37 through 40, sketch the graph of the given function.

37. $f(x) = 3^{x+1}$

38. $g(x) = \log_{10} |x|$

39. $f(x) = \log_{10}(2x - 3)$

40. $g(x) = 2^{4-x}$

In Exercises 41 and 42, find an equation of the tangent line to the graph of the given equation at the specified point.

41. $y = x2^x$ $(3, 24)$

42. $x3^y - y - 7 = 0$ $(1, 2)$

43. The rate of change of the population of a city is considered to be proportional to the population at any time. Its population was $\frac{1}{2}$ million in 1930 and 1 million in 1970. (a) Express the population P as a function of the time t. (b) What would be the estimated population in 1990 to the nearest 100?

44. The number of bacteria in a culture was 6000 at a certain instant and 18,000 4 hours later. Assuming ideal conditions of growth (that is, the rate of growth is proportional to the number present) find the number of bacteria 5 hours after the initial reading (to the nearest 100).

45. The rate of decay of a radioactive substance is proportional to the amount of material present. It takes 4 years for a third of the material to decay. Find, to the nearest day, the half-life of the material, that is, how long it takes for half the material to decay.

In Exercises 46 through 51, prove the given properties where $a > 0$, $b > 0$ and x and y are arbitrary real numbers.

46. $a^x \cdot a^y = a^{x+y}$

47. $(a^x)^y = a^{xy}$

48. $(ab)^x = a^x \cdot b^x$

49. $\log_a xy = \log_a x + \log_a y$ $(x > 0$ and $y > 0)$

50. $\log_a (x^y) = y \log_a x \qquad (x > 0)$

51. $\log_a x = \log_b x \cdot \log_a b \qquad (x > 0)$

52. Prove that $\log_{10} e = \dfrac{1}{\ln 10}$

Review and Miscellaneous Exercises

In Exercises 1 through 14, differentiate the following functions.

1. $f(x) = x^3 \ln 6x$

2. $g(x) = e^{(4x - e^{3x})}$

3. $h(x) = \dfrac{\ln ax}{e^{bx}}$, $\quad (a, b$ constants$)$

4. $F(t) = 8^{-5\sqrt{t+1}}$

5. $G(t) = \ln [\ln (\ln t)]$

6. $f(t) = \ln e^{\sqrt{t^2+1}}$

7. $g(x) = (\log_a x)^m$, $\quad (m$ constant$)$

8. $h(z) = e^{(z^3 + \ln z)}$

9. $F(x) = (x^2 + x + 1)^{x^2}$

10. $G(t) = \dfrac{1}{\sqrt{\pi}} (t + 3\sqrt{t})^e$

11. $H(t) = \displaystyle\int_{t^2}^{0} \log_{10} (x^4 + 1) \, dx$

12. $f(x) = \log_{10} (\ln \sqrt{x})$

13. $g(t) = t^{(t^t)}$

14. $h(t) = t^{(3^t)}$

In Exercises 15 and 16 use logarithmic differentiation to find the derivative of the given function.

15. $f(x) = (x - a)^2(x - b)(x - c) \qquad (a, b, c$ distinct constants$)$

16. $g(x) = \dfrac{u^2}{vw} \qquad (u, v, w$ are differentiable functions of $x)$

In Exercises 17 through 32, evaluate the integrals.

17. $\displaystyle\int \dfrac{x \, dx}{10x^2 + 7}$

18. $\displaystyle\int 2x^2 \, e^{-x^3} \, dx$

19. $\displaystyle\int \dfrac{e^{4x} + 3e^{3x} - 1}{e^{2x}} \, dx$

20. $\displaystyle\int (e^{2x} + e^{-x})^3 \, e^{-x} \, dx$

21. $\displaystyle\int \dfrac{5x^4 - 10x^3 + 7x - 19}{x^2} \, dx$

22. $\displaystyle\int \dfrac{e^{2x} + e^{-2x}}{e^{2x} - e^{-2x}} \, dx$

23. $\displaystyle\int (3 + 5e)^x \, dx$

24. $\displaystyle\int \dfrac{(4e)^x}{3^x} \, dx$

25. $\displaystyle\int 6^{5t+3} \, dt$

26. $\displaystyle\int (3 + 4^{-x})^2 \, dx$

27. $\displaystyle\int_{-2}^{-1} 3^{-t} \, dt$

28. $\displaystyle\int_{0}^{1} 4^{-x^2} x \, dx$

29. $\displaystyle\int \dfrac{x^2}{x + a} \, dx$

30. $\displaystyle\int \dfrac{e^{-2x} - 1}{e^{-2x} + 1} \, dx$

31. $\displaystyle\int e^x 3^{(e^x)} \, dx$

32. $\displaystyle\int x^2 4^{(2x^3 + 5)} \, dx$

33. If $x^2 e^{3y} + \ln (x - y) = 2$, find $D_x y$.

34. (a) Given only that $\log_a 1 = 0$ and that $\log_a rs = \log_a r + \log_a s$, prove that $\log_a \left(\dfrac{1}{r}\right) = -\log_a r$.

(b) Then prove that $\log_a \dfrac{r}{s^2} = \log_a r - 2 \log_a s$.

35. In how many years would a population increase by 50% if it increased at a uniform rate of 3% per year.

36. A substance in a chemical reaction is used up at a rate proportional to the amount of substance present at any time t. Suppose that $\frac{1}{3}$ of the substance is used up in $2\frac{1}{2}$ hours, how much of the substance is left at the end of 4 hours?

37. Find the area bounded by the curves $y = 3^x$, $y = 9$, and the x-axis.

38. Find the volume of the solid generated by rotating the region bounded by $y = 2e^{-x}$, $x = 0$, $x = 4$, and $y = 0$ about the x-axis.

39. Show that the equation $e^{-x} - x = 0$ has only one root. Find the root to the nearest 0.01. (*Hint:* Consider the function $f(x) = e^{-x} - x$.)

40. If $G(x) = \int_0^x e^{-t^2}\, dt$, find $G'(2)$ and $G''(2)$.

41. Solve the differential equation $e^{2x+y}\dfrac{dy}{dx} - 1 = 0$ subject to $y = 0$ when $x = 0$.

In Exercises 42 through 45, determine whether or not the function has an inverse over the given domain. Determine the inverse function and its domain, where possible.

42. $f(x) = -3x + 7$ $(-\infty < x < \infty)$ 43. $f(x) = e^{-x}$ $(0 \le x \le 3)$

44. $f(x) = \frac{1}{2}(e^{3x} + e^{-3x})$ $(-1 \le x \le 3)$ 45. $f(x) = x^2 - 2x + 3$ $(2 \le x \le 4)$

46. The demand law for a particular commodity is given by $p = 1000e^{-x/30}$, where x is the number of items sold and p is the unit price in dollars. Determine the maximum revenue.

47. Solve the equation $e^x + e^{-x} = 3$, $(x > 0)$, correct to two decimal places.

48. A woman places $4000 in a bank which pays 6.5% compounded continuously. What will the value of this deposit be in 7 years?

49. (a) We seek a formula for $\int xe^x\, dx$ in the form $e^x P(x)$ where $P(x)$ is a polynomial in x. Find $P(x)$.

   (b) Find $\int x^2 e^x\, dx$.

50. Suppose that a function f is defined for all real values of x, and satisfies the relation

$$f(x + y) = f(x) \cdot f(y)$$

for all real values of x and y.

   (a) Show that $f(0) = 0$ or $f(0) = 1$ must hold.

   (b) If $f(0) = 0$, deduce that $f(x) = 0$ for all x.

   (c) If f is differentiable at 0, then f must be differentiable for all x, and, furthermore, $f'(x) = f(x)\,f'(0)$. [*Hint:* Use the definition of the derivative and hypothesis that $f(x + h) = f(x) \cdot f(h)$.]

9

Trigonometric, Inverse Trigonometric, and Hyperbolic Functions

---

9.1   REVIEW OF TRIGONOMETRY

In this chapter we continue the study of the elementary transcendental functions started in Chapter 8. The emphasis in the chapter is on the trigonometric functions. It is the analytic portions of plane trigonometry rather than the solutions of triangles that will be our concern. Although it is assumed that the reader is somewhat familiar with elementary trigonometry, we shall begin by reviewing those parts that will be of use in calculus.

The radian rather than degrees may be used to measure the size of angles. Recall that the radian measure of an angle is defined to be the length s of the arc of a circle subtended by the angle (Figure 9.1.1) divided by the radius r of the circle, that is,

$$\theta = \frac{s}{r} \tag{1}$$

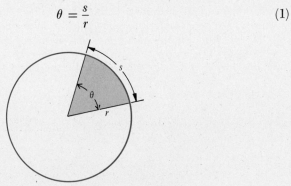

Figure 9.1.1

EXAMPLE 1   Find the radian measure of a straight angle AOB (Figure 9.1.2).

Solution   The circumference C of a circle of radius r is $C = 2\pi r$. The straight angle AOB intercepts one half of the circumference. Therefore $s = \pi r$ and

$$\theta = \frac{s}{r} = \pi$$ ●

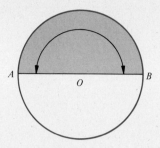

Figure 9.1.2

Example 1 tells us that a straight angle has a measure π radians (approximately 3.1416 radians). Thus π radians corresponds to 180° (the number of degrees in a straight angle). Therefore, the following proportion enables us to go from degrees to radians and conversely from radians to degrees:

$$\frac{\text{number of degrees}}{180°} = \frac{\text{number of radians}}{\pi} \tag{2}$$

EXAMPLE 2 Determine the measure in radians of a 45° angle.

Solution From (2),

$$\frac{45°}{180°} = \frac{\text{number of radians}}{\pi}$$

so that

$$\text{number of radians} = \frac{\pi}{4} \doteq 0.7854$$

This means that

$$45° \doteq 0.7854 \text{ radians} \qquad \bullet$$

EXAMPLE 3 How many degrees are there in 1 radian?

Solution From (2)

$$\frac{\text{number of degrees}}{180°} = \frac{1}{\pi}$$

Therefore, 1 radian corresponds to $\dfrac{180°}{\pi} \doteq 57.30°$ $\qquad \bullet$

Roughly speaking, 1 radian is a little less than 60° (see Figure 9.1.3). Notice that the central angle is 1 radian if the length of the intercepted arc is equal to the radius.

From the fact that 360° is equivalent to 2π radians, we can find a formula for the area of a sector (a pie shaped region). If the central angle is 2π radians, the sector becomes a circle with area πr^2. However, the area of a sector is proportional to the measure of the central angle. Thus if θ is the *radian* measure of the

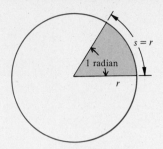

Figure 9.1.3

central angle of the sector, then

$$\frac{\theta}{2\pi} = \frac{A}{\pi r^2}$$

where A is the area of the sector. Therefore

$$A = \frac{r^2\theta}{2} \tag{3}$$

is the required area.

There is no reason to restrict the angle θ to the interval $[0, 2\pi]$. Instead we can associate an angle with *any* number θ. To this end we consider a unit circle centered at the origin, with the ***initial side*** or ***arm*** OA of the angle placed along the positive x-axis. The ***second arm*** or ***terminal side*** OP of the angle θ may be obtained by pivoting about O a distance θ in the *counterclockwise direction* (this follows from (1) since $s = \theta$ when $r = 1$). In this way any positive angle θ may be interpreted geometrically (see Figure 9.1.4). Note, in particular, that if $\theta = \dfrac{9\pi}{2}$ it is necessary to "wrap around" the circle twice and then reach the point P above the center O of the circle $\left(\text{since } \dfrac{9\pi}{2} = 2\pi + 2\pi + \dfrac{\pi}{2}\right)$.

In a similar manner we may associate angles with the negative number θ by proceeding in a *clockwise direction* around the unit circle. Thus, for example, to display the angle $-\dfrac{3\pi}{4}$ we start with the initial side along the x-axis and proceed clockwise through the angle of $\dfrac{3\pi}{4}$ radians or $135°$ as shown in Figure 9.1.4.

We now define the cosine and sine functions. Let θ be an arbitrary real number. Draw the angle of θ radians in ***standard position*** as shown in Figure 9.1.5, where P is the intersection of the terminal side of the angle with the unit circle having its center at the origin. If P is the point (x, y) then the ***cosine*** and ***sine*** functions are defined by

$$\begin{cases} x = \cos\theta \\ y = \sin\theta \end{cases} \tag{4}$$

From the definitions (4) it follows that the $\cos\theta$ and the $\sin\theta$ are defined for all real values of θ. This means that the domain of the cosine and sine functions is the set of all real numbers. We can thus speak of the sine or cosine of a number, rather than the sine or cosine of an angle. We then call θ the ***argument*** of the trigonometric function.

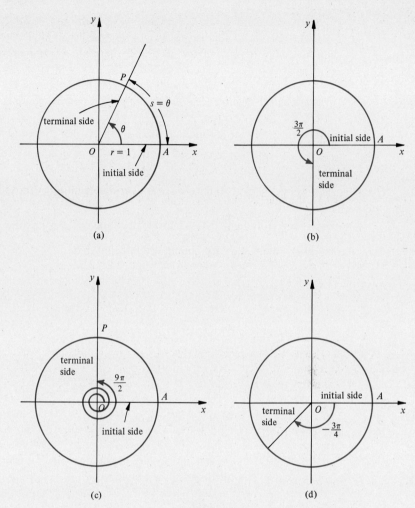

Figure 9.1.4

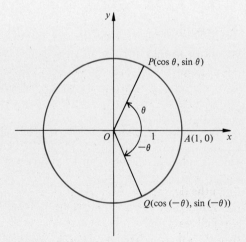

Figure 9.1.5

Since on the unit circle $|x| \le 1$ and $|y| \le 1$, the largest value of either function is 1, and the least value is -1. It will be established later that the functions are differentiable for all values of θ and therefore continuous everywhere. This implies that the range of the two functions is $[-1, 1]$.

The reader can verify without difficulty that, if θ is an acute angle, the definitions in (4) correspond to the standard right triangle definitions of $\sin \theta$ and $\cos \theta$.

The cosine and sine functions are easily obtained for certain values of θ. Thus $\cos 0 = 1$ and $\sin 0 = 0$, $\cos \dfrac{\pi}{2} = 0$ and $\sin \dfrac{\pi}{2} = 1$, $\cos \pi = -1$ and $\sin \pi = 0$.

Also elementary geometry enables us to determine that $\cos \dfrac{\pi}{4} = \sin \dfrac{\pi}{4} = \dfrac{\sqrt{2}}{2}$, $\cos \dfrac{\pi}{3} = \dfrac{1}{2}$ and $\sin \dfrac{\pi}{3} = \dfrac{\sqrt{3}}{2}$, and so on. Table 9.1.1 gives these values in addition to other values that are commonly encountered.

Table 9.1.1

θ	0	$\dfrac{\pi}{6}$	$\dfrac{\pi}{4}$	$\dfrac{\pi}{3}$	$\dfrac{\pi}{2}$	$\dfrac{2\pi}{3}$	$\dfrac{3\pi}{4}$	$\dfrac{5\pi}{6}$	π	$\dfrac{3\pi}{2}$	2π
$\sin \theta$	0	$\dfrac{1}{2}$	$\dfrac{\sqrt{2}}{2}$	$\dfrac{\sqrt{3}}{2}$	1	$\dfrac{\sqrt{3}}{2}$	$\dfrac{\sqrt{2}}{2}$	$\dfrac{1}{2}$	0	-1	0
$\cos \theta$	1	$\dfrac{\sqrt{3}}{2}$	$\dfrac{\sqrt{2}}{2}$	$\dfrac{1}{2}$	0	$\dfrac{-1}{2}$	$\dfrac{-\sqrt{2}}{2}$	$\dfrac{-\sqrt{3}}{2}$	-1	0	1

Note from the definitions (4) that $\cos \theta$ is positive in the first and fourth quadrants (where x is positive), whereas it is negative in the second and third. Also $\sin \theta$ is positive in the first and second quadrants (where y is positive), whereas it is negative in the third and fourth quadrants.

An equation of the unit circle having its center at the origin is $x^2 + y^2 = 1$. Substitution of (4) into this relation yields the significant *identity*

$$\cos^2 \theta + \sin^2 \theta = 1 \tag{5}$$

This relation is true for all values of θ. Note that $\cos^2 \theta$ and $\sin^2 \theta$ stand, respectively, for $(\cos \theta)^2$ and $(\sin \theta)^2$.

From Figure 9.1.5 we see that the points P and Q corresponding to θ and $-\theta$ are mirror images of each other with the x-axis as the line of symmetry. Thus

$$\cos(-\theta) = \cos \theta \tag{6}$$

and

$$\sin(-\theta) = -\sin \theta \tag{7}$$

which implies that $\cos \theta$ is an even function of θ while $\sin \theta$ is odd. Also since $\theta + 2\pi$ and θ yield the same point P for any value of θ we have

$$\cos(\theta + 2\pi) = \cos \theta \tag{8}$$

and

$$\sin(\theta + 2\pi) = \sin \theta \tag{9}$$

Equations (8) and (9) expresses the fact that cosine and sine are periodic functions of θ with period 2π in accordance with the following definition.

Definition   A function $f(x)$ is said to be ***periodic*** with a period $p \neq 0$ if whenever x is in the domain of f, then $x + p$ is also in the domain of f and

$$f(x + p) = f(x) \qquad (10)$$

In the case of $\cos x$ and $\sin x$ we have seen that $p = 2\pi$ is a suitable value. Other possible values for p are $4\pi, 6\pi, 8\pi, \ldots, -2\pi, -4\pi, \ldots$. The smallest positive value of p that satisfies (10) is called ***the period*** of f. On the graph of the cosine or sine functions it is the distance from one wave crest to the next (Figures 9.1.6 and 9.1.7). We conjecture from geometric considerations that 2π is the period of the cosine function—let us prove this analytically in the next illustration.

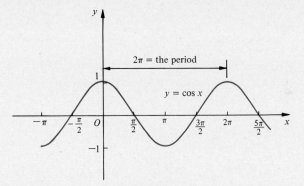

Figure 9.1.6

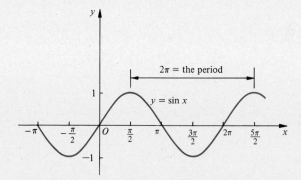

Figure 9.1.7

EXAMPLE 4   Prove that 2π is the period of the function $\cos x$.

Proof   If p is the period of $\cos x$ then $\cos (x + p) = \cos x$ for all p.
     This must be true for all x. Thus, let $x = 0$ so that $\cos p = 1$, which implies for $p > 0$, that $p = 2n\pi$ ($n = 1, 2, \ldots$). Thus $p = 2\pi$ is the period.   ∎

There are many important relationships or identities involving the sine and cosine functions. One of them is

$$\cos(\theta_1 - \theta_2) = \cos\theta_1 \cos\theta_2 + \sin\theta_1 \sin\theta_2 \qquad (11)$$

This identity is known as a difference formula and leads to the rapid derivation of other identities that we will need. We will prove (11) for the case where $\theta_1 > \theta_2 > 0$ (Figure 9.1.8). If the points P_1, P_2, P, and A on the unit circle have coordinates as shown in Figure 9.1.8 then $|\overset{\frown}{P_1P_2}| = |\overset{\frown}{AP}| = \theta_1 - \theta_2$. But from elementary geometry we know that, in the same or in equal circles, equal arcs have equal chords, which yields

$$|\overline{P_1P_2}| = |\overline{AP}| \qquad (12)$$

The distance formula between two points requires that

$$\sqrt{[(\cos\theta_1 - \cos\theta_2)^2 + (\sin\theta_1 - \sin\theta_2)^2} =$$
$$\sqrt{[\cos(\theta_1 - \theta_2) - 1]^2 + \sin^2(\theta_1 - \theta_2)}$$

Square both sides and expand both sides,

$$\cos^2\theta_1 - 2\cos\theta_1\cos\theta_2 + \cos^2\theta_2 + \sin^2\theta_1 - 2\sin\theta_1\sin\theta_2 + \sin^2\theta_2$$
$$= \cos^2(\theta_1 - \theta_2) - 2\cos(\theta_1 - \theta_2) + 1 + \sin^2(\theta_1 - \theta_2)$$

From (5), we have, after simplification

$$2 - 2(\cos\theta_1\cos\theta_2 + \sin\theta_1\sin\theta_2) = 2 - 2\cos(\theta_1 - \theta_2)$$

from which

$$\cos(\theta_1 - \theta_2) = \cos\theta_1\cos\theta_2 + \sin\theta_1\sin\theta_2 \qquad (11)$$

Similar derivations can be given for the other cases involving the arguments θ_1 and θ_2. It is recommended that the reader try some other cases.

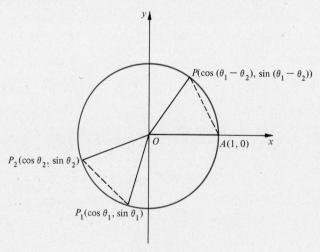

Figure 9.1.8

From properties (4), (5), (6), (7), (11), and the known values for the cosine and sine of special angles we can derive a number of useful identities. Some of the identities below are derived by utilizing earlier identities in the list.

$$\cos\left(\frac{\pi}{2} - \theta\right) = \sin\theta \tag{13}$$

$$\sin\left(\frac{\pi}{2} - \theta\right) = \cos\theta \tag{14}$$

$$\cos(\theta_1 + \theta_2) = \cos\theta_1 \cos\theta_2 - \sin\theta_1 \sin\theta_2 \tag{15}$$

$$\sin(\theta_1 - \theta_2) = \sin\theta_1 \cos\theta_2 - \cos\theta_1 \sin\theta_2 \tag{16}$$

$$\sin(\theta_1 + \theta_2) = \sin\theta_1 \cos\theta_2 + \cos\theta_1 \sin\theta_2 \tag{17}$$

$$\sin 2\theta = 2 \sin\theta \cos\theta \tag{18}$$

$$\cos 2\theta = \cos^2\theta - \sin^2\theta \tag{19}$$

$$\cos 2\theta = 2 \cos^2\theta - 1 \tag{20}$$

$$\cos 2\theta = 1 - 2 \sin^2\theta \tag{21}$$

$$\sin^2\frac{\theta}{2} = \frac{1 - \cos\theta}{2} \tag{22}$$

$$\cos^2\frac{\theta}{2} = \frac{1 + \cos\theta}{2} \tag{23}$$

$$\cos\theta_1 \cos\theta_2 = \tfrac{1}{2}[\cos(\theta_1 - \theta_2) + \cos(\theta_1 + \theta_2)] \tag{24}$$

$$\sin\theta_1 \cos\theta_2 = \tfrac{1}{2}[\sin(\theta_1 - \theta_2) + \sin(\theta_1 + \theta_2)] \tag{25}$$

$$\sin\theta_1 \sin\theta_2 = \tfrac{1}{2}[\cos(\theta_1 - \theta_2) - \cos(\theta_1 + \theta_2)] \tag{26}$$

The identities (18) through (21) are the double argument formulas; (22) and (23) are the half argument formulas.

We shall prove properties (15) and (17) and leave the proofs of the others to the reader.

Proof   Proof of (15): From (11) with θ_2 replaced by $-\theta_2$,

$$\cos(\theta_1 + \theta_2) = \cos\theta_1 \cos(-\theta_2) + \sin\theta_1 \sin(-\theta_2)$$
$$= \cos\theta_1 \cos\theta_2 - \sin\theta_1 \sin\theta_2$$

from (6) and (7). ∎

Proof   Proof of (17):

$$\sin(\theta_1 + \theta_2) = \cos\left[\frac{\pi}{2} - (\theta_1 + \theta_2)\right] \qquad \text{from (13)}$$

$$= \cos\left(\frac{\pi}{2} - \theta_1 - \theta_2\right) \qquad \text{and from (11)}$$

$$= \cos\left(\frac{\pi}{2} - \theta_1\right)\cos\theta_2 + \sin\left(\frac{\pi}{2} - \theta_1\right)\sin\theta_2$$

$$\sin(\theta_1 + \theta_2) = \sin\theta_1 \cos\theta_2 + \cos\theta_1 \sin\theta_2 \qquad \text{from (13) and (14)} \quad ∎$$

There are four additional trigonometric functions, namely, tangent, cotangent, secant, and cosecant. These functions are defined in terms of cosine and sine:

$$\tan \theta = \frac{\sin \theta}{\cos \theta} \qquad \cot \theta = \frac{\cos \theta}{\sin \theta} = \frac{1}{\tan \theta} \tag{27}$$

$$\sec \theta = \frac{1}{\cos \theta} \qquad \csc \theta = \frac{1}{\sin \theta} \tag{28}$$

From (27) and (28) it follows that $\tan \theta$ and $\sec \theta$ are not defined when $\cos \theta = 0$, that is, when $\theta = \frac{(2k + 1)\pi}{2}$ where k is an arbitrary integer. Figures 9.1.9 and 9.1.10 show the graphs of $y = \tan x$ and $y = \sec x$. Similarly $\cot \theta$ and $\csc \theta$ are not defined when $\sin \theta = 0$, that is if $\theta = k\pi$. The graphs of $y = \cot x$ and $y = \csc x$ are given, respectively, in Figures 9.1.11 and 9.1.12.

From the definitions and some of the previous identities of this section the following additional identities can be obtained

$$1 + \tan^2 \theta = \sec^2 \theta \tag{29}$$

$$1 + \cot^2 \theta = \csc^2 \theta \tag{30}$$

$$\tan (-\theta) = -\tan \theta \qquad \cot (-\theta) = -\cot \theta$$
$$\sec (-\theta) = \sec \theta \qquad \csc (-\theta) = -\csc \theta \tag{31}$$

$$\tan (\theta_1 + \theta_2) = \frac{\tan \theta_1 + \tan \theta_2}{1 - \tan \theta_1 \tan \theta_2} \tag{32}$$

$$\tan (\theta_1 - \theta_2) = \frac{\tan \theta_1 - \tan \theta_2}{1 + \tan \theta_1 \tan \theta_2} \tag{33}$$

and the following double and half argument formulas

$$\tan 2\theta = \frac{2 \tan \theta}{1 - \tan^2 \theta} \tag{34}$$

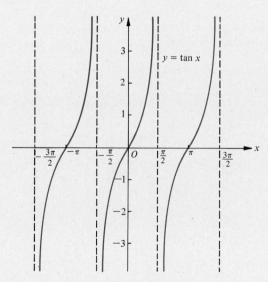

Figure 9.1.9

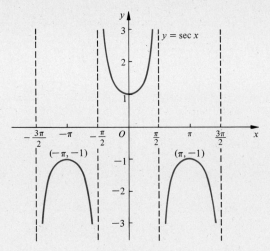

Figure 9.1.10

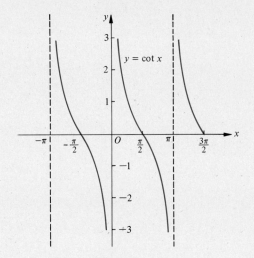

Figure 9.1.11

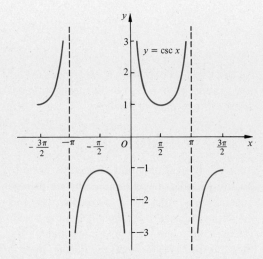

Figure 9.1.12

$$\tan \frac{\theta}{2} = \frac{1 - \cos \theta}{\sin \theta} \qquad (35)$$

$$\tan \frac{\theta}{2} = \frac{\sin \theta}{1 + \cos \theta} \qquad (36)$$

In our next examples two of the above identities are proved.

EXAMPLE 5 Prove that $1 + \cot^2 \theta = \csc^2 \theta$

Proof
$$1 + \cot^2 \theta = 1 + \frac{\cos^2 \theta}{\sin^2 \theta} = \frac{\sin^2 \theta + \cos^2 \theta}{\sin^2 \theta}$$

and from (5)

$$1 + \cot^2 \theta = \frac{1}{\sin^2 \theta}$$

$$= \csc^2 \theta \qquad \blacksquare$$

EXAMPLE 6 Prove that

$$\tan (\theta_1 + \theta_2) = \frac{\tan \theta_1 + \tan \theta_2}{1 - \tan \theta_1 \tan \theta_2}$$

Proof
$$\sin (\theta_1 + \theta_2) = \sin \theta_1 \cos \theta_2 + \cos \theta_1 \sin \theta_2 \qquad (17)$$
$$\cos (\theta_1 + \theta_2) = \cos \theta_1 \cos \theta_2 - \sin \theta_1 \sin \theta_2 \qquad (15)$$

and thus

$$\tan (\theta_1 + \theta_2) = \frac{\sin (\theta_1 + \theta_2)}{\cos (\theta_1 + \theta_2)} = \frac{\sin \theta_1 \cos \theta_2 + \cos \theta_1 \sin \theta_2}{\cos \theta_1 \cos \theta_2 - \sin \theta_1 \sin \theta_2}$$

Divide numerator and denominator by $\cos \theta_1 \cos \theta_2$ and obtain

$$\tan (\theta_1 + \theta_2) = \frac{\dfrac{\sin \theta_1}{\cos \theta_1} + \dfrac{\sin \theta_2}{\cos \theta_2}}{1 - \dfrac{\sin \theta_1}{\cos \theta_1} \dfrac{\sin \theta_2}{\cos \theta_2}}$$

$$= \frac{\tan \theta_1 + \tan \theta_2}{1 - \tan \theta_1 \tan \theta_2} \qquad \blacksquare$$

The tangent, cotangent, secant, and cosecant functions are all periodic. From their graphs, we see that tangent and cotangent have period π, whereas secant and cosecant have period 2π. Thus

$$\tan (\theta + \pi) = \tan \theta \qquad \cot (\theta + \pi) = \cot \theta$$
$$\sec (\theta + 2\pi) = \sec \theta \qquad \csc (\theta + 2\pi) = \csc \theta$$

Many physical phenomena involve periodic or wave motion. Some examples are the swinging of a pendulum, the vibration of a spring, beam or rod, alternating current, the motion of the planets around the sun, sound, water, light and radio waves. Because of their wavelike and periodic character the cosine and the sine functions are significant mathematical tools in the analytic treatment of these phenomena. Thus the small oscillations of a simple pendulum are often described by $x = x_0 \cos \omega t$ where x describes its displacement as a function of the time t.

The quantities x_0 and ω are positive constants, the latter depending upon the length of the pendulum. Note that in this instance the period of oscillation is $\dfrac{2\pi}{\omega}$ while the range of the function is $[-x_0, x_0]$. We call x_0 the **amplitude** of the vibration that is, the absolute maximum of the displacement. More generally, if the displacement of an object is given by

$$A \cos(\omega t + \alpha), \qquad A \sin(\omega t + \alpha) \tag{37}$$

or, equivalently, by

$$c_1 \cos \omega t + c_2 \sin \omega t \tag{38}$$

where A, ω, α, c_1, and c_2 are constants then the oscillatory motion is said to be **simple harmonic motion.** The three expressions (37) and (38) are equivalent in the sense that each can be expressed in the other two forms (Exercises 53 and 54). The graphs are similar to that of the cosine and sine function shown in Figures 9.1.6 and 9.1.7. The number $A > 0$ is called the amplitude of the function, the period is $\dfrac{2\pi}{\omega}$ while α is referred to as the **phase angle.**

Exercises 9.1

In Exercises 1 through 6 draw a unit circle, locate the point P with the given angular coordinate θ, and find $(\cos \theta, \sin \theta)$

1. $\theta = \dfrac{\pi}{2}$

2. $\theta = \dfrac{7\pi}{4}$

3. $\theta = -4\pi$

4. $\theta = \dfrac{7\pi}{3}$

5. $\theta = 5\pi$

6. $\theta = \dfrac{11\pi}{2}$

In Exercises 7 through 23 derive the given formula.

7. Formula (13) $\left[\textit{Hint: Let } \theta_1 = \dfrac{\pi}{2} \text{ and } \theta_2 = \theta \text{ in (11).} \right]$

8. Formula (14) [*Hint:* Use (13).]   9. Formula (16)
10. Formula (18)   11. Formula (19)
12. Formula (20) [*Hint:* Use (19) and (5).]   13. Formula (21)
14. Formula (22)   15. Formula (23)
16. Formula (24)   17. Formula (25)
18. Formula (26)   19. Formula (29)
20. Formula (31)   21. Formula (33)
22. Formula (34)   23. Formula (35)

In Exercises 24 through 32 derive the following identities.

24. $\sin\left(\dfrac{\pi}{2} + \theta\right) = \cos \theta$

25. $\sin(\theta + \pi) = -\sin \theta$

26. $\cos\left(\dfrac{3\pi}{2} - \theta\right) = -\sin \theta$

27. $\cos(\theta + \pi) = -\cos \theta$

28. $\sin\left(\theta + \dfrac{\pi}{4}\right) = \dfrac{\sqrt{2}}{2}(\sin\theta + \cos\theta)$

29. $\cos\left(\theta + \dfrac{\pi}{4}\right) = \dfrac{\sqrt{2}}{2}(\cos\theta - \sin\theta)$

30. $\sin\left(\theta + \dfrac{\pi}{6}\right) = \dfrac{1}{2}(\sqrt{3}\sin\theta + \cos\theta)$

31. $\sin k\theta = 2\sin(k-1)\theta\cos\theta - \sin(k-2)\theta$

32. $\cos k\theta = 2\cos(k-1)\theta\cos\theta - \cos(k-2)\theta$

33. Derive a formula for $\sin 3\theta$ in terms of $\sin\theta$. [*Hint:* $\sin 3\theta = \sin(2\theta + \theta)$, then use (17).]

34. Derive a formula for $\cos 3\theta$ in terms of $\cos\theta$.

35. Derive the **law of cosines** for an arbitrary triangle, namely that

$$c^2 = a^2 + b^2 - 2ab\cos\theta$$

where a, b, and c are the lengths of the sides and angle θ is opposite side c. [*Hint:* Place the triangle so that one vertex is at the origin, a second is on the x-axis with coordinates $(b, 0)$, and the third vertex has the coordinates $(a\cos\theta,\ a\sin\theta)$.]

36. Derive the formula for the area K of a triangle, namely that

$$K = \frac{ab\sin\theta}{2} = \frac{bc\sin\alpha}{2} = \frac{ca\sin\beta}{2}$$

where θ, α, β are the included angles between sides a and b, b and c, c and a, respectively. (*Hint:* See the hint of Exercise 35.)

37. Using the notation of the previous exercise, prove that

$$\frac{c}{\sin\theta} = \frac{a}{\sin\alpha} = \frac{b}{\sin\beta}$$

This is the **law of sines** for an arbitrary triangle. (*Hint:* Use the result of Exercise 36.)

In Exercises 38 through 43, find the solution sets of the given equations

38. $5\sin t + 4 = 3\sin t + 5$

39. $\tan^2 t = 1$

40. $2\sin\theta\cos\theta = 3$

41. $\cos 2\theta + \cos\theta = 0$

42. $\sin 2x\cos x = 1 - \cos 2x\sin x$

43. $\sin\dfrac{\theta}{2} + \cos\dfrac{\theta}{2} = 1$

In Exercises 44 through 49, determine which of the functions are even, odd or neither even nor odd.

44. $f(\theta) = \sin 10\theta$

45. $g(\theta) = 3\sin\theta + 2\cos\theta$

46. $h(\theta) = \cos^2 3\theta + \sin^2 3\theta - 4$

47. $F(\theta) = 2\sin\theta + 8\tan^3\theta$

48. $G(\theta) = \sqrt{\cos^2\theta + 3}$

49. $H(\theta) = |\sin\theta|$

50. If $f(x)$ is a function with period p, then show analytically that $f(x)$ must possess a period $2p$; that is, $f(x + 2p) = f(x)$ must hold. [*Hint:* Write $x + 2p = (x + p) + p$ and use the hypothesis.]

51. This is a generalization of Exercise 50. If $f(x)$ is a function with period p, then show that $f(x)$ is a function with period np where n is an arbitrary positive integer. (*Hint:* Use mathematical induction.) Does it also follow that $f(x - np) = f(x)$?

52. If f and g are periodic functions with periods p and q, respectively, show that $af + bg$, where a and b are constants is also periodic. Find its period. (*Hint:* Use the previous exercise.)

53. Without calculus, find the extrema of a function f defined by $f(t) = a\cos\omega t + b\sin\omega t$ (a, b, and ω are nonzero constants). $\left(\textit{Hint: Multiply and divide by } \sqrt{a^2 + b^2} \textit{ and let } \dfrac{a}{\sqrt{a^2 + b^2}} = \cos\alpha, \textit{ and so on}\right).$

9.2 DERIVATIVES OF TRIGONOMETRIC FUNCTIONS

From their graphs, it is clear that $\sin \theta$ and $\cos \theta$ are continuous for all values of θ. [This can be proved using definition (4) of section 9.1.] Similarly, $\tan \theta$, $\cot \theta$, $\sec \theta$, and $\csc \theta$ are continuous for all values of θ in their domains $\Big($for example, $\tan \theta$ is continuous at $\theta = 0$ and $\theta = \dfrac{\pi}{4}$ but is not continuous at $\theta = \dfrac{\pi}{2}\Big)$.

The determination of the derivatives of the trigonometric functions depends upon our finding the derivatives of the sine function. The derivative of the sine function, in turn, can be found once certain limits are evaluated. In fact, the whole sequence of calculations depends upon the determination of $\lim\limits_{x \to 0} \dfrac{\sin x}{x}$.

Theorem 1

$$\lim_{x \to 0} \frac{\sin x}{x} = 1 \qquad (x \text{ in radians}) \qquad (1)$$

It is important to notice that both the numerator and denominator are approaching zero as $x \to 0$. Thus we *cannot* use the theorem that the limit of a quotient equals the quotient of the limits. Also, there is no way to manipulate this algebraically so that a common factor can be cancelled out. Therefore we must try something different.

Proof   Temporarily we restrict the values of x to the open interval $0 < x < \dfrac{\pi}{2}$. Consider the unit circle in Figure 9.2.1. Since $|\overline{OQ}| = \cos x$, $|\overline{PQ}| = \sin x$ we have from

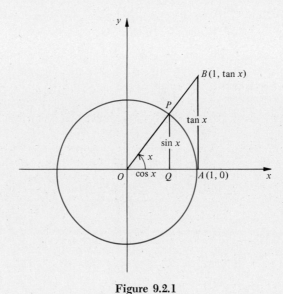

Figure 9.2.1

similar triangles

$$\frac{|\overline{PQ}|}{|\overline{OQ}|} = \frac{|\overline{AB}|}{|\overline{OA}|} = |\overline{AB}|$$

so that

$$|\overline{AB}| = \frac{\sin x}{\cos x} = \tan x$$

Comparison of areas yields

area of triangle $OPQ <$ area of sector $OAP <$ area of triangle OAB

which means that

$$\frac{\sin x \cos x}{2} < \frac{x}{2} < \frac{\tan x}{2}$$

Multiplication of the three members of the double inequality by $\dfrac{2}{\sin x}$ (which is positive) yields

$$\cos x < \frac{x}{\sin x} < \frac{1}{\cos x}$$

Take the reciprocals of each member to obtain

$$\frac{1}{\cos x} > \frac{\sin x}{x} > \cos x$$

(Note the reversal of the inequality signs.)

By continuity, we know that $\lim\limits_{x \to 0^+} \cos x = 1$ so that

$$\lim_{x \to 0^+} \frac{1}{\cos x} = \frac{1}{\lim\limits_{x \to 0^+} \cos x} = \frac{1}{1} = 1$$

Thus $\dfrac{\sin x}{x}$ must approach 1 as $x \to 0^+$, or

$$\lim_{x \to 0^+} \frac{\sin x}{x} = 1$$

Also because $\sin x$ and x are odd functions,

$$\lim_{x \to 0^-} \frac{\sin x}{x} = \lim_{x \to 0^+} \frac{\sin (-x)}{-x} \qquad \text{(if the latter limit exists)}$$

$$= \lim_{x \to 0^+} \frac{-\sin x}{-x}$$

$$= \lim_{x \to 0^+} \frac{\sin x}{x}$$

$$= 1$$

Since the two one-sided limits are equal,

$$\lim_{x \to 0^-} \frac{\sin x}{x} = \lim_{x \to 0^+} \frac{\sin x}{x} = 1$$

then (1) is established. ∎

We next establish the

Corollary

$$\lim_{x \to 0} \frac{\cos x - 1}{x} = 0 \tag{2}$$

Proof   We shall make use of (1).

$$\begin{aligned}
\lim_{x \to 0} \frac{\cos x - 1}{x} &= \lim_{x \to 0} \frac{(\cos x - 1)(\cos x + 1)}{x(\cos x + 1)} \\
&= \lim_{x \to 0} \frac{\cos^2 x - 1}{x(\cos x + 1)} \\
&= \lim_{x \to 0} \frac{-\sin^2 x}{x(\cos x + 1)} \\
&= \lim_{x \to 0} \frac{\sin x}{x} \lim_{x \to 0} \frac{-\sin x}{\cos x + 1} \\
&= (1)\left(\frac{0}{2}\right) = (1)(0) \\
&= 0 \qquad\blacksquare
\end{aligned}$$

Of course, any other letter could be substituted for x in equations (1) and (2). Thus if x is replaced by h,

$$\lim_{h \to 0} \frac{\sin h}{h} = 1 \qquad \text{and} \qquad \lim_{h \to 0} \frac{\cos h - 1}{h} = 0 \tag{3}$$

The results (3) will now be applied directly to the determination of the derivative of the sine function.

Theorem 2   If $f(x) = \sin x$, then $f'(x) = \cos x$ $\tag{4}$

Proof   We utilize the definition of the derivative.

$$\begin{aligned}
f'(x) &= \lim_{h \to 0} \frac{f(x + h) - f(x)}{h} \\
&= \lim_{h \to 0} \frac{\sin(x + h) - \sin x}{h}
\end{aligned}$$

and from (17) of Section 9.1

$$\begin{aligned}
f'(x) &= \lim_{h \to 0} \frac{\sin x \cos h + \cos x \sin h - \sin x}{h} \\
&= \lim_{h \to 0} \frac{\sin x (\cos h - 1)}{h} + \lim_{h \to 0} \frac{\cos x \sin h}{h}
\end{aligned}$$

since x is independent of h

$$f'(x) = \sin x \lim_{h \to 0} \frac{\cos h - 1}{h} + \cos x \lim_{h \to 0} \frac{\sin h}{h}$$

and from (3)

$$\begin{aligned}
f'(x) &= \sin x\,(0) + \cos x\,(1) \\
&= \cos x \qquad\blacksquare
\end{aligned}$$

If the argument of sine is a function of x as in $\sin 2x$ or $\sin (3x^2 - 1)$, we find the derivative by the chain rule as follows:

Let $y = \sin u$ where $u = u(x)$ is a differentiable function of x. Then $D_x y = D_u y\, D_x u = (\cos u)\, D_x u$. We have thus shown that

$$D_x(\sin u) = (\cos u)\, D_x u \tag{5}$$

[Note that we used u instead of x in (4).]

EXAMPLE 1   Differentiate $y = \sin (3x + 1)$.

Solution   In this instance, $u(x) = 3x + 1$ so that $D_x u = 3$. Therefore from (5) with $D_u y$ $= \cos u = \cos (3x + 1)$

$$y' = 3 \cos (3x + 1)$$ ●

EXAMPLE 2   Differentiate $y = \sin^3 x$.

Solution $y = \sin^3 x = (\sin x)^3$. Let $u = \sin x$, which implies that $y = u^3$. Therefore, the chain rule and (4) yield

$$y' = 3u^2\, u'$$
$$= 3 \sin^2 x \cos x$$ ●

Theorem 3   If $f(x) = \cos x$ then $f'(x) = -\sin x$

Proof   From (14), Section 9.1, we have $\cos x = \sin \left(\dfrac{\pi}{2} - x\right)$. If $u = \dfrac{\pi}{2} - x$, then $f(x) = \cos x = \sin u$. From (5),

$$f'(x) = (\cos u)\, D_x u = (\cos u)(-1) = -\cos \left(\frac{\pi}{2} - x\right)$$

By formula (13) of Section 9.1, $\sin x = \cos \left(\dfrac{\pi}{2} - x\right)$ so

$$f'(x) = -\sin x$$ ∎

When the argument is a function of x, $u = u(x)$, we have (by the chain rule)

$$D_x(\cos u) = -(\sin u)\, D_x u \tag{6}$$

EXAMPLE 3   Differentiate $f(x) = \cos \sqrt{x^2 + 4}$.

Solution   We set $u = \sqrt{x^2 + 4}$ so that application of (6) with $u' = \dfrac{x}{\sqrt{x^2 + 4}}$ by the chain rule, there results

$$f'(x) = -\frac{x}{\sqrt{x^2 + 4}} \sin \sqrt{x^2 + 4}$$ ●

EXAMPLE 4   Differentiate $y = \cos^2 (x^3 - 4)$.

Solution   Let $u = \cos (x^3 - 4)$. Then $y = u^2$.
By the chain rule

$$D_x y = 2u \, D_x u.$$

To find $D_x u$, let $v = x^3 - 4$ so that $u = \cos v$. Then

$$D_x u = D_v u \, D_x v = (-\sin v)(3x^2)$$

Thus

$$D_x y = (2u)(-\sin v)(3x^2) = 2 \cos (x^3 - 4)[-\sin (x^3 - 4)]3x^2$$

or

$$D_x y = -6x^2 \cos (x^3 - 4) \sin (x^3 - 4) \qquad \bullet$$

Theorem 4   If $f(x) = \tan x$, then $f'(x) = \sec^2 x$ for all x in the domain of $\tan x$.

Proof $f(x) = \tan x = \dfrac{\sin x}{\cos x}$. By the quotient rule

$$f'(x) = \frac{\cos x \, (\cos x) - \sin x \, (-\sin x)}{\cos^2 x}$$

$$= \frac{\cos^2 x + \sin^2 x}{\cos^2 x}$$

$$= \frac{1}{\cos^2 x}$$

$$= \sec^2 x \qquad\qquad \blacksquare$$

The following three theorems can be proven in similar manner. Their proofs are left as exercises for the reader.

Theorem 5   If $f(x) = \cot x$, then $f'(x) = -\csc^2 x$

Theorem 6   If $f(x) = \sec x$, then $f'(x) = \sec x \tan x$

Theorem 7   If $f(x) = \csc x$, then $f'(x) = -\csc x \cot x$

In Theorems 5 through 7, the values of x are to be restricted to the domains of the functions.

The results of Theorems 4 through 7 must be used with the chain rule if the argument is a function of x. We then have the following:

$$\text{If } f(x) = \tan u, \text{ then } f'(x) = \sec^2 u \, D_x u \qquad (7)$$
$$\text{If } f(x) = \cot u, \text{ then } f'(x) = -\csc^2 u \, D_x u \qquad (8)$$
$$\text{If } f(x) = \sec u, \text{ then } f'(x) = \sec u \tan u \, D_x u \qquad (9)$$
$$\text{If } f(x) = \csc u, \text{ then } f'(x) = -\csc u \cot u \, D_x u \qquad (10)$$

EXAMPLE 5   Differentiate $y = \sec \dfrac{1}{x}$.

Solution   With $u = \dfrac{1}{x}$ and $D_x u = -\dfrac{1}{x^2}$, we have from (9)

$$y' = -\frac{1}{x^2} \sec \frac{1}{x} \tan \frac{1}{x} \qquad \bullet$$

We shall now summarize our results by listing the "standard forms" for the derivatives of the six trigonometric functions.

$$D_x \sin u = \cos u \cdot u' \qquad\qquad D_x \cos u = -\sin u \cdot u'$$
$$D_x \tan u = \sec^2 u \cdot u' \qquad\qquad D_x \cot u = -\csc^2 u \cdot u'$$
$$D_x \sec u = \sec u \tan u \cdot u' \qquad\qquad D_x \csc u = -\csc u \cot u \cdot u'$$

These formulas are placed side by side to aid the memorization of the results.

EXAMPLE 6   Differentiate $y = \csc^3 6x$.

Solution   Let $v = \csc 6x$, then $y = v^3$. By the chain rule

$$D_x y = D_v y \, D_x v$$

Furthermore,

$$D_v y = 3v^2 = 3 \csc^2 6x$$
$$D_x v = -6 \csc 6x \cot 6x \qquad \text{from (10)}$$

so that

$$D_x \csc^3 6x = -18 \csc^3 6x \cot 6x \qquad \bullet$$

EXAMPLE 7   Differentiate $y = \ln \tan 4x$.

Solution   Let $u = \tan 4x$, then $y = \ln u$.

$$D_x y = D_u y \, D_x u$$
$$= \frac{1}{u} 4 \sec^2 4x$$
$$= \frac{4}{\tan 4x} \sec^2 4x$$
$$= \frac{4}{\sin 4x \cos 4x}$$
$$= \frac{8}{\sin 8x}$$
$$= 8 \csc 8x \qquad \bullet$$

EXAMPLE 8   Study the graph of $y = \sec x$.

Solution   Although this graph was already given in Section 9.1, we are now going to illustrate how the techniques we have learned can be used to draw the sketch.

We have

$$y = \sec x = \frac{1}{\cos x}$$

The domain of $\sec x$ is all values of x except those for which $\cos x = 0$; that is $x = \frac{\pi}{2} + n\pi$ $(n = 0, \pm 1, \pm 2, \ldots)$. These values of x correspond to the vertical asymptotes of the curve. For all x in the domain, $\sec x$ is continuous.

If $0 < \cos x \leq 1$, then $\sec x = \frac{1}{\cos x} \geq \frac{1}{1} = 1$; and if $-1 \leq \cos x < 0$, then $\sec x = \frac{1}{\cos x} \leq \frac{1}{-1} = -1$. The range of $\sec x$ is then $y \geq 1$ or $y \leq -1$.

$$y' = \sec x \tan x$$

For all x in the domain, the graph is smooth since y' exists for these values.

Setting $y' = 0$ to find the critical values yields $\sec x \tan x = 0$. But $\sec x$ is never zero, so $y' = 0$ if and only if $\tan x = 0$. Then $x = n\pi$ (where n is any integer). Corresponding to this we have

$$y = \sec n\pi = \begin{cases} 1 & \text{if } n \text{ is even} \\ -1 & \text{if } n \text{ is odd} \end{cases}$$

$$y'' = \sec x \,(\sec^2 x) + \tan x \,(\sec x \tan x) = \sec x \,(\sec^2 x + \tan^2 x)$$

Therefore, y'' has the same sign as $\sec x$. The graph is concave upward whenever $\sec x > 0$ (and hence $\cos x > 0$) and concave downward whenever $\sec x < 0$ (and hence $\cos x < 0$). It then follows that the points $(n\pi, 1)$ are relative minima if n is even and the points $(n\pi, -1)$ are relative maxima if n is odd.

The graph is drawn in Figure 9.1.10. ●

EXAMPLE 9 Find y' if $\cot (2x + y) = y$.

Solution Form the derivative with respect to x of both sides

$$-\csc^2 (2x + y) \cdot (2 + y') = y'$$

Collection of the coefficient of y' yields

$$y'[1 + \csc^2 (2x + y)] = -2 \csc^2 (2x + y)$$
$$y' = \frac{-2 \csc^2 (2x + y)}{1 + \csc^2 (2x + y)}$$ ●

EXAMPLE 10 Find an equation of the line tangent to $y = 3x \tan^2 x$ at $\left(\frac{\pi}{4}, \frac{3\pi}{4}\right)$.

Solution

$$y' = 3[x\, 2 \tan x \sec^2 x + \tan^2 x]$$
$$y' \Big|_{x = \pi/4} = 3\left[\frac{\pi}{2} (1)(\sqrt{2})^2 + 1\right] = 3(\pi + 1)$$

An equation of the tangent line at the given point is

$$y - \frac{3\pi}{4} = 3(\pi + 1)\left(x - \frac{\pi}{4}\right)$$ ●

Exercises 9.2

In Exercises 1 through 16, evaluate the limit, if it exists

1. $\lim\limits_{x \to 0} \dfrac{\sin 3x}{x}$

2. $\lim\limits_{x \to 0} \dfrac{\sin 5x}{3x}$

3. $\lim\limits_{x \to 0} \dfrac{\sin 4x}{x \cos x}$

4. $\lim\limits_{x \to 0^+} \dfrac{\sin 3x}{2\sqrt{x}}$

5. $\lim\limits_{x \to 0} \dfrac{7x}{\sin^2 x}$

6. $\lim\limits_{t \to 0} \dfrac{1 - \cos 2t}{t^2}$

7. $\lim\limits_{u \to \pi} \dfrac{\pi - u}{\sin u}$

8. $\lim\limits_{u \to \pi/2} \dfrac{\cos u}{u - \dfrac{\pi}{2}}$

9. $\lim\limits_{x \to 0} \dfrac{1 - \cos^2 2x}{x^2}$

10. $\lim\limits_{x \to 0} \dfrac{1 - \cos^4 3x}{7x^2}$
    (*Hint:* Factor the numerator.)

11. $\lim\limits_{h \to 0} \dfrac{\cos (a + h) - \cos a}{h}$

12. $\lim\limits_{\Delta x \to 0} \dfrac{\sin (a + 2\,\Delta x) - \sin a}{\Delta x}$ (*Hint:* Let $2\,\Delta x = h$.)

13. $\lim\limits_{t \to 0} \dfrac{\tan 5t}{t}$

14. $\lim\limits_{t \to 0} \dfrac{\tan 3t}{\sin 5t}$

15. $\lim\limits_{t \to 0} \cot 2t \sin^2 4t$

16. $\lim\limits_{t \to 0} \dfrac{3 \tan 3t - 2 \tan t}{t}$

In Exercises 17 through 36 find the derivative of the given function with respect to the indicated independent variable.

17. $y = \sin x \cos 2x$

18. $y = \dfrac{x}{\sin x}$

19. $z = \sin^2 3t + \cos^2 3t$

20. $y = (\cos t + \sin t)^2$

21. $u = \dfrac{\sin 4\theta}{2 \sin 2\theta}$

22. $v = \cos (\sin t)$

23. $w = a(1 - \cos \theta)$

24. $u = x^2 \sin \dfrac{1}{x}$

25. $z = \sqrt{\cos x^2}$

26. $y = \sin^2 \sqrt{t}$

27. $y = e^{3x}(4 \cos 2x + 3 \sin 2x)$

28. $u = \sin^3 t \cos^2 t$

29. $u = \ln \sec bt,$ (b constant)

30. $u = \ln \sqrt[3]{\dfrac{1 + \sin t}{1 - \sin t}}$

31. $u = e^{\sin b\theta} \cos c\theta,$ (b and c constants)

32. $g = t^{\sin t}$

33. $h = (\sin \theta)^{\sin \theta}$

34. $z = e^{\sin mx} \cos nx,$ (m and n constants)

35. $w = \sin kx \sin^k x,$ (k constant)

36. $u = \dfrac{\tan nx}{nx},$ (n constant)

37. Find a formula for the nth derivative of
    (a) $\sin 3x$ (b) $\cos 2x$

In Exercises 38 through 41, find y'.

38. $\cos x = \sin y$

39. $\cot (x + y) = 2y$

40. $\sec (xy) = \tan 3x$

41. $e^{2x+y} = \cos \left(x + \dfrac{\pi}{4}\right)$

42. Verify that the functions $\sin kt$ and $\cos kt$ satisfy the differential equation $y'' + k^2 y = 0$, (k constant). Verify also that $A \sin kt + B \cos kt$ also satisfies the equation where A and B are arbitrary constants. Differential equations of this type are significant in the theory of socalled **undamped free vibrations** of systems. If $y(t)$ is the displacement of a particle then the motion as a function of time is said to be **simple harmonic.**

43. Verify that the functions $e^{-t} \cos t$ and $e^{-t} \sin t$ satisfy the differential equation $y'' + 2y' + 2y = 0$. Does $Ae^{-t} \cos t + Be^{-t} \sin t = e^{-t} (A \cos t + B \sin t)$, where A and B are arbitrary constants, also satisfy the differential equation? Equations of this type are significant in the analysis of **damped free vibrations** of systems.

44. A particle moves back and forth along the x-axis, its position at any time being given by $x = A \cos kt + B \sin kt$, where k, A, and B are constants. Show that its acceleration is proportional to its position x but oppositely directed. (*Hint:* $x''(t)$ is the acceleration of the particle. This is simple harmonic motion once again.)

In Exercises 45 through 48, find the absolute maximum and minimum values of the following functions in the given interval

45. $f(x) = 3x + \cos 2x$, $[0, \pi]$ **46.** $g(x) = 1 + \cos x - \cos^2 x$, $[0, 2\pi]$

47. $h(x) = \tan x - 8 \sin x$, $\left[0, \dfrac{\pi}{2}\right)$ **48.** $F(x) = \sqrt{3} \sin x + \cos x$, $[0, 2\pi]$

49. (a) Show that $y = \tan x$ satisfies the differential equation $y' = 1 + y^2$.
(b) Does $y = 2 \tan x$ also satisfy it?
(c) Does $y = \tan(x + k)$, where k is an arbitrary constant, satisfy the differential equation?

50. By differentiating both sides of the identity $\sin(x + \alpha) = \sin x \cos \alpha + \cos x \sin \alpha$ where α is an arbitrary constant. Derive a formula for $\cos(x + \alpha)$.

51. Let $\sin \theta°$ and $\cos \theta°$ denote the sine and cosine of an angle of θ degrees. Find a formula for $D_\theta \sin \theta°$.

52. Find equations of the tangent and normal lines to the curve $y = \sin 2x$ at $(0, 0)$.

53. (a) Find the points of the graph of $y = x + \cot x$ at which the tangent line is horizontal.
(b) Are these points relative extrema for the given function?

54. Given the function

$$f(x) = \begin{cases} \dfrac{\sin x}{x} & x \neq 0 \\ 1 & x = 0 \end{cases}$$

(a) Prove that $f(x)$ is continuous at $x = 0$.
(b) What kind of symmetry (if any) does $f(x)$ have?

55. Let $f(x) = x^2 \sin \dfrac{1}{x}$ and $f(0) = 0$. Prove that $f'(0)$ exists and find it.

56. (a) Prove that the area A_n of a regular polygon with n sides that is inscribed in a circle of radius r is given by

$$A_n = \frac{nr^2}{2} \sin \frac{2\pi}{n}$$

(b) Show that $\lim\limits_{n \to \infty} A_n = \pi r^2$.
(c) Does this prove that the area of a circle is πr^2?

INTEGRALS INVOLVING TRIGONOMETRIC FUNCTIONS

From the formulas for the derivatives of the six trigonometric functions we immediately have corresponding indefinite integrals. We summarize these results as follows.

$$\int \sin u \, du = -\cos u + C \qquad (1)$$

$$\int \cos u \, du = \sin u + C \qquad (2)$$

$$\int \sec^2 u \, du = \tan u + C \qquad (3)$$

$$\int \csc^2 u \, du = -\cot u + C \qquad (4)$$

$$\int \sec u \tan u \, du = \sec u + C \qquad (5)$$

$$\int \csc u \cot u \, du = -\csc u + C \qquad (6)$$

For example, (2) follows from the fact that $D_u(\sin u) = \cos u$. Similarly, (4) follows from $D_u(\cot u) = -\csc^2 u$.

EXAMPLE 1   Evaluate $\int 5 \sin 5x \, dx$.

Solution   With $u = 5x$, $du = 5 \, dx$, and the integral is in the form $\int \sin u \, du$. Thus from (1),

$$\int 5 \sin 5x \, dx = \int \sin u \, du = -\cos u + C = -\cos 5x + C \qquad \bullet$$

EXAMPLE 2   Evaluate $\int \cos 3x \, dx$.

Solution   Let $u = 3x$ so that $du = 3 \, dx$ or $dx = \dfrac{du}{3}$

Therefore

$$\int \cos 3x \, dx = \int \cos u \, \frac{du}{3} = \frac{1}{3} \int \cos u \, du$$

$$= \tfrac{1}{3} \sin u + C \qquad \text{from (2)}$$

$$= \tfrac{1}{3} \sin 3x + C$$

CHECK:

$$D_x(\tfrac{1}{3}\sin 3x + C) = \tfrac{1}{3} \cdot 3 \cos 3x = \cos 3x$$

and our computation is verified. ●

EXAMPLE 3   Evaluate $\displaystyle\int_0^{\pi/6} \cos 3x \, dx$.

Solution   **Method 1:**   Follow the procedure used in Example 2 to obtain

$$\int \cos 3x \, dx = \tfrac{1}{3}\sin 3x + C$$

Then

$$\int_0^{\pi/6} \cos 3x \, dx = \frac{1}{3}\sin 3x \Big|_0^{\pi/6} = \frac{1}{3}\sin\frac{\pi}{2} - \frac{1}{3}\sin 0 = \frac{1}{3}$$

Method 2:   Again letting $u = 3x$ we have $du = 3\,dx$ or $dx = \tfrac{1}{3}\,du$. We change the limits of integration to correspond to the new variable u by noting that when $x = 0$ then $u = 0$ and when $x = \dfrac{\pi}{6}$ then $u = \dfrac{\pi}{2}$. Thus

$$\int_0^{\pi/6} \cos 3x \, dx = \tfrac{1}{3}\int_0^{\pi/2} \cos u \, du = \tfrac{1}{3}\sin u \Big|_0^{\pi/2} = \tfrac{1}{3}\sin\frac{\pi}{2} - \tfrac{1}{3}\sin 0 = \tfrac{1}{3} \quad ●$$

EXAMPLE 4   Evaluate $\displaystyle\int x \cos x^2 \, dx$.

Solution   This looks as if we may be able to apply (2) where $u = x^2$. Then $du = 2x\,dx$ or $x\,dx = \dfrac{du}{2}$ so that

$$\int x \cos x^2 \, dx = \int \cos u \, \frac{du}{2} = \frac{1}{2}\int \cos u \, du$$

$$= \frac{\sin u}{2} + C \quad \text{from (2)}$$

$$= \frac{\sin x^2}{2} + C \qquad ●$$

EXAMPLE 5   Evaluate $\displaystyle\int \sin^5 x \cos x \, dx$.

Solution   None of the given formulas can be applied directly to this integral. However, if we substitute $u = \sin x$, then with $du = \cos x \, dx$ our integral becomes

$$\int \sin^5 x \cos x \, dx = \int u^5 \, du = \frac{u^6}{6} + C = \frac{\sin^6 x}{6} + C \qquad ●$$

EXAMPLE 6   Evaluate $\displaystyle\int \cot^3 4x \csc^2 4x \, dx$.

Solution   Let $u = \cot 4x$ so that $du = -4\csc^2 4x \, dx$. Therefore

$$\int \cot^3 4x \csc^2 4x \, dx = -\frac{1}{4} \int u^3 \, du = -\frac{u^4}{16} + C$$

$$= -\frac{\cot^4 4x}{16} + C \qquad \bullet$$

So far, all our trigonometric integration examples were performed by a simple substitution. In Examples 1 through 4, the substitution put the integral into a "standard form" [formulas (1) through (6)] and, in Examples 5 and 6, the substitution gave us the generalized power rule of Section 5.3. Other integrals of trigonometric functions may require more manipulation. Some involve use of one of the identities given in Section 9.1; others are more complicated. We shall look at a few of the easier manipulations now. In Sections 10.2 and 10.3, we shall study these integrals in more depth.

EXAMPLE 7   Evaluate $\int \cot^2 x \, dx$.

Solution   None of the given formulas allows us to evaluate this integral directly. However, this problem can be handled easily if we utilize the identity $\cot^2 x = \csc^2 x - 1$. Therefore

$$\int \cot^2 x \, dx = \int (\csc^2 x - 1) \, dx = \int \csc^2 x \, dx - \int 1 \, dx$$

then from (4) with $u = x$

$$\int \cot^2 x \, dx = -\cot x - x + C \qquad \bullet$$

EXAMPLE 8   Evaluate $\int \tan u \, du$.

Solution $\int \tan u \, du = \int \frac{\sin u}{\cos u} \, du$

Letting $v = \cos u$, we have $dv = -\sin u \, du$. Thus

$$\int \tan u \, du = \int -\frac{dv}{v} = -\ln |v| + C = -\ln |\cos u| + C \qquad (7)$$

Our result can be put in another form if we note that

$$|\sec u| = \frac{1}{|\cos u|}$$

which implies $\ln |\sec u| = \ln 1 - \ln |\cos u| = -\ln |\cos u|$. Thus (7) may be rewritten in the form

$$\int \tan u \, du = \ln |\sec u| + C \qquad \bullet \qquad (8)$$

EXAMPLE 9 Evaluate $\int e^{4x} \tan(e^{4x} + 3)\, dx$.

Solution Let $u = e^{4x} + 3$. Then $du = 4e^{4x}\, dx$. From (8), we have

$$\int e^{4x} \tan(e^{4x} + 3)\, dx = \tfrac{1}{4} \int \tan u\, du$$

$$= \tfrac{1}{4} \ln \sec |e^{4x} + 3| + C$$
$$= \tfrac{1}{4} \ln \sec (e^{4x} + 3) + C$$

The absolute value sign is not needed since $e^{4x} + 3$ is positive for all x. ●

To find $\int \sec^2 u\, du$ we need only apply formula (3). But how about $\int \sec u\, du$?

This requires a rather ingenious derivation. We multiply both numerator and denominator by $\sec u + \tan u$. Then

$$\int \sec u\, du = \int \frac{(\sec^2 u + \sec u \tan u)}{\sec u + \tan u}\, du$$

If $v = \sec u + \tan u$ then $dv = (\sec u \tan u + \sec^2 u)\, du$, which is the numerator of the integrand. Thus

$$\int \sec u\, du = \int \frac{dv}{v} = \ln |v| + C$$

$$= \ln |\sec u + \tan u| + C \qquad (9)$$

The following four formulas may be added to our list of integrals. We have already derived two of them. The others can be derived by similar methods.

$$\int \tan u\, du = -\ln |\cos u| + C = \ln |\sec u| + C \qquad (10)$$

$$\int \cot u\, du = \ln |\sin u| + C \qquad (11)$$

$$\int \sec u\, du = \ln |\sec u + \tan u| + C \qquad (12)$$

$$\int \csc u\, du = -\ln |\csc u + \cot u| + C \qquad (13)$$

Formula (10) illustrates that in trigonometric integration problems, two apparently different answers may be equivalent. In Exercise 49, this idea is again demonstrated.

EXAMPLE 10 Find the area of the region bounded by $y = \sec x$, $y = \sin x$, the y-axis, and the line $x = \dfrac{\pi}{3}$.

Solution   Referring to Figure 9.3.1 and using the method of Section 7.1, we see that

$$A = \int_0^{\pi/3} [\sec x - \sin x]\, dx$$
$$= [\ln |\sec x + \tan x| + \cos x]_0^{\pi/3}$$
$$= [\ln(2 + \sqrt{3}) + \tfrac{1}{2}] - [\ln 1 + 1]$$
$$= \ln(2 + \sqrt{3}) - \tfrac{1}{2}$$

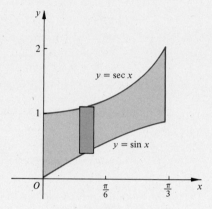

Figure 9.3.1

The identities

$$\sin^2 u = \frac{1 - \cos 2u}{2} \tag{14}$$

$$\cos^2 u = \frac{1 + \cos 2u}{2} \tag{15}$$

are very useful in integrating even powers of $\sin u$ and $\cos u$. These applications will be treated in more detail in Section 10.2, which is devoted to an extensive treatment of integrals involving powers of trigonometric functions. We shall look at one illustration now.

EXAMPLE 11   Evaluate $\int \cos^2 3x\, dx$.

Solution   From (15) with $u = 3x$ and $du = 3\, dx$ we have

$$\int \cos^2 3x\, dx = \frac{1}{3} \int \cos^2 u\, du = \frac{1}{3} \int \frac{1 + \cos 2u}{2}\, du$$
$$= \frac{1}{6} \int du + \frac{1}{6} \int \cos 2u\, du$$
$$= \frac{1}{6} u + \frac{1}{12} \int \cos 2u\, (2\, du)$$
$$= \frac{u}{6} + \frac{1}{12} \sin 2u + C$$
$$= \frac{x}{2} + \frac{\sin 6x}{12} + C$$

Exercises 9.3

Evaluate the integrals in Exercises 1 through 42.

1. $\int \cos 3x \, dx$

2. $\int \sin \frac{\theta}{2} \, d\theta$

3. $\int \sec 5\theta \tan 5\theta \, d\theta$

4. $\int \csc^2 \left(\frac{t}{3} \right) dt$

5. $\int \csc \frac{\theta}{3} \cot \frac{\theta}{3} \, d\theta$

6. $\int t \sec^2 t^2 \, dt$

7. $\int \tan^2 v \, dv$

8. $\int (\sin \theta + \cos \theta)^2 \, d\theta$

9. $\int (\sin 3t - \cos 3t)^2 \, dt$

10. $\int (\cos^2 3x - \sin^2 3x) \, dx$

11. $\int \cot^2 5\theta \, d\theta$

12. $\int \frac{\cos y}{10 + \sin y} \, dy$

13. $\int \sin^5 2t \cos 2t \, dt$

14. $\int \cos^4 3t \sin 3t \, dt$

15. $\int \frac{1 + \sin 3t}{\sin 3t} \, dt$

16. $\int \sqrt{3 + 2 \tan \theta} \, \sec^2 \theta \, d\theta$

17. $\int \frac{\csc^2 x}{\cot^3 x} \, dx$

18. $\int \frac{\cos t}{\cos^2 t - 1} \, dt$

19. $\int \frac{\sin 2y}{\sin^2 y - 1} \, dy$

20. $\int e^{\cot 2x} \csc^2 2x \, dx$

21. $\int \frac{1}{x^2} \cos \left(\frac{1}{x} \right) dx$

22. $\int \frac{a \cos x + b \sin x}{\sin x} \, dx$,
(a and b are constants)

23. $\int (\sec t - \tan t)^2 \, dt$

24. $\int \csc^4 3\theta \cot 3\theta \, d\theta$

25. $\int \sin^2 7x \, dx$

26. $\int \frac{3 + \tan t}{\cos t} \, dt$

27. $\int_0^{\pi/3} \cos 3x \, dx$

28. $\int_0^{\pi} \sin^2 x \, dx$

29. $\int_{-\pi/4}^{\pi/4} \sec^2 x \, dx$

30. $\int_0^{\pi/4} \sin x \cos x \, dx$

31. $\int_0^{\pi} (1 - \cos t)^2 \, dt$

32. $\int_0^{\pi} \cos^3 \theta \sin \theta \, d\theta$

33. $\int_0^{\pi/4} \tan^2 t \, dt$

34. $\int_0^{\pi/4} \sec \theta \, d\theta$

35. $\int_{-\pi/4}^{\pi/3} \tan^3 u \sec^2 u \, du$

36. $\int_{\pi}^{\pi/2} (1 - \cos \theta)(1 - \sin \theta) \, d\theta$

37. $\int_{\pi/8}^{\pi/4} \frac{dt}{\sin 2t}$

38. $\int_{2\pi/3}^{\pi/2} \frac{2 \sin x - 3 \cos x}{\sin x} \, dx$

39. $\int_0^{2\pi} \sin x \cos 3x \, dx$ [*Hint:* Use (25) of Section 9.1.]

40. $\int_0^{\pi/4} e^{\cos 2x} \sin 2x \, dx$

41. $\int_0^{\pi/3} \sin 2\theta \cos \theta \, d\theta$

42. $\int_0^{\pi} \sin mx \cos nx \, dx$,   ($m$ and n are integers)

43. Find the area of the region bounded by $y = \sin x$ and $y = \cos x$, between $x = 0$ and $x = \dfrac{\pi}{4}$.

44. Find the area of the region bounded by $y = \sin^2 x$ and the x-axis between $x = 0$ and $x = \pi$.

45. Find the area of the region bounded by $y = \cos 2x$ and $y = \frac{1}{2}$ between $x = -\dfrac{\pi}{6}$ and $x = \dfrac{\pi}{6}$.

46. Find the area of one of the arches bounded by $y = \sin x$ and the x-axis.

47. Suppose that the area bounded by the arch and the x-axis of the previous exercise is revolved about the x-axis. Find the volume of the solid generated.

48. Evaluate $\displaystyle\int_0^{\pi} |\cos x - \sin x|\, dx$.

49. The $\displaystyle\int \sin x \cos x\, dx$ may be looked at from two points of view:

   (a) $\displaystyle\int \sin x \cos x\, dx = \int \sin x\, d(\sin x) = \frac{\sin^2 x}{2} + C$

   (b) $\displaystyle\int \sin x \cos x\, dx = \frac{1}{2} \int \sin 2x\, dx = \frac{-\cos 2x}{4} + C$

   Which (if either) of (a) or (b) is correct?

50. Evaluate $\displaystyle\int_0^{\pi/2} \sin^3 x\, dx$. (*Hint:* Write $\sin^3 x = \sin^2 x \sin x$ and replace $\sin^2 x$ by $1 - \cos^2 x$.) More complicated trigonometric integrals will be discussed more fully later on where systematic formal methods will be developed.

51. Comment on the calculation

$$\int_0^{\pi} \sec^2 x\, dx = \tan x \bigg|_0^{\pi} = 0$$

   Note that $\sec^2 x > 0$ (where $\sec x$ is defined).

52. If $F(x) = \displaystyle\int_0^x \sin^3 t\, dt$, find

   (a) $F\left(\dfrac{\pi}{2}\right)$ and (b) $F'\left(\dfrac{\pi}{2}\right)$

53. Find $\displaystyle\int_{-\pi/2}^{\pi} |\sin 3t|\, dt$

54. Find the area bounded by one arch of $y = \sin^2 t$ and the line $y = \frac{1}{2}$.

9.4   INVERSE TRIGONOMETRIC FUNCTIONS

We know that the sine function $x = \sin y$ is differentiable everywhere and therefore also continuous everywhere. Furthermore its domain is unlimited while its range is the closed interval $[-1, 1]$. From the periodicity of the sine function shown in Figure 9.4.1, it follows that for each number x in $[-1, 1]$ there exists infinitely many numbers y such that $x = \sin y$. For example some of the roots of $\sin y = \frac{1}{2}$ are $y = \dfrac{\pi}{6}, \dfrac{5\pi}{6}, \dfrac{13\pi}{6}$, and so on. This means that the equation $x = \sin y$ does define x as a function of y; however, it does *not* define y as a function of x.

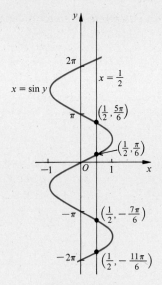

Figure 9.4.1

Note that for each x in $[-1, 1]$ there exists a unique number y in each of the intervals

$$\ldots, \left[-\frac{3\pi}{2}, -\frac{\pi}{2}\right], \left[-\frac{\pi}{2}, \frac{\pi}{2}\right], \left[\frac{\pi}{2}, \frac{3\pi}{2}\right], \ldots \qquad (1)$$

such that $x = \sin y$ is satisfied.

Refer now to the graph of $y = \sin x$, which is shown in Figure 9.1.7. Let us restrict the sine function to $\left[-\dfrac{\pi}{2}, \dfrac{\pi}{2}\right]$ where $D_x\,(\sin x) = \cos x \geq 0$ the equality holding only at $\pm\dfrac{\pi}{2}$. The function $f(x) = \sin x$ is strictly increasing in $\left[-\dfrac{\pi}{2}, \dfrac{\pi}{2}\right]$ and hence must have an inverse. The domain of f is $\left[-\dfrac{\pi}{2}, \dfrac{\pi}{2}\right]$, while its range is $[-1, 1]$. The graph of f is sketched in Figure 9.4.2. The inverse of

$$f(x) = \sin x \qquad -\frac{\pi}{2} \leq x \leq \frac{\pi}{2} \qquad (2)$$

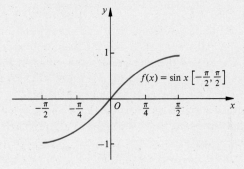

Figure 9.4.2

is called the ***inverse sine function*** and is denoted by the symbol $\sin^{-1}$. Formally, we have the following definition:

Definition $\qquad\qquad y = \sin^{-1} x$ if and only if $x = \sin y$ and $-\dfrac{\pi}{2} \le y \le \dfrac{\pi}{2}.$ (2a)

From this definition the function $\sin^{-1}$ has domain $[-1, 1]$ and range $\left[-\dfrac{\pi}{2}, \dfrac{\pi}{2}\right]$ (that is, its domain is the range of $f(x) = \sin x$ while its range is the domain of f). A sketch of $y = \sin^{-1} x$ is given in Figure 9.4.3 based upon the data in Table 9.4.1.

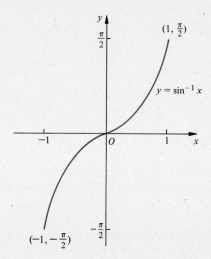

Figure 9.4.3

Table 9.4.1

x	-1	$-\dfrac{\sqrt{3}}{2}$	$-\dfrac{1}{\sqrt{2}}$	$-\dfrac{1}{2}$	0	$\dfrac{1}{2}$	$\dfrac{1}{\sqrt{2}}$	$\dfrac{\sqrt{3}}{2}$	1
$y = \sin^{-1} x$	$-\dfrac{\pi}{2}$	$-\dfrac{\pi}{3}$	$-\dfrac{\pi}{4}$	$-\dfrac{\pi}{6}$	0	$\dfrac{\pi}{6}$	$\dfrac{\pi}{4}$	$\dfrac{\pi}{3}$	$\dfrac{\pi}{2}$

In our definition of $y = \sin^{-1} x$, we restricted y to the closed interval $\left[-\dfrac{\pi}{2}, \dfrac{\pi}{2}\right]$. This interval is referred to as the ***principal value range*** of the inverse sine function. We could instead have chosen any of the other intervals in (1) as our principal value range and the definition would also have made sense. However, once one of these intervals is specified, it must always be used.

Note that our choice of $\left[-\dfrac{\pi}{2}, \dfrac{\pi}{2}\right]$ for the principal values means that y is either a fourth or a first quadrant angle. If $x > 0$, then y is an acute angle in the first quadrant. If $x < 0$ then y is in the fourth quadrant and is the negative of an acute angle.

It should be noted that $\sin^{-1} x \ne (\sin x)^{-1}$; $\sin^{-1} x$ denotes the inverse of the sine function, whereas $(\sin x)^{-1}$ is the reciprocal of the sine function, that is, $\dfrac{1}{\sin x}$.

From the definition of $\sin^{-1}$ we must have

$$\sin(\sin^{-1} x) = x \quad \text{for any } x \text{ in } [-1, 1] \tag{3}$$

and

$$\sin^{-1}(\sin y) = y \quad \text{for any } y \text{ in } \left[-\frac{\pi}{2}, \frac{\pi}{2}\right] \tag{4}$$

EXAMPLE 1 $\sin^{-1}\left(\dfrac{1}{\sqrt{2}}\right) = \dfrac{\pi}{4}$ because $-\dfrac{\pi}{2} \leq \dfrac{\pi}{4} \leq \dfrac{\pi}{2}$,   and $\sin\dfrac{\pi}{4} = \dfrac{1}{\sqrt{2}}$.   Also,

$\sin^{-1}\left(-\dfrac{\sqrt{3}}{2}\right) = -\dfrac{\pi}{3}$ since $-\dfrac{\pi}{2} \leq -\dfrac{\pi}{3} \leq \dfrac{\pi}{2}$ and $\sin\left(-\dfrac{\pi}{3}\right) = -\dfrac{\sqrt{3}}{2}$.

However, $\sin^{-1}\left(-\dfrac{1}{2}\right) \neq \dfrac{7\pi}{6}$ even though $\sin\left(\dfrac{7\pi}{6}\right) = -\dfrac{1}{2}$ because $\dfrac{7\pi}{6}$ is *not* in the range of the inverse sine function.   ●

In a similar manner, we must restrict the domain of the cosine function in order that the modified function possess an inverse. To this end, we choose the interval $[0, \pi]$ on which the cosine is decreasing. $\left(\text{Note that we cannot choose} \left[-\dfrac{\pi}{2}, \dfrac{\pi}{2}\right].\right)$ Thus, if

$$F(x) = \cos x \quad \text{where } 0 \leq x \leq \pi \tag{5}$$

then $[0, \pi]$ is the domain and $[-1, 1]$ is the range of the function (see Figure 9.4.4). Furthermore $F(x)$ must possess an inverse function defined as follows.

Definition   The ***inverse cosine function,*** denoted $\cos^{-1}$, is defined by $y = \cos^{-1} x$ if and only if $x = \cos y$ and $0 \leq y \leq \pi$.   5(a)

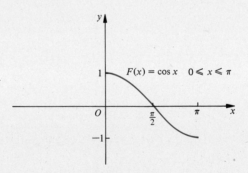

Figure 9.4.4

The domain of $\cos^{-1} x$ is the closed interval $[-1, 1]$ and the range is the closed interval $[0, \pi]$. We call $[0, \pi]$ the ***principal value range*** of $y = \cos^{-1} x$. y corresponds to an angle in either the first or second quadrant. The graph of $y = \cos^{-1} x$ is shown in Figure 9.4.5.

From the definition, it follows that:

$$\cos(\cos^{-1} x) = x \quad \text{for } x \text{ in } [-1, 1] \tag{6}$$

$$\cos^{-1}(\cos y) = y \quad \text{for } y \text{ in } [0, \pi] \tag{7}$$

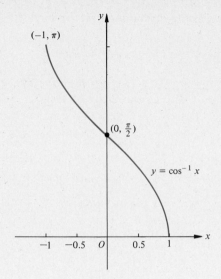

Figure 9.4.5

EXAMPLE 2   Find the values of $\cos^{-1} x$ for $x = -1, -\frac{1}{2}, 0, \frac{1}{2}$ and 1.

Solution   The reader is referred to Table 9.4.2 where $y = \cos^{-1} x$ if and only if $x = \cos y, \ 0 \le y \le \pi$.   ●

Table 9.4.2

x	-1	$-\dfrac{1}{2}$	0	$\dfrac{1}{2}$	1
$y = \cos^{-1} x$	π	$\dfrac{2\pi}{3}$	$\dfrac{\pi}{2}$	$\dfrac{\pi}{3}$	0

EXAMPLE 3   Find $\cos(\sin^{-1} k)$ where $|k| \le 1$. In particular, evaluate $\cos \sin^{-1}(-\frac{5}{13})$.

Solution   Let $\sin^{-1} k = c$; that is, c must be a number in $-\dfrac{\pi}{2} \le c \le \dfrac{\pi}{2}$ for which $\sin c = k$. Now

$$\cos c = \pm\sqrt{1 - \sin^2 c} = \pm\sqrt{1 - k^2}$$

and we must choose a plus sign because $\cos c \ge 0$ when $-\dfrac{\pi}{2} \le c \le \dfrac{\pi}{2}$. Therefore $\cos(\sin^{-1} k) = \sqrt{1 - k^2}$ where $|k| \le 1$. In particular,

$$\begin{aligned}\cos(\sin^{-1} -\tfrac{5}{13}) &= \sqrt{1 - (-\tfrac{5}{13})^2} \\ &= \sqrt{1 - \tfrac{25}{169}} = \sqrt{\tfrac{144}{169}} \\ &= \tfrac{12}{13}\end{aligned}$$ ●

In order to obtain the inverse tangent function we first restrict the tangent function to the open interval $\left(-\dfrac{\pi}{2}, \dfrac{\pi}{2}\right)$ where it is both continuous and increasing. If we let

$$F(x) = \tan x \qquad -\frac{\pi}{2} < x < \frac{\pi}{2}$$

(see Figure 9.4.6), then $F(x)$ takes on all real values exactly once. Furthermore $F(x)$ has an inverse function that is called the inverse tangent function denoted by $\tan^{-1}$. Therefore we have the following definition.

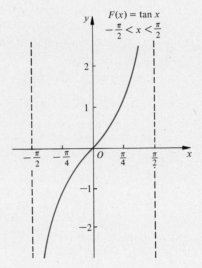

Figure 9.4.6

Definition   The **inverse tangent function** denoted by $\tan^{-1}$ is defined by

$$y = \tan^{-1} x \text{ if and only if } x = \tan y \text{ and } -\frac{\pi}{2} < y < \frac{\pi}{2} \qquad (8)$$

Thus the domain of $y = \tan^{-1} x$ is the set of all real numbers and the **principal value range** is the open interval $\left(-\frac{\pi}{2}, \frac{\pi}{2}\right)$. A sketch of the graph of $y = \tan^{-1} x$ is given in Figure 9.4.7.

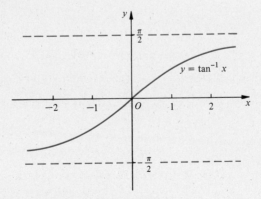

Figure 9.4.7

EXAMPLE 4 Evaluate $\tan (2 \cos^{-1} \frac{3}{5})$.

Solution Let $\theta = \cos^{-1} \frac{3}{5}$ so that $\cos \theta = \frac{3}{5}$, where $0 < \theta < \pi/2$. Therefore $\sin \theta = \frac{4}{5}$ and $\tan \theta = \frac{4}{3}$. Now $\tan 2\theta = \dfrac{2 \tan \theta}{1 - \tan^2 \theta}$ is required. Thus

$$\tan 2\theta = \frac{2(\frac{4}{3})}{1 - (\frac{4}{3})^2} = \frac{\frac{8}{3}}{-\frac{7}{9}} = -\frac{24}{7} \qquad \bullet$$

EXAMPLE 5 Evaluate $\tan [\cos^{-1} (-\frac{3}{5}) + \tan^{-1} (\frac{3}{4})]$.

Solution Let $\alpha = \cos^{-1} (-\frac{3}{5})$ and $\beta = \tan^{-1} (\frac{3}{4})$.

Then $\cos \alpha = -\frac{3}{5}$ and $\dfrac{\pi}{2} < \alpha < \pi$. Therefore $\sin \alpha = \frac{4}{5}$ and $\tan \alpha = -\frac{4}{3}$.

Also $\tan \beta = \frac{3}{4}$. Now $\tan (\alpha + \beta) = \dfrac{\tan \alpha + \tan \beta}{1 - \tan \alpha \tan \beta}$ so that

$$\tan (\alpha + \beta) = \frac{-\dfrac{4}{3} + \dfrac{3}{4}}{1 - \left(-\dfrac{4}{3} \cdot \dfrac{3}{4}\right)} = -\frac{7}{24} \qquad \bullet$$

EXAMPLE 6 Next we will establish the important result

$$\sin^{-1} x + \cos^{-1} x = \frac{\pi}{2} \qquad \text{if } x \in [-1, 1] \tag{9}$$

Proof of (9) Let $y = \cos^{-1} x$ which implies that $0 \le y \le \pi$. Therefore (multiplication of the double inequality by -1) $-\pi \le -y \le 0$ and $\left(\text{addition of } \dfrac{\pi}{2}\right)$ $-\dfrac{\pi}{2} \le \dfrac{\pi}{2} - y \le \dfrac{\pi}{2}$. Now

$$\sin \left(\frac{\pi}{2} - y\right) = \cos y = \cos (\cos^{-1} x) = x$$

so that $\dfrac{\pi}{2} - y = \sin^{-1} x$ since $\dfrac{\pi}{2} - y$ is in $\left[-\dfrac{\pi}{2}, \dfrac{\pi}{2}\right]$.
Thus (9) follows. ∎

We could have, alternatively, defined $\cos^{-1} x$ by (9) and, knowing the range of $\sin^{-1} x$ to be $\left[-\dfrac{\pi}{2}, \dfrac{\pi}{2}\right]$, deduced that the range of $\cos^{-1} x$ is $[0, \pi]$. This kind of procedure may be used to define the inverse cotangent function.

Definition The **inverse cotangent function** which is denoted by $\cot^{-1} x$ is defined by

$$\cot^{-1} x = \frac{\pi}{2} - \tan^{-1} x \tag{10}$$

Since the domain of $\tan^{-1} x$ is all real values of x, the domain of $\cot^{-1} x$ is also all real values of x. Also $-\frac{\pi}{2} < \tan^{-1} x < \frac{\pi}{2}$ so that $\frac{\pi}{2} > -\tan^{-1} x > -\frac{\pi}{2}$ and addition of $\frac{\pi}{2}$ yields

$$\pi > \cot^{-1} x > 0$$

Therefore the **principal value range** of the inverse cotangent function is the open interval $(0, \pi)$. This function is sketched in Figure 9.4.8. From this figure we observe that for $x < 0$, $\cot^{-1} x$ exceeds $\frac{\pi}{2}$ by as much as $\frac{\pi}{2}$ exceeds $\cot^{-1}(-x)$. Analytically this suggests the relation

$$\cot^{-1} x - \frac{\pi}{2} = \frac{\pi}{2} - \cot^{-1}(-x)$$

or

$$\cot^{-1} x + \cot^{-1}(-x) = \pi \tag{11}$$

The identity (11) follows from the definition (10) if we establish that $\tan^{-1} x$ is an odd function (see Exercise 20).

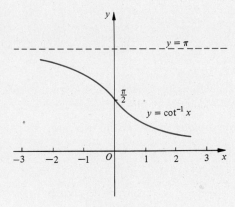

Figure 9.4.8

In an analogous manner to the inverse cotangent function the inverse secant and inverse cosecant functions are defined in terms of $\cos^{-1}$ and $\sin^{-1}$, respectively.

Definition   The **inverse secant function** denoted by $\sec^{-1}$ is defined by

$$\sec^{-1} x = \cos^{-1}\left(\frac{1}{x}\right) \qquad \text{for } |x| \geq 1 \tag{12}$$

This function is sketched in Figure 9.4.9. We observe that from (12) the domain of $\sec^{-1}$ consists of the two intervals $(-\infty, -1]$ and $[1, \infty)$. Also the **principal value range** consists of the two intervals $\left[0, \frac{\pi}{2}\right)$ and $\left(\frac{\pi}{2}, \pi\right]$. The value $\frac{\pi}{2}$ is not

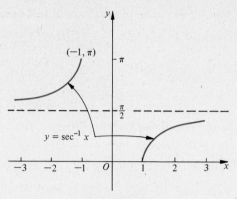

Figure 9.4.9

taken on by $\sec^{-1}$ because $\sec \dfrac{\pi}{2}$ is not defined. Note also that the following formula is suggested

$$\sec^{-1} x + \sec^{-1}(-x) = \pi \qquad |x| \geq 1 \tag{13}$$

Definition   The **inverse cosecant function** denoted by $\csc^{-1}$ is defined by

$$\csc^{-1} x = \sin^{-1}\left(\frac{1}{x}\right) \qquad \text{for } |x| \geq 1 \tag{14}$$

This function is sketched in Figure 9.4.10. From this definition the domain of $\csc^{-1}$, consists of the two intervals $(-\infty, -1]$ and $[1, \infty)$ while the **principal value range** consists of the two intervals $\left[-\dfrac{\pi}{2}, 0\right)$ and $\left(0, \dfrac{\pi}{2}\right]$. We observe that 0 is not in the range because $\csc 0$ is not defined. Also the following formula is suggested

$$\csc^{-1}(-x) = -\csc^{-1} x \qquad |x| \geq 1 \tag{15}$$

This last relation is easily proved from the fact that $\sin^{-1} x$ is an odd function (see Exercise 17).

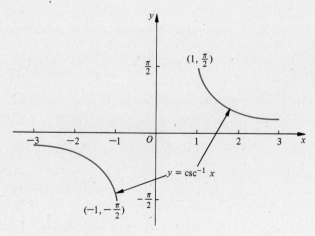

Figure 9.4.10

Note: Some of the notations and definitions given in this section, although fairly common, are not standard. In some books, the authors use arcsin x for $\sin^{-1} x$, arccos x for $\cos^{-1} x$, and so on. Also, whereas the definitions and choices of principal value ranges for $\sin^{-1} x$, $\cos^{-1} x$, $\tan^{-1} x$, and $\cot^{-1} x$ are standard, some books adopt a different definition for $\sec^{-1} x$ and $\csc^{-1} x$. For these, they choose, as the principal value ranges for both functions, angles in the first and third quadrants. There are advantages and disadvantages to each method of defining $\sec^{-1} x$ and $\csc^{-1} x$, and we shall not go into them here.

Exercises 9.4

In Exercises 1 through 8, evaluate the following:

1. (a) $\sin^{-1} \frac{1}{2}$ (b) $\tan^{-1}(-1)$ (c) $\cos^{-1} \frac{1}{\sqrt{2}}$

2. (a) $\sin^{-1} 0$ (b) $\tan^{-1} \sqrt{3}$ (c) $\cos^{-1}(-\frac{1}{2})$

3. (a) $\cos^{-1} \frac{1}{2}$ (b) $\cot^{-1} \frac{1}{\sqrt{3}}$ (c) $\csc^{-1} 2$

4. (a) $\cos(\tan^{-1}(-1))$ (b) $\cot(\cos^{-1} \frac{12}{13})$
5. (a) $3 \sec(\tan^{-1} 2)$ (b) $\sin(2 \sin^{-1} \frac{8}{17})$

6. (a) $\sin^{-1}\left[\sin\left(6\pi + \frac{\pi}{9}\right)\right]$ (b) $\cot^{-1}\left[\sin^2\left(\frac{\pi}{8}\right) + \cos^2\left(\frac{\pi}{8}\right)\right]$

7. (a) $\csc 2 \sin^{-1} \frac{1}{\sqrt{10}}$ (b) $\sin(\tan^{-1} \frac{1}{3} + \tan^{-1} \frac{1}{2})$

8. (a) $\tan(\pi + \sin^{-1}(-\frac{3}{5}))$ (b) $\cos(\tan^{-1} \frac{12}{5} + \cos^{-1} \frac{3}{5})$

In Exercises 9 through 12 simplify.

9. $\tan \sin^{-1} 2x$

10. $\tan(\tan^{-1} x + \tan^{-1} y)$

11. $\sec \sin^{-1} \sqrt{9 - x^2}$

12. $\cos(2 \cos^{-1} x)$

13. Prove that $\tan^{-1} x + \tan^{-1} y = \tan^{-1} \dfrac{x + y}{1 - xy}$, $xy \neq 1$, when $|\tan^{-1} x + \tan^{-1} y| < \dfrac{\pi}{2}$

14. Prove that $\tan^{-1} \frac{1}{2} + \tan^{-1} \frac{1}{3} = \dfrac{\pi}{4}$

15. Prove that $\tan^{-1} \frac{1}{2} = \tan^{-1} \frac{1}{3} + \tan^{-1} \frac{1}{7}$

(*Note:* The formula $\dfrac{\pi}{4} = 2\tan^{-1} \frac{1}{3} + \tan^{-1} \frac{1}{7}$ which follows from the results of Exercises 14 and 15 was used in the calculation of π to 140 places by application of an "infinite series" representation of $\tan^{-1} x$.)

16. Prove that $\tan^{-1} k^2 + \tan^{-1} \dfrac{1}{k^2} = \dfrac{\pi}{2}$

17. Prove the identity $\sin^{-1}(-u) = -\sin^{-1} u$ $-1 \leq u \leq 1$. This implies that $\sin^{-1} u$ is an odd function.

18. If $u = \sin^{-1} \dfrac{\sqrt{3}}{2}$ and $v = \sin^{-1} \frac{1}{2}$, find $\cos(u + v)$, $\cos(u - v)$, and $\tan 2u$.

19. If $u = \sin^{-1} \frac{1}{4}$ and $v = \tan^{-1}(-2)$, find $\sin(u + v)$, $\cos(u + v)$, and $\cos 2u$.

20. Prove the identity $\tan^{-1}(-u) = -\tan^{-1} u$ for all real values of u. From this we conclude that $\tan^{-1} u$ is an odd function.

21. Find $\lim\limits_{u \to 0} \dfrac{\sin^{-1} 3u}{u}$.

22. Find $\lim\limits_{u \to 0} \dfrac{\tan^{-1} ku}{u}$.

DIFFERENTIATION OF INVERSE TRIGONOMETRIC FUNCTIONS

From Theorem 3 of Section 8.3, we can conclude that each of the inverse trigonometric functions is differentiable (except possibly at the end points of the interval of definition). It is simplest to apply the method of implicit differentiation to find these derivatives. Let $y = \sin^{-1} x$. This implies that

$$x = \sin y \qquad -\frac{\pi}{2} \leq y \leq \frac{\pi}{2} \tag{1}$$

Differentiation of both sides of (1) with respect to x yields

$$1 = \cos y \, D_x y$$

or

$$D_x y = \frac{1}{\cos y} \tag{2}$$

provided that $\cos y \neq 0$. Now in general $\cos y = \pm\sqrt{1 - \sin^2 y}$ and for y in $\left[-\frac{\pi}{2}, \frac{\pi}{2} \right]$, $\cos y \geq 0$, so $\cos y = \sqrt{1 - \sin^2 y}$. Therefore

$$D_x y = \frac{1}{\sqrt{1 - \sin^2 y}}$$

$$= \frac{1}{\sqrt{1 - x^2}} \qquad -1 < x < 1$$

From $y = \sin^{-1} x$ it follows that

$$D_x \sin^{-1} x = \frac{1}{\sqrt{1 - x^2}} \qquad -1 < x < 1 \tag{3}$$

If u is a differentiable function of x, then from (3) and the chain rule we have

$$D_x(\sin^{-1} u) = \frac{D_x u}{\sqrt{1 - u^2}} \qquad -1 < u < 1 \tag{4}$$

EXAMPLE 1 Differentiate $y = \sin^{-1} 3x$, $-1 < 3x < 1$.

Solution With $u = 3x$, $D_x u = 3$ we have

$$y' = \frac{3}{\sqrt{1 - (3x)^2}} = \frac{3}{\sqrt{1 - 9x^2}} \qquad -\frac{1}{3} < x < \frac{1}{3} \qquad \bullet$$

To derive the formula for the derivative of the inverse cosine function we could proceed in a completely analogous fashion to the development of the derivative of the inverse sine function. This method is left as an exercise for the reader. Instead we use formula (9) of Example 6 from Section 9.4, namely

$$\cos^{-1} x = \frac{\pi}{2} - \sin^{-1} x$$

Differentiation of both sides with respect to x yields

$$D_x(\cos^{-1} x) = D_x\left(\frac{\pi}{2} - \sin^{-1} x\right)$$

$$= -D_x \sin^{-1} x$$

$$= -\frac{1}{\sqrt{1-x^2}} \qquad \text{where } -1 < x < 1 \qquad (5)$$

If u is a differentiable function of x then from (5) and the chain rule we obtain

$$D_x(\cos^{-1} u) = -\frac{D_x u}{\sqrt{1-u^2}} \qquad \text{where } -1 < u < 1 \qquad (6)$$

Next we derive the formula for the derivative of the inverse tangent function. If

$$y = \tan^{-1} x \qquad -\infty < x < \infty$$

then

$$x = \tan y \qquad \text{where } -\frac{\pi}{2} < y < \frac{\pi}{2}$$

Differentiation of both sides with respect to y yields

$$D_y x = \sec^2 y \qquad -\frac{\pi}{2} < y < \frac{\pi}{2}$$

$$= 1 + \tan^2 y$$

$$= 1 + x^2$$

Therefore

$$D_x y = \frac{1}{1+x^2}$$

or

$$D_x(\tan^{-1} x) = \frac{1}{1+x^2} \qquad -\infty < x < \infty \qquad (7)$$

If u is a differentiable function of x, then from (7) and the chain rule we have

$$D_x(\tan^{-1} u) = \frac{D_x u}{1+u^2} \qquad -\infty < u < \infty \qquad (8)$$

EXAMPLE 2 Differentiate $y = \tan^{-1} \sqrt{x}, \quad x > 0$.

Solution From (8), with $u = \sqrt{x}$, $u^2 = x$, and $D_x u = \dfrac{1}{2\sqrt{x}}$ we have

$$D_x(\tan^{-1} \sqrt{x}) = \frac{1}{2\sqrt{x}(1+x)} \qquad x > 0 \qquad \bullet$$

From the definition of the inverse cotangent function given in Section 9.4, namely that

$$\cot^{-1} x = \frac{\pi}{2} - \tan^{-1} x$$

we have by differentiation of both sides with respect to x,

$$D_x(\cot^{-1} x) = -D_x(\tan^{-1} x) = -\frac{1}{1 + x^2} \qquad -\infty < x < \infty \qquad (9)$$

More generally if u is a differentiable function of x, then from (9) and the chain rule,

$$D_x(\cot^{-1} u) = -\frac{D_x u}{1 + u^2} \qquad -\infty < u < \infty \qquad (10)$$

To find $D_x(\sec^{-1} x)$ we start with its definition

$$\sec^{-1} x = \cos^{-1}\left(\frac{1}{x}\right) \qquad \text{for } |x| \geq 1$$

then

$$\begin{aligned}
D_x \sec^{-1} x &= D_x \cos^{-1}\left(\frac{1}{x}\right) \\[2mm]
&= -\frac{1}{\sqrt{1 - \dfrac{1}{x^2}}}\left(-\frac{1}{x^2}\right) \\[2mm]
&= \frac{\sqrt{x^2}}{x^2 \sqrt{x^2 - 1}} \\[2mm]
&= \frac{|x|}{x^2 \sqrt{x^2 - 1}} \\[2mm]
&= \frac{1}{|x| \sqrt{x^2 - 1}} \qquad \text{where } |x| > 1 \qquad (11)
\end{aligned}$$

In a similar manner $D_x(\csc^{-1} x)$ is obtained from the definition

$$\csc^{-1} x = \sin^{-1}\left(\frac{1}{x}\right) \qquad \text{for } |x| \geq 1.$$

It is readily established that

$$D_x \csc^{-1} x = -\frac{1}{|x| \sqrt{x^2 - 1}} \qquad \text{where } |x| > 1 \qquad (12)$$

Equations (11) and (12) can be generalized by application of the chain rule. We obtain

$$D_x \sec^{-1} u = \frac{D_x u}{|u| \sqrt{u^2 - 1}} \qquad |u| > 1 \qquad (13)$$

and

$$D_x \csc^{-1} u = -\frac{D_x u}{|u| \sqrt{u^2 - 1}} \qquad |u| > 1 \qquad (14)$$

where u is any differentiable function of x.

Our six differentiation formulas are summarized below. They are paired to help in their memorization.

$$D_x(\sin^{-1} u) = \frac{1}{\sqrt{1 - u^2}} D_x u \qquad D_x(\cos^{-1} u) = -\frac{1}{\sqrt{1 - u^2}} D_x u$$

$$D_x(\tan^{-1} u) = \frac{1}{1 + u^2} D_x u \qquad D_x(\cot^{-1} u) = -\frac{1}{1 + u^2} D_x u$$

$$D_x(\sec^{-1} u) = \frac{1}{|u| \sqrt{u^2 - 1}} D_x u \qquad D_x(\csc^{-1} u) = -\frac{1}{|u| \sqrt{u^2 - 1}} D_x u$$

EXAMPLE 3   If $y = \sec^{-1}\left(\dfrac{x + 1}{x}\right)$ find $D_x y$

Solution   Formula (13) is to be used where $u = \dfrac{x + 1}{x} = 1 + \dfrac{1}{x}$ and $D_x u = -\dfrac{1}{x^2}$. Also

$$\sqrt{u^2 - 1} = \sqrt{1 + \frac{2}{x} + \frac{1}{x^2} - 1} = \sqrt{\frac{2x + 1}{x^2}} = \frac{\sqrt{2x + 1}}{|x|}$$

and

$$|u| = \left|1 + \frac{1}{x}\right| = \frac{|x + 1|}{|x|}$$

Substitution into (13) yields

$$D_x \sec^{-1}\left(\frac{x + 1}{x}\right) = \frac{-\dfrac{1}{x^2}}{\dfrac{|x + 1|}{|x|} \dfrac{\sqrt{2x + 1}}{|x|}} = \frac{-1}{|x + 1| \sqrt{2x + 1}} \qquad \bullet$$

EXAMPLE 4   A motion picture screen is 30 ft high. The bottom edge of the screen is 10 ft above eye level. At what distance from the vertical plane of the screen should an observer sit for optimum vision?

Solution   Optimum vision is obtained when the angle subtended at the eye by the screen is a maximum. Thus if we refer to Figure 9.5.1 the angle θ must be an absolute maximum. We assume that θ, α and β in the figure are all measured in radians.

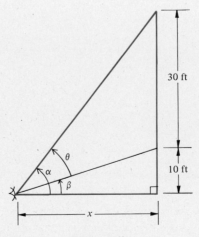

Figure 9.5.1

From the figure,

$$\theta = \alpha - \beta \qquad \alpha = \cot^{-1}\frac{x}{40} \qquad \beta = \cot^{-1}\frac{x}{10}$$

so that

$$\theta = \cot^{-1}\frac{x}{40} - \cot^{-1}\frac{x}{10} \qquad x > 0$$

From (10),

$$D_x\theta = \frac{-\dfrac{1}{40}}{1 + \dfrac{x^2}{1600}} + \frac{\dfrac{1}{10}}{1 + \dfrac{x^2}{100}}$$

$$= -\frac{40}{1600 + x^2} + \frac{10}{100 + x^2}$$

Set $D_x\theta = 0$ to obtain

$$40(100 + x^2) = 10(1600 + x^2)$$
$$30x^2 = 12000$$
$$x = 20 \text{ ft} \qquad (\text{reject } x = -20)$$

If $0 < x < 20$ then $D_x\theta > 0$, whereas if $x > 20$, $D_x\theta < 0$ so that $x = 20$ yields both a relative and an absolute maximum for θ

$$\theta_{\max} = \theta\big|_{x=20} = \cot^{-1}\tfrac{1}{2} - \cot^{-1}2 \doteq 0.645 \doteq 37° \qquad \bullet$$

EXAMPLE 5 A balloon is released at point B which is 600 ft horizontally from a fixed point O (see Figure 9.5.2). If the balloon rises at a uniform rate of 40 ft/sec find the rate at which the angle θ is changing 20 sec after the balloon is released.

Solution From $\theta = \tan^{-1}\left(\dfrac{40t}{600}\right) = \tan^{-1}\left(\dfrac{t}{15}\right)$

$$D_t\theta = \frac{\dfrac{1}{15}}{1 + \left(\dfrac{t}{15}\right)^2} = \frac{15}{225 + t^2}$$

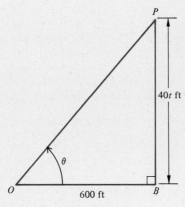

Figure 9.5.2

and therefore

$$D_t\theta \Big|_{t=20} = \frac{15}{225 + 400} = \frac{15}{625} = \frac{3}{125} \text{ rad/sec} \qquad \bullet$$

Exercises 9.5

In Exercises 1 through 20, find the derivative of the given function.

1. $y = \sin^{-1} 5x$

2. $y = \sin^{-1} \dfrac{x}{3}$

3. $y = \cos^{-1} 4x$

4. $y = \tan^{-1} x^3$

5. $y = \tan^{-1} \dfrac{1}{x}$

6. $y = \cos^{-1} (2x - 1)$

7. $y = \sec^{-1} 3x$

8. $y = \csc^{-1} 4x$

9. $y = x \sin^{-1} x$

10. $y = \ln \tan^{-1} 2x$

11. $y = \cos^{-1} x + x\sqrt{1 - x^2}$

12. $y = (\tan^{-1} kx)^2$

13. $y = \tan^{-1} \left(\dfrac{x - a}{x + a} \right)$

14. $y = \cot^{-1} \left(\dfrac{2x}{1 - x^2} \right)$

15. $y = \dfrac{1}{3} \cos^{-1} \dfrac{1}{x} + 2 \sec^{-1} x$

16. $y = \sin^{-1} (3x + 1) + \cos^{-1} (3x + 1)$

17. $y = a \sin^{-1} \dfrac{1}{x} + b \csc^{-1} x$

18. $y = \sin^{-1} (e^{ax})$ For what values of ax is your formula valid?

19. $y = x\sqrt{9 - x^2} + 9 \sin^{-1} \dfrac{x}{3}$

20. $y = \sin (\cos^{-1} x)$

In Exercises 21 through 24 find $D_x y$ by implicit differentiation.

21. $xy = \tan^{-1} \dfrac{y}{x}$

22. $\ln \sqrt{x^2 + y^2} + \tan^{-1} \dfrac{y}{x} = 0$

23. $e^{x+y} = \sin^{-1} x$

24. $\tan^{-1} y = 2 \tan^{-1} \dfrac{x}{2}$

25. A high tower stands at the end of a level road. A man drives toward the tower at a uniform speed of 30 mi/hr (30 mi/hr = 44 ft/sec). The tower rises 800 ft above the level of the man's eyes. At what rate is the angle subtended by the tower at the man's eye increasing when he is 1500 ft from the base of the tower?

26. A picture 4 ft in height is hung on a wall with the lower edge 8 ft above the level of the observer's eye. How far from the wall should the observer stand in order to obtain the most favorable view?

27. A revolving light 8 mi from a straight shore has a constant angular velocity. With what angular velocity does the light revolve if the spot of light moves along the shore at the rate of $\frac{1}{4}$ mi/sec when the beam makes an angle of 30° with the shore line?

28. A particle P starts from rest at the point $(a, 0)$ on the x-axis and freely falls parallel to the y-axis under the influence of gravity. Therefore its distance from the x-axis is given by $s = \dfrac{gt^2}{2}$. If θ is the angle between OP and the x-axis, find the angular velocity of the line OP.

29. A picture a ft in height is hung on a wall with its lower edge b ft above the level of the observer's eye. How far from the wall should the observer stand in order to obtain the most favorable view?

30. A direction indicator located at the origin is following an airplane which is flying toward the y-axis along the line $y = 400$ ft. If the speed of the airplane is constant at 600 ft/sec, how fast is the indicator turning **(a)** when the airplane is 200 ft from the y-axis? **(b)** when the plane crosses the y-axis?

In Exercises 31 and 32 find $f'(x)$

31. $f(x) = \displaystyle\int_2^{x^3} \tan^{-1} t \, dt$ **32.** $f(x) = \tan(2 \tan^{-1} x)$

33. Show that $y = (\sin^{-1} x)^2$ is a solution of the differential equation:

$$(1 - x^2)y'' - xy' - 2 = 0$$

34. Find $\displaystyle\lim_{n \to \infty} \sum_{k=1}^{n} \frac{n}{n^2 + k^2}$ (*Hint:* Express as a definite integral and evaluate.)

35. Find the integral

$$\int \frac{x}{\sqrt{1 - x^4}} \, dx$$

 (*Hint:* Let $x^2 = u$)

36. (a) Find the derivative of $\sin^{-1}\left(\dfrac{x}{\sqrt{1 + x^2}}\right)$

  (b) What is the relationship between this function and $\tan^{-1} x$?

9.6   INTEGRATIONS YIELDING INVERSE TRIGONOMETRIC FUNCTIONS

From the formulas for the derivatives of the inverse trigonometric functions we have the following integration formulas:

$$\int \frac{du}{\sqrt{1 - u^2}} = \sin^{-1} u + C \tag{1}$$

$$\int \frac{du}{1 + u^2} = \tan^{-1} u + C \tag{2}$$

$$\int \frac{du}{u\sqrt{u^2 - 1}} = \sec^{-1}|u| + C \tag{3}$$

The proofs of (1) and (2) follow directly by differentiating $\sin^{-1} u$ and $\tan^{-1} u$. In order to establish (3), we consider $u > 0$ and $u < 0$ separately.

Case i   If $u > 0$, then $u = |u|$ so that

$$\int \frac{du}{u\sqrt{u^2 - 1}} = \int \frac{du}{|u|\sqrt{u^2 - 1}} = \sec^{-1} u + C$$

$$= \sec^{-1}|u| + C$$

Case ii If $u < 0$, then $u = -|u|$

$$\int \frac{du}{u\sqrt{u^2 - 1}} = -\int \frac{du}{|u|\sqrt{u^2 - 1}} = \int \frac{d(-u)}{|-u|\sqrt{(-u)^2 - 1}}$$
$$= \sec^{-1}(-u) + C = \sec^{-1}|u| + C$$

and (3) is verified. ∎

Formulas (1) through (3) can be generalized as follows

$$\int \frac{du}{\sqrt{a^2 - u^2}} = \sin^{-1}\frac{u}{a} + C \qquad a > 0 \tag{4}$$

$$\int \frac{du}{a^2 + u^2} = \frac{1}{a}\tan^{-1}\frac{u}{a} + C \tag{5}$$

$$\int \frac{du}{u\sqrt{u^2 - a^2}} = \frac{1}{a}\sec^{-1}\left|\frac{u}{a}\right| + C \qquad a > 0 \tag{6}$$

Proof of (4) Differentiate the right side to obtain

$$D_u\left(\sin^{-1}\frac{u}{a} + C\right) = \frac{1}{\sqrt{1 - \dfrac{u^2}{a^2}}} D_u\left(\frac{u}{a}\right)$$

$$= \frac{a}{\sqrt{a^2 - u^2}} \cdot \frac{1}{a} \qquad \text{since } a = \sqrt{a^2} \text{ if } a > 0$$

$$= \frac{1}{\sqrt{a^2 - u^2}} \qquad \text{if } a > 0 \qquad ∎$$

The proof of (5) follows in the same way and the proof of (6) is done analogously to the proof of (3). These are left as exercises for the reader.

EXAMPLE 1   Find $\displaystyle\int \frac{dx}{9 + x^2}$.

Solution   Compare with (5) where $a = 3$ and $u = x$. Therefore

$$\int \frac{dx}{9 + x^2} = \frac{1}{3}\tan^{-1}\left(\frac{x}{3}\right) + C \qquad ●$$

EXAMPLE 2   Evaluate $\displaystyle\int \frac{dx}{\sqrt{49 - 16x^2}}$.

Solution

$$\int \frac{dx}{\sqrt{49 - 16x^2}} = \int \frac{dx}{\sqrt{(7)^2 - (4x)^2}}$$

$$= \frac{1}{4}\int \frac{4\,dx}{\sqrt{(7)^2 - (4x)^2}} = \frac{1}{4}\int \frac{d(4x)}{\sqrt{(7)^2 - (4x)^2}}$$

and the last integral can be identified with the left side of (4) with $a = 7$ and $u = 4x$. Therefore from formula (4)

$$\int \frac{dx}{\sqrt{49 - 16x^2}} = \frac{1}{4}\sin^{-1}\left(\frac{4x}{7}\right) + C \qquad \bullet$$

EXAMPLE 3 Find $\int \frac{dx}{x^2 + 4x + 6}$.

Solution We can employ (5) if the method of completing the square is used. Now,

$$\int \frac{dx}{x^2 + 4x + 6} = \int \frac{dx}{(x + 2)^2 + 2} = \int \frac{dx}{(x + 2)^2 + (\sqrt{2})^2}$$

$$= \int \frac{d(x + 2)}{(x + 2)^2 + (\sqrt{2})^2}$$

and from (5)

$$\int \frac{dx}{x^2 + 4x + 6} = \frac{1}{\sqrt{2}}\tan^{-1}\left(\frac{x + 2}{\sqrt{2}}\right) + C \qquad \bullet$$

EXAMPLE 4 Evaluate $\int \frac{dx}{\sqrt{-9x^2 + 30x - 21}}$.

Solution
$$-9x^2 + 30x - 21 = -(9x^2 - 30x + 21)$$
$$= -(3x - 5)^2 + 4$$
$$= 2^2 - (3x - 5)^2$$

$$\int \frac{dx}{\sqrt{-9x^2 + 30x - 21}} = \frac{1}{3}\int \frac{3\,dx}{\sqrt{2^2 - (3x - 5)^2}} = \frac{1}{3}\int \frac{d(3x - 5)}{\sqrt{2^2 - (3x - 5)^2}}$$

$$= \frac{1}{3}\sin^{-1}\left(\frac{3x - 5}{2}\right) + C \qquad \bullet$$

EXAMPLE 5 Evaluate $\int_{2/\sqrt{3}}^{2} \frac{dx}{x\sqrt{x^2 - 1}}$.

Solution Formula (3) can be directly applied [or apply formula (6) with $a = 1$].

$$\int_{2/\sqrt{3}}^{2} \frac{dx}{x\sqrt{x^2 - 1}} = \sec^{-1}|x| \Big|_{2/\sqrt{3}}^{2}$$

$$= \sec^{-1} 2 - \sec^{-1}\frac{2}{\sqrt{3}}$$

$$= \cos^{-1}\frac{1}{2} - \cos^{-1}\frac{\sqrt{3}}{2}$$

$$= \frac{\pi}{3} - \frac{\pi}{6} = \frac{\pi}{6} \qquad \bullet$$

Example 6 Find $\displaystyle\int \frac{x\,dx}{\sqrt{1 + x - x^2}}$.

Solution   This is a more complicated problem than any of the previous examples in this section. Even if we complete the square, the expression will still not be in one of the standard forms given in (4) through (6). Note that if the numerator had been $(-2x + 1)\,dx$, the integral would have been in the form $\int u^{-1/2}\,du$ and the generalized power rule could be used. We must employ here a combination of these techniques.

$$\int \frac{x\,dx}{\sqrt{1 + x - x^2}} = -\frac{1}{2} \int \frac{(-2x\,dx)}{\sqrt{1 + x - x^2}}$$

$$= -\frac{1}{2} \int \frac{(-2x + 1)\,dx}{\sqrt{1 + x - x^2}} + \frac{1}{2} \int \frac{dx}{\sqrt{1 + x - x^2}}$$

The first integral is now in the integrable form $-\dfrac{1}{2} \int u^{-1/2}\,du$.

To evaluate the second integral, we complete the square as follows:

$$1 + x - x^2 = 1 - (x^2 - x) = 1 - (x^2 - x + \tfrac{1}{4}) + \tfrac{1}{4}$$
$$= \tfrac{5}{4} - (x - \tfrac{1}{2})^2$$

Then

$$\int \frac{x\,dx}{\sqrt{1 + x - x^2}} = -\frac{1}{2} \int \frac{(-2x + 1)\,dx}{\sqrt{1 + x - x^2}} + \frac{1}{2} \int \frac{dx}{\sqrt{\dfrac{5}{4} - \left(x - \dfrac{1}{2}\right)^2}}$$

$$= -\frac{1}{2} \cdot 2\sqrt{1 + x - x^2} + \frac{1}{2} \sin^{-1}\left(\frac{x - \dfrac{1}{2}}{\dfrac{\sqrt{5}}{2}}\right) + C$$

$$= -\sqrt{1 + x - x^2} + \frac{1}{2} \sin^{-1}\left(\frac{2x - 1}{\sqrt{5}}\right) + C \qquad \bullet$$

Exercises 9.6

In Exercises 1 through 22 evaluate the indefinite integrals.

1. $\displaystyle\int \frac{dx}{x^2 + 16}$

2. $\displaystyle\int \frac{dx}{2x^2 + 10}$

3. $\displaystyle\int \frac{dx}{\sqrt{25 - x^2}}$

4. $\displaystyle\int \frac{x\,dx}{x^4 + 49}$

5. $\displaystyle\int \frac{x^2 + 5}{x^2 + 4}\,dx$

6. $\displaystyle\int \frac{x\,dx}{x^2 + 64}$

7. $\displaystyle\int \frac{dx}{x\sqrt{4x^2 - 1}}$

8. $\displaystyle\int \frac{x\,dx}{\sqrt{9 - x^2}}$

9. $\displaystyle\int \frac{\sqrt{3}\,dx}{x\sqrt{x^2 - 4}}$

10. $\displaystyle\int \frac{dx}{\sqrt{25 - 4x^2}}$

11. $\int \dfrac{dt}{7 + 5t^2}$

12. $\int \dfrac{dw}{a^2 + b^2w^2}, \quad ab \neq 0$

13. $\int \dfrac{dx}{\sqrt{-x^2 + 6x - 8}}$

14. $\int \dfrac{e^z \, dz}{\sqrt{1 - e^{2z}}}$

15. $\int \dfrac{dx}{(x - 3)\sqrt{x^2 - 6x + 5}}$

16. $\int \dfrac{\sin x}{1 + \cos^2 x} \, dx$

17. $\int \dfrac{\tan^{-1} x}{1 + x^2} \, dx$

18. $\int \dfrac{dx}{\sqrt{21 - 7x^2}}$

19. $\int \dfrac{\sec^2 x}{9 + \tan^2 x} \, dx$

20. $\int \dfrac{dx}{x^2 + 8x + 20}$

21. $\int \dfrac{dx}{\sqrt{8 + 4x - 4x^2}}$

22. $\int \dfrac{dx}{\sqrt{e^{2x} - 1}}$ (*Hint:* multiply numerator and denominator by e^{-x})

In Exercises 23 through 38 evaluate the definite integrals (if possible)

23. $\displaystyle\int_0^{1/2} \dfrac{dx}{\sqrt{1 - x^2}}$

24. $\displaystyle\int_{-\sqrt{3}}^{\sqrt{3}} \dfrac{dx}{1 + x^2}$

25. $\displaystyle\int_0^{1/2} \dfrac{dx}{1 + 4x^2}$

26. $\displaystyle\int_{3\sqrt{2}}^{6} \dfrac{dx}{x\sqrt{x^2 - 9}}$

27. $\displaystyle\int_0^{3} \dfrac{dx}{\sqrt{4 - x^2}}$

28. $\displaystyle\int_0^{1} \dfrac{x^2 \, dx}{x^6 + 1}$

29. $\displaystyle\int_{-5}^{5} \dfrac{x \, dx}{(x^2 + 16)^2}$

30. $\displaystyle\int_0^{-2} \dfrac{dx}{x^2 + 4x + 8}$

31. $\displaystyle\int_{3/8}^{3/4} \dfrac{dx}{\sqrt{9 - 16x^2}}$

32. $\displaystyle\int_{1/2}^{3/2} \dfrac{dx}{\sqrt{2x - x^2}}$

33. $\displaystyle\int_1^{4} \dfrac{dx}{16x^2 - 8x + 10}$

34. $\displaystyle\int_0^{3/\sqrt{2}} \dfrac{t \, dt}{\sqrt{81 - t^4}}$

35. $\displaystyle\int_0^{b/\sqrt{2}} \dfrac{y \, dy}{\sqrt{b^4 - y^4}}$

36. $\displaystyle\int_0^{1} \dfrac{t^2 - 1}{t^2 + 1} \, dt$ (*Hint:* Divide numerator into denominator)

37. $\displaystyle\int_0^{2} \dfrac{2x + 1}{x^2 + 1} \, dx$

38. $\displaystyle\int_0^{2} \dfrac{dx}{e^x + e^{-x}}$

In Exercises 39 through 42 find the area of the given region

39. The region bounded by $y = \dfrac{1}{1 + x^2}$ and

   (a) $y = 0$, $x = -3$, and $x = 3$
   (b) $y = 0$, $x = -k$, and $x = k$, where k is a positive constant
   (c) To what number does the area approach as $k \to \infty$ (if any)? What does the area represent?

40. The region bounded by $y = \dfrac{1}{1 + x^2}$ and the line $y = \dfrac{1}{3}$.

41. The region bounded by $y = \dfrac{1}{\sqrt{1 - x^2}}$, $y = 2$, and the y-axis.

42. The region bounded by the curve $x^2 y = 4a^2 (2a - y)$, the x-axis and the vertical lines $x = \pm k$ where $a > 0$ and $k > 0$. Sketch the curve (known as the **witch of Agnesi**). To what number (if any) does the area approach as $k \to \infty$?

Some simple combinations of the exponential functions appear so frequently, both in the application of mathematics and in theory, that it is worthwhile to give them special names. We shall note that these functions satisfy identities that are quite similar to the trigonometric identities. It is this similarity to the trigonometric functions that accounts for the names that are attached to the functions. These functions are called hyperbolic sine, hyperbolic cosine, and so on, and are abbreviated sinh, cosh, and so forth, respectively. (The term "hyperbolic" is used because of some relationships between the functions and the hyperbola $x^2 - y^2 = 1$).

Definition   The six hyperbolic functions are defined by

$$\text{(a) } \sinh x = \frac{e^x - e^{-x}}{2} \qquad\qquad \text{(b) } \cosh x = \frac{e^x + e^{-x}}{2}$$

$$\text{(c) } \tanh x = \frac{\sinh x}{\cosh x} = \frac{e^x - e^{-x}}{e^x + e^{-x}} \qquad \text{(d) } \coth x = \frac{\cosh x}{\sinh x} = \frac{e^x + e^{-x}}{e^x - e^{-x}} \qquad (1)$$

$$\text{(e) } \operatorname{sech} x = \frac{1}{\cosh x} = \frac{2}{e^x + e^{-x}} \qquad \text{(f) } \operatorname{csch} x = \frac{1}{\sinh x} = \frac{2}{e^x - e^{-x}}$$

We will now establish some identities by way of examples.

EXAMPLE 1   Prove that for all values of x,

$$\cosh^2 x - \sinh^2 x = 1 \qquad (2)$$

Solution

$$\cosh^2 x - \sinh^2 x = \frac{(e^x + e^{-x})^2}{2} - \frac{(e^x - e^{-x})^2}{2}$$

$$= \frac{e^{2x} + 2 + e^{-2x}}{4} - \frac{e^{2x} - 2 + e^{-2x}}{4}$$

$$= \frac{2 - (-2)}{4} = \frac{4}{4} = 1 \qquad \bullet$$

EXAMPLE 2   Prove that for all values of x and y,

$$\sinh (x + y) = \sinh x \cosh y + \cosh x \sinh y \qquad (3)$$

Solution   To prove (3) we work with the right side. By definition

$$\sinh x \cosh y + \cosh x \sinh y = \frac{e^x - e^{-x}}{2} \cdot \frac{e^y + e^{-y}}{2} + \frac{e^x + e^{-x}}{2} \cdot \frac{e^y - e^{-y}}{2}$$

$$= \frac{e^{x+y} - e^{-x+y} + e^{x-y} - e^{-x-y}}{4} + \frac{e^{x+y} + e^{-x+y} - e^{x-y} - e^{-x-y}}{4}$$

$$= \frac{2e^{x+y} - 2e^{-(x+y)}}{4} = \frac{e^{x+y} - e^{-(x+y)}}{2} = \sinh (x + y) \qquad \bullet$$

EXAMPLE 3   Prove that for all values of x,

$$\sinh 2x = 2 \sinh x \cosh x \qquad (4)$$

Solution   This follows from (3) if we replace y by x,

$$\sinh 2x = \sinh(x + x) = \sinh x \cosh x + \cosh x \sinh x$$
$$= 2 \sinh x \cosh x \qquad \bullet$$

EXAMPLE 4   Prove that for all values of x,

$$\operatorname{sech}^2 x + \tanh^2 x = 1 \qquad (5)$$

Solution   This follows from (2), by dividing both sides by $\cosh^2 x$ (which is never zero). Thus

$$\frac{\cosh^2 x - \sinh^2 x}{\cosh^2 x} = \frac{1}{\cosh^2 x}$$

or

$$\frac{\cosh^2 x}{\cosh^2 x} - \frac{\sinh^2 x}{\cosh^2 x} = \frac{1}{\cosh^2 x}$$

Then from the definitions of $\tanh x$ and $\operatorname{sech} x$ we have

$$1 - \tanh^2 x = \operatorname{sech}^2 x$$

which is equivalent to (5). $\bullet$

By this time, the reader should be impressed with the similarities and differences between the trigonometric functions and their counterpart hyperbolic functions. Thus (2) is similar to $\cos^2 x + \sin^2 x = 1$ and (5) is similar to $\sec^2 x - \tan^2 x = 1$, but note the changes of sign. However, (3) coincides completely with its trigonometric counterpart, $\sin(x + y) = \sin x \cos y + \cos x \sin y$.

In a similar manner to (3) we can prove that

$$\cosh(x + y) = \cosh x \cosh y + \sinh x \sinh y \qquad (6)$$

Then from (6) with $y = x$ and (2) we obtain three forms for $\cosh 2x$ namely

$$\cosh 2x = \cosh^2 x + \sinh^2 x \qquad (7a)$$

$$\cosh 2x = 2 \cosh^2 x - 1 \qquad (7b)$$

$$\cosh 2x = 2 \sinh^2 x + 1 \qquad (7c)$$

If we solve (7b) and (7c) for $\cosh x$ and $\sinh x$, respectively, and replace x by $\dfrac{x}{2}$, there results

$$\sinh \frac{x}{2} = \pm \sqrt{\frac{\cosh x - 1}{2}} \qquad (8a)$$

$$\cosh \frac{x}{2} = \sqrt{\frac{\cosh x + 1}{2}} \qquad (8b)$$

Note that there is no $\pm$ sign on the right side of (8b) because the range of the hyperbolic cosine function is the set of numbers $[1, \infty)$.

EXAMPLE 5   Prove that sinh x is an odd function and cosh x is an even function.

Proof   We must show that sinh $(-x) = -\sinh x$. Now

$$\sinh(-x) = \frac{e^{-x} - e^{-(-x)}}{2} = \frac{e^{-x} - e^{x}}{2} = -\frac{e^{x} - e^{-x}}{2} = -\sinh x$$

Similarly,

$$\cosh(-x) = \frac{e^{-x} + e^{-(-x)}}{2} = \frac{e^{-x} + e^{x}}{2} = \cosh x \qquad \blacksquare$$

Thus it follows that tanh x, coth x and csch x are odd while sech x is even. If we replace y by $-y$ in (3) and (6) and utilize the results of Example 5, then

$$\sinh(x - y) = \sinh x \cosh y - \cosh x \sinh y \qquad (9a)$$

and

$$\cosh(x - y) = \cosh x \cosh y - \sinh x \sinh y \qquad (9b)$$

The graphs of the hyperbolic sine and cosine functions are easily obtained from their definitions, properties, and numerical data from Table 8 in the appendix. For example, sinh x is an odd function that is 0 if and only if $x = 0$. Its graph is shown in Figure 9.7.1. Its domain and range are $(-\infty, \infty)$. The hyperbolic cosine function is an even function, with domain $(-\infty, \infty)$ and range $[1, \infty)$ and is sketched in Figure 9.7.2. Note that in the graphs of these functions the similarities to the trigonometric functions break down since (1) the hyperbolic sine and cosine are not periodic while the sine and cosine are periodic and (2) the functions sin x and cos x are bounded functions, in particular $|\sin x| \leq 1$ and $|\cos x| \leq 1$, while their hyperbolic counterparts are unbounded.

Similarly, it is easy to show that $|\tanh x| \leq 1$ and $|\text{sech } x| \leq 1$ for all real x,

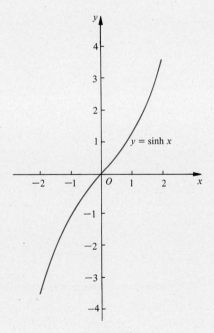

Figure 9.7.1

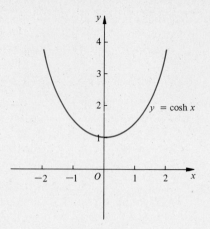

Figure 9.7.2

whereas the ranges of tan x and sec x are unbounded. The graphs of the other four hyperbolic functions are shown in Figures 9.7.3 through 9.7.6.

Next let us turn to the **inverse hyperbolic functions.** From the graph of $y = \sinh x$ it follows that the equation $x = \sinh y$ has only one solution for all x in $(-\infty, \infty)$. Now the inverse hyperbolic sine written

$$y = \sinh^{-1} x \tag{10}$$

is equivalent to $x = \sinh y$ and therefore

$$x = \sinh y = \frac{e^y - e^{-y}}{2} \tag{11}$$

Multiplication of both sides of (11) by $2e^y$ yields

$$2xe^y = e^{2y} - 1$$

or

$$e^{2y} - 2xe^y - 1 = 0 \tag{12}$$

Equation (12) may be treated as a quadratic in e^y (since $e^{2y} = (e^y)^2$) and the quadratic formula yields

$$e^y = x \pm \sqrt{x^2 + 1}$$

However $e^y > 0$ and $x - \sqrt{x^2 + 1}$ is negative. We reject the negative sign and write

$$e^y = x + \sqrt{x^2 + 1}$$

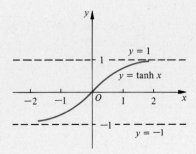

Figure 9.7.3

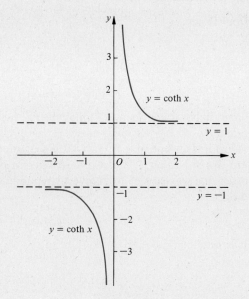

Figure 9.7.4

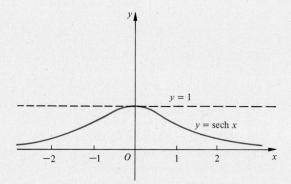

Figure 9.7.5

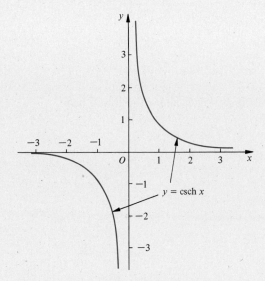

Figure 9.7.6

from which

$$y = \ln(x + \sqrt{x^2 + 1}) \qquad (13)$$

Equations (10) and (13) yield

$$\sinh^{-1} x = \ln(x + \sqrt{x^2 + 1}) \qquad (14)$$

Since $\sinh x$ is continuous, monotonically increasing, and has both domain and range $(-\infty, \infty)$, the same must be true for $\sinh^{-1} x$ defined by (14).

From the graph of $y = \cosh x$ it is suggested that to each value of $y > 1$ there are two values of x. Thus we must select a **principal value** for this function in order that we may define an inverse hyperbolic cosine function. To this end, if we write

$$y = \cosh^{-1} x \qquad (15)$$

it is implied that

$$x = \cosh y = \frac{e^y + e^{-y}}{2}$$

thus

$$2xe^y = e^{2y} + 1$$

or

$$e^{2y} - 2xe^y + 1 = 0 \qquad (16)$$

and the solution of this quadratic equation in e^y is

$$e^y = x \pm \sqrt{x^2 - 1} \qquad (17)$$

If $x < 1$, there is no value of y that satisfies this equation since $\sqrt{x^2 - 1}$ is not real; if $x = 1$, then $y = 0$; and, if $x > 1$, there are two values of y for each value of x. Now from the quadratic equation (16), the product of its two roots is 1; that is, the roots are reciprocals of each other. We choose for the principal value the larger of the two roots, namely, the one that corresponds to $x + \sqrt{x^2 - 1}$. Therefore from (17) (rejecting the minus sign)

$$y = \ln(x + \sqrt{x^2 - 1}) \qquad (18)$$

Comparison of (15) and (18) yields the definition

$$\cosh^{-1} x = \ln(x + \sqrt{x^2 - 1}) \qquad (x \geq 1) \qquad (19)$$

Note that the range of $\cosh^{-1} x$ is $[0, \infty)$. Geometrically this implies that the graph of (19) may be obtained by reflecting the right half of the curve $y = \cosh x$ (Figure 9.7.2) about the line $y = x$

EXAMPLE 6   Prove the identity

$$\tanh^{-1} x = \frac{1}{2} \ln \frac{1 + x}{1 - x} \qquad |x| < 1$$

Proof   Let $y = \tanh^{-1} x$, $-1 < x < 1$. Then,

$$x = \tanh y = \frac{\sinh y}{\cosh y} = \frac{\frac{1}{2}(e^y - e^{-y})}{\frac{1}{2}(e^y + e^{-y})} = \frac{e^{2y} - 1}{e^{2y} + 1}$$

This last equation can be solved for e^{2y}.

$$xe^{2y} + x = e^{2y} - 1$$

or

$$e^{2y}(1 - x) = 1 + x$$

Therefore

$$e^{2y} = \frac{1 + x}{1 - x}$$

from which the required identity follows

$$y = \tanh^{-1} x = \frac{1}{2} \ln \left(\frac{1 + x}{1 - x} \right) \qquad |x| < 1 \qquad \blacksquare \qquad (20)$$

Note that the domain of the function defined by (20) is $-1 < x < 1$, while the range is $(-\infty, \infty)$ or, equivalently, all real values of y.

In a similar manner, the expressions for the other inverse hyperbolic functions in terms of the logarithms are readily obtained. We summarize these formulas:

$$
\begin{aligned}
\sinh^{-1} x &= \ln(x + \sqrt{x^2 + 1}) & &-\infty < x < \infty \\
\cosh^{-1} x &= \ln(x + \sqrt{x^2 - 1}) & &x \geq 1 \\
\tanh^{-1} x &= \frac{1}{2} \ln\left(\frac{1 + x}{1 - x}\right) & &|x| < 1 \\
\coth^{-1} x &= \frac{1}{2} \ln\left(\frac{x + 1}{x - 1}\right) = \tanh^{-1}\left(\frac{1}{x}\right) & &|x| > 1 \\
\operatorname{sech}^{-1} x &= \ln\left(\frac{1 + \sqrt{1 - x^2}}{x}\right) = \cosh^{-1}\left(\frac{1}{x}\right) & &0 < x \leq 1 \\
\operatorname{csch}^{-1} x &= \ln\left(\frac{1}{x} + \frac{\sqrt{1 + x^2}}{|x|}\right) = \sinh^{-1}\left(\frac{1}{x}\right) & &x \neq 0
\end{aligned}
\qquad (21)
$$

Exercises 9.7

In Exercises 1 through 10 prove the following assertions about the hyperbolic functions. Check these assertions with the corresponding graphs.

1. $f(x) = \tanh x$ is an odd function
2. $g(x) = \operatorname{sech} x$ is an even function
3. $\cosh x \geq 1$ for all x
4. $|\tanh 2x| < 1$ for all x
5. $0 < \operatorname{sech} x \leq 1$ for all x
6. $\lim\limits_{x \to -\infty} \operatorname{sech} x = 0$
7. $\cosh x > \sinh x$ for all x
8. $y = \cosh x$, $-\infty < x < \infty$ does not possess an inverse function
9. $\lim\limits_{x \to \infty} (\cosh x - \sinh x) = 0$
10. $h(x) = x \operatorname{csch} x$ is an even function

In Exercises 11 through 20 prove the following identities and state the trigonometric counterparts.

11. $\coth^2 x - 1 = \operatorname{csch}^2 x$
12. $\cosh^2(3x + 1) - \sinh^2(3x + 1) = 1$
13. $\cosh(x + y) = \cosh x \cosh y + \sinh x \sinh y$

14. $\sinh (x - y) = \sinh x \cosh y - \cosh x \sinh y$

15. $\tanh (x + y) = \dfrac{\tanh x + \tanh y}{1 + \tanh x \tanh y}$

16. $\tanh 2x = \dfrac{2 \tanh x}{1 + \tanh^2 x}$

17. $\tanh \dfrac{x}{2} = \dfrac{\sinh x}{1 + \cosh x}$

18. $\sinh x + \sinh y = 2 \sinh \frac{1}{2}(x + y) \cosh \frac{1}{2}(x - y)$

19. $\cosh x + \cosh y = 2 \cosh \frac{1}{2}(x + y) \cosh \frac{1}{2}(x - y)$

20. $\sinh 3x = 3 \sinh x + 4 \sinh^3 x$

21. Given that $\cosh x_1 = \frac{17}{8}$ and $x_1 > 0$, find the values of the other five hyperbolic functions at x_1.

22. Given that $\sinh x_1 = -\frac{9}{40}$, find the values of the other five hyperbolic functions at x_1.

23. Prove that $\cosh x + \sinh x = e^x$ and that $\cosh x - \sinh x = e^{-x}$. From these identities prove the identity

$$\cosh^2 x - \sinh^2 x = 1$$

24. Prove the identity $(\cosh x + \sinh x)^n = \cosh nx + \sinh nx$ where n is an arbitrary positive integer. (*Hint:* It is not necessary to use mathematical induction; instead use a result from Exercise 23.)

25. Does the identity $(\cosh x + \sinh x)^n = \cosh nx + \sinh nx$ hold for nonpositive integers n? Prove your assertion.

9.8 DIFFERENTIATION AND INTEGRATION OF HYPERBOLIC FUNCTIONS

The formulas for the differentiation and integration of the hyperbolic function are again very similar to the corresponding formulas for the trigonometric functions. Let us start with

$$D_x(\sinh x) = D_x \left(\frac{e^x - e^{-x}}{2} \right) = \frac{e^x + e^{-x}}{2} = \cosh x \qquad (1)$$

Also

$$D_x(\cosh x) = D_x \left(\frac{e^x + e^{-x}}{2} \right) = \frac{e^x - e^{-x}}{2} = \sinh x \qquad (2)$$

More generally, if $u(x)$ is any differentiable function of x, the chain rule yields

$$D_x(\sinh u) = \cosh u \, D_x u \qquad (3)$$

and

$$D_x(\cosh u) = \sinh u \, D_x u \qquad (4)$$

Continuing, by the quotient rule

$$D_x(\tanh x) = D_x \frac{\sinh x}{\cosh x}$$

$$= \frac{\cosh x \, D_x \sinh x - \sinh x \, D_x \cosh x}{\cosh^2 x}$$

$$= \frac{\cosh^2 x - \sinh^2 x}{\cosh^2 x} = \frac{1}{\cosh^2 x}$$

$$= \operatorname{sech}^2 x \qquad\qquad (5)$$

By the chain rule,

$$D_x(\tanh u) = \operatorname{sech}^2 u \, D_x u \qquad\qquad (6)$$

In a similar manner, it is left to the reader to verify the following formulas

$$D_x \coth u = -\operatorname{csch}^2 u \, D_x u \qquad\qquad (7)$$

$$D_x \operatorname{sech} u = -\operatorname{sech} u \tanh u \, D_x u \qquad\qquad (8)$$

$$D_x \operatorname{csch} u = -\operatorname{csch} u \coth u \, D_x u \qquad\qquad (9)$$

where u is any differentiable function of x.

EXAMPLE 1   Prove that $y = \sinh x$ is a monotonically increasing function of x for all x.

Proof   The result follows from (1) because

$$D_x(\sinh x) = \cosh x > 0 \qquad\qquad ■$$

EXAMPLE 2   Discuss the concavity of $y = \sinh x$.

Solution   From (1) and (2) by repeated differentiation,

$$y' = \cosh x \qquad \text{and} \qquad y'' = \sinh x$$

Therefore

$$(\sinh x)'' = \sinh x \text{ is} \begin{cases} \text{negative} & \text{for } x < 0 \\ \text{zero} & \text{for } x = 0 \\ \text{positive} & \text{for } x > 0 \end{cases}$$

and we conclude that the graph is concave downward for $x < 0$ and concave upward for $x > 0$. Furthermore the point $(0, \sinh 0) = (0, 0)$ is an inflection point, since it separates the part of the curve which is concave downward from the portion which is concave upward.   ●

The formulas for the derivatives of the inverse hyperbolic functions may be readily obtained in a straightforward manner. By way of illustration, consider $y = \sinh^{-1} u$, where u is a differentiable function of x. We seek $D_x y$. If $y = \sinh^{-1} u$, then

$$u = \sinh y$$

Form D_x of both sides

$$D_x u = \cosh y \, D_x y$$

and, solving for $D_x y$, recalling that $\cosh y \geq 1$, we have

$$D_x y = \frac{D_x u}{\cosh y} = \frac{D_x u}{\sqrt{1 + \sinh^2 y}} = \frac{D_x u}{\sqrt{1 + u^2}}$$

Therefore we have derived the formula

$$D_x \sinh^{-1} u = \frac{D_x u}{\sqrt{1 + u^2}} \qquad -\infty < u < \infty \qquad (10)$$

Alternatively we may start with

$$y = \sinh^{-1} u = \ln\left(u + \sqrt{u^2 + 1}\right)$$

then

$$D_x y = \frac{1}{u + \sqrt{u^2 + 1}}\left(1 + \frac{u}{\sqrt{u^2 + 1}}\right) D_x u$$

$$= \frac{1}{u + \sqrt{u^2 + 1}} \frac{\sqrt{u^2 + 1} + u}{\sqrt{u^2 + 1}} D_x u$$

$$= \frac{D_x u}{\sqrt{1 + u^2}}$$

in agreement with (10)

EXAMPLE 3   Find $D_x \sinh^{-1}(5x - 2)$.

Solution   In this instance, $u = 5x - 2$ and $D_x u = 5$; therefore, from (10),

$$D_x \sinh^{-1}(5x - 2) = \frac{5}{\sqrt{1 + (5x - 2)^2}} = \frac{5}{\sqrt{25x^2 - 20x + 5}}$$

$$= \frac{\sqrt{5}}{\sqrt{5x^2 - 4x + 1}}$$

It is left to the reader to prove the additional formulas (11) through (15).

$$D_x \cosh^{-1} u = \frac{D_x u}{\sqrt{u^2 - 1}} \qquad u > 1 \qquad (11)$$

$$D_x \tanh^{-1} u = \frac{D_x u}{1 - u^2} \qquad |u| < 1 \qquad (12)$$

$$D_x \coth^{-1} u = \frac{D_x u}{1 - u^2} \qquad |u| > 1 \qquad (13)$$

$$D_x \operatorname{sech}^{-1} u = \frac{-D_x u}{u \sqrt{1 - u^2}} \qquad 0 < u < 1 \tag{14}$$

$$D_x \operatorname{csch}^{-1} u = \frac{-D_x u}{|u| \sqrt{u^2 + 1}} \qquad u \neq 0 \tag{15}$$

From each differentiation formula we can obtain a corresponding integration formula, and these integration formulas are important.

$$\int \sinh u \, du = \cosh u + C \tag{16}$$

$$\int \cosh u \, du = \sinh u + C \tag{17}$$

$$\int \operatorname{sech}^2 u \, du = \tanh u + C \tag{18}$$

$$\int \operatorname{csch}^2 u \, du = -\coth u + C \tag{19}$$

$$\int \operatorname{sech} u \tanh u \, du = -\operatorname{sech} u + C \tag{20}$$

$$\int \operatorname{csch} u \coth u \, du = -\operatorname{csch} u + C \tag{21}$$

$$\int \frac{du}{\sqrt{u^2 + 1}} = \sinh^{-1} u + C = \ln\left(u + \sqrt{u^2 + 1}\right) + C \tag{22}$$

$$\int \frac{du}{\sqrt{u^2 - 1}} = \cosh^{-1} u + C = \ln\left(u + \sqrt{u^2 - 1}\right) + C \qquad u > 1 \tag{23}$$

$$\int \frac{du}{1 - u^2} = \tanh^{-1} u + C = \frac{1}{2} \ln \frac{1 + u}{1 - u} + C \qquad -1 < u < 1 \tag{24}$$

$$\int \frac{du}{1 - u^2} = \coth^{-1} u + C = \frac{1}{2} \ln \frac{u + 1}{u - 1} + C \qquad |u| > 1 \tag{25}$$

$$\int \frac{du}{u \sqrt{1 - u^2}} = -\operatorname{sech}^{-1} u + C = -\ln\left[\frac{1 + \sqrt{1 - u^2}}{u}\right] + C \qquad 0 < u < 1 \tag{26}$$

$$\int \frac{du}{|u|\sqrt{u^2+1}} = -\operatorname{csch}^{-1} u + C = -\ln\left[\frac{1}{u} + \frac{\sqrt{1+u^2}}{|u|}\right]$$
$$+ C \qquad u \neq 0 \qquad (27)$$

EXAMPLE 4   Find $\int \tanh^2 3x \, dx$.

Solution   A survey of our list of integrals in this section indicates that our problem is *not* on it. However,

$$\tanh^2 3x = 1 - \operatorname{sech}^2 3x \qquad (i)$$

and $\int \operatorname{sech}^2 u \, du$ is the left side of (18). Thus we have

$$\int \tanh^2 3x \, dx = \int (1 - \operatorname{sech}^2 3x) \, dx$$

$$= \int dx - \tfrac{1}{3}\int \operatorname{sech}^2 3x \, 3 \, dx$$

$$= x - \tfrac{1}{3}\tanh 3x + C$$

This may be verified by differentiating the expression on the right side and using (i) to recover the integrand.   ●

EXAMPLE 5   Find $\int \dfrac{dx}{\sqrt{x^2 - 49}}$.

Solution   Divide numerator and denominator by 7, which results in

$$\int \frac{dx}{\sqrt{x^2-49}} = \int \frac{d\left(\dfrac{x}{7}\right)}{\sqrt{\left(\dfrac{x}{7}\right)^2 - 1}} = \int \frac{du}{\sqrt{u^2 - 1}}$$

where $u = \dfrac{x}{7}$. Thus, (23) may be applied to yield

$$\int \frac{dx}{\sqrt{x^2-49}} = \int \frac{du}{\sqrt{u^2 - 1}} = \cosh^{-1} u + C$$

$$= \cosh^{-1}\left(\frac{x}{7}\right) + C \qquad ●$$

EXAMPLE 6   Find the value of the definite integral $\displaystyle\int_{-3}^{3} \frac{dx}{\sqrt{x^2 + 9}}$.

Solution   Since the integrand is an even function of x,

$$\int_{-3}^{3} \frac{dx}{\sqrt{x^2 + 9}} = 2 \int_{0}^{3} \frac{dx}{\sqrt{x^2 + 9}}$$

$$= 2 \int_{0}^{3} \frac{d\left(\dfrac{x}{3}\right)}{\sqrt{\left(\dfrac{x}{3}\right)^2 + 1}}$$

and from (22),

$$\int_{-3}^{3} \frac{dx}{\sqrt{x^2 + 9}} = 2 \ln\left(\frac{x}{3} + \sqrt{\left(\frac{x}{3}\right)^2 + 1}\right)\Bigg|_{0}^{3}$$

$$= 2\left[\ln\left(1 + \sqrt{2}\right) - \ln 1\right]$$

$$= 2\left[\ln\left(1 + \sqrt{2}\right)\right] \qquad \bullet$$

EXAMPLE 7   A particle of mass m is falling through a medium such that the resistive force to its motion is proportional to the square of its velocity. If the particle falls from rest, determine its velocity as a function of time and, in particular, deduce the behavior of its velocity with increase in time.

Solution   There are two forces acting on the particle: (i) its weight $= mg$ and, if v is the velocity, (ii) a resistive force kv^2 where $k > 0$ is a constant of proportionality. Newton's second law of motion yields (for downward direction positive and t representing time)

$$m\, D_t v = mg - kv^2 \qquad (28)$$

or, solving for the acceleration,

$$D_t v = g - \frac{k}{m} v^2 = g\left(1 - \frac{k}{mg} v^2\right)$$

$$D_t v = g(1 - c^2 v^2) \qquad (29)$$

where we have defined c by $c = \sqrt{\dfrac{k}{mg}}$. Therefore, separation of variables and multiplication of both sides by c yields

$$\frac{d(cv)}{1 - (cv)^2} = cg\, dt \qquad (30)$$

Integration of both sides of (30) using (24) results in

$$\tanh^{-1}(cv) = cgt + C$$

However $v|_{t=0} = 0$ and $\tanh^{-1} 0 = 0$, from which $C = 0$. Therefore

$$cv = \tanh(cgt)$$

or

$$v = \frac{1}{c} \tanh(cgt)$$

From the definition of c,

$$v = \sqrt{\frac{mg}{k}} \tanh\left(\sqrt{\frac{kg}{m}}\, t\right) \qquad (31)$$

Equation (31) is the required velocity function in terms of t. Note that the velocity is an increasing function of t that approaches its terminal velocity as $t \to \infty$, namely

$$\lim_{t \to \infty} v = \sqrt{\frac{mg}{k}}$$

●

EXAMPLE 8   This section will be concluded with an application of hyperbolic functions to the solution of a physical problem called the **hanging cable** or **catenary** problem. We seek to determine the shape of a flexible inextensible cable hanging under its own weight with its endpoints at B_1 and B_2 (see Figure 9.8.1). Let $y = y(x)$ be the equation of the curve we seek, and also let $P(x, y)$ be any point on the curve.

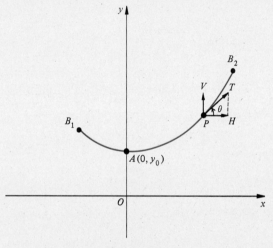

Figure 9.8.1

The forces acting on a portion of the cable are shown in Figure 9.8.2. The assumption of flexibility means that at any section P we have a resultant force denoted by tension T in the cable along the tangent (that is, there are no shearing

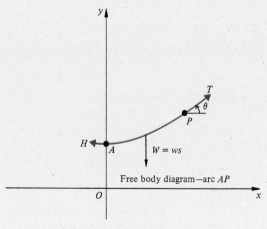

Free body diagram—arc AP

Figure 9.8.2

components). If H denotes the horizontal component of T then

$$H = T \cos \theta \tag{32}$$

where θ is the inclination of the tangent with the horizontal. For equilibrium in the horizontal direction it is necessary that H be constant. The weight W of the portion of the cable is given by $W = ws$, where w is the constant weight per unit of length and s is the arc length from the low point A to a variable point P. For equilibrium, the algebraic sum of the vertical forces must equal zero or if V denotes the vertical component of T,

$$V = W = ws = T \sin \theta \tag{33}$$

From (32) and (33),

$$\frac{V}{H} = \frac{W}{H} = \frac{ws}{H} = \frac{T \sin \theta}{T \cos \theta} = \tan \theta$$

But $\tan \theta = y'$ and we have

$$y' = \frac{ws}{H} \tag{34}$$

Differentiation of both sides of (34) with respect to x yields [since $s' = \sqrt{1 + (y')^2}$]

$$y'' = \frac{ws'}{H} = \frac{w}{H} \sqrt{1 + (y')^2} \tag{35}$$

which is a second order differential equation. Equation (35) must be solved subject to two initial conditions. These conditions are needed to enable us to determine the two constants of integration that are introduced when (35) is solved. Let the y-axis pass through the lowest point A on the cable (Figure 9.8.1). To find the other initial condition we observe that $y = y_0$ when $x = 0$ and y_0 will be selected in order to put the equation of the cable in simplest form.

In order to solve (35) we observe that y is not explicitly present. We take advantage of this by letting $p = y'$ and $p' = y''$ which reduces (35) to a first order differential equation

$$p' = \frac{w}{H} \sqrt{1 + p^2}$$

Separation of variables yields

$$\frac{w}{H} \, dx = \frac{dp}{\sqrt{1 + p^2}} \tag{36}$$

and integration of both sides using (22) results in

$$\frac{wx}{H} + C_1 = \sinh^{-1} p \tag{37}$$

The constant of integration C_1 is found from the requirement that $p = y' = 0$ when $x = 0$. Thus $\frac{w0}{H} + C_1 = \sinh^{-1} 0 = 0$ or $C_1 = 0$. From (37) we then have

$$p = y' = \sinh \frac{wx}{H} \tag{38}$$

Integration of both sides of (38) yields

$$y = \frac{H}{w} \cosh \frac{wx}{H} + C_2$$

and

$$y \Big|_{x=0} = \frac{H}{w} + C_2 = y_0$$

Therefore

$$C_2 = y_0 - \frac{H}{w}$$

and we choose $y_0 = \dfrac{H}{w}$ so that $C_2 = 0$. Therefore the equation of the curve that describes the hanging cable is

$$y = \frac{H}{w} \cosh \frac{w}{H} x$$

or finally

$$y = a \cosh \frac{x}{a} \tag{39}$$

where

$$a = \frac{H}{w} \tag{40}$$

has the units of length. The curve defined by (39) is called a **catenary**. In most applications, H (and therefore a) is not known and must be determined from the given geometry and given weight of the cable.

Exercises 9.8

In Exercises 1 through 10 find the derivative of the given function.

1. $f(x) = \sinh(10x + 9)$
2. $g(x) = \tanh ax$
3. $h(x) = \ln \cosh x$
4. $F(x) = \cosh^2(3x + 4) - \sinh^2(3x + 4)$
5. $G(x) = \cosh \sqrt{x^2 + 1}$
6. $H(x) = \operatorname{sech}^2 3x$
7. $f(x) = \coth(2x^3 - 5x)$
8. $g(x) = \operatorname{csch} \sqrt{3x + 4}$
9. $F(t) = \dfrac{1 - \cosh t}{1 + \cosh t}$
10. $G(t) = \tan^{-1}(\sinh t)$

In Exercises 11 through 15 prove the indicated formulas of this section

11. Formula (7)
12. Formula (9)
13. Formula (11)
14. Formula (14)
15. Formula (15)

In Exercises 16 through 20 find the derivative of the given function

16. $\sinh^{-1}(3x^2 + 1)$
17. $\cosh^{-1}(x^4)$
18. $\tanh^{-1}(1 - x^2)$
19. $\tanh^{-1}(\tan ax)$
20. $\coth^{-1}(e^{2x})$

In Exercises 21 through 30, find the indicated integral

21. $\int \dfrac{dx}{\sqrt{x^2 + 4}}$

22. $\int \dfrac{dx}{\sqrt{16x^2 - 1}}$

23. $\int x \coth x^2 \, dx$

24. $\int \cosh^2 x \, dx$ (*Hint:* Use an identity.)

25. $\int \sinh^2 3x \, dx$

26. $\int \dfrac{dx}{\sqrt{x^2 - 6x + 10}}$

27. $\int \dfrac{dx}{\sqrt{x^2 + 8x + 12}}$

28. $\int \dfrac{3x \, dx}{1 - x^4}$

29. $\int \dfrac{\tanh t}{\tanh^2 t - 1} \, dt$

30. $\int \dfrac{3 \, dt}{t\sqrt{1 - 4t^2}}$

In Exercises 31 through 34 evaluate the definite integral

31. $\displaystyle\int_0^{1/2} \dfrac{dx}{1 - x^2}$

32. $\displaystyle\int_2^3 \dfrac{dx}{\sqrt{x^2 - 1}}$

33. $\displaystyle\int_2^4 \dfrac{x \, dx}{\sqrt{x^4 - 1}}$

34. $\displaystyle\int_{1/8}^{1/6} \dfrac{dx}{x\sqrt{1 - 16x^2}}$

35. Prove that

$$\int \frac{du}{\sqrt{u^2 + a^2}} = \ln\left(u + \sqrt{u^2 + a^2}\right) + C$$

and also verify the result by differentiation. This generalizes equation (22) of this section.

36. Prove that

$$\int \frac{du}{\sqrt{u^2 - a^2}} = \ln\left(u + \sqrt{u^2 - a^2}\right) \qquad u > a > 0$$

and also verify the result by differentiation. This generalizes equation (23) of this section.

37. The *tractrix* is an important curve in both pure and applied mathematics. Its equation may be shown to be

$$x = a \operatorname{sech}^{-1}\left(\frac{y}{a}\right) - \sqrt{a^2 - y^2} \qquad 0 < y \le a$$

Show that

$$D_x y = \frac{-y}{\sqrt{a^2 - y^2}} \qquad 0 < y < a$$

Use this information to sketch the curve.

Exercises 38 through 41 are amplifications of the hanging cable or catenary problem (see Example 8 of this section).

38. Find the length of arc of the catenary $y = a \cosh \dfrac{x}{a}$ from the point $A(0, a)$ to point $P_1(x_1, y_1)$, where $x_1 > 0$.

39. Let d ft be the dip of the hanging cable suspended from two points at the same level. Let L ft be the span, and s_1 ft the entire length of the cable (Figure 9.8.3). Prove that

$$s_1 = 2a \sinh \frac{L}{2a} \qquad d = a\left(\cosh \frac{L}{2a} - 1\right)$$

and that the tension T in the cable at any point is given by $T = wy$ where w is the uniform weight density of the cable and y is the ordinate of the point in question.

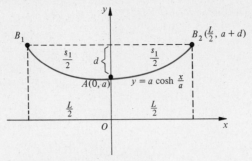

Figure 9.8.3

40. A cable is fastened at the two posts 100 ft apart and the dip is 20 ft. (a) Find the tension at the lowest point of the cable. (b) Find the tension T at the highest point if the weight density of the cable is $w = 0.30$ lb/ft. (*Hint:* There are two unknowns a and H; the quantity a is found by solving the transcendental equation (of Exercise 39) by *trial and error*. Then H is obtained from $H = aw$ and $T = wy$.)

41. A cable 40 ft long, weighing 3 lb/ft hangs under its own weight between two supports 30 ft apart. Find (a) the equation of the curve, (b) the tension at the lowest point, and (c) the sag.

42. A particle moves along the x-axis in accordance with one of the following laws

$$\text{(i)} \quad x = a \cos kt + b \sin kt$$
$$\text{(ii)} \quad x = A \cosh kt + B \sinh kt$$

(a) Show in each instance that the acceleration $x''(t)$ is proportional to the displacement x.

(b) However, show that in case (i) the proportionality constant is negative, whereas in case (ii) it is positive. This means that in case (i) the acceleration is directed toward the origin whereas in case (ii) it is directed away from the origin.

Review and Miscellaneous Exercises

1. Find the period of the following functions (if any):

    (a) $f(\theta) = \sin 3\theta$

    (b) $g(\theta) = 5 \cos \left(\dfrac{\theta}{4} \right)$

    (c) $h(\theta) = \sin^2 3\theta$

    (d) $F(\theta) = \sinh 2\theta$

    (e) $G(\theta) = 2 \sin \theta - 3 \cos 2\theta + \sqrt{\pi}$

2. For what values of θ is the following equation satisfied

    $$2 \sin^2 \theta + 3 \cos \theta = 7?$$

    Justify your conclusion.

3. If $\cos \theta = -\dfrac{3}{5}$ and $\sin \theta < 0$, find $\cos \dfrac{\theta}{2}$.

4. Prove that the following is an identity:

    $$\frac{\sin 3\theta}{\cos \theta} + \frac{\cos 3\theta}{\sin \theta} = 2 \cot 2\theta.$$

In each of the Exercises 5 through 10, evaluate the given limit, if it exists.

5. $\displaystyle \lim_{x \to 0} \frac{\sin 3x}{7x}$

6. $\displaystyle \lim_{x \to 0} \frac{x}{\sin 4x}$

7. $\lim\limits_{x\to 0} \dfrac{\sin^2 6x}{x}$

8. $\lim\limits_{x\to 0} x \cot 5x$

9. $\lim\limits_{x\to 0} \dfrac{\tan^3 4x}{x^3}$

10. $\lim\limits_{x\to 0^+} \dfrac{\sqrt{x + x^2}}{\sin 2x}$

In Exercises 11 through 21, differentiate each of the functions.

11. $f(x) = \sin^2 5x$

12. $g(x) = (\cos 3x - \sin 3x)^2$

13. $h(x) = \sqrt{\cos x}$

14. $F(x) = \ln \cos 2x$

15. $g(t) = \dfrac{\ln 3t}{\csc t}$

16. $h(t) = \ln \sqrt[3]{8 + \tan^2 4t}$ (*Hint:* First simplify and then differentiate.)

17. $F(t) = \ln \left(\dfrac{1 - \cos t}{1 + \cos t} \right)$

18. $G(t) = e^{\sqrt{\sin 4t}}$

19. $f(x) = \sec^4 (ax^2 + b)$, (*a* and *b* are constants)

20. $g(x) = \sin^{-1} \sqrt{x}$

21. $h(x) = \dfrac{\sqrt{a^2 - x^2}}{x} + \sin^{-1} \dfrac{x}{a}$

22. If the identity $\cos (x - b) = \cos x \cos b + \sin x \sin b$ is differentiated with respect to x (where b is independent of x), show that the resulting relation is also an identity. Would you apply this same technique to $\tan x = x - \dfrac{x^3}{3}$? Explain.

23. Is it valid to differentiate both sides of $\sin 4x = 2 \sin 2x \cos 2x$? What is the result of doing this?

In Exercises 24 through 31, differentiate each of the functions.

24. $f(t) = (\sin^{-1} at)^2$ (*a* is constant)

25. $g(\theta) = (\sin a\theta)^{\sin a\theta}$ (*a* is constant)

26. $h(t) = \cos (\sinh bt)$ (*b* is constant)

27. $F(\theta) = \tan^{-1} \left(\dfrac{1 - \cos \theta}{1 + \cos \theta} \right)^{1/2}, \ 0 < \theta < \pi$

Determine $F(\theta)$ in *simplest form* as a function of θ.

28. $u(x) = \tan^{-1} \dfrac{x}{\sqrt{1 - x^2}} - \sin^{-1} x$ when $-1 < x < 1$ and show that $u(x) \equiv 0$ in this interval.

29. $G(x) = 3 \cos^{-1} \left(1 - \dfrac{x}{3} \right) - \sqrt{6x - x^2}$

30. $v(x) = \sinh^{-1} (ax + b)$ (*a* and *b* are constants)

31. $w(x) = (\cos 2x)^{3x}$

In Exercises 32 through 59, determine the following integrals:

32. $\displaystyle\int \dfrac{dx}{\sqrt{9x^2 + 49}}$

33. $\displaystyle\int \dfrac{dx}{\sqrt{9 - 4x^2}}$

34. $\displaystyle\int \dfrac{dx}{\sqrt{4x^2 - 20x + 21}}$

35. $\displaystyle\int \dfrac{e^{3t}}{5 + e^{6t}} dt$

36. $\displaystyle\int \dfrac{x}{x^4 + 100} dx$

37. $\displaystyle\int \dfrac{dx}{\sqrt{-4x^2 + 12x + 7}}$

38. $\displaystyle\int \dfrac{x^2 + 4}{x^2 + a^2} dx$, (*a* is constant)
(*Hint:* divide first)

39. $\displaystyle\int \dfrac{x^2 - x + 1}{x^2 + 2x + 3} dx$

40. $\displaystyle\int \frac{dt}{e^{2t} + 16e^{-2t}}$

41. $\displaystyle\int \frac{x}{\sqrt{4 - 4x - x^2}}\, dx$

42. $\displaystyle\int \cosh^3 x\, dx$

43. $\displaystyle\int \tanh^3 ax\, dx$, (*a* is a nonzero constant)

44. $\displaystyle\int \sinh^2 4x \cosh^5 4x\, dx$

45. $\displaystyle\int \sinh^2 t \cosh^2 t\, dt$

46. $\displaystyle\int_0^{2/3} \frac{dx}{4 + 9x^2}$

47. $\displaystyle\int_0^2 \frac{dx}{\sqrt{3 + 2x - x^2}}$

48. $\displaystyle\int_0^2 \frac{dx}{\sqrt{3 + 2x - x^2}}$

49. $\displaystyle\int_{1/3}^{2/3} \frac{dx}{\sqrt{1 + 2x - 3x^2}}$

50. $\displaystyle\int_0^3 \frac{x^2 + 2}{x^2 + 9}\, dx$

51. $\displaystyle\int_{-1}^1 \frac{2x + 9}{x^2 + 2x + 5}\, dx$

52. $\displaystyle\int_1^3 \frac{5x - 18}{x^2 - 6x + 10}\, dx$

53. $\displaystyle\int_0^{\pi/2} (3 - \sin x)^3 \cos x\, dx$

54. $\displaystyle\int_0^{\pi/4} \frac{9 \tan^2 x + 4}{\cos^2 x}\, dx$

55. $\displaystyle\int_0^1 \frac{\sinh 4x}{\cosh^3 4x}\, dx$

56. $\displaystyle\int_{-1/4}^{1/4} (\cosh x + \sinh x)^3 e^x\, dx$

57. $\displaystyle\int_1^3 \cosh 2x \cosh x\, dx$

58. $\displaystyle\int_0^{\pi/3} \sec^4 z\, dz$

59. $\displaystyle\int_{\pi/6}^{\pi/2} \csc^{10} v \cot v\, dv$

60. Describe the behavior as $x \to \infty$, for each of the following:

(a) $\dfrac{\cosh x}{e^x}$

(b) $e^x(\cosh 2x - \sinh 2x)$

(c) $\dfrac{\sin x}{x}$

61. Express $ae^{mx} + be^{-mx}$, (*a*, *b* and *m* constants), in terms of cosh *mx* and sinh *mx*.
62. Find an equation of the line tangent to $y = x \cosh x + 3$ at $(0, 3)$.
63. Find an equation of the line normal to $y = 3x \cos^2 x - \sinh x$ at $(0, 0)$.
64. Find all the solutions of the differential equation $(\cosh^2 x)y' = 2 \sinh x$.
65. Find the arc length of the catenary $y = \cosh x$ from $(0, 1)$ to $(a, \cosh a)$, where $a > 0$.
66. Find the volume of the solid generated if the region bounded by the catenary $y = \cosh x$, the line $x = a$, $(a > 0)$, and the *x*- and *y*-axes, is revolved about the *x*-axis.

67. Find the average value of $|\cos 2x|$ in $\left[0, \dfrac{\pi}{2}\right]$.

68. A body of weight W is to be moved along a rough horizontal plane by means of a force P whose line of action makes an angle of θ with the plane. Simple mechanical considerations show that the required force to move the body with constant velocity is given by

$$P = \frac{\mu W}{\mu \sin \theta + \cos \theta}$$

where μ is a constant coefficient of friction. Determine θ so that the required force P is least.
69. If a projectile is fired from O so as to strike an inclined plane which makes a constant angle α with the horizontal axis at O, the range R is given by

$$R = \frac{2v_0^2 \cos \theta \sin (\theta - \alpha)}{g \cos^2 \alpha}$$

where v_0^2 is the square of the initial speed and is constant. Also g is the constant gravitational acceleration, and θ is the angle of elevation and the only variable. Determine the value of θ that maximizes the range up the plane.

In Exercises 70 through 73, use logarithmic differentiation to find the indicated derivative.

70. If $f(x) = \sin x \sin 2x \sin 3x$, find $f'(x)$.

71. If $g(x) = (3x)^{\sin x}$, $x > 0$, find $g'(x)$.

72. If $F(x) = \dfrac{e^x \cos x}{1 + x^4}$, find $F'(x)$ and $F'(0)$.

73. If $G(x) = (x^2 + 2)^{\cos x}$, find $G'(x)$.

10
Methods of Integration

10.1   INTRODUCTION AND SOME BASIC FORMULAS

In our earlier development we learned how to determine the derivative of any combination of standard functions such as polynomials, rational functions, more general algebraic functions, exponential, logarithmic, trigonometric, inverse trigonometric, hyperbolic and inverse hyperbolic functions. With the help of the chain rule, derivatives of products and quotients, and the establishment of particular limits, considerable success was achieved.

We have also obtained the indefinite integrals of many of these functions, and, in fact, it is desirable to summarize the important formulas obtained up to this point:

$$\int a\, f(u)\, du = a \int f(u)\, du \quad (a \text{ constant}) \tag{1}$$

$$\int (f(u) + g(u))\, du = \int f(u)\, du + \int g(u)\, du \tag{2}$$

$$\int u^n\, du = \frac{u^{n+1}}{n+1} + C \qquad n \neq -1 \tag{3}$$

$$\int \frac{du}{u} = \ln |u| + C \tag{4}$$

$$\int e^u\, du = e^u + C \tag{5}$$

$$\int a^u\, du = \frac{a^u}{\ln a} + C \qquad a > 0 \tag{6}$$

$$\int \sin u\, du = -\cos u + C \tag{7}$$

$$\int \cos u\, du = \sin u + C \tag{8}$$

$$\int \tan u\, du = -\ln |\cos u| + C \tag{9}$$

$$\int \cot u \, du = \ln |\sin u| + C \qquad (10)$$

$$\int \sec u \, du = \ln |\sec u + \tan u| + C \qquad (11)$$

$$\int \csc u \, du = -\ln |\csc u + \cot u| + C \qquad (12)$$

$$\int \sec^2 u \, du = \tan u + C \qquad (13)$$

$$\int \csc^2 u \, du = -\cot u + C \qquad (14)$$

$$\int \sec u \tan u \, du = \sec u + C \qquad (15)$$

$$\int \csc u \cot u \, du = -\csc u + C \qquad (16)$$

$$\int \sinh u \, du = \cosh u + C \qquad (17)$$

$$\int \cosh u \, du = \sinh u + C \qquad (18)$$

$$\int \operatorname{sech}^2 u \, du = \tanh u + C \qquad (19)$$

$$\int \operatorname{csch}^2 u \, du = -\coth u + C \qquad (20)$$

$$\int \operatorname{sech} u \tanh u \, du = -\operatorname{sech} u + C \qquad (21)$$

$$\int \operatorname{csch} u \coth u \, du = -\operatorname{csch} u + C \qquad (22)$$

$$\int \frac{du}{a^2 + u^2} = \frac{1}{a} \tan^{-1} \frac{u}{a} + C \qquad a \neq 0 \qquad (23)$$

$$\int \frac{du}{\sqrt{a^2 - u^2}} = \sin^{-1} \frac{u}{a} + C \qquad a > |u| > 0 \qquad (24)$$

$$\int \frac{du}{\sqrt{u^2 + a^2}} = \ln (u + \sqrt{u^2 + a^2}) + C \qquad (25)$$

$$\int \frac{du}{\sqrt{u^2 - a^2}} = \ln (u + \sqrt{u^2 - a^2}) + C \qquad |u| > 0 > 0 \qquad (26)$$

We have already seen that some of these formulas can be derived from others. Formula (9) was derived from (4) after an appropriate substitution. Formulas (25) and (26) followed from results on the inverse hyperbolic functions, but they can also be derived using the technique of trigonometric substitution that will be covered in Section 10.4.

EXAMPLE 1   Find $\int \dfrac{e^{2x} \, dx}{e^{2x} + 9}$.

Solution   Clearly this integral is not on our list as it stands. However, if we let $u = e^{2x} + 9$ so that $du = 2e^{2x} \, dx$ or $e^{2x} \, dx = \dfrac{du}{2}$ then from (4)

$$\int \frac{e^{2x}\,dx}{e^{2x}+9} = \frac{1}{2}\int \frac{du}{u} = \frac{1}{2}\ln|u| + C$$

$$= \frac{1}{2}\ln(e^{2x}+9) + C$$

This is easily verified by differentiating the result.   ●

EXAMPLE 2 Find $\displaystyle\int \frac{\sinh x\,dx}{(3+\cosh x)^3}$.

Solution If we let $u = 3 + \cosh x$ then $du = \sinh x\,dx$ and therefore from (3)

$$\int \frac{\sinh x\,dx}{(3+\cosh x)^3} = \int (3+\cosh x)^{-3}\,d(3+\cosh x)$$

$$= \frac{(3+\cosh x)^{-2}}{-2} + C$$

$$= -\frac{1}{2(3+\cosh x)^2} + C$$ ●

EXAMPLE 3 Find $\displaystyle\int \frac{2x^3 + 11x + 5}{x^2 + 4}\,dx$.

Solution Since the degree of the numerator is greater than the degree of the denominator we divide until the quotient is a polynomial in x and the remainder is a polynomial of degree less than the denominator $x^2 + 4$. In this instance, we have

$$\frac{2x^3 + 11x + 5}{x^2 + 4} = 2x + \frac{3x + 5}{x^2 + 4}$$

$$= 2x + \frac{3x}{x^2 + 4} + \frac{5}{x^2 + 4}$$

Each term is easily integrated. Specifically from (4),

$$\int \frac{3x\,dx}{x^2 + 4} = \frac{3}{2}\int \frac{2x\,dx}{x^2 + 4} = \frac{3}{2}\ln(x^2 + 4)$$

while (23) is directly applicable to the third term with $a = 2$,

$$\int \frac{5\,dx}{x^2 + 4} = \frac{5}{2}\tan^{-1}\frac{x}{2}$$

Therefore

$$\int \frac{2x^3 + 11x + 5}{x^2 + 4}\,dx = x^2 + \frac{3}{2}\ln(x^2 + 4) + \frac{5}{2}\tan^{-1}\frac{x}{2} + C$$ ●

EXAMPLE 4 Find $I = \displaystyle\int (2\sec 3t + 5\tan 3t)^2\,dt$.

Solution We expand the integrand to obtain

$$I = \int (2 \sec 3t + 5 \tan 3t)^2 \, dt = \int (4 \sec^2 3t + 20 \sec 3t \tan 3t + 25 \tan^2 3t) \, dt$$

However, $\tan^2 3t = \sec^2 3t - 1$ and our third term becomes $25 (\sec^2 3t - 1)$ so that

$$I = \int (29 \sec^2 3t - 25 + 20 \sec 3t \tan 3t) \, dt$$

Now we apply (13) and (15) with $u = 3t$ and $du = 3 \, dt$ to obtain

$$I = \tfrac{29}{3} \tan 3t - 25t + \tfrac{20}{3} \sec 3t + C \qquad \bullet$$

EXAMPLE 5 Find $I = \int x \sqrt{2x + 3} \, dx$.

Solution This is not in our basic list. However, we recall that the algebraic substitution technique (Section 5.3) is often useful for problems of this type. Thus, let

$$u = 2x + 3 \qquad du = 2 \, dx \qquad \text{then} \quad x = \frac{u - 3}{2}$$

and

$$I = \int \frac{u - 3}{2} \sqrt{u} \, \frac{du}{2} = \frac{1}{4} \int (u^{3/2} - 3u^{1/2}) \, du$$

$$I = \tfrac{1}{4}(\tfrac{2}{5})u^{5/2} - \tfrac{3}{4}(\tfrac{2}{3})u^{3/2} + C$$
$$= \tfrac{1}{10}u^{5/2} - \tfrac{1}{2}u^{3/2} + C$$

and in terms of x,

$$I = \frac{1}{10} (2x + 3)^{5/2} - \frac{1}{2} (2x + 3)^{3/2} + C.$$

Alternative Solution Let $v = \sqrt{2x + 3}$, $x = \dfrac{v^2 - 3}{2}$ and $dx = v \, dv$

$$I = \int \frac{v^2 - 3}{2} v^2 \, dv$$

$$= \frac{1}{2} \int (v^4 - 3v^2) \, dv = \frac{v^5}{10} - \frac{v^3}{2} + C$$

$$= \frac{1}{10} (2x + 3)^{5/2} - \frac{1}{2} (2x + 3)^{3/2} + C$$

in agreement with our previous solution. $\qquad \bullet$

Before we tackle the exercises, there are several important comments that are in order. Certainly the basic list can be augmented and the results can be verified by differentiation. However, this approach is not intellectually satisfying and is too haphazard. Rather we seek to develop additional procedures that show how these formulas are derived. It is the objective of the remaining sections of this chapter to develop these formal techniques. We shall first treat the problems of trigonometric integrals more generally. Then we shall establish the methods of

(1) trigonometric substitution (2) partial fractions, and (3) integration by parts. These are the most powerful formal techniques of integration.

Exercises 10.1

In Exercises 1–38, evaluate the indefinite integrals.

1. $\int \dfrac{x^2}{x^3 + 5} \, dx$

2. $\int \dfrac{x}{x^2 + 100} \, dx$

3. $\int \dfrac{e^{1/x}}{x^2} \, dx$

4. $\int e^{\cos x} \sin x \, dx$

5. $\int \sqrt{3x + 7} \, dx$

6. $\int \dfrac{e^{\sqrt{x}}}{\sqrt{x}} \, dx$

7. $\int 3^{2x} \, dx$

8. $\int \csc^2 \left(\dfrac{x}{8} \right) dx$

9. $\int \dfrac{dx}{4 + 25x^2}$

10. $\int \dfrac{dx}{\sqrt{1 - 64x^2}}$

11. $\int \dfrac{dx}{\sqrt{3x^2 - 10}}$

12. $\int \dfrac{\sin 2x}{4 + 7 \cos 2x} \, dx$

13. $\int \dfrac{\sin ax}{\cos^3 ax} \, dx$

14. $\int \dfrac{e^{-x}}{e^{-x} + 1} \, dx$

15. $\int \dfrac{e^x}{e^{2x} + 4} \, dx$

16. $\int \dfrac{x^2}{x + 1} \, dx$

17. $\int \dfrac{x^2}{x^2 + 10} \, dx$

18. $\int \tan^2 ax \, dx$

19. $\int (\sin^2 x - \cos^2 x) \, dx$

20. $\int \dfrac{\cos^3 x}{\sin^2 x} \, dx$ (*Hint:* Write $\cos^3 x = \cos^2 x \cdot \cos x$ and then use a basic identity.)

21. $\int \cos^3 x \, dx$

22. $\int \dfrac{1 - 2 \cos x}{3 \sin x} \, dx$

23. $\int \dfrac{dx}{x^2 + 6x + 13}$

24. $\int \dfrac{dt}{(1 - 3t)^6}$

25. $\int \dfrac{e^{2x} + 4}{e^x} \, dx$

26. $\int \dfrac{\sinh 3t}{1 + \cosh 3t} \, dt$

27. $\int (1 + \tan \theta)^2 \, d\theta$

28. $\int \dfrac{a + b \ln x}{x} \, dx$

29. $\int \dfrac{x^2}{x^6 + 36} \, dx$

30. $\int \dfrac{\sin 2\theta}{\sin^2 \theta - \cos^2 \theta} \, d\theta$

31. $\int \dfrac{\sin \theta \cos \theta}{\sin^2 \theta + 3} \, d\theta$

32. $\int \dfrac{4x^2 + 2x + 1}{x + 1} \, dx$

33. $\int \dfrac{dx}{\sqrt{x^2 - 10x + 61}}$

34. $\int \dfrac{dx}{\sqrt{7 - 6x + x^2}}$

35. $\int t \sqrt[3]{t^2 + 4} \, dt$

36. $\int x^3 \sqrt{x^2 + 100} \, dx$ (*Hint:* Let $x^2 + 100 = u$.)

37. $\int (10 + \sin at)^5 \cos at \, dt$ (*Hint:* Although the binomial may be expanded, it is unnecessary.)

38. $\int \sin^5 t \, dt$ $\left[\text{\textit{Hint:} The answer is \textit{not} } \dfrac{\sin^6 t}{6} + C. \text{ Why not? Can you do anything with factoring } \sin^5 t \text{ into } (\sin^4 t)(\sin t)? \right]$

39. Find a value of n for which $\int e^{-x^2} x^n \, dx$ can be readily integrated. Evaluate the integral for this value of n. (*Warning:* $n = 0$ is *not* the answer.)

40. Find a value of n for which each of the following is easily integrated. Evaluate each integral for this choice

(a) $\int (a + bx^2)^{-1/2} x^n \, dx$

(b) $\int x^n \cos (a + bx^3) \, dx$

(c) $\int x^n e^{\sqrt[3]{x}} \, dx$

41. Evaluate $\int \dfrac{x}{(4 + x)^{1/3}} \, dx$

42. Find $\int \dfrac{dx}{1 + 5e^{-2x}}$

43. Find $\int \dfrac{dx}{2 + 3e^{4x}}$

10.2   TRIGONOMETRIC INTEGRALS—PART I

The integration of the basic trigonometric functions is by now well known to us (Section 9.3). Furthermore we have been treating problems such as $\int \sin^2 x \, dx$ or $\int \cos^3 x \, dx$, which are in reality special cases of what will be developed in this section. We begin by first working out a particular example.

EXAMPLE 1   Find $\int \sin^4 x \cos^3 x \, dx$.

Solution   We write $\cos^3 x = \cos^2 x \cos x$ and note that $\cos x = D_x \sin x$ and $\cos^2 x = 1 - \sin^2 x$. Therefore our given integral may be written in the form

$$\int \sin^4 x \cos^2 x \cos x \, dx = \int \sin^4 x (1 - \sin^2 x) \, d(\sin x)$$

$$= \int (\sin^4 x - \sin^6 x) \, d(\sin x)$$

$$= \frac{\sin^5 x}{5} - \frac{\sin^7 x}{7} + C$$

by application of the generalized power formula.   ●

More generally we now prove the following

Case 1. If m and n are rational numbers and either m or n is a positive odd integer, then

$$\int \sin^m x \cos^n x \, dx$$

is readily integrable.

Proof Suppose that n is a positive odd integer of the form $n = 2k + 1$ where k is a nonnegative integer. Then it follows that

$$\int \sin^m x \cos^n x \, dx = \int \sin^m x \cos^{n-1} x \, (\cos x \, dx)$$

$$= \int \sin^m x \cos^{2k} x \, d(\sin x)$$

$$= \int \sin^m x \, (1 - \sin^2 x)^k \, d(\sin x)$$

$$= \int \sin^m x \, P_{2k} \, (\sin x) \, d(\sin x)$$

where $P_{2k} \, (\sin x)$ is a polynomial of degree $2k$ in $\sin x$. The integration then may be performed term by term.

The proof for the case when m is a positive odd integer follows analogously and is left to the reader as an exercise. ∎

EXAMPLE 2 Find $\int \sqrt{\cos^3 x} \, \sin^5 x \, dx$.

Solution We factor $\sin^5 x$ into $\sin x \sin^4 x$ and write

$$\sin^4 x = (\sin^2 x)^2 = (1 - \cos^2 x)^2.$$

Therefore

$$\int \sqrt{\cos^3 x} \, \sin^5 x \, dx = \int \sqrt{\cos^3 x} \, (1 - \cos^2 x)^2 \sin x \, dx$$

and, since $d(\cos x) = -\sin x \, dx$,

$$\int \sqrt{\cos^3 x} \, \sin^5 x \, dx = \int -\sqrt{\cos^3 x} \, (1 - 2\cos^2 x + \cos^4 x) \, d(\cos x)$$

$$= \int (-\cos^{3/2} x + 2\cos^{7/2} x - \cos^{11/2} x) \, d(\cos x)$$

$$= -\tfrac{2}{5} \cos^{5/2} x + \tfrac{4}{9} \cos^{9/2} x - \tfrac{2}{13} \cos^{13/2} x + C \qquad ●$$

Case 2. If m and n are *both* nonnegative even integers then

$$\int \sin^m x \cos^n x \, dx$$

may be integrated by application of the identities

$$\sin u \cos u = \tfrac{1}{2} \sin 2u$$

$$\sin^2 u = \frac{1 - \cos 2u}{2}$$

$$\cos^2 u = \frac{1 + \cos 2u}{2}$$

These identities are used until at least one of the exponents of sine or cosine becomes odd.

EXAMPLE 3   Determine $I = \int \sin^2 5x \cos^2 5x \, dx$

Solution
$$I = \int \frac{(1 - \cos 10x)}{2} \cdot \frac{(1 + \cos 10x)}{2} \, dx$$

$$= \frac{1}{4} \int (1 - \cos^2 10x) \, dx$$

However, $\cos^2 10x = \frac{1}{2} + \frac{1}{2} \cos 20x$ and therefore

$$I = \frac{1}{4} \int \left(1 - \frac{1}{2} - \frac{1}{2} \cos 20x\right) dx$$

$$= \frac{1}{8} \int (1 - \cos 20x) \, dx = \frac{1}{8} \left(x - \frac{\sin 20x}{20}\right) + C$$

Alternative
Solution
We start with the identity
$$\sin 5x \cos 5x = \frac{1}{2} \sin 10x$$

and therefore
$$\sin^2 5x \cos^2 5x = \frac{1}{4} \sin^2 10x = \frac{1}{4}(\frac{1}{2} - \frac{1}{2} \cos 20x)$$

and
$$\int \sin^2 5x \cos^2 5x \, dx = \frac{1}{8} \int (1 - \cos 20x) \, dx$$

$$= \frac{1}{8} \left(x - \frac{\sin 20x}{20}\right) + C$$

which is in agreement with the former solution.   ●

EXAMPLE 4   Find $\int \sin^4 x \cos^2 x \, dx$.

Solution
$$\int \sin^4 x \cos^2 x \, dx = \int \sin^2 x \cos^2 x \sin^2 x \, dx$$

$$= \int (\sin x \cos x)^2 \sin^2 x \, dx$$

$$= \int \left(\frac{\sin 2x}{2}\right)^2 \left(\frac{1}{2} - \frac{1}{2} \cos 2x\right) dx$$

$$= \frac{1}{8} \int \sin^2 2x \, dx - \frac{1}{8} \int \sin^2 2x \cos 2x \, dx$$

$$= \frac{1}{16} \int (1 - \cos 4x) \, dx - \frac{1}{16} \int \sin^2 2x \, d(\sin 2x)$$

$$= \frac{x}{16} - \frac{\sin 4x}{64} - \frac{\sin^3 2x}{48} + C$$ ●

Notice that in the first three lines we took advantage of the identity $\sin x \cos x = \dfrac{\sin 2x}{2}$, thereby providing a rather compact solution. Now we give a second solution following perhaps a more routine procedure.

Alternative Solution Since $\sin^2 x = \dfrac{1 - \cos 2x}{2}$ and $\cos^2 x = \dfrac{1 + \cos 2x}{2}$,

$$\int \sin^4 x \cos^2 x \, dx = \int (\sin^2 x)^2 \cos^2 x \, dx$$

$$= \int \left(\frac{1 - \cos 2x}{2}\right)^2 \left(\frac{1 + \cos 2x}{2}\right) dx$$

$$= \frac{1}{8} \int (1 - \cos 2x - \cos^2 2x + \cos^3 2x) \, dx$$

$$= \frac{1}{8} \int dx - \frac{1}{8} \int \cos 2x \, dx - \frac{1}{8} \int \cos^2 2x \, dx + \frac{1}{8} \int \cos^3 2x \, dx$$

The first two integrals may be found immediately.

$$\int \cos^2 2x \, dx = \int \left(\frac{1}{2} + \frac{1}{2}\cos 4x\right) dx = \frac{x}{2} + \frac{\sin 4x}{8} + \text{constant}$$

and

$$\int \cos^3 2x \, dx = \int \cos^2 2x \cos 2x \, dx = \frac{1}{2} \int (1 - \sin^2 2x) \cos 2x \, d(2x)$$

$$= \frac{1}{2} \int (1 - \sin^2 2x) \, d(\sin 2x)$$

$$= \frac{\sin 2x}{2} - \frac{\sin^3 2x}{6} + \text{constant}$$

Therefore by combination of terms

$$\int \sin^4 x \cos^2 x \, dx = \frac{x}{8} - \frac{\sin 2x}{16} - \left(\frac{x}{16} + \frac{\sin 4x}{64}\right) + \frac{\sin 2x}{16} - \frac{\sin^3 2x}{48} + C$$

or

$$\int \sin^4 x \cos^2 x \, dx = \frac{x}{16} - \frac{\sin 4x}{64} - \frac{\sin^3 2x}{48} + C$$

Clearly the first method is the neater of the two solutions.   ●

Example 5 Find $I = \int \sin^4 3x \, dx$.

Solution Let $u = 3x$ and $du = 3 \, dx$ from which

$$I = \frac{1}{3} \int \sin^4 u \, du$$

$$= \frac{1}{3} \int (\sin^2 u)^2 \, du$$

$$= \frac{1}{12} \int (1 - \cos 2u)^2 \, du$$

$$= \frac{1}{12} \int (1 - 2 \cos 2u + \cos^2 2u) \, du$$

$$= \frac{1}{12} \int \left(1 - 2 \cos 2u + \frac{1}{2} + \frac{1}{2} \cos 4u \right) du$$

$$= \frac{1}{12} \int \left(\frac{3}{2} - 2 \cos 2u + \frac{1}{2} \cos 4u \right) du$$

$$= \frac{1}{12} \left(\frac{3}{2} u - \sin 2u + \frac{\sin 4u}{8} \right) + C$$

$$= \frac{u}{8} - \frac{\sin 2u}{12} + \frac{\sin 4u}{96} + C$$

and in terms of x,

$$I = \frac{3x}{8} - \frac{\sin 6x}{12} + \frac{\sin 12x}{96} + C \qquad \bullet$$

EXAMPLE 6 Find $I = \int \dfrac{\tan^5 4x}{\sec^6 4x} \, dx$.

Solution Express in terms of sines and cosines,

$$I = \int \frac{\tan^5 4x}{\sec^6 4x} \, dx = \int \frac{\sin^5 4x}{\cos^5 4x} \cos^6 4x \, dx$$

$$= \int \sin^5 4x \cos 4x \, dx$$

$$= \frac{1}{4} \int \sin^5 4x \, d(\sin 4x)$$

$$= \frac{\sin^6 4x}{24} + C \qquad \bullet$$

In the next section we will discuss in more depth integrals of the form $\int \tan^m x \sec^n x \, dx$ or $\int \cot^m x \csc^n x \, dx$.

Notice that in the integrals of this section the arguments have been the same. Suppose then that the arguments are not the same. This problem can be treated with the aid of the following basic identities (24) through (26) of Section 9.1:

$$\cos \theta_1 \cos \theta_2 = \tfrac{1}{2}[\cos(\theta_1 - \theta_2) + \cos(\theta_1 + \theta_2)] \qquad \text{(i)}$$

$$\sin \theta_1 \cos \theta_2 = \tfrac{1}{2}[\sin(\theta_1 - \theta_2) + \sin(\theta_1 + \theta_2)] \qquad \text{(ii)}$$

$$\sin \theta_1 \sin \theta_2 = \tfrac{1}{2}[\cos(\theta_1 - \theta_2) - \cos(\theta_1 + \theta_2)] \qquad \text{(iii)}$$

EXAMPLE 7 Find $\int \cos 4x \cos 7x \, dx$.

Solution From (i) with $\theta_1 = 4x$ and $\theta_2 = 7x$

$$\int \cos 4x \cos 7x \, dx = \frac{1}{2} \int [\cos(-3x) + \cos 11x] \, dx$$

$$= \frac{1}{2} \left(\frac{\sin 3x}{3} + \frac{\sin 11x}{11} \right) + C$$

since $\cos(-3x) = \cos 3x$. $\qquad \bullet$

EXAMPLE 8 Find $\int \sin x \sin 2x \cos 5x \, dx$

Solution From (iii)

$$\sin x \sin 2x = \tfrac{1}{2}(\cos(-x) - \cos 3x)$$
$$= \tfrac{1}{2}(\cos x - \cos 3x)$$

and thus

$$\int \sin x \sin 2x \cos 5x \, dx = \frac{1}{2}\int \cos x \cos 5x \, dx - \frac{1}{2}\int \cos 3x \cos 5x \, dx$$

$$= \frac{1}{4}\int (\cos(-4x)$$

$$+ \cos 6x) \, dx - \frac{1}{4}\int (\cos(-2x) + \cos 8x) \, dx$$

$$= \frac{\sin 4x}{16} + \frac{\sin 6x}{24} - \frac{\sin 2x}{8} - \frac{\sin 8x}{32} + C \qquad \bullet$$

By now it should be apparent that no one method will enable us to evaluate all integrals involving sine and cosine. We may try one or a combination of the following:

1. Standard forms
2. Generalized power rule
3. Substitution (either for the argument or for a trigonometric function)
4. Case 1 or case 2 for $\int \sin^m x \cos^n x \, dx$
5. Trigonometric identities to change the integrand.

There are other procedures (which will be covered in future sections) that also can be used. Unfortunately, there are some integrals that cannot be evaluated by any of the techniques that we shall learn.

Exercises 10.2

In Exercises 1 through 27 evaluate the integrals.

1. $\int \sin^2 x \cos x \, dx$

2. $\int \cos^4 x \sin x \, dx$

3. $\int \cos^3 2t \sin 2t \, dt$

4. $\int \sqrt{\sin \theta} \cos \theta \, d\theta$

5. $\int \sqrt{\cos t} \sin^3 t \, dt$

6. $\int \cos^2 3x \, dx$

7. $\int (\cos^2 4x - \sin^2 4x) \, dx$

8. $\int \sin^2 t \cos^2 t \, dt$

9. $\int \sin t \cos^k t \, dt, \quad (k = \text{constant})$

10. $\int \cos^4 \theta \, d\theta$

11. $\int \sin^5 x \, dx$

12. $\int \sin^3 t \cos^2 t \, dt$

13. $\int \sin^2 \theta \cos^4 \theta \, d\theta$

14. $\int \cos^7 x \, dx$

15. $\int \sin^3 t \cos^{20} t \, dt$

16. $\int (\sqrt{\sin t} \cos t)^5 \, dt$

17. $\int \sin 3x \cos x \, dx$

18. $\int \sin 5\theta \sin 3\theta \, d\theta$

19. $\int \sin^2 2x \cos x \, dx$

20. $\int (\sin t + 2 \cos t)^2 \, dt$

21. $\int \dfrac{\tan^3 \theta}{\sec^5 \theta} \, d\theta$

22. $\int \dfrac{\sin^2 2t}{\cos 2t} \, dt,$

23. $\int \left(a \cos \dfrac{\theta}{2} + b \sin \dfrac{\theta}{2}\right)^2 \, d\theta$ $(a, b \text{ constant})$

24. $\int \sin^6 u \, du$

25. $\int \dfrac{d\theta}{1 + \sin \theta}$ $\left(\textit{Hint: }\text{multiply the integrand by } \dfrac{1 - \sin \theta}{1 - \sin \theta}\right)$

26. $\int \dfrac{d\theta}{1 - \cos \theta}$

27. $\int t^2 \sin^2 t^3 \cos^3 t^3 \, dt$

28. Verify that for any positive integer n

$$\int_0^\pi \cos^2 nx \, dx = \frac{\pi}{2}$$

    (*Hint:* $\cos^2 u = \frac{1}{2}(1 + \cos 2u)$)

29. Find the value of $\displaystyle\int_0^\pi \sin^2 nx \, dx$, where n is an arbitrary positive integer.

30. Find the value of $\displaystyle\int_0^{\pi/3} \sin 3x \cos 3x \, dx$.

31. Prove that for any positive integer n,

$$\int_0^{\pi/n} \sin nx \cos nx \, dx = 0.$$

    This is a generalization of Exercise 30.

32. Find the value of $\displaystyle\int_0^{\pi/2} \sin^4 \theta \cos^3 \theta \, d\theta$.

33. Find the value of $\displaystyle\int_0^{\pi/4} \sin^2 \theta \cos^2 \theta \, d\theta$.

34. Find the integral: $\int \sin ax \sin bx \cos cx \, dx$. $(a, b, \text{ and } c \text{ constant})$

TRIGONOMETRIC INTEGRALS—PART II

This section is first devoted to some integrals of the form

$$\int \tan^m x \sec^n x \, dx \qquad (1)$$

or

$$\int \cot^m x \csc^n x \, dx \qquad (2)$$

where m is a positive odd integer or n is a positive even integer. Let us initially consider some examples.

EXAMPLE 1   Find $\int \sqrt{\tan x} \sec^4 x \, dx$.

Solution   We note that n is an even integer and that

$$\sec^4 x = \sec^2 x \, (\sec^2 x)$$

$$= (1 + \tan^2 x) \sec^2 x = (1 + \tan^2 x) \frac{d}{dx} (\tan x)$$

Therefore

$$\int \sqrt{\tan x} \sec^4 x \, dx = \int \sqrt{\tan x} \, (1 + \tan^2 x) \, d(\tan x)$$

$$= \int [(\tan x)^{1/2} + (\tan x)^{5/2}] \, d(\tan x)$$

$$= \tfrac{2}{3}(\tan x)^{3/2} + \tfrac{2}{7}(\tan x)^{7/2} + C \qquad \bullet$$

EXAMPLE 2   Find $\int \tan^5 x \sec^3 x \, dx$.

Solution   We recall that $\frac{d}{dx}(\sec x) = \sec x \tan x$ and therefore

$$\int \tan^5 x \sec^3 x \, dx = \int \tan^4 x \sec^2 x \tan x \sec x \, dx$$

$$= \int (\sec^2 x - 1)^2 \sec^2 x \, d(\sec x)$$

$$= \int (\sec^6 x - 2 \sec^4 x + \sec^2 x) \, d(\sec x)$$

$$= \frac{\sec^7 x}{7} - \frac{2 \sec^5 x}{5} + \frac{\sec^3 x}{3} + C \qquad \bullet$$

EXAMPLE 3 Find $\int \tan^3 x\, dx$

Solution
$$\int \tan^3 x\, dx = \int \tan^2 x \tan x\, dx$$

We would like to factor out sec x tan x, however, there is no sec x in the integrand. We can introduce one by multiplying and dividing by sec x. Therefore

$$\int \tan^3 x\, dx = \int \frac{\tan^2 x}{\sec x} \sec x \tan x\, dx$$

$$= \int \frac{\sec^2 x - 1}{\sec x}\, d(\sec x)$$

$$= \int \sec x\, d(\sec x) - \int \frac{d(\sec x)}{\sec x}$$

$$= \frac{\sec^2 x}{2} - \ln |\sec x| + C$$

Alternative Solution
$$\int \tan^3 x\, dx = \int \tan^2 x \tan x\, dx$$

$$= \int (\sec^2 x - 1) \tan x\, dx$$

$$= \int \tan x \sec^2 x\, dx - \int \tan x\, dx$$

$$= \int \tan x\, d(\tan x) - \int \tan x\, dx$$

$$= \frac{\tan^2 x}{2} + \ln |\cos x| + C$$

The equivalence of the two solutions is easily verified and is left to the reader.

         ⬤

For a general proof of the integrability of (1) where n is a positive even integer (as in Example 1) we proceed as follows.

$$\int \tan^m x \sec^n x\, dx = \int \tan^m x \sec^{n-2} x \sec^2 x\, dx$$

$$= \int \tan^m x\, (1 + \tan^2 x)^{\frac{n-2}{2}} \sec^2 x\, dx$$

$$= \int f(\tan x)\, d(\tan x)$$

where $f(\tan x)$ stands for a function of $(\tan x)$ and the integral is readily determined.

         ■

A similar method can be used for $\int \cot^m x \csc^n x\, dx$ where n is again a positive even integer. This is illustrated in the next example.

EXAMPLE 4 Find $\int \cot^5 x \csc^2 x\, dx$.

Solution   **Method 1:**

$$\int \cot^5 x \csc^2 x \, dx = -\int \cot^5 x \, d(\cot x)$$

$$= -\frac{\cot^6 x}{6} + C$$

Method 2: $\displaystyle\int \cot^5 x \csc^2 x \, dx = \int \cot^4 x \csc x \, (\csc x \cot x \, dx)$

$$= \int (\csc^2 x - 1)^2 \csc x \, (\csc x \cot x \, dx)$$

$$= -\int (\csc^5 x - 2 \csc^3 x + \csc x) \, d(\csc x)$$

$$= -\left(\frac{\csc^6 x}{6} - \frac{\csc^4 x}{2} + \frac{\csc^2 x}{2} \right) + C$$

The two apparently different answers can be shown to be equivalent.   ●

Next we give a general proof of the integrability of (1) when m is a positive odd integer. In this instance,

$$\int \tan^m x \sec^n x \, dx = \int \tan^{m-1} x \sec^{n-1} x \sec x \tan x \, dx$$

$$= \int (\tan^2 x)^{\frac{m-1}{2}} \sec^{n-1} x \, d(\sec x)$$

$$= \int (\sec^2 x - 1)^{\frac{m-1}{2}} \sec^{n-1} x \, d(\sec x)$$

$$= \int f(\sec x) \, d(\sec x)$$

Similarly the same method can be used for (2) in this instance.   ∎

Now let us focus on

$$\int \tan^m x \, dx \tag{3}$$

or

$$\int \cot^m x \, dx \tag{4}$$

If m is a positive odd integer we can proceed as previously developed. However, a method can be given which applies for arbitrary positive integer m. To this end we observe that $\int \tan x \, dx$ and $\int \tan^2 x \, dx$ are easily determined. Let m be an arbitrary positive integer > 2 then

$$\int \tan^m x \, dx = \int \tan^{m-2} x \tan^2 x \, dx$$

$$= \int \tan^{m-2} x \, (\sec^2 x - 1) \, dx$$

$$= \int \tan^{m-2} x \, d(\tan x) - \int \tan^{m-2} x \, dx$$

$$= \frac{\tan^{m-1} x}{m-1} - \int \tan^{m-2} x \, dx \qquad (5)$$

Equation (5) is an example of a **reduction** formula because the exponent of the second term is two less than the original exponent. If $m - 2 > 2$, this reduction formula can be repeated until the exponent is reduced to 1 or 2.

EXAMPLE 5   Evaluate $\int \tan^8 x \, dx$.

Solution   Substitute into (5) with $m = 8$

$$\int \tan^8 x \, dx = \frac{\tan^7 x}{7} - \int \tan^6 x \, dx$$

From (5) with $m = 6$,

$$\int \tan^6 x \, dx = \frac{\tan^5 x}{5} - \int \tan^4 x \, dx$$

Again from (5) with $m = 4$,

$$\int \tan^4 x \, dx = \frac{\tan^3 x}{3} - \int \tan^2 x \, dx$$

$$= \frac{\tan^3 x}{3} - \int (\sec^2 x - 1) \, dx$$

$$= \frac{\tan^3 x}{3} - \tan x + x + C$$

Therefore by successive substitution,

$$\int \tan^8 x \, dx = \frac{\tan^7 x}{7} - \frac{\tan^5 x}{5} + \frac{\tan^3 x}{3} - \tan x + x + C \qquad \bullet$$

EXAMPLE 6   Evaluate $\int \cot^5 x \, dx$.

Solution   **Method 1:** $\int \cot^5 x \, dx = \int \cot^3 x \cot^2 x \, dx$

$$= \int \cot^3 x \, (\csc^2 x - 1) \, dx$$

$$= -\int \cot^3 x \, d(\cot x) - \int \cot^3 x \, dx$$

since $d(\cot x) = -\csc^2 x \, dx$. Therefore

$$\int \cot^5 x \, dx = - \frac{\cot^4 x}{4} - \int \cot^3 x \, dx \qquad (i)$$

Continuing,

$$\int \cot^3 x \, dx = \int \cot x \cot^2 x \, dx$$

$$= \int \cot x \, (\csc^2 x - 1) \, dx$$

$$\int \cot^3 x \, dx = -\int \cot x \, d \cot x - \int \cot x \, dx$$

$$= -\frac{\cot^2 x}{2} - \ln |\sin x| + \text{constant} \qquad \text{(ii)}$$

Combination of (i) and (ii), yields

$$\int \cot^5 x \, dx = -\frac{\cot^4 x}{4} + \frac{\cot^2 x}{2} + \ln |\sin x| + C \qquad \text{(iii)}$$

Method 2: $\displaystyle \int \cot^5 x \, dx = \int \cot^4 x \cot x \, dx$

$$= \int (\csc^2 x - 1)^2 \cot x \, dx$$

$$= \int (\csc^4 x - 2 \csc^2 x + 1) \cot x \, dx$$

$$= \int (\csc^3 x - 2 \csc x) \csc x \cot x \, dx + \int \cot x \, dx$$

However, $d(\csc x) = -\csc x \cot x \, dx$

and thus

$$\int \cot^5 x \, dx = -\frac{\csc^4 x}{4} + \csc^2 x + \ln |\sin x| + C \qquad \text{(iv)}$$

We note that the results (iii) and (iv) appear different. However, we can easily verify that the right sides of these equations differ by a constant. To this end, the difference between the two solutions (excluding the constants) is

$$\frac{-\cot^4 x}{4} + \frac{\csc^4 x}{4} + \frac{\cot^2 x}{2} - \csc^2 x$$

$$= \frac{-\cot^4 x + (\cot^2 x + 1)^2}{4} + \frac{\cot^2 x}{2} - (\cot^2 x + 1)$$

$$= \frac{-\cot^4 x + \cot^4 x + 2 \cot^2 x + 1}{4} + \frac{\cot^2 x}{2} - \cot^2 x - 1$$

$$= \frac{\cot^2 x}{2} + \frac{1}{4} + \frac{\cot^2 x}{2} - \cot^2 x - 1 = -\frac{3}{4}$$

and the verification of the equivalence of the two solutions is completed.

●

Exercises 10.3

In Exercises 1 through 32 evaluate the integrals.

1. $\displaystyle \int \tan x \sec^2 x \, dx$

2. $\displaystyle \int \sec^2 x \tan^3 x \, dx$

3. $\displaystyle \int \sec^2 x (\tan x + 1) \, dx$

4. $\displaystyle \int \tan x \sec^3 x \, dx$

5. $\displaystyle \int \cot 3x \csc^3 3x \, dx$

6. $\displaystyle \int \cot^3 x \csc x \, dx$

7. $\int \cot^3 5\theta \csc^2 5\theta\, d\theta$

8. $\int \cot 2y \csc 2y\, dy$

9. $\int \cot^3 \theta\, d\theta$

10. $\int \cot^4 2u\, du$

11. $\int \sec^4 \left(\dfrac{t}{2}\right) dt$

12. $\int \tan^3 (2x + 1)\, dx$

13. $\int \tan^5 (a\theta)\, d\theta \quad a = \text{constant} \neq 0$

14. $\int \tan^3 t \sec t\, dt$

15. $\int \tan^3 v \sec^3 v\, dv$

16. $\int \tan^7 3x \sec^4 3x\, dx$

17. $\int \dfrac{\cos^3 x}{\sin^5 x}\, dx$ (*Hint:* Express in terms of other trigonometric functions and then integrate.)

18. $\int \dfrac{\cos^4 x + \sin^2 x}{\sin^6 x}\, dx$

19. $\int \tan^3 (a\theta + b)\, d\theta,$ where a and b are constants and $a \neq 0$.

20. $\int (\tan x + \cot x)^2\, dx$

21. $\int (\tan 2x - \cot 2x)^2\, dx$

22. $\int (\sec x + \csc x)^2\, dx$

23. $\int \tan^3 w \sec^{3/2} w\, dw$

24. $\int \cot^5 v \csc^{3/2} v\, dv$

25. $\int (\tan^2 \theta + 3) \sec^4 \theta\, d\theta$

26. $\int \dfrac{d\theta}{\sin^6 \theta}$

27. $\int_0^{\pi/4} \tan \theta \sec^3 \theta\, d\theta$

28. $\int_0^{\pi/4} \tan^5 x\, dx$

29. $\int_{-\pi/3}^{\pi/3} \tan^3 x \sec x\, dx$

30. $\int_{\pi/4}^{\pi/2} \csc^4 x\, dx$

31. $\int \sec^n \theta \tan^5 \theta\, d\theta, \quad n = \text{positive integer}$

32. $\int \csc^n \theta \cot^3 \theta\, d\theta, \quad n = \text{nonnegative integer}$

10.4   TRIGONOMETRIC SUBSTITUTION

In Section 5.3 we applied the algebraic substitution technique to integrals in order to eliminate radicals that are present in the integrand. Thus, for example, when faced with $\int x^2 \sqrt{4 + x}\, dx$ we found that the substitution $u = \sqrt{4 + x}$ transformed the integral into another integral in u that contained no radicals. Note however that the term under the radical is linear. Furthermore a little experimentation will convince the reader that the algebraic substitution technique is often of no avail when the radical is quadratic. We will now show that the trigonometric substitution method can be effectively used to eliminate the radical when dealing with an integral that involves the square root of a quadratic expression.

If the integrand contains:

(i) $\sqrt{a^2 - u^2}$ and $a > 0$, set $u = a \sin \theta$ where $-\frac{\pi}{2} \leq \theta \leq \frac{\pi}{2}$. Then

$-a \leq u \leq a$ and $\theta = \sin^{-1}\left(\frac{u}{a}\right)$.

(ii) $\sqrt{a^2 + u^2}$ and $a > 0$, set $u = a \tan \theta$ where $-\frac{\pi}{2} < \theta < \frac{\pi}{2}$. Then

$\theta = \tan^{-1}\left(\frac{u}{a}\right)$ and u can be any real number.

(iii) $\sqrt{u^2 - a^2}$ and $u \geq a > 0$, set $u = a \sec \theta$ where $0 \leq \theta < \frac{\pi}{2}$. Then

$\theta = \sec^{-1}\left(\frac{u}{a}\right)$.

Notice that in all three cases, the θ values are restricted to the principal value range of the corresponding inverse trigonometric function.

When the integrand contains $\sqrt{u^2 - a^2}$, as in (iii), there is also a second possibility, namely $u \leq -a < 0$. In this instance, we shall change the variable by letting $u = -y$. Then $\sqrt{u^2 - a^2} = \sqrt{y^2 - a^2}$ and $y \geq a > 0$. We now proceed as outlined in (iii), letting $y = a \sec \theta$ where $0 \leq \theta < \frac{\pi}{2}$ (see Example 6 below).†

Useful devices for recalling substitutions (i), (ii), and (iii) are the right triangles shown in Figures 10.4.1 through 10.4.3. The sides of these triangles are a, u, and the related radical. Note that the positions of a and u depend on which case is being considered. In (i) since $\sin \theta = \frac{u}{a}$, a is the hypotenuse and u is the side opposite the angle θ (Figure 10.4.1). In (ii) because $\tan \theta = \frac{u}{a}$, u is the side opposite and a is the side adjacent to angle θ (Figure 10.4.2). In (iii) we have $\sec \theta = \frac{u}{a}$ and therefore u is the hypotenuse and a is the side adjacent to angle θ (Figure 10.4.3). The objective in each case is to make the expression under the radical a perfect square so that the radical is eliminated. The examples below illustrate the procedure.

EXAMPLE 1   Find $\int \frac{\sqrt{a^2 - x^2}}{x^2} \, dx$, $a > 0$, $x \neq 0$, and $|x| \leq a$.

Solution   We set $x = a \sin \theta$ where $0 < |\theta| < \frac{\pi}{2}$. Then $\theta = \sin^{-1} \frac{x}{a}$ and in this interval, $\cos \theta = \sqrt{\cos^2 \theta}$ and $dx = a \cos \theta \, d\theta$.‡

---

† A second method of handling the situation where $u \leq -a < 0$ is to set $u = a \sec \theta$ where $\pi \leq \theta < \frac{3\pi}{2}$. However, this choice of θ is not part of the principal value range of $\sec^{-1}\left(\frac{u}{a}\right)$ and the reason for this seemingly artificial choice of θ values is not immediately obvious. Although some books use this method, we shall use the $u = -y$ substitution when necessary.

‡ Note that $dx \neq d\theta$.

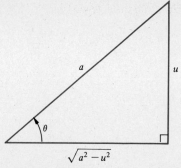

$u = a \sin \theta$

$\sqrt{a^2 - u^2} = \sqrt{a^2 - a^2 \sin^2 \theta} = a \cos \theta \qquad \left(-\dfrac{\pi}{2} \leqslant \theta \leqslant \dfrac{\pi}{2}\right)$

$du = a \cos \theta \, d\theta$

Figure 10.4.1

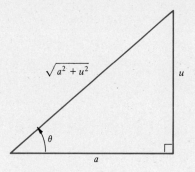

$u = a \tan \theta$

$\sqrt{a^2 + u^2} = \sqrt{a^2 + a^2 \tan^2 \theta} = a \sec \theta \qquad \left(-\dfrac{\pi}{2} < \theta < \dfrac{\pi}{2}\right)$

$du = a \sec^2 \theta \, d\theta$

Figure 10.4.2

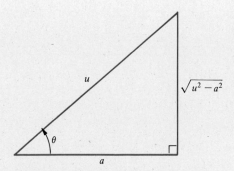

$u = a \sec \theta$

$\sqrt{u^2 - a^2} = \sqrt{a^2 \sec^2 \theta - a^2} = a \tan \theta \qquad \left(0 \leqslant \theta < \dfrac{\pi}{2}\right)$

$du = a \sec \theta \tan \theta \, d\theta$

Figure 10.4.3

Therefore

$$\int \frac{\sqrt{a^2 - x^2}}{x^2}\, dx = \int \frac{\sqrt{a^2 - a^2 \sin^2 \theta}\,(a \cos \theta\, d\theta)}{a^2 \sin^2 \theta} = \int \frac{a \cos \theta\,(a \cos \theta\, d\theta)}{a^2 \sin^2 \theta}$$

$$= \int \frac{\cos^2 \theta}{\sin^2 \theta}\, d\theta = \int \cot^2 \theta\, d\theta$$

$$= \int (\csc^2 \theta - 1)\, d\theta$$

$$= -\cot \theta - \theta + C$$

Now if $\sin \theta = \dfrac{x}{a}$ then $\cot \theta = \sqrt{a^2 - x^2}/x$ (see Figure 10.4.4). This relation is defined for $0 < \theta < \dfrac{\pi}{2}$ when $x > 0$ and is also still valid in $-\dfrac{\pi}{2} < \theta < 0$ (when $x < 0$). Therefore

$$\int \frac{\sqrt{a^2 - x^2}}{x^2}\, dx = -\frac{\sqrt{a^2 - x^2}}{x} - \sin^{-1}\left(\frac{x}{a}\right) + C \qquad \bullet$$

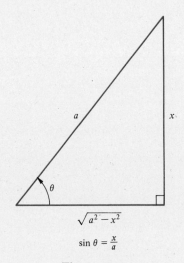

$$\sin \theta = \frac{x}{a}$$

Figure 10.4.4

EXAMPLE 2   Evaluate $\displaystyle\int \frac{x^2}{(x^2 + 16)^2}\, dx$.

Solution   This is not precisely of the form of case (ii) but it does contain $\sqrt{x^2 + 16}$ in the sense that $(x^2 + 16)^2 = \sqrt{(x^2 + 16)^4}$. Therefore we let $x = 4 \tan \theta$, $-\dfrac{\pi}{2} < \theta < \dfrac{\pi}{2}$, so that $\theta = \tan^{-1}\left(\dfrac{x}{4}\right)$ for all real x. Then it follows that $dx = 4 \sec^2 \theta\, d\theta$ and $(x^2 + 16)^2 = (16 \tan^2 \theta + 16)^2 = (16)^2 \sec^4 \theta$. With these results,

$$\int \frac{x^2}{(x^2 + 16)^2}\, dx = \int \frac{16 \tan^2 \theta\, 4 \sec^2 \theta\, d\theta}{(16)^2 \sec^4 \theta} = \frac{1}{4} \int \frac{\tan^2 \theta}{\sec^2 \theta}\, d\theta = \frac{1}{4} \int \frac{\dfrac{\sin^2 \theta}{\cos^2 \theta}}{\dfrac{1}{\cos^2 \theta}}\, d\theta$$

$$= \frac{1}{4} \int \sin^2 \theta \, d\theta = \frac{1}{4} \int \left(\frac{1}{2} - \frac{1}{2} \cos 2\theta \right) d\theta$$

$$= \frac{1}{8} \left(\theta - \frac{1}{2} \sin 2\theta \right) + C \qquad (1)$$

$$= \frac{1}{8} \left(\theta - \sin \theta \cos \theta \right) + C$$

Since $\tan \theta = \dfrac{x}{4}$ we may draw Figure 10.4.5. Then $\cos \theta = \dfrac{4}{\sqrt{x^2 + 16}}$ and $\sin \theta = \dfrac{x}{\sqrt{x^2 + 16}}$. Note that in (1) we had to express $\sin 2\theta$ as $2 \sin \theta \cos \theta$ because we only have information about the argument (angle) θ. We now have

$$\int \frac{x^2}{(x^2 + 16)^2} \, dx = \frac{1}{8} \left(\tan^{-1} \left(\frac{x}{4} \right) - \frac{4x}{x^2 + 16} \right) + C \qquad \bullet$$

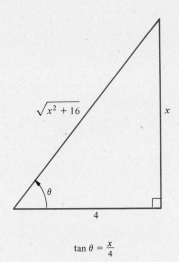

$$\tan \theta = \frac{x}{4}$$

Figure 10.4.5

EXAMPLE 3 Evaluate $\displaystyle \int_0^4 \frac{x^2}{(x^2 + 16)^2} \, dx$.

Solution The integrand is the same as in Example 2 except that now a definite integral must be evaluated. As in Example 2, we let $x = 4 \tan \theta$. Since $0 \leq x \leq 4$, we have $0 \leq \theta \leq \dfrac{\pi}{4}$. Therefore, from (1) since $\theta = 0$ corresponds to $x = 0$ while $\theta = \dfrac{\pi}{4}$ when $x = 4$,

$$\int_0^4 \frac{x^2}{(x^2 + 16)^2} \, dx = \frac{1}{8} \left(\theta - \frac{\sin 2\theta}{2} \right) \Bigg|_{\theta=0}^{\theta=\pi/4}$$

$$= \frac{1}{8} \left[\left(\frac{\pi}{4} - \frac{1}{2} \right) - (0 - 0) \right]$$

$$= \frac{1}{32} (\pi - 2) \qquad \bullet$$

EXAMPLE 4   Evaluate $\int \sqrt{x^2 + a^2}\, dx$, $a > 0$.

Solution   Again we use the substitution $x = a \tan \theta$, $-\frac{\pi}{2} < \theta < \frac{\pi}{2}$. Therefore $dx = a \sec^2 \theta\, d\theta$ and $\sqrt{x^2 + a^2} = a \sec \theta$. Our integral

$$\int \sqrt{x^2 + a^2}\, dx = a^2 \int \sec^3 \theta\, d\theta \qquad (2)$$

and it appears as if the problem is now routine. Certainly we have extensive experience with this kind of problem as developed in Section 10.3. We could write $\int \sec^3 \theta\, d\theta = \int \sec \theta\, (\sec^2 \theta\, d\theta) = \int \sec \theta\, (d \tan \theta)$. However, how is this integrated? If we use the substitution $\sec \theta = \sqrt{1 + \tan^2 \theta}$, then our problem is essentially our initial one and we have run full circle. This problem requires a different technique which will be developed in Section 10.7 (integration by parts). At this stage, we must wait for further developments and go on to the next illustration.

EXAMPLE 5   Evaluate $\int \dfrac{dx}{\sqrt{x^2 - 49}}$, where $x > 7$.

Solution   This is an instance of (iii). Therefore let $x = 7 \sec \theta$ where $0 < \theta < \frac{\pi}{2}$. Consequently
$$dx = 7 \sec \theta \tan \theta\, d\theta$$
and
$$\sqrt{x^2 - 49} = \sqrt{49 \sec^2 \theta - 49} = \sqrt{49(\sec^2 \theta - 1)} = 7 \tan \theta$$
because $\tan \theta > 0$ in the first quadrant.
  With these relations
$$\int \frac{dx}{\sqrt{x^2 - 49}} = \int \frac{7 \sec \theta \tan \theta\, d\theta}{7 \tan \theta} = \int \sec \theta\, d\theta$$
$$= \ln |\sec \theta + \tan \theta| + \text{constant}$$
$$= \ln \left| \frac{x}{7} + \frac{\sqrt{x^2 - 49}}{7} \right| + \text{constant} \qquad \text{(see Figure 10.4.6)}$$
$$= \ln |x + \sqrt{x^2 - 49}| - \ln 7 + \text{constant}$$
$$= \ln |x + \sqrt{x^2 - 49}| + C \qquad \bullet$$

EXAMPLE 6   Evaluate $\displaystyle\int_{-14}^{-7\sqrt{2}} \dfrac{dx}{\sqrt{x^2 - 49}}$.

Solution   This is the same integrand as in Example 5 except that now $x < -7$. Letting $y = -x$ we have
$$\int_{-14}^{-7\sqrt{2}} \frac{dx}{\sqrt{x^2 - 49}} = \int_{14}^{7\sqrt{2}} \frac{-dy}{\sqrt{y^2 - 49}} = \int_{7\sqrt{2}}^{14} \frac{dy}{\sqrt{y^2 - 49}}$$

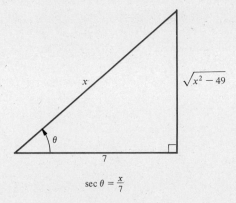

$$\sec \theta = \frac{x}{7}$$

Figure 10.4.6

By the result of Example 5, we have

$$\int_{7\sqrt{2}}^{14} \frac{dy}{\sqrt{y^2 - 49}} = [\ln |y + \sqrt{y^2 - 49}|]_{7\sqrt{2}}^{14}$$

$$= \ln (14 + \sqrt{147}) - \ln 7(\sqrt{2} + 1)$$

$$= \ln \frac{7(2 + \sqrt{3})}{7(\sqrt{2} + 1)} = \ln \left(\frac{2 + \sqrt{3}}{1 + \sqrt{2}}\right)$$

●

EXAMPLE 7 Evaluate $\displaystyle\int \frac{x}{\sqrt{8 - x^2 + 2x}} \, dx.$

Solution The integrand is not in one of the basic forms (i), (ii), or (iii) as it stands. However, if we complete the square

$$8 - x^2 + 2x = 9 - (x^2 - 2x + 1) = (3)^2 - (x - 1)^2$$

it is recognized that this integral is of type (i).
We set

$$x - 1 = 3 \sin \theta \qquad \text{where } -\frac{\pi}{2} < \theta < \frac{\pi}{2}$$

Then

$$dx = 3 \cos \theta \, d\theta \text{ and } x = 3 \sin \theta + 1$$

$$\int \frac{x}{\sqrt{8 - x^2 + 2x}} \, dx = \int \frac{x}{\sqrt{9 - (x - 1)^2}} \, dx$$

$$= \int \frac{3 \sin \theta + 1}{\sqrt{9 - 9 \sin^2 \theta}} 3 \cos \theta \, d\theta$$

$$= \int \frac{3 \sin \theta + 1}{3 \cos \theta} 3 \cos \theta \, d\theta$$

$$= \int (3 \sin \theta + 1) \, d\theta$$

$$= -3 \cos \theta + \theta + C$$

From $\sin \theta = \dfrac{x-1}{3}$, $-\dfrac{\pi}{2} < \theta < \dfrac{\pi}{2}$, we have $\theta = \sin^{-1}\left(\dfrac{x-1}{3}\right)$. Also, we may draw Figure 10.4.7, showing us that

$$\cos \theta = \frac{\sqrt{9 - (x-1)^2}}{3}$$

Then

$$\int \frac{x}{\sqrt{8 - x^2 + 2x}}\, dx = -3\,\frac{\sqrt{9 - (x-1)^2}}{3} + \sin^{-1}\left(\frac{x-1}{3}\right) + C$$

$$= -\sqrt{8 - x^2 + 2x} + \sin^{-1}\left(\frac{x-1}{3}\right) + C \qquad \bullet$$

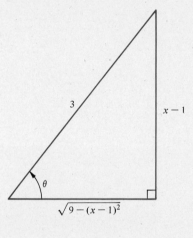

$$\sin \theta = \tfrac{x-1}{3}$$

Figure 10.4.7

EXAMPLE 8 Evaluate $I = \displaystyle\int \frac{dx}{\sqrt{9x^2 + 6x + 26}}$.

Solution Here the integrand is also not in a basic form as it stands. However, if we complete squares there results

$$9x^2 + 6x + 26 = (3x + 1)^2 + (5)^2$$

and we see that this integral is of type (ii).
We set

$$3x + 1 = 5 \tan \theta \qquad -\frac{\pi}{2} < \theta < \frac{\pi}{2}$$

from which

$$3\, dx = 5 \sec^2 \theta\, d\theta \qquad \text{and} \qquad dx = \tfrac{5}{3} \sec^2 \theta\, d\theta$$

$$9x^2 + 6x + 26 = (5 \tan \theta)^2 + 5^2 = 5^2(\tan^2 \theta + 1) = 5^2 \sec^2 \theta$$

Therefore

$$I = \int \frac{\frac{5}{3}\sec^2\theta\, d\theta}{5\sec\theta} = \frac{1}{3}\int \sec\theta\, d\theta$$

$$= \frac{1}{3}\ln|\sec\theta + \tan\theta| + C_1 \qquad (C_1 \text{ is an arbitrary constant})$$

$$= \frac{1}{3}\ln\left|\sqrt{\frac{(3x+1)^2 + 25}{5}} + \frac{3x+1}{5}\right| + C_1 \qquad (\text{see Figure 10.4.8})$$

$$= \frac{1}{3}\ln\left|\frac{\sqrt{9x^2 + 6x + 26} + 3x + 1}{5}\right| + C_1$$

$$= \frac{1}{3}\ln\left(\sqrt{9x^2 + 6x + 26} + 3x + 1\right) + C$$

where the absolute value sign has been removed since the argument of the logarithmic function is positive and C is a new arbitrary constant ($C = C_1 - \frac{1}{3}\ln 5$). ●

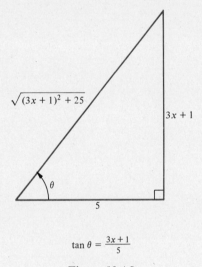

$$\tan\theta = \frac{3x+1}{5}$$

Figure 10.4.8

It is interesting to note that substitutions using hyperbolic functions rather than trigonometric functions may be utilized for cases (i), (ii), and (iii). The two identities $\cosh^2\theta - \sinh^2\theta = 1$ and $\tanh^2\theta + \operatorname{sech}^2\theta = 1$ are used in this analogous development. Therefore, if the integrand contains

(I) $\sqrt{a^2 - u^2}$, set $u = a\tanh\theta$
(II) $\sqrt{a^2 + u^2}$, set $u = a\sinh\theta$
(III) $\sqrt{u^2 - a^2}$, set $u = a\cosh\theta$

The application of this procedure is given in the exercises. The fact that these problems already have been solved with the aid of trigonometric substitution may appear to reduce their significance. However at this stage in our development of techniques of integration there are problems which may be treated more effectively through the usage of hyperbolic functions (see for example Exercise 15 of this section).

Exercises 10.4

Evaluate the following integrals.

1. $\int \sqrt{49 - x^2}\, dx$

2. $\int \dfrac{dx}{\sqrt{36 - x^2}}$

3. $\int \dfrac{x}{\sqrt{x^2 + 9}}\, dx$

4. $\int (x^2 + 4)^{-3/2}\, x\, dx$

5. $\int \dfrac{dx}{x\sqrt{5 + x^2}}$

6. $\int \dfrac{dx}{(9 + x^2)^{3/2}}$

7. $\int \dfrac{\sqrt{x^2 - 4}}{x}\, dx$

8. $\int \dfrac{\sqrt{25x^2 - 1}}{x}\, dx$

9. $\int \dfrac{dx}{(16 + x^2)^2}$

10. $\int \dfrac{x\, dx}{(x^2 - 7)^3}$

11. $\int \dfrac{dx}{x^4\sqrt{3 - x^2}}$

12. $\int \dfrac{dx}{x^4\sqrt{x^2 - 16}}$

13. $\int \dfrac{x^2}{\sqrt{4 - x^2}}\, dx$

14. $\int \dfrac{x^3\, dx}{\sqrt{25 - x^2}}$

15. $\int \dfrac{x^2}{\sqrt{x^2 + 9}}\, dx$ [*Hint:* By trigonometric substitution our procedure is to let $x = 3 \tan \theta$. Show that the

problem involves $\int \sec^3 \theta\, d\theta$ (which will be done by the method of integration by parts). Then start

again using an appropriate substitution involving hyperbolic functions and complete the problem.]

16. $\int \dfrac{dx}{x^2 + 6x + 13}$

17. $\int \dfrac{dx}{4x^2 + 12x + 34}$

18. $\int \dfrac{dx}{\sqrt{1 + x - x^2}}$

19. $\int \dfrac{dx}{\sqrt{4x - x^2}}$

20. $\int \dfrac{x\, dx}{\sqrt{x^4 - 1}}$

10.5   INTEGRATION OF RATIONAL FUNCTIONS, LINEAR FACTORS

A basic problem in elementary algebra is to combine fractions by finding a common denominator. For example

$$\frac{3}{x} - \frac{4}{x + 1} = \frac{3 - x}{x(x + 1)} \tag{1}$$

The expression on the left side of (1) is easily integrated term by term and therefore the indefinite integral of the right side can be found as follows:

$$\int \frac{3 - x}{x(x + 1)}\, dx = \int \left(\frac{3}{x} - \frac{4}{x + 1} \right) dx = 3 \ln |x| - 4 \ln |x + 1| + C \tag{2}$$

However, had we just been given the fraction $\dfrac{3 - x}{x(x + 1)}$, how would we have known to express it as $\dfrac{3}{x} - \dfrac{4}{x+1}$? We must *reverse* the preceding process. Given a proper rational fraction (that is a fraction where the degree of the polynomial in the numerator is less than the degree of the polynomial in the denominator), we would like to write it as a sum of "partial fractions," each of which can then be integrated.

EXAMPLE 1   Find the partial fraction expansion of $\dfrac{3-x}{x(x+1)}$.

Solution   We conjecture that constants A and B exist such that

$$\frac{3-x}{x(x+1)} = \frac{A}{x} + \frac{B}{x+1} \tag{3}$$

is an identity (that is, holds for all values of x for which the expression is defined). Multiplication of both sides of (3) by $x(x+1)$ yields

$$3 - x = A(x+1) + Bx \tag{4}$$

There are two methods we can use to find A and B.

Method 1 (Substitution): Since (4) is true for all values of x, we may substitute particular values.

Letting $x = 0$,   gives $3 = A$
Letting $x = -1$,   gives $4 = -B$ or $B = -4$

Thus, we have obtained the result in (1).

Method 2 (Regrouping): Rewrite (4) so that the various powers of x are grouped together. Then we have

$$3 - x = (A + B)x + A$$

Since this is an identity, both sides of the equation must be the same. So the coefficients of corresponding powers of x must be equal. Therefore

$$A + B = -1 \quad \text{and} \quad A = 3$$

Then $B = -4$ and we have again obtained the result in (1).   ●

EXAMPLE 2   Find $\displaystyle\int \frac{10x^2 - 13x - 59}{(x+1)(x-2)(x+3)}\,dx$.

Solution   As in Example 1, we assume that we have an identity

$$\frac{10x^2 - 13x - 59}{(x+1)(x-2)(x+3)} = \frac{A}{x+1} + \frac{B}{x-2} + \frac{C}{x+3} \tag{5}$$

Multiplication by $(x+1)(x-2)(x+3)$ yields

$$10x^2 - 13x - 59 = A(x-2)(x+3) + B(x+1)(x+3) + C(x+1)(x-2) \tag{6}$$

Using the substitution method, we have

$$\text{if } x = -1, \quad \text{then } -36 = -6A; \quad \text{so } A = 6$$
$$\text{if } x = 2, \quad \text{then } -45 = 15B; \quad \text{so } B = -3$$
$$\text{if } x = -3, \quad \text{then } 70 = 10C; \quad \text{so } C = 7$$

Then

$$\int \frac{10x^2 - 13x - 59}{(x+1)(x-2)(x+3)}\, dx = \int \left(\frac{6}{x+1} + \frac{-3}{x-2} + \frac{7}{x+3} \right) dx$$
$$= 6 \ln |x + 1| - 3 \ln |x - 2| + 7 \ln |x + 3| + C. \qquad \bullet$$

Now let us generalize our procedure. Suppose that we want to integrate a real rational function that is, the quotient of two polynomials. *If the numerator is of degree greater than or equal to the degree of the denominator then we first divide out until the remainder divided by the original divisor is in proper form;* that is, the degree of the numerator is less than the degree of the denominator. This reduces the problem of integrating any rational function to that of integrating a proper rational function (since the integration of any polynomial is routine). Next, as shown in algebra, the denominator of this proper fraction can be factored so that every factor is either linear or an irreducible quadratic function (that is, the discriminant is negative) with real coefficients. (Although this can generally be done, we must realize that it may sometimes be a difficult task to determine these factors.) Furthermore, the proper rational function can always be expressed as a sum of partial fractions, and the form of these fractions depends upon the factors in the denominator. We consider four separate cases:

    I. nonrepeated linear factors
    II. repeated linear factors
    III. nonrepeated quadratic factors
    IV. repeated quadratic factors

This section will be concerned with cases I and II, and the next section will be devoted to Cases III and IV.

10.5.1 *Case I. Nonrepeated Linear Factors*

Examples 1 and 2 exhibit special cases of this situation. More generally if $F(x)$ is a polynomial of degree less than n, and $G(x)$ is given by $G(x) = (x + a_1)(x + a_2)(x + a_3) \cdots (x + a_n)$, then constants $A_1, A_2, \ldots, A_n$ exist such that

$$\frac{F(x)}{G(x)} = \frac{A_1}{x + a_1} + \frac{A_2}{x + a_2} + \cdots + \frac{A_n}{x + a_n} \qquad (7)$$

Note that if the coefficient of x^n in $G(x)$ is not 1, we can divide numerator and denominator by it and consider the original numerator divided by this coefficient as the numerator $F(x)$. The equality in (7) is to be interpreted to mean identically equal.

EXAMPLE 3   Find the indefinite integral

$$I = \int \frac{2x^4 - 9x^3 - 7x^2 + 40x + 15}{x^3 - 5x^2 - 2x + 24}\, dx$$

Solution   Here the numerator is the fourth degree and the denominator is the third degree, so our first step is to divide. We obtain, after long division

$$I = \int \left(2x + 1 + \frac{2x^2 - 6x - 9}{x^3 - 5x^2 - 2x + 24}\right) dx$$

$$= x^2 + x + \int \frac{2x^2 - 6x - 9}{x^3 - 5x^2 - 2x + 24} dx \qquad (8)$$

We factor the denominator

$$\frac{2x^2 - 6x - 9}{x^3 - 5x^2 - 2x + 24} = \frac{2x^2 - 6x - 9}{(x - 3)(x + 2)(x - 4)}$$

and from (7), constants A, B, and C must be found so that

$$\frac{2x^2 - 6x - 9}{(x - 3)(x + 2)(x - 4)} = \frac{A}{x - 3} + \frac{B}{x + 2} + \frac{C}{x - 4} \qquad (9)$$

$$2x^2 - 6x - 9 = A(x + 2)(x - 4) + B(x - 3)(x - 4) + C(x - 3)(x + 2) \qquad (10)$$

We shall find A, B, and C by the substitution method:

$$\text{if } x = 3 \qquad \text{then } -9 = -5A; \quad \text{so } A = \tfrac{9}{5}$$
$$\text{if } x = -2 \quad \text{then } 11 = 30B; \qquad \text{so } B = \tfrac{11}{30}$$
$$\text{if } x = 4 \qquad \text{then } -1 = 6C; \qquad \text{so } C = -\tfrac{1}{6}$$

Substituting into (9) and then (8) we have

$$I = x^2 + x + \int \left(\frac{\tfrac{9}{5}}{x - 3} + \frac{\tfrac{11}{30}}{x + 2} + \frac{-\tfrac{1}{6}}{x - 4}\right) dx$$

$$= x^2 + x + \tfrac{9}{5}\ln|x - 3| + \tfrac{11}{30}\ln|x + 2| - \tfrac{1}{6}\ln|x - 4| + C \qquad \bullet$$

Notice that in the substitution method, we have chosen the values of x to be the ones that make the various factors zero. Although this is not necessary, it greatly simplifies the computation in that the solution of simultaneous linear equations is avoided.

The reader is invited to find A, B, and C in Examples 2 and 3 by the regrouping method. It turns out to be considerably more difficult and involves the simultaneous solution of three linear equations in three unknowns.

10.5.2   *Case II. Repeated Linear Factors*

If the denominator of the integrand contains a factor $(ax + b)$ repeated m times then in the partial decomposition of this rational function the corresponding term is

$$\frac{A_1}{ax + b} + \frac{A_2}{(ax + b)^2} + \cdots + \frac{A_m}{(ax + b)^m} \qquad (11)$$

where the numerators are constants to be determined. Any nonrepeated linear factor is treated as before in Case I.

EXAMPLE 4   Evaluate $I = \int \dfrac{x^2 - 2}{x(x - 1)^3} dx$.

Solution   The integrand is a proper rational fraction and we utilize (11) with $m = 3$ in writing

$$\frac{x^2 - 2}{x(x - 1)^3} = \frac{A}{x} + \frac{B}{x - 1} + \frac{C}{(x - 1)^2} + \frac{D}{(x - 1)^3}$$

Clearing fractions, there results the identity

$$x^2 - 2 = A(x - 1)^3 + Bx(x - 1)^2 + Cx(x - 1) + Dx \tag{12}$$

Let $x = 0$ be substituted into (12) and obtain $A = 2$. Next, if we take $x = 1$, then $D = -1$. If we select $x = 2$, there results

$$2 = 2 + 2B + 2C - 2$$

or

$$B + C = 1 \tag{i}$$

Also for $x = -1$, substituted into (12)

$$-1 = -16 - 4B + 2C + 1$$

or

$$-4B + 2C = 14 \tag{ii}$$

Simultaneous solution of (i) and (ii) yields $B = -2$ and $C = 3$. Therefore

$$I = \int \left(\frac{2}{x} - \frac{2}{x - 1} + \frac{3}{(x - 1)^2} - \frac{1}{(x - 1)^3} \right) dx$$

$$= 2 \ln |x| - 2 \ln |x - 1| - 3(x - 1)^{-1} + \tfrac{1}{2}(x - 1)^{-2} + C$$

$$= \ln \left(\frac{x}{x - 1} \right)^2 - 3(x - 1)^{-1} + \tfrac{1}{2}(x - 1)^{-2} + C \qquad \bullet$$

Exercises 10.5

In Exercises 1 through 24 evaluate the following integrals.

1. $\int \dfrac{dx}{x^2 - 9}$

2. $\int \dfrac{3\, dx}{x^2 - x - 6}$

3. $\int \dfrac{x\, dx}{x^2 - 5x + 4}$

4. $\int \dfrac{(x + 2)\, dx}{x^2 + 4x - 5}$

5. $\int \dfrac{x^2 - 5x - 3}{x^3 - 4x}\, dx$

6. $\int \dfrac{7x - 4}{x^3 + x^2 - 2x}\, dx$

7. $\int \dfrac{x}{(x + 1)(x + 2)(x + 3)}\, dx$

8. $\int \dfrac{x^2 + 3x + 5}{x^2 + 2x}\, dx$

9. $\int \dfrac{x^4 - 7x^3 + 2x + 9}{x}\, dx$

10. $\int \dfrac{x^3 + x^2 - 12x - 69}{x^2 - 3x - 10}\, dx$

11. $\int \dfrac{x^4 - 4x^3 + x^2 - 9x - 24}{x^2 - 4x}\, dx$

12. $\int \dfrac{x^4 + 12x^2 - 7x - 18}{x^3}\, dx$

13. $\int \dfrac{11x^2 - x - 4}{x^3 + x^2}\, dx$

14. $\int \dfrac{7x^2 - 39x + 45}{x^3 - 6x^2 + 9x}\, dx$

15. $\int \dfrac{4x^2 - 15x + 17}{(x - 1)^3}\, dx$

16. $\int \dfrac{13x^3 - 89x^2 + 160x - 128}{(x - 4)^2 x^2}\, dx$

17. $\int \dfrac{x + 1}{(x + 3)^4}\, dx$

18. $\int \dfrac{3x}{(2x + 5)^3}\, dx$

19. $\int \dfrac{10x^2 - 3x - 4}{4x^3 - 3x + 1}\, dx$

20. $\int \dfrac{3x^2 + 40x + 104}{x^3 + 12x^2 + 45x + 50}\, dx$

21. $\int \dfrac{dx}{x^4 - 3x^3}$

22. $\int \dfrac{9x + 7}{(3x + 2)^5}\, dx$

23. $\int \dfrac{\cos t}{\sin^3 t + \sin^2 t - 2\sin t}\, dt$
(*Hint:* let $x = \sin t$.)

24. $\int \dfrac{e^{2t}}{e^{4t} - 10e^{2t} + 9}\, dt$

In Exercises 25 through 30 evaluate the definite integral.

25. $\displaystyle\int_{-1}^{1} \dfrac{dx}{(x - 2)(x + 4)}$

26. $\displaystyle\int_{2}^{5} \dfrac{x}{(x + 1)^2}\, dx$

27. $\displaystyle\int_{1}^{4} \dfrac{dx}{x^3 - x^2 - 6x}$

28. $\displaystyle\int_{3}^{4} \dfrac{x^3 + 2}{x^3 - x}\, dx$

29. $\displaystyle\int_{0}^{1} \dfrac{6x^2 - 15}{x^3 - 3x^2 + 4}\, dx$

30. $\displaystyle\int_{0}^{1} \dfrac{x}{(4x + 1)^3}\, dx$

In Exercises 31 through 36 evaluate the following integrals.

31. $\int \dfrac{dx}{x^2 - a^2}, \quad a \neq 0$

32. $\int \dfrac{dx}{x^2 - (a + b)x + ab}, \quad a \neq b$

33. $\int \dfrac{dx}{(x - a)(x - b)(x - c)}, \quad a \neq b \neq c \text{ and } abc \neq 0$

34. $\int \dfrac{x^2}{(x - a)^2}\, dx, \quad a \neq 0 \quad (\textit{Hint: } x^2 = [(x - a) + a]^2)$

35. $\int \dfrac{x^3}{(x - a)^2}\, dx, \quad a \neq 0$

36. $\int \dfrac{dx}{(ax + b)(cx + d)}, \quad ad - bc \neq 0$

10.6   INTEGRATION OF RATIONAL FUNCTIONS, QUADRATIC FACTORS

In Section 10.5 the method of partial fraction expansion for finding antiderivatives of rational functions was developed. Our considerations were restricted to rational functions whose denominators can be factored into *linear* polynomials with real coefficients. To complete our discussion of this problem we now turn to the instance of irreducible quadratic factors in the denominator. Again from algebra we have the following procedure.

10.6.1   *Case III. Nonrepeated Irreducible Quadratic Factors*

The factors of the denominator $G(x)$ are linear and quadratic; however, it is assumed that none of the quadratic factors is repeated.

To each irreducible quadratic factor of the form $ax^2 + bx + c$ in the denominator the corresponding term in the partial fraction decomposition is

$$\frac{Ax + B}{ax^2 + bx + c} \qquad (1)$$

where A and B are unknowns to be determined. Linear factors are still treated as before.

EXAMPLE 1   Evaluate $\int \dfrac{3x + 2}{(x + 1)(x^2 + 4)}\, dx$.

Solution   From (1) we have the identity

$$\frac{3x + 2}{(x + 1)(x^2 + 4)} = \frac{A}{x + 1} + \frac{Bx + D}{x^2 + 4}$$

where A, B and D must be determined.

As done in the previous section we write (using the regrouping method)

$$3x + 2 = A(x^2 + 4) + (Bx + D)(x + 1)$$
$$= (A + B)x^2 + (B + D)x + 4A + D$$

which implies that

$$A + B = 0 \qquad B + D = 3 \qquad 4A + D = 2$$

The solution of the system is

$$A = -\tfrac{1}{5} \qquad B = \tfrac{1}{5} \qquad D = \tfrac{14}{5}$$

Therefore

$$\frac{3x + 2}{(x + 1)(x^2 + 4)} = \frac{-\tfrac{1}{5}}{x + 1} + \frac{14 + x}{5(x^2 + 4)}$$

and observing that the derivative of $x^2 + 4$ is $2x$ we rewrite the right side as

$$-\frac{1}{5(x + 1)} + \frac{28 + 2x}{10(x^2 + 4)}$$

Thus by rearrangement of terms

$$\frac{3x + 2}{(x + 1)(x^2 + 4)} = -\frac{1}{5(x + 1)} + \frac{1}{10}\left(\frac{2x}{x^2 + 4}\right) + \frac{14}{5(x^2 + 4)}$$

Finally integration term by term results in

$$\int \frac{3x + 2}{(x + 1)(x^2 + 4)}\, dx = -\frac{1}{5}\ln |x + 1| + \frac{1}{10}\ln (x^2 + 4) + \frac{7}{5}\tan^{-1}\left(\frac{x}{2}\right) + C$$

We could also have found A, B, and D by the substitution method. However, with irreducible quadratic factors, this method is no longer as easy to use as with linear factors.   ●

EXAMPLE 2   Evaluate $\int \dfrac{dx}{x^3 + 8}$.

Solution   The denominator is the sum of two cubes $x^3 + 2^3$, which readily factors into the product $(x + 2)$ and $(x^2 - 2x + 4)$. The discriminant of the quadratic form is

$(-2)^2 - 4(1)(4) = -12$ so that the quadratic is not reducible in terms of real linear factors. Thus we set

$$\frac{1}{x^3 + 8} = \frac{A}{x + 2} + \frac{Bx + D}{x^2 - 2x + 4}$$

and this implies that

$$1 = A(x^2 - 2x + 4) + (Bx + D)(x + 2)$$

or

$$1 = (A + B)x^2 + (2B + D - 2A)x + 4A + 2D$$

From this identity we obtain the following system of equations

$$A + B = 0 \qquad 2B + D - 2A = 0 \qquad 4A + 2D = 1$$

from which $A = \frac{1}{12}$, $B = -\frac{1}{12}$, and $D = \frac{1}{3}$. Therefore

$$\frac{1}{x^3 + 8} = \frac{\frac{1}{12}}{x + 2} + \frac{-\frac{1}{12}x + \frac{1}{3}}{x^2 - 2x + 4}$$

In order to integrate the second term we seek to make the numerator the derivative of the denominator. To do this we rewrite

$$-\frac{x}{12} + \frac{1}{3} = \frac{2x - 2}{-24} + \frac{1}{4}$$

which implies that

$$\frac{1}{x^3 + 8} = \frac{\frac{1}{12}}{x + 2} - \frac{1}{24}\frac{2x - 2}{x^2 - 2x + 4} + \frac{1}{4}\frac{1}{x^2 - 2x + 4}$$

$$= \frac{\frac{1}{12}}{x + 2} - \frac{1}{24}\frac{2x - 2}{x^2 - 2x + 4} + \frac{1}{4}\frac{1}{(x - 1)^2 + (\sqrt{3})^2}$$

Hence

$$\int \frac{dx}{x^3 + 8} = \frac{1}{12}\ln|x + 2| - \frac{1}{24}\ln(x^2 - 2x + 4) + \frac{1}{4\sqrt{3}}\tan^{-1}\left(\frac{x - 1}{\sqrt{3}}\right) + C$$

where the argument of the second term is positive because $x^2 - 2x + 4 = (x - 1)^2 + 3$ and the absolute value sign is not necessary. ●

EXAMPLE 3   Evaluate $\int \frac{dx}{x^4 - 1}$.

Solution   We factor the denominator and express as a sum of fractions

$$\frac{1}{x^4 - 1} = \frac{1}{(x - 1)(x + 1)(x^2 + 1)}$$

$$= \frac{A}{x - 1} + \frac{B}{x + 1} + \frac{Dx + E}{x^2 + 1}$$

from which

$$1 = A(x + 1)(x^2 + 1) + B(x - 1)(x^2 + 1) + (Dx + E)(x^2 - 1)$$

$$\text{Let } x = 1, \qquad 4A = 1 \text{ or } A = \tfrac{1}{4}$$

$$\text{Let } x = -1, \quad -4B = 1 \text{ or } B = -\tfrac{1}{4}$$

If we set $x = 0$, $A - B - E = 1$ or $E = -\frac{1}{2}$; and if we use $x = 2$, $1 = \frac{15}{4} - \frac{5}{4} + (2D - \frac{1}{2})3$, or $D = 0$. Therefore

$$\frac{1}{x^4 - 1} = \frac{\frac{1}{4}}{x - 1} - \frac{\frac{1}{4}}{x + 1} - \frac{\frac{1}{2}}{x^2 + 1}$$

and integration term by term, followed by combining the natural logarithm terms, yields

$$\int \frac{dx}{x^4 - 1} = \frac{1}{4} \ln \left| \frac{x - 1}{x + 1} \right| - \frac{1}{2} \tan^{-1} x + C \qquad \bullet$$

The only case remaining to be discussed is that of repeated quadratic factors.

10.6.2 Case IV. Repeated Irreducible Quadratic Factors

If $(ax^2 + bx + c)^n$, where $n > 1$ is a positive integer, is a factor in the denominator and $ax^2 + bx + c$ is irreducible, then corresponding to this factor we have the sum of the n partial fractions

$$\frac{A_1 x + B_1}{ax^2 + bx + c} + \frac{A_2 x + B_2}{(ax^2 + bx + c)^2} + \cdots + \frac{A_n x + B_n}{(ax^2 + bx + c)^n} \qquad (2)$$

where $A_1, B_1, A_2, \ldots, B_n$ are constants to be determined.

EXAMPLE 4 Evaluate $\int \dfrac{3x^4 + 3x^3 + 15x^2 + 12x + 16}{x^5 + 8x^3 + 16x} \, dx$.

Solution We factor the denominator and use (2) to obtain

$$\frac{3x^4 + 3x^3 + 15x^2 + 12x + 16}{x^5 + 8x^3 + 16x} = \frac{3x^4 + 3x^3 + 15x^2 + 12x + 16}{x(x^2 + 4)^2}$$

$$= \frac{A}{x} + \frac{Bx + D}{x^2 + 4} + \frac{Ex + F}{(x^2 + 4)^2}$$

Therefore

$$3x^4 + 3x^3 + 15x^2 + 12x + 16 = A(x^2 + 4)^2 + (Bx + D)(x^2 + 4)x +$$
$$(Ex + F)x = (A + B)x^4 + Dx^3 + (8A + 4B + E)x^2 + (4D + F)x + 16A$$

and equating coefficients of like powers of x yields

$$A + B = 3$$
$$D = 3$$
$$8A + 4B + E = 15$$
$$4D + F = 12$$
$$16A = 16$$

Thus $A = 1$, $B = 2$, $D = 3$, $E = -1$ and $F = 0$. Consequently

$$\frac{3x^4 + 3x^3 + 15x^2 + 12x + 16}{x(x^2 + 4)^2} = \frac{1}{x} + \frac{2x + 3}{x^2 + 4} - \frac{x}{(x^2 + 4)^2}$$

$$= \frac{1}{x} + \frac{2x}{x^2 + 4} + \frac{3}{x^2 + 4} - \frac{x}{(x^2 + 4)^2}$$

Then integration term by term yields

$$\int \frac{3x^4 + 3x^3 + 15x^2 + 12x + 16}{x^5 + 8x^3 + 16x} \, dx$$

$$= \ln|x| + \ln(x^2 + 4) + \frac{3}{2}\tan^{-1}\left(\frac{x}{2}\right) + \frac{1}{2(x^2 + 4)} + C$$

$$= \ln|x(x^2 + 4)| + \frac{3}{2}\tan^{-1}\left(\frac{x}{2}\right) + \frac{1}{2(x^2 + 4)} + C \qquad \bullet$$

Exercises 10.6

In Exercises 1 through 22 find the indicated indefinite integral.

1. $\int \dfrac{3x^2 + 4x - 1}{(1 + x^2)(1 + x)} \, dx$

2. $\int \dfrac{x^2 - 3x + 32}{(x^2 + 16)(x + 3)} \, dx$

3. $\int \dfrac{x^3 - 2}{x^2 + x + 1} \, dx$

4. $\int \dfrac{x^3 - 5}{x^2 + 2x + 4} \, dx$

5. $\int \dfrac{4x^2 + 17x + 23}{(x^2 + 4x + 5)(x + 1)} \, dx$

6. $\int \dfrac{5x^2 - 5x - 17}{(x^2 + 2x + 2)(4x - 1)} \, dx$

7. $\int \dfrac{dx}{16x^2 + 8x + 1}$

8. $\int \dfrac{x \, dx}{x^4 + 2x^2 + 1}$

9. $\int \dfrac{x}{(x^2 + 4)(x^2 + 9)} \, dx$

10. $\int \dfrac{dx}{x^4 + 4x^2 + 3}$

11. $\int \dfrac{2x^2 - x + 2}{x^5 + 2x^3 + x} \, dx$

12. $\int \dfrac{2x^2 + 3x + 8}{x^3 + 4x} \, dx$

13. $\int \dfrac{dx}{x^3 - x^2 + x - 1}$

14. $\int \dfrac{7x^2 - 16x + 11}{x^3 - 3x^2 + 4x - 12} \, dx$

15. $\int \dfrac{x + 3}{x^4 - 1} \, dx$

16. $\int \dfrac{x^4}{x^2 + 16} \, dx$

17. $\int \dfrac{x^4 + 10x^2 + 3x + 1}{x^2 + 9} \, dx$

18. $\int \dfrac{x^7}{x^4 + 81} \, dx$

19. $\int \dfrac{x^5}{x^3 + 10} \, dx$

20. $\int \dfrac{x^5}{(x^2 + 1)^2} \, dx$

21. $\int \dfrac{2x^4 - 3x^3 + 10x^2 - 6x + 16}{(x - 3)(x^2 + 4)^2} \, dx$

22. $\int \dfrac{6 \, dx}{x^4 + x^3 - x - 1}$

In Exercises 23 through 28, evaluate the definite integral.

23. $\displaystyle\int_1^2 \dfrac{dx}{x(1 + x^2)}$

24. $\displaystyle\int_0^2 \dfrac{x^2 - x + 8}{(x^2 + 4)(x + 1)} \, dx$

25. $\displaystyle\int_1^3 \dfrac{dx}{x^3 + x^2 + x + 1}$

26. $\displaystyle\int_1^4 \dfrac{x^2}{x^3 + 2x^2 - 3x - 10} \, dx$

27. $\displaystyle\int_{-3}^{-2} \dfrac{x}{x^4 - 1} \, dx$

28. $\displaystyle\int_{-2}^2 \dfrac{x^3 + x + 1}{x^2 + 4} \, dx$

29. The integral $\int \dfrac{dx}{x^n(1 + x^2)}$, where n is a positive integer may be simplified by introducing the reciprocal substitution $x = \dfrac{1}{u}$. Transform this integral in general and work out the cases $n = 3$ and $n = 4$ explicitly. Note that in this special case we can bypass the method of partial fractions.

10.7

The formula for the differential of a product of two functions is

$$d(uv) = u\,dv + v\,du \tag{1}$$

and by transposition

$$u\,dv = d(uv) - v\,du.$$

If we then integrate both sides there results

$$\int u\,dv = uv - \int v\,du \tag{2}$$

$\left(\text{since } \int d(uv) = uv + \text{constant}\right).$

Equation (2) is the basic formula for **integration by parts.** It expresses the $\int u\,dv$ in terms of uv and $\int v\,du$. The selection of the parts for u and for dv requires care—essentially what we do is replace one integration problem with another. Obviously our objective will be to choose u and dv so that a complicated problem $\int u\,dv$ is replaced by a more manageable one, namely $\int v\,du$.

For definite integrals (2) is replaced by

$$\int_a^b u\,dv = uv \Big|_a^b - \int_a^b v\,du \tag{3}$$

EXAMPLE 1   Find $\int xe^x\,dx$.

Solution   Let $u = x$ and $dv = e^x\,dx$. Then $du = dx$ and $v = e^x$. (Note that v could be $e^x + C$, where C is an arbitrary constant; however, we select the simplest v.) From (2) we obtain

$$\int \underset{u}{x}\,\underset{dv}{e^x\,dx} = \underset{u\ \ v}{x\,e^x} - \int \underset{v\ \ du}{e^x\,dx} \tag{4}$$

Is $\int e^x\,dx$ simpler than $\int xe^x\,dx$? The answers is clearly yes and in fact $\int e^x\,dx = e^x + C$, so that

$$\int xe^x\,dx = xe^x - e^x + C = e^x(x-1) + C \tag{5}$$

This result is easily verified by differentiation of the right side to obtain xe^x.

●

Methods of Integration / Ch. 10

There is a second (and just as natural) way to choose our parts; namely, let $u = e^x$ and $dv = x\,dx$. Then $du = e^x\,dx$ and $v = \dfrac{x^2}{2}$ from which (2) yields

$$\int e^x x\,dx = e^x \frac{x^2}{2} - \int \frac{x^2}{2} e^x\,dx \tag{6}$$

The integral on the right side of (6) is more complicated than the left side, that is, $\int x^2 e^x\,dx$ is less manageable than our given integral. So this choice of u and dv would not have helped us.

EXAMPLE 2   Find $\int x^2 \ln x\,dx$.

Solution   Suppose that we choose the parts $u = x^2$ and $dv = \ln x\,dx$, then we cannot proceed because $\int \ln x\,dx$ presents a problem (see Example 3). Instead let $u = \ln x$ and $dv = x^2\,dx$, from which $du = \dfrac{dx}{x}$ and $v = \dfrac{x^3}{3}$. Therefore from (2),

$$\int \underbrace{(\ln x)}_{u}\, \underbrace{x^2\,dx}_{dv} = \underbrace{(\ln x)\frac{x^3}{3}}_{uv} - \int \underbrace{\frac{x^3}{3}}_{v}\, \underbrace{\frac{dx}{x}}_{du}$$

$$= \frac{x^3}{3}\ln x - \frac{1}{3}\int x^2\,dx$$

$$= \frac{x^3}{3}\ln x - \frac{x^3}{9} + C$$

$$= \frac{x^3}{9}(3\ln x - 1) + C \qquad \bullet$$

EXAMPLE 3   Find $\int \ln x\,dx$

Solution   The reader must be wondering—where are the parts? There is only one factor in the integrand. However, we can easily choose two, namely $\ln x$ and 1. If u is taken to be 1 and $dv = \ln x\,dx$ then $uv = \int 1 \ln x\,dx = \int \ln x\,dx$ and this is our original problem. Instead we select

$$u = \ln x \qquad \text{and} \qquad dv = dx$$

from which $du = \dfrac{dx}{x}$ and $v = x$ so that from (2)

$$\int \ln x\,dx = x \ln x - \int x\frac{dx}{x} = x \ln x - \int dx$$

$$= x \ln x - x + C$$
$$= x(\ln x - 1) + C \qquad \bullet$$

EXAMPLE 4 Find $\int_1^2 \ln x \, dx$.

Solution From Example 3 and (3) we have

$$\int_1^2 \ln x \, dx = (x \ln x)\Big|_1^2 - \int_1^2 dx$$

$$= x(\ln x - 1)\Big|_1^2$$

$$= 2(\ln 2 - 1) - [1(\ln 1) - 1]$$
$$= 2(\ln 2 - 1) + 1$$
$$= 2 \ln 2 - 1 \qquad \bullet$$

EXAMPLE 5 Find $\int \sin^{-1} x \, dx$.

Solution Let $u = \sin^{-1} x$ and $dv = 1 \, dx$ so that $du = \dfrac{1}{\sqrt{1 - x^2}} dx$ and $v = x$. Therefore from (2)

$$\int \sin^{-1} x \, dx = x \sin^{-1} x - \int \frac{x}{\sqrt{1 - x^2}} \, dx$$

$$= x \sin^{-1} x - \int (1 - x^2)^{-1/2} x \, dx$$

$$= x \sin^{-1} x + \sqrt{1 - x^2} + C \qquad \bullet$$

EXAMPLE 6 Find $\int x^2 \sin x \, dx$.

Solution Let $u = x^2$ and $dv = \sin x \, dx$. Then $du = 2x \, dx$ and $v = -\cos x$. From (2) there results

$$\int x^2 \sin x \, dx = -x^2 \cos x - \int (-\cos x) \, 2x \, dx$$

$$= -x^2 \cos x + \int 2x \cos x \, dx \qquad (7)$$

We cannot evaluate this new integral immediately, but at least it is a simpler form than our original one (that is, the power of x in the integrand is smaller). We try integration by parts again, this time letting $u = 2x$ and $dv = \cos x \, dx$ in the new integral. Then $du = 2 \, dx$ and $v = \sin x$, giving

$$\int x^2 \sin x \, dx = -x^2 \cos x + 2x \sin x - \int 2 \sin x \, dx$$

$$= -x^2 \cos x + 2x \sin x + 2 \cos x + C$$

In the second integration by parts, if we had chosen $u = \cos x$ and $dv = 2x \, dx$, then $du = -\sin x \, dx$ and $v = x^2$. From (7) we would then have

$$\int x^2 \sin x \, dx = -x^2 \cos x + x^2 \cos x - \int x^2 (-\sin x \, dx)$$

which simplifies to $\int x^2 \sin x \, dx = \int x^2 \sin x \, dx$

In other words, we have come full circle and accomplished nothing. The reason for this is that our second choice of the parts is the opposite of the original choice and just brought us back to the starting point. In general, if it is necessary to use integration by parts more than once in a problem, *never reverse the roles of the parts* when making successive choices. ●

EXAMPLE 7   Find $\int e^x \cos x \, dx$.

Solution   Let $u = e^x$ and $dv = \cos x \, dx$, from which $du = e^x \, dx$ and $v = \sin x$. Therefore (2) implies that

$$\int e^x \cos x \, dx = e^x \sin x - \int e^x \sin x \, dx$$

This new integral is no easier to evaluate than the old one, so we try integration by parts again. Let $u = e^x$ and $dv = \sin x \, dx$ (observe that we chose the parts in the same manner as before–we did not reverse the roles). Then $du = e^x \, dx$ and $v = -\cos x$, thus giving

$$\int e^x \cos x \, dx = e^x \sin x - \left[e^x (-\cos x) - \int (-\cos x) e^x \, dx \right]$$

or

$$\int e^x \cos x \, dx = e^x \sin x + e^x \cos x - \int e^x \cos x \, dx \qquad (8)$$

Notice that the integral on the right is the same integral we started with. However, this time our work can be utilized. If we add $\int e^x \cos dx$ to both sides of Equation (8), we obtain

$$2 \int e^x \cos x \, dx = e^x \sin x + e^x \cos x$$

Dividing by 2 and putting in the constant term C (since this is an indefinite integral), we have the solution

$$\int e^x \cos x \, dx = \frac{e^x}{2} (\sin x + \cos x) + C$$

This example could also have been done if the original choice of the parts had been $u = \cos x$ and $dv = e^x \, dx$. Again, a second integration by parts would be necessary and we would finally have to solve an equation for the integral. ●

EXAMPLE 8   Establish the following formula: If n is a positive integer greater than 1,

$$\int \sec^n x \, dx = \frac{1}{n-1} \left[\sec^{n-2} x \tan x + (n-2) \int \sec^{n-2} x \, dx \right] \qquad (9)$$

Solution   We write $\sec^n x = \sec^{n-2} x \sec^2 x$ and set $u = \sec^{n-2} x$ and $dv = \sec^2 x\, dx$ $\Big($because we know $\int \sec^2 x\, dx = \tan x\Big)$. Therefore

$$du = (n-2)\sec^{n-3} x\, d(\sec x) = (n-2)\sec^{n-3} x \sec x \tan x\, dx$$
$$= (n-2)\sec^{n-2} x \tan x\, dx$$

and $v = \tan x$. Consequently from (2)

$$\int \sec^n x\, dx = \sec^{n-2} x \tan x - (n-2)\int \sec^{n-2} x \tan^2 x\, dx \qquad (10)$$

In order to obtain (9) from (10), we replace $\tan^2 x$ by $\sec^2 x - 1$ in the integral on the right side of (10). Thus

$$\int \sec^{n-2} x \tan^2 x\, dx = \int (\sec^n x - \sec^{n-2} x)\, dx$$

and (10) becomes

$$\int \sec^n x\, dx = \sec^{n-2} x \tan x - (n-2)\int \sec^n x\, dx + (n-2)\int \sec^{n-2} x\, dx$$

and if we solve this equation for $\int \sec^n x\, dx$, Equation (9) results. ●

Formula (9) is known as a ***reduction formula*** because it expresses the integral of the nth power of the secant in terms of the integral of a lower power of the secant. In particular if $n = 3$, we obtain from (9)

$$\int \sec^3 x\, dx = \tfrac{1}{2}\Big(\sec x \tan x + \int \sec x\, dx\Big)$$
$$= \tfrac{1}{2}(\sec x \tan x + \ln |\sec x + \tan x|) + C$$

By repetition of (9) we may integrate $\sec x$ raised to any positive integral power.

Exercises 10.7

In Exercises 1 through 28 evaluate the indefinite integral.

1. $\int x \cos x\, dx$

2. $\int x \sin 2x\, dx$

3. $\int x e^{-x}\, dx$

4. $\int x \ln x\, dx$

5. $\int \tan^{-1} x\, dx$

6. $\int x \tan^{-1} x\, dx$

7. $\int x \sec^2 x\, dx$

8. $\int x^2 \cos x\, dx$

9. $\int x^2 \ln x\, dx$

10. $\int (\ln x)^2\, dx$

11. $\int \sin^2 x \, dx$, by applying integration by parts. (*Hint:* Write $u = \sin x$ and $dv = \sin x \, dx$, integrate by parts, and use a basic identity.)

12. $\int \cos^3 x \, dx$, by applying integration by parts. (*Hint:* Write $u = \cos^2 x$ and $dv = \cos x \, dx$.)

13. $\int \sin^3 x \, dx$, by applying integration by parts.

14. $\int \sin (\ln x) \, dx$

15. $\int e^x \sin x \, dx$

16. $\int e^{ax} \cos bx \, dx$

17. $\int e^{ax} \sin bx \, dx$

18. $\int x^2 e^{-x} \, dx$

19. $\int x^3 e^{-x^2} \, dx$ $\left(\text{*Hint:* If } u = x^3 \text{ is chosen, then } \int e^{-x^2} \, dx \text{ must be evaluated. However this integral } \textit{cannot} \text{ be expressed in terms of elementary functions; that is, there is no elementary function whose derivative is } e^{-x^2}. \text{ Therefore try to choose } u \text{ so that } v \text{ can easily be determined.}\right)$

20. $\int x^3 \cos (x^2) \, dx$

21. $\int x(x + 4)^{20} \, dx$

22. $\int (4x + 1)(3x + 2)^{12} \, dx$

23. $\int x^3 \sqrt{9 - x^2} \, dx$

24. $\int \sec^4 6x \, dx$

25. $\int \sin 3x \sin x \, dx$, using integration by parts. (*Hint:* Integrate by parts twice.)

26. $\int x^n \ln ax \, dx, \quad n \neq -1$

27. $\int x^2 \tan^{-1} x \, dx$

28. $\int x^2 \cosh bx \, dx$

In Exercises 29 through 34 calculate the definite integral.

29. $\int_0^{\pi/2} x \sin 3x \, dx$

30. $\int_{-1/2}^{1/2} \sin^{-1} x \, dx$

31. $\int_{-1}^1 xe^{-x} \, dx$

32. $\int_0^{\pi/4} \cos^3 x \, dx$

33. $\int_2^3 (x - 3)(x - 2)^8 \, dx$

34. $\int_0^3 \frac{x}{\sqrt{x + 1}} \, dx$

35. Use integration by parts to establish the reduction (or recursion) formula

$$\int \sin^n x \, dx = -\frac{\sin^{n-1} x \cos x}{n} + \frac{n-1}{n} \int \sin^{n-2} x \, dx$$

$\left(\text{*Hint:* let } u = \sin^{n-1} x \text{ and } dv = \sin x \, dx, \text{ then replace } \cos^2 x \text{ by } 1 - \sin^2 x \text{ and solve for } \int \sin^n x \, dx.\right)$

36. Establish the reduction formula

$$\int x^n e^{ax} \, dx = \frac{x^n e^{ax}}{a} - \frac{n}{a} \int x^{n-1} e^{ax} \, dx \qquad a \neq 0$$

37. Establish the reduction formula

$$\int \cos^n x \, dx = \frac{\cos^{n-1} x \sin x}{n} + \frac{n-1}{n} \int \cos^{n-2} x \, dx$$

38. Use the results of Exercises 35 and 37 to derive the following formulas:

(a) $\int \sin^3 x \, dx = \frac{1}{3}(\cos^3 x - 3 \cos x) + C$

(b) $\int \cos^4 x \, dx = \frac{\cos^3 x \sin x}{4} + \frac{3}{8} \cos x \sin x + \frac{3}{8}x + C$

(c) $\int_0^{\pi/2} \sin^4 x \, dx = \frac{3\pi}{16}$

(d) $\int_0^{\pi/2} \cos^5 x \, dx = \frac{8}{15}$

10.8 RATIONAL FUNCTIONS OF TRIGONOMETRIC FUNCTIONS

The problem of the integration of rational functions of trigonometric functions is equivalent to the integration of rational functions of the two basic trigonometric functions $\sin x$ and $\cos x$. There is a substitution that enables us to integrate any rational expression in $\sin x$ and $\cos x$. In fact we shall prove that the substitution

$$z = \tan \frac{x}{2} \tag{1}$$

converts the problem of integrating a rational function of $\sin x$ and $\cos x$ into the integration of a rational function of z.

Proof We must show that with the application of (1), $\sin x$ and $\cos x$ and $\dfrac{dx}{dz}$ are rational functions of z. Observe that

$$dz = \sec^2 \frac{x}{2} \frac{dx}{2} = \left(1 + \tan^2 \frac{x}{2}\right) \frac{dx}{2} = (1 + z^2) \frac{dx}{2}$$

or

$$dx = \frac{2 \, dz}{1 + z^2} \tag{2}$$

Next

$$\cos x = 2 \cos^2 \left(\frac{x}{2}\right) - 1 = \frac{2}{\sec^2 \dfrac{x}{2}} - 1$$

$$= \frac{2}{1 + \tan^2 \dfrac{x}{2}} - 1 = \frac{2}{1 + z^2} - 1$$

or

$$\cos x = \frac{1 - z^2}{1 + z^2} \tag{3}$$

Finally

$$\sin x = 2 \sin \frac{x}{2} \cos \frac{x}{2} = 2 \frac{\sin \dfrac{x}{2}}{\cos \dfrac{x}{2}} \cos^2 \frac{x}{2}$$

$$= 2 \frac{\tan \dfrac{x}{2}}{\sec^2 \dfrac{x}{2}} = 2 \frac{\tan \dfrac{x}{2}}{1 + \tan^2 \dfrac{x}{2}}$$

$$\sin x = \frac{2z}{1 + z^2} \tag{4}$$

From (2), (3), and (4) it follows that the integration of any rational function of $\sin x$ and $\cos x$ becomes a rational function of z under the transformation (1). ∎

EXAMPLE 1 Find $\displaystyle\int \frac{dx}{1 + \cos x}$.

Solution From (2) and (3)

$$\int \frac{dx}{1 + \cos x} = \int \frac{\dfrac{2\,dz}{1 + z^2}}{1 + \dfrac{1 - z^2}{1 + z^2}} = \int \frac{\dfrac{2}{1 + z^2}}{\dfrac{2}{1 + z^2}}\,dz$$

$$= \int 1\,dz = z + C$$

$$= \tan \frac{x}{2} + C$$

EXAMPLE 2 Find $\displaystyle\int \frac{dx}{\sin x + \tan x}$.

Solution $\tan x = \dfrac{\sin x}{\cos x} = \dfrac{2z}{1 - z^2}$ from (3) and (4)

Therefore

$$\int \frac{dx}{\sin x + \tan x} = \int \frac{\dfrac{2\,dz}{1 + z^2}}{\dfrac{2z}{1 + z^2} + \dfrac{2z}{1 - z^2}}$$

$$= \int \frac{2}{1 + z^2} \frac{(1 + z^2)(1 - z^2)}{4z}\,dz$$

$$= \frac{1}{2} \int \frac{1 - z^2}{z}\,dz$$

$$= \frac{1}{2} \left(\int \frac{dz}{z} - \int z \, dz \right)$$

$$= \frac{1}{2} \left(\ln |z| - \frac{z^2}{2} \right) + C$$

$$= \frac{1}{2} \left(\ln \left| \tan \frac{x}{2} \right| - \frac{1}{2} \tan^2 \frac{x}{2} \right) + C$$

●

EXAMPLE 3   Find $\int \dfrac{dx}{3 + 2 \sin x - \cos x}$.

Solution   From (2), (3), and (4)

$$\int \frac{dx}{3 + 2 \sin x - \cos x} = \int \frac{\dfrac{2 \, dz}{1 + z^2}}{3 + \dfrac{4z}{1 + z^2} - \dfrac{1 - z^2}{1 + z^2}}$$

$$= \int \frac{\dfrac{2 \, dz}{1 + z^2}}{\dfrac{3 + 3z^2 + 4z - 1 + z^2}{1 + z^2}}$$

$$= \int \frac{dz}{2z^2 + 2z + 1}$$

$$= \frac{1}{2} \int \frac{dz}{z^2 + z + \frac{1}{2}}$$

and completion of the square in the denominator yields

$$\int \frac{dx}{3 + 2 \sin x - \cos x} = \frac{1}{2} \int \frac{dz}{(z + \frac{1}{2})^2 + (\frac{1}{2})^2}$$

$$= \frac{1}{2} \cdot 2 \tan^{-1} \left(\frac{z + \frac{1}{2}}{\frac{1}{2}} \right) + C$$

$$= \tan^{-1}(2z + 1) + C$$

$$= \tan^{-1} \left(2 \tan \frac{x}{2} + 1 \right) + C$$

●

EXAMPLE 4   Find $\int \csc x \, dx$.

Solution

$$\int \csc x \, dx = \int \frac{dx}{\sin x}$$

and by (2) and (4) of this section

$$\int \csc x \, dx = \int \frac{\dfrac{2 \, dz}{1 + z^2}}{\dfrac{2z}{1 + z^2}} = \int \frac{dz}{z} = \ln |z| + C$$

$$= \ln \left| \tan \frac{x}{2} \right| + C$$

$$= \ln \left| \frac{1 - \cos x}{\sin x} \right| + C$$

$$= \ln |\csc x - \cot x| + C\dagger$$

which is in agreement with our formula developed previously in a less direct manner. ●

Exercises 10.8

In each of Exercises 1 through 10, find the indefinite integral by the method developed in this section.

1. $\int \dfrac{dx}{1 - \cos x}$

2. $\int \dfrac{dx}{2 + \cos x}$

3. $\int \dfrac{dx}{1 + \sin x}$

4. $\int \dfrac{dx}{\cos x + \sin x}$

5. $\int \sec x \, dx$, and reconcile with the standard formula for this integral.

6. $\int \dfrac{\sec t}{1 + \tan t + \sec t} \, dt$

7. $\int \dfrac{\csc \theta}{1 + 2 \csc \theta - \cot \theta} \, d\theta$

8. $\int \dfrac{d\theta}{4 + 5 \cos \theta}$

9. $\int \dfrac{d\theta}{\sqrt{2} + \sin \theta}$

10. $\int \dfrac{2 \sec t}{3 \tan t + 4} \, dt$

In Exercises 11 through 16, evaluate the definite integrals by the method of this section.

11. $\int_0^{\pi/2} \dfrac{dx}{2 + \cos x}$

12. $\int_0^{\pi} \dfrac{dx}{2 + 3 \cos x}$

13. $\int_0^{\pi/4} \dfrac{dt}{\cos t}$

14. $\int_0^{\pi/2} \dfrac{d\theta}{1 + \sin \theta + \cos \theta}$

15. $\int_{-\pi/2}^{0} \dfrac{dt}{2 + \sin t}$

16. $\int_0^{-\pi/2} \dfrac{\sin x}{1 + \sin^2 x} \, dx$

10.9   ADDITIONAL TECHNIQUES

The methods of Sections 10.5 and 10.6 show that the indefinite integrals of rational functions result in elementary functions. In this section we consider other integrals that can be transformed into integrals of rational functions. To this end, we shall examine some particular examples.

$\dagger \ln |\csc x - \cot x| = -\ln |\csc x + \cot x|$

since

$$\ln |\csc x - \cot x| + \ln |\csc x + \cot x| = \ln (|\csc x - \cot x| \, |\csc x + \cot x|)$$
$$= \ln |\csc^2 x - \cot^2 x| = \ln 1 = 0$$

EXAMPLE 1   Evaluate $\int \dfrac{dx}{x + x^{4/3}}$.

Solution   The fractional power of x can be eliminated by the substitution of $x^{1/3} = u$ or $x = u^3$. The integral then becomes (since $dx = 3u^2\,du$)

$$\int \frac{dx}{x + x^{4/3}} = \int \frac{3u^2\,du}{u^3 + u^4} = 3\int \frac{du}{u(1 + u)}$$

$$= 3\int \left(\frac{1}{u} - \frac{1}{1 + u}\right) du$$

$$= 3\ln \left|\frac{u}{1 + u}\right| + C$$

$$= 3\ln \left|\frac{x^{1/3}}{1 + x^{1/3}}\right| + C \qquad \bullet$$

EXAMPLE 2   Evaluate $\int \dfrac{dx}{\sqrt{x}(1 + \sqrt[3]{x})}$.

Solution   All radicals can be eliminated by letting $\sqrt[6]{x} = u$ (we choose the sixth root since, if we write $\sqrt{x} = x^{1/2}$ and $\sqrt[3]{x} = x^{1/3}$, then 6 is the least common multiple of the denominators of the fractional exponents). Therefore $x = u^6$, $dx = 6u^5\,du$, $\sqrt{x} = u^3$, and $\sqrt[3]{x} = u^2$. Consequently

$$\int \frac{dx}{\sqrt{x}(1 + \sqrt[3]{x})} = 6\int \frac{u^5\,du}{u^3(1 + u^2)}$$

$$= 6\int \frac{u^2}{1 + u^2}\,du$$

$$= 6\int \left(1 - \frac{1}{1 + u^2}\right) du$$

$$= 6(u - \tan^{-1} u) + C$$

$$= 6(\sqrt[6]{x} - \tan^{-1}\sqrt[6]{x}) + C \qquad \bullet$$

EXAMPLE 3   Find $\displaystyle\int_1^2 \dfrac{x^5}{(8 + x^3)^{1/2}}\,dx$.

Solution   First we note that the integrand is continuous if $8 + x^3 > 0$ or $x > -2$. Therefore our integral exists, since the interval is from $x = 1$ to $x = 2$. Let us try the substitution $u = (8 + x^3)^{1/2}$ in the hope that the resulting integral with respect to u will involve a rational integrand. With our choice of u, $u^2 = 8 + x^3$ and $2u\,du = 3x^2\,dx$. Also $x^3 = u^2 - 8$ and, when $x = 2$, $u = 4$; whereas $u = 3$ when $x = 1$. Therefore

$$\int_1^2 \frac{x^5}{(8 + x^3)^{1/2}}\,dx = \int_1^2 \frac{x^3 \cdot x^2}{(8 + x^3)^{1/2}}\,dx$$

$$= \int_3^4 \frac{u^2 - 8}{u}\,\frac{2u\,du}{3}$$

$$= \frac{2}{3} \int_3^4 (u^2 - 8)\, du$$

$$= \frac{2}{3} \left(\frac{u^3}{3} - 8u \right) \Big|_3^4$$

$$= \frac{26}{9} \qquad \bullet$$

EXAMPLE 4   Find $\displaystyle\int_1^3 \frac{dx}{\sqrt{x} + \sqrt{x+1}}$.

Solution   The integrand is continuous for $x > -1$ and therefore our integral exists. Our first impulse (it was mine) is to let $\sqrt{x} = u$ and eliminate one square root, then to use a second substitution to eliminate the second radical in u. Instead (in this special case) we observe that the numerator and denominator of the integrand can be multiplied by $\sqrt{x+1} - \sqrt{x}$. Thus

$$\int_1^3 \frac{dx}{\sqrt{x} + \sqrt{x+1}} = \int_1^3 \frac{1}{\sqrt{x+1} + \sqrt{x}} \frac{\sqrt{x+1} - \sqrt{x}}{\sqrt{x+1} - \sqrt{x}} dx$$

$$= \int_1^3 (\sqrt{x+1} - \sqrt{x})\, dx$$

$$= (\tfrac{2}{3}(x+1)^{3/2} - \tfrac{2}{3} x^{3/2}) \Big|_1^3$$

$$= (\tfrac{16}{3} - \tfrac{2}{3} 3\sqrt{3}) - (\tfrac{2}{3} 2^{3/2} - \tfrac{2}{3})$$

$$= \tfrac{2}{3}(9 - 2^{3/2} - 3^{3/2}) \qquad \bullet$$

There are many other procedures and substitutions that can be used to evaluate integrals. We have only illustrated some of the simpler ones. A more detailed discussion of these methods would not be of sufficient benefit at this point to warrant going through the detailed computation that accompanies them. Additional substitutions are given in the exercises.

Exercises 10.9

1. $\displaystyle\int \frac{dx}{1 + \sqrt{x}}$

2. $\displaystyle\int \frac{dx}{x - \sqrt[3]{x}}$

3. $\displaystyle\int \frac{dx}{x(1 + \sqrt[5]{x})}$

4. $\displaystyle\int \frac{\sqrt{t+2} + 1}{\sqrt{t+2} - 1}\, dt$

5. $\displaystyle\int \frac{e^{2t}}{e^t + 2}\, dt$

6. $\displaystyle\int \sqrt{e^z + 3}\, dz$

7. $\displaystyle\int \frac{e^{3t}}{e^t + 1}\, dt$

8. $\displaystyle\int \frac{w\sqrt{w}}{1 + \sqrt{w}}\, dw$

9. $\displaystyle\int \frac{\sqrt{t}}{1 + \sqrt[4]{t}}\, dt$

10. $\displaystyle\int \frac{t^{1/3}}{1 + t^{2/3}}\, dt$

11. $\displaystyle\int \frac{x^5}{(1 + x^3)^{3/2}}$

12. $\displaystyle\int x^3(5 + 4x^2)^2\, dx$ *(Hint: Let $u = 5 + 4x^2$.)*

13. Show that for

$$\int x^m (a + bx^n)^{r/s}\, dx \qquad (s > 0,\ b \neq 0,\ \text{and } n \neq 0)$$

where m, n, r, and s are integers, the substitution $u^s = a + bx^n$ will yield an integrand in u which is rational provided that $\dfrac{m + 1}{n}$ is an integer. Identify this exercise with Exercises 11 and 12.

14. $\displaystyle\int x^2 (3 + 4x)^{1/3}\, dx$

15. $\displaystyle\int x^{-1}(2 + 7x^3)^{1/2}\, dx$

16. $\displaystyle\int \frac{dx}{\sqrt{x} + \sqrt[3]{x}}$

17. $\displaystyle\int \frac{dx}{\sqrt{2 + \sqrt{3 + x}}}$

18. $\displaystyle\int \frac{dx}{\sqrt[4]{1 + \sqrt{1 - x}}}$

19. Find $\displaystyle\int \frac{dx}{x\sqrt{x^2 - a^2}}$ by letting $x = \dfrac{u^2 + a^2}{2u}, \quad u > 0,\ a > 0$

20. Find $\displaystyle\int \frac{dx}{x^2 - a^2}$ by using the same substitution as in Exercise 19

21. Find $\displaystyle\int \frac{dx}{x\sqrt{a^2 - x^2}}$ by letting $x = \dfrac{a(1 - u^2)}{1 + u^2}, \quad a > 0$

22. Find $\displaystyle\int \frac{dx}{\sqrt{x^2 + x}}$ by letting $\sqrt{x^2 + x} = x + u$

23. Find $\displaystyle\int \frac{dx}{\sqrt{x^2 + 3x + 2}}$ by letting $\sqrt{(x + 1)(x + 2)} = (x + 2)u$

In Exercises 24 through 32, compute each of the following definite integrals.

24. $\displaystyle\int_0^{\pi^2/4} \cos \sqrt{x}\, dx$

25. $\displaystyle\int_1^9 \frac{\sqrt{x}}{1 + \sqrt{x}}\, dx$

26. $\displaystyle\int_0^1 \frac{\sqrt{x + 2}}{\sqrt{x + 1}}\, dx$

27. $\displaystyle\int_0^8 e^{\sqrt[3]{x}}\, dx$

28. $\displaystyle\int_2^4 \frac{\sqrt{x - 1}}{1 + \sqrt{x}}\, dx$

29. $\displaystyle\int_0^2 \frac{\sqrt{2 - x} + \sqrt{1 - x}}{\sqrt{2 - x} + 1}\, dx$

30. $\displaystyle\int_0^a \frac{x^3}{(a^2 + x^2)^{5/2}}\, dx$ (*Hint:* Use a trigonometric substitution.)

31. $\displaystyle\int_1^4 \frac{dx}{x\sqrt{2x^2 - 2x + 1}}$ $\left(Hint:\ \text{Let } x = \dfrac{1}{u}.\right)$

32. $\displaystyle\int_{1/3}^{1/\sqrt{3}} \frac{\sqrt{1 + 9y^2}}{y^2}\, dy$ (*Hint:* Let $3y = x$, then let $\sinh x = u$.)

10.10 INTEGRAL TABLES

In this chapter we have developed a number of methods for finding indefinite integrals, such as integration by parts, partial fractions, algebraic, trigonometric, and hyperbolic substitutions, reduction formulas, and so on. However, it must be admitted that a scientist generally does not recall all these special techniques so that when an integral must be evaluated we generally depend upon integral tables. Clearly, an accurate set of tables saves a considerable amount of time and helps eliminate errors. To aid the reader, a table of indefinite integrals or antiderivatives is given in the front and back inside covers. More extensive tables of integrals may be found, for example, in (1) *Burington's Handbook of Mathematical Tables and Formulas* (1933); (2) Chemical Rubber Company (or CRC) *Standard Mathematical Tables;* (3) *Pierce* or *Pierce-Foster Table of Integrals.* It is suggested that the reader obtain at least one of these mathematical tables, which generally has other information of importance to a practicing scientist, such as some significant definite integrals, derivatives, fundamental identities, logarithms, trigonometric, exponential, and hyperbolic functions, and so on.

It must be emphasized that intelligent use of integral tables is greatly enhanced when the user has a good understanding of the various techniques of integration. For one thing, not every integral can be found directly in a table. This will be understood better through illustrative examples.

EXAMPLE 1   Find $\int x^4 e^x \, dx$.

Solution   This integral cannot be found in the previously mentioned tables. However, all of these tables list the reduction formula

$$\int x^n e^{ax} \, dx = \frac{x^n e^{ax}}{a} - \frac{n}{a} \int x^{n-1} e^{ax} \, dx \tag{1}$$

found by integrating by parts.

In our case $n = 4$ and $a = 1$ so that (1) becomes

$$\int x^4 e^x \, dx = x^4 e^x - 4 \int x^3 e^x \, dx$$

Then we use (1) again with $n = 3$ and $a = 1$ to obtain

$$\int x^3 e^x \, dx = x^3 e^x - 3 \int x^2 e^x \, dx$$

and similarly

$$\int x^2 e^x \, dx = x^2 e^x - 2 \int x e^x \, dx$$

$$\int x e^x \, dx = x e^x - \int e^x \, dx$$

$$= x e^x - e^x + C$$

By simple substitution

$$\int x^4 e^x \, dx = x^4 e^x - 4x^3 e^x + (4)(3)x^2 e^x - 4(3)(2)x e^x + 4(3)(2)e^x + C$$

or

$$\int x^4 e^x \, dx = e^x(x^4 - 4x^3 + 12x^2 - 24x + 24) + C \tag{2}$$

This result may (and should) be verified by differentiating the right side of (2) to obtain $x^4 e^x$.   ●

EXAMPLE 2 Find $\int (\ln 5x)^3 \, dx$.

Solution This integral cannot be found in the standard tables. Instead we can find the reduction or recursion formula

$$\int (\ln ax)^n \, dx = x(\ln ax)^n - n \int (\ln ax)^{n-1} \, dx \qquad (3)$$

obtained from integration by parts. (In many tables $\log x$ means $\log_e x$ or $\ln x$ in our notation.)

From (3) with $a = 5$ and $n = 3$,

$$\int (\ln 5x)^3 \, dx = x(\ln 5x)^3 - 3 \int (\ln 5x)^2 \, dx$$

From (3) again,

$$\int (\ln 5x)^2 \, dx = x(\ln 5x)^2 - 2 \int \ln 5x \, dx$$

and again from the tables

$$\int \ln 5x \, dx = x \ln 5x - x + C$$

Therefore by successive substitution

$$\int (\ln 5x)^3 \, dx = x(\ln 5x)^3 - 3x(\ln 5x)^2 + 6(x \ln 5x - x) + C \qquad (4)$$

and our result is easily verified. ●

EXAMPLE 3 Find the following three integrals with the use of a table of integrals:

$$\int \frac{dx}{x^2 - 2x - 1} \qquad (5)$$

$$\int \frac{dx}{4x^2 + 4x + 5} \qquad (6)$$

$$\int \frac{dx}{9x^2 - 12x + 4} \qquad (7)$$

Solution In the table by Burington (1933) under "expressions containing $(ax^2 + bx + c)$" we have three formulas listed (depending upon the value of the discriminant)

$$\int \frac{dx}{ax^2 + bx + c} = \frac{1}{\sqrt{b^2 - 4ac}} \ln \frac{2ax + b - \sqrt{b^2 - 4ac}}{2ax + b + \sqrt{b^2 - 4ac}} \qquad b^2 > 4ac \qquad (8)$$

$$\int \frac{dx}{ax^2 + bx + c} = \frac{2}{\sqrt{4ac - b^2}} \tan^{-1} \frac{2ax + b}{\sqrt{4ac - b^2}} \qquad b^2 < 4ac \qquad (9)$$

$$\int \frac{dx}{ax^2 + bx + c} = -\frac{2}{2ax + b} \qquad b^2 = 4ac \qquad (10)$$

In (5), $a = 1$, $b = -2$, and $c = -1$, $b^2 - 4ac = 4 - 4(1)(-1) = 8$ so that (8) applies and

$$\int \frac{dx}{x^2 - 2x - 1} = \frac{1}{\sqrt{8}} \ln \frac{2x - 2 - \sqrt{8}}{2x - 2 + \sqrt{8}} + C$$

In (6), $a = 4$, $b = 4$, and $c = 5$, $b^2 - 4ac = 16 - 4(4)(5) = -64$, and (9) yields

$$\int \frac{dx}{4x^2 + 4x + 5} = \frac{2}{8} \tan^{-1} \frac{8x + 4}{8} + C$$

$$= \frac{1}{4} \tan^{-1} \left(x + \frac{1}{2} \right) + C$$

In (7), $a = 9$, $b = -12$, and $c = 4$, $b^2 - 4ac = 144 - 4(9)(4) = 0$, and (10) yields

$$\int \frac{dx}{9x^2 - 12x + 4} = \frac{-2}{18x - 12} + C = -\frac{1}{9x - 6} + C$$

Again the reader is encouraged to check the result by differentiating back. ●

EXAMPLE 4 Find $\displaystyle\int \frac{3x^4 + x^3 + 24x^2 + 2x + 47}{x^4 + 8x^2 + 16} \, dx$.

Solution We do not expect to find this integral in any table as it stands. Rather, we know that as a first step the denominator must be divided into the numerator and partial fraction expansion is to be used. Now

$$x^4 + 8x^2 + 16 = (x^2 + 4)^2$$

and we have

$$\frac{3x^4 + x^3 + 24x^2 + 2x + 47}{(x^2 + 4)^2} = 3 + \frac{x^3 + 2x - 1}{(x^2 + 4)^2}$$

$$= 3 + \frac{Ax + B}{x^2 + 4} + \frac{Dx + E}{(x^2 + 4)^2}$$

$$3x^4 + x^3 + 24x^2 + 2x + 47 = 3(x^2 + 4)^2 + (Ax + B)(x^2 + 4) + (Dx + E)$$
$$= 3x^4 + Ax^3 + (24 + B)x^2 + (4A + D)x + 48 + 4B + E$$

from which

$$A = 1 \qquad 24 + B = 24 \quad \text{or} \quad B = 0 \qquad 4A + D = 2 \quad \text{or} \quad D = -2$$
$$4B + E = -1 \quad \text{or} \quad E = -1$$

Thus

$$\frac{3x^4 + x^3 + 24x^2 + 2x + 47}{(x^2 + 4)^2} = 3 + \frac{x}{x^2 + 4} - \frac{2x + 1}{(x^2 + 4)^2} \qquad \text{(i)}$$

Formula (90) in Burington is

$$\int \frac{x \, dx}{ax^2 + c} = \frac{1}{2a} \ln(ax^2 + c)$$

so that with $a = 1$ and $c = 4$

$$\int \frac{x}{x^2 + 4} \, dx = \frac{1}{2} \ln(x^2 + 4) \qquad \text{(ii)}$$

[(ii) is an easy enough integral for us to have found directly; however, our purpose here is to illustrate the use of integral tables].

Next we write

$$-\int \frac{2x + 1}{(x^2 + 4)^2} \, dx = -2 \int \frac{x}{(x^2 + 4)^2} \, dx - \int \frac{dx}{(x^2 + 4)^2}$$

and formula (89) states that

$$\int x(ax^2 + c)^n \, dx = \frac{1}{2a} \frac{(ax^2 + c)^{n+1}}{n + 1} \qquad n \neq -1$$

In our case

$$\int x(x^2 + 4)^{-2} \, dx = \frac{1}{2} \frac{(x^2 + 4)^{-1}}{-1} = -\frac{1}{2} \frac{1}{(x^2 + 4)} \qquad \text{(iii)}$$

Finally the $\int \dfrac{dx}{(x^2 + 4)^2}$ must be found. We use the reduction formula (88)

$$\int \frac{dx}{(ax^2 + c)^n} = \frac{1}{2(n - 1)c} \frac{x}{(ax^2 + c)^{n-1}} + \frac{2n - 3}{2(n - 1)c} \int \frac{dx}{(ax^2 + c)^{n-1}}$$

which in our case ($a = 1$, $c = 4$, and $n = 2$) becomes

$$\int \frac{dx}{(x^2 + 4)^2} = \frac{1}{8} \frac{x}{x^2 + 4} + \frac{1}{8} \int \frac{dx}{x^2 + 4}$$

$$= \frac{1}{8} \frac{x}{x^2 + 4} + \frac{1}{16} \tan^{-1}\left(\frac{x}{2}\right) \qquad \text{(iv)}$$

[from formula (84)]:

$$\int \frac{dx}{p^2 + x^2} = \frac{1}{p} \tan^{-1}\left(\frac{x}{p}\right)$$

Therefore from (i) through (iv)

$$\int \frac{3x^4 + x^3 + 24x^2 + 2x + 47}{x^4 + 8x^2 + 16} \, dx = 3x + \frac{1}{2} \ln (x^2 + 4)$$

$$+ \frac{1}{x^2 + 4} - \frac{x}{8(x^2 + 4)} - \frac{1}{16} \tan^{-1}\left(\frac{x}{2}\right) + C$$

$$= 3x + \frac{1}{2} \ln (x^2 + 4) + \frac{1 - \dfrac{x}{8}}{x^2 + 4} - \frac{1}{16} \tan^{-1}\left(\frac{x}{2}\right) + C$$

Exercises 10.10

Find the following integrals using a table of integrals as well as any standard method (if necessary):

1. $\displaystyle\int \frac{dx}{x^2(ax + b)}, \quad b \neq 0$

2. $\displaystyle\int \frac{x \, dx}{x^2 - x + 2}$

3. $\displaystyle\int \sec^3 2\theta \, d\theta$

4. $\displaystyle\int x^2 \cos 3x \, dx$

5. $\int \sin^{-1} 6x \, dx$

6. $\int e^x \cos^2 x \, dx$

7. $\int (\ln 3x)^3 \, dx$

8. $\int \dfrac{\sin^4 ax}{\cos ax} \, dx, \quad a \neq 0$

9. $\int \dfrac{x \, dx}{\sqrt{3x^2 + x - 2}}$

10. $\int \sqrt{\dfrac{x+3}{x+1}} \, dx$

11. $\int \dfrac{dx}{x\sqrt{x^5 + 2}}$

12. $\int \sin^8 x \, dx$

13. $\int \dfrac{x \, dx}{x^4 + 3x^2 + 2}$

14. $\int \dfrac{x^2}{(3x - 2)^2} \, dx$

15. $\int \dfrac{dx}{\sqrt{-x^2 + 7x - 12}}$

16. $\int \sin 2x \cos 6x \, dx$

17. $\int \csc^4 5x \, dx$

18. $\int x^4 (\ln 2x)^2 \, dx$

19. $\int (x^2 + 8x + 4) \cos^2 x \, dx$

20. $\int e^{2x} \sin^3 x \, dx$

21. $\int (\cot x + \csc^2 x)^2 \, dx$

22. $\int \dfrac{dx}{x^5 \sqrt{x^2 + 2}}$

23. $\int \dfrac{x}{(4 - x^2)^{7/2}} \, dx$

24. $\int \dfrac{2x^2 + 9x - 4}{x^3 + 3x^2 + 4x + 12} \, dx$

25. $\int (x^2 + 3x - 1) \cosh x \, dx$

26. $\int \sinh^3 x \, dx$

27. $\int (2 + x - x^3) \sqrt{25 - x^2} \, dx$

28. $\int \dfrac{6x - 5}{81 - x^4} \, dx$

29. $\int \dfrac{x}{x^4 + 1} \, dx$

30. $\int \dfrac{x^3}{2x^8 + 3} \, dx$

31. $\int \dfrac{x \, dx}{x^6 - 16x^3 + 64}$

32. $\int \dfrac{z^5 \, dz}{z^8 + 81}$

10.11

TRAPEZOIDAL RULE FOR APPROXIMATING AN INTEGRAL

Our procedure for evaluating a definite integral $\int_a^b f(x) \, dx$ has been to use the fundamental theorem of calculus. This means that we must determine a function $F(x)$ such that $F'(x) = f(x)$. However, there are functions f such as $f(x) = e^{-x^2}$ and $f(x) = \dfrac{\sin x}{x}$ for which no elementary antiderivative exists, and a direct numerical evaluation is our only recourse. Also, there are many instances when such a numerical procedure is preferred over working with a complicated function $F(x)$. For these reasons, several methods of approximate integration have been developed. We shall consider two of them in this and the next section. These procedures are applicable even if the integrand is known only at a number of selected values of the variable x.

Since a definite integral for a continuous function $f(x) \geq 0$ defined on a given interval $[a, b]$ may be interpreted as an area under the curve $y = f(x)$, any method for finding this area can be used to approximate the value of this integral $\int_a^b f(x)\, dx$. Let us partition $[a, b]$ into n equal parts of width $h = \dfrac{b-a}{n}$ and the coordinates of the subdivision are

$$x_0 = a,\; x_1 = a + h,\; x_2 = a + 2h, \ldots,\; x_k = a + kh, \ldots,\; x_n = a + nh = b$$

Draw the vertical lines $x = x_k$, $k = 0, 1, 2, \ldots, n$, and let these lines intersect the curve at the points $P_0, P_1, P_2, \ldots, P_n$, respectively. Next, draw the line segments $P_{k-1}P_k$ for each k. We have thus constructed n trapezoids (see Figure 10.11.1).

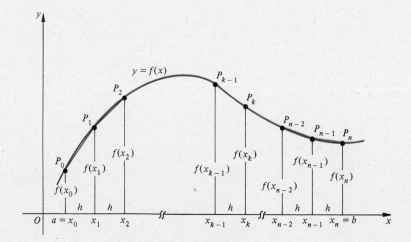

Figure 10.11.1

The area of the kth trapezoid is

$$\frac{h}{2}\,[f(x_{k-1}) + f(x_k)]$$

(that is, $\frac{1}{2}$ altitude · sum of bases). The sum of the areas of the n trapezoids is

$$T_n = \frac{h}{2}\,\{[f(x_0) + f(x_1)] + [f(x_1) + f(x_2)] + \cdots + [f(x_{n-1}) + f(x_n)]\}$$

or by combining like terms

$$T_n = \frac{h}{2}\,[f(x_0) + 2f(x_1) + 2f(x_2) + \cdots + 2f(x_{n-1}) + f(x_n)] \qquad (1)$$

Equation (1) is called the ***trapezoidal rule*** for approximating the integral

$$\int_a^b f(x)\, dx$$

that is

$$\int_a^b f(x)\,dx \approx \frac{h}{2}\left[f(x_0) + 2f(x_1) + 2f(x_2) + \cdots + 2f(x_{n-1}) + f(x_n)\right] \qquad (2)$$

where the symbol $\approx$ means "approximately equal to." More generally (2) may be applied even when $f(x)$ changes sign on $[a, b]$ (or $f(x) \leq 0$ on $[a, b]$) and may be interpreted geometrically if we use the concept of "signed area."

In order to enhance the usefulness of an approximation method such as the trapezoidal rule, it is important to estimate the accuracy of our approximation (2). It can be shown that the error, E_T, made by using the trapezoidal rule with n trapezoids to approximate $\int_a^b f(x)\,dx$ must satisfy the equation

$$E_T = \int_a^b f(x)\,dx - T_n = -\frac{(b-a)}{12}f''(c)h^2 \qquad (3)$$

where c is some number on $a < x < b$. This is proved in advanced calculus texts by using an extension of the mean value theorem, where it is assumed that $f(x)$ is continuous on $[a, b]$ and twice differentiable on (a, b). This proof will not be given here. Note that, since the value of c is generally not known, this only estimates the error. Actually in many practical situations if $f''(x)$ is continuous on $[a, b]$ we replace (3) by

$$|E_T| \leq \frac{h^2(b-a)}{12}\max|f''(x)| \qquad (4)$$

where $\max|f''(x)|$ denotes the maximum value of $|f''(x)|$ on $[a, b]$.

EXAMPLE 1   Use the trapezoidal rule with $n = 6$ to estimate $\int_0^{3/2} x^2\,dx$ and compare this approximation with the exact value of the integral.

Solution   The exact value of the integral is

$$\int_0^{3/2} x^2\,dx = \frac{x^3}{3}\bigg|_0^{3/2} = \frac{1}{3}\left(\frac{3}{2}\right)^3 - 0 = \frac{9}{8} = 1.125$$

For the trapezoidal approximation we have

$$x_0 = a = 0 \qquad x_n = b = \tfrac{3}{2} \qquad n = 6$$

$$h = \frac{b-a}{n} = \frac{\tfrac{3}{2}}{6} = \frac{1}{4}$$

$$
\begin{aligned}
x_0 &= a = 0 & y_0 &= f(x_0) = f(0) = 0^2 = 0 \\
x_1 &= a + h = \tfrac{1}{4} & y_1 &= f(x_1) = (\tfrac{1}{4})^2 = \tfrac{1}{16} \\
x_2 &= a + 2h = \tfrac{1}{2} & y_2 &= f(x_2) = (\tfrac{1}{2})^2 = \tfrac{1}{4} \\
x_3 &= a + 3h = \tfrac{3}{4} & y_3 &= f(x_3) = (\tfrac{3}{4})^2 = \tfrac{9}{16} \\
x_4 &= a + 4h = 1 & y_4 &= f(x_4) = (1)^2 = 1 \\
x_5 &= a + 5h = \tfrac{5}{4} & y_5 &= f(x_5) = (\tfrac{5}{4})^2 = \tfrac{25}{16} \\
x_6 &= a + 6h = \tfrac{6}{4} = \tfrac{3}{2} & y_6 &= f(x_6) = (\tfrac{3}{2})^2 = \tfrac{9}{4}
\end{aligned}
$$

and from (1)

$$T_6 = \frac{h}{2} [f(x_0) + 2f(x_1) + 2f(x_2) + 2f(x_3) + 2f(x_4) + 2f(x_5) + f(x_6)]$$

$$= \tfrac{1}{8}[0 + \tfrac{1}{8} + \tfrac{1}{2} + \tfrac{9}{8} + 2 + \tfrac{25}{8} + \tfrac{9}{4}] = \tfrac{73}{64} = 1\tfrac{9}{64} \doteq 1.141$$

Thus the approximation is too large by one part in 72 or about 1.4%.

In this particular case, $f''(x) = 2 = f''(c)$ for all c in $[a, b]$ so that from (3)

$$E_T = -\frac{\tfrac{3}{2} - 0}{12} (2) \left(\frac{1}{16}\right) = -\frac{1}{64} = \int_0^{3/2} x^2\, dx - T_6$$

Therefore T_6 exceeds $\int_0^{3/2} x^2\, dx$ by $\tfrac{1}{64}$ as found numerically.

Of course the only reason why the exact value for the error is obtained is that $f''(x)$ is constant. In general, the best we can do is *estimate* the error. ●

EXAMPLE 2   Use the trapezoidal rule with $n = 5$ to estimate $\int_1^2 \frac{dx}{x}$. Then find an upper bound for $|E_T|$ and a range of values for $\int_1^2 \frac{dx}{x}$. Compare with the exact value.

Solution   For the trapezoidal approximation we have

$$x_0 = a = 1 \qquad x_n = b = 2 \qquad n = 5 \qquad h = \frac{b - a}{5} = \frac{1}{5} = 0.2$$

Therefore,

$$x_0 = a = 1 \qquad\qquad y_0 = f(x_0) = f(1) = \tfrac{1}{1} = 1$$
$$x_1 = a + h = 1.2 \qquad\quad y_1 = f(x_1) = f(1.2) = \tfrac{1}{1.2} \doteq 0.83333$$
$$x_2 = a + 2h = 1.4 \qquad\; y_2 = f(x_2) = f(1.4) = \tfrac{1}{1.4} \doteq 0.71429$$
$$x_3 = a + 3h = 1.6 \qquad\; y_3 = f(x_3) = f(1.6) = \tfrac{1}{1.6} = 0.62500$$
$$x_4 = a + 4h = 1.8 \qquad\; y_4 = f(x_4) = f(1.8) = \tfrac{1}{1.8} \doteq 0.55556$$
$$x_5 = a + 5h = b = 2.0 \quad y_5 = f(x_5) = f(2) = \tfrac{1}{2} = 0.50000$$

and from (1)

$$T_5 = \frac{h}{2} [f(x_0) + 2f(x_1) + 2f(x_2) + 2f(x_3) + 2f(x_4) + f(x_5)]$$

$$\doteq 0.1[1 + 2(0.83333) + 2(0.71429) + 2(0.62500) + 2(0.55556) + 0.50000]$$

$$\doteq 0.69564$$

In order to obtain an upper bound for $|E_T|$, we observe that $f(x) = \dfrac{1}{x}, f'(x) = -\dfrac{1}{x^2}$, and $f''(x) = \dfrac{2}{x^3}$ and $|f''(x)| \le 2$ in $[1, 2]$. Because $f''(x) \ge 0$ in $[1, 2]$, this implies that our estimate by the trapezoidal rule is too large. From (4) we obtain a crude upper bound for $|E_T|$:

$$|E_T| \le \frac{(0.2)^2}{12} (2 - 1)(2) \le 0.0067$$

Thus a range for $\int_1^2 \dfrac{dx}{x}$ is determined, namely

$$0.69564 - 0.0067 \le \int_1^2 \frac{dx}{x} \le 0.69564 + 0.0067$$

or

$$0.68894 \le \int_1^2 \frac{dx}{x} \le 0.70234$$

In fact, the exact value of $\int_1^2 \dfrac{dx}{x}$ is

$$\int_1^2 \frac{dx}{x} = \ln 2 = 0.69314 \cdots$$

and the *actual* $|E_T| \doteq 0.0025$. Our approximation obtained by the trapezoidal formula is too large by about 0.36%. ●

Exercises 10.11

In Exercises 1 through 16 compute the approximate value of the definite integral obtained by the trapezoidal rule for the given value of n. Express the result to three decimal places. Compare with the exact result.

1. $\displaystyle\int_0^2 (x + 3)\, dx, \quad n = 4$

2. $\displaystyle\int_0^1 x^2\, dx, \quad n = 5$

3. $\displaystyle\int_0^1 x^3\, dx, \quad n = 10$

4. $\displaystyle\int_1^3 \sqrt{x + 1}\, dx, \quad n = 4$

5. $\displaystyle\int_0^2 \frac{x}{x + 1}\, dx, \quad n = 4$

6. $\displaystyle\int_1^2 x(x - 1)\, dx, \quad n = 5$

7. $\displaystyle\int_0^1 \frac{dx}{x^2 + 1}, \quad n = 5$

8. $\displaystyle\int_0^1 \frac{x}{x^2 + 1}\, dx, \quad n = 5$

9. $\displaystyle\int_0^2 \cos x\, dx, \quad n = 4$

10. $\displaystyle\int_0^3 e^{-x}\, dx, \quad n = 6$

11. $\displaystyle\int_0^3 xe^{-x}\, dx, \quad n = 6$

12. $\displaystyle\int_2^4 \sqrt{x^2 + 1}\, x\, dx, \quad n = 4$

13. $\displaystyle\int_1^2 x \sin x\, dx, \quad n = 4$

14. $\displaystyle\int_1^3 \frac{x^2 + 3x}{x + 1}\, dx, \quad n = 5$

15. $\displaystyle\int_0^3 \frac{x^2}{x^2 + 4}\, dx, \quad n = 6$

16. $\displaystyle\int_1^2 \frac{dx}{x(x - 3)}, \quad n = 5$

In Exercises 17 through 22 find the maximum value of $|E_T|$ and the range of values of the integral.

17. Exercise 1
18. Exercise 2
19. Exercise 3
20. Exercise 4
21. Exercise 5
22. Exercise 9
23. The function F defined by

$$F(x) = \frac{1}{\sqrt{2\pi}} \int_0^x e^{-t^2/2}\, dt$$

is significant in mathematical statistics. In particular $F(x)$ is extensively tabulated. Find $F(1)$ approximately by applying the trapezoidal rule with $n = 5$. Express your answer to three decimal places.

$$\left(\textit{Hint:}\ \frac{1}{\sqrt{2\pi}} \doteq 0.3989\right)$$

24. The definite integral $\int_{1}^{3} \dfrac{dx}{x}$ is approximated by application of the trapezoidal rule. How large must n be chosen so that $|E_T| < 0.004$?

10.12   SIMPSON'S RULE

The method known as **Simpson's rule** usually yields an even better approximation to a definite integral with very little increase in labor. We subdivide the interval $[a, b]$ into an *even* number of subintervals of equal length. Then parabolic arcs are made to fit through sets of three given noncollinear points (Figure 10.12.1).

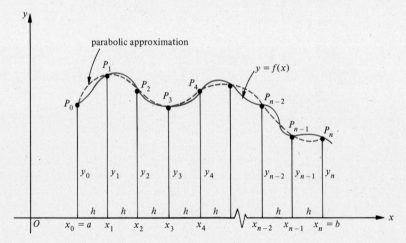

Figure 10.12.1

Suppose for convenience that the coordinates of three specified points are $(-h, y_0)$, $(0, y_1)$ and (h, y_2), where y_0, y_1, and y_2 are taken to be nonnegative (Figure 10.12.2). There is a unique parabola with equation of the form

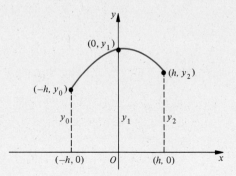

Figure 10.12.2

$$y = ax^2 + bx + c \tag{1}$$

passing through the three points. This follows from the fact that since $(-h, y_0)$, $(0, y_1)$, and (h, y_2) lie on (1),

$$\begin{aligned} y_0 &= a(-h)^2 + b(-h) + c \\ y_1 &= c \\ y_2 &= a(h)^2 + bh + c \end{aligned} \tag{2}$$

and the system (2) has a unique solution for a, b, and c. Actually

$$c = y_1$$

and

$$a = \frac{y_0 + y_2 - 2y_1}{2h^2} \tag{3}$$

and the formula for b is not needed for what follows.

From (1) we obtain, for the area under the parabola,

$$\int_{-h}^{h} (ax^2 + bx + c)\, dx = 2ch + \frac{2a}{3}h^3 \tag{4}$$

and substitution of (3) into (4) yields

$$\int_{-h}^{h} (ax^2 + bx + c)\, dx = \frac{h}{3}(y_0 + 4y_1 + y_2) \tag{5}$$

If a congruent parabola is located elsewhere in the plane, as in Figure 10.12.3, the form $y = ax^2 + bx + c$ applies again. If the curve passes through the three points (x_0, y_0), (x_1, y_1) and (x_2, y_2), where $h = x_2 - x_1 = x_1 - x_0$, then, since the area under the parabola is the same as before,

$$\int_{x_0}^{x_2} (ax^2 + bx + c)\, dx = \frac{h}{3}(y_0 + 4y_1 + y_2) \tag{6}$$

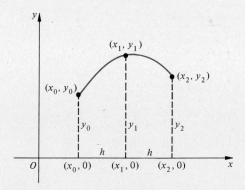

Figure 10.12.3

Repeating this process for pairs of intervals (Figure 10.12.1) from $x = a$ to $x = b$ and adding, we obtain (where the number of intervals must be even)

$$\int_{a}^{b} f(x)\, dx \approx \frac{h}{3}[(y_0 + 4y_1 + y_2) + (y_2 + 4y_3 + y_4) + \cdots$$

$$+ (y_{n-2} + 4y_{n-1} + y_n)] \tag{7}$$

or

$$\int_a^b f(x)\, dx \approx \frac{h}{3}[y_0 + 4y_1 + 2y_2 + 4y_3 + 2y_4 + \cdots + 2y_{n-2}$$
$$+ 4y_{n-1} + y_n] \qquad (8)$$

This is **Simpson's rule.** Furthermore, if I_S represents the right side of (8), we may write

$$\int_a^b f(x)\, dx = I_S + E_S \qquad (9)$$

and E_S is the error made in using I_S to approximate $\int_a^b f(x)\, dx$. By an extension of the mean value theorem of differential calculus it can be shown that, if f is continuous on $[a, b]$ and four times differentiable on $a < x < b$, then

$$\int_a^b f(x)\, dx = I_S - \frac{b-a}{180} f^{(4)}(c) \cdot h^4 \qquad (10)$$

where c is some number in $a < x < b$. As a corollary to (10), if $f^{(4)}(x)$ is continuous on the closed interval $a \le x \le b$, an upper bound on the error $|E_S|$ made by application of Simpson's rule in approximating $\int_a^b f(x)\, dx$ is given by

$$|E_S| \le \frac{b-a}{180} M \cdot h^4 \qquad (11)$$

where M is the maximum value of $|f^{(4)}(x)|$ on $[a, b]$. The proof of (10) will not be given.

EXAMPLE 1   Use Simpson's rule with $n = 4$ to estimate $\int_1^3 \dfrac{dx}{x}$. Then find an upper bound for $|E_S|$ and a range of values for $\int_1^3 \dfrac{dx}{x}$. Compare with the exact value.

Solution   For the approximation by Simpson's rule, $x_0 = a = 1$, $n = 4$, $x_n = x_4 = b = 3$,
$h = \dfrac{b-a}{n} = \dfrac{3-1}{4} = \dfrac{1}{2} = 0.5$

In tabular form we have

x_i	$f(x_i)$	Weighting Factor	Product
1	1	1	1
1.5	$\frac{2}{3}$	4	$\frac{8}{3}$
2.0	$\frac{1}{2}$	2	1
2.5	$\frac{2}{5}$	4	$\frac{8}{5}$
3.0	$\frac{1}{3}$	1	$\frac{1}{3}$
			sum $= 6\frac{3}{5} = \frac{33}{5}$

Since $h = \frac{1}{2}$, formula (8) yields

$$\int_1^3 \frac{dx}{x} \approx \frac{1}{6}\left(\frac{33}{5}\right) = 1.1$$

$$f(x) = x^{-1} \qquad f'(x) = -x^{-2} \qquad f''(x) = 2x^{-3} \qquad f'''(x) = -6x^{-4}$$

$$f^{(4)}(x) = 24x^{-5}$$

and in $[1, 3]$,

$$|f^{(4)}(x)|_{\max} = 24$$

Substitution into (11) yields,

$$|E_S| \le \frac{3-1}{180}(24)\left(\frac{1}{2}\right)^4 = \frac{1}{60}$$

$$1.1 - \frac{1}{60} \le \int_1^3 \frac{dx}{x} \le 1.1 + \frac{1}{60}$$

$$1.0833 \le \int_1^3 \frac{dx}{x} \le 1.1167$$

gives us a range of possible values for $\int_1^3 \frac{dx}{x}$. Actually

$$\int_1^3 \frac{dx}{x} = \ln 3 \doteq 1.09861$$

●

EXAMPLE 2   Find $\int_1^{1.6} \frac{\sin x}{x}\, dx$ by using Simpson's rule with $n = 6$.

Solution   There is no elementary function $F(x)$ such that $F'(x) = \dfrac{\sin x}{x}$ and therefore the second fundamental theorem cannot be applied. We proceed to Simpson's rule with $x_0 = a = 1$, $x_n = x_6 = b = 1.6$, $h = \dfrac{b-a}{n} = \dfrac{1.6-1}{6} = 0.1$. In tabular form we have

x_i	$\sin x_i$	$f(x_i)$	Weighting Factor	Product
1	0.84147	0.84147	1	0.84147
1.1	0.89121	0.81019	4	3.24076
1.2	0.93204	0.77670	2	1.55340
1.3	0.96356	0.74120	4	2.96480
1.4	0.98545	0.70389	2	1.40778
1.5	0.99749	0.66499	4	2.65996
1.6	0.99957	0.62473	1	0.62473
			sum =	13.29290

Since $h = 0.1$, formula (8) yields

$$\int_1^{1.6} \frac{\sin x}{x}\, dx \approx \frac{0.1}{3}(13.29290) \doteq 0.44310$$

From CRC *Standard Mathematical Table* we obtain, with the notation $\int_0^z \frac{\sin x}{x}\, dx = Si(z)$, that

$$\int_1^{1.6} \frac{\sin x}{x} = Si(1.6) - Si(1) = 1.38918 - 0.94608 = 0.44310$$

Therefore our calculations apparently agree with results obtained by more exact procedures to five decimal places. ●

Exercises 10.12

In each of Exercises 1 through 16, approximate the definite integral by applying Simpson's rule to the given integrand for the prescribed value of n. Express the answer to three decimal places and compare with the exact result. A calculator and tables are required for Exercises 7 through 16.

1. $\int_0^4 (3x^2 - 5)\, dx, \quad n = 4$

2. $\int_{-1}^3 (3x - 2)(x + 1)\, dx, \quad n = 4$

3. $\int_1^3 \frac{dx}{x}, \quad n = 4$

4. $\int_{-1}^2 \frac{dx}{x + 2}, \quad n = 6$

5. $\int_1^5 x(x - 1)(x - 2)\, dx, \quad n = 4$

6. $\int_{-2}^4 x^2(x - 3)\, dx, \quad n = 6$

7. $\int_0^2 \frac{dx}{\sqrt{x + 1}}, \quad n = 4$

8. $\int_0^2 \frac{dx}{\sqrt{x + 1}}, \quad n = 8$

9. $\int_0^1 \frac{dx}{x^2 + 1}, \quad n = 4$

10. $\int_0^1 \frac{dx}{x^2 + 1}, \quad n = 8$

11. $\int_0^1 \cos x\, dx, \quad n = 4$

12. $\int_0^{\pi/4} \tan x\, dx, \quad n = 6$

13. $\int_1^{5/2} e^{-2x}\, dx, \quad n = 6$

14. $\int_0^3 xe^x\, dx, \quad n = 6$

15. $\int_{-1}^2 \sinh^2 x\, dx, \quad n = 6$

16. $\int_0^3 e^{2x} \sin x\, dx, \quad n = 6$

17. For the function F defined by

$$F(x) = \frac{1}{\sqrt{2\pi}} \int_0^x e^{-t^2/2}\, dt$$

find $F(1.2)$ approximately by an application of Simpson's rule with $n = 6$. Express your answer to three decimal places.

18. Find $\int_{0.5}^{1.5} \frac{\sin t}{t}\, dt$ by an application of Simpson's rule with $n = 10$. Express your answer to four decimal places.

19. Find $\int_0^1 \frac{1 + x^2}{1 + x^4}\, dx$ by an application of Simpson's rule with $n = 10$. Express your answer to four decimal places.

20. Find $\int_0^1 \sqrt{1 + x^4}\, dx$ by an application of Simpson's rule with $n = 10$. Express your answer to three decimal places.

21. Find an approximate value of the arc length of the graph of $y = \dfrac{1}{x}$ between the points $(1, 1)$ and $(4, \frac{1}{4})$ by using Simpson's rule with $n = 6$. Express your answer to three decimal places.

22. Evaluate $\displaystyle\int_0^4 \sqrt{16 - x^2}\, dx$ by an application of Simpson's rule with $n = 8$. Express your answer to three decimal places, and interpret the result as an area.

23. Evaluate $\displaystyle\int_0^a \sqrt{a^2 - x^2}\, dx$, $(a > 0)$, by an application of Simpson's rule with $n = 10$. Express your answer to three decimal places, and interpret the result as an area. (*Hint:* Let $x = au$ and then use Simpson's rule on the resulting integral with respect to u.)

24. Prove that Simpson's rule with $n = 2$, namely

$$\int_a^b f(x)\, dx \approx \frac{b-a}{6}\left[f(a) + 4f\left(\frac{a+b}{2}\right) + f(b) \right]$$

yields the exact result if $f(x)$ is a polynomial of degree 3. (*Hint:* For algebraic convenience write $f(x) = A + 2Bx + 3Cx^2 + 4Dx^3$ and verify the result.)

Review and Miscellaneous Exercises

In Exercises 1 through 64, find the indefinite integral, or evaluate the definite integral.

1. $\displaystyle\int (e^x + 3)^2\, dx$

2. $\displaystyle\int \frac{\sec^2 \theta}{3 + \tan \theta}\, d\theta$

3. $\displaystyle\int 5^{3x+4}\, dx$

4. $\displaystyle\int \frac{dt}{\sqrt{9t^2 + 4}}$

5. $\displaystyle\int \frac{z\, dz}{z^4 + 3}$

6. $\displaystyle\int \left(x + \frac{1}{x} \right)^3 dx$

7. $\displaystyle\int \frac{w\, dw}{\sqrt{w + 5}}$

8. $\displaystyle\int (z^3 + 9)^{1/3} z^5\, dz$

9. $\displaystyle\int \frac{(3t + 11)\, dt}{\sqrt{t^2 + t + 1}}$

10. $\displaystyle\int \frac{z^2\, dz}{(z + 1)^{2/3}}$

11. $\displaystyle\int \frac{x + 3}{x^2 + 8x + 20}\, dx$

12. $\displaystyle\int \frac{dx}{\sqrt{x^2 + 12x}}$

13. $\displaystyle\int \frac{e^{ax}\, dx}{\sqrt{e^{2ax} + \pi}}$

14. $\displaystyle\int \frac{dt}{3t^2 - 2t - 1}$

15. $\displaystyle\int \cos^6 2\theta \sin 2\theta\, d\theta$

16. $\displaystyle\int \sin^2 2z \cos z\, dz$

17. $\displaystyle\int \left(\sin \frac{t}{3} + \cos \frac{t}{3} \right)^2 dt$

18. $\displaystyle\int (\sec t - 2 \csc t)^2\, dt$

19. $\displaystyle\int \cos^4 3\theta\, d\theta$

20. $\displaystyle\int \sin 3\theta \cos 7\theta\, d\theta$

21. $\displaystyle\int \frac{\sinh t}{(3 + \cosh t)^7}\, dt$

22. $\displaystyle\int \frac{e^{8x}}{\sqrt{e^{8x} + 5}}\, dx$

23. $\displaystyle\int \sinh^5 z \cosh^3 z\, dz$

24. $\displaystyle\int \sqrt{\tan^5 t}\, \sec^4 t\, dt$

25. $\displaystyle\int \frac{\cot^3 3x}{\csc^6 3x}\, dx$

26. $\displaystyle\int \frac{11x - 29}{x^2 - 5x + 4}\, dx$

27. $\int \dfrac{x}{(x-4)^2}\,dx$

28. $\int \dfrac{x^3 + 7x^2 + 11x - 3}{x^2 + 6x + 5}\,dx$

29. $\displaystyle\int_0^{\pi/2} \dfrac{dx}{1 + \cos x}$

30. $\displaystyle\int_{-\pi/4}^{\pi/4} \sin^4 x\,dx$

31. $\displaystyle\int_1^2 x^3 \ln x\,dx$

32. $\displaystyle\int_0^1 te^{-t}\,dt$

33. $\int e^x \sinh x\,dx$

34. $\int \operatorname{csch} v \coth v\,dv$

35. $\int e^{ax} \cos^2 bx\,dx$, $(a, b$ are nonzero constants$)$

36. $\int \sinh^{-1} bx\,dx$, $(b$ is a nonzero constant$)$

37. $\displaystyle\int_{-\pi/3}^{\pi/3} \dfrac{dx}{1 - \sin x}$

38. $\int \dfrac{1 + \sin t}{1 + \cos t}\,dt$

39. $\int \dfrac{x^2 + 4}{x^3 - 4x}\,dx$

40. $\int \dfrac{3x + 4}{(x - 2)^3}\,dx$

41. $\int \dfrac{x^3 - x^2 + 3}{x(x + 2)(x^2 - 9)}\,dx$

42. $\int \dfrac{dx}{16 - x^4}$

43. $\int \dfrac{2t^4 + 3t^3 + 4t^2 + 2t + 2}{t^5 + 2t^3 + t}\,dt$

44. $\displaystyle\int_1^3 \dfrac{dt}{t^2\sqrt{1 + t^2}}$

45. $\displaystyle\int_0^{\pi} \dfrac{\sin t}{(2 + \cos t)^2}\,dt$

46. $\displaystyle\int_0^{\pi/2} \dfrac{\sin t}{2 + \sin t}\,dt$

47. $\displaystyle\int_0^{\pi/2} (\cos^6 x - \sin^6 x)\,dx$

48. $\displaystyle\int_0^{\pi/2} \sin^4 2x\,dx$

49. $\displaystyle\int_0^{\pi/2} |\cos^3 x - \sin^3 x|\,dx$

50. $\int \sec^5 \theta\,d\theta$

51. $\int t^3 e^{-t^2}\,dt$

52. $\int \sinh^{-1} bt\,dt$, $(b$ is a nonzero constant$)$

53. $\int \dfrac{dx}{1 + \cos ax}$, $(a \neq 0)$

54. $\displaystyle\int_0^{\pi/2} \sin^4 x \cos^5 x\,dx$

55. $\int \dfrac{x\,dx}{2 + \sqrt{x + 1}}$

56. $\int \dfrac{\sec (3e^x)}{e^{-x}}\,dx$

57. $\int \dfrac{x^{3/2}}{1 + x^{1/2}}\,dx$

58. $\int x(a - x)^{2/3}\,dx$, $(a = \text{constant})$

59. $\displaystyle\int_0^3 \dfrac{5x^2 + 2x + 1}{x^3 + x^2 + x + 1}\,dx$

60. $\int \dfrac{dx}{\sqrt[3]{2 + \sqrt{x}}}$

61. $\int \dfrac{x^2}{c^4 - x^4}\,dx$, $(c$ is a nonzero constant$)$

62. $\int \dfrac{dx}{x(a + bx)^2}$, $(a$ and b are nonzero constants$)$

63. $\displaystyle\int_0^{\pi/2} \sin^{2n}x\,dx$, where n is a positive integer. (*Hint:* Refer to the formula given in Exercise 35 in Section 10.7)

64. $\displaystyle\int_0^{\pi/2} \cos^{2n+1} x\,dx$, where n is a positive integer. (*Hint:* Refer to the formula given in Exercise 37 in Section 10.7)

65. (a) Establish the following reduction formula by the method of integration by parts:

$$\int (a^2 - x^2)^n\,dx = \dfrac{x(a^2 - x^2)^n}{2n + 1} + \dfrac{2a^2 n}{2n + 1} \int (a^2 - x^2)^{n-1}\,dx \qquad (n \neq -\tfrac{1}{2})$$

(b) Use (a) to find $\int (a^2 - x^2)^{3/2}\,dx$.

66. (a) If m and n are positive integers, show that

$$\int_0^1 x^m(1-x)^n \, dx = \int_0^1 x^n(1-x)^n \, dx$$

(b) Evaluate $\displaystyle\int_0^1 x^3(1-x)^{10} \, dx$.

67. Evaluate $\displaystyle\int \frac{dx}{x^4+1}$. [*Hint:* $x^4 + 1 = (x^2+1)^2 - 2x^2 = (x^2 + \sqrt{2}x + 1)(x^2 - \sqrt{2}x + 1)$.]

68. Evaluate $\displaystyle\int \frac{dy}{3y^4+48}$ (see *Hint*, Exercise 67).

69. (a) Derive the reduction formula

$$\int x^m(\ln x)^n \, dx = \frac{x^{m+1}(\ln x)^n}{m+1} - \frac{n}{m+1}\int x^m(\ln x)^{n-1} \, dx$$

where m and n are arbitrary positive integers.

(b) Evaluate $\displaystyle\int x^3(\ln x)^2 \, dx$.

70. Evaluate $\displaystyle\int \frac{dx}{x(x+1)(x+2)\cdots(x+m)}$.   (*Hint:* Use partial fractions by writing

$$\frac{1}{x(x+1)\cdots(x+m)} = \sum_{r=0}^{m} \frac{A_r}{x+r}$$ and then clearing fractions; determine A_r.)

71. Given that $f(x)$ has a continuous second derivative on $[0, a]$, evaluate

$$\int_0^a (x+2)f''(x) \, dx$$

knowing that $f(0) = 1$, $f'(0) = -2$, $f(a) = 4$, and $f'(a) = 3$.

72. Prove that for any constant k,

$$\int_a^b [f(x) + f'(x)(x-k)] \, dx = (b-k)f(b) - (a-k)f(a)$$

if f' is continuous on $[a, b]$.

73. Prove that

$$\int_a^b [f(x)g''(x) - g(x)f''(x)] \, dx = (f(x)g'(x) - g(x)f'(x))\,|_a^b$$

where f and g have continuous derivatives of the second order on $[a, b]$.

74. It is reasonable to conjecture that constants A and B can be determined so that

$$\int e^{ax}\cos bx \, dx = e^{ax}[A\sin bx + B\cos bx] + C$$

Why? Find A and B in terms of the constants a and b. (*Hint:* Differentiate both sides with respect to x.)

75. Evaluate $\displaystyle\int xe^x \cos x \, dx$ by the "undetermined coefficients" method of Exercise 74.

76. Evaluate $\displaystyle\int_0^2 \sqrt{9+8x} \, dx$ approximately by the trapezoidal rule with $n = 4$. Express your answer to two decimal places and compare with the exact result.

77. Evaluate $\displaystyle\int_0^3 \sqrt{9+x^3} \, dx$ approximately by **(a)** the trapezoidal rule, and **(b)** Simpson's rule. Take six subintervals and express the results to three decimal places.

78. Evaluate $\int_{-1}^{7} \sqrt{5 + x^2}\, dx$ approximately by Simpson's rule with $n = 8$. Express your answer to two decimal places and compare with the exact result.

79. Find the approximate length of one arch of the curve $y = |\sin x|$ by application of Simpson's rule with $n = 6$. Express your answer to two decimal places.

80. Sketch the graph of $\dfrac{x^2}{4} + \dfrac{y^2}{1} = 1$ in the first quadrant connecting the two points $\left(1, \dfrac{\sqrt{3}}{2}\right)$ and $(0, 1)$ on the curve. Find an approximate length of this curve using Simpson's rule with $n = 4$. Express your answer to three decimal places.

81. The polynomials $B_n(x)$ (known as Bernoulli polynomials) are defined by the relations:

$$B_1(x) = x - \tfrac{1}{2} \qquad B_n'(x) = nB_{n-1}(x) \qquad \int_0^1 B_n(x)\, dx = 0$$

Find $B_2(x)$ and $B_3(x)$.

82. (a) Find a polynomial $P(x)$ such that

$$P'(x) + P(x) = x^2 - x - 2$$

(b) Prove that the polynomial found in (a) is the only polynomial satisfying the given differential equation. [*Hint:* Assume that $P(x)$ and $Q(x)$ are two polynomials satisfying the given equation and show that $P(x) \equiv Q(x)$.]

83. Evaluate: $\displaystyle\lim_{h \to 0} \frac{1}{h} \int_3^{3+h} \frac{e^{-t}}{t}\, dt$.

84. Obtain a recurrence formula for $\int x^n\, e^{ax} \cos bx\, dx$.

85. A function $F(x)$ is defined by the integral

$$F(x) = \int_0^x \frac{dt}{t^3 + 1}$$

Find $F(1)$.

11

Vectors in the Plane and Parametric Representation

11.1   VECTORS—AN INTRODUCTION

Many concepts in physics and engineering include both magnitude and direction. For example, we have position, velocity, acceleration, force, electric fields, and so on. It has been found that vector analysis is very useful for the development of these concepts in a straightforward and yet elegant manner. In fact, an introduction to vectors and their combination is routine subject matter in physics courses such as mechanics, electricity and magnetism, and others. The applied mathematician finds vector (and tensor) analysis an indispensible tool in, for example, the development of advanced structural analysis—plates, shells, and elasticity. Subjects such as differential geometry and fluid dynamics have been simplified considerably by the application of vector (and tensor) analysis.

Vector analysis was originally suggested by the work of a brilliant Irish mathematician, William Rowan Hamilton (1805–1865), and independently by the great German mathematician, Hermann Günther Grassman (1809–1877). However, neither the system of Hamilton nor that of Grassman met the requirements of the physicists and applied mathematicians of that day, being both too general and too complex. The development of vector analysis in its present form is primarily due to an American mathematician, Josiah Willard Gibbs (1839–1903), a distinguished professor at Yale, and an ingenious English scientist, Oliver Heaviside (1850–1925).

11.2   BASIC DEFINITIONS OF A VECTOR

There are two distinct ways in which to define vectors: (i) geometrically and (ii) analytically. In accordance with (i), a vector is a directed line segment. Thus a vector has an initial point A and a final or terminal point B. Such a vector is denoted by $\overrightarrow{AB}$. Two vectors $\overrightarrow{AB}$ and $\overrightarrow{CD}$ are said to be **equal** if and only if they have the *same length* and *direction*, and then we write $\overrightarrow{AB} = \overrightarrow{CD}$. Thus $\overrightarrow{AB} = \overrightarrow{CD}$ if the quadrilateral $ABCD$ is a parallelogram (assuming that points A, B, C, and D are not collinear—Figure 11.2.1). Therefore a vector is considered unchanged if it is moved parallel to itself. Such vectors are sometimes called *free vectors,* whereas the term *bound vectors* is used for those vectors with a fixed initial point. In this text we shall omit adjectives such as free and bound and, if nothing is said to the contrary, we shall always assume the vectors to be free.

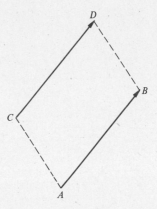

Figure 11.2.1

A vector is also denoted by a single letter set in bold face type such as **A**, **B**, **a**, **b**, and so on.

In this chapter it will be assumed that all vectors lie in a rectangular coordinate plane. If **A** is a vector, we can draw it with initial point at the origin, $\mathbf{A} = \overrightarrow{OP}$. Then the vector is uniquely determined by its endpoint $P(x, y)$. Thus, the ordered pair of real numbers $\langle x, y \rangle$ can be thought of as another way of representing **A**. This leads to the following definition, which is part of the analytic approach (ii) referred to above.

Definition   A *two dimensional vector,* or a *vector in the plane* is an ordered pair of real numbers $\langle x, y \rangle$. The numbers x and y are called the *components* of the vector.

For example, $\langle 3, 2 \rangle$, $\langle -1, -4 \rangle$, $\langle \pi, \ln 3 \rangle$ and $\langle 0, -1 \rangle$ are vectors. We use the notation $\langle x, y \rangle$ rather than (x, y) to avoid confusing the notation for a vector with the notation for a point.

The vector $\langle 0, 0 \rangle$, both of whose components are zero, is called the *zero vector* and is denoted by the special symbol **0**, that is

$$\mathbf{0} = \langle 0, 0 \rangle \qquad \text{is the zero vector}$$

Note that any point is a representation of the zero vector. Also, the zero vector has *no* direction.

Geometrically, we have said that the vector $\mathbf{A} = \langle a_1, a_2 \rangle$ corresponds to the vector from the origin to the point (a_1, a_2). The same vector can be represented with any starting point. Thus, if $\mathbf{A} = \overrightarrow{PQ}$ and $P = (x, y)$, then $Q = (x + a_1, y + a_2)$ (see Figure 11.2.2a).

Conversely, if $\mathbf{A} = \overrightarrow{PQ}$ where $P = (x_1, y_1)$ and $Q = (x_2, y_2)$, then $\mathbf{A}$ can be represented as a vector from the origin to the point $(x_2 - x_1, y_2 - y_1)$. Thus, $\mathbf{A} = \langle x_2 - x_1, y_2 - y_1 \rangle$ (see Figure 11.2.2b).

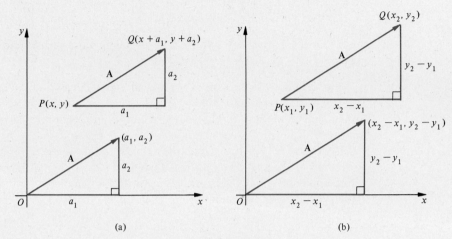

(a)   (b)

Figure 11.2.2

Definition   Two vectors $\mathbf{a} = \langle a_1, a_2 \rangle$ and $\mathbf{b} = \langle b_1, b_2 \rangle$ are said to be **equal** if and only if $a_1 = b_1$ and $a_2 = b_2$. Thus

$$\mathbf{a} = \mathbf{b} \quad \leftrightarrow \quad a_1 = b_1 \quad \text{and} \quad a_2 = b_2 \qquad (1)$$

(where the notation $\leftrightarrow$ means "if and only if").

Definition   The **negative** of a vector $\mathbf{a} = \langle a_1, a_2 \rangle$, written $-\mathbf{a}$, is the vector $\langle -a_1, -a_2 \rangle$. Thus, for example, the negative of the vector $\mathbf{c} = \langle 2, -9 \rangle$ is $-\mathbf{c} = \langle -2, 9 \rangle$. Note that the zero vector is the negative of itself.

Definition   The **sum** of two vectors $\mathbf{a} = \langle a_1, a_2 \rangle$ and $\mathbf{b} = \langle b_1, b_2 \rangle$ is defined to be the vector

$$\mathbf{a} + \mathbf{b} = \langle a_1 + b_1, a_2 + b_2 \rangle \qquad (2)$$

For example, $\langle 3, 4 \rangle + \langle -5, 2 \rangle = \langle 3 + (-5), 4 + 2 \rangle = \langle -2, 6 \rangle$.
From the above definitions the following rules are readily derived:

Theorem   Let $\mathbf{a}$, $\mathbf{b}$, and $\mathbf{c}$ be any three vectors, then

$$
\begin{array}{lll}
\mathbf{a} + \mathbf{b} = \mathbf{b} + \mathbf{a} & \text{(commutative law of addition)} & (3) \\
\mathbf{a} + (\mathbf{b} + \mathbf{c}) = (\mathbf{a} + \mathbf{b}) + \mathbf{c} & \text{(associative law of addition)} & (4) \\
\mathbf{a} + (-\mathbf{a}) = \mathbf{0} & & (5) \\
\mathbf{a} + \mathbf{0} = \mathbf{0} + \mathbf{a} = \mathbf{a} & & (6)
\end{array}
$$

Proof For example, we prove the associative law of addition—Equation (4). If $\mathbf{a} = \langle a_1, a_2 \rangle$, $\mathbf{b} = \langle b_1, b_2 \rangle$ and $\mathbf{c} = \langle c_1, c_2 \rangle$ then

$$\mathbf{a} + (\mathbf{b} + \mathbf{c}) = \langle a_1, a_2 \rangle + \langle b_1 + c_1, b_2 + c_2 \rangle$$
$$= \langle a_1 + b_1 + c_1, a_2 + b_2 + c_2 \rangle$$

while

$$(\mathbf{a} + \mathbf{b}) + \mathbf{c} = \langle a_1 + b_1, a_2 + b_2 \rangle + \langle c_1, c_2 \rangle$$
$$= \langle a_1 + b_1 + c_1, a_2 + b_2 + c_2 \rangle$$

and the associative law follows. The proofs of rules (3), (5) and (6) are left as exercises. ∎

The definition of the sum of two planar vectors (equation (2)) has a significant geometrical interpretation. Figure 11.2.3 shows vectors $\overrightarrow{OP} = \mathbf{a}$ and $\overrightarrow{PQ} = \mathbf{b}$ and their sum $\overrightarrow{OQ} = \mathbf{a} + \mathbf{b}$. The vectors $\mathbf{a}$ and $\mathbf{b}$ can be interpreted as vectors along adjacent sides of a parallelogram, while $\mathbf{a} + \mathbf{b}$ is a vector along the diagonal (Figure 11.2.4). For this reason, this is referred to as the *parallelogram law* of addition of vectors. The fact that displacements, velocities, accelerations, and forces may be combined in accordance with the parallelogram law is one of the basic axioms of mechanics. Note that Figure 11.2.4 is a verification of the commutative law of addition.

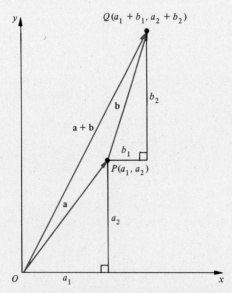

Figure 11.2.4

Figure 11.2.3

In a similar fashion, Figure 11.2.5 shows the sum of three vectors $\mathbf{a} + \mathbf{b} + \mathbf{c}$. We find $\mathbf{a} + \mathbf{b}$ by sliding the vector $\mathbf{b}$ parallel to itself until the foot of $\mathbf{b}$ coincides with the head of $\mathbf{a}$. Then $\mathbf{c}$ is moved parallel to itself until its foot coincides with the head of $\mathbf{a} + \mathbf{b}$.

Next, we define the *difference* $\mathbf{a} - \mathbf{b}$ to be the sum of the vectors $\mathbf{a}$ and $-\mathbf{b}$. Thus

$$\mathbf{a} - \mathbf{b} = \langle a_1 - b_1, a_2 - b_2 \rangle$$

Note that if $\mathbf{c}$ is the vector $\mathbf{b} - \mathbf{a}$, then $\mathbf{a} - \mathbf{b} = -\mathbf{c}$ and that

$$(\mathbf{a} - \mathbf{b}) + \mathbf{b} = (\mathbf{a} + -\mathbf{b}) + \mathbf{b} = \mathbf{a} + (-\mathbf{b} + \mathbf{b})$$
$$= \mathbf{a} + 0 = \mathbf{a}$$

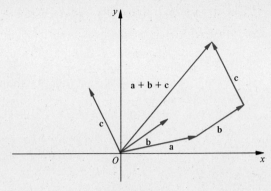

Figure 11.2.5

Therefore if **a** and **b** are vectors with initial points at the origin, then **a** − **b** is the vector from the terminal of **b** to the terminal of **a** (Figure 11.2.6). This is easily verified.

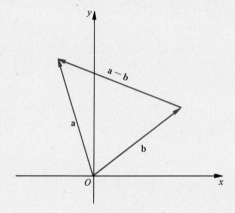

Figure 11.2.6

The *length* or *magnitude* of a vector **a** is denoted by |**a**| and is defined by the equation

$$|\mathbf{a}| = \sqrt{a_1{}^2 + a_2{}^2} \qquad \text{if } \mathbf{a} = \langle a_1, a_2 \rangle \tag{7}$$

(|**a**| is thus a nonnegative number). Thus, for example, the length of the vector $\langle -2, 5 \rangle$ is $\sqrt{(-2)^2 + 5^2} = \sqrt{29}$. We note that the length of a vector is the length of the line segment representing it. It is positive for all nonzero vectors and is zero for the zero vector.

Another important operation with vectors is that of *scalar* multiplication.

Definition For each vector $\mathbf{a} = \langle a_1, a_2 \rangle$ and each real number k (called a scalar) the *scalar multiple* of **a** by k is the vector $k\mathbf{a}$ defined as follows:

$$k\mathbf{a} = k\langle a_1, a_2 \rangle = \langle ka_1, ka_2 \rangle \tag{8}$$

Thus, for example,

$$4\langle -7, -3 \rangle = \langle -28, -12 \rangle$$

The geometric interpretation of $k\mathbf{a}$ is as follows:

(i)   If $k > 0$, $k\mathbf{a}$ has the same direction as $\mathbf{a}$, but its length is k times the length of $\mathbf{a}$.

(ii) If $k = 0$, then $k\mathbf{a} = \mathbf{0}$.

(iii) If $k < 0$, $k\mathbf{a}$ has the direction opposite to that of $\mathbf{a}$ and its length is $|k|$ times the length of $\mathbf{a}$.

The following rules hold for scalar multiplication (k and m scalars)

$$
\begin{aligned}
k(\mathbf{a} + \mathbf{b}) &= k\mathbf{a} + k\mathbf{b} &\text{(i)} \\
(k + m)\mathbf{a} &= k\mathbf{a} + m\mathbf{a} &\text{(ii)} \\
k(m\mathbf{a}) &= (km)\mathbf{a} &\text{(iii)} \\
(-k)\mathbf{a} &= -(k\mathbf{a}) &\text{(iv)} \\
1\mathbf{a} &= \mathbf{a} &\text{(v)} \\
0\mathbf{a} &= \mathbf{0} &\text{(vi)} \\
k\mathbf{0} &= \mathbf{0} &\text{(vii)} \\
|k\mathbf{a}| &= |k||\mathbf{a}| &\text{(viii)}
\end{aligned}
\qquad (9)
$$

Proof of (i)
$$k(\mathbf{a} + \mathbf{b}) = k\mathbf{a} + k\mathbf{b}$$

With $\mathbf{a} = \langle a_1, a_2 \rangle$, $\mathbf{b} = \langle b_1, b_2 \rangle$, we have

$$
\begin{aligned}
k(\mathbf{a} + \mathbf{b}) &= k(\langle a_1, a_2 \rangle + \langle b_1, b_2 \rangle) = k(\langle a_1 + b_1, a_2 + b_2 \rangle) \\
&= \langle k(a_1 + b_1), k(a_2 + b_2) \rangle = \langle ka_1 + kb_1, ka_2 + kb_2 \rangle \\
&= \langle ka_1, ka_2 \rangle + \langle kb_1, kb_2 \rangle \\
&= k\langle a_1, a_2 \rangle + k\langle b_1, b_2 \rangle = k\mathbf{a} + k\mathbf{b}
\end{aligned}
$$

The proofs of the other rules, (ii) through (viii) are left as exercises. ∎

EXAMPLE 1   Solve for $\mathbf{a}$: $\langle 2, 4 \rangle + \mathbf{a} = \langle -3, 0 \rangle$

Solution   Add $\langle -2, -4 \rangle$ to both sides. Thus

$$\langle -2, -4 \rangle + (\langle 2, 4 \rangle + \mathbf{a}) = \langle -2, -4 \rangle + \langle -3, 0 \rangle$$

and by the associative law of addition

$$(\langle -2, -4 \rangle + \langle 2, 4 \rangle) + \mathbf{a} = \langle -2 - 3, -4 + 0 \rangle$$

and

$$\mathbf{0} + \mathbf{a} = \langle -5, -4 \rangle$$

thus

$$\mathbf{a} = \langle -5, -4 \rangle \qquad \bullet$$

EXAMPLE 2   Find the magnitude of the vector $\mathbf{a}$ from the point $(-1, 2)$ to the point $(4, 9)$

Solution $$\mathbf{a} = \langle 4 - (-1), 9 - 2 \rangle = \langle 5, 7 \rangle$$

and the magnitude of $\mathbf{a}$

$$|\mathbf{a}| = \sqrt{5^2 + 7^2} = \sqrt{25 + 49} = \sqrt{74} \qquad \bullet$$

EXAMPLE 3   Find the slope of the vector $\mathbf{b}$ joining the points $(-3, -5)$ and $(2, 6)$

Solution The slope of **b** is by definition the slope of the line segment connecting the given points. Thus the slope of **b** is given by

$$\frac{6 - (-5)}{2 - (-3)} = \frac{11}{5}$$

●

EXAMPLE 4 Find scalars k and m, such that

$$k\langle 3, 2 \rangle + m\langle 1, 7 \rangle = \langle -5, 22 \rangle$$

Solution By Definition (8)—multiplication of a vector by a scalar, we have

$$\langle 3k + m, 2k + 7m \rangle = \langle -5, 22 \rangle$$

from which we obtain the system of equations

$$\begin{cases} 3k + m = -5 \\ 2k + 7m = 22 \end{cases}$$

The simultaneous solution is $k = -3$ and $m = 4$.

●

Exercises 11.2

In Exercises 1 through 8, express each of the following as a single vector.

1. $\langle 4, 3 \rangle + \langle 2, 5 \rangle$
2. $\langle 0, 0 \rangle + \langle 4, 7 \rangle$
3. $\langle 2, 5 \rangle - \langle -3, -7 \rangle$
4. $(-2)\langle -1, 10 \rangle$
5. $4(\langle 2, -3 \rangle + \langle 5, -1 \rangle)$
6. $(3 - 7)\langle 4, 0 \rangle$
7. $\langle -2, -1 \rangle - \langle 5, 3 \rangle + \langle 2, 6 \rangle$
8. $9(\langle -1, 2 \rangle + \langle -3, 6 \rangle)$

In Exercises 9 through 12 determine the vector **a**.

9. $\langle 5, 3 \rangle + \mathbf{a} = \langle -2, 0 \rangle$
10. $6\mathbf{a} - 12\mathbf{a} = \langle -3, -1 \rangle$
11. $\frac{2}{5}\mathbf{a} = \langle 1, -2 \rangle$
12. $-4\mathbf{a} - \langle 2, 7 \rangle = -2\mathbf{a} + \langle 6, 5 \rangle$

In Exercises 13 through 15, solve for **a** and **b**.

13. $2\mathbf{a} + \mathbf{b} = \langle 8, 4 \rangle$
    $\mathbf{a} - \mathbf{b} = \langle -2, 5 \rangle$
14. $\mathbf{a} - 3\mathbf{b} = \mathbf{b}$
    $-\mathbf{a} + 2\mathbf{b} = \langle 6, -8 \rangle$
15. $3\mathbf{a} - 4\mathbf{b} = -2\mathbf{a} + 3\mathbf{b} = \langle 1, 2 \rangle$

In Exercises 16 and 17, find scalars k and m such that:

16. $k\langle 3, 1 \rangle + m\langle 1, -1 \rangle = \langle 1, 7 \rangle$

17. $\frac{k}{3}\langle -2, 5 \rangle + m\langle 1, 3 \rangle = \langle -2, -17 \rangle$

18. Represent graphically
    (a) A force of 75 lb in a direction of 30° north of east.
    (b) A force of 100 lb in a direction of 40° east of north.
    (c) The negative of a force of 50 lb in a direction of 45° east of north.
19. A truck travels 10 mi due south and then 8 mi 60° east of north. Represent these displacements graphically and determine the resultant displacement (a) graphically and (b) analytically.

In Exercises 20 through 25, let $\mathbf{a} = \langle 2, 3 \rangle$, $\mathbf{b} = \langle -1, 4 \rangle$ *and* $\mathbf{c} = \langle 5, 6 \rangle$. *Find*

20. $3\mathbf{a} - 2\mathbf{b}$ **21.** $5\mathbf{a} + 3\mathbf{b} + 2\mathbf{c}$

22. $-(\mathbf{a} - 4\mathbf{c})$ **23.** $|\mathbf{b}|$

24. $|2\mathbf{b} + \mathbf{c}|$ **25.** $|3\mathbf{a} + \mathbf{b} - \mathbf{c}|$

26. Simplify $3(\mathbf{a} - \mathbf{b}) + 3(\mathbf{b} + \mathbf{a})$ **27.** Show that $|\mathbf{b} - \mathbf{a}| = |\mathbf{a} - \mathbf{b}|$

28. In any triangle it is known that the length of one side is less than or equal to the sum of the lengths of the other two sides. **(a)** Why is this true? **(b)** Use this to prove that for any vectors $\mathbf{a}$ and $\mathbf{b}$,

$$|\mathbf{a} + \mathbf{b}| \leq |\mathbf{a}| + |\mathbf{b}|$$

29. By the same technique utilized in the previous exercise show that

$$|\mathbf{a} - \mathbf{b}| \geq |\mathbf{a}| - |\mathbf{b}|$$

30. Prove (ii) of (9) in the text, namely

$$(k + m)\mathbf{a} = k\mathbf{a} + m\mathbf{a}$$

where k and m are scalars.

31. Prove (iii) of (9) in the text, namely

$$k(m\mathbf{a}) = (km)\mathbf{a}$$

32. Prove (viii) of (9) in the text, namely

$$|k\mathbf{a}| = |k||\mathbf{a}|$$

11.3   DOT PRODUCTS AND BASIS VECTORS

In two dimensions there is only one product of two vectors called the *dot product.*

Definition   If $\mathbf{a} = \langle a_1, a_2 \rangle$ and $\mathbf{b} = \langle b_1, b_2 \rangle$ are any two vectors, we define the **dot product** written $\mathbf{a} \cdot \mathbf{b}$ to be the number

$$\mathbf{a} \cdot \mathbf{b} = \langle a_1, a_2 \rangle \cdot \langle b_1, b_2 \rangle = a_1 b_1 + a_2 b_2 \qquad (1)$$

It is important to emphasize that the dot product is a number (or scalar) rather than a vector. For example, from (1)

$$\langle 3, -4 \rangle \cdot \langle -5, 2 \rangle = (3)(-5) + (-4)(2) = -23$$
$$\langle 4, 3 \rangle \cdot \langle 6, -8 \rangle = (4)(6) + (3)(-8) = 0$$

The dot product is also often called the *scalar product* and less frequently the *inner product.*

The following rules may be readily established from (1).

Theorem 1   If $\mathbf{a}$, $\mathbf{b}$, and $\mathbf{c}$ are vectors and k is a scalar then

$$\mathbf{a} \cdot \mathbf{b} = \mathbf{b} \cdot \mathbf{a} \qquad \text{(commutative law)} \qquad (2)$$

$$\left.\begin{array}{l} \mathbf{a} \cdot (\mathbf{b} + \mathbf{c}) = \mathbf{a} \cdot \mathbf{b} + \mathbf{a} \cdot \mathbf{c} \\ (\mathbf{a} + \mathbf{b}) \cdot \mathbf{c} = \mathbf{a} \cdot \mathbf{c} + \mathbf{b} \cdot \mathbf{c} \end{array}\right\} \quad \text{(distributive laws)}$$

(3a)

(3b)

$$k(\mathbf{a} \cdot \mathbf{b}) = (k\mathbf{a}) \cdot \mathbf{b} = \mathbf{a} \cdot (k\mathbf{b}) \tag{4}$$

$$\mathbf{0} \cdot \mathbf{a} = 0 \tag{5}$$

$$\mathbf{a} \cdot \mathbf{a} = |\mathbf{a}|^2 = a^2 \qquad \text{where } a \text{ denotes } |\mathbf{a}| \tag{6}$$

We illustrate the proofs of (2) and (3a).

Proof To prove the commutative law, (2):

$$\begin{aligned} \mathbf{a} \cdot \mathbf{b} &= \langle a_1, a_2 \rangle \cdot \langle b_1, b_2 \rangle = a_1 b_1 + a_2 b_2 \\ &= b_1 a_1 + b_2 a_2 = \langle b_1, b_2 \rangle \cdot \langle a_1, a_2 \rangle = \mathbf{b} \cdot \mathbf{a} \end{aligned}$$ ∎

Proof To prove the distributive law (3a):

$$\begin{aligned} \mathbf{a} \cdot (\mathbf{b} + \mathbf{c}) &= \langle a_1, a_2 \rangle \cdot (\langle b_1, b_2 \rangle + \langle c_1, c_2 \rangle) \\ &= \langle a_1, a_2 \rangle \cdot \langle b_1 + c_1, b_2 + c_2 \rangle \\ &= a_1(b_1 + c_1) + a_2(b_2 + c_2) \\ &= a_1 b_1 + a_1 c_1 + a_2 b_2 + a_2 c_2 \\ &= \langle a_1, a_2 \rangle \cdot \langle b_1, b_2 \rangle + \langle a_1, a_2 \rangle \cdot \langle c_1, c_2 \rangle \\ &= \mathbf{a} \cdot \mathbf{b} + \mathbf{a} \cdot \mathbf{c} \end{aligned}$$ ∎

The remaining proofs are left as exercises.

Definition Let $\mathbf{a}$ and $\mathbf{b}$ be two nonzero vectors, represented by directed line segments $\overrightarrow{OA}$ and $\overrightarrow{OB}$, respectively. By the angle between $\mathbf{a}$ and $\mathbf{b}$ is meant the smallest angle θ formed by vectors $\mathbf{a}$ and $\mathbf{b}$.

Therefore in all cases $0 \leq \theta \leq \pi$ (see Figure 11.3.1).

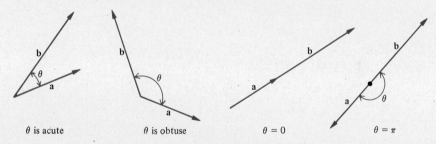

θ is acute $\qquad$ θ is obtuse $\qquad$ $\theta = 0$ $\qquad$ $\theta = \pi$

Figure 11.3.1

Next, we establish the most important result concerning the dot product.

Theorem 2 If $\mathbf{a} = \langle a_1, a_2 \rangle$ and $\mathbf{b} = \langle b_1, b_2 \rangle$ are nonzero vectors, then

$$\mathbf{a} \cdot \mathbf{b} = |\mathbf{a}||\mathbf{b}| \cos \theta \tag{7}$$

Proof We refer to Figure 11.3.2. From the law of cosines,

$$|\mathbf{b} - \mathbf{a}|^2 = |\mathbf{a}|^2 + |\mathbf{b}|^2 - 2|\mathbf{a}|\,|\mathbf{b}| \cos \theta \tag{i}$$

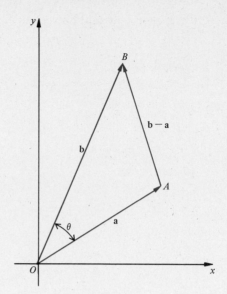

Figure 11.3.2

But,

$$|\mathbf{b} - \mathbf{a}|^2 = (\mathbf{b} - \mathbf{a}) \cdot (\mathbf{b} - \mathbf{a})$$
$$= \mathbf{b} \cdot (\mathbf{b} - \mathbf{a}) - \mathbf{a} \cdot (\mathbf{b} - \mathbf{a})$$
$$= \mathbf{b} \cdot \mathbf{b} - \mathbf{b} \cdot \mathbf{a} - \mathbf{a} \cdot \mathbf{b} + \mathbf{a} \cdot \mathbf{a}$$
$$= |\mathbf{b}|^2 - 2(\mathbf{a} \cdot \mathbf{b}) + |\mathbf{a}|^2 \qquad \text{(ii)}$$

From (i) and (ii), we obtain

$$|\mathbf{a}|^2 + |\mathbf{b}|^2 - 2|\mathbf{a}| \, |\mathbf{b}| \cos \theta = |\mathbf{b}|^2 + |\mathbf{a}|^2 - 2(\mathbf{a} \cdot \mathbf{b})$$

and therefore

$$\mathbf{a} \cdot \mathbf{b} = |\mathbf{a}| \, |\mathbf{b}| \cos \theta \qquad \blacksquare \qquad (7)$$

The angle θ between two nonzero vectors can then be determined from:

$$\cos \theta = \frac{\mathbf{a} \cdot \mathbf{b}}{|\mathbf{a}| \, |\mathbf{b}|} \qquad 0 \le \theta \le \pi \qquad (8)$$

Theorem 3   Two nonzero vectors $\mathbf{a}$ and $\mathbf{b}$ are perpendicular (orthogonal) if and only if $\mathbf{a} \cdot \mathbf{b} = 0$.

Proof   From (8), $\mathbf{a} \cdot \mathbf{b} = 0$ if and only if $\cos \theta = 0$. Since $0 \le \theta \le \pi$, $\cos \theta = 0$ if and only if $\theta = \dfrac{\pi}{2}$. $\qquad \blacksquare$

From (8), we can also observe the following:

(i) $\mathbf{a} \cdot \mathbf{b} > 0$ if and only if $0 \le \theta < \dfrac{\pi}{2}$.

(ii) $\mathbf{a} \cdot \mathbf{b} < 0$ if and only if $\dfrac{\pi}{2} < \theta \le \pi$.

EXAMPLE 1   Find the angle θ between the vectors $\langle 2, 5 \rangle$ and $\langle -1, 3 \rangle$.

Solution   From (8),

$$\cos \theta = \frac{2(-1) + 5(3)}{\sqrt{2^2 + 5^2}\sqrt{(-1)^2 + 3^2}} = \frac{13}{\sqrt{29}\sqrt{10}}$$

Using six place approximations to the square roots, we obtain

$$\cos \theta \doteq \frac{13}{(5.385164)(3.162278)} \doteq \frac{13}{17.029386}$$

$$\doteq 0.76339$$

and from the tables, $\theta \doteq 40° 14'$. The vectors $\langle 2, 5 \rangle$, $\langle -1, 3 \rangle$ and the angle θ are shown in Figure 11.3.3.   ●

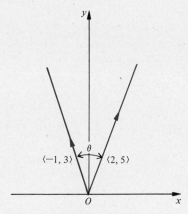

Figure 11.3.3

EXAMPLE 2   Find a vector of length 5 that is perpendicular to the given vector, $\langle 6, 8 \rangle$.

Solution   A sketch shows that there should be two answers to this problem (Figure 11.3.4).
Let $\langle a_1, a_2 \rangle$ be the required vector. The condition of perpendicularity requires that

$$6a_1 + 8a_2 = 0 \tag{i}$$

and the fact that the length of the required vector is 5 implies that

$$\sqrt{a_1^2 + a_2^2} = 5 \tag{ii}$$

From (i) we obtain $a_2 = -\frac{3}{4}a_1$ and substitution into (ii) yields

$$\sqrt{a_1^2 + \frac{9}{16}a_1^2} = 5 \qquad \text{or} \qquad \pm \frac{5}{4}a_1 = 5$$

and hence $a_1 = \mp 4$, and it follows that

$$a_2 = \pm 3$$

Hence, we have two answers (as shown in Figure 11.3.4): $\langle -4, 3 \rangle$ and $\langle 4, -3 \rangle$.   ●

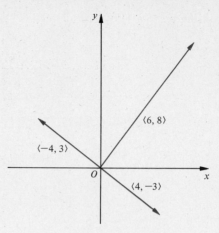

Figure 11.3.4

We recall that if $\mathbf{a} = \langle a_1, a_2 \rangle$, then the length of the vector $\mathbf{a}$ is defined to be $\sqrt{a_1{}^2 + a_2{}^2}$. In particular a vector of length 1 is said to be a **unit vector.** For example, $\left\langle \dfrac{1}{2}, \dfrac{\sqrt{3}}{2} \right\rangle$ is a unit vector because

$$\sqrt{\left(\frac{1}{2}\right)^2 + \left(\frac{\sqrt{3}}{2}\right)^2} = \sqrt{\frac{1}{4} + \frac{3}{4}} = \sqrt{1} = 1$$

Theorem 4   If $\mathbf{a}$ is an arbitrary nonzero vector, then $\dfrac{\mathbf{a}}{|\mathbf{a}|}$ is the **unit vector in the same direction as** $\mathbf{a}$.

For example, if $\mathbf{a} = \langle -6, 8 \rangle$, then $|\mathbf{a}| = \sqrt{(-6)^2 + 8^2} = 10$ and $\langle -\frac{6}{10}, \frac{8}{10} \rangle$ is the unit vector in the same direction as $\mathbf{a}$. The proof of Theorem 4 is left as an exercise.

Of specific importance are two unit vectors along the coordinate axes. They are the **basis vectors** $\mathbf{i}$ and $\mathbf{j}$ defined by $\mathbf{i} = \langle 1, 0 \rangle$ and $\mathbf{j} = \langle 0, 1 \rangle$. Thus a representation of $\mathbf{i}$ is a unit vector from the origin to the point $(1, 0)$. Similarly, a representation of $\mathbf{j}$ is a unit vector from the origin to the point $(0, 1)$—Figure 11.3.5.

Note that the fact that $\mathbf{i}$ and $\mathbf{j}$ are perpendicular to each other follows from $\mathbf{i} \cdot \mathbf{j} = 0$, and the relations $\mathbf{i} \cdot \mathbf{i} = 1$ and $\mathbf{j} \cdot \mathbf{j} = 1$ verify that they are unit vectors.

The vectors $\mathbf{i}$ and $\mathbf{j}$ are called basis vectors because any vector $\mathbf{a}$ in the plane can be expressed as a linear combination of $\mathbf{i}$ and $\mathbf{j}$ in one and only one way.

Theorem 5   If $\mathbf{a} = \langle a_1, a_2 \rangle$ is an arbitrary vector in the plane, then

$$\mathbf{a} = a_1 \mathbf{i} + a_2 \mathbf{j}$$

and this representation is unique.

Proof   We write

$$\begin{aligned}\mathbf{a} = \langle a_1, a_2 \rangle &= a_1 \langle 1, 0 \rangle + a_2 \langle 0, 1 \rangle \\ &= a_1 \mathbf{i} + a_2 \mathbf{j}\end{aligned} \tag{9}$$

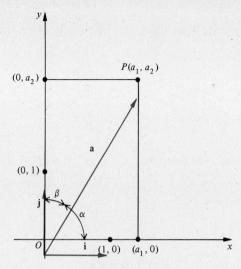

Figure 11.3.5

Thus the existence of a linear combination if $\mathbf{i}$ and $\mathbf{j}$ has been obtained. We must next show uniqueness. To this end, suppose that

$$\mathbf{a} = b_1\mathbf{i} + b_2\mathbf{j}$$

we must prove that $b_1 = a_1$ and $b_2 = a_2$.

$$\mathbf{a} = b_1\mathbf{i} + b_2\mathbf{j} = b_1\langle 1, 0 \rangle + b_2\langle 0, 1 \rangle$$
$$= \langle b_1, 0 \rangle + \langle 0, b_2 \rangle = \langle b_1, b_2 \rangle$$

Thus $\langle a_1, a_2 \rangle = \langle b_1, b_2 \rangle$ from which $b_1 = a_1$ and $b_2 = a_2$, and the uniqueness is established. ∎

For example, $\langle 6, -4 \rangle = 6\mathbf{i} - 4\mathbf{j}$, whereas $\langle -3, 0 \rangle = -3\mathbf{i} + 0\mathbf{j} = -3\mathbf{i}$. Theorem 4 states that any vector $\mathbf{a}$ may be expressed in two ways, namely, $\mathbf{a} = \langle a_1, a_2 \rangle$ or $\mathbf{a} = a_1\mathbf{i} + a_2\mathbf{j}$, and both representations have their applications as will be seen. In taking a dot product, Definition (1) becomes

$$(a_1\mathbf{i} + a_2\mathbf{j}) \cdot (b_1\mathbf{i} + b_2\mathbf{j}) = a_1b_1 + a_2b_2$$

We say that a_1 and a_2 are the ***components*** of $\mathbf{a}$ or, more precisely, the ***scalar components*** of $\mathbf{a}$ in the x and y directions. Furthermore $a_1\mathbf{i}$ and $a_2\mathbf{j}$ are the ***orthogonal vector components*** of $\mathbf{a}$; that is, these vector components are perpendicular to each other.

Given a vector $\mathbf{a} = a_1\mathbf{i} + a_2\mathbf{j}$, we seek expressions for a_1 and a_2. To find a_1, take the scalar product of both sides with $\mathbf{i}$. Thus

$$\mathbf{a} \cdot \mathbf{i} = (a_1\mathbf{i} + a_2\mathbf{j}) \cdot \mathbf{i} = a_1(\mathbf{i} \cdot \mathbf{i}) + a_2(\mathbf{j} \cdot \mathbf{i}) = a_1$$

and similarly, taking the scalar product of both sides with $\mathbf{j}$ yields,

$$\mathbf{a} \cdot \mathbf{j} = (a_1\mathbf{i} + a_2\mathbf{j}) \cdot \mathbf{j} = a_1(\mathbf{i} \cdot \mathbf{j}) + a_2(\mathbf{j} \cdot \mathbf{j}) = a_2$$

From

$$a_1 = \mathbf{a} \cdot \mathbf{i} \quad \text{and} \quad a_2 = \mathbf{a} \cdot \mathbf{j} \tag{10}$$

we have the following representation of $\mathbf{a}$:

$$\mathbf{a} = (\mathbf{a} \cdot \mathbf{i})\mathbf{i} + (\mathbf{a} \cdot \mathbf{j})\mathbf{j} \tag{11}$$

If α and β are the angles that **a** makes with the x and y-axes (Figure 11.3.5), then

$$\mathbf{a} = |\mathbf{a}| (\cos \alpha)\mathbf{i} + |\mathbf{a}| (\cos \beta)\mathbf{j}$$
$$= |\mathbf{a}| [(\cos \alpha)\mathbf{i} + (\cos \beta)\mathbf{j}] \qquad (12)$$

From the fact that

$$(\cos \alpha)\mathbf{i} + (\cos \beta)\mathbf{j} = \frac{\mathbf{a}}{|\mathbf{a}|} \qquad (\mathbf{a} \neq \mathbf{0}) \qquad (13)$$

it follows that $(\cos \alpha)\mathbf{i} + (\cos \beta)\mathbf{j}$ is a unit vector. This is readily verified since

$$|(\cos \alpha)\mathbf{i} + (\cos \beta)\mathbf{j}| = \sqrt{\cos^2 \alpha + \cos^2 \beta}$$
$$= \sqrt{\cos^2 \alpha + \sin^2 \alpha} \qquad \left(\cos \beta = \sin \alpha \text{ since } \alpha + \beta = \frac{\pi}{2}\right)$$
$$= 1$$

Equation (12) is a third and also very useful way to represent an arbitrary vector in the plane.

We can generalize the preceding discussion by defining the scalar and vector projection of a vector **a** onto a second vector **b**. The two vectors are shown in Figure 11.3.6. By the *scalar projection* of **a** onto **b** we mean the scalar product of **a** with a unit vector in the direction of **b**. Thus the scalar is found by forming

$$\mathbf{a} \cdot \left(\frac{\mathbf{b}}{|\mathbf{b}|}\right) = \frac{\mathbf{a} \cdot \mathbf{b}}{|\mathbf{b}|} = \frac{|\mathbf{a}| \, |\mathbf{b}| \cos \theta}{|\mathbf{b}|} = |\mathbf{a}| \cos \theta \qquad (14)$$

Thus the scalar projection is positive if $0 \leq \theta < \dfrac{\pi}{2}$, zero if $\theta = \dfrac{\pi}{2}$ and negative if $\dfrac{\pi}{2} < \theta \leq \pi$.

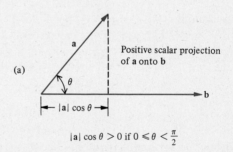

(a)

Positive scalar projection
of **a** onto **b**

$|\mathbf{a}| \cos \theta > 0$ if $0 \leqslant \theta < \frac{\pi}{2}$

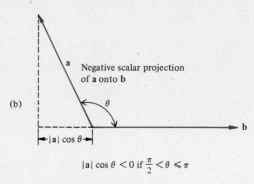

(b)

Negative scalar projection
of **a** onto **b**

$|\mathbf{a}| \cos \theta < 0$ if $\frac{\pi}{2} < \theta \leqslant \pi$

Figure 11.3.6

Similarly, the *vector projection* of **a** onto **b** is defined as the scalar projection of **a** onto **b** times a unit vector in the direction of **b**. Thus it is given by

$$\left(\mathbf{a} \cdot \frac{\mathbf{b}}{|\mathbf{b}|}\right)\frac{\mathbf{b}}{|\mathbf{b}|} = \left[\frac{\mathbf{a} \cdot \mathbf{b}}{|\mathbf{b}|^2}\right]\mathbf{b} = |\mathbf{a}|\,(\cos\theta)\frac{\mathbf{b}}{|\mathbf{b}|} \tag{15}$$

(Figure 11.3.7 shows $\overrightarrow{OC} = \mathbf{c}$ which is a vector representation of the vector projection.)

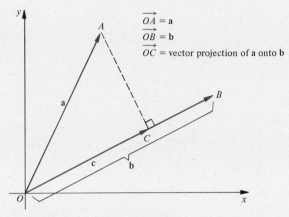

$\overrightarrow{OA} = \mathbf{a}$
$\overrightarrow{OB} = \mathbf{b}$
$\overrightarrow{OC} =$ vector projection of **a** onto **b**

Figure 11.3.7

EXAMPLE 3 Given $\mathbf{a} = 3\mathbf{i} - 4\mathbf{j}$ and $\mathbf{b} = 2\mathbf{i} + \mathbf{j}$, find the scalar and vector projections of **a** onto **b**.

Solution From (14), the scalar projection of **a** onto **b** is

$$\mathbf{a} \cdot \frac{\mathbf{b}}{|\mathbf{b}|} = \frac{3(2) - 4(1)}{\sqrt{2^2 + 1^2}} = \frac{2}{\sqrt{5}}$$

while from (15), the vector projection of **a** and **b** is

$$\left[\frac{\mathbf{a} \cdot \mathbf{b}}{|\mathbf{b}|^2}\right]\mathbf{b} = \frac{3(2) - 4(1)}{(\sqrt{5})^2}(2\mathbf{i} + \mathbf{j}) = \frac{2}{5}(2\mathbf{i} + \mathbf{j}) \qquad\bullet$$

In Chapter 7 Section 7.5 we defined the work done by a force F during a displacement s in the direction of the force to be Fs. More generally if **F** and **s** are vectors representing the force and the displacement, respectively, inclined at an angle θ to one another (Figure 11.3.8), then the work done is defined to be

$$\mathbf{F} \cdot \mathbf{s} = |\mathbf{F}|\,|\mathbf{s}|\cos\theta = Fs\cos\theta \tag{16}$$

Thus the work done is a scalar quantity that is given by the magnitude of the force times the scalar projection of the displacement **s** onto the direction of the force. We observe that the work done is positive if $0 \le \theta < \frac{\pi}{2}$ and negative if $\frac{\pi}{2} < \theta \le \pi$. Furthermore, the work done is zero only when **s** is perpendicular to **F**. For example, the work done by the gravitational force acting on a block is zero

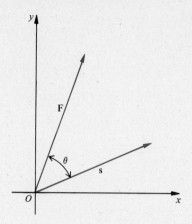

Figure 11.3.8

when the block is dragged along a horizontal line. This is because the displacement vector is at right angles to the vector representing the weight of the block.

Suppose next that a particle (or point) is acted upon by n forces $\mathbf{F}_1, \mathbf{F}_2, \ldots, \mathbf{F}_n$. Then, during a displacement $\mathbf{s}$ of the particle, the separate forces do work $\mathbf{F}_1 \cdot \mathbf{s}, \mathbf{F}_2 \cdot \mathbf{s}, \ldots, \mathbf{F}_n \cdot \mathbf{s}$. Therefore the total work W is given by

$$W = \mathbf{F}_1 \cdot \mathbf{s} + \mathbf{F}_2 \cdot \mathbf{s} + \mathbf{F}_3 \cdot \mathbf{s} + \cdots + \mathbf{F}_n \cdot \mathbf{s}$$

or more compactly by

$$W = \sum_{i=1}^{n} \mathbf{F}_i \cdot \mathbf{s} = \mathbf{s} \cdot \left(\sum_{i=1}^{n} \mathbf{F}_i \right) = \mathbf{s} \cdot \mathbf{R} \tag{17}$$

and is therefore the same as if the system of forces were replaced by its resultant $\mathbf{R}$. (The **resultant** $\mathbf{R}$ of any system of vectors $\mathbf{v}_1, \mathbf{v}_2, \ldots, \mathbf{v}_n$ is *defined* to be their sum, $\mathbf{v}_1 + \mathbf{v}_2 + \cdots + \mathbf{v}_n$.)

EXAMPLE 4   A particle is acted upon by two forces $\mathbf{F}_1 = 100(3\mathbf{i} + \mathbf{j})$ and $\mathbf{F}_2 = 60(-\mathbf{i} + 2\mathbf{j})$ where the magnitude of force is measured in pounds. Find the total work done by these forces in moving the particle in a straight line from the origin to the point $(5, 6)$. The distance is measured in feet.

Solution $\mathbf{F}_1 = 300\mathbf{i} + 100\mathbf{j}$ and $\mathbf{F}_2 = -60\mathbf{i} + 120\mathbf{j}$ so that the resultant force is

$$\mathbf{R} = \mathbf{F}_1 + \mathbf{F}_2 = (300\mathbf{i} + 100\mathbf{j}) + (-60\mathbf{i} + 120\mathbf{j})$$
$$= 240\mathbf{i} + 220\mathbf{j}$$

The displacement $\mathbf{s}$ is given by $5\mathbf{i} + 6\mathbf{j}$ and therefore from (17)

$$W = \mathbf{R} \cdot \mathbf{s} = 240(5) + 220(6) = 2520 \text{ ft-lb} \qquad \bullet$$

We conclude this section with some interesting geometric applications of the dot product showing the potency of the vector method.

EXAMPLE 5   Prove that the median to the base of an isosceles triangle is also an altitude to that base.

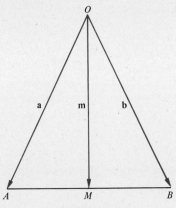

Figure 11.3.9

Solution   The isosceles triangle is shown in Figure 11.3.9 where $|\mathbf{a}| = |\mathbf{b}|$ and $\overrightarrow{OM} = \mathbf{m}$ is the median from point O to side AB. Now, $\overrightarrow{AB} = \mathbf{b} - \mathbf{a}$ so $\overrightarrow{AM} = \frac{1}{2}(\mathbf{b} - \mathbf{a})$. Then,

$$\mathbf{m} = \mathbf{a} + \tfrac{1}{2}(\mathbf{b} - \mathbf{a}) = \tfrac{1}{2}(\mathbf{a} + \mathbf{b})$$

and therefore

$$\mathbf{m} \cdot (\mathbf{b} - \mathbf{a}) = \tfrac{1}{2}(\mathbf{b} + \mathbf{a}) \cdot (\mathbf{b} - \mathbf{a})$$
$$= \tfrac{1}{2}(\mathbf{b} \cdot \mathbf{b} - \mathbf{a} \cdot \mathbf{a}) = \tfrac{1}{2}(|\mathbf{b}|^2 - |\mathbf{a}|^2) = 0$$

since $|\mathbf{a}| = |\mathbf{b}|$.

Hence $\overrightarrow{OM}$ is perpendicular to $\overrightarrow{AB}$.   ●

EXAMPLE 6   Verify the law of cosines by an application of vector analysis methods.

Solution   An arbitrary triangle is shown in Figure 11.3.10. With the directions of the vectors shown there, we have

$$\mathbf{c} = \mathbf{b} - \mathbf{a}$$

from which

$$\mathbf{c} \cdot \mathbf{c} = (\mathbf{b} - \mathbf{a}) \cdot (\mathbf{b} - \mathbf{a}) = |\mathbf{b}|^2 + |\mathbf{a}|^2 - 2(\mathbf{a} \cdot \mathbf{b})$$

so that

$$|\mathbf{c}|^2 = |\mathbf{b}|^2 + |\mathbf{a}|^2 - 2|\mathbf{a}|\,|\mathbf{b}|\cos\theta$$

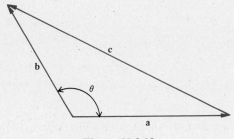

Figure 11.3.10

or, equivalently,

$$c^2 = a^2 + b^2 - 2ab\,(\cos\theta)$$

where $a = |\mathbf{a}|$, $b = |\mathbf{b}|$, and $c = |\mathbf{c}|$.

(It should be remarked that this technique does not prove the law of cosines, since the law of cosines was used in obtaining the relation $\mathbf{a}\cdot\mathbf{b} = |\mathbf{a}|\,|\mathbf{b}|\cos\theta$, which, in turn, was used in the proof above. The technique does illustrate the power of vector methods.) ●

EXAMPLE 7 Prove that the three altitudes of a triangle meet in a point (that is, are concurrent).

Solution Refer to Figure 11.3.11. Let O be the intersection of two altitudes BD and CE. (*Note:* the triangle in the figure is acute so that point O is inside the triangle. An obtuse triangle can be treated in a similar fashion, and the result is apparent for a right triangle.)

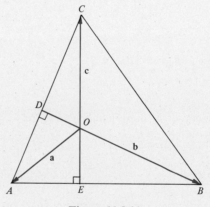

Figure 11.3.11

Let $\overrightarrow{OA} = \mathbf{a}$, $\overrightarrow{OB} = \mathbf{b}$ and $\overrightarrow{OC} = \mathbf{c}$, then

$$\overrightarrow{AB} = \mathbf{b} - \mathbf{a} \qquad \overrightarrow{BC} = \mathbf{c} - \mathbf{b} \qquad \overrightarrow{CA} = \mathbf{a} - \mathbf{c}$$

But $\overrightarrow{OB} \perp \overrightarrow{AC}$ and therefore

$$\mathbf{b}\cdot(\mathbf{a} - \mathbf{c}) = 0$$

or, equivalently

$$\mathbf{b}\cdot\mathbf{a} = \mathbf{b}\cdot\mathbf{c}$$

Also $\overrightarrow{OC} \perp \overrightarrow{AB}$ from which

$$\mathbf{c}\cdot(\mathbf{b} - \mathbf{a}) = 0 \qquad \text{or} \qquad \mathbf{c}\cdot\mathbf{b} = \mathbf{c}\cdot\mathbf{a}$$

Hence

$$\mathbf{b}\cdot\mathbf{a} = \mathbf{b}\cdot\mathbf{c} = \mathbf{c}\cdot\mathbf{a} \qquad \text{or} \qquad (\mathbf{b} - \mathbf{c})\cdot\mathbf{a} = 0$$

Therefore $\overrightarrow{OA}$ is perpendicular to $\overrightarrow{BC}$; thus the three altitudes are concurrent. ●

Exercises 11.3

In Exercises 1 through 4, **a**, **b** and **c** are vectors and k is a scalar. Prove the following:

1. $(\mathbf{a} + \mathbf{b}) \cdot \mathbf{c} = \mathbf{a} \cdot \mathbf{c} + \mathbf{b} \cdot \mathbf{c}$ [Formula (3b) of the text]
2. $(k\mathbf{a}) \cdot \mathbf{b} = \mathbf{a} \cdot (k\mathbf{b})$ [Formula (4)]
3. $\mathbf{0} \cdot \mathbf{a} = 0$ [Formula (5)]
4. $\mathbf{a} \cdot \mathbf{a} = |\mathbf{a}|^2 = a^2$ [Formula (6)]

In Exercises 5 through 10, find the dot (or scalar) product of the following vectors:

5. $\langle 1, 3 \rangle \cdot \langle 2, -1 \rangle$
6. $\langle 0, 2 \rangle \cdot \langle -3, 0 \rangle$
7. $\langle -1, -2 \rangle \cdot \langle 3, 4 \rangle$
8. $\langle a, b \rangle \cdot \langle b, a \rangle$
9. $\langle a, b \rangle \cdot \langle -3b, 3a \rangle$
10. $3\langle 2, -4 \rangle \cdot \langle 4, 0 \rangle$

In Exercises 11 through 14, determine whether **a** and **b** are perpendicular.

11. $\mathbf{a} = 6\mathbf{i} - 10\mathbf{j}, \quad \mathbf{b} = 5\mathbf{i} + 3\mathbf{j}$
12. $\mathbf{a} = -\mathbf{i} + \mathbf{j}, \quad \mathbf{b} = -2\mathbf{i} + 2\mathbf{j}$
13. $\mathbf{a} = -3\mathbf{i} - 4\mathbf{j}, \quad \mathbf{b} = -10\mathbf{i} + 15\mathbf{j}$
14. $\mathbf{a} = (\cos\theta)\mathbf{i} - (\sin\theta)\mathbf{j}, \quad \mathbf{b} = (\sin\theta)\mathbf{i} + (\cos\theta)\mathbf{j}$

In Exercises 15 through 18, find m so that $\langle 3, m \rangle$ is perpendicular to:

15. $\langle 0, 2 \rangle$
16. $\langle -1, 2 \rangle$
17. $\langle 1, m \rangle$
18. $\langle m, 2m \rangle$
19. Determine a vector **a** such that (a) $\mathbf{a} \cdot \mathbf{a} = -1$, (b) $\mathbf{a} \cdot \mathbf{a} = -\frac{1}{4}$.
20. Determine a vector **a** such that (a) $\mathbf{a} \cdot \mathbf{a} = 0$,   (b) $\mathbf{a} \cdot \mathbf{a} = 16$,   (c) describe the answer to part (b) geometrically.
21. If $\mathbf{a} = \langle -2, 1 \rangle$, $\mathbf{b} = \langle 1, 3 \rangle$, and $\mathbf{c} = \langle 5, 10 \rangle$, find
    (a) the angle between vectors **a** and **b**
    (b) the angle between vectors **a** and **c**
22. Given the vector $2\mathbf{i} - \mathbf{j}$, find the vectors of length 3 that are orthogonal (perpendicular) to it.
23. Find the totality of vectors of length 2 orthogonal (perpendicular) to $b_1\mathbf{i} + b_2\mathbf{j}$ where $b_2 \neq 0$.
24. Given that $\mathbf{a} \cdot \mathbf{b} = \mathbf{a} \cdot \mathbf{c}$ where $\mathbf{a} \neq \mathbf{0}$, does this imply that $\mathbf{b} = \mathbf{c}$. If not, what does it imply?
25. When does $\mathbf{a} \cdot (\mathbf{b} \cdot \mathbf{c}) = (\mathbf{a} \cdot \mathbf{b}) \cdot \mathbf{c}$?
26. (a) Prove that $|\mathbf{a} - \mathbf{b}|^2 + |\mathbf{a} + \mathbf{b}|^2 = 2(|\mathbf{a}|^2 + |\mathbf{b}|^2)$ (this equality is frequently referred to as the *parallelogram law*.)
    (b) If **a** and **b** represent two adjacent sides of a parallelogram give a geometrical interpretation of the result in part (a).
27. Prove that the diagonals of a rhombus are perpendicular to each other.
28. Prove by vector methods the *Cauchy-Schwarz inequality,* namely that

$$(a_1 b_1 + a_2 b_2)^2 \leq (a_1^2 + a_2^2)(b_1^2 + b_2^2)$$

for arbitrary real numbers a_1, a_2, b_1, and b_2. (*Hint:* Show that $(\mathbf{a} \cdot \mathbf{b})^2 \leq |\mathbf{a}|^2 |\mathbf{b}|^2$)
29. Show that the vectors $\mathbf{i} + \mathbf{j}$ and $\mathbf{i} - 2\mathbf{j}$ may be used as basis vectors for any vector in the plane; that is, every vector in the plane can be expressed as a linear combination of the two given vectors in one and only one way.
30. Prove by vector methods that an angle inscribed in a semicircle is a right angle.
31. Given $\mathbf{a} = 3\mathbf{i} - 2\mathbf{j}$ and $\mathbf{b} = 2\mathbf{i} + \mathbf{j}$, find the
    (a) vector projection of **a** onto **b**
    (b) vector projection of **b** onto **a**.

32. Two forces denoted by $\mathbf{F}_1$ and $\mathbf{F}_2$ are acting together on a particle and cause it to move from a point $(3, 4)$ to a point $(5, 9)$. If $\mathbf{F}_1 = 60(4\mathbf{i} - \mathbf{j})$ and $\mathbf{F}_2 = 80(2\mathbf{i} + 3\mathbf{j})$, and the magnitudes of the forces are in pounds while the distance is measured in feet, find the work done by the two forces acting simultaneously.

33. By vector methods prove that the diagonals of a parallelogram bisect each other.

11.4   VECTOR VALUED FUNCTIONS; PARAMETRIC EQUATIONS

We will now consider functions whose domains are sets of numbers and whose ranges are sets of vectors. Such functions are called *vector valued functions*. More precisely we have the following definition:

Definition   Let f and g be two real valued functions of a parameter (that is, a variable) t. Furthermore for every number t in the domain I common to f and g, there is a vector $\mathbf{r}$ defined by

$$\mathbf{r}(t) = f(t)\mathbf{i} + g(t)\mathbf{j} \qquad t \in I$$

and $\mathbf{r}$ is said to be a *vector valued function*. Thus if $\mathbf{r}(t)$ is a vector valued function of t with domain I, then we write

$$\mathbf{r}(t) = \langle f(t), g(t) \rangle \qquad t \in I \tag{1}$$

EXAMPLE 1   Determine the domain of the function

$$\mathbf{r}(t) = \frac{t + 1}{t - 1}\mathbf{i} + \sqrt{t^2 - 1}\,\mathbf{j}$$

Solution   The domain of $\mathbf{r}(t)$ are those values for which both $f(t)$ and $g(t)$ are defined. In this instance $f(t) = \dfrac{t + 1}{t - 1}$ is defined for all t, where $t \neq 1$, and $g(t) = \sqrt{t^2 - 1}$ is defined for all t such that $|t| \geq 1$. Therefore the domain of $\mathbf{r}$ is defined by $|t| \geq 1$ and $t \neq 1$.   ●

If the left side of (1) is expressed in component form namely, $\mathbf{r} = x\mathbf{i} + y\mathbf{j}$, then (1) may be replaced by

$$x = f(t) \qquad \text{and} \qquad y = g(t) \tag{2}$$

Generally as t assumes the values in the domain of $\mathbf{r}$, the endpoint of $\mathbf{r}$ traces out a curve C in the xy-plane. Equations (2) are said to be the *parametric equations* of C. The curve C is the *graph* of the vector function $\mathbf{r}$.

If t increases monotonically then (1) or (2) implies that the curve C is being traced in a particular direction at each point. If the curve is interpreted as being traced by a particle, it is often convenient to consider the positive direction to be coincident with the direction obtained by following the motion of the particle as the parameter increases.

In many cases the parameter t may be eliminated between the pair of equations (2). The resulting equation is called a **Cartesian equation** of C. Occasionally the graph of the Cartesian equation (not properly restricted) will consist of more points than the graph of (2). Therefore, the graph of (2) is always a subset of the graph of the corresponding graph of the Cartesian equation and the two graphs may coincide.

EXAMPLE 2 Sketch the graph of the two parametric equations

$$x = 2t + 3 \quad \text{and} \quad y = \sqrt{t}$$

Find the corresponding Cartesian equation.

Solution We note that the domain of $\mathbf{r}(t) = x(t)\,\mathbf{i} + y(t)\,\mathbf{j}$ is $t \geq 0$ since $x(t)$ is defined for all t, and $y(t)$ is defined for $t \geq 0$. We calculate Table 11.4.1.

Table 11.4.1

t	0	1	2	3	4	5	6
x	3	5	7	9	11	13	15
y	0	1	$\sqrt{2} \doteq 1.41$	$\sqrt{3} \doteq 1.73$	2	$\sqrt{5} \doteq 2.24$	$\sqrt{6} \doteq 2.45$

A sketch of the graph is shown in Figure 11.4.1.

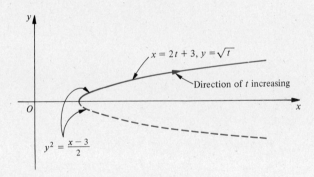

Figure 11.4.1

We may eliminate t by writing $y^2 = t = \dfrac{x-3}{2}$. The graph of $y^2 = \dfrac{x-3}{2}$ is a parabola with the x-axis as an axis of symmetry. The graph of the given equations in this example is the *upper half* of the parabola (since $t \geq 0$, $y \geq 0$ and $x \geq 3$).

EXAMPLE 3 Sketch the graph of the vector function $\mathbf{r}(t) = 5\,(\cos 2t)\mathbf{i} + 5\,(\sin 2t)\mathbf{j}$, $0 \leq t \leq \pi$.

Solution The magnitude of the vector $\mathbf{r}$ is given by

$$|\mathbf{r}| = \sqrt{(5\cos 2t)^2 + (5\sin 2t)^2} \quad 0 \leq t \leq \pi$$
$$= \sqrt{25\,(\cos^2 2t + \sin^2 2t)} = 5$$

since $\cos^2\theta + \sin^2\theta = 1$. Thus as t increases from $t = 0$ to $t = \pi$, the circle of radius 5 with center at the origin is traversed once in a counterclockwise direction (Figure 11.4.2).

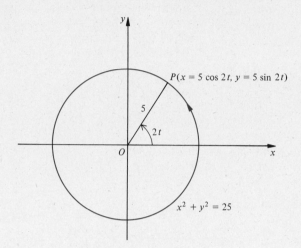

Figure 11.4.2

Parametric equations of the graph are

$$x = 5\cos 2t \qquad y = 5\sin 2t \qquad 0 \le t \le \pi$$

Thus, $\dfrac{x}{5} = \cos 2t$ and $\dfrac{y}{5} = \sin 2t$, from which $\left(\dfrac{x}{5}\right)^2 + \left(\dfrac{y}{5}\right)^2 = \cos^2 2t + \sin^2 2t = 1$ or $x^2 + y^2 = 25$. The graphs obtained from the parametric and Cartesian equations are identical.

The parametric equations

$$x = a\cos\theta, \qquad y = a\sin\theta \qquad 0 \le \theta \le 2\pi$$

are a very useful representation of a circle of radius a centered at the origin. Example 3 is a special case of this with $a = 5$ and $\theta = 2t$. ●

EXAMPLE 4   Find a system of parametric equations whose graph is the graph of the equation

$$x^{2/3} + y^{2/3} = a^{2/3} \tag{3}$$

Solution   There is no systematic way for doing this in general. In this instance, we divide both sides of the equation by $a^{2/3}$ to obtain:

$$\left(\frac{x}{a}\right)^{2/3} + \left(\frac{y}{a}\right)^{2/3} = 1$$

which may be rewritten in the form

$$\left[\left(\frac{x}{a}\right)^{1/3}\right]^2 + \left[\left(\frac{y}{a}\right)^{1/3}\right]^2 = 1$$

Thus we have the sum of two squares is 1. This suggests the introduction of the trigonometric functions $\cos t$ and $\sin t$ by setting

$$\left(\frac{x}{a}\right)^{1/3} = \cos t \qquad \text{and} \qquad \left(\frac{y}{a}\right)^{1/3} = \sin t$$

since $\cos^2 t + \sin^2 t = 1$. Hence

$$x = a \cos^3 t \qquad \text{and} \qquad y = a \sin^3 t \qquad\qquad (4)$$

and it may be verified that the graph of (4) for $0 \leq t \leq 2\pi$ is identical with the graph of the original relation in Cartesian coordinates. The curve defined by (3) or (4) is known as the **hypocycloid of four cusps** or **astroid**. We shall return to it in the exercises (see Exercises 37 and 38). ●

EXAMPLE 5　A wheel of radius a is rolling along a straight line. Find the parametric equations of the path traced out by a given point on the circumference of the wheel, assuming that it is initially in contact with the straight line.

Solution　Let us suppose that the given point P originally is at the origin of a Cartesian coordinate system and that the disk is rolling (without slipping) along the x-axis. In Figure 11.4.3, we have the wheel in a displaced position after it has rolled through an angle θ (in radians). We must find a formula for the position vector **OP** in terms of θ. From the figure it follows that

$$\mathbf{OP} = \mathbf{OT} + \mathbf{TQ} + \mathbf{QP} \qquad\qquad (i)$$

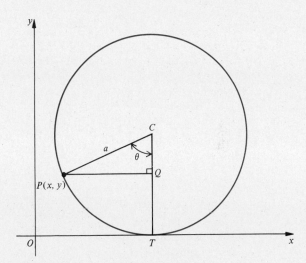

Figure 11.4.3

Because $|\mathbf{OT}| = \text{arc } TP = a\theta$, we have $\mathbf{OT} = a\theta\mathbf{i}$. Furthermore, from triangle CQP there results

$$\mathbf{TQ} = (a - a \cos \theta)\mathbf{j} = a(1 - \cos \theta)\mathbf{j}$$

and

$$\mathbf{QP} = -\mathbf{PQ} = -a\,(\sin \theta)\mathbf{i}$$

Thus by substitution into (i)

$$\mathbf{OP} = a\theta\mathbf{i} + a(1 - \cos\theta)\mathbf{j} - a\,(\sin\theta)\mathbf{i}$$
$$= a(\theta - \sin\theta)\mathbf{i} + a(1 - \cos\theta)\mathbf{j}$$
$$= x\mathbf{i} + y\mathbf{j} \qquad\qquad\qquad (ii)$$

This equation has been derived in the case where θ is an acute angle. It can be shown that we obtain the same result no matter how θ is drawn.

From the vector equation (ii), we obtain the parametric equations:

$$x = a(\theta - \sin\theta) \qquad \text{and} \qquad y = a(1 - \cos\theta) \qquad (5)$$

where θ is in $[0, \infty)$. This curve is called a *cycloid* and a sketch of part of its graph is given in Figure 11.4.4. ●

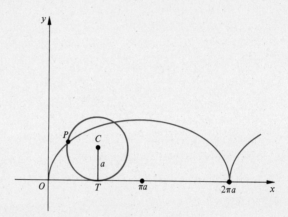

Figure 11.4.4

Let $\mathbf{r}$ be a vector function defined by $\mathbf{r}(t) = \langle f(t), g(t) \rangle$. Let t_0 be a value of t in the domain of $\mathbf{r}$. Then the *limit* of $\mathbf{r}(t)$ as t approaches t_0 is defined by

$$\lim_{t \to t_0} \mathbf{r}(t) = \langle \lim_{t \to t_0} f(t), \lim_{t \to t_0} g(t) \rangle \qquad (6)$$

provided that the two latter limits exist. For example, if

$$\mathbf{r}(t) = \langle \cos t, \sin t \rangle$$

then

$$\lim_{t \to 0} \mathbf{r}(t) = \langle \lim_{t \to 0} \cos t, \lim_{t \to 0} \sin t \rangle = \langle 1, 0 \rangle$$

Hence the limit is the vector $\langle 1, 0 \rangle$.

The vector function $\mathbf{r}$ is said to be *continuous* at t_0, if

(i) $\mathbf{r}$ is defined at t_0

(ii) $\lim_{t \to t_0} \mathbf{r}(t)$ exists

(iii) $\lim_{t \to t_0} \mathbf{r}(t) = \mathbf{r}(t_0)$

From the preceding definitions, a necessary and sufficient condition that $\mathbf{r}(t) = \langle f(t), g(t) \rangle$ be continuous at an interior point t_0 of an interval I is that *both* $f(t)$ and $g(t)$ be continuous at t_0.

In the definition of the derivative that is to follow, we must consider the expression

$$\frac{\mathbf{r}(t + \Delta t) - \mathbf{r}(t)}{\Delta t} \qquad (\Delta t \neq 0)$$

By the division of a vector by a nonzero scalar we mean the multiplication by its reciprocal, that is,

$$\frac{1}{\Delta t}[\mathbf{r}(t + \Delta t) - \mathbf{r}(t)]$$

Thus if $\mathbf{r}(t)$ is a vector valued function, then the **_derivative_** of $\mathbf{r}$ is another vector valued function denoted by $\mathbf{r}'$ and defined by

$$\mathbf{r}'(t) = \lim_{\Delta t \to 0} \frac{\mathbf{r}(t + \Delta t) - \mathbf{r}(t)}{\Delta t} \qquad (7)$$

provided that the limit exists. Alternatively, we may use the notation $D_t \mathbf{r}(t)$ for the derivative of $\mathbf{r}(t)$.

The interpretation of $\mathbf{r}'(t)$ is obtained geometrically by considering representations of the vectors $\mathbf{r}(t)$, $\mathbf{r}(t + \Delta t)$ and thus $\mathbf{r}(t + \Delta t) - \mathbf{r}(t)$. In Figure 11.4.5, $\overrightarrow{OP}$ and $\overrightarrow{OQ}$ are the position representations of $\mathbf{r}(t)$ and $\mathbf{r}(t + \Delta t)$, respectively. Therefore $\overrightarrow{PQ}$ is a representation of $\mathbf{r}(t + \Delta t) - \mathbf{r}(t)$. Division by Δt, or equivalently, multiplication by $\frac{1}{\Delta t}$ yields a vector in the same or opposite direction to $\overrightarrow{PQ}$ and magnified by a factor $\left|\frac{1}{\Delta t}\right|$. Hence if we let $\Delta t \to 0$ $(\Delta t \neq 0)$ the limiting position of the secants is the tangent and thus $\mathbf{r}'(t)$ is a vector in the direction of the tangent to the curve C at the point P.

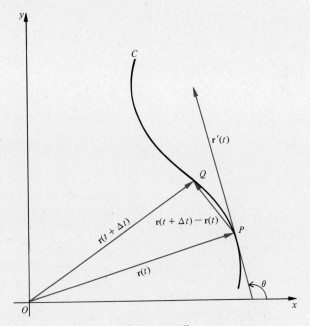

Figure 11.4.5

The following theorem is derived from definition (7) and the definition of the derivative of a real valued function.

Theorem 1 If **r** is a vector valued function given by

$$\mathbf{r}(t) = f(t)\mathbf{i} + g(t)\mathbf{j}$$

then

$$\mathbf{r}'(t) = f'(t)\mathbf{i} + g'(t)\mathbf{j} \qquad (8)$$

provided that $f'(t)$ and $g'(t)$ both exist.

Proof From (7),

$$\mathbf{r}'(t) = \lim_{\Delta t \to 0} \frac{\mathbf{r}(t + \Delta t) - \mathbf{r}(t)}{\Delta t}$$

$$= \lim_{\Delta t \to 0} \frac{[f(t + \Delta t)\mathbf{i} + g(t + \Delta t)\mathbf{j}] - [f(t)\mathbf{i} + g(t)\mathbf{j}]}{\Delta t}$$

$$= \lim_{\Delta t \to 0} \frac{f(t + \Delta t) - f(t)}{\Delta t}\mathbf{i} + \lim_{\Delta t \to 0} \frac{g(t + \Delta t) - g(t)}{\Delta t}\mathbf{j}$$

$$= f'(t)\mathbf{i} + g'(t)\mathbf{j} \qquad \blacksquare$$

The **second derivative r″** is defined as the derivative of the first derivative. Hence it is the vector function given by

$$\mathbf{r}''(t) = \langle f''(t), g''(t) \rangle \qquad (9)$$

assuming both $f''(t)$ and $g''(t)$ exist. In expanded form if $\mathbf{r}(t) = f(t)\mathbf{i} + g(t)\mathbf{j}$, then $\mathbf{r}''(t) = f''(t)\mathbf{i} + g''(t)\mathbf{j}$ if $f''(t)$ and $g''(t)$ exist.

EXAMPLE 6 If $\mathbf{r}(t) = (3 + \cos 2t)\mathbf{i} + (6 + \sin 2t)\mathbf{j}$, find

(a) $\mathbf{r}'(t)$; (b) $\mathbf{r}'\left(\dfrac{\pi}{2}\right)$; (c) $|\mathbf{r}'(t)|$; (d) $\mathbf{r}''(t)$; (e) $\mathbf{r}'(t) \cdot \mathbf{r}''(t)$.

Solution (a)
$$\mathbf{r}'(t) = (-2 \sin 2t)\mathbf{i} + (2 \cos 2t)\mathbf{j}$$

(b)
$$\mathbf{r}'\left(\frac{\pi}{2}\right) = (-2 \sin \pi)\mathbf{i} + (2 \cos \pi)\mathbf{j} = -2\mathbf{j}$$

(c)
$$|\mathbf{r}'(t)| = \sqrt{(-2 \sin 2t)^2 + (2 \cos 2t)^2}$$
$$= \sqrt{4(\sin^2 2t + \cos^2 2t)} = 2$$

(d)
$$\mathbf{r}''(t) = (-4 \cos 2t)\mathbf{i} + (-4 \sin 2t)\mathbf{j}$$

(e)
$$\mathbf{r}'(t) \cdot \mathbf{r}''(t) = (-2 \sin 2t)(-4 \cos 2t) + (2 \cos 2t)(-4 \sin 2t) = 0$$

This implies that the nonzero vectors $\mathbf{r}'(t)$ and $\mathbf{r}''(t)$ are orthogonal (perpendicular) to each other for all t. ●

There are a number of rules for differentiating combinations of vector valued functions and scalar functions. We will prove Theorem 3 and the remaining proofs are left for the exercises. These rules are all results that we should expect in that they correspond to the rules of differentiation that we learned for scalar functions.

Theorem 2   If **r** and **u** are differentiable vector valued functions on an interval (that is, $D_t\mathbf{r}(t)$ and $D_t\mathbf{u}(t)$ exist for all values of t in this interval) then

$$D_t[\mathbf{r}(t) + \mathbf{u}(t)] = D_t\mathbf{r}(t) + D_t\mathbf{u}(t) \tag{10}$$

The proof of (10) is left for the Exercises.

Theorem 3   If **r** and **u** are differentiable vector valued functions of t, then

$$D_t[\mathbf{r}(t) \cdot \mathbf{u}(t)] = \mathbf{r}(t) \cdot D_t\mathbf{u}(t) + D_t\mathbf{r}(t) \cdot \mathbf{u}(t) \tag{11}$$

Proof   Let $\mathbf{r}(t) = \langle\, f(t), g(t)\rangle$ and $\mathbf{u}(t) = \langle F(t), G(t)\rangle$, then

$$\mathbf{r}(t) \cdot \mathbf{u}(t) = f(t)\,F(t) + g(t)\,G(t)$$

Therefore

$$
\begin{aligned}
D_t[\mathbf{r}(t) \cdot \mathbf{u}(t)] &= f(t)\,F'(t) + g(t)\,G'(t) + f'(t)\,F(t) + g'(t)\,G(t)\\
&= \langle\, f(t), g(t)\rangle \cdot \langle F'(t), G'(t)\rangle + \langle\, f'(t), g'(t)\rangle \cdot \langle F(t), G(t)\rangle\\
&= \mathbf{r}(t) \cdot D_t\mathbf{u}(t) + D_t\mathbf{r}(t) \cdot \mathbf{u}(t) \qquad\blacksquare
\end{aligned}
$$

Theorem 4   If a scalar valued function h and a vector valued function **r** are differentiable functions of t on an interval I, then for all t in this interval

$$D_t[(h(t)\,\mathbf{r}(t)] = h(t)\,\mathbf{r}'(t) + h'(t)\,\mathbf{r}(t) \tag{12}$$

EXAMPLE 7   If $\mathbf{r}(t) = (3t^2 + 2)\mathbf{i} + e^{4t}\mathbf{j}$, find $D_t(\mathbf{r}(t) \cdot \mathbf{r}'(t))$

Solution   From (11) with $\mathbf{u}(t) = \mathbf{r}'(t)$, we have

$$D_t[\mathbf{r}(t) \cdot \mathbf{r}'(t)] = \mathbf{r}(t) \cdot \mathbf{r}''(t) + \mathbf{r}'(t) \cdot \mathbf{r}'(t) \tag{i}$$

In our case, $\mathbf{r}(t) = (3t^2 + 2)\mathbf{i} + e^{4t}\mathbf{j}$, hence

$$\mathbf{r}'(t) = (6t)\mathbf{i} + 4e^{4t}\mathbf{j} \qquad \text{and} \qquad \mathbf{r}''(t) = 6\mathbf{i} + 16e^{4t}\mathbf{j}$$

Therefore from (i),

$$
\begin{aligned}
D_t[\mathbf{r}(t) \cdot \mathbf{r}'(t)] &= 6(3t^2 + 2) + 16e^{8t} + 36t^2 + 16e^{8t}\\
&= 54t^2 + 12 + 32e^{8t}
\end{aligned}
$$

Alternative Solution

$$\mathbf{r}(t) \cdot \mathbf{r}'(t) = 6t(3t^2 + 2) + 4e^{8t}$$

and by direct calculation we obtain

$$D_t[\mathbf{r}(t) \cdot \mathbf{r}'(t)] = 54t^2 + 12 + 32e^{8t} \qquad\bullet$$

which agrees with the first solution.

It should be noted that the alternative solution (in example 7) gave a shorter solution than that obtained using the general formula for the derivative of a dot product. However, the former is a systematic method for attacking such problems that, in other cases, may yield more efficient solutions.

We shall now verify analytically our earlier geometric interpretation of the direction of $\mathbf{r}'(t)$.

From (8), the direction of the vector $\mathbf{r}'(t)$ is readily found. If θ is the angle between the tangent to the curve and the positive x-axis where $0 \leq \theta < 2\pi$

$$\tan \theta = \frac{dy}{dx} = \frac{\dfrac{dy}{dt}}{\dfrac{dx}{dt}} = \frac{g'(t)}{f'(t)} \qquad \text{if } f'(t) \neq 0 \tag{13}$$

Therefore if $\mathbf{r} = \mathbf{r}(t)$ is given by $\langle f(t), g(t) \rangle$ and traces out a curve C in the xy-plane, then the direction of $\mathbf{r}'(t)$ is along the *tangent line*.

EXAMPLE 8   Given $x = a(\theta - \sin\theta)$ and $y = a(1 - \cos\theta)$, which are the parametric equations of the cycloid, find $D_x y$ and $D_x^2 y$.

Solution

$$D_x y = \frac{D_\theta y}{D_\theta x} = \frac{a \sin\theta}{a(1 - \cos\theta)} = \frac{\sin\theta}{1 - \cos\theta}$$

$$D_x^2 y = D_x(D_x y) = \frac{D_\theta(D_x y)}{D_\theta x}$$

$$= \frac{(1 - \cos\theta)\cos\theta - \sin\theta\,(\sin\theta)}{(1 - \cos\theta)^2} \frac{1}{a(1 - \cos\theta)}$$

$$= \frac{\cos\theta - 1}{a(1 - \cos\theta)^3} = -\frac{1}{a(1 - \cos\theta)^2} \qquad \bullet$$

It is useful to replace Formula (13) by

$$D_x[\;\;] = \frac{D_t[\;\;]}{D_t x} \tag{14}$$

with the understanding that any function of t may be placed in the brackets, provided that the same function of t is placed in both brackets. Therefore if y and x are functions of t then $D_x y$ is again a function of t. Therefore from (14), with $D_x y$ placed in the brackets, we have

$$D_x^2 y = \frac{D_t[D_x y]}{D_t x} = \frac{D_t\left[\dfrac{D_t y}{D_t x}\right]}{D_t x}$$

and if the numerator is differentiated by the quotient rule:

$$D_x^2 y = \frac{\dfrac{D_t x \cdot D_t^2 y - D_t y \cdot D_t^2 x}{(D_t x)^2}}{D_t x} = \frac{D_t x \cdot D_t^2 y - D_t y \cdot D_t^2 x}{(D_t x)^3} \tag{15}$$

or, in prime notation

$$D_x^2 y = \frac{x'(t)y''(t) - y'(t)x''(t)}{[x'(t)]^3} \tag{16}$$

Although $D_y x$ and $D_x y$ are reciprocals of each other, in general $D_y^2 x$ and $D_x^2 y$ are *not*. This is easily shown by constructing specific examples which illustrate that $D_x^2 y$ and $\dfrac{1}{D_y^2 x}$ need not be the same.

EXAMPLE 9   Given $x = a(\theta - \sin \theta)$ and $y = a(1 - \cos \theta)$, find $D_x^2 y$ by an application of (15). This is the same given data as in Example 8.

Solution

$$D_\theta x = a(1 - \cos \theta) \qquad D_\theta^2 x = a \sin \theta$$
$$D_\theta y = a \sin \theta, \qquad D_\theta^2 y = a \cos \theta$$

Formula (15) yields, (with t replaced by θ)

$$D_x^2 y = \frac{a(1 - \cos \theta)a \cos \theta - (a \sin \theta)(a \sin \theta)}{[a(1 - \cos \theta)]^3}$$

$$= \frac{a^2(\cos \theta - \cos^2 \theta - \sin^2 \theta)}{a^3(1 - \cos \theta)^3} = \frac{\cos \theta - 1}{a(1 - \cos \theta)^3}$$

$$= -\frac{1}{a(1 - \cos \theta)^2}$$

in agreement with the result of Example 8.   ●

EXAMPLE 10   Find a parametric form for the circle $x^2 + y^2 = a^2$ $(a > 0)$, by using the slope m of a line through the origin as a parameter. Are all points on the circle obtained by this procedure?

Solution   The equation of the line through the origin with slope m is $y = mx$. Substitution of $y = mx$ into $x^2 + y^2 = a^2$, yields $x^2 + m^2 x^2 = a^2$, from which

$$x = \pm \frac{a}{\sqrt{1 + m^2}} \qquad \text{and} \qquad y = \pm \frac{ma}{\sqrt{1 + m^2}}$$

For each value of m, two sets of number pairs are determined that represent the intersections of the line with the given circle. The points $(0, \pm a)$ are not obtained because the slope is not defined for vertical lines.   ●

Exercises 11.4

In Exercises 1 through 6, find the domain of each of the following vector functions.

1. $\mathbf{r}(t) = \sqrt{9 - t^2}\,\mathbf{i} + (t^3 + 1)\mathbf{j}$
2. $\mathbf{r}(t) = \sqrt{t^2 + 4}\,\mathbf{i} + t[t]\mathbf{j}$ ([ ] = greatest integer function)
3. $\mathbf{r}(t) = |t|\mathbf{i} + \sqrt{t - 5}\,\mathbf{j}$
4. $\mathbf{r}(t) = \dfrac{e^{-3t}}{2t - 1}\mathbf{i} + \dfrac{\sqrt{t}}{t - 1}\mathbf{j}$
5. $\mathbf{r}(t) = (\sin^{-1} t)\mathbf{i} + \ln (t + 1)\mathbf{j}$
6. $\mathbf{r}(t) = (\cos^{-1} 3t)\mathbf{i} + (t - 3)^{2/3}\mathbf{j}$

In Exercises 7 through 12, we are given a vector equation. Draw a sketch of its graph and determine the corresponding Cartesian equation.

7. $\mathbf{r} = t^2\mathbf{i} + 4t\mathbf{j}$

8. $\mathbf{r} = 3t\mathbf{i} + 4(t - 1)\mathbf{j}, \quad 0 \leq t \leq 2$

9. $\mathbf{r} = (4 \cos 3t)\mathbf{i} + (4 \sin 3t)\mathbf{j}, \quad 0 \leq t \leq \dfrac{\pi}{3}$

10. $\mathbf{r} = (7 \sin 4t)\mathbf{i} + (7 \cos 4t)\mathbf{j}, \quad 0 \leq t \leq \dfrac{\pi}{8}$

11. $\mathbf{r} = (3 \cos t)\mathbf{i} + (2 \sin t)\mathbf{j}, \quad 0 \leq t \leq \dfrac{\pi}{2}$

12. $\mathbf{r} = (2 \sinh t)\mathbf{i} + (\cosh t)\mathbf{j}$

13. Find a Cartesian equation of the cycloid
$$x = a(\theta - \sin \theta) \qquad y = a(1 - \cos \theta)$$

14. Identify the curve defined parametrically by
$$x = 3 \cos t + 4 \sin t \qquad \text{and} \qquad y = 3 \sin t - 4 \cos t$$
where $0 \leq t \leq 2\pi$.

15. A straight line passes through the points $A(a_1, a_2)$ and $B(b_1, b_2)$ in the xy-plane. Let $\mathbf{a} = \langle a_1, a_2 \rangle$ and $\mathbf{b} = \langle b_1, b_2 \rangle$ be the two vectors from O to A and O to B, respectively. Let $\mathbf{r} = \langle x, y \rangle$ be a vector from O to any point P on the line. Find parametric equations of the line.

16. Let $\mathbf{r}_1$ and $\mathbf{r}_2$ be the position vectors of two masses m_1 and m_2, respectively, relative to a point O. The position vector of the **center of mass** is defined to be
$$\mathbf{r} = \frac{m_1\mathbf{r}_1 + m_2\mathbf{r}_2}{m_1 + m_2}$$
Show that the center of mass is on the line connecting the two point masses and divides the segment connecting m_2 to m_1 in the ratio m_1/m_2.

17. Show that the position of the center of mass defined in the previous exercise is independent of the origin of coordinates. (*Hint:* Let O_1 be a second origin, $\overrightarrow{OO_1} = \mathbf{a}$, and $\mathbf{s}$ be the position vector of the center of mass relative to O_1. Show that $\mathbf{s} = \mathbf{r} - \mathbf{a}$.

In Exercises 18 through 22, find $\mathbf{r}'(t)$ and $\mathbf{r}''(t)$.

18. $\mathbf{r}(t) = (t^2 + 2)\mathbf{i} + (1 - t)^2\mathbf{j}$

19. $\mathbf{r}(t) = (1 + 2t)^2\mathbf{i} + e^{-6t}\mathbf{j}$

20. $\mathbf{r}(t) = \dfrac{t}{1 + t}\mathbf{i} + (\ln 3t^2)\mathbf{j}$

21. $\mathbf{r}(t) = (\sin 10t)\mathbf{i} + (\cosh t^2)\mathbf{j}$

22. $\mathbf{r}(t) = (\cos^3 t)\mathbf{i} + [\sinh^2(3t + 5)]\mathbf{j}$

In Exercises 23 through 26, find vectors and unit vectors tangent to the graphs of the following equations at the point corresponding to the given parametric value.

23. $\mathbf{r} = 3t^2\mathbf{i} + 4t^3\mathbf{j} \qquad t = 2$

24. $\mathbf{r} = a[(\cos t)\mathbf{i} + (\sin t)\mathbf{j}] \qquad (a > 0), t = \dfrac{\pi}{4}$

25. $\mathbf{r} = a(t - \sin t)\mathbf{i} + a(1 - \cos t)\mathbf{j} \qquad (a > 0), t = \dfrac{\pi}{3}$

26. $\mathbf{r} = a(\cos^3 t)\mathbf{i} + a(\sin^3 t)\mathbf{j}$ $(a > 0), t = \dfrac{\pi}{6}$

27. If $\mathbf{r} = (4 + \sin 5t)\mathbf{i} + (7 + \cos 5t)\mathbf{j}$, find
   (a) $\mathbf{r}'(t)$, (b) $\mathbf{r}''(t)$, (c) $|\mathbf{r}''(t)|$, and (d) $D_t[\mathbf{r}'(t) \cdot \mathbf{r}''(t)]$

28. Find a parametric form for the circle $x^2 + y^2 = a^2$, by using the slope m of a line through the point $(0, a)$ on the circle as a parameter. Are all points on the circle obtained by this procedure?

29. Find a parametric form for the parabola $y^2 = 4px$ by using as a parameter the slope m of the straight line through the origin. Are all points on the parabola obtained by this procedure?

30. Discuss and graph the curve known as the *folium of Descartes,* which has the equation $x^3 + y^3 - 3axy = 0, a > 0$, by introducing the slope m of a line through the origin as a parameter. (a) What are the parametric equations? Construct a table of values, that is, assign m and find x and y. (b) Why is the parametric form a virtual necessity in this instance? (*Hint:* Note that $y = x$ is an axis of symmetry. Set $y = x$ and solve for x.)

31. Let $\mathbf{r}(t)$ be a nonconstant vector valued function such that $|\mathbf{r}(t)| = $ positive constant for all t in $[t_0, t_1]$. Show that $\mathbf{r}(t)$ and $\mathbf{r}'(t)$ are perpendicular vectors for all t in the given interval. Interpret the result geometrically.

32. Find $D_x y$ and $D_x^2 y$ if $x = t^3$ and $y = \dfrac{t}{4}$ by (a) an application of formulas (13) and (14) of the text; (b) then by expressing y in terms of x and differentiating.

33. Find $D_x^2 y$ if $x = \sin t - t$ and $y = t + \cos t$.

34. Given that $x = x(t)$ and $y = y(t)$, find formulas for $D_y x$ and $D_y^2 x$. (*Hint:* Interchange the roles of x and y.)

35. Show that $D_y^2 x$ need not be the same as $\dfrac{1}{D_x^2 y}$. (*Hint:* One example such as $x = \cos t$, $y = \sin t$, should prove the assertion.)

36. A circle of radius b rolls without slipping on the *outside* of a circle of radius a with center at the origin. Let P be the point on the circumference of the rolling circle that was at the point $(a, 0)$ when the circles were tangent at $(a, 0)$. Find parametric equations of the path of P. The curve described by the point P is called an *epicycloid* (Figure 11.4.6).

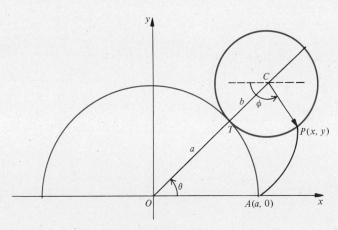

Figure 11.4.6

37. A circle of radius b rolls without slipping on the *inside* of a circle of radius a with center at the origin $(b < a)$. Let P be the point of the circumference of the rolling circle that was at the point $(a, 0)$ when the circles were tangent at $(a, 0)$. Find parametric equations of the path of P. The curve described by the point P is called an *hypocycloid* (Figure 11.4.7).

38. If in Exercise 37, we set $b = \dfrac{a}{4}$, then the hypocycloid of four cusps (or astroid) is obtained. Show that the parametric equations of the curve are $x = a \cos^3 \theta$ and $y = a \sin^3 \theta$.

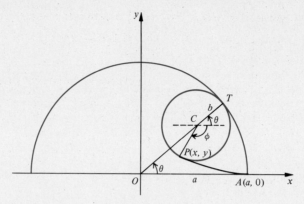

Figure 11.4.7

39. The epicycloid and hypocycloid (see Exercises 36 and 37, respectively) are used in large gear design. The teeth of such gears are sometimes cut in the shape of epicycloids or hypocycloids to avoid the binding and friction that occur between straight teeth. Of special interest is that of the hypocycloid when $a = 2b$. Find the parametric equations in this case and explain why it is of particular interest.
40. Prove Theorem 2 of the text.
41. Prove Theorem 4 of the text.

11.5   MOTION IN THE PLANE; VELOCITY AND ACCELERATION

As a preliminary to our main objective, namely, the analysis of motion along a curved path in the plane, we introduce some definitions of antiderivatives or integrals of vector valued functions.

Definition   Let $\mathbf{r}(t)$ be a vector function given by

$$\mathbf{r}(t) = f(t)\mathbf{i} + g(t)\mathbf{j}$$

then

$$\int \mathbf{r}(t)\,dt = \left[\int f(t)\,dt\right]\mathbf{i} + \left[\int g(t)\,dt\right]\mathbf{j} \tag{1}$$

assuming that the integrals on the right side exist.

Definition (1) is consistent with the definition of integrals of real valued functions since

$$D_t\left(\left[\int f(t)\,dt\right]\mathbf{i} + \left[\int g(t)\,dt\right]\mathbf{j}\right) = D_t\left(\int f(t)\,dt\right)\mathbf{i} + D_t\left(\int g(t)\,dt\right)\mathbf{j}$$

$$= f(t)\mathbf{i} + g(t)\mathbf{j} = \mathbf{r}(t)$$

that is,

$$D_t\left(\int \mathbf{r}(t)\,dt\right) = \mathbf{r}(t)$$

In general, if $\mathbf{w}(t)$ is *any* vector function for which

$$D_t\mathbf{w}(t) = \mathbf{r}(t) \tag{2}$$

then

$$\int \mathbf{r}(t)\, dt = \mathbf{w}(t) + \mathbf{C} \tag{3}$$

where $\mathbf{C}$ is an arbitrary *constant vector;* that is, $\mathbf{C}$ must be independent of t. Equation (3) is the totality of solutions of (2).

EXAMPLE 1   Find the most general vector valued function whose derivative is $\mathbf{r}(t) = t^2\mathbf{i} + 7e^{4t}\mathbf{j}$.

Solution

$$\int \mathbf{r}(t)\, dt = \left[\int t^2\, dt\right]\mathbf{i} + \left[7\int e^{4t}\, dt\right]\mathbf{j}$$

$$= \left[\frac{t^3}{3} + C_1\right]\mathbf{i} + 7\left[\frac{e^{4t}}{4} + C_2\right]\mathbf{j}$$

$$= \frac{t^3}{3}\mathbf{i} + \frac{7e^{4t}}{4}\mathbf{j} + (C_1\mathbf{i} + 7C_2\mathbf{j})$$

$$= \frac{t^3}{3}\mathbf{i} + \frac{7e^{4t}}{4}\mathbf{j} + \mathbf{C}$$

where $\mathbf{C}$ is an arbitrary constant vector.   ●

EXAMPLE 2   Find the vector function for which

$$D_t\mathbf{r}(t) = (\cos 3t)\mathbf{i} + (e^{-2t})\mathbf{j}$$

such that

$$\mathbf{r}(0) = 2\mathbf{i} - \mathbf{j}$$

Solution   From (1) and (3),

$$\mathbf{r}(t) = \left(\frac{\sin 3t}{3}\right)\mathbf{i} - \frac{e^{-2t}}{2}\mathbf{j} + \mathbf{C}$$

where $\mathbf{C}$ is a constant vector (to be found from the initial conditions). Now,

$$\mathbf{r}(0) = 0\mathbf{i} - \tfrac{1}{2}\mathbf{j} + \mathbf{C} = 2\mathbf{i} - \mathbf{j}$$

so that

$$\mathbf{C} = 2\mathbf{i} - \tfrac{1}{2}\mathbf{j}.$$

Therefore

$$\mathbf{r}(t) = \left(\frac{\sin 3t}{3} + 2\right)\mathbf{i} - \frac{1}{2}(e^{-2t} + 1)\mathbf{j}$$

is the vector solution of this initial value problem.   ●

Consider a particle P moving along a plane curve C described parametrically by $x = f(t)$, $y = g(t)$ where now t represents the time. Then the position vector from the origin O to point P is given by

$$\overrightarrow{OP} = \mathbf{r} = x\mathbf{i} + y\mathbf{j} = f(t)\mathbf{i} + g(t)\mathbf{j} \tag{4}$$

The tangent vector $\mathbf{r}'(t) = D_t\mathbf{r}(t)$ is defined to be the *velocity* $\mathbf{v}(t)$ of the particle and is found by

$$\mathbf{v}(t) = \mathbf{r}'(t) = f'(t)\mathbf{i} + g'(t)\mathbf{j} \tag{5}$$

where $x'(t) = f'(t)$ and $y'(t) = g'(t)$ are the scalar components of the velocity in the x- and y-directions, respectively. The magnitude of the velocity vector is called the *speed* and is a scalar determined by

$$|\mathbf{v}(t)| = \sqrt{[f'(t)]^2 + [g'(t)]^2} \tag{6}$$

Note that from (6) the speed is always nonnegative and is zero if and only if $\mathbf{v} = \mathbf{0}$.

The *acceleration* vector is defined as the derivative of the velocity vector. Thus if $\mathbf{a}(t)$ denotes the acceleration vector at time t, we have

$$\mathbf{a}(t) = \mathbf{v}'(t) = \mathbf{r}''(t) = f''(t)\mathbf{i} + g''(t)\mathbf{j} \tag{7}$$

EXAMPLE 3   A particle moves along the path governed by

$$\mathbf{r}(t) = (1 - t)\mathbf{i} + (t^2 + 2)\mathbf{j}$$

Find (a) The Cartesian equation of the path and sketch it
(b) The velocity and acceleration vectors at times $t = 0$ and $t = 1$.
(c) The speed and magnitude of the acceleration at these times.
(d) Sketch representations of the velocity and acceleration vectors at the two times $t = 0$ and $t = 1$.

Solution   (a) $x(t) = 1 - t$, $y(t) = t^2 + 2$, so that if t is eliminated we obtain

$$y = (1 - x)^2 + 2 = x^2 - 2x + 3$$

which is the equation of the parabola with $x = 1$ as an axis. Furthermore, y has an absolute minimum at $x = 1$, namely $y_{\min} = 2$. The curve is sketched in Figure 11.5.1.

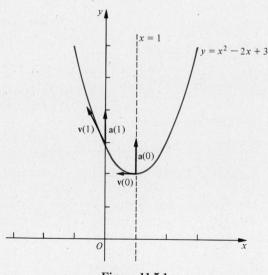

Figure 11.5.1

(b) The velocity and acceleration vectors are given by

$$\mathbf{v}(t) = -\mathbf{i} + 2t\mathbf{j} \qquad \text{and} \qquad \mathbf{a}(t) = 2\mathbf{j}.$$

In particular,

$$\mathbf{v}(0) = -\mathbf{i} \qquad \text{and} \qquad \mathbf{v}(1) = -\mathbf{i} + 2\mathbf{j}$$

while

$$\mathbf{a}(0) = 2\mathbf{j} \qquad \text{and} \qquad \mathbf{a}(1) = 2\mathbf{j}$$

In fact, the acceleration vector is always in the y-direction.

(c) $\text{speed} = |\mathbf{v}(t)| = \sqrt{(-1)^2 + (2t)^2} = \sqrt{1 + 4t^2}$

$$|\mathbf{v}(0)| = 1 \qquad \text{and} \qquad |\mathbf{v}(1)| = \sqrt{5}$$

$$|\mathbf{a}(t)| = |2\mathbf{j}| = 2 \qquad \text{for all } t$$

(d) Refer to Figure 11.5.1. Note that at $t = 0$, $\mathbf{a}(0)$ is orthogonal to $\mathbf{v}(0)$, whereas at $t = 1$, $\mathbf{a}(1)$ is not orthogonal to $\mathbf{v}(1)$. In fact, $\mathbf{a}(t) \cdot \mathbf{v}(t) = 4t$ and is zero if and only if $t = 0$. ●

EXAMPLE 4   The position vector of a particle of mass m is given by the expression

$$r = (b \cos \omega t)\mathbf{i} + (b \sin \omega t)\mathbf{j} \qquad (b > 0 \text{ and } \omega > 0) \qquad (8(\text{a}))$$

Describe the motion by finding the Cartesian equation of the path, velocity, and acceleration.

Solution   The mass m is moving in the xy-plane, where

$$x = b \cos \omega t \qquad y = b \sin \omega t \qquad (b \text{ and } \omega \text{ are constants}) \qquad (8(\text{b}))$$

The Cartesian equation of its path is easily obtained by squaring each of the equations (8(b)) and adding

$$x^2 + y^2 = b^2(\cos^2 \omega t + \sin^2 \omega t) = b^2$$

The mass m moves in a circular path of radius b and center at the origin of coordinates (Figure 11.5.2). Initially (when $t = 0$) the mass is located at $(b, 0)$. It

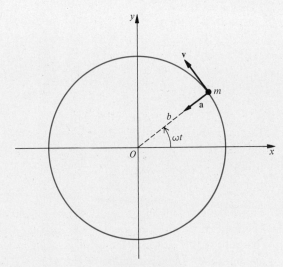

Figure 11.5.2

returns to this point when

$$t = \frac{2\pi}{\omega}, t = \frac{4\pi}{\omega}, t = \frac{6\pi}{\omega}, \ldots$$

The motion is said to be periodic because it repeats itself. Since the mass goes through a complete cycle of motion every $\frac{2\pi}{\omega}$ sec (we assume that t is measured in seconds), it is said to have *period* $T = \frac{2\pi}{\omega}$ sec/cycle and a *frequency* $f = \frac{1}{T} = \frac{\omega}{2\pi}$ cycles/sec. Note that $\omega = 2\pi f$ is the *angular frequency* in radians per second. The velocity and acceleration of the mass are obtained by successive differentiation of (8(a)):

$$\mathbf{v} = D_t\mathbf{r} = (-b\omega \sin \omega t)\mathbf{i} + (b\omega \cos \omega t)\mathbf{j} \tag{9}$$

and

$$\mathbf{a} = D_t\mathbf{v} = D_t^2\mathbf{r} = (-b\omega^2 \cos \omega t)\mathbf{i} + (-b\omega^2 \sin \omega t)\mathbf{j} \tag{10}$$

The speed of the moving point is

$$|\mathbf{v}| = v = [(-b\omega \sin \omega t)^2 + (-b\omega \cos \omega t)^2]^{1/2} = b\omega \tag{11}$$

In this instance, the speed is constant. The acceleration vector, however, is not zero (the acceleration is the derivative of the velocity and *not* the derivative of the speed). From (10),

$$\mathbf{a} = -\omega^2[(b \cos \omega t)\mathbf{i} + (b \sin \omega t)\mathbf{j}] = -\omega^2\mathbf{r} \tag{12}$$

Therefore the acceleration vector is always directed toward the center of the circular path (Figure 11.5.2) and is called the *centripetal acceleration.* Centripetal means "seeking a center."

From Newton's second law of motion, the force $\mathbf{F}$ on the mass m is given by $\mathbf{F} = m\mathbf{a}$ and therefore

$$\mathbf{F} = mD_t^2\mathbf{r} = -m\omega^2\mathbf{r} \tag{13}$$

This is the resultant force required to keep the mass m on a circular path, and it acts toward the center of the circle. Hence, it is called the centripetal force.

Note that the magnitudes of the acceleration and force [as obtained from (12), (13) and (11)] are given by

$$a = \omega^2 b = \frac{v^2}{b} \qquad (\text{since } |\mathbf{r}| = b) \tag{14}$$

and

$$F = \frac{mv^2}{b} \tag{15}$$

These are special cases of a more general result to be established in Section 7 of this chapter. ●

EXAMPLE 5   The Moon revolves about the Earth, making a complete revolution in 27.3 days. Assume that the orbit of the Moon is circular and that the speed of the Moon is constant. If the radius of the orbit is taken to be 239,000 miles, determine the magnitude of the acceleration of the Moon.

Solution In this instance, $b = 239{,}000$ miles $\doteq 1.262 \times 10^9$ ft. The period $T = 27.3$ days $\doteq 2.359 \times 10^6$ sec, from which the speed v is found from

$$v = \frac{2\pi b}{T} \doteq 3{,}360 \text{ ft/sec}$$

The magnitude of the centripetal acceleration is found from (14),

$$a = \frac{(3.360 \times 10^3)^2}{1.262 \times 10^9} \doteq 0.00896 \text{ ft/sec}^2 \qquad\qquad \bullet$$

Our last development in this section is concerned with the determination of the motion of a projectile in the vertical plane. It is assumed that the projectile is fired from the ground with known initial velocity which is the muzzle velocity of the weapon (cannon, rifle, and so on). Furthermore, it is assumed that air resistance is negligible so that the only force acting on the projectile in flight is the gravitational force.

The coordinate system to be used is shown in Figure 11.5.3 where x and y are the components of the displacement in the horizontal and vertical directions, respectively. Let α be the angle of elevation in radians of the projectile initially, and let $\mathbf{v}_0 = v_0[(\cos \alpha)\mathbf{i} + (\sin \alpha)\mathbf{j}]$ be the given initial velocity. With our choice of coordinates, the initial displacement $\mathbf{r}_0 = \mathbf{0}$ since $x = y = 0$ when the time t is zero.

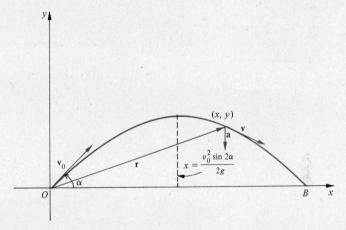

Figure 11.5.3

The only force acting on the mass m is the gravitational force, which is acting towards the center of the Earth and therefore in the $-y$-direction. Thus if $\mathbf{F}$ denotes the force,

$$\mathbf{F} = -mg\mathbf{j} \qquad\qquad (16)$$

Now, from Newton's second law of motion

$$\mathbf{F} = m\mathbf{a} \qquad\qquad (17)$$

where $\mathbf{a}$ is the acceleration of the particle. From (16) and (17),

$$\mathbf{a} = -g\mathbf{j} \qquad\qquad (18)$$

However, $\mathbf{a} = \mathbf{v}'(t)$, where $\mathbf{v}$ is the velocity vector, and integration of both sides of (18) with respect to t yields

$$\mathbf{v} = -gt\mathbf{j} + \mathbf{C}_1 \qquad (19)$$

where $\mathbf{C}_1$ is a vector constant. We determine $\mathbf{C}_1$ from $\mathbf{v}(0) = \mathbf{v}_0$ and therefore $\mathbf{C}_1 = \mathbf{v}_0$. Thus

$$\mathbf{v}(t) = \mathbf{v}_0 - gt\mathbf{j} \qquad (20)$$

and from $\mathbf{v} = \mathbf{r}'(t)$ and integration of both sides with respect to t,

$$\mathbf{r}(t) = \mathbf{v}_0 t - \frac{gt^2}{2}\mathbf{j} + \mathbf{C}_2$$

where $\mathbf{C}_2$ is a second vector constant. However, $\mathbf{r}(0) = \mathbf{0}$ by our choice of coordinates, which implies that $\mathbf{C}_2 = \mathbf{0}$.

Hence
$$\mathbf{r}(t) = \mathbf{v}_0 t - \frac{gt^2}{2}\mathbf{j}$$

$$= v_0 t(\cos\alpha)\mathbf{i} + \left(v_0 t \sin\alpha - \frac{gt^2}{2}\right)\mathbf{j} \qquad (21)$$

Equation (21) is the vector equation of the displacement $\mathbf{r}(t)$ as a function of the time t. This yields the component equations

$$x(t) = v_0 t \cos\alpha \qquad \text{and} \qquad y(t) = v_0 t \sin\alpha - \frac{gt^2}{2} \qquad (22)$$

It is easy to eliminate the parameter t between the two equations (22) to obtain

$$y(x) = x \tan\alpha - \frac{gx^2}{2v_0^2 \cos^2\alpha} \qquad \left(0 < \alpha < \frac{\pi}{2}\right) \qquad (23)$$

which is the equation of a parabola with a vertical axis of symmetry,

$$x = \frac{v_0^2 \sin\alpha \cos\alpha}{g} = \frac{v_0^2 \sin 2\alpha}{2g} \qquad (24)$$

The reader should verify this last result (equation (24)).

Exercises 11.5

1. If $\mathbf{r}'(t) = 3e^{-t}\mathbf{i} - e^{2t}\mathbf{j}$, find $\mathbf{r}(t)$ if $\mathbf{r}(0) = -2\mathbf{i} + 3\mathbf{j}$.

2. If $\mathbf{r}'(t) = (5 \cos 3t)\mathbf{i} - (2 \sin t)\mathbf{j}$, find $\mathbf{r}(t)$ if $\mathbf{r}\left(\frac{\pi}{2}\right) = 4\mathbf{i} + \mathbf{j}$.

3. If $\mathbf{r}''(t) = (t^2 + 4)\mathbf{i} + t^{-2}\mathbf{j}$, find $\mathbf{r}(t)$ if $\mathbf{r}'(1) = \mathbf{i} + \mathbf{j}$ and $\mathbf{r}(1) = \mathbf{0}$.
4. If $\mathbf{r}''(t) = t\mathbf{i} + (3t^2 + 1)\mathbf{j}$, find $\mathbf{r}(t)$ if $\mathbf{r}'(0) = \mathbf{i}$ and $\mathbf{r}(0) = \mathbf{i} - 2\mathbf{j}$.

In Exercises 5 through 8, a particle moves so that at time t its position vector is given by $\mathbf{r}(t) = \langle x(t), y(t) \rangle$. (a) Find its velocity, speed, and acceleration at time t. (b) Sketch the curve along which the particle moves. (c) Find the velocity and acceleration vectors at the indicated time t_1 and draw them so that their initial point corresponds to the position at $t = t_1$.

5. $\mathbf{r}(t) = \langle 10 \sin 3t, 10 \cos 3t \rangle, \quad t_1 = \dfrac{\pi}{6}$

6. $\mathbf{r}(t) = \langle e^{t-3}, e^2 \rangle, \quad t_1 = 1$

7. $\mathbf{r}(t) = \langle 7t - 3, 2t^2 + 5 \rangle, \quad t_1 = 0$

8. $\mathbf{r}(t) = \langle 3 \cos 2t - 5, 6 \sin 2t \rangle, \quad t_1 = \dfrac{\pi}{8}$

In Exercises 9 through 12, determine the displacement $\mathbf{r}(t)$ of the particle given the following information about its displacement $\mathbf{r}(t)$, its velocity $\mathbf{v}(t)$, and its acceleration $\mathbf{a}(t)$.

9. $\mathbf{v}(t) = \langle t^2, t^3 \rangle; \quad \mathbf{r}(0) = \langle 0, 2 \rangle$

10. $\mathbf{v}(t) = \langle (3t + 1)^2, t^{-2} \rangle; \quad \mathbf{r}(1) = \langle -1, 1 \rangle$

11. $\mathbf{a}(t) = \langle 4, 3 \rangle; \quad \mathbf{v}(0) = \langle -1, 0 \rangle; \quad \mathbf{r}(0) = \langle 2, 5 \rangle$

12. $\mathbf{a}(t) = \langle 6, t \rangle; \quad \mathbf{v}(1) = \langle 0, 1 \rangle; \quad \mathbf{r}(1) = \langle 2, 0 \rangle$

13. A particle moves on the straight line $y = 3x - 1$ in the direction in which x is increasing and its speed is uniform at 5 units/sec. Sketch the path, and the velocity and acceleration vectors at $(1, 2)$.

14. A particle moves on the curve $x = y^2$ in the direction in which y is increasing and its speed is uniform at 4 units/sec. Sketch the path and the velocity vector at $(4, 2)$.

15. The planar motion of a particle of constant mass m is given by $\mathbf{r}(t) = \mathbf{r}(0) + \mathbf{C}t$ where $\mathbf{C}$ is a constant vector. **(a)** Find the velocity, speed, and the acceleration. **(b)** What is the force acting on the particle? **(c)** Describe the motion.

16. Prove that if $\mathbf{r}(t)$ and $\mathbf{r}'(t)$ are orthogonal, then $|\mathbf{r}(t)| = $ a constant. Interpret your result.

17. A train rounds a curve in a level road at 40 mi/hr. The curve is in the form of a circular arc of 1600 ft radius. One of the cars together with the load it is carrying weighs 30 tons. Determine the total lateral force that is exerted by the rails on the wheel flanges of this car to keep it moving in its circular path.

(The projectile problem referred to in Exercises 18 through 20 is discussed at the end of Section 11.5)

18. For the projectile problem, determine **(a)** the impact speed and **(b)** the maximum altitude.

19. For the projectile problem, find its range.

20. For the projectile problem, determine the angle α that maximizes its range. What is this maximum range?

11.6   ARC LENGTH-PARAMETRIC FORM

In Section 7.10 we developed formulas for arc length of a plane curve between two points on it. This was derived for curves of the form $y = f(x)$ (for which $f'(x)$ is continuous), which means that a line parallel to the y-axis intersects the graph of f in at most one point.

We shall now derive a method and formula that will enable us to find the length of other kinds of curves, such as that pictured in Figure 11.6.1 (which is not the graph of a function).

Let a curve C be defined parametrically by

$$x = f(t) \quad \text{and} \quad y = g(t) \quad a \leq t \leq b, \tag{1}$$

where f' and g' are continuous on $[a, b]$. From formula (16), Section 7.10, we have

$$(ds)^2 = (dx)^2 + (dy)^2$$

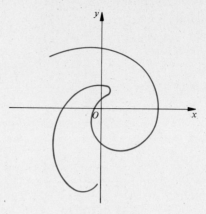

Figure 11.6.1

where ds is the differential of arc length. Then assuming s is an increasing function of t,

$$\frac{(ds)^2}{(dt)^2} = \frac{(dx)^2}{(dt)^2} + \frac{(dy)^2}{(dt)^2}$$

and

$$\frac{ds}{dt} = \sqrt{\left(\frac{dx}{dt}\right)^2 + \left(\frac{dy}{dt}\right)^2}$$

Let L be the length of C. As s goes from O to L, t goes from a to b. Thus

$$L = \int_O^L ds = \int_a^b \frac{ds}{dt}\, dt = \int_a^b \sqrt{\left(\frac{dx}{dt}\right)^2 + \left(\frac{dy}{dt}\right)^2}\, dt$$

$$= \int_a^b \sqrt{[f'(t)]^2 + [g'(t)]^2}\, dt \qquad (2)$$

We have arrived at the following result.

Theorem 1   Let f and g be functions of t having continuous derivatives on a closed interval $a \leq t \leq b$. Then the length of arc of the curve C expressed by

$$x = f(t) \qquad \text{and} \qquad y = g(t) \qquad \text{on} \qquad [a, b] \qquad (1)$$

from the point $(f(a), g(a))$ to the point $(f(b), g(b))$ is given by the formula

$$L = \int_a^b \sqrt{[f'(t)]^2 + [g'(t)]^2}\, dt \qquad (2)$$

Our notational manipulation does not constitute a proof of this theorem. We shall, however, delay giving a formal proof until we first look at some examples.

EXAMPLE 1   Find the length of the plane curve with the parametric equations $x = t + 4$ and $y = 2t - 5$ from $t = 1$ to $t = 3$.

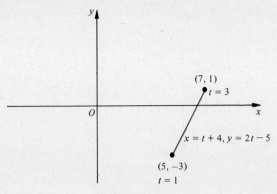

Figure 11.6.2

Solution The curve is a segment of a straight line shown in Figure 11.6.2. We find $D_t x = 1$ and $D_t y = 2$, so that from (2),

$$L = \int_1^3 \sqrt{(1)^2 + (2)^2}\, dt = 2\sqrt{5}$$

This result is easily verified because when $t = 1$, $x = 5$ and $y = -3$; and when $t = 3$, $x = 7$ and $y = 1$. The length of the segment joining points $(5, -3)$ and $(7, 1)$ is

$$\sqrt{(7 - 5)^2 + (1 - (-3))^2} = \sqrt{20} = 2\sqrt{5} \qquad \bullet$$

EXAMPLE 2 Find the length of the curve with parametric equations $x = t^3$ and $y = t^2$ from $t = 0$ to $t = 2$ (Figure 11.6.3).

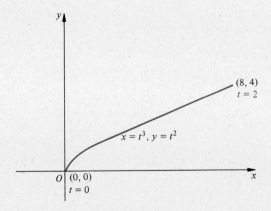

Figure 11.6.3

Solution From the given parametric equations

$$D_t x = 3t^2 \qquad \text{and} \qquad D_t y = 2t$$

and hence from (2),

$$L = \int_0^2 \sqrt{(3t^2)^2 + (2t)^2}\, dt = \int_0^2 t\sqrt{9t^2 + 4}\, dt$$

since $\sqrt{t^2} = t$ when $t \geq 0$. By the power formula

$$L = \int_0^2 (9t^2 + 4)^{1/2} \frac{d(9t^2 + 4)}{18}$$

$$= \tfrac{1}{18} \tfrac{2}{3} (9t^2 + 4)^{3/2} \big|_0^2 = \tfrac{1}{27}[(40)^{3/2} - 8]$$

$$= \tfrac{8}{27}[(10)^{3/2} - 1] \doteq 9.07$$

●

EXAMPLE 3   Determine the length of one arch of the cycloid

$$x = a(\theta - \sin\theta), y = a(1 - \cos\theta) \qquad a > 0$$

Solution   By differentiation, we have

$$D_\theta x = a(1 - \cos\theta) \qquad \text{and} \qquad D_\theta y = a(\sin\theta)$$

from which

$$L = \int_0^{2\pi} \sqrt{a^2(1 - \cos\theta)^2 + a^2 \sin^2\theta} \; d\theta$$

$$= a \int_0^{2\pi} \sqrt{1 - 2\cos\theta + \cos^2\theta + \sin^2\theta} \; d\theta$$

$$= a \int_0^{2\pi} \sqrt{2(1 - \cos\theta)} \; d\theta$$

$$= a \int_0^{2\pi} \sqrt{2\left(2\sin^2\frac{\theta}{2}\right)} \; d\theta$$

$$= 2a \int_0^{2\pi} \sin\frac{\theta}{2} \; d\theta$$

since $\sin\dfrac{\theta}{2} \geq 0$ when $0 \leq \theta \leq 2\pi$

$$L = -4a\left(\cos\frac{\theta}{2}\right)\bigg|_0^{2\pi} = -4a(-1 - 1) = 8a$$

●

Proof of   Introduce a partition P of $[a, b]$ namely $a = t_0 < t_1 < t_2 < \cdots < t_{n-1} < t_n = b$
Theorem 1   into n subintervals with norm $|P|$. Thus if $\Delta t_i = t_i - t_{i-1}$, we have $|\Delta t_i| \leq |P|$ for
$i = 1, 2, \ldots, n$.

Corresponding to each of the numbers t_i are the points $P_i(f(t_i), g(t_i))$ on the curve (1). Furthermore, P_0 denotes the point $(f(a), g(a))$ as shown in Figure 11.6.4. Join P_{i-1} to P_i with a straight line segment or chord of the curve and form the sum

$\displaystyle\sum_{i=1}^{n} |\overline{P_{i-1}P_i}|$ of the lengths of the n chords. Then the **length** of the curve (1) from the

point $P_0(f(a), g(a))$ to $P_n(f(b), g(b))$ is defined to be

$$L = \lim_{|P| \to 0} \sum_{i=1}^{n} |\overline{P_{i-1}P_i}| \tag{3}$$

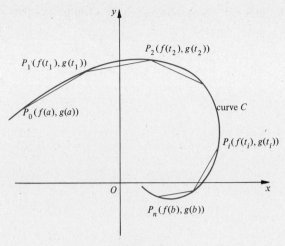

Figure 11.6.4

provided that the limit exists. This means that corresponding to any positive number ε there exists a second positive number δ such that

$$\left| L - \sum_{i=1}^{n} |\overline{P_{i-1}P_i}| \right| < \varepsilon$$

provided that $0 < |P| < \delta$ (for *all* partitions P).

From the distance formula between two points in a plane, we have

$$\sum_{i=1}^{n} |\overline{P_{i-1}P_i}| = \sum_{i=1}^{n} \sqrt{[f(t_i) - f(t_{i-1})]^2 + [g(t_i) - g(t_{i-1})]^2} \qquad (4)$$

Since $f'(t)$ and $g'(t)$ are continuous on $[a, b]$, they must also be continuous on any subinterval $[t_{i-1}, t_i]$ and therefore the hypotheses of the basic mean value theorem of calculus (Section 4.6) are satisfied. Hence there exist numbers z_i and w_i between t_{i-1} and t_i such that for $i = 1, 2, \ldots, n$,

$$f(t_i) - f(t_{i-1}) = f'(z_i)\,\Delta t_i \qquad (5)$$

and

$$g(t_i) - g(t_{i-1}) = g'(w_i)\,\Delta t_i \qquad (6)$$

hold.

Substitution of (5) and (6) into (4) yields,

$$\sum_{i=1}^{n} |\overline{P_{i-1}P_i}| = \sum_{i=1}^{n} \sqrt{[f'(z_i)]^2 + [g'(w_i)]^2}\,\Delta t_i \qquad (7)$$

where $t_{i-1} < z_i < t_i$ and $t_{i-1} < w_i < t_i$. Thus if the limit of the sum exists, we have

$$L = \lim_{|P| \to 0} \sum_{i=1}^{n} \sqrt{[f'(z_i)]^2 + [g'(w_i)]^2}\,\Delta t_i \qquad (8)$$

The sum in (8) is not a Riemann sum because z_i and w_i are not necessarily the same numbers in (t_{i-1}, t_i). However, it can be shown that the value of (8) is precisely the same as would occur if $z_i = w_i$ for each i, $i = 1, 2, 3, \ldots, n$ and we can conclude that

$$L = \int_a^b \sqrt{[f'(t)]^2 + [g'(t)]^2}\, dt \qquad (2) \qquad \blacksquare$$

We return to the curve C, which has parametric equations (1). Let $s(t)$ be the units of length from the point $(f(t_0), g(t_0))$ to a point $(f(t), g(t))$, where it is assumed that s is a nonnegative increasing function of t; that is, as t increases s increases. Then it follows that we may write

$$s(t) = \int_{t_0}^t \sqrt{[f'(u)]^2 + [g'(u)]^2}\, du \qquad (9)$$

from which

$$D_t s = \sqrt{[f'(t)]^2 + [g'(t)]^2} \qquad (10)$$

A vector equation of the curve C is

$$\mathbf{r}(t) = f(t)\mathbf{i} + g(t)\mathbf{j} \qquad (11)$$

so that

$$\mathbf{r}'(t) = f'(t)\mathbf{i} + g'(t)\mathbf{j} \qquad (12)$$

and

$$|\mathbf{r}'(t)| = \sqrt{\mathbf{r}'(t) \cdot \mathbf{r}'(t)} = \sqrt{[f'(t)]^2 + [g'(t)]^2} \qquad (13)$$

From (10) and (13) it follows that

$$|\mathbf{r}'(t)| = D_t s \qquad (14)$$

and the length of a curve for which $D_t s \geq 0$ may be expressed as

$$L = \int_a^b |\mathbf{r}'(t)|\, dt \qquad (15)$$

assuming also that $\mathbf{r}'(t)$ is continuous on $[a, b]$.

EXAMPLE 4   Find the length of arc traced out by the terminal point of the position representation of $\mathbf{r}(t)$ given by

$$\mathbf{r}(t) = a[(\cos^3 t)\mathbf{i} + (\sin^3 t)\mathbf{j}] \qquad 0 \leq t \leq \frac{\pi}{2}, \quad a > 0$$

Solution

$$\mathbf{r}'(t) = 3a[\cos^2 t\,(-\sin t)\mathbf{i} + \sin^2 t\,(\cos t)\mathbf{j}]$$

$$\begin{aligned}
|\mathbf{r}'(t)|^2 = \mathbf{r}'(t) \cdot \mathbf{r}'(t) &= 9a^2[\cos^4 t \sin^2 t + \sin^4 t \cos^2 t] \\
&= 9a^2[\sin^2 t \cos^2 t(\cos^2 t + \sin^2 t)] \\
&= 9a^2 \sin^2 t \cos^2 t
\end{aligned}$$

Thus $|\mathbf{r}'(t)| = 3a \sin t \cos t$ since $a > 0$ and $\sin t$ and $\cos t$ are nonnegative in $\left[0, \frac{\pi}{2}\right]$. Hence the arc length from Equation (15) is given by

$$L = 3a \int_0^{\pi/2} \sin t \cos t\, dt = 3a \left. \frac{\sin^2 t}{2} \right|_0^{\pi/2} = \frac{3a}{2}$$

This is one fourth the total length of a hypocycloid of four cusps (or astroid) obtained by the peripheral trace of a circle of radius $\dfrac{a}{4}$ rolling without slipping on the inside of a circle of radius a (see Figure 11.4.6 and Exercise 38 of that section). The total length of the astroid is $6a$. ●

Exercises 11.6

In Exercises 1 through 12, determine the lengths of the following curves. For the evaluation of more complicated definite integrals use the methods of Chapter 10 or integral tables.

1. $x = 2t + 3$, $y = -t + 2$ from $t = 1$ to $t = 4$.

2. $x = 4 \cos 3t$, $y = 4 \sin 3t$ from $t = 0$ to $t = \dfrac{\pi}{8}$.

3. $x = t^3$, $y = t^2 + 2$ from $t = 0$ to $t = 3$.
4. $x = \frac{4}{3}t^3$, $y = 4t^2$ from $t = -\sqrt{3}$ to $t = 0$.

5. $x = 3 + 5 \sin t$, $y = 4 - 5 \cos t$ from $t = \dfrac{\pi}{8}$ to $t = \dfrac{\pi}{3}$

6. $x = t + 2$, $y = \dfrac{t^2}{2}$ from $t = 0$ to $t = 2$.

7. $x = e^t \sin t$, $y = e^t \cos t$ from $t = -3$ to $t = 5$.
8. $x = a \cos 3t$, $y = a \sin 3t$, $(a > 0)$, from $t = 0$ to $t = 3\pi$; explain your answer.

9. $\mathbf{r}(t) = 10(\cos t + t \sin t)\mathbf{i} + 10(\sin t - t \cos t)\mathbf{j}$, from $t = 0$ to $t = \dfrac{\pi}{4}$.

10. $\mathbf{r}(t) = t^4\mathbf{i} + t^6\mathbf{j}$ from $t = 0$ to $t = 1$.
11. $\mathbf{r}(t) = e^{3t}\mathbf{i} + e^{2t}\mathbf{j}$ from $t = 0$ to $t = 3$.

12. $\mathbf{r}(t) = (\sin t)\mathbf{i} + (\sin^2 t)\mathbf{j}$ from $t = 0$ to $t = \dfrac{\pi}{2}$.

In Exercises 13 through 16 express the arc length of the given arc by means of a definite integral. Do not evaluate the definite integral.

13. $x = t + 5$, $y = t^3$ from $t = -1$ to $t = 3$.
14. $x = e^{3t}$, $y = e^t$ from $t = 0$ to $t = 4$.
15. $x = a \cos t$, $y = b \sin t$ from $t = 0$ to $t = \pi$.
16. $x = \sqrt{1 + t}$, $y = t^2$ from $t = 1$ to $t = 3$.

The *centroid* of a curve C given by $\mathbf{r}(t) = \langle x(t), y(t) \rangle$ where $\mathbf{r}'(t)$ is continuous on $[t_0, t_1]$ is defined by

$$\bar{x} = \frac{1}{L} \int_{t_0}^{t_1} x(t) \, |\mathbf{r}'(t)| \, dt \qquad \bar{y} = \frac{1}{L} \int_{t_0}^{t_1} y(t) \, |\mathbf{r}'(t)| \, dt \tag{16}$$

where L is the length of the curve.

In each of Exercises 17 through 19, find the centroid of the given curve defined on the given domain.

17. $\mathbf{r}(t) = 2t\mathbf{i} + (t - 3)\mathbf{j}$, $\quad -1 \le t \le 5$
18. $\mathbf{r}(t) = (b \cos t)\mathbf{i} + (b \sin t)\mathbf{j}$, $\quad b > 0$ and constant for $0 \le t \le \pi$. What is the anticipated value of $\bar{x}$?
19. $\mathbf{r}(t) = t\mathbf{i} + (\cosh t)\mathbf{j}$, $\quad -a \le t \le a$
20. Derive the formulas (16) for $\bar{x}$ and $\bar{y}$ by using the concept of moment of arc length with respect to an axis.

Let

$$\mathbf{r}(t) = f(t)\mathbf{i} + g(t)\mathbf{j} \qquad (1)$$

be a vector valued function for which $f'(t)$ and $g'(t)$ are continuous in $[a, b]$. Also we assume that $(f'(t))^2 + (g'(t))^2 > 0$; that is, $f'(t)$ and $g'(t)$ are not zero simultaneously. Then the curve C traced by the tip of $\mathbf{r}(t)$ is said to be a *smooth* curve.

The velocity vector $\mathbf{v}$ at $P(t) = (f(t), g(t))$, $(\mathbf{v} \neq \mathbf{0})$, is directed tangentially to C for $a \leq t \leq b$. Let

$$\mathbf{T}(t) = \frac{\mathbf{v}(t)}{|\mathbf{v}(t)|} \qquad (2)$$

We call $\mathbf{T}(t)$ the *unit tangent vector* at $P(f(t), g(t))$. It is so named because its length is 1 and its direction coincides with that of $\mathbf{v}(t)$.

Let $s(t)$ be the length of arc of C measured from $P(a)$ to $P(t)$. Then from the last section, we have

$$s(t) = \int_a^t |\mathbf{r}'(u)| \, du \qquad a \leq t \leq b$$

where it is assumed that $s(t)$ increases with increase in t. Then the speed of the moving point is given by

$$D_t s = |\mathbf{r}'(t)| = |\mathbf{v}(t)| \qquad (3)$$

Next, we define the *curvature vector* $\mathbf{K}(t)$ at P. It is the rate of change of the unit tangent vector $\mathbf{T}(t)$ with respect to the arc length s. Hence the curvature vector $\mathbf{K}(t)$ is defined by

$$\mathbf{K}(t) = D_s \mathbf{T}(t) \qquad (4)$$

Since the length of $\mathbf{T}(t)$ is 1 it follows that the change in $\mathbf{T}$ occurs because of the change in its direction. Thus from (4), *the curvature vector measures the rate of change of the direction of the tangent to the curve with respect to arc length traversed during this change.*

By means of the chain rule, we can express $\mathbf{K}(t)$ in terms of t. We have from (4),

$$\mathbf{K}(t) = D_t \mathbf{T}(t) D_s t = \frac{\mathbf{T}'(t)}{D_t s} = \frac{\mathbf{T}'(t)}{|\mathbf{v}(t)|} \qquad (5)$$

since

$$D_s t = \frac{1}{D_t s} \qquad \text{and} \qquad D_t s = |\mathbf{v}(t)| \text{ from (3)}.$$

The **curvature** $K(t)$ of the arc $x = f(t)$, $y = g(t)$, $a \leq t \leq b$ at a point P is defined to be the magnitude of the curvature vector. Therefore from (5),

$$K(t) = |\mathbf{K}(t)| = \left| \frac{\mathbf{T}'(t)}{|\mathbf{v}(t)|} \right| \qquad (6)$$

The curvature tells us how sharply the curve is turning.

EXAMPLE 1   Find the curvature at each point of a straight line.

Solution   In the case of a straight line the unit tangent vector $\mathbf{T}(t)$ lies along that line at all points on it. Thus, the direction of $\mathbf{T}(t)$ does not change as we move from point to point. Therefore the curvature vector

$$\mathbf{K}(t) = D_s\mathbf{T}(t) = \mathbf{0}$$

and the curvature $K = 0$. Hence the curvature of a straight line is zero at all its points (which is to be expected since a straight line is not turning).   ●

EXAMPLE 2   Find the curvature vector and the curvature of a circle of radius a with center at the origin.

Solution   The points on the boundary of the circle may be expressed by the vector equation

$$\mathbf{r}(t) = (a \cos t)\mathbf{i} + (a \sin t)\mathbf{j} \qquad (a > 0 \text{ and constant}) \qquad (7)$$

where t is the angle between the radius and the x-direction.
Then

$$\mathbf{v}(t) = D_t\mathbf{r} = (-a \sin t)\mathbf{i} + (a \cos t)\mathbf{j}$$

and

$$|\mathbf{v}(t)| = \sqrt{(-a \sin t)^2 + (a \cos t)^2} = a$$

Therefore from (2),

$$\mathbf{T}(t) = (-\sin t)\mathbf{i} + (\cos t)\mathbf{j}$$
$$\mathbf{T}'(t) = (-\cos t)\mathbf{i} + (-\sin t)\mathbf{j}$$

and

$$\mathbf{K}(t) = \frac{\mathbf{T}'(t)}{|\mathbf{v}(t)|} = -\frac{1}{a}((\cos t)\mathbf{i} + (\sin t)\mathbf{j})$$

From (7),

$$\mathbf{K}(t) = -\frac{\mathbf{r}}{a^2}$$

Therefore $\mathbf{K}(t)$ is a vector pointing in the radially inward direction (opposite to $\mathbf{r}$) and

$$K(t) = |\mathbf{K}(t)| = \left|\frac{-\mathbf{r}}{a^2}\right| = \frac{1}{a^2}|\mathbf{r}| = \frac{a}{a^2} = \frac{1}{a} \qquad (8)$$

Hence, the curvature of a circle is constant and is the reciprocal of the radius.
●

Consider once again the unit tangent vector $\mathbf{T}(t)$, which is directed tangentially to a given curve. Since $\mathbf{T}(t)$ is a unit vector, we have

$$\mathbf{T}(t) \cdot \mathbf{T}(t) = 1 \qquad (9)$$

If both sides of (9) are differentiated with respect to t, then after simplifying we have

$$\mathbf{T}(t) \cdot \mathbf{T}'(t) = 0 \qquad (10)$$

Equation (10) implies that the vectors $\mathbf{T}(t)$ and $\mathbf{T}'(t)$ are perpendicular to one another provided that $|\mathbf{T}'(t)| \neq 0$ at the point in question. Let us assume that $|\mathbf{T}'(t)| \neq 0$ and therefore that $\mathbf{T}'(t)$ must be normal to the curve. We introduce a second unit vector $\mathbf{N}(t)$ defined by

$$\mathbf{N}(t) = \frac{\mathbf{T}'(t)}{|\mathbf{T}'(t)|} \tag{11}$$

$\mathbf{N}(t)$ is called the ***unit normal vector*** and it can be verified that $\mathbf{N}(t)$ is always on the concave side of the curve as shown in Figure 11.7.1. Note that the direction of $\mathbf{N}(t)$ is approximately in the direction of

$$\frac{\mathbf{T}(t + \Delta t) - \mathbf{T}(t)}{\Delta t}$$

and that their directions coincide when the limit is taken as $\Delta t \to 0$. From (5) and (11), it follows that the curvature vector $\mathbf{K}(t)$ points in the same direction as $\mathbf{N}(t)$; that is, $\mathbf{K}(t)$ is on the normal to curve C pointing in the direction of the concave side.

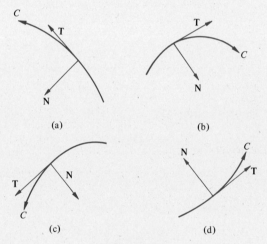

Figure 11.7.1

Let P be a particular point on C for which $\mathbf{T}(t)$ and $\mathbf{N}(t)$ are defined and, furthermore, assume that the curvature $K(t) \neq 0$. We seek a circle that is tangent to C at P and has the curvature $K(t)$ at P. Such a circle must have its center on the line perpendicular to the tangent line in the direction of $\mathbf{N}(t)$. This circle is called the ***circle of curvature*** and its radius is called the ***radius of curvature.*** This circle is also called the ***osculating circle*** in that it approximates the curve more closely than any other circle. The ordinates, slopes, and curvatures for the osculating circle are equal to the corresponding quantities of the curve at the given point. Also the center of the osculating circle is on the concave side of the curve (Figure 11.7.2).

If $K(t)$ is the curvature at a point P on a curve C where it is assumed that $K(t) \neq 0$, then we define

$$\rho(t) = \frac{1}{K(t)} \tag{12}$$

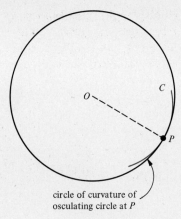

circle of curvature of
osculating circle at P

Figure 11.7.2

to be the *radius of curvature* of the curve at P. It is the radius of the circle of curvature, that is, the radius of the circle of "closest fit" in the sense described above.

Let α be the angle of inclination of the unit tangent vector $\mathbf{T}$ to a given curve C at a point P. Then

$$\mathbf{T} = (\cos \alpha)\mathbf{i} + (\sin \alpha)\mathbf{j} \tag{13}$$

(where α, and hence $\mathbf{T}$ are functions of t.) If we differentiate both sides of (13) with respect to the arc length s, measured from an arbitrary initial point on C, there results

$$D_s\mathbf{T} = -\sin \alpha D_s\alpha\mathbf{i} + \cos \alpha D_s\alpha\mathbf{j}$$

Therefore

$$|D_s\mathbf{T}| = |D_s\alpha| \, |(-\sin \alpha)\mathbf{i} + (\cos \alpha)\mathbf{j}| = |D_s\alpha| \tag{14}$$

From (4) and (14) it follows that the curvature $K(t)$ may be calculated from

$$K = |D_s\alpha| \tag{15}$$

From (15) we can find a formula for the curvature K directly from the parametric equations of the curve C given by $x = f(t)$ and $y = g(t)$.

$$\tan \alpha = \frac{g'(t)}{f'(t)} \tag{16}$$

Differentiation of both sides of (16) with respect to t yields

$$\sec^2 \alpha D_t\alpha = \frac{f'(t)\,g''(t) - g'(t)\,f''(t)}{[f'(t)]^2} \tag{17}$$

But

$$\sec^2 \alpha = 1 + \tan^2 \alpha = 1 + \left[\frac{g'(t)}{f'(t)}\right]^2 \tag{18}$$

and substitution of (18) into (17) yields

$$D_t\alpha = \frac{f'(t)\,g''(t) - g'(t)\,f''(t)}{[f'(t)]^2 + [g'(t)]^2} \tag{19}$$

and, if s increases as t increases

$$D_t s = [(f')^2 + (g')^2]^{1/2} \tag{20}$$

Therefore, by the chain rule

$$K = |D_s \alpha| = \left| \frac{D_t \alpha}{D_t s} \right| \tag{21}$$

and we have from (19) through (21),

$$K(t) = \frac{|f'(t)\, g''(t) - g'(t)\, f''(t)|}{[f'(t))^2 + (g'(t))^2]^{3/2}} \tag{22}$$

EXAMPLE 3 Apply (22) to show that the curvature of a circle of radius a expressed parametrically by

$$x = f(t) = a \cos t \qquad \text{and} \qquad y = g(t) = a \sin t \qquad 0 \le t < 2\pi$$

is $K = \dfrac{1}{a}$.

Solution

$$\begin{aligned} f'(t) &= -a \sin t & f''(t) &= -a \cos t \\ g'(t) &= a \cos t & g''(t) &= -a \sin t \end{aligned}$$

From (22),

$$\begin{aligned} K(t) &= \frac{|(-a \sin t)(-a \sin t) - (a \cos t)(-a \cos t)|}{[(-a \sin t)^2 + (a \cos t)^2]^{3/2}} \\[2mm] &= \frac{a^2}{a^3} = \frac{1}{a} \end{aligned}$$

Note that $\rho(t) = \dfrac{1}{K(t)} = a$.

Thus, the radius of curvature is the same as the radius of the circle. This is to be expected since the circle of curvature for a circle should be the circle itself.

●

EXAMPLE 4 Find the curvature of the parabola given parametrically by

$$x = f(t) = t - 1 \qquad \text{and} \qquad y = g(t) = t^2 + 1$$

(a) in general (for an arbitrary value of t); (b) when $t = 0$; (c) find the maximum value of $K(t)$; (d) find the circle of curvature when $t = 0$.

Solution (a)

$$\begin{aligned} f'(t) &= 1 & f''(t) &= 0 \\ g'(t) &= 2t & g''(t) &= 2 \end{aligned}$$

From (22),

$$K(t) = \frac{|1(2) - 0(2t)|}{[1^2 + (2t)^2]^{3/2}} = \frac{2}{[1 + 4t^2]^{3/2}}$$

(b) $K(0) = 2$

(c) $K(t)$ is largest when $1 + 4t^2$ is least. This occurs when $t = 0$ and $K_{max} = K(0) = 2$.

(d) If $t = 0$ then $(x, y) = (-1, 1)$. $\rho(0) = \dfrac{1}{K(0)} = \dfrac{1}{2}$.

$$\mathbf{r}'(t) = \mathbf{i} + 2t\mathbf{j} \qquad \text{so} \qquad \mathbf{r}'(0) = \mathbf{i} = \mathbf{T}(0)$$

Then

$$\mathbf{N}(0) = \mathbf{j}$$

Therefore, the center of curvature is in the positive y-direction from $(-1, 1)$. The coordinates of the center of curvature are thus $(-1, 1 + \frac{1}{2}) = (-1, \frac{3}{2})$.

An equation of the circle of curvature is

$$(x + 1)^2 + (y - \tfrac{3}{2})^2 = \tfrac{1}{4}$$

Figure 11.7.3 shows the parabola and the circle of curvature. ●

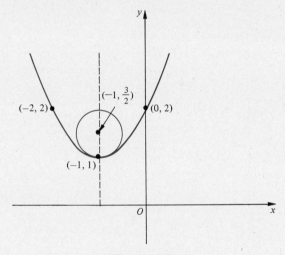

Figure 11.7.3

An important special case of (22) occurs when the given curve is expressed in Cartesian form, say, $y = g(x)$. Then we can apply (22) by setting $x = f(t) = t$ and $y = g(t)$. Then $f'(t) = 1$ and $f''(t) = 0$ and (22) yields

$$K = \frac{|g''(t)|}{[1 + (g'(t))^2]^{3/2}} = \frac{|g''(x)|}{[1 + (g'(x))^2]^{3/2}} \tag{23}$$

Similarly, if $x = f(y)$ we find with $y = g(t) = t$ that

$$K = \frac{|f''(t)|}{[1 + (f'(t))^2]^{3/2}} = \frac{|f''(y)|}{[1 + (f'(y))^2]^{3/2}} \tag{24}$$

EXAMPLE 5 Find the curvature of $y = \ln x$, $x > 0$. What is the limit of the curvature (i) as $x \to 0^+$ and (ii) as $x \to \infty$?

Solution Substitute $g(x) = \ln x$, $g'(x) = \dfrac{1}{x}$, $g''(x) = -\dfrac{1}{x^2}$ into (23), and we obtain

$$K(x) = \frac{\left| -\dfrac{1}{x^2} \right|}{\left(1 + \dfrac{1}{x^2} \right)^{3/2}} = \frac{x}{(x^2 + 1)^{3/2}}$$

Therefore

(i) $K(x) \to 0$ as $x \to 0^+$

(ii) $K(x) < \dfrac{1}{x^2}$ since $(x^2 + 1)^{3/2} < x^3$

Hence, $K(x) \to 0$ as $x \to \infty$. ●

The velocity vector at any point P on a curve C is given by

$$\mathbf{v}(t) = D_t s \, \mathbf{T}(t) \tag{25}$$

where $\mathbf{T}(t)$ is a unit vector tangent to the curve in the direction in which s is increasing. Thus $D_t s = |\mathbf{v}(t)| = $ speed at a point P. We now seek a formula for the acceleration $\mathbf{a}(t)$ at P. In general, $\mathbf{a}(t)$ can be expressed as the sum of a tangential vector component and a normal vector component; that is, the former will be a multiple of $\mathbf{T}(t)$ and the latter a multiple of $\mathbf{N}(t)$. If we differentiate both sides of (25) with respect to t, there results

$$\mathbf{a}(t) = D_t^2 s \, \mathbf{T}(t) + D_t s \, D_t \mathbf{T}(t)$$

But from (5)

$$D_t \mathbf{T}(t) = |\mathbf{v}(t)| \, \mathbf{K}(t)$$
$$= D_t s \, \mathbf{K}(t)$$

and we have

$$\mathbf{a}(t) = D_t^2 s \, \mathbf{T}(t) + (D_t s)^2 \, \mathbf{K}(t)$$

$$\mathbf{a}(t) = D_t^2 s \, \mathbf{T}(t) + (D_t s)^2 \, K(t) \, \mathbf{N}(t) \tag{26}$$

[That $\mathbf{K}(t) = K(t) \, \mathbf{N}(t)$ follows from equations (5), (6) and (11).]

Equation (26) expresses the acceleration vector $\mathbf{a}(t)$ at $P(t)$ on the curve C as the sum of its *tangential vector component* $D_t^2 s \, \mathbf{T}(t)$ and its *normal vector component* $(D_t s)^2 \, K(t) \, \mathbf{N}(t)$. The corresponding scalar components are

$$a_T = D_t^2 s = \text{tangential scalar component of acceleration} \tag{27}$$

$$a_N = (D_t s)^2 K(t) = \frac{v^2}{\rho} = \text{normal scalar component of acceleration} \tag{28}$$

EXAMPLE 6 A particle moves along a curved path given by the equations

$$x = a \cos t \qquad y = b \sin t \qquad (a > 0, b > 0 \text{ and constant})$$

Find the tangential and normal scalar components of acceleration.

Solution

$$v = D_t s = [(D_t x)^2 + (D_t y)^2]^{1/2} = (a^2 \sin^2 t + b^2 \cos^2 t)^{1/2}$$

$$a_T = D_t v = \frac{(a^2 - b^2) \sin 2t}{2[a^2 \sin^2 t + b^2 \cos^2 t]^{1/2}}$$

To find a_N, we first find the curvature $K(t)$. From (22) with $f(t) = a \cos t$ and $g(t) = b \sin t$, we have

$$K(t) = \frac{|f'(t) g''(t) - f''(t) g'(t)|}{[(f'(t))^2 + (g'(t))^2]^{3/2}}$$

$$= \frac{|(-a \sin t)(-b \sin t) + (a \cos t)(b \cos t)|}{[a^2 \sin^2 t + b^2 \cos^2 t]^{3/2}}$$

$$= \frac{ab}{(a^2 \sin^2 t + b^2 \cos^2 t)^{3/2}}$$

Hence

$$a_N = \frac{ab(a^2 \sin^2 t + b^2 \cos^2 t)}{(a^2 \sin^2 t + b^2 \cos^2 t)^{3/2}} = \frac{ab}{(a^2 \sin^2 t + b^2 \cos^2 t)^{1/2}} \qquad \bullet$$

Exercises 11.7

In each of Exercises 1 through 6, find the unit tangent vector $\mathbf{T}$ and the unit normal vector $\mathbf{N}$ to the given curve.

1. $\mathbf{r}(t) = t^2\mathbf{i} + t\mathbf{j}$
2. $\mathbf{r}(t) = (A + b \cos t)\mathbf{i} + (B + b \sin t)\mathbf{j}$,　(A, B, and b are constants, $b > 0$)
3. $\mathbf{r}(t) = e^t\mathbf{i} + t\mathbf{j}$
4. $\mathbf{r}(t) = t\mathbf{i} + (\cosh t)\mathbf{j}$
5. $\mathbf{r}(t) = (\cos t + t \sin t)\mathbf{i} + (\sin t - t \cos t)\mathbf{j}$,　$t > 0$
6. $\mathbf{r}(t) = (\ln t)\mathbf{i} + (t - 1)\mathbf{j}$,　$t > 0$

In Exercises 7 through 10, (a) find the curvature at any point on the indicated curve; (b) find where the radius of curvature is least and its minimum value; (c) draw a diagram of the curve and show the circle of curvature for which the radius of curvature is a minimum.

7. $y = 3 + 2x^2$
8. $y = \cos x$,　$-\pi \leq x < \pi$
9. $xy = 1$,　$x > 0$
10. $y = \ln x$,　$x > 0$

11. Find the curvature of $y = \frac{a}{2}(e^{x/a} + e^{-x/a})$, $a > 0$, at any point on the curve. In particular find $K(a)$.

12. If $[y'(x)]^2 \ll 1$, show that $|y''(x)|$ can be used as an approximation for the curvature of a curve whose equation is $y = f(x)$. This result is used extensively in the analysis of beam deflections and their corresponding bending stresses. (The symbol $\ll$ means much less than.)

13. Find the curvature of $y^2 = 4px$ at any point on the curve. In particular, find K at $(0, 0)$. [*Hint:* Use (24).]

14. Find the curvature of $x^2 + y^2 = 100$ at $(-6, 8)$ by using implicit differentiation.

15. Find the radius of curvature at the point $(0, b)$ on the curve $\dfrac{x^2}{a^2} + \dfrac{y^2}{b^2} = 1$, where $a > 0$ and $b > 0$.

In Exercises 16 through 19, use formula (22) to find the curvature $K(t)$.

16. $x = a(t - \sin t)$, $y = a(1 - \cos t)$,　$a > 0, 0 < t < 2\pi$.
17. $x = e^t \cos t$, $y = e^t \sin t$
18. $x = a \cos^3 t$, $y = a \sin^3 t$,　$a > 0$

19. $x = 2 \cos t + 5$, $y = 3 \sin t - 8$

20. Show that the radius of curvature of the curve $b^2x^2 - a^2y^2 = a^2b^2$, $(ab \neq 0)$, at any point (x, y) on it is given by $\dfrac{(a^4y^2 + b^4x^2)^{3/2}}{a^4b^4}$.

21. At any time $t \geq 0$, the position of a particle is given parametrically by

$$x = \sin t - t \cos t \qquad \text{and} \qquad y = \cos t + t \sin t$$

   (a) Find the speed $v = |\mathbf{v}|$ and the magnitude of the acceleration.
   (b) Find the radius of curvature.
   (c) Find the scalar tangential and normal components of the acceleration vector.

22. At any time t, the position of a particle is given by

$$\mathbf{r}(t) = 3t\mathbf{i} + t^2\mathbf{j}$$

   (a) Find the velocity $\mathbf{v}$ and the acceleration $\mathbf{a}$.
   (b) Find the speed v and $D_t v$.
   (c) Find the radius of curvature.
   (d) Find the scalar tangential and normal components of the acceleration vector.

23. An automobile weighs 4000 lb and is moving with uniform speed of 40 mi/hr and makes a circular turn on a flat road. If the radius of the circular turn is 132 ft, what frictional force exerted by the road on the bottom of the tires is required to prevent the car from skidding? [*Hint:* The frictional force F must act radially inward, and it is required to keep the automobile moving on a circular path with constant speed. Also, $F = (\text{mass}) \times (\text{normal component of acceleration}).$]

24. If the speed of the vehicle in Exercise 23 were 20 mi/hr rather than 40 mi/hr, what is the required frictional force to prevent the car from skidding?

25. A particle moves along the curve $y - 2x^2 = 0$ with uniform speed $v = 10$. Find the tangential and normal components of acceleration.

26. A particle moves along the curve $x = y^2 + 3$ with the y component of velocity constant at 5. Find the tangential and normal components of acceleration. What is the magnitude of the acceleration?

27. The normal component of acceleration a_N is often the most difficult to obtain. Show how to determine a_N knowing the x, y, and tangential components of acceleration a_x, a_y, and a_T.

28. The locus of the centers of the circles of curvature for a given curve C is a curve called the *evolute* E of C. Show that if (X, Y) denotes the Cartesian coordinates of E corresponding to (x, y) on C that

$$X = x - R(D_s y), \qquad Y = y + R(D_s x)$$

where R is the radius of curvature of C at (x, y) and s is the arc length measured positively. [*Hint:* $\mathbf{N} = (-D_s y)\mathbf{i} + (D_s x)\mathbf{j}$.]

29. Find the equation of the evolute of the parabola whose equation is $y = x^2$. Sketch the parabola and its evolute.

30. From Exercise 28 show that if x and y are twice differentiable functions of a general parameter t, say, $x = f(t)$ and $y = g(t)$, the Cartesian coordinates of the evolute are determined from

$$X(t) = f(t) - g'(t) \frac{[f'(t)]^2 + [g'(t)]^2}{|f'(t)\,g''(t) - f''(t)\,g'(t)|}$$

$$Y(t) = g(t) + f'(t) \frac{[f'(t)]^2 + [g'(t)]^2}{|f'(t)\,g''(t) - f''(t)\,g'(t)|}$$

31. From Exercise 30, show that the evolute of $x = a \cos t$, $y = b \sin t$, where $a > 0$ and $b > 0$, is given parametrically by

$$X(t) = \frac{a^2 - b^2}{a} \cos^3 t \qquad Y(t) = -\frac{a^2 - b^2}{a} \sin^3 t$$

Find the Cartesian form for the evolute.

Review and Miscellaneous Exercises

In Exercises 1, 2, and 3, express each of the following as a single vector.

1. $3\langle -1, -4 \rangle - 2\langle 5, 0 \rangle$
2. $6[\langle 2, 5 \rangle - 3\langle 0, 7 \rangle + 4\langle -1, 4 \rangle]$
3. $(12 - 8 - 1)\langle k, -k \rangle + k\langle -1, 2 \rangle$
4. Solve for **a**: $6\mathbf{a} - \langle 2, 5 \rangle = -4\mathbf{a} + \langle -1, 3 \rangle$
5. Solve for **a** and **b**:

$$\begin{cases} 8\mathbf{a} - 5\mathbf{b} = \langle 55, -76 \rangle \\ 4\mathbf{b} + 3\mathbf{a} = \langle 3, -5 \rangle \end{cases}$$

6. Find a vector of length 20 that is orthogonal to the vector $3\mathbf{i} + 4\mathbf{j}$.

In Exercises 7 and 8, evaluate the limits.

7. $\lim\limits_{t \to 2} \left\langle \dfrac{t^2 - 4}{t - 2}, \dfrac{t^3 - 8}{t - 2} \right\rangle$

8. $\lim\limits_{t \to 0} \left\langle \dfrac{\sin 3t}{t}, \dfrac{1 - \cos^2 t}{t} \right\rangle$

9. Find k so that $-3\mathbf{i} + k\mathbf{j}$ is orthogonal to $-3\mathbf{i} + 7\mathbf{j}$.
10. Prove that $\mathbf{a} - \mathbf{b}$ is orthogonal to $\mathbf{a} + \mathbf{b}$ if and only if $|\mathbf{a}| = |\mathbf{b}|$, where it is understood that $\mathbf{a}$ and $\mathbf{b}$ are nontrivial vectors. Give a geometric interpretation of this result.
11. It has been observed that if $\mathbf{a} \cdot \mathbf{b} = \mathbf{a} \cdot \mathbf{c}$ then $\mathbf{b} = \mathbf{c}$ is not necessarily true. Show, however, that if $\mathbf{a} \cdot \mathbf{b} = \mathbf{a} \cdot \mathbf{c}$ for all $\mathbf{a}$, then $\mathbf{b} = \mathbf{c}$ must hold.
12. Find a linear combination of the vectors $\langle 3, 4 \rangle$, $\langle 4, 6 \rangle$ and $\langle 1, -2 \rangle$ with coefficients not all zero that equals zero.
13. Let $\mathbf{a} = \langle a_1, a_2 \rangle$ and $\mathbf{b} = \langle b_1, b_2 \rangle$ be two noncollinear vectors, but otherwise arbitrary. Show that *any* vector in the plane may be expressed as a linear combination of $\mathbf{a}$ and $\mathbf{b}$. (a) Would you say that $\mathbf{a}$ and $\mathbf{b}$ serve as basis vectors for vectors in the plane? (b) What happens if the two vectors are collinear?
14. $\overrightarrow{OA} = \mathbf{a}$ and $\overrightarrow{OB} = \mathbf{b}$ be two arbitrary vectors with initial point O. On the segment joining A and B is a point whose distance from A is 4 times its distance from B. Find $\overrightarrow{OP}$ in terms of $\mathbf{a}$ and $\mathbf{b}$.
15. Find a vector form of the equation of a straight line passing through (a_1, a_2) with slope m.
16. Prove by vector methods that the line segments joining the midpoints of the opposite sides of a plane quadrilateral bisect each other.
17. Prove that the medians of a triangle intersect in a point two thirds of the distance from each vertex to the midpoint of the opposite side. (This is to be done by vector methods.)
18. Show that the bisector of an angle of a triangle intersects the opposite side dividing it into two segments, the ratio of whose lengths is the same as the ratio of the lengths of the other two sides. (*Hint:* Introduce vectors and unit vectors along the sides of the angle which is bisected.)
19. A man travelling southward at 15 mi/hr observed that the wind appears to be coming from the west. On increasing his speed to 35 mi/hr it appears to come from 45° south of west. Find the direction and speed of the wind. (*Hint:* Let $x\mathbf{i} + y\mathbf{j}$ be the velocity of the wind where $\mathbf{i}$ represents 5 mi/hr eastward, whereas $\mathbf{j}$ represents 5 mi/hr in the southward direction. The velocity of the wind relative to the man is the vector difference of the velocity of the wind and the velocity of the man.)
20. If $x = 3t^2 - 7t + 1$ and $y = 2t^2 + 5t + 2$, find $D_x y$ and $D_x^2 y$.
21. Extend the triangular inequality to three vectors, then generalize to n vectors, namely that

$$|\mathbf{r}_1 + \mathbf{r}_2 + \cdots + \mathbf{r}_n| \leq |\mathbf{r}_1| + |\mathbf{r}_2| + \cdots + |\mathbf{r}_n|$$

for arbitrary vectors in the plane.

22. The resultant of a horizontal force and a force inclined at 45° to the vertical balances a force of 200 lb acting downward. Find the magnitude of the two forces.
23. Show that the length of the segment of each tangent line to $\mathbf{r}(t) = a[(\cos^3 t)\mathbf{i} + (\sin^3 t)\mathbf{j}]$, $a > 0$, cut off

by the coordinate axes is constant. What is the value of the constant? Note that the curve is the hypocycloid of four cusps (or astroid).

24. Find the vector projection of $2\mathbf{i} - 3\mathbf{j}$ onto $3\mathbf{i} + 4\mathbf{j}$.

25. A man wishes to row directly across a river in an easterly direction. There is a constant current of 3 mi/hr which is flowing in a southerly direction. If the man's rowing speed (in still water) is 5 mi/hr, find the direction along which the boat should be aimed. If the width of the river is 6 mi, how long does it take to row across? Use vector methods to solve this problem.

26. A particle moves on a plane curve and its position vector is given by $\mathbf{r}(t) = (t^2 - 3t + 1)\mathbf{i} + (t^3 - 10t + 2)\mathbf{j}$. Find the velocity, speed, acceleration, and magnitude of acceleration functions.

27. If $\mathbf{r}(t) = \dfrac{t}{\sqrt{t^2 + 9}}\mathbf{i} + \dfrac{t}{2 + t}\mathbf{j}$, find $\int \mathbf{r}(t)\,dt$.

28. Prove that $\mathbf{r}(t) \cdot \mathbf{r}'(t) = \frac{1}{2}D_t|r(t)|^2$.

29. If $\mathbf{r}''(t) = t(3\mathbf{i} - \mathbf{j})$, find $\mathbf{r}(t)$ if $\mathbf{r}'(2) = \mathbf{i} - 2\mathbf{j}$ and $\mathbf{r}(2) = \mathbf{0}$.

30. The position vector of a moving point is given by $\mathbf{r}(t) = (A \cos \omega t)\mathbf{i} + (B \sin \omega t)\mathbf{j}$ where A, B, and ω are constants.
    (a) Find the Cartesian equation of its path.
    (b) Show that the magnitude of the acceleration is proportional to the distance of the point from the origin.
    (c) What is the direction of the acceleration vector?

31. If $\mathbf{r}(t) = (\cosh \omega t)\mathbf{a} + (\sinh \omega t)\mathbf{b}$ where $\mathbf{a}$, $\mathbf{b}$ and ω are constants, show that $D_t^2\mathbf{r} - \omega^2\mathbf{r} = \mathbf{0}$.

32. A particle moves on the boundary of the circle $x^2 + y^2 = 100$ (x and y are measured in feet) in a counterclockwise direction, with uniform speed of 20 ft/sec. Find the velocity and acceleration vectors at the point $(6, 8)$.

33. A rock is thrown horizontally from the top of a cliff 144 ft high with an initial speed of 64 ft/sec. Find an equation for the path of the projectile, the time to reach the ground, and the distance from the base of the cliff to where it lands. Use vector methods.

34. A rock is thrown horizontally from the top of a cliff h ft high with an initial speed of v_0 ft/sec. Find an equation for the path of the projectile, the time to reach the ground, and the distance from the base of cliff to where it lands. Does the time to reach the ground depend on v_0? Use vector methods.

35. Find the length of the curve with parametric equations $x = 5t - 3$ and $y = t + 2$ from $t = 0$ to $t = 7$.

36. Find the length of the curve with parametric equations $x = 4t^2 + 1$ and $y = t^3 - 10$ from $t = 0$ to $t = 5$.

37. Let $\mathbf{u} = (\sin \alpha)\mathbf{i} + (\cos \alpha)\mathbf{j}$, and $\mathbf{v} = (\cos \alpha)\mathbf{i} - (\sin \alpha)\mathbf{j}$.
    (a) Find the lengths of $\mathbf{u}$ and $\mathbf{v}$
    (b) Find the angle between $\mathbf{u}$ and $\mathbf{v}$
    (c) Find $\mathbf{i}$ and $\mathbf{j}$ in terms of $\mathbf{u}$ and $\mathbf{v}$.

38. Find the curvature, radius of curvature and the coordinates of the center of curvature for $y = ax^2$ at $(0, 0)$ where a is a positive constant.

39. Find the curvature, radius of curvature and the coordinates of the center of curvature for $y^2 = x^3$ at the point $(1, 1)$.

40. Show that the curvature of a plane curve $\mathbf{r} = \mathbf{r}(s)$ where s is arc length, may be expressed by
$$K = |D_s^2\mathbf{r}| = [(D_s^2x)^2 + (D_s^2y)^2]^{1/2}$$

41. If two plane curves with Cartesian equations $y = f(x)$ and $y = g(x)$ have the same tangent and curvature at point (x_0, y_0) on both curves, prove that $|f''(x_0)| = |g''(x_0)|$ or, equivalently, $f''(x_0) = \pm g''(x_0)$.

42. Find the curvature of the graph of $y = f(x)$ at an inflection point, where it is assumed that $f''(x)$ exists.

43. Find the radius of curvature of the graph of $x^{2/3} + y^{2/3} = a^{2/3}$, $(a > 0)$, at the point $P(-2^{-3/2}a, 2^{-3/2}a)$.

44. Find the radius of curvature of the cycloid
$$x = a(t - \sin t), \; y = a(1 - \cos t) \qquad a > 0,\, 0 < t < 2\pi$$

45. Prove that if the curvature of the graph of a function $y = f(x)$ is identically equal to zero, then the graph is a straight line.

46. If a particle of mass m moves along a curve with constant speed, show that a nonzero resultant force must act at right angles to the path.

47. Find the position vector $\mathbf{r}(t)$ if the acceleration vector $\mathbf{a}(t) = 6t\mathbf{i} + 2\mathbf{j}$ subject to the initial conditions $\mathbf{v}(0) = \mathbf{i}$ and $\mathbf{r}(0) = -\mathbf{i} + 5\mathbf{j}$.

48. Find the parametric equations of the evolute of the curve defined by

$$x = a(\cos t + t \sin t)$$
$$a > 0, t > 0$$
$$y = a(\sin t - t \cos t)$$

49. The position of a particle at time t is given parametrically by the equations

$$x(t) = \int_0^t \cos (u^2)\, du \qquad \text{and} \qquad y(t) = \int_0^t \sin (u^2)\, du$$

Find the curvature as a function of the arc length s where s is an increasing function of t measured from $(0, 0)$.

50. It is determined empirically that the maximum range of a gun is R mi. Find the initial speed (or muzzle speed) as a function of R in miles per second.

12

Plane Polar Coordinates

12.1   POLAR COORDINATES OF A POINT

Thus far we have located points in the plane by means of ordered pairs of numbers (x, y) which are directed distances from the y- and x-axes, respectively.

There are, however, other ways to establish such a correspondence. The only other coordinate system for the plane of sufficient significance for us to treat in this text is the system of **polar coordinates.** In this system, a fixed point O, called the **pole** or **origin,** is chosen and from this fixed point a fixed ray (or half-line) Ox, called the **polar axis** or **initial line,** is drawn (Figure 12.1.1). We have (for convenience) selected the pole and polar axis to correspond to the nonnegative x-axis.

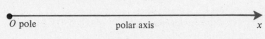

O pole   polar axis x

Figure 12.1.1

Now, let P be any point in the plane and draw a half-line from O through P as shown in Figure 12.1.2. Let r be the undirected distance from O to P and let θ be the angle between OA and OP (where A is a point on the positive x-axis). Thus $r = |\overline{OP}|$ and (r, θ) are said to be the polar coordinates of P. If we artificially restrict r and θ so that $r \geq 0$ and (in radians) $0 \leq \theta < 2\pi$, then, with the exception of the origin, to each point P in the plane there corresponds a unique pair of coordinates (the origin is excluded because $(0, \theta)$ describes the origin for all values of θ). Conversely, to each r and θ satisfying the given restrictions there is a unique point in the plane.

We wish now to remove the artificial restrictions imposed on r and θ to obtain greater flexibility and convenience. Let r and θ be any real numbers. We first locate a ray OM by rotating from the positive x-axis through an angle $|\theta|$. Then θ is taken to be positive if the rotation is counterclockwise and negative if clock-

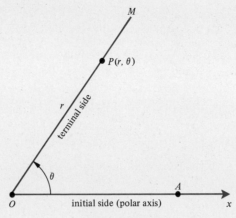

Figure 12.1.2

wise. To the point P on the ray OM, we assign the r value $r = |\overline{OP}| \geq 0$. If the ray is extended backward through the pole to reach P, we then assign the value $r = -|\overline{OP}| < 0$. Figure 12.1.3 shows several points and a polar coordinate representation of the given points.

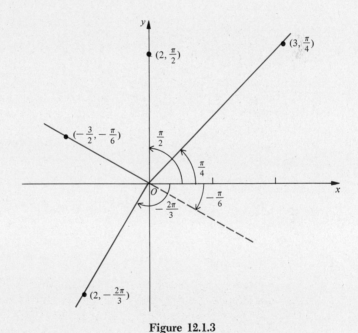

Figure 12.1.3

In broadening the values that r and θ can assume, we recognize that polar coordinates present a problem because each point in the plane $(r \neq 0)$ has an infinite number of representations. For example, $\left(3, \dfrac{\pi}{4}\right)$, $\left(-3, \dfrac{5\pi}{4}\right)$ and $\left(3, \dfrac{9\pi}{4}\right)$ represent the same point. In fact, if (r, θ) is one representation of a point, then $(r, \theta + 2n\pi)$ and $(-r, \theta + (2n + 1)\pi)$, where n is an arbitrary integer, are also representations of the same point. The pole may be represented by $(0, \theta)$ where θ is arbitrary.

EXAMPLE 1   (a) Plot the point whose polar coordinates are $\left(-2, -\dfrac{\pi}{4}\right)$. (b) Find two other sets of polar coordinates for the same point such that (i) $r < 0$ and $0 < \theta < 2\pi$ and (ii) $r > 0$ and $4\pi < \theta < 6\pi$.

Solution   (a) The point $\left(-2, -\dfrac{\pi}{4}\right)$ is shown in Figure 12.1.4(a). It is found by drawing the angle $\dfrac{\pi}{4}$ in a clockwise direction and then extending the ray of the terminal side backward through the pole and marking off two units.

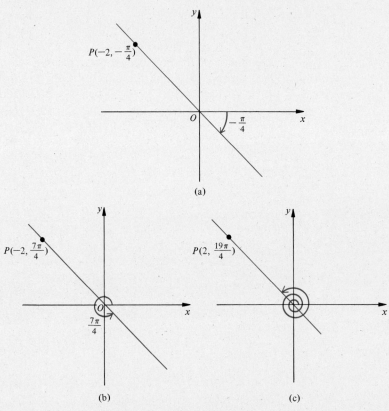

(a)

(b)   (c)

Figure 12.1.4

(b) (i) Any other representation of the same point where $r < 0$ is of the form $\left(-2, -\dfrac{\pi}{4} + 2n\pi\right)$ where n is an arbitrary integer. Since $0 < \theta < 2\pi$ then $0 < -\dfrac{\pi}{4} + 2n\pi < 2\pi$ is satisfied if and only if $n = 1$. Therefore the polar coordinate representation is $\left(-2, \dfrac{7\pi}{4}\right)$. This is illustrated in Figure 12.1.4(b).

(ii) Any representation where $r > 0$ and thus equal to 2 is of the form $\left(2, -\dfrac{\pi}{4} + (2n + 1)\pi\right)$. Since $4\pi < \theta < 6\pi$ must hold then

$4\pi < -\dfrac{\pi}{4} + (2n + 1)\pi < 6\pi$ is required. The only solution of this double inequality is $n = 2$. Therefore the polar coordinate representation of the point is given by $\left(2, \dfrac{19\pi}{4}\right)$. Figure 12.1.4(c) shows this representation.

●

With the introduction of a new set of coordinates, it is desirable and useful to relate this system to the old system. In short, we seek the rectangular coordinates in terms of the polar coordinates and conversely. To accomplish this, we super-impose the positive x-axis of the Cartesian system on the polar axis of the polar system so that their origins coincide. Also, the ray corresponding to $\theta = \dfrac{\pi}{2}$ coincides with the positive y-axis. Then any point P in the plane has polar coordinates (r, θ) and also Cartesian coordinates (x, y).

There are two cases to be considered, namely, $r > 0$ and $r < 0$. If r is positive, say $r = |\overline{OP}|$ then P is on the terminal side of θ (Figure 12.1.5(a)). Then

$$\cos\theta = \frac{x}{r} \qquad \text{and} \qquad \sin\theta = \frac{y}{r}$$

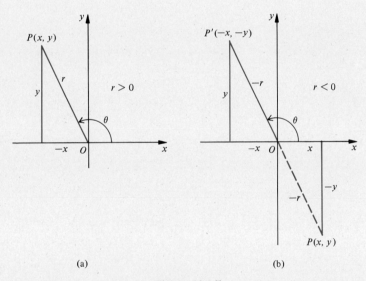

(a)   (b)

Figure 12.1.5

Note that in the figures, the letters next to the sides of the triangle denote the actual lengths of the segments. Thus, since in Figure 12.1.5(a), the x coordinate of P is negative and the length of the side is given by $-x$, and so on. In the same figure, $\cos\theta < 0$ as $\dfrac{\pi}{2} < \theta < \pi$ and $\cos\theta = -\cos(\pi - \theta) = -\left(\dfrac{-x}{r}\right) = \dfrac{x}{r}$. Angles in other quadrants can be treated similarly. It is readily verified in all instances where $r > 0$ that

$$x = r\cos\theta \qquad \text{and} \qquad y = r\sin\theta \qquad (1)$$

Next, we consider $r < 0$ so that P is on the extension of the terminal side and $r = -|\overline{OP}|$ (see Figure 12.1.5(b)). Now, if $P = P(x, y)$ then the corresponding coordinates of P' are $P'(-x, -y)$ and

$$\cos \theta = \frac{-x}{|\overline{OP'}|} = \frac{-x}{|\overline{OP}|} = \frac{-x}{-r} = \frac{x}{r}$$

and hence

$$x = r \cos \theta$$

Also,

$$\sin \theta = \frac{-y}{|\overline{OP'}|} = \frac{-y}{|\overline{OP}|} = \frac{-y}{-r} = \frac{y}{r}$$

and we have

$$y = r \sin \theta$$

Other angles can be treated analogously, and therefore formulas (1) hold in all cases.

Consequently we can determine r and θ (not uniquely) when x and y are known. From (1), by squaring both sides of the two equations, we obtain

$$x^2 = r^2 \cos^2 \theta \quad \text{and} \quad y^2 = r^2 \sin^2 \theta$$

Addition of these two equations yields

$$x^2 + y^2 = r^2(\cos^2 \theta + \sin^2 \theta) = r^2$$

Hence

$$r = \pm \sqrt{x^2 + y^2} \tag{2}$$

Similarly, from the equations (1) division yields

$$\frac{r \sin \theta}{r \cos \theta} = \frac{y}{x}$$

or, equivalently,

$$\tan \theta = \frac{y}{x} \tag{3}$$

EXAMPLE 2   Given the point whose rectangular coordinates are $(\sqrt{3}, 1)$, determine all the possible sets of polar coordinates representing this point.

Solution   Since $r^2 = x^2 + y^2$, it follows that

$$r^2 = (\sqrt{3})^2 + 1^2 = 3 + 1 = 4$$

and thus $r = \pm 2$

$$\tan \theta = \frac{y}{x} = \frac{1}{\sqrt{3}} \quad \text{(see Equation (3))}$$

The point $(\sqrt{3}, 1)$ lies in the first quadrant, and we choose $\theta = \dfrac{\pi}{6}$, since this is the unique angle in the first quadrant for which $\tan \theta = \dfrac{1}{\sqrt{3}}$. [In general, we determine θ to satisfy equation (3) in such a way that θ is also an angle in the same quadrant as the given point.] Therefore, if $r = 2$, $\theta = \dfrac{\pi}{6}$, but adding any integer multiple of 2π to $\dfrac{\pi}{6}$ will still yield an angle that represents the point in question.

Thus, $\left(2, \dfrac{\pi}{6} + 2n\pi\right)$, where n is any integer is the complete set of polar coordinates representing $(\sqrt{3}, 1)$ in the case where $r = 2$. If $r = -2$, $\theta = \dfrac{\pi}{6} + (2n + 1)\pi$, where n is any integer is the set of polar coordinate representatives in this case.

To summarize, if n is any integer

$$\left(2, \frac{\pi}{6} + 2n\pi\right) \qquad \text{and} \qquad \left(-2, \frac{\pi}{6} + (2n + 1)\pi\right)$$

is the complete set of polar representations of $(\sqrt{3}, 1)$. ●

EXAMPLE 3   Graph the point whose polar coordinates are $P\left(3, \dfrac{7\pi}{6}\right)$ and find its corresponding rectangular Cartesian coordinates.

Solution   The point $P\left(3, \dfrac{7\pi}{6}\right)$ is graphed in Figure 12.1.6. From (1) it follows that

$$x = r \cos \theta = 3 \cos \frac{7\pi}{6} = 3\left(-\frac{\sqrt{3}}{2}\right) = -\frac{3\sqrt{3}}{2}$$

and

$$y = r \sin \theta = 3 \sin \frac{7\pi}{6} = 3\left(-\frac{1}{2}\right) = -\frac{3}{2}$$

Therefore the rectangular Cartesian coordinates are $\left(-\dfrac{3\sqrt{3}}{2}, -\dfrac{3}{2}\right)$. ●

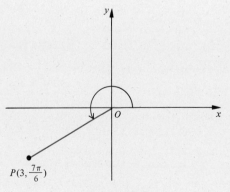

$P(3, \frac{7\pi}{6})$

Figure 12.1.6

EXAMPLE 4   Sketch the graph of the straight line whose Cartesian coordinates satisfy the relation $x - y = 2$ and find the equation in polar coordinates.

Solution   The graph of $x - y = 2$ is shown in Figure 12.1.7. From $x = r \cos \theta$ and $y = r \sin \theta$ we have

$$r \cos \theta - r \sin \theta = 2 \text{ or equivalently,}$$
$$r(\cos \theta - \sin \theta) = 2$$

The polar coordinates of a point P on the line are shown in the figure.   ●

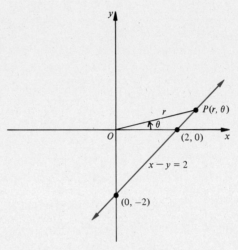

Figure 12.1.7

EXAMPLE 5   A circle has its center at $(a, 0)$, $(a > 0)$, and its radius is a. Find its Cartesian and polar equations.

Solution   From the given data an equation of the circle in Cartesian coordinates is

$$(x - a)^2 + y^2 = a^2 \qquad (4)$$

or, equivalently,

$$x^2 - 2ax + y^2 = 0 \qquad (5)$$

However, $x^2 + y^2 = r^2$ and $x = r \cos \theta$ and the equation becomes

$$r^2 - 2ar \cos \theta = 0 \qquad (6)$$

Now, by factoring we have

$$r = 0 \qquad \text{or} \qquad r = 2a \cos \theta$$

Note that the pole $(r = 0)$ is also on $r = 2a \cos \theta$ $\left(\text{when } \theta = \dfrac{\pi}{2}\right)$. We therefore conclude that if the point is on the circle, then

$$r = 2a \cos \theta \qquad (7)$$

Conversely, the steps in the above argument can be reversed and we conclude that if (7) is satisfied then (5) and (4) must also hold. Thus the circle with center at $(a, 0)$ and radius a has (7) as an equation.   ●

Exercises 12.1

In Exercises 1 through 10 a point is given by a pair of polar coordinates. Plot the point and determine its rectangular coordinates.

1. $(2, 0)$

2. $(5, \pi)$

3. $(-3, 0)$

4. $\left(4, -\dfrac{\pi}{2}\right)$

5. $\left(\dfrac{3}{2}, \dfrac{\pi}{4}\right)$

6. $\left(1, -\dfrac{\pi}{6}\right)$

7. $\left(-2, \dfrac{5\pi}{4}\right)$

8. $\left(\dfrac{3}{4}, -\dfrac{5\pi}{3}\right)$

9. $\left(\dfrac{3}{2}, \dfrac{11\pi}{4}\right)$

10. $\left(2, \dfrac{16\pi}{3}\right)$

In Exercises 11 through 16 the rectangular coordinates of a point are given. Determine all the possible sets of polar coordinates representing the given point.

11. $(3, 0)$

12. $(2, 2)$

13. $(1, \sqrt{3})$

14. $(-1, -1)$

15. $(4, 3)$

16. $(-7, 24)$

In Exercises 17 through 24, transform the given equations in Cartesian form into polar form.

17. $x = 3$
18. $x = a$ (a is a constant)
19. $y = b$ (b is a constant)
20. $x^2 + y^2 = 16$
21. $x^2 - y^2 = a^2$ (a is a constant)
22. $ax + by + c = 0$ (a, b and c are constants)
23. $(x - h)^2 + (y - k)^2 = a^2$ (h, k, and a are constants)

24. $y = \dfrac{3x}{x + 2}$

In Exercises 25 through 36 express the given equations in Cartesian coordinates.

25. $r = 7$

26. $\theta = \dfrac{3\pi}{4}$

27. $\theta = \dfrac{2\pi}{3}$

28. $r = 5 \sin \theta$

29. $2r = \sin \theta$

30. $r^2 = 3 \cos \theta$

31. $r(3 \cos \theta - 2 \sin \theta) = 1$

32. $r = 5 \cos^2 \theta$

33. $r^2 = 6 \sin 2\theta$

34. $r^2 = \cos 2\theta$

35. $r = \dfrac{3}{1 + \cos \theta}$

36. $r = \dfrac{4}{1 + 2 \sin \theta}$

THE GRAPH OF A POLAR EQUATION

By the graph of a polar equation

$$F(r, \theta) = 0 \tag{1}$$

we mean the collection of all points $P(r, \theta)$ whose polar coordinates satisfy the equation. The point P has many different pairs of polar coordinates, however P is on the graph if just one of its pairs of coordinates satisfies the given equation. Frequently the given equation may be solved explicitly for r in terms of θ say

$$r = f(\theta) \tag{2}$$

and less often for θ as a function of r,

$$\theta = g(r) \tag{3}$$

Thus the graph of (2) can be obtained by selecting a series of values of θ and then computing the corresponding values of r. The points are then plotted (using an appropriate scale) and a smooth curve is drawn through these points.

EXAMPLE 1   Sketch the graph of $r = 3 \sin \theta$.

Solution   We shall plot points and then connect them by a (hopefully) smooth curve. We choose convenient angles such as $\theta = 0, \dfrac{\pi}{6}$, etc. and then determine the corresponding value of r. We have Table 12.2.1(a) if $0 \le \theta \le \pi$, and Table 12.2.1(b) if $\pi \le \theta \le 2\pi$.

Table 12.2.1(a)

θ	0	$\dfrac{\pi}{6}$	$\dfrac{\pi}{4}$	$\dfrac{\pi}{3}$	$\dfrac{\pi}{2}$	$\dfrac{2\pi}{3}$	$\dfrac{3\pi}{4}$	$\dfrac{5\pi}{6}$	π
r	0	$\dfrac{3}{2}$	$\dfrac{3\sqrt{2}}{2}$	$\dfrac{3\sqrt{3}}{2}$	3	$\dfrac{3\sqrt{3}}{2}$	$\dfrac{3\sqrt{2}}{2}$	$\dfrac{3}{2}$	0

Table 12.2.1(b)

θ	π	$\dfrac{7\pi}{6}$	$\dfrac{5\pi}{4}$	$\dfrac{4\pi}{3}$	$\dfrac{3\pi}{2}$	$\dfrac{5\pi}{3}$	$\dfrac{7\pi}{4}$	$\dfrac{11\pi}{6}$	2π
r	0	$-\dfrac{3}{2}$	$-\dfrac{3\sqrt{2}}{2}$	$-\dfrac{3\sqrt{3}}{2}$	-3	$-\dfrac{3\sqrt{3}}{2}$	$-\dfrac{3\sqrt{2}}{2}$	$-\dfrac{3}{2}$	0

If we compare the values in these tables, we find that the same points are repeating although with different representations. For example, $(r, \theta) = \left(\dfrac{3}{2}, \dfrac{\pi}{6}\right)$ and $(r, \theta) = \left(-\dfrac{3}{2}, \dfrac{7\pi}{6}\right)$ determine the same point (the other points in the tables can be paired similarly).

The resulting graph is drawn in Figure 12.2.1. The curve looks like a circle with

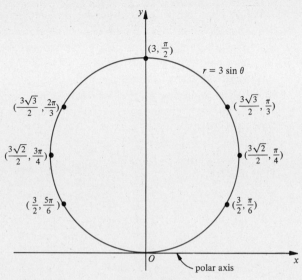

Figure 12.2.1

diameter 3 and center at $(x, y) = (0, \frac{3}{2})$. To see that it is, in fact, this circle is easily verified if we change $r = 3 \sin \theta$ to rectangular coordinates. Multiplying by r we have $r^2 = 3r \sin \theta$. But $r^2 = x^2 + y^2$ and $y = r \sin \theta$. So

$$x^2 + y^2 = 3y \qquad \text{or} \qquad x^2 + y^2 - 3y = 0.$$

Completing the square, we have

$$x^2 + (y - \tfrac{3}{2})^2 = (\tfrac{3}{2})^2 \qquad\qquad \bullet$$

In the preceding example, it took a lot of work to sketch a curve that we could have drawn very quickly had we started with the rectangular equation. However, this is not always the case. Many curves can be drawn much more easily if we work from a polar representation.

To sketch the graph of a polar equation, it is not necessary to plot as many points as we did in Example 1. A reasonably accurate sketch can be obtained if one just observes how the function varies—in particular, when does it increase and when does it decrease. In Example 1, we could have used only a few points in Table 12.2.1(a) along with the observation that $r = 3 \sin \theta$ increases for $0 \leq \theta \leq \frac{\pi}{2}$ and decreases for $\frac{\pi}{2} \leq \theta \leq \pi$.

EXAMPLE 2 Sketch the graph of the equation $r = \dfrac{2}{1 + \cos \theta}$.

Solution The graph is symmetrical with respect to the polar axis since $\cos(-\theta) = \cos \theta$ and therefore, if (r, θ) is on the curve, its reflection in the polar axis $(r, -\theta)$ will also be on the curve. Hence it is only necessary to consider values of $\theta \geq 0$. In this instance, our interval of interest is $0 \leq \theta \leq \pi$ since these values of θ will determine all points on or above the polar axis. In the table of values, if θ determines r we can then indicate that $\pm\theta$ determine r.

We observe that $r \to \infty$ as $\theta \to \pi$ since $\cos \pi = -1$ makes the value of the denominator zero. Also r can never be zero since the numerator equals 2. Furthermore the least value of r occurs when $1 + \cos \theta$ is largest, that is, when $\cos \theta = 1$ or $\theta = 0$. Hence $r_{\min} = 1$.

In Table 12.2.2 we list the coordinates of some points on the graph (Figure 12.2.2). The points are plotted for convenience on polar coordinate paper, which displays lines at various angles passing through the pole O and concentric circles with centers at O.

Table 12.2.2

θ	0	$\pm\dfrac{\pi}{6}$	$\pm\dfrac{\pi}{4}$	$\pm\dfrac{\pi}{3}$	$\pm\dfrac{\pi}{2}$	$\pm\dfrac{2}{3}\pi$	$\pm\dfrac{3}{4}\pi$	$\pm\dfrac{5}{6}\pi$
r	1	$\dfrac{4}{2+\sqrt{3}}$	$\dfrac{4}{2+\sqrt{2}}$	$\dfrac{4}{3}$	2	4	$\dfrac{4}{2-\sqrt{2}}$	$\dfrac{4}{2-\sqrt{3}}$

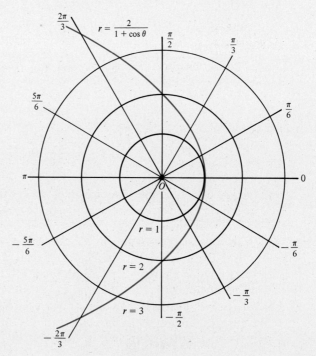

Figure 12.2.2

The graph is reminiscent of a parabola. That the curve actually is a parabola is easily verified by transforming to rectangular coordinates. We rewrite the given equation in the form

$$r + r \cos \theta = 2$$

and replace $r \cos \theta$ by x to obtain

$$r = 2 - x.$$

If we square both sides, we obtain, $r^2 = 4 - 4x + x^2$ or, using $r^2 = x^2 + y^2$, this simplifies to

$$y^2 = 4 - 4x$$

which is indeed a parabola with the x-axis as its axis of symmetry.   ●

EXAMPLE 3   Sketch the graph of the equation $r = 2(1 + \cos\theta)$.

Solution   Once again, since $\cos\theta = \cos(-\theta)$, the graph is symmetric with respect to the polar axis. As θ increases from 0 to π, $\cos\theta$ decreases from 1 to -1. Therefore, correspondingly, r decreases from $r = 4$ to $r = 0$. From the given equation we obtain Table 12.2.3.

Table 12.2.3

θ	0	$\pm\dfrac{\pi}{6}$	$\pm\dfrac{\pi}{4}$	$\pm\dfrac{\pi}{3}$	$\pm\dfrac{\pi}{2}$	$\pm\dfrac{2}{3}\pi$	$\pm\dfrac{3}{4}\pi$	$\pm\dfrac{5}{6}\pi$	$\pm\pi$
r	4	$2+\sqrt{3}$	$2+\sqrt{2}$	3	2	1	$2-\sqrt{2}$	$2-\sqrt{3}$	0

The graph is shown in Figure 12.2.3 and is heart shaped. We call this curve a *cardioid.* In general, a cardioid is a graph of an equation of the form

$$r = a(1 \pm \cos\theta) \qquad \text{or} \qquad r = a(1 \pm \sin\theta) \qquad\qquad ●$$

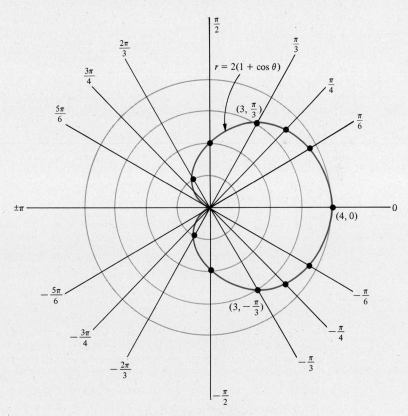

Figure 12.2.3

EXAMPLE 4   Sketch the graph of the equation $r = \dfrac{\theta}{3}$ where $\theta \geq 0$.

Solution   The polar coordinate r is a monotonically increasing function of θ (observe that $D_\theta r = \frac{1}{3}$). We tabulate r versus θ (see Table 12.2.4) and the values of r are rounded off to two decimal places.

The graph is shown in Figure 12.2.4 and is called a ***spiral of Archimedes.*** ●

Table 12.2.4

θ	0	$\dfrac{\pi}{6}$	$\dfrac{\pi}{3}$	$\dfrac{\pi}{2}$	π	$\dfrac{3}{2}\pi$	2π	$\dfrac{5}{2}\pi$	3π	$\dfrac{7}{2}\pi$
r	0	0.17	0.35	0.52	1.05	1.57	2.09	2.62	3.14	3.67

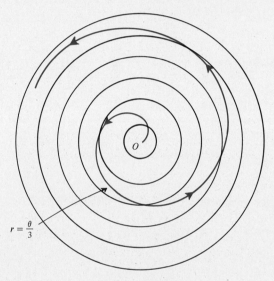

$r = \dfrac{\theta}{3}$

Figure 12.2.4

In Examples 2 and 3, our curve sketching was simplified by observing that the graphs were symmetric with respect to the polar axis. More generally, Figure 12.2.5 shows that:

(i)   The points (r, θ) and $(r, -\theta)$ are symmetric with respect to the polar axis (the x-axis).

(ii)   The points (r, θ) and $(r, \pi - \theta)$ are symmetric with respect to the line $\theta = \dfrac{\pi}{2}$ (the y-axis).

(iii)   The points (r, θ) and $(-r, \theta)$ are symmetric with respect to the pole (the origin).

Unfortunately, the tests for symmetry are not as simple as the preceding cases seem to indicate. For example, to test for symmetry with respect to the polar axis it is not sufficient to replace (r, θ) by $(r, -\theta)$ and see if an equivalent equation is obtained. The reason for this is that the point $(r, -\theta)$ has many other possible representations—either $(r, -\theta + 2n\pi)$ or $(-r, (2n + 1)\pi - \theta)$ where n is an integer. Similar comments apply to the other two symmetries. Taking this into account, we have the following rules:

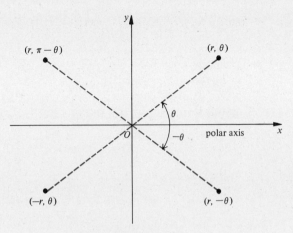

Figure 12.2.5

Rule I: If the same or an equivalent equation is obtained when (r, θ) is replaced by $(r, -\theta + 2n\pi)$ or when (r, θ) is replaced by $(-r, (2n + 1)\pi - \theta)$, where n is an arbitrary integer, the graph of the equation is symmetric with respect to the polar axis (that is, $\theta = 0$) or x-axis.

Rule II: If the same or an equivalent equation is obtained when (r, θ) is replaced by $(-r, 2n\pi - \theta)$ or when (r, θ) is replaced by $(r, (2n + 1)\pi - \theta)$, where n is an arbitrary integer, the graph of the equation is symmetric with respect to the $\theta = \dfrac{\pi}{2}$ or y-axis.

Rule III: If the same or an equivalent equation is obtained when (r, θ) is replaced by $(-r, 2n\pi + \theta)$ or when (r, θ) is replaced by $(r, (2n + 1)\pi + \theta)$, where n is an arbitrary integer, the graph is symmetric with respect to the origin or pole.

In many practical applications it is necessary to utilize only the special case $n = 0$ as will be demonstrated in the examples below.

EXAMPLE 5 Test the graph of the equation

$$r = 3 \sin \theta \tag{4}$$

for symmetry with respect to the polar axis, the $\theta = \dfrac{\pi}{2}$ axis and the pole.

Solution From Rule I we replace r by r and θ with $-\theta + 2n\pi$ in (4) to obtain

$$r = 3 \sin (-\theta + 2n\pi) = 3 \sin (-\theta) = -3 \sin \theta$$

which is not equivalent to the original equation. Therefore the polar axis is not an axis of symmetry of (4).

From Rule II with $n = 0$ we replace r by $-r$ and θ by $-\theta$ in (4) and we have

$$-r = 3 \sin (-\theta) \qquad \text{or, equivalently,} \qquad -r = -3 \sin \theta$$

which implies that $r = 3 \sin \theta$. Therefore the graph of (4) is symmetrical with respect to the $\theta = \dfrac{\pi}{2}$ or y-axis.

To test for symmetry with respect to the pole, we replace r by $-r$ and θ by $2n\pi + \theta$ in (4). This results in

$$-r = 3\sin(2n\pi + \theta) = 3\sin\theta$$

which implies that $r = -3\sin\theta$. This is not equivalent to (4) and therefore the graph of (4) is not symmetrical with respect to the pole.

This curve was sketched in Figure 12.2.1. Our results on symmetry can be seen in that diagram. ●

EXAMPLE 6   Examine the curve

$$r = 3\sin 2\theta \qquad\qquad (5)$$

for symmetry and sketch the graph.

Solution   If we replace (r, θ) by $(r, -\theta)$ in Rule I, we obtain $r = 3\sin 2(-\theta) = -3\sin 2\theta$, and this is not equivalent to the original equation. However, if we replace (r, θ) by $(-r, \pi - \theta)$ also in Rule I, then equation (5) becomes

$$-r = 3\sin 2(\pi - \theta) = 3\sin(2\pi - 2\theta) = -3\sin 2\theta$$

or $r = 3\sin 2\theta$. Therefore (5) is indeed symmetrical about the polar or x-axis. This result also implies that any one of the above tests is *sufficient* to ensure the particular symmetry involved, but it is not *necessary*.

If (r, θ) is replaced by $(-r, -\theta)$, Rule II with $n = 0$ in (5), there results

$$-r = 3\sin(-2\theta) = -3\sin 2\theta$$

which yields (5). Therefore the graph of (5) is symmetric with respect to the y-axis. Now any curve that is symmetric with respect to the x- and y-axes must also be symmetric with respect to the origin.

We leave it to the reader to show that (5) is also symmetric with respect to the line $\theta = \dfrac{\pi}{4}$ or $y = x$ by devising a test for this kind of symmetry and verifying the invariance of the equation under the appropriate replacements of r and θ. Because of all these symmetries it is only necessary to sketch the curve in the sector between the lines $\theta = 0$ and $\theta = \dfrac{\pi}{4}$ and then the remainder of the curve may be obtained by reflections in the lines of symmetry. The curve is sketched in Figure 12.2.6 and is called a **four leaved rose.**

A curve having a polar equation of the form $r = a\sin n\theta$ or $r = a\cos n\theta$ where n is a positive integer is called a **rose** or a **rose leaved curve**. A rose has n leaves if n is odd and $2n$ leaves when n is even. ●

EXAMPLE 7   Sketch the graph of $r = 1 + 2\sin\theta$

Solution   Since $\sin(\pi - \theta) = \sin\theta$, the curve is symmetrical with respect to the $\dfrac{\pi}{2}$ or y-axis. It has no other symmetries.

The maximum value of r is 3, when $\theta = \dfrac{\pi}{2}$ and the least is -1, when $\theta = -\dfrac{\pi}{2}$. It is also helpful to determine when $r = 0$. This occurs if and only if $1 +$

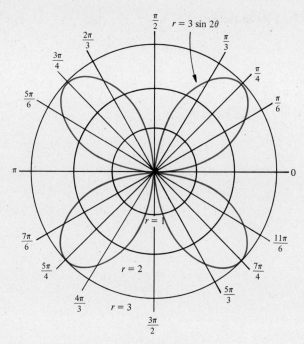

Figure 12.2.6

$2 \sin \theta = 0$ or $\sin \theta = -\frac{1}{2}$. Hence the curve passes through the pole when $\theta = -\frac{5\pi}{6}$ or $-\frac{\pi}{6}$. It will be rigorously established in the next section that these are the angles that the tangents to the curve make with the x-axis. These are two of the radial lines shown in Figure 12.2.7. We compute a table of values (Table

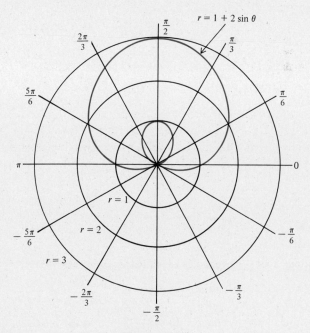

Figure 12.2.7

Table 12.2.5

θ	$-\dfrac{\pi}{2}$	$-\dfrac{\pi}{3}$	$-\dfrac{\pi}{6}$	0	$\dfrac{\pi}{6}$	$\dfrac{\pi}{3}$	$\dfrac{\pi}{2}$
r	-1	$1-\sqrt{3}$	0	1	2	$1+\sqrt{3}$	3

12.2.5). The rest of the graph is quickly sketched because of its symmetry about the $\theta = \dfrac{\pi}{2}$ axis. This curve is called a *limacon*. In general *limacons* are represented by equations of the form

$$r = a + b\cos\theta \qquad \text{or} \qquad r = a + b\sin\theta$$

where a and b are arbitrary constants. The special form when $a = \pm b$ is the cardioid shown in Figure 12.2.3. ●

Exercises 12.2

In Exercises 1 through 20 sketch the graph of the given curve.

1. $r\sin\theta = 4$

2. $r = 6$

3. $r(\cos\theta - \sin\theta) = 5$

4. $\theta = \dfrac{\pi}{4}$

5. $r = -7$

6. $r\cos\theta = -3$

7. $r\sin\theta = 0$

8. $r(3\cos\theta + \sin\theta) = 0$

9. $\theta^2 + \dfrac{\pi}{6}\theta = 0$

10. $r = 10\cos\theta$

11. $r = 12\sin\theta$

12. $r(a\cos\theta + b\sin\theta) = 0$, ($a$ and b are positive constants)

13. $r + 6\sin\theta + 8\cos\theta = 0$

14. $r\cos\left(\theta - \dfrac{\pi}{4}\right) = 6$

15. $r^2 = 9\cos 2\theta$, (lemniscate)

16. $r = 3(1 - \cos\theta)$, (cardioid)

17. $r = 2 - \cos\theta$, (limaçon)

18. $r = 1 - 2\cos\theta$, (limaçon)

19. $r = 8\sin^2\theta\cos^2\theta$

20. $r^2 = 4\sin 2\theta$, (lemniscate)

21. Sketch the graph of $r = \sqrt{\theta}$, $0 \leq \theta \leq 2\pi$ by tabulating r_k versus θ_k for $\theta_k = \dfrac{k\pi}{4} (k = 0, 1, 2, \ldots, 8)$ to two decimal places and passing a smooth curve through the points (r_k, θ_k).

22. Graph $\sin 3\theta = 0$.

23. Find the maximum and minimum of the function f defined by $r = f(\theta) = a\cos\theta + b$ where $a < 0$ and $b > 0$.

24. Identify the curve defined by the equation

$$(x^2 + y^2 - 2ax)^2 = 4a^2(x^2 + y^2)$$

(*Hint:* Transform to polar coordinates.)

25. Identify the curve defined by the equation

$$x = r - 1$$

and sketch it.

26. A circle of diameter a is tangent to the polar axis at the pole. Find its equation in polar coordinates *without* using rectangular coordinates.

27. Find the equation in polar coordinates of a vertex P of a triangle which is opposite a fixed base AB of length $2a$ and is subject to the constraint that $|\overline{PA}| \cdot |\overline{PB}| = a^2$. Identify the curve. (*Hint:* First find an equation in rectangular coordinates by placing the origin at the midpoint of the base and then transform to polar coordinates.)

28. Show that the distance between two points with polar coordinates (r_1, θ_1) and (r_2, θ_2) is given by $\sqrt{r_1{}^2 + r_2{}^2 - 2r_1 r_2 \cos(\theta_1 - \theta_2)}$. Find a formula for the special case when the line connecting the two points passes through the origin.

12.3   INTERSECTION OF CURVES IN POLAR COORDINATES

In rectangular coordinates all the points of intersection of two curves are obtained by the simultaneous solution of the equations that define them. This follows from the fact that to each point P in the plane we associate a unique pair of rectangular coordinates, and conversely. However, in the case of polar coordinates the problem is more bothersome because each point has infinitely many sets of polar coordinates. Thus it may happen (and often does) that two curves intersect at a point P with polar coordinates (r_1, θ_1) that satisfy the equation of one curve while the same point P has another pair of coordinates (r_2, θ_2) that satisfy the equation of the second curve. This implies that the simultaneous solution of the two equations may fail to reveal all the points of intersection.

One very useful procedure is to sketch the two curves and determine the number of intersection points to be expected. Then solve the equations simultaneously and determine whether any solution or solutions have been lost. This is best illustrated by means of examples.

EXAMPLE 1   Find the points of intersection of the two curves $r = a \cos \theta$ and $r = a \sin \theta$, $(a > 0)$.

Solution   The two curves are circles of diameter a with centers on the polar and $\theta = \dfrac{\pi}{2}$ axes, respectively, as shown in Figure 12.3.1. The two curves clearly intersect at two points.

The simultaneous solution, however, yields

$$\cos \theta = \sin \theta$$

from which we obtain $\theta = \dfrac{\pi}{4} + 2n\pi$ and $\theta = \dfrac{5\pi}{4} + 2n\pi$ where n is an integer. These yield only one point $\left(\dfrac{a}{\sqrt{2}}, \dfrac{\pi}{4}\right)$ since $\left(-\dfrac{a}{\sqrt{2}}, \dfrac{5\pi}{4}\right)$ represents the same point.

We have lost one point of intersection, namely the origin. On the circle $r = a \cos \theta$, if we set $r = 0$, there follows $\theta_0 = \pm\dfrac{\pi}{2} + 2n\pi$, whereas for the second circle $r = a \sin \theta$, if we set $r = 0$, there results $\theta_1 = n\pi$ and the angle

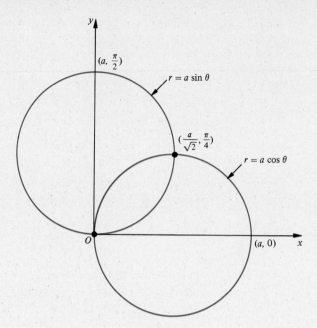

Figure 12.3.1

$\theta_1 \neq \theta_0$. Although the origin is on both curves, the coordinates $(0, \theta_0)$ and $(0, \theta_1)$ are different. Thus the simultaneous solution in this case fails to yield this point.

●

In more complicated situations it is preferable to proceed in an analytical fashion. Given a function f, the equations

$$r = f(\theta) \qquad \text{and} \qquad r = (-1)^n f(\theta + n\pi) \tag{1}$$

have the same graphs where n is an arbitrary integer; that is, a graph as well as a point may have alternative designations.

EXAMPLE 2   Find the different equations of the graph, given one of its equations $r = \cos \dfrac{\theta}{2}$.

Solution   From (1), possible equations of this graph are

$$r = (-1)^n \cos \left[\tfrac{1}{2}(\theta + n\pi)\right] \qquad (n \text{ is an arbitrary integer})$$

For $n = 0, 1, 2, 3$ these equations are

$$r = (-1)^0 \cos \left[\tfrac{1}{2}(\theta + 0\pi)\right] = \cos \frac{\theta}{2}$$

$$r = (-1)^1 \cos \left[\tfrac{1}{2}(\theta + \pi)\right] = -\cos \left(\frac{\theta}{2} + \frac{\pi}{2}\right) = \sin \frac{\theta}{2}$$

$$r = (-1)^2 \cos \left[\tfrac{1}{2}(\theta + 2\pi)\right] = \cos \left(\frac{\theta}{2} + \pi\right) = -\cos \frac{\theta}{2}$$

$$r = (-1)^3 \cos \left[\tfrac{1}{2}(\theta + 3\pi)\right] = -\cos \left(\frac{\theta}{2} + \frac{3\pi}{2}\right) = -\sin \frac{\theta}{2}$$

For $n = 4, 5, 6, \ldots$ or $n = -1, -2, -3, \ldots$ one of these four preceding equations is obtained. Therefore $r = \cos\dfrac{\theta}{2}$, $r = \sin\dfrac{\theta}{2}$, $r = -\cos\dfrac{\theta}{2}$, and $r = -\sin\dfrac{\theta}{2}$ are all the equations of this graph. ●

To find the intersection of two graphs expressed in polar coordinates

(i) Solve each equation of one graph simultaneously with each equation of the other graph.
(ii) Determine whether the pole lies on the two graphs.

This is done for each equation by setting $r = 0$ and determining whether the resulting equation has one or more solutions.

EXAMPLE 3   Find the points of intersection of the graphs of

$$r = 1 - \cos 2\theta \tag{2}$$

and

$$r = \sin\theta - 1 \tag{3}$$

Solution   From (1) we see that (2) has the general form

$$r = (-1)^n[1 - \cos 2(\theta + n\pi)]$$

which yields only one other equation for the graph of (2). This occurs when n is an odd integer and we have

$$r = -(1 - \cos 2\theta) \tag{2a}$$

Again, from (1) we find the general form for (3) to be

$$r = (-1)^n[\sin(\theta + n\pi) - 1]$$

which yields, for odd n, the only other additional equation for the graph of (3)

$$r = \sin\theta + 1 \tag{3a}$$

First we solve (2) and (3) simultaneously by equating

$$1 - \cos 2\theta = \sin\theta - 1$$

Now $1 - \cos 2\theta = 2\sin^2\theta$ and we have

$$2\sin^2\theta - \sin\theta + 1 = 0$$

This equation has no real solutions since its discriminant < 0. Next we solve (3) and (2a) simultaneously

$$-2\sin^2\theta = \sin\theta - 1$$

or, equivalently,

$$2\sin^2\theta + \sin\theta - 1 = 0$$

This may be factored into

$$(2\sin\theta - 1)(\sin\theta + 1) = 0$$

which implies that

$$\sin\theta = \tfrac{1}{2} \text{ or } \sin\theta = -1$$

This yields

$$\theta = \frac{\pi}{6} + 2n\pi \qquad \theta = \frac{5\pi}{6} + 2n\pi \qquad \theta = -\frac{\pi}{2} + 2n\pi$$

and the points of intersection

$$\left(-\frac{1}{2}, \frac{\pi}{6} + 2n\pi\right) \qquad \left(-\frac{1}{2}, \frac{5\pi}{6} + 2n\pi\right) \qquad \left(-2, -\frac{\pi}{2} + 2n\pi\right) \qquad (4)$$

The simultaneous solution of (2) and (3a) implies that

$$2 \sin^2 \theta = \sin \theta + 1$$

or, equivalently,

$$(2 \sin \theta + 1)(\sin \theta - 1) = 0$$

This results in the points of intersection

$$\left(\frac{1}{2}, -\frac{\pi}{6} + 2n\pi\right) \qquad \left(\frac{1}{2}, -\frac{5\pi}{6} + 2n\pi\right) \qquad \left(2, \frac{\pi}{2} + 2n\pi\right) \qquad (5)$$

Actually the points of (5) are the same as the points of (4).

Next, we solve (2a) and (3a) simultaneously and obtain the quadratic equation $2 \sin^2 \theta + \sin \theta + 1 = 0$, which has a negative discriminant and hence no real solutions.

It only remains to check whether both curves pass through the pole by setting $r = 0$ in, say, (2) and (3) to obtain $\cos 2\theta = 1$ and $\sin \theta = 1$ from which $\theta = 0$ and $\theta = \frac{\pi}{2}$, respectively.

Therefore the curves obtained from (2) and (3) intersect at the four points, namely, the pole and the three points given by (4) (see Figure 12.3.2).   ●

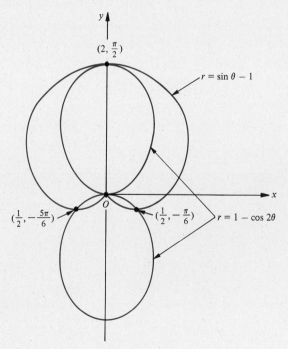

Figure 12.3.2

REMINDER: In most simple cases, points of intersection are found by solving the *given* equations simultaneously and investigating the pole separately. A sketch will indicate whether any more extensive analytical work (as shown in Example 3) is necessary.

Exercises 12.3

In Exercises 1 through 14 find all of the points of intersection of the graphs of the given pair of curves. Sketch the curves.

1. $r = 4\cos\theta, \quad \theta = \dfrac{\pi}{4}$

2. $r = 2\sin\theta, \quad \theta = \dfrac{\pi}{2}$

3. $r = \frac{3}{2}, \quad r = 2\cos\theta$

4. $r = 2, \quad r = 4\sin\theta$

5. $r = 1 - \sin\theta, \quad r = 2\sin\theta$

6. $r = \sqrt{3}, \quad r = \tan\theta$

7. $r = \sin\theta, \quad r = \sin 2\theta$

8. $r\cos\theta = 2, \quad r = 4\cos\theta$

9. $r = 3, \quad r = e^{(\frac{\theta}{2}+1)}$ (*Hint:* Use tables.)

10. $r = \cos\theta, \quad r = \sin\theta\tan\theta$

11. $r = 2 - \sin\theta, \quad r = \cos\theta$

12. $r = 2(1 - 2\sin\theta), \quad r = 2(1 - \sin\theta)$

13. $r = 2 - \sin\theta, \quad r = 1 + \cos\theta$

14. $r = 2\cos 2\theta, \quad r = 1$

(*Hint:* there are eight points of intersection)

12.4

DIFFERENTIATION IN POLAR COORDINATES

Given a function f in polar coordinates

$$r = f(\theta) \tag{1}$$

then the derivative $f'(\theta)$ is defined by

$$f'(\theta) = \lim_{\Delta\theta\to 0} \frac{f(\theta + \Delta\theta) - f(\theta)}{\Delta\theta} \tag{2}$$

In other words the definition of the derivative is not altered. However, the geometric interpretation of the derivative must be altered; that is, the slope of a curve cannot be equated to $f'(\theta)$. Therefore it is of interest to see what is the expression for the slope of a line tangent to a curve $r = f(\theta)$. This is given in the following theorem.

Theorem 1   If m is the slope of the line tangent to a curve $r = f(\theta)$ at a point $P_1(r_1, \theta_1)$ on it, then

$$m|_{P_1} = \frac{f(\theta_1)\cos\theta_1 + f'(\theta_1)\sin\theta_1}{f'(\theta_1)\cos\theta_1 - f(\theta_1)\sin\theta_1} \tag{3}$$

provided that the denominator is not zero.

Proof The rectangular coordinates x and y are given by the differentiable functions of a parameter θ, namely

$$x = r\cos\theta = f(\theta)\cos\theta$$
$$y = r\sin\theta = f(\theta)\sin\theta \qquad (4)$$

Then from (4) we have

$$m = D_x y = \frac{D_\theta y}{D_\theta x} = \frac{f(\theta)\cos\theta + f'(\theta)\sin\theta}{f'(\theta)\cos\theta - f(\theta)\sin\theta} \qquad (5)$$

and this is (3) when θ is replaced by θ_1, provided that the denominator is not zero at $\theta = \theta_1$. The particular case when the denominator is zero and the numerator is different from zero implies that the tangent line to the curve is vertical. ∎

Formula (3) leads to another formula which is used more frequently because of its simplicity. We introduce an angle ψ between the radius vector OP_1 and the tangent line at P_1, measured counterclockwise from OP_1 (see Figure 12.4.1). We derive a formula for $\tan\psi$ given in the following theorem.

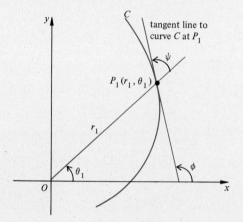

Figure 12.4.1

Theorem 2 Let P_1 be a given point on the curve $r = f(\theta)$ and let ψ be the angle from the radial line OP_1 extended to the tangent line to the curve $r = f(\theta)$ at P_1 (as shown in Figure 12.4.1), then

$$\tan\psi = \frac{f(\theta_1)}{f'(\theta_1)} \qquad (6)$$

provided that $f'(\theta_1)$ and $r_1 \neq 0$.

Proof If $\cos\theta_1 \neq 0$ then division of numerator and denominator of (3) by $\cos\theta_1$ yields

$$m\big|_{P_1} = \tan\phi = \frac{f(\theta_1) + f'(\theta_1)\tan\theta_1}{f'(\theta_1) - f(\theta_1)\tan\theta_1} \qquad (7)$$

If we solve (7) for $f(\theta_1)$, we have

$$f(\theta_1) = \frac{\tan\phi - \tan\theta_1}{1 + \tan\phi\tan\theta_1}f'(\theta_1) = \tan(\phi - \theta_1)f'(\theta_1) \qquad (8)$$

Since $\psi = \phi - \theta_1$ and thus $\tan \psi = \tan(\phi - \theta_1)$ we obtain (6) from (8) by division of both sides by $f'(\theta_1) \neq 0$.

The proof of (6) when $\cos \theta_1 = 0$ is left as an exercise for the reader. ■

With our knowledge of ψ, we can find ϕ from

$$\phi = \psi + \theta_1$$

and the slope $m = \tan \phi$ is readily obtained.

EXAMPLE 1   Given the circle $r = 3 \sin \theta$, find the angle ψ at $\left(\dfrac{3}{2}, \dfrac{\pi}{6}\right)$. (See Figure 12.4.2.)

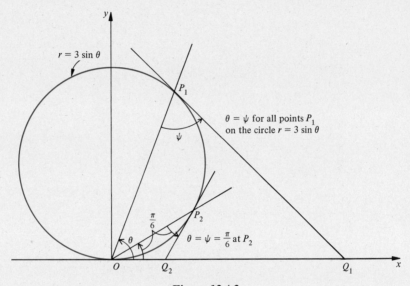

Figure 12.4.2

Solution   From (6) with $f(\theta) = 3 \sin \theta$, $f'(\theta) = 3 \cos \theta$ we find that

$$\tan \psi = \frac{f(\theta)}{f'(\theta)} = \tan \theta$$

Therefore $\psi = \theta$. In particular, at $\left(\dfrac{3}{2}, \dfrac{\pi}{6}\right)$, $\psi = \dfrac{\pi}{6}$.

The result that $\psi = \theta$ follows from elementary geometry since from Figure 12.4.2 OQ_1 and P_1Q_1 are tangents to the circle from an outside point. Thus triangle OP_1Q_1 is isosceles and $\theta = \psi$ because base angles of an isosceles triangle are equal. Triangle OP_2Q_2 shows the particular case $\theta = \psi = \dfrac{\pi}{6}$. ●

EXAMPLE 2   Find ψ for the cardioid $r = a(1 + \cos \theta)$, $a > 0$ at the points $\left(a, \dfrac{\pi}{2}\right)$ and $\left(\dfrac{3a}{2}, \dfrac{\pi}{3}\right)$.

Solution The cardioid $r = f(\theta) = a(1 + \cos\theta)$ is shown in Figure 12.4.3. We have

$$f'(\theta) = -a\sin\theta$$

and from (6),

$$\tan\psi = \frac{f(\theta)}{f'(\theta)} = \frac{1 + \cos\theta}{-\sin\theta}$$

If $\theta = \dfrac{\pi}{2}$,

$$\tan\psi_1 = \frac{1 + 0}{-1} = -1$$

from which $\psi_1 = \dfrac{3\pi}{4}$.

If $\theta = \dfrac{\pi}{3}$,

$$\tan\psi_2 = \frac{1 + \dfrac{1}{2}}{-\dfrac{\sqrt{3}}{2}} = -\sqrt{3}$$

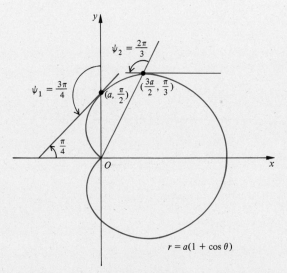

$$r = a(1 + \cos\theta)$$

Figure 12.4.3

Therefore $\psi_2 = \dfrac{2\pi}{3}$. The results are illustrated in Figure 12.4.3. Note that the tangent line to the curve $r = a(1 + \cos\theta)$ at the point $\left(a, \dfrac{\pi}{2}\right)$ intersects the polar axis at an angle $= \dfrac{\pi}{4}$. Also, the tangent line at the point $\left(\dfrac{3a}{2}, \dfrac{\pi}{3}\right)$ is parallel to the polar axis. ●

Proof In the last section, we claimed and will now prove that if *a curve $r = f(\theta)$ passes through the pole, the slope (s) of the curve at that point may be obtained by* *solving the equation $f(\theta) = 0$ and then finding $\tan \theta$ for each solution.* To see this, consider an arc C of a curve $r = f(\theta)$ given in rectangular coordinates by $y = F(x)$ (Figure 12.4.4). Let f be differentiable at $\theta = \theta_1$ where $f(\theta_1) = 0$ whereas $f'(\theta_1) \neq 0$; then, if m denotes the slope at the pole, equation (3) implies that

$$m = D_x y \big|_{(0,0)} = \tan \theta_1$$

Hence the line $\theta = \theta_1$ is tangent to the given arc C at the pole. ∎

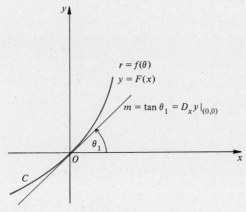

Figure 12.4.4

A formula for the length of an arc of a curve expressed in polar coordinates is easily obtained. To this end, we establish the following theorem.

Theorem 3 If $r = f(\theta)$ possesses a continuous derivative in $[\theta_0, \theta_1]$, then the length s of the arc of the curve between the points $P_0(r_0, \theta_0)$ and $P_1(r_1, \theta_1)$ is given by

$$s = \int_{\theta_0}^{\theta_1} \sqrt{[f(\theta)]^2 + [f'(\theta)]^2} \, d\theta \qquad (9)$$

Proof If θ is regarded as a parameter for the curve in rectangular coordinates, then, from Equation (2) of Section 11.6,

$$s = \int_{\theta_0}^{\theta_1} \sqrt{(D_\theta x)^2 + (D_\theta y)^2} \, d\theta \qquad (10)$$

But, $x = f(\theta) \cos \theta$ and $y = f(\theta) \sin \theta$, so that

$$
\begin{aligned}
(D_\theta x)^2 + (D_\theta y)^2 &= [-f(\theta) \sin \theta + f'(\theta) \cos \theta]^2 + [f(\theta) \cos \theta + f'(\theta) \sin \theta]^2 \\
&= [f(\theta)]^2(\sin^2 \theta + \cos^2 \theta) + [f'(\theta)]^2(\cos^2 \theta + \sin^2 \theta) \\
&= [f(\theta)]^2 + [f'(\theta)]^2
\end{aligned}
$$

Substitution of this into (10) yields (9). ∎

Corollary As a corollary to (9), if we have

$$s(\theta) = \int_{\theta_0}^{\theta} \sqrt{[f(\theta)]^2 + [f'(\theta)]^2} \, d\theta$$

then
$$D_\theta s = \sqrt{[f(\theta)]^2 + [f'(\theta)]^2}$$
(11)

or, equivalently, in differential form,

$$(ds)^2 = [f(\theta)]^2 (d\theta)^2 + [f'(\theta)]^2 (d\theta)^2$$

$$(ds)^2 = r^2 (d\theta)^2 + (dr)^2$$
(12)

since $(dr)^2 = [f'(\theta)]^2 (d\theta)^2$.

The formulas for $\tan \psi$ and $(ds)^2$ can be obtained by the following crude argument, which does *not* constitute a proof. We form a curved right triangle as shown in Figure 12.4.5. The following relations are suggested by the figure

$$\tan \psi = \frac{r\, d\theta}{dr} = r/(dr/d\theta), \text{ where } r = f(\theta)$$
(6)

$$(ds)^2 = (dr)^2 + r^2 (d\theta)^2$$
(12)

and

$$\sin \psi = r \frac{d\theta}{ds} \qquad \cos \psi = \frac{dr}{ds}$$
(13)

if s increases with θ.

A rigorous proof of (13) is left for the reader (see Exercise 26).

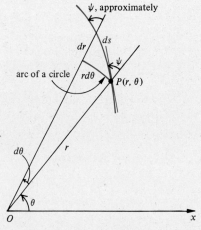

Figure 12.4.5

EXAMPLE 3   Find the total length of the cardioid $r = a(1 + \cos \theta)$, $a > 0$.

Solution   The curve is shown in Figure 12.4.3. From (9) with $f(\theta) = a(1 + \cos \theta)$ we obtain

$$s = \int_0^{2\pi} \sqrt{a^2(1 + \cos \theta)^2 + a^2(-\sin \theta)^2}\, d\theta$$

$$= a \int_0^{2\pi} \sqrt{1 + 2\cos \theta + \cos^2 \theta + \sin^2 \theta}\, d\theta$$

$$= a \int_0^{2\pi} \sqrt{2 + 2\cos \theta}\, d\theta$$

$$= 2a \int_0^{2\pi} \left| \cos \frac{\theta}{2} \right|\, d\theta$$

From the symmetry of the curve about the polar axis,

$$s = 4a \int_0^\pi \left| \cos \frac{\theta}{2} \right| d\theta$$

Now, $\cos \dfrac{\theta}{2} \geq 0$ in $[0, \pi]$ and thus the absolute value signs may be dropped.

$$s = 4a \int_0^\pi \cos \frac{\theta}{2} \, d\theta = 8a \sin \frac{\theta}{2} \Big|_0^\pi = 8a \qquad \bullet$$

In some cases it is advantageous to use r as the independent variable. We have for this situation

$$s = \int_{r_0}^{r_1} \sqrt{1 + r^2 (D_r \theta)^2} \, dr \qquad (14)$$

This is left as an exercise for the reader.

Exercises 12.4

In Exercises 1 through 8, find ψ at the point indicated.

1. $r = a \sin \theta \quad \left(\dfrac{\sqrt{3}a}{2}, \dfrac{\pi}{3} \right), a > 0$

2. $r = 2\theta \quad \left(\dfrac{\pi}{4}, \dfrac{\pi}{8} \right)$

3. $r = a(1 - \cos \theta) \quad \left(\dfrac{a}{2}, \dfrac{\pi}{3} \right), a > 0$

4. $r = a(2 + \cos \theta) \quad \left(2a, \dfrac{\pi}{2} \right), a > 0$

5. $r = 2 + 3 \cos \theta \quad \left(\dfrac{7}{2}, \dfrac{\pi}{3} \right)$

6. $r = a\theta^2 \quad (a\pi^2, \pi), a < 0$

7. $r = \dfrac{a}{1 - \cos \theta} \quad \left(a, \dfrac{\pi}{2} \right), a > 0$

8. $r = a^2 \cos 2\theta \quad \left(\dfrac{a^2}{2}, \dfrac{\pi}{6} \right), a \neq 0$

9. For the parabola $r = a \sec^2 \dfrac{\theta}{2}$, where $a \neq 0$, show that $\psi + \phi = \pi$.

10. Prove that the angle $\psi = $ constant for the **logarithmic spiral** $r = ae^{b\theta}$ (a and b nonzero constants). Since the angle between the radius vector and the tangent to the curve is constant, this curve is also called the **equiangular spiral**.

11. Find equations of all curves such that all rays from the pole make a *constant* angle ψ in $(0, \pi)$ with the curve. $\left(Hint: \text{ Use } \cot \psi = \dfrac{D_\theta r}{r} \text{ and integrate.} \right)$

In Exercises 12 through 15, find the slope of the tangent line to the given curve at the specified point using equation (3).

12. $r = a \cos \theta \quad \left(\dfrac{a\sqrt{2}}{2}, \dfrac{\pi}{4} \right), a > 0$

13. $r = 2\theta$ $\left(\pi, \dfrac{\pi}{2}\right)$

14. $r = a \sin 2\theta$ $\left(a, \dfrac{\pi}{4}\right)$

15. $r = 1 - 2 \cos \theta$ $\left(1, \dfrac{\pi}{2}\right)$

16. Let α be the angle of intersection between two curves $r = f_1(\theta)$ and $r = f_2(\theta)$ at a point P. If α is measured from the first to the second curve, prove that

$$\tan \alpha = \frac{\tan \psi_2 - \tan \psi_1}{1 + \tan \psi_2 \tan \psi_1}$$

As a corollary, deduce a condition that implies that the two curves intersect at right angles.

In Exercises 17 through 20, find the angle between the tangent lines of the given pair of curves at the specified point of intersection.

17. $r = 4$, $r \cos \theta = 2$ $\left(4, \dfrac{\pi}{3}\right)$

18. $r = a \cos \theta$, $r = a \sin \theta$ $\left(\dfrac{\sqrt{2}a}{2}, \dfrac{\pi}{4}\right), a > 0$

19. $r = a(1 + \cos \theta)$, $r = a(1 - \cos \theta)$ $\left(a, \dfrac{\pi}{2}\right), a > 0$

20. $r \sin \theta = 2a$, $r = a \sec^2 \dfrac{\theta}{2}$ $\left(2a, \dfrac{\pi}{2}\right), a > 0$

In Exercises 21 through 25, find the length of the indicated arc.

21. $r = a \cos \theta$, $a > 0, \theta = 0$ to $\theta = \dfrac{\pi}{4}$

22. $r = 5\theta^2$, $\theta = 1$ to $\theta = 2$

23. $r = 3e^{2\theta}$, $\theta = \theta_0$ to $\theta = \theta_1$.

24. $r = \cos^2 \dfrac{\theta}{2}$, $\theta = 0$ to $\theta = \dfrac{\pi}{2}$

25. $r = \dfrac{p}{1 - \cos \theta}$, $p > 0, \theta = \dfrac{\pi}{2}$ to $\theta = \pi$

26. Prove equations (13) of the text; that is,

$$\sin \psi = r \frac{d\theta}{ds}, \qquad \cos \psi = \frac{dr}{ds}$$

(*Hint:* Express $\sin^2 \psi$ in terms of $\tan^2 \psi$ and use the formula for $\tan \psi$.)

27. Starting with $x = r \cos \theta$ and $y = r \sin \theta$, where $r = f(\theta)$ is a twice differentiable function of θ as a parameter, show that the curvature K is given by

$$K = \frac{|f^2 + 2(f')^2 - ff''|}{[f^2 + (f')^2]^{3/2}}$$

(*Hint:* Use (22) of Section 11.7 with t replaced by θ.)

28. Use the result of Exercise 27 to find the curvature of the curve

$$r = a \cos \theta + b \sin \theta$$

at any point (r, θ) on it. Explain your result by finding the Cartesian representation.

29. (a) Use the result of Exercise 27 to find the curvature of the cardioid

$$r = a(1 - \cos \theta), \qquad a > 0 \text{ for } 0 < \theta < 2\pi$$

(b) Find the location and the corresponding value of the minimum curvature.

12.5   MOTION IN POLAR COORDINATES

Suppose that a particle is moving along a planar curve given by $r = f(\theta)$. It is our objective in this section to develop expressions for its velocity **v** and acceleration **a**. It turns out that this can be conveniently done if we refer these quantities to orthogonal unit vectors in the radial and transverse directions, respectively. Thus we let $\mathbf{e}_r$ be a vector in the increasing radial direction and $\mathbf{e}_\theta$ be a unit vector in the direction of increasing θ (Figure 12.5.1), and these unit vectors serve as a basis for any vector in the plane. However, these vectors are different from the constant unit vectors **i** and **j** of the Cartesian coordinate system in that $\mathbf{e}_r$ and $\mathbf{e}_\theta$ *are not constant vectors* in general. At different points their direction may vary. This fact must be accommodated in our endeavor to find the formulas for the velocity **v** and acceleration **a**.

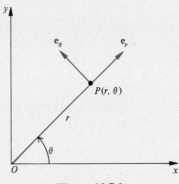

Figure 12.5.1

The position vector to a point $P(r, \theta)$ is given by

$$\mathbf{r} = r\mathbf{e}_r \tag{1}$$

where $r = |\mathbf{r}|$. This follows from the fact that the vector **r** is in the same direction as $\mathbf{e}_r$ and its magnitude is given by r. The velocity **v** is given by

$$\mathbf{v} = D_t\mathbf{r} = D_t(r\mathbf{e}_r) = rD_t\mathbf{e}_r + (D_t r)\mathbf{e}_r \tag{2}$$

We must find $D_t\mathbf{e}_r$. This is accomplished by first using the chain rule

$$D_t\mathbf{e}_r = (D_\theta\mathbf{e}_r)(D_t\theta) \tag{3}$$

and calculating $D_\theta\mathbf{e}_r$.

Now,

$$\mathbf{e}_r = (\mathbf{e}_r \cdot \mathbf{i})\mathbf{i} + (\mathbf{e}_r \cdot \mathbf{j})\mathbf{j}$$

and therefore

$$\mathbf{e}_r = (\cos\theta)\mathbf{i} + (\sin\theta)\mathbf{j} \tag{4}$$

Since later we will also need $D_t\mathbf{e}_\theta$, we write

$$\mathbf{e}_\theta = (\mathbf{e}_\theta \cdot \mathbf{i})\mathbf{i} + (\mathbf{e}_\theta \cdot \mathbf{j})\mathbf{j}$$

$$= \left[\cos\left(\frac{\pi}{2} + \theta\right)\right]\mathbf{i} + \left[\sin\left(\frac{\pi}{2} + \theta\right)\right]\mathbf{j}$$

and hence,

$$\mathbf{e}_\theta = (-\sin\theta)\mathbf{i} + (\cos\theta)\mathbf{j} \qquad (5)$$

Differentiation of (4) and (5) with respect to θ yields the useful formulas

$$D_\theta\mathbf{e}_r = (-\sin\theta)\mathbf{i} + (\cos\theta)\mathbf{j} = \mathbf{e}_\theta \qquad (6)$$

and

$$D_\theta\mathbf{e}_\theta = (-\cos\theta)\mathbf{i} - (\sin\theta)\mathbf{j} = -\mathbf{e}_r \qquad (7)$$

Substitution of (6) and (7) into (3) results in

$$D_t\mathbf{e}_r = (D_\theta\mathbf{e}_r)(D_t\theta) = \mathbf{e}_\theta\, D_t\theta \qquad (8)$$

and

$$D_t\mathbf{e}_\theta = (D_\theta\mathbf{e}_\theta)(D_t\theta) = -\mathbf{e}_r\, D_t\theta \qquad (9)$$

Substitution of (8) into (2) gives the velocity $\mathbf{v}$,

$$\mathbf{v} = (D_t r)\mathbf{e}_r + r(D_t\theta)\mathbf{e}_\theta \qquad (10)$$

in terms of its radial and transverse (or circumferential) components. If we use the Newtonian notation for the derivative, $\cdot \equiv D_t$ (so that $\dot{r} = D_t r$, and so on), there results

$$\mathbf{v} = \dot{r}\mathbf{e}_r + r\dot{\theta}\mathbf{e}_\theta \qquad (11)$$

We denote the radial and transverse components of velocity by v_r and v_θ so that

$$\mathbf{v} = v_r\mathbf{e}_r + v_\theta\mathbf{e}_\theta \qquad (12)$$

where

$$v_r = \dot{r} \quad\text{and}\quad v_\theta = r\dot{\theta} \qquad (13)$$

The speed $v = |\mathbf{v}|$ is given by

$$v = (\dot{r}^2 + r^2\dot{\theta}^2)^{1/2} \qquad (14)$$

If expression (11) is examined from a geometrical viewpoint, it is not surprising. The radial component of velocity is $\dot{r}$ as expected, whereas the transverse component is $r\dot{\theta}$, which is what would be obtained in the case of motion on a circular path.

These results are readily recalled if a differential triangle is shown as in Figure 12.5.2.

EXAMPLE 1   A particle P moves on the circle $r = a\cos\theta$, $a > 0$, so that the radius vector OP turns counterclockwise at a constant rate of 3 radians/sec. Find v_r, v_θ, and the speed of the particle.

Solution $r = a\cos\theta$, $v_r = \dot{r} = (-a\sin\theta)\dot{\theta} = -3a\sin\theta$ since $\dot{\theta} = 3$ rad/sec. Also,

$$v_\theta = r\dot{\theta} = 3a\cos\theta$$

and

$$v = \sqrt{v_r^2 + v_\theta^2} = 3a$$

This says that the speed is constant.   ●

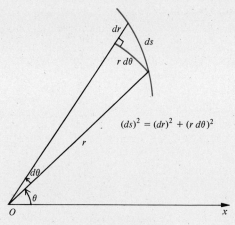

$$(ds)^2 = (dr)^2 + (r\,d\theta)^2$$

Figure 12.5.2

EXAMPLE 2 A particle moves on the cardioid $r = a(1 + \sin\theta)$, $a > 0$, with constant speed $v = b$. Find v_r and v_θ if $\dot\theta \geq 0$.

Solution
$$v_r = \dot r = a(\cos\theta)\dot\theta \qquad v_\theta = r\dot\theta = a(1 + \sin\theta)\dot\theta$$

The square of the speed is given by

$$v^2 = v_r{}^2 + v_\theta{}^2 = a^2[\cos^2\theta + (1 + \sin\theta)^2]\dot\theta^2$$

Thus,

$$v^2 = b^2 = 2a^2(1 + \sin\theta)\dot\theta^2$$

$$\dot\theta^2 = \frac{b^2}{2a^2(1 + \sin\theta)} \qquad \dot\theta = \frac{b}{a\sqrt{2}\sqrt{1 + \sin\theta}} \qquad (\text{since } \dot\theta \geq 0)$$

Hence

$$v_r = \frac{b\cos\theta}{\sqrt{2}\sqrt{1 + \sin\theta}} \qquad \text{and} \qquad v_\theta = \frac{b\sqrt{1 + \sin\theta}}{\sqrt{2}} \qquad \bullet$$

The acceleration vector **a** is obtained by differentiation of **v** with respect to time:

$$\mathbf{a} = \dot{\mathbf{v}} = D_t(\dot r \mathbf{e}_r + r\dot\theta\,\mathbf{e}_\theta)$$
$$= \dot r D_t \mathbf{e}_r + \ddot r \mathbf{e}_r + r\dot\theta\,D_t\mathbf{e}_\theta + D_t(r\dot\theta)\mathbf{e}_\theta$$

Substitution of (8) and (9) and differentiation of the last term yields

$$\mathbf{a} = \dot r\dot\theta\,\mathbf{e}_\theta + \ddot r \mathbf{e}_r - r\dot\theta^2\,\mathbf{e}_r + (r\ddot\theta + \dot r\dot\theta)\mathbf{e}_\theta$$

hence

$$\mathbf{a} = (\ddot r - r\dot\theta^2)\mathbf{e}_r + (r\ddot\theta + 2\dot r\dot\theta)\mathbf{e}_\theta \tag{15}$$

and we have relatively complicated expressions for the radial and transverse components, a_r and a_θ. These are given by

$$a_r = \ddot r - r\dot\theta^2 \qquad \text{and} \qquad a_\theta = r\ddot\theta + 2\dot r\dot\theta \tag{16}$$

EXAMPLE 3 A particle moves counterclockwise on a circular path of radius b. Find the radial and transverse components of velocity and acceleration.

Solution Let the circle have polar equation $r = b$, so that $\dot{r} = 0$ and $\ddot{r} = 0$. Hence from (13) (with $r = b$)

$$v_r = 0 \qquad v_\theta = b\dot{\theta}$$

and the speed

$$v = b\dot{\theta},$$

and from (16),

$$a_r = -b\dot{\theta}^2 \qquad \text{and} \qquad a_\theta = b\ddot{\theta}$$

In the particular case when the angular velocity is constant, $\ddot{\theta} = 0$ and $a_\theta = 0$. Furthermore,

$$a_r = -b\dot{\theta}^2 = \text{constant}$$

which implies that the acceleration $\mathbf{a}$ is constant and is directed toward the center of the circle. Since $v = b\dot{\theta}$ we have for the magnitude of the acceleration,

$$a = \sqrt{a_r^2 + a_\theta^2} = \frac{v^2}{b} \qquad \bullet$$

EXAMPLE 4 A particle moves on the curve $r = a(1 - \cos\theta)$ so that its angular position is given by $\theta = kt$ ($k > 0$ is constant). Find its radial and transverse components of velocity and acceleration.

Solution We differentiate with respect to t,

$$\dot{r} = a(\sin\theta)\dot{\theta} \qquad \dot{\theta} = k$$

Therefore,

$$v_r = \dot{r} = ka\sin\theta \qquad \text{and} \qquad v_\theta = r\dot{\theta} = ka(1 - \cos\theta)$$

Furthermore,

$$\ddot{\theta} = 0 \qquad \text{and} \qquad \ddot{r} = k^2 a\cos\theta$$

Hence,

$$a_r = \ddot{r} - r(\dot{\theta})^2 = k^2 a\cos\theta - k^2 a(1 - \cos\theta)$$
$$= k^2 a(2\cos\theta - 1)$$

and

$$a_\theta = r\ddot{\theta} + 2\dot{r}\dot{\theta} = 2k^2 a\sin\theta \qquad \bullet$$

EXAMPLE 5 **Motion of a Simple Pendulum.** Consider a simple pendulum consisting of a mass m attached to a rod of fixed length and suspended from a fixed point O (Figure 12.5.3). It is assumed the mass of the pendulum and all friction (at pivot point, atmospheric, and so on) are negligible. We seek the equations of motion for the bob of the pendulum using polar coordinates.

Solution For convenience, the polar axis (pointing downward) coincides with the rest position of the pendulum. The bob of the pendulum is acted on by two forces

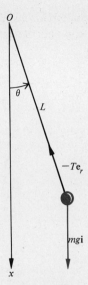

Figure 12.5.3

1. The tension in the rod given by $-Te_r$.
2. The gravitational force or weight $mg\mathbf{i}$, where g is the gravitational constant.

Now,

$$\mathbf{i} = (\mathbf{i} \cdot \mathbf{e}_r)\mathbf{e}_r + (\mathbf{i} \cdot \mathbf{e}_\theta)\mathbf{e}_\theta = (\cos \theta)\mathbf{e}_r - (\sin \theta)\mathbf{e}_\theta$$

and with this decomposition of $\mathbf{i}$ in the radial and transverse directions the resultant force is

$$(mg \cos \theta - T)\mathbf{e}_r - mg(\sin \theta)\mathbf{e}_\theta \qquad (17)$$

In accordance with Newton's law, we have from (16)

$$(mg \cos \theta - T)\mathbf{e}_r - mg(\sin \theta)\mathbf{e}_\theta = m(\ddot{r} - r\dot{\theta}^2)\mathbf{e}_r + m(r\ddot{\theta} + 2\dot{r}\dot{\theta})\mathbf{e}_\theta$$

and since $\mathbf{e}_r$ and $\mathbf{e}_\theta$ are orthogonal to each other their respective coefficients must be equal. Therefore,

$$\begin{aligned} mg \cos \theta - T &= m(\ddot{r} - r\dot{\theta}^2) \\ -mg \sin \theta &= m(r\ddot{\theta} + 2\dot{r}\dot{\theta}) \end{aligned} \qquad (18)$$

This system of equations can be simplified by observing that in the case of the pendulum $r = L$, and since L is fixed,

$$\dot{r} = \dot{L} = 0 \qquad \text{and} \qquad \ddot{r} = \ddot{L} = 0$$

Hence (18) reduces to

$$g \cos \theta - \frac{T}{m} = -L\dot{\theta}^2 \qquad (19a)$$

$$-g \sin \theta = L\ddot{\theta} \qquad (19b)$$

For small oscillations, θ is small and $\sin \theta \approx \theta$, thus (19b) becomes

$$\ddot{\theta} + \frac{g}{L}\theta = 0 \qquad (20)$$

which is the equation for a simple harmonic motion of angular frequency $\sqrt{g/L}$. Once $\theta(t)$ is determined, the unknown tension $T(t)$ is readily obtained algebraically from (19a). ●

Exercises 12.5

In Exercises 1 through 4, find the radial and transverse components of the velocity and acceleration of a particle moving along each of the curves. Express answers in terms of θ, ω, $\dot\omega$ where $\omega = \dot\theta$ is the angular velocity.

1. $r = 3\theta$
2. $r = b \sin 2\theta$, b is constant
3. $r = b \cos\theta + c \sin\theta$, b and c constants
4. $r = \frac{1}{2}(e^{b\theta} + e^{-b\theta})$, b constant
5. An automobile moves along the path $r = a(1 - \cos\theta)$, $0 \le \theta \le \pi$, so that $\theta = bt$ where t is the time. Find its velocity and acceleration in the radial and transverse directions as funtions of θ. It is assumed that a and b are constants.
6. A point moves on the limaçon $r = 3 + 2 \sin\theta$. Find the radial and transverse components of the velocity at $\theta = \dfrac{\pi}{2}$ if its speed $|\mathbf{v}| = 30$ and $\dot\theta < 0$.
7. A point moves on the cardioid $r = a(1 + \cos\theta)$, $a > 0$. Find v_r and v_θ at $\theta = \dfrac{\pi}{3}$ if $|\mathbf{v}| = 5$ units/sec and $\dot\theta > 0$.
8. A particle moves on the logarithmic spiral $r = e^\theta$ with constant speed $v = v_0$. Find formulas for a_r, a_θ and $|\mathbf{a}|$. Show that $|\mathbf{a}|$ varies inversely with r.
9. A particle moves counterclockwise on a circle $r = b \cos\theta$, $b > 0$, with constant speed v_0. Find a_r, a_θ, and $|\mathbf{a}|$.
10. Assume a particle in motion is *attracted* to a fixed point O, by a force $\mathbf{F}$. In such cases, the particle is said to move subject to a **central force.** It can be established that the particle must move in a plane containing point O. Accepting this, prove that if $P(r, \theta)$ denotes the polar coordinates of P relative to O and an arbitrary polar axis, that the motion must satisfy the condition that $r^2\dot\theta = $ constant. See Exercise 25 of the next section for an important application of this result. [*Hint:* The transverse component of force and the corresponding component of acceleration must be zero. Show that $D_t(r^2\dot\theta) = 0$.]
11. Prove that if $A\mathbf{e}_r + B\mathbf{e}_\theta = C\mathbf{e}_r + D\mathbf{e}_\theta$, where $\mathbf{e}_r$ and $\mathbf{e}_\theta$ are the orthogonal unit vectors in the radial and transverse directions, respectively, then $A = C$ and $B = D$. (*Hint:* Use dot product of both sides with the unit vectors $\mathbf{e}_r$ and $\mathbf{e}_\theta$.)
12. Show that $\theta = A \cos pt + B \sin pt$ is a solution of (20) for arbitrary constants A and B, provided that $p = \sqrt{g/L}$ if $p > 0$. This expression can be shown to be the totality of solutions of (20) where constants A and B are determined from initial conditions $\theta|_{t=0} = \theta_0$ and $\dot\theta|_{t=0} = \dot\theta_0$ and where θ_0 and $\dot\theta_0$ are the initial angular displacement and velocity, respectively. Find A and B in terms of θ_0 and $\dot\theta_0$.

12.6   AREAS IN POLAR COORDINATES

In this section, we shall determine the area A of a region R bounded by the polar curve $r = f(\theta)$, and the rays $\theta = \alpha$ and $\theta = \beta$ where $f(\theta) \ge 0$ and is continuous, (Figure 12.6.1). We take $\beta > \alpha$ and angles are measured in radians.

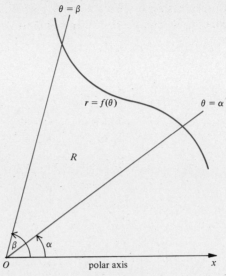

$\theta = \beta$

$r = f(\theta)$ $\theta = \alpha$

R

β α

O polar axis x

Figure 12.6.1

We are going to approximate the area of R by using sectors of circles. Recall that the area of a circular sector is $\frac{1}{2}r^2 \Delta\theta$ where r is the radius and $\Delta\theta$ is the central angle. If we add these sector areas together we obtain

$$A \approx \Sigma \tfrac{1}{2} r^2 \, \Delta\theta$$

Taking the limit as $\Delta\theta \to 0$ we can expect the result

$$A = \tfrac{1}{2} \int_\alpha^\beta r^2 \, d\theta = \tfrac{1}{2} \int_\alpha^\beta [f(\theta)]^2 \, d\theta$$

This is, in fact, correct, and we now derive it formally.

Let P denote a partition of $[\alpha, \beta]$ determined by

$$\alpha = \theta_0 < \theta_1 < \theta_2 < \cdots < \theta_n = \beta$$

and also set $\Delta\theta_i = \theta_i - \theta_{i-1}$. Therefore, we have n intervals of the form $[\theta_{i-1}, \theta_i]$, $i = 1, 2, \ldots, n$. The lines with $\theta = \theta_i$, $i = 1, 2, \ldots, n$ divide the given region into n wedgeshaped subregions. We set

$$f(u_i) = \text{minimum value of } r \text{ on } [\theta_{i-1}, \theta_i]$$

and

$$f(v_i) = \text{maximum value of } r \text{ on } [\theta_{i-1}, \theta_i]$$

The portion of the region R that lies between θ_{i-1} and θ_i contains a circular sector of radius $f(u_i)$ and central angle $\Delta\theta_i = \theta_i - \theta_{i-1}$ and is contained in a circular sector of radius $f(v_i)$ and central angle $\Delta\theta_i = \theta_i - \theta_{i-1}$. Thus, the area A_i, of the portion of R lying between θ_{i-1} and θ_i, must satisfy the double inequality

$$\tfrac{1}{2}[f(u_i)]^2 \, \Delta\theta_i \leq A_i \leq \tfrac{1}{2}[f(v_i)]^2 \, \Delta\theta_i$$

Therefore, there exists a ξ_i in $[\theta_{i-1}, \theta_i]$ such that

$$A_i = \tfrac{1}{2}[f(\xi_i)]^2 \, \Delta\theta_i$$

(see Figure 12.6.2), and $A_i = $ area of this specific circular sector of radius $f(\xi_i)$.

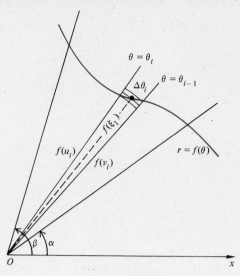

Figure 12.6.2

There is a specific ξ_i for each of the intervals so that the total area A is given by

$$A = \sum_{i=1}^{n} \frac{1}{2}[f(\xi_i)]^2 \, \Delta\theta_i \qquad \theta_{i-1} \leq \xi_i \leq \theta_i \tag{1}$$

and if we let the norm of the partition $|P|$ tend to zero the Riemann sum becomes the Riemann integral. Therefore

$$A = \frac{1}{2}\int_{\alpha}^{\beta} [f(\theta)]^2 \, d\theta \tag{2}$$

EXAMPLE 1   Find the area of the region shown in Figure 12.6.3, bounded by the polar axis and a portion of the spiral $r = 3\theta$.

Solution $$r = 3\theta, \text{ so } f(\theta) = 3\theta.$$

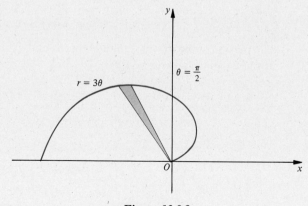

Figure 12.6.3

$$\text{area of } R = \frac{1}{2} \int_0^\pi [f(\theta)]^2 \, d\theta = \frac{1}{2} \int_0^\pi (3\theta)^2 \, d\theta$$

$$= \frac{9}{2} \int_0^\pi \theta^2 \, d\theta = \frac{9}{2} \frac{\theta^3}{3} \Big|_0^\pi = \frac{3}{2} \pi^3 \text{ square units.} \quad \bullet$$

EXAMPLE 2 Find the area of the circle $r = a \sin \theta$, $a > 0$.

Solution The circle is shown in Figure 12.6.4. The region inside the circle is swept out by a ray rotating from $\theta = 0$ to $\theta = \pi$. Hence the area of the circle is given by $(f(\theta) = a \sin \theta)$

$$A = \frac{1}{2} \int_0^\pi [f(\theta)]^2 \, d\theta = \frac{a^2}{2} \int_0^\pi \sin^2 \theta \, d\theta$$

$$= \frac{a^2}{2} \int_0^\pi \left(\frac{1}{2} - \frac{1}{2} \cos 2\theta \right) d\theta = \frac{a^2}{2} \left(\frac{\theta}{2} - \frac{\sin 2\theta}{4} \right) \Big|_0^\pi$$

$$= \frac{\pi a^2}{4} \text{ square units}$$

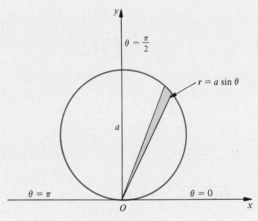

Figure 12.6.4

The result is easily verified because a is the diameter of the circle. The area could also have been obtained by finding the area of the semicircle by integration from $\theta = 0$ to $\theta = \pi/2$ and doubling the result. $\quad \bullet$

EXAMPLE 3 Find the area enclosed by the cardioid $r = 2(1 + \cos \theta)$.

Solution The graph is sketched in Figure 12.2.3. Because of the symmetry about the polar axis, it is only necessary to use (2) to find the area of the upper half and double the result, $(f(\theta) = 2(1 + \cos \theta))$.

$$\frac{A}{2} = \frac{1}{2} \int_0^\pi [f(\theta)]^2 \, d\theta = 2 \int_0^\pi (1 + \cos \theta)^2 \, d\theta$$

$$= 2 \int_0^\pi (1 + 2 \cos \theta + \cos^2 \theta) \, d\theta$$

and using the identity $\cos^2\theta = \frac{1}{2} + \frac{1}{2}\cos 2\theta$, there results

$$\frac{A}{2} = 2\int_0^\pi \left(\frac{3}{2} + 2\cos\theta + \frac{\cos 2\theta}{2}\right)d\theta$$

$$= 2\left(\frac{3}{2}\theta + 2\sin\theta + \frac{\sin 2\theta}{4}\right)\bigg|_0^\pi = 3\pi$$

Hence, $A = 6\pi$ square units. ●

Consider next the region R bounded by $r = f(\theta)$, $r = g(\theta)$, $\theta = \alpha$ and $\theta = \beta$ (Figure 12.6.5). It is assumed that $f(\theta) \geq g(\theta) \geq 0$ and f and g are continuous for all θ in the interval $[\alpha, \beta]$. Clearly, the area A may be determined by subtracting the areas of the two regions of the type considered previously. Therefore

$$A = \frac{1}{2}\int_\alpha^\beta [f(\theta)]^2\,d\theta - \frac{1}{2}\int_\alpha^\beta [g(\theta)]^2\,d\theta$$

or

$$A = \frac{1}{2}\int_\alpha^\beta ([f(\theta)]^2 - [g(\theta)]^2)\,d\theta \tag{3}$$

The last integral (3) can be expressed as a limit of a Riemann sum as follows:

$$A = \lim_{|P|\to 0} \sum_i \frac{1}{2}([f(\xi_i)]^2 - [g(\xi_i)]^2)\,\Delta\theta_i \tag{4}$$

where $\theta_{i-1} \leq \xi_i \leq \theta_i$.

The sum (4) may be obtained directly by using an element of area as sketched in Figure 12.6.5 and then sweeping out the region by letting θ vary from α to β.

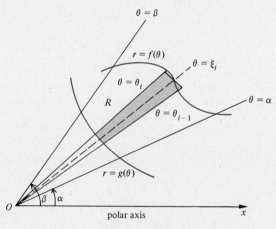

Figure 12.6.5

EXAMPLE 4   Find the area that is inside the cardioid $r = 2a(1 + \cos\theta)$ and outside the circle $r = 3a$ where $a > 0$.

Solution   The region is sketched in Figure 12.6.6 where a typical polar element of area is shown. Simultaneous solution yields the two points of intersection $(3a, \pm\pi/3)$.

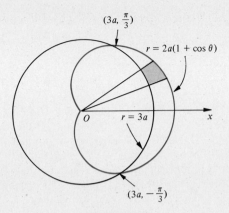

Figure 12.6.6

Again we have symmetry about the polar axis, and consequently, with $f(\theta) = 2a(1 + \cos\theta)$ and $g(\theta) = 3a$ substituted into (3), we obtain

$$\frac{A}{2} = \frac{a^2}{2} \int_0^{\pi/3} [2^2(1 + \cos\theta)^2 - 3^2]\, d\theta$$

$$= \frac{a^2}{2} \int_0^{\pi/3} [4\cos^2\theta + 8\cos\theta - 5]\, d\theta$$

$$= \frac{a^2}{2} \int_0^{\pi/3} [2\cos 2\theta + 8\cos\theta - 3]\, d\theta$$

$$= \frac{a^2}{2}(\sin 2\theta + 8\sin\theta - 3\theta)\big|_0^{\pi/3}$$

$$= \frac{a^2}{2}\left(\frac{\sqrt{3}}{2} + 4\sqrt{3} - \pi - 0\right) = \frac{a^2}{2}\left(\frac{9}{2}\sqrt{3} - \pi\right)$$

Thus $A = a^2(\frac{9}{2}\sqrt{3} - \pi)$ square units. $\qquad\bullet$

Exercises 12.6

In Exercises 1 through 12, find the area of the region bounded by the graph of the given equation. Sketch the curve.

1. $r = b$, $b > 0$
2. $r = 2b\cos\theta$, $b > 0$
3. $r = 1 + \sin\theta$
4. $r = a(1 - \cos\theta)$, $a > 0$
5. $r = b\sin 2\theta$, $b > 0$
6. $r = a\sin 3\theta$, $a > 0$
7. $r = a\sin n\theta$, $a > 0$, n is an odd integer
8. $r = a\sin n\theta$, $a > 0$, n is an even integer

9. $r^2 = a^2\cos 2\theta$, $a > 0$
10. $r = a\sin^2\frac{\theta}{2}$, $a > 0$

11. $r = 3 + 2\cos\theta$
12. $r = a + b\cos\theta$, $a > b > 0$

In Exercises 13 through 16, calculate the area of the region bounded by the graphs of the following:

13. $r = a\cos\theta$, $r = a\sin\theta$, and the rays $\theta = 0$, $\theta = \frac{\pi}{4}$

14. $r = 2 + \cos\theta$, $r = \cos\theta$, and the rays $\theta = 0$, $\theta = \frac{\pi}{2}$

15. $r = e^{2\theta}$, $0 \le \theta \le \pi$; $r = \theta$, $0 \le \theta \le \pi$; the rays $\theta = 0$ and $\theta = \pi$

16. The parabola $r = \dfrac{a}{2} \sec^2 \dfrac{\theta}{2}$ and the vertical line through the origin.

In Exercises 17 through 20, find the area of the region common to the two given regions.

17. $r = 3 \cos \theta$, $r = 3 \sin \theta$ 18. $r = 1 - \cos \theta$, $r = \sin \theta$

19. $r = \sqrt{2} \sin \theta$, $r^2 = \cos 2\theta$ 20. $r = 3 - \sin \theta$, $r = 5 \sin \theta$

21. Determine the area enclosed by the small loop of the limaçon $r = a(1 + 2 \cos \theta)$, $a > 0$.

22. Find the area that is inside the circle $r = 2a \cos \theta$ and outside the circle $r = a$, where $a > 0$.

23. Find the area that is inside the graph of the region $r = 6 \sin \theta$ and outside the graph $r = 2(1 + \sin \theta)$.

24. Find the area of the region between the inner and outer loops of the curve $r = a(1 - 2 \sin \theta)$, $a > 0$.

25. Kepler deduced three laws of planetary motion assuming that the planet may be treated as a particle acted upon by a central force exerted by the sun. The second law states that the radius vector drawn from the sun to the planet sweeps out equal areas in equal times. Prove this by referring to the result of Exercise 10 of Section 12.5 where the sun is the fixed point 0. (Hint: Set up the expression for the area $A(t)$ swept out by the planet in time $[t_0, t]$ and show that $\dot{A} = $ constant).

Review and Miscellaneous Exercises

In Exercises 1 through 10 sketch and identify the graph of the following equations in polar coordinates.

1. $r = 4$ 2. $\theta = -\dfrac{\pi}{3}$

3. $r \cos \theta = -4$ 4. $r(3 \cos \theta + 4 \sin \theta) = 0$

5. $r = a(1 + \cos \theta)$, $\quad a > 0$ 6. $r^2 + 3r = 0$

7. $r^2 = 4 \sin 2\theta$ 8. $r = \cos (\theta + \pi)$

9. $r = \dfrac{3}{1 - \sin \theta}$ 10. $\theta^2 = \dfrac{\pi^2}{16}$

In Exercises 11 through 14, find a polar equation of the graph having the given Cartesian equation.

11. $x^2 - y^2 = a^2$ (*a* is constant)

12. $(x - 3)^2 + y^2 = 4$

13. $(x^2 + y^2)^2 = a^2(x^2 - y^2)$

14. $y^2 = x^2 \left(\dfrac{a + x}{a - x} \right)$ (strophoid)   (*a* is a constant)

In Exercises 15 through 18, find a Cartesian equation of the graph having the given polar equation.

15. $r = 2 \cos \theta - \sin \theta$ 16. $r \cos \left(\theta - \dfrac{\pi}{3} \right) = 3$

17. $r^2 \sin 2\theta = 2$ 18. $r = \dfrac{10}{1 + 5 \cos \theta}$

19. Find the equation in polar coordinates of a parabola with focus at the origin and vertex at $(2, 0)$.

20. Find the equation in polar coordinates of a parabola with focus at the origin and vertex at $\left(3, \dfrac{3\pi}{2} \right)$.

21. Show that the equations $r = \cos \theta + 2$ and $r = \cos \theta - 2$ possess the same graph.

22. What kind of symmetry occurs if an equation in r and θ is unchanged when
   (a) r is replaced by $-r$
   (b) θ is replaced by $\pi - \theta$
   (c) r and θ are replaced by $-r$ and $-\theta$, respectively.
23. Find all points of intersection of $r = a(1 - \cos\theta)$ and $r = a(1 - \sin\theta)$, $a > 0$.
24. Find the total length of the curve $r = a(1 - \cos\theta)$, $a > 0$.
25. The limaçon $r = 1 + 2\sin\theta$ has a small inner loop. Find its area.
26. Find the area of the region inside the curve $r = 4a\sin\theta$ and outside the curve $r = 2a$, where $a > 0$.
27. Show that the graph of $r = 1/\theta, \theta > 0$ has a horizontal asymptote and find it. $\left(\text{Hint: } y = r\sin\theta = \dfrac{\sin\theta}{\theta}.\right)$
28. The curve $y^2 = \dfrac{x^3}{2a - x}$, where a is positive constant, is known as the *cissoid of Diocles*. Find its polar coordinate representation and sketch it.
29. Find the area of one loop of the curve $r = a\cos 3\theta$.
30. Find the area of a right triangle with legs a and b by using polar coordinates.
31. By proceeding geometrically identify the curve $r\cos\left(\theta - \dfrac{\pi}{6}\right) = p$, where p is a positive constant. Verify your assertion analytically.
32. Given $r = 1 + 2\cos\theta$, find the values of θ in $0 \le \theta < 2\pi$ at which the slope is zero. (*Hint:* Find $D_x y$ and set it equal to zero.)
33. Find the slope of the tangent line to the curve $r^2 \sin 2\theta = a$ at $\theta = \theta_0$, where $a > 0$, $r > 0$ and $0 < \theta_0 < \dfrac{\pi}{2}$. Transform from polar to rectangular coordinates and check your result.
34. Find the length of the curve $\theta = \dfrac{1}{2}\left(r + \dfrac{1}{r}\right)$ from $r = 1$ to $r = 4$.
35. Find the length of the cissoid $r = 2a\tan\theta\sec\theta$ from $\theta = 0$ to $\theta = \dfrac{\pi}{4}$.
36. A circle of diameter $2b$, $(b > 0)$, passes through the pole O and has its center on the line $\theta = \dfrac{\pi}{2}$. Find an equation representing the locus of midpoints of chords through the origin and identify the curve.
37. Find the curvature of the curve $r = a\sec^2\dfrac{\theta}{2}$. (Use the result of Exercise 27 of Section 12.4.)
38. The kinetic energy T of a particle of mass m is by definition $T = \dfrac{m}{2}|\mathbf{v}|^2$. What is the expression for the kinetic energy in plane polar coordinates?
39. For the simple pendulum (see Example 5 of Section 12.5) we had deduced the equation $-g\sin\theta = L\ddot\theta$. Show that this differential equation can be integrated to obtain

$$\frac{m(L\dot\theta)^2}{2} - mgL\cos\theta = C$$

or equivalently,

$$\frac{m(L\dot\theta)^2}{2} + mgL(1 - \cos\theta) = \text{constant}$$

The quantity $V = mgL(1 - \cos\theta)$ is the potential energy due to position of the mass relative to the rest position. Thus $T + V = C$. This expresses the law of conservation of energy in the absence of friction; that is, the total energy is constant. (*Hint:* multiply both sides of $-g\sin\theta = L\ddot\theta$ by $\dot\theta$ and integrate.)

13

Conic Sections

13.1   INTRODUCTION TO CONIC SECTIONS, THE ELLIPSE

13.1.1 *Introduction*

We have already worked with cones, in particular, right circular cones, when we studied volume and surface area. We now consider a right circular cone of two nappes (as illustrated in Figure 13.1.1) extending infinitely far in both directions. A *generator* of the cone is any line through the vertex lying on the surface of the cone (in Chapter 16 we shall give a formal definition of "cone" and see why these lines "generate" the surface).

The term *conic section* is used to describe the curves obtained when a right

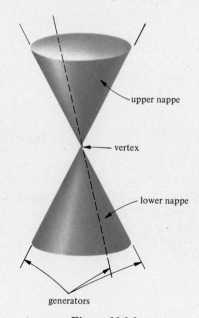

upper nappe

vertex

lower nappe

generators

Figure 13.1.1

circular cone of two nappes is intersected by a plane. If the plane does not pass through the vertex of the cone, we obtain either an ellipse (or a circle), a parabola, or a hyperbola. Conic sections were studied extensively in ancient Greece where the most notable contributions to this subject were made by Apollonius (born 262 B.C.). The Greek mathematicians utilized a plane perpendicular to a generator of the cone, and depending upon whether the vertex angle of the cone was acute, right, or obtuse, either an ellipse, a parabola, or a hyperbola is obtained.

We can instead consider a fixed cone with a given vertex angle and vary the angle at which the plane intersects the axis of the cone. Let α denote the fixed vertex angle and θ be the angle between the plane and the axis. We obtain, respectively, an ellipse, a parabola, and a hyperbola when $\alpha < \theta < \frac{\pi}{2}, \theta = \alpha$ and $0 < \theta < \alpha$ (Figure 13.1.2). A parabola occurs if and only if the plane is parallel to a generator. If the sectioning plane makes an angle $\frac{\pi}{2}$ with the cone axis, the curve of intersection is a circle. In addition to these curves, if the plane passes through the vertex of the cone we obtain either a point, one line, or two intersecting lines. We sometimes refer to these as "degenerate conics."

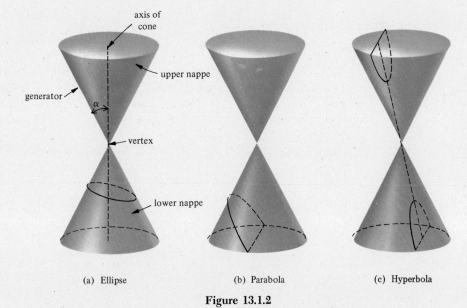

(a) Ellipse   (b) Parabola   (c) Hyperbola

Figure 13.1.2

In addition to their interesting mathematical properties, there are many applications of conic sections in such fields as rocketry, architecture, astronomy, mechanical and electrical design, and military science. We have already studied the parabola in Chapter 1. We shall now also study the ellipse and the hyperbola, but instead of using the geometry of a cone in three dimensions, we shall provide analytic definitions that are easier to use when studying plane curves.

13.1.2 The Ellipse

Definition An **ellipse** is the set of points P, the sum of whose distances from distinct fixed points F and F', called **foci,** is a constant.

Let $2a$ be a given positive number such that $2a$ is greater than the distance between the foci, then the set of all points P for which the sum of the undirected distances

$$|\overline{F'P}| + |\overline{FP}| = 2a \tag{1}$$

is an ellipse.

A line segment drawn from a point on the ellipse to the focus is called a *focal radius.* Thus Definition 1 requires that the sum of the two lengths of the focal radii, drawn to an arbitrary point P, must be equal to $2a$.

The ellipse can be constructed by using an inextensible piece of string of length $2a$. Place the ends of the string at the foci F and F'. If a pencil or marker is held taut, the possible positions of the point P determine the set of points of the ellipse (see Figure 13.1.3). Let $2c$ denote the distance between the foci so that

$$|\overline{F'P}| + |\overline{FP}| = 2a > 2c \tag{2}$$

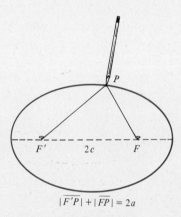

$$|\overline{F'P}| + |\overline{FP}| = 2a$$

Figure 13.1.3

A simple equation of an ellipse is obtained by placing the foci of the ellipse at the two points $F'(-c, 0)$ and $F(c, 0)$. Let $P(x, y)$ be an arbitrary point on the ellipse as shown in Figure 13.1.4.

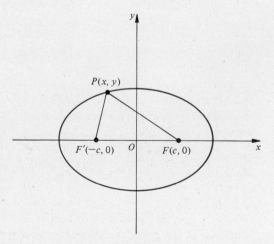

Figure 13.1.4

If we use the distance formula between two points the requirement (1) implies that

$$\sqrt{(x + c)^2 + (y - 0)^2} + \sqrt{(x - c)^2 + (y - 0)^2} = 2a \qquad (3)$$

Transpose the second radical (thereby isolating the first) and square both sides. We obtain

$$\sqrt{(x + c)^2 + y^2} = 2a - \sqrt{(x - c)^2 + y^2}$$
$$(x + c)^2 + y^2 = 4a^2 - 4a\sqrt{(x - c)^2 + y^2} + (x - c)^2 + y^2$$

and by elementary simplification

$$4cx - 4a^2 = -4a\sqrt{(x - c)^2 + y^2}$$

Division by -4 yields

$$a^2 - cx = a\sqrt{(x - c)^2 + y^2} \qquad (4)$$

We square both sides once more and combine terms to find that x and y satisfy

$$x^2(a^2 - c^2) + a^2y^2 = a^2(a^2 - c^2) \qquad (5)$$

Noting that $a^2 - c^2 > 0$ since $a > c > 0$ we find that if we let

$$b^2 = a^2 - c^2 \qquad (b > 0) \qquad (6)$$

then division by a^2b^2 yields

$$\frac{x^2}{a^2} + \frac{y^2}{b^2} = 1 \qquad (7)$$

Conversely, if $P_1(x, y)$ is a point on the curve defined by (7), then it is left to the reader to show that it must satisfy (5), (4), and (3). Therefore $|\overline{F'P_1}| + |\overline{FP_1}| = 2a$ and P_1 is a point on the ellipse. Hence (7) is an equation of the ellipse.

Because of its importance, we summarize this result as a theorem.

Theorem 1   Let the distance between the two foci be $2c$ and take the sum of the focal radii to be $2a$ where $a > c$. Choose the coordinate axes so that the origin is at the midpoint of the segment joining the foci of the ellipse and the x-axis contains both foci. Then an equation of the ellipse is given by

$$\frac{x^2}{a^2} + \frac{y^2}{b^2} = 1 \qquad (7)$$

where
$$b^2 = a^2 - c^2.$$

For any ellipse, the *center* is defined to be the midpoint of the line segment joining the foci. The *major axis* of an ellipse is the chord of an ellipse that contains both foci; the *minor axis* is the chord through the center perpendicular to the major axis.

Equation (7) may now be utilized to deduce information about the ellipse. Because $(x, -y)$ and $(-x, y)$ both satisfy (7) whenever (x, y) does, the ellipse is symmetrical with respect to the x-axis, the y-axis, and the origin. If we set $y = 0$ in (7) then $x = \pm a$. Therefore $V(a, 0)$ and $V'(-a, 0)$ are x-intercepts. Similarly,

$B(0, b)$ and $B'(0, -b)$ are the y-intercepts. The points V' and V are called the major vertices (for they are at the ends of the major axis). The length of the major axis is $2a$. The points B and B' are called the minor vertices (for they are the ends of the minor axis). The length of the minor axis is $2b$. The origin O is the center C (of symmetry) of the ellipse.

If (7) is solved for y, we obtain

$$y = \pm \frac{b}{a} \sqrt{a^2 - x^2} \tag{8}$$

Equation (8) implies that each of the two functions is defined on the domain $-a \le x \le a$. Similarly if we solve for x in terms of y, we obtain $-b \le y \le b$ for the extent of y.

If we join a focus F to an end B of the minor axis we obtain a right triangle OBF with BF the hypotenuse (Figure 13.1.5). The length of $BF = a$ because $b^2 + c^2 = a^2$.

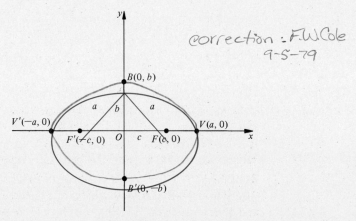

correction : F.W.Cole
9-5-79

Figure 13.1.5

In the preceding development, we chose to place the x-axis along the major axis of the ellipse. Of course, we could choose the y-axis to contain the major axis, and our equation of the ellipse would analogously be

$$\frac{x^2}{b^2} + \frac{y^2}{a^2} = 1 \tag{9}$$

where the roles of a and b are the same as before $(a > b)$. Equations (7) and (9) represent the ellipse in **standard position.**

EXAMPLE 1   Find the major vertices and foci of the ellipse with equation $9x^2 + 16y^2 = 144$. Sketch the graph of the ellipse.

Solution   The given equation may be written in the form

$$\frac{x^2}{16} + \frac{y^2}{9} = 1$$

Comparison with (7) yields $a^2 = 16$, $b^2 = 9$, from which $a = 4$ and $b = 3$. Now, $c^2 = a^2 - b^2 = 16 - 9 = 7$ so $c = \sqrt{7}$. The major axis lies along the x-axis. The foci are at $(\pm \sqrt{7}, 0)$ and the vertices are at $(\pm 4, 0)$ and the ends of the minor axis are at $(0, \pm 3)$ (see Figure 13.1.6). ●

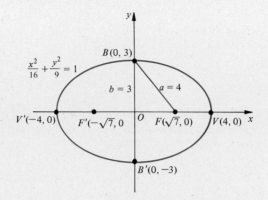

Figure 13.1.6

EXAMPLE 2   Find the vertices and foci of the ellipse with equation $\dfrac{x^2}{12} + \dfrac{y^2}{16} = 1$ and sketch the graph.

Solution   In this case the major axis lies along the y-axis (since the larger denominator is attached to the term with y in it). The quantities $a^2 = 16$ and $b^2 = 12$, from which $a = 4$ and $b = 2\sqrt{3}$. Also, $c^2 = 16 - 12 = 4$, and $c = 2$. Therefore the vertices are at $(0, \pm 4)$ and the ends of the minor axis are at $(\pm 2\sqrt{3}, 0)$. The coordinates of the foci are at $(0, \pm 2)$—see Figure 13.1.7. ●

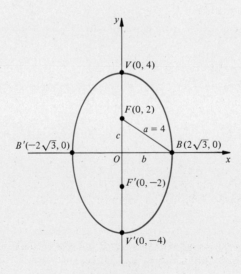

Figure 13.1.7

EXAMPLE 3   Find an equation of an ellipse with foci at $(0, 3)$ and $(0, -3)$ and passing through the point $(1, 4)$.

Solution   Since the foci are at $(0, \pm 3)$ we know that the y-axis contains the major axis of the ellipse, the center is at the origin, and $c = 3$. If we write an equation of the required ellipse in the form

$$\frac{x^2}{b^2} + \frac{y^2}{a^2} = 1 \tag{i}$$

and since $(1, 4)$ is a point on the ellipse, substitution into (i) yields

$$\frac{1}{b^2} + \frac{16}{a^2} = 1 \tag{ii}$$

But

$$a^2 - b^2 = c^2 = 9 \tag{iii}$$

The simultaneous solution of (ii) and (iii) for the two unknowns a^2 and b^2 is $a^2 = 18$ and $b^2 = 9$. Hence,

$$\frac{x^2}{9} + \frac{y^2}{18} = 1 \qquad \text{or, equivalently,} \qquad 2x^2 + y^2 = 18$$

is an equation for the ellipse.

Alternative Solution The sum of the focal radii is $2a$. Thus using the distance formula between the foci and the given point, we have

$$\sqrt{(1 - 0)^2 + (4 - 3)^2} + \sqrt{(1 - 0)^2 + (4 - (-3))^2}$$
$$= \sqrt{2} + \sqrt{50} = 6\sqrt{2} = 2a$$

or

$$a = 3\sqrt{2}$$

Therefore,

$$a^2 = 18 \qquad \text{and} \qquad b^2 = a^2 - c^2 = 18 - 9 = 9 \quad \text{or} \quad b = 3.$$

Substituting these values of a and b into (i) yields the same equation as we obtained by the previous solution. ●

The shape of an ellipse is determined by a (dimensionless) ratio known as the eccentricity.

Definition The quantity $e = \dfrac{c}{a}$ is called the ***eccentricity*** of the ellipse. Note that $0 < e < 1$ for an ellipse.

If two ellipses have the same eccentricity, they are said to be ***similar.*** The constants a, b, and c must be proportional in this instance. If the two ellipses have the same center at the origin and if their major axes are in the same direction, then it is possible to associate with each point P_1 on the first ellipse a corresponding point P_2 on the second ellipse so that $\overrightarrow{OP_2} = k(\overrightarrow{OP_1})$. Furthermore, this can be done for all points P on the ellipse where the proportionality constant k is the same for all such points. The proof of this is left to the reader as an exercise.

EXAMPLE 4 Two ellipses E_1 and E_2 have the same eccentricity $3/4$, the same center O and their major axes are of lengths 8 and 16 along the x-axis, respectively. Find equations for E_1 and E_2 and sketch them relative to the same Cartesian reference frame. Verify that, if P_1 is an arbitrary point on E_1, there is a corresponding point P_2 on E_2 such that $\overrightarrow{OP_2} = 2(\overrightarrow{OP_1})$.

Solution For the ellipse E_1: $2a_1 = 8$, $a_1 = 4$, and therefore $c_1 = 3$, since $e_1 = c_1/a_1 = 3/4$. Also, $b_1 = \sqrt{a_1^2 - c_1^2} = \sqrt{7}$.

For ellipse E_2: $2a_2 = 16$, $a_2 = 8$, and consequently, $c_2 = 6$. Also, $b_2 = \sqrt{a_2^2 - c_2^2} = 2\sqrt{7}$.

An equation for E_1 is $\dfrac{x^2}{16} + \dfrac{y^2}{7} = 1$ and a corresponding equation for E_2 is

$$\frac{x^2}{64} + \frac{y^2}{28} = 1.$$

The reader may verify that if $P_1(r, s)$ satisfies the equation for E_1, then $P_2(2r, 2s)$ must satisfy the equation for E_2, and $\overrightarrow{OP_2} = 2r\mathbf{i} + 2s\mathbf{j} = 2(r\mathbf{i} + s\mathbf{j}) = 2(\overrightarrow{OP_1})$ (see Figure 13.1.8). ●

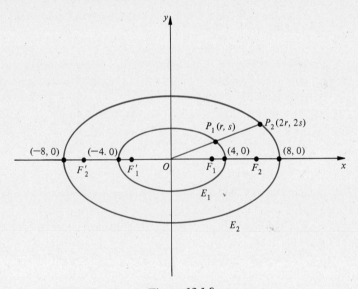

Figure 13.1.8

The eccentricity is a measure of the flatness of the ellipse and as we know $0 < e < 1$. For small values of e in comparison with 1, the ellipse is nearly circular, whereas if e is close to 1 the ellipse is relatively long and narrow.

As $e \to 0$, the two foci F and F' of the ellipse approach coincidence and in the limit the ellipse becomes a circle. In this sense, a circle may be regarded as an ellipse with zero eccentricity. At the other extreme, if we allow e to become 1 then $a = c$ and the ellipse degenerates to a straight line segment connecting points F and F'.

Planets such as the Earth travel in elliptical orbits about the Sun with the Sun at one of the focal points. The Earth travels in an elliptic path about the Sun and since $e \doteq 0.0166$ the trajectory is almost circular. On the contrary, Halley's comet travels on a long, narrow elliptical orbit where $e \doteq 0.98$.

Next, we state as a second theorem a very important property known as the **reflection property** of an ellipse.

Theorem 2   At each point on an ellipse the tangent line to the ellipse makes equal angles with the focal radii drawn to that point.

Proof   We refer to Figure 13.1.9. In order to prove Theorem 2, it must be established that angle FPV = angle $F'PW$. It is left as an exercise for the reader (see Exercise 34).   ∎

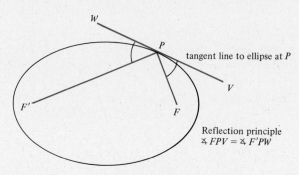

Figure 13.1.9

This result implies that a sound or light wave originating at point F will be reflected toward F'. One important application is to the field of acoustics in the design and analysis of elliptic whispering galleries. A sound emanating from focal point F will be heard at focal point F' (and conversely) even though the sound may be inaudible at points closer to F than F'.

Exercises 13.1

In Exercises 1 through 10, find the coordinates of the vertices, ends of the minor axis, foci; also find the eccentricity and sketch the graph of each ellipse.

1. $\dfrac{x^2}{9} + \dfrac{y^2}{4} = 1$

2. $\dfrac{x^2}{36} + \dfrac{y^2}{25} = 1$

3. $\dfrac{x^2}{4} + \dfrac{y^2}{16} = 1$

4. $x^2 + \dfrac{y^2}{4} = 1$

5. $9x^2 + 25y^2 = 225$

6. $25x^2 + 4y^2 = 100$

7. $2x^2 + 3y^2 = 6$

8. $9x^2 + 20y^2 = 45$

9. $400x^2 + 36y^2 = 225$

10. $9x^2 + 900y^2 = 25$

In Exercises 11 through 16, find an equation of an ellipse with center at the origin and with

11. Foci at $(\pm 3, 0)$ and major axis of length 10.

12. Focus at $(0, 8)$ and passing through $(-6, 0)$.

13. Focus at $(0, -6)$ and passing through $(5, 0)$.

14. Focus at $(5, 0)$ and eccentricity $\tfrac{1}{4}$.

15. Foci on the x-axis and passing through $(4, -3)$ with eccentricity $\tfrac{3}{4}$.

16. Foci on the y-axis and passing through $(2, 5)$ with eccentricity $1/\sqrt{5}$.

17. Find an equation of the tangent line to the ellipse $3x^2 + 4y^2 = 48$ at $(2, 3)$.

18. Show that an equation of the tangent line to the ellipse $\dfrac{x^2}{a^2} + \dfrac{y^2}{b^2} = 1$ at the point $P_1(x_1, y_1)$ on it is

$$\frac{xx_1}{a^2} + \frac{yy_1}{b^2} = 1.$$

19. Given the equation $\dfrac{x^2}{a^2} + \dfrac{y^2}{b^2} = 1$ show by implicit differentiation, that $y' = -\dfrac{b^2x}{a^2y}$ and $y'' = -\dfrac{b^4}{a^2y^3}$.

20. A *latus rectum* of an ellipse is defined to be a chord through a focus perpendicular to the major axis (so called **right focal chord**). Show that a latus rectum of the ellipse $\dfrac{x^2}{a^2} + \dfrac{y^2}{b^2} = 1$ has the length $\dfrac{2b^2}{a}$.

21. Find the slope of $\dfrac{x^2}{a^2} + \dfrac{y^2}{b^2} = 1$ at the fourth quadrant end of a latus rectum. Express your answer in terms of the eccentricity of the ellipse.

22. Find the volume of the *prolate spheroid* generated by rotating an ellipse $b^2x^2 + a^2y^2 = a^2b^2$ $(a > b)$ about its major axis, that is, the x-axis.

23. Find the volume of the *oblate spheroid* generated by rotating an ellipse $b^2x^2 + a^2y^2 = a^2b^2$, $(a > b)$, about its minor axis, that is, the y-axis.

24. Find the area bounded by an ellipse $\dfrac{x^2}{a^2} + \dfrac{y^2}{b^2} = 1$. (*Hint:* Consider the area in the first quadrant and relate it to the area of a circle.)

25. Find an equation of the line containing the point $(4, -1)$ and tangent to the ellipse $x^2 + 2y^2 = 6$.

26. Find an equation of the line with slope -6 and tangent to the ellipse $3x^2 + y^2 = 13$.

27. Find an equation of the line with slope m and tangent to the ellipse $b^2x^2 + a^2y^2 = a^2b^2$.

28. A rectangle is inscribed in an ellipse so that (i) the sides of the rectangle are parallel to the axes of the ellipse and (ii) the ratio of the length of the sides is equal to the ratio of the length of the axes of the ellipse. Find the area of the rectangle. (Assume that the equation of the ellipse is $b^2x^2 + a^2y^2 = a^2b^2$.)

29. Find the rectangle of largest area that may be inscribed in an ellipse $\dfrac{x^2}{a^2} + \dfrac{y^2}{b^2} = 1$ so that its sides are parallel to the axes of the ellipse.

30. The following method is used by draftsmen to construct an ellipse. Draw the concentric circles of radii a and b with $a > b > 0$ (Figure 13.1.10). Draw a line from the center of the circles making an angle θ with the horizontal axis and intersecting the circles at A and B, respectively. Let $P(x, y)$ denote the intersection of the line through B parallel to the x-axis and the line through A parallel to the y-axis. Find x and y in terms of the parameter θ and verify that P traces out an ellipse as θ varies from $0 \leq \theta < 2\pi$. *Note:* By varying θ from $0 \leq \theta < 2\pi$, as many points as desired can be constructed in this manner.

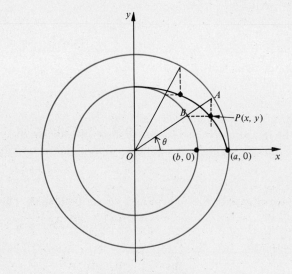

Figure 13.1.10

31. Find the area of an ellipse using the parametric equations $x = a \cos \theta$ and $y = b \sin \theta$, $0 \leq \theta < 2\pi$. (See Exercise 30.)

32. By a diameter of an ellipse, $\dfrac{x^2}{a^2} + \dfrac{y^2}{b^2} = 1$, we mean a chord passing through the center. Determine the largest and smallest chord lengths assuming that $a > b > 0$.

33. The orbit of the Earth about the Sun is elliptical with the Sun at one focus. If m and M are, respectively, the least and greatest distances from the Earth to the Sun, find the eccentricity e of the ellipse in terms of m and M. Find e numerically if $m = 91,466,000$ mi and $M = 94,560,000$ mi.

34. Prove the reflection principle for the ellipse, namely, that the focal radii to a point P (on the ellipse) make equal angles with the tangent line at P. This implies that a sound wave emanating from one focus of the ellipse will be reflected from the ellipse to the other focus.

13.2   THE HYPERBOLA

Definition   A **hyperbola** is the set of points P the difference of whose distances from two distinct fixed points F and F', called the *foci*, is a constant.

Let $2a$ be a fixed positive number which is less than the distance $2c$ between the foci. Then the set of points P such that

$$\left|\,|\overline{FP}| - |\overline{F'P}|\,\right| = \left|\,|\overline{F'P}| - |\overline{FP}|\,\right| = 2a \tag{1}$$

is a hyperbola. A line segment drawn from a point on the hyperbola to a focus is called a *focal radius.* Our definition requires that the absolute value of the difference in the lengths of the focal radii be the same for all points on the hyperbola, and $2a$ is to represent this common difference. Furthermore, from the fact that the difference between the lengths of two sides of triangle $FF'P$ must be less than the length of the third side (in essence, the triangular inequality) we have

$$\left|\,|\overline{FP}| - |\overline{F'P}|\,\right| = 2a < 2c$$

and hence

$$c > a \tag{2}$$

A simple equation for the hyperbola may be found by placing the coordinates of the foci at $F(c, 0)$ and $F'(-c, 0)$, respectively (see Figure 13.2.1). Let $P(x, y)$ be an arbitrary point on the hyperbola. Since $|\overline{F'P}| - |\overline{FP}| = \pm 2a$ we have

$$\sqrt{(x + c)^2 + y^2} - \sqrt{(x - c)^2 + y^2} = \pm 2a$$

Again we isolate the radical and square both sides. Thus

$$\sqrt{(x + c)^2 + y^2} = \sqrt{(x - c)^2 + y^2} \pm 2a$$
$$x^2 + 2cx + c^2 + y^2 = x^2 - 2cx + c^2 + y^2 \pm 4a\sqrt{(x - c)^2 + y^2} + 4a^2$$
$$4cx - 4a^2 = \pm 4a\sqrt{(x - c)^2 + y^2}$$
$$(cx - a^2)^2 = a^2[(x - c)^2 + y^2]$$

and simplification yields

$$(c^2 - a^2)x^2 - a^2y^2 = a^2c^2 - a^4 = a^2(c^2 - a^2).$$

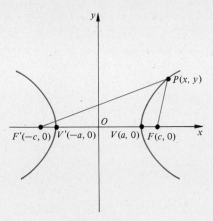

Figure 13.2.1

Then if we set $b^2 = c^2 - a^2$ $(b > 0)$ and divide both sides of the equation by the right side we have

$$\frac{x^2}{a^2} - \frac{y^2}{b^2} = 1 \qquad \text{where } b^2 = c^2 - a^2 \tag{3}$$

[Note that since $c > a > 0$, $c^2 - a^2 > 0$, and thus b can be chosen so that $b^2 = c^2 - a^2$.]

Conversely, if $P(x, y)$ is a point on the curve defined by (3), it can be shown that the absolute value of the difference between the distances from this point to the foci $F(c, 0)$ and $F'(-c, 0)$ is $2a$ where $c^2 = a^2 + b^2$. The proof of this nontrivial assertion is left to the reader as an exercise.

We summarize our conclusions in the form of a theorem.

Theorem 1 If the distance between the two foci is $2c$, if the absolute value of the difference between the focal radii is $2a$, where $a < c$, and if, furthermore, the origin is at the midpoint of the segment joining the foci and the foci lie on the x-axis, then an equation of the hyperbola is

$$\frac{x^2}{a^2} - \frac{y^2}{b^2} = 1 \qquad \text{where } b^2 = c^2 - a^2 \tag{3}$$

To draw the graph of a hyperbola it is only necessary to locate points in one quadrant since the curve from (3) is symmetric with respect to the x-axis, the y-axis, and the origin. Consequently, the origin is the **center of symmetry** or simply referred to as the **center** of the hyperbola. The line containing the foci is called the **transverse axis** of the hyperbola. The intersection points of the transverse axis with the hyperbola are called the **vertices** of the hyperbola. These points are labeled V and V' in Figure 13.2.1. The line through the center O, which is perpendicular to the transverse axis, is called the **conjugate axis** of the hyperbola. This is the y-axis in Figure 13.2.1.

Note that the curve in Figure 13.2.1 corresponding to (3) falls into two pieces called **branches.** The right hand branch corresponds to $|\overline{F'P}| - |\overline{FP}| = 2a$, and the left hand branch corresponds to $|\overline{FP}| - |\overline{F'P}| = 2a$.

We proceed now to examine Equation (3) in algebraic detail.

If we set $y = 0$ in (3) we find that $x = \pm a$. Therefore $(\pm a, 0)$ are the x-intercepts. However, the curve does not have y-intercepts because $x = 0$ yields the contradictory result $-\dfrac{y^2}{b^2} = 1$ for real values of y. If Equation (3) is solved for y^2, we have

$$y^2 = \frac{b^2}{a^2}(x^2 - a^2) \qquad (4)$$

and since $y^2 \geq 0$ it follows that $x^2 - a^2 \geq 0$ from which $|x| \geq a$. Thus the extent for x is $x \geq a$ or $x \leq -a$. If (4) is solved for y, we have

$$y = \pm\frac{b}{a}\sqrt{x^2 - a^2} = \pm\frac{b}{a}x\sqrt{1 - \frac{a^2}{x^2}}$$

Because $1 - \dfrac{a^2}{x^2}$ tends to 1 as $x \to \infty$, it is suggested that y approaches $\pm\dfrac{bx}{a}$ which in turn would imply that $y = \pm\dfrac{bx}{a}$ are **oblique asymptotes** of the hyperbola according to the following definition. (Theorem 2 will rigorously prove that these lines passing through O are indeed asymptotes.)

Definition The line $y = Mx + B$ is an **asymptote** of the graph of $y = f(x)$ if either

$$\lim_{x \to \infty} |Mx + B - f(x)| = 0 \qquad \text{or} \qquad \lim_{x \to -\infty} |Mx + B - f(x)| = 0$$

If $M \neq 0$, then $y = Mx + B$ is called an **oblique asymptote**.

skip

Thus, to draw the graph of (3), we draw a rectangle with sides $2a$ and $2b$ parallel to the x-axis and y-axis, respectively, and with the center at the origin of coordinates (see Figure 13.2.2). After drawing the diagonals of the rectangle, which are segments of the asymptotes $y = \pm\dfrac{bx}{a}$, the hyperbola is easily drawn. Furthermore, since $c^2 = a^2 + b^2$ a circle with center at O and radius c passes through the vertices of the rectangle and intersects the x-axis at the foci $(\pm c, 0)$.

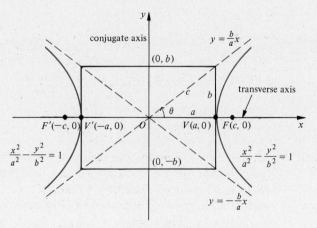

Figure 13.2.2

Theorem 2 The hyperbola with equation

$$\frac{x^2}{a^2} - \frac{y^2}{b^2} = 1 \tag{3}$$

Know

has the lines with equations

$$y = \pm \frac{bx}{a} \tag{5}$$

as asymptotes.

Proof From symmetry, it is sufficient to show that $y = \dfrac{bx}{a}$ is an asymptote of the branch

of the hyperbola defined by $y = \dfrac{b}{a}\sqrt{x^2 - a^2}$. We must show that

$$\lim_{x \to \infty} \left| \frac{bx}{a} - \frac{b}{a}\sqrt{x^2 - a^2} \right| = 0 \tag{6}$$

By rationalizing, we have for $x > a$,

$$\left| \frac{bx}{a} - \frac{b}{a}\sqrt{x^2 - a^2} \right| = \left| \frac{b}{a}(x - \sqrt{x^2 - a^2})\frac{(x + \sqrt{x^2 - a^2})}{(x + \sqrt{x^2 - a^2})} \right|$$

$$= \frac{b}{a}\left| \frac{x^2 - (x^2 - a^2)}{x + \sqrt{x^2 - a^2}} \right| = \frac{ab}{|x + \sqrt{x^2 - a^2}|} < \frac{ab}{x}$$

SKIP

Since $\dfrac{ab}{x} \to 0$ as $x \to \infty$, it follows that (6) is proved. ∎

If the foci of the hyperbola are on the y-axis and the center is at the origin, it is easy to show that the equation of the hyperbola is

$$\frac{y^2}{a^2} - \frac{x^2}{b^2} = 1 \tag{7}$$

and the asymptotes are

$$y = \pm \frac{a}{b}x \tag{8}$$

The y-axis is now the transverse axis and the x-axis is the conjugate axis.

EXAMPLE 1 Discuss and sketch the hyperbola defined by

$$\frac{x^2}{64} - \frac{y^2}{36} = 1.$$

Solution If we set $y = 0$, there results $x = \pm 8$. Therefore $(\pm 8, 0)$ are the vertices of the hyperbola with the x-axis the transverse axis and the y-axis the conjugate axis. Also, comparison with equation (3) yields $a^2 = 64$ and $b^2 = 36$, so $a = 8$ and $b = 6$. Furthermore, $c^2 = a^2 + b^2 = 100$ and $c = 10$; that is, the foci are at $(\pm 10, 0)$. The equations of the asymptotes are $y = \pm \dfrac{3x}{4}$ since $\dfrac{b}{a} = \dfrac{6}{8} = \dfrac{3}{4}$. The graph is shown in Figure 13.2.2 where a, b, and c are replaced, respectively, by 8, 6, and 10. ●

EXAMPLE 2 Discuss and sketch the hyperbola defined by

$$9y^2 - 4x^2 = 36$$

Solution We divide by 36 to put this equation into standard form

$$\frac{y^2}{4} - \frac{x^2}{9} = 1$$

Let $x = 0$ and find $y = \pm 2$. Therefore $(0, \pm 2)$ are the vertices of the hyperbola with the y-axis as the transverse axis and the x-axis the conjugate axis. Comparison with (7) yields $a^2 = 4$, $a = 2$, and $b^2 = 9$, $b = 3$. The asymptotes are the lines $y = \pm \frac{2}{3}x$. The quantity c is found from $c = \sqrt{a^2 + b^2} = \sqrt{13}$, that is, $(0, \pm \sqrt{13})$ are the coordinates of the foci. The curve is sketched in Figure 13.2.3. ●

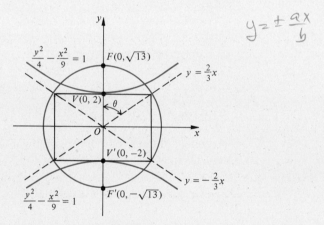

Figure 13.2.3

EXAMPLE 3 Find an equation of a hyperbola with transverse axis on a coordinate axis and center at the origin which passes through the points $(-4, 0)$ and $(8, -6)$. Find the equations of its asymptotes.

Solution Because the hyperbola contains the point $(-4, 0)$ its equation must be of the form

$$\frac{x^2}{a^2} - \frac{y^2}{b^2} = 1$$

Substitution of $x = -4$, $y = 0$ [since $(-4, 0)$ lies on the curve] yields $a = 4$ and the equation

$$\frac{x^2}{16} - \frac{y^2}{b^2} = 1$$

If we then utilize the point $(8, -6)$ and substitute $x = 8$ and $y = -6$, we obtain

$b^2 = 12$ or $b = 2\sqrt{3}$. Hence the required equation is

$$\frac{x^2}{16} - \frac{y^2}{12} = 1$$

and $y = \pm\dfrac{\sqrt{3}}{2}x$ are the equations of its asymptotes.   ●

We again define the **eccentricity** e by the ratio $e = \dfrac{c}{a}$ and it is noted that, as in the case of the ellipse, the eccentricity is related to shape rather than size. Reference to Figure 13.2.2 shows that e is a measure of the secant of the acute angle θ between the asymptote $y = \dfrac{b}{a}x$ and the transverse axis which is the x-axis there. Also, $c > a$ or, equivalently, $e > 1$. When e is close to 1, the angle is small, whereas increasing eccentricity implies an increase in the angle. To illustrate, in Example 1, $e = \frac{10}{8} = \frac{5}{4} = 1.25$, whereas in Example 2, $e = \dfrac{\sqrt{13}}{2} \doteq 1.80$. Note that in Example 2, e again is the $\sec\theta$, where θ is the angle between the transverse axis (the y-axis) and the asymptote $y = \frac{2}{3}x$. The greater eccentricity in Example 2 is manifested by the larger value of θ.

When the asymptotes are orthogonal to each other $\left(\theta = \dfrac{\pi}{4}\right)$ the hyperbola is said to be **equilateral** or **rectangular.** This occurs if and only if $a = b$. When $a = b$, $c^2 = a^2 + b^2 = 2a^2$, or $c = a\sqrt{2}$, and $e = \dfrac{c}{a} = \sqrt{2}$. If the transverse axis is the x-axis its equation is $x^2 - y^2 = a^2$. If the y-axis is the transverse axis, the equilateral hyperbola's equation is $y^2 - x^2 = a^2$.

EXAMPLE 4   A hyperbola has its center at the origin and the coordinate axes are axes of symmetry. Find its equation if a vertex is at $(-3, 0)$ and its eccentricity is $\frac{4}{3}$.

Solution   Since its vertex is on the x-axis we have the equation

$$\frac{x^2}{a^2} - \frac{y^2}{b^2} = 1$$

Furthermore, $x = -3$ when $y = 0$, so that $a = 3$. Also, the eccentricity $= \dfrac{c}{a} = \dfrac{4}{3}$ from which $c = 4$. Hence $b^2 = c^2 - a^2 = 16 - 9 = 7$ and our equation is

$$\frac{x^2}{9} - \frac{y^2}{7} = 1$$ ●

EXAMPLE 5   Find an equation of a hyperbola and its eccentricity if its center is at the origin, its transverse axis on the y-axis, asymptotes $y = \pm\dfrac{x}{3}$ and it passes through the point $(3, 6)$.

Solution   From the fact that the transverse axis is on the y-axis, our equation must be of the form

$$\frac{y^2}{a^2} - \frac{x^2}{b^2} = 1 \qquad \text{(i)}$$

with asymptotes $y = \pm\frac{a}{b}x$. Thus $\frac{a}{b} = \frac{1}{3}$ or $b = 3a$. From $c^2 = a^2 + b^2$, we obtain $c^2 = a^2 + (3a)^2 = 10a^2$ or the eccentricity $e = \frac{c}{a} = \sqrt{10}$. Using $b = 3a$ in (i) we have

$$\frac{y^2}{a^2} - \frac{x^2}{(3a)^2} = 1 \qquad \text{(ii)}$$

as an equation for the hyperbola. Since $(3, 6)$ lies on the curve, substituting the coordinates $(3, 6)$ into (ii) yields $a^2 = 35$ and $b^2 = 315$ and our equation is

$$\frac{y^2}{35} - \frac{x^2}{315} = 1$$

or, equivalently, $9y^2 - x^2 = 315$.   ●

   We often encounter rectangular hyperbolas with the coordinate axes as asymptotes. The equation of the hyperbola in such a case is of the form

$$xy = k \qquad (k \neq 0) \qquad \text{(9)}$$

where k is a constant. If the constant is positive, the branches of the hyperbola are in the first and third quadrants. If the constant is negative, the branches of the hyperbola are in the second and fourth quadrants. Curves corresponding to (9) for specific values of k are shown in Figures 13.2.4 and 13.2.5.

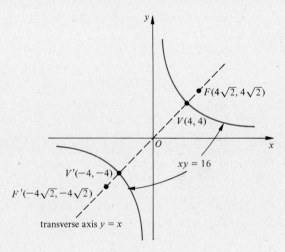

Figure 13.2.4

   At this point, the reader should be wondering how we know that (9) is a hyperbola since the form of (9) is different from (3) and (7). To show this, we must

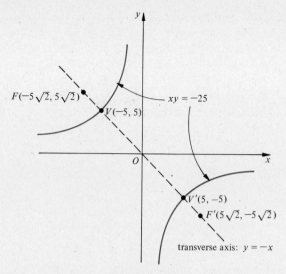

Figure 13.2.5

revert back to the definition of a hyperbola. Let the coordinates of the foci be at

$$\left(\frac{c}{\sqrt{2}}, \frac{c}{\sqrt{2}}\right) \quad \text{and} \quad \left(-\frac{c}{\sqrt{2}}, -\frac{c}{\sqrt{2}}\right) \quad \text{on the line } y = x.$$

Since $a = b$ we have $a = \dfrac{c}{\sqrt{2}}$ or $2a = c\sqrt{2}$. Then by definition, we have

$$\sqrt{\left(x - \frac{c}{\sqrt{2}}\right)^2 + \left(y - \frac{c}{\sqrt{2}}\right)^2} - \sqrt{\left(x + \frac{c}{\sqrt{2}}\right)^2 + \left(y + \frac{c}{\sqrt{2}}\right)^2} = \pm\sqrt{2}c$$

By the usual process of isolating the radical and squaring we obtain, after simplification

$$xy = \frac{c^2}{4} \tag{10}$$

The detailed derivation of (10) is left as an exercise for the reader. The case xy equals a negative constant may be handled analogously.

A number of significant applications of the hyperbola occur. It can be shown that a particle moving with sufficient speed and attracted by another body in accordance with the inverse square law (force varies as the inverse square of the distance between the bodies) moves in a hyperbolic path. In this instance, the body would eventually move away from the attracting body, never to return.

The compressibility of gases was studied by Robert Boyle (1627–1694). It is found empirically that the pressure of a gas P varies inversely with its volume V, provided that its temperature T remains constant; that is, $PV = $ constant. Hence for constant T, the graph of P versus V is one branch of a hyperbola since P and V must be positive.

A third application of the hyperbola is in range-finding work. Suppose that we wish to locate the position of an enemy artillery piece—say, a cannon fired at a point P. The report is heard at station S_1 and, say, t_0 sec later it is heard at a second station S_2. Assuming that sound travels at a rate of r ft/sec in air, point P must be

rt_0 ft further from S_2 than from S_1. This places P on one branch of a hyperbola with foci at S_1 and S_2 since $|\overline{PS_2}| - |\overline{PS_1}| = rt_0 =$ constant. If a third station say S_3 also heard this report t_1 sec after station S_1, then P is also on the branch of a second hyperbola because $|\overline{PS_3}| - |\overline{PS_1}| = rt_1 =$ constant. This implies that P may now be located at the intersection point of the two hyperbolas. It is this principle that is the basis for a system of navigation using radio signals named **LORAN** (which stands for LOng RAnge Navigation). Loran helps ships and aircraft to locate their positions with great accuracy despite the large distances involved.

Exercises 13.2

In each of Exercises 1 through 10, find the vertices, foci, eccentricity, and asymptotes of the hyperbola. Sketch the graph.

1. $\dfrac{x^2}{9} - \dfrac{y^2}{4} = 1$

2. $\dfrac{x^2}{49} - \dfrac{y^2}{16} = 1$

3. $\dfrac{y^2}{16} - \dfrac{x^2}{4} = 1$

4. $\dfrac{y^2}{1} - \dfrac{x^2}{25} = 1$

5. $x^2 - 4y^2 = 4$

6. $25y^2 - 64x^2 = 1600$

7. $4x^2 - 7y^2 = 28$

8. $x^2 - y^2 = 3$

9. $xy = 4$

10. $4x^2 - y^2 = 1$

*In each of Exercises 11 through 16 a hyperbola is in **standard position**; that is, its center is at the origin and the axes of symmetry are along the coordinate axes. Find its equation from the given data.*

11. Vertex is at $(8, 0)$ and focus at $(10, 0)$.
12. Vertex is at $(5, 0)$ and focus at $(13, 0)$.
13. Focus at $(0, 6)$ and eccentricity 2.

14. Vertex is at $(0, 1)$ and asymptotes $y = \pm \dfrac{4x}{3}$.

15. Focus on x-axis, passing through the point $(3, 4)$, and asymptotes $y = \pm 2x$.
16. Vertex at $(-\tfrac{3}{2}, 0)$ with eccentricity 2.

In Exercises 17 through 20, find by implicit differentiation an equation of the tangent line to the given hyperbola at the point indicated.

17. $2x^2 - y^2 = 14$, $(3, 2)$
18. $5x^2 - 3y^2 = 17$, $(-2, -1)$
19. $xy = -40$, $(8, -5)$

20. $\dfrac{x^2}{a^2} - \dfrac{y^2}{b^2} = 1$ at any point (x_0, y_0) on it. Simplify your answer by using the fact that the coordinates of the given point must satisfy the equation.

21. Given the hyperbola $\dfrac{x^2}{a^2} - \dfrac{y^2}{b^2} = 1$ deduce by implicit differentiation that

$$D_x y = \frac{b^2 x}{a^2 y} \quad \text{and} \quad D_x^2 y = -\frac{b^4}{a^2 y^3}$$

22. A *latus rectum* of a hyperbola is defined to be a chord through a focus perpendicular to the transverse axis. Show that the length of a latus rectum of $\dfrac{x^2}{a^2} - \dfrac{y^2}{b^2} = 1$ is given by $\dfrac{2b^2}{a}$.

23. Find the point on the curve $xy = a^2$ that is closest to the origin, where x and a are positive. What is the minimum distance?

24. Show that the hyperbola $xy = 10$ is symmetric with respect to the lines $y = x$, $y = -x$ and the origin. Find the coordinates of the vertices and foci.

25. Find the distance between a focus of the hyperbola $\dfrac{x^2}{a^2} - \dfrac{y^2}{b^2} = 1$ and an asymptote.

26. Prove that the tangent to the hyperbola $\dfrac{x^2}{a^2} - \dfrac{y^2}{b^2} = 1$ at any point $P_1(x_1, y_1)$ on it bisects the angle at P_1 formed by the focal radii.

27. Identify the curves defined by the relation $x^3 y^3 - 100xy = 0$.

28. Find the parametrization of the right branch of the hyperbola

$$\frac{x^2}{a^2} - \frac{y^2}{b^2} = 1$$

in terms of trigonometric functions.

29. Sketch the two hyperbolas $x^2 - y^2 = 5$ and $xy = 6$ on the same set of axes and prove that they intersect at right angles.

30. For what values of r is the equation

$$\frac{x^2}{14 - r} + \frac{y^2}{6 - r} = 1$$

a hyperbola. Find the coordinates of the foci and show that they are independent of r. For this reason, these hyperbolas are called *confocal* hyperbolas.

13.3   TRANSLATION OF AXES

Suppose that we have a circle of radius 5 in the xy-plane. The circumference C of the circle is given by $C = 2\pi(5) = 10\pi$ and the area $A = \pi(5)^2 = 25\pi$. Furthermore, these factors are independent of where the xy-axes are placed relative to the circle. However, the equation of the circle is dependent upon the location of the coordinate axes. For example, in the case of the given circle its equation is $x^2 + y^2 = 25$ if the center of the circle is at the origin. On the other hand, relative to the center at $(2, 3)$ the equation is $(x - 2)^2 + (y - 3)^2 = 25$ or, in expanded form,

$$x^2 + y^2 - 4x - 6y - 12 = 0$$

In general, the coordinate system may be chosen for our convenience; therefore, it is usually selected so that the equation is simplest. This is significant because we have learned to recognize "standard forms" of equations for curves such as circles, parabolas, ellipses, and hyperbolas.

Definition   In the Cartesian plane with given coordinates x and y, if new coordinate axes are chosen parallel to the given ones, then the axes are said to be *translated* in the plane.

Let the given x- and y-axes be translated to x' and y' so that the origin of the $x'y'$ system is at (h, k) of the xy system; that is, $x' = y' = 0$, when $x = h$ and $y = k$ (Figure 13.3.1). We shall now find a more general relation between the primed and unprimed variables.

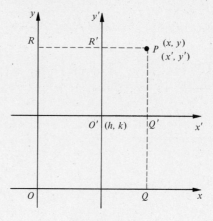

Figure 13.3.1

Consider an arbitrary point P that has coordinates (x, y) with respect to the original Cartesian system and (x', y') with respect to the "new" translated axes. If we draw parallels to the y- and y'-axes and to the x- and x'-axes suppose that these projections intersect the x'-axis in Q', the x-axis in Q, the y'-axis in R' and the y-axis in R. Then with respect to the x- and y-axes the coordinates of P are (x, y), the coordinates of Q are $(x, 0)$ and the coordinates of Q' are (x, k). However, $\overline{Q'P} = \overline{QP} - \overline{QQ'}$ so that

$$y' = y - k \qquad \text{or} \qquad y = y' + k$$

Similarly, with respect to the xy-axes the coordinates of R are $(0, y)$ and the coordinates of R' are (h, y). But $\overline{R'P} = \overline{RP} - \overline{RR'}$ and therefore

$$x' = x - h \qquad \text{or} \qquad x = x' + h$$

To summarize, we have the following theorem.

Theorem 1   Let (x, y) be the coordinates of a point P relative to a given Cartesian reference frame and (x', y') be the coordinates of a second, translated Cartesian reference frame so that the new origin has coordinates (h, k) relative to the given axes. Then the following relations must hold in general

$$x = x' + h \qquad \text{and} \qquad y = y' + k \tag{1}$$

or, equivalently,

$$x' = x - h \qquad \text{and} \qquad y' = y - k \tag{2}$$

Equations (1) or (2) are called the equations of *translation of axes* and their application will now follow.

EXAMPLE 1   Given the equation $(x - 3)^2 + (y - 2)^2 = 9$, find an equation of the graph with respect to the $x'y'$-axes after translating to the point $(3, 2)$.

Solution   In this case $h = 3$ and $k = 2$ so that from (2),

$$x' = x - 3 \quad \text{and} \quad y' = y - 2.$$

The equation in the primed system becomes

$$x'^2 + y'^2 = 9$$

which is the equation of a circle with center at O' the origin of the primed system. The circle is shown relative to the two sets of coordinates in Figure 13.3.2.

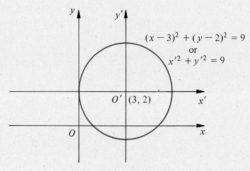

Figure 13.3.2

EXAMPLE 2   Given the equation $x^2 - 4x - 8y - 4 = 0$, translate the axes so that the new origin is at $(2, -1)$. Find the equation relative to the $x'y'$-axes and graph the curve relative to both sets of axes.

Solution   In our case $h = 2$ and $k = -1$ so that from (1), $x = x' + 2$ and $y = y' - 1$. Substitution into the original equation yields

$$(x' + 2)^2 - 4(x' + 2) - 8(y' - 1) - 4 = 0$$

or expansion yields after simplification

$$x'^2 = 8y'$$

which is a parabola with vertex at $x' = y' = 0$ and focus at $x' = 0$ and $y' = 2$, since $4p = 8$. The axis of the parabola is the y'-axis. The graph with respect to the xy-axes is therefore a parabola with vertex at $(2, 1)$, focus at $(2, 3)$, axis $x = 2$ and directrix $y = -1$ (Figure 13.3.3).

Actually a more realistic formulation of this example is to start with the given equation $x^2 - 4x - 8y - 4 = 0$ and request that a translation of axes be obtained so that the equation is simplified in the prime system. In other words, the quantities h and k are **unknowns** and must be determined. Let $x = x' + h$, $y = y' + k$ then substitution into the given equation yields

$$x'^2 + 2hx' + h^2 - 4x' - 4h - 8y' - 8k - 4 = 0$$

or collecting terms,

$$x'^2 + (2h - 4)x' - 8y' + h^2 - 4h - 8k - 4 = 0$$

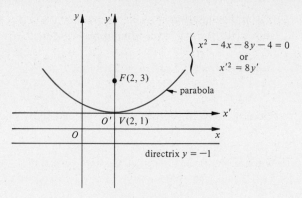

Figure 13.3.3

Setting the coefficient of x' equal to zero yields $h = 2$ and then setting the constant term $h^2 - 4h - 8k - 4 = 0$ (and using $h = 2$) yields $k = -1$. Thus the equation in the primed system reduces to $x'^2 - 8y' = 0$, as before.

Another method is to use completing the squares. With this in mind we rewrite the given equation as

$$x^2 - 4x + 4 - 8y - 8 = 0$$

or

$$(x - 2)^2 - 8(y + 1) = 0$$

and thus the indicated substitution is

$$x - 2 = x' \qquad \text{and} \qquad y + 1 = y'$$

or

$$x = x' + 2 \qquad \text{and} \qquad y = y' - 1$$

as found previously. ●

EXAMPLE 3 Suppose that we have a parabola with vertex $V(h, k)$ and opening upward (that is, $p > 0$). The equation of the parabola relative to the $x'y'$-axes is

$$x'^2 = 4py'$$

and from (2) this becomes

$$(x - h)^2 = 4p(y - k) \tag{3}$$

The axis of symmetry is $x = h$, the focus is at $(h, k + p)$, and the directrix is $y = k - p$.

If $p < 0$ then (3) represents a parabola still symmetric about $x = h$ and opening downward.

Analogously,

$$(y - k)^2 = 4p(x - h) \tag{4}$$

is an equation of a parabola with the line $y = k$ as an axis of symmetry. If $p > 0$ the parabola opens to the right (that is, $x \geq h$) while if $p < 0$ the parabola opens to the left ($x \leq h$).

It is important to note that *the clue to a parabola with axis parallel to one of the coordinate axes is that it is quadratic in one of the coordinates and linear in the other.* Whenever such an equation occurs, it may be reduced to one of the standard forms (3) or (4) by completing the square of the coordinate, which appears quadratically. ●

EXAMPLE 4   Remove the first degree terms from $xy - 3x - 2y + 1 = 0$ by a translation of axes.

Solution   Replace x by $x' + h$, y by $y' + k$ and thus the equation becomes

$$(x' + h)(y' + k) - 3(x' + h) - 2(y' + k) + 1 = 0$$

or collecting coefficients of x' and y'.

$$x'y' + (k - 3)x' + (h - 2)y' + hk - 3h - 2k + 1 = 0$$

Set the coefficient of $x' = 0$ and the coefficient of $y' = 0$, so that

$$k - 3 = 0 \quad \text{and} \quad h - 2 = 0$$

which yields $k = 3$ and $h = 2$. Substitution of these values for h and k results in $x'y' = 5$. The equation in the primed variables is considerably simpler than the original xy equation, and represents a rectangular hyperbola (Figure 13.3.4). ●

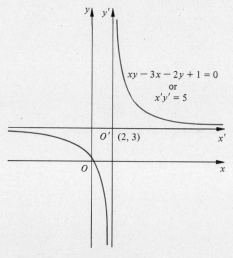

$$xy - 3x - 2y + 1 = 0$$
$$\text{or}$$
$$x'y' = 5$$

Figure 13.3.4

EXAMPLE 5   Simplify the equation $16x^2 + 9y^2 - 64x + 54y + 1 = 0$ by means of a translation of axes to the $x'y'$-system. Determine the center, vertices, foci, and axes of the curve and sketch it.

Solution   We regroup the given equation and complete the square:

$$16(x^2 - 4x \quad) + 9(y^2 + 6y \quad) = -1$$

which is equivalent to

$$16(x^2 - 4x + 4) + 9(y^2 + 6y + 9) = 64 + 81 - 1 = 144$$

or

$$16(x - 2)^2 + 9(y + 3)^2 = 144$$

Therefore we set

$$x - 2 = x' \quad \text{or} \quad x = x' + 2$$

and

$$y + 3 = y' \quad \text{or} \quad y = y' - 3$$

(that is, $h = 2$ and $k = -3$) to obtain an ellipse in standard position (relative to the x' and y' axes)

$$16x'^2 + 9y'^2 = 144$$

or, equivalently

$$\frac{x'^2}{9} + \frac{y'^2}{16} = 1$$

The center of the ellipse is at the point $(2, -3)$. Its major axis has length 8 and is part of the y'-axis and is parallel to the y-axis, whereas the minor axis is of length 6 along the x'-axis and is parallel to the x-axis. The coordinates of the vertices are $V(2, 1)$ and $V'(2, -7)$. Since $a = 4$ and $b = 3$ we find $c = \sqrt{a^2 - b^2} = \sqrt{7}$ and the coordinates of the foci are $F(2, \sqrt{7} - 3)$ and $F'(2, -\sqrt{7} - 3)$. The curve is sketched relative to the primed and unprimed coordinate axes in Figure 13.3.5. ●

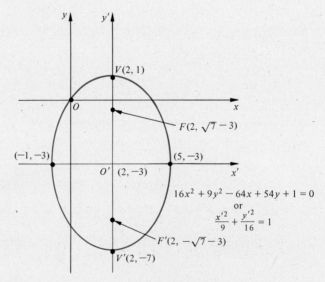

Figure 13.3.5

As a result of the developments of this section we now generalize our previous "standard forms" for the conics. The only restriction is that, for the parabola, ellipse, and hyperbola, the principal, major, and transverse axes are, respectively, parallel to an axis of the coordinate system. We have the following

PARABOLA: (i) With vertex at $V(h, k)$, axis $y = k$ and focus at $F(h + p, k)$:

$$(y - k)^2 = 4p(x - h) \tag{4}$$

(ii) With vertex at $V(h, k)$, axis $x = h$, and focus at $F(h, k + p)$:

$$(x - h)^2 = 4p(y - k) \tag{3}$$

ELLIPSE: (i) With center at (h, k), major axis on $y = k$, vertices $V(h + a, k)$, $V'(h - a, k)$, and foci at $F(h + c, k)$ and $F'(h - c, k)$:

$$\frac{(x - h)^2}{a^2} + \frac{(y - k)^2}{b^2} = 1 \tag{5}$$

(ii) With center at (h, k), transverse axis on $x = h$, vertices $V(h, k + a)$, $V'(h, k - a)$, and foci at $F(h, k + c)$ and $F'(h, k - c)$:

$$\frac{(y - k)^2}{a^2} + \frac{(x - h)^2}{b^2} = 1 \tag{6}$$

For the ellipse, $c^2 = a^2 - b^2$.

HYPERBOLA: (i) With center at (h, k), transverse axis on $y = k$, vertices $V(h + a, k)$, $V'(h - a, k)$, and foci at $F(h + c, k)$ and $F'(h - c, k)$:

$$\frac{(x - h)^2}{a^2} - \frac{(y - k)^2}{b^2} = 1 \tag{7}$$

The asymptotes are the lines $y - k = \pm \dfrac{b}{a}(x - h)$

(ii) With center at (h, k), transverse axis on $x = h$, vertices $V(h, k + a)$ and $V'(h, k - a)$, and foci at $F(h, k + c)$ and $F'(h, k - c)$:

$$\frac{(y - k)^2}{a^2} - \frac{(x - h)^2}{b^2} = 1 \tag{8}$$

The asymptotes are the lines $y - k = \pm \dfrac{a}{b}(x - h)$. For the hyperbola, $c^2 = a^2 + b^2$.

These forms, (3) through (8), are easy to recall, that is, we keep in mind that they are, respectively, generalizations of $y^2 = 4px$, $x^2 = 4py$, $\dfrac{x^2}{a^2} + \dfrac{y^2}{b^2} = 1$, $\dfrac{y^2}{a^2} + \dfrac{x^2}{b^2} = 1$, $\dfrac{x^2}{a^2} - \dfrac{y^2}{b^2} = 1$ and $\dfrac{y^2}{a^2} - \dfrac{x^2}{b^2} = 1$.

EXAMPLE 6   Simplify the equation $x^2 - 4y^2 + 2x + 16y + 1 = 0$ by means of a translation of axes. Draw the graph.

Solution   In order to complete the square, group terms as follows:

$$(x^2 + 2x \qquad) - 4(y^2 - 4y \qquad) = -1$$

Then

$$(x^2 + 2x + 1) - 4(y^2 - 4y + 4) = -1 + 1 - 16$$

or

$$(x + 1)^2 - 4(y - 2)^2 = -16$$

Dividing by -16 gives

$$\frac{(y - 2)^2}{4} - \frac{(x + 1)^2}{16} = 1$$

This is a hyperbola having the form of equation (8) where $(h, k) = (-1, 2)$, $a = 2$, and $b = 4$.

Letting $x' = x + 1$ and $y' = y - 2$, the translated equation is

$$\frac{(y')^2}{4} - \frac{(x')^2}{16} = 1$$

The asymptotes are $y - 2 = \pm\frac{1}{2}(x + 1)$ or, in the $x'y'$-system, $y' = \pm\frac{1}{2}x'$. Figure 13.3.6 illustrates these results.

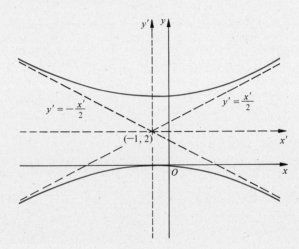

Figure 13.3.6

Exercises 13.3

In Exercises 1 through 8, simplify each equation by a translation of axes. Draw a graph of each curve relative to both the xy- and x'y'-axes.

1. $y = (x - 3)^2$

3. $y = (x - 1)^2 - 3$ Draw

5. $\dfrac{(x + 4)^2}{9} + (y - 7)^2 = 1$

7. $(x - 3)y - 10 = 0$

2. $(y + 6)^2 = -3x$

4. $(y + 2)(x - 5) = 4$

6. $x^2 - \dfrac{(y + 4)^2}{16} = 1$

8. $(x - 4)(y + 7) + 8 = 0$

In Exercises 9 through 12, simplify the following equations by completing the squares and translating axes. Identify the curve and describe it in detail.

9. $x^2 + y^2 - 4x - 6y + 9 = 0$

11. $4x^2 + y^2 + 8x - 10y + 25 = 0$

10. $x^2 + y^2 + 4x + 2y = 0$

12. $x^2 - 9y^2 - 6x + 18y - 9 = 0$

13. A translation of coordinates moves the origin to the point of intersection of the lines $2x - 3y - 1 = 0$

and $5x - 7y - 3 = 0$. Find the translation of coordinates and the equations of these lines in the new system. Use the method of Example 6.

14. Show that the distance between two points in a plane is unchanged (or invariant) under a translation of axes.

In Exercises 15 through 18, translate the coordinates so as to eliminate the first degree terms in the cases for which both x and y appear quadratically. In the cases where one variable occurs only to the first degree, eliminate one first degree term and the constant term. Describe the principal properties and sketch the curve.

15. $x^2 + 4y^2 - 6x - 16y + 21 = 0$ 16. $x^2 - y^2 - 2x + 6y - 9 = 0$

17. $y = x^2 - 4x + 1$ 18. $y^2 + 4x + 4y = 4$

*In Exercises 19 through 22, eliminate the terms indicated by translation of axes. (**Hint:** Use the substitution of Example 4 in Exercises 19, 20, and 22 and set the coefficients equal to zero.)*

19. $x^2 + 3xy + y^2 - 4x - y - 5 = 0$; first degree terms.

20. $y = x^3 + 3x^2 + x + 3$; x^2 and constant term.

21. $y = x^4 + x^3 - x^2 - x - 7$; x and constant term. (*Hint:* regroup the terms.)

22. $x^2y + x^2 + 2xy - x - 3y + 2 = 0$; second degree terms.

23. Show that every quadratic $y = ax^2 + bx + c$, where $a \neq 0$ represents a parabola with a vertical axis of symmetry. Find the axis, the coordinates of the vertex, the focus and its directrix.

24. Remove the first degree terms from $xy + ax + by + c = 0$, $(a, b,$ and c are constants), and identify the curve.

25. Suppose that under a translation of axes, the equation

$$Ax^2 + Bxy + Cy^2 + Dx + Ey + F = 0$$

becomes

$$A'x'^2 + B'x'y' + C'y'^2 + D'x' + E'y' + F' = 0.$$

Verify that $A' = A$, $B' = B$, and $C' = C$ for any translation of axes; that is, A, B, and C are invariant under translation.

26. Find the equation of translation of axes that transforms the equation $Ax^2 + Cy^2 + Dx + Ey + F = 0$, $(AC \neq 0)$, into the form $A'x'^2 + C'y'^2 + F' = 0$, where the linear terms are absent. What is the formula for F'?

In Exercises 27 through 34, graph the following equations using the methods of this section.

27. $x^2 + y^2 - 8x - 6y - 75 = 0$ 28. $3x^2 + 4y^2 - 40y + 88 = 0$

29. $4y^2 - 9x^2 - 18x - 24y - 9 = 0$ 30. $3x^2 + y^2 + 12x + 10y + 47 = 0$

31. $5x^2 + 2y^2 + 30x + 4y + 37 = 0$ 32. $x^2 + y^2 + 14x - 6y + 56 = 0$

33. $x^2 - y^2 - 4x + 2y + 1 = 0$ 34. $4x^2 - 3y - 24x + 36 = 0$

13.4 THE CONIC SECTIONS— A UNIFIED APPROACH

Our treatment of the parabola, ellipse, and hyperbola has thus far been on an individual basis. In Chapter 1 the parabola was introduced as the locus of points equidistant from a fixed point called a focus and a fixed line called the directrix. The term eccentricity was not mentioned in conjunction with the parabola. The ellipse and hyperbola were defined in terms of sums and differences of distances from two fixed points called the foci and the concept of eccentricity as a shape factor was significant. The term directrix was never mentioned in connection with either of these curves. We shall now give an alternative definition of the conic sections that will reveal that the ellipse and hyperbola also have a directrix (actually they have two directrices) and the parabola will be a conic section with eccentricity equal to 1.

Definition A **conic** is the collection of all points whose undirected distances from a fixed point called a **focus** are in constant ratio to their undirected distances from a fixed line (not passing through the focus) called the **directrix.**

The constant ratio e of the distances is called the **eccentricity** of the conic. Thus if F denotes the focus, P is a point and $|\overline{PQ}|$ is the undirected distance from P to the directrix (Figure 13.4.1). Then P is on the conic if and only if

$$|\overline{FP}| = e|\overline{QP}| \qquad (1)$$

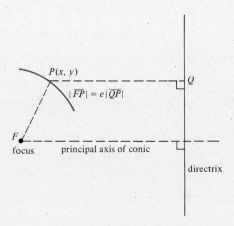

Figure 13.4.1

The line through the focus F perpendicular to the directrix d is called the **principal axis** of the conic. The point (or points) of intersection of the principal axis with the conic is called a **vertex** (or vertices) of the conic. Since e is the ratio of undirected distances it follows that e must be positive.

With this definition we shall now establish the following theorem.

Theorem 1 If $e = 1$, the conic is a parabola; when $e < 1$, the conic is an ellipse; if $e > 1$, the conic is a hyperbola.

Proof Let $x = d$ be the equation of the directrix of the conic and place the focus F on the x-axis where $F(c, 0)$ (Figure 13.4.2) and $c > 0$. Let $P(x, y)$ be any point of the conic distance $|\overline{QP}|$ from d. Then from $|\overline{FP}| = \sqrt{(x - c)^2 + y^2}$ and $|\overline{QP}| = |d - x|$ and (1) we have

$$\sqrt{(x - c)^2 + y^2} = e|d - x|$$

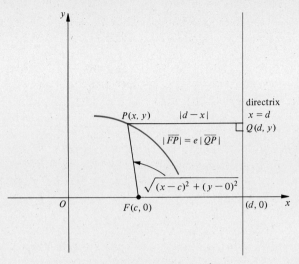

Figure 13.4.2

If we square both sides and note that $(e|d - x|)^2 = e^2(d - x)^2$, collection of like terms yields

$$(1 - e^2)x^2 + y^2 + (2e^2d - 2c)x + c^2 - e^2d^2 = 0 \qquad (2)$$

Inspection of (2) yields

(i)   If $0 < e < 1$, then $1 - e^2 > 0$ and the coefficients of x^2 and y^2 are both positive and unequal. Therefore the conic is an ellipse.

(ii)   If $e > 1$, then $1 - e^2 < 0$ and the coefficient of x^2 is negative while the coefficient of y^2 is positive. Hence the conic defined by (2) is a hyperbola.

(iii)   If $e = 1$, then $1 - e^2 = 0$ and (2) becomes

$$y^2 + 2(d - c)x + c^2 - d^2 = 0, \ (d \neq c) \qquad (3)$$

which is a parabola since (3) is of the second degree in y and the first degree in x. ∎

Conversely, we can establish that each of the conics conforms to the property defined by (1). In the case of a parabola the result follows directly from the definition. However, the result is not obvious for the ellipse or hyperbola. In order to establish this for the ellipse we place the origin of coordinates at the center of symmetry with the ellipse in standard position.

Theorem 2   Let $a > b > 0$ and set

$$c = \sqrt{a^2 - b^2} \qquad e = \frac{c}{a} \qquad d = \frac{a}{e} \qquad (4)$$

If F denotes the point $(c, 0)$ and $x = d$ is an equation of the directrix, then for all points on the ellipse

$$\frac{x^2}{a^2} + \frac{y^2}{b^2} = 1 \qquad (5)$$

we must also have (1) where $0 < e < 1$.

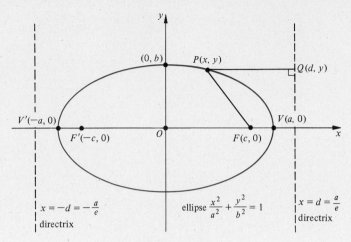

Figure 13.4.3

Proof Let $P(x, y)$ be an arbitrary point on the ellipse defined by (5)—see Figure 13.4.3. Thus

$$|\overline{PF}|^2 = (x - c)^2 + y^2 = x^2 - 2cx + c^2 + y^2$$

From (5) for all points P, $y^2 = b^2\left(1 - \dfrac{x^2}{a^2}\right)$, so that

$$|\overline{PF}|^2 = x^2 - 2cx + c^2 + b^2\left(1 - \frac{x^2}{a^2}\right)$$

$$= x^2\left(1 - \frac{b^2}{a^2}\right) - 2cx + c^2 + b^2$$

$$= \frac{c^2}{a^2}x^2 - 2cx + a^2 = \frac{c^2}{a^2}\left(x^2 - \frac{2a^2x}{c} + \frac{a^4}{c^2}\right)$$

$$= \frac{c^2}{a^2}\left(x - \frac{a^2}{c}\right)^2$$

from (4). If the square root of each side is taken, we have

$$|\overline{PF}| = \frac{c}{a}\left|x - \frac{a^2}{c}\right| = e\left|x - \frac{a}{e}\right| = e|x - d|$$

Hence $|\overline{PF}| = e|\overline{QP}|$, where $e = \dfrac{c}{a} < 1$. ∎

From the symmetry of (5) with respect to the y-axis it follows that $F'(-c, 0)$ also is a focus and $x = -d = -\dfrac{a}{e}$ is a corresponding directrix.

It is left to the reader to prove the corresponding theorem about the graph $\dfrac{x^2}{b^2} + \dfrac{y^2}{a^2} = 1$, where $a > b > 0$; that is, the major axis is along the y-axis. If the foci are $(0, \pm c)$, then the directrices are $y = \pm\dfrac{a}{e}$, where once again $c = \sqrt{a^2 - b^2}$ and $e = \dfrac{c}{a}$.

Next, we state and prove an analogous theorem about the hyperbola.

Theorem 3   Let a and b be positive numbers and set

$$c = \sqrt{a^2 + b^2} \qquad e = \frac{c}{a} \qquad d = \frac{a}{e} \tag{6}$$

If F denotes the point $(c, 0)$ and $x = d$ is an equation of the directrix, then for all points on the hyperbola

$$\frac{x^2}{a^2} - \frac{y^2}{b^2} = 1 \tag{7}$$

we must also have (1) where $e > 1$.

Proof   Let $P(x, y)$ be an arbitrary point on the hyperbola defined by (7) and shown in Figure 13.4.4. From (7) we obtain

$$y^2 = b^2\left(\frac{x^2}{a^2} - 1\right)$$

so that

$$|\overline{PF}|^2 = (x - c)^2 + y^2 = x^2 - 2xc + c^2 + \frac{b^2 x^2}{a^2} - b^2$$

$$= x^2\left(\frac{a^2 + b^2}{a^2}\right) - 2cx + a^2 \qquad \text{(from (6))}$$

$$= \frac{c^2}{a^2}\left(x^2 - \frac{2a^2 x}{c} + \frac{a^4}{c^2}\right) = \frac{c^2}{a^2}\left(x - \frac{a^2}{c}\right)^2$$

and if we take the square root of both sides

$$|\overline{PF}| = \frac{c}{a}\left|x - \frac{a^2}{c}\right| = e\left|x - \frac{a}{e}\right| = e|x - d|$$

Hence $|\overline{PF}| = e|\overline{QP}|$ where $e = \frac{c}{a} > 1$. ∎

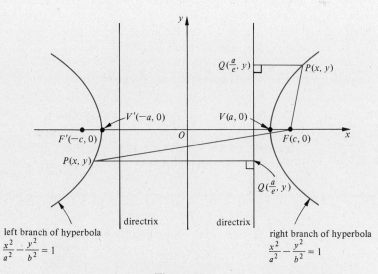

Figure 13.4.4

Note that since $e > 1$ the directrix $x = d = \dfrac{a}{e}$ is to the left of the focus $F(c, 0)$ and also the vertex $V(a, 0)$. Also, because the graph is symmetric with respect to the y-axis, the point $F'(-c, 0)$ is also a focus and the line $x = -d$ is a directrix.

It is left to the reader to show analogously that for the hyperbola $\dfrac{y^2}{a^2} - \dfrac{x^2}{b^2} = 1$ with transverse axis along the y-axis, the foci are at $(0, \pm c)$, where $c = \sqrt{a^2 + b^2}$ and the directrices are given by $y = \pm d = \pm \dfrac{a}{e}$, where $e = \dfrac{c}{a} > 1$.

The ellipse and hyperbola are called **central conics** because they have a center of symmetry, namely, the midpoint of the segment joining their two vertices. On the other hand, the parabola has only one focus, one directrix, and one vertex and is not a central conic.

EXAMPLE 1   Given the central conic defined by $4x^2 + 25y^2 = 100$, (a) identify it and (b) find its vertices, foci, eccentricity, and directrices.

Solution   (a) Division of both sides of the given equation by 100 yields

$$\frac{x^2}{25} + \frac{y^2}{4} = 1$$

so that we have an ellipse in standard position with semimajor and semiminor axes $a = 5$ and $b = 2$ on the x- and y-axes respectively.

(b) The vertices are at $(\pm a, 0) = (\pm 5, 0)$; the foci $(\pm c, 0)$, where $c = \sqrt{a^2 - b^2}$ are given by $(\pm \sqrt{21}, 0)$; the eccentricity $e = \dfrac{c}{a} = \dfrac{\sqrt{21}}{5} \doteq 0.92$ while the directrices are the parallel lines

$$x = \pm \frac{a}{e} = \pm \frac{25}{\sqrt{21}} \doteq \pm 5.46 \qquad \bullet$$

EXAMPLE 2   Given the central conic defined by $9y^2 - 36x^2 = 324$, (a) identify it and (b) find its foci, eccentricity, and directrices.

Solution   (a) Division of both sides of the given equation by 324 yields

$$\frac{y^2}{36} - \frac{x^2}{9} = 1$$

so that we have a hyperbola in standard position. The transverse axis is along the y-axis, $a = 6$ and $b = 3$.

(b) The vertices are at $(0, \pm a) = (0, \pm 6)$; the foci $(0, \pm c)$ are given by $(0, \pm \sqrt{a^2 + b^2}) = (0, \pm 3\sqrt{5}) \doteq (0, \pm 6.71)$; the eccentricity $e = \dfrac{c}{a} = \dfrac{\sqrt{5}}{2} \doteq 1.12$ and the directrices are the parallel lines $y = \pm \dfrac{a}{e} = \pm 6/(\sqrt{5}/2) = \pm \dfrac{12}{\sqrt{5}} \doteq \pm 5.37$. $\qquad \bullet$

EXAMPLE 3   Discuss the central conic $16x^2 - 25y^2 + 64x + 50y - 361 = 0$ giving its center, vertices, foci, and eccentricity. If it is a hyperbola, what are its asymptotes? Sketch the graph of the curve.

Solution   Collect the terms in x and y to obtain

$$16x^2 + 64x - 25y^2 + 50y = 361$$

or, equivalently,

$$16(x + 2)^2 - 25(y - 1)^2 = 361 + 64 - 25 = 400$$

Let $x + 2 = X$ and $y - 1 = Y$, hence

$$\frac{X^2}{25} - \frac{Y^2}{16} = 1$$

which is the equation of a hyperbola in standard position relative to the X- and Y-axes. Its center is at $X = Y = 0$, and its transverse axis is on the X-axis. Also, $a^2 = 25$, $b^2 = 16$, and $c^2 = a^2 + b^2 = 41$, from which $a = 5$, $b = 4$, and $c = \sqrt{41} \doteq 6.40$. The coordinates of the vertices, foci, and directrices relative to the XY-axes are, respectively, $(\pm 5, 0)$, $(\pm \sqrt{41}, 0)$ and $X = \pm \dfrac{a^2}{c} = \pm \dfrac{25}{\sqrt{41}} =$ ± 3.90. The center relative to the x- and y-axes is at $(-2, 1)$, the vertices are at $(3, 1)$ and $(-7, 1)$, the foci are at $(-2 + \sqrt{41}, 1)$ and $(-2 - \sqrt{41}, 1)$, its eccentricity $\doteq \dfrac{6.40}{5} = 1.28$. The asymptotes of the hyperbola are $Y = \pm \frac{4}{5}X$ or $y - 1 = \pm \frac{4}{5}(x + 2)$. The graph of the hyperbola with its outstanding properties is given in Figure 13.4.5.   ●

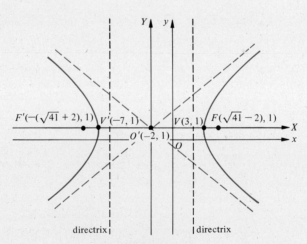

Figure 13.4.5

EXAMPLE 4   Discuss the central conic $25x^2 - 150x + 169y^2 = 4000$, giving its center, vertices, foci, eccentricity, and directrices. If its a hyperbola, what are its asymptotes?

Solution   By completion of the square on x, we have

$$25(x - 3)^2 + 169y^2 = 4225$$

or, equivalently, division of both sides by 4225 yields

$$\frac{(x-3)^2}{169} + \frac{y^2}{25} = 1$$

Let $x - 3 = X$ and $y = Y$, hence we have

$$\frac{X^2}{169} + \frac{Y^2}{25} = 1$$

which is the equation of an ellipse in standard position relative to the X- and Y-axes. Its center is at $X = Y = 0$, and its major axis is on the X-axis. Also, $a^2 = 169$, $b^2 = 25$, and $c^2 = a^2 - b^2 = 144$, from which $a = 13$, $b = 5$, and $c = 12$. The coordinates of the vertices, foci, and directrices relative to the XY-axes are, respectively, $(\pm 13, 0)$, $(\pm 12, 0)$ and $X = \pm 13/(12/13) = \pm 169/12$. Its eccentricity is $12/13$. Since $x = X + 3$ and $y = Y$, the center relative to the x- and y-axes is at $(3, 0)$. The vertices are at $(16, 0)$ and $(-10, 0)$; the foci are at $(15, 0)$ and $(-9, 0)$; and its directrices are $x = 3 \pm \frac{169}{12}$, or, equivalently, $x = \frac{205}{12}$, and $x = -\frac{133}{12}$. Because the curve is an ellipse there are no asymptotes. $\quad \bullet$

Conic sections that have an oblique directrix will be studied in Section 13.5. The equation of such a conic will always include an xy-term.

Exercises 13.4

In Exercises 1 through 6, find the center, vertices, foci, eccentricity, and directrices of the given central conic. If the curve is also a hyperbola find its asymptotes.

1. $16x^2 + 81y^2 - 1296 = 0$
3. $9x^2 - 18x - 16y^2 - 135 = 0$
5. $25x^2 + 16y^2 - 100x - 128y - 44 = 0$

2. $4x^2 - y^2 = 36$
4. $9x^2 + y^2 - 8y + 7 = 0$
6. $7x(x + 6) + 117 = 3y(10 - y)$

In Exercises 7 through 14, the central conic has its center of symmetry at the origin of coordinates and is in standard position. Find its equation satisfying the additional conditions:

7. Major axis is of length 8 and a focus is at $(-3, 0)$.
8. Vertex is at $(8, 0)$ and a focus is at $(10, 0)$.
9. Focus is at $(6, 0)$ and eccentricity is 2.
10. Focus is at $(0, \sqrt{10})$ and asymptotes $3y = \pm x$.
11. Directrix $y = -26$ and eccentricity $\frac{5}{13}$.
12. Eccentricity $\frac{2}{3}$ and directrix $x = 15$.
13. Focus at $(13, 0)$ and asymptotes $12y = \pm 5x$.
14. Major axis is of length 11 on the x-axis and passing through $(2, 1)$.

In each of Exercises 15 through 20 obtain the equation of the conic directly from its definition. Identify the curve and sketch it.

15. Focus at $(0, 3)$; directrix $y = -5$ and eccentricity $\frac{1}{3}$.
16. Focus at $(4, 0)$, directrix $x = -2$ and eccentricity 2.
17. Focus at $(2, 1)$, directrix $x = 6$ and eccentricity 3.
18. Focus at $(-1, 4)$, directrix $y = -4$ and eccentricity $\frac{1}{3}$.
19. Focus at $(-6, 3)$, directrix $y = 2x$ and eccentricity $\frac{1}{2}$.
20. Focus at $(-2, 6)$, directrix $x - 3y = 0$ and eccentricity 5.

ROTATION OF AXES

We have previously shown how a translation of axes can simplify equations. However, translation of axes (that is, moving axes parallel to themselves), does *not* affect the second degree terms. We shall now demonstrate that by rotating coordinates a further simplification can be made.

To this end, suppose that we have two rectangular coordinate systems with the same origin (Figure 13.5.1). Let xy be one system and $x'y'$ be the rotated system

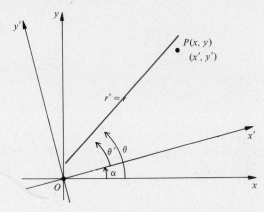

Figure 13.5.1

such that the x'-axis makes an angle α with the x-axis measured positively in the counterclockwise direction. It then follows that angle α is also the angle between the y- and y'-axes. A point P in the plane will have coordinates (x, y) with respect to the xy-axes and coordinates (x', y') with respect to the $x'y'$-axes. It is our objective to find a relationship between the two sets of coordinates involving the angle α as well. To determine this, we introduce two polar coordinate systems (r, θ) and (r', θ') relative to the positive x-axis and positive x'-axis as polar axes, respectively. Thus point P has two sets of polar coordinates (r, θ) and (r', θ') such that

$$r' = r \qquad \text{and} \qquad \theta' = \theta - \alpha \tag{1}$$

Furthermore, for the primed system the relationship between the polar and rectangular coordinates is

$$x' = r' \cos \theta' \qquad \text{and} \qquad y' = r' \sin \theta' \tag{2}$$

so that from (1) and (2),

$$x' = r \cos (\theta - \alpha) \qquad \text{and} \qquad y' = r \sin (\theta - \alpha) \tag{3}$$

But

$$\cos (\theta - \alpha) = \cos \theta \cos \alpha + \sin \theta \sin \alpha$$

and

$$\sin (\theta - \alpha) = \sin \theta \cos \alpha - \cos \theta \sin \alpha$$

so that equations (3) become

$$x' = r \cos \theta \cos \alpha + r \sin \theta \sin \alpha$$

and

$$y' = r \sin \theta \cos \alpha - r \cos \theta \sin \alpha$$

However, $x = r \cos \theta$ and $y = r \sin \theta$. Therefore it follows that

$$x' = x \cos \alpha + y \sin \alpha \qquad \text{and} \qquad y' = -x \sin \alpha + y \cos \alpha \qquad (4)$$

Equations (4) give the primed rectangular coordinates in terms of the unprimed ones for a given angle α. Conversely, we can solve equations (4) for x and y in terms of x' and y'. This yields

$$x = x' \cos \alpha - y' \sin \alpha \qquad \text{and} \qquad y = x' \sin \alpha + y' \cos \alpha \qquad (5)$$

More expeditiously we could obtain (5) from (4) by replacing α by $-\alpha$; $\cos \alpha$ by $\cos(-\alpha) = \cos \alpha$, $\sin \alpha$ by $\sin(-\alpha) = -\sin \alpha$ and interchanging the primed and unprimed quantities. The reader is urged to establish (5) from (4) by both methods.

EXAMPLE 1 Find a new representation of the given equation $3x - 2y + 6 = 0$ after rotating counterclockwise through the acute angle α such that $\tan \alpha = \frac{3}{2}$.

Solution $\tan \alpha = \dfrac{3}{2}$ and thus $\cos \alpha = \dfrac{2}{\sqrt{13}}$ and $\sin \alpha = \dfrac{3}{\sqrt{13}}$.

From (5),

$$x = \frac{1}{\sqrt{13}}(2x' - 3y') \qquad \text{and} \qquad y = \frac{1}{\sqrt{13}}(3x' + 2y')$$

Therefore the given equation becomes

$$\frac{3}{\sqrt{13}}(2x' - 3y') - \frac{2}{\sqrt{13}}(3x' + 2y') + 6 = 0$$

Simplification yields

$$-\frac{13y'}{\sqrt{13}} + 6 = 0 \qquad \text{or} \qquad y' = \frac{6}{\sqrt{13}}$$

which is a line parallel to the x'-axis, $\dfrac{6}{\sqrt{13}}$ units in the positive y'-direction (Figure 13.5.2). ●

EXAMPLE 2 Find a new representation of the given equation $x^2 - y^2 = 4$ after rotating the x- and y-axes counterclockwise through an angle of $\dfrac{\pi}{4}$ radians (or 45°).

Solution From (5), with $\cos \alpha = \cos \dfrac{\pi}{4} = \dfrac{1}{\sqrt{2}}$ and $\sin \alpha = \sin \dfrac{\pi}{4} = \dfrac{1}{\sqrt{2}}$, there results

$$x = \frac{1}{\sqrt{2}}(x' - y') \qquad \text{and} \qquad y = \frac{1}{\sqrt{2}}(x' + y')$$

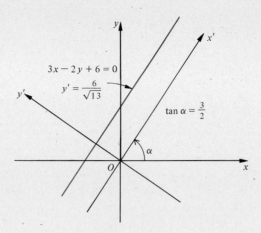

Figure 13.5.2

Then $x^2 - y^2 = 4$ becomes

$$\left[\frac{1}{\sqrt{2}}(x' - y')\right]^2 - \left[\frac{1}{\sqrt{2}}(x' + y')\right]^2 = 4$$

or

$$\tfrac{1}{2}[(x'^2 - 2x'y' + y'^2) - (x'^2 + 2x'y' + y'^2)] = 4$$

so that

$$x'y' = -2$$

Both $x^2 - y^2 = 4$ and $x'y' = -2$ are equations of an equilateral hyperbola (Figure 13.5.3). Note that the asymptotes $y = \pm x$ become the coordinate axes $x' = 0$ and $y' = 0$ in the rotated Cartesian $x'y'$-system of coordinates. ●

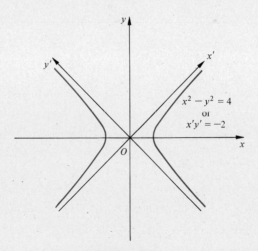

Figure 13.5.3

EXAMPLE 3   Discuss the graph of

$$3x^2 + 4xy + 3y^2 = 36$$

by rotating coordinates so as to eliminate the xy (or product term). Sketch the curve relative to both sets of axes.

Solution From (5) we obtain

$$3(x' \cos \alpha - y' \sin \alpha)^2 + 4(x' \cos \alpha - y' \sin \alpha)(x' \sin \alpha + y' \cos \alpha)$$
$$+ 3(x' \sin \alpha + y' \cos \alpha)^2 = 36$$

Collecting coefficients results in

$$(3 \cos^2 \alpha + 4 \sin \alpha \cos \alpha + 3 \sin^2 \alpha)x'^2$$
$$+ [-6 \sin \alpha \cos \alpha + 4(\cos^2 \alpha - \sin^2 \alpha) + 6 \sin \alpha \cos \alpha]x'y'$$
$$+ (3 \sin^2 \alpha - 4 \sin \alpha \cos \alpha + 3 \cos^2 \alpha)y'^2 = 36$$

Using elementary trigonometric identities we have

$$(3 + 2 \sin 2\alpha)x'^2 + (4 \cos 2\alpha)x'y' + (3 - 2 \sin 2\alpha)y'^2 = 36$$

In order to eliminate the $x'y'$-term we set its coefficient $4 \cos 2\alpha = 0$ to obtain $\alpha = \dfrac{\pi}{4}$ radians or $45°$. With this choice of α our equation reduces to

$$5x'^2 + y'^2 = 36$$

or, equivalently

$$\frac{x'^2}{\left(\dfrac{6}{\sqrt{5}}\right)^2} + \frac{y'^2}{6^2} = 1$$

This is an ellipse with major axis along the y'-axis. Furthermore, the semimajor and semiminor axes are $a = 6$ and $b = \dfrac{6\sqrt{5}}{5}$, respectively, and the distance from the center to the focus is $c = \dfrac{12\sqrt{5}}{5}$ (see Figure 13.5.4). ●

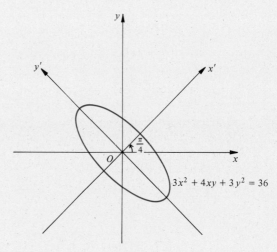

Figure 13.5.4

The method used in Example 3 may be applied for removing the xy-term in the general quadratic expression. To show that this is indeed true, suppose that we have the general quadratic equation

$$Ax^2 + Bxy + Cy^2 + Dx + Ey + F = 0 \tag{6}$$

where the coefficient of xy, $B \neq 0$. If the axes are rotated through an angle α then substitution of Equations (5) and simplifying yields an equation of the form

$$A'x'^2 + B'x'y' + C'y'^2 + D'x' + E'y' + F' = 0 \tag{7}$$

where

$$
\begin{aligned}
A' &= A\cos^2\alpha + B\cos\alpha\sin\alpha + C\sin^2\alpha \\
B' &= B(\cos^2\alpha - \sin^2\alpha) + 2(C - A)\sin\alpha\cos\alpha \\
C' &= A\sin^2\alpha - B\sin\alpha\cos\alpha + C\cos^2\alpha \\
D' &= D\cos\alpha + E\sin\alpha \\
E' &= -D\sin\alpha + E\cos\alpha \\
F' &= F
\end{aligned}
\tag{8}
$$

To remove the $x'y'$-term it is necessary to set $B' = 0$. Thus

$$B\cos 2\alpha + (C - A)\sin 2\alpha = 0$$

from which

$$\cot 2\alpha = \frac{A - C}{B} \qquad B \neq 0 \tag{9}$$

For arbitrary A, B, and C subject to the above restrictions a unique value of α satisfying $0 < 2\alpha < \pi$ or, equivalently, $0 < \alpha < \frac{\pi}{2}$ is determined that will remove the $x'y'$-term. Once α is known, then $\cos\alpha$ and $\sin\alpha$ are also known and equations (5) can be used.

Alternatively, from our knowledge of $\cot 2\alpha$ we can directly find $\cos\alpha$ and $\sin\alpha$ by application of the trigonometric identities

$$\cos\alpha = \sqrt{\frac{1 + \cos 2\alpha}{2}} \quad \text{and} \quad \sin\alpha = \sqrt{\frac{1 - \cos 2\alpha}{2}} \tag{10}$$

and then apply (5). Actually, it is not necessary to find α—only $\cos\alpha$ and $\sin\alpha$ are needed.

We have proved that an equation of the form (6) with $B \neq 0$ can always be transformed into (7), where $B' = 0$. Also we can easily show that the coefficients A' and C' cannot *both* be zero. To establish this, we suppose to the contrary that an α has been found such that $A' = B' = C' = 0$. This implies that (7) is an equation that is linear in x' and y'. Hence (6) can be obtained from (7) by using the rotation $-\alpha$, from the primed to the unprimed quantities. From (4) we would have

$$D'(x\cos\alpha + y\sin\alpha) + E'(-x\sin\alpha + y\cos\alpha) + F' = 0$$

which is an equation of the first degree in x and y and therefore must be different from (6), where at least $B \neq 0$. With this contradiction we conclude that at least one of A' or C' must be different from zero. Therefore, we have proved the following theorem.

Theorem 1 An equation of the form

$$Ax^2 + Bxy + Cy^2 + Dx + Ey + F = 0$$

for which $B \neq 0$, can always be transformed into an equation

$$A'x'^2 + C'y'^2 + D'x' + E'y' + F' = 0$$

where A' and C' are not both zero by application of a counterclockwise rotation of axes through an angle α such that $\cot 2\alpha = \dfrac{A - C}{B}$.

The quantity $B^2 - 4AC$ is called the **discriminant** of the quadratic expression (6). It is interesting and very useful that the discriminant is *unchanged* (or *invariant*) under *any* rotation of coordinates. This means that for arbitrary rotation angle α,

$$B^2 - 4AC = B'^2 - 4A'C' \tag{11}$$

The invariance of $B^2 - 4AC$ may be verified by direct calculation using the formulas (8) and is left as an exercise for the reader (see Exercise 19).

Now, suppose that α is chosen so that $B' = 0$. Thus from (11),

$$B^2 - 4AC = -4A'C' \tag{12}$$

If the graph of

$$A'x'^2 + C'y'^2 + D'x' + E'y' + F' = 0 \tag{13}$$

is nondegenerate, it is a parabola if A' or C' is zero, it is an ellipse if A' and C' have the same sign, and it is a hyperbola if A' and C' have opposite signs. This comment in conjunction with equation (12) implies that if the graph of (6) is nondegenerate it is a parabola, ellipse, or hyperbola if the discriminant $B^2 - 4AC$ is zero, negative or positive, respectively. We state this as a second theorem.

Theorem 2 An equation of the form

$$Ax^2 + Bxy + Cy^2 + Dx + Ey + F = 0$$

is either a conic section or a degenerate form. If it is a conic then

(i)   It is a parabola if $B^2 - 4AC = 0$
(ii) It is an ellipse if $B^2 - 4AC < 0$
(iii) It is a hyperbola if $B^2 - 4AC > 0$

Let us briefly look at the **degenerate forms.** Suppose that the axes are rotated and the $x'y'$-term is removed. Consider then the resulting equation

$$A'x'^2 + C'y'^2 + D'x' + E'y' + F' = 0 \tag{13}$$

Now if $A'C' = 0$ (that is, A' or C' or both are zero), then the degenerate forms corresponding to a parabola (since $B'^2 - 4A'C' = 0$) yields either a pair of parallel lines, a single line, or no points. If A' and C' possess the same sign, the degenerate forms corresponding to an ellipse are a circle, a single point, or no points. If A' and C' are of different signs, the degenerate form corresponding to a hyperbola is two intersecting lines.

EXAMPLE 4   Discuss the locus of

$$2x^2 + xy - y^2 + 5x - y + 2 = 0.$$

Solution $B^2 - 4AC = 1^2 - 4(2)(-1) = 9$. The curve is either a hyperbola or a pair of intersecting lines. It is a pair of intersecting straight lines only if the left side factors into the product of two linear expressions

$$(ax + by + c)(mx + ny + p) = 0$$

We now attempt to factor the left side. The second degree terms factor as

$$(2x - y)(x + y)$$

The constant term may be determined by trial and error testing $1, 2; 2, 1; -1, -2$ and $-2, -1$. We find that the left side does factor into

$$(2x - y + 1)(x + y + 2)$$

and therefore we have

$$(2x - y + 1)(x + y + 2) = 0$$

The solution of this is the pair of intersecting lines

$$2x - y + 1 = 0 \quad \text{and} \quad x + y + 2 = 0.$$

We could also first rotate the axes to eliminate the xy term and then factor (the factoring would then have been slightly easier).

Alternative   If the given locus is a pair of intersecting lines, we look for the form $y = mx + b$.
Solution   Substitution of this into the original equation must yield an *identity* in x. Hence

$$2x^2 + x(mx + b) - (mx + b)^2 + 5x - (mx + b) + 2 = 0$$

and collection of like powers yields

$$(2 + m - m^2)x^2 + (b - 2bm + 5 - m)x - (b^2 + b - 2) = 0$$

Thus

$$2 + m - m^2 = 0 \qquad \qquad \text{(i)}$$
$$b - 2bm + 5 - m = 0 \qquad \qquad \text{(ii)}$$
$$b^2 + b - 2 = 0 \qquad \qquad \text{(iii)}$$

Equation (i) implies that $m = -1$ or $m = 2$, whereas (iii) implies that $b = 1$ or $b = -2$. In order to satisfy (ii), if $b = 1$, $m = 2$; and if $b = -2$, then $m = -1$. Therefore $y = 2x + 1$ and $y = -x - 2$ are the equations of the two lines, or, equivalently, $2x - y + 1 = 0$ or $x + y + 2 = 0$ are their equations—in agreement with the previously developed solution.   ●

EXAMPLE 5   Simplify the equation

$$13x^2 - 6\sqrt{3}xy + 7y^2 + 52x - 12\sqrt{3}y - 92 = 0 \qquad \qquad \text{(a)}$$

by translating and rotating axes.

Solution Since $B^2 - 4AC = (-6\sqrt{3})^2 - 4(13)(7) = 108 - 364 = -256$ we have an ellipse or a degenerate locus. Let us translate axes first by substituting $x = x' + h$ and $y = y' + k$ into (a) and we obtain

$$13x'^2 - 6\sqrt{3}x'y' + 7y'^2 + (26h - 6\sqrt{3}k + 52)x'$$
$$+ (-6\sqrt{3}h + 14k - 12\sqrt{3})y'$$
$$+ 13h^2 - 6\sqrt{3}hk + 7k^2 + 52h - 12\sqrt{3}k - 92 = 0$$

We set the coefficients of x' and y' equal to zero in order to eliminate them. Hence

$$26h - 6\sqrt{3}k + 52 \quad\ = 0$$
$$-6\sqrt{3}h + 14k \quad - 12\sqrt{3} = 0 \tag{b}$$

which yields $h = -2$ and $k = 0$. Therefore $x = x' - 2$ and $y = y'$ reduces (a) to

$$13x'^2 - 6\sqrt{3}x'y' + 7y'^2 - 144 = 0 \tag{c}$$

Next, we seek to rotate axes so that the $x''y''$ term is eliminated. To accomplish this we choose

$$\cot 2\alpha = \frac{A - C}{B} = \frac{13 - 7}{-6\sqrt{3}} = -\frac{1}{\sqrt{3}}$$

or $2\alpha = 120°$ and $\alpha = 60°$. Hence

$$x' = x'' \cos 60° - y'' \sin 60° = \frac{x'' - \sqrt{3}y''}{2}$$

and

$$y' = x'' \sin 60° + y'' \cos 60° = \frac{\sqrt{3}x'' + y''}{2} \tag{d}$$

Substitution of (d) into (c) yields

$$13\left[\frac{(x'' - \sqrt{3}y'')^2}{4}\right] - \frac{6\sqrt{3}}{4}(x'' - \sqrt{3}y'')(\sqrt{3}x'' + y'')$$
$$+ \frac{7}{4}(\sqrt{3}x'' + y'')^2 - 144 = 0$$

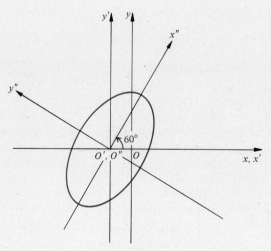

Figure 13.5.5

which simplifies to

$$4x''^2 + 16y''^2 = 144 \quad \text{or} \quad \frac{x''^2}{36} + \frac{y''^2}{9} = 1$$

This is an ellipse with major axis along the x''-axis. Also $a = 6$, $b = 3$ and $c = 3\sqrt{3}$ so that its eccentricity $e = \dfrac{c}{a} = \dfrac{\sqrt{3}}{2}$. The ellipse is sketched in Figure 13.5.5 relative to the three sets of axes. ●

Exercises 13.5

1. Find the coordinates $(2, 2)$ in the $x'y'$-system obtained by rotating the xy-axes counterclockwise through an angle of (a) $\frac{\pi}{4}$, (b) $\frac{\pi}{2}$. Illustrate graphically.

2. Find an equation of the curve $xy = 8$, referred to $x'y'$-axes obtained by rotating the x- and y-axes counterclockwise through an angle of $\frac{\pi}{4}$. Illustrate graphically.

3. What does the equation $y^2 - x^2 = 6$ become under a $45°$ rotation of axes?
4. Show that the equation $x^2 + y^2 = a^2$ becomes $x'^2 + y'^2 = a^2$ for every choice of the angle α in the equations for rotation of axes.
5. Find the sine and cosine of the angle α through which the axes must be rotated so as to eliminate the product term from $9x^2 + 24xy + 2y^2 - 7 = 0$.
6. Find the sine and cosine of the angle α through which the axes must be rotated so as to eliminate the product term from $2x^2 + 4\sqrt{3}xy - 2y^2 = 16$. Rotate through this angle and find the simplified equation in the $x'y'$-system. Identify the curve.
7. Discuss the graph of $2x^2 + 3xy + 2y^2 = 7$ by rotating coordinates so as to eliminate the product term.

In Exercises 8 through 11 remove the product term from the following equations by rotating the axes. Graph the curve showing both sets of axes.

8. $3x^2 - 4\sqrt{3}xy - y^2 = 15$
10. $5x^2 + 4xy + 8y^2 = 36$

9. $x^2 + 2xy + y^2 = 32$
11. $41x^2 - 24xy + 34y^2 + 15x + 20y = 25$

In Exercises 12 through 14, simplify the given equation by rotation and translation of axes. Draw a sketch of the graph showing all three sets of axes.

12. $x^2 - 2xy + y^2 - 2\sqrt{2}x - 6\sqrt{2}y = 14$

13. $6x^2 + 3xy + 2y^2 + 81x + 17y + 236 = 0$ $\left(\textit{Hint: } x = \dfrac{3x' - y'}{\sqrt{10}} \text{ and } y = \dfrac{x' + 3y'}{\sqrt{10}} \text{ then complete the squares.}\right)$

14. $7x^2 - 48xy - 7y^2 - 10x + 70y + 25 = 0$
15. Can rotation of axes be used to simplify the general equation $ax + by + c = 0$ of a straight line?
16. The slope m of a line $y = mx + b$ is invariant under translation of axes. Is this also true if the axes are rotated?
17. Prove that the distance between two points is an invariant under a rotation of axes.
18. Prove that $A + C$ is invariant under a rotation of axes.
19. Establish the invariance of $B^2 - 4AC$. (*Hint:* First establish that

$$\begin{aligned} B' &= B\cos 2\alpha + (C - A)\sin 2\alpha \\ A' + C' &= A + C \\ A' - C' &= (A - C)\cos 2\alpha + B\sin 2\alpha \end{aligned}$$

(i)

Then use the identity

$$B'^2 - 4A'C' = B'^2 + (A' - C')^2 - (A' + C')^2 \qquad \text{(ii)}$$

in conjunction with (i) to establish that

$$B'^2 - 4A'C' = B^2 - 4AC.)$$

20. Discuss the following degenerate conics
    (a) $x^2 + 2xy + y^2 - 4x - 4y + 4 = 0$
    (b) $3x^2 + 6xy + 3y^2 - 5x - 5y + 2 = 0$
    (c) $x^2 + y^2 - 4x - 6y + 13 = 0$
    (d) $8x^2 + 10xy - 3y^2 - 6x - 16y - 5 = 0$

In Exercises 21 through 24, describe and sketch the graphs of the equations.

21. $xy - 3x + 2y + 10 = 0$
22. $11x^2 - 24xy + 4y^2 + 70x - 40y + 55 = 0$
23. $5x^2 + 4xy + 8y^2 - 8x - 32y + 16 = 0$
24. $x^2 - 2xy + y^2 - 4x - 2y = 12$

13.6   POLAR EQUATIONS OF A CONIC

Applications of conics in space mechanics are often treated and simplified through the use of polar coordinates. It is convenient to let a focus coincide with the pole and to take the polar axis and its extension along the principal axis of the conic. To illustrate the derivation of the polar equation of a conic, we select the directrix to be (i) to the left of the focus F and (ii) perpendicular to the principal axis of the conic passing through the point (p, π) which is on the extension of the polar axis (Figure 13.6.1). By the general definition of a conic section, $P(r, \theta)$ is a point on the conic if and only if

$$\frac{|\overline{OP}|}{|\overline{QP}|} = e \qquad \text{(1)}$$

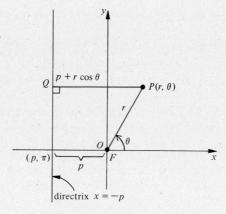

Figure 13.6.1

where $e > 0$ is the eccentricity of the conic and Q is a point on the directrix such that PQ is perpendicular to the directrix. Since P is to the right of the directrix, we have the directed segment $\overline{QP} > 0$ and $|\overline{OP}| = r$, where $r > 0$. Hence

$$\frac{r}{p + r\cos\theta} = e \qquad (p > 0, r > 0) \tag{2}$$

which when solved for r yields

$$r = \frac{ep}{1 - e\cos\theta} \tag{3}$$

In a similar manner, an equation of a conic for the case where the directrix is to the right of the focus (that is, $x = p$) is easily derived. It is left as an exercise for the reader (see Exercise 11) to show that the directrix is perpendicular to the principal axis, which, in turn, contains the polar axis and its extension, and that an equation of a conic assumes the form

$$r = \frac{ep}{1 + e\cos\theta} \tag{4}$$

Similarly, if again the focus is at the pole and if the polar axis is parallel to the corresponding directrix, then the line $\theta = \frac{\pi}{2}$ and its extension must coincide with the principal axis of the conic. The following equations for the conics are obtained

$$r = \frac{ep}{1 \pm e\sin\theta} \tag{5}$$

where e and p, respectively, are the eccentricity and the undirected distance from the focus and its corresponding directrix. The minus sign in (5) is taken when the directrix corresponding to the given focus at the pole is below it (that is, $y = -p$); the plus sign is used when the directrix is above the focus (that is, $y = p$). The derivations of equations (5) are also left as exercises for the reader (see Exercises 12 and 13).

EXAMPLE 1   Find an equation of the conic with focus at the origin and directrix $r\cos\theta = -4$ whose eccentricity is $\frac{2}{3}$.

Solution   The equation for the directrix $r\cos\theta = -4$ yields $x = -4$. Therefore $p = 4$ and the directrix is 4 units to the left of the pole. Our equation is from (3)

$$r = \frac{ep}{1 - e\cos\theta}$$

where $p = 4$ and $e = \frac{2}{3}$. Hence the conic is an ellipse and has an equation

$$r = \frac{\frac{8}{3}}{1 - \frac{2}{3}\cos\theta} \tag{6}$$

The graph of (6) is shown in Figure 13.6.2.   ●

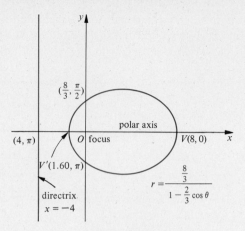

Figure 13.6.2

EXAMPLE 2   Find an equation of the conic with focus at the origin and directrix $r \sin \theta = 4$ whose eccentricity is 1.

Solution   The eccentricity is 1 so that the curve is a parabola. The equation for the directrix is $y = 4$; therefore $p = 4$ and equation (5) applies with the plus sign in the denominator. Hence

$$r = \frac{4}{1 + \sin \theta} \tag{7}$$

is an equation of the parabola. Its axis of symmetry is $\theta = \frac{\pi}{2}$ and its extension since $\sin \left(\frac{\pi}{2} + \theta \right) = \sin \left(\frac{\pi}{2} - \theta \right)$; r is least when $1 + \sin \theta$ is largest, that is, when $\theta = \frac{\pi}{2}$. We have

$$r_{\min} = r|_{\theta = \pi/2} = 2$$

and this is an absolute minimum. A sketch of the parabola with its directrix is shown in Figure 13.6.3.   ●

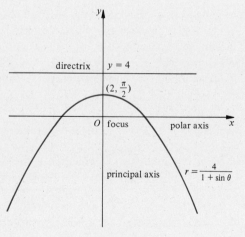

Figure 13.6.3

EXAMPLE 3   Identify and graph $r = \dfrac{6}{1 - 2\cos\theta}$.

Solution   The equation is of the form $r = \dfrac{ep}{1 - e\cos\theta}$ so that $e = 2$, which implies that the graph is a hyperbola. Furthermore $ep = 6$, from which $p = 3$. Hence, for this hyperbola one focus is at the origin and its associated directrix is a vertical line (parallel to the $\theta = \dfrac{\pi}{2}$ axis) three units to the left of the focus at the origin.

As θ increases from $\theta = 0$ to $\theta = \dfrac{\pi}{3}$, r decreases from -6 to $-\infty$ (see Figure 13.6.4).

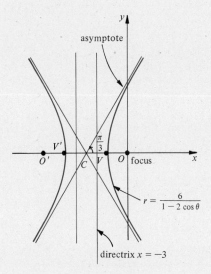

asymptote

$r = \dfrac{6}{1 - 2\cos\theta}$

directrix $x = -3$

Figure 13.6.4

If $r < 0$ is *not* permitted, as in our derivation of (3), then the left branch of the hyperbola cannot be generated from $r = \dfrac{6}{1 - 2\cos\theta}$. If we replace r by $-r$ and θ by $\theta + \pi$, then we have an equation $r = \dfrac{-6}{1 + 2\cos\theta}$ [since $\cos(\theta + \pi) = -\cos\theta$] which will yield the same graph when $\dfrac{2\pi}{3} < \theta < \dfrac{4\pi}{3}$ as the original equation does for $-\dfrac{\pi}{3} < \theta < \dfrac{\pi}{3}$. Thus, for example, we can denote vertex V' by $(6, \pi)$ rather than $(-6, 0)$. The center C of the hyperbola is 4 units to the left of the pole on the polar axis and the asymptotes must pass through C. It may be shown that the angles of inclination of the asymptotes are $\pm\dfrac{\pi}{3}$ with the x-axis. These angles are obtained by setting the denominator equal to zero; that is, $1 - 2\cos\theta = 0$ or $\theta = \pm\dfrac{\pi}{3}$. The asymptotes are lines through C parallel to these two lines.   ●

EXAMPLE 4 Show by transformation to rectangular coordinates that if $0 < e < 1$, then

$$r = \frac{1}{1 + e \sin \theta}$$

is an equation of an ellipse.

Solution We have $r \sin \theta = y$ so that by cross multiplication

$$r + ey = 1$$
$$r = 1 - ey$$
$$r^2 = (1 - ey)^2 = 1 - 2ey + e^2 y^2$$

Therefore

$$x^2 + y^2 = 1 - 2ey + e^2 y^2$$
$$x^2 + y^2(1 - e^2) + 2ey = 1$$
$$x^2 + (1 - e^2)\left(y^2 + \frac{2e}{1 - e^2} y\right) = 1$$

and, by completing the square

$$x^2 + (1 - e^2)\left(y + \frac{e}{1 - e^2}\right)^2 = 1 + \frac{e^2}{1 - e^2} = \frac{1}{1 - e^2}$$

which is an ellipse because $1 - e^2$, the coefficient of y^2 and its reciprocal are both positive. Note that $\left(0, -\dfrac{e}{1 - e^2}\right)$ are the coordinates of the center of the ellipse.

●

Exercises 13.6

In Exercises 1 through 10, the equation of a conic with its focus at the origin is given. Find (a) its eccentricity, (b) identify the particular conic, (c) write an equation for the directrix that corresponds to the given focus, and (d) sketch the given curve.

1. $r = \dfrac{10}{1 - \cos \theta}$

2. $r = \dfrac{2}{1 - \frac{1}{2} \cos \theta}$

3. $r = \dfrac{7}{1 + 3 \cos \theta}$

4. $r = \dfrac{5}{3 + 2 \cos \theta}$

5. $r = \dfrac{9}{1 + 2 \sin \theta}$

6. $r = \dfrac{10}{7 - 3 \sin \theta}$

7. $r = \dfrac{100}{3 - 15 \sin \theta}$

8. $r = \dfrac{24}{7(1 + \sin \theta)}$

9. $r = \dfrac{a}{b + c \cos \theta}$, $\quad 0 < a < b < c$

10. $r = \dfrac{a}{b - a \cos \theta}$, $\quad 0 < a < b$

11. Derive an equation of a conic having its principal axis along the polar axis and its extension, with one focus at its pole and its associated directrix given by $x = p > 0$.

12. Derive an equation of a conic having its principal axis along the $\dfrac{\pi}{2}$ – axis and its extension, with one focus at its pole and its associated directrix given by $y = -p$, where $p > 0$.

13. Derive an equation of a conic having its principal axis along the $\dfrac{\pi}{2}$ – axis and its extension, with one focus at its pole and its associated directrix given by $y = p$, where $p > 0$.

In Exercises 14 through 17, find in polar coordinates an equation of the conic with focus at the pole and satisfying the additional condition:

14. parabola, directrix $r \cos \theta = -7$
15. $e = \frac{1}{3}$, directrix $r \sin \theta = 4$
16. $e = \frac{2}{3}$, directrix $r \cos \left(\theta - \frac{\pi}{2} \right) = 5$
17. $e = \frac{3}{2}$, vertex at $\left(\frac{12}{5}, \frac{\pi}{2} \right)$

18. Prove that the graph of $r = a \sec^2 \frac{\theta}{2}$, $a > 0$, is a parabola. Find the positions of the focus, the directrix and the vertex. Sketch the graph.

19. The equation $r = \dfrac{ep}{1 - e \cos \theta}$, where $0 < e < 1$ and $p > 0$, is a polar coordinate representation of an ellipse. Apply differential calculus to determine the points on it that are closest and furthest from the origin. What is the length of the major axis?

20. Derive the polar equation of a conic where its focus is at the origin O and its directrix is at distance p from O and intersects the polar axis or its extension in an angle α. This is a generalization of the cases discussed in the text where $\alpha = \frac{\pi}{2}$ and $\alpha = 0$ yields equations (3) and one of (5). Verify that your formula simplifies correctly in these two instances.

21. Prove that

$$r = \frac{6}{1 - [(0.3) \cos \theta - (0.4) \sin \theta]}$$

is a conic. Find its eccentricity and identify it. (*Hint:* Refer to Exercise 20.)

22. Given the hyperbola

$$r = \frac{ep}{1 - e \cos \theta} \qquad e > 1$$

Find its equation in rectangular coordinates. Locate the rectangular coordinates of its center in terms of e and p. Find its asymptotes and verify that their inclinations with the polar axis is given by $\pm \cos^{-1} \dfrac{1}{e}$ (which may be obtained by equating the denominator of the given expression equal to zero).

Review and Miscellaneous Exercises

In Exercises 1 through 3 write the equation of the ellipse in standard form. Locate the center, foci, and vertices. Determine the eccentricity and the lengths of the major and minor axes. Sketch the graph.

1. $4x^2 + y^2 = 36$
2. $4x^2 + 9y^2 - 16x - 18y - 11 = 0$
3. $36x^2 + 4y^2 + 108x + 45 = 0$
4. Find an equation in standard form of an ellipse with foci $F(2, -4)$ and $F'(2, 2)$ and with its major axis having length 10.
5. Identify the graph of the equation $2x^2 + y^2 = 2(2x + 3y - 6)$.
6. Find an equation of an ellipse in standard position passing through the two points $(5, 3\sqrt{3})$ and $(5\sqrt{3}, -3)$.
7. Show that any equation of the form $\dfrac{1}{r} = k + n \cos \theta$ where k and n are positive constants, represents a conic section. Find the eccentricity e and the quantity p in terms of k and n.

8. Find an equation of the set of points such that the difference of the distances from any point P of the set to the points $(6, -3)$ and $(-4, -3)$ is 8. Identify the conic and its asymptotes (if any).
9. Find a Cartesian equation for the conic consisting of all points $P(x, y)$ whose distance from the point $(4, 1)$ is $\frac{3}{2}$ the distance from the line $x = 5$. Identify the conic.
10. Find a Cartesian equation for the conic consisting of all points $P(x, y)$ equidistant from the point $(0, 0)$ and the line $5x - 12y + 39 = 0$. Identify the conic.

Each of the equations in Exercises 11 through 15 represents a hyperbola. Find the coordinates of the center, the foci, and the vertices. Also determine the eccentricity and the equations of the asymptotes. Sketch the curve.

11. $\dfrac{x^2}{49} - \dfrac{y^2}{16} = 1$

12. $3x^2 - 7y^2 + 21 = 0$

13. $\dfrac{(x - 5)^2}{9} - \dfrac{y^2}{25} = 1$

14. $\dfrac{(y - 4)^2}{4} - \dfrac{(x - 1)^2}{9} = 1$

15. $x^2 - 4y^2 - 6x - 40y - 95 = 0$

In Exercises 16 through 20, find a Cartesian equation for the given conic that satisfies the conditions given. Identify the conic.

16. Directrix $x = -6$, focus at $(4, 0)$, and eccentricity 1.
17. Center at $(0, 0)$, one vertex at $(-5, 0)$, and asymptotes $y = \pm \frac{3}{5}x$.
18. Center at $(-10, 0)$, eccentricity $\dfrac{\sqrt{40}}{7}$, the length of its minor axis is 6, and its major axis is on the x-axis.
19. Noncentral conic with directrix $y = x$ and focus at $(3, -3)$.
20. Conic with eccentricity 2, directrix $y = x$, and focus at $(3, -3)$.
21. If $k > 0$ and $9x^2 + ky^2 = 9k$, then for each constant k we have an ellipse. What common property is possessed by all these ellipses?
22. Discuss the family of curves $kx^2 + (1 - k)y^2 = 1$ where k is a constant. Point out any common properties, if
    (a) $0 < k < 1$ (b) $k = 1$ (c) $k > 1$
23. The general conic contains five essential constants. Verify this and find the conic that passes through $(0, 0)$, $(3, 1)$, $(5, -5)$, $(7, -1)$, and $(8, -4)$.
24. Find an equation of the conic that is symmetrical with respect to the origin and passes through the points $(4, -6)$, $(8, 0)$, and $(16, 6)$. (*Hint:* If the conic is symmetrical with respect to the origin, then $D = E = 0$—why?)
25. Find an equation of a parabola that passes through the points with coordinates $(4, -2)$, $(2, -3)$, and $(6, -4)$ and has its axis parallel to the x-axis.
26. Prove that each equilateral hyperbola $x^2 - y^2 = a$ is orthogonal to each hyperbola $xy = b$, where $ab \neq 0$. (This generalizes Exercise 29 of Section 13.2.)
27. Eliminate the xy-term in $xy = -1$. What equation results?
28. Find a rotation of axes α that eliminates the xy-term in $x^2 + xy + Dx + Ey + F = 0$. Find $\cos \alpha$ and $\sin \alpha$ explicitly.
29. By definition, a diameter of an ellipse is a chord passing through the center. Show that the diameter of maximum length of the ellipse $\dfrac{x^2}{a^2} + \dfrac{y^2}{b^2} = 1$, where $a > b > 0$, is $2a$; that is, it is its major axis.
30. Prove that the locus of the middle points of a system of parallel chords with slope m of an ellipse $\dfrac{x^2}{a^2} + \dfrac{y^2}{b^2} = 1$ must be a diameter of the conic. What is the equation of the diametral line?
31. The equation $r = \dfrac{9}{2 + \cos \theta}$, $0 \leq \theta < 2\pi$, defines a particular conic with a focus at the pole. Find its eccentricity, directrix associated with its focus, and the length of the latus rectum. Identify the conic.
32. A particle moves in the xy-plane so that its position P at any time t is given by

$$\mathbf{R}(t) = \mathbf{a}(\cos t) + \mathbf{b}(\sin t) + \mathbf{c}$$

where **a**, **b**, and **c** are constant vectors in the plane, such that $\mathbf{a} \cdot \mathbf{b} = 0$, where $\mathbf{a} \neq \mathbf{0}$ and $\mathbf{b} \neq \mathbf{0}$. Show that the path of the particle is an ellipse.

33. Show that the length L of an ellipse $\dfrac{x^2}{a^2} + \dfrac{y^2}{b^2} = 1$ may be expressed by the definite integral

$$L = 4a \int_0^{\pi/2} (1 - e^2 \cos^2 \theta)^{1/2} \, d\theta$$

where $e = \sqrt{a^2 - b^2}/a$ is the eccentricity. (*Hint:* Write $x = a \cos \theta$ and $y = b \sin \theta$, and so on.)

34. Show that the length L of a hyperbola $\dfrac{x^2}{a^2} - \dfrac{y^2}{b^2} = 1$ between $x = a$ and $x = a \cosh \lambda$ where λ is a positive constant, is given by the definite integral

$$L = 2a \int_0^{\lambda} (e^2 \cosh^2 u - 1)^{1/2} \, du$$

where $e = \sqrt{a^2 + b^2}/a > 1$ is the eccentricity of the hyperbola. (*Hint:* Write $x = a \cosh u$ and $y = b \sinh u$, $u \geq 0$.)

35. Find the shortest distance from the point $(4, 2)$ to the graph of $x^2 - 4y^2 - 6x + 16y - 11 = 0$. (*Hint:* Write the given curve in standard form by completing the squares.)

36. Compute the area A of the shaded sector OBP of an ellipse as shown in the following diagram.

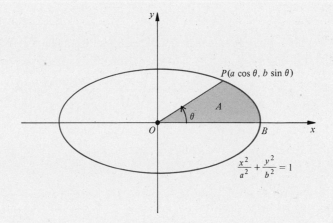

37. A hyperbola may be defined as the locus of points such that the product of the directed distances from the two intersecting lines $y = \pm \dfrac{bx}{a}$ is equal to a constant given by $\dfrac{a^2 b^2}{a^2 + b^2}$, where $ab \neq 0$. Verify this.

38. Identify the central conic defined by $3xy - 6x + 8y = 20$. Find the center and, if the conic has asymptotes, locate them.

39. The graph $x^{1/2} + y^{1/2} = a^{1/2}$, where a is a positive constant, lies entirely within the square $0 \leq x \leq a$ and $0 \leq y \leq a$. Show that the graph is an arc of a parabola.

40. Sketch the loci $(xy + y - x - 1)(x^2 - y^2 - 4) = 0$.

41. Sketch the loci $x^3 + y^3 + xy^2 + yx^2 + 3x + 3y < 0$ (*Hint:* Factor the expression.)

42. Prove that no tangent to the equilateral hyperbola $xy = 10$ can pass through the origin.

43. Show that any tangent line to the equilateral hyperbola $xy = a^2$ determines with its asymptotes a triangle of constant area. What is this area?

44. Find the equation of the locus of points $P(x, y)$ such that the ratio of the undirected distances from two fixed points $(a, 0)$ and $(b, 0)$ is a positive constant k. Identify the locus.

45. On a level plane the crack of a rifle and the sound of the bullet hitting its target are heard simultaneously by a fixed observer. Show that the position of the observer must be located on a particular hyperbola.

14

Indeterminate Forms, Taylor's Formula, and Improper Integrals

14.1   THE INDETERMINATE FORM 0/0

We have considerable experience with the problem of evaluating $\lim\limits_{x \to a} \dfrac{f(x)}{g(x)}$. One rule that has been utilized is as follows:

$$\lim_{x \to a} \frac{f(x)}{g(x)} = \frac{\lim\limits_{x \to a} f(x)}{\lim\limits_{x \to a} g(x)}$$

provided that $\lim\limits_{x \to a} f(x)$ and $\lim\limits_{x \to a} g(x)$ exist and, furthermore, $\lim\limits_{x \to a} g(x) \neq 0$. For example, let $\dfrac{f(x)}{g(x)} = \dfrac{\sin 4x}{\cos x}$, then

$$\lim_{x \to 0} \frac{\sin 4x}{\cos x} = \frac{\lim\limits_{x \to 0} \sin 4x}{\lim\limits_{x \to 0} \cos x} = \frac{0}{1} = 0$$

There are many important cases where the hypothesis of the given rule is not satisfied. To illustrate, let us find $\lim\limits_{x \to 2} \dfrac{x^3 - 8}{x - 2}$. Here $f(x) = x^3 - 8$ and $g(x) = x - 2$, so that $\lim\limits_{x \to 2} g(x) = 0$ and, thus, the preceding rule is not applicable. Instead, we recall that, since $f(2) = 2^3 - 8 = 0$, it follows that $x - 2$ is a factor of $x^3 - 8$. In fact, $x^3 - 8 = (x - 2)(x^2 + 2x + 4)$ which means that

$$\lim_{x \to 2} \frac{x^3 - 8}{x - 2} = \lim_{x \to 2} \frac{(x - 2)(x^2 + 2x + 4)}{x - 2} = \lim_{x \to 2} (x^2 + 2x + 4)$$
$$= 4 + 4 + 4 = 12$$

However, when we sought to derive rules for the differentiation of trigonometric functions our difficulties increased. The key result, $\lim\limits_{x \to 0} \dfrac{\sin x}{x}$, was derived geometrically with some effort. From our derivations of $\lim_{x\to 0} \dfrac{\sin x}{x} = 1$ and $\lim_{x\to 0} \dfrac{1 - \cos x}{x} = 0$ we then proved that $D_x(\sin x) = \cos x$. Once this was obtained the main obstacle to the determination of the differentiation of the trigonometric functions was removed.

It is our objective in this section to develop a general method for determining such limits. This procedure was first published by the affluent Marquis de l'Hospital who wrote the first calculus book in 1696. In developing this rule, l'Hospital received a strong assist from his young and struggling teacher Jakob Bernoulli who actually derived it. However, with the passage of time this procedure has simply become known as l'Hospital's rule.

Theorem 1   Let f and g be differentiable functions of x in an open interval I containing $x = a$, except possibly at the number a itself. If $\lim_{x\to a} f(x) = 0$ and $\lim_{x\to a} g(x) = 0$ and if $g'(x) \neq 0$ for all $x \neq a$ and if $\lim_{x\to a} \dfrac{f'(x)}{g'(x)} = L$, then $\lim_{x\to a} \dfrac{f(x)}{g(x)} = L$ where L is a real number, ∞ or $-\infty$.

L'Hospital's rule is valid if all limits are taken as $x \to a^+$ (in this case, the open interval I need only have a as a left end point). The theorem is also valid if all limits are taken as $x \to a^-$ (and I has a as a right end point). Furthermore, we shall see in Section 14.2 that if $a = \infty$ or $a = -\infty$ the theorem is valid provided $I = (M, \infty)$ or $I = (-\infty, M)$ for some number M. Thus, under the differentiability hypotheses, we may use l'Hospital's rule to try to evaluate all limits of the "form $\dfrac{0}{0}$."

The proof of Theorem 1 will be given after some examples showing its use.

EXAMPLE 1   Apply l'Hospital's rule to find
$$\lim_{x\to 2} \frac{x^2 - 3x + 2}{x^2 - 4}$$

Solution   We observe first that
$$\lim_{x\to 2} (x^2 - 3x + 2) = 0 \qquad \text{and} \qquad \lim_{x\to 2} (x^2 - 4) = 0$$

and since the numerator and denominator are differentiable for all x (and, in particular, in any interval containing $x = 2$), l'Hospital's rule applies. Thus
$$\lim_{x\to 2} \frac{x^2 - 3x + 2}{x^2 - 4} = \lim_{x\to 2} \frac{2x - 3}{2x} = \frac{1}{4} \qquad \bullet$$

EXAMPLE 2   Find $\lim_{x\to 0^+} \dfrac{\sin 3x}{x}$.

Solution Since

$$\lim_{x \to 0^+} \sin 3x = 0 \qquad \text{and} \qquad \lim_{x \to 0^+} x = 0$$

and both functions are differentiable for all x, we may apply l'Hospital's rule to obtain

$$\lim_{x \to 0^+} \frac{\sin 3x}{x} = \lim_{x \to 0^+} \frac{3 \cos 3x}{1} = 3(\lim_{x \to 0^+} \cos 3x) = 3(1) = 3.$$ ●

EXAMPLE 3 Find $\lim_{x \to 0} \dfrac{1 - \cos bx}{x^2}$ where b is an arbitrary nonzero constant.

Solution We may use l'Hospital's rule because $\lim_{x \to 0} (1 - \cos bx) = 0$ and $\lim_{x \to 0} x^2 = 0$, and the differentiability requirements are met. Hence

$$\lim_{x \to 0} \frac{1 - \cos bx}{x^2} = \lim_{x \to 0} \frac{b \sin bx}{2x} \tag{i}$$

Since $\lim_{x \to 0} b \sin bx = 0$ and $\lim_{x \to 0} 2x = 0$, we apply l'Hospital's rule again to obtain

$$\lim_{x \to 0} \frac{b \sin bx}{2x} = \lim_{x \to 0} \frac{b^2 \cos bx}{2} = \frac{b^2}{2} \tag{ii}$$

Hence, from (i) and (ii),

$$\lim_{x \to 0} \frac{1 - \cos bx}{x^2} = \frac{b^2}{2}$$

The reader may verify that if $b = 0$ then

$$\lim_{x \to 0} \frac{1 - \cos bx}{x^2} = \lim_{x \to 0} \frac{0}{x^2} = 0$$

This means that our result is also valid when $b = 0$. ●

EXAMPLE 4 Find $\lim_{t \to 0^-} \dfrac{(1 + 2t)^{1/5} - (1 + t)^{1/3}}{t}$.

Solution Since

$$\lim_{t \to 0^-} [(1 + 2t)^{1/5} - (1 + t)^{1/3}] = \lim_{t \to 0^-} (1 + 2t)^{1/5} - \lim_{t \to 0^-} (1 + t)^{1/3}$$

$$= 1 - 1 = 0$$

and $\lim_{t \to 0^-} t = 0$ and both the numerator and the denominator are differentiable functions of t in $-\frac{1}{2} < t < 0$, for example, then l'Hospital's rule may be applied. Hence

$$\lim_{t \to 0^-} \frac{(1 + 2t)^{1/5} - (1 + t)^{1/3}}{t} = \lim_{t \to 0^-} \left\{ \frac{\frac{1}{5}(1 + 2t)^{-4/5} \cdot 2 - \frac{1}{3}(1 + t)^{-2/3}}{1} \right\}$$

$$= \frac{\frac{2}{5} - \frac{1}{3}}{1} = \frac{1}{15}$$ ●

EXAMPLE 5 Criticize the following argument for the problem: Find $\lim\limits_{x\to 1}\dfrac{2x^2-3x+1}{x^2-1}$.

Solution Since the numerator and denominator both approach zero as x tends to 1 and both functions are differentiable for all x, l'Hospital's rule applies. Consequently, applying this rule twice,

$$\lim_{x\to 1}\frac{2x^2-3x+1}{x^2-1}=\lim_{x\to 1}\frac{4x-3}{2x}=\lim_{x\to 1}\frac{4}{2}=2$$

This solution is *wrong* because l'Hospital's rule is only applicable to finding the limit of the first fraction. It does not apply to $\lim\limits_{x\to 1}\dfrac{4x-3}{2x}$ since, for example,

$\lim\limits_{x\to 1}(4x-3)=1\neq 0$.

The correct solution is

$$\lim_{x\to 1}\frac{2x^2-3x+1}{x^2-1}=\lim_{x\to 1}\frac{4x-3}{2x}=\frac{4(1)-3}{2}=\frac{1}{2}\qquad\bullet$$

EXAMPLE 6 Find $\lim\limits_{t\to 0}\dfrac{t-\tan^{-1}t}{t^5}$.

Solution Since $\lim\limits_{t\to 0}(t-\tan^{-1}t)=0$ and $\lim\limits_{t\to 0}t^5=0$ and the numerator and denominator are differentiable functions of t, l'Hospital's rule is applicable. We have

$$\lim_{t\to 0}\frac{t-\tan^{-1}t}{t^5}=\lim_{t\to 0}\frac{1-\dfrac{1}{1+t^2}}{5t^4}$$

Simplifying algebraically and taking the limit gives

$$\lim_{t\to 0}\frac{\dfrac{t^2}{1+t^2}}{5t^4}=\lim_{t\to 0}\frac{1}{5t^2(1+t^2)}=\infty\qquad\bullet$$

In order to prove Theorem 1, we will need a generalization of the basic mean value theorem of differential calculus. Because this was originally established by the nineteenth century French mathematician Augustin Louis Cauchy (1789–1857), it is known as the Cauchy mean value theorem.

Theorem 2 **(*Cauchy's mean value theorem*).** Let $f(x)$ and $g(x)$ be two functions that are continuous on the closed interval $[a,b]$ and differentiable on the open interval (a,b). Suppose further that $g'(x)$ is not zero for all x in $a<x<b$. Then there exists at least one number c in (a,b) such that

$$\frac{f(b)-f(a)}{g(b)-g(a)}=\frac{f'(c)}{g'(c)}\qquad\qquad(1)$$

Proof First it must be shown that $g(b)\neq g(a)$. To this end, suppose on the contrary that $g(b)=g(a)$. Then from the basic mean value theorem which applies to g (since g

satisfies the continuity and differentiability hypotheses) a number c exists for which $a < c < b$ and

$$\frac{g(b) - g(a)}{b - a} = g'(c)$$

This, together with the hypothesis that $g(b) = g(a)$ implies that $g'(c) = 0$. But, by hypothesis $g'(x) \neq 0$ for all x in (a, b) and the condition $g(b) = g(a)$ has yielded a conclusion that is false. Hence $g(b) \neq g(a)$ and therefore the divisor in the left side of (1) is never zero.

Next, we consider a function ϕ defined by

$$\phi(x) = f(x) - f(a) - \left[\frac{f(b) - f(a)}{g(b) - g(a)}\right][g(x) - g(a)] \qquad (2)$$

We observe that ϕ must be continuous on $[a, b]$ and differentiable on (a, b), because both $f(x)$ and $g(x)$ possess these properties by hypothesis. Also by direct substitution into (2)

$$\phi(a) = f(a) - f(a) - \left[\frac{f(b) - f(a)}{g(b) - g(a)}\right][g(a) - g(a)] = 0$$

and

$$\phi(b) = f(b) - f(a) - \left[\frac{f(b) - f(a)}{g(b) - g(a)}\right][g(b) - g(a)] = 0$$

Therefore $\phi(x)$ satisfies all the hypotheses of Rolle's theorem. Hence, the conclusion of Rolle's theorem applies, which implies that there is at least one c in (a, b) at which $\phi'(c) = 0$. Thus from (2),

$$\phi'(x) = f'(x) - \left[\frac{f(b) - f(a)}{g(b) - g(a)}\right]g'(x)$$

and replacement of x by c yields

$$\phi'(c) = f'(c) - \left[\frac{f(b) - f(a)}{g(b) - g(a)}\right]g'(c) = 0$$

and therefore (since $g'(x) \neq 0$ in (a, b))

$$\frac{f'(c)}{g'(c)} = \frac{f(b) - f(a)}{g(b) - g(a)}$$

so that (1) is established. ∎

In particular, if $g(x) = x$, from which $g'(x) = 1$ for all x and $g(b) - g(a) = b - a \neq 0$, thus, we have for some c in (a, b)

$$\frac{f(b) - f(a)}{b - a} = f'(c) \qquad (3)$$

Hence the Cauchy mean value theorem is indeed a generalization of the basic mean value theorem (3).

Cauchy's mean value theorem has a geometrical interpretation similar to that of the basic mean value theorem. For the sake of this illustration, we let f and g be functions of t. Then $x = g(t)$ and $y = f(t)$ are parametric equations of a smooth

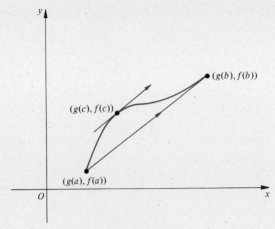

Figure 14.1.1

curve C for $a \leq t \leq b$. As shown in Figure 14.1.1, the end points of C have coordinates $(g(a), f(a))$ and $(g(b), f(b))$. The slope of the line segment joining these points is $\dfrac{f(b) - f(a)}{g(b) - g(a)}$. There is then some point P on the curve at which the tangent line is parallel to this secant line; let $t = c$ be the t-value that determines this point. The slope of the tangent line is

$$\frac{dy}{dx}\bigg|_{t=c} = \frac{\dfrac{dy}{dt}}{\dfrac{dx}{dt}}\bigg|_{t=c} = \frac{f'(c)}{g'(c)}, \, g'(c) \neq 0.$$

Thus, since the slopes are the same, (1) follows.

Now we return to the problem of determining

$$\lim_{x \to a} \frac{f(x)}{g(x)} \tag{4}$$

and l'Hospital's rule for the so-called $\frac{0}{0}$ case. We restate the theorem:

Theorem 1   (*L'Hospital's Rule—$\frac{0}{0}$ Case*).   Let a be an arbitrary real number, and f and g be functions such that
   (i) $f'(x)$ and $g'(x)$ exist for all x in some open interval I containing a, except possibly at $x = a$ itself, and $g'(x) \neq 0$ in that deleted interval I.
   (ii) $\lim\limits_{x \to a} f(x) = 0$ and $\lim\limits_{x \to a} g(x) = 0$

   (iii) $\lim\limits_{x \to a} \dfrac{f'(x)}{g'(x)} = L$, where L is a real number, ∞ or $-\infty$
   Then

$$\lim_{x \to a} \frac{f(x)}{g(x)} = L$$

Proof   In order to establish this we must consider $x \to a^+$ and $x \to a^-$ separately. We restrict our considerations to the right-hand limit only.

The statement of Theorem 1 does not require that either $f(a)$ or $g(a)$ exist [although in many cases in practice $f(a) = g(a) = 0$]. Thus, let us introduce two new functions F and G for which

$$F(x) = \begin{cases} f(x) & x \neq a \\ 0 & x = a \end{cases} \qquad G(x) = \begin{cases} g(x) & x \neq a \\ 0 & x = a \end{cases}$$

Then, by definition of continuity at a point, F and G are continuous at $x = a$. If $x \, \varepsilon \, I$ and $x \geq a$ then F and G are continuous in $[a, x]$. This follows because F and G are continuous at a and differentiable for each $x \, \varepsilon \, I$ where $x > a$ and thus they must be continuous in the same interval. Also $F' = f'$ and $G' = g'$ for each number in (a, x), where $x \, \varepsilon \, I$, and furthermore from (i) $G'(x) \neq 0$ in (a, x). From the Cauchy mean value theorem there is a number c between a and x such that

$$\frac{F(x) - F(a)}{G(x) - G(a)} = \frac{F'(c)}{G'(c)}$$

But $F(a) = 0$ and $G(a) = 0$, consequently

$$\frac{F(x)}{G(x)} = \frac{F'(c)}{G'(c)}$$

Now let $x \to a^+$ so that $c \to a^+$ also (since $a < c < x$). Hence

$$\lim_{x \to a^+} \frac{f(x)}{g(x)} = \lim_{x \to a^+} \frac{F(x)}{G(x)} = \lim_{c \to a^+} \frac{F'(c)}{G'(c)} = L$$

Therefore we have proved the theorem for the one sided approach.

The proof of the theorem for the other one sided approach

$$\lim_{x \to a^-} \frac{f(x)}{g(x)} = L$$

is similar and is left to the reader. As a consequence, the corresponding result for the case

$$\lim_{x \to a} \frac{f(x)}{g(x)}$$

must follow. ∎

Exercises 14.1

In Exercises 1 through 25 find the limit if it exists.

1. $\lim\limits_{x \to 1} \dfrac{x^2 - 1}{x - 1}$

2. $\lim\limits_{x \to 3} \dfrac{x^2 - 9}{x - 3}$

3. $\lim\limits_{x \to 5} \dfrac{x^3 - 11x^2 + 35x - 25}{3x - 15}$

4. $\lim\limits_{x \to -2} \dfrac{x^3 + 9x^2 + 26x + 24}{x^2 - 4}$

5. $\lim\limits_{x \to 0^+} \dfrac{\sin x}{x^2}$

6. $\lim\limits_{x \to 0} \dfrac{e^x - 1}{x}$

7. $\lim\limits_{x \to 0} \dfrac{\tan ax}{x}$ (a is constant)

8. $\lim\limits_{x \to 0} \dfrac{\sin x - x}{\tan x - x}$

9. $\lim\limits_{u \to 2} \dfrac{u - 1 - e^{u-2}}{1 - \cos 2\pi u}$

10. $\lim\limits_{t \to 3} \dfrac{3 - t}{\ln t - \ln 3}$

11. $\lim\limits_{t \to 4} \dfrac{t^{3/2} - 4t^{1/2} - 2t + 8}{t^3 + 64}$

12. $\lim\limits_{u \to 1} \dfrac{u^4 - 1}{\sqrt{u} - 1}$

13. $\lim\limits_{t \to 0} \dfrac{(1 + t)^{1/3} - 1}{t}$

14. $\lim\limits_{x \to 0} \dfrac{5^x - e^x}{x}$

15. $\lim\limits_{x \to 0} \dfrac{e^x - (1 + x) - \dfrac{x^2}{2}}{x^3}$

16. $\lim\limits_{t \to 0} \dfrac{(1 + 3t)^{1/3} - (1 - t)^{1/3}}{t}$

17. $\lim\limits_{x \to 0} \dfrac{\cosh x - (1 + x - \sinh x)}{x^2}$

18. $\lim\limits_{x \to 0} \dfrac{\ln(1 + 7x^2)}{x \ln(1 + 10x)}$

19. $\lim\limits_{\theta \to \pi/2} \dfrac{\sin^2 2\theta}{\cos 3\theta}$

20. $\lim\limits_{t \to 0} \dfrac{e^{at} - at - 1}{at^2}$, $\quad (a \neq 0$ and constant$)$

21. $\lim\limits_{u \to 0} \dfrac{\sin^{-1} u}{\tan^{-1} u}$

22. $\lim\limits_{h \to 0} \dfrac{\sin(x + 3h) - \sin x}{h}$ $\quad$ (*Hint:* Note that x is not affected by variations in h.)

23. $\lim\limits_{h \to 0} \dfrac{\sqrt[3]{x + 2h} - \sqrt[3]{x}}{h}, \quad x \neq 0$

24. $\lim\limits_{h \to 0} \dfrac{e^{x+h} + e^{x-h} - 2e^x}{h^2}$

25. $\lim\limits_{r \to 1} \dfrac{a - ar^n}{1 - r}$, $\quad (a$ is constant and n is an arbitrary positive integer$)$

26. Find a, b, and c such that

$$\lim_{x \to 0} \frac{e^{ax} + bx + c}{x^2} = 2$$

27. Find a, b, and c such that

$$\lim_{x \to 0} \frac{\sinh ax + bx + c}{x^3} = \frac{1}{6}$$

28. Show that

$$\lim_{h \to 0} \frac{f(x + h) + f(x - h) - 2f(x)}{h^2} = f''(x)$$

provided $f''(x)$ is continuous. This generalizes Exercise 24 where $f(x) = e^x$.

29. Find $\lim\limits_{x \to 0^+} \dfrac{(x + \sqrt{x}) \sin^2 2x}{x^{3/2}}$

30. Find $\lim\limits_{x \to 0} \dfrac{e^{3x} - \frac{9}{2}x^3 - \frac{9}{2}x^2 - 3x - 1}{x^4}$

14.2 $\quad$ ADDITIONAL L'HOSPITAL'S RULES

We now state and prove that l'Hospital's rule for the form $\frac{0}{0}$ applies if we are taking the limit as $x \to \infty$. It will be left to the reader to establish the corresponding result when $x \to -\infty$.

Theorem 1   Let f and g be differentiable for all $x > M$, where M is a positive constant. Suppose also that for $x > M$, $g'(x) \neq 0$. Furthermore $\lim_{x \to \infty} f(x) = 0$ and $\lim_{x \to \infty} g(x) = 0$. Then

$$\lim_{x \to \infty} \frac{f(x)}{g(x)} = \lim_{x \to \infty} \frac{f'(x)}{g'(x)} \tag{1}$$

provided that the latter limit exists.

Proof   (We prove the theorem for x becoming positively infinite and leave to the reader to establish the corresponding result when $x \to -\infty$.)

We bring the "neighborhood of infinity" to a "neighborhood of the origin" by the reciprocal substitution. Thus for all $x > M$, we let $x = \dfrac{1}{t}$ or, equivalently, $t = \dfrac{1}{x}$. Then by Theorem 1 of Section 14.1,

$$\lim_{x \to \infty} \frac{f(x)}{g(x)} = \lim_{t \to 0^+} \frac{f\left(\frac{1}{t}\right)}{g\left(\frac{1}{t}\right)} = \lim_{t \to 0^+} \frac{D_t\left[f\left(\frac{1}{t}\right)\right]}{D_t\left[g\left(\frac{1}{t}\right)\right]}$$

By the chain rule of differentiation $\left(\text{using } x = \dfrac{1}{t}\right)$ we have

$$D_t\left[f\left(\frac{1}{t}\right)\right] = D_t[f(x)] = D_x[f(x)] \cdot D_t x = -t^{-2}D_x[f(x)],$$

Similarly

$$D_t\left[g\left(\frac{1}{t}\right)\right] = D_t g(x) = -t^{-2}D_x[g(x)]$$

We then have

$$\lim_{x \to \infty} \frac{f(x)}{g(x)} = \lim_{t \to 0^+} \frac{-t^{-2}D_x[f(x)]}{-t^{-2}D_x[g(x)]} = \lim_{t \to 0^+} \frac{D_x[f(x)]}{D_x[g(x)]} = \lim_{x \to \infty} \frac{f'(x)}{g'(x)} \qquad \blacksquare$$

Next, we state another form of l'Hospital's rule which applies to the limit of a quotient $\dfrac{f(x)}{g(x)}$, where both the numerator and denominator are becoming infinite (either positive or negative infinity) as x approaches a finite limit.

Theorem 2   Let f and g be differentiable in an open interval I, except possibly at $x = a$ in its interior. Furthermore, it is assumed that $g'(x) \neq 0$ for all x in I (except possibly $x = a$). If $\lim_{x \to a} |f(x)| = \infty$ and $\lim_{x \to a} |g(x)| = \infty$, and if $\lim_{x \to a} \dfrac{f'(x)}{g'(x)} = L$, then $\lim_{x \to a} \dfrac{f(x)}{g(x)} = L$ also.

Note: Theorem 2 is valid if all the limits are one sided, that is, right-hand or left-hand limits. This theorem is also true if all limits are taken as $x \to \infty$ or, instead, as $x \to -\infty$, provided an appropriate modification is made for the interval I.

The proof of Theorem 2 and its extensions are omitted.

Besides the forms $\dfrac{0}{0}$ and $\dfrac{\infty}{\infty}$, there are other indeterminate forms that occur frequently, namely, $0 \cdot \infty$, $\infty - \infty$, 0^0, ∞^0 and 1^∞. Thus, for example, the ∞^0 form means that our function is of the form $f(x)^{g(x)}$ where $\lim\limits_{x \to a} f(x) = \infty$ while $\lim\limits_{x \to a} g(x) = 0$. These forms are treated by converting them to $\dfrac{0}{0}$ or $\dfrac{\infty}{\infty}$ forms and then applying l'Hospital's rule. The following examples will illustrate applications of Theorems 1 and 2 and the technique for the treatment of the additional indeterminate forms.

EXAMPLE 1   Find $\lim\limits_{x \to \infty} \dfrac{\sin \dfrac{7}{x^2}}{\dfrac{4}{x^2}}$.

Solution   This is a $\dfrac{0}{0}$ form since $\lim\limits_{x \to \infty} \dfrac{7}{x^2} = 0$ and $\lim\limits_{x \to \infty} \dfrac{4}{x^2} = 0$ (and $\sin \alpha \to 0$ as $\alpha \to 0$) and both functions are differentiable for all $x \neq 0$. Also $D_x \left(\dfrac{4}{x^2} \right) = -\dfrac{8}{x^3}$ is never zero and the hypotheses of Theorem 1 are satisfied. By l'Hospital's rule—Theorem 1,

$$\lim_{x \to \infty} \frac{\sin \dfrac{7}{x^2}}{\dfrac{4}{x^2}} = \lim_{x \to \infty} \frac{\dfrac{-14}{x^3} \cos \dfrac{7}{x^2}}{\dfrac{-8}{x^3}} = \lim_{x \to \infty} \frac{7}{4} \cos \frac{7}{x^2} = \frac{7}{4}(1) = \frac{7}{4}.$$

Alternative Solution   Let $u = \dfrac{4}{x^2}$ so that $\dfrac{7}{x^2} = \dfrac{7}{4}u$ and since $u \to 0^+$ as $x \to \infty$, we have

$$\lim_{x \to \infty} \frac{\sin \dfrac{7}{x^2}}{\dfrac{4}{x^2}} = \lim_{u \to 0^+} \frac{\sin \dfrac{7}{4}u}{u}$$

which is a $\dfrac{0}{0}$ form. By l'Hospital's rule of Section 14.1,

$$\lim_{u \to 0^+} \frac{\sin \dfrac{7}{4}u}{u} = \lim_{u \to 0^+} \frac{\dfrac{7}{4} \cos \dfrac{7}{4}u}{1} = \frac{7}{4} \lim_{u \to 0^+} \cos \frac{7}{4}u = \frac{7}{4}(1) = \frac{7}{4}. \qquad \bullet$$

EXAMPLE 2   Find $\lim\limits_{x \to \infty} \dfrac{\ln 3x}{x + 1}$.

Solution Both the numerator and the denominator tend to ∞ as $x \to \infty$, that is, this is an $\dfrac{\infty}{\infty}$ form. By l'Hospital's rule

$$\lim_{x \to \infty} \frac{\ln 3x}{x+1} = \lim_{x \to \infty} \frac{\dfrac{3}{3x}}{1} = \lim_{x \to \infty} \frac{1}{x} = 0. \qquad \bullet$$

EXAMPLE 3 Find $\lim\limits_{x \to \infty} xe^{-3x}$.

Solution As $x \to \infty$, $e^{-3x} = \dfrac{1}{e^{3x}} \to 0$. Therefore, our limit is of the form $\infty \cdot 0$; $xe^{-3x} = \dfrac{x}{e^{3x}}$

and, as $x \to \infty$, we now have the $\dfrac{\infty}{\infty}$ form. Hence, by l'Hospital's rule,

$$\lim_{x \to \infty} xe^{-3x} = \lim_{x \to \infty} \frac{x}{e^{3x}} = \lim_{x \to \infty} \frac{1}{3e^{3x}} = 0 \qquad \bullet$$

EXAMPLE 4 Find $\lim\limits_{x \to \frac{\pi}{2}} \dfrac{\sec x}{\tan x}$.

Solution Because both $\lim\limits_{x \to \frac{\pi}{2}} \sec x = \infty$ and $\lim\limits_{x \to \frac{\pi}{2}} \tan x = \infty$ we apply l'Hospital's rule and obtain

$$\lim_{x \to \frac{\pi}{2}} \frac{\sec x}{\tan x} = \lim_{x \to \frac{\pi}{2}} \frac{\sec x \tan x}{\sec^2 x} = \lim_{x \to \frac{\pi}{2}} \frac{\tan x}{\sec x}$$

Since $\dfrac{\sec x}{\tan x}$ is the reciprocal of $\dfrac{\tan x}{\sec x}$ and both the numerator and denominator are positive for $0 < x < \dfrac{\pi}{2}$, it follows that *if* the limit exists it must be 1. However, this is not the same as proving that the limit exists.

If l'Hospital's rule is used once more, we get

$$\lim_{x \to \frac{\pi}{2}} \frac{\sec x}{\tan x} = \lim_{x \to \frac{\pi}{2}} \frac{\tan x}{\sec x} = \lim_{x \to \frac{\pi}{2}} \frac{\sec^2 x}{\sec x \tan x} = \lim_{x \to \frac{\pi}{2}} \frac{\sec x}{\tan x}$$

Thus we are now back to the original ratio and the method is a partial failure.

Fortunately, this problem may be treated simply by expressing $\sec x$ and $\tan x$ in terms of $\sin x$ and $\cos x$. Thus

$$\lim_{x \to \frac{\pi}{2}} \frac{\sec x}{\tan x} = \lim_{x \to \frac{\pi}{2}} \frac{\dfrac{1}{\cos x}}{\dfrac{\sin x}{\cos x}} = \lim_{x \to \frac{\pi}{2}} \frac{1}{\sin x} = 1$$

and l'Hospital's rule is obviated. $\qquad \bullet$

EXAMPLE 5 Find $\lim\limits_{x \to 0^+} \left(1 + \dfrac{3}{x}\right)^x$.

Solution As $x \to 0^+$, $1 + \dfrac{3}{x} \to \infty$ so that we have the indeterminate form ∞^0. Let

$$y = \left(1 + \frac{3}{x}\right)^x$$

We seek $\lim\limits_{x \to 0^+} y$. If the natural logarithm of both sides is taken, there results

$$\ln y = x \ln\left(1 + \frac{3}{x}\right) = \frac{\ln\left(1 + \dfrac{3}{x}\right)}{\dfrac{1}{x}}$$

By l'Hospital's rule $\left(\dfrac{\infty}{\infty} \text{ case}\right)$,

$$\lim_{x \to 0^+} \ln y = \lim_{x \to 0^+} \frac{\left(1 + \dfrac{3}{x}\right)^{-1}(-3x^{-2})}{-x^{-2}} = 3 \lim_{x \to 0^+} \frac{1}{1 + \dfrac{3}{x}} = 0$$

Since $\ln y$ is continuous for all $y > 0$ we have

$$\lim_{x \to 0^+} \ln y = \ln\left[\lim_{x \to 0^+} y\right]$$

thus $\ln\left[\lim\limits_{x \to 0^+} y\right] = 0$ or, equivalently, (since $e^{\ln u} = u$)

$$\lim_{x \to 0^+} y = \lim_{x \to 0^+} \left(1 + \frac{3}{x}\right)^x = e^0 = 1 \qquad \bullet$$

EXAMPLE 6 Find $\lim\limits_{x \to 0} (x \csc^3 x - \csc^2 x)$.

Solution $x \csc^3 x = \dfrac{x}{\sin x} \csc^2 x$ and when $x \to 0$, $\dfrac{x}{\sin x} \to 1$ and $\csc^2 x \to \infty$. Hence we may think of this as an indeterminate expression of the form $\infty - \infty$. If we replace $\csc x$ by $\dfrac{1}{\sin x}$ throughout we have

$$\lim_{x \to 0} (x \csc^3 x - \csc^2 x) = \lim_{x \to 0} \left(\frac{x}{\sin^3 x} - \frac{1}{\sin^2 x}\right)$$

$$= \lim_{x \to 0} \frac{x - \sin x}{\sin^3 x}$$

$$= \lim_{x \to 0} \frac{x - \sin x}{x^3} \lim_{x \to 0} \frac{x^3}{\sin^3 x}$$

$$= \lim_{x \to 0} \frac{x - \sin x}{x^3} \lim_{x \to 0} \left(\frac{x}{\sin x}\right)^3$$

provided that the limits on the right side exist. We now prove this by l'Hospital's rule used repeatedly $\left(\dfrac{0}{0}\text{-form}\right)$:

$$\lim_{x\to 0}\frac{x-\sin x}{x^3}=\lim_{x\to 0}\frac{1-\cos x}{3x^2}=\lim_{x\to 0}\frac{\sin x}{6x}=\frac{1}{6}\lim_{x\to 0}\frac{\sin x}{x}$$

$$=\frac{1}{6}\lim_{x\to 0}\frac{\cos x}{1}=\frac{1}{6}(1)=\frac{1}{6}$$

Also, $\displaystyle\lim_{x\to 0}\frac{x}{\sin x}=\lim_{x\to 0}\frac{1}{\cos x}=1$, hence

$$\lim_{x\to 0}(x\csc^3 x-\csc^2 x)=\tfrac{1}{6}(1)^3=\tfrac{1}{6}$$ •

EXAMPLE 7 Show that $\displaystyle\lim_{x\to 0}(1+bx)^{1/2x}=e^{b/2}$, where b is an arbitrary real number.

Solution This is of the form 1^∞. Let

$$y=(1+bx)^{1/2x}\qquad \ln y=\frac{\ln(1+bx)}{2x}$$

By l'Hospital's rule,

$$\lim_{x\to 0}\ln y=\lim_{x\to 0}\frac{\ln(1+bx)}{2x}=\lim_{x\to 0}\frac{\dfrac{b}{1+bx}}{2}=\frac{b}{2}$$

$$\lim_{x\to 0}y=\lim_{x\to 0}e^{\ln y}=e^{\lim_{x\to 0}(\ln y)}=e^{b/2}$$ •

EXAMPLE 8 Find $\displaystyle\lim_{x\to\infty}\frac{\displaystyle\int_3^x e^{t^2}\,dt}{e^{x^2}}$.

Solution This is an $\dfrac{\infty}{\infty}$ form so that, by l'Hospital's rule and the first fundamental theorem of calculus, we have

$$\lim_{x\to\infty}\frac{\displaystyle\int_3^x e^{t^2}\,dt}{e^{x^2}}=\lim_{x\to\infty}\frac{e^{x^2}}{e^{x^2}\cdot 2x}=\lim_{x\to\infty}\frac{1}{2x}=0$$ •

EXAMPLE 9 Let

$$f(x)=\begin{cases}e^{-\frac{1}{x^2}} & \text{if } x\neq 0\\ 0 & \text{if } x=0\end{cases}$$

Find $f'(0)$.

Solution The reader can verify that our usual techniques of differentiation will not help us. We must instead use the definition of derivative (given in Chapter 3).

$$f'(0)=\lim_{h\to 0}\frac{f(h)-f(0)}{h}=\lim_{h\to 0}\frac{e^{-1/h^2}}{h} \tag{2}$$

This limit is of the form $\frac{0}{0}$. Applying l'Hospital's rule (and simplifying) twice, we obtain

$$f'(0) = \lim_{h \to 0} \frac{\dfrac{2}{h^3} e^{-1/h^2}}{1} = \lim_{h \to 0} \frac{2e^{-1/h^2}}{h^3}$$

$$= \lim_{h \to 0} \frac{\dfrac{4}{h^3} e^{-1/h^2}}{3h^2} = \lim_{h \to 0} \frac{4e^{-1/h^2}}{3h^5}$$

Inspection shows that the expressions are becoming more complicated (the numerator still involves e^{-1/h^2} and the power of h in the denominator is increasing). Further application of l'Hospital's rule seems pointless. We return to (2) but rewrite it as

$$f'(0) = \lim_{h \to 0} \frac{\dfrac{1}{h}}{e^{1/h^2}}$$

which is in the form $\frac{\infty}{\infty}$. Now, l'Hospital's rule gives

$$f'(0) = \lim_{h \to 0} \frac{-\dfrac{1}{h^2}}{-\dfrac{2}{h^3} e^{1/h^2}} = \lim_{h \to 0} \frac{h}{2e^{1/h^2}} = 0 \qquad \bullet$$

We thus see that, for some problems, one form of l'Hospital's rule is more useful than another.

In a similar manner, it can be shown that for the function in Example 9 $f''(0) = 0$, $f'''(0) = 0, \ldots, f^{(n)}(0) = 0$ for all positive integers n.

Exercises 14.2

In Exercises 1 through 42, evaluate, if possible, the limits

1. $\displaystyle\lim_{x \to \infty} xe^{-x}$

2. $\displaystyle\lim_{x \to \infty} x^2 e^{-x}$

3. $\displaystyle\lim_{x \to 0} x \cot x$

4. $\displaystyle\lim_{x \to \pi/2} (\sec x - \tan x)$

5. $\displaystyle\lim_{x \to \infty} \frac{10 + 3 \sin x}{x}$

6. $\displaystyle\lim_{x \to \infty} \frac{e^x}{(x^{10} + 3x + 5)^3}$

7. $\displaystyle\lim_{x \to \infty} \frac{x^3 - 4x + 300}{5x^3 + 7x^2 + 100}$

8. $\displaystyle\lim_{t \to \infty} \frac{a + bt + ct^2}{d + ft + gt^2}$, where $a, \ldots, g$ are constants and $cg \neq 0$

9. $\displaystyle\lim_{x \to 0^+} x^x$

10. $\displaystyle\lim_{u \to \infty} \frac{\sqrt{u}}{\ln \ln u}$

11. $\displaystyle\lim_{x \to \infty} \frac{\ln 3x}{\sqrt{x}}$

12. $\displaystyle\lim_{x \to 0^+} x^{100x}$

13. $\displaystyle\lim_{\theta \to \pi/2} (\sec^2 3\theta - \tan^2 3\theta)$

14. $\displaystyle\lim_{\theta \to 0} (1 + \sin \theta)^{1/\theta}$

15. $\lim_{x \to 0} (1 - 3x)^{1/x}$

16. $\lim_{t \to 0^+} t^{\sqrt{3t}}$

17. $\lim_{x \to \infty} \left(1 - \sqrt{\dfrac{3x + 1}{3x}}\right)^{1000}$

18. $\lim_{x \to 0^+} x^r \ln x, \, r > 0$

19. $\lim_{x \to 0} (1 + x^{20})^{1/x}$

20. $\lim_{y \to \infty} \dfrac{(1 + y)^2 - (1 - y)^2}{\sqrt{y^2 + 1}}$

21. $\lim_{x \to \infty} \dfrac{x^4 \ln^2 x}{e^x}$ (*Hint:* Find a convenient function that dominates the given function.)

22. $\lim_{x \to 0} \dfrac{1}{\pi x} \ln \left(\dfrac{1 + 3x}{1 - 3x}\right)$

23. $\lim_{x \to 0} (e^x + x)^{5/x}$

24. $\lim_{x \to 0} (\cos 2x)^{1/x^2}$

25. $\lim_{x \to 0} (x^2 + e^x)^{b/x}, \quad (b \text{ is constant})$

26. $\lim_{x \to 0} x \sqrt{x^2 + x + 4} \cot 3x$ (*Hint:* limit of a product = product of the limits, provided that each limit exists.)

27. $\lim_{x \to \infty} \dfrac{\sqrt{7e^{4x} + 3e^{2x} + 12}}{5e^{2x} + 1}$ (*Hint:* Do not use l'Hospital's rule although it is indeed applicable.)

28. $\lim_{x \to 0} \dfrac{\cot mx}{\cot nx}$, where m and n are positive integers

29. $\lim_{t \to -\infty} t^3 \sin \left(-\dfrac{1}{t^2}\right)$

30. $\lim_{t \to \infty} t^2 \sin^3 \left(\dfrac{1}{t}\right)$

31. $\lim_{n \to \infty} n^{1/n}$

32. $\lim_{t \to 0} \dfrac{(t^2 + t + 1)^4}{3^{2-t}}$

33. $\lim_{h \to \infty} (8^{-h})^{1/(h+1)}$

34. $\lim_{t \to 0} \left(\dfrac{1}{t^2} - \dfrac{1}{\sin^2 t}\right)$

35. $\lim_{x \to \infty} \dfrac{1}{3x + 1} \int_0^x e^{\sqrt{t}} \, dt$

36. $\lim_{n \to \infty} (n^5 + 2n^4)^{1/n}$

37. $\lim_{n \to \infty} \dfrac{n^3}{k^n}, \quad k$ is constant and > 1

38. $\lim_{x \to \infty} x e^{-\sqrt{x}}$

39. $\lim_{x \to 0^+} (\sin 3x)^{2x}$

40. $\lim_{x \to \infty} \dfrac{\sin^2 x \cos^4 x \ln x}{x}$

41. $\lim_{x \to \infty} \dfrac{\displaystyle\int_0^{[\![x]\!]} t \, dt}{x^2 + 3x + 1}$, where $[\![x]\!]$ is the greatest integer function

42. $\lim_{x \to \infty} \dfrac{\displaystyle\int_2^x e^{t^2}(3t^2 + 7t + 4) \, dt}{\displaystyle\int_1^x e^{t^2}(5t^2 + 6y + 3) \, dt}$

43. Show that $e^{bx} = 1 + bx + \phi(x)$, where $\dfrac{\phi(x)}{x} \to 0$ as $x \to 0$ and where b is constant.

44. Show that $\cos bx = 1 - \dfrac{b^2 x^2}{2} + \phi(x)$, where $\dfrac{\phi(x)}{x^2} \to 0$ as $x \to 0$ and where b is constant.

IMPROPER INTEGRALS—INFINITE LIMITS OF INTEGRATION

Until now definite integrals have been defined over finite intervals $[a, b]$ as a limit of a Riemann sum. However, there are other situations that have significant applications. For example, an integral of the type

$$\int_2^\infty \frac{dx}{x^2}$$

is said to be *improper* because one of its limits is infinite. Also an integral of the form

$$\int_{-2}^3 \frac{dx}{x}$$

is said to be improper because the function $f(x) = \dfrac{1}{x}$ becomes infinite at $x = 0$, which is in the interval of integration. More generally, integrals of the following types are called improper:

(i) $\displaystyle\int_a^\infty f(x)\,dx, \ \int_{-\infty}^b f(x)\,dx, \ \text{or} \int_{-\infty}^\infty f(x)\,dx$

(ii) $\displaystyle\int_a^b f(x)\,dx$, where $f(x)$ becomes infinite at one or more values of x in $[a, b]$.

In this section we shall treat case (i) in detail.

For problems defined by (i) and (ii) the definition of the Riemann integral as given in Chapter 6 is inadequate. In fact, there is no way to partition an infinite interval such as $[a, \infty]$ and have the widths of *all* the subintervals approach zero. Also, we may recall that, if a function is unbounded, the Riemann integral as defined previously simply does not exist. Consequently, we seek now to *define* what we mean by saying an improper integral of a function defined on an infinite interval has a value. To this end, consider the following example:

EXAMPLE 1 Find the area under the curve $y = \dfrac{4}{x^2}$ from $x = 2$ to $x = \infty$, that is, to the right of the line $x = 2$ (see Figure 14.3.1).

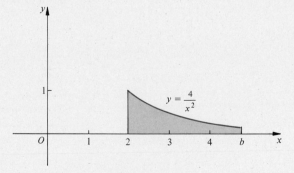

Figure 14.3.1

Solution Certainly we know how to find the area under the curve $y = \dfrac{4}{x^2}$ from $x = 2$ to $x = b$, where b is an arbitrary number greater than 2. The area is given by

$$\int_2^b \frac{4}{x^2}\, dx = -\frac{4}{x}\bigg|_2^b = -\frac{4}{b} - \left(-\frac{4}{2}\right) = 2 - \frac{4}{b} \text{ square units}$$

It is reasonable to *define* the area A from $x = 2$ to $x = \infty$ to be

$$A = \lim_{b \to \infty} \int_2^b \frac{4}{x^2}\, dx = \lim_{b \to \infty}\left(2 - \frac{4}{b}\right) = 2 \text{ square units}$$

In abbreviated form, we say

$$A = \int_2^\infty \frac{4}{x^2}\, dx = \lim_{b \to \infty} \int_2^b \frac{4}{x^2}\, dx \tag{1}$$

provided that the limit on the right of (1) exists. ●

Definition In general, suppose that a function $f(x)$ is continuous for all $x \geq a$.† Taking $b > a$, if $\lim\limits_{b \to \infty} \int_a^b f(x)\, dx = L$ (where L is a finite number), we write

$$\int_a^\infty f(x)\, dx = \lim_{b \to \infty} \int_a^b f(x) = L. \tag{2}$$

If (2) occurs, the improper integral is said to be **convergent.** If the limit in (2) does not exist (as a real number) then the improper integral is said to be **divergent.** Particular cases of divergence occur when L is one of the symbols ∞ or $-\infty$ (remember these are not numbers).

EXAMPLE 2 Test the integral $\displaystyle\int_0^\infty e^{-4x}\, dx$ for convergence.

Solution
$$\int_0^b e^{-4x}\, dx = -\frac{1}{4} e^{-4x}\bigg|_0^b = -\frac{1}{4}(e^{-4b} - 1)$$

$$= \frac{1}{4}(1 - e^{-4b})$$

for any number b. Furthermore,

$$\int_0^\infty e^{-4x}\, dx = \lim_{b \to \infty} \frac{1}{4}(1 - e^{-4b}) = \frac{1}{4}$$

which means that the given integral converges. ●

EXAMPLE 3 Test the integral $\displaystyle\int_0^\infty \sin x\, dx$ for convergence.

---

† We assume for simplicity that $f(x)$ is a continuous function for $x \geq a$, even though we could replace it by the more general integrability requirement, that is, $\int_a^b f(x)\, dx$ exists for all finite b such that $b > a$.

Solution
$$\int_0^b \sin x \, dx = -\cos x \Big|_0^b = 1 - \cos b$$

for any number b. However, because of the oscillatory nature of $\cos b$ for large positive arguments, it follows that $\lim_{b \to \infty} (1 - \cos b)$ does not exist and the improper integral diverges. ●

EXAMPLE 4   Determine the values of k for which $\int_1^\infty \dfrac{dx}{x^k}$ converges.

Solution   We must consider three cases, namely, $k > 1$, $k < 1$, and $k = 1$. For $k \neq 1$, we have by the power rule

$$\int_1^b \frac{1}{x^k} \, dx = \int_1^b x^{-k} \, dx = \frac{x^{1-k}}{1-k} \Big|_1^b = \frac{1}{1-k}(b^{1-k} - 1)$$

and therefore we must determine whether or not $\dfrac{1}{1-k}(\lim_{b \to \infty} [b^{1-k} - 1])$ exists.

(i)   If $k > 1$, $1 - k < 0$ and $\lim_{b \to \infty} [b^{1-k} - 1] = 0 - 1 = -1$ so that

$$\int_1^\infty \frac{1}{x^k} \, dx = \frac{1}{k-1}$$

(ii)   If $k < 1$, $1 - k > 0$ and $\lim_{b \to \infty} [b^{1-k} - 1] = \infty$ so that

$$\int_1^\infty \frac{1}{x^k} \, dx \text{ diverges}$$

(iii)   If $k = 1$, $\displaystyle\int_1^b \frac{1}{x^k} \, dx = \int_1^b x^{-1} \, dx = \ln x \Big|_1^b = \ln b$ and   therefore

$\lim_{b \to \infty} \displaystyle\int_1^b x^{-1} \, dx = \infty$, and again the integral diverges.

To summarize: $\displaystyle\int_1^\infty \frac{1}{x^k} \, dx$ exists if and only if $k > 1$ and for these values of k,

$$\int_1^\infty \frac{1}{x^k} \, dx = \frac{1}{k-1}.$$ ●

Definition   In a similar manner, if b is finite and $f(x)$ is continuous for all $a \leq b$, we define

$$\int_{-\infty}^b f(x) \, dx = \lim_{a \to -\infty} \int_a^b f(x) \, dx$$

provided the limit exists. Again we say that the improper integral is *convergent* or *divergent* according to whether the limit does or does not exist.

EXAMPLE 5   Investigate the improper integral $\displaystyle\int_{-\infty}^{-1} \frac{dx}{x^2}$.

Solution For any negative number a,

$$\int_a^{-1} \frac{dx}{x^2} = -\frac{1}{x}\bigg|_a^{-1} = 1 + \frac{1}{a}$$

Also,

$$\int_{-\infty}^{-1} \frac{dx}{x^2} = \lim_{a \to -\infty} \int_a^{-1} \frac{dx}{x^2} = \lim_{a \to -\infty} \left(1 + \frac{1}{a}\right) = 1 \qquad \bullet$$

Up to this point we have been concerned with integrals for which the integrand is continuous and one of the limits is ∞ or $-\infty$. Sometimes we must integrate functions from $-\infty$ to ∞. If f is continuous at each number x, we define

$$\int_{-\infty}^{\infty} f(x)\, dx = \int_{-\infty}^{0} f(x)\, dx + \int_{0}^{\infty} f(x)\, dx \qquad (3)$$

provided that each integral on the right side of (3) exists. This means that $\lim_{a \to -\infty} \int_a^0 f(x)\, dx$ and $\lim_{b \to \infty} \int_0^b f(x)\, dx$ must exist individually; that is, the passage of the limit as $a \to -\infty$ and $b \to \infty$ must be independently taken. We must be careful about this because, for example, $\lim_{t \to \infty} \int_{-t}^t f(x)\, dx$ can exist even though $\int_{-\infty}^{\infty} f(x)\, dx$ does not exist (see Example 7 of this section).

Definition $\int_{-\infty}^{\infty} f(x)\, dx$ is **convergent** if and only if both integrals on the right in (3) are convergent. Otherwise, $\int_{-\infty}^{\infty} f(x)\, dx$ is **divergent.**

Instead of (3), it would have been correct to define

$$\int_{-\infty}^{\infty} f(x)\, dx = \int_{-\infty}^{c} f(x)\, dx + \int_{c}^{\infty} f(x)\, dx \qquad (4)$$

for some real number c, provided the integrals on the right exist. In practice however, it is usually easiest to work with $c = 0$.

EXAMPLE 6 Find $\displaystyle\int_{-\infty}^{\infty} \frac{dx}{1 + x^2}$.

Solution We shall use (3).

$$\int_0^{\infty} \frac{dx}{1 + x^2} = \lim_{b \to \infty} \int_0^b \frac{dx}{1 + x^2} = \lim_{b \to \infty} \tan^{-1} x \bigg|_0^b$$

$$= \lim_{b \to \infty} (\tan^{-1} b - \tan^{-1} 0) = \frac{\pi}{2}$$

Also,

$$\int_{-\infty}^{0} \frac{dx}{1 + x^2} = \lim_{a \to -\infty} \int_a^0 \frac{dx}{1 + x^2} = \lim_{a \to -\infty} (\tan^{-1} 0 - \tan^{-1} a)$$

$$= -\left(-\frac{\pi}{2}\right) = \frac{\pi}{2}$$

Thus from (3),

$$\int_{-\infty}^{\infty} \frac{dx}{1 + x^2} = \int_{-\infty}^{0} \frac{dx}{1 + x^2} + \int_{0}^{\infty} \frac{dx}{1 + x^2}$$

$$= \frac{\pi}{2} + \frac{\pi}{2} = \pi$$

EXAMPLE 7   Find $\lim\limits_{t \to \infty} \int_{-t}^{t} \sin x \, dx$ and compare this result with $\int_{-\infty}^{\infty} \sin x \, dx$.

Solution

$$\int_{-t}^{t} \sin x \, dx = -\cos x \Big|_{-t}^{t} = -\cos t + \cos(-t)$$

$$= -\cos t + \cos t = 0$$

for each value of t since $\cos(-t) = \cos t$. Hence

$$\lim_{t \to \infty} \int_{-t}^{t} \sin x \, dx = 0$$

However, since $\int_{0}^{\infty} \sin x \, dx$ does not exist (see Example 3 of this section), it follows from (3) that

$$\int_{-\infty}^{\infty} \sin x \, dx$$

does not exist.

Exercises 14.3

In Exercises 1 through 21, evaluate the improper integrals if they are convergent or show that the integrals are divergent.

1. $\int_{3}^{\infty} \frac{dx}{x^2}$

2. $\int_{0}^{\infty} e^{-7x} \, dx$

3. $\int_{1}^{\infty} \frac{dx}{\sqrt{x}}$

4. $\int_{1}^{\infty} \frac{dx}{1 + x^2}$

5. $\int_{1}^{\infty} \frac{\ln x}{x} \, dx$

6. $\int_{-\infty}^{-1} e^{kx} \, dx, \quad k > 0$

7. $\int_{0}^{\infty} \frac{dx}{a^2 + x^2}, \quad a \neq 0$

8. $\int_{2}^{\infty} \frac{dx}{x \ln x}$

9. $\int_{0}^{\infty} x e^{-x} \, dx$

10. $\int_{5}^{\infty} \frac{dx}{x^2 - 16}$

11. $\int_{3}^{\infty} \frac{dx}{(x - 2)^5}$

12. $\int_{1}^{\infty} \frac{x \, dx}{3x^2 - 2}$

13. $\int_{0}^{\infty} \frac{x \, dx}{(5x^2 + 1)^{3/2}}$

14. $\int_{-\infty}^{-1} \frac{x + 1}{x^3} \, dx$

15. $\int_{2}^{\infty} \frac{dx}{x^2(1 + x)}$

16. $\int_{2}^{\infty} \frac{dx}{1 - x^4}$

17. $\displaystyle\int_{1}^{-\infty} e^{-t^2}t\,dt$

18. $\displaystyle\int_{4}^{\infty} \frac{dx}{\ln e^x}$

19. $\displaystyle\int_{-\infty}^{0} 2^{kx}\,dx, \quad k > 0$

20. $\displaystyle\int_{1}^{\infty} \frac{\ln x}{x^2}\,dx$

21. $\displaystyle\int_{1}^{\infty} \frac{\ln x}{x^{1+r}}\,dx$ where $r > 0$. This generalizes Exercise 20.

22. Use mathematical induction to find a formula for $\displaystyle\int_{0}^{\infty} x^n e^{px}\,dx$, where n is a positive integer and $p < 0$.

23. The region under the graph of $y = e^{-2x}$ to the right of the line $x = 0$ is revolved about the x-axis. Find the volume of the solid of revolution. Sketch the volume of interest.

24. Find an expression for the area A under the graph $y = \dfrac{a^2}{\sqrt{a^2 + x^2}}$ from $x = -b$ to $x = b$, where $a > 0$ and $b > 0$. Does $A(b)$ tend to a limit as $b \to \infty$?

25. The region of the previous exercise from $x = -b$ to $x = b$ is revolved about the x-axis. Find an expression for the volume $V(b)$. Does $V(b)$ tend to a limit as $b \to \infty$?

26. The integral

$$\int_{2}^{\infty} \left(\frac{kx}{x^2 - 1} - \frac{1}{x + 1} \right) dx$$

converges for a certain value of k. Find this value of k and evaluate the corresponding integral.

14.4   IMPROPER INTEGRALS — OTHER CASES

Consider the function f defined by $f(x) = x^{-1/3}$ on the interval $(0, 1]$. We seek the determination of the area bounded by the graph of $y = x^{-1/3}$, the x-axis, and the vertical lines $x = 0$ and $x = 1$ (see Figure 14.4.1).

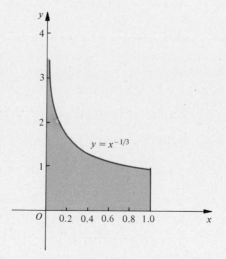

Figure 14.4.1

If the function were continuous in [0, 1], the area would be found by evaluating the Riemann integral from $x = 0$ to $x = 1$. But, in this case the function is becoming unbounded as $x \to 0^+$. Thus the Riemann integral does not exist, and once again we must deal with an improper integral. However, $\int_{\varepsilon}^{1} x^{-1/3} \, dx$ does exist for any positive number ε, no matter how small. Hence, it is reasonable to *define*

$$\int_0^1 x^{-1/3} \, dx = \lim_{\varepsilon \to 0^+} \int_{\varepsilon}^1 x^{-1/3} \, dx \tag{1}$$

provided that the limit on the right side exists.

Definition   More generally we have the definition: if $f(x)$ is continuous in the half open interval $(a, b]$ then

$$\int_a^b f(x) \, dx = \lim_{\varepsilon \to 0^+} \int_{a+\varepsilon}^b f(x) \, dx \tag{2}$$

provided that the limit on the right side of (2) exists. If the limit exists, the integral is said to be **convergent.** If the limit does not exist, the integral is said to be **divergent.**

EXAMPLE 1   Determine whether the improper integral $\int_0^1 x^{-1/3} \, dx$ exists (converges).

Solution   In accordance with our discussion we find

$$\int_{\varepsilon}^1 x^{-1/3} \, dx = \frac{3}{2} x^{2/3} \Big|_{\varepsilon}^1 = \frac{3}{2} (1 - \varepsilon^{2/3})$$

where ε is a positive number. Then from (1),

$$\int_0^1 x^{-1/3} \, dx = \lim_{\varepsilon \to 0^+} \frac{3}{2} (1 - \varepsilon^{2/3}) = \frac{3}{2} (1 - 0) = \frac{3}{2}$$

Hence $\frac{3}{2}$ square units is a measure of the area shown in Figure 14.4.1.   ●

Consider next the problem of determining the area bounded by the graph of $y = \dfrac{1}{(4 - x)^2}$, the x-axis and the vertical lines $x = 1$ and $x = 4$ (Figure 14.4.2). The function f defined by $f(x) = \dfrac{1}{(4 - x)^2}$ is continuous for all x in $(1, 4)$. However, $\lim\limits_{x \to 4^-} f(x) = \infty$ so that once again we are faced with an improper integral. We define

$$\int_1^4 \frac{1}{(4 - x)^2} \, dx = \lim_{\varepsilon \to 0^+} \int_1^{4-\varepsilon} \frac{1}{(4 - x)^2} \, dx \tag{3}$$

More generally, we have the following definition.

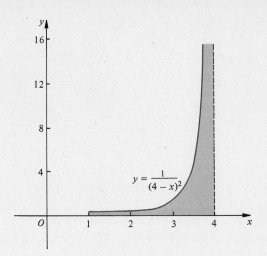

$$y = \frac{1}{(4-x)^2}$$

Figure 14.4.2

Definition   If $f(x)$ is continuous in $[a, b)$ then

$$\int_a^b f(x)\,dx = \lim_{\varepsilon \to 0^+} \int_a^{b-\varepsilon} f(x)\,dx \tag{4}$$

provided that the limit on the right side exists. (The terms **convergent** and **divergent** are defined as before.)

EXAMPLE 2   Compute the integral defined by (3).

Solution

$$\int_1^4 \frac{dx}{(4-x)^2} = \lim_{\varepsilon \to 0^+} \int_1^{4-\varepsilon} \frac{dx}{(4-x)^2} = \lim_{\varepsilon \to 0^+} (4-x)^{-1}\bigg|_1^{4-\varepsilon}$$

$$= \lim_{\varepsilon \to 0^+} \left(\frac{1}{\varepsilon} - \frac{1}{3}\right) = \infty$$

In this case the integral diverges and the associated area is infinite.   ●

Next, suppose that the integrand has an infinite discontinuity at an interior point of a given interval. Then the associated improper integral is defined as follows:

Definition   Let f be a function that is continuous at all points in $[a, b]$ with the exception of $x = c$ in (a, b), then

$$\int_a^b f(x)\,dx = \lim_{\varepsilon \to 0^+} \int_a^{c-\varepsilon} f(x)\,dx + \lim_{\delta \to 0^+} \int_{c+\delta}^b f(x)\,dx \tag{5}$$

provided that each integral on the right side of (5) exists.

EXAMPLE 3   Test the convergence of $\displaystyle\int_0^4 \frac{dx}{\sqrt[3]{x-1}}$ and if it converges find its value. (See Figure 14.4.3.)

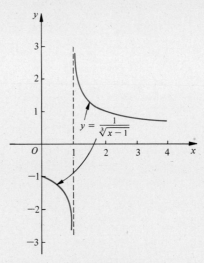

Figure 14.4.3

Solution   The integrand is continuous for all x with the exception of $x = 1$, which is inside the interval of integration. Therefore from (5),

$$\int_0^4 \frac{dx}{\sqrt[3]{x-1}} = \lim_{\varepsilon \to 0^+} \int_0^{1-\varepsilon} (x-1)^{-1/3}\, dx + \lim_{\delta \to 0^+} \int_{1+\delta}^4 (x-1)^{-1/3}\, dx$$

provided each limit exists. Hence

$$\int_0^4 \frac{dx}{\sqrt[3]{x-1}} = \lim_{\varepsilon \to 0^+} \frac{3}{2}(x-1)^{2/3}\bigg|_0^{1-\varepsilon} + \lim_{\delta \to 0^+} \frac{3}{2}(x-1)^{2/3}\bigg|_{1+\delta}^4$$

$$= \lim_{\varepsilon \to 0^+} \frac{3}{2}(-\varepsilon)^{2/3} - \frac{3}{2} + \frac{3}{2}(3^{2/3}) - \lim_{\delta \to 0^+} \frac{3}{2}\delta^{2/3}$$

$$= \frac{3}{2}(3^{2/3} - 1) \qquad\qquad\qquad\qquad \bullet$$

EXAMPLE 4   Find $\displaystyle\int_0^1 \sqrt{x}\, \ln x\, dx$ if it is convergent.

Solution   The integrand is continuous at all x in the given interval except at $x = 0$ where it is not defined.

$$\int_0^1 \sqrt{x}\, \ln x\, dx = \lim_{\varepsilon \to 0^+} \int_\varepsilon^1 \sqrt{x}\, \ln x\, dx$$

Using integration by parts:

let $\qquad u = \ln x \qquad dv = x^{1/2}\, dx \qquad du = \dfrac{dx}{x} \qquad v = \dfrac{2}{3}x^{3/2}$

$$\int_\varepsilon^1 \sqrt{x}\, \ln x\, dx = \frac{2}{3}x^{3/2} \ln x \bigg|_\varepsilon^1 - \frac{2}{3}\int_\varepsilon^1 x^{1/2}\, dx$$

$$= 0 - \frac{2}{3}\varepsilon^{3/2}\ln\varepsilon - \frac{4}{9}x^{3/2}\bigg|_\varepsilon^1$$

$$= -\frac{2}{3}\varepsilon^{3/2}\ln\varepsilon - \frac{4}{9} + \frac{4}{9}\varepsilon^{3/2}$$

Thus $\int_0^1 \sqrt{x} \ln x \, dx = -\frac{4}{9}$, since by l'Hospital's rule, $\lim_{\varepsilon \to 0^+} -\frac{2}{3}\varepsilon^{3/2} \ln \varepsilon = 0$ and clearly $\lim_{\varepsilon \to 0^+} \frac{4}{9}\varepsilon^{3/2} = 0$. (The reader should verify that $\lim_{\varepsilon \to 0^+} \varepsilon^{3/2} \ln \varepsilon = 0$.) ●

In some cases it may be difficult to determine whether or not an improper integral exists because an antiderivative of the integrand cannot be found. For example, consider the integral $\int_3^\infty e^{-x^2} \, dx$. We cannot find an antiderivative of e^{-x^2} and the previous method of first finding $\int_3^b e^{-x^2} \, dx$ and then letting $b \to \infty$ fails because we cannot find $\int_3^b e^{-x^2} \, dx$ analytically. Instead we shall use the technique of comparing the given integral with a known integral. To this end, we need the basic property of real functions, namely, that if $f(x)$ has domain $[a, \infty)$ and $f(x)$ is monotonically increasing then either $\lim_{x \to \infty} f(x)$ exists or $\lim_{x \to \infty} f(x) = \infty$.

The two alternatives are shown in Figure 14.4.4(a) and 14.4.4(b). This property will be established in the next chapter in conjunction with sequences.

We shall now establish the following *comparison test for convergence or divergence:*

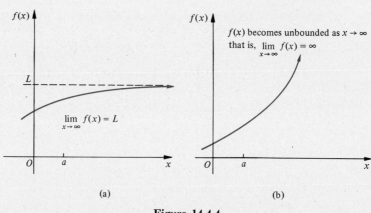

(a)   (b)

Figure 14.4.4

Theorem 1
(Comparison Test)

Let f and g be two continuous functions for all $x \geq a$. Also it is assumed that

$$0 \leq f(x) \leq g(x) \qquad \text{for all } x \geq a$$

It follows that

(I) If $\int_a^\infty g(x) \, dx$ exists, then $\int_a^\infty f(x) \, dx$ must also exist.

(II) If $\int_a^\infty f(x) \, dx$ diverges, then $\int_a^\infty g(x) \, dx$ diverges also.

Proof of (I)   Consider the integrals from a to b, with a fixed and b varying. We define

$$F(b) = \int_a^b f(x)\,dx \quad \text{and} \quad G(b) = \int_a^b g(x)\,dx$$

for any positive b no matter how large. Furthermore, $F(b) \leq G(b)$ since $0 \leq f(x) \leq g(x)$. Now $\int_a^\infty g(x)\,dx$ exists, which implies that $\lim_{b \to \infty} G(b) = k$, where k is a specific positive number. Hence $F(b) \leq k$ for all b, no matter how large and, since $F(b)$ is a monotonically increasing function of b, it follows from the preceding basic property that

$$\lim_{b \to \infty} F(b) = \lim_{b \to \infty} \int_a^b f(x)\,dx$$

exists and is $\leq k$.

The proof of (II) is left to the reader.   ∎

EXAMPLE 5   Investigate the convergence or divergence of

$$\int_3^\infty e^{-x^2}\,dx.$$

Solution   We note that for $x \geq 3$, we have $0 \leq e^{-x^2} \leq e^{-3x}$. Furthermore, for all $b \geq 3$,

$$\int_3^b e^{-3x}\,dx = \frac{e^{-3x}}{-3}\Big|_3^b = -\frac{1}{3}(e^{-3b} - e^{-9}) = \frac{1}{3}(e^{-9} - e^{-3b})$$

Therefore $\int_3^\infty e^{-3x}\,dx$ exists $\left(\text{and equals } \dfrac{e^{-9}}{3}\right)$ so that by Theorem 1, $\int_3^\infty e^{-x^2}\,dx$ converges to a positive number which is $\leq \dfrac{e^{-9}}{3}$.   ●

EXAMPLE 6   Is $\int_0^\infty e^{-x^2}\,dx$ convergent or divergent?

Solution   Unlike the preceding example, we cannot readily find a convenient function with which to compare e^{-x^2}. Instead, note that

$$\int_0^\infty e^{-x^2}\,dx = \int_0^3 e^{-x^2}\,dx + \int_3^\infty e^{-x^2}\,dx \tag{6}$$

provided that both integrals on the right side of (6) exist.

Since e^{-x^2} is continuous on $[0, 3]$, $\int_0^3 e^{-x^2}\,dx$ exists. By Example 5, $\int_3^\infty e^{-x^2}\,dx$ exists and the convergence of $\int_0^\infty e^{-x^2}\,dx$ follows.   ●

EXAMPLE 7   Test $\int_{10}^\infty \dfrac{dx}{\ln x}$ for convergence or divergence.

Solution Again, we cannot find an antiderivative of $\dfrac{1}{\ln x}$. However, it is easy to show that

$$\frac{1}{\ln x} > \frac{1}{x} > 0 \text{ for } x \geq 10. \text{ Also, } \int_{10}^{\infty} \frac{dx}{x} \text{ diverges since } \int_{10}^{b} \frac{dx}{x} = \ln \frac{b}{10} \text{ for } b \geq 10$$

and $\displaystyle\lim_{b \to \infty} \ln \frac{b}{10} = \infty$. Therefore, $\displaystyle\int_{10}^{\infty} \frac{dx}{\ln x}$ diverges by (II) of Theorem 1. ●

Next, we shall state another useful test for the convergence or divergence of integrals.

Theorem 2 **(Limit Comparison Test).** It is assumed that both $\displaystyle\int_{a}^{b} f(x)\,dx$ and $\displaystyle\int_{a}^{b} g(x)\,dx$ exist for arbitrary $b \geq a$ and that for each $x \geq a$, $f(x) \geq 0$ and $g(x) > 0$. Also

$$\lim_{x \to \infty} \frac{f(x)}{g(x)} = k$$

where k is a nonzero constant. Then $\displaystyle\int_{a}^{\infty} f(x)\,dx$ and $\displaystyle\int_{a}^{\infty} g(x)\,dx$ both converge or both diverge.

The proof is left as an exercise.

EXAMPLE 8 Test $\displaystyle\int_{1}^{\infty} \frac{dx}{x\sqrt{3x+2}}$ for convergence or divergence.

Solution Let $f(x) = \dfrac{1}{x\sqrt{3x+2}}$ and choose $g(x) = \dfrac{1}{x\sqrt{x}}$ then it is easily verified that $f(x) > 0$ and $g(x) > 0$ for all x in $[1, \infty)$. Also, since the functions are continuous for all positive x, both integrals $\displaystyle\int_{1}^{b} f(x)\,dx$ and $\displaystyle\int_{1}^{b} g(x)\,dx$ must exist for arbitrary $b \geq 1$. In addition

$$\lim_{x \to \infty} \frac{f(x)}{g(x)} = \lim_{x \to \infty} \frac{x\sqrt{x}}{x\sqrt{3x+2}} = \lim_{x \to \infty} \frac{\sqrt{x}}{\sqrt{x}\sqrt{3 + \dfrac{2}{x}}} = \frac{1}{\sqrt{3}}$$

and

$$\int_{1}^{\infty} \frac{1}{x\sqrt{x}}\,dx = \lim_{b \to \infty} \int_{1}^{b} x^{-3/2}\,dx = \lim_{b \to \infty} -2x^{-1/2}\Big|_{1}^{b}$$

$$= \lim_{b \to \infty} (2 - 2b^{-1/2}) = 2$$

Thus we conclude that the given integral must converge also. ●

Exercises 14.4

In Exercises 1 through 24, evaluate the improper integrals that converge:

1. $\displaystyle\int_0^1 \frac{1}{\sqrt{x}}\,dx$

2. $\displaystyle\int_0^1 \frac{dx}{x\sqrt{x}}$

3. $\displaystyle\int_0^4 \frac{1}{(4-x)^2}\,dx$

4. $\displaystyle\int_0^1 \frac{x}{\sqrt{1-x^2}}\,dx$

5. $\displaystyle\int_0^3 \frac{1}{x^2 - 7x + 12}\,dx$

6. $\displaystyle\int_0^2 \frac{dx}{\sqrt{4-x^2}}$

7. $\displaystyle\int_{-3}^3 \frac{dx}{\sqrt{9-x^2}}$

8. $\displaystyle\int_5^6 \frac{dx}{\sqrt{6-x}}$

9. $\displaystyle\int_{\pi/3}^{\pi/2} \sec x\,dx$

10. $\displaystyle\int_2^0 x \ln x\,dx$

11. $\displaystyle\int_0^{\pi/2} \frac{1}{1-\sin x}\,dx$

12. $\displaystyle\int_{-2}^1 \frac{1}{\sqrt[3]{x}}\,dx$

13. $\displaystyle\int_a^{2a} \frac{dx}{\sqrt{x^2 - a^2}}, \quad$ where $a > 0$

14. $\displaystyle\int_a^{2a} \frac{dx}{(x-a)^{2/3}}$

15. $\displaystyle\int_0^{10} \frac{dx}{100 - x^2}$

16. $\displaystyle\int_{1/2}^1 \frac{dx}{x^3 - x}$

17. $\displaystyle\int_0^a \frac{dx}{\sqrt{a^2 - x^2}}, \quad a > 0$

18. $\displaystyle\int_0^a \frac{x\,dx}{\sqrt{a^2 - x^2}}, \quad a > 0$

19. $\displaystyle\int_1^6 \frac{5}{(x-3)^2}\,dx$

20. $\displaystyle\int_0^3 \frac{dx}{\sqrt{(x-2)(x+1)}}$

21. $\displaystyle\int_0^8 \frac{dx}{(x-5)^{2/3}}$

22. $\displaystyle\int_{-2}^5 \frac{dx}{x^2 - 10x + 16}$

23. $\displaystyle\int_0^\infty \frac{dx}{(x-10)^2}$

24. $\displaystyle\int_4^\infty \frac{dx}{x^2 - 1}$

In Exercises 25 through 34, test each integral for convergence or divergence by comparing it with simpler or known integrals using Theorem 1 or 2 and the techniques illustrated in Examples 5 through 8.

25. $\displaystyle\int_1^\infty \frac{dx}{x^2 + 4x}$

26. $\displaystyle\int_\pi^\infty \frac{\cos^2 3x}{x^2}\,dx$

27. $\displaystyle\int_1^\infty \frac{\sqrt{x+1}}{x}\,dx$

28. $\displaystyle\int_{100}^\infty \frac{dx}{3 + x^3}$

29. $\displaystyle\int_0^\infty \frac{dx}{\sqrt{9 + x^3}}$

30. $\displaystyle\int_e^\infty \frac{4}{x}\sqrt{\frac{x+1}{x+3}}\,dx$

31. $\displaystyle\int_{20}^\infty \frac{\ln x}{x^2}\,dx$

32. $\displaystyle\int_{-1}^\infty \frac{x^5\,dx}{e^x}$

33. $\displaystyle\int_1^\infty \frac{dx}{\sqrt[3]{x^4 + 8}}$

34. $\displaystyle\int_4^\infty e^{-3x}\ln^2 x\,dx$

35. Evaluate $\displaystyle\int_0^1 \ln^2 x\,dx$.

36. Derive the reduction formula

$$\int \ln^n x \, dx = x \ln^n x - n \int \ln^{n-1} x \, dx \qquad (n = \text{positive integer})$$

and compute $\int_0^1 \ln^4 x \, dx$.

37. Find $\int_0^a \dfrac{dx}{\sqrt{ax - x^2}}$, where $a > 0$.

38. Find the length of the quarter circle of radius a:

$$y = \sqrt{a^2 - x^2} \qquad 0 \le x \le a.$$

14.5   EXTENDED LAW OF THE MEAN — TAYLOR'S FORMULA

Many times in mathematics, as well as in its applications, it is useful to have a function approximated by a polynomial. This is due in part to the fact that it is usually easier to work with polynomials than with other types of functions. Computers can easily be programmed to evaluate a polynomial for any desired values of x. Thus, for example, if we have a polynomial approximation for $f(x) = \sin x$, we can have a computer generate the corresponding trigonometric table.

We obtain our polynomial approximations by an extension of the basic mean value theorem of calculus. In our development of that theorem we found that, if f is a continuous function in $[a, b]$ and differentiable in (a, b), then a point c in (a, b) exists such that

$$f(b) = f(a) + f'(c)(b - a) \tag{1}$$

Equation (1) is exact; however, the determination of c in (a, b) such that (1) holds may be a difficult task. If we assume furthermore that $f'(x)$ is continuous also and b is close to a, then $f'(x)$ will be close to $f'(a)$ and (1) may be replaced by the approximation

$$f(b) \approx f(a) + f'(a)(b - a) \tag{2}$$

The result (2) has a simple geometric interpretation. An equation of the tangent line $y = P_1(x)$ to the curve $y = f(x)$ at the point $(a, f(a))$ is given by $P_1(x) = f(a) + f'(a)(x - a)$. Note that $P_1(a) = f(a)$ and $P_1'(a) = f'(a)$ (see Figure 14.5.1).

Consider next the case of a function f such that f and f' are continuous at all points in an interval I containing the point $x = a$. Furthermore $f''(x)$ is assumed to exist in the interior of I and, in particular, $f''(a)$ exists. We seek a polynomial $P_2(x)$ of degree 2, at most, such that

$$P_2(a) = f(a) \qquad P_2'(a) = f'(a) \qquad P_2''(a) = f''(a) \tag{3}$$

It would be anticipated that this quadratic polynomial would generally be a better approximation to $f(x)$ in some neighborhood of $x = a$ than the previously obtained linear form. In fact, it is reasonable to conjecture that $P_2(x)$ is the "best

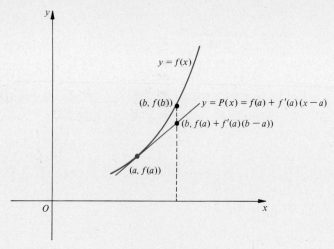

Figure 14.5.1

approximation" to $f(x)$ of all quadratic polynomials in the vicinity of $x = a$. We write

$$P_2(x) = c_0 + c_1(x - a) + c_2(x - a)^2 \qquad (4)$$

where the coefficients c_0, c_1, and c_2 are to be expressed in terms of the given function and its derivatives at $x = a$ [by meeting the requirement that $P_2(a) = f(a)$, $P_2'(a) = f'(a)$ and $P_2''(a) = f''(a)$]. From (4), with x replaced by a, we have $P_2(a) = c_0 = f(a)$. Differentiation of both sides of (4) yields $P_2'(x) = c_1 + 2c_2(x - a)$ so that $P_2'(a) = f'(a) = c_1$. Also, one more differentiation yields $P_2''(x) = 2c_2$ and $P_2''(a) = f''(a) = 2c_2$ or $c_2 = \frac{1}{2}f''(a)$. Summarizing, we have the required quadratic polynomial

$$P_2(x) = f(a) + f'(a)(x - a) + \frac{f''(a)}{2}(x - a)^2 \qquad (5)$$

Note that, due to (3), the curve defined by the quadratic in (5) must possess the same ordinate, slope, and curvature as the given curve $y = f(x)$ at the point $(a, f(a))$.

More generally, we may approximate $f(x)$ in a neighborhood of $x = a$ by a polynomial $P_n(x)$ of degree $\leq n$ which possesses the property that $P_n(a) = f(a)$, $P_n'(a) = f'(a)$, $P_n''(a) = f''(a)$, ..., $P_n^{(n)}(a) = f^{(n)}(a)$. To this end, we write

$$P_n(x) = c_0 + c_1(x - a) + c_2(x - a)^2 + \cdots + c_n(x - a)^n \qquad (6)$$

where the coefficients are easily obtained successively by substituting $x = a$ into (6), then differentiating, substituting $x = a$, differentiating again, and so on. This calculation yields

$$P_n(a) = f(a) = c_0, \ P_n'(a) = f'(a) = c_1, \ P_n''(a) = f''(a) = 2c_2 = 2!c_2,$$
$$P_n'''(a) = f'''(a) = 3!c_3, \ldots, P_n^{(n)}(a) = f^{(n)}(a) = n!c_n$$

Thus we obtain

$$c_k = \frac{P_n^{(k)}(a)}{k!} = \frac{f^{(k)}(a)}{k!} \qquad (k = 0, 1, 2, \ldots, n) \qquad (7)$$

and substitution of (7) into (6) yields the approximating polynomial known as **Taylor's polynomial** †

$$P_n(x) = f(a) + f'(a)(x - a) + \frac{f''(a)}{2!}(x - a)^2 + \cdots + \frac{f^{(n)}(a)}{n!}(x - a)^n \quad (8)$$

By analogy with equations (1) and (2) we may anticipate the following generalization to the basic mean value theorem.

Theorem   If $f(x), f'(x), f''(x), \ldots, f^{(n)}(x)$ are continuous on the closed interval $[a, b]$ and if $f^{(n+1)}(x)$ exists for all x in the open interval (a, b), then there is some point c in the open interval (a, b) such that

$$f(b) = f(a) + \frac{f'(a)}{1!}(b - a) + \frac{f''(a)}{2!}(b - a)^2 + \cdots$$

$$+ \frac{f^{(n)}(a)}{n!}(b - a)^n + \frac{f^{(n+1)}(c)}{(n + 1)!}(b - a)^{n+1} \quad (9)$$

Proof   In order to prove (9) we shall employ a technique similar to that utilized in the proof of the basic mean value theorem of differential calculus—namely, the application of Rolle's theorem. We define a constant K by the equation

$$f(b) = f(a) + \frac{f'(a)}{1!}(b - a) + \frac{f''(a)}{2!}(b - a)^2$$

$$+ \cdots + \frac{f^{(n)}(a)}{n!}(b - a)^n + \frac{K}{(n + 1)!}(b - a)^{n+1} \quad (10)$$

and a function $g(x)$ defined by bringing all terms in (10) onto the same side of the equation and then replacing a by x.

$$g(x) = f(b) - f(x) - \frac{f'(x)}{1!}(b - x) - \frac{f''(x)}{2!}(b - x)^2$$

$$- \cdots - \frac{f^{(n)}(x)}{n!}(b - x)^n - \frac{K}{(n + 1)!}(b - x)^{n+1} \quad (11)$$

The function $g(x)$ is continuous on $[a, b]$. Also $g'(x)$ exists on the open interval (a, b) because $f^{(n+1)}(x)$ exists there. Furthermore, $g(a) = 0$ from (10) and $g(b) = 0$ by direct substitution into (11). Thus $g(x)$ satisfies all the conditions of Rolle's theorem. Therefore $g'(x)$ must be zero for some c in (a, b). If we differentiate (11) term by term and apply the product rule, we obtain

$$g'(x) = -f'(x) + f'(x) - f''(x)(b - x) + f''(x)(b - x)$$

$$- \cdots - \frac{f^{(n+1)}(x)}{n!}(b - x)^n + \frac{K}{n!}(b - x)^n$$

and we observe that the terms will cancel in pairs with the exception of the last two terms. The requirement that $g'(c) = 0$ yields

$$K = f^{(n+1)}(c) \quad (12)$$

and substitution of (12) into (10) gives the required result (9).   ∎

---

† Named in honor of Brook Taylor (1685–1731) an English mathematician who did some important work on infinite series that will be treated in the next chapter.

If in (9), b is replaced by x we obtain **Taylor's theorem** or **Taylor's formula**

$$f(x) = f(a) + \frac{f'(a)}{1!}(x - a) + \frac{f''(a)}{2!}(x - a)^2$$

$$+ \cdots + \frac{f^{(n)}(a)}{n!}(x - a)^n + \frac{f^{(n+1)}(c)}{(n + 1)!}(x - a)^{n+1} \tag{13}$$

where c is some number between a and x.

We may write

$$f(x) = P_n(x) + R_n(x)$$

where $P_n(x)$ is the polynomial of degree at most n defined by (8) and

$$R_n(x) = \frac{f^{(n+1)}(c)}{(n + 1)!}(x - a)^{n+1} \tag{14}$$

where c is between a and x. Hence $f(x)$ is approximated by the **Taylor polynomial** $P_n(x)$ and $R_n(x)$ is the **remainder** or discrepancy between $f(x)$ and $P_n(x)$. The form (14) of the remainder is due to Lagrange and is therefore known as the **Lagrange form** of the remainder.

There are other forms in which the remainder is sometimes written. We shall only mention one of these, **the integral form of the remainder**, which is given by

$$R_n(x) = \frac{1}{n!}\int_a^x (x - t)^n f^{(n+1)}(t)\, dt$$

(See Review and Miscellaneous Exercise 74 at the end of this chapter.)

EXAMPLE 1   Express $x^2 - 5x + 8$ in powers of $x - 3$.

First
Solution   Let $x = (x - 3) + 3$, $x^2 = [(x - 3) + 3]^2$ so that

$$x^2 - 5x + 8 = [(x - 3) + 3]^2 - 5[(x - 3) + 3] + 8$$
$$= (x - 3)^2 + 6(x - 3) + 9 - 5(x - 3) - 15 + 8$$
$$= (x - 3)^2 + (x - 3) + 2$$

Second
Solution   Let $u = x - 3$ so that $x = u + 3$. Hence

$$x^2 - 5x + 8 = (u + 3)^2 - 5(u + 3) + 8$$
$$= u^2 + u + 2 = (x - 3)^2 + (x - 3) + 2$$

Third
Solution   We use (5) directly, where $f(x) = x^2 - 5x + 8$ and $a = 3$. Now, $f(3) = 2$, $f'(x) = 2x - 5$, $f'(3) = 1$, $f''(x) = f''(3) = 2$. Hence from (5)

$$x^2 - 5x + 8 = 2 + 1(x - 3) + \tfrac{2}{2}(x - 3)^2$$
$$= (x - 3)^2 + (x - 3) + 2$$ ●

EXAMPLE 2   Find the Taylor polynomial of nth degree which approximates e^x for x near zero, that is, $a = 0$. Discuss the error function $R_n(x)$ for x near zero.

Solution   We have $f(x) = e^x, f'(x) = e^x, f''(x) = e^x, \ldots, f^{(n)}(x) = e^x$. Thus $f(0) = f'(0) = f''(0) = \cdots = f^{(n)}(0) = e^0 = 1$. Hence the approximating Taylor polynomial is

$$P_n(x) = 1 + \frac{x}{1!} + \frac{x^2}{2!} + \cdots + \frac{x^n}{n!} \qquad (15)$$

By Taylor's theorem (13) and the expression (14) for the remainder $R_n(x)$,

$$R(x) = e^x - P_n(x) = \frac{e^c}{(n+1)!} x^{n+1} \qquad (16)$$

Even though c is not known precisely we can easily estimate $R_n(x)$:

(i) If $x > 0$ then $0 < c < x$ and $e^c < e^x$ (because e^x is an increasing function of x) so that

$$0 < R_n(x) < \frac{e^x x^{n+1}}{(n+1)!} \qquad (17)$$

For example, let us calculate e to within 10^{-5}. In this case, $x = 1$ and if we use 3 as an upper bound for $e^1 = e$ in (17), we seek n such that

$$0 < R_n(1) < \frac{e}{(n+1)!} < \frac{3}{(n+1)!} < 10^{-5}$$

or $3(10^5) < (n+1)!$ and a simple calculation of factorials (or reference to the tables) yields $n \geq 8$. Thus

$$e = 1 + \frac{1}{1!} + \frac{1}{2!} + \frac{1}{3!} + \frac{1}{4!} + \frac{1}{5!} + \frac{1}{6!} + \frac{1}{7!} + \frac{1}{8!} + R_8$$

where $R_8 < 0.00001$. Calculation of this sum to six decimal places yields

$$e = 2.718279 + R_8 \qquad \text{or} \qquad 2.718279 < e < 2.718289$$

(ii) If $x < 0$, then $x < c < 0$ and $e^c < 1$. Hence (16) yields

$$|R_n(x)| < \frac{|x^{n+1}|}{(n+1)!} \qquad (18)$$

Since $R_n(x) = \frac{e^c x^{n+1}}{(n+1)!}$ and $x < 0$, the sign of $R_n(x)$ alternates between negative and positive according as n is even or odd, respectively. This implies that the sequence of polynomials

$$P_0(x) = 1, P_1(x) = 1 + x, P_2(x) = 1 + x + \frac{x^2}{2!}, \ldots$$

will alternately overestimate and underestimate the given function e^x when $x < 0$. ●

EXAMPLE 3   Find the Taylor polynomial of degree $2n$ which approximates $\cos x$ near zero, that is, $a = 0$. Discuss the error function $R_n(x)$.

Solution   We have $f(x) = \cos x$, $f'(x) = -\sin x$, $f''(x) = -\cos x$, $f'''(x) = \sin x$, $f^{(4)}(x) = \cos x$, and then the cycle repeats. Also, $f(0) = 1$, $f'(0) = 0$, $f''(0) = -1$,

$f'''(0) = 0, f^{(4)}(0) = 1$, and so on. Therefore we have from (13)

$$\cos x = 1 - \frac{x^2}{2!} + \frac{x^4}{4!} - \frac{x^6}{6!} + \cdots + \frac{(-1)^n x^{2n}}{(2n)!} + \frac{(-1)^{n+1}(\cos c)x^{2n+2}}{(2n+2)!} \quad (19)$$

By way of an illustration we seek $\cos 1$ to five decimal places. In this case $x = 1$, so that (19) yields

$$\cos 1 = 1 - \frac{1}{2!} + \frac{1}{4!} - \frac{1}{6!} + \cdots + \frac{(-1)^n}{(2n)!} + R_{2n}(1)$$

where if we desire $|R_{2n}(1)| < 10^{-6}$,

$$0 < |R_{2n}(1)| \le \frac{1}{(2n+2)!} < 10^{-6}$$

or equivalently $(2n + 2)! > 10^6$. Evaluation of $(2n + 2)!$ results in $2n + 2 \ge 10$ (since $10! > 10^6 > 9!$, as the reader may readily verify) or $n \ge 4$. Hence

$$\cos 1 = 1 - \frac{1}{2!} + \frac{1}{4!} - \frac{1}{6!} + \frac{1}{8!} + R_8$$

where $|R_8| \le 10^{-6}$. A simple calculation yields $\cos 1 = 0.54030$ to five decimal places. ●

In the same way, it is left for the reader to show that

$$\sin x = x - \frac{x^3}{3!} + \frac{x^5}{5!} - \frac{x^7}{7!} + \cdots + \frac{(-1)^n x^{2n+1}}{(2n+1)!} + R_{2n+1}(x) \quad (20)$$

where

$$|R_{2n+1}(x)| \le \frac{|x|^{2n+3}}{(2n+3)!} \quad (21)$$

EXAMPLE 4 Find the nth degree Taylor polynomial representing $\ln x$ at $x = 10$. Find $R_n(x)$ and determine an upper bound for $R_n(x)$ if $9 \le x \le 11$.

Solution With $f(x) = \ln x$, we have by successive differentiation

$$f'(x) = \frac{1}{x}, f''(x) = -\frac{1}{x^2}, f'''(x) = \frac{2!}{x^3}, f^{(4)}(x) = \frac{-3!}{x^4}$$

and in general

$$f^{(k)}(x) = \frac{(-1)^{k-1}(k-1)!}{x^k}$$

The coefficient of $(x - 10)^k$ in the Taylor polynomial is

$$\frac{f^{(k)}(10)}{k!} = \frac{(-1)^{k-1}(k-1)!}{k!(10^k)} = \frac{(-1)^{k-1}}{k10^k}$$

and we have

$$P_n(x) = \ln 10 + \frac{1}{1(10)}(x - 10) - \frac{1}{2(10)^2}(x - 10)^2$$

$$+ \frac{1}{3(10)^3}(x - 10)^3 + \cdots + \frac{(-1)^{n-1}}{n(10)^n}(x - 10)^n$$

An error estimate in the interval $9 \le x \le 11$ is

$$|R_n(x)| \le \frac{M}{(n + 1)!}|x - 10|^{n+1} \le \frac{M}{(n + 1)!}(1)^{n+1}$$

where M is an upper bound for $|f^{(n+1)}(x)|$ in the given interval. Now, for the given function $f(x) = \ln x$ in $[9, 11]$ we have

$$|f^{(n+1)}(x)| = \left| \pm \frac{n!}{x^{n+1}} \right| \le \frac{n!}{9^{n+1}} \quad \text{in } [9, 11]$$

Thus, with $M = \dfrac{n!}{9^{n+1}}$ we have

$$|R_n(x)| \le \frac{n!}{(n + 1)!(9^{n+1})} = \frac{1}{(n + 1)9^{n+1}}$$

Exercises 14.5

In Exercises 1 through 6, express the given polynomial in powers of x − a.

1. $x^2 - 5x + 11$, $a = 2$
2. $x^2 + x + 6$, $a = 1$
3. x^3, $a = -1$
4. $x^3 - 3x + 5$, $a = -1$
5. $x^3 - 9x^2 + 29x - 13$, $a = 3$
6. x^4, $a = 2$

In Exercises 7 through 21, find the Taylor polynomial of given degree n *for the specified function at the given point. Write the Lagrange form of the remainder.*

7. $f(x) = \cos x$; $a = \dfrac{\pi}{3}, n = 2$
8. $f(x) = \sin x$; $a = \dfrac{\pi}{4}, n = 2$

9. $f(x) = \cos 2x$; $a = 0, n = 2$
10. $f(x) = \sin x$; $a = 0, n = 5$

11. $f(x) = \ln(1 + x)$; $a = 0, n = 5$
12. $f(x) = \sqrt{1 + x}$; $a = 0, n = 3$

13. $f(x) = \dfrac{1}{x}$; $a = 0, n = 2$
14. $f(x) = \tan^{-1} x$; $a = 1, n = 3$

15. $f(x) = \cos x$; $a = \dfrac{\pi}{2}, n = 4$
16. $f(x) = e^{x^2}$; $a = 1, n = 2$

17. $f(x) = \sec x$; $a = \dfrac{\pi}{3}, n = 2$
18. $f(x) = \tan x$; $a = 0, n = 3$

19. $f(x) = x(x - 2)(x - 4)$; $a = 1, n = 3$
20. $f(x) = \sin x$; a arbitrary, $n = 4$

21. $f(x) = \sin^2 x$; $a = 0, n = 4$ $\left(\textit{Hint: } \sin^2 x = \dfrac{1 - \cos 2x}{2} \right)$

22. Find the value of $\sin 5°$ to five decimal places by applying Taylor's formula.
23. Compute the Taylor polynomial of degree 2 and $R_2(x)$ about $x = 0$ for the function $f(x) = (x + 1)^{1/3}$. Find an upper bound to your error in the interval $[0, \frac{1}{2}]$.
24. Use Taylor's formula to compute $\sqrt[3]{28}$ to four decimal places. (*Hint:* $\sqrt[3]{28} = \sqrt[3]{27 + 1} = 3(1 + \frac{1}{27})^{1/3}$).
25. Compute the Taylor polynomial of degree n and $R_n(x)$ at $x = 0$ for the function $\ln(1 - x)$, where $0 \le x < 1$.

26. Use Taylor's formula to compute $(0.99)^8$ correctly to four decimal places.

27. Find the Taylor polynomial $P_3(x)$ at $x = 0$ for the function $e^{ax} \sin bx$, where a and b are constants.

28. Find $\sqrt[4]{e}$ to four-place accuracy by an application of Taylor's formula.

29. Compute $\cos \dfrac{3\pi}{8}$ to four-place accuracy by application of Taylor's formula. (*Hint:* Do not expand at $x = 0$.)

30. Suppose that $f(x)$ is a function such that at $x = a$, $f(a) = f'(a) = \cdots = f^{(n)}(a) = 0$, and $f^{(n+1)}(a) \neq 0$. To be specific, suppose that $f^{(n+1)}(x)$ is continuous and positive in an interval I containing the point $x = a$. Then for any x belonging to I,

$$f(x) - f(a) = f^{(n+1)}(c) \frac{(x - a)^{n+1}}{(n + 1)!}$$

where c is between a and x in I. Thus if $n + 1$ is an even integer, we have $f(x) > f(a)$ if $x \neq a$ in I. This means that $f(a)$ is a relative minimum of $f(x)$ in I. When do we have a relative maximum at $x = a$? When does neither occur?

14.6 ROOT DETERMINATION—APPLICATION OF TAYLOR'S FORMULA

14.6.1 Introduction

The determination of the roots of equations is important in many branches of mathematics and applied mathematics. Thus, for example, we seek a real root or a real solution of the equation $f(x) = 0$. Such a root is called a "zero" of the function. Consider first polynomial equations. If the equation is linear or first degree, the problem is easy. If the equation is quadratic, we can rely on the quadratic formula (which depends on the completion of the square for its derivation) or better still just complete the square in each case. There are also formulas coupled with procedures for polynomial equations of the third and fourth degrees. However these formulas are rarely used in practice because

    (i)   They are cumbersome.
    (ii) There are better ways to proceed.

Quite surprisingly there are no algebraic formulas for the zeros of polynomials of fifth degree or higher. Thus for example, there is no formula for the root(s) of the equation $x^5 - 4x + 2 = 0$.

In addition nonpolynomial or transcendental equations occur very frequently in pure and applied mathematics. Two examples (for which no root finding formulas are available) are

$$3 \sin x - 2x = 0 \qquad \text{and} \qquad e^x - 100x^2 = 0$$

Our main objective in this section is to discuss and develop the Newton–Raphson method for the approximate determination of the roots of algebraic equations and transcendental equations. Because of the computations involved, a calculating machine is indispensible.

14.6.2 Bisection Method

The real roots of $f(x) = 0$ are those values of x at which $f(x)$ crosses the x-axis. Suppose that we find two numbers a and b where $a < b$ such that $f(a) < 0$ and $f(b) > 0$. If furthermore $f(x)$ is continuous in $[a, b]$, then by the intermediate value theorem $f(x)$ must possess a root in this interval (see Figure 14.6.1). Let c be the midpoint of segment $[a, b]$. Compute $f(c)$. Then if $f(c) > 0$ there is a root in $[a, c]$, whereas if $f(c) < 0$ there is a root in $[c, b]$. In either case, let d be the midpoint of the interval in which the function changes sign and repeat the process.

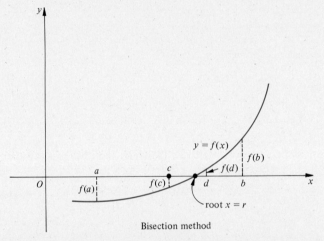

Bisection method

Figure 14.6.1

This procedure is called the **bisection method** because at each stage the root must lie in an interval of width one half that of the preceeding step. Therefore after a sufficient number of steps the root will be known to the required accuracy. In other words any root can be determined by this method; however, the effort involved may be enormous when greater precision is required. This procedure is straightforward and is easily programmed for machine computation.

14.6.3 Newton–Raphson Method

A more efficient method for finding close approximations to the real roots of an equation of the form $f(x) = 0$ is known as the Newton–Raphson method (or, more simply, as the Newton method). Again let us assume that we have found two numbers a and b ($a < b$) such that a and b are "close to each other" and $f(a)$ and $f(b)$ are of opposite signs. If $f(x)$ is continuous in $[a, b]$, then the equation $f(x) = 0$ must possess at least one root, say r, where $a < r < b$. Consider then the graph of $f(x)$ in this interval in the vicinity of r. From the graph (or by simple analytic considerations such as finding the intersection of the chord connecting $(a, f(a))$ and $(b, f(b))$ with the x-axis) we make a first estimate x_0 of r. At the point $(x_0, f(x_0))$ on the graph of the curve $y = f(x)$ draw a tangent line that intersects the x-axis at point $(x_1, 0)$ (Figure 14.6.2). If the value $x = x_0$ is close to the required root $x = r$, Figure 14.6.2 suggests that $x = x_1$ is even closer. The process is then repeated by letting $x = x_1$ play the role that $x = x_0$ played. This means that at $(x_1, f(x_1))$ we

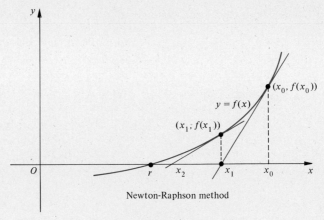

Newton-Raphson method

Figure 14.6.2

draw a tangent line to $y = f(x)$ that intersects the x-axis at $(x_2, 0)$. This iterative process produces a sequence which, in many cases, converges to r with remarkable rapidity.

We next must find the iterative formula that enables us to find x_{i+1} from x_i. Let us start with the point $(x_0, f(x_0))$ on the curve $y = f(x)$. We write an equation of the tangent line to $y = f(x)$ at this point:

$$y - f(x_0) = f'(x_0)(x - x_0)$$

Since $(x_1, 0)$ is on this line, substitution yields

$$0 - f(x_0) = f'(x_0)(x_1 - x_0)$$

from which

$$x_1 = x_0 - \frac{f(x_0)}{f'(x_0)} \tag{1}$$

Since x_{n+1} is obtained from x_n in exactly the same way that x_1 is obtained from x_0, we have in general the recursion formula

$$x_{n+1} = x_n - \frac{f(x_n)}{f'(x_n)} \qquad (n = 0, 1, 2, \ldots) \tag{2}$$

EXAMPLE 1   Find $\sqrt{7}$ to two decimal places.

Solution   We write $f(x) = x^2 - 7$ and consider the curve $y = f(x) = x^2 - 7$. We seek a positive root of $x^2 - 7 = 0$, that is, where the curve crosses the x-axis (and $x > 0$). We know that our root is between 2 and 3 and a little closer to $x = 3$. We therefore choose $x_0 = 3$, a crude first approximation. The quantity x_1 can be calculated from equation (1). In this instance,

$$f(x) = x^2 - 7 \qquad f(3) = 2$$
$$f'(x) = 2x \qquad f'(3) = 6$$

so that

$$x_1 = 3 - \tfrac{2}{6} = \tfrac{8}{3} \doteq 2.67$$

To obtain the next approximation x_2 we use the recurrence formula (2) with $n = 1$:

$$x_2 = x_1 - \frac{f(x_1)}{f'(x_1)}$$

where

$$f(x_1) = f\left(\frac{8}{3}\right) = \left(\frac{8}{3}\right)^2 - 7 = \frac{1}{9}$$

$$f'(x_1) = 2x_1 = 2\left(\frac{8}{3}\right) = \frac{16}{3}$$

Therefore

$$x_2 = \frac{8}{3} - \frac{\frac{1}{9}}{\frac{16}{3}} = \frac{8}{3} - \frac{1}{48} = \frac{127}{48} \doteq 2.6458$$

Now, $(2.645)^2 = 6.996025$ and $(2.646)^2 = 7.001316$; therefore, it follows that the positive root must be between 2.645 and 2.646. Consequently 2.65 is the required root to two decimal places. Actually

$$\sqrt{7} \doteq 2.645751 \quad \text{(to six decimal places)}$$

and our x_2 approximates the root to four significant figures after the decimal point and represents an excellent approximation for most practical purposes. ●

EXAMPLE 2   Show that the equation $x^3 + 3x - 1 = 0$ has only one real root and find it correct to three decimal places.

Solution   Every polynomial equation of odd degree must have at least one real root. If $f(x) = x^3 + 3x - 1$, then $f'(x) = 3x^2 + 3$, which is positive. Therefore $f(x)$ is a monotonically increasing function of x and has precisely one zero.

Now, $f(0) = -1$ whereas $f(1) = 3$ so that $0 < r < 1$, where r is the root we are seeking. Taking into account the values of the function, we choose $x_0 = 0.25$ as our initial guess.

$$f(0.25) = 0.015625 + 0.75 - 1 = -0.234375$$
$$f'(0.25) = 3(0.25)^2 + 3 = 3.1875$$

Therefore

$$x_1 = x_0 - \frac{f(x_0)}{f'(x_0)} = 0.25 + \frac{0.234375}{3.1875}$$

$$\doteq 0.25 + 0.0735 \doteq 0.32 \qquad \text{(to two decimal places)}$$

$$f(0.32) = 0.032768 + 0.96 - 1 = -0.007232$$
$$f'(0.32) = 3(0.32)^2 + 3 = 3.0372$$

$$x_2 = x_1 - \frac{f(x_1)}{f'(x_1)} = 0.32 + \frac{0.007232}{3.0372} \doteq 0.322$$

$$f(0.322) = (0.322)^3 + 3(0.322) - 1$$
$$= 0.033386 + 0.966 - 1 \doteq -0.0006$$
$$f'(0.322) = 3(0.322)^2 + 3 \doteq 3.311$$

$$x_3 = x_2 - \frac{f(x_2)}{f'(x_2)} = 0.322 + \frac{0.0006}{3.311}$$

$$\doteq 0.322 + 0.00018$$

$$= r \doteq 0.322 \qquad \text{(correct to three decimal places)}$$

It is easy to check that $f(0.3215) < 0$ while $f(0.3225) > 0$, which verifies that our solution is indeed correct to three decimal places. ●

EXAMPLE 3 The equation $x^3 - e^x = 0$ has a root in the interval $[1, 2]$. Find it correct to two decimal places.

Solution A root occurs at the intersection of the graph of $y = e^x$ and $y = x^3$. A sketch of these functions shows that we have a root at about $x_0 = 1.9$. Then if

$$f(x) = x^3 - e^x$$

$$f(1.9) = 6.859 - 6.6859 = 0.1731$$

$$f'(x) = 3x^2 - e^x$$

$$f'(1.9) = 10.83 - 6.6859 = 4.1441$$

Therefore

$$x_1 = x_0 - \frac{f(x_0)}{f'(x_0)} = 1.9 - \frac{0.1731}{4.1441} \doteq 1.9 - 0.042$$

$$\doteq 1.858$$

$$\doteq 1.86$$

$$f(1.86) = 6.434856 - 6.4237 \doteq 0.0111$$
$$f'(1.86) \doteq 10.3788 - 6.4237 = 3.9551$$

$$x_2 = x_1 - \frac{f(x_1)}{f'(x_1)} = 1.86 - \frac{0.0111}{3.9551}$$

$$\doteq 1.86 - 0.003 = 1.857$$

so that the root $r = 1.86$ correct to two decimal places. ●

EXAMPLE 4 (A particular function and nonconvergence). In the preceding examples the approximations x_n to the root r improved with increase in n. However, this will not always occur. For instance, if one considers the function

$$f(x) = \frac{4x}{x^2 + 3}$$

which

(i)   is odd
(ii) has a unique zero at $x = 0$
(iii) has an inflection point at $(0, 0)$, and
(iv) $f'(\pm\sqrt{3}) = 0$

then it is easily shown that, if x_0 is chosen to be 1, we have $x_1 = -1$, $x_2 = 1$, $x_3 = -1$, and so on. (Figure 14.6.3(a)). Furthermore, if x_0 is chosen so that

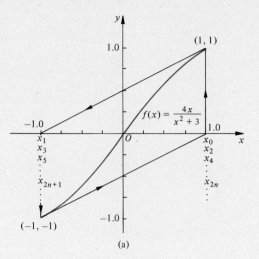

Figure 14.6.3(a)

$1 < x_0 < \sqrt{3}$, we find $|x_1| > |x_0|$ and $|x_2| > |x_1| > |x_0|$, and so on. (Figure 14.6.3(b)).

●

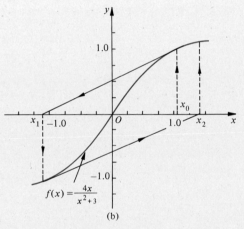

Figure 14.6.3(b)

This disturbing result of Example 4 makes the following theorem all the more significant. This theorem provides a bound for the error at any step in the Newton–Raphson process. Furthermore, it also establishes sufficient conditions for the convergence of this process to the exact root and enables us to estimate this rate of convergence.

Theorem Let $f(x)$ be a twice differentiable function of x in an interval I that contains a point $x = r$ where $f(x) = 0$. Let x_n be an nth approximation to r in the Newton–Raphson method such that x_n is also in I. Furthermore it is assumed that *for all x in I*

(i) $f'(x) \neq 0$
(ii) $f''(x)$ is one-signed, that is either $f''(x) > 0$ or $f''(x) < 0$

(iii) positive numbers m and M exist that are lower and upper bounds for $|f'(x)|$ and $|f''(x)|$, respectively; or, symbolically, $m \le |f'(x)|$ and $|f''(x)| \le M$.

Then

$$|r - x_{n+1}| \le \frac{M}{2m}|r - x_n|^2$$

Proof In accordance with the Newton–Raphson procedure the $(n + 1)$ approximation, x_{n+1}, is obtained from

$$x_{n+1} = x_n - \frac{f(x_n)}{f'(x_n)}$$

Subtract the root r from both sides to obtain

$$x_{n+1} - r = x_n - r - \frac{f(x_n)}{f'(x_n)} \tag{3}$$

From the extended mean value theorem (Section 14.5)

$$f(r) = f(x_n) + (r - x_n)f'(x_n) + \frac{(r - x_n)^2}{2}f''(z) \tag{4}$$

where z is some number between the numbers x_n and r. However, we know that $f(r) = 0$ and therefore relation (4) may be written in the form

$$-f(x_n) = (r - x_n)f'(x_n) + \frac{(r - x_n)^2}{2}f''(z) \tag{5}$$

Substitution of the right member of (5) for $-f(x_n)$ into equation (3) yields

$$x_{n+1} - r = x_n - r + \frac{(r - x_n)f'(x_n) + \dfrac{(r - x_n)^2 f''(z)}{2}}{f'(x_n)}$$

$$= \frac{(r - x_n)^2}{2}\frac{f''(z)}{f'(x_n)} \tag{6}$$

Therefore if we take the absolute value of both sides of (6), the equality remains

$$|r - x_{n+1}| = \frac{|r - x_n|^2}{2}\left|\frac{f''(z)}{f'(x_n)}\right|$$

Application of the two inequalities from assumptions (iii) yields the required result

$$|r - x_{n+1}| \le \frac{M}{2m}|r - x_n|^2 \tag{7}$$

∎

A useful corollary may be obtained by rewriting (6) in the form

$$(x_{n+1} - r) = \frac{x_n - r}{2}\frac{(x_n - r)f''(z)}{f'(x_n)} \tag{8}$$

and from the fact that $f''(x)$ is one-signed, the sign of $f''(x_n)$ is the same as $f''(z)$. This means that $x_{n+1} - r$ and $x_n - r$ have the same or opposite signs when $(x_n - r)f'(x_n)f''(x_n)$ is positive or negative, respectively.

Equation (7) is easily interpreted. The quantity $|r - x_n|$ is the magnitude of the error in the nth approximation, whereas $|r - x_{n+1}|$ is the magnitude of the error in the $(n + 1)$ approximation. Suppose that M and m are such that $\dfrac{M}{2m} < 2$ for all x in I. Then from (7) we deduce that the number of decimal places of accuracy at each stage is at least twice that in the previous approximation once one or more decimal place accuracy has been achieved; that is, $|r - x_n| < 0.05$. As an example, suppose that x_2 is accurate to three decimal places. This implies that $|r - x_2| < 0.0005$ and (7) states that the magnitude of the error in x_3 is less than $2(0.0005)^2 = 0.0000005$.

Exercises 14.6

In Exercises 1 through 4 estimate the root (or roots) to two decimal places by application of the bisection method.

1. $x^2 = 10$
2. $x^3 = 11$
3. $x^3 + x - 12 = 0$
4. $x^3 + 3x^2 + 6x - 1 = 0$

In Exercises 5 through 10 verify that $f(a)$ and $f(b)$ are of opposite sign and use Newton's method to approximate the root of $f(x) = 0$, which lies between a and b, to three decimal places.

5. $f(x) = x^2 - x - 3; \quad a = 2, b = 3$
6. $f(x) = x^2 - 19; \quad a = 4, b = 5$
7. $f(x) = x^3 - 8x - 8; \quad a = 3, b = 4$
8. $f(x) = x^3 + 2x - 4; \quad a = 1, b = 2$
9. $f(x) = 3x^3 - 2x^2 + 6x - 4; \quad a = 0, b = 1$
10. $f(x) = x^3 + x^2 - 4x + 5; \quad a = -3, b = -2$
11. Using Newton's method, find $\sqrt[3]{6}$ to three decimal places.
12. Using Newton's method, find $\sqrt[4]{316}$ to three decimal places.
13. Let b be an arbitrary positive number and suppose that $\sqrt{b}$ is required. Prove that Newton's recurrence formula for estimating $\sqrt{b}$ is given by

$$x_{i+1} = \frac{1}{2}\left(x_i + \frac{b}{x_i}\right)$$

that is, x_{i+1} is the arithmetic mean of x_i and $\dfrac{b}{x_i}$.

14. Use the result of Exercise 13 to estimate $\sqrt{10}$. Start with $x_0 = 3$ and find x_1 and x_2 to three decimal places.
15. Let b be an arbitrary real number and suppose that $\sqrt[3]{b}$ is required. Prove that Newton's recurrence formula for estimating $\sqrt[3]{b}$ is given by

$$x_{i+1} = \frac{2}{3}x_i + \frac{1}{3}\left(\frac{b}{x_i^2}\right)$$

That is, x_{i+1} is the "weighted" average of x_i and $\dfrac{b}{x_i^2}$ with x_i being weighted twice as heavily as $\dfrac{b}{x_i^2}$.

16. Use the result of Exercise 15 to estimate $\sqrt[3]{40}$. Start with $x_0 = 4$ and find x_1 and x_2 to three decimal places.
17. Generalize the procedures of Exercises 13 and 15 to find $b^{1/n}$, where b is a positive number and n is a positive integer > 1. Show that x_{i+1} is a "weighted average" of x_i and $\dfrac{b}{x_i^{n-1}}$; that is, find coefficients k

and g such that

$$x_{i+1} = kx_i + g\left(\frac{b}{x_i^{n-1}}\right)$$

where k and g are positive numbers for which $k + g = 1$. In particular, find x_{i+1} if $x_i = b/x_i^{n-1}$. (*Hint:* $f(x) = x^n - b$.)

18. Estimate $\sqrt[5]{150}$. Start with $x_0 = 3$ and find x_1 and x_2 to three decimal places.

For each of Exercises 19 through 24, use Newton's method to find the indicated root to two decimal places.

19. $2 \cos x = 3x$; real root
20. $e^x = 3 - x$; real root
21. $x \sin x = 0.60$; positive root
22. $2 \tan x = 3x$; smallest positive root
23. $\cosh x = 2x + 3$; negative root, choose $x_0 = -1$
24. $\ln \dfrac{1}{x} = x - 2$; real root

25. Suppose that we are given a function

$$f(x) = \begin{cases} \sqrt{x - r} & \text{for } x \geq r \\ -\sqrt{x - r} & \text{for } x \leq r \end{cases}$$

that has a unique zero at $x = r$. Let $x_0 = r + h$ where $h > 0$ be our first approximation to $x = r$. Find x_1 and x_2 in a neighborhood of $x = r$ using Newton's method. Interpret your result geometrically.

26. Find the area bounded by the two curves $y = e^x$ and $y = 4x^2$. (*Hint:* Find the points of intersection of the two curves to three decimal places and then estimate the area using linear interpolation for the values of e^x.)

In Exercises 27 through 30, find the smallest positive critical number for the given function $g(x)$; that is, the least positive root of $g'(x) = 0$. Accuracy is required to the specified number of decimal places. Use Newton's method.

27. $g(x) = 2x^2 - \sin 2x$; 2 places
28. $g(x) = x^4 - 6x^3 + x^2 + 24x + 17$; 4 places
29. $g(x) = x^6 - 240x + 18$; 3 places
30. $g(x) = x^2 + 20e^{-x} + 8x + 7$; 2 places

Review and Miscellaneous Exercises

In Exercises 1 through 28, evaluate the limit, if it exists.

1. $\displaystyle\lim_{t \to 0} \frac{4 \sin^2 t - t^2}{2t^2}$

2. $\displaystyle\lim_{u \to 0} \frac{\sqrt{49 + 3u} - 7}{u}$

3. $\displaystyle\lim_{x \to \infty} \sqrt{x^2 + 1} \sinh \frac{1}{x}$

4. $\displaystyle\lim_{t \to \infty} \frac{t^5 \ln^3 t}{e^{2t}}$

5. $\displaystyle\lim_{x \to \infty} \left(1 + \frac{10}{x^2}\right)^{x^2}$

6. $\displaystyle\lim_{t \to 3^+} \left(\frac{1}{t - 3} - \frac{1}{\sqrt{t - 3}}\right)$

7. $\displaystyle\lim_{t \to e} \frac{\ln t^2 - 2}{t^2 - e^2}$

8. $\displaystyle\lim_{x \to \infty} \frac{e^{2x} - xe^x}{3e^{2x} + x^2 \ln x}$

9. $\displaystyle\lim_{x \to \infty} \frac{\cosh^2 x}{e^{2x} + x^3}$

10. $\displaystyle\lim_{x \to 1} \frac{1 - x^m}{1 - x^n}$, where $m > 0$ and $n > 0$

11. $\displaystyle\lim_{x \to 4} \frac{48 - 40x + 11x^2 - x^3}{16 + 8x - 7x^2 + x^3}$

12. $\displaystyle\lim_{x \to 0} \frac{\sin (2\pi \cos x)}{\sin^2 x}$

13. $\displaystyle\lim_{x \to 3} \frac{x(x^3 - 7x^2 + 16x - 12)}{x^3 - 7x^2 + 15x - 9}$

14. $\displaystyle\lim_{h \to 0} \frac{\tan (x + 2h) - \tan (x - h)}{h}$

15. $\displaystyle\lim_{h \to 0} \frac{\cos (x - h) - \cos (x + 3h)}{h}$

16. $\displaystyle\lim_{x \to 0} \frac{(e^{2x} - 1) \sin x}{\sinh^2 x}$

17. $\displaystyle\lim_{x \to \pi/2} \frac{(1 - \sin x) \cos^2 x}{(\cos x - 1)^2 \cosh x}$

18. $\displaystyle\lim_{t \to 0} \frac{8^t + 6^t - 2}{e^t - 1}$

19. $\displaystyle\lim_{x \to 0} \frac{e^x + e^{2x} + e^{3x} - (3 + 6x)}{x^2}$

20. $\displaystyle\lim_{x \to 0} \frac{a^{2x} - b^{2x}}{x}$, where a and b are positive constants

21. $\displaystyle\lim_{x \to 0} \left(\frac{\sin kx}{kx}\right)^{1/x}$, where k is a positive constant

22. $\displaystyle\lim_{u \to \infty} (u - \sqrt{u^2 - u + 10})$

23. $\displaystyle\lim_{t \to \infty} \frac{3t^4 - 7t^3 + 9t - 8}{(4t^2 - 1)^2}$

24. $\displaystyle\lim_{u \to \infty} \frac{\sqrt[3]{4u^3 + 2u^2 - u + 3}}{5u - 7}$

25. $\displaystyle\lim_{x \to 0} \left(\frac{\csc x}{x} - \frac{1}{x^2}\right)$

26. $\displaystyle\lim_{x \to 0} x^2(\csc^2 x)(\csc x - \cot x)$

27. $\displaystyle\lim_{x \to \infty} \frac{(\ln x)^n}{x}$, where n is a positive integer

28. $\displaystyle\lim_{h \to 0} \frac{f(x + 3h) - 3f(x + 2h) + 3f(x + h) - f(x)}{h^3}$, assuming that $f'''(x)$ is continuous

29. Prove that $\displaystyle\lim_{x \to \infty} \frac{P_n(x)}{e^x} = 0$ where $P_n(x)$ is a polynomial in x of degree n.

30. The function $g(x) = \dfrac{\sinh 3x - 3x}{x^3}$ is undefined at $x = 0$. Choose $g(0)$ so that $g(x)$ is continuous at $x = 0$.

31. It would be expected that for any $a > 0$,

$$\lim_{k \to 1} \int_1^a \frac{dx}{x^k} = \int_1^a \frac{dx}{x}$$

Prove that this is true.

32. Find $\displaystyle\lim_{x \to \infty} x^2 e^{-x^3} \int_0^x e^{t^3}\, dt$.

In Exercises 33 through 44 find the value of the improper integral if it exists.

33. $\displaystyle\int_{-\infty}^5 \frac{dx}{(x - 8)^3}$

34. $\displaystyle\int_2^\infty \frac{\ln t}{t^3}\, dt$

35. $\displaystyle\int_{-2}^\infty \frac{dz}{z^2 + 4z + 7}$

36. $\displaystyle\int_{-\infty}^\infty \frac{t^3}{t^4 + 25}\, dt$

37. $\displaystyle\int_0^\infty e^{ax} x^2\, dx$, $(a < 0)$

38. $\displaystyle\int_0^\infty e^{ax} \cos bx\, dx$, $(a < 0)$

39. $\displaystyle\int_0^\infty x^4 e^{-x^2}\, dx$, where we may use $\displaystyle\int_0^\infty e^{-x^2}\, dx = \frac{\sqrt{\pi}}{2}$

40. $\displaystyle\int_{-2}^2 \frac{dx}{\sqrt{1 - x^2}}$

41. $\displaystyle\int_0^{\pi/2} \frac{\sin x}{\sqrt{1 - \cos x}}\, dx$

42. $\displaystyle\int_e^1 \frac{dz}{z\sqrt{\ln z}}$

43. $\int_{-1}^{4} f(x)\, dx$, where $f(x) = \begin{cases} x^2 - 3x & \text{if } -1 \leq x < 0 \\ \sqrt{x} & \text{if } 0 \leq x \leq 4 \end{cases}$

44. $\int_{0}^{2} g(x)\, dx$, where $g(x) = \begin{cases} x^{-1/2} & \text{if } 0 \leq x < 1 \\ 3 & \text{if } x = 1 \\ \dfrac{x+1}{x^{1/2}} & \text{if } 1 < x \leq 2 \end{cases}$

*In Exercises 45 through 53 test the integral for convergence or divergence. Do **not** evaluate the convergent integrals.*

45. $\displaystyle\int_{0}^{\pi/2} \frac{dx}{\sqrt[3]{x}\,\sin x}$

46. $\displaystyle\int_{0}^{\infty} \frac{x^2 + x + 3}{e^x}\, dx$

47. $\displaystyle\int_{0}^{\infty} \frac{dx}{\sqrt{x^3 + 5x + 3}}$

48. $\displaystyle\int_{0}^{\infty} \frac{x^2 + 5x - 9}{\cosh x}\, dx$

49. $\displaystyle\int_{2}^{\infty} \frac{1}{\sqrt{x}(x+3)}\, dx$

50. $\displaystyle\int_{1}^{\infty} \frac{\ln x}{\sqrt{x}(x+3)}\, dx$

51. $\displaystyle\int_{0}^{\infty} \frac{e^{2x}\sinh x}{e^{x^2}}\, dx$

52. $\displaystyle\int_{10}^{\infty} \frac{\sqrt{x}}{\sqrt{x^3 + x + 4}}\, dx$

53. $\displaystyle\int_{1}^{\infty} \frac{dx}{x(x+a)(x+b)}$, where $0 < a < b$.

54. The **gamma function** $\Gamma(x)$ is defined by the improper integral

$$\Gamma(x) = \int_{0}^{\infty} t^{x-1} e^{-t}\, dt \qquad \text{for } x > 0$$

(a) Why do we impose the restriction that $x > 0$? (*Hint:* consider $\int_{0}^{1} t^{x-1}e^{-t}\, dt$ and $\int_{1}^{\infty} t^{x-1}e^{-t}\, dt$.)

(b) Show that integration by parts yields the recurrence formula $\Gamma(x+1) = x\Gamma(x)$, for $x > 0$.

(c) Show that $\Gamma(1) = 1$, $\Gamma(2) = 1$, $\Gamma(3) = 2 = 2!$, and in general for arbitrary positive integers n, $\Gamma(n) = (n-1)!$ (recall that $0! = 1$).

55. Clarify the following discussion

$$\int_{-1}^{1} x^{-2}\, dx = \frac{x^{-1}}{-1}\Bigg|_{-1}^{1} = (-1) - (+1) = -2$$

However, the integrand x^{-2} is positive, which implies that $\int_{-1}^{1} x^{-2}\, dx$ should also be positive. What is wrong?

56. Find an upper bound for the value of $\displaystyle\int_{2}^{\infty} \frac{dx}{3x^6 + 1}$.

57. Find the value of $\displaystyle\int_{0}^{\infty} x(1 + x^2)^{-2}\, dx$. What is the value of $\displaystyle\int_{0}^{\infty} x(1 + x^2)^{-1}\, dx$?

58. Evaluate $\displaystyle\int_{0}^{\infty} e^{-x}\sin 2x\, dx$

In Exercises 59 through 61 test the improper integral for convergence.

59. $\displaystyle\int_{0}^{\infty} |e^{-x}\cos x|\, dx$

60. $\displaystyle\int_{1}^{\infty} \frac{\tanh^3 x}{x^{3/2}}\, dx$

61. $\displaystyle\int_{2}^{\infty} \frac{dx}{x - \sqrt{x}}$

62. Find $\sin 1$ to four place accuracy by application of Taylor's formula.

In Exercises 63 through 71, find the Taylor polynomial of nth degree (at most) which approximates the function for the given values of a and n.

63. $f(x) = x^3 - 3x + 1$; $a = -1, n = 1, 2, 3, 6$.

64. $f(x) = \dfrac{1}{x + 2}$; $a = 0, n = 0, 1, 2, 3$.

65. $f(x) = \dfrac{1}{1 + x^2}$; $a = 0, n = 0, 1, 2$.

66. $f(x) = \sec x$; $a = 0, n = 0, 2, 4$.

67. $f(x) = \sinh x$; $a = 0, n = 1, 3, 5$.

68. $f(x) = \cosh bx$; $a = 0, n = 0, 2, 4, 6$.

69. $f(x) = \ln x$; a an arbitrary positive number, $n = 5$.

70. $f(x) = \dfrac{1}{x + b}$; $b \neq 0, a = 0, n = 4$.

71. $f(x) = \dfrac{7 - x}{3 + x}$; $a = 2, n = 4$.

72. The Taylor approximation $P_n(x)$ (of degree $\leq n$) to a function f is often called "the polynomial of best fit" for this function f in the vicinity of $x = a$. This is because it can be shown that the difference $f(x) - P_n(x)$ tends to zero more rapidly than $(x - a)^n$. Furthermore, $P_n(x)$ is the only polynomial of degree $\leq n$ that possesses this property. Show that, if $f(x)$ has a continuous derivative of order $n + 1$ in a closed interval I containing $x = a$ then

$$\lim_{x \to a} \frac{f(x) - P_n(x)}{(x - a)^n} = 0$$

73. Given the function

$$F(x) = \begin{cases} e^{-1/x^2} & x \neq 0 \\ 0 & x = 0 \end{cases}$$

Show that $F'(0) = 0$ since $\lim\limits_{x \to 0} \dfrac{e^{-1/x^2}}{x} = 0$

Furthermore, show that $\lim\limits_{x \to 0} \dfrac{e^{-1/x^2}}{x^n} = 0$, where n is an arbitrary positive integer. Hence show that $F^{(n)}(0) = 0$. How well would Taylor's polynomial of degree 10 approximate $F(x)$?

74. Derive the *integral formula for the remainder* in Taylor's theorem, namely, that if

$$f(x) = P_n(x) + R_n(x)$$

where $P_n(x)$ is the Taylor polynomial of degree n, then

$$R_n(x) = \frac{1}{n!} \int_a^x f^{(n+1)}(t)(x - t)^n \, dt$$

It is assumed that $f^{(n+1)}(x)$ is continuous in an interval I containing the points a and x in its interior. (Hint: Note that $\int_a^x f'(t) \, dt = f(t) \Big|_a^x = f(x) - f(a)$ or $f(x) = f(a) + \int_a^x f'(t) \, dt$. Use integration by parts and establish by mathematical induction, recognizing that we have established the formula for $n = 0$.)

75. Prove as a corollary to Exercise 74 that

$$|R_n(x)| \leq \frac{M}{(n + 1)!} |x - a|^{n+1}$$

where M is the maximum of $|f^{(n+1)}(t)|$ for t in $[a, x]$.

76. The function $f(x) = 1 - \dfrac{x^2}{2} - \cos x$ is such that $f'(0) = 0$. Determine whether or not $f(0)$ is a relative extremum of $f(x)$. (*Hint:* Apply Exercise 30 of Section 14.5.)

77. Find $\sqrt[3]{2}$ to three decimal places by Newton's method using 1.20 as a first approximation.

78. Find to two decimal places the abscissa of the point of intersection of the two curves $2x = e^y$ and $y = 2(1 - x)$.

79. Find the root of the equation $x^3 + 2x - 5 = 0$ to three decimal places by Newton's method. How do you know that there is a root and that it is unique?

80. Find the real root(s) of $x^3 - 3x^2 + 4x - 10 = 0$ to three decimal places.

81. A spherical segment of one base has volume

$$V = \frac{\pi h^2}{3}(3r - h)$$

where r is the radius of the sphere and h is the height of the segment. Determine h to three decimal places if $V = 30 \text{ ft}^3$ and $r = 3$ ft.

82. A function $y = f(x)$ satisfies the initial value problem $y'' + y = 0$, $y(0) = 0$, $y'(0) = 0$ in an arbitrary interval $[a, b]$ containing $x = 0$ in its interior. Find $f(x)$ by applying Taylor's theorem. (*Hint:* Differentiate the given differential equation and show that the required function must satisfy $f^{(n)}(0) = 0$ for all nonnegative integers n.)

15
Sequences and Series

15.1   SEQUENCES

We have been considering functions of a continuous variable defined on an interval or the union of such intervals. However, many cases occur in mathematics where the functions are defined on the set of positive integers. Such functions $a(n)$ are called sequences and, more precisely, infinite sequences, when n ranges over all the positive integers.

Definition 1   A function that takes on real values and whose domain is the set of positive integers is called a **sequence of real numbers** or simply a **sequence.**

Thus the functions defined for each positive integer by

$$a(n) = n^2 + 1, \qquad b(n) = \frac{\cos n\pi}{n}, \qquad c(n) = e^{1/n}, \qquad d(n) = \frac{n^2}{3^n}$$

are illustrations of sequences. For example, $a(1) = 1^2 + 1 = 2$, $a(2) = 2^2 + 1 = 5, \ldots,$ while $b(1) = \frac{\cos \pi}{1} = -1,$ $b(2) = \frac{\cos 2\pi}{2} = \frac{1}{2},$ $b(3) = \frac{\cos 3\pi}{3} = -\frac{1}{3},$ and so forth.

The sum of the first n positive integers

$$S(n) = 1 + 2 + 3 + \cdots + n = \frac{n(n+1)}{2}$$

is a function of n that yields the sequence

$$1, 3, 6, 10, 15, 21, \ldots, \frac{n(n+1)}{2}, \ldots$$

This sequence may be listed conveniently as $\left\{\dfrac{n(n+1)}{2}\right\}$ because the domain of the sequence is always the set of natural numbers. We also find it useful to use subscript notation $\{a_n\}$ to denote the elements of the sequence for which $a(n) = a_n$. Thus $\{a_n\}$ stands for the entire sequence $\{a_1, a_2, a_3, \ldots\}$. We call a_n the "nth term of the sequence".

A sequence is defined by the elements which are ordered in a precise way. Thus the sequence is *not* determined by its elements alone. Hence we distinguish between

$$\{a_n\} = \left\{\frac{1}{2n-1}\right\} = \left\{1, \frac{1}{3}, \frac{1}{5}, \frac{1}{7}, \ldots\right\}$$

and

$$\{b_n\} = \begin{cases} \dfrac{1}{n} \text{ if } n \text{ is odd} \\ 1 \text{ if } n \text{ is even} \end{cases} \quad = \left\{1, 1, \frac{1}{3}, 1, \frac{1}{5}, 1, \ldots\right\}$$

because, even though both sequences contain the same numbers, the second sequence has the terms in a different order and the number 1 keeps repeating.

Definition 2

A sequence $\{a_n\}$ is said to be

(i) **Increasing** if and only if $a_{n+1} > a_n$ for each positive integer n.
(ii) **Nondecreasing** if and only if $a_{n+1} \geq a_n$ for each positive integer n.
(iii) **Decreasing** if and only if $a_{n+1} < a_n$ for each positive integer n.
(iv) **Nonincreasing** if and only if $a_{n+1} \leq a_n$ for each positive integer n.

If a sequence satisfies any one of the above definitions it is said to be **monotonic**.

Hence the sequences $\{1, \frac{1}{3}, \frac{1}{5}, \frac{1}{7}, \ldots\}$, $\{1, 1, \frac{1}{2}, \frac{1}{2}, \frac{1}{3}, \frac{1}{3}, \frac{1}{4}, \frac{1}{4}, \ldots\}$, and $\{1, 4, 9, 16, \ldots\}$ are monotonic, whereas $\{1, 1, \frac{1}{3}, 1, \frac{1}{5}, 1, \ldots\}$ is not.

Definition 3

A sequence $\{a_n\}$ is said to have an **upper bound** M if

$$a_n \leq M \quad \text{for all } n.$$

Similarly, a sequence $\{a_n\}$ is said to have a **lower bound** m if

$$m \leq a_n \quad \text{for all } n.$$

Once an upper bound M is known then any number $\geq M$ is also an upper bound. Analogously, if a lower bound m has been determined then any number $\leq m$ will also serve as a lower bound.

Definition 4

A sequence $\{a_n\}$ is said to be **bounded** if and only if it has an upper and a lower bound.

Now let us consider some less obvious examples.

EXAMPLE 1 Show that the sequence $\{a_n\} = \left\{\dfrac{n}{n+2}\right\}$ is an increasing sequence of positive numbers which is bounded below by $\frac{1}{3}$ and bounded above by 1.

Solution This result appears to be correct if we inspect the terms of the sequence: $\frac{1}{3}, \frac{2}{4}, \frac{3}{5}, \frac{4}{6}, \frac{5}{7}, \ldots$

We prove the result formally by noting that

$$a_n = \frac{n}{n+2} \qquad \text{and} \qquad a_{n+1} = \frac{n+1}{n+3}$$

Thus

$$a_{n+1} - a_n = \frac{n+1}{n+3} - \frac{n}{n+2} = \frac{(n^2 + 3n + 2) - (n^2 + 3n)}{(n+3)(n+2)}$$

$$= \frac{2}{(n+3)(n+2)}$$

and therefore the right side is positive for all natural numbers n. Thus $a_{n+1} - a_n > 0$ or $a_{n+1} > a_n$ for all positive integers n. Furthermore, $a_1 = \dfrac{1}{3}$ whereas $a_n = \dfrac{n}{n+2} < 1$ for all positive integers n. This yields the inequalities

$$\tfrac{1}{3} \le a_n < 1 \qquad \text{for all positive integers } n. \qquad \bullet$$

EXAMPLE 2 Show that the sequence $\{a_n\} = \left\{\dfrac{3^n}{n!}\right\}$ is *not* a monotonic sequence of positive numbers.

Solution
$$a_{n+1} = \frac{3^{n+1}}{(n+1)!} = \frac{3^n}{n!} \frac{3}{n+1} = \frac{3}{n+1} a_n$$

is the recurrence formula relating $\{a_{n+1}\}$ and $\{a_n\}$. Thus by letting $n = 1$ and 2 we find that

$$a_2 = \frac{3}{2} a_1 \qquad a_3 = a_2$$

and from the recurrence formula, $a_{n+1} < a_n$ if $n \ge 3$. Also, $a_n = \dfrac{3^n}{n!} > 0$ for each positive integer. Hence $a_1 < a_2 = a_3 > a_4 > a_5 \ldots$. The specific numbers are

$$\{3, \tfrac{9}{2}, \tfrac{9}{2}, \tfrac{27}{8}, \tfrac{81}{40}, \ldots\} \qquad \bullet$$

EXAMPLE 3 Show that the sequence $\{a_n\} = \{|r|^n\}$ is bounded if $-1 \le r \le 1$ and is unbounded if $|r| > 1$.

Solution If $-1 \le r \le 1$ the sequence is bounded by 0 and 1; that is, $0 \le |r|^n \le 1$.

If $|r| > 1$ we write $|r| = 1 + h$ where $h > 0$. Now,

$$a_n = (1 + h)^n \geq 1 + nh$$

(see Example 4 in Section 1.9 on mathematical induction or apply the binomial theorem). But the expression $1 + nh$ becomes unbounded as n increases so that the same must be true for $\{a_n\}$. ●

Since the sequences are defined on the set of positive integers only, rather than on an interval, the methods of calculus may not be applied directy to them. However, this problem may be surmounted by applying calculus to functions that take on the same functional values as the sequence at the positive integers. This technique will be illustrated in our next example.

EXAMPLE 4 Show that the sequence $\left\{ \dfrac{\ln n}{n} \right\}$ is a decreasing sequence if $n \geq 3$.

Solution We set $f(x) = \dfrac{\ln x}{x}$, $x > 0$ so that

$$f'(x) = \frac{1 - \ln x}{x^2}$$

and the right side is negative if $\ln x > 1$ or, equivalently, $x > e$. Since $2 < e < 3$, it follows that $f'(x) < 0$ when $x \geq 3$. Hence $\left\{ \dfrac{\ln n}{n} \right\}$ is a decreasing sequence for $n \geq 3$. To three decimal places, the numbers of the sequence are

$$\{0, 0.347, 0.366, 0.347, 0.322, 0.299, 0.278, \ldots\}$$ ●

The algebra of sequences is defined in a rather apparent manner. If $\{a_n\}$ and $\{b_n\}$ are two sequences and k and m are constants (that is, independent of n), then a linear combination of the sequences is defined by

$$k\{a_n\} + m\{b_n\} = \{ka_n + mb_n\} \tag{1}$$

Also the product of the two sequences is by definition

$$\{a_n\} \cdot \{b_n\} = \{a_n \cdot b_n\} \tag{2}$$

Furthermore, if $b_n \neq 0$ for any value of n, then the quotient of two sequences is defined by

$$\frac{\{a_n\}}{\{b_n\}} = \left\{ \frac{a_n}{b_n} \right\} \tag{3}$$

Exercises 15.1

In Exercises 1 through 26 discuss the boundedness and monotonicity of the sequences with the given nth term.

1. $\dfrac{10}{n}$

2. $\dfrac{1}{3n}$

3. $\dfrac{(-1)^n}{n}$

4. $\dfrac{n + 1}{n}$

5. $\dfrac{2n - 3}{n}$

6. $\dfrac{3n + 5}{n}$

7. $\dfrac{n^2 + 3}{n^2}$

8. $\dfrac{n^2 + 1}{n}$

9. $\ln \dfrac{n}{n + 1}$

10. $\left(\dfrac{1 + \sqrt{3}}{3}\right)^n$

11. $\left(-\dfrac{1}{\pi}\right)^n$

12. $\dfrac{2n - 1}{2n + 1}$

13. $\dfrac{5^n + 100}{5^n}$

14. $\sin \dfrac{n\pi}{2}$

15. $\dfrac{e^n}{n}$

16. $n^2 e^{-n}$

17. $\dfrac{\sinh n}{n}$

18. $\dfrac{\ln 5n}{n}$

19. $\dfrac{n + 3\sqrt{n}}{n}$

20. $\sqrt{n + 1} - \sqrt{n}$

21. $n \sin \dfrac{1}{n}$

22. $\displaystyle\int_n^{n+1} \dfrac{dt}{t}$

23. $\displaystyle\int_n^{n+1} \dfrac{dt}{t^2}$

24. $\dfrac{1 + 2 + 3 + \cdots + n}{n^2}$

25. $\dfrac{1 + 3 + 5 + 7 + \cdots + (2n - 1)}{n^3}$

26. $\dfrac{1^2 + 2^2 + 3^2 + \cdots + n^2}{n^3}$

27. Show that the sequence

$$a_n = \frac{1}{\sqrt{n}} + \frac{1}{\sqrt{n + 1}} + \frac{1}{\sqrt{n + 2}} + \cdots + \frac{1}{\sqrt{2n}} \qquad (n = 1, 2, 3, \ldots)$$

does *not* have an upper bound.

28. Show that the sequence

$$a_n = \frac{1}{(1)(2)} + \frac{1}{(2)(3)} + \frac{1}{(3)(4)} + \cdots + \frac{1}{n(n + 1)} \qquad (n = 1, 2, 3, \ldots)$$

has an upper bound. Find its least upper bound (that is, the smallest upper bound).

29. Let $\{a_n\}$ be an increasing sequence. Prove that the sequence of arithmetic means $\{b_n\}$ where

$$b_n = \frac{a_1 + a_2 + \cdots + a_n}{n}$$

is also an increasing sequence. (*Hint:* Form $b_{n+1} - b_n$ and show that it must be positive.)

15.2   THE LIMIT OF A SEQUENCE

A sequence $\{a_n\}$ is said to have a limit L if the nonnegative quantity $|a_n - L|$ is arbitrarily small provided that n is sufficiently large. More precisely we have

Definition 1   (*Limit of a Sequence*).   A sequence $\{a_n\}$ is said to have a limit L if for every positive number ε there exists a positive integer N such that for all

$$n > N$$

$$|a_n - L| < \varepsilon$$

We write $\lim\limits_{n \to \infty} a_n = L$.

EXAMPLE 1   Apply Definition 1 to prove that the sequence $\{a_n\} = \left\{ \dfrac{3n + \sqrt{n}}{n} \right\}$ has a limit $L = 3$.

Solution   The fact that $\dfrac{3n + \sqrt{n}}{n} = 3 + \dfrac{1}{\sqrt{n}}$ strongly suggests that $\lim\limits_{n \to \infty} \dfrac{3n + \sqrt{n}}{n} = 3$.

We have $|a_n - 3| = \left| \dfrac{1}{\sqrt{n}} \right| = \dfrac{1}{\sqrt{n}} < \varepsilon$ provided that n is chosen $> \dfrac{1}{\varepsilon^2}$. There-

fore if $N = \left[\!\!\left[\dfrac{1}{\varepsilon^2} \right]\!\!\right]$ where $\left[\!\!\left[\dfrac{1}{\varepsilon^2} \right]\!\!\right]$ is the greatest integer less than or equal to $\dfrac{1}{\varepsilon^2}$,

we have $|a_n - 3| < \varepsilon$ when $n > N$. Thus $\lim\limits_{n \to \infty} \dfrac{3n + \sqrt{n}}{n} = 3$.   ●

Definition 2   A sequence $\{a_n\}$ that has a limit L is said to **converge** to that limit. If the sequence is not convergent it is said to be **divergent**.

EXAMPLE 2   Test the sequence $\{a_n\} = \left\{ \dfrac{1 - (-1)^n}{n} \right\}$ for convergence or divergence.

Solution   We observe that $a_n = \dfrac{2}{n}$ if n is odd and $a_n = 0$ if n is an even integer. Hence the first few terms of our sequence are

$$\{2, 0, \tfrac{2}{3}, 0, \tfrac{2}{5}, 0, \tfrac{2}{7}, \ldots\}$$

Although the sequence is not monotonic, it does appear to be convergent to the number 0. In fact,

$$|a_n - 0| \leq \frac{2}{n} < \varepsilon \qquad \text{provided that } n > \frac{2}{\varepsilon}$$

Hence we choose $N = \left[\!\!\left[\dfrac{2}{\varepsilon} \right]\!\!\right]$ and then

$$|a_n - 0| < \varepsilon \qquad \text{whenever } n > N.$$

Therefore

$$\lim_{n \to \infty} a_n = 0$$ ●

EXAMPLE 3   Determine if the sequence $\left\{ \dfrac{1 - (-1)^n}{2} \right\}$ is convergent or divergent.

Solution With $a_n = \dfrac{1 - (-1)^n}{2}$ we observe that our sequence is

$$\{1, 0, 1, 0, 1, 0, \ldots\}$$

Clearly it does not converge to a limit L. For example, if we choose $L = 0$ then $|a_n - 0| > \frac{1}{2}$ for all odd $n > N$. Similarly, if we choose $L = 1$ then $|a_n - 1| > \frac{1}{2}$ for all even $n > N$. Finally if we choose any other L (not equal to 0 or 1) then $|a_n - L| \geq \min\left(|0 - L|, |1 - L|\right)$ for all $n = 1, 2, 3, \ldots$. ●

Next, we establish the **uniqueness theorem for limits.**

Theorem 1 If $\lim\limits_{n \to \infty} a_n = L_1$ and $\lim\limits_{n \to \infty} a_n = L_2$, then $L_1 = L_2$.

Proof If $L_1 \neq L_2$ then set $\varepsilon = |L_2 - L_1|$. Now, since $\lim\limits_{n \to \infty} a_n = L_1$, we have $|L_1 - a_n| < \dfrac{\varepsilon}{2}$, if $n > N_1$ (for some positive integer N_1) and also, from $\lim\limits_{n \to \infty} a_n = L_2$, $|L_2 - a_n| < \dfrac{\varepsilon}{2}$ if $n > N_2$ (for some positive integer N_2). Thus if $N = \max(N_1, N_2)$, we have for $n > N$

$$|L_2 - L_1| = |(L_2 - a_n) + (a_n - L_1)| \leq |L_2 - a_n| + |a_n - L_1| < \varepsilon$$

Hence

$$\varepsilon = |L_2 - L_1| < \varepsilon$$

which is clearly invalid. We conclude that $L_1 = L_2$, that is, a convergent sequence has a unique limit. ∎

Theorem 2 A convergent sequence is bounded.

The proof of Theorem 2 is left to the reader (see Exercise 27).

The converse of this theorem is not valid; that is, if a sequence is bounded, it may be divergent. The sequence in Example 3 is bounded and divergent whereas the sequences in Examples 1 and 2 are bounded and convergent. Hence boundedness is a necessary but not sufficient condition for convergence. Thus the following sequences

$$\left\{\frac{n+3}{5}\right\} \qquad \left\{\frac{n^2 + 4n}{n+3}\right\} \qquad \left\{\frac{n}{\sin n}\right\} \qquad \left\{\frac{e^n}{n^2}\right\}$$

are all unbounded sequences and therefore are divergent.

Definition 3 Let B be an upper bound of a given sequence $\{a_n\}$, that is, $a_n \leq B$ for all n. If B has the further property that, if M is any upper bound, $B \leq M$, then B is said to be the **least upper bound** of the sequence.

Similarly, let A be a lower bound of a given sequence $\{a_n\}$, that is,

$A \leq a_n$ for all n. If A has the property that, if m is any lower bound, $A \geq m$, then A is said to be the **greatest lower bound** of the sequence.

As an illustration of these definitions, let us look at the sequence

$$\{a_n\} = \left\{\frac{1}{n}\right\} = \left\{1, \frac{1}{2}, \frac{1}{3}, \frac{1}{4}, \ldots\right\}$$

$5, \pi, 1, 27$, and $\sqrt{2}$ are all examples of upper bounds. The smallest upper bound is 1, so it is the least upper bound. Similarly, $-2, 0, -e, -100$, and $-\sqrt{7}$ are all lower bounds, but 0 is the greatest lower bound.

It should be noted that a least upper bound and a greatest lower bound might or might not be part of a sequence. In our illustration, the least upper bound is a term of the sequence but the greatest lower bound is not.

For the proof of the next theorem we will need the following significant property of the real number system.

Axiom of Completeness   Every set of real numbers that has an upper bound must also have a least upper bound. Similarly, every set of real numbers that has a lower bound must also have a greatest lower bound.

Theorem 3   A bounded, nondecreasing sequence converges to its least upper bound, and a bounded, nonincreasing sequence converges to its greatest lower bound.

Proof   Let $\{a_n\}$ be a bounded nondecreasing sequence. Let ε be an arbitrary positive number. Choose B to be its least upper bound, so that $a_n \leq B$. Also, for some $n = N$ we must have

$$a_N > B - \varepsilon$$

otherwise B would not be the *least* upper bound. Furthermore, by the hypothesis of monotonicity

$$a_n \geq a_N \qquad \text{if } n \geq N$$

But this means that for arbitrary positive ε,

$$B - \varepsilon < a_n \leq B \qquad \text{for all } n \geq N$$

Hence $|a_n - B| < \varepsilon$ for all $n \geq N$ or, equivalently,

$$\lim_{n \to \infty} a_n = B$$

The nonincreasing case may be proved similarly and this is left to the reader as an exercise. ∎

EXAMPLE 4   Show that the sequence $\left\{\dfrac{n}{3^n}\right\}$ is convergent.

Solution   First we shall show that the sequence $\{a_n\} = \left\{\dfrac{n}{3^n}\right\}$ is a decreasing sequence; that is, $a_n > a_{n+1}$ or $\dfrac{n}{3^n} > \dfrac{n+1}{3^{n+1}}$. This is equivalent to $3n > n + 1$ or $n > \frac{1}{2}$, which is true for all positive integers. Furthermore, $\left\{\dfrac{n}{3^n}\right\}$ is a sequence of positive numbers and therefore is bounded below by the number zero. Hence from Theorem 3, $\lim\limits_{n \to \infty} a_n = A$ where $A \geq 0$.

It is left to the reader to show that the limit $A = 0$.   ●

Theorem 4   If a sequence $\{a_n\}$ converges to a limit L and a sequence $\{b_n\}$ converges to a limit M, then

    (i) $\lim\limits_{n \to \infty} (a_n \pm b_n) = L \pm M$

    (ii) $\lim\limits_{n \to \infty} (a_n b_n) = LM$

Also, if $b_n \neq 0$ $(n = 1, 2, 3, \ldots)$ and $M \neq 0$, then

    (iii) $\lim\limits_{n \to \infty} \dfrac{a_n}{b_n} = \dfrac{L}{M}$

Proof   The proofs of (i) and (iii) are left as exercises. To prove (ii), since $a_n \to L$ and $b_n \to M$, then, for any positive number ε_1, $|L - a_n| < \varepsilon_1$ provided that $n > N_1(\varepsilon_1)$, and $|M - b_n| < \varepsilon_1$ provided that $n > N_2(\varepsilon_1)$, [N_1 and N_2 are positive integers that depend on ε_1]. Thus, if $n > N = \max(N_1, N_2)$, we have

$$|L - a_n| < \varepsilon_1 \qquad \text{and} \qquad |M - b_n| < \varepsilon_1 \qquad \text{when } n > N.$$

Hence,

$$|LM - a_n b_n| = |M(L - a_n) + a_n(M - b_n)|$$
$$\leq |M|\,|L - a_n| + |a_n|\,|M - b_n| \leq (|M| + |a_n|)\varepsilon_1$$

Now since $\{a_n\}$ converges, it is bounded so that $|a_n| \leq K$ where K is a positive constant for all n. Hence

$$|LM - a_n b_n| \leq (|M| + K)\varepsilon_1 = \varepsilon$$

and the positive quantity ε can be made arbitrarily small. Therefore

$$\lim_{n \to \infty} (a_n b_n) = LM \qquad\qquad ■$$

It follows that we can interchange the rational operation of calculation with that of forming a limit. In other words, the same result is obtained if we first perform a passage to a limit and then perform a rational operation upon the limits or vice versa.

EXAMPLE 5   Find $\lim\limits_{n \to \infty} \dfrac{2n^3 + 7n - 10}{n^3 + n^2 + 3n + 5}$

Solution $\lim\limits_{n \to \infty} \dfrac{2n^3 + 7n - 10}{n^3 + n^2 + 3n + 5} = \lim\limits_{n \to \infty} \dfrac{2 + \dfrac{7}{n^2} - \dfrac{10}{n^3}}{1 + \dfrac{1}{n} + \dfrac{3}{n^2} + \dfrac{5}{n^3}} = 2$

where the numerator and denominator were both divided by n^3 before we passed to the limit. ●

EXAMPLE 6   Find $\lim\limits_{n\to\infty} \dfrac{7n^3 + 2n^2 + 5n + 10}{n^2 + 100n + 4}$

Solution

$$\frac{7n^3 + 2n^2 + 5n + 10}{n^2 + 100n + 4} = \frac{7 + \dfrac{2}{n} + \dfrac{5}{n^2} + \dfrac{10}{n^3}}{\dfrac{1}{n} + \dfrac{100}{n^2} + \dfrac{4}{n^3}}$$

Since the numerator tends to 7 while the denominator tends to zero the sequence becomes unbounded. Therefore the sequence $\left\{ \dfrac{7n^3 + 2n^2 + 5n + 10}{n^2 + 100n + 4} \right\}$ is divergent. ●

Rule: Another rule that is simply stated (and easily proved) is that, if $a_n < b_n$ for all $n > N$ and if $\lim\limits_{n\to\infty} a_n = A$ and $\lim\limits_{n\to\infty} b_n = B$, then $A \leq B$

Note that one might be tempted to say that $A < B$ must hold. However, this is not correct as is seen by the following counterexample. Let $a_n = \dfrac{1}{3n}$ and $b_n = \dfrac{1}{n}$, then $a_n < b_n$ for all positive integers n and $\lim\limits_{n\to\infty} a_n = \lim\limits_{n\to\infty} b_n = 0$.

If a comparison is made between the definition of $\lim\limits_{x\to\infty} f(x)$ in Chapter 2 and $\lim\limits_{n\to\infty} a_n$, one is impressed with the similarity and difference of the two definitions. When we say $\lim\limits_{x\to\infty} f(x) = L$, this function must be defined for all real values of x greater than some fixed value x_0; whereas, when $\lim\limits_{n\to\infty} a_n$ is considered, n is restricted to positive integers. We have, however, the following theorem, which again enables us to apply calculus in dealing with problems concerned with sequences.

Theorem 5   If $\lim\limits_{x\to\infty} f(x) = L$ and f is defined for all real numbers x greater than some number x_0, then $\lim\limits_{n\to\infty} f(n) = L$ where the domain is restricted to the positive integers $> x_0$.

The proof of Theorem 5 is left to the reader as an exercise.

Now that the similarity between limit of a sequence and the limit of a function (as the variable becomes infinite) have been stated, the reader should note that the proofs of some of our theorems are similar to the proofs of the comparable theorems on limits of functions (given in Chapter 2).

EXAMPLE 7   Does the sequence $\left\{ \dfrac{100 \ln n}{e^n} \right\}$ converge?

Solution $f(x) = \dfrac{100 \ln x}{e^x}$ exists for all positive numbers x. Also, $100 \ln x \to \infty$ and $e^x \to \infty$ as $x \to \infty$. Hence l'Hospital's rule yields

$$\lim_{x \to \infty} f(x) = \lim_{x \to \infty} \frac{100}{x e^x} = 0$$

Therefore, $\lim\limits_{n \to \infty} \dfrac{100 \ln n}{e^n} = 0$ when n is restricted to the positive integers. This means that the sequence $\dfrac{100 \ln n}{e^n}$ converges to the limit 0. ●

The sequences

$$\{e^{\sqrt{n}}\} \qquad \left\{ \sin \frac{2\pi}{n} \right\} \qquad \left\{ \ln \frac{n^2 + 3}{n^2 + 1} \right\}$$

are all of the form $f(a_n)$ where f is a continuous function in a domain that includes all a_n. Now suppose that a_n tends to L as $n \to \infty$, then, if f is continuous at L, we have $\lim\limits_{n \to \infty} f(a_n) = f(L)$. More succinctly we have the following theorem:

Theorem 6 Suppose that $\lim\limits_{n \to \infty} a_n = L$ and that for each n, a_n is in the domain of f. Furthermore if f is continuous at L, then $\lim\limits_{n \to \infty} f(a_n) = f(L)$.

Proof It is assumed that f is a continuous function at $x = L$. Choose $\varepsilon > 0$, then from the continuity of f at $x = L$ it follows that a $\delta > 0$ must also exist such that

$$|f(x) - f(L)| < \varepsilon \qquad \text{provided that } |x - L| < \delta$$

However, $\lim\limits_{n \to \infty} a_n = L$ by hypothesis so that $|a_n - L| < \delta$ provided that $n > N(\delta)$ where N is a positive integer. Hence if $n > N$, we have $|f(a_n) - f(L)| < \varepsilon$. ∎

EXAMPLE 8 Find $\lim\limits_{n \to \infty} \ln \dfrac{n^2 + 3}{n^2 + 1}$.

Solution

$$\lim_{n \to \infty} \frac{n^2 + 3}{n^2 + 1} = \lim_{n \to \infty} \frac{1 + \dfrac{3}{n^2}}{1 + \dfrac{1}{n^2}} = 1$$

and $\ln x$ is continuous at $x = 1$. Hence

$$\lim_{n \to \infty} \ln \frac{n^2 + 3}{n^2 + 1} = \ln \left[\lim_{n \to \infty} \frac{n^2 + 3}{n^2 + 1} \right] = \ln 1 = 0 \qquad ●$$

Exercises 15.2

In Exercises 1 through 20, determine whether or not the sequence with the indicated general term converges. If it converges find the limit.

1. $\dfrac{8}{n}$

2. $\dfrac{1}{2n + 1}$

3. $\dfrac{(-1)^n}{3}$

4. $\dfrac{n + 3}{n}$

5. $\dfrac{n + 2}{n^2}$

6. $\dfrac{n + 1}{10\sqrt{n}}$

7. $\dfrac{1}{\pi}\cos n\pi$

8. $\dfrac{\sqrt{n}}{n + 10}$

9. $\dfrac{n^4 + n^3 - 10}{2n^4 + n^2 - 1}$

10. $\ln\dfrac{2n + 1}{n}$

11. $\sin\left(\dfrac{\pi}{8}\left(\dfrac{2n + 7}{n}\right)\right)$

12. $\dfrac{3n^2 + 7n + 10}{n + 100}$

13. $\dfrac{n^3 - 4n^5}{7n^5 + 1}$

14. $\dfrac{n}{(\sqrt{3n} + 4)^2}$

15. $\sqrt{n}(\sqrt{n + 2} - \sqrt{n})$

16. $\dfrac{2^n}{3^n + 4}$

17. $\dfrac{0.66\ldots 6}{n\ 6\text{'s}}$

18. $\dfrac{3^n - 2^n}{4^n}$

19. $\dfrac{(-1)^n\sqrt{n}}{n + 5}$

20. $n^2 - (-1)^n 3n$

In Exercises 21 and 22 use Theorem 3 to show that each sequence is convergent. It is **not** required to find the limit.

21. $a_n = \dfrac{2\cdot 4\cdot 6 \cdots (2n - 2)(2n)}{3\cdot 5\cdot 7 \cdots (2n - 1)(2n + 1)}$

22. $a_n = 10 - \dfrac{2^n}{n!}$

23. Let $\{a_n\}$ be the sequence in which a_1 is arbitrary and $a_n = ka_{n-1}$, where k is a constant (that is, k is independent of n). Find a_2, a_3, and a_4. Find a formula for a_n by inspection and verify it. Show that $\lim\limits_{n\to\infty} a_n = 0$ if $|k| < 1$.

24. The sequence of positive constants $a_1, a_2, \ldots, a_n, \ldots$ is determined by the property that $a_{n+1} = \dfrac{1}{2}\left(a_n + \dfrac{k}{a_n}\right)$ where k is a positive constant. Show that if a_n tends to a limit L as $n \to \infty$ then $L = \sqrt{k}$. If $a_1 = 2$ and $k = 5$ find a_2 and a_3 to four decimal places.

25. Let s_n denote the length of a side of a regular polygon of 2^n sides inscribed in a circle of radius equal to 1. Show that $s_{n+1} = \sqrt{2 - \sqrt{4 - s_n^2}}$. Start with the side of a square $= s_2 = \sqrt{2}$ and write formulas for s_3 and s_4. What is the $\lim\limits_{n\to\infty} 2^n s_n$?

26. Given the sequence $\{a_n\} = \left\{\dfrac{n!}{n^n}\right\}$ find $\lim\limits_{n\to\infty}\dfrac{a_{n+1}}{a_n}$. What does this imply about $\lim\limits_{n\to\infty} a_n$?

27. Prove Theorem 2 of the text, namely, that a convergent sequence is bounded.

In this section we will apply our understanding of the convergence or divergence of sequences to that of the question of summing an infinite number of constant terms. Let $u_1 + u_2 + u_3 + \cdots$ be a given series of real constants, where each term of the series u_k ($k = 1, 2, \ldots$) is obtained by some prescribed rule. From the given series of real numbers, we form a sequence S_n of **partial sums**

$$
\begin{aligned}
S_1 &= u_1 \\
S_2 &= u_1 + u_2 \\
S_3 &= u_1 + u_2 + u_3 \\
&\cdots \\
S_n &= u_1 + u_2 + \cdots + u_n \\
&\cdots
\end{aligned}
\tag{1}
$$

We shall be concerned with various methods for determining whether or not the sequence S_n converges to a limit.

If the sigma notation is used we write

$$
S_n = \sum_{k=1}^{n} u_k
$$

for the given partial sum S_n in (1) and $\displaystyle\sum_{k=1}^{\infty} u_k$ for the given infinite series.

Definition   An infinite series

$$
u_1 + u_2 + u_3 + \cdots
$$

with partial sums $S_1, S_2, S_3, \ldots, S_n, \ldots$ is said to be **convergent** if and only if $\lim\limits_{n \to \infty} S_n$ exists. If the limit does not exist, the series is said to be **divergent** and there is no sum. For a convergent series the number

$$
S = \lim_{n \to \infty} S_n
\tag{2}
$$

is called the **sum** of the series.

EXAMPLE 1   An infinite series of the form

$$
\sum_{k=1}^{\infty} ar^{k-1} = a + ar + ar^2 + ar^3 + \cdots
\tag{3}
$$

is called a **geometric series,** where r, the common ratio, is the ratio of u_n to u_{n-1}; that is, for $n = 2, 3, \ldots$

$$
\frac{u_n}{u_{n-1}} = \frac{ar^{n-1}}{ar^{n-2}} = r
$$

If we use the identity

$$(1 - r^n) = (1 - r)(1 + r + r^2 + \cdots + r^{n-1})$$

it follows that

$$S_n = a + ar + ar^2 + \cdots + ar^{n-1} = a(1 + r + r^2 + \cdots + r^{n-1})$$
$$= \frac{a(1 - r^n)}{1 - r} \qquad \text{if } r \neq 1$$

or

$$S_n = \frac{a}{1 - r} - \frac{ar^n}{1 - r} \qquad (r \neq 1)$$

Since

$$\lim_{n \to \infty} r^n = 0 \text{ if and only if } |r| < 1$$

we have for $|r| < 1$,

$$\lim_{n \to \infty} S_n = \lim_{n \to \infty} \left(\frac{a}{1 - r} - \frac{a}{1 - r} r^n \right)$$
$$= \frac{a}{1 - r} - \frac{a}{1 - r} (\lim_{n \to \infty} r^n) = \frac{a}{1 - r}$$

Therefore it follows that the geometric series (3) converges if $|r| < 1$ and its sum S is given by

$$S = a + ar + ar^2 + \cdots = \frac{a}{1 - r} \tag{4}$$

It is left to the reader to provide the needed steps to show that (3) diverges if $|r| \geq 1$. (We shall also prove this shortly by another means). ●

EXAMPLE 2   Test $1 - \frac{1}{3} + \frac{1}{9} - \frac{1}{27} + \cdots$ for convergence.

Solution   This is an infinite geometric series with $a = 1$ and ratio $r = -\frac{1}{3}$. From (4) the series converges to

$$S = \frac{1}{1 - (-\frac{1}{3})} = \frac{3}{4}$$

It is interesting to note that the partial sums are *not* monotonic. For example, $S_1 = 1$, $S_2 = \frac{2}{3} \doteq 0.67$, $S_3 = \frac{7}{9} \doteq 0.78$, and so on. This suggests, and further calculations will verify, that the partial sums are approaching 0.75 in an oscillatory fashion (much the way an oscillating pendulum will approach its rest position). ●

EXAMPLE 3   Test $1 - 1 + 1 - 1 + \cdots$ for convergence.

Solution   This is also an infinite geometric series with $a = 1$ and ratio $r = -1$. Since $|r| = 1$ it follows that the series diverges. Its partial sums are $S_n = 0$ if n is even and $S_n = 1$ if n is odd. Hence $\lim_{n \to \infty} S_n$ does not exist. ●

EXAMPLE 4   Express the nonterminating and repeating decimal $0.292929\ldots$ as a rational number.

Solution We recognize that the given decimal is expressible in the form $u_1 + u_2 + u_3 + \cdots$ where

$$u_1 = 0.29 = \frac{29}{10^2}, u_2 = 0.0029 = \frac{29}{10^4}, u_3 = 0.000029 = \frac{29}{10^6}, \cdots$$

Hence formula (4) is applicable with $a = \frac{29}{10^2}$ and $r = \frac{1}{10^2}$. Therefore

$$0.292929\ldots = \frac{\dfrac{29}{10^2}}{1 - \dfrac{1}{10^2}} = \frac{\dfrac{29}{10^2}}{\dfrac{99}{10^2}} = \frac{29}{99} \qquad \bullet$$

Theorem 1 If $\sum\limits_{k=0}^{\infty} u_k$ converges, then $\lim\limits_{n\to\infty} u_n = 0$.

Proof With the notation defined in (1)

$$S_n - S_{n-1} = (u_1 + u_2 + \cdots + u_n) - (u_1 + u_2 + \cdots + u_{n-1}) = u_n$$

so that

$$\lim_{n\to\infty} u_n = \lim_{n\to\infty} S_n - \lim_{n\to\infty} S_{n-1} \qquad (5)$$

provided that each limit on the right side of (5) exists. Now the given series converges to some number L, hence

$$\lim_{n\to\infty} S_n = \lim_{n\to\infty} S_{n-1} = L$$

and we have

$$\lim_{n\to\infty} u_n = L - L = 0$$

which is the required result. ∎

The contrapositive of the conclusion of Theorem 1 must also be true, namely, we have

Theorem 2 If $\lim\limits_{n\to\infty} u_n \neq 0$, then the infinite series diverges.

We may use Theorem 2 to show that the geometric series (3) diverges if $a \neq 0$ and $|r| \geq 1$. This follows since

$$|ar^n| \geq |a| \neq 0 \qquad \text{for all } n$$

so that $\lim\limits_{n\to\infty} u_n \neq 0$.

The converse of Theorem 2 is false. There are divergent series $\sum\limits_{k=1}^{\infty} u_k$ for which $\lim\limits_{n\to\infty} u_n = 0$. We demonstrate this by means of a specific example.

EXAMPLE 5   Show that the series $\displaystyle\sum_{k=1}^{\infty} \frac{1}{\sqrt{k}}$ diverges.

Solution   First we note that $u_n = \dfrac{1}{\sqrt{n}}$ and that $\lim\limits_{n\to\infty} u_n = 0$. Hence the necessary condition for convergence is met. However, it is not sufficient because

$$S_n = \frac{1}{1} + \frac{1}{\sqrt{2}} + \frac{1}{\sqrt{3}} + \cdots + \frac{1}{\sqrt{n}} > \frac{1}{\sqrt{n}} + \frac{1}{\sqrt{n}} + \cdots + \frac{1}{\sqrt{n}}$$

$$= \frac{n}{\sqrt{n}} = \sqrt{n}$$

and S_n becomes unbounded with increase in n. The given series diverges.   ●

If the infinite series $\displaystyle\sum_{k=1}^{\infty} u_k$ with nth partial sum S_n converges to a limit S, then we have for any positive ε a number $N(\varepsilon)$ such that

$$|S - S_n| < \varepsilon \qquad \text{if } n > N$$

Also, $|S - S_m| < \varepsilon$ if $m > N$. Then

$$|S_m - S_n| = |(S_m - S) + (S - S_n)| \leq |S_m - S| + |S - S_n| < 2\varepsilon \qquad (6)$$

if $m > N$ and $n > N$. Hence, the difference between any two partial sums S_m and S_n can be made arbitrarily small provided that $m > N$ and $n > N$, where N depends upon the ε chosen.

Relation (6) is known as the **Cauchy intrinsic test** for convergence. Its importance lies in the fact that it can be shown conversely that, if (6) holds, then the sequence of partial sums must also converge. Thus (6) is a test for convergence that involves only the elements of the series and not the limit S—hence the adjective intrinsic.

We turn now to an application of this concept to prove the divergence of an important series called the harmonic series. The **harmonic series** is defined by

$$\sum_{k=1}^{\infty} \frac{1}{k} = 1 + \frac{1}{2} + \frac{1}{3} + \cdots + \frac{1}{n} + \cdots \qquad (7)$$

Clearly, $\lim\limits_{n\to\infty} \dfrac{1}{n} = 0$, however, the series diverges as we shall now show. To this end we form

$$S_{2n} - S_n = \frac{1}{n+1} + \frac{1}{n+2} + \cdots + \frac{1}{2n} > \frac{1}{2n} + \frac{1}{2n} + \cdots + \frac{1}{2n}$$

$$= n\left(\frac{1}{2n}\right) = \frac{1}{2}$$

that is,

$$S_{2n} - S_n > \tfrac{1}{2} \qquad (i)$$

for each positive integer n. Now if $\lim\limits_{n\to\infty} S_n = S$ then for any $\varepsilon > 0$ we can find an N such that

$$|S_{2n} - S_n| = S_{2n} - S_n < \varepsilon \qquad \text{when } n > N$$

In particular, if we choose $\varepsilon = \frac{1}{2}$, then

$$S_{2n} - S_n < \tfrac{1}{2} \tag{ii}$$

for all $n > N$. Hence (ii) contradicts (i), and the harmonic series (7) diverges. In particular, S_n becomes unbounded as n increases.

Theorem 3   If $\displaystyle\sum_{k=1}^{\infty} u_k$ and $\displaystyle\sum_{k=1}^{\infty} v_k$ are convergent series with sums U and V, respectively,

then $\displaystyle\sum_{k=1}^{\infty} (cu_k + dv_k)$, where c and d are arbitrary constants, is a convergent

series and its sum is given by

$$\sum_{k=1}^{\infty} (cu_k + dv_k) = cU + dV \tag{8}$$

The proof of Theorem 3 is left to the reader as an exercise.

Corollary 1   If $\displaystyle\sum_{k=1}^{\infty} u_k$ and $\displaystyle\sum_{k=1}^{\infty} v_k$ are convergent series with sums U and V, respectively,

then their sum or difference is also a convergent series. Furthermore,

(i) $\displaystyle\sum_{k=1}^{\infty} (u_k + v_k) = U + V$

(ii) $\displaystyle\sum_{k=1}^{\infty} (u_k - v_k) = U - V$

The proof of Corollary 1 follows from (8) if we replace (c, d) with $(1, 1)$ and then with $(1, -1)$.

Corollary 2   If the series $\displaystyle\sum_{k=1}^{\infty} u_k$ is convergent and the series $\displaystyle\sum_{k=1}^{\infty} v_k$ is divergent, then the

series $\displaystyle\sum_{k=1}^{\infty} (u_k + v_k)$ is divergent.

Proof   Assume instead that $\displaystyle\sum_{k=1}^{\infty} (u_k + v_k)$ is a convergent series and let S be its sum. Let

$\displaystyle\sum_{k=1}^{\infty} u_k$ be U. Then [since $v_k = (u_k + v_k) - u_k$]

$$\sum_{k=1}^{\infty} v_k = \sum_{k=1}^{\infty} [(u_k + v_k) - u_k] = \sum_{k=1}^{\infty} (u_k + v_k) - \sum_{k=1}^{\infty} u_k$$

from (ii) of Corollary 1. Hence, $\displaystyle\sum_{k=1}^{\infty} v_k$ converges to the sum $S - U$. But this

contradicts the hypothesis that $\displaystyle\sum_{k=1}^{\infty} v_k$ diverges. Therefore the assumption that

$\displaystyle\sum_{k=1}^{\infty} (u_k + v_k)$ converges is false and we conclude that $\displaystyle\sum_{k=1}^{\infty} (u_k + v_k)$ is a divergent series. ∎

Corollary 3 Let c be an arbitrary real number.

(i) If $\displaystyle\sum_{k=1}^{\infty} u_k$ is a convergent series, then $\displaystyle\sum_{k=1}^{\infty} cu_k$ is also convergent.

(ii) If $\displaystyle\sum_{k=1}^{\infty} u_k$ is a divergent series, then $\displaystyle\sum_{k=1}^{\infty} cu_k$ also diverges, providing that $c \neq 0$. (What happens when $c = 0$?)

The proof of Corollary 3 is left as an exercise for the reader.

EXAMPLE 6 Determine whether the infinite series

$$\sum_{k=1}^{\infty} \left[\left(\frac{1}{3}\right)^k + 7\left(-\frac{1}{2}\right)^k \right]$$

converges or diverges.

Solution The series $\displaystyle\sum_{k=1}^{\infty} \left(\frac{1}{3}\right)^k$ is a geometric series with ratio $\frac{1}{3}$ so that it converges.

Also $\displaystyle\sum_{k=1}^{\infty} \left(-\frac{1}{2}\right)^k$ is a geometric series with ratio $-\frac{1}{2}$, which implies that it also converges. From Theorem 3 with $c = 1$ and $d = 7$ the given series converges. We leave it to the reader to find the sum. ●

EXAMPLE 7 Determine whether or not the infinite series

$$\sum_{k=1}^{\infty} \left(\frac{1}{100k} + \frac{1}{5^k} \right)$$

converges or diverges.

Solution The first series diverges (by Corollary 3 to Theorem 3) since it is a nonzero

multiple of the harmonic series. Because the series $\displaystyle\sum_{k=1}^{\infty} \frac{1}{5^k}$ is a geometric series with ratio $r = \frac{1}{5}$, it is a convergent series. Hence from Corollary 2 the given series is divergent. ●

Now suppose that we are given two divergent series $\sum\limits_{k=1}^{\infty} u_k$ and $\sum\limits_{k=1}^{\infty} v_k$. Can we then correctly conclude that $\sum\limits_{k=1}^{\infty} (u_k + v_k)$ diverges? The answer is no—the resulting series may converge. For example, if $v_k = -u_k$ or $u_k + v_k = 0$, the resulting series is the convergent series $\sum\limits_{k=1}^{\infty} 0 = 0$. It is left to the reader to construct other examples where the resulting series converges.

Finally, it may be noted that *any* finite number of terms may be changed or removed from a given series without affecting convergence or divergence. However, the sum of a given convergent series may be altered. The proof of this is straightforward and is left to the reader as an exercise.

EXAMPLE 8   Determine the convergence or divergence of the infinite series

$$\frac{1}{\sqrt{101}} + \frac{1}{\sqrt{102}} + \frac{1}{\sqrt{103}} + \cdots + \frac{1}{\sqrt{n}} + \cdots$$

Solution   In Example 5 we showed that the series $\sum\limits_{k=1}^{\infty} \frac{1}{\sqrt{k}}$ diverges. Since the given series is

the same as $\sum\limits_{k=1}^{\infty} \frac{1}{\sqrt{k}}$ with the first 100 terms removed, that is, $\sum\limits_{k=101}^{\infty} \frac{1}{\sqrt{k}}$, it follows

that the series also diverges. In fact, more generally, $\sum\limits_{k=N}^{\infty} \frac{1}{\sqrt{k}}$ diverges for any

fixed positive integer N. ●

Exercises 15.3

In Exercises 1 through 9, find the sum of each of the following convergent series.

1. $3 + \dfrac{3}{8} + \dfrac{3}{8^2} + \cdots + \dfrac{3}{8^{n-1}} + \cdots$

2. $-1 + \frac{1}{4} - \frac{1}{16} + \frac{1}{64} - \cdots + (-1)^n (\frac{1}{4})^{n-1} + \cdots$

3. $\dfrac{1}{\sqrt{3}} + \dfrac{1}{3} + \dfrac{1}{3\sqrt{3}} + \cdots + \dfrac{1}{(\sqrt{3})^n} + \cdots$

4. $1 + x^2 + x^4 + x^6 + \cdots + x^{2(n-1)} + \cdots$,   where $0 \leq x^2 < 1$

5. $(xy)^2 + (xy)^3 + \cdots + (xy)^{n+1} + \cdots$,   where $|xy| < 1$

6. $3.\overline{27}$ (where the bar over 27 means that the number 27 is repeated indefinitely, i.e., $3.272727\ldots$)

7. $0.\overline{012}$

8. $0.63\overline{421}$

9. $4.3\overline{214}$

In Exercises 10 through 16, S_n represents the nth partial sum of an infinite series $\sum_{i=1}^{\infty} u_i$. Find the nth term of the series, and determine whether or not the series converges. If it converges, find its sum. (Hint: $u_n = S_n - S_{n-1}$, $n > 1$; $u_1 = S_1$.)

10. $S_n = \dfrac{n}{n+1}$

11. $S_n = \dfrac{1}{2}\left(1 - \dfrac{1}{3^n}\right)$

12. $S_n = \dfrac{n}{3(2n+3)}$

13. $S_n = \dfrac{(-1)^n}{3n}$

14. $S_n = 5 + (-1)^n$

15. $S_n = \ln(n+1)$

16. $S_n = \dfrac{1 - x^{3n}}{1 - x^3}$, $\quad -1 < x < 1$

In Exercises 17 through 32, determine whether or not the given infinite series converges. Find the sum of each convergent series.

17. $1 + \dfrac{1}{7} + \dfrac{1}{7^2} + \dfrac{1}{7^3} + \cdots$

18. $1 - \dfrac{3}{2} + \left(\dfrac{3}{2}\right)^2 - \left(\dfrac{3}{2}\right)^3 + \cdots$

19. $\dfrac{2}{1} + \dfrac{3}{2} + \dfrac{4}{3} + \dfrac{5}{4} + \cdots$

20. $\dfrac{1}{5} + \dfrac{1}{10} + \dfrac{1}{15} + \dfrac{1}{20} + \cdots$

21. $\dfrac{1}{8} + \dfrac{1}{8\sqrt{8}} + \dfrac{1}{8^2} + \dfrac{1}{8^2\sqrt{8}} + \cdots$

22. $\dfrac{1}{1} + \dfrac{2}{6} + \dfrac{3}{11} + \dfrac{4}{16} + \dfrac{5}{21} + \cdots$

23. $1 - 2x + 4x^2 - 8x^3 + \cdots$

24. $r + r^5 + r^9 + r^{13} + \cdots$

25. $1 + \dfrac{1}{1+k^2} + \dfrac{1}{(1+k^2)^2} + \dfrac{1}{(1+k^2)^3} + \cdots$

26. $\sum_{k=0}^{\infty} \left[\dfrac{1}{4^k} + \dfrac{1}{(-3)^k}\right]$

27. $\sum_{k=1}^{\infty} \left(\dfrac{1}{2^k} + \dfrac{1}{3k}\right)$

28. $\sum_{k=1}^{\infty} \left(\dfrac{1}{\sqrt{k}} - \dfrac{1}{k}\right)$ $\left(\text{Hint: Compare with } \sum \dfrac{1}{k}.\right)$

29. $\sum_{k=1}^{\infty} \dfrac{\cosh k}{k^4 + 1}$

30. $\sum_{k=0}^{\infty} \dfrac{5^k + 3^k}{7^k}$

31. $\dfrac{1}{1 \cdot 5} + \dfrac{1}{5 \cdot 9} + \dfrac{1}{9 \cdot 13} + \cdots + \dfrac{1}{(4n-3)(4n+1)} + \cdots$ (Hint: First decompose the general term of the series into partial fractions.)

32. $\dfrac{1}{1 \cdot 2 \cdot 3} + \dfrac{1}{2 \cdot 3 \cdot 4} + \dfrac{1}{3 \cdot 4 \cdot 5} + \cdots + \dfrac{1}{n(n+1)(n+2)} + \cdots$

33. A ball is dropped from a height of a ft above the ground. The impact of the ball with the ground is not "perfectly elastic" so that the ball rises to a height of $\frac{2}{3}a$ ft and after its next impact to a height of $\frac{2}{3}(\frac{2}{3}a) = \frac{4}{9}a$ ft, and so on. Find the total distance traveled before the ball comes to rest.

34. Construct an example where $\sum\limits_{k=1}^{\infty} u_k$ diverges and yet the infinite series

$$(u_1 + u_2) + (u_3 + u_4) + \cdots$$

converges. What does this indicate?

35. Find S_n and $\lim\limits_{n \to \infty} S_n$ for the infinite series

$$\frac{1}{1 \cdot 3 \cdot 5} + \frac{1}{3 \cdot 5 \cdot 7} + \cdots + \frac{1}{(2n-1)(2n+1)(2n+3)} + \cdots$$

36. Prove that all periodic decimals are rational numbers.
37. Find a formula for $S_n = r + 2r^2 + 3r^3 + \cdots + nr^n$ (*Hint:* Multiply both sides by r and subtract—use the formula for the sum of a geometric series). Show that $\sum\limits_{k=1}^{\infty} kr^k$ converges if and only if $|r| < 1$ and find its sum.

15.4   INFINITE SERIES WITH NONNEGATIVE TERMS

The terms of the infinite series that we have considered thus far have been arbitrary real numbers. This implies that the definitions and theorems of the last section hold for general real numbers, that is, positive, negative, or zero. Now we wish to restrict our considerations to infinite series with positive terms or, more generally, to nonnegative terms.

If u_k is nonnegative for all positive integers k, then its sequence of partial sums

$$S_n = \sum_{k=1}^{n} u_k \text{ must be nondecreasing. We now easily prove the following theorem.}$$

Theorem 1   A series with nonnegative terms converges if and only if its partial sums are bounded.

Proof   If the sequence of partial sums is not bounded, then it must diverge (since it cannot tend to a finite limit). On the contrary, if it is bounded, then the nondecreasing sequence of partial sums must converge (see Theorem 3 of Section 15.2)   ∎

The main problem with applying Theorem 1 directly is the difficulty in finding the expression for the partial sum of the given series. However, in many instances we can establish convergence or divergence of a given series by making term by term comparison with a known infinite series.

Theorem 2   (**Comparison Test**).   Let $\sum\limits_{k=1}^{\infty} u_k$ be a given series with nonnegative terms.

(i)   If $\displaystyle\sum_{k=1}^{\infty} v_k$ is a series with nonnegative terms that is known to be convergent and, furthermore, $u_k \le v_k$ for all k sufficiently large, then $\displaystyle\sum_{k=1}^{\infty} u_k$ must converge also.

(ii) If $\displaystyle\sum_{k=1}^{\infty} v_k$ is a series with nonnegative terms that is known to be divergent and $u_k \ge v_k$ for all k sufficiently large, then $\displaystyle\sum_{k=1}^{\infty} u_k$ must diverge also.

The proof follows by noting that in case (i) both series are bounded and hence convergent whereas in (ii) both series are unbounded and therefore divergent. The reader is asked to complete the details.

If $u_k \le v_k$ for all k, we say that the series $\displaystyle\sum_{k=1}^{\infty} v_k$ **dominates** the series $\displaystyle\sum_{k=1}^{\infty} u_k$.

EXAMPLE 1   Determine whether the infinite series

$$\sum_{k=1}^{\infty} \frac{1}{k3^k}$$

is convergent or divergent.

Solution   The given series is

$$\frac{1}{3} + \frac{1}{2(3)^2} + \frac{1}{3(3)^3} + \frac{1}{4(3)^4} + \cdots + \frac{1}{n(3)^n} + \cdots$$

If a comparison is made of the nth term of this series with the nth term of the convergent series

$$\frac{1}{3} + \frac{1}{3^2} + \frac{1}{3^3} + \cdots + \frac{1}{3^n} + \cdots$$

(which is a geometric series with ratio $r = \frac{1}{3} < 1$), we have

$$\frac{1}{n3^n} \le \frac{1}{3^n} \qquad (n = 1, 2, 3, \ldots)$$

Hence by Theorem 2(i) the given series must converge.   ●

EXAMPLE 2   Determine whether the infinite series

$$\sum_{k=1}^{\infty} \frac{1}{\sqrt[3]{k}}$$

is convergent or divergent.

Solution   The given series is

$$\frac{1}{\sqrt[3]{1}} + \frac{1}{\sqrt[3]{2}} + \frac{1}{\sqrt[3]{3}} + \cdots + \frac{1}{\sqrt[3]{n}} + \cdots$$

If a comparison is made of the nth term of this series with the nth term of the divergent harmonic series

$$\frac{1}{1} + \frac{1}{2} + \frac{1}{3} + \cdots + \frac{1}{n} + \cdots$$

we have (since $n \geq \sqrt[3]{n} > 0$ when $n = 1, 2, 3, \ldots$ and taking reciprocals of both sides of this inequality)

$$\frac{1}{n} \leq \frac{1}{\sqrt[3]{n}} \qquad (n = 1, 2, 3, \ldots)$$

Hence by Theorem 2(ii) the given series diverges.   ●

Our next theorem is often easier to apply than Theorem 2 upon which it is based.

Theorem 3   (*Limit Comparison Test*). Let $\displaystyle\sum_{k=1}^{\infty} u_k$ and $\displaystyle\sum_{k=1}^{\infty} v_k$ be two positive series such that

$$\lim_{n \to \infty} \frac{u_n}{v_n} = L > 0$$

then either both series converge or both series diverge.

Proof   Since $\displaystyle\lim_{n \to \infty} \frac{u_n}{v_n} = L > 0$, there is a positive number N such that for any positive ε,

$$\left| \frac{u_n}{v_n} - L \right| < \varepsilon \qquad \text{if } n > N$$

Now, choose $\varepsilon = \dfrac{L}{2}$ so that

$$\left| \frac{u_n}{v_n} - L \right| < \frac{L}{2}$$

which is equivalent to

$$-\frac{L}{2} < \frac{u_n}{v_n} - L < \frac{L}{2}$$

or upon adding L to all members of the inequality

$$\frac{L}{2} < \frac{u_n}{v_n} < \frac{3L}{2} \qquad \text{for } n > N$$

This implies that (since $u_n > 0, v_n > 0$)

$$\frac{L}{2} v_n < u_n \qquad \text{and} \qquad \frac{2u_n}{3L} < v_n \qquad \text{for } n > N$$

Hence the series $\displaystyle\sum_{k=N+1}^{\infty} u_k$ dominates the series $\dfrac{L}{2}\displaystyle\sum_{k=N+1}^{\infty} v_k$, and the series $\displaystyle\sum_{k=N+1}^{\infty} v_k$

dominates the series $\dfrac{2}{3L}\displaystyle\sum_{k=N+1}^{\infty} u_k$. Thus, from Theorem 2, either both series

converge or both diverge. The same must be true for $\displaystyle\sum_{k=1}^{\infty} u_k$ and $\displaystyle\sum_{k=1}^{\infty} v_k$. ∎

Corollary   If $\displaystyle\sum_{k=1}^{\infty} u_k$ and $\displaystyle\sum_{k=1}^{\infty} v_k$ are series of positive terms and if $\displaystyle\lim_{n\to\infty}\dfrac{u_n}{v_n}=0$ and $\displaystyle\sum_{k=1}^{\infty} v_k$

converges then $\displaystyle\sum_{k=1}^{\infty} u_k$ also converges.

Proof   Since $\displaystyle\lim_{n\to\infty}\dfrac{u_n}{v_n}=0$ then a positive integer N must exist such that $\dfrac{u_n}{v_n}<1$ for all

$n>N$. Hence $0<u_n<v_n$ for all $n>N$. Thus the series $\displaystyle\sum_{k=N+1}^{\infty} v_k$ dominates the

series $\displaystyle\sum_{k=N+1}^{\infty} u_k$ and since $\displaystyle\sum_{k=N+1}^{\infty} v_k$ converges, $\displaystyle\sum_{k=N+1}^{\infty} u_k$ converges also. ∎

EXAMPLE 3   Test the series

$$\sum_{k=1}^{\infty}\frac{k+5}{3k(k+\sqrt{k})}$$

for convergence.

Solution
$$u_n=\frac{n+5}{3n(n+\sqrt{n})}=\frac{1+\dfrac{5}{n}}{3n\left(1+\dfrac{1}{\sqrt{n}}\right)}$$

where the numerator and denominator have been divided by n. Therefore if we
introduce the nth term of the harmonic series $v_n=\dfrac{1}{n}$ and divide u_n by v_n there
results

$$\frac{u_n}{v_n}=\frac{1+\dfrac{5}{n}}{3\left(1+\dfrac{1}{\sqrt{n}}\right)}$$

from which

$$\lim_{n\to\infty}\frac{u_n}{v_n}=\frac{1}{3}$$

From the divergence of the harmonic series it follows that the given series diverges also. ●

EXAMPLE 4 Test the series

$$\sum_{k=1}^{\infty} \frac{3k+2}{(2k+1)5^k}$$

for convergence.

Solution

$$u_n = \frac{3n+2}{(2n+1)5^n} = \frac{3+\dfrac{2}{n}}{\left(2+\dfrac{1}{n}\right)5^n}$$

and if we let $v_n = \dfrac{1}{5^n}$ then

$$\lim_{n\to\infty} \frac{u_n}{v_n} = \frac{3}{2}$$

Since the geometric series $\displaystyle\sum_{n=1}^{\infty} \frac{1}{5^n}$ converges (it has ratio $r = \frac{1}{5} < 1$) it follows that the given series must also converge. ●

The effectiveness of the preceding comparison tests depends upon our ability to furnish a simple series whose convergence or divergence is known. The following test known as the **integral test** will assist us in finding these simple series.

Theorem 4 **(Cauchy Integral Test).** Let f be a continuous, decreasing, and positive valued function of x that is defined for all $x \geq 1$. Let $u_k = f(k)$ for all positive integers k. Then the infinite series

$$\sum_{k=1}^{\infty} u_k = f(1) + f(2) + f(3) + \cdots + f(n) + \cdots$$

will converge if the improper integral

$$\int_1^{\infty} f(x)\, dx$$

exists and the infinite series will diverge if the improper integral increases without bound.

Proof To visualize the proof, refer to Figure 15.4.1. Since f decreases on the interval $[1, n]$,

$$f(2) + f(3) + \cdots + f(n-1) + f(n)$$

is a lower sum for $\displaystyle\int_1^n f(x)\, dx$, whereas

$$f(1) + f(2) + \cdots + f(n-1)$$

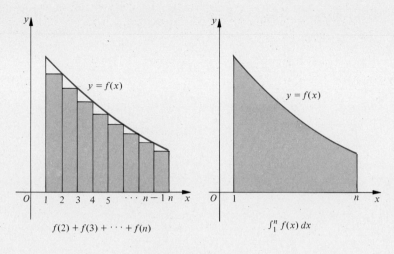

$$f(2) + f(3) + \cdots + f(n) \qquad\qquad \int_1^n f(x)\, dx$$

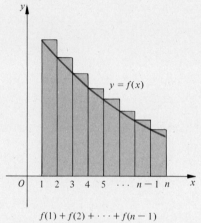

$$f(1) + f(2) + \cdots + f(n-1)$$

Figure 15.4.1

is an upper sum for $\displaystyle\int_1^n f(x)\, dx$. Hence, we have the simultaneous inequalities

$$f(2) + f(3) + \cdots + f(n) < \int_1^n f(x)\, dx < f(1) + f(2) + \cdots + f(n-1) \quad (1)$$

If the sequence

$$\left\{ \int_1^n f(x)\, dx \right\}$$

converges as n becomes infinite, it must be a bounded function of n and $\displaystyle\int_1^\infty f(x)\, dx$

exists. Thus, from the left inequality of (1), $\displaystyle\sum_{k=1}^\infty f(k)$ converges, which implies that

$\displaystyle\sum_{k=1}^\infty u_k$ (the given series) must also converge. On the other hand, if $\displaystyle\lim_{n\to\infty} \int_1^n f(x)\, dx$

does not exist, that is, $\displaystyle\int_1^\infty f(x)\, dx$ diverges, then $\displaystyle\int_1^n f(x)\, dx$ is an unbounded function of n. From the right inequality of (1), $\sum\limits_{k=1}^{\infty} f(k)$ must also be divergent. Hence $\sum\limits_{k=1}^{\infty} u_k$ diverges also. ∎

EXAMPLE 5   Use the integral test to show that the harmonic series

$$\sum_{k=1}^{\infty} \frac{1}{k}$$

diverges.

Solution   Let $f(x) = \dfrac{1}{x}$ so that $f(k) = \dfrac{1}{k}$. Also, $f(x)$ satisfies the requirements of the integral test, that is, it is continuous, decreasing and a positive valued function for all $x \geq 1$.

$$\int_{1}^{\infty} f(x)\, dx = \lim_{b \to \infty} \int_{1}^{b} \frac{dx}{x} = \lim_{b \to \infty} \ln b$$

$\ln b \to \infty$ as $b \to \infty$. Therefore the harmonic series diverges. ●

EXAMPLE 6   The series

$$\sum_{k=1}^{\infty} \frac{1}{k^p} = \frac{1}{1^p} + \frac{1}{2^p} + \frac{1}{3^p} + \cdots$$

is called the *p series.* Prove that the series converges if $p > 1$ and diverges if $p \leq 1$ by applying the integral test.

Solution   Let $f(x) = \dfrac{1}{x^p}$ so that $f(k) = \dfrac{1}{k^p}$. For $p \neq 1$, we have

$$\int_{1}^{\infty} f(x)\, dx = \lim_{b \to \infty} \int_{1}^{b} \frac{1}{x^p}\, dx = \lim_{b \to \infty} \int_{1}^{b} x^{-p}\, dx$$

$$= \lim_{b \to \infty} \frac{x^{1-p}}{1-p}\bigg|_{1}^{b} = \lim_{b \to \infty} \frac{1}{1-p}(b^{1-p} - 1)$$

If $p > 1$, $\int_{1}^{\infty} f(x)\, dx = \dfrac{1}{p-1}$ so that the corresponding infinite series must converge.

If $p < 1$, $\int_{1}^{\infty} f(x)$ does not exist, since $\lim\limits_{b \to \infty} b^{1-p}$ does not exist and the series diverges.

If $p = 1$, the series is the harmonic series treated in Example 5 which we know diverges. Hence by the integral test we have established that the p series converges if and only if $p > 1$. ●

EXAMPLE 7   Determine whether or not the infinite series

$$\sum_{k=2}^{\infty} \frac{1}{k \ln k}$$

converges.

Solution   This is a particularly interesting series because $\frac{1}{k^2} < \frac{1}{k \ln k} < \frac{1}{k}$ for $k = 3, 4, 5, \ldots$. However, the comparison test *fails* to give us information since, $\sum_{k=2}^{\infty} \frac{1}{k^2}$ converges (p series with $p = 2$), whereas $\sum_{k=2}^{\infty} \frac{1}{k}$ diverges since it is the harmonic series. Thus our series dominates a convergent series whereas a divergent series dominates it.

We turn now to the integral test. Let $f(x) = \frac{1}{x \ln x}$ then

$$f'(x) = -\frac{1 + \ln x}{(x \ln x)^2}$$

Since $f(x) > 0$ and $f'(x) < 0$ if $x > 2$, the integral test may be utilized to test the convergence of our given series. Now since $d(\ln x) = \frac{dx}{x}$, we have

$$\int_2^\infty \frac{dx}{x \ln x} = \lim_{b \to \infty} \int_2^b \frac{dx}{x \ln x} = \lim_{b \to \infty} \ln(\ln x) \Big|_2^b$$
$$= \lim_{b \to \infty} [\ln(\ln b) - \ln(\ln 2)]$$

which clearly diverges. Therefore $\sum_{k=2}^{\infty} \frac{1}{k \ln k}$ must diverge also.   ●

Under the hypotheses of the integral test, if $\int_1^\infty f(x)\, dx = L$ (where L is a real number) then the series $\sum_{k=1}^{\infty} u_k$ converges. If S is the sum of the series, it follows from (1) that

$$L \leq S \leq L + u_1 \qquad \text{(where } u_1 = f(1)\text{)}.$$

Thus, the integral test also gives us upper and lower bounds on the sum of a series.

As noted earlier, if a finite number of terms are eliminated from a series, it will not change the convergence or divergence of the series (although, if the series is convergent, the sum will clearly be affected). Thus $\sum_{k=1}^{\infty} u_k$ converges if and only if $\sum_{k=r}^{\infty} u_k$ also converges, where r is any positive integer. If the hypotheses of the integral test are satisfied, then to determine the convergence or divergence of $\sum_{k=r}^{\infty} u_k$ we need only examine $\int_r^\infty f(x)\, dx$.

Exercises 15.4

In each of Exercises 1 through 26, determine whether the series is convergent or divergent.

1. $1 + \dfrac{1}{2^4} + \dfrac{1}{3^4} + \cdots$

2. $\dfrac{1}{1 \cdot 3} + \dfrac{1}{2 \cdot 4} + \dfrac{1}{3 \cdot 5} + \cdots$

3. $\displaystyle\sum_{k=1}^{\infty} \dfrac{2^k}{3^k + 4}$

4. $\frac{1}{5} + \frac{1}{9} + \frac{1}{13} + \cdots$

5. $\dfrac{2}{1^2} + \dfrac{3}{2^2} + \dfrac{4}{3^2} + \cdots$

6. $\displaystyle\sum_{k=1}^{\infty} \dfrac{1}{k(2k - 1)}$

7. $\displaystyle\sum_{k=1}^{\infty} \dfrac{2k + 1}{k} e^{-k}$

8. $\displaystyle\sum_{k=1}^{\infty} \dfrac{k}{k^2 + 10^3}$

9. $\displaystyle\sum_{k=1}^{\infty} \dfrac{k^2}{k^3 + 10^3}$ by the integral test

10. $\displaystyle\sum_{k=1}^{\infty} \dfrac{k}{k^2 + a^2}$ by the integral test

11. $\displaystyle\sum_{k=1}^{\infty} (\tfrac{3}{4})^k \, |\cos k|$

12. $\displaystyle\sum_{k=1}^{\infty} \dfrac{7k^2 + 10}{k^3 + k^2 + 1}$

13. $\displaystyle\sum_{k=1}^{\infty} \dfrac{\ln k}{k^2}$ (*Hint:* Use the integral test)

14. $\displaystyle\sum_{k=2}^{\infty} \dfrac{1}{k \ln^2 k}$

15. $\displaystyle\sum_{k=1}^{\infty} \dfrac{k}{(2k + 1)(5k - 1)(7k - 3)}$

16. $\displaystyle\sum_{k=2}^{\infty} \dfrac{1}{\ln^2 k}$

17. $\displaystyle\sum_{k=1}^{\infty} \dfrac{3^k}{k^2}$ $\left(\textit{Hint: Find } \dfrac{u_{n+1}}{u_n}\right)$

18. $\displaystyle\sum_{k=1}^{\infty} \dfrac{k^{1/2}}{(k + 100)^{3/2}}$

19. $\displaystyle\sum_{k=1}^{\infty} \dfrac{k^{2/3} + 3k^{1/2}}{3k^2 + 5k - 1}$

20. $\displaystyle\sum_{k=1}^{\infty} \sin \dfrac{\pi}{2k}$

21. $\displaystyle\sum_{k=1}^{\infty} \dfrac{7^{k+10}}{2^{3k}}$

22. $\displaystyle\sum_{k=1}^{\infty} \dfrac{10^{k-2}}{3^k 5^{k+1}}$

23. $\displaystyle\sum_{k=1}^{\infty} \dfrac{(k + 2)(k + 4)(k + 6)}{k^{3/2}(k + 1)(k + 3)}$

24. $\dfrac{1^{3/2}}{1 \cdot 3 \cdot 5} + \dfrac{2^{3/2}}{5 \cdot 7 \cdot 9} + \dfrac{3^{3/2}}{9 \cdot 11 \cdot 13} + \cdots$

25. $\dfrac{\ln 3}{1^2} + \dfrac{\ln 4}{2^2} + \dfrac{\ln 5}{3^2} + \cdots$

26. $\displaystyle\sum_{k=1}^{\infty} \dfrac{\sqrt{k}}{\sqrt{k + 1}\sqrt{k^2 + 3k + 8}}$

27. Prove that if $\displaystyle\sum_{k=1}^{\infty} a_k$ converges, where $a_k \geq 0$ for all positive integers k, then $\displaystyle\sum_{k=1}^{\infty} a_k^2$ must also converge.

28. If $a_k \geq 0$ and $\displaystyle\sum_{k=1}^{\infty} a_k$ converges, prove that $\displaystyle\sum_{k=1}^{\infty} \dfrac{a_k}{\sqrt{1 + a_k}}$ must also converge.

29. Does the series $\displaystyle\sum_{k=1}^{\infty} \dfrac{\tanh k}{k}$ converge or diverge? Why?

30. Establish the convergence or divergence of the series $\displaystyle\sum_{k=1}^{\infty} \dfrac{\sqrt{k + 3} - \sqrt{k}}{k}$.

THE RATIO AND ROOT TESTS

Theorem 1 (*The Ratio Test*). Let $\sum\limits_{k=1}^{\infty} u_k$ be a series with positive terms and suppose

that $\lim\limits_{k\to\infty} \dfrac{u_{k+1}}{u_k} = \rho$. Then

  (i)   If $\rho < 1$, the series $\sum\limits_{k=1}^{\infty} u_k$ converges.

  (ii) If $\rho > 1$ or $\rho = \infty$, the series $\sum\limits_{k=1}^{\infty} u_k$ diverges.

  (iii) If $\rho = 1$, the test fails to determine whether or not the series converges.

Proof **Case (i), $\rho < 1$.** Choose $\varepsilon > 0$ so small that $\rho + \varepsilon < 1$. Therefore an $N(\varepsilon)$ exists such that

$$\frac{u_{k+1}}{u_k} < \rho + \varepsilon \qquad \text{for all } k \geq N$$

This implies that $u_{N+1} < (\rho + \varepsilon)u_N$, $u_{N+2} < (\rho + \varepsilon)u_{N+1} < (\rho + \varepsilon)^2 u_N$ and more generally,

$$u_{N+j} < (\rho + \varepsilon)^j \, u_N \qquad (j = 1, 2, 3, \ldots)$$

Therefore the series $\sum\limits_{k=N}^{\infty} u_k$ converges since it is dominated by a convergent

geometric series

$$u_N[1 + (\rho + \varepsilon) + (\rho + \varepsilon)^2 + \cdots] \qquad \text{where } \rho + \varepsilon < 1$$

Hence the given series converges.

Case (ii), $\rho > 1$. Choose $\varepsilon > 0$ so small that $\rho - \varepsilon > 1$. Therefore an $N(\varepsilon)$ exists such that

$$\frac{u_{k+1}}{u_k} > \rho - \varepsilon \qquad \text{for all } k \geq N$$

This implies that $u_{N+1} > (\rho - \varepsilon)u_N$, $u_{N+2} > (\rho - \varepsilon)u_{N+1} > (\rho - \varepsilon)^2 u_N$, and, more generally,

$$u_{N+j} > (\rho - \varepsilon)^j \, u_N \qquad (j = 1, 2, 3, \ldots)$$

Therefore the series $\sum\limits_{k=N}^{\infty} u_k$ diverges since it dominates the divergent series

$$u_N[1 + (\rho - \varepsilon) + (\rho - \varepsilon)^2 + \cdots] \qquad \text{where } \rho - \varepsilon > 1$$

The proof of the case $\rho = \infty$ is left for the reader.

Case (iii), $\rho = 1$. This is the inconclusive case. This simply means that $\rho = 1$ can hold for both convergent and divergent series. For example, consider the harmonic series $\sum\limits_{k=1}^{\infty} \dfrac{1}{k}$ which *diverges*. For this series

$$\lim_{k\to\infty} \frac{u_{k+1}}{u_k} = \lim_{k\to\infty} \frac{\dfrac{1}{k+1}}{\dfrac{1}{k}} = \lim_{k\to\infty} \frac{k}{k+1} = \lim_{k\to\infty} \frac{1}{1+\dfrac{1}{k}} = 1$$

Next, consider the *convergent* series $\sum\limits_{k=1}^{\infty} \dfrac{1}{k^2}$. For this series

$$\lim_{k\to\infty} \frac{u_{k+1}}{u_k} = \lim_{k\to\infty} \frac{\dfrac{1}{(k+1)^2}}{\dfrac{1}{k^2}} = \lim_{k\to\infty} \frac{k^2}{(k+1)^2}$$

$$= \lim_{k\to\infty} \left(\frac{1}{1+\dfrac{1}{k}}\right)^2 = 1$$

Hence the ratio test fails when $\rho = 1$. ∎

EXAMPLE 1 Test the series $\sum\limits_{k=1}^{\infty} \dfrac{3^k}{k!}$ for convergence.

Solution
$$\frac{u_{k+1}}{u_k} = \frac{\dfrac{3^{k+1}}{(k+1)!}}{\dfrac{3^k}{k!}} = \frac{3^{k+1}}{(k+1)!} \cdot \frac{k!}{3^k} = \frac{3}{k+1}$$

so that

$$\lim_{k\to\infty} \frac{u_{k+1}}{u_k} = \lim_{k\to\infty} \frac{3}{k+1} = 0$$

The series therefore converges.

EXAMPLE 2 Test the series $\dfrac{1^3}{7} + \dfrac{2^3}{7^2} + \dfrac{3^3}{7^3} + \cdots$ for convergence.

Solution In this instance, $u_k = \dfrac{k^3}{7^k}$ so that

$$\lim_{k\to\infty} \frac{u_{k+1}}{u_k} = \lim_{k\to\infty} \frac{(k+1)^3}{7^{k+1}} \cdot \frac{7^k}{k^3} = \lim_{k\to\infty} \left(\frac{k+1}{k}\right)^3\left(\frac{1}{7}\right) = \frac{1}{7}$$

Hence the series converges.

EXAMPLE 3   Test the series $\displaystyle\sum_{k=1}^{\infty} \frac{k^k}{(k+2)!}$

Solution   We have $u_k = \dfrac{k^k}{(k+2)!}$ from which

$$\lim_{k\to\infty} \frac{u_{k+1}}{u_k} = \lim_{k\to\infty} \frac{(k+1)^{k+1}}{(k+3)!} \cdot \frac{(k+2)!}{k^k} = \lim_{k\to\infty} \left(\frac{k+1}{k}\right)^k \left(\frac{k+1}{k+3}\right)$$

$$= \lim_{k\to\infty} \left(1 + \frac{1}{k}\right)^k \left(\frac{k+1}{k+3}\right) = \lim_{k\to\infty} \left(1 + \frac{1}{k}\right)^k \cdot \lim_{k\to\infty} \frac{k+1}{k+3}$$

$$= e \cdot 1 = e > 1$$

Therefore the given series diverges.   ●

EXAMPLE 4   Test the series $\dfrac{1}{1\cdot 3} + \dfrac{1}{3\cdot 5} + \dfrac{1}{5\cdot 7} + \cdots$ for convergence.

Solution $$u_k = \frac{1}{(2k-1)(2k+1)} \qquad \text{and} \qquad u_{k+1} = \frac{1}{(2k+1)(2k+3)}$$

so that

$$\lim_{k\to\infty} \frac{u_{k+1}}{u_k} = \lim_{k\to\infty} \frac{2k-1}{2k+3} = 1$$

Therefore the ratio test fails to give us any information. Actually the series converges by application of the comparison test. It is left as an exercise for the reader to verify this.   ●

REMARK: After using the ratio test on a number of examples, the following pattern seems to arise:

(i)   If the terms of the series are algebraic expressions, then the ratio test usually fails to determine a result (that is, $\rho = 1$),

(ii)   If the terms of the series involve factorials or exponentials, then the ratio test usually will determine whether the series converges or diverges.

Theorem 2   (**The Root Test**).   Let $\displaystyle\sum_{k=1}^{\infty} u_k$ be a series with nonnegative terms for which

$$\lim_{k\to\infty} u_k^{1/k} = \rho$$

Then

(i)   If $\rho < 1$, the series $\displaystyle\sum_{k=1}^{\infty} u_k$ converges.

(ii)   If $\rho > 1$, the series $\displaystyle\sum_{k=1}^{\infty} u_k$ diverges.

(iii)   If $\rho = 1$, the test fails to determine whether or not the series converges.

Proof **Case (i), $\rho < 1$.** Choose $\varepsilon > 0$ so small that $\rho + \varepsilon < 1$. Since $\lim\limits_{k \to \infty} u_k^{1/k} = \rho$, an $N(\varepsilon)$ exists such that

$$u_k^{1/k} < \rho + \varepsilon < 1 \qquad \text{for all } k > N$$

We have

$$u_k < (\rho + \varepsilon)^k \qquad k > N$$

But, $\sum\limits_{k=N+1}^{\infty} (\rho + \varepsilon)^k$ converges because it is a geometric series with $r = \rho + \varepsilon$ and

$0 < \rho + \varepsilon < 1$. Then $\sum\limits_{k=N+1}^{\infty} u_k$ converges by the comparison test. Therefore $\sum\limits_{k=1}^{\infty} u_k$ also must converge.

The proof of (ii) and the verification of the failure in the case of (iii) is left as exercises for the reader. ∎

EXAMPLE 5 Test the series

$$\sum_{k=1}^{\infty} \frac{k}{10^k}$$

for convergence by application of the root test.

Solution $$u_k = \frac{k}{10^k} \qquad \text{and} \qquad u_k^{1/k} = \frac{k^{1/k}}{10}$$

so that

$$\lim_{k \to \infty} u_k^{1/k} = \lim_{k \to \infty} \frac{k^{1/k}}{10} = \frac{1}{10}.$$

(The reader should verify that $\lim\limits_{k \to \infty} k^{1/k} = 1$). Therefore the given series converges. ●

EXAMPLE 6 Test the series

$$\sum_{k=1}^{\infty} kr^k \qquad (0 \le r < 1)$$

for convergence.

Solution $$u_k = kr^k \qquad \text{and} \qquad u_k^{1/k} = k^{1/k}r$$

so that

$$\lim_{k \to \infty} u_k^{1/k} = r \qquad \text{where } 0 \le r < 1$$

and thus the given series converges. ●

Exercises 15.5

In Exercises 1 through 16 use the ratio or root test to determine whether the given series converges or diverges. If neither test answers the question, resolve by another procedure.

1. $1 + \dfrac{1}{10} + \dfrac{1}{10^2} + \cdots$

2. $\dfrac{1}{4} + \dfrac{4}{4^2} + \dfrac{7}{4^3} + \dfrac{10}{4^4} + \cdots$

3. $\dfrac{2}{1!} + \dfrac{4}{2!} + \dfrac{6}{3!} + \cdots$

4. $\dfrac{3}{\sqrt{5}} + \dfrac{3^2}{5} + \dfrac{3^3}{5\sqrt{5}} + \cdots$

5. $\dfrac{1}{4} + \dfrac{1 \cdot 2}{4 \cdot 8} + \dfrac{1 \cdot 2 \cdot 3}{4 \cdot 8 \cdot 12} + \cdots$

6. $7 + \dfrac{7^2}{11} + \dfrac{7^3}{11^2} + \dfrac{7^4}{11^3} + \cdots$

7. $\displaystyle\sum_{k=1}^{\infty} \dfrac{e^{100k}}{k!}$

8. $\dfrac{1}{1 \cdot 2} + \dfrac{4}{2 \cdot 5} + \dfrac{16}{3 \cdot 10} + \cdots + \dfrac{4^{n-1}}{n(n^2 + 1)} + \cdots$

9. $\displaystyle\sum_{k=1}^{\infty} \dfrac{k!}{10^{3k}}$

10. $\displaystyle\sum_{k=1}^{\infty} \dfrac{(k+1)^2 e^{k+2}}{\sqrt{k} 3^{k+1}}$

11. $1 + 2r + r^2 + 2r^3 + r^4 + \cdots$ where $0 < r < 1$

12. $\displaystyle\sum_{k=1}^{\infty} \dfrac{(k+1)!}{(k+2)!}$

13. $\displaystyle\sum_{k=1}^{\infty} \dfrac{k^k}{k!}$

14. $\displaystyle\sum_{n=2}^{\infty} \dfrac{1}{(\ln n)^n}$

15. $\displaystyle\sum_{k=1}^{\infty} \dfrac{k^2 + 1}{k^2 + k} \cdot \dfrac{10^{k-1}}{5^{k+2}}$

16. $\displaystyle\sum_{n=1}^{\infty} \dfrac{4^n n!}{n^n}$

17. Show that $\displaystyle\lim_{n \to \infty} \dfrac{r^n}{n!} = 0$ for all values of r. (*Hint:* Consider the infinite series $\displaystyle\sum_{n=1}^{\infty} \dfrac{r^n}{n!}$.)

18. Show that the positive series $\displaystyle\sum_{k=1}^{\infty} u_k$ may or may not converge if $\dfrac{u_{k+1}}{u_k} < 1$ for all positive integers k. Does this contradict the ratio test?

19. Prove the following theorem: Suppose $\displaystyle\sum_{k=1}^{\infty} u_k$ is an infinite series of positive terms. Then this series will converge if there is a number $\rho < 1$ and an integer N such that

$$\dfrac{u_{k+1}}{u_k} \leq \rho \qquad \text{if } k \geq N$$

(Note that this does not require that $\displaystyle\lim_{k \to \infty} \dfrac{u_{k+1}}{u_k}$ exists and in this sense is a generalization of the ratio test in the text.)

20. Prove the theorem: Let $\displaystyle\sum_{k=1}^{\infty} u_k$ be an infinite series of positive terms. Then this series will diverge if there is an integer N such that

$$\dfrac{u_{k+1}}{u_k} \geq 1 \qquad \text{if } k \geq N$$

15.6 ABSOLUTE CONVERGENCE— ALTERNATING SERIES

In this section we shall treat infinite series consisting of an infinite number of positive and negative numbers.

Let $\displaystyle\sum_{k=1}^{\infty} u_k$ consist of an infinite number of positive and negative terms. We shall prove the following

Theorem 1 If $\displaystyle\sum_{k=1}^{\infty} |u_k|$ converges then $\displaystyle\sum_{k=1}^{\infty} u_k$ must also converge.

Proof For each positive integer k

$$-|u_k| \le u_k \le |u_k| \qquad (k = 1, 2, 3, \ldots)$$

so that if we add $|u_k|$ to each member of the double inequality, we obtain

$$0 \le |u_k| + u_k \le 2|u_k|$$

Now $\displaystyle\sum_{k=1}^{n} (|u_k| + u_k)$ is a nondecreasing function of n that is bounded above by

$2\displaystyle\sum_{k=1}^{n} |u_k| \le 2 \sum_{k=1}^{\infty} |u_k|$. Therefore $\displaystyle\sum_{k=1}^{\infty} (|u_k| + u_k)$ converges. Now

$$u_k = (|u_k| + u_k) - |u_k|$$

and since both series $\displaystyle\sum_{k=1}^{\infty} (|u_k| + u_k)$ and $\displaystyle\sum_{k=1}^{\infty} |u_k|$ converge, then $\displaystyle\sum_{k=1}^{\infty} u_k$ must also converge. ∎

Definition 1 A series $\displaystyle\sum_{k=1}^{\infty} u_k$ for which $\displaystyle\sum_{k=1}^{\infty} |u_k|$ converges is said to be *absolutely convergent.*

Theorem 1 states that if a series converges absolutely it must also converge.

EXAMPLE 1 Test $1 - \dfrac{1}{2} - \dfrac{1}{2^2} + \dfrac{1}{2^3} + \dfrac{1}{2^4} - \cdots$ for convergence.

Solution If we replace each of the terms by its absolute value, there results the series

$$1 + \frac{1}{2} + \frac{1}{2^2} + \frac{1}{2^3} + \cdots$$

This series is a geometric series with ratio $\frac{1}{2}$ and therefore it is convergent. Thus the original series is absolutely convergent and by Theorem 1 is also convergent.

●

EXAMPLE 2   Test $\dfrac{1}{1\cdot 2} - \dfrac{1}{2\cdot 3} + \dfrac{1}{3\cdot 4} - \cdots$ for convergence.

Solution   If each of the terms is replaced by its absolute value, we have the series

$$\frac{1}{1\cdot 2} + \frac{1}{2\cdot 3} + \frac{1}{3\cdot 4} + \cdots + \frac{1}{n(n+1)} + \cdots$$

But this series is dominated by the convergent p series $(p = 2)$

$$\frac{1}{1^2} + \frac{1}{2^2} + \frac{1}{3^2} + \cdots + \frac{1}{n^2} + \cdots$$

Hence the original series is absolutely convergent and thus convergent.   ●

We shall see that the converse of Theorem 1 is *false;* that is, a convergent series might not converge absolutely.

EXAMPLE 3   The series

$$1 - \tfrac{1}{2} + \tfrac{1}{3} - \tfrac{1}{4} + \cdots \quad \text{(alternating harmonic series)}$$

is such that if we replace each term by its absolute value there results

$$1 + \tfrac{1}{2} + \tfrac{1}{3} + \tfrac{1}{4} + \cdots$$

which is the divergent harmonic series. However, Theorem 4 (proved below) shows that the original series is convergent. The series $\displaystyle\sum_{k=1}^{\infty} \frac{(-1)^{k+1}}{k}$ is an example of a *conditionally convergent* series.   ●

Definition 2   A series $\displaystyle\sum_{k=1}^{\infty} u_k$, which is convergent but not absolutely convergent (that is, $\displaystyle\sum_{k=1}^{\infty} |u_k|$ diverges) is said to be ***conditionally convergent.***

Theorem 2   (***The Ratio Test for Absolute Convergence—and Thus Convergence***).

Let $\displaystyle\sum_{k=1}^{\infty} u_k$ be a given infinite series for which $u_k \neq 0, k = 1, 2, 3, \ldots$. Then

(i)   If $\displaystyle\lim_{k\to\infty} \left| \frac{u_{k+1}}{u_k} \right| = \rho < 1$, the given series converges absolutely.

(ii)   If $\displaystyle\lim_{k\to\infty} \left| \frac{u_{k+1}}{u_k} \right| = \rho > 1$ or if $\rho = \infty$, the given series diverges.

(iii)   If $\displaystyle\lim_{k\to\infty} \left| \frac{u_{k+1}}{u_k} \right| = \rho = 1$, the test fails.

The proof of Theorem 2 follows readily from the proof of Theorem 1 of Section 15.5 and is left to the reader.

Theorem 3 **(*The Root Test for Absolute Convergence—and Thus Convergence*).**

Let $\displaystyle\sum_{k=1}^{\infty} u_k$ be a series for which $\displaystyle\lim_{k\to\infty} |u_k|^{1/k}$ exists or is infinite

(i)   If $\displaystyle\lim_{k\to\infty} |u_k|^{1/k} = \rho < 1$, the given series converges absolutely.

(ii)   If $\displaystyle\lim_{k\to\infty} |u_k|^{1/k} = \rho > 1$ or if $\rho = \infty$, the given series diverges.

(iii)   If $\displaystyle\lim_{k\to\infty} |u_k|^{1/k} = \rho = 1$, the test fails.

The proof of Theorem 3 follows easily from the proof of Theorem 2 of Section 15.5 and is left to the reader.

Definition 3   A series $u_1 - u_2 + u_3 - \cdots$, where $u_k > 0$ for all k, in which the successive terms have opposite signs is called an ***alternating series.***

Some examples of alternating series are

$$1 - \frac{1}{3} + \frac{1}{5} - \frac{1}{7} + \cdots$$

$$-\frac{1}{1!} + \frac{1}{2!} - \frac{1}{3!} + \frac{1}{4!} - \cdots$$

$$\frac{1}{2\ln 2} - \frac{1}{3\ln 3} + \frac{1}{4\ln 4} - \cdots$$

whereas

$$1 + \frac{1}{2^2} - \frac{1}{3^2} + \frac{1}{4^2} + \frac{1}{5^2} - \frac{1}{6^2} + \cdots$$

is *not* an alternating series because there are consecutive terms with the same sign.

Theorem 4   **(*Alternating Series Test—or Leibnitz Test*).** If $0 < u_{k+1} \leq u_k$ for each $k = 1, 2, 3, \ldots$, that is $\{u_k\}$ is a nonincreasing sequence of positive numbers such that $\displaystyle\lim_{k\to\infty} u_k = 0$, then $\displaystyle\sum_{k=1}^{\infty} (-1)^{k-1} u_k$ converges.†

Proof   Let us examine first the even partial sums,

$$S_2 = u_1 - u_2, \quad S_4 = (u_1 - u_2) + (u_3 - u_4)$$

---

† We assume for simplicity that the first term of the alternating series is positive. If the first term is negative, factor out -1 and then proceed as in Theorem 4.

and in general

$$S_{2n} = (u_1 - u_2) + (u_3 - u_4) + \cdots + (u_{2n-1} - u_{2n})$$

By assumption, $u_1 - u_2 \geq 0, u_3 - u_4 \geq 0$, and in general, $u_{2n-1} - u_{2n} \geq 0$, from which it follows that

$$0 \leq S_2 \leq S_4 \leq \cdots \leq S_{2n} \leq \cdots$$

We may also write S_{2n} in the form

$$S_{2n} = u_1 - (u_2 - u_3) - (u_4 - u_5) - \cdots - (u_{2n-2} - u_{2n-1}) - u_{2n}$$

from which it follows that

$$S_{2n} \leq u_1$$

Therefore, the sequence of even partial sums $\{S_{2n}\}$ must converge to a limit $\leq u_1$.
   Now, $S_{2n+1} = S_{2n} + u_{2n+1}$ so that

$$\lim_{n \to \infty} S_{2n+1} = \lim_{n \to \infty} S_{2n} + \lim_{n \to \infty} u_{2n+1} = \lim_{n \to \infty} S_{2n} + 0$$

from which we have

$$\lim_{n \to \infty} S_n = \lim_{n \to \infty} S_{2n} = \lim_{n \to \infty} S_{2n+1} = S \text{ where } S \leq u_1 \qquad \blacksquare$$

From Theorem 4 it follows that the alternating series

$$1 - \frac{1}{3} + \frac{1}{5} - \frac{1}{7} + \cdots \qquad \text{(i)}$$

$$-\frac{1}{1!} + \frac{1}{2!} - \frac{1}{3!} + \frac{1}{4!} - \cdots \qquad \text{(ii)}$$

$$\frac{1}{2 \ln 2} - \frac{1}{3 \ln 3} + \frac{1}{4 \ln 4} - \cdots \qquad \text{(iii)}$$

are convergent; however, it is left to the reader to apply the ratio and the integral tests and show that (ii) is the only infinite series (in the preceding list) that is also absolutely convergent.
   Let us consider again the alternating series satisfying the hypotheses of Theorem 4. Let R_n denote the remainder of this series after n term. Thus

$$R_n = S - S_n = (-1)^n(u_{n+1} - u_{n+2} + u_{n+3} - \cdots)$$

and the expression on the right side is again an alternating series. Since $u_{n+1} \geq u_{n+2} \geq u_{n+3}$, and so on, and $|(-1)^n| = 1$, we have

$$|R_n| = u_{n+1} - u_{n+2} + u_{n+3} - \cdots$$

so that $|R_n| \leq u_{n+1}$ (as proved analogously to Theorem 4). Therefore

$$|R_n| = |S - S_n| \leq u_{n+1} \qquad \text{(1)}$$

and we have the very useful result

Theorem 5   If $S = u_1 - u_2 + u_3 - u_4 + \cdots$ is a convergent alternating series for which $0 < u_{k+1} \leq u_k, (k = 1, 2, 3, \ldots)$, and $\lim_{k \to \infty} u_k = 0$, then, if $S = S_n + R_n$

where S_n is the nth partial sum of the series, we have $|S - S_n| \leq u_{n+1}$. This means that the magnitude of the error if S_n is used to approximate S is less than or equal to the first term not included in the partial summation.

It is important to note that in many applications $u_{n+1} \leq u_n$ may not hold true for all positive integers n but rather for $n \geq n_0$, where n_0 is some fixed positive integer. This, of course, does not affect convergence.

Also, one may find that the sign alternation takes place only if $n \geq n_1$, where n_1 is some fixed integer. Again convergence is not affected.

EXAMPLE 4   Test the series

$$1 - \tfrac{1}{2} + \tfrac{1}{3} - \tfrac{1}{4} + \cdots \tag{2}$$

for convergence.

Solution   If $u_k = \dfrac{1}{k}$ then the series is of the form $\displaystyle\sum_{k=1}^{\infty} \dfrac{(-1)^{k-1}}{k}$. We must make sure that *all* the hypotheses of Theorem 4 are met. The series is alternating. Also, $u_k = \dfrac{1}{k} > \dfrac{1}{k+1} = u_{k+1}$ and finally $\displaystyle\lim_{k \to \infty} u_k = \lim_{k \to \infty} \dfrac{1}{k} = 0$. Therefore the series converges.   ●

EXAMPLE 5   For the series (2) of Example 4 find a positive integer N such that S_n approaches S, the sum of the series, with at least three decimal place accuracy when $n \geq N$.

Solution   The problem requires that $|S - S_n| < 0.0005$. From (1) this implies that

$$u_{n+1} = \frac{1}{n+1} \leq 0.0005$$

or equivalently, $n \geq 1999$.

Hence, S_{1999} yields the required accuracy, in fact, $S_{1999} > S$. This implies that the series is a very slowly convergent one in that almost 2000 terms are needed for at least three decimal place accuracy.   ●

EXAMPLE 6   (a) Show that the series

$$1 - \frac{1}{2!} + \frac{1}{3!} - \frac{1}{4!} + \frac{1}{5!} - \cdots + \frac{(-1)^k}{k!} + \cdots$$

is convergent. (b) Is the series absolutely convergent? (c) Find n such that $|S_n - S| < 0.0001$.

Solution   (a) The series is alternating and $u_{k+1} = \dfrac{1}{(k+1)!} < \dfrac{1}{k!} = u_k$. Also, $\displaystyle\lim_{k \to \infty} u_k = \lim_{k \to \infty} \dfrac{1}{k!} = 0$. Therefore the series converges.

(b) The series $\sum\limits_{k=1}^{\infty} \dfrac{1}{k!}$ converges by application of the ratio test—

$$\lim_{k \to \infty} \frac{u_{k+1}}{u_k} = \lim_{k \to \infty} \frac{1}{(k+1)!} \cdot \frac{k!}{1} = \lim_{k \to \infty} \frac{1}{k+1} = 0$$

Hence the given series converges absolutely.

(c) $u_{n+1} = \dfrac{1}{(n+1)!} < 0.0001$ is required. Direct calculation yields $n = 7$, since

$$\frac{1}{8!} = \frac{1}{40,320} < 0.00003. \text{ Since from}$$

$$S_7 = 1 - \frac{1}{2!} + \frac{1}{3!} - \frac{1}{4!} + \frac{1}{5!} - \frac{1}{6!} + \frac{1}{7!} \doteq 0.63212$$

and $|S - S_7| < 0.00003$ we have that 0.6321 approximates S correctly to four decimal places. In fact, it will presently be verified that $1 - \dfrac{1}{2!} + \dfrac{1}{3!} - \cdots = 1 - e^{-1}$. ●

Exercises 15.6

In Exercises 1 through 12, determine whether or not the given series is convergent.

1. $1 - \dfrac{1}{4} + \dfrac{1}{7} - \cdots + \dfrac{(-1)^{n-1}}{3n-2} + \cdots$

2. $-1 + \dfrac{1}{2^2} - \dfrac{1}{2^4} + \dfrac{1}{2^6} - \cdots + \dfrac{(-1)^n}{2^{2(n-1)}} + \cdots$

3. $2 - \frac{3}{2} + \frac{4}{3} - \frac{5}{4} + \cdots + (-1)^{n-1}\left(1 + \dfrac{1}{n}\right) + \cdots$

4. $\sum\limits_{k=1}^{\infty} (-1)^k \dfrac{4^k}{k^2}$

5. $\sum\limits_{k=2}^{\infty} \dfrac{(-1)^k}{k} \ln k$

6. $\sum\limits_{k=2}^{\infty} \dfrac{(-1)^{k-1}}{k \ln k}$

7. $\sum\limits_{k=1}^{\infty} \dfrac{(-1)^k k}{k^2 + 1}$ (*Hint:* let $f(x) = \dfrac{x}{x^2 + 1}$ and check that $f'(x) < 0$ for $x > 1$.)

8. $\sum\limits_{k=1}^{\infty} \dfrac{(-1)^k}{k^2 - 2k + 3}$

9. $\sum\limits_{k=1}^{\infty} \dfrac{(-1)^k(k-2)}{k^2 + 3}$

10. $\sum\limits_{k=1}^{\infty} \dfrac{(-1)^{k+1}\sqrt{k^2 + 1}}{k^{3/2}}$

11. $\displaystyle\sum_{k=1}^{\infty} \frac{(-1)^{k-1}(2k-1)}{100(2k+1)}$

12. $\displaystyle\sum_{k=2}^{\infty} \frac{(-1)^{k-1}\ln k}{k^r} \qquad \frac{1}{2} \leq r < 1$

In Exercises 13 through 16, find the sum of each of the series to three decimal places.

13. $1 - \dfrac{1}{3^2} + \dfrac{1}{3^4} - \dfrac{1}{3^6} + \cdots + \dfrac{(-1)^{n-1}}{3^{2(n-1)}} + \cdots$

14. $1 - \dfrac{1}{2(1!)^2} + \dfrac{1}{2^2(2!)^2} + \cdots + \dfrac{(-1)^{n-1}}{2^{n-1}[(n-1)!]^2} + \cdots$

15. $1 - \dfrac{1}{2!} + \dfrac{1}{4!} - \dfrac{1}{6!} + \cdots + \dfrac{(-1)^{n-1}}{[2(n-1)]!} + \cdots$

16. $1 - \dfrac{1}{3^2} + \dfrac{1}{15^2} - \dfrac{1}{35^2} + \cdots + \dfrac{(-1)^{n-1}}{[4(n-1)^2-1]^2} + \cdots$

In Exercises 17 through 32, determine whether the given series is absolutely convergent, conditionally convergent, or divergent.

17. $\displaystyle\sum_{k=1}^{\infty} \frac{(-1)^k}{\sqrt{k}}$

18. $\displaystyle\sum_{k=1}^{\infty} \frac{(-1)^k}{k^2 + 3k + 4}$

19. $\displaystyle\sum_{k=1}^{\infty} \frac{(-1)^k k}{\sqrt{k^2 + 100}}$

20. $\displaystyle\sum_{k=1}^{\infty} \frac{(-1)^k}{\sqrt[3]{k}(k+7)}$

21. $\displaystyle\sum_{k=2}^{\infty} \frac{\sin k}{k(k-1)}$

22. $\displaystyle\sum_{k=1}^{\infty} \frac{(-1)^k}{\sqrt[3]{k(k+1)(k+2)}}$

23. $\displaystyle\sum_{k=1}^{\infty} \frac{(-1)^k k!}{k^4 + 1}$

24. $\displaystyle\sum_{k=2}^{\infty} \frac{(-1)^k \ln k}{k}$

25. $\displaystyle\sum_{k=1}^{\infty} \frac{(-1)^k \cos(3k\pi)}{k}$

26. $\displaystyle\sum_{k=2}^{\infty} \frac{(-1)^k \ln \sqrt[k]{k}}{k}$

27. $\displaystyle\sum_{k=1}^{\infty} \frac{(-1)^k \sin \dfrac{1}{k}}{k}$

28. $\displaystyle\sum_{k=1}^{\infty} (-1)^k \frac{3\cdot5\cdot7\cdots(2k+1)}{2\cdot4\cdot6\cdots(2k)}$

29. $\displaystyle\sum_{k=1}^{\infty} (-1)^k \frac{1\cdot3\cdot5\cdots(2k-1)}{3\cdot6\cdot9\cdots(3k)}$

30. $\displaystyle\sum_{k=1}^{\infty} (-1)^k \frac{1\cdot4\cdot7\cdots(3k-2)}{(3k)!}$

31. $\displaystyle\sum_{k=1}^{\infty} (-1)^k k(k+1)(\ln 2)^k$

32. $\displaystyle\sum_{k=1}^{\infty} (-1)^k \left(1 - \frac{1}{3k}\right)^k$

33. We know that the alternating harmonic series $\displaystyle\sum_{k=1}^{\infty} \frac{(-1)^{k+1}}{k}$ is a conditionally convergent series. Show that the positive series taken alone diverges. Also show that the negative terms taken alone diverge. In fact, it can be shown more generally that the positive and negative component series of any conditionally convergent series must diverge.

34. Using the result stated in Exercise 33, show that the terms of a conditionally convergent series can be rearranged to obtain *any* desired sum, say π. Describe how this can be done.

35. Find the fallacy: Let

$$A = 1 - \tfrac{1}{2} + \tfrac{1}{3} - \tfrac{1}{4} + \tfrac{1}{5} - \tfrac{1}{6} + \tfrac{1}{7} - \tfrac{1}{8} + \cdots$$

$$\tfrac{1}{2}A = \quad \tfrac{1}{2} \quad - \tfrac{1}{4} \quad + \tfrac{1}{6} \quad - \tfrac{1}{8} + \cdots$$

sum both sides of the equalities

$$A + \tfrac{1}{2}A = 1 + \tfrac{1}{3} - \tfrac{1}{2} + \tfrac{1}{5} + \tfrac{1}{7} - \tfrac{1}{4} + \cdots$$

But the right side is just a rearrangement of the terms in the series for A. Hence its sum is again A. Thus

$$A + \tfrac{1}{2}A = A$$

or $A = 0$
But,

$$A = (1 - \tfrac{1}{2}) + (\tfrac{1}{3} - \tfrac{1}{4}) + (\tfrac{1}{5} - \tfrac{1}{6}) + \cdots$$

so that $A > 0$. What is wrong?

15.7 POWER SERIES

An important extension of infinite series is obtained by considering series whose terms are x-dependent. In particular, we shall now consider series of the form

$$\sum_{n=0}^{\infty} a_n x^n = a_0 + a_1 x + a_2 x^2 + \cdots + a_n x^n + \cdots \tag{1}$$

Definition   A series of this form is called a **_power series in x_** or, more briefly, a **_power series._**

Examples of power series are

$$\sum_{k=0}^{\infty} x^k = 1 + x + x^2 + x^3 + \cdots \tag{2}$$

$$\sum_{k=0}^{\infty} \frac{x^k}{k!} = 1 + x + \frac{x^2}{2!} + \frac{x^3}{3!} + \cdots \tag{3}$$

$$\sum_{k=1}^{\infty} \frac{x^k}{k3^k} = \frac{x}{3} + \frac{x^2}{2(3^2)} + \frac{x^3}{3(3^3)} + \cdots \tag{4}$$

$$\sum_{k=1}^{\infty} \frac{(-1)^{k-1} x^k}{k} = x - \frac{x^2}{2} + \frac{x^3}{3} - \cdots \tag{5}$$

If the variable in a power series is replaced by a specific value, an infinite series of constant terms is obtained. Thus, for example, in (2) if the variable x is replaced by $\frac{1}{3}$, we have the series

$$1 + \frac{1}{3} + \frac{1}{3^2} + \frac{1}{3^3} + \cdots \qquad (6)$$

Therefore a power series can be viewed as representing an infinite collection of constant series—one series for each value of x. The power series may converge for some values of x and diverge for others. For example, the series (6) converges since it is a geometric series with ratio $\frac{1}{3}$, whereas the series

$$1 + 3 + 3^2 + 3^3 + \cdots \qquad (7)$$

obtained by replacing in (2) the variable x by 3, is a divergent series. In fact, we know that the series (2) converges if $-1 < x < 1$ and diverges for all other values of x, since (2) is a geometric series.

EXAMPLE 1   Determine the values of x for which the series

$$\frac{x}{3} + \frac{x^2}{2(3^2)} + \frac{x^3}{3(3^3)} + \cdots + \frac{x^n}{n(3^n)} + \cdots \qquad (8)$$

of (4) converges.

Solution   As shown, the nth term of the series is $u_n = \dfrac{x^n}{n(3^n)}$. By the ratio test for absolute convergence we have

$$\lim_{n \to \infty} \left| \frac{u_{n+1}}{u_n} \right| = \lim_{n \to \infty} \frac{|x|^{n+1}}{(n+1)3^{n+1}} \cdot \frac{n3^n}{|x|^n} = \lim_{n \to \infty} \frac{n}{3(n+1)} |x| = \frac{|x|}{3}.$$

Hence series (8) converges absolutely if $|x| < 3$, that is, $-3 < x < 3$, and diverges if $|x| > 3$. The ratio test fails to give information if $x = \pm 3$. When $x = 3$, series (8) reduces to the harmonic series

$$1 + \frac{1}{2} + \frac{1}{3} + \cdots$$

which diverges. When $x = -3$, the series (8) becomes

$$-1 + \frac{1}{2} - \frac{1}{3} + \frac{1}{4} - \cdots$$

which is the negative of the alternating harmonic series and converges. To summarize, series (8) converges if and only if $-3 \leq x < 3$. We call this internal $[-3, 3)$ the *interval of convergence* of the power series (8).   ●

EXAMPLE 2   Find the values of x for which the series

$$1 + x + \frac{x^2}{2!} + \frac{x^3}{3!} + \cdots + \frac{x^{n-1}}{(n-1)!} + \cdots$$

in (3) converges.

Solution   The nth term of the series is $u_n = \dfrac{x^{n-1}}{(n-1)!}$ and again we apply the ratio test:

(for $x \neq 0$)

$$\lim_{n \to \infty} \left| \frac{u_{n+1}}{u_n} \right| = \lim_{n \to \infty} \frac{|x|^n}{n!} \cdot \frac{(n-1)!}{|x|^{n-1}} = \lim_{n \to \infty} \frac{|x|}{n} = 0$$

Hence the series converges absolutely for all values of x, ●

EXAMPLE 3　Find the values of x for which the series

$$1 + 1!x + 2!x^2 + 3!x^3 + \cdots + (n-1)!x^{n-1} + \cdots \tag{9}$$

converges.

Solution　The nth term of the series is $u_n = (n-1)!x^{n-1}$ and by the ratio test

$$\lim_{n \to \infty} \left| \frac{u_{n+1}}{u_n} \right| = \lim_{n \to \infty} \frac{n!|x|^n}{(n-1)!|x|^{n-1}} = \lim_{n \to \infty} n|x| = \infty$$

for all $x \neq 0$. Thus the power series (9) diverges when $x \neq 0$. However, this power series (as well as any other power series) must converge when $x = 0$. Hence series (9) converges if and only if $x = 0$. ●

The following theorem is fundamental:

Theorem 1　If power series (1) is convergent for $x = x_0$, then it must converge absolutely for all values of x such that $|x| < |x_0|$.

Proof　Since $\sum_{k=0}^{\infty} a_k x_0{}^k$ converges (by hypothesis), it follows that $a_k x_0{}^k$ must approach 0 as $k \to \infty$. Then certainly there must exist a positive constant M such that $|a_k x_0{}^k| \leq M \ (k = 0, 1, 2, \ldots)$. Now consider any value of x for which $|x| < |x_0|$. For this x we must have

$$|a_k x^k| = \left| a_k \left(\frac{x}{x_0} \right)^k \right| |x_0{}^k| = |a_k x_0{}^k| \left| \frac{x}{x_0} \right|^k \leq M \left| \frac{x}{x_0} \right|^k$$

$(k = 0, 1, 2, \ldots)$, and therefore

$$\sum_{k=0}^{\infty} |a_k x^k| \leq \sum_{k=0}^{\infty} M \left| \frac{x}{x_0} \right|^k \tag{10}$$

The series on the right side of (10) is a convergent geometric series since $\left| \frac{x}{x_0} \right| < 1$. Therefore the series $\sum_{k=0}^{\infty} |a_k x^k|$ converges and the given series $\sum_{k=0}^{\infty} a_k x^k$ converges absolutely. ∎

Corollary 1　If power series (1) is divergent for $x = x_1$, then it must also diverge for all values of x such that $|x| > |x_1|$.

Proof　Suppose that the series converges at $x = x_0$ where $|x_0| > |x_1|$. Then by Theorem 1, it must also converge at $x = x_1$ since $|x_1| < |x_0|$, contradicting the hypothesis. ∎

Corollary 2   Either the power series is absolutely convergent for all values of x, for $x = 0$ only, or there exists a positive number R such that the series is absolutely convergent if $-R < x < R$ (or $|x| < R$) and divergent for $|x| > R$.

Proof   Suppose that the power series is convergent when $x = x_0$ where $x_0 \neq 0$. Furthermore it is assumed that the series does not converge for all values of x, say it diverges when $x = x_1$. Then from Corollary 1, $|x_0| \leq |x_1|$, or, equivalently, $-|x_1| \leq |x_0| \leq |x_1|$. Hence, the values of x for which the series converges are bounded and therefore must possess a least upper bound. Let R be the least upper bound of the convergence set. This means that the series converges if $|x| < R$ (remember that the least upper bound of a set may or may not belong to the set), and diverges if $|x| > R$ (see Figure 15.7.1).   ∎

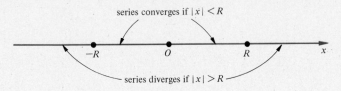

series converges if $|x| < R$

series diverges if $|x| > R$

(The points $x = \pm R$ must be treated separately.)

Figure 15.7.1

The interval $-R < x < R$, together with none, one, or both of the end points (R and $-R$), is called the *interval of convergence,* and the number R is called the *radius of convergence.* The series must be tested separately for convergence at each of the points $x = R$ and $x = -R$. If the power series converges only at $x = 0$ (and all power series must converge at $x = 0$) then $R = 0$, whereas $R = \infty$ corresponds to the case when the power series converges for all values of x.

Knowing the interval of convergence of the series $\sum\limits_{k=0}^{\infty} a_k x^k$ gives us immediately the interval of convergence of the series $\sum\limits_{k=0}^{\infty} a_k (x - c)^k$, which is centered about $x = c$ (Figure 15.7.2). If, for example, $\sum\limits_{k=0}^{\infty} a_k x^k$ has interval of convergence $-R < x < R$ then the series $\sum\limits_{k=0}^{\infty} a_k (x - c)^k$ must converge in the interval

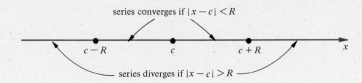

series converges if $|x - c| < R$

series diverges if $|x - c| > R$

(The points $x = c \pm R$ must be treated separately.)

Figure 15.7.2

$(c - R, c + R)$. This interval is called the interval of convergence for the series

$$\sum_{k=0}^{\infty} a_k(x - c)^k.$$

In most cases the ratio test yields the interval of convergence (except for the end points) of a power series. Now we consider additional examples.

EXAMPLE 4   Find the radius of convergence R for the binomial series

$$1 + mx + \frac{m(m - 1)}{2!}x^2 + \cdots + \frac{m(m - 1) \cdots (m - n + 1)}{n!}x^n + \cdots$$

where $m \neq$ positive integer.

Solution
$$\lim_{n \to \infty} \left| \frac{u_{n+1}}{u_n} \right| = \lim_{n \to \infty} \left| \frac{m(m - 1) \cdots (m - n + 1)(m - n) \cdot n!}{(n + 1)! \cdot m(m - 1) \cdots (m - n + 1)} \right| \left| \frac{x^{n+1}}{x^n} \right|$$

$$= \lim_{n \to \infty} \left| \frac{m - n}{n + 1} \right| |x| = |x|$$

Therefore the radius of convergence is $R = 1$.   ●

EXAMPLE 5   Find the interval of convergence of the power series

$$\frac{1}{2} + \frac{x - 5}{16} + \frac{(x - 5)^2}{72} + \frac{(x - 5)^3}{256} + \cdots + \frac{(x - 5)^{n-1}}{2^n n^2} + \cdots$$

Solution
$$\lim_{n \to \infty} \left| \frac{u_{n+1}}{u_n} \right| = \lim_{n \to \infty} \left| \frac{(x - 5)^n}{2^{n+1}(n + 1)^2} \cdot \frac{2^n n^2}{(x - 5)^{n-1}} \right|$$

$$= \lim_{n \to \infty} \left| \frac{x - 5}{2} \right| \left(\frac{n}{n + 1} \right)^2 = \left| \frac{x - 5}{2} \right|$$

Consequently, the series converges if $\dfrac{|x - 5|}{2} < 1$, or, equivalently, if $3 < x < 7$. It diverges when $|x - 5| > 2$, that is, if $x > 7$ or if $x < 3$.

To test the end points of the interval of convergence, let $x = 7$. Then $u_n = \dfrac{1}{2n^2}$ and the series converges since it is $\frac{1}{2}$ times a p-series with $p = 2$. Next, let $x = 3$, $u_n = \dfrac{(-1)^{n-1}}{2n^2}$ and the alternating series converges since $|u_{n+1}| < |u_n|$ and $\lim_{n \to \infty} u_n = 0$. Hence the given series converges in the closed interval $[3, 7]$.   ●

Exercises 15.7

In each of Exercises 1 through 22 find the interval of convergence of the given power series.

1. $1 + \dfrac{x}{2} + \dfrac{x^2}{2^2} + \dfrac{x^3}{2^3} + \cdots$

2. $1 - \dfrac{x}{5} + \dfrac{x^2}{5^2} - \dfrac{x^3}{5^3} + \cdots$

3. $1 - \dfrac{x^2}{2!} + \dfrac{x^4}{4!} - \dfrac{x^6}{6!} + \cdots$

4. $x - \dfrac{x^2}{2} + \dfrac{x^3}{3} - \dfrac{x^4}{4} + \cdots$

5. $1 - 2x + 4x^2 - 8x^3 + \cdots$

6. $1 + x + \dfrac{x^2}{2} + \dfrac{x^3}{3} + \cdots$

7. $1 - x + \dfrac{x^2}{\sqrt{2}} - \dfrac{x^3}{\sqrt{3}} + \dfrac{x^4}{\sqrt{4}} - \cdots$

8. $\dfrac{x}{3} + \dfrac{2x^2}{3^2} + \dfrac{3x^3}{3^3} + \cdots + \dfrac{nx^n}{3^n} + \cdots$

9. $\dfrac{x}{b} + \dfrac{x^2}{2b^2} + \dfrac{x^3}{3b^3} + \cdots + \dfrac{x^n}{nb^n} + \cdots, \quad b > 0$

10. $(x - 2) + \dfrac{(x - 2)^2}{\sqrt{2}} + \dfrac{(x - 2)^3}{\sqrt{3}} + \cdots + \dfrac{(x - 2)^n}{\sqrt{n}} + \cdots$

11. $\displaystyle\sum_{k=1}^{\infty} \dfrac{(-1)^{k-1}x^{2k}}{k^2}$

12. $\displaystyle\sum_{k=0}^{\infty} \dfrac{k!x^k}{3^k}$

13. $\displaystyle\sum_{k=1}^{\infty} \dfrac{(4z)^k}{10^k}$

14. $\displaystyle\sum_{k=1}^{\infty} \dfrac{(2t - 1)^k}{k(k + 1)}$

15. $\displaystyle\sum_{k=0}^{\infty} \dfrac{(3x + 2)^k}{5^k}$

16. $\displaystyle\sum_{k=2}^{\infty} \dfrac{x^k}{\ln k}$

17. $\displaystyle\sum_{k=1}^{\infty} \dfrac{k^k}{k!}(7x)^k$ (omit consideration of the end points of interval of convergence)

18. $\displaystyle\sum_{k=1}^{\infty} \dfrac{(k + 2)(5t - 1)^k}{k^2}$

19. $\displaystyle\sum_{k=0}^{\infty} \dfrac{(y - 2)^k}{4^k \sqrt{k + 1}}$

20. $\displaystyle\sum_{k=0}^{\infty} \dfrac{(t + 10)^k}{3^k + 2^k}$

21. $\displaystyle\sum_{k=1}^{\infty} 2^{k-1}(z - 1)^{4(k-1)}$

22. $\displaystyle\sum_{k=1}^{\infty} \dfrac{(3x)^{k!}}{k}$

In Exercises 23 through 26, find the values of x for which the given series converges.

23. $\displaystyle\sum_{k=0}^{\infty} (x^2 - 2x + 2)^k$ (*Hint:* replace $x^2 - 2x + 2$ by t and find the minimum value of $x^2 - 2x + 2$.)

24. $\displaystyle\sum_{k=100}^{\infty} k^k(2x - 3)^k$

25. $\displaystyle\sum_{k=0}^{\infty} \dfrac{\left(\dfrac{3x - 4}{x}\right)^k}{2^k}$

26. $\displaystyle\sum_{k=1}^{\infty} \dfrac{(x^2 - 4)^k}{10^k}$

15.8 DIFFERENTIATION AND INTEGRATION OF POWER SERIES

Consider the power series

$$1 - x + x^2 - x^3 + x^4 + \cdots \tag{1}$$

This is a geometric series with ratio $-x$ and it represents its sum function $\dfrac{1}{1 + x}$ inside its interval of convergence, that is

$$\dfrac{1}{1 + x} = 1 - x + x^2 - x^3 + x^4 + \cdots \qquad \text{if } |x| < 1 \tag{2}$$

More generally, a power series $\sum_{k=0}^{\infty} a_k x^k$ represents a function $f(x)$ within its

interval of convergence. If the radius of convergence is R then we have

$$f(x) = \sum_{k=0}^{\infty} a_k x^k \qquad |x| < R \qquad\qquad (3)$$

If this had been a finite sum, we know that term by term differentiation and integration would have been permissible. We shall show in this section that such term by term operations are valid when x is confined to the interior of its interval of convergence.

Theorem 1 If power series (3) converges absolutely for $|x| < R$ then the series $\sum_{k=1}^{\infty} k a_k x^{k-1}$

also converges absolutely for $|x| < R$.

Proof The series $\sum_{k=1}^{\infty} k a_k x^{k-1}$ converges for $x = 0$. Consider $x \neq 0$ and select x_0 such that

$|x| < |x_0| < R$. Thus with $\left| \dfrac{x}{x_0} \right| = r < 1$, we have

$$|a_k x^k| = |a_k x_0{}^k| \left| \frac{x}{x_0} \right|^k = |a_k x_0{}^k| r^k$$

But, $\sum_{k=0}^{\infty} |a_k x_0{}^k|$ is convergent so that an integer N must exist so that $|a_k x_0{}^k| < 1$ for $k > N$. Hence

$$|a_k x^k| < r^k$$

from which

$$|k a_k x^{k-1}| = \frac{k}{|x|} |a_k x^k| < \frac{k}{|x|} r^k \qquad \text{when } k > N.$$

But, $\sum_{k=1}^{\infty} \dfrac{k r^k}{|x|}$ converges by application of the ratio test. Thus, $\left| \sum_{k=1}^{\infty} k a_k x^{k-1} \right|$ converges for all x, $|x| < R$. ∎

An immediate application of Theorem 1 is to establish the continuity of the function $f(x)$ represented by a convergent power series.

Theorem 2 If $\sum_{k=0}^{\infty} a_k x^k = f(x)$ in $|x| < R$, then $f(x)$ is continuous in $|x| < R$.

Proof To prove this we go back to the definition of continuity of a function at a point $x = x_0$ for which $|x_0| < R$. We must show that $\lim_{x \to x_0} (f(x) - f(x_0)) = 0$ or, equivalently, that

$$\lim_{x \to x_0} |f(x) - f(x_0)| = 0 \qquad (4)$$

To this end, we form

$$f(x) - f(x_0) = \sum_{k=1}^{\infty} a_k(x^k - x_0{}^k) \qquad (5)$$

Now since $x^k - x_0{}^k$ possesses $x - x_0$ as a factor and the fact that the absolute value of a product equals the product of their absolute values,

$$|x^k - x_0{}^k| = |x - x_0|\,|x^{k-1} + x^{k-2}x_0 + \cdots + x_0{}^{k-1}| \qquad (6)$$

Next we choose r such that $0 < r < R$ and also $|x| < r$, $|x_0| < r$. Then equations (5) and (6) yield

$$|f(x) - f(x_0)| \le \sum_{k=1}^{\infty} |a_k|\,|x^k - x_0{}^k| < |x - x_0| \sum_{k=1}^{\infty} k|a_k|r^{k-1} \qquad (7)$$

But $0 < r < R$ so that the infinite series $\sum_{k=1}^{\infty} k|a_k|r^{k-1} = B = $ constant. Hence,

$$|f(x) - f(x_0)| < B|x - x_0|$$

which implies that

$$\lim_{x \to x_0} |f(x) - f(x_0)| = 0 \text{ or, equivalently,}$$

$$\lim_{x \to x_0} (f(x) - f(x_0)) = 0$$

and the continuity of $f(x)$ at $x = x_0$ is proved. ∎

We are now ready to prove a theorem regarding the validity of term by term integration of a power series within its interval of convergence.

Theorem 3 Let $f(x) = \sum_{k=0}^{\infty} a_k x^k$ in $|x| < R$ and let A and B be any two numbers in $|x| < R$. Then

$$\int_A^B f(x)\,dx = \sum_{k=0}^{\infty} \int_A^B a_k x^k\,dx = \sum_{k=0}^{\infty} a_k \frac{B^{k+1} - A^{k+1}}{k+1} \qquad (8)$$

Proof From Theorem 2 we know that $f(x)$ is continuous in $|x| < R$ and therefore $\int_A^B f(x)\,dx$ exists. Let $s_n = \sum_{k=0}^{n} \int_A^B a_k x^k\,dx$ and it must be shown that $\int_A^B f(x)\,dx = \lim_{n \to \infty} s_n$ to complete the proof. To this end we write (for $|x| < R$),

$$f(x) = \sum_{k=0}^{n} a_k x^k + R_n(x)$$

where $R_n(x) = \sum\limits_{k=n+1}^{\infty} a_k x^k$ and we must show that

$$\lim_{n \to \infty} \int_A^B R_n(x)\, dx = 0$$

For $A < B$,

$$\left| \int_A^B R_n(x)\, dx \right| \leq \int_A^B |R_n(x)|\, dx \leq \int_A^B \sum_{k=n+1}^{\infty} |a_k x^k|\, dx$$

Choose a quantity x_0 such that $|A| < |x_0| < R$ and $|B| < |x_0| < R$. Then if we let $\left| \dfrac{x}{x_0} \right| = r < 1$, as in the proof of Theorem 1,

$$|a_k x^k| = |a_k x_0{}^k| \left| \frac{x}{x_0} \right|^k = |a_k x_0{}^k| r^k$$

and, since $\sum\limits_{k=0}^{\infty} a_k x^k$ converges for $x = x_0$, there exists an integer N such that $|a_k x_0{}^k| < 1$ for $k > N$, thus

$$|a_k x^k| < r^k \qquad k > N, A \leq x \leq B, \text{ where } r < 1$$

Therefore

$$\sum_{k=N+1}^{\infty} |a_k x^k| < \sum_{k=N+1}^{\infty} r^k = \frac{r^{N+1}}{1-r}$$

and

$$\int_A^B \sum_{k=N+1}^{\infty} |a_k x^k|\, dx < \frac{r^{N+1}}{1-r}(B - A)$$

But the number on the right side tends to zero as $N \to \infty$ (since $r < 1$), hence

$$\lim_{n \to \infty} \int_A^B R_n(x)\, dx = 0$$

If $A > B$, the same conclusion holds and the proof is complete. ∎

We state the following corollary to Theorem 3 without proof.

Corollary If $f(x) = \sum\limits_{k=0}^{\infty} a_k x^k$ and the series converges in $|x| < R$ then the power series

for $\int f(x)\, dx$ in $|x| < R$ is

$$\int f(x)\, dx = \sum_{k=0}^{\infty} \int a_k x^k\, dx = \sum_{k=0}^{\infty} a_k \frac{x^{k+1}}{k+1} + C$$

Our next theorem justifies term by term differentiation of a power series within its interval of convergence.

Theorem 4   If $f(x) = \displaystyle\sum_{k=0}^{\infty} a_k x^k$ in $|x| < R$ then

$$f'(x) = \sum_{k=1}^{\infty} k a_k x^{k-1} \qquad \text{in } |x| < R \tag{9}$$

Proof   Let $h(x) = \displaystyle\sum_{k=1}^{\infty} k a_k x^{k-1}$; then from Theorem 1 it is known that this series converges in $(-R, R)$ and Theorem 2 yields the fact that $h(x)$ is continuous. Hence $\displaystyle\int_0^u h(x)\, dx$ exists for $|u| < R$ and Theorem 3 yields

$$\int_0^u h(x)\, dx = \int_0^u \sum_{k=1}^{\infty} k a_k x^{k-1}\, dx = \sum_{k=1}^{\infty} \int_0^u k a_k x^{k-1}\, dx$$

$$= \sum_{k=1}^{\infty} a_k u^k$$

Now,

$$f(u) = \sum_{k=0}^{\infty} a_k u^k = a_0 + \sum_{k=1}^{\infty} a_k u^k = a_0 + \int_0^u h(x)\, dx$$

Differentiation of both sides with respect to u and application of the first fundamental theorem of the calculus yields

$$f'(u) = h(u) \qquad \text{or} \qquad f'(x) = h(x) \qquad\qquad \blacksquare$$

Theorem 5   If two power series are equal in an interval $|x| < R$ then the coefficients of like powers must be equal. Hence if

$$a_0 + a_1 x + a_2 x^2 + \cdots = b_0 + b_1 x + b_2 x^2 + \cdots \qquad \text{in } |x| < R \tag{10}$$

then

$$a_k = b_k \qquad (k = 0, 1, 2, \ldots) \tag{11}$$

that is, the two power series must be identical.

Proof   Let $x = 0$ in (10) so that $a_0 = b_0$. Next differentiate both sides of (10) so that

$$a_1 + 2a_2 x + \cdots = b_1 + 2b_2 x + \cdots \tag{12}$$

Set $x = 0$ and obtain $a_1 = b_1$. Differentiate both sides of (12) again and set $x = 0$, and so on. Hence (11) follows. $\qquad\qquad \blacksquare$

The theorems of this section may be restated easily in terms of the series $\displaystyle\sum_{k=0}^{\infty} a_k (x - c)^k$, $|x - c| < R$, where R is the radius of convergence of the series.

Notice that all of our theorems in this section apply only to the interior of the interval of convergence. This restriction is due in part to the fact that there are examples where a given series converges at an end point of the interval, but the series obtained using term by term differentiation does not converge at that end point.

EXAMPLE 1 Prove that for $-1 < x < 1$

$$\frac{1}{(1-x)^2} = \sum_{k=1}^{\infty} kx^{k-1}$$

Solution $$\frac{1}{1-x} = 1 + x + x^2 + x^3 + \cdots + x^k + \cdots \qquad \text{in } |x| < 1$$

Thus by differentiation of both sides with respect to x we obtain

$$\frac{1}{(1-x)^2} = 1 + 2x + 3x^2 + \cdots + kx^{k-1} + \cdots \qquad \text{if } |x| < 1$$

or, more compactly,

$$\frac{1}{(1-x)^2} = \sum_{k=1}^{\infty} kx^{k-1} \qquad -1 < x < 1 \qquad \bullet \quad (13)$$

EXAMPLE 2 Find the function represented by the infinite series $\sum_{k=0}^{\infty} \dfrac{x^{k+1}}{k+1}$ in its interval of convergence.

Solution Let f be the function defined by

$$f(x) = \sum_{k=0}^{\infty} \frac{x^{k+1}}{k+1} \qquad (14)$$

The radius of convergence is $R = 1$ as shown by the ratio test, so the series converges if $|x| < 1$.

From Equation (9) and the sum of the infinite geometric progression,

$$f'(x) = \sum_{k=0}^{\infty} x^k = \frac{1}{1-x} \qquad |x| < 1$$

Therefore integration results in

$$f(x) = \int \frac{dx}{1-x} = -\ln(1-x) + C \qquad (15)$$

However, from (14) we know that $f(0) = 0$. Thus, substituting $x = 0$ and $f(0) = 0$ into (15) yields $C = 0$, since $\ln 1 = 0$. Hence

$$f(x) = -\ln(1-x) = \sum_{k=0}^{\infty} \frac{x^{k+1}}{k+1} \qquad |x| < 1 \qquad \bullet \quad (16)$$

EXAMPLE 3 Prove that the function represented by the infinite series $\sum_{k=0}^{\infty} \dfrac{(-1)^k x^k}{k!}$, $-\infty < x < \infty$, is e^{-x}.

Solution The ratio test verifies that the series is absolutely convergent for all real values of x. Hence if f is the function defined by

$$f(x) = \sum_{k=0}^{\infty} \frac{(-1)^k x^k}{k!} = 1 - x + \frac{x^2}{2!} - \frac{x^3}{3!} + \cdots \qquad (17)$$

for all values of x, from (9), it follows that

$$f'(x) = \sum_{k=1}^{\infty} \frac{(-1)^k k x^{k-1}}{k!} = \sum_{k=1}^{\infty} \frac{(-1)^k x^{k-1}}{(k-1)!}$$

since $k! = k(k-1)!$. Changing the index of summation to start at $k = 0$, we have

$$f'(x) = \sum_{k=0}^{\infty} \frac{(-1)^{k+1} x^k}{k!} = -\sum_{k=0}^{\infty} \frac{(-1)^k x^k}{k!} \qquad (18)$$

From (17) and (18),

$$f'(x) = -f(x) \qquad -\infty < x < \infty \qquad (19)$$

and from (17), $f(0) = 1$.

Now the function $y = e^{-x}$ is such that $y' = -e^{-x} = -y$. In fact, the most general solution of this equation is $y = Ce^{-x}$ and if $y = 1$ when $x = 0$, then $C = 1$. Thus $y = e^{-x}$ is the *only* solution of (19) subject to the initial condition that $y(0) = 1$.

Hence $e^{-x} = f(x)$ for all x, that is

$$e^{-x} = \sum_{k=0}^{\infty} \frac{(-1)^k x^k}{k!} \qquad \bullet \qquad (20)$$

EXAMPLE 4 A function $f(x)$ is defined by the differential equation $f'(x) = f(x)$ subject to the condition $f(0) = 1$. Find a power series representation of the function and identify it.

Solution We assume tentatively that the function sought has a power series representation. Thus we write in $|x| < R$,

$$f(x) = a_0 + a_1 x + a_2 x^2 + a_3 x^3 + \cdots + a_{n-1} x^{n-1} + a_n x^n + \cdots$$

by differentiation

$$f'(x) = a_1 + 2a_2 x + 3a_3 x^2 + \cdots + na_n x^{n-1} + \cdots$$

Since $f'(x) = f(x)$ by hypothesis, we may equate like powers of x (in accordance with Theorem 5)

$$na_n = a_{n-1}$$

or

$$a_n = \frac{a_{n-1}}{n} \qquad (n = 1, 2, 3, \ldots)$$

Hence

$$a_1 = a_0 \qquad a_2 = \frac{a_1}{2} = \frac{a_0}{2!} \qquad a_3 = \frac{a_2}{3} = \frac{a_0}{3!}$$

and in general,

$$a_n = \frac{a_0}{n!} \qquad (n = 1, 2, 3, \ldots)$$

Therefore

$$f(x) = a_0 + a_0 x + \frac{a_0}{2!} x^2 + \cdots + \frac{a_0}{n!} x^n + \cdots$$

$$f(x) = a_0 \left(1 + x + \frac{x^2}{2!} + \cdots + \frac{x^n}{n!} + \cdots \right)$$

But $f(0) = 1$ so that $a_0 = 1$ (which could have been determined in our first step). Thus

$$f(x) = 1 + x + \frac{x^2}{2!} + \frac{x^3}{3!} + \cdots + \frac{x^n}{n!} + \cdots = \sum_{k=0}^{\infty} \frac{x^k}{k!} \qquad (21)$$

By the ratio test the series in (21) converges for all values of x (that is, $R = \infty$) and therefore represents a function for which $f'(x) = f(x)$ and $f(0) = 1$. In fact, the only function that satisfies these conditions is e^x and we have

$$e^x = \sum_{k=0}^{\infty} \frac{x^k}{k!} \qquad \bullet \qquad (22)$$

EXAMPLE 5   Find $\displaystyle\int_0^{0.5} \frac{dx}{1 + x^4}$ approximately to four decimal places.

Solution

$$(1 + u)^{-1} = 1 - u + u^2 - u^3 + u^4 - \cdots \qquad \text{if } |u| < 1$$

so that with u replaced by x^4,

$$(1 + x^4)^{-1} = 1 - x^4 + x^8 - x^{12} + x^{16} - \cdots \qquad \text{if } |x| < 1$$

Thus

$$\int_0^{0.5} \frac{dx}{1 + x^4} = \int_0^{0.5} (1 - x^4 + x^8 - x^{12} + x^{16} - \cdots)\, dx$$

and integration term by term yields

$$\int_0^{0.5} \frac{dx}{1 + x^4} = \left(x - \frac{x^5}{5} + \frac{x^9}{9} - \frac{x^{13}}{13} + \frac{x^{17}}{17} - \cdots \right) \Bigg|_0^{0.5}$$

$$= \frac{1}{2} - \frac{1}{2^5 \cdot 5} + \frac{1}{2^9 \cdot 9} - \frac{1}{2^{13} \cdot 13} + \cdots$$

where the right side is a rapidly convergent alternating series. The sum of the first three terms will gives us the definite integral to four decimal places (since the fourth term of the alternating series is less than 0.00005). We have

$$\int_0^{0.5} \frac{dx}{1 + x^4} \doteq 0.4940 \qquad \bullet$$

Exercises 15.8

In Exercises 1 through 8, a function $f(x)$ is defined by a given power series. Find the radius of convergence for the given series. Determine the power series for $\int f(x)\,dx$ and for $f'(x)$ and verify that the radius of convergence R is the same for the three power series.

1. $f(x) = \sum_{k=0}^{\infty} x^k$

2. $f(x) = \sum_{k=1}^{\infty} \frac{x^k}{k}$

3. $f(x) = \sum_{k=1}^{\infty} \frac{x^k}{k^2}$

4. $f(x) = \sum_{k=0}^{\infty} (-1)^k x^{2k}$

5. $f(x) = \sum_{k=0}^{\infty} (-1)^k \frac{x^{2k+1}}{(2k+1)!}$

6. $f(x) = \sum_{k=0}^{\infty} \frac{x^{2k}}{(2k)!}$

7. $f(x) = \sum_{k=1}^{\infty} \frac{2^k(x-3)^k}{k}$

8. $f(x) = \sum_{k=1}^{\infty} \frac{10^k(x-b)^k}{k(k+1)}, \quad (b = \text{constant})$

9. Obtain a power series expression for $\dfrac{1}{1-x^2}$. What is its interval of convergence?

10. Obtain a power series expansion for $\ln(1+x)$. Determine its interval of convergence. Approximate $\ln 1.1$ to three decimal places.

11. Obtain a power series representation for $f(x) = \dfrac{1}{1000 + x^3}$. What is its interval of convergence?

12. By combining the power series for $\ln(1+x)$ and $\ln(1-x)$ (assuming this operation is valid) verify that

$$\ln\frac{1+x}{1-x} = 2\left(x + \frac{x^3}{3} + \frac{x^5}{5} + \cdots\right) \quad \text{when } |x| < 1$$

Use this formula to approximate $\ln 2$ to three decimal places.

13. Find the power series expansion of $\int_0^x e^{-t^2}\,dt$ and determine its interval of convergence.

14. Find $\int_0^1 e^{-t^2}\,dt$ approximately to three decimal places.

15. Find $\int_0^{1/2} \dfrac{dx}{1 + x^5}$ approximately to four decimal places.

16. Find a power series for $\tan^{-1} x$. (*Hint:* $D_x(\tan^{-1} x) = \dfrac{1}{1+x^2}$)

17. If $F(x) = e^{-x^2}$, find $F^{(6)}(0)$. (*Hint:* Expand e^{-x^2} in a power series.)

18. If $f(x) = \sum_{k=0}^{\infty} \dfrac{b^k x^k}{k!}$, where b is an arbitrary constant, show that $f'(x) = bf(x)$ for all values of x.

19. A function f is defined by the differential equation $f'(x) = bf(x)$ where b is an arbitrary constant subject to the condition $f(0) = 1$. Find the power series representation of $f(x)$ and identify the function.

20. If $f(x) = \sum_{k=0}^{\infty} \dfrac{x^{2k}}{(2k)!}$ and $g(x) = \sum_{k=0}^{\infty} \dfrac{x^{2k+1}}{(2k+1)!}$ for all values of x. Show that

   (a) $f'(x) = g(x)$ (b) $g'(x) = f(x)$ (c) $f''(x) = f(x)$

21. If an *even* function $f(x)$ is expressible by a power series $\sum_{k=0}^{\infty} a_k x^k$ show that all $a_n = 0$ if n is an odd integer. This means that only the even powers of x will appear. Recall that a function $f(x)$ is even if and only if $f(x) = f(-x)$.

22. As in Exercise 21 state and prove a corresponding result for *odd* functions $f(x)$.

23. A function f is expressible by a power series $\sum_{k=0}^{\infty} a_k x^k$ for $|x| < R$, where $R > 0$. Furthermore $f''(x) + f(x) = 0$, $f(0) = 1$ and $f'(0) = 0$. Find the power series and the domain of the function.

24. Find a power series expansion for $\dfrac{1}{(1-x)^3}$ and show that the series represents the function in $(-1, 1)$.

(*Hint:* Start with the expansion for $\dfrac{1}{1-x}$ and differentiate twice.)

25. Evaluate $\displaystyle\int_0^1 \dfrac{1 - e^{-t}}{t}\, dt$ to three decimal places by starting with $e^u = \sum_{k=0}^{\infty} \dfrac{u^k}{k!}$ for all values of u. Does

$\displaystyle\int_0^1 \dfrac{1 - e^{-t}}{t}\, dt = \int_0^1 \dfrac{dt}{t} - \int_0^1 \dfrac{e^{-t}}{t}\, dt$? Explain your conclusion.

26. Evaluate the sum of the series

$$F(x) = \frac{x^2}{1 \cdot 2} - \frac{x^3}{2 \cdot 3} + \frac{x^4}{3 \cdot 4} - \frac{x^5}{4 \cdot 5} + \cdots + (-1)^{n-1} \frac{x^{n+1}}{n(n+1)} + \cdots$$

where $|x| < 1$, by first finding $F'(x)$ and utilizing the series for $\ln(1 + t) = \displaystyle\int_0^t \dfrac{du}{1 + u}$.

15.9   TAYLOR SERIES

We know that a power series $\sum_{k=0}^{\infty} a_k (x - c)^k$ defines a function $f(x)$ for each value of x in the interior of its interval of convergence $|x - c| < R$. Furthermore the function possesses derivatives of all orders, and these may be obtained by termwise differentiation of the series with the same interval of validity. Conversely, given a function $f(x)$ we may ask: If the function is differentiable to all orders, is the function representable by a power series? Although this is not always true, it is true for most of the elementary functions with which we will be concerned in this book.

Suppose then that in $|x - c| < R$,

$$f(x) = \sum_{k=0}^{\infty} a_k (x - c)^k$$

or, in expanded form

$$f(x) = a_0 + a_1(x - c) + a_2(x - c)^2 + \cdots + a_n(x - c)^n + \cdots \qquad (1)$$

We assume that $f(x)$ is infinitely differentiable and seek a formula for the coefficients a_k. To this end, let us differentiate both sides of (1) with respect to x successively, and find

$$f'(x) = a_1 + 2a_2(x - c) + 3a_3(x - c)^2 + 4a_4(x - c)^3$$
$$+ \cdots + na_n(x - c)^{n-1} + \cdots \qquad (2)$$

$$f''(x) = 2a_2 + 3 \cdot 2a_3(x - c) + 4 \cdot 3a_4(x - c)^2 + \cdots$$
$$+ n(n - 1)a_n(x - c)^{n-2} + \cdots \qquad (3)$$

$$f'''(x) = 3 \cdot 2a_3 + 4 \cdot 3 \cdot 2a_4(x - c) + \cdots$$
$$+ n(n - 1)(n - 2)a_n(x - c)^{n-3} + \cdots \quad (4)$$

and so on.

Evaluation of the function f and its derivatives at the number c yields, from (1) through (4), and so on,

$$f(c) = a_0 \qquad f'(c) = a_1 \qquad f''(c) = 2a_2 = 2!a_2$$
$$f'''(c) = 3 \cdot 2a_3 = 3!a_3, \ldots, f^{(n)}(c) = n!a_n, \ldots$$

Thus

$$a_0 = f(c) \qquad a_1 = f'(c) \qquad a_2 = \frac{f''(c)}{2!} \qquad a_3 = \frac{f'''(c)}{3!}$$

and, in general, for each $n = 0, 1, 2, \ldots$

$$a_n = \frac{f^{(n)}(c)}{n!} \quad (5)$$

Substitution of (5) into (1) yields the power series

$$f(x) = f(c) + f'(c)(x - c) + \frac{f''(c)}{2!}(x - c)^2 + \cdots$$
$$+ \frac{f^{(n)}(c)}{n!}(x - c)^n + \cdots \quad (6)$$

Series (6) is called the **Taylor series** of f about $x = c$.

In particular, if $c = 0$ then we obtain the **Maclaurin series** for f, namely,

$$f(x) = f(0) + f'(0)x + \frac{f''(0)}{2!}x^2 + \cdots + \frac{f^{(n)}(0)}{n!}x^n + \cdots \quad (7)$$

EXAMPLE 1   Find the Maclaurin series for e^x and its interval of convergence. Find the series for $\sqrt{e}$.

Solution   Set $f(x) = e^x$ and observe that $f^{(n)}(x) = e^x$ for all x and that $f^{(n)}(0) = e^0 = 1$. Hence the Maclaurin series for e^x from (7) is given by

$$e^x = 1 + x + \frac{x^2}{2!} + \frac{x^3}{3!} + \cdots + \frac{x^n}{n!} + \cdots \quad (8)$$

and the power series on the right side of (8) converges absolutely for all values of x by the ratio test:

$$\lim_{n \to \infty} \left| \frac{x^{n+1}}{(n+1)!} \cdot \frac{n!}{x^n} \right| = \lim_{n \to \infty} \frac{|x|}{n+1} = 0$$

To illustrate, if $x = \frac{1}{2}$ is substituted into (8), we obtain

$$\sqrt{e} = e^{1/2} = 1 + \frac{1}{2} + \frac{\left(\frac{1}{2}\right)^2}{2!} + \frac{\left(\frac{1}{2}\right)^3}{3!} + \cdots + \frac{\left(\frac{1}{2}\right)^n}{n!} + \cdots \quad \bullet \quad (9)$$

EXAMPLE 2   Find the Taylor series for $\cos x$ about $x = c$. In particular, find the Maclaurin series (Taylor series about $x = 0$).

Solution   If $f(x) = \cos x$ then $f'(x) = -\sin x, f''(x) = -\cos x, f'''(x) = \sin x, f^{(4)}(x) = \cos x$, and so forth. Therefore $f(c) = \cos c$, $f'(c) = -\sin c$, $f''(c) = -\cos c$, $f'''(c) = \sin c, f^{(4)}(c) = \cos c, \ldots$ and substitution of these numbers into (6) yields, for all x,

$$\cos x = \cos c - (\sin c)(x - c) - \cos c \frac{(x-c)^2}{2!} + \sin c \frac{(x-c)^3}{3!} + \cdots \qquad (10)$$

If we substitute $c = 0$ into (10), there results (since $\cos 0 = 1$ and $\sin 0 = 0$) for arbitrary x,

$$\cos x = 1 - \frac{x^2}{2!} + \frac{x^4}{4!} - \cdots + (-1)^n \frac{x^{2n}}{(2n)!} + \cdots \qquad \bullet \qquad (11)$$

The equality signs in Equations (8) through (11) have not been validated. Actually we have found the power series *associated* with the functions e^x and $\cos x$—it still has to be shown that the series converge to the functions on the left.

The nth partial sum of (6) is the Taylor polynomial (13) of Section 14.5 (taken about $x = c$ rather than $x = a$). Thus

$$f(x) = P_n(x) + R_n(x) \qquad (12)$$

where

$$P_n(x) = f(c) + f'(c)(x - c) + \cdots + \frac{f^{(n)}(c)}{n!}(x - c)^n \qquad (13)$$

and

$$R_n(x) = \frac{f^{(n+1)}(z)}{(n+1)!}(x - c)^{n+1} \qquad (14)$$

where z is some number between c and x.

The fundamental theorem for proving that the power series represents the function is as follows:

Theorem 1   If the function f has derivatives of all orders in an interval containing the number c, then a necessary and sufficient condition that (6) holds is that

$$\lim_{n \to \infty} R_n(x) = 0 \qquad (15)$$

Proof   From (12), $P_n(x) = f(x) - R_n(x)$, we have

$$\lim_{n \to \infty} P_n(x) = \lim_{n \to \infty} f(x) - \lim_{n \to \infty} R_n(x) = f(x) - \lim_{n \to \infty} R_n(x)$$

Thus $\lim_{n \to \infty} P_n(x) = f(x)$ if and only if $\lim_{n \to \infty} R_n(x) = 0$. ∎

EXAMPLE 3   Prove that the Maclaurin series for e^x represents e^x for all values of x.

Solution From (14) with $c = 0$, $f(x) = e^x$, $f^{(n+1)}(x) = e^x$ we have

$$R_n(x) = \frac{e^z}{(n+1)!}x^{n+1}$$

where z is a number between 0 and x and thus $e^{|z|} \leq e^{|x|}$. Hence

$$|R_n(x)| \leq \frac{e^{|x|}|x|^{n+1}}{(n+1)!}$$

But, $\lim\limits_{n\to\infty} \dfrac{|x|^{n+1}}{(n+1)!} = 0$ for any given x (why?) and therefore $\lim\limits_{n\to\infty} R_n(x) = 0$. ●

EXAMPLE 4 Prove that the Maclaurin series for $\cos x$ represents the function.

Solution $f(x) = \cos x$ and from Example 2 its nth derivative is one of the four functions $\pm\sin x$ and $\pm\cos x$ depending on n. In all cases, $|f^{(n)}(x)| \leq 1$ for all n, hence

$$\lim_{n\to\infty}|R_n(x)| = \lim_{n\to\infty}|f^{(n+1)}(z)|\frac{|x^{n+1}|}{(n+1)!} = 0$$

Therefore the Maclaurin series for $\cos x$ is valid for all x. ●

EXAMPLE 5 Find the Taylor series for $\ln x$ in powers of $x - 1$ and its interval of convergence. What is the Maclaurin expansion of this function?

Solution We have $f(x) = \ln x$ and $c = 1$. Hence we must obtain the values of f and its successive derivatives at 1.

$$f(x) = \ln x \qquad f(1) = 0$$
$$f'(x) = \frac{1}{x} \qquad f'(1) = 1$$
$$f''(x) = -\frac{1}{x^2} \qquad f''(1) = -1$$
$$f'''(x) = \frac{2}{x^3} \qquad f'''(1) = 2$$

and in general, for all positive integers

$$f^{(n)}(x) = \frac{(-1)^{n-1}(n-1)!}{x^n} \qquad f^{(n)}(1) = (-1)^{n-1}(n-1)!$$

Therefore the Taylor series for $\ln x$ about 1 is

$$\ln x = (x-1) - \frac{(x-1)^2}{2} + \frac{(x-1)^3}{3} - \cdots + \frac{(-1)^{n-1}(x-1)^n}{n} + \cdots \quad (16)$$

By the ratio test,

$$\lim_{n\to\infty}\left|\frac{u_{n+1}}{u_n}\right| = \lim_{n\to\infty}\left|-\frac{(x-1)^{n+1}}{n+1}\cdot\frac{n}{(x-1)^n}\right| = |x-1|$$

and the series converges for $0 < x < 2$. It also converges when $x = 2$ and it diverges when $x = 0$. We next show that if x is in $(1, 2]$ then $\lim_{n \to \infty} R_n(x) = 0$.

$$R_n(x) = \frac{(-1)^n n!}{z^{n+1}} \cdot \frac{(x - 1)^{n+1}}{(n + 1)!} = \frac{(-1)^n}{n + 1} \left(\frac{x - 1}{z} \right)^{n+1}$$

where z is between 1 and x. Now for x in $(1, 2]$, $0 < x - 1 \le 1 < z$ so that $\dfrac{x - 1}{z} < 1$ and

$$\lim_{n \to \infty} R_n(x) = 0$$

It may also be shown that the series (16) converges if $0 < x \le 1$.

Finally the Maclaurin series for $\ln x$ does not exist since $\ln 0$ is not defined. ●

EXAMPLE 6   Find $\cos 56°$ correct to four decimal places.

Solution   Rather than expand $\cos x$ about $x = 0$, we shall use a Taylor expansion about $x = c$, which is close to $56°$. The number c must be chosen so that the function and its derivatives are known there. We choose $c = \dfrac{\pi}{3}$, (that is, $60°$) which is the number cloest to $56°$ meeting our requirements.

$$\cos x = \cos \frac{\pi}{3} - \left(\sin \frac{\pi}{3} \right)\left(x - \frac{\pi}{3} \right) - \frac{\cos \dfrac{\pi}{3}}{2!} \left(x - \frac{\pi}{3} \right)^2$$

$$+ \frac{\sin \dfrac{\pi}{3}}{3!} \left(x - \frac{\pi}{3} \right)^3 + \frac{\cos \dfrac{\pi}{3}}{4!} \left(x - \frac{\pi}{3} \right)^4 + \cdots$$

$$= \frac{1}{2} - \frac{\sqrt{3}}{2}\left(x - \frac{\pi}{3} \right) - \frac{1}{2!2}\left(x - \frac{\pi}{3} \right)^2 + \frac{\sqrt{3}}{3!2}\left(x - \frac{\pi}{3} \right)^3$$

$$+ \frac{1}{4!2}\left(x - \frac{\pi}{3} \right)^4 + \cdots$$

Now, $56° = \dfrac{56\pi}{180}$ radians $= \dfrac{14\pi}{45}$ radians and $\dfrac{14\pi}{45} - \dfrac{\pi}{3} = -\dfrac{\pi}{45}$ so that

$$\cos \frac{14\pi}{45} = \frac{1}{2} - \frac{\sqrt{3}}{2}\left(-\frac{\pi}{45} \right) - \frac{1}{2!2}\left(-\frac{\pi}{45} \right)^2 + \frac{\sqrt{3}}{3!2}\left(-\frac{\pi}{45} \right)^3$$

$$+ \frac{1}{4!2}\left(-\frac{\pi}{45} \right)^4 + \cdots$$

$$= 0.500000 + 0.060460 - 0.001218 - 0.000049 + 0.0000005 + \cdots$$

The sum of the first four terms $(n = 3)$ is

$$\cos \frac{14\pi}{45} \doteq 0.559193 \doteq 0.5592$$

We determine that this is correct to four decimal places as follows

$$R_n(x) = \frac{f^{(n+1)}(z)}{(n + 1)!}\left(x - \frac{\pi}{3} \right)^{n+1}$$

and since $|f^{(n+1)}(z)| \leq 1$, we have

$$|R_n(x)| \leq \frac{1}{(n+1)!} \left| \left(x - \frac{\pi}{3}\right) \right|^{n+1}$$

In particular,

$$\left| R_3 \left(\frac{14\pi}{45}\right) \right| \leq \frac{1}{4!} \left(\frac{\pi}{45}\right)^4 \doteq 0.000001$$

In fact, $\cos \dfrac{14\pi}{45} \doteq 0.55919$ to five decimal places.

EXAMPLE 7 Find $\displaystyle\int_0^1 \frac{\sin x}{x}\, dx$ to four decimal places.

Solution The Maclaurin series for $\sin x$ is

$$\sin x = x - \frac{x^3}{3!} + \frac{x^5}{5!} - \frac{x^7}{7!} + \cdots + \frac{(-1)^{n-1} x^{2n-1}}{(2n-1)!} + \cdots$$

for all values of x (the reader should verify this). Thus for all nonzero x,

$$\frac{\sin x}{x} = 1 - \frac{x^2}{3!} + \frac{x^4}{5!} - \frac{x^6}{7!} \cdots + \frac{(-1)^{n-1} x^{2(n-1)}}{(2n-1)!} + \cdots$$

and $\displaystyle\lim_{x \to 0} \frac{\sin x}{x} = 1$. Hence

$$\int_0^1 \frac{\sin x}{x}\, dx = \int_0^1 \left(1 - \frac{x^2}{3!} + \frac{x^4}{5!} - \frac{x^6}{7!} + \cdots \right) dx$$

and integration term by term implies that

$$\int_0^1 \frac{\sin x}{x}\, dx = \left(x - \frac{x^3}{3(3!)} + \frac{x^5}{5(5!)} - \frac{x^7}{7(7!)} + \cdots \right) \Bigg|_0^1$$

$$= 1 - \frac{1}{3(3!)} + \frac{1}{5(5!)} - \frac{1}{7(7!)} + E$$

where $0 < E < \dfrac{1}{9(9!)} < \dfrac{10^{-6}}{3}$. The addition of the four terms yields

$$\int_0^1 \frac{\sin x}{x}\, dx \doteq 0.9461$$

to four decimal places.

Theorem 1 tells us when the Taylor series (and the Maclaurin series as a special case) represents a function inside the interval of convergence of the series. The following example shows that there are times when the power series expansion does not represent the function.

EXAMPLE 8 Let

$$f(x) = \begin{cases} e^{-1/x^2} & \text{if } x \neq 0 \\ 0 & \text{if } x = 0 \end{cases}$$

Find the Maclaurin series for $f(x)$. Show that if $x \neq 0$ then $f(x)$ is not equal to its Maclaurin series.

Solution   In Section 14.2, Example 9, we showed that, for this function, $f'(0) = 0$. It can similarly be shown that $f^{(n)}(0) = 0$ for all positive integers n. Thus, the Maclaurin series is

$$\sum_{n=0}^{\infty} \frac{1}{n!} f^{(n)}(0) x^n = 0 + 0 + 0 + \cdots + 0 + \cdots = 0$$

and this series converges for all values of x.

However, if $x \neq 0$ then $f(x) = e^{-1/x^2} \neq 0$, so the function does not equal its Maclaurin series.

Exercises 15.9

In Exercises 1 through 12, find the Maclaurin series for the given function, assuming convergence to the function in a suitable interval.

1. $\sin x$
3. $\sinh x$
5. $(1 + x)^{-1}$
7. $(x - 1)^3$

2. $\cosh x$
4. e^{3x}
6. $\ln(1 - x)$
8. $(x^2 + 1)^{-1}$ (*Hint:* Use the function of Exercise 5)

9. $\cos bx$, (b constant)

10. $\cos\left(x + \dfrac{\pi}{4}\right)$

11. $\sin^2 x$ (*Hint:* Use double argument formula.)

12. $\sin x \cos x$ (*Hint:* Use double argument formula.)

In Exercises 13 through 20, find the Taylor series for the given function at the specified value of c. State the radius of convergence.

13. e^x, $c = a$

14. e^{bx}, $c = a$, ($b = $ constant)

15. $\ln x$, $c = 2$

16. $\sin x$, $c = \dfrac{\pi}{6}$

17. $\sin x$, $c = a$ [*Hint:* Write $x = a + (x - a)$ and use $\sin(A + B) = \sin A \cos B + \cos A \sin B$.]

18. $\dfrac{1}{x}$, $c = 1$

19. $\tan^{-1} x$, $c = 0$ $\left[\textit{Hint:} \text{ Form } \dfrac{d}{dx}(\tan^{-1} x). \right]$

20. $\dfrac{1}{x + 2}$, $c = 3$

In Exercises 21 through 26, compute each of the following to the stated degree of accuracy by application of Taylor series.

21. $e^{0.2}$, four decimal places

22. $\sin 12°$, four decimal places $\left(\textit{Hint:} \ 12° = \dfrac{\pi}{15} \text{ radians.} \right)$

23. $\cosh 1$, five decimal places
24. $\sin 32°$, four decimal places (*Hint:* $1° \doteq 0.017453$ radians.)

25. $\displaystyle\int_0^1 \sin t^2\, dt,$ five decimal places

26. $\displaystyle\int_0^{1/2} e^{-t^2}\, dt,$ five decimal places

In Exercises 27 through 30, find the terms of the Maclaurin series up to the specified degree for the given function. In some cases it may be difficult to find an explicit formula for the nth derivative.

27. $e^x \cos 2x,$ degree 3

28. $\tanh x,$ degree 3

29. $e^{\cos x},$ degree 4

30. $\displaystyle\int_0^x \frac{dt}{1+t^3},$ degree 10

15.10   FURTHER APPLICATIONS OF TAYLOR SERIES—BINOMIAL SERIES

In some of the exercises of the earlier sections of this text it has been assumed that certain algebraic operations with power series are valid. We now state and apply some important rules concerning the manipulation of power series.

Theorem 1   *(Multiplication of Power Series Representation by a Constant).* If

$$f(x) = \sum_{k=0}^{\infty} a_k x^k = a_0 + a_1 x + a_2 x^2 + \cdots \qquad \text{for } -R < x < R$$

then

$$cf(x) = \sum_{k=0}^{\infty} c a_k x^k = c a_0 + c a_1 x + c a_2 x^2 + \cdots \qquad \text{for } -R < x < R.$$

Hence if a power series represents a function f in a given interval $|x| < R$, then the power series may be multiplied term by term by a constant and the new power series will represent the function cf. Of course, this statement includes division by any nonzero constant.

Theorem 2   *(Addition of Power Series Representations).* If

$$f(x) = \sum_{k=0}^{\infty} a_k x^k = a_0 + a_1 x + a_2 x^2 + \cdots \qquad \text{for } |x| < R_1$$

and

$$g(x) = \sum_{k=0}^{\infty} b_k x^k = b_0 + b_1 x + b_2 x^2 + \cdots \qquad \text{for } |x| < R_2$$

then

$$f(x) + g(x) = \sum_{k=0}^{\infty} (a_k + b_k)x^k$$

$$= (a_0 + b_0) + (a_1 + b_1)x + (a_2 + b_2)x^2 + \cdots \qquad -R < x < R \quad (1)$$

where $R = \min (R_1, R_2)$.

This means that we may add the terms of the same degree representing f and g respectively to obtain the series that represents $f + g$ in the common interval of convergence.

From Theorems 1 and 2 we have

Corollary 1   If $f(x) = \sum_{k=0}^{\infty} a_k x^k$ in $-R_1 < x < R_1$ and $g(x) = \sum_{k=0}^{\infty} b_k x^k$ in $-R_2 < x < R_2$,

then for arbitrary constants c_1 and c_2, $c_1 f(x) + c_2 g(x) = \sum_{k=0}^{\infty} (c_1 a_k + c_2 b_k)x^k$

in $-R < x < R$ where $R = \min (R_1, R_2)$.

This result is readily extended to n functions and the proofs of the theorems and the corollary are left to the reader as exercises.

EXAMPLE 1   Find the Maclaurin series for the function $\dfrac{x}{(1 - 2x)(1 - x)}$.

Solution   By partial fractions we write

$$\frac{x}{(1 - 2x)(1 - x)} = \frac{A}{1 - 2x} + \frac{B}{1 - x} \qquad (A \text{ and } B \text{ are constants})$$

or

$$x = A(1 - x) + B(1 - 2x) = A + B - (A + 2B)x$$

Hence $A + B = 0$ and $-(A + 2B) = 1$ from which

$$A = 1 \qquad \text{and} \qquad B = -1$$

Therefore

$$\frac{x}{(1 - 2x)(1 - x)} = \frac{1}{1 - 2x} - \frac{1}{1 - x} \qquad (2)$$

From

$$\frac{1}{1 - r} = \sum_{k=0}^{\infty} r^k, \ |r| < 1 \text{ we have}$$

$$\frac{1}{1 - 2x} = \sum_{k=0}^{\infty} (2x)^k = \sum_{k=0}^{\infty} 2^k x^k \qquad \text{for } |x| < \tfrac{1}{2} \qquad (3)$$

while

$$\frac{1}{1 - x} = \sum_{k=0}^{\infty} x^k \qquad \text{for } |x| < 1 \tag{4}$$

But, $\frac{1}{2} = \min(\frac{1}{2}, 1)$ so that, from Corollary 1, and (2), (3), and (4) we have the required Maclaurin series representation:

$$\frac{x}{(1 - 2x)(1 - x)} = \sum_{k=0}^{\infty} (2^k - 1)x^k \qquad \text{if } |x| < \tfrac{1}{2} \tag{5}$$

Note that $\dfrac{x}{(1 - 2x)(1 - x)}$ is not defined at $x = \frac{1}{2}$ and $x = 1$ so that $x = \frac{1}{2}$ is its nearest "singular point" to $x = 0$. Furthermore, $x = \frac{1}{2}$ is an end point of the interval of convergence. ●

Theorem 3 **(Multiplication of Power Series Representations).** If

$$f(x) = \sum_{k=0}^{\infty} a_k x^k = a_0 + a_1 x + a_2 x^2 + \cdots \qquad \text{for } |x| < R_1$$

and

$$g(x) = \sum_{k=0}^{\infty} b_k x^k = b_0 + b_1 x + b_2 x^2 + \cdots \qquad \text{for } |x| < R_2$$

Let $R = \min(R_1, R_2)$, then

$$f(x) \cdot g(x) = \sum_{k=0}^{\infty} c_k x^k \qquad \text{for } |x| < R, \tag{6}$$

where

$$\begin{aligned}
c_0 &= a_0 b_0 \\
c_1 &= a_0 b_1 + a_1 b_0 \\
c_2 &= a_0 b_2 + a_1 b_1 + a_2 b_0 \\
&\cdots \\
c_n &= a_0 b_n + a_1 b_{n-1} + a_2 b_{n-2} + \cdots + a_n b_0
\end{aligned} \tag{7}$$

In other words, we simply collect together the terms of the same degree.

The proofs of Theorem 3 and Theorem 4 (which follows) are more difficult than the earlier theorems and are best relegated to a course in advanced calculus.

EXAMPLE 2   Find the Maclaurin expansion of $\dfrac{\ln(1 + x)}{1 + x}$.

Solution   We write the given function as a product $(1 + x)^{-1} \ln(1 + x)$. It is known that

$$(1 + x)^{-1} = 1 - x + x^2 - x^3 + \cdots \text{ if } |x| < 1$$

and

$$\ln(1 + x) = x - \frac{x^2}{2} + \frac{x^3}{3} - \frac{x^4}{4} + \cdots \text{ if } |x| < 1$$

so that we obtain the interesting formula

$$(1 + x)^{-1} \ln(1 + x) = x - (1 + \tfrac{1}{2})x^2 + (1 + \tfrac{1}{2} + \tfrac{1}{3})x^3 - (1 + \tfrac{1}{2} + \tfrac{1}{3} + \tfrac{1}{4})x^4$$
$$+ \cdots \quad (8)$$

for $|x| < 1$. ●

Theorem 4 **(*Division of Power Series Representations*).** If $f(x) = \sum\limits_{k=0}^{\infty} a_k x^k$ for $|x| < R_1$ and $g(x) = \sum\limits_{k=0}^{\infty} b_k x^k$ for $|x| < R_2$ and we let $g(0) = b_0 \neq 0$. Also let R_3 be the smallest absolute value of the solutions of the equation $g(x) = 0$, where all solutions real or complex must be considered. Then the series obtained by dividing $\sum\limits_{k=0}^{\infty} a_k x^k$ by $\sum\limits_{k=0}^{\infty} b_k x^k$ represents the function $\dfrac{f(x)}{g(x)}$ in $|x| < R$, where $R = \min(R_1, R_2, R_3)$. Furthermore, if the power series for the quotient is $\sum\limits_{k=0}^{\infty} c_k x^k$, then c_k may be found from

$$\begin{aligned}
a_0 &= c_0 b_0 \qquad (b_0 \neq 0) \\
a_1 &= c_0 b_1 + c_1 b_0 \\
a_2 &= c_0 b_2 + c_1 b_1 + c_2 b_0 \\
&\cdots \\
a_n &= c_0 b_n + c_1 b_{n-1} + \cdots + c_n b_0
\end{aligned} \qquad (9)$$

where the first equation of (9) yields c_0, the second c_1, and so on.

EXAMPLE 3   Find the first four nonzero terms of the power series representation for $\tan x$.

Solution

$$\tan x = \frac{\sin x}{\cos x} = \frac{x - \dfrac{x^3}{3!} + \dfrac{x^5}{5!} - \cdots}{1 - \dfrac{x^2}{2!} + \dfrac{x^4}{4!} - \cdots}$$

where both power series represent the functions for all values of x. Since $\tan x$ is an odd function, that is, $\tan(-x) = -\tan x$, we write

$$\tan x = c_1 x + c_3 x^3 + c_5 x^5 + c_7 x^7 + \cdots$$

Hence

$$x - \frac{x^3}{3!} + \frac{x^5}{5!} - \frac{x^7}{7!} + \cdots = \left(1 - \frac{x^2}{2!} + \frac{x^4}{4!} - \frac{x^6}{6!} + \cdots\right)$$
$$(c_1 x + c_3 x^3 + c_5 x^5 + \cdots)$$

from which, equating like coefficients yields

$$1 = c_1$$

$$-\frac{1}{3!} = c_3 - \frac{c_1}{2!}$$

$$\frac{1}{5!} = c_5 - \frac{c_3}{2!} + \frac{c_1}{4!}$$

$$-\frac{1}{7!} = c_7 - \frac{c_5}{2!} + \frac{c_3}{4!} - \frac{c_1}{6!}$$

and so on.

Thus the first four coefficients are

$$c_1 = 1 \qquad c_3 = \tfrac{1}{3} \qquad c_5 = \tfrac{2}{15} \qquad c_7 = \tfrac{17}{315}$$

and we have

$$\tan x = x + \frac{x^3}{3} + \frac{2}{15}x^5 + \frac{17}{315}x^7 + \cdots \quad -\frac{\pi}{2} < x < \frac{\pi}{2} \tag{10}$$

Alternative Solution   Proceed directly by long division. This solution is left to the reader as an exercise. ●

We next turn to a discussion of the binomial series. Consider the function $f(x) = (1 + x)^m$, where m is a real number. The Maclaurin series for $(1 + x)^m$ is easily obtained. We have

$$f(x) = (1 + x)^m \qquad\qquad f(0) = 1$$
$$f'(x) = m(1 + x)^{m-1} \qquad\qquad f'(0) = m$$
$$f''(x) = m(m - 1)(1 + x)^{m-2} \qquad\qquad f''(0) = m(m - 1)$$
$$\cdots$$

$$f^{(k)}(x) = m(m - 1)(m - 2) \cdots [m - (k - 1)](1 + x)^{m-k}$$
$$f^{(k)}(0) = m(m - 1)(m - 2) \cdots [m - (k - 1)]$$

Therefore the Maclaurin series for $(1 + x)^m$ is†

$$(1 + x)^m = 1 + mx + \frac{m(m - 1)}{2!}x^2 + \frac{m(m - 1)(m - 2)}{3!}x^3 + \cdots$$
$$+ \frac{m(m - 1)(m - 2) \cdots [m - (k - 1)]}{k!}x^k + \cdots \tag{11}$$

The series on the right side of (11) is called the **binomial series** for the function $(1 + x)^m$. It still remains to show that the series converges to $(1 + x)^m$. The behavior of the right side when m is a positive integer is vastly different from the case when $m \neq$ positive integer.

If $m = n$, a positive integer, then $(1 + x)^n$ is a polynomial of degree n and the

---

† The justification of the equality in (11) is given later in this section.

Maclaurin series terminates. We have

$$(1 + x)^n = 1 + nx + \frac{n(n - 1)}{2!}x^2 + \frac{n(n - 1)(n - 2)}{3!}x^3 + \cdots +$$
$$\frac{n(n - 1)(n - 2) \cdots [n - (k - 1)]}{k!}x^k + \cdots + x^n \tag{12}$$

$$= \sum_{k=0}^{n} \binom{n}{k} x^k \tag{13}$$

where

$$\binom{n}{k} = \frac{n(n - 1)(n - 2) \cdots (n - k + 1)}{k!} \tag{14}$$

Formula (12) is called the **binomial theorem** and was originally developed by Newton. Its coefficients $\binom{n}{k}$ defined by (14) are integers and are called the **binomial coefficients.** The quantity $\binom{n}{k}$ is the number of ways k objects can be selected from n objects where order is not significant. Another useful form for $\binom{n}{k}$ is

$$\binom{n}{k} = \frac{n!}{k!(n - k)!} \qquad (k = 0, 1, 2, \ldots, n) \tag{15}$$

and it follows immediately that

$$\binom{n}{k} = \binom{n}{n - k} \tag{16}$$

This shows that the coefficients in (12) are symmetrical. The verification of (15) and (16) is left as an exercise for the reader.

If m is not a positive integer, the series in (11) does not terminate; that is, it is an infinite Maclaurin series. The radius of convergence of the series in (11) is easily obtained by the ratio test. We have

$$\lim_{n \to \infty} \left| \frac{u_{n+1}}{u_n} \right| = \lim_{n \to \infty} \frac{\dfrac{m(m - 1) \cdots (m - n + 1)(m - n)}{(n + 1)!}x^{n+1}}{\dfrac{m(m - 1) \cdots (m - n + 1)}{n!}x^n}$$

$$= \lim_{n \to \infty} \left| \frac{m - n}{n + 1} \right| |x| = \lim_{n \to \infty} \left| \frac{\dfrac{m}{n} - 1}{1 + \dfrac{1}{n}} \right| |x| = |x|$$

Consequently the Maclaurin series for $(1 + x)^m$ converges for $|x| < 1$. However, it still remains to show that series (11) represents the function $(1 + x)^m$ in $(-1, 1)$. This would follow if we could show that $\lim_{n \to \infty} R_n(x) = 0$, but this is difficult.

Fortunately we can use instead the following method.

Proof We let $f(x)$ equal the binomial series (the right side of equation (11)). It must be shown that $f(x) = (1 + x)^m$. We have

$$f(x) = 1 + mx + \frac{m(m-1)}{2!}x^2 + \cdots$$

$$+ \frac{m(m-1)(m-2)\cdots(m-k+1)}{k!}x^k + \cdots \quad (17)$$

Term by term differentiation yields

$$f'(x) = m + m(m-1)x + \cdots$$

$$+ \frac{m(m-1)(m-2)\cdots(m-k+1)}{(k-1)!}x^{k-1} + \cdots \quad (18)$$

We now multiply (18) by $1 + x$ and regroup. The constant term is m. The x^k term in $(1 + x)f'(x)$ is obtained by multiplying the x^{k-1} term in (18) by x and the x^k term in (18) by 1 and then adding. This term is

$$\frac{m(m-1)(m-2)\cdots(m-k+1)}{(k-1)!}x^{k-1} \cdot x$$

$$+ \frac{m(m-1)(m-2)\cdots(m-k+1)(m-k)}{k!}x^k$$

$$= \frac{m(m-1)(m-2)\cdots(m-k+1)}{(k-1)!}\left[1 + \frac{m-k}{k}\right]x^k$$

$$= \frac{m(m-1)(m-2)\cdots(m-k+1)}{(k-1)!} \cdot \frac{m}{k} \cdot x^k$$

$$= m\frac{m(m-1)(m-2)\cdots(m-k+1)}{k!}x^k \quad (19)$$

Note that (19) is m times the coefficient of x^k in (17). Thus

$$(1 + x)f'(x) = m\left[1 + mx + \frac{m(m-1)}{2!}x^2 + \cdots\right.$$

$$\left. + \frac{m(m-1)(m-2)\cdots(m-k+1)}{k!}x^k + \cdots\right] = mf(x) \quad (20)$$

Letting $y = f(x)$, we must solve the differential equation

$$(1 + x)\frac{dy}{dx} = my \quad \text{where } y = 1 \text{ if } x = 0 \text{ (because } f(0) = 1) \quad (21)$$

We have $\dfrac{dy}{y} = m\dfrac{dx}{1 + x}$ so that

$$\ln y = m \ln(1 + x) + C \quad (22)$$

Substituting $y = 1$ and $x = 0$ into (22) gives $C = 0$ and (22) becomes

$$\ln y = m \ln(1 + x) = \ln(1 + x)^m$$

Therefore $f(x) = y = (1 + x)^m$ as desired. We rewrite this as

$$(1 + x)^m = \sum_{k=0}^{\infty} \binom{m}{k} x^k \qquad |x| < 1 \tag{23}$$

which completes our proof. ∎

When m is not a positive integer (23) is valid if $|x| < 1$, whereas the Maclaurin series diverges if $|x| > 1$. Proving convergence at the end points $|x| = 1$ is a more difficult task. Thus, we state without proof that

(i)   If $m \le -1$, the series diverges for $x = \pm 1$.
(ii) If $-1 < m < 0$, the series diverges for $x = -1$ and represents the function for $x = 1$; that is, (23) is valid.
(iii) If $m > 0$, (23) is correct for $x = \pm 1$ also.

EXAMPLE 4   Find the binomial series for $(1 + x)^{1/3}$ and use it to approximate $1.1^{1/3}$ to five decimal places.

Solution   From (23) with $m = \frac{1}{3}$, we have

$$(1 + x)^{1/3} = 1 + \frac{1}{3}x + \frac{\frac{1}{3}(-\frac{2}{3})}{2!}x^2 + \frac{\frac{1}{3}(-\frac{2}{3})(-\frac{5}{3})}{3!}x^3 + \frac{\frac{1}{3}(-\frac{2}{3})(-\frac{5}{3})(-\frac{8}{3})}{4!}x^4$$

$$+ \cdots + (-1)^{n-1}\frac{2 \cdot 5 \cdot 8 \cdots (3n - 4)}{3^n n!}x^n + \cdots$$

$$= 1 + \frac{1}{3}x - \frac{1}{9}x^2 + \frac{5}{81}x^3 - \frac{10}{243}x^4 + \frac{22}{729}x^5 + \cdots \quad \text{if } |x| < 1$$

In particular,

$$(1 + 0.1)^{1/3} = 1 + \frac{0.1}{3} - \frac{0.01}{9} + \frac{5}{81}(0.001) - \frac{10}{243}(0.0001) + E$$

where because the series alternates after the first term

$$|E| < \frac{22}{729}(10^{-5}) < \frac{1}{3}(10^{-6})$$

or by addition of the first five terms,

$$(1.1)^{1/3} \doteq 1.03228 \text{ to five decimal places.} \qquad \bullet$$

EXAMPLE 5   Find the power series for the function $\sin^{-1} x$.

Solution   We note that

$$D_x(\sin^{-1} x) = \frac{1}{\sqrt{1 - x^2}}$$

so that the required series may be obtained by integrating the power series of

$(1 - t^2)^{-1/2}$. Now,

$$(1 - t^2)^{-1/2} = 1 - \frac{1}{2}(-t^2) + \frac{(-\frac{1}{2})(-\frac{3}{2})}{2!}(-t^2)^2$$

$$+ \frac{(-\frac{1}{2})(-\frac{3}{2})(-\frac{5}{2})}{3!}(-t^2)^3 + \frac{(-\frac{1}{2})(-\frac{3}{2})(-\frac{5}{2})(-\frac{7}{2})}{4!}(-t^2)^4 + \cdots$$

$$= 1 + \frac{t^2}{2} + \frac{1 \cdot 3}{2 \cdot 4}t^4 + \frac{1 \cdot 3 \cdot 5}{2 \cdot 4 \cdot 6}t^6 + \frac{1 \cdot 3 \cdot 5 \cdot 7}{2 \cdot 4 \cdot 6 \cdot 8}t^8 + \cdots$$

and

$$\sin^{-1} x = \int_0^x (1 - t^2)^{-1/2}\, dt$$

$$= x + \frac{1}{2}\frac{x^3}{3} + \frac{1 \cdot 3}{2 \cdot 4}\frac{x^5}{5} + \frac{1 \cdot 3 \cdot 5}{2 \cdot 4 \cdot 6}\frac{x^7}{7} + \frac{1 \cdot 3 \cdot 5 \cdot 7}{2 \cdot 4 \cdot 6 \cdot 8}\frac{x^9}{9} + \cdots$$

$$-1 < x < 1 \qquad \bullet \quad (24)$$

EXAMPLE 6   Find $\int_2^4 \sqrt{1 + x^4}\, dx$ correct to three decimal places.

Solution   If we apply the binomial expansion to the function $(1 + x^4)^{1/2}$ the interval of convergence is $|x| < 1$ which is outside the interval of integration. This difficulty can be surmounted by writing

$$\sqrt{1 + x^4} = x^2 \sqrt{1 + \frac{1}{x^4}}$$

since $x > 0$, and therefore

$$\int_2^4 (1 + x^4)^{1/2} = \int_2^4 x^2 \left(1 + \frac{1}{x^4}\right)^{1/2} dx$$

Expansion of the binomial in powers of $\frac{1}{x^4}$ yields

$$\int_2^4 (1 + x^4)^{1/2}\, dx = \int_2^4 x^2 \left(1 + \frac{1}{2x^4} + \frac{\frac{1}{2}(-\frac{1}{2})}{2!}\frac{1}{x^8} + \frac{\frac{1}{2}(-\frac{1}{2})(-\frac{3}{2})}{3!}\frac{1}{x^{12}} + \cdots\right) dx$$

so that from integration term by term

$$\int_2^4 (1 + x^4)^{1/2}\, dx = \int_2^4 (x^2 + \tfrac{1}{2}x^{-2} - \tfrac{1}{8}x^{-6} + \tfrac{1}{16}x^{-10} + \cdots)\, dx$$

$$= \left(\frac{x^3}{3} - \frac{x^{-1}}{2} + \frac{x^{-5}}{40} - \frac{x^{-9}}{144} + \cdots\right) \Big|_2^4$$

and after the first term we have a convergent alternating series. To within the accuracy required, it is only necessay to use three terms. We have

$$\int_2^4 (1 + x^4)^{1/2}\, dx \doteq 18.791$$

to three decimal places. $\bullet$

EXAMPLE 7 Find the Maclaurin series for $f(x) = \ln(1 + \sin x)$ up through the term x^4. Is the function even or odd?

Solution Let $u = \sin x$, so that

$$\ln(1 + \sin x) = \ln(1 + u) = u - \frac{u^2}{2} + \frac{u^3}{3} - \cdots \qquad |u| < 1$$

$$= \sin x - \frac{\sin^2 x}{2} + \frac{\sin^3 x}{3} - \cdots$$

$$= \left(x - \frac{x^3}{3!} + \cdots\right) - \frac{1}{2}\left(x - \frac{x^3}{3!} + \cdots\right)^2$$

$$+ \frac{1}{3}\left(x - \frac{x^3}{3!} + \cdots\right)^3 + \cdots$$

If we start squaring and cubing as indicated and rearrange the terms in ascending powers of x, we have

$$\ln(1 + \sin x) = x - \frac{x^2}{2} + \frac{x^3}{6} + \cdots$$

The function is neither even nor odd. Why?

This is known as the **substitution method.** ●

More generally, if we have a composite function h defined by $h(x) = f(g(x))$, we can formally find its Maclaurin expansion as follows: If

$$f(u) = a_0 + a_1 u + a_2 u^2 + \cdots \qquad \text{and} \qquad g(x) = b_0 + b_1 x + b_2 x^2 + \cdots$$

then we write

$$f(g(x)) = a_0 + a_1(b_0 + b_1 x + b_2 x^2 + \cdots)$$

$$+ a_2(b_0 + b_1 x + b_2 x^2 + \cdots)^2 + \cdots$$

We then collect the constant terms, the terms involving x, the terms involving x^2, and so forth, and the resultant series is the Maclaurin series for $f(g(x))$ if certain conditions of convergence are met.

Exercises 15.10

1. Derive the power series representation

$$\cosh x = \sum_{k=0}^{\infty} \frac{x^{2k}}{(2k)!} = 1 + \frac{x^2}{2!} + \frac{x^4}{4!} + \cdots \qquad \text{for all } x$$

from those for e^x and e^{-x} by applying the corollary to Theorem 2 and the basic definition of $\cosh x$.

2. Find the Maclaurin series for $(1 - x)^{-2}$ by multiplying the Maclaurin series for $(1 - x)^{-1}$ by itself. Verify the expansion

$$(1 - x)^{-2} = \sum_{k=0}^{\infty} (k + 1)x^k \qquad |x| < 1$$

3. Obtain the expansion of Exercise 2 by differentiating the Maclaurin series for $(1 - x)^{-1}$.

4. Find the Maclaurin series for $\dfrac{1}{1-x} \ln \dfrac{1}{1-x}$ and determine its interval of absolute convergence.

5. Find the first four nonzero terms of the Maclaurin series for $e^x \cos x$. What is the interval of convergence?

6. Find the Maclaurin series for e^{b-bx}, where b is a constant.

7. Find the Taylor series for $\dfrac{1}{x}$ about $x = 1$. (*Hint:* let $x = 1 + u$ and find the Maclaurin series for $\dfrac{1}{1+u}$ about $u = 0$.) What is the radius of convergence of your series?

8. Find the first four nonzero terms of the Maclaurin series of sec x, from the identity $\sec x = \dfrac{1}{\cos x}$ and the power series for $\cos x$. (*Hint:* sec x is an even function of x.) What would you conjecture to be the interval of convergence?

9. Find the first four terms of the Taylor series for $x^{3/2}$ about $x = 1$.

10. Find the Taylor series for $\dfrac{1}{x}$ about $x = 4$ and its domain of convergence.

11. By starting with the power series expansions of sinh x and cosh x, find the first four terms of the Maclaurin series for tanh x.

12. Find the function $f(x)$ that uniquely satisfies the initial value problem

$$f'(x) = \frac{1}{1 + x^2} \qquad f(0) = 0$$

Deduce the Maclaurin series for this function.

In Exercises 13 through 16, use the substitution method of Example 7 to find the Maclaurin expansion for the given function through the indicated degree.

13. $e^{\sin x}$, degree 4

14. $\cos (\sin x)$, degree 4

15. $e^{x(1+x)}$, degree 3

16. $\dfrac{1}{1 - x^2 \sin x}$, degree 6

17. Find the Maclaurin series expansion for $f(x) = \ln (x + \sqrt{1 + x^2})$ (*Hint:* Show that $f'(x) = \dfrac{1}{\sqrt{1 + x^2}}$, apply the binomial expansion for this function, and then integrate.)

18. Determine the Maclaurin series for $\dfrac{1}{1 - x + x^2}$. (*Hint:* Use elementary algebra).

19. If $f(x) = (1 - x)^{-2}$, find $f^{(n)}(0)$.

In Exercises 20 through 23, use a binomial series to find the Maclaurin series for each given function. Determine the radius of convergence.

20. $f(x) = (1 + x^2)^{-1/2}$
22. $F(x) = \sqrt{1 - 6x}$

21. $g(x) = (1 - x^3)^{-1/2}$
23. $f(x) = (27 + x)^{1/3}$

In Exercises 24, 25, and 26, find the value of the given quantity accurate to three decimal places by application of the binomial series.

24. $\sqrt{10}$ (*Hint:* $9 + x = 3\sqrt{1 + \dfrac{x}{9}} = 3\left(1 + \dfrac{x}{9}\right)^{1/2}$, and so on.)

25. $\sqrt[3]{29}$

26. $\sqrt[5]{996}$

In Exercises 27 through 30 find the value of the definite integral to the specified accuracy by application of the binomial series.

27. $\displaystyle\int_0^{1/2} \sqrt{1 + x^2}\, dx$, three decimal places

28. $\displaystyle\int_0^{0.8} \sqrt{4 - x^2}\, dx$, three decimal places

29. $\int_0^{0.6} (1 + x^2)^{1/3} \, dx$, four decimal places

30. $\int_0^{1/3} x(1 + x)^{1/2} \, dx$, three decimal places

Review and Miscellaneous Exercises

In Exercises 1 through 12, the nth term of the sequence is given. If it converges, find the limit. If it diverges, does it become infinite or does it remain bounded? Justify your response.

1. $a_n = \dfrac{n^2 - n}{8n^2 + 1}$

2. $a_n = \left(10 - \dfrac{n^3}{7^n}\right)^2$

3. $a_n = \begin{cases} 5 + \dfrac{\ln n}{n} & 1 \le n \le N \\[2mm] \dfrac{(-1)^n \ln n}{\sqrt{n}} & n > N \end{cases}$

4. $a_n = n - \ln n$

5. $a_n = \dfrac{e^n - e^{-n}}{2n^2}$

6. $a_n = \dfrac{(-1)^n n + 20}{2n + \pi}$

7. $a_n = \sin^2\left(\dfrac{n\pi}{2}\right)$

8. $a_n = \sqrt{n}(\sqrt{n + 10} - \sqrt{n})$

9. $a_n = \dfrac{n^2}{n + 2} - \dfrac{n^2 + 1}{n}$

10. $a_n = \sum_{k=1}^{n} \ln\left(1 + \dfrac{1}{k}\right)$ $\left[\text{Hint: } \ln\left(1 + \dfrac{1}{k}\right) = \ln(k + 1) - \ln k.\right]$

11. $a_n = \dfrac{P_m(n)}{2^{n/2}}$, where $P_m(n)$ is an arbitrary polynomial in n of degree m.

12. $a_n = \dfrac{6^n + 2(n!)}{10^n + 5(n!)}$

In Exercises 13 through 16, find the sum of the given convergent infinite series.

13. $100 - 10 + 1 - 10^{-1} + \cdots$

14. $0.\overline{359}$ (see Exercise 15.3, Problem 6 for the meaning of the bar.)

15. $\dfrac{1}{1 \cdot 3} + \dfrac{1}{3 \cdot 5} + \dfrac{1}{5 \cdot 7} + \dfrac{1}{7 \cdot 9} + \cdots$ (*Hint:* Write the nth term as a difference and obtain a telescoping series.)

16. $\dfrac{1}{10} + \dfrac{\left(\frac{1}{10}\right)^2}{2!} + \dfrac{\left(\frac{1}{10}\right)^3}{3!} + \cdots + \dfrac{\left(\frac{1}{10}\right)^n}{n!} + \cdots$

In Exercises 17 through 30, determine whether or not the given series converges. Do not attempt to find the sum in case of convergence.

17. $\sum_{k=1}^{\infty} \dfrac{1}{k^2 + 3k}$

18. $\sum_{k=1}^{\infty} \dfrac{k^2}{100k^2 + 1}$

19. $\sum_{k=2}^{\infty} \dfrac{1}{\ln^2 k}$

20. $\sum_{k=1}^{\infty} \dfrac{1}{k + 3\sqrt{k}}$

21. $\displaystyle\sum_{k=3}^{\infty} \frac{k^2}{k!}$

22. $\displaystyle\sum_{k=100}^{\infty} \frac{10k}{k^4 - 100},$ by application of the integral test

23. $\displaystyle\sum_{k=2}^{\infty} k^2 e^{-k},$ by application of the integral test

24. $\displaystyle\sum_{k=1}^{\infty} \frac{7k + \pi}{k^3 - k^2 + k + 3}$

25. $\displaystyle\sum_{k=1}^{\infty} \frac{k!}{b^k},$ (b is an arbitrary nonzero constant)

26. $\displaystyle\sum_{k=1}^{\infty} \frac{k!}{1 \cdot 3 \cdot 5 \cdots (2k - 1)}$

27. $\displaystyle\sum_{k=1}^{\infty} \tan \frac{1}{k}$

28. $\displaystyle\sum_{k=1}^{\infty} \frac{1}{k(1 + \ln k)^2}$

29. $\displaystyle\sum_{k=2}^{\infty} \frac{1}{k(\ln k)^p},$ (p is constant)

30. $\displaystyle\sum_{k=1}^{\infty} \frac{2^{3k} + k^3}{3^{2k}}$

In Exercises 31 through 42, determine if the series is absolutely convergent, conditionally convergent, or divergent. Justify your answer.

31. $\displaystyle\sum_{k=2}^{\infty} (-1)^k \frac{\ln k^2}{k}$

32. $\displaystyle\sum_{k=1}^{\infty} \frac{(-1)^k}{k + 2\sqrt{k}}$

33. $\displaystyle\sum_{k=2}^{\infty} \frac{(-1)^k(k - \sqrt{k})}{1000k}$

34. $\displaystyle\sum_{k=1}^{\infty} \frac{\sin k \cos k}{k\sqrt{k}}$

35. $\displaystyle\sum_{k=1}^{\infty} \frac{(-1)^k(k^3 - k^2 + 10)}{k^3 + k + 1}$

36. $\displaystyle\sum_{k=1}^{\infty} \frac{(-1)^k 6^{3k}}{(2k - 1)!}$

37. $\displaystyle\sum_{k=1}^{\infty} (-1)^k \frac{(k + 1)^{3/4}}{k^2}$

38. $\displaystyle\sum_{k=1}^{\infty} \frac{(-1)^{k-1}}{k^p},$ $p > 0$

39. $\displaystyle\sum_{k=2}^{\infty} \frac{(-1)^k k!}{3 \cdot 5 \cdot 7 \cdots (2k - 1)}$

40. $\displaystyle\sum_{k=1}^{\infty} \frac{(-1)^k}{k - \ln k}$

41. $\displaystyle\sum_{k=2}^{\infty} \frac{(-1)^k k}{k^2 - 1}$

42. $\displaystyle\sum_{k=1}^{\infty} \frac{(-1)^k k^{100}}{(\sqrt{2})^k}$

In Exercises 43 through 52, determine the interval of convergence of the given series.

43. $\displaystyle\sum_{k=2}^{\infty} x^{3k}$

44. $\displaystyle\sum_{k=0}^{\infty} \frac{x^k}{10^{k+2}}$

45. $\displaystyle\sum_{k=1}^{\infty} \frac{(x + 3)^k}{2k}$

46. $\displaystyle\sum_{k=1}^{\infty} \frac{k^3(x - 2)^k}{k!}$

47. $\displaystyle\sum_{k=1}^{\infty} \frac{kx^k}{k + 2}$

48. $\displaystyle\sum_{k=1}^{\infty} b^k x^k,$ $b \neq 0$

49. $\displaystyle\sum_{k=1}^{\infty} (\sqrt[k]{k} - 1)^k x^k$

50. $\displaystyle\sum_{k=1}^{\infty} \frac{x^k}{10 + b^k},$ $b > 1$

51. $\displaystyle\sum_{k=2}^{\infty} \frac{(-1)^k x^k}{k(\ln k)^3}$

52. $\displaystyle\sum_{k=0}^{\infty} \frac{e^k(x - \pi)^k}{k!}$

In Exercises 53, 54, and 55, determine the set of values of x for which the given series converges.

53. $\displaystyle\sum_{k=1}^{\infty} (-1)^k \frac{\sin kx \cos kx}{k^2}$

54. $\displaystyle\sum_{k=1}^{\infty} e^{-3kx^2} \cos kx$

55. $\displaystyle\sum_{k=1}^{\infty} \frac{7^k}{k^2 x^k}$

In Exercises 56 through 62, find the Taylor series for the given function at the indicated value of c. Determine the interval of convergence.

56. $e^{-bx}, \quad c = 1$

57. $\ln x^3, \quad c = 1$

58. $\dfrac{1}{1 - 36x^2}, \quad c = 0$

59. $x|x|, \quad c = 0$

60. $x|x|, \quad c = -1$

61. $\dfrac{1}{x}, \quad c$ arbitrary $\neq 0$

62. $\dfrac{1}{3 + 4x}, \quad c = 0$

In Exercises 63 through 66, find the exact sum of the series by using the known series for elementary functions.

63. $\dfrac{1}{2!} - \dfrac{1}{3!} + \dfrac{1}{4!} - \cdots$

64. $1 + 2(\tfrac{1}{3}) + 3(\tfrac{1}{3})^2 + 4(\tfrac{1}{3})^3 + \cdots$

65. $1 + \dfrac{(\tfrac{3}{2})^2}{2!} + \dfrac{(\tfrac{3}{2})^4}{4!} + \dfrac{(\tfrac{3}{2})^6}{6!} + \cdots$

66. $\dfrac{1}{3!} - \dfrac{1}{5!} + \dfrac{1}{7!} - \dfrac{1}{9!} + \cdots$

In Exercises 67 through 72 apply power series to compute each of the following to the indicated accuracy.

67. $8.5^{1/3}$, four decimal places
68. $\cosh 0.90$, four decimal places
69. $\cos 0.30$, five decimal places
70. $990^{1/5}$, five decimal places

71. If $erf(x) = \dfrac{2}{\sqrt{\pi}} \displaystyle\int_0^x e^{-t^2}\, dt$, find $erf(1)$ to three decimal places.

72. If $Si(x) = \displaystyle\int_0^x \frac{\sin t}{t}\, dt$, find $Si(0.3)$ to four decimal places.

73. Prove that $\displaystyle\sum_{k=1}^{\infty} kx^{2k-1} = \frac{x}{1 - 2x^2 + x^4}$ in $|x| < 1$. Does the series represent the function on the right side when $x = \pm 1$?

74. A function satisfies the following initial value problem
$$f'(x) + 2xf(x) = 0 \qquad f(0) = 1$$
Find $f(x)$ in the form of a power series $\displaystyle\sum_{k=0}^{\infty} a_k x^k$ and identify the function. In what interval does the series represent the function?

75. Consider the function $f(x) = \dfrac{1}{x}$ from $x = 1$ to $x = n$, where n is a positive integer > 1. Show that

(a) $1 + \displaystyle\int_2^{n+1} \frac{dx}{x} < \sum_{k=1}^{n} \frac{1}{k} < 1 + \int_1^n \frac{dx}{x}$

(b) The sequence $F(n) = \sum\limits_{k=1}^{n} \dfrac{1}{k} - \ln n$ is bounded below.

(c) $F(n+1) - F(n) < 0$, so that $F(n)$ is a decreasing sequence which then yields

$$\lim_{n \to \infty} F(n) \text{ exists and } = \gamma \qquad \text{where } 0 < \gamma < 1$$

In fact,

$$\gamma = \lim_{n \to \infty} F(n) = \text{Euler's constant} \doteq 0.5772\ldots$$

(d) What does this imply about the divergence of the harmonic series $\sum\limits_{k=1}^{\infty} \dfrac{1}{k}$?

76. Find $\lim\limits_{n \to \infty} \dfrac{\sum\limits_{k=1}^{n} \dfrac{1}{k}}{\sqrt{n}}$ (*Hint:* see Exercise 75.)

77. Prove that $\int_0^\infty \dfrac{\sin x}{x}\, dx$ exists. $\left(Hint:\ \int_0^\infty \dfrac{\sin x}{x}\, dx\text{ exists if each of the integrals }\int_0^{\pi/2} \dfrac{\sin x}{x}\, dx\text{ and}\right.$

$\int_{\pi/2}^\infty \dfrac{\sin x}{x}\, dx$ exist. Use integration by parts to show that

$$\int_{\pi/2}^{b} \frac{\sin x}{x}\, dx = -\frac{\cos b}{b} - \int_{\pi/2}^{b} \frac{\cos x}{x^2}\, dx\Bigg)$$

78. Prove that $\int_0^\infty \left| \dfrac{\sin x}{x} \right| dx$ diverges. (*Hint:* Compare

$$\int_0^{n\pi} \frac{|\sin x|}{x}\, dx = \sum_{k=1}^{n} \int_{(k-1)\pi}^{k\pi} \frac{|\sin x|}{x}\, dx$$

with the harmonic series). Hence, from this and the previous exercise, the $\int_0^\infty \dfrac{\sin x}{x}\, dx$ is a "conditionally convergent improper integral."

In the next exercise, we will prove a theorem that will yield a technique to establish the convergence or divergence of typical series usually treated by the more detailed integral test.

79. Let $\sum\limits_{n=1}^{\infty} u_n$ be an infinite series with $u_n > 0$ and $u_{n+1} \leq u_n$ for all n. Then $\sum\limits_{n=1}^{\infty} u_n$ converges if and only if

$\sum\limits_{k=0}^{\infty} 2^k u_{2^k}$ converges. (*Hint:* Let $s_n = \sum\limits_{j=1}^{n} u_j$ and $t_k = u_1 + 2u_2 + \cdots + 2^k u_{2^k}$ and show that if $n < 2^k$,

$s_n \leq t_k$, whereas if $n > 2^k$, $2s_n \geq t_k$. Hence the partial sums s_n and t_k are either bounded or unbounded simultaneously.)

80. Test for convergence using the result of Exercise 79.

(a) $\sum\limits_{n=2}^{\infty} \dfrac{1}{n \ln n}$ **(b)** $\sum\limits_{n=2}^{\infty} \dfrac{1}{n(\ln n)^2}$

81. Test for convergence using the results of Exercises 79 and 80

(a) $\sum\limits_{n=2}^{\infty} \dfrac{1}{n(\ln n)^p}$, $p > 0$ **(b)** $\sum\limits_{n=2}^{\infty} \dfrac{1}{n \ln n \ln (\ln n)}$

82. Prove that $\dbinom{n}{k} = \dbinom{n-1}{k-1} + \dbinom{n-1}{k}$, where $\dbinom{n}{k} = \dfrac{n!}{k!(n-k)!}$.

83. Prove by mathematical induction the binomial theorem

$$(a + b)^n = \sum_{r=0}^{n} \binom{n}{r} a^{n-r} b^r$$

where n is a positive integer and a and b are real numbers.

84. Prove that if $\sum_{k=1}^{\infty} a_k$ is a divergent series of positive numbers, then $\sum_{k=1}^{\infty} \frac{a_k}{s_k^2}$ must converge, where

$s_n = \sum_{k=1}^{n} a_k.$ $\left(\textit{Hint:} \text{ Show that } \frac{a_n}{s_n^2} \leq \frac{1}{s_{n-1}} - \frac{1}{s_n}, \text{ which is a term of a convergent telescoping series.} \right)$

85. Prove that if $\sum_{k=1}^{\infty} a_k$ is a divergent series of positive numbers, then $\sum_{k=1}^{\infty} \frac{a_k}{s_k}$ must diverge, where

$s_n = \sum_{k=1}^{n} a_k.$ (*Hint:* Form $\frac{a_{n+1}}{s_{n+1}} + \frac{a_{n+2}}{s_{n+2}} + \cdots + \frac{a_{n+p}}{s_{n+p}}$ and show that p can be chosen sufficiently large so that this sum does not tend to zero.)

16

Solid Analytic Geometry— Vectors in Space

The objective in this chapter is to extend our considerations to the study of vectors in three-dimensional space and solid analytic geometry. We shall mention noteworthy geometrical and physical applications of these mathematical concepts. As the material is developed, it will become apparent that there are many striking analogies between problems in three-dimensional space and in the plane as, for example,

(i)   the sphere in three-dimensional space and the circle in the plane
(ii) the plane in space and the line in the plane and, in particular, tangent plane and tangent line
(iii) the formulas for the distance between two points in space and in the plane

Consequently our understanding of the previous concepts in the plane will be most useful in this transition from two- to three-dimensional space.

16.1   CARTESIAN SPACE COORDINATES

It is natural to introduce rectangular coordinates in space as an extension of the rectangular coordinates in the plane. To this end, suppose that we have three mutually perpendicular lines called the x, y, and z-axes, with common unit of length, meeting at an origin of coordinates O (Figure 16.1.1) where $x = y = z = 0$. The axes shown in Figure 16.1.1 form a *right-handed system of coordinates.* This means that if the index and middle fingers of the right hand point in the direction of the positive x- and y-axes then the thumb points in the direction of the positive z-axis. If the x- and y-axes of Figure 16.1.1 are interchanged and the z-axis remains unaltered, the three axes would form a left-handed system. We shall primarily utilize the right-handed system of coordinates. Two intersecting lines in space determine a plane. Three *coordinate planes* are formed by the coordinate axes, namely, the xy-plane containing the x- and y-axes, the yz-plane, and the zx-plane.

A point $P(x, y, z)$ is shown in Figure 16.1.2. A plane through P and perpendic-

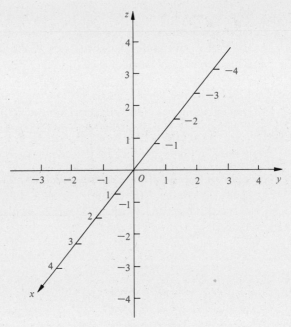

Figure 16.1.1

ular to the x-axis will intersect the x-axis at $(x, 0, 0)$. Similarly, a plane through P and perpendicular to the y-axis will intersect the y-axis at $(0, y, 0)$ and a plane through P and perpendicular to the z-axis will intersect the z-axis at $(0, 0, z)$.

For each point P in space there is a unique ordered triple of coordinates and, conversely, for each ordered triple of numbers, there is a unique point in space having these numbers as coordinates. The association of ordered triples of numbers with a point in space is called a ***rectangular Cartesian coordinate system in***

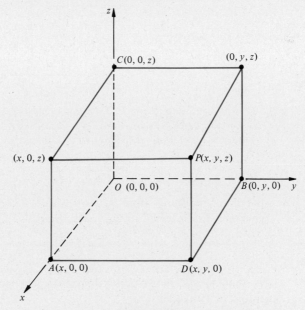

Figure 16.1.2

space. The set of all ordered triples of numbers is called the ***three-dimensional number space*** and is denoted by R_3.

The three coordinate planes separate space into eight parts called ***octants.*** Although we shall not number all of them, the one in which each coordinate is positive is called the ***first octant,*** that is, $P(x, y, z)$ is a point in the first octant if and only if $x > 0$, $y > 0$, and $z > 0$.

The equation $x = a$ with no restrictions on y and z, represents a plane that is parallel to the yz-plane and therefore perpendicular to the x-axis. Similarly, $y = b$ represents a plane parallel to the zx-plane. Their intersection $\begin{cases} x = a \\ y = b \end{cases}$ is the equation of a straight line that is parallel to the intersection of the yz-plane and zx-plane, that is, the z-axis (Figure 16.1.3). Let $A_1(a, b, z_1)$ and $A_2(a, b, z_2)$ be two points on this line; then we define the ***directed distance*** from A_1 to A_2 denoted by $\overline{A_1A_2}$ to be

$$\overline{A_1A_2} = z_2 - z_1 \tag{1a}$$

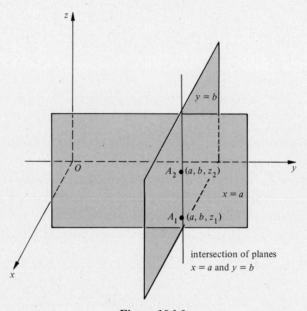

Figure 16.1.3

Similarly, if $B_1(x_1, b, c)$ and $B_2(x_2, b, c)$ are two points on a line parallel to the x-axis, then

$$\overline{B_1B_2} = x_2 - x_1 \tag{1b}$$

is the directed distance from B_1 to B_2. Also, if $C_1(a, y_1, c)$ and $C_2(a, y_2, c)$ are two points on a line parallel to the y-axis, then

$$\overline{C_1C_2} = y_2 - y_1 \tag{1c}$$

is the directed distance from C_1 to C_2.

Distances between two points in space may be found by two applications of the theorem of Pythagoras.

Theorem 1   **_Distance Formula._** The undirected distance between the points $P_1(x_1, y_1, z_1)$ and $P_2(x_2, y_2, z_2)$ is given by

$$|\overline{P_1P_2}| = \sqrt{(x_2 - x_1)^2 + (y_2 - y_1)^2 + (z_2 - z_1)^2} \qquad (2)$$

Proof   Construct a rectangular parallelepiped (a box) with faces parallel to the coordinate planes and having P_1 and P_2 as opposite vertices (Figure 16.1.4). With points A and B chosen as shown in the figure, $|\overline{P_1A}| = x_2 - x_1$, $|\overline{AB}| = y_2 - y_1$, and $|\overline{BP_2}| = z_2 - z_1$. From right triangle P_1BP_2 with right angle at B and hypotenuse P_1P_2, we have

$$|\overline{P_1P_2}|^2 = |\overline{P_1B}|^2 + |\overline{BP_2}|^2 \qquad (i)$$

But P_1B is the hypotenuse of triangle P_1AB in the upper rectangle. Hence

$$|\overline{P_1B}|^2 = |\overline{P_1A}|^2 + |\overline{AB}|^2 \qquad (ii)$$

Combination of (i) and (ii) yields

$$|\overline{P_1P_2}|^2 = |\overline{P_1A}|^2 + |\overline{AB}|^2 + |\overline{BP_2}|^2$$
$$= (x_2 - x_1)^2 + (y_2 - y_1)^2 + (z_2 - z_1)^2$$

and if we take the square root of each side equation (2) follows. ∎

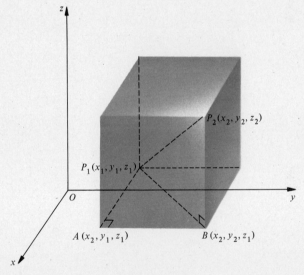

Figure 16.1.4

EXAMPLE 1   Find the undirected distance d between the points $P(-2, 3, -5)$ and $Q(4, -1, -9)$.

Solution   From (2),

$$d = |\overline{PQ}| = \sqrt{(4 + 2)^2 + (-1 - 3)^2 + (-9 + 5)^2} = \sqrt{68} \qquad \bullet$$

EXAMPLE 2   Show that the points $P(0, 1, 5)$, $Q(6, 5, 3)$ and $R(9, 7, 2)$ lie on a straight line.

Solution   It is sufficient to show that the length of one of the segments formed by joining the given points in pairs is equal to the sum of the other two. From (2), we have

$$|\overline{PQ}| = \sqrt{(6-0)^2 + (5-1)^2 + (3-5)^2} = \sqrt{56} = 2\sqrt{14}$$
$$|\overline{QR}| = \sqrt{(9-6)^2 + (7-5)^2 + (2-3)^2} = \sqrt{14}$$
$$|\overline{PR}| = \sqrt{(9-0)^2 + (7-1)^2 + (2-5)^2} = \sqrt{126} = 3\sqrt{14}$$

Hence $|\overline{PR}| = |\overline{PQ}| + |\overline{QR}|$, which implies that the points P, Q, and R lie on a straight line; Q is between P and R.   ●

EXAMPLE 3   Show that the points $P(1,4,6)$, $Q(3,7,5)$ and $R(5,2,8)$ are vertices of a right triangle.

Solution   We use the converse of the theorem of Pythagoras:

$$|\overline{PQ}|^2 = (3-1)^2 + (7-4)^2 + (5-6)^2 = 14$$
$$|\overline{QR}|^2 = (5-3)^2 + (2-7)^2 + (8-5)^2 = 38$$
$$|\overline{PR}|^2 = (5-1)^2 + (2-4)^2 + (8-6)^2 = 24$$

Hence $|\overline{QR}|^2 = |\overline{PQ}|^2 + |\overline{PR}|^2$, and the triangle is a right triangle with the right angle at P.   ●

Definition   The **graph** of an equation in R_3 are those points and only those points which satisfy the equation.

Definition   A graph of an equation in R_3 is called a **surface**†. (For example, planes are surfaces that we will examine in more detail later in this chapter.)

    With the distance formula as a tool, we now define the surface known as the sphere.

Definition   A **sphere** is the set of points that are at a fixed distance r from a given point (h, k, m) in R_3. The point (h, k, m) is said to be the **center** of the sphere and r is its **radius** (Figure 16.1.5).

Theorem 2   An equation of the sphere with center at $C(h, k, m)$ and radius r is given by

$$(x - h)^2 + (y - k)^2 + (z - m)^2 = r^2 \tag{3}$$

Proof   A point $P(x, y, z)$ is on the surface of the sphere if and only if $|\overline{CP}| = r$, or, equivalently, from (2)

$$\sqrt{(x-h)^2 + (y-k)^2 + (z-m)^2} = r$$

Equation (3) is obtained by squaring both sides of this last equation.   ■

---

† Degenerate graphs such as the empty set, points, lines, and so on are excluded in the definition.

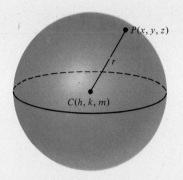

$P(x, y, z)$

r

$C(h, k, m)$

Figure 16.1.5

In particular, when the center of the sphere is at the origin of coordinates, we obtain

$$x^2 + y^2 + z^2 = r^2 \qquad (4)$$

If equation (3) is expanded and terms combined, there results

$$x^2 + y^2 + z^2 - 2hx - 2ky - 2mz + h^2 + k^2 + m^2 - r^2 = 0 \qquad (5)$$

This may be written in the form

$$x^2 + y^2 + z^2 + Dx + Ey + Fz + G = 0 \qquad (6)$$

where

$$D = -2h \qquad E = -2k \qquad F = -2m \qquad G = h^2 + k^2 + m^2 - r^2 \qquad (7)$$

Equation (6) is called the **general** or **expanded** form of an equation of a sphere.

The system of equations (7) may be solved for h, k, m, and r. We obtain

$$h = -\frac{D}{2} \qquad k = -\frac{E}{2} \qquad m = -\frac{F}{2} \qquad r^2 = \frac{D^2 + E^2 + F^2 - 4G}{4} \qquad (8)$$

Since any sphere has a center and a radius, its equation can always be expressed in form (3) or in its expanded form (6). Conversely, by the method of completion of the squares and relations (8), it follows that an equation of form (6) represents a sphere with center (h, k, m) given by (8) and radius $r = \dfrac{\sqrt{D^2 + E^2 + F^2 - 4G}}{2}$ provided that $D^2 + E^2 + F^2 - 4G$ is positive. If this quantity is zero, the graph of the equation is a point, (h, k, m), and if it is negative there is no locus.

EXAMPLE 4   Discuss the graph of the equation

$$x^2 + y^2 + z^2 - 2x + 4y - 14z + 28 = 0$$

Solution   By regrouping and completion of the squares

$$x^2 - 2x + 1 + y^2 + 4y + 4 + z^2 - 14z + 49 = 1 + 4 + 49 - 28$$

or, equivalently,

$$(x - 1)^2 + (y + 2)^2 + (z - 7)^2 = 26$$

Hence the graph is a sphere with center at $(1, -2, 7)$ and radius $\sqrt{26}$. It is recommended that the reader follow this procedure in lieu of memorizing (8).

●

Theorem 3   (*Midpoint Formula*).   The segment with endpoints (x_1, y_1, z_1) and (x_2, y_2, z_2) has midpoint

$$\left(\frac{x_1 + x_2}{2}, \frac{y_1 + y_2}{2}, \frac{z_1 + z_2}{2} \right) \tag{9}$$

Its proof involves congruent triangles and proceeds in a manner similar to the two-dimensional case. It is left to the reader to establish (9)—see Exercise 31.

EXAMPLE 5   Find an equation of the sphere having the points $P(-2, -8, 1)$ and $Q(4, 2, 3)$ as end points of a diameter.

Solution   The center of the sphere will be the midpoint of the line segment PQ. Hence from (9) the center is the point

$$C\left(\frac{-2 + 4}{2}, \frac{-8 + 2}{2}, \frac{1 + 3}{2} \right) \quad \text{or} \quad C(1, -3, 2)$$

Also, the radius of the sphere is given by $|\overline{CP}|$ (or, equivalently, $|\overline{CQ}|$), and we have

$$r = |\overline{CP}| = \sqrt{(-2 - 1)^2 + (-8 + 3)^2 + (1 - 2)^2} = \sqrt{35}$$

Therefore from (3), an equation of the sphere is

$$(x - 1)^2 + (y + 3)^2 + (z - 2)^2 = 35$$

or, in expanded form,

$$x^2 + y^2 + z^2 - 2x + 6y - 4z - 21 = 0 \qquad \bullet$$

Exercises 16.1

1. Sketch a right-handed coordinate system and plot the following points
   - (a) $(2, 0, 0)$
   - (b) $(-1, 0, 1)$
   - (c) $(1, 4, -2)$
   - (d) $(0, 0, 3)$
   - (e) $(-1, -2, -1)$
   - (f) $(-2, -1, 1)$
2. Given a right-handed coordinate system, suppose that we permute the axes in accordance with the scheme $x \to y \to z \to x$, that is, the x-axis is replaced by the y-axis, the y-axis by the z-axis, and the z-axis by the x-axis. Is the resulting system left-handed or right-handed?

In Exercises 3 through 6, A and B are the opposite vertices of a rectangular parallelepiped, having its faces parallel to the coordinate planes. Sketch the parallelepiped and determine the coordinates of the other six vertices.

3. $A(0, 0, 0)$; $B(1, 2, 4)$
5. $A(-2, 1, 3)$; $B(1, 2, 5)$
4. $A(1, 0, 1)$; $B(3, 3, 2)$
6. $A(2, 0, 4)$; $B(0, -1, 3)$

In Exercises 7 through 12, find the distances between the pairs of points.

7. $P(2, 3, 4)$ and $Q(6, 4, 5)$
8. $P(-1, -2, 3)$ and $Q(0, 4, -2)$
9. $P(3, -3, 2)$ and $Q(4, \frac{1}{2}, 0)$
10. $P(a, b, a)$ and $Q(-a, -a, b)$
11. $P(b, -b, 3)$ and $Q(-b, b, -3)$
12. $P(a, b, c)$ and $Q(b, c, a)$
13. Prove that the three points $A(0, 0, 5)$, $B(1, 2, 3)$ and $C(3, 5, 7)$ are the vertices of a right triangle and find its area.

14. Find an equation of the plane through the point (a, b, c) which is (a) parallel to the zx-plane; (b) parallel to the yz-plane.
15. Prove, by an application of the distance formula, that the three points $P(7, 10, 1)$, $Q(6, 12, 5)$ and $R(9, 6, -7)$ are collinear.

In each of Exercises 16 through 22, determine the graph of the equation.

16. $x^2 + y^2 + z^2 - 2x - 4y - 8z + 20 = 0$ 17. $x^2 + y^2 + z^2 - 6x - 16 = 0$
18. $x^2 + y^2 + z^2 + 4y + 6z + 11 = 0$ 19. $x^2 + y^2 + z^2 - 2x + 6y + 4z + 18 = 0$
20. $x^2 + y^2 + z^2 - 8x + 6y - 4z + 29 = 0$ 21. $4x^2 + 4y^2 + 4z^2 - 4x + 8y - 8z - 7 = 0$
22. $x^2 + y^2 + z^2 - 8x + 4y - 10z = 0$
23. Find an equation of the sphere with center at $(-3, -4, 7)$ and passing through the origin.
24. Find an equation of the sphere having the points $P(-3, 7, 2)$ and $Q(1, 9, -6)$ as end points of a diameter.
25. Find an equation of the graph of all points equidistant from the points $(1, 0, 0)$ and $(0, 1, 0)$. What is the graph?
26. Find the distance between an arbitrary point $P(x_0, y_0, z_0)$ and each of the coordinate axes. Sketch the graph showing the coordinates and lengths involved.
27. Let $P(x_0, y_0, z_0)$ be an arbitrary point in xyz-space. Find the coordinates of Q, given that the line segment PQ is symmetric
    (a) about the xy-plane   (b) about the yz-plane
    (c) about the zx-plane   (d) about the origin O
    Sketch the graphs and interpret the results.
28. Under what conditions on the function† f is the graph of the equation $f(x, y, z) = 0$ symmetric with respect to the
    (a) xy-plane?   (b) yz-plane?
    (c) zx-plane?   (d) the origin?
29. Under what conditions on the function f is the graph of the equation $f(x, y, z) = 0$ symmetric with respect to the
    (a) x-axis?   (b) y-axis?   (c) z-axis?
30. Let P, Q, R, and S be four points in three-dimensional space with no three points collinear. If A, B, C, and D are the midpoints of PQ, QR, RS, and SP, respectively, prove that $ABCD$ is a parallelogram.
31. Prove Theorem 3.

16.2   THREE-DIMENSIONAL VECTORS

In Chapter 11, we introduced vectors in the plane or two-dimensional vectors. Many of the concepts introduced there are easily extended to three-dimensional space. This will lead to a number of simplifications of and applications to problems in solid analytic geometry.

Definition   A **three-dimensional vector** or a **vector in three-dimensional space** is an ordered triple of real numbers $\langle x, y, z \rangle$ that satisfies the following rules:

---

† The concept of functions of two or more variables will be formally treated in Chapter 17.

If $\mathbf{a} = \langle a_1, a_2, a_3 \rangle$ and $\mathbf{b} = \langle b_1, b_2, b_3 \rangle$ are vectors and k is a scalar, then

$$\mathbf{a} = \mathbf{b} \text{ if and only if } a_1 = b_1, a_2 = b_2, \text{ and } a_3 = b_3 \tag{1}$$

$$k\mathbf{a} = \langle ka_1, ka_2, ka_3 \rangle \tag{2}$$

$$\mathbf{a} + \mathbf{b} = \langle a_1 + b_1, a_2 + b_2, a_3 + b_3 \rangle \tag{3}$$

Statements (1), (2), and (3) are simply definitions of equality, multiplication by a scalar, and addition of vectors.

The numbers x, y, and z that appear in the vector $\langle x, y, z \rangle$ are defined to be the *components* of the vector.

The set of all triples of numbers defined by (2) and (3) form a vector space $\mathbf{V}_3$ of three dimensions.

Three-dimensional vectors are frequently represented by a directed line segment $\overrightarrow{PQ}$ in R_3. Thus if $\mathbf{a} = \langle a_1, a_2, a_3 \rangle$ is an arbitrary vector, we may choose $P(x, y, z)$ and $Q(x + a_1, y + a_2, z + a_3)$. Hence $\overrightarrow{PQ}$ is a representative directed line segment (Figure 16.2.1). If, in particular, the initial point of $\mathbf{a}$ is the origin $(0, 0, 0)$, then the end point of $\mathbf{a}$ is simply the point (a_1, a_2, a_3). Note that the direction of a vector is determined by the orientation of any of its representative directed line segments.

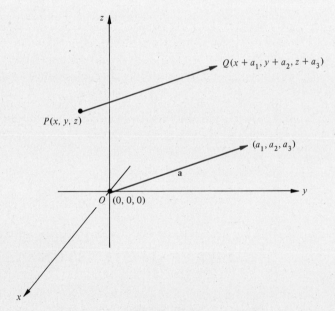

Figure 16.2.1

The *length* or *magnitude* of a vector $\mathbf{a}$ is denoted by $|\mathbf{a}|$ and is the number

$$|\mathbf{a}| = \sqrt{a_1{}^2 + a_2{}^2 + a_3{}^2} \tag{4}$$

The vector $\mathbf{0} = \langle 0, 0, 0 \rangle$ is called the *zero vector*. It may be represented by a point. As in the two dimensional case, the zero vector has no direction.

The negative of vector $\mathbf{a} = \langle a_1, a_2, a_3 \rangle$ written $-\mathbf{a}$, is the vector

$$-\mathbf{a} = \langle -a_1, -a_2, -a_3 \rangle \tag{5}$$

It has the reversed direction and same magnitude as $\mathbf{a}$.

Next, we define the difference $\mathbf{a} - \mathbf{b}$ to be the sum of the vectors $\mathbf{a}$ and $-\mathbf{b}$:

$$\mathbf{a} - \mathbf{b} = \langle a_1 - b_1, a_2 - b_2, a_3 - b_3 \rangle \tag{6}$$

From the preceding definitions, the following rules are readily derived:

Theorem 1   Let $\mathbf{a}$, $\mathbf{b}$, and $\mathbf{c}$ be vectors; k, m, and 1 be scalars, then

$\mathbf{a} + \mathbf{b} = \mathbf{b} + \mathbf{a}$	(commutative law of addition)	(7)
$\mathbf{a} + (\mathbf{b} + \mathbf{c}) = (\mathbf{a} + \mathbf{b}) + \mathbf{c}$	(associative law of addition)	(8)
$\mathbf{a} + (-\mathbf{a}) = \mathbf{0}$		(9)
$\mathbf{a} + \mathbf{0} = \mathbf{a}$		(10)
$k(\mathbf{a} + \mathbf{b}) = k\mathbf{a} + k\mathbf{b}$	$(k + m)\mathbf{a} = k\mathbf{a} + m\mathbf{a}$	(11)
$(km)\mathbf{a} = k(m\mathbf{a})$	$(-k)\mathbf{a} = -(k\mathbf{a})$	(12)
$0\mathbf{a} = \mathbf{0}$	$1\mathbf{a} = \mathbf{a}$ $k\mathbf{0} = \mathbf{0}$	(13)
$|k\mathbf{a}| = |k|\,|\mathbf{a}|$		(14)

The proofs of the various parts of Theorem 1 are straightforward applications of the basic definitions given in equations (1) through (6) and are left as exercises. ∎

The addition of vectors in space may be readily interpreted geometrically in terms of the parallelogram law. Without loss in generality, consider two vectors $\mathbf{a} = \langle a_1, a_2, a_3 \rangle$ and $\mathbf{b} = \langle b_1, b_2, b_3 \rangle$ with the same initial point O and terminal points at P and Q respectively (Figure 16.2.2). Then the vector

$$\mathbf{a} + \mathbf{b} = \langle a_1 + b_1, a_2 + b_2, a_3 + b_3 \rangle$$

is represented by the directed line segment $\overrightarrow{OR}$ along the diagonal of the parallelogram.

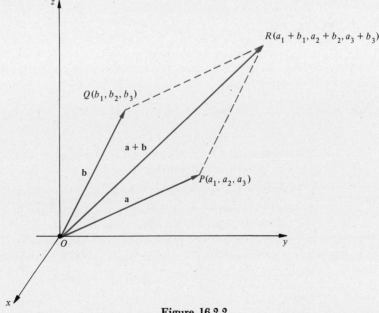

Figure 16.2.2

Figure 16.2.3 illustrates a geometrical picture of $\mathbf{a} - \mathbf{b}$. If $\mathbf{a}$ and $\mathbf{b}$ are, respectively, represented by the directed line segments $\overrightarrow{OR}$ and $\overrightarrow{OP}$ then $\overrightarrow{PR}$ represents $\mathbf{a} - \mathbf{b}$. This follows from the fact that $\mathbf{b} + (\mathbf{a} - \mathbf{b}) = \mathbf{a}$.

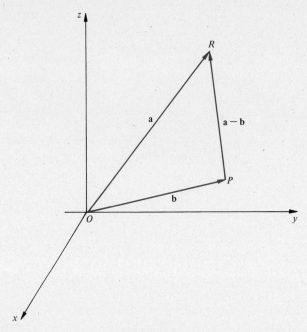

Figure 16.2.3

Definition   Two vectors $\mathbf{a}$ and $\mathbf{b}$ are said to be *parallel* provided that

$$\mathbf{b} = k\mathbf{a} \qquad \text{from some real number } k \neq 0$$

If $k > 0$, then $\mathbf{a}$ and $\mathbf{b}$ have the *same direction*, whereas if $k < 0$, they are said to have *opposite directions*.

In the case of

$$\mathbf{a} = \langle 6, -6, 12 \rangle \qquad \mathbf{b} = \langle 3, -3, 6 \rangle \qquad \mathbf{c} = \langle -1, 1, -2 \rangle$$

we have

$$\mathbf{a} = 2\mathbf{b} \qquad \text{and} \qquad \mathbf{a} = -6\mathbf{c}$$

This means that $\mathbf{a}$ and $\mathbf{b}$ are parallel and have the same direction, whereas $\mathbf{a}$ and $\mathbf{c}$ are also parallel but oppositely directed.

EXAMPLE 1   Let $\mathbf{a}$ be any nonzero vector. Find a *unit vector,* that is, a vector of length 1, that is parallel to $\mathbf{a}$ and has the same direction as $\mathbf{a}$.

Solution   Let $\mathbf{b}$ be the required vector. Then a scalar k is sought such that

$$\mathbf{b} = k\mathbf{a} \qquad (k > 0) \tag{i}$$

and

$$|\mathbf{b}| = 1 \tag{ii}$$

If we take the magnitude of both sides of (i),

$$|\mathbf{b}| = |k\mathbf{a}|$$ (iii)

then use of (ii) and (14) gives

$$1 = |k|\,|\mathbf{a}| = k|\mathbf{a}| \qquad \text{since } k > 0$$

Hence $k = \dfrac{1}{|\mathbf{a}|}$, and therefore

$$\mathbf{b} = \frac{\mathbf{a}}{|\mathbf{a}|}$$ (15)

is the required vector. ●

Note that although $\dfrac{\mathbf{a}}{|\mathbf{a}|}$ is a unit vector in the same direction as $\mathbf{a}$, $-\dfrac{\mathbf{a}}{|\mathbf{a}|}$ is a unit vector in the opposite direction.

There are three vectors that will require our special attention. These are the vectors

$$\mathbf{i} = \langle 1, 0, 0 \rangle \qquad \mathbf{j} = \langle 0, 1, 0 \rangle \qquad \mathbf{k} = \langle 0, 0, 1 \rangle$$

These vectors are unit vectors and if we draw arrows emanating from the origin O, the vectors $\mathbf{i}$, $\mathbf{j}$, and $\mathbf{k}$ coincide with the x-, y- and z-axes, respectively (Figure 16.2.4).†

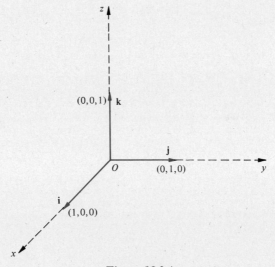

Figure 16.2.4

It is now an easy matter to establish a second theorem.

Theorem 2   *Any* vector in three-dimensional space can be expressed as a linear combination of the unit coordinate vectors $\mathbf{i}$, $\mathbf{j}$, and $\mathbf{k}$.

---

† Recall that in two dimensions, $\mathbf{i} = \langle 1, 0 \rangle$ and $\mathbf{j} = \langle 0, 1 \rangle$.

Proof   Let $\mathbf{a} = \langle a_1, a_2, a_3 \rangle$ be an arbitrary vector. Then

$$\begin{aligned}
\mathbf{a} = \langle a_1, a_2, a_3 \rangle &= \langle a_1, 0, 0 \rangle + \langle 0, a_2, 0 \rangle + \langle 0, 0, a_3 \rangle \\
&= a_1 \langle 1, 0, 0 \rangle + a_2 \langle 0, 1, 0 \rangle + a_3 \langle 0, 0, 1 \rangle \\
&= a_1 \mathbf{i} + a_2 \mathbf{j} + a_3 \mathbf{k} \qquad\qquad\qquad\blacksquare \quad (16)
\end{aligned}$$

If $\mathbf{a} = a_1 \mathbf{i} + a_2 \mathbf{j} + a_3 \mathbf{k}$, then the numbers a_1, a_2, and a_3 are the *scalar components* of $\mathbf{a}$, whereas $a_1 \mathbf{i}$, $a_2 \mathbf{j}$, and $a_3 \mathbf{k}$ are the *vector components* of $\mathbf{a}$ in the x-, y- and z-directions, respectively. The vectors $\mathbf{i}$, $\mathbf{j}$, and $\mathbf{k}$ form a *basis* for vectors in three-dimensional space; that is, any vector in R_3 may be expressed uniquely as a linear combination of $\mathbf{i}$, $\mathbf{j}$, and $\mathbf{k}$.

Note that from the definition of equality of vectors that the vector equation $\mathbf{a} = \mathbf{b}$ (Equation (1)) becomes

$$a_1 \mathbf{i} + a_2 \mathbf{j} + a_3 \mathbf{k} = b_1 \mathbf{i} + b_2 \mathbf{j} + b_3 \mathbf{k}$$

This implies three scalar equations

$$a_1 = b_1 \qquad a_2 = b_2 \qquad a_3 = b_3$$

EXAMPLE 2   Given that $\mathbf{a} = 3\mathbf{i} - 2\mathbf{j} - \mathbf{k}$ and $\mathbf{b} = -\mathbf{i} + 4\mathbf{j} + 5\mathbf{k}$

   (i)   express $3\mathbf{a} - 4\mathbf{b}$ as a linear combination of $\mathbf{i}$, $\mathbf{j}$, and $\mathbf{k}$
   (ii)   compute $|3\mathbf{a} - 4\mathbf{b}|$
   (iii) find a unit vector $\mathbf{u}$ in the direction of $\mathbf{a} + \mathbf{b}$

Solution   (i)

$$\begin{aligned}
3\mathbf{a} - 4\mathbf{b} &= 3(3\mathbf{i} - 2\mathbf{j} - \mathbf{k}) - 4(-\mathbf{i} + 4\mathbf{j} + 5\mathbf{k}) \\
&= 9\mathbf{i} - 6\mathbf{j} - 3\mathbf{k} + 4\mathbf{i} - 16\mathbf{j} - 20\mathbf{k} \\
&= 13\mathbf{i} - 22\mathbf{j} - 23\mathbf{k}
\end{aligned}$$

   (ii)

$$\begin{aligned}
|3\mathbf{a} - 4\mathbf{b}| &= \sqrt{13^2 + (-22)^2 + (-23)^2} \\
&= \sqrt{1182} \doteq 34.38
\end{aligned}$$

   (iii) $\mathbf{a} + \mathbf{b} = 2\mathbf{i} + 2\mathbf{j} + 4\mathbf{k}$ and its magnitude is

$$|\mathbf{a} + \mathbf{b}| = \sqrt{2^2 + 2^2 + 4^2} = 2\sqrt{6}$$

Hence for the unit vector $\mathbf{u}$ in the direction of $\mathbf{a} + \mathbf{b}$, (15) yields

$$\mathbf{u} = \frac{\mathbf{a} + \mathbf{b}}{|\mathbf{a} + \mathbf{b}|} = \frac{2\mathbf{i} + 2\mathbf{j} + 4\mathbf{k}}{2\sqrt{6}} = \frac{\mathbf{i} + \mathbf{j} + 2\mathbf{k}}{\sqrt{6}} \qquad\qquad\bullet$$

EXAMPLE 3   Express the vector $\mathbf{a} = a_1 \mathbf{i} + a_2 \mathbf{j} + a_3 \mathbf{k}$ as a linear combination of the vectors $\mathbf{i} + \mathbf{j} + \mathbf{k}$, $\mathbf{i} - \mathbf{k}$ and $\mathbf{j} + \mathbf{k}$. What is implied by this result?

Solution   We seek scalars m, n, and p such that

$$\mathbf{a} = m(\mathbf{i} + \mathbf{j} + \mathbf{k}) + n(\mathbf{i} - \mathbf{k}) + p(\mathbf{j} + \mathbf{k})$$

or, equivalently, by collecting coefficients of the vectors $\mathbf{i}$, $\mathbf{j}$ and $\mathbf{k}$

$$a_1 \mathbf{i} + a_2 \mathbf{j} + a_3 \mathbf{k} = (m + n)\mathbf{i} + (m + p)\mathbf{j} + (m - n + p)\mathbf{k}$$

Therefore

$$m + n = a_1$$
$$m + p = a_2$$
$$m - n + p = a_3$$

which yields

$$m = a_1 - a_2 + a_3 \qquad n = a_2 - a_3 \qquad p = -a_1 + 2a_2 - a_3$$

This implies that the vectors $\mathbf{i} + \mathbf{j} + \mathbf{k}$, $\mathbf{i} - \mathbf{k}$ and $\mathbf{j} + \mathbf{k}$ (as well as $\mathbf{i}, \mathbf{j}$, and $\mathbf{k}$) also form a basis for vectors in three-dimensional space. ●

Exercises 16.2

In Exercises 1 through 4, simplify the following linear combinations

1. $3(\mathbf{i} - \mathbf{j} - \mathbf{k}) - 2(\mathbf{i} + 4\mathbf{j} - 3\mathbf{k})$
2. $10(2\mathbf{i} - \mathbf{j} - 3\mathbf{k}) - 4(\mathbf{i} - 7\mathbf{k})$
3. $-(-\mathbf{i} + \mathbf{j} + 2\mathbf{k}) + \frac{1}{2}(3\mathbf{i} + 2\mathbf{j} + 5\mathbf{k})$
4. $-(\mathbf{i} - \mathbf{j}) + 3(\mathbf{j} - \mathbf{k}) - (\mathbf{i} - \mathbf{j} - \mathbf{k})$

In Exercises 5 through 8, find the magnitude of the vectors.

5. $2\mathbf{i} - \mathbf{j} + 2\mathbf{k}$
6. $\frac{1}{3}(\mathbf{i} - \mathbf{j} + \mathbf{k})$
7. $4(\mathbf{i} - \mathbf{j} - \mathbf{k}) - 3(2\mathbf{i} + 3\mathbf{j} + \mathbf{k}) + 2\mathbf{i} + 13\mathbf{j} + 7\mathbf{k}$
8. $\mathbf{i} - \mathbf{j} - 3(\mathbf{j} - \mathbf{k}) - (3\mathbf{i} - \mathbf{j} + 2\mathbf{k})$

In Exercises 9 through 14, let $\mathbf{a} = \langle 3, 1, 2 \rangle$, $\mathbf{b} = \langle -1, 2, 1 \rangle$, $\mathbf{c} = \langle -2, -1, 4 \rangle$, and $\mathbf{d} = \langle 2, 3, 5 \rangle$, find

9. $3\mathbf{a} - \mathbf{b} - 2\mathbf{c}$
10. $\mathbf{a} + 2\mathbf{b} - \mathbf{c} + \mathbf{d}$
11. $|3\mathbf{b}| - |4\mathbf{d}|$
12. $|\mathbf{b} - \frac{1}{2}\mathbf{c}|$
13. $|\mathbf{a}| - (\mathbf{b} + \mathbf{c} + \mathbf{d})$
14. $|\mathbf{a}| \, |\mathbf{c}|(\mathbf{d} - 3\mathbf{b})$
15. Find a unit vector in the opposite direction to $3\mathbf{i} - 2\mathbf{j} + 4\mathbf{k}$.
16. For what values of m and p is $\mathbf{k} = m\mathbf{i} + p\mathbf{j}$?
17. Find m so that $|m\mathbf{i} + (m + 2)\mathbf{j} + (m + 4)\mathbf{k}|^2 = 200$
18. If $\mathbf{a} = -2\mathbf{i} + 3\mathbf{j} - \mathbf{k}$ and $\mathbf{b} = -4\mathbf{i} + 5\mathbf{j} + 3\mathbf{k}$, find (i) $3\mathbf{a} - 2\mathbf{b}$; (ii) $|3\mathbf{a} - 2\mathbf{b}|$; and (iii) a unit vector $\mathbf{u}$ in the direction of $3\mathbf{a} - 2\mathbf{b}$.
19. If $\mathbf{a} = 3\mathbf{i} - 4\mathbf{j} + 7\mathbf{k}$, find scalars m and n such that $\mathbf{b} = (m + n)\mathbf{i} - 16\mathbf{j} + (3m + 2n - 1)\mathbf{k}$ is parallel to $\mathbf{a}$.
20. Let $\mathbf{a}$ and $\mathbf{b}$ be two nonzero vectors in three dimensional space with initial point O and terminal points A and B, respectively. Determine the position vector from O to the midpoint M of the segment AB.
21. Find vectors of magnitude 7 that are parallel to the vector $5\mathbf{i} - 3\mathbf{j} + 2\mathbf{k}$.

Exercises 22 through 26 refer to Theorem 1

22. Prove (7) and (8)
23. Prove (9) and (10)
24. Prove (11)
25. Prove (12)
26. Prove (13) and (14)

DOT PRODUCT, DIRECTION ANGLES, AND DIRECTION COSINES

The dot or scalar product is widely applicable to geometrical and physical problems. Its definition parallels and generalizes our previous definition for vectors in the plane.

Definition If $\mathbf{a} = \langle a_1, a_2, a_3 \rangle$ and $\mathbf{b} = \langle b_1, b_2, b_3 \rangle$ are vectors, we define

$$\mathbf{a} \cdot \mathbf{b} = a_1 b_1 + a_2 b_2 + a_3 b_3 \tag{1}$$

to be the **dot** or **scalar product** of $\mathbf{a}$ and $\mathbf{b}$. Note that the dot product is a number, not a vector, and is read "a dot b."

EXAMPLE 1   If $\mathbf{a} = \langle 2, 4, -1 \rangle$ and $\mathbf{b} = \langle -3, 1, 8 \rangle$, find $\mathbf{a} \cdot \mathbf{b}$.

Solution   From (1),

$$\mathbf{a} \cdot \mathbf{b} = \langle 2, 4, -1 \rangle \cdot \langle -3, 1, 8 \rangle$$
$$= (2)(-3) + (4)(1) + (-1)(8) = -10 \qquad \bullet$$

EXAMPLE 2   If $\mathbf{a} = 3\mathbf{i} + 5\mathbf{j} - 2\mathbf{k}$, $\mathbf{b} = 4\mathbf{i} + 7\mathbf{j} + 6\mathbf{k}$, and $\mathbf{c} = 2\mathbf{i} - 3\mathbf{k}$, find $\mathbf{a} \cdot \mathbf{b}$ and $\mathbf{a} \cdot \mathbf{c}$.

Solution   We may write $\mathbf{a} = \langle 3, 5, -2 \rangle$, $\mathbf{b} = \langle 4, 7, 6 \rangle$ and $\mathbf{c} = \langle 2, 0, -3 \rangle$. Therefore

$$\mathbf{a} \cdot \mathbf{b} = (3)(4) + (5)(7) + (-2)(6) = 35$$
$$\mathbf{a} \cdot \mathbf{c} = (3)(2) + (5)(0) + (-2)(-3) = 12 \qquad \bullet$$

If we take the scalar product of a vector $\mathbf{a}$ with itself we obtain from (1)

$$\mathbf{a} \cdot \mathbf{a} = a_1 a_1 + a_2 a_2 + a_3 a_3 = a_1^2 + a_2^2 + a_3^2 \tag{2}$$

But $|\mathbf{a}| = \sqrt{a_1^2 + a_2^2 + a_3^2}$ and therefore

$$\mathbf{a} \cdot \mathbf{a} = |\mathbf{a}|^2 \tag{3}$$

Consequently, $\mathbf{a} \cdot \mathbf{a} \geq 0$ where the equality holds if and only if $|\mathbf{a}| = 0$ or, equivalently, $\mathbf{a} = \mathbf{0}$.

Definition†   Let $\mathbf{a}$ and $\mathbf{b}$ be two nonzero vectors represented in Figure 16.3.1 by the directed line segments $\overrightarrow{PQ}$ and $\overrightarrow{PR}$, respectively. The **angle** between $\mathbf{a}$ and $\mathbf{b}$ is the smallest nonnegative angle θ formed by $\overrightarrow{PQ}$ and $\overrightarrow{PR}$. In all instances, $0 \leq \theta \leq \pi$.

---

† Since it is always possible to choose representatives of $\mathbf{a}$ and $\mathbf{b}$ that have a common initial point and then pass a plane through these representatives, the situation in Figure 16.3.1 holds—θ is unique.

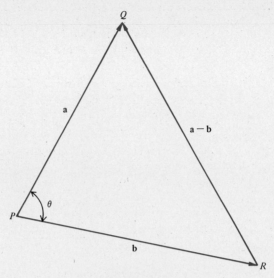

Figure 16.3.1

Theorem 1 If **a** and **b** are nonzero vectors, the angle θ is given by

$$\cos \theta = \frac{\mathbf{a} \cdot \mathbf{b}}{|\mathbf{a}|\,|\mathbf{b}|} \qquad 0 \le \theta \le \pi \tag{4}$$

If either **a** or **b** is zero, the angle θ is not defined.

Proof Refer to Figure 16.3.1. The sides of the triangle have lengths $|\mathbf{a}|$, $|\mathbf{b}|$, and $|\mathbf{a} - \mathbf{b}|$, and application of the law of cosines yields

$$|\mathbf{a} - \mathbf{b}|^2 = |\mathbf{a}|^2 + |\mathbf{b}|^2 - 2|\mathbf{a}|\,|\mathbf{b}|\cos\theta$$

from which

$$\begin{aligned}
2|\mathbf{a}|\,|\mathbf{b}|\cos\theta &= |\mathbf{a}|^2 + |\mathbf{b}|^2 - |\mathbf{a} - \mathbf{b}|^2 \\
&= a_1{}^2 + a_2{}^2 + a_3{}^2 + b_1{}^2 + b_2{}^2 + b_3{}^2 \\
&\quad - (a_1 - b_1)^2 - (a_2 - b_2)^2 - (a_3 - b_3)^2 \\
&= 2(a_1b_1 + a_2b_2 + a_3b_3)
\end{aligned}$$

Hence

$$\cos\theta = \frac{a_1b_1 + a_2b_2 + a_3b_3}{|\mathbf{a}|\,|\mathbf{b}|} = \frac{\mathbf{a} \cdot \mathbf{b}}{|\mathbf{a}|\,|\mathbf{b}|}$$

or, equivalently,

$$\mathbf{a} \cdot \mathbf{b} = |\mathbf{a}|\,|\mathbf{b}|\cos\theta \qquad\qquad \blacksquare \tag{5}$$

Note that for nonzero vectors **a** and **b**, (5) implies that $\mathbf{a} \cdot \mathbf{b} = 0$ if and only if the angle between the vectors is $\frac{\pi}{2}$ because $\cos\frac{\pi}{2} = 0$. Thus we have a corollary.

Corollary Two nonzero vectors are perpendicular (or orthogonal) if and only if their dot product is zero.

From (5), it follows that $\mathbf{a} \cdot \mathbf{b} > 0$ if and only if $0 \le \theta < \frac{\pi}{2}$ (that is, θ is acute) while $\mathbf{a} \cdot \mathbf{b} < 0$ if and only if $\frac{\pi}{2} < \theta \le \pi$ (that is, θ is obtuse).

EXAMPLE 3   Find the angle θ between the two vectors $\mathbf{a} = \langle 2, 5, 1 \rangle$ and $\mathbf{b} = \langle -3, -4, 5 \rangle$.

Solution

$$\cos \theta = \frac{\mathbf{a} \cdot \mathbf{b}}{|\mathbf{a}| \, |\mathbf{b}|} = \frac{(2)(-3) + (5)(-4) + (1)(5)}{\sqrt{2^2 + 5^2 + 1^2} \sqrt{(-3)^2 + (-4)^2 + 5^2}}$$

$$= \frac{-21}{\sqrt{30}\sqrt{50}} = -\frac{2.1}{\sqrt{15}} \doteq -0.5422$$

$$\theta \doteq 122°50'$$

●

EXAMPLE 4   Given the four points in space, $P(1, 3, 2)$, $Q(5, 2, 4)$, $R(-2, 5, 1)$, and $S(-3, 3, 2)$, show that the line connecting P and Q is perpendicular to the line through R and S.

Solution   Let $\overrightarrow{PQ}$ and $\overrightarrow{RS}$ denote vectors connecting pairs of points P and Q, and R and S, respectively, $\overrightarrow{PQ} = \overrightarrow{OQ} - \overrightarrow{OP}$ where O is the origin of coordinates so that

$$\overrightarrow{PQ} = \langle 5, 2, 4 \rangle - \langle 1, 3, 2 \rangle = \langle 4, -1, 2 \rangle$$

Similarly,

$$\overrightarrow{RS} = \overrightarrow{OS} - \overrightarrow{OR} = \langle -3, 3, 2 \rangle - \langle -2, 5, 1 \rangle = \langle -1, -2, 1 \rangle$$

hence

$$\overrightarrow{PQ} \cdot \overrightarrow{RS} = (4)(-1) + (-1)(-2) + (2)(1) = 0$$

Since the dot product is zero, the two vectors are perpendicular.   ●

The unit vectors $\mathbf{i}$, $\mathbf{j}$, and $\mathbf{k}$ are mutually perpendicular, which means that the dot product of any two distinct vectors in the coordinate directions must be zero. Also, the dot product of any one of these three unit vectors with itself must be 1. (Why?) Hence we have the array

$$
\begin{array}{lll}
\mathbf{i} \cdot \mathbf{i} = 1 & \mathbf{i} \cdot \mathbf{j} = 0 & \mathbf{i} \cdot \mathbf{k} = 0 \\
\mathbf{j} \cdot \mathbf{i} = 0 & \mathbf{j} \cdot \mathbf{j} = 1 & \mathbf{j} \cdot \mathbf{k} = 0 \\
\mathbf{k} \cdot \mathbf{i} = 0 & \mathbf{k} \cdot \mathbf{j} = 0 & \mathbf{k} \cdot \mathbf{k} = 1
\end{array}
\tag{6}
$$

From the definition of a scalar product it is straightforward to verify the following theorem.

Theorem 2   The dot or scalar product has the following properties (m and p are scalars)

$$\mathbf{a} \cdot \mathbf{b} = \mathbf{b} \cdot \mathbf{a} \qquad \text{(commutative law for dot products)} \tag{7}$$

$$\mathbf{a} \cdot (\mathbf{b} + \mathbf{c}) = \mathbf{a} \cdot \mathbf{b} + \mathbf{a} \cdot \mathbf{c} \quad \text{(dot distribution over addition)} \tag{8}$$

$$(m\mathbf{a}) \cdot (p\mathbf{b}) = mp(\mathbf{a} \cdot \mathbf{b}) \quad \text{(scalars can be factored out)} \tag{9}$$

$$|\mathbf{a} \cdot \mathbf{b}| \le |\mathbf{a}| \, |\mathbf{b}| \qquad \text{(Cauchy-Schwarz inequality)} \tag{10}$$

Proof   To illustrate the manner of proof of (7), (8), and (9), we shall establish (8). Let $\mathbf{a} = \langle a_1, a_2, a_3 \rangle$, $\mathbf{b} = \langle b_1, b_2, b_3 \rangle$ and $\mathbf{c} = \langle c_1, c_2, c_3 \rangle$. Therefore

$$\begin{aligned} \mathbf{a} \cdot (\mathbf{b} + \mathbf{c}) &= \langle a_1, a_2, a_3 \rangle \cdot \langle b_1 + c_1, b_2 + c_2, b_3 + c_3 \rangle \\ &= a_1(b_1 + c_1) + a_2(b_2 + c_2) + a_3(b_3 + c_3) \\ &= (a_1 b_1 + a_2 b_2 + a_3 b_3) + (a_1 c_1 + a_2 c_2 + a_3 c_3) \\ &= \mathbf{a} \cdot \mathbf{b} + \mathbf{a} \cdot \mathbf{c} \end{aligned}$$

To prove (10):

$$\begin{aligned} |\mathbf{a} \cdot \mathbf{b}| &= |\,|\mathbf{a}|\,|\mathbf{b}| \cos \theta\,| \\ &= |\mathbf{a}|\,|\mathbf{b}|\,|\cos \theta| \\ &\leq |\mathbf{a}|\,|\mathbf{b}| \end{aligned}$$

because $|\cos \theta| \leq 1$.

The proofs of (7) and (9) are left to the reader.   ∎

EXAMPLE 5   Show that $\mathbf{a} \cdot (\mathbf{b} + \mathbf{c} + \mathbf{d}) = \mathbf{a} \cdot \mathbf{b} + \mathbf{a} \cdot \mathbf{c} + \mathbf{a} \cdot \mathbf{d}$ (11)

Solution   Let $\mathbf{c} + \mathbf{d} = \mathbf{s}$ then from (8), employed twice

$$\begin{aligned} \mathbf{a} \cdot (\mathbf{b} + \mathbf{s}) &= \mathbf{a} \cdot \mathbf{b} + \mathbf{a} \cdot \mathbf{s} \\ &= \mathbf{a} \cdot \mathbf{b} + \mathbf{a} \cdot (\mathbf{c} + \mathbf{d}) \\ \mathbf{a} \cdot (\mathbf{b} + \mathbf{c} + \mathbf{d}) &= \mathbf{a} \cdot \mathbf{b} + \mathbf{a} \cdot \mathbf{c} + \mathbf{a} \cdot \mathbf{d} \end{aligned}$$ ●

EXAMPLE 6   Show that $(\mathbf{a} + \mathbf{b}) \cdot \mathbf{c} = \mathbf{a} \cdot \mathbf{c} + \mathbf{b} \cdot \mathbf{c}$ (12)

Solution
$$\begin{aligned} (\mathbf{a} + \mathbf{b}) \cdot \mathbf{c} &= \mathbf{c} \cdot (\mathbf{a} + \mathbf{b}) && \text{from (7)} \\ &= \mathbf{c} \cdot \mathbf{a} + \mathbf{c} \cdot \mathbf{b} && \text{from (8)} \\ &= \mathbf{a} \cdot \mathbf{c} + \mathbf{b} \cdot \mathbf{c} && \text{from (7)} \end{aligned}$$ ●

Theorem 3   **(Triangular Inequality).**   If $\mathbf{a}$ and $\mathbf{b}$ are any two vectors in three dimensional space, then

$$|\mathbf{a} + \mathbf{b}| \leq |\mathbf{a}| + |\mathbf{b}|$$

Proof   From the distributive and commutative properties of the scalar product, we have

$$\begin{aligned} |\mathbf{a} + \mathbf{b}|^2 &= (\mathbf{a} + \mathbf{b}) \cdot (\mathbf{a} + \mathbf{b}) \\ &= \mathbf{a} \cdot \mathbf{a} + \mathbf{b} \cdot \mathbf{a} + \mathbf{a} \cdot \mathbf{b} + \mathbf{b} \cdot \mathbf{b} \\ &= |\mathbf{a}|^2 + 2(\mathbf{a} \cdot \mathbf{b}) + |\mathbf{b}|^2 \\ &\leq |\mathbf{a}|^2 + 2|\mathbf{a}|\,|\mathbf{b}| + |\mathbf{b}|^2 = (|\mathbf{a}| + |\mathbf{b}|)^2 \end{aligned}$$

Therefore

$$|\mathbf{a} + \mathbf{b}| \leq |\mathbf{a}| + |\mathbf{b}|$$ ∎   (13)

From the formula $\mathbf{a} \cdot \mathbf{b} = |\mathbf{a}|\,|\mathbf{b}| \cos \theta$, division by $|\mathbf{b}|$, (if $|\mathbf{b}| \neq 0$), yields

$$\mathbf{a} \cdot \frac{\mathbf{b}}{|\mathbf{b}|} = |\mathbf{a}| \cos \theta$$ (14)

Equation (14) states that the scalar product of **a** with a unit vector in the direction of **b** is given by $|\mathbf{a}|\cos\theta$. Either side of (14) is called the *scalar projection* of a vector **a** on a vector **b** where θ is the angle between **a** and **b**. Note that this number is positive, zero, or negative according as θ is acute, right, or obtuse (Figure 16.3.2).

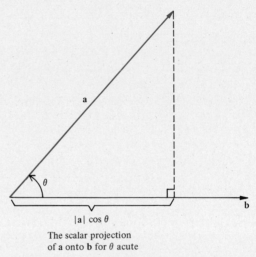

The scalar projection
of a onto b for θ acute

Figure 16.3.2

The *vector projection* of a vector **a** on a vector **b** is the vector

$$|\mathbf{a}|\cos\theta\,\frac{\mathbf{b}}{|\mathbf{b}|} \tag{15}$$

where $|\mathbf{a}|\cos\theta$ is the scalar projection of **a** on **b** and $\dfrac{\mathbf{b}}{|\mathbf{b}|}$ is a unit vector in the direction of **b**. The vector projection of **a** onto **b** is also called the *vector component* of **a** along **b**. The vector component of **a** in the direction of **b** is also given by

$$\frac{\mathbf{a}\cdot\mathbf{b}}{|\mathbf{b}|^2}\,\mathbf{b} \tag{16}$$

EXAMPLE 7   Given $\mathbf{a} = 2\mathbf{i} - \mathbf{j} - 4\mathbf{k}$ and $\mathbf{b} = \mathbf{i} + 3\mathbf{j} + \mathbf{k}$
Find (a) the scalar projection of **a** onto **b** and (b) the vector component of **a** along **b**.

Solution   (a) We compute $\mathbf{a}\cdot\mathbf{b} = (2)(1) + (-1)(3) + (-4)(1) = -5$ and $|\mathbf{b}| = \sqrt{1^2 + 3^2 + 1^2} = \sqrt{11}$. Application of (14) yields the scalar projection of **a** onto $\mathbf{b} = \dfrac{-5}{\sqrt{11}}$.

(b) $\dfrac{\mathbf{a}\cdot\mathbf{b}}{|\mathbf{b}|^2} = -\dfrac{5}{11}$ and from (16), the vector component of **a** along **b** is $-\dfrac{5}{11}(\mathbf{i} + 3\mathbf{j} + \mathbf{k})$.   ●

In Figure 16.3.3, we are given a nonzero vector **a** and the unit vectors **i, j,** and **k** in the positive coordinate directions. The angles α, β, and γ between **a** and the positive coordinate directions x, y, and z, respectively, are called the *direction angles* of the vector **a**. The numbers $\cos\alpha$, $\cos\beta$, and $\cos\gamma$ are called the *direction cosines* of **a**.

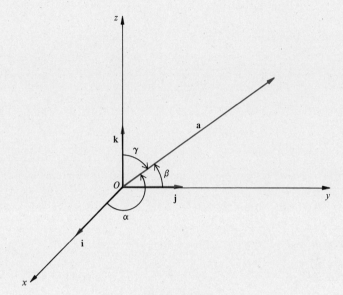

Figure 16.3.3

If $\mathbf{a} = \langle a_1, a_2, a_3 \rangle$ then

$$\cos\alpha = \frac{\mathbf{a} \cdot \mathbf{i}}{|\mathbf{a}|\,|\mathbf{i}|} = \frac{a_1}{|\mathbf{a}|}$$

$$\cos\beta = \frac{\mathbf{a} \cdot \mathbf{j}}{|\mathbf{a}|\,|\mathbf{j}|} = \frac{a_2}{|\mathbf{a}|} \qquad (17)$$

$$\cos\gamma = \frac{\mathbf{a} \cdot \mathbf{k}}{|\mathbf{a}|\,|\mathbf{k}|} = \frac{a_3}{|\mathbf{a}|}$$

from which

$$a_1 = |\mathbf{a}|\cos\alpha \qquad a_2 = |\mathbf{a}|\cos\beta \qquad a_3 = |\mathbf{a}|\cos\gamma$$

Hence

$$\mathbf{a} = a_1\mathbf{i} + a_2\mathbf{j} + a_3\mathbf{k} = |\mathbf{a}|[(\cos\alpha)\mathbf{i} + (\cos\beta)\mathbf{j} + (\cos\gamma)\mathbf{k}] \qquad (18)$$

If we take the absolute value of both sides there results

$$|\mathbf{a}|\,|(\cos\alpha)\mathbf{i} + (\cos\beta)\mathbf{j} + (\cos\gamma)\mathbf{k}| = |\mathbf{a}|$$

or, equivalently,

$$\sqrt{\cos^2\alpha + \cos^2\beta + \cos^2\gamma} = 1$$

Thus the angles α, β and γ may not be chosen independently of each other, rather we must have

$$\cos^2\alpha + \cos^2\beta + \cos^2\gamma = 1 \qquad (19)$$

Furthermore, if **u** is any *unit* vector, then

$$\mathbf{u} = (\cos \alpha)\mathbf{i} + (\cos \beta)\mathbf{j} + (\cos \gamma)\mathbf{k} \tag{20}$$

EXAMPLE 8 Find the direction cosines of the vector represented by the directed line segment whose initial point is $P(-2, -1, 3)$ and whose terminal point is $Q(1, -5, 8)$.

Solution The vector $\overrightarrow{PQ}$ is $\langle 3, -4, 5 \rangle$. Its length is

$$\sqrt{3^2 + (-4)^2 + 5^2} = \sqrt{50} = 5\sqrt{2}$$

From (17) its direction cosines are

$$\cos \alpha = \frac{3}{5\sqrt{2}} = \frac{3\sqrt{2}}{10}$$

$$\cos \beta = -\frac{4}{5\sqrt{2}} = -\frac{2\sqrt{2}}{5}$$

$$\cos \gamma = \frac{5}{5\sqrt{2}} = \frac{\sqrt{2}}{2}$$

Exercises 16.3

In Exercises 1 through 6, find the dot product of the two given vectors.

1. $\mathbf{i} - \mathbf{j}$ and $\mathbf{i} + \mathbf{j} - 2\mathbf{k}$
2. $\mathbf{j}$ and $\mathbf{k} - \mathbf{j}$
3. $2\mathbf{i} + \mathbf{j} - \mathbf{k}$ and $\mathbf{i} - 5\mathbf{j} + 6\mathbf{k}$
4. $3\mathbf{i} - 2\mathbf{j} + 7\mathbf{k}$ and $-3\mathbf{j}$
5. $\langle -1, 0, 5 \rangle$ and $\langle 0, 2, 8 \rangle$
6. $\frac{1}{3}\langle 2, 6, -4 \rangle$ and $\langle 9, -2, \frac{3}{8} \rangle$

In Exercises 7 through 10, find the cosine of the angle between the two given vectors.

7. $\langle 2, -2, -1 \rangle$ and $\langle 1, -1, 0 \rangle$
8. $\langle -3, 6, 2 \rangle$ and $\langle 2, -1, 6 \rangle$
9. $3\mathbf{i} - 2\mathbf{j} - \mathbf{k}$ and $\mathbf{j} - \mathbf{i} + 3\mathbf{k}$
10. $2\mathbf{k} - 2\mathbf{j} + \mathbf{i}$ and $2\mathbf{i} + \mathbf{j} - 2\mathbf{k}$
11. Show in two ways that the triangle with vertices $A(-1, 3, 5)$, $B(0, 1, 3)$ and $C(2, 1, 4)$ is a right triangle.
12. Find $\cos C$ in the triangle with vertices $A(0, 1, 2)$, $B(-1, 1, 3)$ and $C(2, 0, 1)$.

In Exercises 13 through 18, let $\mathbf{a} = 4\mathbf{i} - 3\mathbf{j} + \mathbf{k}$, $\mathbf{b} = 2\mathbf{i} - 5\mathbf{k}$, and $\mathbf{c} = \mathbf{i} + 2\mathbf{j} - 4\mathbf{k}$

13. Compute $\mathbf{a} \cdot (\mathbf{b} - \mathbf{c})$
14. Find $3\mathbf{a} + \mathbf{b} - 2\mathbf{c}$
15. Find $(\mathbf{a} - \mathbf{b}) \cdot (\mathbf{c} - \mathbf{b})$
16. Calculate $(\mathbf{a} \cdot \mathbf{b})\mathbf{c} - (\mathbf{b} \cdot \mathbf{c})\mathbf{a}$
17. Find the scalar components of **a** in the **b** and **c** directions.
18. Find the vector component of **c** in the **b** direction.
19. Does the "cancellation law" hold true for scalar products; namely, if $\mathbf{a} \cdot \mathbf{b} = \mathbf{c} \cdot \mathbf{b}$, $\mathbf{b} \neq \mathbf{0}$, then does $\mathbf{a} = \mathbf{c}$ follow?

In Exercises 20 through 24, find the direction cosines of the given line or line segment.

20. Positively directed y-axis.
21. From the origin into the first octant making equal angles with the coordinate axes.
22. From the origin into the first quadrant of the yz-plane, making equal angles with the y- and z-axes.
23. From $(2, 3, 7)$ to $(1, -5, 3)$.
24. From (a, b, c) to (b, c, a).

In Exercises 25 and 26, tell whether the three numbers could be the direction cosines of a vector. If not, state why not.

25. $\frac{1}{3}, \frac{1}{3}, \frac{1}{3}$

26. $\frac{3}{\sqrt{21}}, \frac{-2}{\sqrt{21}}, \sqrt{\frac{8}{21}}$

27. Prove that for any vector **a**, $\sqrt{(\mathbf{a} \cdot \mathbf{a})} = |\mathbf{a}|$.

28. Given that $3\mathbf{a} + \mathbf{b}$ is perpendicular to $3\mathbf{a} - \mathbf{b}$, find a relationship between $|\mathbf{a}|$ and $|\mathbf{b}|$.

29. Find the cosine of the angle between the diagonal of a cube and the diagonal of one of its faces. (*Hint:* refer the cube to a convenient set of rectangular coordinates in space.)

30. Let **a** and **b** be two vectors of unit length. Show that $\mathbf{a} + \mathbf{b}$ bisects the angle between the two vectors. Show also that $\mathbf{a} - \mathbf{b}$ is perpendicular to $\mathbf{a} + \mathbf{b}$.

31. Find a vector that bisects the angle between the vectors $2\mathbf{i} - \mathbf{j} - \mathbf{k}$ and $\mathbf{i} + 2\mathbf{j} + \mathbf{k}$. (*Hint:* utilize the result of Exercise 30.)

32. (i)   Express $-2\mathbf{j} + 3\mathbf{k}$ as a linear combination of the vectors $3\mathbf{i} + \mathbf{j} + 2\mathbf{k}$, $-\mathbf{i} - \mathbf{j} + \mathbf{k}$ and $\mathbf{i} + \mathbf{j} + \mathbf{k}$.
   (ii) Generalize by showing that any vector $\mathbf{a} = a_1\mathbf{i} + a_2\mathbf{j} + a_3\mathbf{k}$ may be expressed as a linear combination of these vectors.
   (iii) What does this imply about the three vectors?

33. Three vectors $\mathbf{a} = a_1\mathbf{i} + a_2\mathbf{j} + a_3\mathbf{k}$, $\mathbf{b} = b_1\mathbf{i} + b_2\mathbf{j} + b_3\mathbf{k}$, and $\mathbf{c} = c_1\mathbf{i} + c_2\mathbf{j} + c_3\mathbf{k}$ are said to be **linearly independent** if and only if the only scalars that satisfy an equation of the form

$$\lambda_1\mathbf{a} + \lambda_2\mathbf{b} + \lambda_3\mathbf{c} = \mathbf{0}$$

are $\lambda_1 = \lambda_2 = \lambda_3 = 0$. It can be proved that any three linearly independent vectors in three-dimensional space form a basis in that *any* vector in the space can be expressed as a linear combination of them. Show that the vectors of the preceeding exercise $3\mathbf{i} + \mathbf{j} + 2\mathbf{k}$, $-\mathbf{i} - \mathbf{j} + \mathbf{k}$, and $\mathbf{i} + \mathbf{j} + \mathbf{k}$ are linearly independent.

34. Prove equations (7) and (9) of the text.

16.4   Lines in Space

Through two points in space there passes one and only one straight line. This means that a straight line in space is uniquely determined by any two points on it. Also, a straight line is uniquely prescribed by giving the coordinates of one point on it and the direction of the line. This direction may be specified by giving a vector parallel to the given line. Suppose that two points $P_0(x_0, y_0, z_0)$ and $P_1(x_1, y_1, z_1)$ are on line m (Figure 16.4.1). The vector

$$\overrightarrow{P_0P_1} = \overrightarrow{OP_1} - \overrightarrow{OP_0} = (x_1 - x_0)\mathbf{i} + (y_1 - y_0)\mathbf{j} + (z_1 - z_0)\mathbf{k} \tag{1}$$

connecting the two points yields the direction of the given line. We seek the vector equation of this line.

Refer to Figure 16.4.1. Any point on the line may be reached by first moving from the origin O to point P_0 and then from P_0 to P. But $\overrightarrow{P_0P}$ is parallel to $\overrightarrow{P_0P_1}$; thus, a scalar t must exist so that

$$\overrightarrow{P_0P} = t(\overrightarrow{P_0P_1}) \tag{2}$$

Therefore, from

$$\overrightarrow{OP} = \overrightarrow{OP_0} + \overrightarrow{P_0P} \tag{3}$$

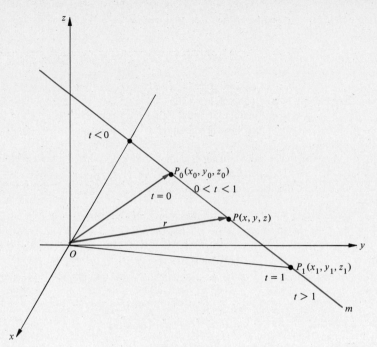

Figure 16.4.1

and (2), we have

$$\overrightarrow{OP} = \overrightarrow{OP_0} + t(\overrightarrow{P_0P_1}) \qquad (4)$$

If we denote the position vector $\overrightarrow{OP}$ by **r** then

$$\mathbf{r} = \overrightarrow{OP} = \overrightarrow{OP_0} + t(\overrightarrow{P_0P_1}) \qquad (5)$$

Hence (5) is the vector equation of the line through the points P_0 and P_1. Note that $t = 0$ corresponds to the point P_0 and $t = 1$ yields the point P_1. As t varies from 0 to 1, the segment joining P_0 and P_1 is generated. Other values of t are shown in the figure. If $t > 1$ then the points are in the order P_0P_1P, while if $t < 0$, the points are in the order PP_0P_1. The reader should verify these results.

If we let $\mathbf{r} = x\mathbf{i} + y\mathbf{j} + z\mathbf{k}$, then (5) may be written

$$x\mathbf{i} + y\mathbf{j} + z\mathbf{k} = x_0\mathbf{i} + y_0\mathbf{j} + z_0\mathbf{k}$$
$$+ t[(x_1 - x_0)\mathbf{i} + (y_1 - y_0)\mathbf{j} + (z_1 - z_0)\mathbf{k}] \qquad (6)$$

and equating corresponding coefficients of **i, j,** and **k** yields

$$\begin{aligned} x &= x_0 + t(x_1 - x_0) \\ y &= y_0 + t(y_1 - y_0) \\ z &= z_0 + t(z_1 - z_0) \end{aligned} \qquad (7)$$

Equations (7) are three scalar equations that correspond to the vector equation (5) and are called the **parametric equations** for the line with parameter t. If we let $a = x_1 - x_0$, $b = y_1 - y_0$, and $c = z_1 - z_0$, then the vector $a\mathbf{i} + b\mathbf{j} + c\mathbf{k}$ is *parallel* to the given line. The triplet of numbers $\langle a, b, c \rangle$ is said to be a set of **direction numbers** for the line. Two lines are said to be **parallel** if and only if their direction numbers are proportional. Note that any set of direction numbers $\langle a, b, c \rangle$ of a line (for which $a^2 + b^2 + c^2 > 0$) are proportional to their direction cosines:

$$\frac{a}{\sqrt{a^2 + b^2 + c^2}}, \frac{b}{\sqrt{a^2 + b^2 + c^2}}, \frac{c}{\sqrt{a^2 + b^2 + c^2}}$$

Equations (7) may be rewritten in the form

$$x = x_0 + at \qquad y = y_0 + bt \qquad z = z_0 + ct \tag{8}$$

If $a, b,$ and c are nonzero, then by solving each equation of (8) for t and equating the resulting equations we obtain

$$\frac{x - x_0}{a} = \frac{y - y_0}{b} = \frac{z - z_0}{c} \tag{9}$$

Equations (9) are called **symmetric equations** of the line.

The first equation of (9), $\dfrac{x - x_0}{a} = \dfrac{y - y_0}{b}$ is an equation of a plane perpendicular to the xy-plane (Section (16.5).† Similarly, $\dfrac{y - y_0}{b} = \dfrac{z - z_0}{c}$, the second equation of (9), is an equation of a plane perpendicular to the yz-plane. Thus the two equations (9) define the straight line that is the intersection of these two planes.

EXAMPLE 1 Find vector, parametric, and symmetric equations of a straight line that passes through $P_0(-2, 1, 4)$ and $P_1(3, 5, 11)$. At what point does the line intersect the zx-plane?

Solution $\overrightarrow{P_0P_1} = 5\mathbf{i} + 4\mathbf{j} + 7\mathbf{k}$ and a vector equation of the line from (5) is

$$\mathbf{r} = -2\mathbf{i} + \mathbf{j} + 4\mathbf{k} + t(5\mathbf{i} + 4\mathbf{j} + 7\mathbf{k}) \tag{i}$$

or (with $\mathbf{r} = x\mathbf{i} + y\mathbf{j} + z\mathbf{k}$)

$$x = -2 + 5t \qquad y = 1 + 4t \qquad z = 4 + 7t \tag{ii}$$

are the parametric equations of the given line. In symmetric form, we have

$$\frac{x + 2}{5} = \frac{y - 1}{4} = \frac{z - 4}{7} \tag{iii}$$

To find the intersection of (iii) with the zx-plane, we set $y = 0$ in (iii) to obtain $x = -\frac{13}{4}$ and $z = \frac{9}{4}$. Therefore the line pierces the zx-plane at $(-\frac{13}{4}, 0, \frac{9}{4})$. ●

EXAMPLE 2 Find the coordinates of the point $P*$ on the line joining P_0 and P_1 and such that $\overrightarrow{P_0P*} = \frac{2}{3}(\overrightarrow{P_0P_1})$

Solution
$$\overrightarrow{OP*} = \overrightarrow{OP_0} + \overrightarrow{P_0P*}$$
$$= \overrightarrow{OP_0} + \tfrac{2}{3}(\overrightarrow{P_0P_1})$$

---

† It will be established in Section 16.5 that the linear equation $Ax + By + Cz = D$ is an equation of a plane if $A^2 + B^2 + C^2 > 0$. For example, $x = z$ is an equation of a plane and not a line in three-dimensional space.

Hence

$$x*\mathbf{i} + y*\mathbf{j} + z*\mathbf{k} = x_0\mathbf{i} + y_0\mathbf{j} + z_0\mathbf{k} + \tfrac{2}{3}[(x_1 - x_0)\mathbf{i} + (y_1 - y_0)\mathbf{j} + (z_1 - z_0)\mathbf{k}]$$

and, therefore,

$$x* = x_0 + \tfrac{2}{3}(x_1 - x_0) = \tfrac{1}{3}x_0 + \tfrac{2}{3}x_1 = \tfrac{1}{3}(x_0 + 2x_1)$$
$$y* = \tfrac{1}{3}(y_0 + 2y_1), \quad z* = \tfrac{1}{3}(z_0 + 2z_1)$$

Note that the result also follows directly from (7) by letting $t = \tfrac{2}{3}$. ●

Definition Two lines in R_3 are said to be *skew* if they do not intersect and are nonparallel.

EXAMPLE 3 Prove that the line of Example 1 and the z-axis are skew lines.

Solution If the line given by (iii) did intersect the z-axis then $(0, 0, z)$ must satisfy (iii) for some value of z. If $x = y = 0$ are substituted into the first equation of (iii), we obtain a contradiction, namely, $\tfrac{2}{5} = -\tfrac{1}{4}$. Therefore the lines do not intersect.

A set of direction numbers for the z-axis are $\langle 0, 0, 1 \rangle$ which are clearly not proportional to the direction numbers $\langle 5, 4, 7 \rangle$ of Example 1. Thus the given line and the z-axis are nonparallel. Hence the lines are skew to each other. ●

EXAMPLE 4 Find the vector equation of the straight line m which is the intersection of the two planes

$$3x - 5y + 2z = 2 \quad \text{and} \quad 2x - y + 4z = 6$$

Solution We seek two points on the required line. If we set $z = 0$, there results

$$3x - 5y = 2 \quad \text{and} \quad 2x - y = 6$$

which has the solution $x = 4$ and $y = 2$. Hence $P_0(4, 2, 0)$ is a point on the line m. If we set $y = 0$, then

$$3x + 2z = 2 \quad \text{and} \quad 2x + 4z = 6$$

which yields $x = -\tfrac{1}{2}$, $z = \tfrac{7}{4}$. Hence $P_1(-\tfrac{1}{2}, 0, \tfrac{7}{4})$ is a second point on the line.

A vector equation for m, from (5), is

$$\mathbf{r} = \overrightarrow{OP_0} + t(\overrightarrow{P_0P_1})$$
$$= 4\mathbf{i} + 2\mathbf{j} + t[(-\tfrac{1}{2} - 4)\mathbf{i} + (0 - 2)\mathbf{j} + (\tfrac{7}{4} - 0)\mathbf{k}]$$
$$= 4\mathbf{i} + 2\mathbf{j} + t[-\tfrac{9}{2}\mathbf{i} - 2\mathbf{j} + \tfrac{7}{4}\mathbf{k}]$$ ●

EXAMPLE 5 Find the point at which the lines

$$m_1: \mathbf{r}_1(t) = \mathbf{i} - 2\mathbf{j} + 2\mathbf{k} + t(2\mathbf{i} + \mathbf{j} + \mathbf{k})$$

and

$$m_2: \mathbf{r}_2(u) = 3\mathbf{i} - 5\mathbf{j} - \mathbf{k} + u(2\mathbf{i} + 3\mathbf{j} + 3\mathbf{k})$$

intersect.

Solution In order to have a point of intersection, there must be values of t and u for which $\mathbf{r}_1(t) = \mathbf{r}_2(u)$. Hence, we equate $\mathbf{r}_1(t) = \mathbf{r}_2(u)$ and solve for t and u. Therefore

$$\mathbf{i} - 2\mathbf{j} + 2\mathbf{k} + t(2\mathbf{i} + \mathbf{j} + \mathbf{k}) = 3\mathbf{i} - 5\mathbf{j} - \mathbf{k} + u(2\mathbf{i} + 3\mathbf{j} + 3\mathbf{k})$$

from which

$$(1 + 2t - 3 - 2u)\mathbf{i} + (-2 + t + 5 - 3u)\mathbf{j} + (2 + t + 1 - 3u)\mathbf{k} = \mathbf{0}$$

This yields

$$t - u = 1 \qquad t - 3u = -3 \qquad t - 3u = -3$$

The last two equations are the same. We obtain $t = 3$ and $u = 2$. For these numbers,

$$\mathbf{r}_1(3) = \mathbf{r}_2(2) = 7\mathbf{i} + \mathbf{j} + 5\mathbf{k}$$

as is easily verified. Therefore $(7, 1, 5)$ is the point of intersection. ●

Exercises 16.4

In Exercises 1 through 4, find vector and scalar equations of a straight line joining the two given points.

1. $(4, 4, 0)$ and $(7, 10, -3)$ 2. $(2, -3, -1)$ and $(4, -1, 5)$
3. $(-4, 1, -2)$ and $(3, -2, 6)$ 4. $(-7, 2, -5)$ and $(7, -2, 5)$
5. Where does the line of Exercise 1 meet the xy-plane?
6. Where does the line of Exercise 2 meet the plane $x = -8$?
7. Where does the line of Exercise 3 meet the plane $x = z$?
8. In Exercises 1 through 4, which line, if any, passes through the origin?

In Exercises 9 through 12, find the line that contains the given point and is parallel to the given line.

9. $(3, -4, 6), \mathbf{r} = (t + 5)\mathbf{i} + (2t - 1)\mathbf{j} + (4 - t)\mathbf{k}$ (*Hint:* Find a set of direction numbers of $\mathbf{r}$ by choosing two convenient values of t.)
10. $(0, -1, 2), \mathbf{r} = (3t - 1)\mathbf{i} + (t + 5)\mathbf{j} + (2t + 1)\mathbf{k}$
11. $(0, 0, 0), \mathbf{r} = (2 - t)\mathbf{i} + (3 - 2t)\mathbf{j} + (5 - 3t)\mathbf{k}$
12. $(0, 0, 1), \mathbf{r} = (-2 + 7t)\mathbf{i} + (4 - 5t)\mathbf{j} + 6t\mathbf{k}$

In Exercises 13 through 16, find a vector equation of the line of intersection of the two given planes.

13. $x + 2y = 3, y + z = 1$ 14. $x - z = 4, y + 2z = 2$
15. $x + 2y + z = 3, 3x - y + z = 2$ 16. $3x + y - z = 5, 2x + y + z = 3$

In Exercises 17 through 20, a line passes through points P_0, P_1, and P_2 in that order.

17. Find the coordinates of P_2 if $\overrightarrow{P_0P_2} = 3(\overrightarrow{P_0P_1})$ and P_0 and P_1 are $(1, 2, 4)$ and $(2, 4, 6)$, respectively.
18. Find the coordinates of P_1 if $\overrightarrow{P_0P_1} = \frac{3}{4}(\overrightarrow{P_0P_2})$ and P_0 and P_2 are $(0, 1, -1)$ and $(4, 13, 7)$, respectively.
19. Find the coordinates of P_1 if $\overrightarrow{P_0P_1} = 4(\overrightarrow{P_1P_2})$ and P_0 and P_2 are $(-1, 2, 5)$ and $(14, 12, 10)$, respectively.
20. Find the coordinates of P_0 if $\overrightarrow{P_0P_1} = 2(\overrightarrow{P_1P_2})$, $P_1(5, 2, 8)$, and $P_2(8, -3, 4)$.
21. Prove that the straight line

$$\frac{x + 1}{2} = y - 3 = \frac{z + 7}{5}$$

lies in the plane $2x + y - z = 8$.

22. Prove that the two sets of equations

$$\frac{x + 6}{4} = y - 1 = \frac{z + 9}{3} \quad \text{and} \quad \frac{x - 6}{8} = \frac{y - 4}{2} = \frac{z}{6}$$

are equations of the same line.

23. Give a vector parametrization for a line that passes through a point $P_0(4, -2, 3)$ and is parallel to the line

$$\frac{x - 8}{13} = \frac{y + 9}{7} = z - 5$$

24. Find the point at which the lines

$$m_1: \mathbf{r}_1(t) = \mathbf{i} - 5\mathbf{j} + \mathbf{k} + t(3\mathbf{i} + \mathbf{j} - \mathbf{k})$$
$$m_2: \mathbf{r}_2(u) = 3\mathbf{i} + 5\mathbf{j} + 3\mathbf{k} + u(\mathbf{i} - 2\mathbf{j} - \mathbf{k})$$

intersect.

25. Find the point of intersection of the line containing points $(2, 1, -1)$ and $(10, -2, 6)$ with the plane $x + 2y - z = 10$.

In Exercises 26 and 27, find the point where the lines m_1 and m_2 intersect and determine the angle of intersection.

26. $m_1: \mathbf{r}_1(t) = (t + 2)\mathbf{i} + (t + 1)\mathbf{j} + 3t\mathbf{k}$
$m_2: \mathbf{r}_2(u) = (2u + 4)\mathbf{i} + (u + 2)\mathbf{j} + (-u - 1)\mathbf{k}$

27. $m_1: \mathbf{r}_1(t) = (-t - 3)\mathbf{i} + (3t + 10)\mathbf{j} + (5t + 12)\mathbf{k}$
$m_2: \mathbf{r}_2(u) = (2u - 3)\mathbf{i} + (u + 3)\mathbf{j} + (u + 1)\mathbf{k}$

28. Prove that the lines m_1 and m_2 are skew to each other.

$$m_1: \mathbf{r}_1(t) = (2t - 1)\mathbf{i} + 3t\mathbf{j} + (t + 1)\mathbf{k}$$
$$m_2: \mathbf{r}_2(u) = u\mathbf{i} + (u - 1)\mathbf{j} + (2u + 3)\mathbf{k}$$

(Hint: Recall that it is sufficient to show that the lines are not parallel and do not intersect.)

29. Find the perpendicular distance from the origin to the line

$$\mathbf{r} = (1 + t)\mathbf{i} + t\mathbf{j} + (t - 2)\mathbf{k}$$

(Hint: Write a formula for the square of the distance $s^2(t)$ from the origin to any point on the given line. Then minimize s^2.)

30. Find the perpendicular distance from the point $(1, 2, 3)$ to the line

$$\mathbf{r} = (3 + t)\mathbf{i} + 2t\mathbf{j} + (3t - 1)\mathbf{k}$$

16.5   PLANES

Our objective is to determine an equation of a plane from given data. A plane is uniquely determined if we know a point on it and its orientation in space. This orientation may be prescribed by giving the direction numbers of a vector perpendicular to the plane.

Let $\mathbf{n} = A\mathbf{i} + B\mathbf{j} + C\mathbf{k}$ be the known nonzero vector which is perpendicular to the given plane so that $\langle A, B, C \rangle$ are a set of direction numbers of $\mathbf{n}$. Let $P_0(x_0, y_0, z_0)$ be the fixed point on the plane Π with position vector $\mathbf{r} = x_0\mathbf{i} + y_0\mathbf{j} + z_0\mathbf{k}$ (Figure 16.5.1). From the definition of a line being perpendicular to a

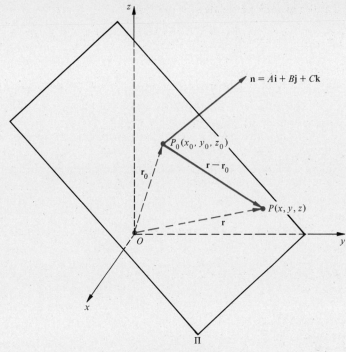

$\mathbf{n} = A\mathbf{i} + B\mathbf{j} + C\mathbf{k}$

$P_0(x_0, y_0, z_0)$

$\mathbf{r} - \mathbf{r}_0$

$P(x, y, z)$

$\mathbf{r}_0$

$\mathbf{r}$

O

Π

Figure 16.5.1

plane, it follows that every line passing through P_0 and in the plane Π must be perpendicular to $\mathbf{n}$. Thus a point $P(x, y, z)$, where P is distinct from P_0, with position vector $\mathbf{r} = x\mathbf{i} + y\mathbf{j} + z\mathbf{k}$ is on the plane if and only if $\mathbf{r} - \mathbf{r}_0$ is perpendicular to $\mathbf{n}$. Hence,

$$\mathbf{n} \cdot (\mathbf{r} - \mathbf{r}_0) = 0 \tag{1}$$

or

$$(A\mathbf{i} + B\mathbf{j} + C\mathbf{k}) \cdot [(x - x_0)\mathbf{i} + (y - y_0)\mathbf{j} + (z - z_0)\mathbf{k}] = 0 \tag{2}$$

Expansion of (2) yields

$$A(x - x_0) + B(y - y_0) + C(z - z_0) = 0 \tag{3}$$

which is an equation of the plane Π.

EXAMPLE 1   Find an equation of a plane which passes through a point $(4, 5, -1)$ and is perpendicular to the vector $\mathbf{n} = 2\mathbf{i} - 3\mathbf{j} + 6\mathbf{k}$.

Solution   We apply (3) directly with $(x_0, y_0, z_0) = (4, 5, -1)$ and $\langle A, B, C \rangle = \langle 2, -3, 6 \rangle$ to obtain

$$2(x - 4) - 3(y - 5) + 6(z + 1) = 0$$

or, equivalently,

$$2x - 3y + 6z + 13 = 0 \qquad \bullet$$

Equation (3) is equivalent to

$$Ax + By + Cz = Ax_0 + By_0 + Cz_0 \tag{4}$$

and may be written in the form

$$Ax + By + Cz = D \qquad (5)$$

Consider now any equation of the form (5) where at least one of the coefficients A, B, or C is not zero. Then one can always determine a triplet of numbers (x_0, y_0, z_0) such that $Ax_0 + By_0 + Cz_0 = D$. For example, if $C \neq 0$ then we may choose x_0 and y_0 arbitrarily and solve for z_0. Thus (5) can always be written in form (4) or, equivalently, in form (3). This means that (5) represents a plane through the point (x_0, y_0, z_0) and perpendicular to a vector with direction numbers A, B, and C.

Therefore we have established the following theorem.

Theorem 1 In three-dimensional space, every plane has a Cartesian equation of the first degree and, conversely, the graph of every first degree equation is a plane.

In symbols, if A, B, or C are not all zero, the graph of

$$Ax + By + Cz = D$$

is a plane where A, B, and C are direction numbers of a normal vector to the plane.

From this theorem, we see the analogy between the straight line in two-dimensional space and the plane in three-dimensional space.

Definition Two planes are **parallel** if and only if a normal vector of one is parallel to a normal vector of the other (Figure 16.5.2).

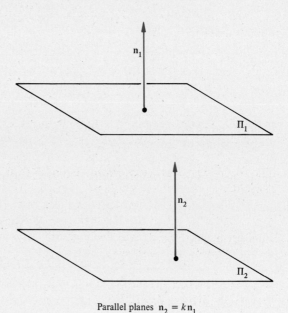

Parallel planes $n_2 = k n_1$

Figure 16.5.2

Theorem 2 Two planes

$$A_1x + B_1y + C_1z + D_1 = 0$$

and

$$A_2x + B_2y + C_2z + D_2 = 0$$

are parallel if and only if the normal vectors $\mathbf{n}_1 = \langle A_1, B_1, C_1 \rangle$ and $\mathbf{n}_2 = \langle A_2, B_2, C_2 \rangle$ are proportional; that is, a nonzero constant k exists so that

$$\mathbf{n}_2 = k\mathbf{n}_1 \tag{6}$$

Definition Two planes are **perpendicular** if and only if a normal vector of one is perpendicular to a normal vector of the other.

Thus if $\mathbf{n}_1$ and $\mathbf{n}_2$ are such normal vectors, the two planes are perpendicular if and only if

$$\mathbf{n}_1 \cdot \mathbf{n}_2 = 0 \tag{7}$$

(See Figure 16.5.3.)

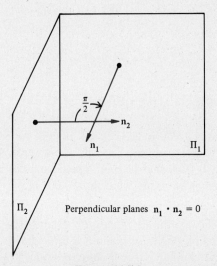

Perpendicular planes $\mathbf{n}_1 \cdot \mathbf{n}_2 = 0$

Figure 16.5.3

Definition The **angle between two planes** is defined to be the angle between their normals.

EXAMPLE 2 Find an equation of the plane that passes through the point $(2, 3, 5)$ and is parallel to the plane $3x - y - 2z = 4$.

Solution We may use (3) directly with $\langle 3, -1, -2 \rangle$ as a normal vector and $(2, 3\ 5)$ as the given point. Hence from (3),

$$3(x - 2) - 1(y - 3) - 2(z - 5) = 0$$

or, multiplying out,

$$3x - y - 2z + 7 = 0.$$

Alternative
Solution The required plane has an equation of the form

$$3x - y - 2z = D$$

But, $(2, 3, 5)$ must satisfy its equation so that

$$3(2) - 3 - 2(5) = D \qquad \text{or} \qquad D = -7$$

Therefore an equation of the plane is

$$3x - y - 2z + 7 = 0 \qquad\qquad \bullet$$

EXAMPLE 3 Find an equation of the plane through $P_0(7, 1, -5)$ and perpendicular to the line passing through points $P_1(-1, 9, 2)$ and $P_2(3, 6, 10)$.

Solution A normal $\mathbf{n}$ to the plane is the vector

$$\mathbf{n} = 4\mathbf{i} - 3\mathbf{j} + 8\mathbf{k}$$

from point P_1 to point P_2. From (3), an equation of the plane is

$$4(x - 7) - 3(y - 1) + 8(z + 5) = 0$$

or, equivalently,

$$4x - 3y + 8z + 15 = 0 \qquad\qquad \bullet$$

EXAMPLE 4 Sketch the plane $3x + 4y + 6z = 24$.

Solution Setting $y = 0$ and $z = 0$, we find that the plane has x-intercept 8. Similarly, the y- and z-intercepts are 6 and 4, respectively. Since the three points $(8, 0, 0)$, $(0, 6, 0)$ and $(0, 0, 4)$ must lie on the plane and determine it, the graph of a portion of the plane is easily drawn (Figure 16.5.4). $\qquad\qquad \bullet$

The *coordinate planes* are the yz-plane, the zx-plane and the xy-plane. The equation of the yz-plane is $x = 0$; that is, $(0, y, z)$ is an arbitrary point in this plane. Similarly, $y = 0$ is the equation of the zx-plane and $z = 0$ is the equation of the xy-plane. The equations of planes parallel to the yz-plane are of the form $x = c = $ constant, where $c \neq 0$. Planes parallel to the zx- and xy-planes are defined analogously by equations $y = c_1$ and $z = c_2$, respectively, where c_1 and c_2 are arbitrary nonzero constants.

The lines of intersection of a plane with the coordinate planes are called its *traces*. These traces are very helpful in sketching planes not through the origin nor parallel to a coordinate axis, as shown in Figure 16.5.4.

Planes through the origin are also easily sketched. Their analytic representation is $Ax + By + Cz = 0$, where the coefficients A, B, and C are direction numbers of vectors normal to the plane. The plane is obtained if we determine any three noncollinear points on it. To illustrate, suppose that we are given an equation of a plane $3x - y - 2z = 0$ that passes through the origin. If we choose (for convenience) $x = 0$ and $y = -2$, substitution into the equation yields $z = 1$;

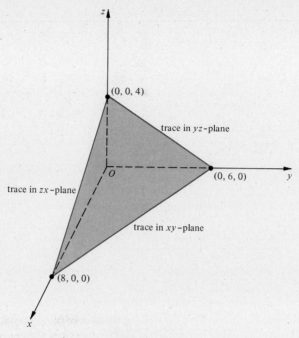

(0, 0, 4)

trace in yz-plane

O

(0, 6, 0) y

trace in zx-plane

trace in xy-plane

(8, 0, 0)

x

Figure 16.5.4

hence, $P_1(0, -2, 1)$ is a point on the plane. Also the line connecting O and P_1 lies on the plane. Next we choose $y = 0$ and $x = 2$ to obtain $z = 3$. Hence $P_2(2, 0, 3)$ and the line connecting O and P_2 is also on the plane. The three points O, P_1, and P_2 uniquely determine the plane, which is now easily drawn.

EXAMPLE 5   Find an equation of the plane through the three points $P_0(3, -1, -5)$, $P_1(1, 4, 6)$, and $P_2(7, 2, 12)$.

Solution   Let the equation of the plane be

$$Ax + By + Cz = D \qquad (8)$$

where A, B, C, and D are to be determined.

The plane passes through the three given points if and only if the coordinates of these points satisfy (8). From the given data, by successive substitution into (8), we obtain

$$3A - B - 5C = D \qquad (9)$$
$$A + 4B + 6C = D \qquad (10)$$
$$7A + 2B + 12C = D \qquad (11)$$

This is a system of three homogeneous linear equations in four unknowns. This system can be solved for the ratios of the unknowns (and this is all that is necessary—why?). The quantity D may be eliminated by subtracting (10) from (9) and (10) from (11) to obtain

$$2A - 5B - 11C = 0 \qquad (12)$$
$$6A - 2B + 6C = 0 \qquad (13)$$

If C is eliminated between (12) and (13), we obtain

$$3A = 2B \tag{14}$$

If we *choose* $A = 2$, then $B = 3$ in order to satisfy (14) and $C = -1$ from (12) or (13). Substitution into any one of equations (9), (10), or (11) yields $D = 8$. Hence

$$2x + 3y - z = 8 \tag{15}$$

is the desired equation of the plane through the three given points. It is left to the reader to verify that the coordinates of the three points actually satisfy (15). ●

EXAMPLE 6   Find the distance between the plane $8x - y + 4z = 12$ and the point $P(-1, 2, 1)$.

Solution   Setting $x = 0$ and $y = 0$, we find $Q(0, 0, 3)$ is a point on the given plane. The vector from point P to point Q is $\mathbf{v} = \mathbf{i} - 2\mathbf{j} + 2\mathbf{k}$. Now, $(8, -1, 4)$ are a set of direction numbers of a normal to the plane and, since $\sqrt{8^2 + (-1)^2 + 4^2} = 9$, we have a unit normal vector $\mathbf{n} = \frac{8}{9}\mathbf{i} - \frac{1}{9}\mathbf{j} + \frac{4}{9}\mathbf{k}$. The required distance d is the absolute value of the scalar component of $\mathbf{v}$ in the direction of $\mathbf{n}$ and is given by (see Figure 16.5.5 for the general case)

$$d = |\mathbf{v} \cdot \mathbf{n}| = |(1)(\tfrac{8}{9}) + (-2)(-\tfrac{1}{9}) + (2)(\tfrac{4}{9})|$$

or

$$d = 2 \qquad\qquad ●$$

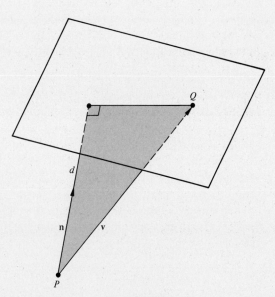

Figure 16.5.5

The method of Example 6 is easily generalized to find the distance d between the point $P(x_0, y_0, z_0)$ and the plane $Ax + By + Cz = D$.

Refer to Figure 16.5.5. Set $x = 0$ and $y = 0$ to find that $Q\left(0, 0, \dfrac{D}{C}\right)$ is a point on the plane assuming that $C \neq 0$. $\left[\text{If } C = 0, \text{ we may use } \left(0, \dfrac{D}{B}, 0\right) \text{ or}\right.$

$\left(\dfrac{D}{A}, 0, 0\right).\,\Big]$ The vector from point P to point Q is $-x_0\mathbf{i} - y_0\mathbf{j} + \left(\dfrac{D}{C} - z_0\right)\mathbf{k}.$

Since $A\mathbf{i} + B\mathbf{j} + C\mathbf{k}$ is a vector normal to the plane, a unit normal $\mathbf{n}$ is given by

$$\mathbf{n} = \frac{1}{\sqrt{A^2 + B^2 + C^2}}(A\mathbf{i} + B\mathbf{j} + C\mathbf{k})$$

and

$$d = \left| -\left(x_0\mathbf{i} + y_0\mathbf{j} + (z_0 - \frac{D}{C})\mathbf{k}\right) \cdot \left(\frac{A\mathbf{i} + B\mathbf{j} + C\mathbf{k}}{\sqrt{A^2 + B^2 + C^2}}\right) \right|$$

$$= \frac{|Ax_0 + By_0 + Cz_0 - D|}{\sqrt{A^2 + B^2 + C^2}} \tag{16}$$

In particular, if the given point is the origin, then $x_0 = y_0 = z_0 = 0$ and the distance d between the origin and the plane is

$$d = \frac{|D|}{\sqrt{A^2 + B^2 + C^2}} \tag{17}$$

As a corollary to (16), we have the formula for the distance d from a point $P_0(x_0, y_0)$ to a line $Ax + By = C$, both in the xy-plane, given by

$$d = \frac{|Ax_0 + By_0 - C|}{\sqrt{A^2 + B^2}} \tag{18}$$

Exercises 16.5

In Exercises 1 through 6, find an equation of a plane that passes through a given point P and is perpendicular to a given vector **n**.

1. $P(0, 0, 0)$, $\mathbf{n} = \langle 2, 5, 7\rangle$
2. $P(-1, 3, 4)$, $\mathbf{n} = \langle 1, 0, 2\rangle$
3. $P(-5, 6, 2)$, $\mathbf{n} = \langle -3, -1, 7\rangle$
4. $P(0, 3, 0)$, $\mathbf{n} = 4\mathbf{i} + 5\mathbf{j} + 8\mathbf{k}$
5. $P(0, 0, 5)$, $\mathbf{n} = 7\mathbf{i} - \mathbf{j} - 2\mathbf{k}$
6. $P(-3, -6, 8)$, $\mathbf{n} = 2\mathbf{j} - 7\mathbf{k}$

In Exercises 7 through 10, sketch the plane and give a unit vector normal to it.

7. $x + y + 2z = 4$
8. $3y - 7 = 0$
9. $4x - y = 0$
10. $6x - 3z - 2y - 40 = 0$

In Exercises 11 through 14, find an equation of the plane which passes through the three given points.

11. $(1, 0, 0), (0, 1, 0), (0, 0, 2)$
12. $(8, 0, 0), (1, 4, 2), (0, 3, 7)$
13. $(2, -5, 1), (4, -1, -3), (3, -2, -4)$
14. $(2a, 0, -c), (0, -b, 2c), (3a, -b, -c)$, where $abc \neq 0$

In Exercises 15 through 20, find an equation of the plane.

15. The plane through $(3, -1, 4)$ and parallel to the plane $x - 2y - 3z = 8$.
16. The plane that is the perpendicular bisector of the line segment joining $(-3, -1, 2)$ and $(5, 7, 4)$.
17. The plane that contains the point $(7, -2, 5)$ and is parallel to the zx-plane.
18. The plane through the origin and perpendicular to the vector $\langle 2, -6, 5\rangle$.
19. The plane at distance 12 from the origin and perpendicular to the vector $\langle 2, 2, -1\rangle$.

20. The plane perpendicular to each of the planes $x - 5y - 2z = 1$ and $2x + 2y - z = 5$ and containing the point $(3, 2, -1)$.

21. Find the distance between the parallel planes

$$4x + 2y + 5z = 10 \quad \text{and} \quad 4x + 2y + 5z = 25$$

[*Hint:* Find a convenient point on one of the planes and use the distance formula (16).]

22. Prove that the distance between the two parallel planes $Ax + By + Cz = D_1$ and $Ax + By + Cz = D_2$ is given by

$$\frac{|D_1 - D_2|}{\sqrt{A^2 + B^2 + C^2}}$$

23. Find the cosine of the angle between the planes $7x - 2y + 3z - 10 = 0$ and $3x + 5y - 6z - 3 = 0$.

24. Find the angle between the planes $z = 0$ and $2x + 2y + z = 100$.

25. Find an equation of the plane containing the line

$$\frac{x - 4}{-4} = \frac{y - 1}{-2} = \frac{z}{5}$$

and the point $(3, 1, 2)$.

26. Find all the solutions of the simultaneous linear equations

$$\begin{cases} x + y + z = 5 \\ 3x - y - 2z = -1 \\ 2x + 5y + 3z = 11 \end{cases}$$

Interpret your result geometrically.

27. Find all the solutions of the simultaneous linear equations

$$\begin{cases} x + 2y - z = 4 \\ 2x + y + 2z = 3 \\ x + 5y - 5z = 9 \end{cases}$$

Interpret your result geometrically.

28. Show that the plane that passes through the three points $(a, 0, 0)$, $(0, b, 0)$ and $(0, 0, c)$ where $abc \neq 0$ has an equation $\frac{x}{a} + \frac{y}{b} + \frac{z}{c} = 1$. This is known as the *intercept* form of an equation of a plane because a, b, and c are the x-, y- and z-intercepts of the plane, respectively.

29. Prove that the plane $Ax + By + Cz = D$ passes through the origin if and only if $D = 0$.

30. Find the volume of the tetrahedron bounded by the three coordinate planes and the portion of the plane $4x + 9y + 6z = 36$ in the first octant.

16.6   CROSS PRODUCT OF TWO VECTORS

In order to motivate the definition of the cross product, we consider two vectors $\mathbf{A} = a_1\mathbf{i} + a_2\mathbf{j} + a_3\mathbf{k}$ and $\mathbf{B} = b_1\mathbf{i} + b_2\mathbf{j} + b_3\mathbf{k}$. We seek a third vector, say, $\mathbf{u} = x\mathbf{i} + y\mathbf{j} + z\mathbf{k}$ that is perpendicular to both $\mathbf{A}$ and $\mathbf{B}$. Hence

$$\mathbf{u} \cdot \mathbf{A} = 0 \quad \text{and} \quad \mathbf{u} \cdot \mathbf{B} = 0$$

must hold simultaneously. In terms of components, the quantities x, y, and z must satisfy

$$a_1 x + a_2 y + a_3 z = 0 \qquad b_1 x + b_2 y + b_3 z = 0 \tag{1}$$

If it is assumed that the vectors $\mathbf{A}$ and $\mathbf{B}$ are not parallel, then, for example, if $a_1 b_2 \neq a_2 b_1$ we can solve for x and y in terms of z to obtain

$$x = \frac{a_2 b_3 - a_3 b_2}{a_1 b_2 - a_2 b_1} z \qquad y = \frac{a_3 b_1 - a_1 b_3}{a_1 b_2 - a_2 b_1} z \tag{2}$$

Thus any vector of the form

$$s[(a_2 b_3 - a_3 b_2)\mathbf{i} + (a_3 b_1 - a_1 b_3)\mathbf{j} + (a_1 b_2 - a_2 b_1)\mathbf{k}] \tag{3}$$

where $s \neq 0$, is an arbitrary scalar, is orthogonal to the vectors $\mathbf{A}$ and $\mathbf{B}$. It is left to the reader to verify this result.

Definition   The **cross product** of two vectors $\mathbf{A} = \langle a_1, a_2, a_3 \rangle$ and $\mathbf{B} = \langle b_1, b_2, b_3 \rangle$ denoted by $\mathbf{A} \times \mathbf{B}$ is given by

$$\mathbf{A} \times \mathbf{B} = \langle a_2 b_3 - a_3 b_2, a_3 b_1 - a_1 b_3, a_1 b_2 - a_2 b_1 \rangle \tag{4}$$

The cross product $\mathbf{A} \times \mathbf{B}$ is a vector and therefore it is often called the **vector product.**

EXAMPLE 1   Given $\mathbf{A} = \langle 3, 5, -1 \rangle$ and $\mathbf{B} = \langle 2, -4, 7 \rangle$, find $\mathbf{A} \times \mathbf{B}$.

Solution   From the Definition (4),

$$
\begin{aligned}
\mathbf{A} \times \mathbf{B} &= \langle 3, 5, -1 \rangle \times \langle 2, -4, 7 \rangle \\
&= \langle (5)(7) - (-1)(-4), (-1)(2) - (3)(7), (3)(-4) - (5)(2) \rangle \\
&= \langle 31, -23, -22 \rangle
\end{aligned}
$$ ●

EXAMPLE 2   Using the data of Example 1, show that $\mathbf{A} \times \mathbf{B}$ is perpendicular to $\mathbf{A}$ and $\mathbf{B}$.

Solution   We must verify that $\mathbf{A} \times \mathbf{B}$ is perpendicular to $\mathbf{A}$ and to $\mathbf{B}$.

$$
\begin{aligned}
(\mathbf{A} \times \mathbf{B}) \cdot \mathbf{A} &= \langle 31, -23, -22 \rangle \cdot \langle 3, 5, -1 \rangle \\
&= (31)(3) + (-23)(5) + (-22)(-1) \\
&= 93 - 115 + 22 = 0
\end{aligned}
$$

$$
\begin{aligned}
(\mathbf{A} \times \mathbf{B}) \cdot \mathbf{B} &= \langle 31, -23, -22 \rangle \cdot \langle 2, -4, 7 \rangle \\
&= (31)(2) + (-23)(-4) + (-22)(7) \\
&= 62 + 92 - 154 = 0
\end{aligned}
$$

and the required result is established. The satisfaction of this property also serves as an excellent check on the determination of $\mathbf{A} \times \mathbf{B}$.   ●

Equation (4) is not easy to memorize, but fortunately the formula for $\mathbf{A} \times \mathbf{B}$ can be expressed in terms of a determinant of third order, namely

$$
\begin{vmatrix} \mathbf{i} & \mathbf{j} & \mathbf{k} \\ a_1 & a_2 & a_3 \\ b_1 & b_2 & b_3 \end{vmatrix} = \begin{vmatrix} a_2 & a_3 \\ b_2 & b_3 \end{vmatrix} \mathbf{i} - \begin{vmatrix} a_1 & a_3 \\ b_1 & b_3 \end{vmatrix} \mathbf{j} + \begin{vmatrix} a_1 & a_2 \\ b_1 & b_2 \end{vmatrix} \mathbf{k} \tag{5}
$$

where, in general,

$$\begin{vmatrix} a & b \\ c & d \end{vmatrix} = ad - bc$$

From this observation we have the following theorem.

Theorem 1 If $\mathbf{A} = a_1\mathbf{i} + a_2\mathbf{j} + a_3\mathbf{k}$ and $\mathbf{B} = b_1\mathbf{i} + b_2\mathbf{j} + b_3\mathbf{k}$, then

$$\mathbf{A} \times \mathbf{B} = \begin{vmatrix} \mathbf{i} & \mathbf{j} & \mathbf{k} \\ a_1 & a_2 & a_3 \\ b_1 & b_2 & b_3 \end{vmatrix} \tag{6}$$

EXAMPLE 3 Apply (6) to find $\mathbf{A} \times \mathbf{B}$ for the vectors $\mathbf{A}$ and $\mathbf{B}$ of Example 1.

Solution
$$\mathbf{A} \times \mathbf{B} = \begin{vmatrix} \mathbf{i} & \mathbf{j} & \mathbf{k} \\ 3 & 5 & -1 \\ 2 & -4 & 7 \end{vmatrix} = \begin{vmatrix} 5 & -1 \\ -4 & 7 \end{vmatrix}\mathbf{i} - \begin{vmatrix} 3 & -1 \\ 2 & 7 \end{vmatrix}\mathbf{j} + \begin{vmatrix} 3 & 5 \\ 2 & -4 \end{vmatrix}\mathbf{k}$$

$$= [(5)(7) - (-1)(-4)]\mathbf{i} - [(3)(7) - (-1)(2)]\mathbf{j} + [(3)(-4) - (5)(2)]\mathbf{k}$$

$$= 31\mathbf{i} - 23\mathbf{j} - 22\mathbf{k} \qquad \bullet$$

Theorem 2 For any two vectors $\mathbf{A}$ and $\mathbf{B}$ in three-dimensional space

$$\mathbf{A} \times \mathbf{B} = -(\mathbf{B} \times \mathbf{A}) \tag{7}$$

In other words, the cross product is not commutative, but is rather anti-commutative.

Proof By Definition (4)

$$\mathbf{A} \times \mathbf{B} = \langle a_2 b_3 - a_3 b_2, a_3 b_1 - a_1 b_3, a_1 b_2 - a_2 b_1 \rangle$$

while formal interchanges $a_i \to b_i$ and $b_i \to a_i$ $(i = 1, 2, 3)$ yield

$$\mathbf{B} \times \mathbf{A} = \langle b_2 a_3 - b_3 a_2, b_3 a_1 - b_1 a_3, b_1 a_2 - b_2 a_1 \rangle$$

from which (7) follows. ∎

Corollary For any vector $\mathbf{A}$, $\mathbf{A} \times \mathbf{A} = 0$

Proof Replace $\mathbf{B}$ by $\mathbf{A}$ in (7), so that

$$\mathbf{A} \times \mathbf{A} = -(\mathbf{A} \times \mathbf{A})$$

or

$$\mathbf{A} \times \mathbf{A} = 0 \qquad ∎$$

Theorem 3 For any scalar s and any pair of vectors $\mathbf{A}$ and $\mathbf{B}$

$$s(\mathbf{A} \times \mathbf{B}) = (s\mathbf{A}) \times \mathbf{B} = \mathbf{A} \times (s\mathbf{B}) \tag{8}$$

The proof of Theorem 3 is left for the reader (Exercise 15).

Theorem 4 **(The Distributive Law).**
For any three vectors **A**, **B** and **C** in three-dimensional space,

$$\mathbf{A} \times (\mathbf{B} + \mathbf{C}) = \mathbf{A} \times \mathbf{B} + \mathbf{A} \times \mathbf{C} \tag{9}$$

Proof Let $\mathbf{A} = a_1\mathbf{i} + a_2\mathbf{j} + a_3\mathbf{k}$, $\mathbf{B} = b_1\mathbf{i} + b_2\mathbf{j} + b_3\mathbf{k}$, and $\mathbf{C} = c_1\mathbf{i} + c_2\mathbf{j} + c_3\mathbf{k}$. Then since

$$\mathbf{B} + \mathbf{C} = (b_1 + c_1)\mathbf{i} + (b_2 + c_2)\mathbf{j} + (b_3 + c_3)\mathbf{k}$$

and using (4), we have

$$\mathbf{A} \times (\mathbf{B} + \mathbf{C}) = \langle a_2(b_3 + c_3) - a_3(b_2 + c_2), a_3(b_1 + c_1) - a_1(b_3 + c_3),$$
$$a_1(b_2 + c_2) - a_2(b_1 + c_1)\rangle$$
$$= \langle a_2b_3 - a_3b_2 + a_2c_3 - a_3c_2, a_3b_1 - a_1b_3 + a_3c_1 - a_1c_3,$$
$$a_1b_2 - a_2b_1 + a_1c_2 - a_2c_1\rangle$$
$$= \langle a_2b_3 - a_3b_2, a_3b_1 - a_1b_3, a_1b_2 - a_2b_1\rangle + \langle a_2c_3 - a_3c_2,$$
$$a_3c_1 - a_1c_3, a_1c_2 - a_2c_1\rangle$$
$$= \mathbf{A} \times \mathbf{B} + \mathbf{A} \times \mathbf{C} \qquad \blacksquare$$

If the definition (4) is applied, in particular, to the cross product of the unit coordinate vectors **i**, **j** and **k**, we obtain an interesting array

$$
\begin{array}{lll}
\mathbf{i} \times \mathbf{i} = \mathbf{0} & \mathbf{i} \times \mathbf{j} = \mathbf{k} & \mathbf{i} \times \mathbf{k} = -\mathbf{j} \\
\mathbf{j} \times \mathbf{i} = -\mathbf{k} & \mathbf{j} \times \mathbf{j} = \mathbf{0} & \mathbf{j} \times \mathbf{k} = \mathbf{i} \\
\mathbf{k} \times \mathbf{i} = \mathbf{j} & \mathbf{k} \times \mathbf{j} = -\mathbf{i} & \mathbf{k} \times \mathbf{k} = \mathbf{0}
\end{array}
\tag{10}
$$

To illustrate, we verify that $\mathbf{j} \times \mathbf{k} = \mathbf{i}$

$$\mathbf{j} \times \mathbf{k} = \begin{vmatrix} \mathbf{i} & \mathbf{j} & \mathbf{k} \\ 0 & 1 & 0 \\ 0 & 0 & 1 \end{vmatrix} = (1 - 0)\mathbf{i} + (0 - 0)\mathbf{j} + (0 - 0)\mathbf{k} = i$$

Formulas (10) are easily remembered. The cross product of each of the unit vectors with itself is the zero vector. The cross product of two consecutive unit coordinate vectors, taken in the order of the arrows indicated in Figure 16.6.1

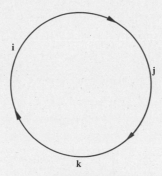

Figure 16.6.1

(that is, clockwise), is equal to the third unit coordinate vector. The cross product of two consecutive unit coordinate vectors taken in the counterclockwise direction is equal to the negative of the third unit coordinate vector.

EXAMPLE 4   Simplify $(\mathbf{i} \times \mathbf{j}) \times (\mathbf{j} + \mathbf{k})$

Solution $\mathbf{i} \times \mathbf{j} = \mathbf{k}$ and from the distributive law (9),

$$(\mathbf{i} \times \mathbf{j}) \times (\mathbf{j} + \mathbf{k}) = \mathbf{k} \times (\mathbf{j} + \mathbf{k}) = \mathbf{k} \times \mathbf{j} + \mathbf{k} \times \mathbf{k}$$
$$= -\mathbf{i} + 0 = -\mathbf{i}$$
●

EXAMPLE 5   Show that $(\mathbf{A} + \mathbf{B}) \times \mathbf{C} = \mathbf{A} \times \mathbf{C} + \mathbf{B} \times \mathbf{C}$.

Solution
$$\begin{aligned}(\mathbf{A} + \mathbf{B}) \times \mathbf{C} &= -\mathbf{C} \times (\mathbf{A} + \mathbf{B}) &&\text{from Theorem 2}\\ &= -\mathbf{C} \times \mathbf{A} + -\mathbf{C} \times \mathbf{B} &&\text{from (9)}\\ &= \mathbf{A} \times \mathbf{C} + \mathbf{B} \times \mathbf{C}\end{aligned}$$

Thus the distributive law has been established both ways (on the left and right).
●

Cross products of vectors are *not* associative in general. For example,

$$(\mathbf{i} \times \mathbf{j}) \times \mathbf{j} = \mathbf{k} \times \mathbf{j} = -\mathbf{i}$$

whereas

$$\mathbf{i} \times (\mathbf{j} \times \mathbf{j}) = \mathbf{i} \times 0 = 0$$

Hence $(\mathbf{i} \times \mathbf{j}) \times \mathbf{j} \neq \mathbf{i} \times (\mathbf{j} \times \mathbf{j})$ and for this operation, parentheses are essential.

We know that the cross product of $\mathbf{A}$ and $\mathbf{B}$ is a vector that is perpendicular to each of the vectors $\mathbf{A}$ and $\mathbf{B}$. But what about its magnitude? This question is answered in the next theorem.

Theorem 5   If $\mathbf{A}$ and $\mathbf{B}$ are any two vectors in three-dimensional space, then

$$|\mathbf{A} \times \mathbf{B}|^2 = |\mathbf{A}|^2 \, |\mathbf{B}|^2 - (\mathbf{A} \cdot \mathbf{B})^2 \tag{11}$$

Proof   We demonstrate this by expanding both sides of (11) and showing that they are equal.

Let $\mathbf{A} = \langle a_1, a_2, a_3 \rangle$ and $\mathbf{B} = \langle b_1, b_2, b_3 \rangle$ so that (4) yields

$$\begin{aligned}|\mathbf{A} \times \mathbf{B}|^2 &= (a_2 b_3 - a_3 b_2)^2 + (a_3 b_1 - a_1 b_3)^2 + (a_1 b_2 - a_2 b_1)^2\\ &= a_2{}^2 b_3{}^2 - 2a_2 a_3 b_2 b_3 + a_3{}^2 b_2{}^2 + a_3{}^2 b_1{}^2 - 2a_1 a_3 b_1 b_3\\ &\qquad + a_1{}^2 b_3{}^2 + a_1{}^2 b_2{}^2 - 2a_1 a_2 b_1 b_2 + a_2{}^2 b_1{}^2 \end{aligned} \tag{i}$$

and for the right side of (11), we obtain

$$\begin{aligned}|\mathbf{A}|^2 \, |\mathbf{B}|^2 - (\mathbf{A} \cdot \mathbf{B})^2 &= (a_1{}^2 + a_2{}^2 + a_3{}^2)(b_1{}^2 + b_2{}^2 + b_3{}^2)\\ &\qquad - (a_1 b_1 + a_2 b_2 + a_3 b_3)^2\\ &= a_1{}^2 b_2{}^2 + a_1{}^2 b_3{}^2 + a_2{}^2 b_1{}^2 + a_2{}^2 b_3{}^2 + a_3{}^2 b_1{}^2 + a_3{}^2 b_2{}^2\\ &\qquad - 2a_1 a_2 b_1 b_2 - 2a_2 a_3 b_2 b_3 - 2a_1 a_3 b_1 b_3 \end{aligned} \tag{ii}$$

and comparison of (i) and (ii) yields (11):

$$|\mathbf{A} \times \mathbf{B}|^2 = |\mathbf{A}|^2 \, |\mathbf{B}|^2 - (\mathbf{A} \cdot \mathbf{B})^2$$
∎

Equation (11) leads to an interesting geometric interpretation of the cross product.

Theorem 6   If **A** and **B** are two vectors in three-dimensional space and θ is the angle between the two vectors, then

$$|\mathbf{A} \times \mathbf{B}| = |\mathbf{A}|\,|\mathbf{B}| \sin \theta \qquad (12)$$

Proof   From (11), we have

$$|\mathbf{A} \times \mathbf{B}|^2 = |\mathbf{A}|^2\,|\mathbf{B}|^2 - (\mathbf{A} \cdot \mathbf{B})^2$$

But we have proved that $\mathbf{A} \cdot \mathbf{B} = |\mathbf{A}|\,|\mathbf{B}| \cos \theta$, so that

$$\begin{aligned}
|\mathbf{A} \times \mathbf{B}|^2 &= |\mathbf{A}|^2\,|\mathbf{B}|^2 - (|\mathbf{A}|\,|\mathbf{B}| \cos \theta)^2 \\
&= |\mathbf{A}|^2\,|\mathbf{B}|^2 - |\mathbf{A}|^2\,|\mathbf{B}|^2 \cos^2 \theta \\
&= |\mathbf{A}|^2\,|\mathbf{B}|^2(1 - \cos^2 \theta) \\
&= |\mathbf{A}|^2\,|\mathbf{B}|^2 \sin^2 \theta
\end{aligned}$$

Since $0 \le \theta \le \pi$, $\sin \theta \ge 0$ must hold, and we obtain (12) (by taking square roots of both sides)

$$|\mathbf{A} \times \mathbf{B}| = |\mathbf{A}|\,|\mathbf{B}| \sin \theta \qquad \blacksquare$$

Consider Figure 16.6.2 which shows a parallelogram that is determined by vectors **A** and **B**, with θ as the angle of inclusion. The quantity $|\mathbf{A}|$ is the base length and the altitude to this base is $|\mathbf{B}| \sin \theta$. Hence $|\mathbf{A}|\,|\mathbf{B}| \sin \theta$ is the area of the parallelogram, and, from (12), we can interpret $|\mathbf{A} \times \mathbf{B}|$ as the area of the parallelogram determined by **A** and **B**. The vector $\mathbf{A} \times \mathbf{B}$ is perpendicular to the plane determined by **A** and **B**, and it can be shown that the vectors **A**, **B**, and $\mathbf{A} \times \mathbf{B}$ form a right-handed system in that order.

From (12), we easily have the following result

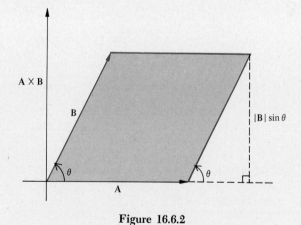

Figure 16.6.2

Theorem 7   Two nonzero vectors **A** and **B** are parallel if and only if $\mathbf{A} \times \mathbf{B} = 0$.

The proof of Theorem 7 is left to reader. (See Exercise 19.)   ∎

Consider next a special product of three vectors. Since $\mathbf{B} \times \mathbf{C}$ is a vector, it is meaningful to form the dot product of it with another vector $\mathbf{A}$. Hence we obtain $\mathbf{A} \cdot (\mathbf{B} \times \mathbf{C})$. Furthermore, it is noted that the parentheses are not necessary because $\mathbf{A} \cdot \mathbf{B}$ is a scalar and the product $(\mathbf{A} \cdot \mathbf{B}) \times \mathbf{C}$ would be meaningless. This implies that $\mathbf{A} \cdot \mathbf{B} \times \mathbf{C}$ can be interpreted in one and only one way.

Definition   The product $\mathbf{A} \cdot \mathbf{B} \times \mathbf{C}$ is a scalar called the ***triple scalar product.***

Theorem 8   If $\mathbf{A}$, $\mathbf{B}$, and $\mathbf{C}$ are vectors in three dimensional space, then

$$\mathbf{A} \cdot \mathbf{B} \times \mathbf{C} = \mathbf{A} \times \mathbf{B} \cdot \mathbf{C} \tag{13}$$

Proof   If $\mathbf{A} = a_1\mathbf{i} + a_2\mathbf{j} + a_3\mathbf{k}$, $\mathbf{B} = b_1\mathbf{i} + b_2\mathbf{j} + b_3\mathbf{k}$ and $\mathbf{C} = c_1\mathbf{i} + c_2\mathbf{j} + c_3\mathbf{k}$, then

$$\mathbf{B} \times \mathbf{C} = \begin{vmatrix} \mathbf{i} & \mathbf{j} & \mathbf{k} \\ b_1 & b_2 & b_3 \\ c_1 & c_2 & c_3 \end{vmatrix}$$

$$= \begin{vmatrix} b_2 & b_3 \\ c_2 & c_3 \end{vmatrix} \mathbf{i} - \begin{vmatrix} b_1 & b_3 \\ c_1 & c_3 \end{vmatrix} \mathbf{j} + \begin{vmatrix} b_1 & b_2 \\ c_1 & c_2 \end{vmatrix} \mathbf{k}$$

Thus

$$\mathbf{A} \cdot \mathbf{B} \times \mathbf{C} = a_1 \begin{vmatrix} b_2 & b_3 \\ c_2 & c_3 \end{vmatrix} - a_2 \begin{vmatrix} b_1 & b_3 \\ c_1 & c_3 \end{vmatrix} + a_3 \begin{vmatrix} b_1 & b_2 \\ c_1 & c_2 \end{vmatrix}$$

which yields upon expansion,

$$\mathbf{A} \cdot \mathbf{B} \times \mathbf{C} = a_1(b_2c_3 - b_3c_2) - a_2(b_1c_3 - b_3c_1) + a_3(b_1c_2 - b_2c_1) \tag{i}$$

For the right side of (13), if we write out the expansion for $\mathbf{A} \times \mathbf{B}$ and then for $\mathbf{A} \times \mathbf{B} \cdot \mathbf{C}$, we obtain an identical expansion with (i). These details are left to the reader.   ∎

Note that right side of (i) is the expansion of the third-order determinant

$$\begin{vmatrix} a_1 & a_2 & a_3 \\ b_1 & b_2 & b_3 \\ c_1 & c_2 & c_3 \end{vmatrix} \tag{ii}$$

where as the right side of (13) is the expansion of the third-order determinant

$$\begin{vmatrix} c_1 & c_2 & c_3 \\ a_1 & a_2 & a_3 \\ b_1 & b_2 & b_3 \end{vmatrix} \tag{iii}$$

Determinant (iii) equals determinant (ii) because it may be obtained from (ii) by two successive interchanges of two rows, each interchange multiplying the value of the determinant by -1. The reader interested in determinants may wish to verify this remark.

Definition   Vectors $\mathbf{A}$, $\mathbf{B}$, and $\mathbf{C}$ form a ***right-handed system*** of vectors (noncoplanar) if the angle α between $\mathbf{B} \times \mathbf{C}$ and $\mathbf{A}$ satisfies the inequalities $0 \leq \alpha < \dfrac{\pi}{2}$.

If $\frac{\pi}{2} < \alpha \le \pi$ then the vectors **A, B,** and **C** form a *left-handed system.*

Theorem 9   Let **A, B,** and **C** be a right-handed system of vectors. Then $\mathbf{A} \cdot \mathbf{B} \times \mathbf{C}$ is the volume of the parallelepiped that has **A, B,** and **C** as coterminal edges, as shown in Figure 16.6.3.

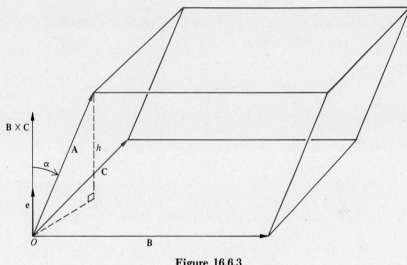

Figure 16.6.3

Proof   We choose as a base the parallelogram that has coterminal edges **B** and **C**. The area of this base is $|\mathbf{B} \times \mathbf{C}|$ and $\mathbf{B} \times \mathbf{C} = |\mathbf{B} \times \mathbf{C}|\mathbf{e}$ where **e** is the unit vector shown in the figure. The height h of the parallelepiped is given by

$$h = |\mathbf{A}| \cos \alpha = \mathbf{A} \cdot \mathbf{e}$$

since $|\mathbf{e}| = 1$ and α is the angle between $\mathbf{B} \times \mathbf{C}$ and **A**. Now, by definition of the dot product,

$$\mathbf{A} \cdot (\mathbf{B} \times \mathbf{C}) = |\mathbf{A}|(|\mathbf{B} \times \mathbf{C}| \cos \alpha) = |\mathbf{A}|(\cos \alpha)|\mathbf{B} \times \mathbf{C}|$$
$$= h \text{ (area of the base)} = V$$

If **A**, **B**, and **C** form a left handed system, then $\mathbf{A} \cdot \mathbf{B} \times \mathbf{C} = -V$ (since $\frac{\pi}{2} < \alpha \le \pi$ and $\mathbf{A} \cdot \mathbf{B} \times \mathbf{C}$ is negative). If **A**, **B**, and **C** are coplanar then $\mathbf{A} \cdot (\mathbf{B} \times \mathbf{C}) = V = 0$, and the parallelpiped has collapsed ($h = 0$).   ■

EXAMPLE 6   Find the volume V of the parallelepiped if three of its edges are $\mathbf{A} = 2\mathbf{i} - \mathbf{j} - \mathbf{k}$, $\mathbf{B} = \mathbf{i} + 2\mathbf{j} + \mathbf{k}$, and $\mathbf{C} = \mathbf{i} - 3\mathbf{j} + 2\mathbf{k}$.

Solution   From Theorem 9, $|\mathbf{A} \cdot \mathbf{B} \times \mathbf{C}| = V$

$$\mathbf{B} \times \mathbf{C} = (\mathbf{i} + 2\mathbf{j} + \mathbf{k}) \times (\mathbf{i} - 3\mathbf{j} + 2\mathbf{k})$$
$$= 7\mathbf{i} - \mathbf{j} - 5\mathbf{k}$$

from which

$$\mathbf{A} \cdot (\mathbf{B} \times \mathbf{C}) = (2)(7) + (-1)(-1) + (-1)(-5) = 20$$

Hence $V = 20$ cubic units.   ●

EXAMPLE 7 Find an equation in x, y, and z for the plane that contains the points $P_1(3, 2, -3)$, $P_2(0, -2, 7)$ and $P_3(4, 1, 3)$.

Solution Let $P(x, y, z)$ be any point in the plane. Then since vectors $\overrightarrow{P_1P}$, $\overrightarrow{P_1P_2}$, and $\overrightarrow{P_1P_3}$ are coplanar, we must have

$$\overrightarrow{P_1P} \cdot \overrightarrow{P_1P_2} \times \overrightarrow{P_1P_3} = 0$$

Now, by subtraction of corresponding coordinates

$$\overrightarrow{P_1P} = \langle x - 3, y - 2, z + 3 \rangle$$
$$\overrightarrow{P_1P_2} = \langle -3, -4, 10 \rangle$$
$$\overrightarrow{P_1P_3} = \langle 1, -1, 6 \rangle$$

and therefore

$$\overrightarrow{P_1P_2} \times \overrightarrow{P_1P_3} = (-3\mathbf{i} - 4\mathbf{j} + 10\mathbf{k}) \times (\mathbf{i} - \mathbf{j} + 6\mathbf{k})$$
$$= -14\mathbf{i} + 28\mathbf{j} + 7\mathbf{k}$$

Hence

$$\overrightarrow{P_1P} \cdot \overrightarrow{P_1P_2} \times \overrightarrow{P_1P_3} = -14(x - 3) + 28(y - 2) + 7(z + 3)$$
$$= -14x + 28y + 7z + 7 = 0$$

or, equivalently, $2x - 4y - z = 1$.   ●

EXAMPLE 8 Find the distance between the two skew lines L_1 and L_2, where L_1 passes through the points $P_1(1, 2, 4)$ and $Q_1(2, 5, 6)$ and L_2 goes through $P_2(3, 5, 7)$ and $Q_2(6, 10, 11)$.

Solution Let $\overrightarrow{P_1Q_1}$ and $\overrightarrow{P_2Q_2}$ denote the vectors joining P_1 and Q_1, and P_2 and Q_2, respectively (see Figure 16.6.4). The vector $\mathbf{N} = \overrightarrow{P_1Q_1} \times \overrightarrow{P_2Q_2}$ is a vector perpendicular to each of the vectors $\overrightarrow{P_1Q_1}$ and $\overrightarrow{P_2Q_2}$ and therefore to the lines L_1 and L_2, respectively. There exist parallel planes Π_1 and Π_2 containing lines L_1 and L_2, which are perpendicular to the vector $\mathbf{N}$ (the vector $\mathbf{N}$ is perpendicular to Π_1 and Π_2 and therefore to L_1 and L_2). If we choose any two points, one in each plane such as P_1 and Q_2 of Figure 16.6.4, then the scalar projection of $\overrightarrow{P_1Q_2}$ onto $\mathbf{N}$ is the distance s between Π_1 and Π_2. But this is also the required distance between the two skew lines. Hence since $\dfrac{\mathbf{N}}{|\mathbf{N}|}$ is a unit vector in the direction of $\mathbf{N}$, we have

$$s = \left| \overrightarrow{P_1Q_2} \cdot \frac{\mathbf{N}}{|\mathbf{N}|} \right| = \left| \frac{\overrightarrow{P_1Q_2} \cdot \overrightarrow{P_1Q_1} \times \overrightarrow{P_2Q_2}}{|\overrightarrow{P_1Q_1} \times \overrightarrow{P_2Q_2}|} \right| \qquad (14)$$

For the particular case,

$$\overrightarrow{P_1Q_1} = \mathbf{i} + 3\mathbf{j} + 2\mathbf{k} \qquad \overrightarrow{P_2Q_2} = 3\mathbf{i} + 5\mathbf{j} + 4\mathbf{k}$$

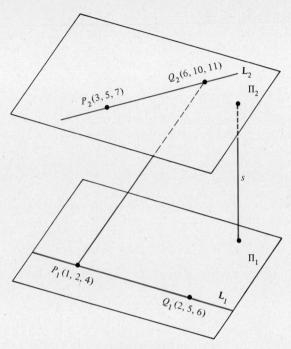

Figure 16.6.4

and

$$\overrightarrow{P_1Q_1} \times \overrightarrow{P_2Q_2} = 2\mathbf{i} + 2\mathbf{j} - 4\mathbf{k}$$

$$|\overrightarrow{P_1Q_1} \times \overrightarrow{P_2Q_2}| = \sqrt{2^2 + 2^2 + (-4)^2} = \sqrt{24}$$

$$\overrightarrow{P_1Q_2} = 5\mathbf{i} + 8\mathbf{j} + 7\mathbf{k}$$

and therefore from (14),

$$s = \left| \frac{(5)(2) + (8)(2) + (7)(-4)}{\sqrt{24}} \right| = \frac{2}{\sqrt{24}} = \frac{\sqrt{6}}{6}$$

Note that in formula (14), $\overrightarrow{P_1Q_2}$ could be replaced by $\overrightarrow{P_1P_2}$, $\overrightarrow{Q_1P_2}$, or $\overrightarrow{Q_1Q_2}$ and the same result would be obtained. It is left to the reader to verify this. ●

Exercises 16.6

In Exercises 1 through 10, if $\mathbf{A} = 2\mathbf{i} - \mathbf{j} + 3\mathbf{k}$, $\mathbf{B} = \mathbf{i} + 4\mathbf{k}$, *and* $\mathbf{C} = 2\mathbf{i} + 3\mathbf{j} - 5\mathbf{k}$, *calculate the given quantity.*

1. $\mathbf{A} \times \mathbf{B}$
2. $\mathbf{A} \times \mathbf{C}$
3. $\mathbf{A} \times (\mathbf{B} + \mathbf{C})$
4. $\mathbf{B} \times (\mathbf{C} - \mathbf{B})$
5. $(\mathbf{B} \times \mathbf{C}) \times \mathbf{A}$
6. $\mathbf{B} \times (\mathbf{C} \times \mathbf{A})$
7. $\mathbf{C} \cdot (\mathbf{B} \times \mathbf{A})$
8. $(\mathbf{C} \times \mathbf{B}) \cdot \mathbf{A}$
9. $(\mathbf{C} \times \mathbf{A}) \cdot (\mathbf{C} \times \mathbf{B})$
10. $(2\mathbf{B} \times 3\mathbf{C}) \cdot (\mathbf{B} + \mathbf{C} - \mathbf{A})$
11. Show that (a) $\mathbf{k} \times \mathbf{i} = \mathbf{j}$ and (b) $\mathbf{k} \times (\mathbf{j} - \mathbf{k}) = -\mathbf{i}$.
12. Simplify: $(\mathbf{i} \times \mathbf{j}) \times [(\mathbf{j} \times \mathbf{k}) \times (\mathbf{k} \times \mathbf{i})]$.
13. Find a vector of length 5 that is perpendicular to both of the vectors $2\mathbf{i} - 3\mathbf{j} - \mathbf{k}$ and $\mathbf{i} + 3\mathbf{j} + 4\mathbf{k}$.
14. Find all vectors that are perpendicular to both of the vectors $3\mathbf{i} + 4\mathbf{j} + 5\mathbf{k}$ and $2\mathbf{i} + \mathbf{j} + 2\mathbf{k}$.

15. Prove Theorem 3 of the text.

16. Find the sine of the angle θ between the unit vectors $\langle \frac{2}{3}, -\frac{1}{3}, \frac{2}{3} \rangle$ and $\langle \frac{4}{9}, \frac{7}{9}, -\frac{4}{9} \rangle$ by two procedures.
    (a) Use the dot product.
    (b) Use the cross product.

17. Find the sine of the angle θ between the unit vectors $\mathbf{u} = \langle u_1, u_2, u_3 \rangle$ and $\mathbf{v} = \langle v_1, v_2, v_3 \rangle$ by the two methods of Exercise 16.

18. Express $(\mathbf{A} + \mathbf{B}) \times (\mathbf{A} - 3\mathbf{B})$ as a scalar multiple of $\mathbf{A} \times \mathbf{B}$.

19. Prove that $\mathbf{A} \times \mathbf{B} = 0$ if and only if $\mathbf{A}$ and $\mathbf{B}$ are parallel, where it is assumed that $\mathbf{A} \neq 0$ and $\mathbf{B} \neq 0$.

20. If $\mathbf{A} \times \mathbf{C} = \mathbf{A} \times \mathbf{B}$ then does it necessarily follow that $\mathbf{C} = \mathbf{B}$? If not, what may be concluded?

In Exercises 21 through 24, by application of the triple scalar product, find the volume V of the parallelepiped whose coterminal edges are $\mathbf{A}$, $\mathbf{B}$, and $\mathbf{C}$.

21. $\mathbf{A} = 2\mathbf{i} - \mathbf{k}$, $\mathbf{B} = \mathbf{i} + 3\mathbf{j}$, and $\mathbf{C} = \mathbf{i} - \mathbf{j} + \mathbf{k}$
22. $\mathbf{A} = \mathbf{i} + \mathbf{j} + \mathbf{k}$, $\mathbf{B} = \mathbf{i} - \mathbf{j} - \mathbf{k}$, and $\mathbf{C} = 2\mathbf{j} + \mathbf{k}$
23. $\mathbf{A} = \mathbf{i} - 2\mathbf{j} - 3\mathbf{k}$, $\mathbf{B} = 3\mathbf{i} + \mathbf{j} + 5\mathbf{k}$, and $\mathbf{C} = \mathbf{i} + 5\mathbf{j} + 10\mathbf{k}$
24. $\mathbf{A} = 3\mathbf{i} - 5\mathbf{j} - 7\mathbf{k}$, $\mathbf{B} = 8\mathbf{i} + 2\mathbf{j} + 6\mathbf{k}$, and $\mathbf{C} = 4\mathbf{i} - \mathbf{j} - 5\mathbf{k}$

In Exercises 25 and 26, find, by application of the triple scalar product, an equation in x, y, and z of the plane that contains the three given points.

25. $P_1(1, 1, 0)$, $P_2(2, -1, -3)$, $P_3(5, 3, -7)$
26. $P_1(1, -3, -2)$, $P_2(5, 2, 2)$, $P_3(9, 3, 2)$
27. Use a cross product to find the area of the triangle with vertices at $P(2, 3, 5)$, $Q(3, 1, 4)$ and $R(-1, 2, 6)$.
28. Let $\mathbf{A}$, $\mathbf{B}$, and $\mathbf{C}$ be position vectors of the vertices of a triangle. Show that the area of the triangle is determined by

$$\tfrac{1}{2}|(\mathbf{B} - \mathbf{A}) \times (\mathbf{C} - \mathbf{A})|$$

In Exercises 29 and 30, find the distance between the line through the points P_1 and Q_1 and the line through the points P_2 and Q_2.

29. $P_1(-1, 0, 3)$, $Q_1(2, 1, 4)$, $P_2(0, 1, 2)$, $Q_2(-2, 5, 0)$
30. $P_1(-4, -2, 1)$, $Q_1(-1, 3, 2)$, $P_2(1, 2, 4)$, $Q_2(5, 5, 7)$

In Exercises 31 and 32, find the distance between the two skew lines.

31. $\dfrac{x-3}{2} = \dfrac{y-1}{3} = \dfrac{z-2}{1}$ and $\dfrac{x+4}{1} = \dfrac{y}{2} = \dfrac{z+1}{2}$

32. $\dfrac{x+1}{-1} = \dfrac{y+2}{2} = \dfrac{z-3}{4}$ and $\dfrac{x}{3} = \dfrac{y-1}{1} = \dfrac{z+2}{5}$

33. In triangle ABC show that $\dfrac{\sin A}{a} = \dfrac{\sin B}{b} = \dfrac{\sin C}{c}$. This is the **law of sines**. (*Hint:* Let $\mathbf{a} = \overrightarrow{BC}$, $\mathbf{b} = \overrightarrow{CA}$, and $\mathbf{c} = \overrightarrow{AB}$, show that $\mathbf{a} + \mathbf{b} + \mathbf{c} = 0$ and form $\mathbf{a} \times (\mathbf{a} + \mathbf{b} + \mathbf{c})$, and so on.)

34. Prove that $\mathbf{A} \times (\mathbf{B} \times \mathbf{C}) = (\mathbf{A} \cdot \mathbf{C})\mathbf{B} - (\mathbf{A} \cdot \mathbf{B})\mathbf{C}$. (*Hint:* Write out both sides in component form and compare.)

16.7 CYLINDERS AND SURFACES OF REVOLUTION

The plane is the simplest surface in three-dimensional space. Next to this comes the cylinder, which is actually a more general surface than we usually envisage when the word is mentioned.

Definition   Let C be a plane curve and L a line which cuts the plane at one point. Then the surface made up of all lines parallel to L, each intersecting C at one point is said to be a ***cylinder.*** The curve C is called the ***directrix*** of the cylinder and the set of straight lines are its ***elements*** (***generators*** or ***rulings***).

We may also say that a cylinder is generated by a straight line moving parallel to itself which intersects a specified plane curve.

The type of cylinder we ordinarily think of is a ***right circular cylinder*** because its directrix is a circle while its elements are ***perpendicular*** to the plane of the circle. An example of such a cylinder is shown in Figure 16.7.1 where the directrix is the circle $x^2 + y^2 = 100$ in the xy-plane.

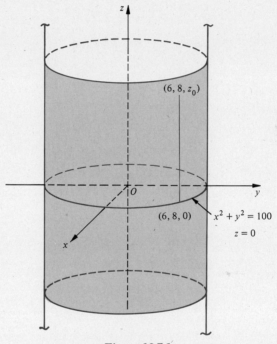

Figure 16.7.1

In order to obtain analytic simplicity, we shall confine our attention to cylinders with directrix in one of the coordinate planes and with generators parallel to a coordinate axis. Equations of such curves are easily recognizable in that a cylinder with elements parallel to a particular coordinate axis must have that coordinate missing. This means, for example, if an equation is of the form $f(x, y) = 0$, then if $(x_0, y_0, 0)$ satisfies the equation and lies on the surface so also does (x_0, y_0, z) for arbitrary z. The surface contains the lines defined by $x = x_0$ and $y = y_0$ which are parallel to the z-axis.

EXAMPLE 1   Discuss and sketch the surface $x^2 + y^2 = 100$.

Solution   The equation $x^2 + y^2 = 100$ is satisfied by all points of the circle of radius 10 about the origin in the xy-plane. This is the directrix. However, since z does not enter into the relation, the equation is satisfied by all points vertically above or below the circle. Also, these are the only points that satisfy the equation, and therefore the surface is a right circular cylinder of radius 10 and whose axis is the z-axis (Figure 16.7.1).　●

EXAMPLE 2　Discuss and sketch $y^2 = 16z$.

Solution　The equation $y^2 = 16z$ is satisfied by the coordinates of the points $(0, y, z)$ of a parabola in the yz-plane. Since x is absent from the equation, the surface is a cylinder with elements parallel to the x-axis. Hence the locus is a parabolic cylinder (Figure 16.7.2).　●

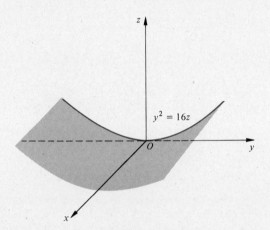

Figure 16.7.2

Other examples of cylinders are

(i)　An *elliptic cylinder* illustrated by $3z^2 + x^2 = 1$, where the elements are parallel to the y-axis and the directrix is the ellipse $3z^2 + x^2 = 1$ in the zx-plane.

(ii)　An *hyperbolic cylinder* as, for example, $4x^2 - 9y^2 = 36$, where the directrix is the hyperbola in the xy-plane and the rulings are parallel to the z-axis.

(iii)　A *plane* exemplified by $y - 2z = 1$ for which the directrix is the straight line $y - 2z = 1$ in the yz-plane and the generators are parallel to the x-axis. This plane is perpendicular to the yz-plane. We note that, in general, a plane is a particular cylinder which has a straight line for a directrix.

The reader is invited to sketch each of the above cylinders referred to the rectangular coordinate system in R_3 showing the directrix and the elements.

We consider next another simple surface that has many applications in applied mathematics and is described as follows:

Definition　A *surface of revolution* is a surface that is formed by rotating a plane curve about a fixed line lying in its plane. The fixed line is called the *axis* of the surface of revolution, and the curve is said to *generate the surface.*

A right circular cylinder is an example of a surface of revolution for which the generating curve is a straight line parallel to the axis of revolution. A sphere can be generated by rotating a semicircle about its diameter. Note that all plane sections of a surface of revolution which are perpendicular to its axis are circles.

Let $y = f(z)$, $x = 0$, for which $f(z) \geq 0$,† be a plane curve in the yz-plane (Figure 16.7.3). We seek an equation of the surface of revolution obtained by rotating this curve about the z-axis.

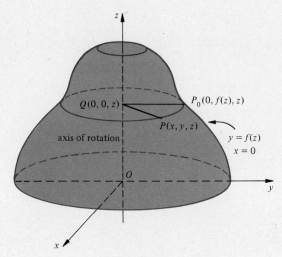

Figure 16.7.3

Suppose that $P(x, y, z)$ is any point on the resulting surface of revolution and consider the cross section made by a plane passing through P and perpendicular to the z-axis. The plane intersects the surface in a circle whose center is at $Q(0, 0, z)$. This plane also intersects the curve $y = f(z)$, $x = 0$ at the point $P_0(0, f(z), z)$. Since P and P_0 are points on a circle, we must have

$$|\overline{QP_0}| = |\overline{QP}|$$

But $|QP| = \sqrt{x^2 + y^2}$ and $|QP_0| = f(z)$ so that we have

$$\sqrt{x^2 + y^2} = f(z)$$

or, by squaring each side

$$x^2 + y^2 = [f(z)]^2 \tag{1}$$

Equation (1) is a surface of revolution.

By cyclical change of variables, it is left to the reader to show analogously that the equation

$$y^2 + z^2 = [g(x)]^2 \tag{2}$$

describes a surface of revolution obtained by rotating $z = g(x)$, $y = 0$, where $g(x) \geq 0$,† about the x-axis.

Similarly, it may be verified that

$$z^2 + x^2 = [h(y)]^2 \tag{3}$$

---

† See the footnote on the next page.

is an equation for a surface of revolution obtained by rotating $x = h(y)$, $z = 0$, where $h(y) \geq 0$,† about the y-axis.

To illustrate, suppose that we have a surface of revolution about the x-axis. This implies that the traces of the surface in planes perpendicular to the x-axis must be circles with centers on the x-axis, and conversely. This in turn means that analytically the equation of the surface is of the form

$$f(x, y^2 + z^2) = 0 ‡ \tag{4}$$

The surface of revolution about the y-axis or about the z-axis is handled analogously.

EXAMPLE 3   The line $z = \dfrac{x}{2}$, $y = 0$ is revolved about the x-axis. Find an equation of the surface thus generated.

Solution   Substitute $g(x) = \dfrac{x}{2}$ into (2) to obtain

$$y^2 + z^2 = \left(\frac{x}{2}\right)^2$$

or, equivalently,

$$4(y^2 + z^2) = x^2 \tag{5}$$

This is the required surface of revolution.

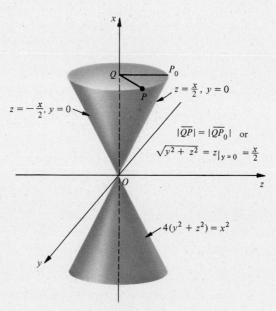

Figure 16.7.4

---

† More generally, $f(z)$, $g(x)$, or $h(y)$ must be nonnegative or nonpositive. If there is a change in sign then, for example, $f(z) = f(-z)$ must be satisfied.
‡ The concept of functions of two or more variables will be treated in detail in Chapter 17. In (4), the two variables may be regarded as x and $y^2 + z^2$.

Note that this same surface would be obtained if we revolve $y = \dfrac{x}{2}, z = 0$ about the x-axis. The reader should verify this observation by setting $z = 0$ in (5) to obtain the trace of the surface of revolution in the xy-plane. The surface given by (5) is a *circular cone* with vertex at the origin of coordinates (Figure 16.7.4). ●

EXAMPLE 4 Sketch the surface $z - 9(x^2 + y^2) = 0$.

Solution This is a surface of revolution with the z-axis as its axis of revolution (special case of $f(x^2 + y^2, z) = 0$). If we set $y = 0$, we obtain the parabola $z = 9x^2$ for the trace in the zx-plane. This parabola generates a paraboloid of revolution (Figure 16.7.5). ●

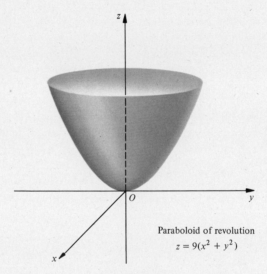

Paraboloid of revolution
$z = 9(x^2 + y^2)$

Figure 16.7.5

Exercises 16.7

For each of Exercises 1 through 12, sketch and describe the surface represented by the given equation.

1. $x^2 + z^2 = 4$
2. $x^2 + y^2 = 49, \quad z \geq 0$
3. $z = x^3$
4. $y = e^{-x}$
5. $x = \sin y, \quad -\pi \leq y \leq \pi$
6. $|x| - |y| = 0$
7. $x^2 + z^2 - y^2 = 0, \quad y \geq 0$
8. $x^2 + 4z^2 = 4$
9. $yz - 10 = 0, \quad y > 0$
10. $y^2 + 2z^2 = 0$
11. $y - x^2 - z^2 = 0$
12. $|x + y| = 1, \quad z \geq 0$

In Exercises 13 through 18, find an equation of the surface that is generated by revolving the given plane curve about the specified coordinate axis. Sketch each surface.

13. $y = x, z = 0$ about the y-axis
14. $y = x, z = 0$ about the x-axis
15. $x = 0, y = 4$ about the z-axis
16. $4z^2 + 9y^2 = 36, x = 0$ about the y-axis
17. $z = e^x, y = 0$ about the x-axis
18. $x - 3y = 9, z = 0$ about the x-axis

In Exercises 19 through 22, we are given a surface of revolution. Find a generating curve and the axis of revolution. Sketch the surface.

19. $y^2 + z^2 = 4(1 - x^2)$

20. $\dfrac{x^2}{a^2} + \dfrac{y^2}{a^2} + \dfrac{z^2}{b^2} = 1, \quad a \neq b$

21. $x^2 + y^2 = 4z^2$

22. $x^{2/3} + (y^2 + z^2)^{1/3} = 1$

16.8   QUADRIC SURFACES

In our development of plane analytic geometry in Chapter 13, we considered the analytic and geometric treatment of quadratic equations in two variables. We learned that by translating and rotating axes ellipses, hyperbolas or parabolas are obtained (or, perhaps, certain degenerate forms). Similarly, the graph of an equation of second degree or quadratic equation in x, y, and z is called a *quadric surface.* It can be shown that, except for degenerate loci, a quadric surface is one of the types listed (when referred to a suitable choice of orthogonal coordinate axes). The complete analysis of the general quadratic in space is best considered in a more specialized course in solid analytic geometry.

1. ELLIPSOID. The ellipsoid when referred to a particular set of coordinate axes (Figure 16.8.1) has the standard equation

$$\frac{x^2}{a^2} + \frac{y^2}{b^2} + \frac{z^2}{c^2} = 1 \tag{1}$$

where $a > 0$, $b > 0$, and $c > 0$.

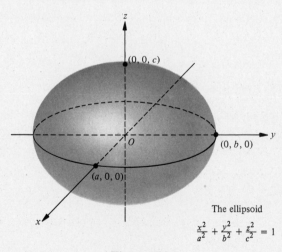

The ellipsoid

$$\frac{x^2}{a^2} + \frac{y^2}{b^2} + \frac{z^2}{c^2} = 1$$

Figure 16.8.1

The surface is symmetrical with respect to the xy-plane for, if a point with coordinates (x, y, z) lies on the surface, so does the point with coordinates $(x, y, -z)$. (Note that these points are located symmetrically with respect to the

xy-plane.) Similarly, the surface is symmetrical with respect to the other coordinate planes. This symmetry follows from the observation that only even powers of x, y, and z occur in (1). Also, the surface is symmetrical with respect to each coordinate line and the origin. Its trace in the xy-plane is determined by setting $z = 0$ to obtain the ellipse $\dfrac{x^2}{a^2} + \dfrac{y^2}{b^2} = 1$. More generally, the cross sections of the surface made with the planes $z = k$ for which $|k| < c$ are

$$\frac{x^2}{a^2} + \frac{y^2}{b^2} = 1 - \frac{k^2}{c^2}$$

which again are ellipses. When the planes $z = \pm c$ are substituted, we obtain the single points $(0, 0, c)$ and $(0, 0, -c)$. If $|k| > c$, then the plane $z = k$ does not intersect the ellipsoid. Therefore the surface lies between the planes $z = \pm c$. Similarly, the traces in planes parallel to either the yz-plane or the zx-plane are also ellipses. Furthermore, the reader may verify that the extent of (1) in the x, y and z directions is determined by

$$|x| \le a \qquad |y| \le b \qquad |z| \le c$$

that is, the surface is also bounded by the planes $x = \pm a$ and $y = \pm b$. The line segments on the coordinate axes with the intercepts as end points have lengths $2a$, $2b$, and $2c$, and are called the **axes** of the ellipsoid. Furthermore, a, b, and c are called the **semiaxes.**

If the three axes are equal, the ellipsoid is a **sphere.**

If two of the axes are equal and different from the third, the ellipsoid is an **ellipsoid of revolution** or a **spheroid.** If the third axis is shorter than the other two (as occurs on the Earth) then we have an **oblate spheroid.** On the contrary, if the third axis is longer than the other two (as in the case of a football) then the ellipsoid is said to be a **prolate spheroid.**

2. ELLIPTIC HYPERBOLOID OF ONE SHEET. The elliptic hyperboloid of one sheet is governed by an equation of the form

$$\frac{x^2}{a^2} + \frac{y^2}{b^2} - \frac{z^2}{c^2} = 1 \qquad a > 0, b > 0, c > 0 \tag{2}$$

The surface is symmetric with respect to the coordinate planes, the coordinate axes, and the origin. If we set $z = 0$, we obtain the standard ellipse $\dfrac{x^2}{a^2} + \dfrac{y^2}{b^2} = 1$. More generally if $z = k$, there results an ellipse $\dfrac{x^2}{a^2} + \dfrac{y^2}{b^2} = 1 + \dfrac{k^2}{c^2}$, and the size of the ellipse increases with increase in $|k|$. This means that the size of the ellipse becomes larger as the cutting plane moves away from the xy-plane (Figure 16.8.2).

The zx-trace and the yz-trace are hyperbolas. For example, consider sections $y = k$ with equations

$$\frac{x^2}{a^2} - \frac{z^2}{c^2} = 1 - \frac{k^2}{b^2} \tag{3}$$

If $k^2 < b^2$, the hyperbolic section (3) has as its transverse axis a line parallel to the x-axis, whereas if $k^2 > b^2$, the transverse axis is parallel to the z-axis. If $k^2 = b^2$

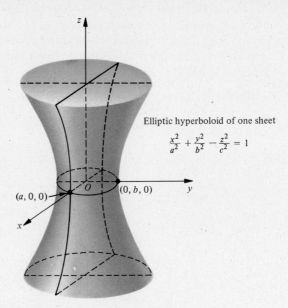

Elliptic hyperboloid of one sheet

$$\frac{x^2}{a^2} + \frac{y^2}{b^2} - \frac{z^2}{c^2} = 1$$

$(a, 0, 0)$

$(0, b, 0)$

Figure 16.8.2

then (3) reduces to $\dfrac{x^2}{a^2} - \dfrac{z^2}{c^2} = 0$ or, equivalently, $cx = \pm az$ which are two straight lines.

If $a = b$ the hyperboloid of one sheet is also a surface of revolution about the z-axis. If the negative sign in (2) had been before the first or second term rather than the third, the surface would still be a hyperboloid of one sheet.

Note that the hyperboloid of one sheet is unbounded in the x, y, and z directions.

3. ELLIPTIC HYPERBOLOID OF TWO SHEETS. The graph of the equation

$$-\frac{x^2}{a^2} - \frac{y^2}{b^2} + \frac{z^2}{c^2} = 1 \qquad a > 0, b > 0, c > 0 \tag{4}$$

is called an ***elliptic hyperboloid of two sheets*** (see Figure 16.8.3). The zx-trace and the yz-trace are hyperbolas; however, there is no xy-trace. In fact, if we set $z = k$, we obtain

$$\frac{x^2}{a^2} + \frac{y^2}{b^2} = \frac{k^2}{c^2} - 1 \tag{5}$$

Hence an ellipse parallel to the xy-plane is obtained if and only if $\dfrac{k^2}{c^2} > 1$ or, equivalently, if $|k| > c$. The size of the ellipse increases as $|k|$ increases, that is, as we move away from the xy-plane. If $k^2 = c^2$ or $k = \pm c$ the intersections are the points $(0, 0, \pm c)$. There is no point of intersection if $|k| < c$, that is, there is no locus between the planes $z = -c$ and $z = c$. Thus the surface consists of two separate parts, and this is the reason for the term "two sheets."

When a plane $x = k$ intersects the surface (4) we obtain the hyperbola

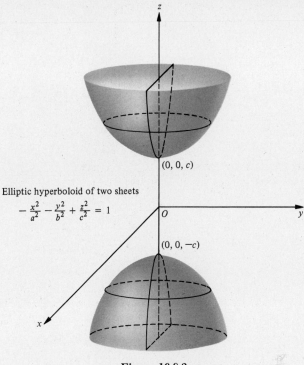

Elliptic hyperboloid of two sheets
$-\dfrac{x^2}{a^2} - \dfrac{y^2}{b^2} + \dfrac{z^2}{c^2} = 1$

$(0, 0, c)$

O

$(0, 0, -c)$

Figure 16.8.3

$\dfrac{z^2}{c^2} - \dfrac{y^2}{b^2} = 1 + \dfrac{k^2}{a^2}$ whose transverse axis is parallel to the z-axis. Similarly, the section made by a plane $y = k$ is a hyperbola $\dfrac{z^2}{c^2} - \dfrac{x^2}{a^2} = 1 + \dfrac{k^2}{b^2}$ for which the transverse axis is also parallel to the z-axis.

If $a = b$, the hyperboloid of two sheets is also a surface of revolution with the z-axis as its axis of revolution.

4. ELLIPTIC PARABOLOID. The elliptic paraboloid is defined by the equation

$$\frac{x^2}{a^2} + \frac{y^2}{b^2} = cz \qquad (6)$$

where a and b are positive and $c \neq 0$.

This surface is symmetrical with respect to the yz-plane, the zx-plane, and the z-axis (Figure 16.8.4 exhibits the surface for $c > 0$). The origin is called the *vertex* of the paraboloid. The section of the surface by the plane $z = k$ is an ellipse if $kc > 0$; that is, k and c have the same sign. Also, the size of the ellipse increases with increase of kc. If $k = 0$, then the section degenerates to a single point $(0, 0, 0)$. If $kc < 0$, that is, k and c are of opposite signs, then there is no intersection. Thus in Figure 16.8.4, the section of the surface by the plane $z = k$ is an ellipse if and only if $k > 0$. Sections by planes parallel to the other coordinate planes, that is, $x = k$ or $y = k$, are parabolas with vertical axes opening upward when $c > 0$ (as in the figure) and, on the other hand, opening downward when $c < 0$. If $a = b$ then the paraboloid becomes a *paraboloid of revolution.*

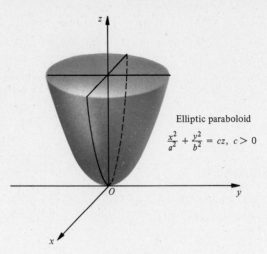

Elliptic paraboloid

$$\frac{x^2}{a^2} + \frac{y^2}{b^2} = cz, \ c > 0$$

Figure 16.8.4

5. HYPERBOLIC PARABOLOID. The **hyperbolic paraboloid**

$$\frac{x^2}{a^2} - \frac{y^2}{b^2} = cz \tag{7}$$

where a and b are positive and $c \neq 0$.

The graph of this surface is given in Figure 16.8.5 for $c > 0$ (where the xy-axes have been rotated for convenience of display). There is symmetry with respect to the yz-plane and the zx-plane. Sections parallel to the yz-plane, that is, $x = k$, are parabolas opening downward if $c > 0$ (as in the figure) and opening upward if $c < 0$. Sections parallel to the zx-plane, that is, $y = k$, are parabolas opening upward if $c > 0$ (as shown) and parabolas opening downward if $c < 0$. The sections $z = k$ are hyperbolas with transverse axes parallel to the y-axis if k and c have opposite signs (Figure 16.8.5, $z = k < 0$) and hyperbolas with transverse axes parallel to the x-axis when $z = k > 0$.

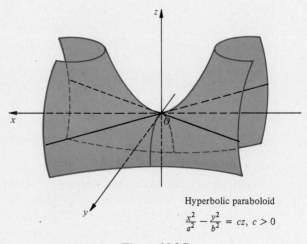

Hyperbolic paraboloid

$$\frac{x^2}{a^2} - \frac{y^2}{b^2} = cz, \ c > 0$$

Figure 16.8.5

6. ELLIPTIC CONE. The *elliptic cone*

$$\frac{x^2}{a^2} + \frac{y^2}{b^2} - \frac{z^2}{c^2} = 0 \tag{8}$$

where a, b, and c are positive (Figure 16.8.6).

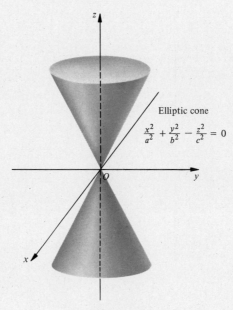

Elliptic cone

$$\frac{x^2}{a^2} + \frac{y^2}{b^2} - \frac{z^2}{c^2} = 0$$

Figure 16.8.6

The graph of (8) is symmetrical with respect to the three coordinate planes, the coordinate axes, and the origin. The surface is unbounded in all directions. The xy-trace (obtained by setting $z = 0$) is the single point, the origin. The origin is the vertex of the cone. The cross sections made by $z = k \neq 0$ are ellipses that increase in size as $|k|$ increases. Cross sections in the planes $x = 0$ (yz-plane) or $y = 0$ (zx-plane) are two intersecting lines at the origin. In the planes $x = k \neq 0$ or $y = k \neq 0$ the cross sections are hyperbolas. If $a = b$ then (8) becomes the usual right circular cone.

7. PARABOLIC, ELLIPTIC, AND HYPERBOLIC CYLINDERS. Representative equations are

$$y^2 = cx \qquad \text{parabolic cylinder} \tag{9}$$

$$\frac{x^2}{a^2} + \frac{y^2}{b^2} = 1 \qquad \text{elliptic cylinder} \tag{10}$$

$$\frac{x^2}{a^2} - \frac{y^2}{b^2} = 1 \qquad \text{hyperbolic cylinder} \tag{11}$$

The sketches of these three surfaces are left to the reader.

EXAMPLE 1 Discuss and sketch the surface

$$24x^2 + 9y^2 + 16z^2 = 144$$

Solution If we divide by 144, there results

$$\frac{x^2}{6} + \frac{y^2}{16} + \frac{z^2}{9} = 1$$

which by comparison with (1) is an ellipsoid for which $a = \sqrt{6}, b = 4$, and $c = 3$. Since only even powers of x, y, and z are present, the surface is symmetrical with respect to the xy-, yz- and zx-planes. It cuts the coordinate axes at $(\pm \sqrt{6}, 0, 0)$, $(0, \pm 4, 0)$ and $(0, 0, \pm 3)$ and is symmetrical with respect to the three coordinate axes and the origin. It is limited in extent to lie inside the rectangular box where

$$|x| \leq \sqrt{6} \qquad |y| \leq 4 \qquad |z| \leq 3$$

The sections cut out by the coordinate planes are ellipses. The upper half of the ellipsoid is sketched in Figure 16.8.7. ●

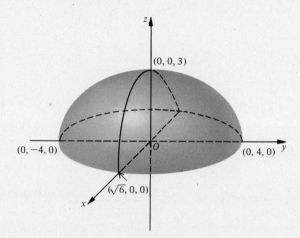

Figure 16.8.7

EXAMPLE 2 Identify the quadric surface in variables X, Y, and Z defined by

$$Z^2 = X^2 + 4Y^2 - 6X + 8Y + 4Z$$

Solution We collect terms and utilize the method of completion of squares

$$Z^2 - 4Z = (X^2 - 6X) + 4(Y^2 + 2Y)$$

$$(Z^2 - 4Z + 4) + 9 = (X^2 - 6X + 9) + 4(Y^2 + 2Y + 1)$$

$$(Z - 2)^2 + 9 = (X - 3)^2 + 4(Y + 1)^2$$

or

$$(X - 3)^2 + 4(Y + 1)^2 - (Z - 2)^2 = 9$$

Therefore if we use the translation of axes

$$x = X - 3 \qquad y = Y + 1 \qquad z = Z - 2$$

there results

$$x^2 + 4y^2 - z^2 = 9$$

or, equivalently,

$$\frac{x^2}{3^2} + \frac{y^2}{(\frac{3}{2})^2} - \frac{z^2}{3^2} = 1$$

This is of the form (2), and the given surface is an elliptic hyperboloid of one sheet with center of symmetry at $(0, 0, 0)$ in the xyz-system or $(3, -1, 2)$ in the XYZ-system. The sketch is left to the reader. ●

EXAMPLE 3   Discuss the graph of

$$9x^2 - 16y^2 - 36z^2 = 144 \qquad (12)$$

Solution   We divide both sides of (12) by 144 to obtain the standard form

$$\frac{x^2}{16} - \frac{y^2}{9} - \frac{z^2}{4} = 1 \qquad (13)$$

Comparison of (13) with (4) shows that (with z and x changing roles) we have an elliptic hyperboloid of two sheets.

If we set $y = z = 0$, then $x = \pm 4$ and therefore $(\pm 4, 0, 0)$ are the x-intercepts. No real value of y is obtained corresponding to $z = x = 0$. Similarly, if we set $x = y = 0$, no real value of z is determined. Thus there are no y- and z-intercepts.

The elliptic sections are obtained when $|x| = R$ where R is a constant greater than 4. Furthermore, the size of the ellipses increase monotonically with increase in $|x|$. The sketch of the graph is left to the reader. ●

EXAMPLE 4   Show that $z = xy$ is a hyperbolic paraboloid.

Solution   We refer to the rotation in the plane development of Chapter 13, Section 5. From (9) of (13.5) the equation $\cot 2\alpha = \dfrac{A - C}{B}$ may be used in order to eliminate the xy-term where $A = C = 0$ and $B = 1$. Hence $\alpha = \dfrac{\pi}{4}$ and since $\cos\dfrac{\pi}{4} = \sin\dfrac{\pi}{4} = \dfrac{1}{\sqrt{2}}$, we have

$$x = \frac{x' - y'}{\sqrt{2}} \qquad \text{and} \qquad y = \frac{x' + y'}{\sqrt{2}}$$

Hence, by multiplication,

$$xy = \frac{x'^2 - y'^2}{2}$$

and the given equation becomes

$$2z = x'^2 - y'^2$$

which is an equation of a hyperbolic paraboloid. ●

EXAMPLE 5   Identify the quadric $x^2 - y^2 + z^2 - 2xz = 0$.

Solution   We rearrange the terms on the left side as follows

$$x^2 - 2xz + z^2 - y^2 = 0$$

or, equivalently,

$$(x - z)^2 - y^2 = 0$$

Next, we factor as the difference of two squares

$$(x - z - y)(x - z + y) = 0$$

which yields two planes passing through the origin

$$x - y - z = 0 \qquad \text{and} \qquad x + y - z = 0$$

Exercises 16.8

In Exercises 1 and 2, determine the semiaxes of the ellipsoid. Sketch the upper half of the ellipsoid.

1. $16x^2 + 36y^2 + 9z^2 = 144$

2. $22x^2 + 16y^2 + 88z^2 = 176$

In Exercises 3 and 4, show that the ellipsoid is a spheroid (that is, an ellipsoid of revolution). Determine whether the spheroid is oblate or prolate and name the axis of revolution. Sketch the figure.

3. $x^2 + 4y^2 + 4z^2 = 4$

4. $5x^2 + 5y^2 + 7z^2 = 35$

In Exercises 5 through 16, identify the following surfaces. If the surface is a surface of revolution, name the axis. Sketch the surface.

5. $4x^2 - y^2 + 4z^2 = 0$

6. $4x^2 + 6y^2 - 9z^2 = 36$

7. $9x^2 + 16y^2 - 36z^2 + 144 = 0$

8. $x^2 - 6x + y^2 + z^2 = 0$

9. $15y - 5x^2 - 3z^2 = 0$

10. $x^2 + y^2 + z^2 + 4x - 2y + 9 = 0$

11. $y^2 + z^2 = 10$

12. $3x^2 - y^2 = 0$

13. $6x^2 + 11z^2 = 0$

14. $y^2 + 4z^2 - 16 = 0$

15. $(z + x)(z - x) = 1$

16. $x^2 + y^2 + z^2 + 2xy - 2yz - 2zx = 1$

In Exercises 17 through 20, write an equation of the prescribed locus of a point P and identify the locus.

17. The point P is equidistant from the point $(2, 0, 0)$ and the plane $x = -2$.

18. The distance of P from the origin is four times its distance from P to the zx-plane.

19. The sum of the squares of the distances of P from the origin and the point $(0, 0, 1)$ equals 5.

20. The sum of the squares of the distances of P from the origin and the point (a, b, c) equals $a^2 + b^2 + c^2$.

In Exercises 21 and 22, find numbers A, B, C, and D such that the quadric surface $Ax^2 + By^2 + Cz^2 + D = 0$ contains the three points and identify the surface.

21. $(1, 1, 1), (-1, 3, -3), (0, 4, 3)$

22. $(2, 1, 0), (3, 0, 4), (2\sqrt{3}, 3, 4)$

16.9   DIFFERENTIATION OF VECTOR FUNCTIONS—SPACE CURVES

Our objective is to extend our discussion of vector functions in two-dimensional space (Chapter 11, Section 11.4) to that of vector-valued functions in three-dimensional space. Because of the similarity to the treatment in two dimensions, most of the results will be stated and their proofs will be left to the reader.

Definition   Let $f(t)$, $g(t)$, and $h(t)$ be three scalar real-valued functions of a real variable t. Then in their common domain of definition, we form

$$\mathbf{r}(t) = f(t)\mathbf{i} + g(t)\mathbf{j} + h(t)\mathbf{k} \qquad (1)$$

which is said to be a ***vector-valued function*** of t.

If $\mathbf{r}(t) = x\mathbf{i} + y\mathbf{j} + z\mathbf{k}$ is the position vector of the point (x, y, z) in three-dimensional space, then (1) is equivalent to three scalar equations

$$x = f(t) \qquad y = g(t) \qquad z = h(t) \qquad (2)$$

The graph of a vector function in space may be determined analogously to its two-dimensional counterpart. When t takes on all its values in the domain of $\mathbf{r}$, the tip of $\mathbf{r}$ traces out a curve C in space which is called the graph of $\mathbf{r}(t)$ defined by (1). The scalar equations (2) are ***parametric equations*** of the curve C.

If the parameter t can be eliminated between, say, $x = f(t)$ and $y = g(t)$ and then between $y = g(t)$ and $z = h(t)$, we obtain two cylinders with rulings parallel to the z- and x-axes, respectively:

$$F(x, y) = 0 \qquad \text{and} \qquad G(y, z) = 0 \qquad (3)$$

These cylinders intersect in curve C and are said to be Cartesian equations of C.

More generally, if t is eliminated from equations (2) and two surfaces are determined in the form

$$F(x, y, z) = 0 \qquad \text{and} \qquad G(x, y, z) = 0 \qquad (4)$$

then these surfaces intersect in C and are the ***Cartesian equations of C.***

EXAMPLE 1   Sketch the space curve defined by

$$\mathbf{r}(t) = a(\cos t)\mathbf{i} + b(\sin t)\mathbf{j} + t\mathbf{k} \qquad (5)$$

where t is in $[0, \infty)$.

Solution   Parametric equations of the path are

$$x = a(\cos t) \qquad y = b(\sin t) \qquad z = t \qquad (6)$$

Table 16.9.1 gives x, y, and z for special values of t in $[0, 2\pi]$ in increments of $\dfrac{\pi}{4}$.

By plotting the coordinates (x, y, z) of point P, we see that the curve is as shown in Figure 16.9.1. It is called an ***elliptic helix.*** The parameter t is easily eliminated by solving (6) for $\cos t$ and $\sin t$; that is

$$\cos t = \frac{x}{a} \qquad \sin t = \frac{y}{b}$$

From the trigonometric identity $\cos^2 t + \sin^2 t = 1$, we have

$$\frac{x^2}{a^2} + \frac{y^2}{b^2} = 1 \qquad (7)$$

Therefore the helix lies entirely on the elliptic cylinder (7) with rulings parallel to the z-axis. In particular, if $a = b$, the curve is called a ***circular helix.*** It winds

Table 16.9.1

t	x	y	z
0	a	0	0
$\dfrac{\pi}{4}$	$\dfrac{a}{\sqrt{2}}$	$\dfrac{b}{\sqrt{2}}$	$\dfrac{\pi}{4}$
$\dfrac{\pi}{2}$	0	b	$\dfrac{\pi}{2}$
$\dfrac{3\pi}{4}$	$-\dfrac{a}{\sqrt{2}}$	$\dfrac{b}{\sqrt{2}}$	$\dfrac{3\pi}{4}$
π	$-a$	0	π
$\dfrac{5\pi}{4}$	$-\dfrac{a}{\sqrt{2}}$	$-\dfrac{b}{\sqrt{2}}$	$\dfrac{5\pi}{4}$
$\dfrac{3\pi}{2}$	0	$-b$	$\dfrac{3\pi}{2}$
$\dfrac{7\pi}{4}$	$\dfrac{a}{\sqrt{2}}$	$-\dfrac{b}{\sqrt{2}}$	$\dfrac{7\pi}{4}$
2π	a	0	2π

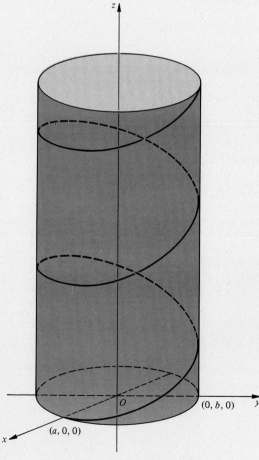

Figure 16.9.1

around a right circular cylinder and is similar to the advance of the threads of a right hand screw. ●

Definition If $\mathbf{r}(t) = f(t)\mathbf{i} + g(t)\mathbf{j} + h(t)\mathbf{k}$ then

$$\lim_{t \to t_0} \mathbf{r}(t) = (\lim_{t \to t_0} f(t))\mathbf{i} + (\lim_{t \to t_0} g(t))\mathbf{j} + (\lim_{t \to t_0} h(t))\mathbf{k}$$

provided that $\lim_{t \to t_0} f(t)$, $\lim_{t \to t_0} g(t)$, and $\lim_{t \to t_0} h(t)$ all exist.

Definition The vector function $\mathbf{r}(t)$ is **continuous** at $t = t_0$ if and only if $\lim_{t \to t_0} \mathbf{r}(t) = \mathbf{r}(t_0)$

Definition The vector function $\mathbf{r}(t)$ possesses a **derivative** $\mathbf{r}'(t)$ and we write

$$\mathbf{r}'(t) = \lim_{\Delta t \to 0} \frac{\mathbf{r}(t + \Delta t) - \mathbf{r}(t)}{\Delta t}$$

provided that the indicated limit exists.

Theorem 1 Let $\mathbf{r}(t)$ be defined by (1) then

(i) $\lim_{t \to t_0} \mathbf{r}(t) = a\mathbf{i} + b\mathbf{j} + c\mathbf{k}$

   if and only if all of the following hold

$$\lim_{t \to t_0} f(t) = a \qquad \lim_{t \to t_0} g(t) = b \qquad \lim_{t \to t_0} g(t) = c$$

(ii) $\mathbf{r}(t)$ is continuous at t_0 if and only if all of the functions $f(t)$, $g(t)$, and $h(t)$ are continuous at t_0.

(iii) $\mathbf{r}'(t)$ exists if and only if all the functions $f'(t)$, $g'(t)$, and $h'(t)$ exist. In this case, we have

$$\mathbf{r}'(t) = f'(t)\mathbf{i} + g'(t)\mathbf{j} + h'(t)\mathbf{k} \tag{8}$$

Higher derivatives are found in exactly the same way, that is, for each positive integer n,

$$\mathbf{r}^{(n)}(t) = f^{(n)}(t)\mathbf{i} + g^{(n)}(t)\mathbf{j} + h^{(n)}(t)\mathbf{k} \tag{9}$$

provided that each of the functions $f^{(n)}(t)$, $g^{(n)}(t)$, and $h^{(n)}(t)$ exists.

Theorem 2 (i) $D_t(b\mathbf{r}(t)) = b\mathbf{r}'(t) \qquad b$ constant $\tag{10}$
(ii) $D_t(\mathbf{r}(t) + \mathbf{u}(t)) = \mathbf{r}'(t) + \mathbf{u}'(t)$ $\tag{11}$
(iii) $D_t(b(t)\,\mathbf{r}(t)) = b(t)\,\mathbf{r}'(t) + b'(t)\,\mathbf{r}(t)$ $\tag{12}$

provided that each of the derivatives in the right members of (10), (11), and (12) exists.

Theorem 3

$$(i) \quad D_t(\mathbf{r}(t) \cdot \mathbf{u}(t)) = \mathbf{r}(t) \cdot \mathbf{u}'(t) + \mathbf{r}'(t) \cdot \mathbf{u}(t) \qquad (13)$$

$$(ii) \quad D_t(\mathbf{r}(t) \times \mathbf{u}(t)) = \mathbf{r}(t) \times \mathbf{u}'(t) + \mathbf{r}'(t) \times \mathbf{u}(t) \qquad (14)$$

where it is assumed that $\mathbf{r}'(t)$ and $\mathbf{u}'(t)$ exist. Note that the cross product must be expressed in correct order.

Proof of (ii)   Let $\mathbf{w} = \mathbf{r} \times \mathbf{u}$ and suppose that if t changes by an amount Δt, $\mathbf{r}$, $\mathbf{u}$, and $\mathbf{w}$ change by amounts $\Delta \mathbf{r}$, $\Delta \mathbf{u}$, and $\Delta \mathbf{w}$, respectively. Hence

$$\mathbf{w} + \Delta \mathbf{w} = (\mathbf{r} + \Delta \mathbf{r}) \times (\mathbf{u} + \Delta \mathbf{u})$$
$$= \mathbf{r} \times \mathbf{u} + \mathbf{r} \times \Delta \mathbf{u} + \Delta \mathbf{r} \times \mathbf{u} + \Delta \mathbf{r} \times \Delta \mathbf{u}$$

Subtract $\mathbf{w} = \mathbf{r} \times \mathbf{u}$ from both sides and divide by Δt to obtain

$$\frac{\Delta \mathbf{w}}{\Delta t} = \mathbf{r} \times \frac{\Delta \mathbf{u}}{\Delta t} + \frac{\Delta \mathbf{r}}{\Delta t} \times \mathbf{u} + \frac{\Delta \mathbf{r}}{\Delta t} \times \Delta \mathbf{u}$$

If we let $\Delta t \to 0$, then $\Delta \mathbf{u} \to 0$ while $\lim_{\Delta t \to 0} \dfrac{\Delta \mathbf{u}}{\Delta t} = \mathbf{u}'(t)$ and $\lim_{\Delta t \to 0} \dfrac{\Delta \mathbf{r}}{\Delta t} = \mathbf{r}'(t)$. Hence (14) is established. The proof of (13) follows in a similar manner and is left to the reader. ∎

EXAMPLE 2   Find the derivative of $\mathbf{F}(t) = \mathbf{r}(t) \cdot \mathbf{u}(t)$ when

$$\mathbf{r}(t) = (3t + 5)\mathbf{i} + (t^3 + t - 1)\mathbf{j} + t^2\mathbf{k}$$

and

$$\mathbf{u}(t) = 4t\mathbf{i} - (t^2 + 1)\mathbf{j} + t^3\mathbf{k}$$

Solution   By formula (13),

$$F'(t) = [(3t + 5)\mathbf{i} + (t^3 + t - 1)\mathbf{j} + t^2\mathbf{k}] \cdot [4\mathbf{i} - 2t\mathbf{j} + 3t^2\mathbf{k}]$$
$$+ [3\mathbf{i} + (3t^2 + 1)\mathbf{j} + 2t\mathbf{k}] \cdot [4t\mathbf{i} - (t^2 + 1)\mathbf{j} + t^3\mathbf{k}]$$
$$= 4(3t + 5) - 2t(t^3 + t - 1) + 3t^4 + 12t - (3t^2 + 1)(t^2 + 1) + 2t^4$$
$$= -6t^2 + 26t + 19$$

Alternative Solution   Take the dot product and obtain

$$F(t) = (3t + 5)(4t) - (t^3 + t - 1)(t^2 + 1) + t^5$$
$$= 12t^2 + 20t - t^5 - 2t^3 + t^2 - t + 1 + t^5$$
$$= -2t^3 + 13t^2 + 19t + 1$$

then $F'(t) = -6t^2 + 26t + 19$

in agreement with our previous calculation. ●

EXAMPLE 3   Prove that if the length of the vector function $\mathbf{r}(t)$ is constant, then the vectors $\mathbf{r}(t)$ and $\mathbf{r}'(t)$ are perpendicular, provided that neither of the vectors is the zero vector.

Solution   By hypothesis, $\mathbf{r} \cdot \mathbf{r} = |\mathbf{r}|^2 = $ constant, that is $\mathbf{r} \cdot \mathbf{r}$ is independent of t. Therefore

$$D_t(\mathbf{r} \cdot \mathbf{r}) = 0$$

and from (13) with $\mathbf{u}$ replaced by $\mathbf{r}$,

$$\mathbf{r} \cdot \mathbf{r}' + \mathbf{r}' \cdot \mathbf{r} = 0$$

or, equivalently,

$$\mathbf{r} \cdot \mathbf{r}' = 0$$

Thus $\mathbf{r}$ and $\mathbf{r}'$ are orthogonal provided that $\mathbf{r}$ and $\mathbf{r}'$ are nonzero vectors. ●

The geometric interpretation of the derivative $\mathbf{r}'(t)$ of a vector function $\mathbf{r}(t)$ follows identically with the two-dimensional case. Thus consider a portion of the space curve C which is the graph of $\mathbf{r}(t)$ (Figure 16.9.2). If $\overrightarrow{OP} = \mathbf{r}(t)$ and $\overrightarrow{OQ} = \mathbf{r}(t + \Delta t)$ then $\overrightarrow{PQ}$ is the chord given by $\overrightarrow{PQ} = \mathbf{r}(t + \Delta t) - \mathbf{r}(t)$. As $\Delta t \to 0$ the vector $\dfrac{\mathbf{r}(t + \Delta t) - \mathbf{r}(t)}{\Delta t}$ approaches a position of tangency to the curve C at P. Thus the vector $\mathbf{r}'(t)$ is a vector tangent to the curve C at the point P, provided that $\mathbf{r}'(t) \neq 0$. Hence the unit vector

$$\mathbf{T}(t) = \frac{\mathbf{r}'(t)}{|\mathbf{r}'(t)|} \qquad \mathbf{r}'(t) \neq 0 \tag{15}$$

is also tangent to the curve.

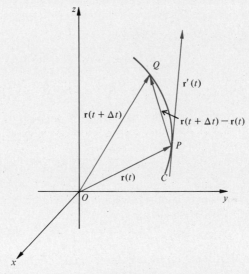

Figure 16.9.2

Definition   The vector $\mathbf{T}(t)$ defined by (15) is the **unit tangent vector** to the curve C defined by $\mathbf{r}(t)$.

EXAMPLE 4   Show that the tangent to the curve

$$\mathbf{r}(t) = t^2\mathbf{i} + (3 + t)\mathbf{j} + (t - 1)\mathbf{k}$$

at the point where it intersects the plane $x - y - 3z + 4 = 0$, lies in the plane.

Solution   First we determine the intersection(s) of the curve with the plane. We substitute the scalar equations

$$x = t^2 \qquad y = 3 + t \qquad z = t - 1$$

into the equation of the plane, and obtain

$$t^2 - (3 + t) - 3(t - 1) + 4 = 0$$

or $t^2 - 4t + 4 = (t - 2)^2 = 0$. Thus there is one repeated solution $t = 2$ and the point of intersection is $(4, 5, 1)$. Since $\mathbf{r}'(t) = 2t\mathbf{i} + \mathbf{j} + \mathbf{k}$, a tangent vector at this point is $\mathbf{r}'(2) = 4\mathbf{i} + \mathbf{j} + \mathbf{k}$. Now this will lie in the plane if and only if it is perpendicular to a vector $\mathbf{N}$ that is normal to the plane. From the coefficients of x, y, and z in the equation of the plane, we choose $\mathbf{N} = \mathbf{i} - \mathbf{j} - 3\mathbf{k}$. But then

$$\mathbf{N} \cdot \mathbf{r}'(2) = (1)(4) + (-1)(1) + (-3)(1) = 0$$

so that $\mathbf{r}'(2)$ is indeed orthogonal to $\mathbf{N}$ and, consequently, the tangent vector to the curve at $t = 2$ lies in the plane.　●

The formula for the length of an arc in three-dimensional space may be derived in the same manner as its two-dimensional counterpart. We state the result without proof.

Theorem 4　Let $\mathbf{r}(t) = f(t)\mathbf{i} + g(t)\mathbf{j} + h(t)\mathbf{k}$ represent a curve C in space. Furthermore, $\mathbf{r}'(t)$ is assumed to be continuous in a closed interval $a \leq t \leq b$. The length L of an arc of the curve C for the points corresponding to $t = a$ and $t = b$ is given by

$$L = \int_a^b |\mathbf{r}'(t)|\, dt = \int_a^b \sqrt{(f'(t))^2 + (g'(t))^2 + (h'(t))^2}\, dt \qquad (16)$$

If $s(t)$ represents the length of an arc of C from $t = a$ to a variable point $(f(t), g(t), h(t))$ on C, then, under the same hypothesis as Theorem 4, it follows that

$$s(t) = \int_a^t \sqrt{(f'(u))^2 + (g'(u))^2 + (h'(u))^2}\, du \qquad (17)$$

where u is simply a dummy letter. From the first fundamental theorem of calculus, we have

$$s'(t) = \sqrt{(f'(t))^2 + (g'(t))^2 + (h'(t))^2} = |\mathbf{r}'(t)| \qquad (18)$$

From (15) and (18),

$$\mathbf{T}(t) = \frac{\mathbf{r}'(t)}{|s'(t)|} = \frac{d\mathbf{r}}{ds} \qquad (19)$$

If $\mathbf{T}(t)$ makes angles α, β, and γ, respectively, with the x-, y- and z-axes, then

$$\mathbf{T} = \cos \alpha \mathbf{i} + \cos \beta \mathbf{j} + \cos \gamma \mathbf{k} = \frac{dx}{ds}\mathbf{i} + \frac{dy}{ds}\mathbf{j} + \frac{dz}{ds}\mathbf{k}$$

from which the direction cosines of $\mathbf{T}$ are given by

$$\cos \alpha = \frac{dx}{ds} \qquad \cos \beta = \frac{dy}{ds} \qquad \cos \gamma = \frac{dz}{ds} \qquad (20)$$

EXAMPLE 5　For the circular helix

$$\mathbf{r}(t) = a(\cos t)\mathbf{i} + a(\sin t)\mathbf{j} + t\mathbf{k}$$

(i)   Find the length of arc of that part of the helix from $t = 0$ to $t = 2\pi$.

(ii) Show that the helix makes a constant angle with the z-axis.

Solution   (i)   From $x = f(t) = a \cos t$, $y = g(t) = a \sin t$, and $z = h(t) = t$, we obtain $f'(t) = -a \sin t$, $g'(t) = a \cos t$, and $h'(t) = 1$ and certainly these derivatives are continuous for all t. Therefore (16) yields

$$L = \int_0^{2\pi} \sqrt{a^2 \sin^2 t + a^2 \cos^2 t + 1}\, dt$$

$$= \int_0^{2\pi} \sqrt{a^2 + 1}\, dt = 2\pi \sqrt{a^2 + 1}$$

(ii) $\mathbf{r}'(t) = -a(\sin t)\mathbf{i} + a(\cos t)\mathbf{j} + \mathbf{k}$ is a vector directed along the tangent to the given curve. The unit tangent vector $\mathbf{T}$ is given by

$$\mathbf{T}(t) = \frac{\mathbf{r}'(t)}{|\mathbf{r}'(t)|} = \frac{1}{\sqrt{a^2 + 1}}(-a(\sin t)\mathbf{i} + a(\cos t)\mathbf{j} + \mathbf{k})$$

Now, $\mathbf{T} \cdot \mathbf{k} = \cos \phi$, where ϕ is the angle between $\mathbf{T}$ and $\mathbf{k}$ and we obtain

$$\cos \phi = \frac{1}{\sqrt{a^2 + 1}}$$

which is constant (independent of t) and (ii) is established.   ●

Exercises 16.9

1. Sketch the arc whose vector equation is

$$\mathbf{r}(t) = t\mathbf{i} + 2t\mathbf{j} + 3\mathbf{k} \qquad (-\infty < t < \infty)$$

Describe the arc.

2. Sketch the arc having parametric equations

$$x = 3 \cos \omega t \qquad y = 3 \sin \omega t \qquad z = \frac{t}{2} \qquad 0 \le t \le \frac{2\pi}{\omega}$$

where ω is a positive constant. Describe the arc.

3. Sketch the arc having parametric equations

$$x = 3t \qquad y = 4t \qquad z = t^2 \qquad t \ge 0$$

Describe the arc.

4. Sketch the arc whose vector equation is

$$\mathbf{r}(t) = t\mathbf{i} + t^2\mathbf{j} + t^3\mathbf{k} \qquad 0 \le t \le 2$$

(*Hint:* It is useful to find the parametric equations and obtain the cylinder relating x to y. Then the curve known as the *twisted cubic* must lie on this cylinder.)

In each of Exercises 5 through 10, find the first and second derivatives $\mathbf{r}'(t)$ and $\mathbf{r}''(t)$.

5. $\mathbf{r}(t) = t\mathbf{i} + (t^2 - 1)\mathbf{j} + (3 - 5t)\mathbf{k}$
6. $\mathbf{r}(t) = t^2\mathbf{i} - (3t^2 - t - 7)\mathbf{j} + t^3\mathbf{k}$
7. $\mathbf{r}(t) = (\ln t)\mathbf{i} - t^{-2}\mathbf{j} + t^{-3}\mathbf{k} \qquad t > 0$
8. $\mathbf{r}(t) = te^t\mathbf{i} - e^{-t}\mathbf{j} + e^t(\sin t)\mathbf{k}$
9. $\mathbf{r}(t) = t(\sin t)\mathbf{i} + e^{-t^2}\mathbf{j} + (\tan^{-1} t)\mathbf{k}$
10. $\mathbf{r}(t) = (\sinh^2 t)\mathbf{i} - [\cosh(2t - 3)]\mathbf{j} - (\tanh 2t)\mathbf{k}$

In Exercises 11 through 14, find the unit tangent vector to the given space curve defined by the vector equation

11. $\mathbf{r}(t) = (3t + 2)\mathbf{i} + (4t - 6)\mathbf{j} + (8t - 5)\mathbf{k}$
12. $\mathbf{r}(t) = t\mathbf{i} + t^2\mathbf{j} + t^3\mathbf{k}$
13. $\mathbf{r}(t) = t(\sin t)\mathbf{i} + t(\cos t)\mathbf{j} + t\mathbf{k}$
14. $\mathbf{r}(t) = e^t\mathbf{i} + (t - 3)\mathbf{j} + e^{-t}\mathbf{k}$

In Exercises 15 through 18, find the length of the given space curve between the indicated limits.

15. $\mathbf{r}(t) = (3t - 1)\mathbf{i} + 5t\mathbf{j} + (t - 2)\mathbf{k}$, for $1 \leq t \leq 4$

16. $\mathbf{r}(t) = a(t - \sin t)\mathbf{i} + a(1 - \cos t)\mathbf{j} + 4a\left(\sin\dfrac{t}{2}\right)\mathbf{k}$, for $0 \leq t \leq 2\pi$, where a is a positive constant. (*Hint:* You will need a trigonometric identity.)

17. $\mathbf{r}(t) = t\mathbf{i} + 2t^2\mathbf{j} + (3t - 4)\mathbf{k}$, for $0 \leq t \leq 1$

18. $\mathbf{r}(t) = a(\sin t)\mathbf{i} + a(\cos t)\mathbf{j} + (\cosh at)\mathbf{k}$, where a is a positive constant, for $t_0 \leq t \leq t_1$.

19. Find an angle between the twisted cubic $\mathbf{r}(t) = t\mathbf{i} + t^2\mathbf{j} + t^3\mathbf{k}$ and the y-axis at their point of intersection.

20. Find the point at which the curves

$$\mathbf{r}_1(t) = t\mathbf{i} + (t^2 - 3)\mathbf{j} - (t - 1)\mathbf{k}$$

$$\mathbf{r}_2(u) = (u - 2)\mathbf{i} + (u - 3)\mathbf{j} + \left(\frac{u^2}{4} - 5\right)\mathbf{k}$$

intersect and then find the cosine of an angle θ between the two curves.

21. Show that the tangent to the curve

$$\mathbf{r}(t) = t\mathbf{i} + (t^2 - 2)\mathbf{j} + (t - 2)\mathbf{k}$$

at the point where it intersects the plane $x + y - z = 0$, lies in the plane. Where does it intersect the plane?

22. If $\mathbf{g}(t) = \mathbf{r}(t) \times \mathbf{u}(t)$ where

$$\mathbf{r}(t) = (\cos t)\mathbf{i} + 2(\sin t)\mathbf{j} - t\mathbf{k}$$

and

$$\mathbf{u}(t) = (\cos t)\mathbf{i} + (\sin t)\mathbf{j} + t\mathbf{k}$$

Find $D_t\mathbf{g}(t)$ by two methods.

23. If $\mathbf{w}(t) = 9(\sin at)\mathbf{i} - 40(\sin at)\mathbf{j} + 41(\cos at)\mathbf{k}$, where a is a constant, show that

$$\mathbf{w}(t) \cdot \mathbf{w}'(t) = 0$$

What development in the text suggests this result?

24. Prove that $D_t(\phi(t)\,\mathbf{r}(t)) = \phi(t)\,\mathbf{r}'(t) + \phi'(t)\,\mathbf{r}(t)$ where $\phi(t)$ and $\mathbf{r}(t)$ are differentiable scalar and vector functions, respectively.

25. Prove that $D_t(\mathbf{r}(t) \cdot \mathbf{u}(t)) = \mathbf{r}(t) \cdot \mathbf{u}'(t) + \mathbf{r}'(t) \cdot \mathbf{u}(t)$ where $\mathbf{r}(t)$ and $\mathbf{u}(t)$ are differentiable vector functions.

26. Prove that

$$D_t|\mathbf{r}| = \frac{\mathbf{r} \cdot \mathbf{r}'}{|\mathbf{r}|}$$

where $\mathbf{r}(t)$ is a differentiable function of t for which $|\mathbf{r}|$ is not zero.

27. Prove the *chain rule* for vector functions:

$$D_t[(\mathbf{f} \circ u)(t)] = D_u(\mathbf{f} \circ u)u'(t)$$

assuming that $\mathbf{f}$ is a differentiable function of u and u is a differentiable function of t. (*Hint:* Write $(\mathbf{f} \circ u)(t) = f_1(u(t))\mathbf{i} + f_2(u(t))\mathbf{j} + f_3(u(t))\mathbf{k}$ and use the chain rule on the scalar functions $f_i(u(t))$, $(i = 1, 2, 3)$.

28. If $\mathbf{f}(t) = (\cosh t)\mathbf{i} + (\sinh t)\mathbf{j} + t\mathbf{k}$, and $u = t \ln t$, $t > 0$, find $D_t\mathbf{f}(u(t))$

16.10 Velocity, Acceleration, Osculating Plane, Curvature, Torsion, and the Frenet-Serret Formulas

Consider a particle that is moving along a given curve C in three-dimensional space. The motion is defined by the position vector $\overrightarrow{OP} = \mathbf{r}$ where

$$\mathbf{r}(t) = f(t)\mathbf{i} + g(t)\mathbf{j} + h(t)\mathbf{k} \tag{1}$$

and t in (1) now measures time in appropriate units. It is assumed that the origin O and the Cartesian reference frame are fixed in space during the entire motion. Furthermore, we also postulate that $\mathbf{r}(t)$, $\mathbf{r}'(t)$, and $\mathbf{r}''(t)$ exist in what follows. Analogous to the planar case, we have the

Definition The *velocity* $\mathbf{v}(t)$ and the *acceleration* $\mathbf{a}(t)$ are defined by

$$\mathbf{v}(t) = \mathbf{r}'(t) \tag{2a}$$

and

$$\mathbf{a}(t) = \mathbf{v}'(t) = \mathbf{r}''(t) \tag{2b}$$

This means that the velocity $\mathbf{v}$ is the time rate of change of the position vector $\mathbf{r}$, and the acceleration $\mathbf{a}$ is the time rate of change of the velocity $\mathbf{v}$. Also we note that the velocity vector, when not $\mathbf{0}$, is tangent to the path of the motion; that is, the direction of $\mathbf{v}$ yields the direction of motion.

The magnitude of $\mathbf{v}$, that is, $|\mathbf{v}|$ tells us how fast a particle is moving. This is the *speed* of the object. In fact, the speed is given by

$$|\mathbf{v}| = |\mathbf{r}'(t)| = |D_t s| \tag{3}$$

or by

$$|\mathbf{v}| = \sqrt{(f'(t))^2 + (g'(t))^2 + (h'(t))^2} \tag{4}$$

EXAMPLE 1 A particle moves along a circular helix

$$\mathbf{r}(t) = a(\cos \omega t)\mathbf{i} + a(\sin \omega t)\mathbf{j} + b\omega t\mathbf{k}$$

where a, b, and ω are positive constants. Find expressions at any time t for

(i)   the velocity vector and the speed
(ii)   the angle between the velocity and the position vectors
(iii) the acceleration vector and its magnitude
(iv)   the angle between the acceleration vector and the velocity vector.

Solution (i) $\mathbf{v} = \mathbf{r}'(t) = -a\omega(\sin \omega t)\mathbf{i} + a\omega(\cos \omega t)\mathbf{j} + b\omega\mathbf{k}$ is the velocity vector. The speed is given by

$$|\mathbf{v}| = \sqrt{a^2\omega^2 \sin^2 \omega t + a^2\omega^2 \cos^2 \omega t + b^2\omega^2}$$
$$= \sqrt{a^2\omega^2 + b^2\omega^2} = \omega\sqrt{a^2 + b^2}$$

Therefore the speed is constant.

(ii) Let θ be the angle between the velocity and position vectors. Therefore

$$\cos\theta = \frac{\mathbf{r}(t)\cdot\mathbf{v}(t)}{|\mathbf{r}(t)|\,|\mathbf{v}(t)|}$$

and from (i),

$$\cos\theta = \frac{[a(\cos\omega t)\mathbf{i} + a(\sin\omega t)\mathbf{j} + b\omega t\mathbf{k}]\cdot[-a\omega(\sin\omega t)\mathbf{i} + a\omega(\cos\omega t)\mathbf{j} + b\omega\mathbf{k}]}{\sqrt{a^2\cos^2\omega t + a^2\sin^2\omega t + (b\omega t)^2}(\omega\sqrt{a^2+b^2})}$$

$$= \frac{b^2\omega t}{\sqrt{a^2+b^2}\sqrt{a^2+b^2\omega^2 t^2}}$$

(iii) $\mathbf{a} = \mathbf{v}'(t) = -a\omega^2(\cos\omega t)\mathbf{i} - a\omega^2(\sin\omega t)\mathbf{j}$ and its magnitude is given by

$$|\mathbf{a}| = a\omega^2$$

(iv) If ϕ is the angle between the velocity and acceleration vectors,

$$\cos\phi = \frac{\mathbf{v}(t)\cdot\mathbf{a}(t)}{|\mathbf{v}(t)|\,|\mathbf{a}(t)|} = \frac{a^2\omega^3\sin\omega t\cos\omega t - a^2\omega^3\sin\omega t\cos\omega t}{(\omega\sqrt{a^2+b^2})a\omega^2}$$

$$= 0$$

Therefore $\phi = \dfrac{\pi}{2}$ ●

In Section 16.9, Equation (15), the unit tangent vector $\mathbf{T}(t)$ to the curve $\mathbf{r}(t)$ is defined by

$$\mathbf{T}(t) = \frac{\mathbf{r}'(t)}{|\mathbf{r}'(t)|}$$

provided that $|\mathbf{r}'(t)| \neq 0$. This means that $|\mathbf{T}(t)| = 1$ and, from Example 3 of Section 16.9, $\mathbf{T}'(t)$ is orthogonal to $\mathbf{T}(t)$. Therefore the vector

$$\mathbf{N}(t) = \frac{\mathbf{T}'(t)}{|\mathbf{T}'(t)|} \qquad \mathbf{T}'(t) \neq \mathbf{0} \tag{5}$$

is also a unit vector that is perpendicular to $\mathbf{T}(t)$.

Definition The vector $\mathbf{N}(t)$ defined by (5) is the ***principal unit normal vector*** for the curve $\mathbf{r}(t)$.

Since $|\mathbf{r}'(t)| = s'(t)$ from (18) of Section 16.9, we may write

$$\mathbf{r}'(t) = s'(t)\,\mathbf{T}(t) \tag{6}$$

Then differentiation of both sides of (6) with respect to t yields

$$\mathbf{r}''(t) = s''(t)\,\mathbf{T}(t) + s'(t)\,\mathbf{T}'(t)$$

and from (5) this may be expressed as

$$\mathbf{r}''(t) = s''(t)\,\mathbf{T}(t) + s'(t)\,|\mathbf{T}'(t)|\,\mathbf{N}(t) \tag{7}$$

Therefore $\mathbf{r}''(t)$ is a linear combination of $\mathbf{T}(t)$ and $\mathbf{N}(t)$ with scalar coefficients $s''(t)$ and $s'(t)|\mathbf{T}'(t)|$, respectively. Hence $\mathbf{r}''(t)$ is parallel to the plane determined by $\mathbf{T}(t)$ and $\mathbf{N}(t)$.

Definition The plane of $\mathbf{T}(t)$ and $\mathbf{N}(t)$ at a point P on a curve $\mathbf{r}(t)$ is called the *osculating plane* at that point.

EXAMPLE 2 Given the space curve

$$\mathbf{r}(t) = e^t\mathbf{i} + e^t(\cos t)\mathbf{j} + e^t(\sin t)\mathbf{k}$$

(a) Find the unit tangent vector $\mathbf{T}(t)$ and unit normal vector $\mathbf{N}(t)$.

(b) Find an equation of the osculating plane at $t = \dfrac{\pi}{2}$.

Solution (a) $\mathbf{r}'(t) = e^t\mathbf{i} + e^t(\cos t - \sin t)\mathbf{j} + e^t(\cos t + \sin t)\mathbf{k}$

$$|\mathbf{r}'(t)| = e^t\sqrt{1 + (\cos t - \sin t)^2 + (\cos t + \sin t)^2}$$

which upon expansion yields

$$|\mathbf{r}'(t)| = \sqrt{3}e^t$$

from which the unit tangent vector is

$$\mathbf{T}(t) = \frac{\mathbf{r}'(t)}{|\mathbf{r}'(t)|} = \frac{1}{\sqrt{3}}[\mathbf{i} + (\cos t - \sin t)\mathbf{j} + (\cos t + \sin t)\mathbf{k}].$$

Differentiation yields

$$\mathbf{T}'(t) = \frac{1}{\sqrt{3}}[-(\sin t + \cos t)\mathbf{j} + (\cos t - \sin t)\mathbf{k}]$$

and

$$|\mathbf{T}'(t)| = \frac{\sqrt{2}}{\sqrt{3}}$$

Hence

$$\mathbf{N}(t) = \frac{\mathbf{T}'(t)}{|\mathbf{T}'(t)|} = \frac{1}{\sqrt{2}}[-(\sin t + \cos t)\mathbf{j} + (\cos t - \sin t)\mathbf{k}]$$

(b) At $t = \dfrac{\pi}{2}$, $\mathbf{T}\left(\dfrac{\pi}{2}\right) = \dfrac{1}{\sqrt{3}}(\mathbf{i} - \mathbf{j} + \mathbf{k})$, and $\mathbf{N}\left(\dfrac{\pi}{2}\right) = \dfrac{1}{\sqrt{2}}(-\mathbf{j} - \mathbf{k})$. Then $\mathbf{T}\left(\dfrac{\pi}{2}\right) \times \mathbf{N}\left(\dfrac{\pi}{2}\right)$ is a vector perpendicular to $\mathbf{T}\left(\dfrac{\pi}{2}\right)$ and $\mathbf{N}\left(\dfrac{\pi}{2}\right)$ and consequently to the osculating plane determined by these vectors. Also, $(\mathbf{i} - \mathbf{j} + \mathbf{k}) \times (-\mathbf{j} - \mathbf{k}) = 2\mathbf{i} + \mathbf{j} - \mathbf{k}$ is orthogonal to the two vectors. Therefore $\langle 2, 1, -1\rangle$ is a set of direction numbers of vectors normal to the plane. The point at which $t = \dfrac{\pi}{2}$ is $\mathbf{r}\left(\dfrac{\pi}{2}\right) = e^{\pi/2}(\mathbf{i} + \mathbf{k})$ or $(e^{\pi/2}, 0, e^{\pi/2})$. Hence an equation of the osculating plane is

$$2(x - e^{\pi/2}) + 1(y - 0) + (-1)(z - e^{\pi/2}) = 0$$

or

$$2x + y - z = e^{\pi/2} \qquad \bullet$$

The definitions of the *curvature vector* $\mathbf{K}(t)$ and the curvature K are the same as for the two-dimensional case given in Section 11.7. We have

Definition   The vector $\mathbf{K}(t) = D_s\mathbf{T}(t)$ is the ***curvature vector*** at a point P on C where
(i) $\mathbf{T}$ is the unit tangent vector to C at P.
(ii) s is arc length from an arbitrarily chosen point on C to point P such that s increases with increase in t (that is, $s'(t) > 0$).
The nonnegative scalar $K(t) = |\mathbf{K}(t)|$ is defined to be the ***curvature***. It is the magnitude of the curvature vector.

We have by application of the chain rule that

$$\mathbf{K}(t) = \frac{D_t\mathbf{T}(t)}{|D_t\mathbf{r}(t)|} \tag{8}$$

and

$$K(t) = |D_s\mathbf{T}(t)| = \left|\frac{\mathbf{T}'(t)}{|\mathbf{r}'(t)|}\right| \tag{9}$$

where $' = \dfrac{d}{dt}$ as usual.

From (8) and $s'(t) = |\mathbf{r}'(t)|$, we have

$$\mathbf{T}'(t) = \mathbf{K}(t)\,s'(t) \tag{10}$$

and from (10) substituted into (7)

$$\mathbf{r}''(t) = s''(t)\,\mathbf{T}(t) + K(t)[s'(t)]^2\,\mathbf{N}(t) \tag{11}$$

Let us now return to the problem of motion of a particle. In particular, we seek to resolve the acceleration $\mathbf{a}(t)$ in terms of its tangential and normal components. The velocity $\mathbf{v}$ of a particle is given from (6) by

$$\mathbf{v}(t) = \mathbf{r}'(t) = s'(t)\,\mathbf{T}(t) \qquad \text{where } t \text{ is the time}$$

Now from (7) and (10),

$$\begin{aligned}
\mathbf{a}(t) = \mathbf{v}'(t) &= s''(t)\,\mathbf{T}(t) + s'(t)|\mathbf{T}'(t)|\,\mathbf{N}(t) \\
&= s''(t)\,\mathbf{T}(t) + K(t)[s'(t)]^2\,\mathbf{N}(t)
\end{aligned} \tag{12}$$

This shows that the acceleration vector always lies in the osculating plane, that is, the plane containing $\mathbf{T}$ and $\mathbf{N}$.

Let us express $\mathbf{a}(t)$ in the form

$$\mathbf{a} = a_T\mathbf{T} + a_N\mathbf{N} \tag{13(a)}$$

where

$$\begin{aligned}
a_T(t) &= s''(t) \\
a_N(t) &= K(t)[s'(t)]^2
\end{aligned} \tag{13(b)}$$

are the ***tangential*** and ***normal*** components of acceleration, respectively.
Note that since $\mathbf{N}$ and $\mathbf{T}$ are perpendicular unit vectors we have

$$a_T^2 + a_N^2 = |\mathbf{a}|^2 \tag{14}$$

and (14) is often used to find a_N and K when a_T and $|\mathbf{a}|$ are more easily obtained.

EXAMPLE 3   The path of a moving particle is defined by

$$\mathbf{r}(t) = t\mathbf{i} + t^2\mathbf{j} + \tfrac{2}{3}t^3\mathbf{k}$$

Find the tangential and normal components of the acceleration and the curvature.

Solution

$$\mathbf{v}(t) = \mathbf{r}'(t) = \mathbf{i} + 2t\mathbf{j} + 2t^2\mathbf{k}$$
$$s'(t) = |\mathbf{v}(t)| = \sqrt{1^2 + (2t)^2 + (2t^2)^2}$$

thus

$$s'(t) = 1 + 2t^2$$

and

$$s''(t) = 4t$$

Therefore

$$a_T = s''(t) = 4t$$

Now,

$$\mathbf{T}(t) = \frac{\mathbf{r}'(t)}{s'(t)} = \frac{1}{1 + 2t^2}(\mathbf{i} + 2t\mathbf{j} + 2t^2\mathbf{k})$$

and

$$\mathbf{T}'(t) = \frac{(1 + 2t^2)(2\mathbf{j} + 4t\mathbf{k}) - 4t(\mathbf{i} + 2t\mathbf{j} + 2t^2\mathbf{k})}{(1 + 2t^2)^2}$$

$$= \frac{-4t\mathbf{i} + (2 - 4t^2)\mathbf{j} + 4t\mathbf{k}}{(1 + 2t^2)^2}$$

Thus the curvature vector

$$\mathbf{K}(t) = D_s\mathbf{T}(t) = \frac{\mathbf{T}'(t)}{s'(t)} = \frac{-4t\mathbf{i} + (2 - 4t^2)\mathbf{j} + 4t\mathbf{k}}{(1 + 2t^2)^3}$$

from which the curvature

$$K(t) = \frac{\sqrt{(-4t)^2 + (2 - 4t^2)^2 + (4t)^2}}{(1 + 2t^2)^3} = \frac{2(1 + 2t^2)}{(1 + 2t^2)^3} = \frac{2}{(1 + 2t^2)^2}$$

From (13(b)),

$$a_N(t) = \frac{2}{(1 + 2t^2)^2}(1 + 2t^2)^2 = 2$$

CHECK:
$$a_T^2 + a_N^2 = 16t^2 + 4 \qquad \text{(i)}$$

$$\mathbf{a}(t) = \mathbf{r}''(t) = 2\mathbf{j} + 4t\mathbf{k}$$

$$|\mathbf{a}(t)|^2 = 4 + 16t^2 \qquad \text{(ii)}$$

and the right sides of (i) and (ii) are identical. ●

With the introduction of the two unit vectors $\mathbf{T}$ and $\mathbf{N}$, it is a simple matter to introduce a third vector orthogonal to both $\mathbf{T}$ and $\mathbf{N}$. We define a vector $\mathbf{B}$ by the equation

Definition
$$\mathbf{B} = \mathbf{T} \times \mathbf{N} \qquad \text{(15)}$$
(See Figure 16.10.1.)

It follows that $\mathbf{B}$ is a unit vector since $|\mathbf{B}| = |\mathbf{T}| \, |\mathbf{N}| \sin\frac{\pi}{2} = 1$ and that $\mathbf{B}$ is perpendicular to the osculating plane containing $\mathbf{T}$ and $\mathbf{N}$. The vector $\mathbf{B}$ is called the *binormal* at P. These three unit vectors $\mathbf{T}$, $\mathbf{N}$, and $\mathbf{B}$ form a set of three mutually perpendicular unit vectors forming a right-handed system in the order given. This is reminiscent of the Cartesian $\mathbf{i}$, $\mathbf{j}$, and $\mathbf{k}$ system ($\mathbf{k} = \mathbf{i} \times \mathbf{j}$). However, there is an important difference in that the $\mathbf{i}$, $\mathbf{j}$, and $\mathbf{k}$ system has always been fixed

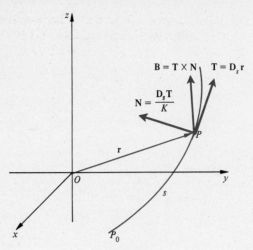

Figure 16.10.1

whereas the **T**, **N**, and **B** system moves from point to point. Thus this latter system of three vectors is known as the *moving trihedral.* It is significant in the study of space curves and is treated at length in the subject called differential geometry. We shall now develop one set of interesting relations known as the *Frenet-Serret* formulas.

We seek formulas for $D_s\mathbf{T}$, $D_s\mathbf{N}$ and $D_s\mathbf{B}$. By definition of **N** and curvature K,

$$D_s\mathbf{T} = |D_s\mathbf{T}|\,\mathbf{N} = K\mathbf{N} \tag{16}$$

and this is one of the three formulas we seek. Since **N** is an orthogonal unit vector to **B** and **T**, it follows that $D_s\mathbf{N}$ lies in the plane of **B** and **T**. Accordingly, we write

$$D_s\mathbf{N} = c\mathbf{T} + \tau\mathbf{B} \tag{17}$$

where c and τ are scalars. But, $\mathbf{N} = \mathbf{B} \times \mathbf{T}$ so that

$$D_s\mathbf{N} = \mathbf{B} \times D_s\mathbf{T} + (D_s\mathbf{B}) \times \mathbf{T} \tag{18}$$

Now, $\mathbf{B} = \mathbf{T} \times \mathbf{N}$ and forming D_s of both sides

$$D_s\mathbf{B} = \mathbf{T} \times D_s\mathbf{N} + (D_s\mathbf{T}) \times \mathbf{N}$$

and from (17), (16), and (15),

$$D_s\mathbf{B} = \mathbf{T} \times (c\mathbf{T} + \tau\mathbf{B}) + k\mathbf{N} \times \mathbf{N}$$
$$D_s\mathbf{B} = \tau\mathbf{T} \times \mathbf{B} = -\tau\mathbf{N} \tag{19}$$

Substitution of (16) and (19) into (18) yields

$$D_s\mathbf{N} = \mathbf{B} \times (K\mathbf{N}) - \tau\mathbf{N} \times \mathbf{T} = -K\mathbf{T} + \tau\mathbf{B} \tag{20}$$

that is, from (17) and (20), $c = -K$.

Formulas (16), (20), and (19) are the Frenet-Serret formulas, collected as follows:

$$D_s\mathbf{T} = K\mathbf{N} \qquad D_s\mathbf{N} = -K\mathbf{T} + \tau\mathbf{B} \qquad D_s\mathbf{B} = -\tau\mathbf{N} \tag{21}$$

The coefficients of **T**, **N**, and **B** involve the curvature K and a function τ, which is called the *torsion* of a space curve. It is a measure of the deviation of the space curve from a plane curve at the point P. The torsion is zero for a plane curve at all points for which τ is defined because $\mathbf{B}$ always has the same direction and $D_s\mathbf{B} = \mathbf{0}$. The development of formulas for the curvature K and the torsion τ are left for the reader (see Exercises 13 and 18, respectively).

EXAMPLE 4   Find the torsion τ for the space curve of Example 3.

Solution   From Example 3,

$$\mathbf{N} = \frac{D_s\mathbf{T}}{K} = \frac{-2t\mathbf{i} + (1 - 2t^2)\mathbf{j} + 2t\mathbf{k}}{1 + 2t^2} \tag{i}$$

and the reader may verify that

$$\mathbf{B} = \mathbf{T} \times \mathbf{N} = \frac{2t^2\mathbf{i} - 2t\mathbf{j} + \mathbf{k}}{1 + 2t^2}$$

Now,

$$D_t\mathbf{B} = \frac{4t\mathbf{i} + (4t^2 - 2)\mathbf{j} - 4t\mathbf{k}}{(1 + 2t^2)^2}$$

and

$$D_s\mathbf{B} = \frac{D_t\mathbf{B}}{D_t s} = \frac{4t\mathbf{i} + (4t^2 - 2)\mathbf{j} - 4t\mathbf{k}}{(1 + 2t^2)^3} \tag{ii}$$

But, $D_s\mathbf{B} = -\tau\mathbf{N}$ so that from (i) and (ii),

$$\tau = \frac{2}{(1 + 2t^2)^2}$$

For this special curve, $\tau = K$.   ●

Exercises 16.10

In Exercises 1 through 6, a particle moves along a curve whose vector equation is given where the parameter t is the time.
(a) Determine its velocity and acceleration at any time.
(b) Find the magnitudes of its velocity and acceleration at any time.
1. $\mathbf{r}(t) = e^{-2t}\mathbf{i} + 5(\cos 3t)\mathbf{j} + 5(\sin 3t)\mathbf{k}$
2. $\mathbf{r}(t) = (\cos t)\mathbf{i} + (\sin t)\mathbf{j} + 2t\mathbf{k}$
3. $\mathbf{r}(t) = b(\sin \omega t)\mathbf{i} + b(\cos \omega t)\mathbf{j} + ct\mathbf{k}$, where b, c, and ω are constants and b is positive.
4. $\mathbf{r}(t) = (3t - 4)\mathbf{i} + (2t + 7)\mathbf{j} - (t - 8)\mathbf{k}$
5. $\mathbf{r}(t) = t\mathbf{i} + t^2\mathbf{j} + t^3\mathbf{k}$
6. $\mathbf{r}(t) = (3t^2 - 5)\mathbf{i} + (\ln t)\mathbf{j} + e^{3t}\mathbf{k}, t > 0$

In Exercises 7 and 8, given the space curve $\mathbf{r}(t)$
(a) Find the unit tangent vector $\mathbf{T}(t)$ and the principal normal vector $\mathbf{N}(t)$.
(b) Find an equation of the osculating plane at $t = t_1$.

7. $\mathbf{r}(t) = (\cos t)\mathbf{i} - (\sin t)\mathbf{j} + t\mathbf{k}, \quad t_1 = \frac{\pi}{2}$

8. $\mathbf{r}(t) = a(\cos t)\mathbf{i} + a(\sin t)\mathbf{j} + bt\mathbf{k}, \quad t_1 = \frac{\pi}{3}$, where a and b are constants, $a > 0$ and $b \neq 0$

*In Exercises 9 and 10, find the curvature and the tangential and normal components of the acceleration vector of a particle moving on the given curve at the specified time t. (**Hint:** Use $\mathbf{a} - a_T\mathbf{T} = a_N\mathbf{N}$.)*

9. $x = t + 1, y = t^2 - 1, z = t^3 + 2, \quad t = 1$

10. $\mathbf{r}(t) = (\sin t)\mathbf{i} + (\cos t)\mathbf{j} + (\sin t)\mathbf{k}, \quad t = \dfrac{\pi}{4}$

In Exercises 11 and 12, find the vectors of the moving trihedral; that is, $\mathbf{T}$, $\mathbf{N}$, and $\mathbf{B}$ and the curvature K for any value of t. Verify that $\mathbf{T} \cdot \mathbf{N} = 0$, $\mathbf{T} \cdot \mathbf{B} = 0$ and $\mathbf{N} \cdot \mathbf{B} = 0$.

11. $\mathbf{r}(t) = a(\cos t)\mathbf{i} + a(\sin t)\mathbf{j} + bt\mathbf{k}, \quad$ where $a > 0$.

12. $\mathbf{r}(t) = t\mathbf{i} + t\mathbf{j} + \dfrac{t^2}{2}\mathbf{k}$

13. Starting with the formulas $\mathbf{r}'(t) = s'(t)\,\mathbf{T}(t)$ and $\mathbf{r}''(t) = s''(t)\,\mathbf{T}(t) + [s'(t)]^2 K(t)\,\mathbf{N}(t)$ prove that the curvature K may be determined from

$$K(t) = \frac{|\mathbf{r}'(t) \times \mathbf{r}''(t)|}{|\mathbf{r}'(t)|^3}$$

14. Use Exercise 13 to find the curvature K of the space curve given in Exercise 11.
15. Use Exercise 13 to find the curvature K of the space curve given in Exercise 12.
16. Starting from $\mathbf{r}'' = s''\mathbf{T} + (s')^2 K\mathbf{N}$ (equation (11) of this section), prove that

$$\mathbf{r}''' = (s''' - (s')^3 K^2)\mathbf{T} + (3s's''K + (s')^2 K')\mathbf{N} + (s')^3 K\tau\mathbf{B}$$

(*Hint:* Use $(\quad)' = s'D_s(\quad)$ and the Frenet-Serret formulas.)

17. Apply the formula developed in Exercise 16 to prove that

$$(\mathbf{r}' \times \mathbf{r}'') \cdot \mathbf{r}''' = K^2\tau(s')^6$$

18. Use the results of Exercise 17 and the formula for the curvature K to prove that the torsion is given by

$$\tau = \frac{(\mathbf{r}' \times \mathbf{r}'') \cdot \mathbf{r}'''}{|\mathbf{r}' \times \mathbf{r}''|^2}$$

19. Use the formula of Exercise 18 to find the torsion of the space curve given in Example 3 of the text.
20. Use the formula of Exercise 18 to find the torsion of the space curve

$$\mathbf{r}(t) = (1 - 3\sin^2 t)\mathbf{i} + (\sin^2 t)\mathbf{j} + (\cos 2t)\mathbf{k}$$

21. Find the torsion of the circular helix $\mathbf{r}(t) = a(\cos t)\mathbf{i} + a(\sin t)\mathbf{j} + bt\mathbf{k}$.
22. Prove that, if a curve lies in the yz-plane, then at all points at which τ is defined, $\tau = 0$ must hold. (*Note:* Essentially the same argument holds for a curve lying in an arbitrary plane.)
23. Reconcile your answer to Exercise 20 with the result given in Exercise 22. (*Hint:* Use a trigonometric identity.)

16.11 CYLINDRICAL AND SPHERICAL COORDINATES

Rectangular coordinates are the most widely used system in the application of mathematics. However, there are two other coordinate systems that are preferred in solving certain kinds of problems in three-dimensional space. These are the spherical coordinates, which are often employed to solve problems with "spherical symmetry," and the cylindrical coordinates, which we will describe now.

16.11.1 *Cylindrical Coordinates*

Problems that have symmetry about an axis (that is, so-called axisymmetrical problems) may be treated most simply by the introduction of ***cylindrical coordinates.*** In this system, the position of a point P in space is given by (r, θ, z) where r and θ are the usual polar coordinates in the xy-plane and z is the third rectangular coordinate (Figure 16.11.1). As discussed in Chapter 12, unless otherwise stated r may be negative and θ is not restricted. This means that to each point P in space, although we have a unique set of rectangular coordinates, there correspond an infinite number of interrelated cylindrical coordinates.

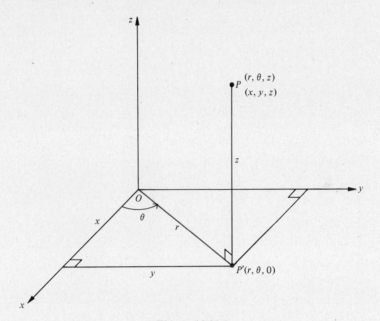

Figure 16.11.1

The equations connecting the cylindrical and rectangular coordinates are

$$x = r\cos\theta \qquad y = r\sin\theta \qquad z = z \tag{1}$$

for the rectangular coordinates in terms of the cylindrical coordinates and, conversely,

$$r = \pm\sqrt{x^2 + y^2} \qquad \tan\theta = \frac{y}{x} \qquad z = z \tag{2}$$

for the cylindrical coordinates in terms of the rectangular coordinates.

EXAMPLE 1   Find cylindrical coordinates of a point P with rectangular coordinates $(3, -3\sqrt{3}, 5)$.

Solution   From (2), we may take

$$r = \sqrt{3^2 + (-3\sqrt{3})^2} = \sqrt{9 + 27} = \sqrt{36} = 6 \qquad \tan\theta = \frac{-3\sqrt{3}}{3} = -\sqrt{3}$$

and $\theta = \frac{5\pi}{3}$, where an angle in the fourth quadrant is used since x is positive and y is negative $\left(\text{recall that } \tan \frac{\pi}{3} = \sqrt{3} \text{ and } \tan\left(2\pi - \frac{\pi}{3}\right) = -\tan \frac{\pi}{3} = -\sqrt{3}\right)$. Hence, one set of cylindrical coordinates is $\left(6, \frac{5\pi}{3}, 5\right)$. For review purposes, the reader is invited to supply two more sets of cylindrical coordinates. ●

If c_i, $(i = 1, 2, 3)$ are constants, then the surfaces $x = c_1$, $y = c_2$, $z = c_3$ are planes parallel to the yz-, the zx-, and the xy-planes, respectively. The point $P(c_1, c_2, c_3)$ is the intersection of the three planes. Corresponding surfaces in cylindrical coordinates are $r = k_1$, a family of right circular cylinders with the z-axis as axis; $\theta = k_2$, a family of planes containing the z-axis; and again $z = k_3$, a family of planes parallel to the xy-plane. The point $P(k_1, k_2, k_3)$ is the intersection of three surfaces $r = k_1$, $\theta = k_2$, and $z = k_3$ (Figure 16.11.2). Our next example discusses some of this in more detail.

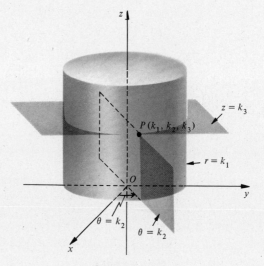

Figure 16.11.2

EXAMPLE 2   Describe each of the surfaces (a) $r = 3$, (b) $\theta = \frac{\pi}{4}$, (c) $r(3 \cos \theta - 5 \sin \theta) - z = 10$, (d) $z = r^2$.

Solution   (a) The points $P(r, \theta, z)$ for which $r = 3$ are just a right circular cylinder of radius 3 with the z-axis as its axis of symmetry.

(b) This is a plane that contains the z-axis and that intersects the xy-plane in a line that forms an angle of $\frac{\pi}{4}$ with the polar line. It may appear that we have only a half-plane; however, our convention is that r can be both positive or negative and it is the negative values that yield the other half-plane.

(c) If we multiply out, there results

$$3r \cos \theta - 5r \sin \theta - z = 10$$

and this becomes

$$3x - 5y - z = 10$$

which is a plane with intercepts $\frac{10}{3}$, -2, and -10, respectively, on the x- y-, and z-axes.

(d) $z = r^2 = x^2 + y^2$ is a paraboloid of revolution. For example, if we set $y = 0$, then we obtain the parabola $z = x^2$. This is the intersection of the surface with the zx-plane. This parabola (as well as $z = y^2$) generates the paraboloid of revolution when rotated about the z-axis. ●

16.11.2 Spherical Coordinates

In the **spherical coordinate system,** denoted by (ρ, θ, ϕ) the first coordinate ρ represents the distance of the point P from the origin of coordinates; that is, $\rho = |\overrightarrow{OP}|$ and $\rho \geq 0$ (Figure 16.11.3). The other two coordinates are angles

(i) θ is the angle in the xy-plane (measured counterclockwise as viewed from the positive z-axis) from the positive half of the x-axis to the projection on the xy-plane (or polar plane) of OP restricted to $0 \leq \theta < 2\pi$† and

(ii) ϕ is the smallest nonnegative angle between the positive direction on the z-axis and $\overrightarrow{OP}$. Therefore $0 \leq \phi \leq \pi$, where $\phi = 0$ corresponds to points on the positive z-axis and $\phi = \pi$ to points on the negative z-axis.

Note that the origin has spherical coordinates $(0, \theta, \phi)$ where θ and ϕ may have any values subject to the restrictions imposed.

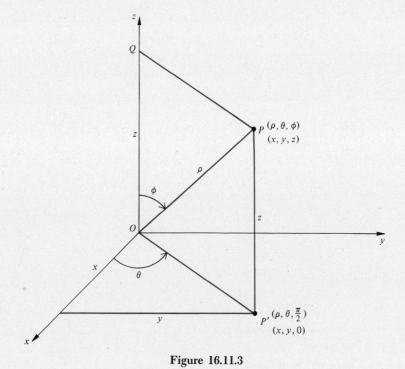

Figure 16.11.3

---

† The θ as used in spherical coordinates is the same as the θ used in cylindrical coordinates with the restriction that $0 \leq \theta < 2\pi$.

From Figure 16.11.3, we obtain

$$x = |\overline{OP'}| \cos \theta \qquad y = |\overline{OP'}| \sin \theta \qquad z = \rho \cos \phi$$

and

$$|\overline{OP'}| = \rho \sin \phi$$

Hence we have

$$x = \rho \sin \phi \cos \theta \qquad y = \rho \sin \phi \sin \theta \qquad z = \rho \cos \phi \qquad (3)$$

for the rectangular coordinates expressed in terms of its spherical coordinates. Using the distance formula from O to P, we have

$$\rho^2 = x^2 + y^2 + z^2 \qquad (4a)$$

from which since $\rho \geq 0$,

$$\rho = \sqrt{x^2 + y^2 + z^2} \qquad (4b)$$

Application of (3) yields

$$\cos \theta = \frac{x}{\sqrt{x^2 + y^2}} \qquad \tan \theta = \frac{y}{x}\dagger \qquad \cos \phi = \frac{z}{\sqrt{x^2 + y^2 + z^2}} \qquad (4c)$$

and equations (4) are formulas for the spherical coordinates in terms of the rectangular coordinates.

The basis surfaces for the spherical coordinate system are

$\rho = $ constant: concentric spheres with centers at the origin

$\theta = $ constant: halfplanes bounded by the z-axis and perpendicular to the xy-plane

$\phi = $ constant: one half of a right circular cone whose axis is the z-axis $\left(\text{except when } \phi = 0, \frac{\pi}{2}, \pi\right)$

The spherical coordinates θ and ϕ correspond to the **longitude** and **colatitude** of the point, respectively. Thus $\phi = \frac{\pi}{2} - $ latitude, where the word latitude is used in the ordinary sense, that is, positive for points north of the equator (xy-plane) and negative for points south of it.

EXAMPLE 3   Find the rectangular coordinates of the point with spherical coordinates P $\left(6, \frac{\pi}{3}, \frac{\pi}{4}\right)$.

Solution   We are given $\rho = 6$, $\theta = \frac{\pi}{3}$, and $\phi = \frac{\pi}{4}$, Hence from (3), we find

$$x = 6\left(\frac{1}{\sqrt{2}}\right)\left(\frac{1}{2}\right) = \frac{3\sqrt{2}}{2} \qquad y = 6\left(\frac{1}{\sqrt{2}}\right)\frac{\sqrt{3}}{2} = \frac{3\sqrt{6}}{2} \qquad z = 3\sqrt{2}$$

The rectangular coordinates of P are $\left(\frac{3\sqrt{2}}{2}, \frac{3\sqrt{6}}{2}, 3\sqrt{2}\right)$.   ●

_____

$\dagger$ Both expressions are needed to determine θ uniquely.

EXAMPLE 4 Find a rectangular equation for and describe the graph of $\rho = 2a \sin \phi \cos \theta$, where a is a positive constant.

Solution Multiply both sides of the given equation by ρ and obtain $\rho^2 = 2a\rho \sin \phi \cos \theta$. Application of (4a) and (3) yields

$$x^2 + y^2 + z^2 = 2ax$$

or, by completing the squares on x,

$$(x - a)^2 + y^2 + z^2 = a^2$$

This is a sphere with center on the x-axis at $(a, 0, 0)$ and radius a (that is, passing through the origin). ●

Exercises 16.11

1. Change the following from cylindrical coordinates to rectangular coordinates
   (a) $\left(7, \frac{\pi}{2}, 2\right)$ (b) $\left(10, \frac{\pi}{3}, -6\right)$

2. Change the following from spherical coordinates to rectangular coordinates
   (a) $\left(8, \frac{\pi}{3}, \frac{\pi}{6}\right)$ (b) $\left(2, \frac{3\pi}{4}, \frac{\pi}{3}\right)$

3. Find a set of cylindrical coordinates having the given rectangular coordinates
   (a) $(3, 3, 4)$ (b) $(-2, 2, -1)$

4. Change the following from rectangular coordinates to spherical coordinates
   (a) $(3, 3, 3)$ (b) $(2, -2, 2\sqrt{2})$

5. Change the following from cylindrical coordinates to spherical coordinates
   (a) $\left(2, \frac{\pi}{7}, 2\right)$ (b) $\left(-8, \frac{\pi}{6}, 6\right)$

6. Find a set of cylindrical coordinates having the given spherical coordinates
   (a) $\left(8, \frac{\pi}{3}, \frac{\pi}{4}\right)$ (b) $\left(6, \frac{\pi}{3}, \frac{2\pi}{3}\right)$

In Exercises 7 through 12, find an equation of the given surface in cylindrical coordinates and identify it.

7. $x^2 + y^2 = 3z^2$ 8. $x^2 = z - y^2$
9. $x^2 - y^2 = 1$ 10. $x^2 + y^2 + 4z^2 = 16$
11. $x^4 + y^4 + 2x^2y^2 - 1 = 0$ 12. $x^2 + y^2 + z^2 - 4y + 6z = 0$

In Exercises 13 through 18, find an equation of the given surface in spherical coordinates and identify it.

13. $x^2 + y^2 + z^2 - a^2 = 0$, $a > 0$ 14. $x^2 + y^2 + z^2 - 10x - 24y = 0$
15. $x^2 + y^2 = 3z^2$ 16. $x^2 + y^2 = a^2$, $a > 0$
17. $3x + 4y + 2z = 24$ 18. $x^2 + y^2 - 6z = 9$

In Exercises 19 through 22, find an equation in rectangular coordinates for the surface whose equation is expressed in cylindrical coordinates. Identify the surface.

19. $r^2 + z^2 = 100$ 20. $r^2 - 2z^2 = 1$
21. $r^2 \cos^2 \theta - z = 3$ 22. $r^2 \sin 2\theta = z$

In Exercises 23 through 26, find an equation in rectangular coordinates for the surface whose equation is expressed in spherical coordinates. Identify the surface.

23. $\cos \phi = \frac{1}{2}$

24. $\rho = \cos \phi$

25. $\rho^2 \sin^2 \phi (\cos^2 \theta - 3 \sin \theta \cos \theta - \sin^2 \theta) = 1$

26. $\rho^2 \sin^2 \phi \cos 2\theta = z$

27. Starting with $s'^2 = x'^2 + y'^2 + z'^2$ show that in cylindrical coordinates

$$s'^2 = r'^2 + r^2\theta'^2 + z'^2$$

and therefore, for a curve with continuous first derivatives $r'(t)$, $\theta'(t)$, and $z'(t)$, its arc length from $t = t_0$ to $t = t_1$ is given by

$$L = \int_{t_0}^{t_1} [r'^2 + r^2\theta'^2 + z'^2]^{1/2} \, dt$$

28. Show that in spherical coordinates

$$s'^2 = \rho'^2 + \rho^2 \sin^2 \phi (\theta')^2 + \rho^2\phi'^2$$

and therefore, for a curve with continuous first derivatives $\rho'(t)$, $\theta'(t)$, and $\phi'(t)$, its arc length from $t = t_0$ to $t = t_1$ is given by

$$L = \int_{t_0}^{t_1} [\rho'^2 + \rho^2 \sin^2 \phi (\theta')^2 + \rho^2\phi'^2]^{1/2} \, dt$$

29. Find the length of arc of the circular helix defined parametrically by

$$x = a \cos t \qquad y = a \sin t \qquad z = t, \quad \text{where } 0 \le t \le 2\pi$$

by application of cylindrical coordinates.

30. Find the length of arc of the space curve defined parametrically by

$$r = 3e^t \qquad \theta = t \qquad z = e^t, \qquad \text{where } a \le t \le b$$

where r, θ, and z are the cylindrical coordinates.

31. Find the length of arc of the space curve defined parametrically by

$$\rho = t^2 \qquad \phi = t \qquad \theta = b, \quad \text{where } 0 \le t \le t_1,$$

b is a constant and ρ, ϕ, and θ are the spherical coordinates.

32. Set up the integral for the determination of the length of arc of the space curve given parametrically by

$$\rho = a \qquad \phi = t \qquad \theta = t + \frac{\pi}{8} \quad \text{where } 0 \le t \le \frac{\pi}{3},$$

a is a positive constant and ρ, ϕ, and θ are spherical coordinates.

Review and Miscellaneous Exercises

In Exercises 1 through 12, sketch the graph in space of the given relation(s). Describe your sketch in words.

1. $y = 4$

2. $x + 2y = 1$

3. $x + y - z = 0$

4. $x = y^2$

5. $x^2 + y^2 + z^2 - 49 = 0$

6. $x^2 - 6y + y^2 + z^2 = 0$

7. $x = 3, \quad y = 2$

8. $y^2 + z^2 \le 4$

9. $3z^2 + 4y^2 = 0$

10. $x^2 + y^2 + z^2 - 6x - 2y + 13 = 0$

11. $z = x^2 + y^2$

12. $z^2 + x^2 - y^2 = 0, \quad y \ge 0$

13. What is the distance from $P(x, y, z)$ to (a) the origin, (b) the x-axis, (c) the point $Q(x, y, -z)$ (d) the yz-plane?

In Exercises 14 and 15, the distance from $P(x, y, z)$ to the origin is s_1 and the distance from P to $Q(4, 0, 0)$ is s_2. Find and identify the locus of P if

14. $s_1 = s_2$ **15.** $3s_1 = s_2$

In Exercises 16 through 20, let $\mathbf{A} = 3\mathbf{i} + 5\mathbf{j} - 2\mathbf{k}$, $\mathbf{B} = -\mathbf{i} + \mathbf{j} + 4\mathbf{k}$, $\mathbf{C} = 4\mathbf{i} - 3\mathbf{j} - \mathbf{k}$, and $\mathbf{D} = 7\mathbf{i} - 10\mathbf{k}$, compute

16. $3\mathbf{A} - \mathbf{B} - 2\mathbf{D}$ **17.** $\mathbf{A} \cdot (\mathbf{B} - \mathbf{C})$

18. $(\mathbf{A} \times \mathbf{B}) \cdot (3\mathbf{A} - 4\mathbf{B})$ **19.** $(\mathbf{A} \times \mathbf{B}) \cdot (\mathbf{C} \times \mathbf{D})$

20. $(\mathbf{A} - \mathbf{B} + 2\mathbf{C} + 3\mathbf{D}) \cdot (\mathbf{B} \times \mathbf{C})$

21. Under what conditions is $(\mathbf{A} + \mathbf{B}) \cdot (\mathbf{A} + \mathbf{B}) = \mathbf{A} \cdot \mathbf{A} + 2\mathbf{A} \cdot \mathbf{B} + \mathbf{B} \cdot \mathbf{B}$?

22. Show that any vector $\mathbf{A}$ in three-dimensional space may be expressed in the form $\mathbf{A} = (\mathbf{A} \cdot \mathbf{i})\mathbf{i} + (\mathbf{A} \cdot \mathbf{j})\mathbf{j} + (\mathbf{A} \cdot \mathbf{k})\mathbf{k}$. This resolves $\mathbf{A}$ into its rectangular components.

23. Show that any vector $\mathbf{A}$ in three-dimensional space may be expressed as a linear combination of the vectors $\mathbf{i}, \mathbf{j}, \mathbf{i} - \mathbf{j} + \mathbf{k}$; that is, these three noncoplanar vectors form a basis for vectors in three-dimensional space. (*Hint:* $\mathbf{A} = A_1\mathbf{i} + A_2\mathbf{j} + A_3\mathbf{k}$ and find constants s_1, s_2, and s_3 so that $\mathbf{A} = s_1\mathbf{i} + s_2\mathbf{j} + s_3(\mathbf{i} - \mathbf{j} + \mathbf{k})$)

24. Find an equation of the plane that passes through the three points $P_1(2, 3, 2)$, $P_2(0, 1, -4)$ and $P_3(-1, 2, -3)$ without using the cross product.

25. Find all vectors of unit length parallel to the plane $3x + 2y - z = 7$ and perpendicular to the vector $\mathbf{i} - \mathbf{j} - 2\mathbf{k}$.

26. Show that the lines

$$\frac{x - 1}{1} = \frac{y - 8}{-4} = \frac{z - 11}{-2} \quad \text{and} \quad \frac{x + 1}{1} = \frac{y + 5}{3} = \frac{z + 3}{4}.$$

intersect. Find the point of intersection.

27. Find the volume V of the parallelepiped having edges AB, AC, and AD, where $A(1, 2, -1)$, $B(3, 5, 0)$, $C(-2, 0, 4)$, and $D(6, 4, -2)$.

28. Simplify the expression

$$\mathbf{i} \times (\mathbf{u} \times \mathbf{i}) + \mathbf{j} \times (\mathbf{u} \times \mathbf{j}) + \mathbf{k} \times (\mathbf{u} \times \mathbf{k})$$

for arbitrary vector $\mathbf{u}$. (*Hint:* The result of Exercise 22 is useful for this exercise.)

29. Find in symmetrical form the line of intersection of the two planes

$$(x - 2) - 2(y - 3) + 4(z - 1) = 0$$

and

$$5(x - 2) + (y - 3) - 3(z - 1) = 0$$

30. Prove that $(\mathbf{A} \times \mathbf{B}) \cdot (\mathbf{C} \times \mathbf{D}) = (\mathbf{A} \cdot \mathbf{C})(\mathbf{B} \cdot \mathbf{D}) - (\mathbf{A} \cdot \mathbf{D})(\mathbf{B} \cdot \mathbf{C})$ (*Hint:* Let $\mathbf{A} \times \mathbf{B} = \mathbf{U}$, interchange dot and cross product in the resulting scalar triple product, and apply the formula $(\mathbf{A} \times \mathbf{B}) \times \mathbf{C} = (\mathbf{A} \cdot \mathbf{C})\mathbf{B} - (\mathbf{B} \cdot \mathbf{C})\mathbf{A}$.)

31. Find the direction cosines of the line joining the points $A(-1, 2, 5)$ and $B(4, 9, 16)$. What relation must exist among the direction cosines?

32. Find the perpendicular distance from the point $P(3, 5, 6)$ to the plane $4x + 5y - z = 50$.

33. Determine an equation of the plane which contains a point (x_0, y_0, z_0) and the line given parametrically by

$$x = x_1 + at \qquad y = y_1 + bt \qquad z = z_1 + ct$$

(*Hint:* Choose three convenient points which lie in the plane.)

34. Find the distance between the skew lines

$$\frac{x - 2}{4} = \frac{y + 1}{2} = \frac{z}{1} \quad \text{and} \quad \frac{x - 1}{3} = \frac{y - 4}{1} = \frac{z + 1}{2}.$$

How do you know that the lines are skew?

35. Find the acute angle between the planes $3x - 4y + 5z = 10$ and $2x + y - 2z = 3$.

36. (a) Show that the three vectors $\mathbf{i} + \mathbf{j} - 2\mathbf{k}, \mathbf{i} - \mathbf{j} + \mathbf{k}$, and $2\mathbf{i} + 3\mathbf{j} - \mathbf{k}$ are **linearly independent**; that is, if we form $c_1(\mathbf{i} + \mathbf{j} - 2\mathbf{k}) + c_2(\mathbf{i} - \mathbf{j} + \mathbf{k}) + c_3(2\mathbf{i} + 3\mathbf{j} - \mathbf{k}) = \mathbf{0}$ where c_1, c_2, and c_3 are scalars (real numbers) then $c_1 = c_2 = c_3 = 0$ must hold. This means that no one of these three vectors may be expressed as a linear combination of the other two; that is, the vectors are noncoplanar.

    (b) Express an arbitrary vector $\mathbf{r} = x\mathbf{i} + y\mathbf{j} + z\mathbf{k}$ as a linear combination of the three vectors: What does this imply about the three vectors?

37. (a) Show that if $\mathbf{A} \neq \mathbf{0}$ and both of the conditions (i) $\mathbf{A} \cdot \mathbf{B} = \mathbf{A} \cdot \mathbf{C}$ and (ii) $\mathbf{A} \times \mathbf{B} = \mathbf{A} \times \mathbf{C}$ are satisfied, then it must follow that $\mathbf{B} = \mathbf{C}$.

    (b) If, on the other hand, only one of the conditions (i) or (ii) is satisfied, is $\mathbf{B} = \mathbf{C}$ a necessary consequence?

38. Let

$$\begin{cases} Ax + By + Cz = D \\ A'x + B'y + C'z = D' \end{cases}$$

be the equations of a line L. Show that for any constant k, the equation

$$(A + kA')x + (B + kB')y + (C + kC')z = D + kD'$$

is a plane passing through L. Show that by varying k we may obtain the totality of planes (or pencil of planes) through L with one exception. What is the exception?

39. Find an equation of the plane through the line

$$\begin{cases} 2x + 7y + 8z = 10 \\ x - 2y - 4z = 9 \end{cases} \quad \text{and the point } (4, 5, -3)$$

using the method of the previous exercise.

40. If $\mathbf{r}(t) = e^t(\sin t)\mathbf{i} + e^t(\cos t)\mathbf{j} + e^t\mathbf{k}$ denotes the position vector of a particle, where t is the time, find

    (a) the velocity at any time t

    (b) the speed at any time t

    (c) the distance travelled in the time interval $[0, 2]$

41. Find the unit tangent vector $\mathbf{T}$ and the curvature K of the space curve

$$\mathbf{r}(t) = (a_1 t + b_1)\mathbf{i} + (a_2 t + b_2)\mathbf{j} + (a_3 t + b_3)\mathbf{k}$$

at any time t, where a_i and b_i are constants ($i = 1, 2, 3$).

42. A curve has curvature K identically zero. Determine it.

43. Find the point at which the curves

$$\mathbf{r}_1(t) = e^t\mathbf{i} + (1 + \sin t)\mathbf{j} - 3t\mathbf{k}$$

and

$$\mathbf{r}_2(u) = (u - 2)\mathbf{i} + (2u - 5)\mathbf{j} + (u - 3)\mathbf{k}$$

intersect, and then find the angle of intersection.

44. A force $\mathbf{F}$ and a unit vector $\mathbf{e}$ in space have initial point O. Show how to express $\mathbf{F}$ as the sum of two vector components $\mathbf{F}_1$ parallel to $\mathbf{e}$ and $\mathbf{F}_2$ perpendicular to $\mathbf{e}$. (*Hint:* $\mathbf{F} = \mathbf{F}_1 + \mathbf{F}_2$.)

45. **Newton's second law of motion** states that the unbalanced force $\mathbf{F}$ acting on a particle of mass m is given by $\mathbf{F} = D_t(m\mathbf{v})$ where $m\mathbf{v}$ is defined to be the **linear momentum** of the particle. The quantity m may vary with the time t as in the case of a rocket that is burning fuel.

    (a) Write Newton's second law in expanded form by calculating the right side in the formula for $\mathbf{F}$.

    (b) What results if $\mathbf{F} = \mathbf{0}$?

46. The moment $\mathbf{M}$ of a force $\mathbf{F}$ acting at a point P about a point O is defined by $\mathbf{M} = \mathbf{r} \times \mathbf{F}$ where $\mathbf{r} = \overrightarrow{OP}$. Show that if $\mathbf{F}_1$ and $\mathbf{F}_2$ are two forces acting at P, that the moment of $\mathbf{F}_1 + \mathbf{F}_2$ about the point O is the vector sum of the moments due to $\mathbf{F}_1$ and $\mathbf{F}_2$. (Any point can be chosen as an origin of coordinates.)

47. Suppose that three forces $\mathbf{F}_1, \mathbf{F}_2$, and $\mathbf{F}_3$ acting on a mass m at a point P are such that $\mathbf{F}_1 + \mathbf{F}_2 + \mathbf{F}_3 = \mathbf{0}$; that is, the forces are self-equilibrating. Show that the vector sum of the moments of these three forces about any point O must also be $\mathbf{0}$. This is a very useful principle in mechanics.

48. The angular momentum (or moment of linear momentum) $\mathbf{L}$ about a point O is defined by

$$\mathbf{L}(t) = \mathbf{r}(t) \times m\mathbf{v}(t)$$

where $\mathbf{r} = \overrightarrow{OP}$ is the position vector of the particle of mass m moving with velocity $\mathbf{v}$ at point P. Show that

$$\mathbf{L}'(t) = \mathbf{r} \times D_t(m\mathbf{v}) = \mathbf{r} \times \mathbf{F}$$

This says that *the time rate of change of angular momentum about a point* O *is equal to the moment of the resultant force acting on* m *about* O.

49. Find the torsion of the space curve

$$\mathbf{r}(t) = (1 - \cos t)\mathbf{i} + 4(\cos t)\mathbf{j} + (5 - 2 \cos t)\mathbf{k}$$

and interpret your result.

50. If $\mathbf{r}(t) = f(t)\mathbf{i} + g(t)\mathbf{j} + h(t)\mathbf{k}$, where f, g, and h possess third order derivatives with respect to t, show that

$$D_t[\mathbf{r} \cdot (\mathbf{r}' \times \mathbf{r}'')] = \mathbf{r} \cdot (\mathbf{r}' \times \mathbf{r}''')$$

51. The space curve whose vector equation is

$$\mathbf{r}(t) = a\sqrt{t}(\sin t)\mathbf{i} + b\sqrt{t}(\cos t)\mathbf{j} + t\mathbf{k} \qquad t \geq 0$$

where a and b are nonzero constants lies on a surface. Find an equation of the surface in rectangular form and describe it.

52. Describe the graph of each of the following
   (a) $z^2 = 2r^2$
   (b) $\sin 4\theta = 0$
   (c) $(\cos \pi r)(\sec \theta)(\sin z) = 0$
   where r, θ, and z are cylindrical coordinates.

53. Describe the graph of the surfaces defined by
   (a) $f(y, z) = 0$ (rectangular coordinates)
   (b) $g(z, r) = 0$ (cylindrical coordinates)

54. Express the following equations in cylindrical and spherical coordinates
   (a) $x^2 + y^2 = 3z$
   (b) $x^2 + y^2 + z^2 - 8z = 0$
   (c) $3x - 2y - 4z = 0$

17

Differential Calculus of Functions of Two or More Variables

Functions of several variables occur very naturally in scientific investigations. Suppose, for example, that we are given a triangle with sides of lengths x and y and with included angle θ. The area A of the triangle is determined by the formula $A = \dfrac{xy \sin \theta}{2}$. Furthermore, the variables x, y, and θ are restricted in that $x > 0$, $y > 0$ and $0 < \theta < \pi$. Clearly the value of A is dependent on the values of the three variables.

Another illustration involving several independent variables is furnished by the following question: Of all rectangular boxes of given volume, find the one that has the least surface area. We note that the lengths of two of the edges may be chosen arbitrarily. However, the length of the third edge is dependent on the other two by the condition that the product of the three edge lengths is prescribed. From this, the surface area is easily expressed in terms of the first two variables. Our problem is to minimize this function of two independent variables. This will be solved later on in this chapter.

Rather than to extend our list of examples (which is virtually endless), we shall turn now to a detailed study of the calculus of functions of several variables. In particular, we will learn some useful procedures for the resolution of these problems. Also, it will be observed that many of the ideas are analogous to and are strongly dependent upon the calculus of functions of one variable.

17.1 FUNCTIONS OF TWO OR MORE VARIABLES

Real valued functions of several real variables occur very frequently. Formulas such as

$$z = x^2 + 2y^2 \tag{1}$$

$$u = xy^2z \tag{2}$$

$$w = (x + y)z^2t \tag{3}$$

define functions of 2, 3, and 4 variables, respectively.

For the major portion of this section, we restrict our considerations to functions of two variables. The domain of a function of two variables is a set of ***ordered pairs*** of numbers, and a function of two variables associates with each pair of numbers in its domain a unique number in its range.

To illustrate, $z = x^2 + 2y^2$ defines a function of two ***independent variables*** x and y because to each number pair (x, y) (the domain in this instance, is the entire xy-plane, assuming no further restrictions on x and y are imposed) there corresponds one and only one value of z, the ***dependent variable.*** If the ordered pair is $(1, 3)$ then the corresponding number z in its range is $z = 1^2 + 2(3^2) = 19$; if the ordered pair is $(-4, 0)$, then the corresponding value of $z = (-4)^2 + 2(0)^2 = 16$, and so on.

Definition   Let D be a set of ordered pairs of real numbers, and let R be a set of real numbers. A rule of correspondence that assigns to each pair of numbers in D a unique number in R is called a ***function of two variables.*** The set D is called the ***domain*** of the function. The set of all numbers in R that corresponds to the set of elements of D is called the ***range.*** If the domain is not given explicitly or implied, it is understood that D consists of all pairs (x, y) for which $z = f(x, y)$ is a real number. We call x and y the ***independent variables,*** and z the ***dependent*** variable.

EXAMPLE 1   The domain D of the function defined by $z = f(x, y) = \sqrt{9 - x^2 - y^2}$, consists of all number pairs (x, y) satisfying $x^2 + y^2 \leq 9$ (geometrically this is all points on and inside the circle centered at the origin with radius 3). The range R of f consists of the interval $0 \leq z \leq 3$. Figure 17.1.1 shows the shaded domain of the function f.   ●

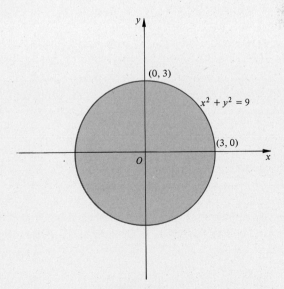

Figure 17.1.1

EXAMPLE 2   The domain of the function defined by $V = \frac{\pi}{3}r^2h$ consists of all number pairs (r, h) of real numbers, and the range consists of all real numbers. However, if it is understood that V is the volume of a right circular cone with r as its base radius and h as its altitude, then r, h, and V are restricted to positive values. Hence the domain is the first quadrant in the rh-plane (with boundary not included).   ●

EXAMPLE 3   Find the domain and range of the function

$$f(x, y) = \ln (x + y - 2)$$

and sketch the domain as a shaded region in the xy-plane.

Solution   Since $\ln t$ is defined if and only if $t > 0$, it follows that the given function is defined in the open half-plane $x + y - 2 > 0$ (boundary $x + y - 2 = 0$ is not included). (See Figure 17.1.2.) The range of f is $(-\infty, \infty)$.   ●

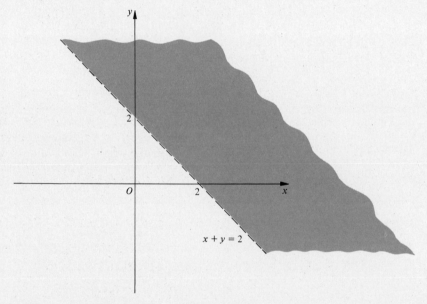

Figure 17.1.2

To summarize, it is convenient geometrically to interpret the domain D of a function $f(x, y)$ as a set of points in the xy-plane where (x_0, y_0) corresponds to the point having rectangular coordinates $x = x_0$ and $y = y_0$.

EXAMPLE 4   The function g of two variables x and y is the set of ordered pairs (x, y) for which

$$z = \frac{\sqrt{x^2 + y^2 - 36}}{2x + y - 4}$$

Find the domain of g and sketch it by showing the set of points in R_2 that correspond to it.

Solution   The domain of g consists of all ordered pairs (x, y) for which $x^2 + y^2 - 36 \geq 0$ and $2x + y - 4 \neq 0$. This is the set of points, exclusive of the straight line $2x + y - 4 = 0$, which are either on the circle $x^2 + y^2 - 36 = 0$ or in the exterior of the region bounded by it. The required sketch is shown in Figure 17.1.3. ●

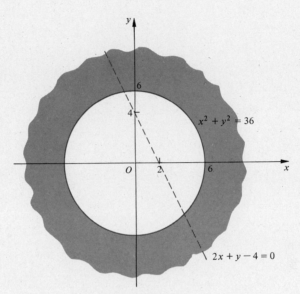

Figure 17.1.3

EXAMPLE 5   Find the domain and range of

$$h(x, y) = \frac{1}{\sqrt{9x^2 - 4y^2}}$$

Solution   A pair (x, y) is in the domain of h if and only if

$$9x^2 - 4y^2 > 0 \tag{i}$$

addition of $4y^2$ to each side of (i) yields, equivalently,

$$9x^2 > 4y^2 \tag{ii}$$

and the solution of this inequality holds if and only if

$$-\frac{3}{2}|x| < y < \frac{3}{2}|x| \tag{iii}$$

The domain of h is shown in Figure 17.1.4. It consists of all points (in the shaded region of the figure) above $y = -\frac{3}{2}|x|$ and below $y = \frac{3}{2}|x|$.

On this set, defined by (iii), $9x^2 - 4y^2$ takes on all positive numbers, and thus the same is true for its reciprocal. Consequently, the range of h is $(0, \infty)$. ●

Definition   The functions f and g of two independent variables are said to be **equal** in a domain D if $f(x, y) = g(x, y)$ for every (x, y) in D.

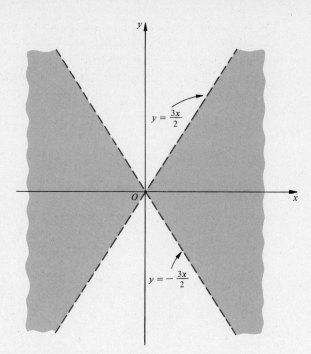

$$y = \frac{3x}{2}$$

$$y = -\frac{3x}{2}$$

Figure 17.1.4

Definition   Let f and g be two functions of two independent variables x and y with domains D_1 and D_2, respectively. Then if D_1 and D_2 have a set D of common points (that is, its intersection), the **sum** of f and g is the new function with domain D, and whose values are $f(x, y) + g(x, y)$. If the sum is denoted by $f + g$, we write

$$(f + g)(x, y) = f(x, y) + g(x, y)$$

Analogous definitions apply to $f - g$, $f \cdot g$ and $\dfrac{f}{g}$ where, of course, the points at which $g(x, y) = 0$ are excluded from the domain of $\dfrac{f}{g}$.

Definition   If f is a function of a single variable and g is a function of two variables then the **composite** (or **compound**) function $f \circ g$ is the function of two variables defined by

$$(f \circ g)(x, y) = f(g(x, y))$$

Note that the domain of the composite function is the set of number pairs (x, y) in the domain of g such that the values of g are in the domain of f.

EXAMPLE 6   Given $f(u) = \cos^{-1} u$ and $g(x, y) = x + 2y - 3$, find the function $f \circ g$ and its domain.

Solution

$$(f \circ g)(x, y) = f(g(x, y))$$
$$= f(x + 2y - 3) = \cos^{-1}(x + 2y - 3)$$

The domain of g is the entire xy-plane, and the domain of f is $[-1, 1]$. Now the domain of $f \circ g$ is the set of points (x, y) such that g (that is, its range) is in the domain of f. Thus the domain of $f \circ g$ is the set (x, y) such that $-1 \le x + 2y - 3 \le 1$ or, equivalently, $2 \le x + 2y \le 4$. The domain is the strip bounded by the two parallel lines $x + 2y = 2$ and $x + 2y = 4$. It is shown shaded in Figure 17.1.5. ●

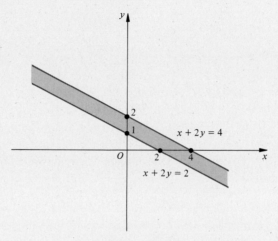

Figure 17.1.5

Definition   The set of points (x, y, z) in R_3 whose coordinates satisfy $z = f(x, y)$ where (x, y) is a point in the domain D of f, is called the **graph** of the function in three-dimensional space.

The graph of a function of two variables is a surface† in three-dimensional space whose Cartesian coordinates are given by the ordered triple of numbers (x, y, z). Since to each (x, y) in the domain D there corresponds one and only one number z, *no line perpendicular to the xy-plane can intersect the graph in more than one point.*

EXAMPLE 7   Graph the function f defined by

$$f(x, y) = \frac{1}{2}\sqrt{4 - 2x^2 - y^2}$$

What is its domain and range?

Solution   With $z = f(x, y) = \frac{1}{2}\sqrt{4 - 2x^2 - y^2}$ the graph is the upper half of the ellipsoid (Figure 17.1.6)

$$\frac{x^2}{2} + \frac{y^2}{4} + z^2 = 1$$

Its domain is all (x, y) such that $2x^2 + y^2 \le 4$ which is an ellipse $2x^2 + y^2 = 4$, center at $(0, 0)$ together with its interior. The range is $0 \le z \le 1$. ●

---

† With the exception of degenerate loci such as the empty set, points, lines, etc.

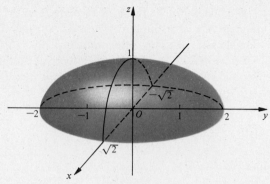

Figure 17.1.6

In many actual instances, it is difficult to draw the graph of a function of two variables. Moreover, even if one is able to draw the surface, its interpretation may be complicated. This problem is circumvented by using a device known by map makers.

In mountainous regions, it is a common practice to sketch curves joining points of common elevation, say 2000 ft above sea level. A collection (or family) of such equielevation curves when properly presented enables one to comprehend the altitude variations in a given region (see Figure 17.1.7).

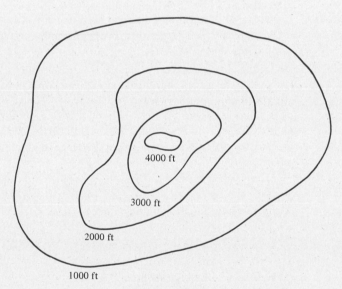

Figure 17.1.7

Definition   If f is a function of x and y and c is a constant, then the set of points where

$$f(x, y) = c$$

is called a ***level curve*** (or ***contour curve***) for f.

EXAMPLE 8   Discuss the level curves for the function $f(x, y) = x^2 + y^2$.

Solution The level curves of f are concentric circles with radii $\sqrt{c}$. Its equations are

$$x^2 + y^2 = c$$

when c is positive. The level curve $x^2 + y^2 = 0$ is just the point $(0, 0)$, and, if $c < 0$, there is no level curve (Figure 17.1.8). ●

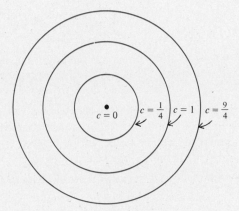

$c = 0$ $c = \frac{1}{4}$ $c = 1$ $c = \frac{9}{4}$

Figure 17.1.8

There are a number of interesting applications of the concept of level curves. One is to the determination of thermal stresses in two-dimensional structures, that is, the computation of the internal forces due to a temperature distribution over its surface. For example, we may have a rectangular plate (made of aluminum, steel, wood, etc.) which is supported in some prescribed fashion subject to a temperature distribution $T(x, y)$. Once $T(x, y)$ is known the internal stresses can be computed. In particular, it is useful to the stress analyst to know where $T(x, y) = c_i =$ constant and to find the associated level curves. These curves are called *isothermal curves* because all points on it have the same temperature. It is also significant for the analyst to look at the spacing between the level curves. This will concern us later in this chapter (Section 17.7). These problems have also been explored optically and have led to the development of the science of "photoelasticity."

To the physicist interested in electrostatics, the determination of the electric potential $V(x, y)$ plays an analogous role. The curves $V(x, y) = c_i$ are known as the *equipotential curves* because all points on such a curve have the same electrostatic potential.

The *isobars* on a weather map ("lines" of constant barometric pressure) are level curves, and aid the meteorologist in predicting weather.

There is no difficulty in extending our definition of a function of two variables to functions of three or more variables.

Definition Let D be a set of ordered n-tuples of real numbers $x_1, x_2, x_3, \ldots, x_n$ where n is *any* positive integer. Also, let R be a set of real numbers. A rule of correspondence that assigns to each ordered n-tuple of D one and only one number in R is called a *function* of n variables. The set D is called the *domain* of the function. The set of all numbers in R that corresponds to the set of elements of D is called the *range*. (It is useful to think of the ordered n-tuple of numbers in D as coordinates of a point in n-dimensional space.)

EXAMPLE 9 Describe the domain and range of the indicated function of three variables

$$f(x, y, z) = \frac{1}{\sqrt{3 - x^2 - y^2 - z^2}}$$

Solution If we rewrite $f(x, y, z)$ in the form

$$f(x, y, z) = \frac{1}{\sqrt{3 - (x^2 + y^2 + z^2)}}$$

it follows that the domain is the interior of the sphere with center at $(0, 0, 0)$ and radius $= \sqrt{3}$, that is, $x^2 + y^2 + z^2 < 3$. The range is any positive number $\geq \dfrac{1}{\sqrt{3}}$ since the least value of f occurs when $x = y = z = 0$ (which makes the denominator largest) and, furthermore, f has no maximum value. This conclusion follows because the denominator can be made as close to zero as we choose when the boundary of the sphere is approached. ●

Definition Let f be a function defined in some region in three-dimensional space (its domain). Let $u = f(x, y, z)$. Then if c is a constant, the equation

$$f(x, y, z) = c$$

is called a ***level surface*** for f.

EXAMPLE 10 Describe the level surfaces defined by

$$f(x, y, z) = x^2 + y^2 + z^2 = c$$

What are the permissible values of c? In particular, determine the level surface that passes through $(3, -4, 5)$.

Solution The level surfaces are concentric spheres with center at $(0, 0, 0)$ and radii $\sqrt{c}$, so that the permissible values of c are $c \geq 0$. Note that $c = 0$ corresponds to a "point sphere" or a single point $(0, 0, 0)$. Substitution of $(3, -4, 5)$ for x, y, and z, respectively, yields $3^2 + (-4)^2 + 5^2 = 50 = c$ and therefore the particular level surface passing through this point is the sphere

$$x^2 + y^2 + z^2 = 50$$

whose radius $= \sqrt{c} = 5\sqrt{2}$. ●

Exercises 17.1

In Exercises 1 through 8, find the values of x and y for which the given function is defined. Describe its domain.

1. $f(x, y) = 3x^2 - 4y + 2$

2. $f(x, y) = \dfrac{1}{x - y}$

3. $f(x, y) = \sqrt{3x + y - 4}$

4. $g(x, y) = \sqrt{-x - y - 1}$

5. $g(x, y) = \ln(16 - 4x^2 - y^2)$

6. $h(x, y) = \dfrac{\sqrt{100 - x^2 - y^2}}{xy}$

7. $h(x, y) = \ln \sin^2 (y + 2x)$

8. $F(x, y) = \sqrt[3]{x^2 - 3y^2}\, e^{x-y}$

9. Given that $f(x, y) = x^2 - 3xy + 4y^2$, determine
   (a) $f(2, -1)$ (b) $f(0, -2)$ (c) $f(y, x)$

10. Given that $f(x, y) = \ln \dfrac{x - y}{x + 2y}$, determine

   (a) $f(1, 0)$ (b) $f(0, 1)$ (c) $f(2x, x)$

11. Given that $g(x, y) = \sqrt{-xy}\, e^{x-y}$, find

   (a) $g(2, -4)$ (b) $g(1, 1)$ (c) $g\left(x, -\dfrac{1}{x}\right)$

12. Given that $h(x, y) = \begin{cases} x + y & \text{if } x \leq 0 \\ \dfrac{x + y}{x - y} & \text{if } x > 0 \text{ and } y < 0 \\ \sqrt{x + y} & \text{if } x > 0 \text{ and } y \geq 0 \end{cases}$

   Find (a) $h(-1, 3)$; (b) $h(2, -5)$ (c) $h(7, 2)$; (d) domain of h.

In Exercises 13 through 18, find the domain and the range of the function. Sketch the domain.

13. $f(x, y) = \ln (3x - 2y - 1)$

14. $f(x, y) = 2 - x^2 - 3y^2$

15. $z = 3 + \sqrt{xy}$

16. $u = \sqrt{49 - x^2 - y^2 - z^2}$

17. $f(x, y, z) = \sqrt{z}\,\sqrt{4x + y - 3}$

18. $g(x, y, z) = e^{x+y-z}$

In Exercises 19 through 26, describe the level curves, that is, $f(x, y) = c$ for each of the following functions. Sketch several level curves and give the range of c (and therefore the range of f).

19. $f(x, y) = x + y$

20. $f(x, y) = 2x - 3y$

21. $f(x, y) = xy$

22. $f(x, y) = x^2 - y^2$

23. $f(x, y) = 3x^2 - y$

24. $f(x, y) = \dfrac{x}{x + y}$

25. $f(x, y) = e^{3xy}$

26. $f(x, y) = \sqrt{x^2 + y^2 - 100}$

In Exercises 27 through 30, identify the level surfaces $f(x, y, z) = c$ and sketch the particular surface at the given number.

27. $f(x, y, z) = 3x + y + 4z, \quad c = 0$

28. $f(x, y, z) = \sqrt{x^2 + y^2 + z^2}, \quad c = 2$

29. $f(x, y, z) = x^2 + y^2 - z, \quad c = 1$

30. $f(x, y, z) = \dfrac{z^2}{x^2 + y^2}, \quad c = 1$

31. If $f(x, y) = x^2 - y^2$, $g(u) = \sqrt{u + 1}$, and $h(t) = t^2 - 1$, find
    (a) $(g \circ f)(3, 2)$ (b) $(g \circ f)(2, 3)$ (c) $(h \circ f)(1, 2)$
    (d) $(h \circ g)(u)$ (e) $(h \circ g)(f(a, b))$

32. If $f(x, y) = \sqrt{1 + x^2 + y^2}$, $g(z) = z^2 - 1$, and $h(\theta) = \sqrt{\theta + 2}$, find
    (a) $(g \circ f)(-1, 4)$ (b) $(h \circ f)(0, 0)$ (c) $(g \circ h)(v)$
    (d) $(g \circ f)(x, x)$ (e) $(h \circ g)(f(x, 2x))$

33. A metal plate occupies the region $-10 \leq x \leq 10$, $-8 \leq y \leq 8$. The temperature distribution is determined to be $T = 2x^2 + y^2 + 3$.
    (a) What are the isothermal curves?
    (b) Where are the maximum and minimum temperatures and what are their values?

34. A potential function of interest in both pure and applied mathematics is

$$V(x, y) = \ln \sqrt{x^2 + y^2}$$

   (a) What is the domain of the function?
   (b) What is the range?

(c) What are the equipotential lines?

(d) What symmetries does the potential function possess?

35. A function $f(x, y)$ is said to be **homogeneous** of degree n (not necessarily an integer) if and only if

$$f(tx, ty) = t^n f(x, y) \qquad \text{(i)}$$

for arbitrary real values of t. If (i) is true for all $t > 0$ only then $f(x, y)$ is said to be **positively homogeneous**. Determine which functions below are homogeneous and state the degree.

(a) $x^3 - 5x^2 y$ (b) $\dfrac{xy}{2x + y}$

(c) $\dfrac{x^2 y}{x^2 + 2y}$ (d) $(x^2 + y^2)^{1/2} - 4x$

(e) $(x - 2y)^{1/3}$

17.2

PARTIAL DERIVATIVES

The concept of a partial derivative is easily explained. Suppose that we are given $f(x, y)$ and that y is held constant. For this particular value of y, $f(x, y)$ becomes a function of a single variable x, and its derivative with respect to x is calculated in the usual way. Such a derivative is called the partial derivative of f with respect to x and is denoted by $f_x(x, y)$ or, alternatively, by $\dfrac{\partial f}{\partial x}(x, y)$.† Also, when it is clear that we are concerned with a function of two variables x and y (or more) the notation f_x and $\dfrac{\partial f}{\partial x}$ may be used. Less frequently, where ambiguities may exist, it is sometimes necessary to emphasize which variable is to be held constant and we write $\left(\dfrac{\partial f}{\partial x}\right)_y$. This means that y is to remain constant during the calculation of the partial derivative with respect to x. Note that f_x gives the rate of change of f with respect to x (y constant) and f_y yields the rate of change of f with respect to y (x constant).

EXAMPLE 1   If $f(x, y) = x^3 - 3x^2 y + 10xy - y^2 + 1$, determine f_x and f_y.

Solution   In the calculation of f_x, y is to be *held constant,* hence

$$f_x = 3x^2 - 6xy + 10y$$

Note that $\dfrac{\partial}{\partial x}(-y^2 + 1) = 0$ since the derivative of a constant with respect to x is zero.

Similarly, in order to find f_y, x is *held constant,* and we obtain

$$f_y = -3x^2 + 10x - 2y \qquad \bullet$$

---

† The notation $\dfrac{\partial}{\partial x} f(x, y)$ is often encountered also. Some textbooks (but not this one) also use the notation $f_1(x, y) = f_x(x, y)$ and $f_2(x, y) = f_y(x, y)$.

EXAMPLE 2   If $f(x, y) = e^{-x} \cos y - y^2 e^{xy}$, calculate f_x and f_y.

Solution   In this instance,

$$f_x = -e^{-x} \cos y - y^3 e^{xy}$$

(since $D_x e^{u(x)} = e^{u(x)} D_x u(x)$)

$$f_y = -e^{-x} \sin y - xy^2 e^{xy} - 2y e^{xy}$$

where the rule for differentiation of a product is utilized.   ●

Definition   Let f be a function of two independent variables x and y. **The partial derivative of f with respect to x** is the function f_x $\left(\text{or } \dfrac{\partial f}{\partial x}\right)$ whose value at any point (x, y) in the domain D of f is

$$f_x(x, y) = \frac{\partial f}{\partial x}(x, y) = \lim_{\Delta x \to 0} \frac{f(x + \Delta x, y) - f(x, y)}{\Delta x} \tag{1}$$

provided that the limit on the right side of (1) exists. Similarly, **the partial derivative of f with respect to y** is the function f_y $\left(\text{or } \dfrac{\partial f}{\partial y}\right)$ whose value at any point in the domain D of f is

$$f_y(x, y) = \frac{\partial f}{\partial y}(x, y) = \lim_{\Delta y \to 0} \frac{f(x, y + \Delta y) - f(x, y)}{\Delta y} \tag{2}$$

provided this latter limit exists.

Partial derivatives as defined by (1) and (2) above have simple geometric interpretations. In Figure 17.2.1 we have sketched a surface $z = f(x, y)$. Consider

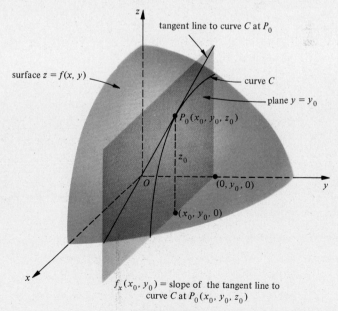

$f_x(x_0, y_0) = $ slope of the tangent line to
curve C at $P_0(x_0, y_0, z_0)$

Figure 17.2.1

then a plane $y = y_0$, which is perpendicular to the y-axis and intersects the surface in a plane curve C as shown and defined by $z = f(x, y_0)$. Then in the plane $y = y_0$, $f_x(x_0, y_0)$ is the slope of the tangent line to the curve C at the point $P(x_0, y_0, z_0)$ of the given surface. The other partial derivative may be interpreted in a similar manner (Figure 17.2.2). We pass a plane $x = x_0$ perpendicular to the x-axis and which also contains the given point $P_0(x_0, y_0, z_0)$ on the surface $z = f(x, y)$. Again, this intersects the surface in a plane curve C defined by $z = f(x_0, y)$ and C together with its tangent line at P_0 are shown in the figure. Also, $f_y(x_0, y_0)$ is the slope of the tangent line to C at P_0.

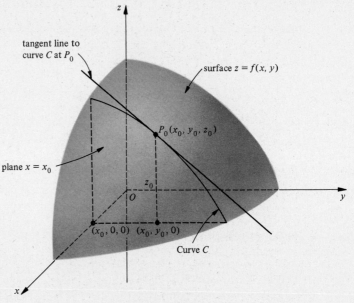

Figure 17.2.2

The notion of partial derivatives may be easily extended to functions of three variables.

Definition Let f be a function of three variables x, y and z, then the ***partial derivatives of f*** are given by

$$f_x(x, y, z) = \frac{\partial f}{\partial x}(x, y, z) = \lim_{\Delta x \to 0} \frac{f(x + \Delta x, y, z) - f(x, y, z)}{\Delta x} \tag{3}$$

$$f_y(x, y, z) = \frac{\partial f}{\partial y}(x, y, z) = \lim_{\Delta y \to 0} \frac{f(x, y + \Delta y, z) - f(x, y, z)}{\Delta y} \tag{4}$$

$$f_z(x, y, z) = \frac{\partial f}{\partial z}(x, y, z) = \lim_{\Delta z \to 0} \frac{f(x, y, z + \Delta z) - f(x, y, z)}{\Delta z} \tag{5}$$

provided that the limits exist. Note that two of the variables are held constant in the calculation of each of the partial derivatives (3), (4), and (5). In (3), y and z are held constant, and so on.

Similarly the notion of partial derivatives is easily extended to functions of n variables, $z = f(x_1, x_2, \ldots, x_n)$.

Definition   Let f be a function of n variables $x_1, x_2, \ldots, x_n$. Then the partial derivatives of f with respect to x_i is

$$f_{x_i}(x_1, x_2, \ldots, x_n) = \lim_{\Delta x_i \to 0} \left[\frac{f(x_1, x_2, \ldots, x_{i-1}, x_i + \Delta x_i, x_{i+1}, \ldots, x_n)}{\Delta x_i} \right.$$
$$\left. - \frac{f(x_1, x_2, \ldots, x_{i-1}, x_i, x_{i+1}, \ldots, x_n)}{\Delta x_i} \right] \quad (6)$$

provided this limit exists. Note that all variables except x_i are held constant.

EXAMPLE 3   If $u(x, y, z) = 2xyz^2 - 3x^2z^2$, show that $xu_x + yu_y + zu_z = 4u$.

Solution

$$u_x = 2yz^2 - 6xz^2$$
$$u_y = 2xz^2$$
$$u_z = 4xyz - 6x^2z$$

Multiplication of u_x by x, u_y by y, and u_z by z gives

$$xu_x + yu_y + zu_z = 2xyz^2 - 6x^2z^2 + 2xyz^2 + 4xyz^2 - 6x^2z^2$$

and combining terms yields

$$xu_x + yu_y + zu_z = 8xyz^2 - 12x^2z^2 = 4(2xyz^2 - 3x^2z^2)$$
$$= 4u$$

and the result is verified.   ●

EXAMPLE 4   If $g(x, y, z) = x^2yz$, find $g_x(2, -1, 3)$ by three methods.

Solution   (a) For constant y and z, $g_x(x, y, z) = 2xyz$ so that, in particular,

$$g_x(2, -1, 3) = 2(2)(-1)(3) = -12$$

(b) If $y = -1$ and $z = 3$, then we have a function of x,

$$g(x, -1, 3) = -3x^2 = h(x)$$

Hence

$$g_x(x, -1, 3) = h'(x) = -6x$$

from which

$$g_x(2, -1, 3) = -6(2) = -12$$

(c) As our last method, we revert back to the definition of the partial derivative and the delta process. Let us find g_x in general (although it is really not necessary)

$$g_x(x, y, z) = \lim_{\Delta x \to 0} \frac{g(x + \Delta x, y, z) - g(x, y, z)}{\Delta x}$$

$$= \lim_{\Delta x \to 0} \frac{(x + \Delta x)^2 yz - x^2 yz}{\Delta x}$$

$$= \lim_{\Delta x \to 0} \frac{[x^2 + 2x\,\Delta x + (\Delta x)^2]yz - x^2 yz}{\Delta x}$$

Subtraction and division by $\Delta x \neq 0$ yields

$$g_x(x, y, z) = \lim_{\Delta x \to 0} (2xyz + (\Delta x)yz)$$

and therefore

$$g_x(x, y, z) = 2xyz$$

Substitution of the given coordinates yields

$$g_x(2, -1, 3) = 2(2)(-1)(3) = -12 \qquad \bullet$$

EXAMPLE 5   If $f(x_1, x_2, x_3, x_4) = x_1 \cos(x_1 + x_2 + 3x_3) + x_3{}^2 e^{x_4} \ln x_1$, find $f_{x_i}(i = 1, 2, 3, 4)$.

Solution $$f_{x_1} = -x_1 \sin(x_1 + x_2 + 3x_3) + \cos(x_1 + x_2 + 3x_3) + \frac{x_3{}^2 e^{x_4}}{x_1}$$

where x_2, x_3, and x_4 were held constant. Similarly,

$$f_{x_2} = -x_1 \sin(x_1 + x_2 + 3x_3)$$
$$f_{x_3} = -3x_1 \sin(x_1 + x_2 + 3x_3) + 2x_3 e^{x_4} \ln x_1$$
$$f_{x_4} = x_3{}^2 e^{x_4} \ln x_1 \qquad \bullet$$

EXAMPLE 6   Partial derivatives can be found when a function is defined implicitly. For example, suppose that $z = z(x, y)$ is defined implicitly by

$$x^3 - y^2 + xz + \ln yz = 0 \qquad (7)$$

and we seek z_x and z_y.

Solution   To find z_x, we hold y constant and differentiate (7) implicitly with respect to x (where x is treated as the independent variable and z is the dependent variable which is a function of x). There results

$$3x^2 + xz_x + z + \frac{1}{yz} yz_x = 0$$

which yields

$$\left(x + \frac{1}{z}\right) z_x = -(3x^2 + z)$$

or

$$z_x = -\frac{z(3x^2 + z)}{xz + 1}$$

Similarly, to find z_y, x is held constant and (7) is differentiated implicitly with respect to y:

$$-2y + xz_y + \frac{1}{yz}(yz_y + z) = 0$$

or

$$-2y + xz_y + \frac{1}{z}z_y + \frac{1}{y} = 0$$

The solution for z_y is

$$z_y = \frac{2y^2 - 1}{y} \cdot \frac{z}{xz + 1}$$

Exercises 17.2

In Exercises 1 through 16, find the partial derivatives.

1. $f(x, y) = xy^2 - x^3y$

2. $g(x, y) = x^3 e^{-y}$

3. $h(x, y) = (2x + y)(x - 3y)$, compute this in two ways in each instance

4. $u = e^{x^2} \cosh{(x + 2y)}$

5. $z = \ln{(x^2 + y^2)}$

6. $z = \cos{(x^2 + xy)}$

7. $u = \sqrt{x^2 + y^2 + 10}$

8. $v = (x^4 + y^2)^{-2}$

9. $f(x, y) = \ln{\dfrac{x}{x + y}}$

10. $g(x, y) = \sin^2{(x - y)}$

11. $u = \tan^{-1}{\dfrac{x}{y}}$

12. $v = (\pi - x)^2 y^2 \sqrt{z}$

13. $u = \dfrac{ax + by}{cx + dy}$, where a, b, c and d are constants. What happens when $ad - bc = 0$? Justify your answer.

14. $\rho = [1 - (x^2 + y^2 + z^2)]^{1/2}$

15. $f(x, y, z) = e^{xyz} + \sqrt{z - x - y}$

16. $g(x, y, z) = \sin^{-1}{(ax + by + cz)}$, where a, b and c are constants

In Exercises 17 through 20, find the indicated partial derivative by using the definition (1) or (2).

17. $f(x, y) = 4x - 3y - 5$, f_x

18. $g(x, y) = xy$, g_y

19. $F(x, y) = x^2 - 2xy + y^2$, F_y

20. $G(x, y) = (2x + y)^{-1}$, G_x

21. For $f(x, y) = e^{3x} \ln{2y}$, find $f_x(0, 1)$ and $f_y(0, e^{-1})$.

22. If $g(x, y) = \dfrac{6x}{y} + \cos{(x - y)}$, find $g_x\left(\dfrac{\pi}{2}, \dfrac{\pi}{3}\right)$ and $g_y(0, \pi)$

23. If $u(x, y) = \left[1 - \dfrac{x^2}{a^2} - \dfrac{y^2}{b^2}\right]^{1/2}$, find $u_x\left(\dfrac{a}{2}, 0\right)$ and $u_y\left(\dfrac{a}{2}, \dfrac{b}{2}\right)$

24. If $T(x, y, z) = \dfrac{x^2}{y + z}$, find $T_x(1, 2, 3)$ and $T_z(1, 2, 3)$

25. Find the rate of change of the volume of a right circular cone of radius r and height h **(a)** first with respect to r and **(b)** then with respect to h. $\left(\text{Hint: } V = \dfrac{\pi}{3}r^2h.\right)$

26. If V is the volume of a rectangular parallelepiped (a box) with sides x, y and z, show that

$$xV_x + yV_y + zV_z = 3V$$

27. Find the slope at $(3, 2, 13)$ of the section of the paraboloid of revolution $z = x^2 + y^2$ made by the plane $x = 3$.

28. Find the slope at $(3, 4, 5)$ of the section of the surface $x^2 + y^2 - z^2 = 0$ made by the plane $y = 4$.

In each of Exercises 29 through 32, show that $u_x = v_y$ and $u_y = -v_x$. These are known as the **Cauchy-Riemann equations in rectangular coordinates.**

29. $u(x, y) = x^2 - y^2$, $v(x, y) = 2xy$

30. $u(x, y) = \dfrac{x}{x^2 + y^2}$, $v(x, y) = -\dfrac{y}{x^2 + y^2}$, where $x^2 + y^2 \neq 0$

31. $u(x, y) = e^x \cos y$, $v(x, y) = e^x \sin y$

32. $u(x, y) = \ln \sqrt{x^2 + y^2}$, $v(x, y) = \tan^{-1} \dfrac{y}{x}$, where $x \neq 0$

In each of Exercises 33 through 36, find F_x and F_y if $F(x, y)$ is defined by

33. $\displaystyle\int_x^y e^{-t}\, dt$

34. $\displaystyle\int_x^y e^t \cos t\, dt$

35. $\displaystyle\int_y^x e^{-t^2/2}\, dt$

36. $\displaystyle\int_{3y}^{2x} \dfrac{\sin t}{t}\, dt$, where x and y are positive

37. If $h = (\cosh^3 \theta - \sinh^2 \phi)^{3/2}$, find h_θ and h_ϕ.

38. If $f(x_1, x_2, x_3, x_4) = x_2 \sin(x_1 + x_2 - x_4) + x_1 \ln x_2 x_3 x_4$ where x_1, x_2, x_3, and x_4 are positive, find f_{x_i}, $(i = 1, 2, 3, 4)$.

39. If $z = z(x, y)$ is defined implicitly by

$$x^2 + y^3 - yz + e^z = 0$$

find z_x and z_y.

40. If $z = z(x, y)$ is defined implicitly by

$$x + y^2 + z + \sin z = \pi$$

find z_x and z_y.

41. (a) If n resistors of resistance x_i are connected in series the total resistance R is given by

$$R = \sum_{i=1}^{n} x_i$$

Find R_{x_k} for $k = 1, 2, \ldots, n$

(b) If n resistors of resistance x_i are connected in parallel the total resistance R is given by

$$\frac{1}{R} = \sum_{i=1}^{n} \frac{1}{x_i}$$

Find R_{x_k} for $k = 1, 2, \ldots, n$.

42. (a) The **arithmetic mean** A of n numbers x_k, $(k = 1, 2, \ldots, n)$ is defined by

$$A = \frac{1}{n} \sum_{k=1}^{n} x_k$$

Find A_{x_k}.

(b) The **geometric mean** G of n positive numbers x_k, $(k = 1, 2, \ldots, n)$ is defined by

$$G = [x_1 x_2 \cdots x_{n-1} x_n]^{1/n}.$$

Find G_{x_k}.

PARTIAL DERIVATIVES OF HIGHER ORDER

The partial derivatives previously considered (in Section 17.2) were *first-order* partial derivatives for functions of two or more variables. Such derivatives are, in general, functions of the same variables and therefore may be differentiated partially with respect to an independent variable once again to obtain *second-order partial derivatives*. From f_x we obtain the second partial derivatives

$$f_{xx} = (f_x)_x \quad \text{or} \quad \frac{\partial}{\partial x}\left(\frac{\partial f}{\partial x}\right) = \frac{\partial^2 f}{\partial x^2} \tag{1}$$

and

$$f_{xy} = (f_x)_y \quad \text{or} \quad \frac{\partial}{\partial y}\left(\frac{\partial f}{\partial x}\right) = \frac{\partial^2 f}{\partial y\, \partial x} \tag{2}$$

In the same way, f_y can be differentiated with respect to x and y to yield

$$f_{yx} = (f_y)_x \quad \text{or} \quad \frac{\partial}{\partial x}\left(\frac{\partial f}{\partial y}\right) = \frac{\partial^2 f}{\partial x\, \partial y} \tag{3}$$

$$f_{yy} = (f_y)_y \quad \text{or} \quad \frac{\partial}{\partial y}\left(\frac{\partial f}{\partial y}\right) = \frac{\partial^2 f}{\partial y^2} \tag{4}$$

In symbols, we have the following definition.

Definition

$$
\begin{aligned}
f_{xx} = (f_x)_x &= \lim_{\Delta x \to 0} \frac{f_x(x + \Delta x, y) - f_x(x, y)}{\Delta x} \\[4pt]
f_{xy} = (f_x)_y &= \lim_{\Delta y \to 0} \frac{f_x(x, y + \Delta y) - f_x(x, y)}{\Delta y} \\[4pt]
f_{yx} = (f_y)_x &= \lim_{\Delta x \to 0} \frac{f_y(x + \Delta x, y) - f_y(x, y)}{\Delta x} \\[4pt]
f_{yy} = (f_y)_y &= \lim_{\Delta y \to 0} \frac{f_y(x, y + \Delta y) - f_y(x, y)}{\Delta y}
\end{aligned}
\tag{5}
$$

provided that the limits exist.

Partial derivatives in which more than one of the independent variables is a variable of differentiation are called *mixed partial derivatives*. For example, f_{xy} and f_{yx} are mixed partial derivatives of the second order.

From (5), it would appear that for a function of two variables, there are four distinct second partial derivatives, namely, f_{xx}, f_{xy}, f_{yx}, and f_{yy}. However, it can be proved under relatively mild conditions on the given function that $f_{xy} = f_{yx}$.[†]

---

† By application of the basic mean value theorem of differential calculus, it can be shown that if f_{xy} and f_{yx} are continuous at an interior point of its domain then $f_{xy} = f_{yx}$ at that point. The concept of continuity of functions of two or more variables is discussed in the next section.

Furthermore, the functions that enter into applied mathematics almost always satisfy the requirements that the mixed partial derivatives are equal.

EXAMPLE 1   If $f(x, y) = x^4 - 6x^2y^2 + y^4$, find f_{xx}, f_{xy}, f_{yx}, and f_{yy} and, in particular, show that $f_{xx} + f_{yy} = 0$. Also, verify that $f_{xy} = f_{yx}$.

Solution   Here

$$f_x = 4x^3 - 12xy^2 \qquad \text{and} \qquad f_y = -12x^2y + 4y^3$$
$$f_{xx} = 12x^2 - 12y^2 \qquad f_{xy} = -24xy$$
$$f_{yx} = -24xy \qquad f_{yy} = -12x^2 + 12y^2$$

We have

$$f_{xy} = f_{yx} = -24xy$$

and by summing the expressions for f_{xx} and f_{yy}:

$$f_{xx} + f_{yy} = 0 \qquad\qquad\qquad \bullet$$

Equations involving partial derivatives of a function are called *partial differential equations*.

EXAMPLE 2   If $f(x, y) = g(x) + h(y)$, find f_{xx}, f_{xy}, f_{yx}, and f_{yy} where g and h are twice differentiable functions of x and y, respectively.

Solution   Differentiation yields

$$f_x = g'(x) \qquad f_{xx} = g''(x)$$
$$f_{xy} = \frac{\partial}{\partial y}(g'(x)) = 0$$

$$f_y = h'(y) \qquad f_{yx} = \frac{\partial}{\partial x}(h'(y)) = 0$$

so that $f_{xy} = f_{yx} = 0$. Also, $f_{yy} = h''(y)$. $\qquad\qquad \bullet$

EXAMPLE 3   If $u(x, t) = A \cos(x + ct) + Be^{x-ct}$, where A, B, and c are constants and $c \neq 0$, show that $u_{xx} = \dfrac{1}{c^2}u_{tt}$ (*wave equation*).

Solution   Form the partial derivatives and obtain

$$u_x = -A \sin(x + ct) + Be^{x-ct}$$
$$u_{xx} = -A \cos(x + ct) + Be^{x-ct} \qquad\qquad (i)$$
$$u_t = -Ac \sin(x + ct) - Bce^{x-ct}$$
$$u_{tt} = -Ac^2 \cos(x + ct) + Bc^2e^{x-ct}$$
$$= c^2[-A \cos(x + ct) + Be^{x-ct}] \qquad\qquad (ii)$$

From (i) and (ii), we have $u_{xx} = \dfrac{1}{c^2}u_{tt}$. $\qquad\qquad \bullet$

Partial derivatives of higher order than the second are defined analogously. In a similar manner, we can define higher order partial derivatives for functions of three or more variables. Thus if f is a function of two independent variables x and y, f_{yxx} means first differentiate f with respect to y and then twice with respect to x. Furthermore, under relatively weak conditions

$$f_{yxx} = f_{xyx} = f_{xxy} \tag{6}$$

where the order of differentiation with respect to adjacent x and y have been interchanged twice to form the two equations (6). Note that x occurs twice and y once as subscripts in each member of (6) indicating two differentiations with respect to x and one differentiation with respect to y. By way of another example, if f is a function of x, y, and z, then $f_{xyzyx} = f_{xzxyy}$ holds for most functions occurring in practice. This follows because the subscript x occurs twice, the y occurs twice, and z once on both sides, indicating the number of differentiations with respect to these variables. This general fact follows from the simpler result $f_{xy} = f_{yx}$ since this latter equality states that two adjacent subscripts can be interchanged, and any permutation of the subscripts can be obtained by a sequence of adjacent interchanges.

EXAMPLE 4   For $u = x^2 e^y \cos z$, obtain all distinct third-order partial derivatives.

Solution

$$u_x = 2xe^y \cos z \qquad u_{xx} = 2e^y \cos z$$

$$u_{xxx} = 0 \qquad u_{xxy} = 2e^y \cos z \qquad u_{xxz} = -2e^y \sin z$$

$$u_y = x^2 e^y \cos z \qquad u_{yy} = x^2 e^y \cos z$$

$$u_{yyy} = x^2 e^y \cos z \qquad u_{yyx} = 2xe^y \cos z \qquad u_{yyz} = -x^2 e^y \sin z$$

$$u_z = -x^2 e^y \sin z \qquad u_{zz} = -x^2 e^y \cos z$$

$$u_{zzz} = x^2 e^y \sin z \qquad u_{zzx} = -2xe^y \cos z \qquad u_{zzy} = -x^2 e^y \cos z$$

$$u_{xy} = 2xe^y \cos z$$

$$u_{xyz} = -2xe^y \sin z \qquad \bullet$$

For a function $u = f(x, y, z)$ there are ten distinct third partial derivatives u_{xxx}, u_{xxy}, u_{xxz}, u_{yyy}, u_{yyx}, u_{yyz}, u_{zzz}, u_{zzx}, u_{zzy}, and u_{xyz}. The proof of this is quite independent of calculus and we will resist any temptation in this matter.

Exercises 17.3

In Exercises 1 through 12, find the specified partial derivatives.

1. $f(x, y) = x^3 - x^2 y^2 + 3y^3$; f_{xx}, f_{xy}, f_{yy}

2. $g(x, y) = x^4 - xy + y^2$; g_{xx}, g_{xxy}, g_{yyx}

3. $h(x, y) = 2xe^y + y^2 e^x$; $\dfrac{\partial^2 h}{\partial x^2}, \dfrac{\partial^2 h}{\partial y\, \partial x}, \dfrac{\partial^3 h}{\partial y^3}$

4. $F(x, z) = \dfrac{x^2}{z}$; F_{xz}, F_{zx}, F_{zzz}

5. $F(u, v) = \sinh^2 u \cosh 3v; \quad F_u, F_{uv}, F_{vvv}$

6. $G(x, y) = e^{xy} \sin ax; \quad \dfrac{\partial^2 G}{\partial x^2}, \dfrac{\partial^2 G}{\partial y\, \partial x}, \dfrac{\partial^2 G}{\partial y^2},$ where a is a constant

7. $f(s, t) = (2s + t)^8 + s^2 - t^2; \quad f_{sst}, f_{sss}, f_{ttt}$

8. $f(x, y) = \ln x^3 y^2;$ where x and y are positive, $f_{xxx}, f_{xyy}, f_{yyy}$

9. $u(x, y) = x[\cos (x + y) + y^2]; \quad u_{xx}, u_{xy}, u_{yy}$

10. $g(x, y) = x^2 y^2 + y \sin xy; \quad g_{xx}, g_{xy}, g_{yy}$

11. $u(x, y, z) = \tan (x^2 + y + z); \quad u_{xzz}, u_{yzz}$

12. $u(x, y) = \tan^{-1} \dfrac{x}{y}; \quad u_{xx}, u_{xy}, u_{yx}, u_{yy}$

In Exercises 13 through 20, show that the given function of x and y satisfies **Laplaces' equation** $u_{xx} + u_{yy} = 0.$ Such functions are called **harmonic functions** in two dimensions.

13. $u = x^3 - 3xy^2 + 10x - y - 4$

14. $u = \dfrac{x}{x^2 + y^2}, \quad x^2 + y^2 \neq 0$

15. $u = \ln \sqrt{x^2 + y^2}, \quad x^2 + y^2 \neq 0$

16. $u = \tan^{-1} \dfrac{y}{x}, \quad x \neq 0$

17. $u = e^x \sin y + e^y \sin x$

18. $u = e^x(x \cos y - y \sin y)$

19. $u = A \sin x \cosh y + B \cos x \sinh y,$ where A and B are arbitrary constants.

20. $u = A \cos x \sinh y + B(3xy^2 - 3y^2 - (x - 1)^3),$ where A and B are arbitrary constants.

21. Show that $f(x, y, z) = (x^2 + y^2 + z^2)^{-1/2},$ where $x^2 + y^2 + z^2 \neq 0$ satisfies **Laplace's equation in three dimensions:**

$$f_{xx} + f_{yy} + f_{zz} = 0$$

22. What relation between constants a and b must exist in order that

$$w(x, y, z) = az^3 + bz(x^2 + y^2)$$

satisfies Laplace's equation: $w_{xx} + w_{yy} + w_{zz} = 0?$

23. What relation between constants a, b and c must exist in order that

$$w(x, y, z) = \sinh ax \sin by \sinh cz$$

satisfies Laplace's equation: $w_{xx} + w_{yy} + w_{zz} = 0?$

24. Show that $u = xye^{3x^2 y^2}$ satisfies the equation $xu_x - yu_y = 0.$

25. Show that

$$w(x, y) = x^4 + y^4 - 2x^2 y^2 + 10 \sin (x^2 - y^2) + 100$$

satisfies the equation

$$yw_x + xw_y = 0$$

26. Verify that

$$u(x, y) = 10x^2 + 4xy + 2y^2 + 3x - 4y - 7$$

satisfies the equation

$$u_{xx} - 2u_{xy} - 3u_{yy} = 0$$

27. Verify that

$$w(x, y) = x^3 + 3x^2 y + 3xy^2 + y^3 + \dfrac{e^x}{e^y}$$

satisfies the equation

$$w_{xxxx} - w_{yyyy} = 0$$

28. Show that

$$z(x, y) = x \sin (x + y) + \cosh (x - y) + x^3 - xy - y^2$$

satisfies the equation

$$z_{xxxx} - 2z_{xxyy} + z_{yyyy} = 0$$

29. If $F(x, y) = \int_y^x e^{-3t^2} dt$, find

    (a) $\dfrac{\partial^2 F}{\partial x^2}$ (b) $\dfrac{\partial^2 F}{\partial y \, \partial x}$ (c) $\dfrac{\partial^2 F}{\partial y^2}$

30. If $G(x, y, z) = (x^5 + x^3 y^2 - y)^{10} + (y + z) \tanh (y^2 - z) + xz^3 \sin z$, find G_{xyz}.

31. If $F(x, y) = y \int_y^x e^{-t^2} dt$, find

    (a) F_x (b) F_y (c) F_{xx} (d) F_{xy} (e) F_{yy}

32. If $f(x, y) = \displaystyle\sum_{n=0}^{\infty} \frac{(x + y)^n}{n!}$, find

    (a) f_x (b) f_y (c) f_{xy} (d) f_{yy}

Assume that term by term differentiation is permissible.

17.4   LIMITS AND CONTINUITY

In this section we will extend the concepts of limit and continuity of a function of one variable to functions of two or more variables.

Let us start with the idea of limit for a function f of two variables. When we say that $\lim\limits_{(x,y)\to(a,b)} f(x, y) = L$, we mean in intuitive terms, that, however close we specify that $f(x, y)$ may come to the constant L there is some two-dimensional region about the point (a, b) in which the desired degree of closeness is attained. Note that (as in the case of a function of one variable) we do not require that the function be defined at (a, b) and, if it is, it need not be equal to L. To make this more precise, we will need some definitions.

Definition   The set of points that are inside a rectangle with sides parallel to the coordinate axes is called an ***open rectangular region*** in two dimensions. Figure 17.4.1 shows the set of all points (x, y) satisfying the inequalities

$$x_0 < x < x_1, \quad y_0 < y < y_1 \tag{1}$$

The ***boundary*** of this point set is the set of points which are on the rectangle itself. Note that the open rectangular region does not include points on the boundary.

A rectangular region consisting of an open rectangular region and its boundary (that is, the union of the two sets) is called a ***closed rectangular region*** (Figure 17.4.2).

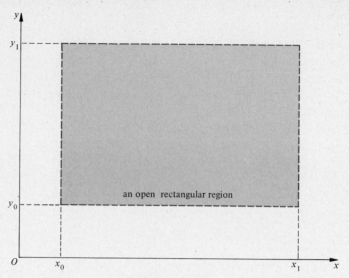

an open rectangular region

Figure 17.4.1

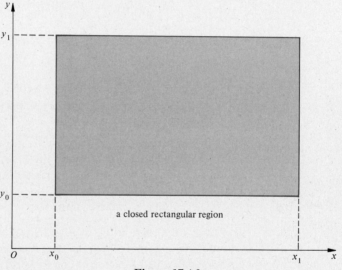

a closed rectangular region

Figure 17.4.2

Definition   A **square neighborhood** of (a, b) is the set of all points interior to some open square region with center at (a, b) and sides parallel to the x- and y-axes such that

$$|x - a| < \delta \quad \text{and} \quad |y - b| < \delta$$

for some $\delta > 0$ (Figure 17.4.3).

Definition   A **circular neighborhood** of (a, b) is the set of all points interior to some circle with center at (a, b); that is, the set of points (x, y) such that

$$(x - a)^2 + (y - b)^2 < \delta^2$$

for some $\delta > 0$ (Figure 17.4.4).

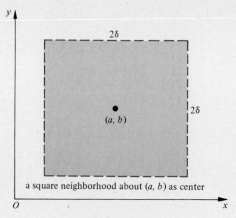

a square neighborhood about (a, b) as center

Figure 17.4.3

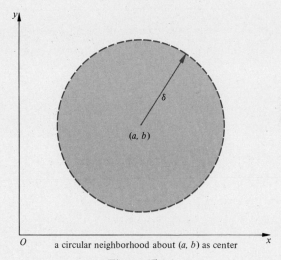

a circular neighborhood about (a, b) as center

Figure 17.4.4

Definition   ***Limit of a Function.*** Let f be a function of two independent variables x and y defined in some circular neighborhood of (a, b) except possibly at (a, b) itself. Then

$$\lim_{(x,y)\to(a,b)} f(x, y) = L$$

if for any given $\varepsilon > 0$, there exists a positive number δ such that

$$|f(x, y) - L| < \varepsilon \quad \text{whenever } 0 < \sqrt{(x - a)^2 + (y - b)^2} < \delta \quad (2)$$

According to this definition, the value of $f(a, b)$, if it exists, is not involved in the limit.

Definition (2) may be recast in terms of a square neighborhood. The reader is invited to do this.

The properties of limits as stated in Section 2.5 for functions of one variable apply equally well to functions of two variables and these details will not be given here.

EXAMPLE 1   Show that $\lim\limits_{(x,y)\to(0,0)} \dfrac{x^2y^2}{x^2+2y^2} = 0$

Solution   Since the limit of the denominator is zero, we cannot use the theorem: limit of the quotient equals quotient of the limits. Now, $x^2 \le x^2 + y^2$ and $y^2 \le x^2 + y^2$ and therefore

$$x^2y^2 \le (x^2 + y^2)^2$$

Also,

$$x^2 + 2y^2 \ge x^2 + y^2$$

so that

$$\left| \frac{x^2y^2}{x^2+2y^2} \right| \le \frac{(x^2+y^2)^2}{x^2+y^2} = x^2 + y^2$$

For any $\varepsilon > 0$, choose $\delta = \varepsilon^{1/2}$. Thus from (2)

$$\left| \frac{x^2y^2}{x^2+2y^2} \right| < \varepsilon \qquad \text{provided that } 0 < \sqrt{x^2+y^2} < \delta$$

and therefore the desired result has been proved.   ●

EXAMPLE 2   Show that $\lim\limits_{(x,y)\to(0,0)} \dfrac{2xy}{x^2+y^2}$ does not exist.

Solution   Let $f(x, y) = \dfrac{2xy}{x^2+y^2}$. Set $x = 0$ so that $f(0, y) = \dfrac{0}{y^2} = 0$ when $y \ne 0$. Thus if we approach $(0, 0)$ along the y-axis, there results $\lim\limits_{y\to0} f(0, y) = \lim\limits_{y\to0} 0 = 0$. Also, if we set $y = 0$, then $f(x, 0) = \dfrac{0}{x^2} = 0$, $(x \ne 0)$. Hence $\lim\limits_{x\to0} f(x, 0) = \lim\limits_{x\to0} 0 = 0$. But suppose we set $y = x$, then $f(x, x) = \dfrac{2x^2}{x^2+x^2} = 1$, for $x \ne 0$. It follows that $\lim\limits_{x\to0} f(x, x) = 1$. Since different limits exist along different paths approaching $(0, 0)$, $\lim\limits_{(x,y)\to(0,0)} f(x, y)$ does not exist.   ●

EXAMPLE 3   Let $f(x, y) = \dfrac{x^2y}{x^4+y^2}$, for $(x, y) \ne (0, 0)$. Show that along *every straight line* through $(0, 0)$, $\lim\limits_{(x,y)\to(0,0)} f(x, y) = 0$. Show, however, that $\lim\limits_{(x,y)\to(0,0)} f(x, y)$ does not exist, by finding a *curved path* passing through the origin for which $\lim\limits_{(x,y)\to(0,0)} f(x, y) \ne 0$.

Solution   Consider $f(x, mx) = \dfrac{mx^3}{x^4+m^2x^2}$. If $m = 0$, $f(x, 0) = \dfrac{0}{x} = 0$, $(x \ne 0)$, hence $\lim\limits_{x\to0} f(x, 0) = 0$. If $m \ne 0$, $f(x, mx) \le \dfrac{mx^3}{m^2x^2} = \dfrac{x}{m}$ from which $\lim\limits_{x\to0} f(x, mx) = 0$. Also, $f(0, y) = \dfrac{0}{y^2} = 0$, $(y \ne 0)$ and $\lim\limits_{y\to0} f(0, y) = 0$. Hence $\lim\limits_{(x,y)\to(0,0)} f(x, y) = 0$ along all straight lines through $(0, 0)$.

However, if we examine the behavior of $f(x, y)$ along the parabola $y = x^2$ then $f(x, x^2) = \dfrac{x^4}{x^4 + x^4} = \dfrac{1}{2}$ for $x \neq 0$. Therefore $\lim\limits_{x \to 0} f(x, x^2) = \dfrac{1}{2}$ and we again have different limits along different paths approaching $(0, 0)$; hence $\lim\limits_{(x,y)\to(0,0)} f(x, y)$ does not exist. ●

Definition   A function $f(x, y)$ is **continuous** at a point (a, b) if

$$\lim_{(x,y)\to(a,b)} f(x, y) = f(a, b) \tag{3}$$

Alternatively, we have the definition that follows.

Definition   A function $f(x, y)$ is **continuous** at a point (a, b) if, given any positive ε, we can find a positive δ such that

$$|f(x, y) - f(a, b)| < \varepsilon \tag{4}$$

for all (x, y) that satisfy

$$(x - a)^2 + (y - b)^2 < \delta^2 \tag{5}$$

Definition   A function is **continuous on a set of points** S if it is continuous at every point of S.

All of the basic theorems on continuity for functions of a single variable can be generalized to functions of two variables. It can be shown (by similar arguments to those in Section 2.5) that the following theorem holds.

Theorem 1   If $f(x, y)$ and $g(x, y)$ are continuous functions of x and y on a set S and c is any constant, then

(i) $f(x, y) \pm g(x, y)$ is continuous on S.
(ii) $cf(x, y)$ is continuous on S.
(iii) $f(x, y) \cdot g(x, y)$ is continuous on S.

(iv) $\dfrac{f(x, y)}{g(x, y)}$ is continuous on S except for those points (x_0, y_0) where $g(x_0, y_0) = 0$

Since every monomial in two variables $cx^r y^s$ is continuous at every point in the plane, it follows from (i) that any polynomial in two variables is also continuous at all points in the plane.

It follows from (iv) that a rational function of two variables, that is, a quotient of two polynomials in two variables is continuous at all points except for those points where the denominator is zero. This implies that a rational function of two variables is continuous at every point in its domain, that is, where it is defined.

We note that even though all definitions and theorems in this section are stated for functions of two variables, we could have considered functions of three or more variables just as well.

EXAMPLE 4 $f(x, y) = \dfrac{x^2 + y^2}{x - 2y}$ is a rational function of x and y that is continuous at all points except for those which lie on the line $x - 2y = 0$. ●

EXAMPLE 5 The function

$$g(x, y) = \begin{cases} \dfrac{2xy}{x^2 + y^2} & \text{if } x^2 + y^2 \neq 0 \\ 0 & \text{if } x = y = 0 \end{cases}$$

Find the points at which g is continuous.

Solution The function g is defined for all x and y. Also, the function $\dfrac{2xy}{x^2 + y^2}$, where $x^2 + y^2 \neq 0$, is a rational function for which the denominator is not zero. Hence g is continuous for all (x, y) such that $x^2 + y^2 \neq 0$. Now, what about $(0, 0)$? From Example 2, $\lim\limits_{(x,y)\to(0,0)} \dfrac{2xy}{x^2 + y^2}$ does not exist. Hence g is not continuous at $(0, 0)$ regardless of its definition at $(0, 0)$. ●

Theorem 2 If a function g of two variables x and y is continuous at (a, b) and a function f of one variable is continuous at the number $g(a, b)$, then the composite function $f \circ g$ defined by $h(x, y) = f(g(x, y))$ is continuous at (a, b).

EXAMPLE 6 Discuss the continuity of the function

$$h(x, y) = \sqrt{x^2 - y}$$

Solution The function $g(x, y) = x^2 - y$ is continuous for all x and y while $f(g(x, y)) = \sqrt{g}$ is continuous if and only if $g > 0$. Hence $h(x, y) = \sqrt{x^2 - y}$ is continuous at all number pairs (x, y) for which $x^2 > y$. This set of points is shown shaded in Figure 17.4.5. ●

EXAMPLE 7 Discuss the continuity of the function

$$h(x, y) = \ln(x^2 + y^2 - 9)$$

Solution The function $g(x, y) = x^2 + y^2 - 9$ is continuous for all x and y while $f(g(x, y)) = \ln g$ is continuous for all $g > 0$. Hence the composite function $h(x, y)$ is continuous for all (x, y) such that $x^2 + y^2 > 9$. This is the set of points (x, y) outside the circle of radius 3 with center at the origin. ●

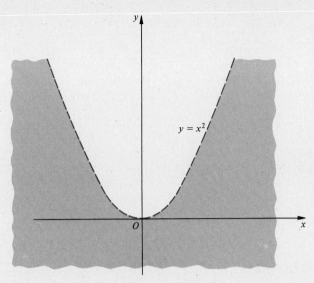

Figure 17.4.5

Exercises 17.4

In each of Exercises 1 through 4, find the limit.

1. $\displaystyle\lim_{(x,y)\to(4,2)} \frac{x^2 - xy}{x + y}$

2. $\displaystyle\lim_{(x,y)\to(1,-2)} \frac{x}{x^2 - y}$

3. $\displaystyle\lim_{(x,y)\to(0,0)} \frac{y^2 + 6y}{xy + 7y}$

4. $\displaystyle\lim_{(x,y)\to(3,3)} \frac{x^2 - y^2}{x^2 - 7xy + 6y^2}$

In Exercises 5 through 12, prove that the limit does not exist.

5. $\displaystyle\lim_{(x,y)\to(0,0)} \frac{y}{x^2 + y^2}$

6. $\displaystyle\lim_{(x,y)\to(0,0)} \frac{2xy}{x^2 - y^2}$

7. $\displaystyle\lim_{(x,y)\to(0,0)} \frac{x}{x + y}$

8. $\displaystyle\lim_{(x,y)\to(0,0)} \frac{y^2}{3(x^2 + y^2)}$

9. $\displaystyle\lim_{(x,y)\to(0,0)} \frac{x^2 - y^2}{x^4 - y^4}$

10. $\displaystyle\lim_{(x,y)\to(0,0)} \sin \frac{1}{x^2 + y^2}$

11. $\displaystyle\lim_{(x,y)\to(0,0)} \frac{x^2 + y^3}{x^2 + y^2}$

12. $\displaystyle\lim_{(x,y)\to(0,0)} \ln (x^2 + y^2)^{1/3}$

In Exercises 13 through 18, determine whether the $\displaystyle\lim_{(x,y)\to(0,0)} f(x, y)$ *exists and discuss the continuity of* $f(x, y)$ *at* $(0, 0)$.

13. $f(x, y) = \begin{cases} \dfrac{\sin 10(x^2 + y^2)}{x^2 + y^2} & \text{if } x^2 + y^2 \neq 0 \\ 10 & \text{if } x = y = 0 \end{cases}$

14. $f(x, y) = \begin{cases} \dfrac{x^2 y^2}{x^2 - 6y^2} & \text{if } x^2 + y^2 \neq 0 \\ -\dfrac{1}{6} & \text{if } x = y = 0 \end{cases}$

15. $f(x, y) = \begin{cases} \dfrac{x^3 + 2y^3}{x^2 + y^2} & \text{if } x^2 + y^2 \neq 0 \\ 0 & \text{if } x = y = 0 \end{cases}$

(*Hint:* Introduce $|f(x, y)|$ and use $|A + B| \leq |A| + |B|$.)

16. $f(x, y) = \begin{cases} \dfrac{\sin(2x^2 + 3y^2)}{x^2 + y^2} & \text{if } x^2 + y^2 \neq 0 \\ 5 & \text{if } x = y = 0 \end{cases}$

17. $f(x, y) = \begin{cases} \dfrac{x^2 + y^6}{x^2 + y^4} & \text{if } x^2 + y^2 \neq 0 \\ 0 & \text{if } x = y = 0 \end{cases}$

18. $f(x, y) = \begin{cases} \dfrac{xy}{\sqrt{x^2 + 3y^2}} & \text{if } x^2 + y^2 \neq 0 \\ 0 & \text{if } x = y = 0 \end{cases}$

In Exercises 19 through 26, determine where the given function is continuous.

19. $f(x, y) = \sin(3 + x^2 y^2 - x^4 y^4)$

20. $g(x, y) = \dfrac{x^2 + 8y^2}{9x^2 - y^2}$

21. $h(x, y) = \dfrac{\cosh(x + y)}{1 + x^2 + y^4}$

22. $F(x, y) = (2x - y) \ln(x + y - 3)$

23. $G(x, y) = \ln|5 + xy|$

24. $f(x, y) = \tan(2x^2 + y^2)$

25. $g(x, y) = \begin{cases} \dfrac{x^2 - y^2}{x^2 + y^2} & \text{if } x^2 + y^2 \neq 0 \\ 0 & \text{if } x = y = 0 \end{cases}$

26. $F(x, y) = \begin{cases} xy(1 + x^2 + y^2)^{-3/2} & \text{if } x^2 + y^2 \neq 0 \\ 0 & \text{if } x = y = 0 \end{cases}$

27. If $f(x, y) = x + 3y^2$, $g(u) = u^2 + 2u$, $h(u) = e^{-u}$, find (a) $g(f(x, y))$, (b) $h(f(x, y))$, and (c) $f(g(u), h(u))$

28. If $f(u, v) = 2u - v + uv$, $g(s, t) = 3s + 2t$, and $h(x, y) = 4x - 3y$, find $f(g(x, y), h(x, y))$.

29. Prove that $f(x, y) = 3x - y$ is continuous at the point $(2, 1)$ by the ε, δ-procedure; that is, choose an arbitrary $\varepsilon > 0$ and find a suitable $\delta(\varepsilon)$. It is important to note that δ is not unique.

30. Prove that $g(x, y) = 7x - 9y$ is continuous at the point $(3, 4)$ by the ε, δ-procedure.

31. (a) Give a definition of continuity at a point (a, b, c) for a function of three variables that parallels the definition in the text for a function of two variables.

(b) Define what is meant by $\displaystyle\lim_{(x, y, z) \to (a, b, c)} f(x, y, z) = L$

In Exercises 32 through 35, use the analog of Theorems 1 and 2, applied to functions of three variables, to discuss the continuity of each of the functions.

32. $f(x, y, z) = \dfrac{x + y + 2z}{\sqrt{81 - x^2 - y^2 - z^2}}$

33. $g(x, y, z) = e^{xyz} + \ln(4 - 2x + y - z)$

34. $h(x, y, z) = \sinh\sqrt{\dfrac{x(1 + y)}{z^2 - 2z + 3}}$

35. $F(x, y, z) = \dfrac{x \tan z}{y^2 - 2y - 2}$

17.5

DIFFERENTIABILITY AND DIFFERENTIALS

We recall that the differential of a function $y = f(x)$ is defined by (Section 5.1)

$$dy = df = f'(x)\,\Delta x = f'(x)\,dx \qquad (1)$$

and that

$$\Delta y = \Delta f = df + \varepsilon\Delta x \qquad (2)$$

where $\varepsilon \to 0$ as $\Delta x \to 0$. From (1) and (2) we have

$$\Delta f = (f'(x) + \varepsilon)\,\Delta x \qquad (3)$$

and in many instances, the "linear part" $f'(x)\,\Delta x$ of Δf is a good approximation to Δf. Also, it is important to remember that $f'(x)$ and Δx may be specified independently of one another.

These ideas can be extended in a perfectly natural way to functions of two or more variables. As in (2), we form

$$\Delta f = f(x + \Delta x, y + \Delta y) - f(x, y) \qquad (4)$$

which gives the difference of the value of a function at two points. A useful expression for Δf is obtained in the following theorem.

Theorem 1 Let $u = f(x, y)$ define a continuous function of x and y for which f_x and f_y are also continuous in a neighborhood of (x, y). Then there are functions $\varepsilon_1(\Delta x, \Delta y)$ and $\varepsilon_2(\Delta x, \Delta y)$ such that

$$\Delta f = f_x(x, y)\,\Delta x + f_y(x, y)\,\Delta y + \varepsilon_1\,\Delta x + \varepsilon_2\,\Delta y \qquad (5)$$

where ε_1 and $\varepsilon_2 \to 0$ as Δx and $\Delta y \to 0$.

Proof $\Delta u = f(x + \Delta x, y + \Delta y) - f(x, y)$ and by subtraction and addition of $f(x, y + \Delta y)$

$$\Delta u = [f(x + \Delta x, y + \Delta y) - f(x, y + \Delta y)] + [f(x, y + \Delta y) - f(x, y)] \qquad (6)$$

Since the quantity in the first bracket involves a change in x only and because of the hypothesis, we may apply the basic mean value theorem of differential calculus to obtain

$$f_x(x + \theta_1\,\Delta x, y + \Delta y)\,\Delta x \qquad (i)$$

where $0 < \theta_1 < 1$. Similarly, the quantity in the second bracket in (6) involves a change in y alone, and, by application of the mean value theorem again, may be written in the form

$$f_y(x, y + \theta_2\,\Delta y)\,\Delta y \qquad (ii)$$

where $0 < \theta_2 < 1$.

Since f_x is continuous in a given neighborhood of (x, y) then

$$f_x(x + \theta_1\,\Delta x, y + \Delta y) = f_x(x, y) + \varepsilon_1 \qquad (iii)$$

where $\varepsilon_1 \to 0$ as Δx and $\Delta y \to 0$. Similarly, due to the continuity of f_y in this same neighborhood,

$$f_y(x, y + \theta_2\,\Delta y) = f_y(x, y) + \varepsilon_2 \qquad (iv)$$

where $\varepsilon_2 \to 0$ as Δx and $\Delta y \to 0$. Substitution of (i) through (iv) into the right side of (6) yields (5). ∎

Definition   A function $f(x, y)$ is said to be **differentiable** at x and y provided that $z = f(x, y)$ can be written in the form (5) where ε_1 and $\varepsilon_2 \to 0$ as Δx and $\Delta y \to 0$.

From (5) it follows that the differentiability of $f(x, y)$ implies that f_x and f_y must exist at (x, y). This condition is necessary for differentiability. However, it can be shown that it is not sufficient. Instead, we have proved that the continuity of f, f_x, and f_y at (x, y) are sufficient to ensure differentiability.

In the case of a function of a single variable the existence of $f'(x)$ alone is sufficient to ensure differentiability at x.

By analogy with the case of a function of one variable, we have the following definition.

Definition   If $u = f(x, y)$ is differentiable, then the **differential** or **total differential** of u is defined by

$$du = u_x \, dx + u_y \, dy \qquad (7)$$

where dx and dy are two independent variables.

The differential du is a function of four independent variables, namely (x, y), the coordinates of the point at which the partial derivatives are calculated, and dx and dy. These latter quantities are called differentials of the *independent* variables. Again we identify these differentials with actual increments. Hence

$$dx = \Delta x \qquad \text{and} \qquad dy = \Delta y \qquad (8)$$

Therefore from (7) and (8) we have

$$du = u_x \, \Delta x + u_y \, \Delta y \qquad (9)$$

and substitution of (9) into (5) yields

$$\Delta u = \Delta f = du + \varepsilon_1 \, \Delta x + \varepsilon_2 \, \Delta y \qquad (10)$$

EXAMPLE 1   Given $u = 2x^2 + y^2$, find du when $x = 3$, $y = 4$, $dx = 0.1$, and $dy = -0.2$.

Solution   From (7)

$$du = 4x \, dx + 2y \, dy$$

from which, when $x = 3$, $y = 4$, $dx = 0.1$, and $dy = -0.2$ we have

$$du = 4(3)(0.1) + 2(4)(-0.2) = -0.4 \qquad \bullet$$

EXAMPLE 2   A right circular cylinder has base radius $r = 5$ in. and height $h = 7$ in. Find the approximate change in the volume of the cylinder if r is increased by 0.2 in. and h is decreased by 0.3 in.

Solution   The volume of a cylinder is a function of the two variables r and h, $V(r, h) = \pi r^2 h$. Therefore

$$\Delta V = V(5.2, 6.7) - V(5, 7)$$

Rather than compute ΔV we shall use

$$dV = V_r(5, 7)\, dr + V_h(5, 7)\, dh$$

as an estimate of ΔV.

We have $V_r = 2\pi rh$, $V_r(5, 7) = 70\pi$, $V_h = \pi r^2$, and $V_h(5, 7) = 25\pi$. Also $dr = 0.2$ and $dh = -0.3$. Hence, upon substitution

$$dV = (70\pi)(0.2) + (25\pi)(-0.3) = 6.5\pi$$

and the volume will increase by approximately 6.5π in.[3]   ●

EXAMPLE 3   If $u = xy$, then

$$\begin{aligned}
\Delta u &= (x + \Delta x)(y + \Delta y) \\
&= xy + x\,\Delta y + y\,\Delta x + \Delta x\,\Delta y
\end{aligned}$$

Now $u = xy$ and $du = y\,dx + x\,dy = x\,\Delta y + y\,\Delta x$ (recall that $dx = \Delta x$ and $dy = \Delta y$). Hence, in this instance,

$$\Delta u - du = \Delta x\,\Delta y$$

so that from (10), we may choose $\varepsilon_1 = \Delta y$ and $\varepsilon_2 = 0$ or $\varepsilon_2 = \Delta x$ and $\varepsilon_1 = 0$.[†] In either case, $\varepsilon_i \to 0$ $(i = 1, 2)$ as Δx and $\Delta y \to 0$.   ●

EXAMPLE 4   A box has a square base with sides of length x and height y. The sides of the base are measured with a possible error of 1% whereas the height is measured with a possible error of 2%. Find the maximum possible error in the volume.

Solution   The maximum relative percent errors have been prescribed. Hence

$$\frac{|\Delta x|}{x} \leq 0.01 \qquad \text{and} \qquad \frac{|\Delta y|}{y} \leq 0.02$$

The volume

$$V = x^2 y$$

and we seek the maximum $\dfrac{|\Delta V|}{V}$. Now,

$$\Delta V \approx dV = 2xy\,\Delta x + x^2\,\Delta y$$

and division of both sides by V yields

$$\frac{dV}{V} = \frac{2xy\,\Delta x}{x^2 y} + \frac{x^2\,\Delta y}{x^2 y} = \frac{2\,\Delta x}{x} + \frac{\Delta y}{y}$$

Hence from $|a + b| \leq |a| + |b|$, we have

$$\frac{|dV|}{V} \leq 2\frac{|\Delta x|}{x} + \frac{|\Delta y|}{y} \leq 2(0.01) + 0.02 = 0.04$$

---

† From this we see that the functions $\varepsilon_1(\Delta x, \Delta y)$ and $\varepsilon_2(\Delta x, \Delta y)$ are generally *not unique*.

This says that the maximum possible error in computing the volume is approximately 4%. Of course, it could in actuality be considerably less if Δx and Δy are of opposite sign. ●

EXAMPLE 5   Use differentials to estimate $\sqrt[3]{28}\sqrt{9990}$.

Solution   We know $\sqrt[3]{27}$ and $\sqrt{10,000}$. If we form

$$f(x, y) = \sqrt[3]{x}\sqrt{y} = x^{1/3}y^{1/2}$$

then we seek the change of Δf that occurs when x is increased from 27 to 28 and y is decreased from 10,000 to 9990. Hence $\Delta x = 1$ and $\Delta y = -10$.

Now

$$df = \tfrac{1}{3}x^{-2/3}y^{1/2}\,dx + \tfrac{1}{2}x^{1/3}y^{-1/2}\,dy$$

Substitution of the given data yields

$$df = \tfrac{1}{3}(27^{-2/3})(10,000)^{1/2}(1) + \tfrac{1}{2}(27^{1/3})(10,000^{-1/2})(-10)$$
$$= \tfrac{1}{27}(100) - \tfrac{3}{20} = \tfrac{1919}{540} \doteq 3.55$$

Now, $f(27, 10,000) = 300$ so that

$$\sqrt[3]{28}\sqrt{9990} \doteq 300 + 3.55 = 303.55$$

A simple calculation (from the tables or a calculator) yields

$$\sqrt[3]{28}\sqrt{9990} \doteq 3.0366 \times 99.9500 \doteq 303.51$$

and it is verified that the error in using df rather than Δf is indeed negligible in this instance. ●

The results of this section are readily extended to functions of more than two variables. However, the details would be repetitious without introducing any new ideas. Therefore we state without proof a second theorem.

Theorem 2   Let $u = f(x, y, z)$ be continuous and have continuous partial derivatives at $P(x, y, z)$ and in a neighborhood of point P. Then

$$\Delta u = f_x\,\Delta x + f_y\,\Delta y + f_z\,\Delta z + \varepsilon_1\,\Delta x + \varepsilon_2\,\Delta y + \varepsilon_3\,\Delta z \qquad (11)$$

where $\varepsilon_1 \to 0$, $\varepsilon_2 \to 0$, and $\varepsilon_3 \to 0$ as Δx, Δy, and $\Delta z \to 0$.

With the definition of the differential we have the following.

Definition
$$du = df = f_x\,dx + f_y\,dy + f_z\,dz$$
$$= f_x\,\Delta x + f_y\,\Delta y + f_z\,\Delta z$$

(11) becomes

$$\Delta u = du + \varepsilon_1\,\Delta x + \varepsilon_2\,\Delta y + \varepsilon_3\,\Delta z \qquad (12)$$

where $\varepsilon_1 \to 0$, $\varepsilon_2 \to 0$, and $\varepsilon_3 \to 0$ as Δx, Δy, and $\Delta z \to 0$.

The extension of (11) and (12) to functions of n variables should now be clear.

Exercises 17.5

In each of Exercises 1 through 12, find du.

1. $u = x^2 - 5y^3$

2. $u = xy$

3. $u = e^{xy}$

4. $u = 4 \cos x - y^2$

5. $u = \dfrac{x}{y}$

6. $u = \sin(x + 3y) + \ln(x^2 + y)$

7. $u = 3x^2 \cos(x + y) - y^3 \sin \pi x$

8. $u = x^3 - y^3 - 3x(x + y) + 2x + 3y - 1$

9. $u = \ln(x^2 + y^2 + z^2)^{1/2}$

10. $u = r^3 s^2 \cos^3 t$

11. $u = x \tan^{-1} \dfrac{z}{y}$

12. $u = (x_1^2 + x_2^2 + \cdots + x_n^2)^{1/2}$

13. Given $u = x^2 + 3y^2$, compute du when $x = 3$, $y = 1$, $dx = -0.2$, and $dy = 0.1$.

14. Given $u = x + xy - y$, compute du when $x = 2$, $y = 3$, $dx = 0.3$, and $dy = 0.2$.

15. Given $u = e^x \sin 2y$, compute du when $x = 0$, $y = \dfrac{\pi}{8}$, $dx = 0.4$, and $dy = -0.3$.

16. Given $u = (2x + y)\sqrt{y - x}$, compute du when $x = 2$, $y = 6$, $dx = 0.3$, and $dy = 0.2$

In Exercises 17 through 20, find (a) Δu; (b) du; and (c) $\Delta u - du$, which yields ε_1 and ε_2; (d) verify that ε_1 and $\varepsilon_2 \to 0$ as Δx and $\Delta y \to 0$.

17. $u = ax + by + c$ where a, b and c are constants

18. $u = x^2 + 2y^2$

19. $u = \dfrac{x}{y}$, $(y \neq 0)$

20. $u = x^2 y$

21. A right circular cone has base radius $r = 6$ in. and height $h = 8$ in. Find the approximate change in the volume of the cone if r is decreased by 0.3 in. and h is increased by 0.4 in. $\left(V = \dfrac{1}{3}\pi r^2 h \right)$

22. The *relative error* in a function $u = f(x, y)$ is given approximately by $\dfrac{du}{u}$. Show that the relative error in the area of a rectangle is approximately the sum of the relative errors in the measurements of the sides.

23. Show that the relative error in the volume V of a box is approximately the sum of the relative errors in the measurements of the sides. (See Exercise 22.)

24. The hypotenuse z of a right triangle is given in terms of its legs x and y by $z = \sqrt{x^2 + y^2}$. Find the approximate maximum possible error in z given that $x = (60 \pm 0.2)$ in. and $y = (80 \pm 0.4)$ in.

25. The period T for small oscillations of a *simple pendulum* of length L is $T = 2\pi \sqrt{\dfrac{L}{g}}$, where g is the acceleration of gravity. If $g = (32.15 \pm 0.1)$ ft/sec^2 and $L = (4 \pm 0.05)$ ft, what is the approximate possible error in the period?

26. The period T for small oscillations of a physical pendulum is given by the formula $T = 2\pi \sqrt{\dfrac{I}{mgL}}$, where I is the moment of inertia of the pendulum with respect to the axis of rotation, m is the mass of the pendulum, g is the acceleration of gravity, and L is the distance between the center of mass of the pendulum and its axis of rotation. If the percent relative error in I, m, g, and L are respectively, 2%, 1%, 0.5%, and 1.5%, find the approximate maximum percentage error in calculating T. (*Hint:* Use logarithmic differentiation.)

27. If three resistances R_1, R_2, R_3 are connected in parallel, the equivalent resistance R is

$$\frac{1}{R} = \frac{1}{R_1} + \frac{1}{R_2} + \frac{1}{R_3}$$

Assume that $R_1 = 300 \pm 3$ ohms, $R_2 = 200 \pm 4$ ohms and $R_3 = 100 \pm 3$ ohms. Find the approximate maximum relative error in the determination of R.

28. A triangle has sides 8.1 in. and 11.95 in., which include an angle of 44°. Approximate the area of the triangle and compare with the exact results. (*Hint: $A = \frac{1}{2}ab \sin C$.*)

29. A closed box with inner dimensions 3 ft, 4 ft, and 5 ft is to be made of aluminum $\frac{1}{10}$ in. thick. Find the approximate volume of the metal.

30. By application of Archimedes' principle, if a body's weight in air is A and in water is W, its specific gravity s is given by

$$s = \frac{A}{A - W}$$

If for a certain body $A = 30$ lb and $W = 20$ lb, and each of these values is subject to a possible error of 2%, compute the specific gravity and find approximately the greatest possible error in the results by application of differentials.

31. Find an approximate value of $\sqrt[4]{627}\sqrt{80}$ by application of differentials. Express your answer to two decimal places.

32. If $u = f(x_1, x_2, \ldots, x_n)$ and $v = g(x_1, x_2, \ldots, x_n)$ are differentiable functions, show that (a) $d(f + g) = df + dg$ (b) $d(f \cdot g) = f \cdot dg + g \cdot df$

17.6 THE CHAIN RULE WITH APPLICATIONS

The chain rule of differentiation for functions of a single variable was developed in Section 3.6. It was proved there that if y is a differentiable function of u and u is a differentiable function of x then the derivative of y with respect to x may be found by the formula

$$D_x y = D_u y \cdot D_x u \tag{1}$$

This provides a method for determining the derivative of a composite function in terms of the derivatives of each of the component functions. A corresponding formula applies to composite functions where functions of more than one variable are involved. We start with a simple case.

Theorem 1 If $x(t)$ and $y(t)$ are differentiable functions of t and if $f(x, y)$ is differentiable at $(x(t), y(t))$ then the function

$$u = f(x(t), y(t))$$

is differentiable and its derivative is given by

$$D_t u = u_x \cdot x'(t) + u_y \cdot y'(t) \tag{2}$$

or, equivalently,

$$D_t u = u_x D_t x + u_y D_t y \tag{2'}$$

Proof From the hypothesis of differentiability we have that, if $u = f(x, y)$ then

$$\Delta u = u_x \, \Delta x + u_y \, \Delta y + \varepsilon_1 \, \Delta x + \varepsilon_2 \, \Delta y$$

Divide both sides by Δt and take the limit as $\Delta t \to 0$. But, as $\Delta t \to 0$, Δx and Δy also tend to zero. Furthermore, ε_1 and $\varepsilon_2 \to 0$ as Δx and $\Delta y \to 0$. Hence (2) follows. ∎

EXAMPLE 1   If $u = 3x^2 - 4xy + 2y$, where $x = e^t$ and $y = e^{-2t}$. Find $D_t u$ in two different ways.

Solution   **Method 1.**   By (2), we have

$$
\begin{aligned}
D_t u &= (6x - 4y)e^t + (2 - 4x)(-2e^{-2t}) \\
&= (6e^t - 4e^{-2t})e^t + (2 - 4e^t)(-2e^{-2t}) \\
&= 6e^{2t} - 4e^{-t} - 4e^{-2t} + 8e^{-t} \\
&= 6e^{2t} + 4e^{-t} - 4e^{-2t}
\end{aligned}
$$

Method 2.   We first substitute and then differentiate the resulting function of t.

$$
u = 3e^{2t} - 4e^{-t} + 2e^{-2t}
$$
$$
D_t u = 6e^{2t} + 4e^{-t} - 4e^{-2t} \qquad\qquad\bullet
$$

Theorem 1 and its associated formula generalize in an apparent manner to the case of n functions of t. If $u = f(x_1, x_2, \ldots, x_n)$ and $x_k = x_k(t)$ $(k = 1, 2, \ldots, n)$ then with corresponding differentiability requirements

$$
D_t u = f_{x_1} \cdot D_t x_1 + f_{x_2} \cdot D_t x_2 + \cdots + f_{x_n} \cdot D_t x_n \tag{3}
$$

or in summation notation

$$
D_t u = \sum_{k=1}^{n} f_{x_k} \cdot D_t x_k \tag{4}
$$

We now turn to a more general case.

Theorem 2   **(Chain Rule).** Let $u = f(x, y)$ be a differentiable function of x and y. Also, let $x = F(r, s)$ and $y = G(r, s)$ be two functions for which x_r, x_s, y_r and y_s exist. Then u is a composite function of r and s and

$$
u_r = u_x x_r + u_y y_r \tag{5\dagger}
$$
$$
u_s = u_x x_s + u_y y_s \tag{6\dagger}
$$

Proof   We shall prove (5) and (6) will follow in a similar manner. If r is changed by an amount Δr and s is held fixed, then

$$
x + \Delta x = F(r + \Delta r, s) \qquad \text{and} \qquad x = F(r, s)
$$

Subtraction yields

$$
\Delta x = F(r + \Delta r, s) - F(r, s) \tag{7}
$$

Similarly,

$$
\Delta y = G(r + \Delta r, s) - G(r, s) \tag{8}
$$

---

† Alternatively, we may write $\dfrac{\partial u}{\partial r} = \dfrac{\partial u}{\partial x}\dfrac{\partial x}{\partial r} + \dfrac{\partial u}{\partial y}\dfrac{\partial y}{\partial r}$ $\qquad$ (5′)

$\dfrac{\partial u}{\partial s} = \dfrac{\partial u}{\partial x}\dfrac{\partial x}{\partial s} + \dfrac{\partial u}{\partial y}\dfrac{\partial y}{\partial s}$ $\qquad$ (6′)

Because f is a differentiable function of x and y,

$$\Delta u(x, y) = u_x(x, y)\,\Delta x + u_y(x, y)\,\Delta y + \varepsilon_1\,\Delta x + \varepsilon_2\,\Delta y \qquad (9)$$

where ε_1 and ε_2 both $\to 0$ as Δx and $\Delta y \to 0$.

Division by Δr $(\Delta r \neq 0)$ yields

$$\frac{\Delta u}{\Delta r} = u_x \frac{\Delta x}{\Delta r} + u_y \frac{\Delta y}{\Delta r} + \varepsilon_1 \frac{\Delta x}{\Delta r} + \varepsilon_2 \frac{\Delta y}{\Delta r}$$

If the limit of both sides is taken as $\Delta r \to 0$, we obtain (5)

$$u_r = u_x x_r + u_y y_r$$

since $\lim\limits_{\Delta r \to 0} \dfrac{\Delta x}{\Delta r} = x_r$, $\lim\limits_{\Delta r \to 0} \dfrac{\Delta y}{\Delta r} = y_r$, $\lim\limits_{\Delta r \to 0} \varepsilon_1 = 0$, and $\lim\limits_{\Delta r \to 0} \varepsilon_2 = 0$. The latter two formulas follow from (7) and (8) because as $\Delta r \to 0$, Δx and $\Delta y \to 0$ (since x and y are differentiable functions of r and s and therefore are continuous functions with respect to r and s, separately.) ∎

EXAMPLE 2 Let $u = f(x, y) = x^3 + 3x^2 y$, $x = 4r + s$, and $y = 2r - s$; find u_r and u_s.

Solution **Method 1. Chain Rule.**

$$u_x = 3x^2 + 6xy \qquad u_y = 3x^2$$
$$x_r = 4 \qquad x_s = 1 \qquad y_r = 2 \qquad y_s = -1$$

Hence from (5),

$$u_r = (3x^2 + 6xy)(4) + (3x^2)(2) = 18x^2 + 24xy$$
$$= 18(4r + s)^2 + 24(4r + s)(2r - s)$$
$$= 480r^2 + 96rs - 6s^2$$
$$= 6(80r^2 + 16rs - s^2)$$

$$u_s = (3x^2 + 6xy)(1) + (3x^2)(-1)$$
$$= 6xy$$
$$= 6(4r + s)(2r - s) = 6(8r^2 - 2rs - s^2)$$

Method 2. Write u in terms of r and s directly and then perform the partial differentiations:

$$u = (4r + s)^3 + 3(4r + s)^2(2r - s)$$
$$= 64r^3 + 48r^2 s + 12rs^2 + s^3 + 96r^3 - 18rs^2 - 3s^3$$
$$= 160r^3 + 48r^2 s - 6rs^2 - 2s^3$$
$$u_r = 480r^2 + 96rs - 6s^2 = 6(80r^2 + 16rs - s^2)$$

and

$$u_s = 48r^2 - 12rs - 6s^2 = 6(8r^2 - 2rs - s^2)$$

in agreement with the previous result. ●

EXAMPLE 3 Prove that if f is any differentiable function then $u = f(x^2 + y^2)$ is a solution of the partial differential equation $y u_x - x u_y = 0$.

Solution   The notation $u = f(x^2 + y^2)$ means that $u = f(z)$ where $z = x^2 + y^2$. Application of the chain rule yields

$$u_x = D_z f \cdot z_x = f'(z) \cdot 2x \qquad \text{(i)}$$

and

$$u_y = D_z f \cdot z_y = f'(z) \cdot 2y \qquad \text{(ii)}$$

where $' = D_z$. If we multiply (i) by y and (ii) by x and subtract, we obtain

$$yu_x - xu_y = (2xy - 2xy)f'(z) = 0$$

which is the required result.   ●

EXAMPLE 4   Prove that if f and g are any differentiable functions, then $u = f(2x + y) + g(5x + y)$ is a solution of

$$u_{xx} - 7u_{xy} + 10u_{yy} = 0$$

Solution   With the notation $v = 2x + y$ and $w = 5x + y$, we have

$$u = f(v) + g(w)$$

By the chain rule, since $v_x = 2$, $v_y = 1$, $w_x = 5$ and $w_y = 1$,

$$u_x = f' \cdot v_x + g' \cdot w_x = 2f' + 5g'$$
$$u_y = f' \cdot 1 + g' \cdot 1 = f' + g'$$

where $f' = f'(v)$ and $g' = g'(w)$, that is, as usual prime means differentiation with respect to the argument. Application of the chain rule again yields the second partial derivatives,†

$$u_{xx} = 4f'' + 25g''$$
$$u_{xy} = 2f'' \cdot 1 + 5g'' \cdot 1 = 2f'' + 5g''$$
$$u_{yy} = f'' + g''$$

$$u_{xx} - 7u_{xy} + 10u_{yy} = (4 - 14 + 10)f'' + (25 - 35 + 10)g''$$
$$= 0$$

and the solution is verified.   ●

Theorem 3   **(The General Chain Rule).** If u is a differentiable function of n independent variables $x_1, x_2, \ldots, x_n$; that is

$$u = f(x_1, x_2, \ldots, x_n)$$

and each x_j, $(j = 1, 2, \ldots, n)$, is a function of m other independent variables r_k, $(k = 1, 2, \ldots, m)$, such that each of the partial derivatives $\dfrac{\partial x_j}{\partial r_k}$ $(j = 1, 2, \ldots, n; k = 1, 2, \ldots, m)$ exists, then f can be considered as a func-

---

† For example, $u_{xx} = \dfrac{\partial}{\partial x} u_x = \dfrac{\partial}{\partial x}(2f' + 5g') = 2f'' \dfrac{\partial}{\partial x}(2x + y) + 5g'' \dfrac{\partial}{\partial x}(5x + y)$

$$= 4f'' + 25g''$$

and similarly for the other second order partial derivatives.

tion of the m variables $r_1, r_2, \ldots, r_m$ and, furthermore,

$$\frac{\partial u}{\partial r_k} = \sum_{j=1}^{n} \frac{\partial f}{\partial x_j} \cdot \frac{\partial x_j}{\partial r_k} \qquad (k = 1, 2, \ldots, m) \tag{10}$$

This is the **generalized chain rule.**

Equations (10) in expanded form are

$$\frac{\partial u}{\partial r_1} = \frac{\partial f}{\partial x_1} \cdot \frac{\partial x_1}{\partial r_1} + \frac{\partial f}{\partial x_2} \cdot \frac{\partial x_2}{\partial r_1} + \cdots + \frac{\partial f}{\partial x_n} \cdot \frac{\partial x_n}{\partial r_1}$$

$$\frac{\partial u}{\partial r_2} = \frac{\partial f}{\partial x_1} \cdot \frac{\partial x_1}{\partial r_2} + \frac{\partial f}{\partial x_2} \cdot \frac{\partial x_2}{\partial r_2} + \cdots + \frac{\partial f}{\partial x_n} \cdot \frac{\partial x_n}{\partial r_2} \tag{11}$$

$$\cdots$$

$$\frac{\partial u}{\partial r_m} = \frac{\partial f}{\partial x_1} \cdot \frac{\partial x_1}{\partial r_m} + \frac{\partial f}{\partial x_2} \cdot \frac{\partial x_2}{\partial r_m} + \cdots + \frac{\partial f}{\partial x_n} \cdot \frac{\partial x_n}{\partial r_m}$$

This constitutes m relations for which the number of terms on the right side of (11) is n, the number of **intermediate variables**, x_j, $(j = 1, 2, \ldots, n)$.

EXAMPLE 5   If $u = x^2 + y^2 + z^2 - 2xy - 2yz$, $x = r + s - t$, $y = r - s - t$,   and $z = r + 3t$, find u_r, u_s, and u_t by (a) application of the chain rule, and (b) expressing u directly in terms of r, s and t and then differentiating.

Solution   (a) $u_x = 2(x - y)$, $u_y = 2(y - x - z)$, $u_z = 2(z - y)$; $x_r = 1$, $x_s = 1$, $x_t = -1$; $y_r = 1$, $y_s = -1$, $y_t = -1$; and $z_r = 1$, $z_s = 0$, $z_t = 3$.
By application of the chain rule,

$$u_r = u_x x_r + u_y y_r + u_z z_r = 2(x - y) + 2(y - x - z) + 2(z - y)$$
$$= -2y = -2(r - s - t)$$

$$u_s = u_x x_s + u_y y_s + u_z z_s = 2(x - y) - 2(y - x - z) + 0$$
$$= 4x - 4y + 2z$$
$$= 4(r + s - t) - 4(r - s - t) + 2(r + 3t)$$
$$= 2r + 8s + 6t = 2(r + 4s + 3t)$$

$$u_t = u_x x_t + u_y y_t + u_z z_t = -2(x - y) - 2(y - x - z) + 6(z - y)$$
$$= -6y + 8z = -6(r - s - t) + 8(r + 3t)$$
$$= 2r + 6s + 30t = 2(r + 3s + 15t)$$

(b) $u = (r + s - t)^2 + (r - s - t)^2 + (r + 3t)^2$

$$- 2(r + s - t)(r - s - t) - 2(r - s - t)(r + 3t)$$

Routine algebraic simplification yields

$$u = -r^2 + 4s^2 + 15t^2 + 2rs + 2rt + 6st$$

from which

$$u_r = -2r + 2s + 2t = -2(r - s - t)$$
$$u_s = 8s + 2r + 6t = 2(r + 4s + 3t)$$

$$u_t = 30t + 2r + 6s = 2(r + 3s + 15t)$$

in agreement with part (a).   ●

EXAMPLE 6   At a certain instant a right circular cylinder has radius of base equal to 7 in. and altitude equal to 18 in. At this instant the radius of the base is decreasing at the rate of 1 in./sec and the altitude is increasing at the rate of 2 in./sec. Find the rate of change of the volume at this moment.

Solution   The volume V of the cylinder is $V = \pi r^2 h$ where r and h are functions of the time t. Application of the chain rule yields

$$D_t V = V_r D_t r + V_h D_t h$$
$$= 2\pi r h \cdot D_t r + \pi r^2 \cdot D_t h$$

At the given instant,

$$D_t V = 2\pi(7)(18)(-1) + \pi(7^2)(2)$$
$$= -252\pi + 98\pi = -154\pi \text{ in.}^3/\text{sec}$$

which means that the rate at which the volume is decreasing is 154π in.^{3}/sec.   ●

Exercises 17.6

In each of Exercises 1 through 6, find $\dfrac{du}{dt}$ by two methods: (a) by using the chain rule, and (b) by first expressing u directly in terms of t and then differentiating.

1. $u = e^{2x+3y}$, $x = \cos t$, $y = \sin t$
2. $u = e^{x^2-y^2}$, $x = \cosh t$, $y = \sinh t$
3. $u = 2xy + yz - zx$, $x = e^t$, $y = e^{-2t}$, $z = e^{3t}$
4. $u = z \sin(x^2 + y)$, $x = 2t$, $y = t^2 + 1$, $z = t$
5. $u = \ln(x^2 + xy)$, $x = t^2$, $y = t - 3$
6. $u = \dfrac{x^2 + y^2}{z}$, $x = e^{-t} \cos t$, $y = e^{-t} \sin t$, $z = e^{-t}$

In Exercises 7 through 16, find the indicated partial derivatives by two methods: (a) usage of the chain rule, and (b) by substitution and partial differentiation.

7. $u = x^4$, $x = r^2 e^s$; u_r, u_s
8. $u = x^2 y$, $x = re^{-s}$, $y = re^s$; u_r, u_s
9. $u = \sinh xyz$, $x = rs$, $y = s$, $z = rs^2$; u_r, u_s
10. $u = \sin^{-1} xy$, $x = r + s$, $y = r - s$; u_r, u_s
11. $u = \cos(x + y - z)$, $x = r + s - t$, $y = r - s + t$, $z = 2r - 3s - t$; u_r, u_s, u_t
12. $u = e^x + y$, $x = \ln(r + s)$, $y = \tan^{-1}\dfrac{r}{s}$; u_r, u_s
13. $u = x^2 + y^2 + z^2 + 2xy - 2xz + 10$, $x = r - s - t$, $y = r + s + 2t$, $z = 3r - t$; u_r, u_s, u_t
14. $u = x^2 + y^2 + z^2 - xy - yz - zx + 100$, $x = 2r + s + t$, $y = r - s$, $z = r + 4t$; u_r, u_s, u_t
15. $u = x^2 + y^2 + z^2 + s^2$, $x = r \cos t$, $y = r \sin t$, $z = 2rt$, $s = r$; u_r, u_t

16. $u = x^2 + y^2 - z^2 - 3s^2, \quad x = r \sin t, \quad y = r \cos t, \quad z = r + t, \quad s = r - 2t; \qquad u_r, u_t$

17. Prove that, if f is any differentiable function, then $u = f(x + y)$ is a solution of the partial differential equation $u_x - u_y = 0$.

18. Prove that, if f is any differentiable function, then $u = f\left(\dfrac{x}{y}\right)$ is a solution of the partial differential equation $xu_x + yu_y = 0$.

19. Prove that, if f is any differentiable function, then $u = f(bx - ay)$ is a solution of the partial differential equation $au_x + bu_y = 0$, where a and b are arbitrary constants.

20. Prove that, if f is any differentiable function, then $u = (x + y)f(x^2 - y^2)$ is a solution of $yu_x + xu_y = u$.

21. Prove that, if f and g are any differentiable functions, then $u = f(4x + y) + g(2x + y)$ is a solution of $u_{xx} - 6u_{xy} + 8u_{yy} = 0$.

22. Prove that, if f and g are any differentiable functions, then $u = f(x + y) + yg(x + y)$ is a solution of $u_{xx} - 2u_{xy} + u_{yy} = 0$.

In Exercises 23 through 28, use the chain rule to find the indicated derivatives at the values given.

23. $u = y^2 - x^2, \quad x = r \cos \theta, \quad y = r \sin \theta; \qquad u_r$ and u_θ at $r = 2, \theta = \dfrac{\pi}{6}$.

24. $u = x^2 + y^2, \quad x = r \cos \theta, \quad y = r \sin \theta; \qquad u_r$ and u_θ at $r = 1, \theta = \dfrac{\pi}{7}$.

25. $u = z^2 - 3z + 5, \quad z = s^2 + t^2 - 2st; \qquad u_s$ and u_t at $s = 2$ and $t = 3$.

26. $u = x^3 + y^3 + z^3, \quad x = st, \quad y = s + t, \quad z = s + t^2; \qquad u_s$ and u_t at $s = 1$ and $t = -1$.

27. $u = x^2 + y^2 - z^2 - r^2, \quad x = 2s + t, \quad y = 2s - t, \quad z = s^2 - t^2, \quad r = st; \qquad u_s$ and u_t at $s = 3$ and $t = 4$.

28. $u = x^3 + y^3 + z^3 - w^2, \quad x = r^2 + s - t, \quad y = r + s^2 + t, \quad z = s + t, \quad w = (s - t)^2; \qquad u_r, u_s$ and u_t at $r = 1, s = 2$ and $t = -1$.

29. An ideal confined gas obeys the gas law $PV = RT$ where P is the pressure in pounds per square unit, V cubic units is the volume, R is a constant, and T is the absolute temperature. At a certain instant while the gas is being compressed, $V = 12$ ft^3, $P = 20$ lb/in.^{2}, V is decreasing at the rate of 2 ft^3/min while P is increasing at the rate of 8 lb/in.^{2} min. Find $D_t T$ in terms of R.

30. At what rate is the volume of a rectangular box changing when its length is a ft and changing at the rate of k_1 ft/sec, its width is b ft and changing at the rate of k_2 ft/sec, and its height is c ft and changing at the rate of k_3 ft/sec?

31. A particle of mass m moves along the surface $z = 3x^2 + 4y^2 + 2x - y$ in such a way that the velocity in the x- and y-directions are given by $D_t x = 4$ and $D_t y = 3$. Find $D_t z$ and the expression for the kinetic energy $\frac{1}{2}m|\mathbf{v}|^2$ when $x = 2$ and $y = -1$. (*Hint:* $|\mathbf{v}|^2 = \dot{x}^2 + \dot{y}^2 + \dot{z}^2$.)

32. A particle of mass m moves along the surface $z = f(x, y)$. If $x(t)$ and $y(t)$ are the horizontal coordinates of the particle of mass m, derive a formula for the kinetic energy of the particle.

In Exercises 33 through 36, it is assumed that the given equation defines z as a differentiable function of x and y. Find z_x and z_y by implicit differentiation.

33. $x^2 y + y^3 z + z^5 x - 8 = 0$

34. $\sin(x^2 + y) + \sin(y + z) = \frac{1}{4}$

35. $\ln(x^2 + y^2) + e^{\pi z} + e^z = 12$

36. $F(x, y, z) = 0$

37. We recall that a function $f(x, y)$ is said to be **homogeneous** of degree n if $f(tx, ty) = t^n f(x, y)$. Prove Euler's theorem, namely, that if $f(x, y)$ has continuous derivatives in a neighborhood of (x, y), then

$$xf_x + yf_y = nf$$

(*Hint:* Differentiate both sides of the equation with respect to t using the chain rule.)

38. If f is a differentiable function of x and y and $u = f(x, y)$ where $x = r \cos \theta$ and $y = r \sin \theta$, show that

$$u_x^2 + u_y^2 = u_r^2 + \frac{1}{r^2}u_\theta^2$$

(*Hint:* Write u_x and u_y as a linear combination of u_r and u_θ.)

THE DIRECTIONAL DERIVATIVE
AND THE GRADIENT

Consider a function f of two variables x and y. The expression $f_x(x, y)$ gives the rate of change of the function f *in the x-direction*, while $f_y(x, y)$ gives the rate of change of f *in the y-direction*. We wish now to generalize this concept by defining and determining the rate of change of a function in *any* direction in the xy-plane (or parallel to it).

Let f be a function of two variables x and y and take $P_0(x_0, y_0)$ to be a fixed point in the domain of f. Introduce a unit vector $\mathbf{u}$ with its initial point at $P_0(x_0, y_0)$ and making an angle θ with the positive x-axis (Figure 17.7.1). Then

$$\mathbf{u} = (\cos \theta)\mathbf{i} + (\sin \theta)\mathbf{j} \tag{1}$$

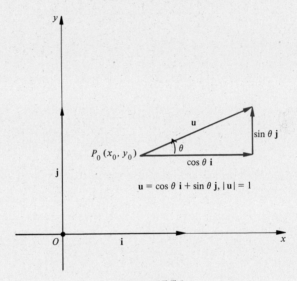

Figure 17.7.1

Select L to be a directed line through P_0 in the direction of $\mathbf{u}$ (Figure 17.7.2).

Suppose that P is any point other than P_0 on L in the domain of f and denote the distance from P_0 to P by Δs. This implies that the x- and y-coordinates of P are $x_0 + \Delta s(\cos \theta)$ and $y_0 + \Delta s(\sin \theta)$, respectively.

Definition   Let f be a function of two independent variables x and y. The **directional derivative** $D_{\mathbf{u}}f$ in the direction $\mathbf{u}$ is defined by

$$D_{\mathbf{u}}f = \lim_{\Delta s \to 0} \frac{f(x_0 + \Delta s(\cos \theta), y_0 + \Delta s(\sin \theta)) - f(x_0, y_0)}{\Delta s} \tag{2}$$

at any point (x_0, y_0) in the domain of f, provided that the limit exists.

In particular, if $\theta = 0$ then $\mathbf{u} = \mathbf{i}$, $\cos \theta = 1$, $\sin \theta = 0$, $\Delta s = \Delta x$ and the directional derivative reduces to the partial derivative with respect to x,

$$f_x(x_0, y_0) = \lim_{\Delta x \to 0} \frac{f(x_0 + \Delta x, y_0) - f(x_0, y_0)}{\Delta x}$$

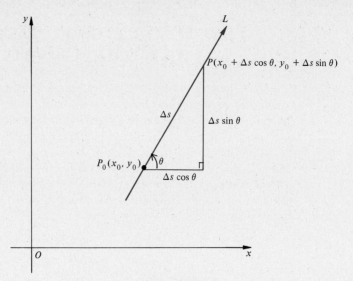

Figure 17.7.2

Similarly, if $\theta = \dfrac{\pi}{2}$, then $\mathbf{u} = \mathbf{j}$, $\cos \theta = 0$, $\sin \theta = 1$, $\Delta s = \Delta y$ and the directional derivative becomes the partial derivative with respect to y,

$$f_y(x_0, y_0) = \lim_{\Delta y \to 0} \frac{f(x_0, y_0 + \Delta y) - f(x_0, y_0)}{\Delta y}$$

In applications of the directional derivative, the following theorem is very important.

Theorem 1   Let f be a differentiable function of two variables x and y, then

$$D_\mathbf{u} f(x_0, y_0) = f_x(x_0, y_0) \cos \theta + f_y(x_0, y_0) \sin \theta \tag{3}$$

Proof   By the definition of the directional derivative

$$D_\mathbf{u} f(x_0, y_0) = \lim_{\Delta s \to 0} \frac{f(x_0 + \Delta s(\cos \theta), y_0 + \Delta s(\sin \theta)) - f(x_0, y_0)}{\Delta s} \tag{2}$$

where $\Delta x = \Delta s(\cos \theta)$ and $\Delta y = \Delta s(\sin \theta)$.

Therefore (2) may be written

$$D_\mathbf{u} f(x_0, y_0) = \lim_{\Delta s \to 0} \frac{f(x_0 + \Delta x, y_0 + \Delta y) - f(x_0, y_0)}{\Delta s} \tag{4}$$

Using the hypothesis of differentiability at (x_0, y_0), we have

$$f(x_0 + \Delta x, y_0 + \Delta y) - f(x_0, y_0) = f_x(x_0, y_0)\, \Delta x + f_y(x_0, y_0)\, \Delta y$$
$$+ \, \varepsilon_1 \, \Delta x + \varepsilon_2 \, \Delta y \tag{5}$$

where ε_1 and $\varepsilon_2 \to 0$ as Δx and $\Delta y \to 0$. Substitution of (5) into (4), yields

$$D_\mathbf{u} f(x_0, y_0) = f_x(x_0, y_0) \lim_{\Delta s \to 0} \frac{\Delta x}{\Delta s} + f_y(x_0, y_0) \lim_{\Delta s \to 0} \frac{\Delta y}{\Delta s}$$

$$+ \lim_{\Delta s \to 0} \varepsilon_1 \cdot \lim_{\Delta s \to 0} \frac{\Delta x}{\Delta s} + \lim_{\Delta s \to 0} \varepsilon_2 \cdot \lim_{\Delta s \to 0} \frac{\Delta y}{\Delta s}$$

Now as $\Delta s \to 0$, Δx and $\Delta y \to 0$ so that ε_1 and $\varepsilon_2 \to 0$. Also, $\dfrac{\Delta x}{\Delta s} = \cos \theta$ and $\dfrac{\Delta y}{\Delta s} = \sin \theta$, where θ is a constant. Hence $\lim\limits_{\Delta s \to 0} \dfrac{\Delta x}{\Delta s} = \cos \theta$ and $\lim\limits_{\Delta s \to 0} \dfrac{\Delta y}{\Delta s} = \sin \theta$, and we have (3). ∎

Analogously to partial derivatives, the directional derivative is easily interpreted geometrically. Let $z = f(x, y)$ be an equation of a surface in three-dimensional space which passes through the point $P_0(x_0, y_0, f(x_0, y_0))$. Draw a vertical plane which contains the points $(x_0, y_0, 0)$ and $(x_0 + \Delta x, y_0 + \Delta y, 0)$ on the directed line L in the xy-plane (Figure 17.7.3). This plane intersects the surface in a plane curve C. The directional derivative of f at (x_0, y_0) gives the slope of the tangent to C at the point $(x_0, y_0, f(x_0, y_0))$.

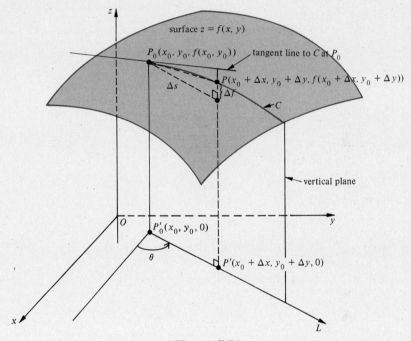

Figure 17.7.3

EXAMPLE 1   Compute the derivative of $f(x, y) = x^2 + 2y^2$ at the point $(3, 2)$ in the direction given by $\theta = \dfrac{\pi}{6}$. Interpret the result if f describes a temperature distribution in the xy-plane.

Solution   We have $f_x(x, y) = 2x$ and $f_y(x, y) = 4y$ and in particular

$$f_x(3, 2) = 6 \qquad \text{and} \qquad f_y(3, 2) = 8$$

Now, $\cos \dfrac{\pi}{6} = \dfrac{\sqrt{3}}{2}$ and $\sin \dfrac{\pi}{6} = \dfrac{1}{2}$ so that from (3), the derivative in the direction $\dfrac{\pi}{6}$ is given by

$$6\left(\dfrac{\sqrt{3}}{2}\right) + 8\left(\dfrac{1}{2}\right) = 3\sqrt{3} + 4$$

If $x^2 + 2y^2$ is the temperature in degrees where x and y are measured in inches, then the space rate of change of temperature at the point $(3, 2)$ and in the direction $\theta = \dfrac{\pi}{6}$ is $(3\sqrt{3} + 4)$ deg/in. ●

EXAMPLE 2 Find the direction θ for which the directional derivative of $f(x, y) = x^2 + 2y^2$ at $(3, 2)$ is a maximum. What is this maximum value?

Solution The directional derivative is

$$f_x(x, y) \cos\theta + f_y(x, y) \sin\theta = 2x(\cos\theta) + 4y(\sin\theta)$$

At $(3, 2)$ the directional derivative is given by

$$F(\theta) = 6\cos\theta + 8\sin\theta \qquad 0 \le \theta \le 2\pi$$

and we want to find the value of θ which maximizes $F(\theta)$. To this end, we form

$$F'(\theta) = -6\sin\theta + 8\cos\theta$$

and if we set $F'(\theta) = 0$, there results

$$-6\sin\theta + 8\cos\theta = 0$$

or

$$\tan\theta = \tfrac{4}{3} \qquad \theta_0 \approx 53° \quad \text{or} \quad \theta_1 \approx 233° \qquad (\theta_1 = 180° + \theta_0)$$

Application of the second derivative test yields

$$F''(\theta_0) < 0 \qquad \text{and} \qquad F''(\theta_1) > 0$$

so that $F(\theta_0)$ is a relative and an absolute maximum, whereas $F(\theta_1)$ is a relative and absolute minimum.

$$F(\theta_0) = 6\cos\theta_0 + 8\sin\theta_0 = 6(\tfrac{3}{5}) + 8(\tfrac{4}{5}) = 10$$

while

$$F(\theta_1) = 6\cos\theta_1 + 8\sin\theta_1 = 6(-\tfrac{3}{5}) + 8(-\tfrac{4}{5}) = -10 \qquad ●$$

The directional derivative is expressible as the dot product of two vectors

$$D_{\mathbf{u}}f(x, y) = [(\cos\theta)\mathbf{i} + (\sin\theta)\mathbf{j}] \cdot [f_x(x, y)\mathbf{i} + f_y(x, y)\mathbf{j}] \qquad (6)$$

The first vector in this dot product is the unit vector $\mathbf{u}$. The second vector is a very important one called the **gradient vector** of f.

Definition If f is a function of two variables having partial derivatives f_x and f_y, then the **gradient vector** of f, denoted by ∇f (read "del f") is defined by

$$\nabla f(x, y) = f_x(x, y)\mathbf{i} + f_y(x, y)\mathbf{j} \qquad (7)$$

Corollary If f is a function of x and y which is differentiable, its directional derivative in the direction of a unit vector $\mathbf{u}$ may be written

$$D_{\mathbf{u}}f(x, y) = \mathbf{u} \cdot \nabla f \qquad (8)$$

where

$\mathbf{u} = (\cos \theta)\mathbf{i} + (\sin \theta)\mathbf{j}$ and ∇f is the gradient vector of f. Note that $\mathbf{u}$ depends only on the angle θ while ∇f depends on the function f and the point (x, y).

Let α be the angle between the unit vector $\mathbf{u}$ and the gradient vector $\nabla f(x, y)$. By (8)

$$D_{\mathbf{u}}f(x, y) = |\mathbf{u}|\,|\nabla f|\cos \alpha = |\nabla f|\cos \alpha \qquad (9)$$

But $-1 \le \cos \alpha \le 1$, so that the directional derivative is greatest when $\alpha = 0$, (that is, when the direction is coincident with the gradient), and the directional derivative is least when $\alpha = \pi$, (that is, when the direction is opposite to that of the gradient). The gradient points in the direction of greatest ascent of the function f, whereas the opposite direction is that of the greatest descent. Hence we have established

Theorem 2 (i) The directional derivative in any direction is obtained by taking the dot product of ∇f with a unit vector in that direction; that is, it is the scalar component of ∇f in that direction. (ii) It is an absolute maximum when its direction coincides with ∇f and it is an absolute minimum when its direction is opposite to ∇f. (iii) Also,

$$D_{\mathbf{u}}f|_{\max} = |\nabla f| = \sqrt{f_x^2 + f_y^2}$$

and

$$D_{\mathbf{u}}f|_{\min} = -|\nabla f| = -\sqrt{f_x^2 + f_y^2}$$

EXAMPLE 3 Given $f(x, y) = x^2 + xy - y^2$, find the maximum value of $D_{\mathbf{u}}f$ at $(4, 3)$.

Solution
$$f_x = 2x + y \qquad f_y = x - 2y$$
from which
$$\nabla f(x, y) = (2x + y)\mathbf{i} + (x - 2y)\mathbf{j}$$
and, in particular,
$$\nabla f(4, 3) = 11\mathbf{i} - 2\mathbf{j}$$

Hence the maximum value of $D_{\mathbf{u}}f$ at $(4, 3)$ is

$$|\nabla f(4, 3)| = \sqrt{11^2 + (-2)^2} = \sqrt{125} \qquad \bullet$$

We recall that the level curves $f(x, y) = k$ are obtained by projecting the intersection of $z = f(x, y)$ and the plane $z = k$ onto the xy-plane. Along such curves $D_{\mathbf{u}}f = 0$, that is, $\nabla f \cdot \mathbf{u} = 0$. Hence the direction of ∇f is perpendicular to the level curve at each point along the curve. This is exhibited in Figure 17.7.4. It has been tacitly assumed that at the point of interest, f_x and f_y are not both zero.

Definition Points at which f_x and f_y are zero are called **critical points**. At such points $\nabla f = \mathbf{0}$ and the directional derivatives in all directions are 0.

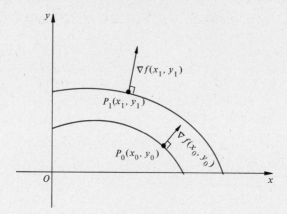

Figure 17.7.4

EXAMPLE 4   Find the critical points of $f(x, y) = x^2 + xy + 3y^2 + x - 2y - 36$

Solution $$f_x(x, y) = 2x + y + 1 \qquad \text{and} \qquad f_y(x, y) = x + 6y - 2$$

and if we set them equal to zero, we have

$$2x + y + 1 = 0 \qquad x + 6y - 2 = 0$$

The simultaneous solution of these equations is $(-\frac{8}{11}, \frac{5}{11})$, the only critical point.

•

The extension of a directional derivative to three (or more) variables is easily done. The proofs of the theorems are similar. In the case of three variables the direction of a unit vector **u** is determined by its direction cosines. Hence from

$$\mathbf{u} = (\mathbf{u} \cdot \mathbf{i})\mathbf{i} + (\mathbf{u} \cdot \mathbf{j})\mathbf{j} + (\mathbf{u} \cdot \mathbf{k})\mathbf{k}$$

we have

$$\mathbf{u} = (\cos \alpha)\mathbf{i} + (\cos \beta)\mathbf{j} + (\cos \gamma)\mathbf{k} \tag{10}$$

Definition   Suppose that f is a function of three variables (x, y, z) and that **u** is given by (10). Then the directional derivative of f in the direction of **u,** denoted by $D_{\mathbf{u}}f$ is given by

$$D_{\mathbf{u}}f(x, y, z) =$$
$$\lim_{\Delta s \to 0} \frac{f(x + \Delta s(\cos \alpha), y + \Delta s(\cos \beta), z + \Delta s(\cos \gamma)) - f(x, y, z)}{\Delta s} \tag{11}$$

if this limit exists.

By the same technique as used in Theorem 1, we can establish a third theorem.

Theorem 3   If f is a differentiable function of (x, y, z) and $\mathbf{u} = (\cos \alpha)\mathbf{i} + (\cos \beta)\mathbf{j} + (\cos \gamma)\mathbf{k}$, then

$$D_{\mathbf{u}}f = f_x(x, y, z) \cos \alpha + f_y(x, y, z) \cos \beta + f_z(x, y, z) \cos \gamma \tag{12}$$

Example 5   Find the directional derivative of the function

$$f(x, y, z) = (x - 2)^2 + 3(y + 3)^2 + 2(z - 1)^2 - 20$$

at the point $(3, -2, 4)$ in the direction of the vector $2\mathbf{i} - \mathbf{j} - 2\mathbf{k}$.

Solution
$$\nabla f(x, y, z) = 2(x - 2)\mathbf{i} + 6(y + 3)\mathbf{j} + 4(z - 1)\mathbf{k}$$

and at $(3, -2, 4)$

$$\nabla f(3, -2, 4) = 2\mathbf{i} + 6\mathbf{j} + 12\mathbf{k}$$

A unit vector in the specified direction is

$$\mathbf{u} = \tfrac{1}{3}(2\mathbf{i} - \mathbf{j} - 2\mathbf{k})$$

so that $\cos \alpha = \tfrac{2}{3}$, $\cos \beta = -\tfrac{1}{3}$, and $\cos \gamma = -\tfrac{2}{3}$ and the required directional derivative is

$$D_\mathbf{u} f = 2(\tfrac{2}{3}) + 6(-\tfrac{1}{3}) + 12(-\tfrac{2}{3}) = -\tfrac{26}{3} \qquad\bullet$$

Definition   If f is a function of three variables (x, y, z) having first partial derivatives f_x, f_y, and f_z, then the **gradient vector** of f denoted by ∇f (read "del f") is defined by

$$\nabla f = f_x(x, y, z)\mathbf{i} + f_y(x, y, z)\mathbf{j} + f_z(x, y, z)\mathbf{k} \qquad (13)$$

From (10), (12), and (13), it follows that if f is a differentiable function of (x, y, z), that

$$D_\mathbf{u} f = \mathbf{u} \cdot \nabla f(x, y, z) \qquad (14)$$

Definition   The **critical points** of $f(x, y, z)$ are the points at which $\operatorname{grad} f = 0$, that is, points at which $f_x = f_y = f_z = 0$, holds simultaneously.

Theorem 4   The directional derivative of a function $f(x, y, z)$ at a point P is a *maximum* in the direction of the gradient of the function at P. It is a *minimum* in the direction of the negative of the gradient of the function at P.

As we have seen in Section 17.1, the surface of the form $f(x, y, z) = c$, where c is a constant is called a **level surface** for f. In Section 17.8 we shall show that *the direction of the gradient vector of f at all points of such a surface must be at right angles to the surface.*

Example 6   The temperature at a point in space is given by $T(x, y, z) = 300 - 2x^2 - 3y^2 - 4z^2$. In what direction should one move from the point $(3, 2, 1)$ in order to cool off as quickly as possible?

Solution   In order to cool off as rapidly as possible, we must choose the direction $-\nabla T$ at any point (x, y, z).

$$-\nabla T = 4x\mathbf{i} + 6y\mathbf{j} + 8z\mathbf{k}$$

and at the given point

$$-\nabla T(3, 2, 1) = 12\mathbf{i} + 12\mathbf{j} + 8\mathbf{k}$$

which yields the required direction has been obtained. ●

Exercises 17.7

In Exercises 1 through 8, find the directional derivative of the indicated function. In particular, find its value at the given point P_0 in the specific direction θ.

1. $f(x, y) = x^2 - 2y^2$; $P_0(2, 3), \theta = \dfrac{\pi}{3}$

2. $g(x, y) = x^2 y$; $P_0(3, 1), \theta = \dfrac{\pi}{6}$

3. $h(x, y) = e^x \cos y$; $P_0\left(0, \dfrac{\pi}{3}\right), \theta = \dfrac{\pi}{4}$

4. $F(x, y) = \ln (2x + y)$; $P_0(2, 1), \theta = \dfrac{\pi}{6}$

5. $G(x, y) = \ln (x^2 + y^2)$; $P_0(-1, 3), \theta = \dfrac{3\pi}{4}$

6. $H(x, y) = 3 \sin x \cos y$; $P_0\left(\dfrac{\pi}{4}, \dfrac{\pi}{6}\right), \theta = \dfrac{\pi}{3}$

7. $f(x, y) = \tan^{-1} \dfrac{y}{x}$; $P_0(3, 2), \theta = \dfrac{2\pi}{3}$

8. $g(x, y) = \dfrac{x^2 + xy}{x^2 - y^2}$; $P_0(5, 3), \theta = -\dfrac{\pi}{3}$

In Exercises 9 through 14, find the direction θ in which the directional derivative of the indicated function at the given point P_0 is a maximum. Find the maximum value.

9. $f(x, y) = 2x^2 - y^2$, $P_0(1, 2)$

10. $g(x, y) = e^x \cos y$, $P_0\left(0, -\dfrac{\pi}{6}\right)$

11. $h(x, y) = x^2 - xy + y^2$, $P_0(3, 2)$

12. $F(x, y) = \sin (x + 2y)$, $P_0\left(\dfrac{\pi}{6}, \dfrac{\pi}{12}\right)$

13. $G(x, y) = \ln (x^2 + 3y^2)$, $P_0(2, -1)$

14. $H(x, y) = e^x \ln (x + y)$, $P_0(0, 1)$

In Exercises 15 through 18, find the directional derivative of the indicated function of three variables in the prescribed direction $\mathbf{u}$ at the given point.

15. $f(x, y, z) = x^2 + 2y^2 + z^3$, $\mathbf{u} = \frac{1}{3}(2\mathbf{i} - \mathbf{j} + 2\mathbf{k})$, $P_0(1, 3, 2)$

16. $g(x, y, z) = xy^2 e^z$, $\mathbf{u} = \frac{1}{5}(3\mathbf{i} + 4\mathbf{j})$, $P_0(4, 2, 0)$

17. $F(x, y, z) = \dfrac{x^2 z}{y}$, $\mathbf{u} = \frac{1}{2}(\mathbf{i} - \mathbf{j} - \sqrt{2}\mathbf{k})$, $P_0(1, 2, 4)$

18. $G(x, y, z) = x^2 e^y z^{1/2}$, $\mathbf{u} = \frac{1}{4}(2\mathbf{i} + \sqrt{3}\mathbf{j} - 3\mathbf{k})$, $P_0(2, 3, 4)$

In Exercises 19 through 22, for the given function, find (a) the gradient, (b) the critical points, (c) the unit vector pointing in the direction of greatest increase at the given point P_0, and (d) the magnitude of the greatest increase at P_0.

19. $f(x, y) = 3x^2 - xy - y^2$, $P_0(1, 2)$
20. $g(x, y) = 2x^2 + xy + y^2 + 3x - 2y + 5$, $P_0(-2, 4)$
21. $f(x, y, z) = x^2 + 2y^2 + xz + z^2 - x - y$, $P_0(4, 2, 3)$
22. $f(x, y, z) = \sqrt{x^2 + y^2 + z^2}$, $(x, y, z) \neq (0, 0, 0)$, $P_0(5, 3, 4)$
23. Suppose that $\nabla f(x, y, z) = \mathbf{0}$ at *all* points in its domain of definition. What conclusion follows?
24. If $f(r, \theta)$, where r and θ are plane polar coordinates, show that

$$\nabla f = f_r \widehat{\mathbf{r}} + \frac{1}{r} f_\theta \widehat{\boldsymbol{\theta}}$$

where $\widehat{\mathbf{r}}$ and $\widehat{\boldsymbol{\theta}}$ are unit vectors in the radial and circumferential directions, respectively. Use this result to find $|\nabla f|^2$ in polar coordinates.

25. Let $\mathbf{r} = x\mathbf{i} + y\mathbf{j} + z\mathbf{k}$ and $|\mathbf{r}| = \sqrt{x^2 + y^2 + z^2} = \rho$ show that (a) $\nabla\rho = \frac{1}{\rho}\mathbf{r}$ and (b) $\nabla\rho^n = n\rho^{n-2}\mathbf{r}$.

26. If f and g are given functions of x, y, and z for which all first partial derivatives exist, show that
   (a) $\nabla(c_1 f + c_2 g) = c_1\nabla f + c_2\nabla g$ (c_1, c_2 constants)
   (b) $\nabla(fg) = f\nabla g + g\nabla f$
27. Show that $\nabla f(u) = f'(u)\nabla u$, where $u = u(x, y, z)$.
28. Let $P(x, y)$ be any point on an ellipse with F_1 and F_2 as its two foci. By definition, the ellipse is defined by the requirement that $r_1 + r_2 = $ constant, where $r_1 = |\overline{PF_1}|$ and $r_2 = |\overline{PF_2}|$. Show that
   (a) $\mathbf{u} \cdot \nabla(r_1 + r_2) = 0$ where $\mathbf{u}$ is a unit vector tangent to the ellipse at P.
   (b) vectors $\overrightarrow{F_1P}$ and $\overrightarrow{PF_2}$ make equal angles with $\mathbf{u}$. (*Hint:* recall from Exercise 25 that ∇r_1 and ∇r_2 are unit vectors in the directions $\overrightarrow{F_1P}$ and $\overrightarrow{F_2P}$ respectively.)

17.8 TANGENT PLANES AND NORMAL LINES TO A SURFACE

In the previous section we introduced the concept of the gradient of a scalar function. In particular, it was established that the vector ∇f is orthogonal to a level curve defined by an equation $f(x, y) = c$ at a point $P_0(x_0, y_0)$. Of course this makes sense only when the level curve is smooth at P_0, that is, a tangent line exists at the point in question. A sufficient condition for ∇f to be orthogonal to a tangent line at P_0 is that $f_x(x, y)$ and $f_y(x, y)$ be continuous in a neighborhood of P_0 and that f_x and f_y are not both zero at P_0, that is, $\nabla f \neq 0$ at P_0.

We now wish to discuss the three-dimensional analog (to the two-dimensional case) more precisely than we did in Section 17.7. Let f be a function defined in some region in space and suppose that

$$u = f(x, y, z)$$

Definition   The set of points where

$$f(x, y, z) = c \qquad (c \text{ constant})$$

is a ***level surface*** for f.

Definition   A vector is said to be a ***normal vector*** to a surface at a point $P_0(x_0, y_0, z_0)$ provided that this vector is perpendicular to any curve on the surface that passes through (x_0, y_0, z_0).

Note that a vector is perpendicular to a curve at a point (x_0, y_0, z_0) provided that it is perpendicular to a tangent vector to the curve at (x_0, y_0, z_0).

Theorem 1   Let $f(x, y, z) = c$ be a level surface. If $f_x, f_y,$ and f_z are continuous at P_0 and not zero simultaneously then the gradient ∇f at (x_0, y_0, z_0) is a normal vector to the level surface of f passing through the point (x_0, y_0, z_0).

Proof   Let $\mathbf{g}(t)$ be the parametrization of a curve C_0 in the level surface of f that passes through the point (x_0, y_0, z_0), Figure 17.8.1. We write $\mathbf{g}(t) = x\mathbf{i} + y\mathbf{j} + z\mathbf{k}$, where $x = x(t), y = y(t), z = z(t),$ and t_0 is the value of t that corresponds to the point (x_0, y_0, z_0); thus $\mathbf{g}(t_0) = x_0\mathbf{i} + y_0\mathbf{j} + z_0\mathbf{k}$. The vector $\mathbf{g}'(t_0) = x'(t_0)\mathbf{i} + y'(t_0)\mathbf{j} + z'(t_0)\mathbf{k}$ is directed along the tangent vector to C_0.

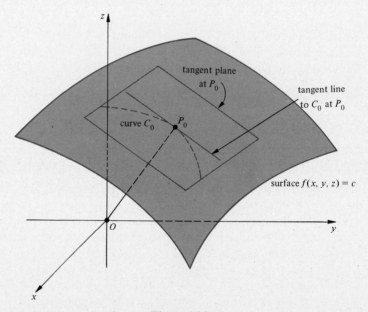

Figure 17.8.1

The gradient of f at P_0 is given by

$$\nabla f|_{P_0} = f_x(x_0, y_0, z_0)\mathbf{i} + f_y(x_0, y_0, z_0)\mathbf{j} + f_z(x_0, y_0, z_0)\mathbf{k}$$

and it must be shown that $\nabla f|_{P_0} \cdot \mathbf{g}'(t_0) = 0.$ We have

$$\nabla f|_{P_0} \cdot \mathbf{g}'(t_0) = f_x(x_0, y_0, z_0)\, x'(t_0) + f_y(x_0, y_0, z_0)\, y'(t_0) + f_z(x_0, y_0, z_0)\, z'(t_0)$$

and by the chain rule this is $f'(t_0)$. But $f = c =$ constant (level surface) so that $f'(t_0) = 0.$ Hence (since neither of these vectors is the zero vector) it follows that ∇f is a normal vector at P_0. ∎

The curve C_0 in Theorem 1 is an arbitrary curve lying in the surface and passing through P_0. This means that ∇f is perpendicular to all such smooth curves; that is, to their tangent lines. Thus all these tangent lines at P_0 lie in a plane which is perpendicular to the gradient at P_0.

Definition   If a surface $f(x, y, z) = c$ has a nonzero gradient vector at a point P_0 on it, then the plane through P_0 that is perpendicular to ∇f is called the **tangent plane** to the surface at P_0.

The components of ∇f at P_0 are $f_x(x_0, y_0, z_0)$, $f_y(x_0, y_0, z_0)$, and $f_z(x_0, y_0, z_0)$. Therefore, an equation of the tangent plane is given by

$$\nabla f(x_0, y_0, z_0) \cdot [(x - x_0)\mathbf{i} + (y - y_0)\mathbf{j} + (z - z_0)\mathbf{k}] = 0 \tag{1}$$

or in expanded form we have

$$(x - x_0)f_x(x_0, y_0, z_0) + (y - y_0)f_y(x_0, y_0, z_0) + (z - z_0)f_z(x_0, y_0, z_0) = 0$$

We summarize the foregoing discussion with the following theorem.

Theorem 2   An equation of the tangent plane to the surface $f(x, y, z) = c$ at any point $P_0(x_0, y_0, z_0)$ on it is

$$(x - x_0)f_x(x_0, y_0, z_0) + (y - y_0)f_y(x_0, y_0, z_0) + (z - z_0)f_z(x_0, y_0, z_0) = 0 \tag{2}$$

provided that f_x, f_y, and f_z are continuous at P_0 and are not all zero simultaneously.

The line through (x_0, y_0, z_0) with a set of direction numbers $f_x(x_0, y_0, z_0)$, $f_y(x_0, y_0, z_0)$ and $f_z(x_0, y_0, z_0)$ is the **normal line** to the surface. It is perpendicular to the tangent plane and its symmetric equations are

$$\frac{x - x_0}{f_x(x_0, y_0, z_0)} = \frac{y - y_0}{f_y(x_0, y_0, z_0)} = \frac{z - z_0}{f_z(x_0, y_0, z_0)} \tag{3}$$

provided that none of the denominators in (3) are zero.
If an equation of the surface is given in the form

$$z = g(x, y) \tag{4}$$

then

$$f(x, y, z) = g(x, y) - z = 0$$

and (since $f_x = g_x$, $f_y = g_y$, and $f_z = -1$) an equation of a tangent plane to a surface defined by (4) at $P_0(x_0, y_0, z_0)$ is from (2)

$$(x - x_0)g_x(x_0, y_0) + (y - y_0)g_y(x_0, y_0) - (z - z_0) = 0 \tag{5}$$

provided that g_x and g_y are continuous at (x_0, y_0).
Similarly, from (3), equations of the **normal line** are found from

$$\frac{x - x_0}{g_x(x_0, y_0)} = \frac{y - y_0}{g_y(x_0, y_0)} = \frac{z - z_0}{-1} \tag{6}$$

EXAMPLE 1 Find an equation of the tangent plane to the sphere $x^2 + y^2 + z^2 = 49$ at the point $(6, 2, 3)$ and write symmetric equations for the normal line to the sphere at the same point. Verify that the normal line passes through the center $(0, 0, 0)$ of the sphere.

Solution
$$f(x, y, z) = x^2 + y^2 + z^2 = 49$$

$$f_x(x, y, z) = 2x \qquad f_y(x, y, z) = 2y \qquad f_z(x, y, z) = 2z$$

and at $(6, 2, 3)$, $f_x(6, 2, 3) = 12$, $f_y(6, 2, 3) = 4$, and $f_z(6, 2, 3) = 6$. Substitution into (2) yields an equation of the tangent plane

$$12(x - 6) + 4(y - 2) + 6(z - 3) = 0$$

or, upon simplification

$$6x + 2y + 3z = 49$$

Since the normal line to the sphere has direction numbers $(6, 2, 3)$, its symmetric equations are

$$\frac{x - 6}{6} = \frac{y - 2}{2} = \frac{z - 3}{3}$$

and if we set these three members equal to t, there results

$$x = 6(t + 1) \qquad y = 2(t + 1) \qquad z = 3(t + 1)$$

Note that when $t = 0$, we recover the point $(6, 2, 3)$, whereas when $t = -1$, we have $x = y = z = 0$. In other words, the normal line to the spherical surface at the given point must pass through the origin. It is left as an exercise for the reader to prove that the normal line to *any* point on the surface of the sphere $x^2 + y^2 + z^2 = 49$ must pass through the origin. ●

EXAMPLE 2 Find equations of the tangent plane and normal line of Example 1 by applying (5) and (6) of this section.

Solution $z = \sqrt{49 - x^2 - y^2} = g(x, y)$ where the positive square root is chosen since $z = 3$ when $x = 6$ and $y = 2$. Thus

$$g_x = \frac{-x}{\sqrt{49 - x^2 - y^2}} \qquad g_x(6, 2) = -\frac{6}{3} = -2$$

and

$$g_y = \frac{-y}{\sqrt{49 - x^2 - y^2}} \qquad g_y(6, 2) = -\frac{2}{3}$$

Substitution into (5) yields an equation of the tangent plane

$$(x - 6)(-2) + (y - 2)(-\tfrac{2}{3}) - (z - 3) = 0$$

which simplifies to: $6x + 2y + 3z = 49$. This is in agreement with the solution of Example 1.

From (6), we obtain the equations of the normal line

$$\frac{x - 6}{-2} = \frac{y - 2}{-\tfrac{2}{3}} = \frac{z - 3}{-1}$$

If each of the members is divided by -3, the system of Example 1 is recovered. ●

Consider two surfaces $f(x, y, z) = 0$ and $g(x, y, z) = 0$ which intersect in a curve C. Let $P_0(x_0, y_0, z_0)$ be a point on each of the surfaces and on curve C. We seek a set of direction numbers of the tangent line to C at P_0 from which symmetric equations of the tangent line are immediately obtained.

A normal vector to the surface

$$f(x, y, z) = 0 \qquad (7)$$

at P_0 is given by

$$\nabla f(x_0, y_0, z_0) = f_x(x_0, y_0, z_0)\mathbf{i} + f_y(x_0, y_0, z_0)\mathbf{j} + f_z(x_0, y_0, z_0)\mathbf{k} \qquad (8)$$

Similarly, a normal vector to the surface

$$g(x, y, z) = 0 \qquad (9)$$

at P_0 is determined by

$$\nabla g(x_0, y_0, z_0) = g_x(x_0, y_0, z_0)\mathbf{i} + g_y(x_0, y_0, z_0)\mathbf{j} + g_z(x_0, y_0, z_0)\mathbf{k} \qquad (10)$$

Each of the normal vectors is perpendicular to C at P_0. Thus, if f and g are not parallel, that is, surfaces $f = 0$ and $g = 0$ are not tangential to each other at P_0, then $\nabla f(x_0, y_0, z_0) \times \nabla g(x_0, y_0, z_0)$ is a vector which is parallel to the tangent vector to C at $P_0(x_0, y_0, z_0)$. Hence a set of direction numbers for the tangent vector is now known and symmetric equations of the tangent line to C at P_0 are determined. The following example illustrates this method.

EXAMPLE 3   Find symmetric equations of the tangent line to the curve of intersection of the surfaces $2x^2 + 9y^2 - z = 0$ and $8x^2 - 27y^2 - 2z + 1 = 0$ at $(1, \frac{1}{3}, 3)$.

Solution   Let $f(x, y, z) = 2x^2 + 9y^2 - z = 0$ and $g(x, y, z) = 8x^2 - 27y^2 - 2z + 1 = 0$. Then

$$\nabla f(x, y, z) = 4x\mathbf{i} + 18y\mathbf{j} - \mathbf{k}$$

and

$$\nabla g(x, y, z) = 16x\mathbf{i} - 54y\mathbf{j} - 2\mathbf{k}$$

Also,

$$\nabla f(1, \tfrac{1}{3}, 3) = 4\mathbf{i} + 6\mathbf{j} - \mathbf{k}$$

and

$$\nabla g(1, \tfrac{1}{3}, 3) = 16\mathbf{i} - 18\mathbf{j} - 2\mathbf{k}$$

At the point $(1, \frac{1}{3}, 3)$

$$\nabla f \times \nabla g = (4\mathbf{i} + 6\mathbf{j} - \mathbf{k}) \times (16\mathbf{i} - 18\mathbf{j} - 2\mathbf{k})$$
$$= -30\mathbf{i} - 8\mathbf{j} - 168\mathbf{k}$$

Division by -2 yields a set of direction numbers $(15, 4, 84)$ and symmetric equations of the tangent line are

$$\frac{x - 1}{15} = \frac{y - \frac{1}{3}}{4} = \frac{z - 3}{84} \qquad \bullet$$

Exercises 17.8

In Exercises 1 through 4, find a vector perpendicular to the surface at the given point.

1. $x^2 + y^2 + z^2 = 9$, $(2, 1, 2)$
2. $3x - y - 4z + 2 = 0$, $(4, 2, 3)$
3. $x^2yz = 2$, $(1, -2, -1)$
4. $xy + yz + zx = 63$, $(3, 5, 6)$

In Exercises 5 throughout 16, find in each case an equation of the tangent plane and equations of the normal line to the given surface at the indicated point.

5. $x^2 + y^2 + z^2 = 50$, $(4, 3, 5)$

6. $4x^2 - y^2 - z = 4$, $(1, 2, -4)$

7. $xy - 2z = 0$, $(3, 4, 6)$

8. $x^2y - 3z = 0$, $(-2, 3, 4)$

9. $3x^2 + y^2 + 2z^2 = 21$, $(2, 1, -2)$

10. $4x^2 - y^2 - 2z^2 = 0$, $(-3, 2, 4)$

11. $e^{2x} \cos y - 3z = 0$, $\left(\dfrac{1}{2}, \dfrac{\pi}{3}, \dfrac{e}{6}\right)$

12. $z^2 - \ln(x^2 + y^2 + 1) = 0$, $(-2, 2, \sqrt{\ln 9})$

13. $x^2 - 3y^2 + z^2 = 13$, $(3, -2, -4)$

14. $xy + yz + zx = 11$, $(2, 3, 1)$

15. $x^{1/2} + y^{1/2} + z^{1/2} = 9$, $(4, 16, 9)$

16. $y = x^2 + 3z^2$, $(-1, 13, 2)$

17. Prove that the plane tangent to the surface $z = x^2 - y^2$ at the point $(2, 1, 3)$ intersects the z-axis at the point for which $z = -3$.

18. Generalize the previous exercise by proving that the plane tangent to the surface $z = x^2 - y^2$ at the point (x_0, y_0, z_0) intersects the z-axis at the point for which $z = -z_0$.

19. Show that the sum of the intercepts on the coordinate axes of any tangent plane to the surface $x^{1/2} + y^{1/2} + z^{1/2} = a^{1/2}$ (where a is a positive constant) is a constant. What is this constant?

20. Show that the sum of the squares of the intercepts on the coordinate axes of any tangent plane to the surface $x^{2/3} + y^{2/3} + z^{2/3} = a^{2/3}$ (where a is a constant) is a constant. What is this constant?

In Exercises 21 through 24, find equations of the tangent line to the curve of intersection of the two given surfaces at the given point.

21. $y - 2x = 0$, $x + 2y + z = 8$; $(1, 2, 3)$

22. $x + 2y - z = 0$, $x^2 + y^2 + z^2 = 14$; $(3, -2, -1)$

23. $x^2 + y^2 + z^2 = 9$, $(x - 1)^2 + (y - 1)^2 + z^2 = 5$; $(2, 1, 2)$

24. $3x^2 + 2y^2 - z^2 = 1$, $2xy - z = 0$; $(1, 1, 2)$

25. Two surfaces $f(x, y, z) = 0$ and $g(x, y, z) = 0$ meet in a space curve C. Let $P_0(x_0, y_0, z_0)$ be a point on C and find equations of the tangent line to C at P_0.

26. Two surfaces $f(x, y, z) = 0$ and $g(x, y, z) = 0$ are orthogonal to each other at a point $P_0(x_0, y_0, z_0)$ of intersection provided that normal vectors to the surfaces are perpendicular to each other at P_0. Show that the two surfaces $2x^2 + 2y^2 + 2z^2 - 33 = 0$ and $2x^2 + 2y^2 + 2x + 2y - z + 1 = 0$ are orthogonal to each other at $P_0(\frac{1}{2}, \frac{1}{2}, 4)$.

27. Show that the acute angle between the surfaces $f(x, y, z) = x^2 + y^2 - 8z - 8 = 0$ and $g(x, y, z) = x^2 + y^2 + 8z - 8 = 0$ is the same at *all* intersection points of the two surfaces. What is the angle to the nearest degree? (*Hint:* Show that the two surfaces, which are paraboloids of revolution, intersect in the circle $z = 0$, $x^2 + y^2 = 8$.)

28. Two surfaces are said to be *tangent* to each other at a point P_0 if they have the same tangent plane at P_0. Show that the surfaces described by $x^2 + z^2 - 2x + 4y + 6z + 10 = 0$ and $x^2 - 2x + y^2 + z^2 = 1$ are tangent to each other at $(1, -1, -1)$. What is an equation of the tangent plane at $(1, -1, -1)$?

17.9 MAXIMA AND MINIMA FOR FUNCTIONS OF TWO VARIABLES

In Chapter 4, we developed the theory and showed diverse applications of maxima and minima for functions of one variable. In this section, we shall extend these considerations to functions of two variables. This again forms one of the most important applications of differentiation. Let us start with some definitions.

Definition   A function f is said to have an ***absolute maximum*** M at a point (x_0, y_0) of a set or region R in the xy-plane if $M = f(x_0, y_0) \geq f(x, y)$ for all points (x, y) belonging to R.

Definition   A function f is said to have an ***absolute minimum*** m at a point (x_0, y_0) of a set or region R in the xy-plane if $m = f(x_0, y_0) \leq f(x, y)$ for all points (x, y) belonging to R.

EXAMPLE 1   Find the absolute maximum and absolute minimum of $f(x, y) = 1 - x^2 - y^2$ if $x^2 + y^2 \leq 4$, defines the region R.

Solution   In this instance, the region R is the interior and boundary of a circle of radius 2 with center at $(0, 0)$. Now

$$f(x, y) = 1 - (x^2 + y^2)$$

and since $x^2 + y^2 \geq 0$ with the equality holding if and only if $x = y = 0$, then

$$M = f(0, 0) = 1$$

is the absolute maximum of f on the given region. Also, since f decreases with increasing $x^2 + y^2$ and if $x^2 + y^2 = r^2 \leq 4$, then $f(x, y) = 1 - r^2$. Hence the minimum value for the function occurs at all points on the boundary of the circle $(r = 2)$ and the absolute minimum is $m = -3$.   ●

EXAMPLE 2   Find the absolute maximum and absolute minimum of $f(x, y) = 1 - x^2 - y^2$ if $x^2 + y^2 < 4$.

Solution   Now the region R consists only of the *interior* of the circle with radius 2. Again, $f(x, y) = 1 - r^2$ where $0 \leq r < 2$, and the absolute maximum of f is

$$M = f(0, 0) = 1.$$

However, there is *no absolute minimum* since there is no smallest number larger than -3. The values of $f > -3$ can be made as close to -3 as desired simply by taking points sufficiently close to the boundary of the circle.   ●

Theorem 1   If a function f of two variables is *continuous* in a ***closed***† ***region*** of the plane (such as a rectangle, a disk, and so on), then the function actually possesses an absolute maximum value M and an absolute minimum value m on that region.

The proof of this theorem will not be given. Note that Theorem 1 guarantees the existence of M and m but does not tell us how they can be determined. The extrema may occur in the interior or on the boundary of the closed region.

---

† ***Closed*** means the region *including* its boundary. Thus, for example, $S = \{(x, y) \mid x^2 + y^2 < 1\}$ is *not* a closed region, whereas $T = \{(x, y) \mid x^2 + y^2 \leq 1\}$ is a closed region, since it includes the boundary $x^2 + y^2 = 1$.

If Theorem 1 is applied to Example 1, we know that M and m exist because $f(x, y) = 1 - x^2 - y^2$ is continuous on the closed region $x^2 + y^2 \leq 4$. On the other hand, the region of Example 2 is not closed and Theorem 1 does not apply. However, this in itself does not mean that an absolute maximum or an absolute minimum (or both) do not exist.

For the determination of the location of maxima and minima, we need once again the concept of relative extrema.

Definition   The function $f(x, y)$ is said to have a ***relative maximum value*** at the point (x_0, y_0) provided that there exists a neighborhood of (x_0, y_0) such that $f(x, y) \leq f(x_0, y_0)$ for all (x, y) in that neighborhood.

Definition   A function $f(x, y)$ is said to have a ***relative minimum value*** at the point (x_0, y_0) provided that there exists a neighborhood of (x_0, y_0) such that $f(x, y) \geq f(x_0, y_0)$ for all (x, y) in that neighborhood.

Theorem 2   If $f(x, y)$ exists at all points in some neighborhood of a point (x_0, y_0) and if f has a relative maximum or a relative minimum at (x_0, y_0), then if $f_x(x_0, y_0)$ and $f_y(x_0, y_0)$ exist, it follows that

$$f_x(x_0, y_0) = 0 \qquad \text{and} \qquad f_y(x_0, y_0) = 0 \tag{1}$$

Proof   We prove that if f has a relative maximum value at (x_0, y_0) and if $f_x(x_0, y_0)$ exists then $f_x(x_0, y_0) = 0$. The argument that follows is the same as the one-variable case because, in proving this, only x varies. By definition,

$$f_x(x_0, y_0) = \lim_{\Delta x \to 0} \frac{f(x_0 + \Delta x, y_0) - f(x_0, y_0)}{\Delta x}$$

Since $f(x, y)$ has a relative maximum value at (x_0, y_0), we have

$$f(x_0 + \Delta x, y_0) - f(x_0, y_0) \leq 0 \tag{i}$$

whenever $|\Delta x|$ is sufficiently small so that $(x_0 + \Delta x, y_0)$ lies in the neighborhood. Now let $\Delta x \to 0$ through positive values. Hence for $\Delta x > 0$, from (i)

$$\frac{f(x_0 + \Delta x, y_0) - f(x_0, y_0)}{\Delta x} \leq 0 \tag{ii}$$

But then

$$\lim_{\Delta x \to 0^+} \frac{f(x_0 + \Delta x, y_0) - f(x_0, y_0)}{\Delta x} \leq 0$$

or

$$f_x(x_0, y_0) \leq 0 \tag{iii}$$

Now let $\Delta x \to 0$ through negative values. Then for $\Delta x < 0$,

$$\frac{f(x_0 + \Delta x, y_0) - f(x_0, y_0)}{\Delta x} \geq 0$$

which implies that

$$\lim_{\Delta x \to 0^-} \frac{f(x_0 + \Delta x, y_0) - f(x_0, y_0)}{\Delta x} \geq 0$$

or

$$f_x(x_0, y_0) \geq 0 \tag{iv}$$

In order for (iii) and (iv) to be consistent, $f_x(x_0, y_0) = 0$ must hold.

The proof that $f_y(x_0, y_0) = 0$ if $f(x_0, y_0)$ is a relative maximum of f and $f_y(x_0, y_0)$ exists, follows in a similar manner (see Exercise 25). The proof of the theorem when $f(x, y)$ has a relative minimum at (x_0, y_0) is also similar (Exercise 26). ∎

Note that the points (x_0, y_0) for which $f_x(x_0, y_0) = f_y(x_0, y_0) = 0$ are critical (or stationary) points of the function f. Conditions (1) are necessary but not sufficient conditions that a function (for which f_x and f_y exist) possesses a relative extremum at (x_0, y_0) (see Figure 17.9.1). Note that if $f_x(x_0, y_0) = f_y(x_0, y_0) = 0$, as shown in the figure, we have a horizontal tangent plane at the maximum point (x_0, y_0, z_0), with $z = z_0$ for its equation. However, just as in the case of functions of one variable, f need not have a maximum nor a minimum at a critical point. Such a point is then called a **saddle point** (or **minimax**).

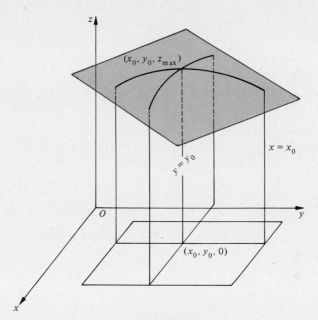

Figure 17.9.1

EXAMPLE 3   Is the origin a relative maximum or a relative minimum for $f(x, y) = x^2 - y^2$?

Solution $f_x = 2x$ and $f_y = -2y$ and the equations $f_x = f_y = 0$ yield $x_0 = y_0 = 0$. Hence $(0, 0)$ is the only critical point. Also, $f(0, 0) = 0$. Now,

$$f(h, k) - f(0, 0) = h^2 - k^2$$

If we let $h = 2k$ then $f(2k, k) = 3k^2$. Also, if $h = \frac{k}{2}$, we have $f\left(\frac{k}{2}, k\right) = -\frac{3}{4}k^2$. What this means is that the function $f(2k, k) > 0$ for $k \neq 0$ and the function $f\left(\frac{k}{2}, k\right) < 0$ for $k \neq 0$ and $(0, 0)$ is neither a maximum nor a minimum of f. Hence, $(0, 0)$ is a saddle point because the surface is saddle shaped near $(0, 0)$ (see Figure 17.9.2).   ●

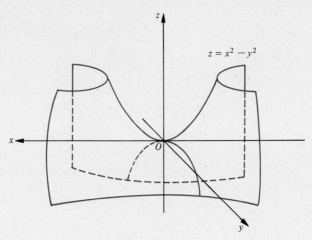

Figure 17.9.2

Next we give a second derivative test for the determination of relative maxima and minima for functions of two variables. After this sufficiency test is stated, it will be illustrated with several examples. We will then close this section by proving this test.

Theorem 3   If f has continuous first and second partial derivatives in a neighborhood of a critical point (x_0, y_0) and if

$$\Delta(x_0, y_0) = f_{xx}(x_0, y_0) \cdot f_{yy}(x_0, y_0) - f_{xy}^2(x_0, y_0) \tag{2}$$

then

(i)   there is a relative minimum at (x_0, y_0) if $\Delta > 0$ and $f_{xx}(x_0, y_0) > 0$
(ii)   there is a relative maximum at (x_0, y_0) if $\Delta > 0$ and $f_{xx}(x_0, y_0) < 0$
(iii)   there is neither a relative minimum nor a relative maximum if $\Delta < 0$
(iv)   the test gives no information if $\Delta = 0$.

EXAMPLE 4   Examine

$$f(x, y) = x^2 - xy + y^2 - 5x + y + 16$$

for relative maxima and minima.

Solution   From the given function,

$$f_x(x, y) = 2x - y - 5$$
$$f_y(x, y) = -x + 2y + 1$$

and if we set $f_x = f_y = 0$, we obtain the critical point $(3, 1)$. Now, $f_{xx}(x, y) = 2$, $f_{xy}(x, y) = -1$, and $f_{yy}(x, y) = 2$, so that

$$\Delta(3, 1) = (2)(2) - (-1)^2 = 3 \quad \text{and} \quad f_{xx}(3, 1) > 0$$

Therefore there is a relative minimum at $(3, 1)$ and $f(3, 1) = 9$.   ●

EXAMPLE 5   Examine

$$f(x, y) = 4y^4 - 5y^2x + x^2$$

for relative maxima and minima.

Solution   From the given function,

$$f_x(x, y) = -5y^2 + 2x$$
$$f_y(x, y) = 16y^3 - 10yx = y(16y^2 - 10x)$$

If $y \neq 0$, the unique solution of

$$-5y^2 + 2x = 0$$
$$16y^2 - 10x = 0$$

is $x = y = 0$ (a contradiction).

Thus $y = 0$ which implies that $x = 0$ and $(0, 0)$ is the only critical point.

$$f_{xx}(x, y) = 2 \qquad f_{xy}(x, y) = -10y \qquad f_{yy}(x, y) = 48y^2 - 10x$$

so that $\Delta(0, 0) = 2(0) - 0^2 = 0$ and the test fails.

Fortunately in this instance, we may write

$$f(x, y) = (y^2 - x)(4y^2 - x)$$

Hence if we let $y = 0$, $f(x, 0) = x^2 \geq 0$, whereas if we set $x = 3y^2$ then

$$f(3y^2, y) = (-2y^2)(y^2) = -2y^4 \leq 0$$

Thus if the point $(0, 0)$ is approached along the x-axis, $f > 0$ whereas if we approach $(0, 0)$ along the parabola $x = 3y^2$, $f < 0$. It follows that $(0, 0)$ is a saddle point of the given function.   ●

EXAMPLE 6   Of all rectangular boxes with volume 27 ft^3, find the one which has least surface area.

Solution   Let x, y and z be the dimensions of the box (in feet) so that

$$xyz = 27 \qquad \text{(fixed volume)} \tag{i}$$

and the surface area S is given by

$$S = 2(xy + yz + zx) \tag{ii}$$

From (i), $z = \dfrac{27}{xy}$ and substitution into (ii) yields

$$S(x, y) = 2\left(xy + \frac{27}{x} + \frac{27}{y}\right) \qquad \text{where } x > 0 \text{ and } y > 0 \tag{iii}$$

$$S_x = 2\left(y - \frac{27}{x^2}\right) \qquad \text{and} \qquad S_y = 2\left(x - \frac{27}{y^2}\right) \tag{iv}$$

Set $S_x = S_y = 0$ to obtain the system of equations

$$y - \frac{27}{x^2} = 0$$
$$x - \frac{27}{y^2} = 0 \tag{v}$$

from which

$$y = \frac{27}{x^2} \quad \text{and} \quad x = \frac{x^4}{27}$$

Hence

$$x^3 = 27 \qquad x = 3 \qquad y = 3$$

Now, let us turn to the second derivative test,

$$S_{xx} = \frac{108}{x^3} \qquad S_{yy} = \frac{108}{y^3} \qquad S_{xy} = 2$$

and at $(3, 3)$,

$$\Delta = (4)(4) - 2^2 = 12 > 0 \qquad S_{xx}(3, 3) = 4$$

which verifies that a relative minimum for S has been obtained. A brief numerical investigation (by choosing values of x less than 3 and greater than 3) yields the fact that the box of absolute minimum surface area must be a cube with edge of length 3 ft. This result is also suggested from geometric considerations. ●

Although thought by many to be peculiar to statistics, curve fitting occurs often in scientific and engineering work. Suppose that we are given data (x_i, y_i), $(i = 1, 2, \ldots, n)$. We seek a function $y = f(x)$ such that (x_i, y_i) lies on the curve, that is, $y_i = f(x_i)$ is satisfied exactly. For example, a polynomial of degree r may be determined that passes through $r + 1$ given points. However, when the number of points is large, the degree of the polynomial is high and the graph contains many wiggles. Any attempt to represent the data exactly is both laborious and foolish since the given data is usually experimental and invariably contains observational errors.

Suppose that we are given a set of experimentally observed points $P_1(x_1, y_1), P_2(x_2, y_2), \ldots, P_n(x_n, y_n)$ and that it is desired to find the straight line of "best fit" to this data (Figure 17.9.3). The method that follows is known as the **method of least squares,** and the line we seek is known as the **line of regression.**

Let $y = mx + b$ be the line we seek to fit to the given data. To each x_i there are two values of y, namely y_i and the value predicted by the straight line, $mx_i + b$. In general, these are different and we introduce the deviation

$$d_i = y_i - (mx_i + b) \qquad (i = 1, 2, \ldots, n) \tag{3}$$

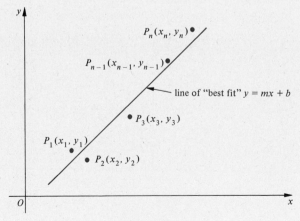

Figure 17.9.3

If the deviation d_i is computed for each point of the given data and the sum of the squares of these quantities is computed, then we have

$$E(m,b) = \sum_{i=1}^{n} d_i^2 = (y_1 - mx_1 - b)^2 + (y_2 - mx_2 - b)^2$$

$$+ \cdots + (y_n - mx_n - b)^2 \qquad (4)$$

Squaring the quantities avoids large positive and negative d_i's from canceling each other and giving an unwarranted impression of accuracy. The quantity E is a measure of how well the line fits the set of points as a whole. We seek m and b such that the sum of the squares of the deviations E is as small as possible, that is, a minimum.

To do this, we must determine the simultaneous solutions of the two equations

$$\frac{\partial E}{\partial b} = 0 \qquad \text{and} \qquad \frac{\partial E}{\partial m} = 0 \qquad (5)$$

From (5) and (4), we obtain

$$\sum_{i=1}^{n} (y_i - b - mx_i) = 0 \qquad \text{and} \qquad \sum_{i=1}^{n} x_i(y_i - b - mx_i) = 0 \qquad (6)$$

The first equation of (6) states that the algebraic sum of the deviations $d_i = 0$ (that is, the positive and negative deviations cancel), whereas the second equation may be interpreted to mean that the algebraic sum of the first moments of the deviations about the y-axis is zero.

If the terms in the unknowns are combined we have the two equations

$$m \sum_{i=1}^{n} x_i + nb = \sum_{i=1}^{n} y_i$$

$$\tag{7}$$

$$m \sum_{i=1}^{n} x_i^2 + \left(\sum_{i=1}^{n} x_i\right)b = \sum_{i=1}^{n} x_i y_i$$

The simultaneous solutions of equations (7) yield

$$m = \frac{N_1}{D} \qquad \text{and} \qquad b = \frac{N_2}{D} \qquad (8)$$

where

$$N_1 = \begin{vmatrix} \sum_{i=1}^{n} y_i & n \\ \sum_{i=1}^{n} x_i y_i & \sum_{i=1}^{n} x_i \end{vmatrix} \qquad N_2 = \begin{vmatrix} \sum_{i=1}^{n} x_i & \sum_{i=1}^{n} y_i \\ \sum_{i=1}^{n} x_i^2 & \sum_{i=1}^{n} x_i y_i \end{vmatrix}$$

and $$\tag{9}$$

$$D = \begin{vmatrix} \sum_{i=1}^{n} x_i & n \\ \sum_{i=1}^{n} x_i^2 & \sum_{i=1}^{n} x_i \end{vmatrix}$$

EXAMPLE 7 If y is the pull (lbs) required to lift a weight x (lbs) by means of a pulley block. We are given the following empirical data

x (lb)	50	70	100	120
y (lb)	12	15	21	25

(a) Find the straight line of best fit in accordance with the least square principle.
(b) Predict y when $x = 150$ lb on the basis of your result in (a).

Solution (a) The given data must be substituted into equations (7) through (9), and we have, with $n = 4$,

$$\sum_{i=1}^{4} x_i = 50 + 70 + 100 + 120 = 340$$

$$\sum_{i=1}^{4} x_i^2 = 50^2 + 70^2 + 100^2 + 120^2 = 31{,}800$$

$$\sum_{i=1}^{4} x_i y_i = (50)(12) + (70)(15) + (100)(21) + (120)(25) = 6{,}750$$

$$\sum_{i=1}^{4} y_i = 12 + 15 + 21 + 25 = 73$$

Hence from (9)

$$D = -11{,}600 \qquad N_1 = -2{,}180 \qquad N_2 = -26{,}400$$

and (8) yields $m \doteq 0.188$ and $b \doteq 2.28$. Thus $y = 0.188x + 2.28$ is the line of best fit.

(b) When $x = 150$, $y \doteq 30.5$ lb. ●

We now complete this section by proving Theorem 3(i).

Proof Let (x_0, y_0) be a critical point of a function f. With $\Delta(x, y) = f_{xx}(x, y)f_{yy}(x, y) - f_{xy}^2(x, y)$ we shall establish that if $\Delta(x_0, y_0) > 0$, if $f_{xx}(x_0, y_0) > 0$, and if f and its first and second partial derivatives are continuous in a neighborhood of (x_0, y_0), then f possesses a relative minimum at (x_0, y_0).

Since f_{xx}, f_{xy}, and f_{yy} are continuous in a neighborhood of (x_0, y_0), then Δ is also continuous in that neighborhood. From $\Delta(x_0, y_0) > 0$, it follows that there exists a neighborhood about (x_0, y_0), perhaps different from the first, such that $\Delta > 0$ and $f_{xx} > 0$ for all points on it. The rest of the discussion is confined to the latter neighborhood, which is denoted by the letter D. Let h and k be constants that are not both zero and whose magnitudes are so small that the point $(x_0 + h, y_0 + k)$ is in D. Then the two equations

$$x = x_0 + ht \qquad y = y_0 + kt \qquad 0 \leq t \leq 1 \tag{i}$$

define a segment of a straight line such that all (x, y) are in D. Let $g(t)$ be a function of t defined by

$$g(t) = f(x_0 + ht, y_0 + kt) \qquad 0 \leq t \leq 1 \tag{ii}$$

Application of Taylor's formula yields

$$g(t) = g(0) + g'(0)t + \frac{g''(c)}{2}t^2 \tag{iii}$$

where c is some number in $[0, t]$. In particular, for $t = 1$, (iii) reduces to

$$g(1) = g(0) + g'(0) + \frac{g''(c)}{2} \qquad \text{(iv)}$$

where $0 < c < 1$. Now, by application of the chain rule of differentiation and (i), we have

$$g'(t) = hf_x(x, y) + kf_y(x, y) \qquad \text{(v)}$$

Also, by application of the chain rule once again, (and then simplifying),

$$g''(t) = h^2 f_{xx}(x, y) + 2hk f_{xy}(x, y) + k^2 f_{yy}(x, y) \qquad \text{(vi)}$$

where $f_{xy} = f_{yx}$ has been utilized. In particular, for $t = 0$

$$g'(0) = hf_x(x_0, y_0) + kf_y(x_0, y_0) \qquad \text{(vii)}$$

and since (x_0, y_0) is a critical point of f,

$$g'(0) = 0 \qquad \text{(viii)}$$

and

$$g''(c) = h^2 f_{xx}(x_0 + hc, y_0 + kc) + 2hk f_{xy}(x_0 + hc, y_0 + kc)$$
$$+ k^2 f_{yy}(x_0 + hc, y_0 + kc) \qquad \text{(ix)}$$

Substitution of (viii) and (ix) into (iv) yields

$$f(x_0 + h, y_0 + k) - f(x_0, y_0) = \tfrac{1}{2}[h^2 f_{xx} + 2hk f_{xy} + k^2 f_{yy}] \qquad \text{(x)}$$

where each second derivative is evaluated at $(x_0 + hc, y_0 + kc)$. Now, the quadratic expression on the right side of (x) may be written (by completion of the squares)

$$h^2 f_{xx} + 2hk f_{xy} + k^2 f_{yy} = f_{xx}\left(h^2 + 2hk\frac{f_{xy}}{f_{xx}} + k^2\frac{f_{yy}}{f_{xx}}\right)$$
$$= f_{xx}\left[\left(h + k\frac{f_{xy}}{f_{xx}}\right)^2 + k^2\left(\frac{f_{yy}}{f_{xx}} - \frac{f_{xy}^2}{f_{xx}^2}\right)\right]$$
$$= f_{xx}\left[\left(h + k\frac{f_{xy}}{f_{xx}}\right)^2 + k^2\left(\frac{f_{xx}f_{yy} - f_{xy}^2}{f_{xx}^2}\right)\right] \qquad \text{(xi)}$$

But in D, $f_{xx} > 0$ and $\Delta = f_{xx}f_{yy} - f_{xy}^2 > 0$, which means that the right side of (xi) and, therefore, (x) is positive. Hence,

$$f(x_0 + h, y_0 + k) - f(x_0, y_0) > 0$$

for all points in D such that $h^2 + k^2 > 0$. This implies that $f(x_0, y_0)$ is a relative minimum.

The proof of the sufficient conditions for a relative maximum and for no relative extrema is left for the exercises (see Exercises 27, 28, and 29). ∎

Exercises 17.9

In Exercises 1 through 10, find the relative maxima and minima, if any, for the given function.

1. $f(x, y) = 2x^2 + 3y^2 - 4x + 12y + 5$
2. $g(x, y) = 3x^2 + 2xy + y^2$
3. $F(x, y) = 3xy + 10$
4. $G(x, y) = 5 + 4x - 12y - 2x^2 - 3y^2$

5. $h(x, y) = e^{-(x^2 + 2y^2)}$

6. $H(x, y) = x^2 + xy - 2y^2 - 3x + 3y - 4$

7. $f(x, y) = x^4 + y^4 - 6x^2 + y^2 + 10$

8. $F(x, y) = x^3 + y^3 - 15xy$

9. $g(x, y) = xy(a - x - y)$, a is a positive constant

10. $G(x, y) = \sin x + \sin y + \sin (x + y)$,   where $0 < x < \frac{\pi}{2}$ and $0 < y < \frac{\pi}{2}$

11. By application of calculus, find the minimum distance of the plane $2x + y - z = 3$ from the origin of coordinates. (*Hint:* Minimize the square of the distance.)

12. Find the minimum distance between the point $(2, -1, 3)$ and the plane $3x + 2y + z = 14$.

13. If x, y, and z are positive numbers, find the maximum value of xyz subject to $2x + y + z = 60$.

14. Find the volume V of the largest box, with sides parallel to the coordinate planes, that can be inscribed in a sphere of radius a. In your solution consider only necessary conditions for an extremum. (*Hint:* Maximize V^2.)

15. Find the volume V of the largest box with sides parallel to the coordinate planes that can be inscribed in an ellipsoid $\dfrac{x^2}{a^2} + \dfrac{y^2}{b^2} + \dfrac{z^2}{c^2} = 1$. Consider only necessary conditions for an extremum. (*Hint:* Maximize V^2—this exercise is a generalization of Exercise 14.)

16. A barometer at various heights (h, 1000 ft) registered pressure (p, in.) as follows

h, 1000 ft	0	2	4	6	8
p, in.	30.00	28.70	27.38	26.10	24.82

Find the straight line of best fit by the least square method. Predict p when $h = 10$.

17. Use the method of least squares to find the straight line which "best" fits the six points $(2, 10.8)$, $(3, 14.1)$, $(4, 16.9)$, $(5, 19.8)$, $(6, 23.2)$, $(7, 26.1)$. Predict y when $x = 10$.

18. A closed rectangular box is to be constructed with a specified surface area S. Find the dimensions of the box of greatest volume and the maximum volume V.

19. Three points P_1, P_2, and P_3 with coordinates (x_1, y_1), (x_2, y_2) and (x_3, y_3) are the vertices of a triangle. Find a fourth point $P(x, y)$ such that the sum of the squares of its distances from the three points is least. Identify your solution.

20. Determine the shortest distance from the origin to the plane $Ax + By + Cz = D$, where $D \neq 0$ and $A^2 + B^2 + C^2 > 0$.

21. Given n fixed points P_i whose coordinates are (x_i, y_i) $(i = 1, 2, \ldots, n)$. Show that the coordinates of the point $P(x, y)$ such that the sum of the squares of the distances from P to the point P_i is a minimum, are given by $\bar{x} = \dfrac{1}{n} \sum_{i=1}^{n} x_i$ and $\bar{y} = \dfrac{1}{n} \sum_{i=1}^{n} y_i$; that is, $(\bar{x}, \bar{y})$ is the centroid of the n points.

22. A one-product company finds that its profit in millions of dollars depends upon its amount x spent on advertising in millions of dollars and the price y charged per item in dollars. The profit function is given by

$$F(x, y) = \left(10 + 2x - \frac{x^2}{10}\right)y - 5y^2 - 5$$

Find the x and y that maximizes F and determine the maximum value of F.

23. A company produces two kinds of radios, which compete with one another. The demand functions for them are as follows

$$q_1 = 80 - 5p_1 - 2p_2$$
$$q_2 = 60 - 3p_1 - 4p_2$$

where p_1, p_2 are the prices of each radio in dollars and q_1, q_2 are the quantities of each radio demanded in hundreds of units.

(a) Find the revenue R as a function of p_1 and p_2.

(b) Determine the price which should be charged to the retailers in order to maximize R.

(c) Find the number of units demanded to the nearest 100.

(d) What is the maximum value of R?

24. Find the triangle of largest area with given perimeter. (*Hint:* If x, y, and z are the sides of the triangle and A is the area, then $A^2 = s(s - x)(s - y)(s - z)$, where s is the fixed semiperimeter $\dfrac{x + y + z}{2}$.)

25. Prove that if $f(x, y)$ exists at all points in a neighborhood of a point (x_0, y_0) and if f has a relative maximum at (x_0, y_0), then, if $f_y(x_0, y_0)$ exists, we must have $f_y(x_0, y_0) = 0$.

26. Prove that if $f(x, y)$ exists at all points in a neighborhood of a point (x_0, y_0) and if f has a relative minimum at (x_0, y_0), then, if $f_x(x_0, y_0)$ and $f_y(x_0, y_0)$ exist, we must have $f_x(x_0, y_0) = f_y(x_0, y_0) = 0$.

27. Prove (ii) of Theorem 3.

28. Prove (iii) of Theorem 3.

29. Give examples of functions f that have a maximum and a minimum at (x_0, y_0) (one of each) and for which $\Delta(x_0, y_0) = 0$, where Δ is defined by (2).

17.10   ABSOLUTE EXTREMA AND LAGRANGE MULTIPLIER METHOD

In our previous section, the emphasis was on the determination of relative extrema, which often yielded absolute extrema as well, although the analytic justification was lacking.

Suppose now that we have a continuous function f of two variables on a closed region such as a triangle, a rectangle, a circle, and so on (interior and boundary points). We seek the largest (or smallest) value of f, on the given closed region R, which we know must exist. If f has continuous partial derivatives, the following procedure may be followed.

1. List all the critical points for f interior to the boundary of R.
2. Eliminate any point that cannot be a relative extremum of f.
3. Find the values of f at these relative extrema.
4. Determine the values of f on the boundary of R and find the largest and smallest values.
5. The largest of all of the values found in 3 and 4 is the absolute maximum and the smallest is the absolute minimum.

EXAMPLE 1   Find the absolute maximum of the function $f(x, y) = 3x + y^2$ on the closed triangular region whose vertices are $(0, 0)$, $(2, 0)$ and $(0, 2)$. (See Figure 17.10.1.)

Solution   The function $f(x, y)$ is continuous over the closed triangular region and therefore the maximum value must be attained at some point in the interior or on the boundary.

For interior extrema,

$$f_x = f_y = 0$$

must hold at these points since $f(x, y)$ is differentiable. But $f_x = 3$, and thus the maximum of f will occur on the *boundary* consisting of (a) the line segment $y = 0$, $0 \le x \le 2$, (b) the line segment $x = 0$, $0 \le y \le 2$, and (c) the line segment $x + y = 2$, $0 \le y \le 2$.

On $y = 0$, $0 \le x \le 2$, $f = 3x$, and $f_{\max} = 6$.

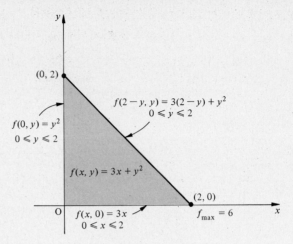

Figure 17.10.1

On $x = 0, 0 \le y \le 2, f(0, y) = y^2, f(0, 2) = 4$ is the largest value of f on this segment.

On $x + y = 2, 0 \le y \le 2, f = 3(2 - y) + y^2 = 6 - 3y + y^2$. For extrema on this line, $D_y f = 0$ must hold. Now, $D_y f = -3 + 2y$ and $D_y f = 0$ yields $y = 3/2$. Also, $D_y^2 f = 2$ and f has a minimum on this boundary.

Hence $f_{max} = 6$ and occurs at $(2, 0)$ and this is the absolute or global maximum of $f(x, y)$ in the given region. ●

EXAMPLE 2   Find the absolute maximum and the absolute minimum of

$$g(x, y) = x^2 - 3xy + y^2 + 4x - 2y$$

on the square with vertices at the points $(0, 0), (2, 0), (2, 2)$ and $(0, 2)$. (See Figure 17.10.2.)

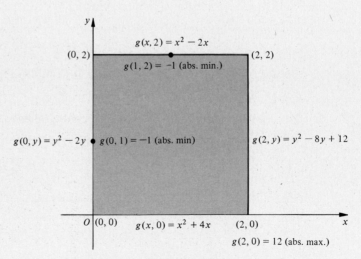

Figure 17.10.2

Solution   The critical points are obtained by setting

$$g_x(x, y) = g_y(x, y) = 0$$

$$g_x(x, y) = 2x - 3y + 4 \qquad \text{and} \qquad g_y(x, y) = -3x + 2y - 2$$

from which the critical point is $(\frac{2}{5}, \frac{8}{5})$. This is inside the square

$$g_{xx}(x, y) = 2 \qquad g_{yy}(x, y) = 2 \qquad g_{xy}(x, y) = -3$$

and $\Delta = (2)(2) - (-3)^2 = -5$. Thus the point $(\frac{2}{5}, \frac{8}{5})$ is a saddle point for g, and this means that both the maximum and minimum of g must occur on the boundary.

The values of g on the boundary of the square are shown in Figure 17.10.2. The absolute maximum is $g(2, 0) = 12$ and the absolute minimum is $g(0, 1) = g(1, 2) = -1$. ●

The rest of this section will be devoted to the problem of determining extrema for functions of two or more variables. Furthermore, only necessary conditions will be treated here. The problem of sufficiency conditions for functions of three or more variables is complicated and is best treated in an advanced calculus text.

The extreme value theorem again applies, namely that every function that is continuous in a closed region R possesses a largest and a smallest value in the interior or on the boundary of the region. Suppose $u = f(x, y, z, \ldots)$ is a differentiable function of n independent variables (that is, the first partial derivatives exist at all interior points of R) then the equations

$$\begin{aligned} f_x(x_0, y_0, z_0, \ldots) &= 0 \\ f_y(x_0, y_0, z_0, \ldots) &= 0 \\ f_z(x_0, y_0, z_0, \ldots) &= 0 \\ \cdots \end{aligned} \qquad (1)$$

are necessary conditions for the occurrence of a relative maximum or a relative minimum of f at an interior point $P_0(x_0, y_0, z_0, \ldots)$.

Equations (1) are easily established by varying the independent variables $x, y, z, \ldots$, one at a time while holding the other variables constant. Thus, for example, if we held $(x, z, \ldots)$ constant and varied only y, we would obtain $f_y(x_0, y_0, z_0, \ldots) = 0$ and the other equations of the system in (1) are determined analogously.

In the system (1) we have n equations in as many unknowns and, as a rule, we can solve the system for the unknowns $(x_0, y_0, \ldots)$. However, the point P_0 obtained in this manner may not be a maximum or a minimum.

It is useful to look at the system (1) from a different viewpoint. If the equations of (1) are multiplied, respectively, by $dx, dy, dz, \ldots$, and added, we obtain

$$df(x_0, y_0, z_0, \ldots) = f_x(x_0, y_0, z_0, \ldots) \, dx + f_y(x_0, y_0, z_0, \ldots) \, dy$$
$$+ f_z(x_0, y_0, z_0, \ldots) \, dz + \cdots = 0 \qquad (2)$$

Therefore at an extreme point the differential, that is, the linear part of the increment f, must be zero for arbitrary values of the differentials $dx, dy, dz, \ldots$, of the independent variables $x, y, z, \ldots$. Conversely, if (2) is satisfied for arbitrary values of $dx, dy, dz, \ldots$, then it is easy to show that system (1) holds. For example, if we take $dx \neq 0$, $dy = dz = \cdots = 0$ then $f_x(x_0, y_0, z_0, \ldots) = 0$ and the other equations of (1) follow in a similar manner. Hence (1) and (2) are equivalent to each other.

Many maximum and minimum problems arise where it is required to find the extremum of a function subject to one or more side or constraint conditions. For example, suppose that we seek to find the shortest distance from a point $(2, 3, 5)$ to the sphere $x^2 + y^2 + z^2 = 4$. Therefore if the square of the distance is used, we seek to minimize

$$f(x, y, z) = (x - 2)^2 + (y - 3)^2 + (z - 5)^2$$

subject to the constraint

$$g(x, y, z) = x^2 + y^2 + z^2 - 4 = 0$$

which means that the required point must be on the given sphere.

To simplify our considerations, we shall first treat the problem of determining the extreme values of a function $f(x, y)$ when the two variables x and y are connected by a subsidiary condition $g(x, y) = 0$. Let us examine the question in a geometrically plausible way. The equation $g(x, y) = 0$ is represented by a curve in the xy-plane (Figure 17.10.3). Also, the equation $f(x, y) = c$ is a family of curves that covers a portion of the plane. Among all the curves $f(x, y) = c$ that intersect $g(x, y) = 0$, we seek the one for which c is the largest or the smallest. In general, in the neighborhood of the point of tangency, as c changes monotonically, the curves $f(x, y) = c$ will intersect $g(x, y) = 0$ in two points, one point, or no points, respectively. Thus it is suspected that the required curve for which f is an extremum is tangent to $g(x, y) = 0$. If $x = x_0$, $y = y_0$ denotes the point of tangency, the condition that the slopes of the curves be the same at this point yields

$$\frac{f_x}{g_x} = \frac{f_y}{g_y} \qquad \text{at } (x_0, y_0) \qquad \text{where } g_x g_y |_{(x_0, y_0)} \neq 0 \qquad (3)$$

If both sides of (3) are equated to a parameter $-\lambda$, the following two equations hold:

$$\begin{aligned} f_x + \lambda g_x &= 0 \\ f_y + \lambda g_y &= 0 \end{aligned} \qquad (4)$$

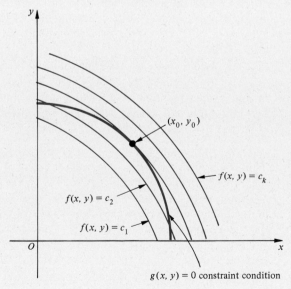

Figure 17.10.3

Equations (4) together with $g(x, y) = 0$ are sufficient, in general, to determine x_0, y_0, and λ. Before proceeding further into this subject, consider the following example.

EXAMPLE 3   Find the rectangle of given perimeter L for which the area is a maximum.

Solution   Let x and y denote the sides of the rectangle $(x > 0, y > 0)$. The area of the rectangle is $f(x, y) = xy$ and f is to be maximized subject to the condition that the perimeter is L. Thus the constraint is

$$g(x, y) = 2x + 2y - L = 0$$

Substitution of the expressions for $f(x, y)$ and $g(x, y)$ into (4) yields

$$y + 2\lambda = 0 \qquad \text{(i)}$$
$$x + 2\lambda = 0 \qquad \text{(ii)}$$

and also,

$$2x + 2y - L = 0 \qquad \text{(iii)}$$

The simultaneous solution of (i) to (iii) results in

$$x = x_0 = \frac{L}{4} \qquad y = y_0 = \frac{L}{4} \qquad \lambda = -\frac{L}{8}$$

Therefore the square $x_0 = y_0$ is a necessary condition that $f(x, y)$ is a maximum. Furthermore, we know the square actually maximizes the area.   ●

Our intuitive investigation suggests the following rule.

Rule: In order that a function $f(x, y)$ attain an extremum subject to the constraint $g(x, y) = 0$, form

$$F(x, y, \lambda) = f(x, y) + \lambda g(x, y)$$

and treat x, y and λ as if they are independent of each other. Then the necessary conditions for making F an extremum are obtained by setting the partial derivatives with respect to each of the variables equal to zero, that is,

$$\begin{aligned} F_x &= 0 = f_x + \lambda g_x \\ F_y &= 0 = f_y + \lambda g_y \\ F_\lambda &= 0 = g \end{aligned} \qquad (5)$$

This is identical with (4) and the constraint condition. This rule is known as *Lagrange's method of undetermined multipliers* and the factor λ is known as *Lagrange's multiplier.*

Suppose next that we seek to extremize $f(x, y, z)$ subject to a constraint condition $g(x, y, z) = 0$. Then *Lagrange's rule* is as follows: Form

$$F(x, y, z, \lambda) = f(x, y, z) + \lambda g(x, y, z)$$

and treat x, y, z, and λ as if they are independent of each other. Then set the partial derivatives of F with respect to each of these variables equal to zero.

Hence

$$F_x = f_x + \lambda g_x = 0$$
$$F_y = f_y + \lambda g_y = 0$$
$$F_z = f_z + \lambda g_z = 0 \tag{6}$$
$$F_\lambda = 0 = g$$

A suggestive geometric argument to obtain the first three equations of (6) is that $f(x, y, z) = c$ is a family of surfaces in xyz-space and $g(x, y, z) = 0$ is a particular surface in space. Among all the surfaces in space, we seek the one for which c is the largest or smallest. Then arguing as in the two-dimensional case it is suggested that tangency of the two surfaces $f(x, y, z) = c$ and $g(x, y, z) = 0$ is required at an extremum $P_0(x_0, y_0, z_0)$. Hence the direction numbers of the normals to f and g must be proportional at P_0; that is, at P_0

$$\frac{f_x}{g_x} = \frac{f_y}{g_y} = \frac{f_z}{g_z} = -\lambda \qquad (g_x g_y g_z \neq 0)$$

from which the first three equations of (6) are obtained.

EXAMPLE 4   Find the triangle with given perimeter $2s$ and largest area A.

Solution   By Heron's formula, the square of the area A^2 is given by

$$A^2 = f(x, y, z) = s(s - x)(s - y)(s - z) \tag{i}$$

where x, y, and z are the sides of the triangle. We seek to maximize f given by (i) subject to the constraint condition

$$g(x, y, z) = x + y + z - 2s = 0 \tag{ii}$$

The variables x, y, and z are restricted by the inequalities

$$x > 0 \qquad y > 0 \qquad z > 0 \qquad x + y > z \qquad y + z > x \qquad z + x > y$$

and the equalities $x = 0$, $y = 0$, ..., $z + x = y$ defines the boundary of the region in space. Furthermore, $f = 0$ at all points on the boundary and, since $f > 0$, a maximum must occur in the interior of the region. By application of Lagrange's method, we form

$$F(x, y, z, \lambda) = s(s - x)(s - y)(s - z) + \lambda(x + y + z - 2s)$$

and substitution into (6) yields

$$-s(s - y)(s - z) + \lambda = 0$$
$$-s(s - x)(s - z) + \lambda = 0 \tag{iii}$$
$$-s(s - x)(s - y) + \lambda = 0$$

$$x + y + z - 2s = 0$$

By subtracting the second equation of (iii) from the first of (iii) λ is eliminated and we obtain $x = y$ (since $s - z \neq 0$). Similarly, from the second and third equations, $y = z$ follows. Hence

$$x = y = z = \frac{2s}{3} \tag{iv}$$

that is, the solution is an equilateral triangle.   ●

More generally, consider the problem of extremizing a function of n variables

$$f(x_1, x_2, \ldots, x_n) \tag{7}$$

subject to constraints

$$g_1(x_1, x_2, \ldots, x_n) = 0$$
$$g_2(x_1, x_2, \ldots, x_n) = 0 \tag{8}$$
$$\cdots$$
$$g_k(x_1, x_2, \ldots, x_n) = 0$$

where $k < n$ must hold. Lagrange's rule is as follows: form

$$F(x_1, x_2, \ldots, x_n, \lambda_1, \lambda_2, \ldots, \lambda_k) = f(x_1, x_2, \ldots, x_n)$$

$$+ \sum_{j=1}^{k} \lambda_j g_j(x_1, x_2, \ldots, x_n) \tag{9}$$

thereby introducing k multipliers $\lambda_1, \lambda_2, \ldots, \lambda_k$ one for each constraint equation. Then set the partial derivatives with respect to the $n + k$ variables $x_1, x_2, \ldots, x_n, \lambda_1, \lambda_2, \ldots, \lambda_k$ equal to zero:

$$\frac{\partial F}{\partial x_i} = 0 \qquad (i = 1, 2, \ldots, n)$$
$$\frac{\partial F}{\partial \lambda_j} = 0 \qquad (j = 1, 2, \ldots, k) \tag{10}$$

and the system (10) is to be solved for the unknowns $x_1, x_2, \ldots, x_n, \lambda_1, \lambda_2, \ldots, \lambda_k$.

EXAMPLE 5   Find the greatest and least values of z on the ellipse formed by the intersection of the plane $x + y + z = 1$ and the ellipsoid $16x^2 + 4y^2 + z^2 = 16$.

Solution   In this instance, $f(x, y, z) = z$ is to be extremized subject to the two constraints

$$g_1(x, y, z) = x + y + z - 1 = 0$$
$$\text{and} \qquad g_2(x, y, z) = 16x^2 + 4y^2 + z^2 - 16 = 0$$

Thus from (9), we have

$$F(x, y, z, \lambda_1, \lambda_2) = z + \lambda_1(x + y + z - 1) + \lambda_2(16x^2 + 4y^2 + z^2 - 16)$$

set the partial derivatives equal to zero:

$$F_z = 1 + \lambda_1 + 2\lambda_2 z = 0 \tag{i}$$
$$F_y = \lambda_1 + 8\lambda_2 y = 0 \tag{ii}$$
$$F_x = \lambda_1 + 32\lambda_2 x = 0 \tag{iii}$$

and

$$x + y + z - 1 = 0 \tag{iv}$$
$$16x^2 + 4y^2 + z^2 - 16 = 0 \tag{v}$$

Instead of solving for the five unknowns, we note that (ii) and (iii) are consistent if $y = 4x$ [$\lambda_1 = \lambda_2 = 0$ is excluded since (i) is violated]. Substitution of $y = 4x$ into

(iv) yields $x = \frac{1}{5}(1 - z)$. Then (v) becomes (in terms of z)

$$\frac{16(1 - z)^2}{25} + \frac{4(16)(1 - z)^2}{25} + z^2 - 16 = 0$$

or, simplifying, this reduces to

$$21z^2 - 32z - 64 = 0$$

Solution of the quadratic yields $z = \frac{8}{3}$ (maximum) and $z = -\frac{8}{7}$ (minimum). ●

EXAMPLE 6   Minimize $x_1^2 + x_2^2 + \cdots + x_n^2$ subject   to   the   condition $a_1 x_1 + a_2 x_2 + \cdots a_n x_n = 1$ where $a_1^2 + a_2^2 + \cdots + a_n^2$ is positive (otherwise the constraint equation is violated).

Solution   Before we solve this problem, we note that if $n = 3$ this has the simple geometric interpretation of determining the point on a given plane (not passing through the origin) which is closest to the origin.

Form

$$F(x_1, x_2, \ldots, x_n) = x_1^2 + x_2^2 + \cdots + x_n^2 + \lambda(a_1 x_1 + a_2 x_2 + \cdots + a_n x_n - 1)$$

$$\frac{\partial F}{\partial x_1} = 2x_1 + \lambda a_1 = 0$$

$$\frac{\partial F}{\partial x_2} = 2x_2 + \lambda a_2 = 0$$

$$\cdots$$

$$\frac{\partial F}{\partial x_n} = 2x_n + \lambda a_n = 0$$

and

$$a_1 x_1 + a_2 x_2 + \cdots + a_n x_n = 1$$

Therefore

$$x_j = -\tfrac{1}{2}(\lambda a_j) \qquad (j = 1, 2, \ldots, n)$$

Thus

$$\sum_{j=1}^{n} a_j x_j = -\frac{\lambda}{2} \sum_{j=1}^{n} a_j^2 = 1$$

or the parameter λ is given by

$$\lambda = -\frac{2}{\displaystyle\sum_{j=1}^{n} a_j^2} \qquad \text{and} \qquad x_j = \frac{a_j}{\displaystyle\sum_{j=1}^{n} a_j^2}$$

The corresponding value of $x_1^2 + x_2^2 + \cdots + x_n^2$ is

$$\sum_{j=1}^{n} x_j^2 = \left(\sum_{j=1}^{n} a_j^2 \right) \left(\sum_{j=1}^{n} a_j^2 \right)^{-2} = \left(\sum_{j=1}^{n} a_j^2 \right)^{-1} \qquad ●$$

Exercises 17.10

In Exercises 1 through 6, locate and find the absolute maximum and the absolute minimum of the given function defined on the square S with opposite vertices at (0, 0) and (1, 1).

1. $f(x, y) = 3x - y$

2. $g(x, y) = x + 2y$

3. $h(x, y) = x^2 + 2y$

4. $F(x, y) = x - 4y^2$

5. $G(x, y) = x^2 + 2y^2 - x - y + 3$

6. $H(x, y) = x^2 + 3y^2 - 3xy - x + 1$

7. Locate and find the absolute maximum and the absolute minimum of $f(x, y) = xy(3 - x - y)$ defined on the closed triangular region with vertices $(0, 0)$, $(3, 0)$, and $(0, 3)$.

8. Locate and find the absolute maximum and the absolute minimum of

$$g(x, y) = x^2 - 2xy + 3y^2 - x - 5y + 6$$

defined on the closed triangular region with vertices $(0, 0)$, $(4, 0)$, and $(0, 2)$.

In Exercises 9 through 22, use the method of Lagrange multipliers involving only necessary conditions for an extremum.

9. Minimize $x^2 + y^2$ subject to the constraint $xy = 4$.

10. Maximize xy subject to $x^2 + y^2 = 25$.

11. Maximize x^2y, where $x^2 + y^2 = 4$.

12. Maximize x^2y such that $\dfrac{x^2}{a^2} + \dfrac{y^2}{b^2} = 1$, where $a > 0$ and $b > 0$.

13. Maximize xyz subject to $x^2 + y^2 + z^2 = 27$.

14. Minimize $(x^2 + y^2 + z^2)^{1/2}$ subject to $4x - y - z = 100$.

15. Find the largest and smallest values of $f(x, y, z) = x^4 + y^4 + z^4$ for points on the surface of the sphere $x^2 + y^2 + z^2 = 25$.

16. Show that of all the triangles inscribed in a circle of radius a, the equilateral triangle has the largest perimeter. (*Hint:* Introduce three central angles θ_1, θ_2, and θ_3 and express the perimeter in terms of θ_1, θ_2, and θ_3. What is the constraint equation?)

17. Find the largest and smallest values of $f(x, y, z) = 2x + y + 3z - 4$ on the unit sphere $x^2 + y^2 + z^2 = 1$.

18. Find the maximum of $(a_1x + a_2y + a_3z)^2$ subject to the constraint that $x^2 + y^2 + z^2 = 1$, where $a_1^2 + a_2^2 + a_3^2 > 0$.

19. Find the maximum of $\left(\displaystyle\sum_{i=1}^{n} a_i x_i\right)^2$ subject to the constraint that $\displaystyle\sum_{i=1}^{n} x_i^2 = 1$, where $\displaystyle\sum_{i=1}^{n} a_i^2 > 0$. This generalizes Exercise 18.

20. Find the maximum of xyz of three positive numbers subject to the constraints that $3x + 6y + 2z = 78$ and $x + 4y + 2z = 58$.

21. Of all tin cans that enclose a given volume V_0, find, by the Lagrange multiplier method, the relative dimensions that require the least metal (that is, minimize the surface area).

22. Find an equation of the plane that contains the point (a, b, c) in the first octant and cuts off the least volume from the first octant. What is the minimum volume?

Review and Miscellaneous Exercises

In each of Exercises 1 through 6, determine the domain of the given function. Interpret your result geometrically.

1. $f(x, y) = \dfrac{x^2 - 2xy - y^2}{16x^2 - y^2}$

2. $g(x, y) = (4 - x^2 - 2y^2)^{-1}$

3. $h(x, y) = e^{\sqrt{x+y-3}} \sin \dfrac{y}{x^2 + \pi}$

4. $F(x, y) = \ln(x + y - 2) + \sqrt{x} + \sqrt{y - 1}$

5. $f(x, y, z) = \sqrt{a^2 - x^2 - y^2 - z^2}$

6. $G(x, y, z) = \dfrac{x + y}{x^2 - y^2 - z^2} + \dfrac{z^2}{\sqrt{y - 3x - z}}$

In Exercises 7 through 12, find the indicated partial derivatives.

7. $u(x, y) = \dfrac{x^4 + 3y^3}{x}$; u_x, u_y, u_{xy}, u_{yx}

8. $v(x, y) = \cos(xy^2) + ye^{xy}$; v_x, v_y, v_{xy}, v_{yx}

9. $F(x, y) = y^2 e^{x+y}$; $F_x, F_y, F_{xx}, F_{xxy}$

10. $G(x, y) = \dfrac{y}{3x + 2}$; $G_x, G_y, G_{xx}, G_{xy}, G_{xxy}$

11. $u(x, y, z) = \cosh xyz$; $u_{xx}, u_{yy}, u_{zz}, u_{xy}, u_{yz}, u_{zx}$

12. $u(x, y, z) = z \tan^{-1}\dfrac{y}{x}$; $u_{xx}, u_{yy}, u_{zz}, u_{xy}, u_{yz}, u_{zx}$

13. Show that $u = e^{kx}(A \cos ky + B \sin ky)$, where k, A, and B are arbitrary constants, satisfies Laplace's equation $u_{xx} + u_{yy} = 0$; that is, u is harmonic.

14. If u and v are functions of x and y that satisfy the Cauchy–Riemann equations $u_x = v_y$, $u_y = -v_x$, and if u and v possess continuous derivatives of the second order in x and y, show that both u and v satisfy Laplace's equation in two dimensions.

In Exercises 15 through 18, find the indicated derivative(s) by (a) the chain rule, and (b) substitution and direct differentiation.

15. $u = x \ln(x + y)$, $x = e^t - e^{-t}$, $y = e^t + e^{-t}$; $D_t u$

16. $u = x^2 + 3xy + y^2$, $x = r + s$, $y = r - 2s$; u_r and u_s

17. $u = (x + 2y - z)^8$, $x = r + s + t$, $y = r + s - t$, $z = 2r - s + t$; $u_r, u_s,$ and u_t

18. $u = \sin(x + y)\cos y \cosh z$, $x = 2r + s - t$, $y = r + 2s + t$, $z = rs$; $u_r, u_s,$ and u_t

In Exercises 19 through 22, find the domain D and the range R of each of the functions.

19. $u = \sqrt{49 - |xy|^2}$

20. $u = \dfrac{\sin(x + y)}{e^{x-y}}$

21. $u = \ln(x^2 + 3y^2 + 5)$

22. $u = \dfrac{x^2 + y^2}{\sqrt{x + y}}$

In Exercises 23 through 26, determine the values of (x, y) at which the given function is continuous.

23. $f(x, y) = \dfrac{xe^{x+y}}{x^2 + y^2 - 4y + 5}$

24. $f(x, y) = \begin{cases} \dfrac{x^2 - y^2}{x - y} & x \neq y \\ 2x & x = y \end{cases}$

25. $h(x, y) = \begin{cases} 2\sqrt{xy} & x \geq 0 \text{ and } y \geq 0 \\ xye^{-(x^2+y^2)} & \text{all other } (x, y) \text{ in } R_2 \end{cases}$

26. $F(x, y) = \begin{cases} \sqrt{x^2 + y^2} & x^2 + y^2 \leq 4 \\ 7 - x^2 - y^2 & x^2 + y^2 > 4 \end{cases}$

27. Given the function

$$u(x, y) = \begin{cases} xy\dfrac{x^2 - y^2}{x^2 + y^2} & (x, y) \neq (0, 0) \\ 0 & (x, y) = (0, 0) \end{cases}$$

show that $u_x(0, y) = -y$ and that $u_y(x, 0) = x$. Therefore we may conclude that $u_{xy}(0, 0) \neq u_{yx}(0, 0)$. (*Hint:* Utilize the definition of the partial derivative.)

28. Given $u = x^y$, where $x > 0$, find u_x, u_y, u_{xy}, and u_{yx}. Does $u_{xy} = u_{yx}$?

29. Prove that if f and g are any differentiable functions, then $u = f(y + 3x) + xg(y + 3x)$ is a solution of

$$u_{xx} - 6u_{xy} + 9u_{yy} = 0.$$

30. If $u = y^2 F\left(\dfrac{y}{x}, \dfrac{z}{x}\right)$, show that

$$xu_x + yu_y + zu_z = 2u.$$

31. If $w = f(x, y)$ and $x = u \cos \lambda - v \sin \lambda$, $y = u \sin \lambda + v \cos \lambda$, where λ is an arbitrary constant, show that

$$w_u^2 + w_v^2 = w_x^2 + w_y^2$$

32. If u and v are functions of x and y satisfying the Cauchy–Riemann equations $u_x = v_y$, $u_y = -v_x$ in a neighborhood about a point (x, y). Show that these equations in polar coordinates are

$$ru_r = v_\theta \qquad -rv_r = u_\theta$$

(*Hint:* Write $u_r = u_x x_r + u_y y_r$, and so on, and use the equations $u_x = v_y$, $u_y = -v_x$.)

33. Prove that the polar form of Laplace's equation $u_{xx} + u_{yy} = 0$ is

$$r^2 u_{rr} + ru_r + u_{\theta\theta} = 0$$

(*Hint:* Use the chain rule to find expressions for u_r, u_{rr}, and $u_{\theta\theta}$.)

34. For an ideal confined gas the pressure p, the volume V, and the absolute temperature T are related by $pV = RT$ where R is a constant. Show that

$$\frac{\partial V}{\partial T} \cdot \frac{\partial T}{\partial p} \cdot \frac{\partial p}{\partial V} = -1$$

What lesson is to be learned from this "surprising" result?

35. Prove that the function

$$f(x, y) = \begin{cases} \dfrac{xy}{x^2 + y^2} & \text{if } x^2 + y^2 \neq 0 \\ 0 & \text{if } (x, y) = (0, 0) \end{cases}$$

is *not* continuous at $(0, 0)$ even though it has partial derivatives at $(0, 0)$ (in fact, the partial derivatives exist at all points in the plane.)

36. It can be proved that if $f(x, y)$ has bounded partial derivatives in a neighborhood of a point (a, b) [that is, a positive constant M exists such that $|f_x| < M$ and $|f_y| < M$ for all (x, y) in that neighborhood] then $f(x, y)$ is continuous at (a, b). Reconcile this with the result of Exercise 35.

In Exercises 37 through 40, find (a) the value of the directional derivative at the particular point P_0 for the given function in the direction of the unit vector $\mathbf{u}$, (b) the maximum rate of change of the function at P_0.

37. $f(x, y) = x^3 e^{2y}$, $\quad \mathbf{u} = \dfrac{\mathbf{i}}{\sqrt{2}} - \dfrac{\mathbf{j}}{\sqrt{2}}$; $\quad P_0(-1, 0)$

38. $g(x, y) = x\sqrt{x + y}$, $\quad \mathbf{u} = \dfrac{1}{2}\mathbf{i} + \dfrac{\sqrt{3}}{2}\mathbf{j}$; $\quad P_0(3, 1)$

39. $f(x, y, z) = 2x^2 + yz + z^2$, $\quad \mathbf{u} = \dfrac{1}{\sqrt{3}}\mathbf{i} - \dfrac{1}{\sqrt{3}}\mathbf{j} + \dfrac{1}{\sqrt{3}}\mathbf{k}$; $\quad P_0(\tfrac{1}{2}, 3, -1)$

40. $g(x, y, z) = \sin x \cos 2y \cosh z$, $\quad \mathbf{u} = \tfrac{1}{3}\mathbf{i} + \tfrac{2}{3}\mathbf{j} + \tfrac{2}{3}\mathbf{k}$; $\quad P_0\left(\dfrac{\pi}{2}, \dfrac{\pi}{4}, 0\right)$

In Exercises 41 and 42, find an equation of the tangent plane and equations of the normal line to the given surface at the indicated point.

41. $x^2 + xy + y^2 - 3z = 16$; $\quad (2, 3, 1)$

42. $x^{2/3} + y^{2/3} + z^{2/3} = 9$; $\quad (1, 8, -8)$

43. Prove that every plane tangent to the surface $2x^2 + 3y^2 - z^2 = 0$ also passes through the origin.
44. Prove that every normal line of a sphere passes through the center of the sphere. (*Hint:* For convenience, but with no loss of generality, place the origin of coordinates at the center of the sphere.)
45. Prove that every normal line to a surface of revolution must intersect the axis of revolution. (*Hint:* An equation of a surface of revolution may be expressed in the form $z = g(x^2 + y^2)$ where the z-axis is the axis of revolution.)

In Exercises 46 through 49, find the relative maxima and minima, if any, for the given function.

46. $f(x, y) = 3x^2 + 4y^2 - 6x - 12y - 7$
47. $g(x, y) = x^2 + y^2 - 4xy + 25$
48. $h(x, y) = x^2 - y^2 - 10x + 2y + 3$
49. $F(x, y) = 2x^2 + 2y^2 - 2xy - 6x - 6y + 7$
50. Determine the triangle for which the product of the sines of the three interior angles is a maximum. What is the maximum value?
51. If $xyz = 1$, where $x > 0$, $y > 0$, and $z > 0$, show that $x + y + z \geq 3$.
52. Find the dimensions of the box of minimum cost if the volume is prescribed at V_0 ft^3, the base costs 30¢/ft^2, the top costs 20¢/ft^2 and the sides 10¢/ft^2.
53. A warehouse is to be built whose cubic footage (or volume) is to be V_0 ft^3. Construction costs are to be C_1 \$/ft^2 for the floor, C_2 \$/ft^2 for the walls and C_3 \$/ft^2 for the ceiling. Find the formula for the dimensions of the building which will minimize the total cost.
54. Consider the following data relating cricket chirps per minute to Fahrenheit temperature

Chirps per minute, x	60	70	82	92
Fahrenheit temperature, y	55°	57°	60°	63°

(a) Find the straight line of "best fit" to the given data by means of the least square criterion.
(b) If a cricket chirps 108 times per minute, predict the temperature by means of the regression line obtained in (a).

55. Suppose that we are given a function f which is continuous in $[-\pi, \pi]$. An approximation to this function is sought employing a three term trigonometric series of the form $A + B \sin x + C \cos x$. The quantities A, B, and C, which are independent of x, are to be found requiring that

$$I = \int_{-\pi}^{\pi} [f(x) - (A + B \sin x + C \cos x)]^2 \, dx$$

be a minimum. Find formulas for A, B, and C.
56. Generalize Exercise 55 by finding the trigonometric series of the nth order (at most)

$$T_n(x) = A_0 + A_1 \cos x + A_2 \cos 2x + \cdots + A_n \cos nx$$
$$+ B_1 \sin x + B_2 \sin 2x + \cdots + B_n \sin nx$$

which approximates a given continuous function $f(x)$ on $[-\pi, \pi]$ "best" in the least square sense, that is,

$$I = \int_{-\pi}^{\pi} [f(x) - T_n(x)]^2 \, dx$$

is a minimum. (*Hint:* Use the fact that for m and n arbitrary integers, $\int_{-\pi}^{\pi} \sin mx \sin nx \, dx = 0$, $m \neq n$;

$\int_{-\pi}^{\pi} \cos mx \cos nx \, dx = 0$, $m \neq n$; and $\int_{-\pi}^{\pi} \sin mx \cos nx \, dx = 0$, with no restriction on the integers m and n. These conditions enable us to determine one coefficient at a time so that the solution of simultaneous equations is not required.)
57. Find the absolute maximum and the absolute minimum of the function

$$f(x, y) = x^2 - xy + y^2 - 3x$$

on the square with vertices $(0, 0)$, $(3, 0)$, $(3, 3)$, $(0, 3)$ and determine the points at which the extrema are achieved.
58. Given a function f which is continuous in $[0, 1]$, we seek a linear function $A + Bx$ (A and B are

independent of x) such that

$$I(A, B) = \int_0^1 [f(x) - (A + Bx)]^2 \, dx$$

is a minimum. In particular, find A, B, and the linear function if $f(x) = x^2$ on $[0, 1]$.

59. An important class of problems occurs when a linear function of two or more independent variables is to be a maximum or a minimum subject to a set of constraints in the form of *linear inequalities*. These are known as *linear programming* problems. An exercise of this type follows: Find the absolute maximum and absolute minimum of $f(x, y) = 10x + 7y$ subject to the constraints

$$2x + y \leq 80$$
$$8x + 3y \leq 300$$
$$x \geq 0 \text{ and } y \geq 0$$

Sketch the region of definition of $f(x, y)$.

60. Find the absolute maximum and absolute minimum of $f(x, y) = 8x + 10y + 270$ subject to the constraints

$$x \leq 60 \qquad x + y \leq 90 \qquad x + 2y \leq 140 \qquad x \geq 0 \qquad y \geq 0$$

Sketch the region of definition of $f(x, y)$.

18

Multiple Integration

We introduced the definite integral in Chapter 6 as a limit of a Riemann sum. The concept of the measurement of area under a curve motivated the development of this topic. In particular, upper and lower bounds were established using, respectively, circumscribed and inscribed rectangles. With the proofs of the two fundamental theorems of calculus, we then demonstrated in Chapters 6 and 7 a diversity of applications to the problems of area, volumes of solids of revolution, work, fluid pressure, and so on, emphasizing the utility of elements.

In this chapter, the concept of volume under a surface is used to motivate the double integral. In general, the double integral may be evaluated by the computation of iterated single integrals; thus, the techniques developed in Chapter 6 and the formal methods of Chapter 10 become prominent once more. The same comments apply to triple and more generally to the n-tuple integral (where n is a positive integer greater than 1). Applications of multiple integrals to the determination of volumes, moments, center of mass, surface area, and moments of inertia are then given, utilizing rectangular, cylindrical, and spherical coordinates.

18.1 DOUBLE AND REPEATED INTEGRATION — SPECIAL CASE

In Chapter 6, we considered the development of the definite integral $\int_a^b f(x)\,dx$, where $a < b$. We partitioned the interval $[a, b]$ with points of division $a = x_0 < x_1 < x_2 < \cdots < x_{n-1} < x_n = b$ into n subintervals of width $\Delta x_i = x_i - x_{i-1}$, $i = 1, 2, \ldots, n$ and formed $\sum_{i=1}^{n} f(\xi_i)\,\Delta x_i$ where $x_{i-1} \le \xi_i \le x_i$ and ξ_i is otherwise arbitrary. Then, with $|P| = \max \Delta x_i$ denoting the norm of the partition, we investigated whether or not $\lim_{|P| \to 0} \sum_{i=1}^{n} f(\xi_i)\,\Delta x_i$ exists. If it does exist, the limit is defined to be the definite integral of $f(x)$ from a to b, that is, $\int_a^b f(x)\,dx$. An application of primary importance is that if $f(x)$ is continuous and nonnegative on $[a, b]$, then $\int_a^b f(x)\,dx$ yields the area between the curve $y = f(x)$, the x-axis, and the lines $x = a$ and $x = b$.

Let $z = f(x, y)$ denote a function of (x, y) that is continuous and nonnegative at all points inside and on the boundary of a closed rectangle R. We seek to determine the volume bounded by the surface $f(x, y)$, the rectangle R, and the four planes $x = a$, $x = b$, $y = c$, and $y = d$ which are perpendicular to the xy-plane (Figure 18.1.1).

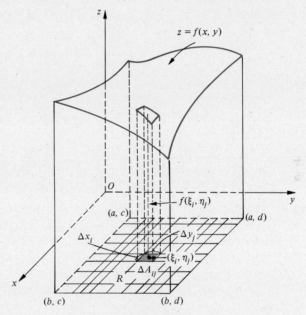

Figure 18.1.1

We subdivide the rectangular region R into $mn = N$ smaller rectangles by partitioning $[a, b]$ and $[c, d]$ so that

$$a = x_0 < x_1 < x_2 < \cdots < x_{m-1} < x_m = b;$$
$$c = y_0 < y_1 < y_2 < \cdots < y_{n-1} < y_n = d$$

and drawing lines $x = x_0, x = x_1, \ldots, x = x_m; y = y_0, y = y_1, \ldots, y = y_n$. The rectangle R_{ij} has area $\Delta A_{ij} = \Delta x_i\,\Delta y_j$ where $\Delta x_i = x_i - x_{i-1}$ and $\Delta y_j = y_j - y_{j-1}$ $(i = 1, 2, \ldots, m; j = 1, 2, \ldots, n)$. Since $z = f(x, y)$ is continuous on R and, therefore, on R_{ij}, z must have an absolute maximum M_{ij} and an absolute minimum m_{ij} on R_{ij}. If V denotes, as usual, the volume we are seeking

$$\sum_{i=1}^{m}\sum_{j=1}^{n} m_{ij}\,\Delta A_{ij} \le V \le \sum_{i=1}^{m}\sum_{j=1}^{n} M_{ij}\,\Delta A_{ij} \tag{1}$$

That is, $\sum_{i=1}^{m}\sum_{j=1}^{n} m_{ij}\,\Delta A_{ij}$ and $\sum_{i=1}^{m}\sum_{j=1}^{n} M_{ij}\,\Delta A_{ij}$ are lower and upper sums, respectively.

If we make the subdivisions finer and finer by letting $m \to \infty$ and $n \to \infty$, while the *greatest diameter* $|P|$ of the regions R_{ij} (that is, the greatest distance between two points of R_{ij}) at the same time must tend to zero, it is suggested that the upper and lower sums must approach one another and therefore the volume is defined as the common limit. Let

$$F(P) = \sum_{i=1}^{m} \sum_{j=1}^{n} f(\xi_i, \eta_j) \, \Delta A_{ij} \tag{2}$$

where (ξ_i, η_j) is any point in R_{ij}, be the **Riemann sum** of f associated with the partition P. By the squeeze theorem of Section 2.5, the same limit is achieved if we form $\lim_{|P| \to 0} \sum_{i=1}^{m} \sum_{j=1}^{n} f(\xi_i, \eta_j) \, \Delta A_{ij}$. Therefore, we *define* V by

$$V = \lim_{|P| \to 0} \sum_{i=1}^{m} \sum_{j=1}^{n} f(\xi_i, \eta_j) \, \Delta A_{ij} \tag{3}$$

Note that as $|P| \to 0$, m and n must $\to \infty$.

It is *not* obvious that the current definition of volume is consistent with the previous definitions given in Chapter 7. However, it can be established that *all* the definitions are consistent; that is, the same result is achieved by the various procedures. This is beyond the scope of the present text.

If we introduce the symbol

$$\iint_R f(x, y) \, dA = \lim_{|P| \to 0} \sum_{i=1}^{m} \sum_{j=1}^{n} f(\xi_i, \eta_j) \, \Delta A_{ij} \tag{4}$$

we call (4) the **double integral** of f over the rectangle R. It can be shown that if $f(x, y)$ is any continuous function over R, not necessarily nonnegative, that the limit of the right side exists. In particular for $f(x, y) \geq 0$ on R, the volume is given by

$$V = \iint_R f(x, y) \, dA \tag{5}$$

Of course we still need a method for evaluating the double integral (5). To this end, we approach volume from a point of view considered in Section 7.4. If we fix x in $[a, b]$, the cross section of the solid is a plane region under the graph $z = f(x, y)$ from $y = c$ to $y = d$, and its area $A(x)$ is given by

$$A(x) = \int_c^d f(x, y) \, dy \tag{6}$$

Since $f(x, y)$ is continuous, $A(x)$ must also be continuous. Therefore the volume V is computed from

$$V = \int_a^b A(x) \, dx = \int_a^b \left[\int_c^d f(x, y) \, dy \right] dx \tag{7}$$

(see Figure 18.1.2).

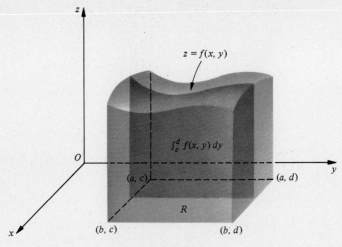

Figure 18.1.2

Alternatively, if we fix y in $[c, d]$ and let (Figure 18.1.3)

$$A(y) = \int_a^b f(x, y)\, dx,$$

then

$$V = \int_c^d A(y)\, dy = \int_c^d \left[\int_a^b f(x, y)\, dx \right] dy \qquad (8)$$

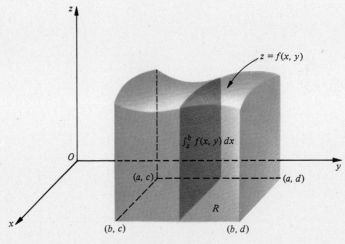

Figure 18.1.3

Combining (7) and (8), there results,

$$V = \int_a^b \left[\int_c^d f(x, y)\, dy \right] dx = \int_c^d \left[\int_a^b f(x, y)\, dx \right] dy \qquad (9)$$

More generally, the following theorem can be proved.

Theorem 1 If $f(x, y)$ is continuous for all values of x and y on a rectangular region R, bounded by the lines $x = a$, $x = b$, $y = c$, and $y = d$, then

$$\iint_R f(x, y)\, dx\, dy = \int_a^b \left[\int_c^d f(x, y)\, dy \right] dx = \int_c^d \left[\int_a^b f(x, y)\, dx \right] dy \quad (10)$$

We shall adopt the following convention: $\displaystyle\int_c^d \int_a^b f(x, y)\, dx\, dy$ is taken to mean that the limits on x are a and b and the limits on y are c and d, hence brackets are no longer required.

EXAMPLE 1 Express the volume under the surface $z = 3x^2y + 2xy^2$ and above the region given by $0 \le x \le 3$, $1 \le y \le 2$ as a double integral and evaluate it in two ways.

Solution If V denotes the required volume, then from (5),

$$V = \iint_R (3x^2y + 2xy^2)\, dA$$

If we use (7),

$$V = \int_0^3 \int_1^2 (3x^2y + 2xy^2)\, dy\, dx$$

But for fixed x,

$$\int_1^2 (3x^2y + 2xy^2)\, dy = \left(3x^2\frac{y^2}{2} + \frac{2xy^3}{3} \right) \Big|_1^2$$

$$= 6x^2 + \frac{16}{3}x - \left(\frac{3}{2}x^2 + \frac{2}{3}x \right)$$

$$= \frac{9}{2}x^2 + \frac{14}{3}x$$

Hence

$$V = \int_0^3 \left(\frac{9}{2}x^2 + \frac{14}{3}x \right) dx$$

$$= \left(\frac{3}{2}x^3 + \frac{7}{3}x^2 \right) \Big|_0^3 = \frac{81}{2} + 21 - 0 = \frac{123}{2} \text{ cubic units}$$

If we change the order of integration, then from (8)

$$V = \int_1^2 \int_0^3 (3x^2y + 2xy^2)\, dx\, dy$$

But for fixed y,

$$\int_0^3 (3x^2y + 2xy^2)\, dx = (x^3y + x^2y^2)|_0^3$$

$$= 27y + 9y^2 - 0$$

so that

$$V = \int_1^2 (27y + 9y^2)\, dy = \left(\frac{27y^2}{2} + 3y^3\right)\Big|_1^2$$

$$= 54 + 24 - \left(\frac{27}{2} + 3\right) = 75 - \frac{27}{2} = \frac{123}{2} \text{ cubic units} \qquad \bullet$$

EXAMPLE 2   Evaluate $\displaystyle\int_R\int \frac{dx\, dy}{x + y + 1}$ over the square $0 \le x \le 2$ and $0 \le y \le 2$.

Solution

$$\int_R\int \frac{dx\, dy}{x + y + 1} = \int_0^2 \int_0^2 \frac{dy}{x + y + 1}\, dx$$

$$= \int_0^2 \ln(x + y + 1)\big|_0^2\, dx$$

$$= \int_0^2 [\ln(x + 3) - \ln(x + 1)]\, dx$$

and by using integration by parts

$$\int_R\int \frac{dx\, dy}{x + y + 1} = [(x + 3)\ln(x + 3) - (x + 3) - (x + 1)\ln(x + 1)$$

$$+ (x + 1)]\big|^2 = 5\ln 5 - 6\ln 3 \qquad \bullet$$

At times the effort required to perform an iterated integration is dependent on the order. This is illustrated by Example 3.

EXAMPLE 3   Evaluate $\displaystyle\int_1^3 \int_0^2 xe^{xy}\, dx\, dy$

Solution   If we first integrate with respect to x (between $x = 0$ and $x = 2$), we must integrate by parts and then the resulting integration with respect to y causes great difficulty. Fortunately we can reverse the order of integration:

$$\int_1^3 \int_0^2 xe^{xy}\, dx\, dy = \int_0^2 \int_1^3 e^{xy} x\, dy\, dx$$

$$= \int_0^2 (e^{xy}\big|_1^3)\, dx$$

$$= \int_0^2 (e^{3x} - e^x)\, dx$$

$$= \left(\frac{e^{3x}}{3} - e^x\right)\Big|_0^2$$

$$= \frac{e^6}{3} - e^2 - \left(\frac{1}{3} - 1\right) = \frac{e^6 + 2}{3} - e^2 \qquad \bullet$$

Suppose that $f(x, y) = F(x) G(y)$ where $F(x)$ and $G(y)$ are continuous functions, respectively, on $[a, b]$ and $[c, d]$. Then, for the rectangular region R,

$$\int_R\!\!\int f(x, y)\, dx\, dy = \int_a^b \int_c^d F(x)\, G(y)\, dy\, dx$$
$$= \int_a^b F(x)\, dx \int_c^d G(y)\, dy \tag{11}$$

EXAMPLE 4   For the rectangular region just described,

$$\int_R\!\!\int x^2 y^3\, dx\, dy = \int_a^b x^2\, dx \int_c^d y^3\, dy$$
$$= \tfrac{1}{12}(b^3 - a^3)(d^4 - c^4) \qquad \bullet$$

Next, we give another physical application of the double integral (4). Suppose we have a lamina or thin rectangular plate R whose mass density, that is, mass per unit area, is a continuous function of x and y given by $\rho(x, y)$. Then the total mass of the plate is given by

$$\text{mass} = \int_R\!\!\int \rho(x, y)\, dA \tag{12}$$

To see this, partition the rectangle into mn subrectangles R_{ij} with area ΔA_{ij} ($i = 1, 2, \ldots, m, j = 1, 2, \ldots, n$). Then if (ξ_i, η_j) is any point in the ijth rectangle, the approximate mass of R_{ij} is given by

$$\rho(\xi_i, \eta_j)\, \Delta A_{ij}$$

and the Riemann sum

$$\sum_{i=1}^m \sum_{j=1}^n \rho(\xi_i, \eta_j)\, \Delta A_{ij}$$

approximates the total mass of the plate. If, as usual, $|P|$ represents the norm or maximum diameter of all the elemental rectangles, we have the definition

$$\text{mass} = \lim_{|P| \to 0} \sum_{i=1}^m \sum_{j=1}^n \rho(\xi_i, \eta_j)\, \Delta A_{ij} = \int_R\!\!\int \rho(x, y)\, dA \tag{13}$$

EXAMPLE 5   Find the mass of the rectangle R bounded by the axes and the lines $x = 2$ and $y = 3$, if the density in slugs per unit area varies as the sum of the distances from the coordinate axes.

Solution   The density is given by the function $\rho = k(x + y)$, where k is a constant of proportionality. Thus from (13),

$$\text{mass} = \int_R\!\!\int k(x + y)\, dA = \int_0^3 \int_0^2 k(x + y)\, dx\, dy$$
$$= \int_0^3 k\left(\frac{x^2}{2} + yx\right)\Big|_0^2 dy = \int_0^3 k(2 + 2y)\, dy$$
$$= k(2y + y^2)|_0^3 = 15k \text{ slugs} \qquad \bullet$$

Exercises 18.1

In Exercises 1 through 6, evaluate each of the following repeated integrals by performing the iterated integration in the order indicated and then by reversing the order of integration.

1. $\int_0^1 \int_0^2 (x + 3y)\, dx\, dy$

2. $\int_{-2}^2 \int_0^1 (x^2 + 2y^2)\, dx\, dy$

3. $\int_{-1}^0 \int_0^1 (x - y)^2\, dx\, dy$

4. $\int_0^1 \int_0^{1/2} (e^x - e^{2y})\, dx\, dy$

5. $\int_0^{\pi/2} \int_0^a (r \sin\theta + \cos\theta)\, dr\, d\theta$
   (a is a constant)

6. $\int_0^{\pi/2} \int_0^1 (r \cos^2\theta - \sin 3\theta)\, dr\, d\theta$

In each of Exercises 7 through 12, sketch the solid whose volume is given by the repeated integral, and compute each volume.

7. $\int_0^4 \int_0^3 y\, dx\, dy$

8. $\int_0^3 \int_1^5 (1 + x + y)\, dx\, dy$

9. $\int_1^3 \int_{1/2}^1 (8 - 2x - y)\, dx\, dy$

10. $\int_{1/2}^{3/2} \int_1^2 (9 - x^2 - y^2)\, dx\, dy$

11. $\int_0^a \int_0^a xy^2\, dx\, dy$,
    a is a positive constant

12. $\int_0^b \int_0^b xye^{-(x+y)}\, dx\, dy$,
    b is a positive constant

13. The *average value* of a function f defined on a rectangle R of area A is defined by

$$\frac{1}{A} \iint_R f\, dA$$

Find the average value of $f(x, y) = k(x^2 + y^2)$, k is a constant, on the square bounded by the lines $x = 1$, $x = 4$, $y = -2$, and $y = 1$.

14. Find the average value of a temperature T, which is defined by $T(x, y) = 3 \sin^2 x + 4 \cos^2 y$, on the rectangle R bounded by the lines $x = 0$, $x = 2\pi$, $y = 0$, and $y = \pi$. (*Hint:* Refer to Exercise 13.)

15. Find the total force against the sides of a completely filled vertical tank, if the dimensions of the rectangular sides are 4 ft in the depth direction by 3 ft, using double integration and the fact that the pressure at a depth y is given by $62.5y$ lb/ft^2.

16. Find the total mass of a square plate of side 8 ft if the mass density ρ in slugs per square foot varies directly as the product of the distances from one vertex.

17. Let R be the rectangle bounded by the lines $x = 0$, $x = 4$, $y = 0$, and $y = 2$. If we partition R into eight congruent squares of side 1, estimate $I = \iint_R e^{-(x+2y)}\, dx\, dy$ by computing upper and lower Riemann sums. Then calculate the exact value of I. Round all answers to three decimal places.

18. Let R be the square bounded by the lines $x = 0$, $x = 2$, $y = 0$, and $y = 2$. If we partition R into four congruent squares of side 1, estimate $I = \iint_R e^{-(x^2+y^2)}\, dx\, dy$, by computing the Riemann sum using the values of the function at the midpoint of each square. Given that $\int_0^2 e^{-x^2}\, dx \doteq 0.8821$, find I and compare the results to two decimal places.

18.2 ITERATED INTEGRALS OVER NONRECTANGULAR REGIONS

In the previous section we restricted our attention to the evaluation of double and repeated integration of a continuous function of x and y defined on a rectangular region in the xy-plane. Suppose more generally that the region R is bounded by the graphs of continuous functions g and h and lies between the x-values $x = a$ and b. This is shown in Figure 18.2.1, where it is assumed that $g(x) \leq h(x)$ on $[a, b]$. One may have analogously the region shown in Figure 18.2.2. We seek the volume of the solid which has R as shown in Figure 18.2.1 as its base and is bounded above or capped by a portion of the surface $z = f(x, y)$ lying over R, where f is a continuous nonnegative function of x and y on R (Figure 18.2.3).

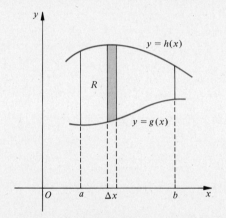

Figure 18.2.1

The cross section of the solid in the plane $x = $ constant is shown in Figure 18.2.3, and the area of this section may be shown to be a continuous function given by

$$A(x) = \int_{g(x)}^{h(x)} f(x, y)\, dy$$

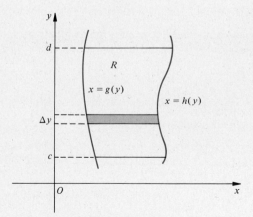

Figure 18.2.2

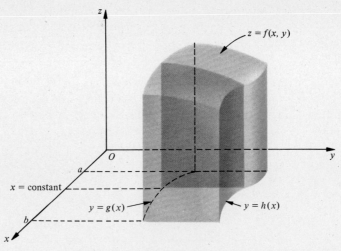

Figure 18.2.3

(where x is held constant for this integration). Therefore, by the method of sections, the volume V is calculated from the iterated integral

$$V = \int_a^b A(x)\,dx = \int_a^b \left[\int_{g(x)}^{h(x)} f(x, y)\,dy \right] dx$$

Therefore we have the following theorem.

Theorem 1 Let S be the solid based on the region R in the xy-plane and capped by the surface $z = f(x, y)$, where $f(x, y)$ is continuous and nonnegative on R. Furthermore, R is a region in the xy-plane bounded by continuous functions g and h and the lines $x = a$ and $x = b$ $(a < b)$ such that $g(x) \leq h(x)$ on $[a, b]$. Then the volume V is determined by the iterated integral

$$V = \int_a^b \left[\int_{g(x)}^{h(x)} f(x, y)\,dy \right] dx \tag{1}$$

Analogously, we have a second theorem.

Theorem 2 Let S be the solid based on the region R in the xy-plane and capped by the surface $z = f(x, y)$, where $f(x, y)$ is continuous and nonnegative on R. Furthermore, R is a region in the xy-plane bounded by continuous functions g and h and the lines $y = c$ and $y = d$, $(c < d)$, such that $g(y) \leq h(y)$ on $[c, d]$. Then the volume is determined by the iterated integral

$$V = \int_c^d \left[\int_{g(y)}^{h(y)} f(x, y)\,dx \right] dy \tag{2}$$

EXAMPLE 1   Evaluate the iterated integral

$$\int_0^2 \left[\int_{x^2}^{2x} 3x^2y \, dy \right] dx$$

and sketch the region R (or domain) of integration.

Solution   First we integrate with respect to y holding x constant. Therefore

$$\int_{x^2}^{2x} 3x^2y \, dy = \frac{3x^2}{2} y^2 \Big|_{x^2}^{2x} = \frac{3x^2}{2}(4x^2 - x^4)$$

$$= 6x^4 - \frac{3}{2}x^6$$

and then the integration with respect to x yields

$$\int_0^2 \left(6x^4 - \frac{3}{2}x^6 \right) dx = \left(\frac{6x^5}{5} - \frac{3}{14}x^7 \right) \Big|_0^2 = \frac{192}{5} - \frac{192}{7} = \frac{384}{35}$$

The region R is found as follows: if $x =$ constant is any number in $[0, 2]$ then we must integrate from $y = x^2$ to $y = 2x$, that is, from the parabola to the straight line as shown in Figure 18.2.4. Then we integrate with respect to x from $x = 0$ to $x = 2$, sweeping out the shaded area R.   ●

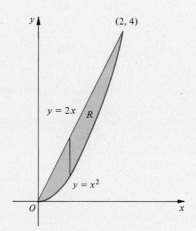

Figure 18.2.4

EXAMPLE 2   Find the volume of the space region under the graph of $z = x^2 + y$ based on the triangular region in the first quadrant bounded by the line $x + 2y = 4$ and the coordinate axes.

Solution   The region R in the xy-plane is shown in Figure 18.2.5(a) and, although a sketch of the surface $z = x^2 + y$ is not difficult, it is actually unnecessary since $z \geq 0$ in the first quadrant and is continuous for all x and y. Hence

$$V = \int_0^4 \int_0^{\frac{4-x}{2}} (x^2 + y) \, dy \, dx$$

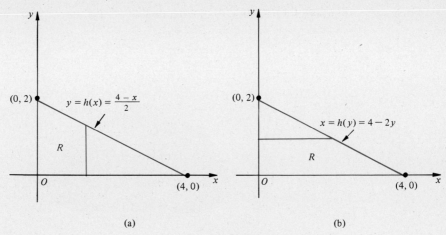

(a)   (b)

Figure 18.2.5

For the inside integral, that is, with respect to y, we have

$$\int_0^{\frac{4-x}{2}} (x^2 + y)\, dy = \left(x^2 y + \frac{y^2}{2}\right)\Big|_0^{\frac{4-x}{2}}$$

$$= \frac{4-x}{2}x^2 + \frac{1}{2}\left(\frac{4-x}{2}\right)^2$$

$$= 2 - x + \frac{17}{8}x^2 - \frac{x^3}{2}$$

Therefore

$$V = \int_0^4 \left(2 - x + \frac{17}{8}x^2 - \frac{x^3}{2}\right) dx$$

$$= \left(2x - \frac{x^2}{2} + \frac{17}{24}x^3 - \frac{x^4}{8}\right)\Big|_0^4$$

$$= \frac{40}{3} \text{ cubic units}$$

Alternatively, we could use an element parallel to the x-axis and integrate from $x = 0$ to $x = h(y) = 4 - 2y$, Figure 18.2.5(b). Hence

$$V = \int_0^2 \int_0^{4-2y} (x^2 + y)\, dx\, dy$$

$$\int_0^{4-2y} (x^2 + y)\, dx = \left(\frac{x^3}{3} + yx\right)\Big|_0^{4-2y} = \frac{(4-2y)^3}{3} + 4y - 2y^2$$

Therefore

$$V = \int_0^2 \left[\frac{(4-2y)^3}{3} + 4y - 2y^2\right] dy$$

$$= \frac{(4-2y)^4}{-24} + 2y^2 - \frac{2y^3}{3}\Big|_0^2$$

$$= \frac{40}{3} \text{ cubic units}$$

  ●

EXAMPLE 3   Find the volume between the paraboloid of revolution $z = a^2 - x^2 - y^2$, $z \geq 0$ and the xy-plane (Figure 18.2.6).

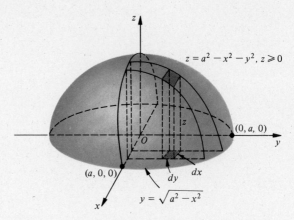

$z = a^2 - x^2 - y^2, z \geqslant 0$

$(0, a, 0)$

$(a, 0, 0)$

dy dx

$y = \sqrt{a^2 - x^2}$

Figure 18.2.6

Solution   By symmetry, we can find the volume in the first octant and multiply by 4. The trace in the xy-plane, obtained by letting $z = 0$ is the circle $x^2 + y^2 = a^2$. The volume V is given by

$$V = 4 \int_0^a \int_0^{\sqrt{a^2-x^2}} z \, dy \, dx = 4 \int_0^a \int_0^{\sqrt{a^2-x^2}} (a^2 - x^2 - y^2) \, dy \, dx \qquad (a > 0)$$

$$= 4 \int_0^a \left[(a^2 - x^2)y - \frac{y^3}{3} \right] \Bigg|_0^{\sqrt{a^2-x^2}} dx$$

$$= 4 \int_0^a \left[(a^2 - x^2)^{3/2} - \frac{1}{3}(a^2 - x^2)^{3/2} \right] dx$$

$$= \frac{8}{3} \int_0^a (a^2 - x^2)^{3/2} \, dx \qquad \text{(i)}$$

If we let $x = a \sin \theta$, $dx = a \cos \theta \, d\theta$, then

$$V = \frac{8}{3} \int_0^{\pi/2} a^3 \cos^3 \theta \, (a \cos \theta) \, d\theta$$

$$= \frac{8a^4}{3} \int_0^{\pi/2} \cos^4 \theta \, d\theta \qquad \text{(ii)}$$

$$\int_0^{\pi/2} \cos^4 \theta \, d\theta = \frac{1}{4} \int_0^{\pi/2} (1 + \cos 2\theta)^2 \, d\theta$$

$$= \frac{1}{4} \int_0^{\pi/2} (1 + 2 \cos 2\theta + \cos^2 2\theta) \, d\theta$$

$$= \frac{1}{4} \int_0^{\pi/2} \left(1 + 2 \cos 2\theta + \frac{1}{2} + \frac{\cos 4\theta}{2} \right) d\theta$$

$$= \frac{1}{4} \left(\frac{3}{2}\theta + \sin 2\theta + \frac{\sin 4\theta}{8} \right) \Bigg|_0^{\pi/2}$$

$$= \frac{3\pi}{16}$$

Hence $V = \dfrac{\pi a^4}{2}$ cubic units

We worked out the integral; however, the reader could easily have used the integral tables to determine the integral defined by (i). Alternatively, after making the trigonometric substitution $x = a \sin \theta$ the definite integral (ii) is given in the tables. ●

Exercises 18.2

In Exercises 1 through 10, evaluate the given iterated integral. Sketch the domain of integration in the xy-plane.

1. $\displaystyle\int_0^1 \int_0^x (2x + 3y)\, dy\, dx$

2. $\displaystyle\int_0^{1/2} \int_0^{2x} xy\, dy\, dx$

3. $\displaystyle\int_0^3 \int_{x^2}^{3x} x^2 y\, dy\, dx$

4. $\displaystyle\int_0^1 \int_{x^2}^{2x} (x + y)\, dy\, dx$

5. $\displaystyle\int_0^2 \int_y^{2y} xy^2\, dx\, dy$

6. $\displaystyle\int_0^1 \int_{y^2}^{4-y} (x + y^2)\, dx\, dy$

7. $\displaystyle\int_0^a \int_1^{e^x} \frac{x}{y}\, dy\, dx,\quad a = \text{positive constant}$

8. $\displaystyle\int_{\pi/6}^{\pi/3} \int_0^{b \sin \theta} \frac{r}{\sqrt{b^2 - r^2}}\, dr\, d\theta,$
$b = \text{positive constant}$

9. $\displaystyle\int_0^2 \int_x^{4x - x^2} (x^2 - 2xy)\, dy\, dx$

10. $\displaystyle\int_{1/2}^1 \int_{1-2y}^{y^2 + 1} \frac{1}{x + 2y}\, dx\, dy$

In Exercises 11 through 16, sketch the solid under the given surface and above the given region in the xy-plane. Find the volume of the solid.

11. $z = 11 - 3x - y;\qquad 0 \le y \le x,\ \ 0 \le x \le 2$

12. $z = 4 + x + y^2;\qquad 0 \le y \le 2x,\ \ 0 \le x \le 1$

13. $z = 3 + y^2;\qquad x^2 \le y \le 4,\ \ 0 \le x \le 2$

14. $z = x^2 + y^2;\qquad 0 \le x \le a - y,\ \ 0 \le y \le a,\ (a \text{ is constant})$

15. $z = 4 + x;\qquad 0 \le x \le \sqrt{9 - y^2},\ 0 \le y \le 3$

16. $z = ax + by;\qquad 0 \le y \le x^2,\ \ 0 \le x \le 2,\ (a \text{ and } b \text{ are positive constants})$

17. Find the volume bounded by the plane $\dfrac{x}{a} + \dfrac{y}{b} + \dfrac{z}{c} = 1$ and the coordinate planes where a, b, and c are positive constants.

18. Find the volume under the paraboloid of revolution $3z = x^2 + y^2$ and above the triangle having vertices $(0, 0)$, $(4, 0)$, and $(4, 3)$ in the xy-plane. Do this two ways.

19. Find the volume of a sphere of radius a by means of an iterated integral. (*Hint:* Use integral tables.)

20. Find the volume under the surface $z = e^{2x+y}$ and above the trapezoid bounded by $x = 0$, $x = \dfrac{y}{2}$, $y = 1$, and $y = 2$.

21. Compute the volume of the ellipsoid

$$\frac{x^2}{a^2} + \frac{y^2}{b^2} + \frac{z^2}{c^2} = 1$$

where a, b, and c are positive constants.

22. Find the volume of the solid capped by $z = y^2$ and above the area bounded by $x - 2y^2 = 0$ and $x - 2y - 4 = 0$.

In this section, we shall first develop the general definition of the double integral without recourse to the particular applications of the concept.

We start with some definitions.

Definition   A **simple arc** in the xy-plane is the graph of the parametric equations

$$x = f(t) \qquad y = g(t) \qquad a \leq t \leq b$$

where f and g are continuous functions of t, such that to distinct values of t in (a, b) correspond distinct points on the graph. This means that a simple arc cannot cross itself.

Definition   If, furthermore, we impose the requirement that the first derivative f' and g' are continuous for all t in $[a, b]$, then the simple arc is said to be **smooth.** This means that the tangent line is turning continuously as the parameter t varies from a to b (Figure 18.3.1(a)).

Definition   A **piecewise smooth curve** in $[a, b]$ is a finite number of smooth simple arcs joined end to end such that distinct values of t in (a, b) correspond to distinct points in its graph (Figure 18.3.1(b)). A piecewise smooth curve is said to be **closed** if the points corresponding to $t = a$ and $t = b$ coincide.

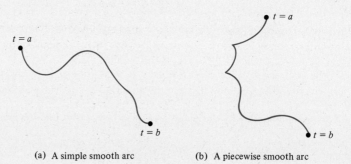

(a) A simple smooth arc   (b) A piecewise smooth arc

Figure 18.3.1

Suppose then that R is a region bounded by a piecewise smooth closed curve. The region R can be partitioned in the following manner. Suppose that we enclose R with a rectangle T (Figure 18.3.2). Now we partition T with horizontal and vertical lines. The totality of closed rectangular regions which lie completely *within* R is called an **inner partition** of R and is shown shaded in the figure. These rectangular regions are denoted by $R_1, R_2, R_3, \ldots, R_N$. Also the length of the longest diagonal of the N rectangles, represented by $|P|$, is the norm of the partition P. Furthermore, ΔA_i is the area of R_i, $(i = 1, 2, \ldots, N)$.

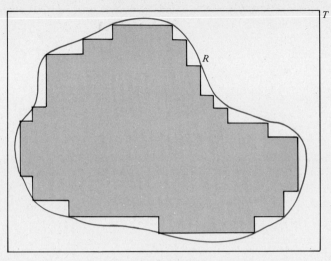

Figure 18.3.2

Definition **Riemann Sum.** Let f be a function of two variables, x and y, defined on a region R, and let (ξ_i, η_i) be an arbitrary point in R_i. Then, any sum of the form

$$\sum_{i=1}^{N} f(\xi_i, \eta_i)\, \Delta A_i$$

is said to be a Riemann sum of f corresponding to the partition P.

EXAMPLE 1   Suppose $f(x, y) = \dfrac{x}{x + y}$ and R is the region defined by $x^2 + y^2 \le 25$, $x \ge 0$, and $y \ge 0$, that is, a region bounded by a quarter circle of radius 5. Let us form an inner partition of R determined by vertical and horizontal lines with integral intercepts as shown in Figure 18.3.3. If (ξ_i, η_i) denotes the centroid of each R_i, calculate the Riemann sum.

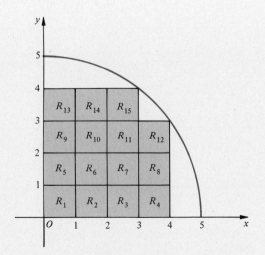

Figure 18.3.3

Solution

i	ξ_i	η_i	$f(\xi_i, \eta_i)$	ΔA_i	$f(\xi_i, \eta_i)\,\Delta A_i$
1	0.5	0.5	0.5	1	0.5
2	1.5	0.5	0.75	1	0.75
3	2.5	0.5	0.833	1	0.833
4	3.5	0.5	0.875	1	0.875
5	0.5	1.5	0.25	1	0.25
6	1.5	1.5	0.50	1	0.50
7	2.5	1.5	0.625	1	0.625
8	3.5	1.5	0.70	1	0.70
9	0.5	2.5	0.167	1	0.167
10	1.5	2.5	0.375	1	0.375
11	2.5	2.5	0.5	1	0.5
12	3.5	2.5	0.583	1	0.583
13	0.5	3.5	0.125	1	0.125
14	1.5	3.5	0.3	1	0.3
15	2.5	3.5	0.417	1	0.417

$$\text{Riemann sum} = \sum_{i=1}^{15} f(\xi_i, \eta_i)\,\Delta A_i = 7.5$$

●

Definition   If a function f of two variables x and y is defined on a region R, then the **double integral** of f over R, denoted by $\int\int_R f(x, y)\, dA$, is given by

$$\int\int_R f(x, y)\, dA = \lim_{|P| \to 0} \sum_{i=1}^{N} f(\xi_i, \eta_i)\,\Delta A_i \qquad (1)$$

provided that the limit of the right side exists.

Definition   A function f is said to be **integrable** over a region R provided that the double integral of f over R exists.

The basic theorem may be expressed as follows.

Theorem 1   If $f(x, y)$ is continuous on a region R, then f is also integrable over R; that is, the double integral

$$\int\int_R f(x, y)\, dA$$

exists.

Just as in the case of a function of one variable, the proof is beyond the scope of the book.

Moreover, for the simple regions of Figures 18.2.1 and 18.2.2, we can establish that the double integral of continuous functions can be evaluated by repeated integration.

Theorem 2 Let $f(x, y)$ be a continuous function on the closed plane region R bounded above and below, respectively, by the graphs of continuous functions $y = h(x)$ and $y = g(x)$ for x in $[a, b]$ then

$$\iint_R f(x, y)\, dA = \int_a^b \int_{g(x)}^{h(x)} f(x, y)\, dy\, dx \tag{2}$$

Similarly, if $f(x, y)$ is a continuous function on the closed plane region bounded on the left and right, respectively, by the graphs of continuous functions $x = g(y)$ and $x = h(y)$ for y in $[c, d]$, then

$$\iint_R f(x, y)\, dA = \int_c^d \int_{g(y)}^{h(y)} f(x, y)\, dx\, dy \tag{3}$$

EXAMPLE 2 Apply (2) to the special case $f(x, y) = 1$ and interpret the result.

Solution If $f(x, y) = 1$, then (2) yields

$$\iint_R 1\, dA = \iint_R dA = \int_a^b \int_{g(x)}^{h(x)} dy\, dx$$

$$= \int_a^b y\, \big|_{g(x)}^{h(x)}\, dx = \int_a^b [h(x) - g(x)]\, dx$$

which is the area of the region of the type depicted in Figure 18.2.1. Hence, if R is the type of region shown in the figure, then

$$A = \iint_R 1\, dA = \text{area of the region } R \tag{4}$$

Similarly, for a region R demonstrated in Figure 18.2.2,

$$A = \iint_R 1\, dA = \int_c^d \int_{g(y)}^{h(y)} 1\, dx\, dy = \int_c^d x\, \big|_{g(y)}^{h(y)}\, dy$$

$$= \int_c^d [h(y) - g(y)]\, dy$$

where A is the area of the region R. ●

EXAMPLE 3 Find the area A bounded by the graphs of $y = 6 - x^2$ and $2x - y - 2 = 0$.

Solution The region lies under the parabola $y = 6 - x^2$ and above the line $y = 2x - 2$ as illustrated in Figure 18.3.4. The intersection points of the two curves are $(-4, -10)$ and $(2, 2)$. Summing first in the y-direction, we obtain using an iterated

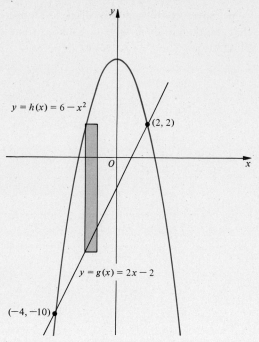

$y = h(x) = 6 - x^2$

$(2, 2)$

O

$y = g(x) = 2x - 2$

$(-4, -10)$

Figure 18.3.4

integral,

$$A = \int_{-4}^{2} \int_{2x-2}^{6-x^2} dy\, dx = \int_{-4}^{2} [(6 - x^2) - (2x - 2)]\, dx$$

$$= \int_{-4}^{2} (8 - 2x - x^2)\, dx = \left(8x - x^2 - \frac{x^3}{3}\right)\bigg|_{-4}^{2}$$

$$= 36 \text{ square units} \qquad\qquad\qquad\qquad\qquad\qquad \bullet$$

We conclude this section by listing without proof some important properties of double integrals that are analogous to the corresponding properties for single integrals in Chapter 6. Proofs are requested in the exercise list.

Theorem 3 If $\iint_R f(x, y)\, dx\, dy$ and $\iint_R g(x, y)\, dx\, dy$ exist over a region R, then so does

$\iint_R [c_1 f(x, y) + c_2 g(x, y)]\, dx\, dy$ where c_1 and c_2 are constants. Furthermore,

$$\iint_R [c_1 f(x, y) + c_2 g(x, y)]\, dx\, dy =$$

$$c_1 \iint_R f(x, y)\, dx\, dy + c_2 \iint_R g(x, y)\, dx\, dy \qquad (5a)$$

In particular, if $c_1 = c_2 = 1$, we have

$$\iint_R [f(x, y) + g(x, y)]\, dx\, dy = \iint_R f(x, y)\, dx\, dy + \iint_R g(x, y)\, dx\, dy \qquad (5b)$$

Also, if $c_1 = c$ and $c_2 = 0$, then

$$\int\int_R cf(x, y)\, dx\, dy = c \int\int_R f(x, y)\, dx\, dy \qquad (5c)$$

The linear superposition (5a) can be readily extended to n functions where n is any positive integer greater than two.

Theorem 4   If $f(x, y) \geq 0$ for all (x, y) in region R, then

$$\int\int_R f(x, y)\, dA \geq 0 \text{ if it exists.}$$

Corollary   If $f(x, y) \geq g(x, y)$ for all (x, y) in region R then

$$\int\int_R f(x, y)\, dA \geq \int\int_R g(x, y)\, dA \qquad (6)$$

assuming both integrals exist.

Theorem 5   If R_1 and R_2 are two nonoverlapping regions having at most boundary points in common and R is the union of R_1 and R_2 then

$$\int\int_R f(x, y)\, dx\, dy = \int\int_{R_1} f(x, y)\, dx\, dy + \int\int_{R_2} f(x, y)\, dx\, dy \qquad (7)$$

provided each double integral on the right side of (7) exists.

Exercises 18.3

1. Evaluate the Riemann sum for $f(x, y) = 2xy$ over the quarter circle $x^2 + y^2 \leq 25$, $x \geq 0$, and $y \geq 0$ using the inner partition of R determined by the vertical and horizontal lines with integral intercepts (as in Example 1) where (ξ_i, η_i) are at the centroid of each square.
2. Evaluate the Riemann sum of the same function, region, and subdivision as in Exercise 1, but with the upper right vertex of each square as evaluation points.
3. Evaluate the double integral for the function and the region defined in Exercise 1.
4. Let R be the trapezoidal region in the first quadrant bounded by the lines $x = 0$, $y = 0$, $y = 3$, and $x + y = 6$. Let P be the inner partition of R determined by the vertical and horizontal lines with integral intercepts. Evaluate the Riemann sum for $f(x, y) = \dfrac{xy}{1 + x}$ with the upper right vertex of each square as an evaluation point.

In Exercises 5 through 12, evaluate the integrals

5. $\displaystyle\int_0^1 \int_x^{\sqrt{x}} (x - 3y)\, dy\, dx$

6. $\displaystyle\int_0^{\pi/2} \int_0^{\pi} (x \sin y - y \cos x)\, dx\, dy$

7. $\displaystyle\int_1^e \int_{x^2}^x \ln x\, dy\, dx$

8. $\displaystyle\int_0^2 \int_x^2 \frac{1}{(1 + y)^2}\, dy\, dx$

9. $\displaystyle\int_{-\pi/2}^{\pi/2}\int_0^{\sin y} e^x \cos y \, dx \, dy$

10. $\displaystyle\int_{\pi/2}^{3\pi/2}\int_{2y}^{y} y \sin{(x-y)} \, dx \, dy$

11. $\displaystyle\int_0^2\int_{x/2}^{x} e^{-x^2} \, dy \, dx$

12. $\displaystyle\int_0^1\int_{-3x}^{-x} e^{-x} \cos{(y+x)} \, dy \, dx$

In Exercises 13 through 16, calculate by double integration the area of the bounded region determined by the given pairs of curves. Sketch the region showing an element of area.

13. $y = x, \quad x = 5y - y^2$

14. $y = 2x, \quad y = 3x - x^2$

15. $y = \sqrt{12x}, \quad 2x^2 - 3y = 0$

16. $x + y - 10 = 0, \quad xy - 21 = 0$

In Exercises 17 through 20, sketch the region of integration and change the order of integration.

17. $\displaystyle\int_0^2\int_{x^3}^{4x} f(x, y) \, dy \, dx$

18. $\displaystyle\int_{-3}^3\int_{x^2}^9 f(x, y) \, dy \, dx$

19. $\displaystyle\int_0^1\int_{x^2}^{3-x} f(x, y) \, dy \, dx$

20. $\displaystyle\int_0^1\int_{v^2}^{\sqrt{v}} F(u, v) \, du \, dv$

21. Evaluate $\displaystyle\int_0^1\int_y^1 \cos{(x^2)} \, dx \, dy$ (Hint: Change the order of integration.)

22. Evaluate $\displaystyle\int_0^1\int_y^1 \frac{x^5}{x^2 + y^2} \, dx \, dy$

23. Evaluate $\displaystyle\iint_R (x + y + \sqrt{xy}) \, dx \, dy$, where R is the region bounded by $x = 0$, $x = y$, and $y = a$, and a is a positive constant.

24. Evaluate $\displaystyle\iint_R y^2 \sqrt{a^2 - x^2} \, dx \, dy$, where R is the region bounded by the circle $x^2 + y^2 = a^2$.

25. Prove that

$$\int_0^a\int_0^x f(x, y) \, dy \, dx = \int_0^a\int_y^a f(x, y) \, dx \, dy$$

assuming that $f(x, y)$ is continuous, where a is a positive constant.

26. As a corollary to Exercise 25, show that if $g(y)$ is continuous

$$\int_0^a\int_0^x g(y) \, dy \, dx = \int_0^a (a - y)g(y) \, dy$$

27. Prove Theorem 3.
28. Prove Theorem 4 and its corollary.
29. Prove Theorem 5.

18.4   MOMENTS, CENTER OF MASS, AND MOMENTS OF INERTIA

The concepts of center of mass (or centroids) for a homogeneous lamina, that is, lamina with constant density was developed in Section 7.8.2. We shall now use double integrals to generalize these concepts to nonhomogeneous laminas.

Consider a lamina that has the shape of a closed region R and suppose that the mass density or mass per unit area is given by a continuous function $\rho = \rho(x, y)$.

Therefore as seen in Section 18.1, the mass M of the lamina is given by

$$M = \lim_{|P| \to 0} \sum_{i=1}^{N} \rho(\xi_i, \eta_i) \, \Delta A_i = \int_R \int \rho(x, y) \, dA \tag{1}$$

where $N = mn$ in the notation of (13) in Section 18.1. If the mass of the ith element of the lamina is assumed to be concentrated at point (ξ_i, η_i), then its moment with respect to the x-axis is given by $\eta_i \rho(\xi_i, \eta_i) \, \Delta A_i$. The total moment of the lamina with respect to the x-axis is defined as the limit of such sums, that is

$$M_x = \lim_{|P| \to 0} \sum_{i=1}^{N} \eta_i \rho(\xi_i, \eta_i) \, \Delta A_i = \int_R \int y \rho(x, y) \, dA \tag{2}$$

Similarly, the moment M_y of the lamina with respect to the y-axis is

$$M_y = \lim_{|P| \to 0} \sum_{i=1}^{N} \xi_i \rho(\xi_i, \eta_i) \, \Delta A_i = \int_R \int x \rho(x, y) \, dA \tag{3}$$

Once M_x, M_y, and M have been determined, the center of mass $(\bar{x}, \bar{y})$ of the lamina is found (by definition) from

$$\bar{x} = \frac{M_y}{M} \qquad \bar{y} = \frac{M_x}{M} \tag{4}$$

EXAMPLE 1   A lamina has the shape of an isosceles right triangle with equal sides of length a, as shown in Figure 18.4.1. Its vertices are at points $(0, 0)$, $(a, 0)$, and $(0, a)$, and in this position its mass density ρ is given by $\rho = \rho_0 xy$ where ρ_0 is a constant. Find a formula for its center of mass.

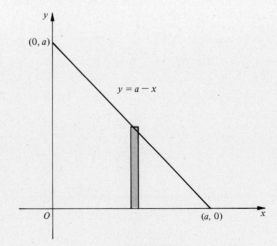

Figure 18.4.1

Solution Its mass is found from

$$M = \rho_0 \iint_R xy \, dA = \rho_0 \int_0^a \int_0^{a-x} xy \, dy \, dx$$

$$= \rho_0 \int_0^a \frac{x}{2} y^2 \Big|_0^{a-x} dx = \frac{1}{2}\rho_0 \int_0^a x(a-x)^2 \, dx$$

$$= \frac{1}{2}\rho_0 \int_0^a (a^2 x - 2ax^2 + x^3) \, dx$$

$$= \frac{1}{2}\rho_0 \left(\frac{a^2 x^2}{2} - \frac{2ax^3}{3} + \frac{x^4}{4}\right) \Big|_0^a = \frac{1}{24}\rho_0 a^4$$

$$M_x = \rho_0 \iint_R xy^2 \, dA = \rho_0 \int_0^a \int_0^{a-x} xy^2 \, dy \, dx \qquad \text{(i)}$$

$$= \rho_0 \int_0^a \frac{xy^3}{3} \Big|_0^{a-x} dx$$

$$= \frac{1}{3}\rho_0 \int_0^a x(a-x)^3 \, dx$$

$$= \frac{1}{3}\rho_0 \int_0^a (a^3 x - 3a^2 x^2 + 3ax^3 - x^4) \, dx$$

$$= \frac{1}{60}\rho_0 a^5$$

$$M_y = \rho_0 \iint_R x^2 y \, dA = \rho_0 \int_0^a \int_0^{a-y} x^2 y \, dx \, dy \qquad \text{(ii)}$$

and from the symmetry of the expressions (i) and (ii)

$$M_y = M_x = \frac{1}{60}\rho_0 a^5$$

Hence from (4), $\bar{x} = \dfrac{2a}{5}$ and $\bar{y} = \dfrac{2a}{5}$. ●

EXAMPLE 2 Find the center of mass of a semicircular lamina that covers the region R: $x^2 + y^2 \le a^2$ and $y \ge 0$, if its density function is given by $\rho(x, y) = \dfrac{1}{a}\rho_0|x|$ where ρ_0 is a constant.

Solution Refer to Figure 18.4.2. The mass M is given by

$$M = \iint_R \frac{1}{a}\rho_0|x| \, dA = \frac{2}{a}\rho_0 \int_0^a \int_0^{\sqrt{a^2-y^2}} x \, dx \, dy$$

$$= \frac{1}{a}\rho_0 \int_0^a (a^2 - y^2) \, dy = \frac{2}{3}\rho_0 a^2$$

$$M_x = \frac{2}{a}\rho_0 \int_0^a \int_0^{\sqrt{a^2-x^2}} xy \, dy \, dx = \frac{1}{a}\rho_0 \int_0^a x(a^2 - x^2) \, dx$$

$$= \frac{1}{4}\rho_0 a^3$$

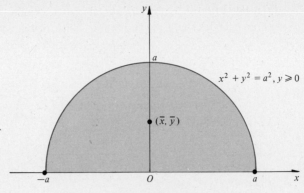

Figure 18.4.2

while $M_y = 0$ by symmetry or direct calculation.† Hence from (4),

$$\bar{x} = 0 \qquad \text{and} \qquad \bar{y} = \tfrac{3}{8}a \qquad\qquad \bullet$$

If n particles of masses $m_1, m_2, \ldots, m_n$ are located at points $(x_1, y_1), (x_2, y_2), \ldots, (x_n, y_n)$, respectively, then the resultant moments with respect to the x- and y-axis were defined in equations (4) and (3) of Section 7.8 by

$$M_x = \sum_{i=1}^{n} m_i y_i \qquad \text{and} \qquad M_y = \sum_{i=1}^{n} m_i x_i \tag{5}$$

These are also called the **first moments** of the system with respect to the coordinate axes. If instead we use the *squares* of the distances from the coordinate axes and form the analogous sum of products, we obtain the **second moments** or **moments of inertia** I_x and I_y with respect to the x- and y-axes, respectively. Hence, by definition,

$$I_x = \sum_{i=1}^{n} m_i y_i^2 \qquad \text{and} \qquad I_y = \sum_{i=1}^{n} m_i x_i^2 \tag{6}$$

The extension of (6) to laminas is straightforward. Suppose that the density $\rho(x, y)$ is a continuous function, then the moments of inertia with respect to the x- and y-axes are defined by

and

$$I_x = \lim_{|P| \to 0} \sum_i \eta_i^2 \rho(\xi_i, \eta_i)\, \Delta A_i = \int\!\!\int_R y^2 \rho(x, y)\, dA$$

$$I_y = \lim_{|P| \to 0} \sum_i \xi_i^2 \rho(\xi_i, \eta_i)\, \Delta A_i = \int\!\!\int_R x^2 \rho(x, y)\, dA \tag{7}$$

If the approximate mass of the element $\rho(\xi_i, \eta_i)\, \Delta A_i$ is multiplied by the square of the distance $\xi_i^2 + \eta_i^2$ from the origin O to the representative point (ξ_i, η_i) and

---

† If the density function $\rho(x, y)$ and the geometry are symmetric about an axis, then the moment with respect to that axis is zero and the center of gravity must be on that axis. The y-axis is the axis of symmetry in the example.

then the limit of a sum of such terms is formed, we obtain (by definition) the moment of inertia I_o of the lamina with respect to the origin, We have

$$I_o = \iint_R (x^2 + y^2)\rho(x, y)\, dA \tag{8}$$

I_o is called the **polar moment of inertia** of the lamina. Furthermore, from (7) and (8), it follows that

$$I_o = I_x + I_y \tag{9}$$

EXAMPLE 3   For the lamina of Example 1, find I_x, I_y, and I_o.

Solution
$$I_x = \iint_R y^2 \rho(x, y)\, dA = \rho_0 \int_0^a \int_0^{a-x} xy^3\, dy\, dx$$

$$= \tfrac{1}{4}\rho_0 \int_0^a xy^4 \Big|_0^{a-x} dx$$

$$= \tfrac{1}{4}\rho_0 \int_0^a x(a - x)^4\, dx$$

$$= \tfrac{1}{4}\rho_0 \int_0^a [a^4 x - 4a^3 x^2 + 6a^2 x^3 - 4ax^4 + x^5]\, dx$$

$$= \tfrac{1}{120}\rho_0 a^6$$

$$I_y = \iint_R x^2 \rho(x, y)\, dA = \rho_0 \int_0^a \int_0^{a-y} x^3 y\, dx\, dy$$

and from the symmetry of the density ρ and the geometry, the reader may verify that

$$I_y = I_x = \tfrac{1}{120}\rho_0 a^6$$

Also,

$$I_o = I_x + I_y = \tfrac{1}{60}\rho_0 a^6 \qquad \bullet$$

EXAMPLE 4   For a homogeneous lamina of constant density ρ_0 in the shape of a circle of radius a, find the polar moment of inertia with respect to the center of the circle O.

Solution   Refer to Figure 18.4.3. From symmetry we have

$$I_x = 4\rho_0 \int_0^a \int_0^{\sqrt{a^2 - x^2}} y^2\, dy\, dx$$

$$= \tfrac{4}{3}\rho_0 \int_0^a (a^2 - x^2)^{3/2}\, dx$$

Let $x = a \sin\theta$, $dx = a \cos\theta\, d\theta$, thus

$$I_x = \frac{4}{3}\rho_0 a^4 \int_0^{\pi/2} \cos^4\theta\, d\theta$$

$$= \frac{4}{3}\rho_0 a^4 \frac{3}{16}\pi = \frac{1}{4}\pi\rho_0 a^4$$

$$= (\pi a^2 \rho_0)\frac{1}{4}a^2 = \frac{1}{4}Ma^2$$

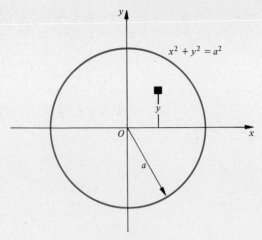

$$x^2 + y^2 = a^2$$

Figure 18.4.3

where M is the total mass of the lamina. But, from the symmetry $I_y = I_x$ so that

$$I_o = \tfrac{1}{2}Ma^2 \qquad \bullet$$

The moment of inertia of a body with respect to an axis is a measure of its resistance to change in the rotational speed about the line. If a wheel is mounted on a smooth shaft as shown in Figure 18.4.4 and a force F is acting tangentially to the lamina as shown, then this force produces a torque $T = Fa$. It is shown in rigid body mechanics that if θ is the angle of rotation, and I_o is the moment of inertia of the wheel about the axis of the shaft containing point O, then

$$T = Fa = I_o\alpha \qquad (10)$$

where

$$\alpha = D_t^2\theta \qquad (11)$$

is the angular acceleration of the wheel or circular lamina. If the force and therefore the torque T is constant, then from (10)

$$\alpha = T/I_o \qquad (12)$$

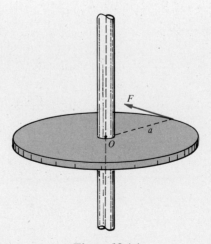

Figure 18.4.4

so that the angular acceleration is inversely proportional to I_o. Therefore I_o is a measure of the resistance of the wheel to being given an angular acceleration in much the same way as mass of a body is a measure of the resistance to being linearly accelerated. Hence the term moment of inertia is used in these dynamic applications.

Also, moment of inertia of area (or the second moment) is significant in the design of beams for buildings and bridges. A greater moment of inertia in general means less deflections and internal forces per unit area for the same loading and support configuration. Hence civil engineers extensively use I beams (wide flange beams) rather than beams of rectangular cross section of the same area (Figure 18.4.5).

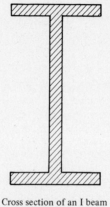

Cross section of an I beam

Figure 18.4.5

In Figure 18.4.6 we are given a lamina and two sets of parallel axes, where one set passes through the centroid $(\overline{x}, \overline{y})$ of the lamina. We seek a relationship between I_x and I_ξ.

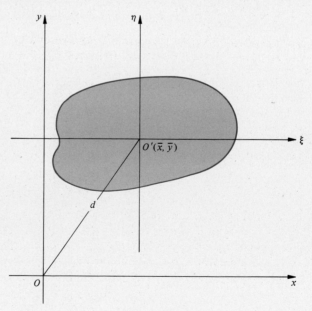

Figure 18.4.6

By definition,

$$I_x = \int\!\!\int_R y^2 \rho(x, y)\, dA$$

$$= \int\!\!\int_R (y - \overline{y} + \overline{y})^2 \rho(x, y)\, dA$$

$$= \int\!\!\int_R (y - \overline{y})^2 \rho(x, y)\, dA + 2\overline{y} \int\!\!\int_R (y - \overline{y})\rho(x, y)\, dA + \overline{y}^2 \int\!\!\int_R \rho(x, y)\, dA$$

Now,

$$\int\!\!\int_R y\rho(x, y)\, dA = M_x \qquad \text{and} \qquad \int\!\!\int_R \overline{y}\rho(x, y)\, dA = \overline{y} \int\!\!\int_R \rho(x, y)\, dA = \overline{y}M$$

and since $\overline{y} = M_x/M$ (by definition) the second double integral is zero. Also

$$I_\xi = \int\!\!\int_R (y - \overline{y})^2 \rho(x, y)\, dA$$

so that we have

$$I_x = I_\xi + \overline{y}^2 M \qquad\qquad (13)$$

Similarly,

$$I_y = I_\eta + \overline{x}^2 M \qquad\qquad (14)$$

and

$$I_o = I_x + I_y = I_\xi + I_\eta + (\overline{x}^2 + \overline{y}^2)M$$

$$I_o = I_{o'} + d^2 M \qquad\qquad (15)$$

where o' is the origin of the $\xi\eta$-system of coordinates and d is the distance between o and o'.

Equations (13), (14), and (15) are known as the **parallel axes or transfer theorems.** From (13) and (14) we have $I_\xi \leq I_x$ and $I_\eta \leq I_y$, which means that, of all parallel axes, the one that passes through the centroid of the lamina yields the least moment of inertia. Also, if we know the moment of inertia with respect to orthogonal axes passing through the centroid, then the parallel axes theorem yields the moments of inertia with respect to axes parallel to the centroidal axes.

Allied to the concept of moment of inertia is that of radius of gyration of a body with respect to a given axis. The distance R defined by the relation

$$I = MR^2 \qquad\qquad (16)$$

is the **radius of gyration** of the body with respect to a particular axis, where I is the moment of inertia with respect to that axis and M is the total mass of the body. It is the distance from the axis in question at which a single particle of mass M must be placed in order to have the same moment of inertia as the distributed mass with respect to that axis. From (16), the radius of gyration is given by

$$R = \sqrt{I/M} \qquad\qquad (17)$$

EXAMPLE 5   Find the moment of inertia of a rectangular lamina of sides a and b and uniform density ρ_0 about the side of length a. Find the radius of gyration with respect to that edge. What is the moment of inertia with respect to an axis through the centroid and parallel to the edges of length a?

Solution   The rectangle is shown in Figure 18.4.7. We seek I_x.

$$I_x = \int_0^b \int_0^a \rho_0 y^2 \, dx \, dy = \int_0^b \rho_0 y^2 x \Big|_0^a \, dy$$

$$= \rho_0 a \int_0^b y^2 \, dy = \tfrac{1}{3}\rho_0 a b^3$$

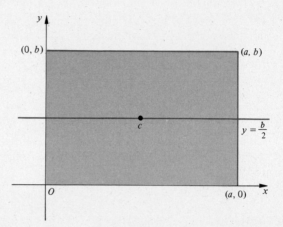

Figure 18.4.7

The total mass M is $\int_0^b \int_0^a \rho_0 \, dx \, dy = \rho_0 ab$, so that the radius of gyration with respect to the x-axis is from (17),

$$R_x = \left(\frac{\rho_0 a b^3/3}{\rho_0 ab}\right)^{1/2} = \frac{b}{\sqrt{3}}$$

From the parallel axis theorem (13),

$$I_{y=b/2} + M\left(\frac{b}{2}\right)^2 = I_x$$

so that

$$I_{y=b/2} = \frac{1}{3}\rho_0 a b^3 - \rho_0 ab\left(\frac{b}{2}\right)^2 = \frac{1}{12}\rho_0 a b^3 \qquad \bullet$$

Exercises 18.4

In Exercises 1 through 4, a metal sheet (or lamina) has the shape of the region bounded by the given lines and has the given surface density $\rho(x, y)$, where ρ_0 and a are positive constants. Sketch the region. Find the coordinates of the center of mass.

1. $x = 0, \quad x = a, \quad y = 0, \quad y = a, \quad \rho = \dfrac{1}{a}\rho_0(x + y)$

2. $y = 0$, $\quad x = a$, $\quad y = x$, $\quad \rho = \dfrac{1}{a}\rho_0 y$

3. $y = x$, $\quad y = a$, $\quad x = 0$, $\quad \rho = \dfrac{1}{a}\rho_0 x$

4. $x = 0$, $\quad y = 0$, $\quad x + y = a$, $\quad \rho = \rho_0\left(1 + \dfrac{x}{a}\right)$

In Exercises 5 through 8, find the center of mass of the given region where ρ_0 and a are positive constants.

5. The region bounded by $y = \dfrac{x^2}{a}$, the x-axis, and $x = a$, with surface density $\rho = \dfrac{1}{a^2}\rho_0 xy$ at (x, y).

6. The region bounded by $y = \dfrac{1}{a}x^2$, the x-axis, and $x = a$ with surface density $\rho = \dfrac{1}{a}\rho_0(x + y)$ at (x, y).

7. The region bounded by $y = \sqrt{a^2 - x^2}$ and the x-axis with surface density $\rho = \dfrac{1}{a^2}\rho_0 y^2$ at (x, y).

8. The region bounded by $ay = x^2$ and $y = 4a$ with surface density $\rho = \dfrac{1}{a^2}\rho_0 x^2$ at (x, y).

9. Find I_x, I_y, and I_o for the lamina of Exercise 1.
10. Find I_x, I_y, and I_o for the lamina of Exercise 2.
11. Find I_x, I_y, and I_o for the lamina of Exercise 3.
12. Find I_x, I_y, and I_o for the lamina of Exercise 4.
13. A plane lamina covers the region R bounded by $y = x^{1/2}$, $y = 0$, and $x = b$, where b is a positive constant. Find I_x, the mass M, and the radius of gyration R_x with respect to the x-axis if the density ρ is proportional to its distance from the y-axis.
14. For the homogeneous lamina ($\rho = \rho_0 = $ constant) defined by $y = 2x - x^2$ and $y = 0$, find M, $\bar{x}$, $\bar{y}$, I_y, and R_y.
15. Find the total mass and the moment of inertia of a semicircular lamina of radius a about its diameter if the density of the lamina at a point is proportional to the distance of the point from the diameter. What is its radius of gyration with respect to the diameter?
16. Find the moment of inertia of a homogeneous circular lamina of radius a (density $\rho = \rho_0 = $ constant) about a tangent to the boundary by two methods. (*Hint:* Use Example 4 in the text and the parallel axis theorem.)
17. Calculate the moments of inertia and radii of gyration about the major and minor axes of a thin homogeneous elliptic lamina of density $\rho = \rho_0 = $ constant and with major and minor axes $2a$ and $2b$, respectively, in the x- and y-directions.
18. Calculate the moment of inertia of a square lamina of side a and constant density ρ_0 with respect to a diagonal.
19. Show that for any plane lamina, the moment of inertia I_θ about the line with inclination θ with the x-axis and passing through the origin is

$$I_\theta = I_x \cos^2\theta - I_{xy}\sin 2\theta + I_y\sin^2\theta$$

where I_{xy} is the **product of inertia** defined by

$$I_{xy} = \int\!\!\int_R xy\rho\,dA$$

(*Hint:* Use a formula for rotation of coordinates in the xy-plane.)
20. Prove that the sum of the moments of inertia of a lamina R about any two perpendicular lines in the plane through a point O is a constant. (*Hint:* Use the result of Exercise 19.)
21. Prove that, at an arbitrary point, two orthogonal axes can be chosen so that $I_{xy} = 0$. Such axes are called **principal axes** at that point and the corresponding moment of inertias are called principal moments of inertia. If x- and y- are principal axes and $I_x > I_y$ what are the maximum and minimum moments of inertia.

DOUBLE INTEGRALS USING POLAR COORDINATES

The double integral using rectangular coordinates and iterated integration has been developed in this chapter. In rectangular coordinates the element is the rectangle with area $\Delta x \, \Delta y$.

For regions described by curves expressed conveniently in polar coordinates another choice for the element of area is desirable. In a manner similar to the rectangular coordinate treatment, we shall deal with regions that are bounded by curves given in polar coordinates by the smooth functions $r = g(\theta)$ and $r = h(\theta)$ defined on the closed interval $\alpha \leq \theta \leq \beta$ over which $0 \leq g(\theta) \leq h(\theta)$ as shown in Figure 18.5.1.

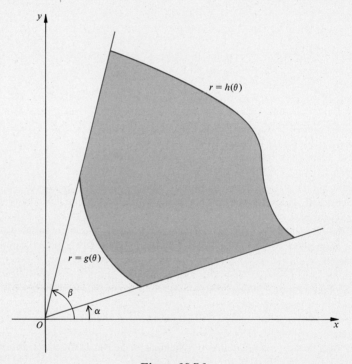

Figure 18.5.1

We partition this region by drawing rays through the pole and circles with center at the pole (not necessarily equally spaced) as shown in Figure 18.5.2. The interior subregions ΔR_{ij} are truncated sectors, that is, the difference of two sectors, with area

$$\Delta A_{ij} = \tfrac{1}{2}[r_i^2(\theta_j - \theta_{j-1}) - r_{i-1}{}^2(\theta_j - \theta_{j-1})]$$

and factoring the difference of two squares yields

$$\Delta A_{ij} = \tfrac{1}{2}(r_i - r_{i-1})(r_i + r_{i-1})(\theta_j - \theta_{j-1})$$

Letting $\bar{r}_i = \tfrac{1}{2}(r_i + r_{i-1})$, $\Delta r_i = r_i - r_{i-1}$ and $\Delta \theta_j = \theta_j - \theta_{j-1}$ we have

$$\Delta A_{ij} = \bar{r}_i \, \Delta r_i \, \Delta \theta_j \qquad (1)$$

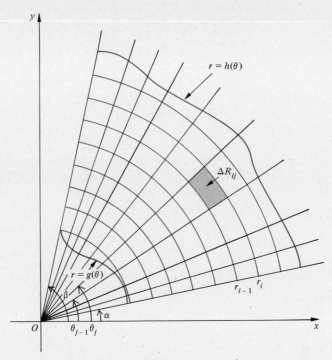

Figure 18.5.2

We then choose $(\bar{r}_i, \bar{\theta}_j)$ in each ijth subregion where $\theta_{j-1} \leq \bar{\theta}_j \leq \theta_j$ and form the Riemann sum over all the interior regions ΔR_{ij}

$$\sum_{i,j} f(\bar{r}_i, \bar{\theta}_j)\, \Delta A_{ij} = \sum_{i,j} f(\bar{r}_i, \bar{\theta}_j)\bar{r}_i \, \Delta r_i \, \Delta \theta_j \qquad (2)$$

It can then be proved that if $f(r, \theta)$ is a continuous function of r and θ in the region R and if $|P|$ is once again the norm of the interior mesh that

$$\lim_{|P| \to 0} \sum_{i,j} f(\bar{r}_i, \bar{\theta}_j)\, \Delta A_{ij} = \int\!\!\int_R f(r, \theta)\, dA \qquad (3)$$

or from (2),

$$\lim_{|P| \to 0} \sum_{i,j} f(\bar{r}_i, \bar{\theta}_j)\, \Delta A_{ij} = \int\!\!\int_R f(r, \theta)r \, dr \, d\theta \qquad (4)$$

Just as with rectangular coordinates, we evaluate the double integral in (4) as an iterated integral,

$$\int\!\!\int_R f(r, \theta)r \, dr \, d\theta = \int_\alpha^\beta \int_{g(\theta)}^{h(\theta)} f(r, \theta)r \, dr \, d\theta \qquad (5)$$

Similarly, if the region is bounded by two circular arcs $r = a$ and $r = b$ and two smooth curves $\theta = g(r)$ and $\theta = h(r)$ where $g(r) \leq h(r)$ (as shown in Figure 18.5.3),

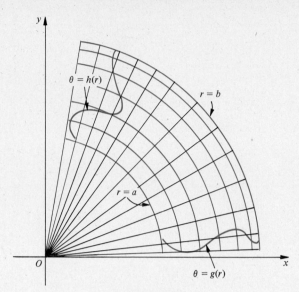

Figure 18.5.3

then

$$\iint_R f(r, \theta)\, dA = \int_a^b \int_{g(r)}^{h(r)} f(r, \theta) r\, d\theta\, dr \qquad (6)$$

EXAMPLE 1   Evaluate $I = \displaystyle\int_0^a \int_0^{\sqrt{a^2 - x^2}} (x^2 + y^2)\, dy\, dx$ where a is a positive constant by transforming to polar coordinates.

Solution   The region of integration is the portion of the circle $x^2 + y^2 = a^2$, which lies in the first quadrant (Figure 18.5.4). We can do this problem by performing the repeated integration in either order.

Method 1. If we integrate first with respect to r (Figure 18.5.4(a))

$$I = \int_0^{\pi/2} \int_0^a r^2 r\, dr\, d\theta$$

$$= \int_0^{\pi/2} \int_0^a r^3\, dr\, d\theta = \int_0^{\pi/2} \left. \frac{r^4}{4} \right|_0^a d\theta$$

$$= \frac{a^4}{4} \int_0^{\pi/2} d\theta = \frac{\pi a^4}{8}$$

Method 2. If we integrate first with respect to θ (Figure 18.5.4(b))

$$I = \int_0^a \int_0^{\pi/2} r^3\, d\theta\, dr$$

$$= \int_0^a \left. r^3 \theta \right|_0^{\pi/2} dr$$

$$= \int_0^a \frac{\pi}{2} r^3\, dr = \left. \frac{\pi}{8} r^4 \right|_0^a = \frac{\pi a^4}{8}$$

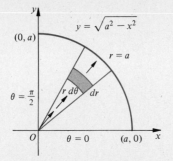

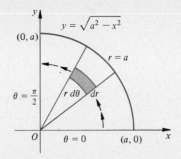

(a) Integrating first with respect to r (b) Integrating first with respect to θ

Figure 18.5.4

EXAMPLE 2   Find the center of mass of a semicircular lamina that covers the region R: $x^2 + y^2 \leq a^2$ and $y \geq 0$ if its density function is given by $\rho(x, y) = \dfrac{1}{a}\rho_0|x|$ where ρ_0 is a constant.

Solution   This is the same example as Example 2 of Section 18.4, where this problem was solved utilizing rectangular coordinates. Hence refer to Figure 18.4.2. From the symmetry with respect to the y-axis, the mass M is given by

$$M = \frac{1}{a}\rho_0 \iint_R |x|\, dA = \frac{2}{a}\rho_0 \int_0^{\pi/2} \int_0^a r\cos\theta \; r\, dr\, d\theta$$

$$= \frac{2}{a}\rho_0 \int_0^{\pi/2} \frac{r^3}{3}\bigg|_0^a \cos\theta \, d\theta = \frac{2}{3}a^2\rho_0 \int_0^{\pi/2} \cos\theta \, d\theta$$

$$= \frac{2}{3}a^2\rho_0 \sin\theta \bigg|_0^{\pi/2} = \frac{2}{3}a^2\rho_0$$

Also, the moment with respect to the x-axis is

$$M_x = \frac{1}{a}\rho_0 \int_0^{\pi/2} \int_0^a r^2 \sin 2\theta \; r\, dr\, d\theta$$

since $xy = (r\cos\theta)(r\sin\theta) = r^2 \sin\theta \cos\theta = \tfrac{1}{2}r^2 \sin 2\theta$

$$M_x = \frac{1}{a}\rho_0 \int_0^{\pi/2} \left(\int_0^a r^3 \, dr\right) \sin 2\theta \, d\theta$$

$$= \frac{1}{4}a^3\rho_0 \int_0^{\pi/2} \sin 2\theta \, d\theta = \frac{1}{8}a^3\rho_0(-\cos 2\theta) \bigg|_0^{\pi/2}$$

$$= \frac{1}{4}a^3\rho_0$$

Since $M_y = 0$, we have

$$\bar{x} = 0 \quad \text{and} \quad \bar{y} = \tfrac{3}{8}a$$ ●

EXAMPLE 3   Find the area which is outside the circle $r = a$ and inside the cardioid $r = a(1 + \cos\theta)$, where $a > 0$.

Solution   Because of symmetry with respect to the polar axis (Figure 18.5.5),

$$A = 2 \int_0^{\pi/2} \int_a^{a(1+\cos\theta)} r \, dr \, d\theta$$

$$= 2 \int_0^{\pi/2} \left[\frac{a^2}{2}(1 + \cos\theta)^2 - \frac{a^2}{2} \right] d\theta$$

$$= a^2 \int_0^{\pi/2} [1 + 2\cos\theta + \cos^2\theta - 1] \, d\theta$$

$$= a^2 \int_0^{\pi/2} [2\cos\theta + (\tfrac{1}{2} + \tfrac{1}{2}\cos 2\theta)] \, d\theta$$

$$= a^2 \left(2\sin\theta + \frac{\theta}{2} + \frac{\sin 2\theta}{4} \right) \Big|_0^{\pi/2}$$

$$= a^2 \left(2 + \frac{\pi}{4} \right) \text{ sq units} \qquad \bullet$$

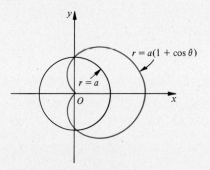

$r = a(1 + \cos\theta)$

$r = a$

Figure 18.5.5

EXAMPLE 4   Evaluate the double integral $I = \iint_R e^{-(x^2+y^2)} \, dx \, dy$ where R is the circle $x^2 + y^2 = a^2$.

Solution   If we try to calculate the integral as a repeated integral in rectangular coordinates, an integral occurs that cannot be performed—namely, $\int e^{-x^2} \, dx$. Fortunately by introducing polar coordinates, we have

$$I = \int_0^{2\pi} \int_0^a e^{-r^2} r \, dr \, d\theta$$

and with the introduction of the r, the integrations are routine,

$$I = \int_0^{2\pi} -\frac{1}{2} e^{-r^2} \Big|_0^a \, d\theta$$

$$= \frac{1}{2}(1 - e^{-a^2}) \int_0^{2\pi} d\theta$$

$$= \pi(1 - e^{-a^2}) \qquad \bullet$$

EXAMPLE 5   Find the volume cut from the upper half of the sphere $x^2 + y^2 + z^2 = 9$ by the cylinder $x^2 + y^2 - 3x = 0$ (Figure 18.5.6).

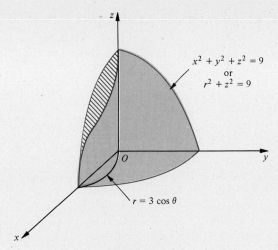

$$x^2 + y^2 + z^2 = 9$$
or
$$r^2 + z^2 = 9$$

$r = 3 \cos \theta$

Figure 18.5.6

Solution   The equation of the sphere in cylindrical coordinates is $r^2 + z^2 = 9$. The desired volume is of the form

$$V = \int\int_R zr \, dr \, d\theta = \int\int_R \sqrt{9 - r^2} \, r \, dr \, d\theta$$

Half of the region R is bounded by the semicircle $x^2 + y^2 - 3x = 0$ in the first quadrant in the xy-plane. This becomes $r = 3 \cos \theta$ in polar coordinates. Then

$$V = 2 \int_0^{\pi/2} \int_0^{3 \cos \theta} (9 - r^2)^{1/2} \, r \, dr \, d\theta$$

$$= 2 \int_0^{\pi/2} -\tfrac{1}{3}(9 - r^2)^{3/2} \big|_0^{3 \cos \theta}$$

$$= -\frac{2}{3} \int_0^{\pi/2} [(9 - 9 \cos^2 \theta)^{3/2} - 27] \, d\theta$$

$$= -18 \int_0^{\pi/2} (\sin^3 \theta - 1) \, d\theta$$

From the tables, we have

$$V = -18[\tfrac{1}{3} \cos^3 \theta - \cos \theta - \theta]\big|_0^{\pi/2}$$
$$= 3(3\pi - 4) \text{ cubic units}$$

●

Exercises 18.5

In Exercises 1 through 6, evaluate the double integrals by changing to polar coordinates (a is a positive constant).

1. $\displaystyle\int_0^a \int_0^{\sqrt{a^2 - x^2}} x \, dy \, dx$

2. $\displaystyle\int_0^4 \int_{-\sqrt{16 - x^2}}^{\sqrt{16 - x^2}} dy \, dx$

3. $\displaystyle\int_0^1 \int_0^x dy \, dx$

4. $\displaystyle\int_0^a \int_0^{\sqrt{ax - x^2}} x \, dy \, dx$

5. $\displaystyle\int_{-a}^a \int_{-\sqrt{a^2 - y^2}}^{\sqrt{a^2 - y^2}} \cos(x^2 + y^2) \, dx \, dy$

6. $\displaystyle\int_0^1 \int_x^{\sqrt{x}} (x^2 + y^2)^{-1/2} \, dy \, dx$

In Exercises 7 through 14, find the areas of the given regions by double integration using polar coordinates.

7. Inside $r = 4$ and outside the parabola $r = 2 \sec^2 \dfrac{\theta}{2}$.

8. Inside the circle $r = 6 \cos \theta$ and outside the circle $r = 3$.
9. Inside the cardioid $r = 10(1 + \sin \theta)$ and outside the circle $r = 10$.
10. One leaf of the curve $r = a \sin 2\theta$.

11. Bounded by $r = 4 \sec \theta$, $\theta = -\dfrac{\pi}{3}$, and $\theta = \dfrac{\pi}{4}$.

12. Bounded by $r = a \cos \theta + b \sin \theta$.
13. Inside the two curves $r = a \cos \theta$ and $r = b \sin \theta$, where a and b are positive constants. In particular, determine the area when $a = b$.
14. One leaf of the region inside $r = 2a \sin 3\theta$ and outside of $r = a$, where a is a positive constant.
15. Find the x and y coordinates of the centroid of the homogeneous lamina that is bounded by the semicircle $r = a$, $a > 0$, in the first and fourth quadrants and the y-axis.
16. Find the x and y coordinates of the centroid of the area bounded by the cardioid $r = a(1 + \cos \theta)$ where a is a positive constant.
17. Find the mass of the lamina inside one leaf of the curve $r = a \sin 2\theta$ if the density at any given point (r, θ) is given by $\rho = \rho_0 a^{-2}(x^2 + y^2)$ where ρ_0 and a are positive constants, using polar coordinates and multiple integration.
18. Find I_x of the lamina covering the region bounded by $r = 2a \cos \theta$ if the density at any point (r, θ) is given by $\rho = \rho_0 a^{-2}(x^2 + y^2)$, where ρ_0 and a are positive constants.
19. Find I_o for the homogeneous lamina ($\rho = \rho_o =$ constant) bounded by the cardioid $r = a(1 + \cos \theta)$.
20. Find the mass of the lamina covering the region bounded by the upper half of the cardioid $r = a(1 + \cos \theta)$ if the density at any point (r, θ) is given by $\rho = \rho_o a^{-2}xy$.

In Exercises 21 through 31, cylindrical coordinates and double integration are to be applied for the determination of the following volumes.

21. The volume of a hemisphere of radius a.
22. The volume of the solid under the surface $z = 9 - r^2$ and above the region R in the xy-plane bounded by the circle $x^2 + y^2 = 4$.
23. The volume of the solid under the surface $z = 4 - r$ and above the region R in the xy-plane bounded by the curve $r = 3 \sin \theta$.
24. The volume of the solid bounded above by the plane $z = x + 5$, below by the xy-plane, and on the sides by the cylinder $x^2 + y^2 = 9$.
25. The volume of the solid bounded above by the spherical surface $x^2 + y^2 + z^2 = a^2$, below by the xy-plane, and on the sides by the cylinder $4(x^2 + y^2) = a^2$, where a is a positive constant.
26. The volume of the solid bounded by the sphere $x^2 + y^2 + z^2 = a^2$ and the cylinder $x^2 + y^2 = 2ax$, where a is a positive constant.
27. The volume of the solid bounded by the cone $z = \sqrt{x^2 + y^2}$ and the paraboloid of revolution $z = \frac{1}{5}(x^2 + y^2)$.
28. The volume of the right circular cone with a as its base radius and h as the altitude. Let the base of radius a lie in the xy-plane and the altitude of the cone coincide with the z-axis. (*Hint:* Use similar triangles to find an equation of the cone.)
29. The volume of the solid capped by the cone $z = h\left(1 - \dfrac{r}{a}\right)$ and based on the circle $r = a \cos \theta$ where h and a are positive constants.
30. The volume of the solid capped by the paraboloid $bz = x^2 + y^2$ and based on the region R which lies inside the circle $r = 2a \cos \theta$ and outside the circle $r = a$, where b and a are positive constants.
31. The volume of the solid drilled out of a sphere of radius a by a bit of radius b if the axis of the bit passes through the center of the sphere.

32. Use Example 4 in the text to find

(i) $\displaystyle\int_0^\infty e^{-x^2}\,dx = \lim_{a\to\infty}\int_0^a e^{-x^2}\,dx$

(ii) $\displaystyle\int_{-\infty}^\infty e^{-x^2}\,dx.$

18.6   SURFACE AREA

In this section we shall intuitively develop a formula for surface area as a double integral. We will consider a surface explicitly defined by an equation $z = f(x, y)$ or implicitly by $F(x, y, z) = 0$ over a domain of definition R in the xy-plane which is suitable for the calculation of a double integral. In any case, it is assumed that all partial derivatives of f and F are continuous and that $f(x, y) \geq 0$ on the region R, while in the implicit form it is further assumed that $F_z \neq 0$ at all interior points of R. Under these conditions the surface area may be defined and computed as a double integral.

Partition the region R into N simple subregions of which ΔR_i, with area ΔA_i, is a typical subregion with (x_i, y_i) as an arbitrary point in ΔR_i (see Figure 18.6.1). This determines a point $P(x_i, y_i, z_i)$, which is on the surface and whose tangent plane at P with normal $\mathbf{n}$ has direction numbers

$$F_x|_P : F_y|_P : F_z|_P$$

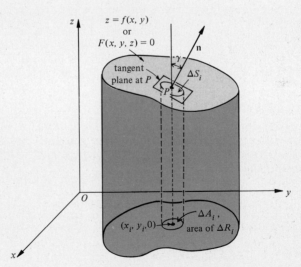

Figure 18.6.1

With ΔR_i as a base, we form a cylindrical column that intersects the tangent plane at P in an area ΔS_i. If γ denotes an angle between the normal to the surface

S at P and the z-axis, then, since ΔA_i is the projection of ΔS_i on the xy-plane, we have

$$\Delta A_i = \Delta S_i \, |\cos \gamma| \tag{1}$$

The surface area we seek is defined by

$$S = \lim_{|P| \to 0} \sum_{i=1}^{N} \Delta S_i \tag{2}$$

where $|P|$ is the maximum diameter for all of the subregions ΔR_i, $(i = 1, 2, 3, \ldots, N)$, that is, the norm of the partition.

Application of the dot product of the two vectors $F_x \mathbf{i} + F_y \mathbf{j} + F_z \mathbf{k}$ and $\mathbf{k}$ yields

$$\cos \gamma = \frac{\mathbf{n} \cdot \mathbf{k}}{|\mathbf{n}|} = \frac{F_z}{\sqrt{F_x^2 + F_y^2 + F_z^2}} \tag{3}$$

and substitution of (3) into (1) and (2) yields

$$\begin{aligned} S &= \lim_{|\Delta P| \to 0} \sum_{i=1}^{N} \frac{\sqrt{F_x^2 + F_y^2 + F_z^2}}{|F_z|} \Delta A_i \\ &= \int_R \int \frac{\sqrt{F_x^2 + F_y^2 + F_z^2}}{|F_z|} \, dA \end{aligned} \tag{4}$$

From (4) with $F(x, y, z) = f(x, y) - z$, we obtain (since $F_x = f_x$, $F_y = f_y$, and $F_z = -1$)

$$S = \int_R \int \sqrt{[f_x(x, y)]^2 + [f_y(x, y)]^2 + 1} \, dA \tag{5}$$

EXAMPLE 1   Let R be the triangular region in the xy-plane with vertices $(0, 0, 0)$, $(0, 1, 0)$ and $(2, 1, 0)$. Compute the surface area of that part of the graph of $z = x + y^2$ which is based on R.

Solution   The region R is bounded by the graphs of $x = 2y$, $y = 1$, and $x = 0$ (Figure 18.6.2). With $f(x, y) = x + y^2$, we have $f_x = 1$, $f_y = 2y$, and substitution into (5) yields

$$\begin{aligned} S &= \int_R \int (2 + 4y^2)^{1/2} \, dA \\ &= \int_0^1 \int_0^{2y} (2 + 4y^2)^{1/2} \, dx \, dy \\ &= \int_0^1 (2 + 4y^2)^{1/2} \, x \big|_0^{2y} \, dy \\ &= \int_0^1 (2 + 4y^2)^{1/2} \, 2y \, dy \\ &= \tfrac{1}{4} \cdot \tfrac{2}{3} (2 + 4y^2)^{3/2} \big|_0^1 = \tfrac{1}{6}(6^{3/2} - 2^{3/2}) \text{ sq units} \qquad \bullet \end{aligned}$$

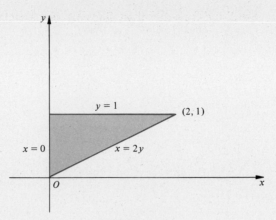

y = 1

(2, 1)

x = 0

x = 2y

O

Figure 18.6.2

EXAMPLE 2 Find the area of the surface $z = 1 + x^2 + \sqrt{3}y$ that lies over the rectangle $0 \le x \le 2, 0 \le y \le 3$.

Solution Here, $f_x = 2x$ and $f_y = \sqrt{3}$, so that from (5),

$$S = \int_0^2 \int_0^3 (4x^2 + 3 + 1)^{1/2} \, dy \, dx$$

$$= 6 \int_0^2 (1 + x^2)^{1/2} \, dx$$

$$= 3\{x(1 + x^2)^{1/2} + \ln[x + (1 + x^2)^{1/2}]\}|_0^2$$

$$= 3[2\sqrt{5} + \ln(2 + \sqrt{5})] \text{ sq units}$$

●

EXAMPLE 3 Calculate the surface area of the portion of a sphere $z = f(x, y) = \sqrt{a^2 - x^2 - y^2}$ in the first quadrant over a circle in the xy-plane $x^2 + y^2 = b^2$, where $0 < b < a$ (see Figure 18.6.3). Use this result to find the surface of the sphere of radius a.

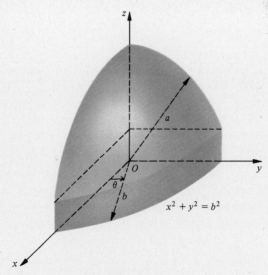

z

a

O

y

θ

b

$x^2 + y^2 = b^2$

x

Figure 18.6.3

Solution $f_x = -\dfrac{x}{z}$ and $f_y = -\dfrac{y}{z}$ so that

$$\sqrt{1 + f_x^2 + f_y^2} = \sqrt{1 + \frac{x^2 + y^2}{z^2}} = \frac{a}{z}$$

since $x^2 + y^2 = a^2 - z^2$. Hence

$$S = \iint_R \frac{a}{z}\, dA$$

and with the introduction of polar coordinates

$$S = \int_0^{\pi/2} \int_0^b \frac{a}{\sqrt{a^2 - r^2}}\, r\, dr\, d\theta$$

$$= \int_0^{\pi/2} -a(a^2 - r^2)^{1/2}\big|_0^b\, d\theta$$

$$= -a[(a^2 - b^2)^{1/2} - a]\frac{\pi}{2}$$

$$= \frac{\pi a^2}{2} - \frac{\pi a}{2}(a^2 - b^2)^{1/2}$$

Now if we let $b \to a^-$ then we obtain the surface of one eighth of a sphere, namely, $\dfrac{\pi a^2}{2}$ so that

$$S = 4\pi a^2 = \text{surface area of a sphere} \qquad \bullet$$

Exercises 18.6

In Exercises 1 through 10, find the area of the portion of the given surface which lies over the specified region in the xy-plane.

1. $z = 3 + 2x + 5y$, $0 \le y \le x$, $0 \le x \le 4$
2. $z = a + bx + cy$, $0 \le y \le x^3$, $0 \le x \le 2$, where a, b and c are positive constants.
3. $z = \dfrac{x^2}{2}$, $0 \le y \le x$, $0 \le x \le 3$
4. $x + 2y + 3z - 10 = 0$, $x = 1$, $y = 1$, $x + y = 3$
5. $z - y^2 - 2 = 0$, $0 \le x \le 2y$, $1 \le y \le 2$
6. $z = x^{3/2} + y^{3/2}$, $0 \le x \le 1$, $0 \le y \le 2$
7. $z = 2 + y + x^2$, $0 \le y \le 3x$, $0 \le x \le 4$
8. $z = x + y^{5/2}$, $0 \le x \le y^2$, $1 \le y \le 2$
9. $z - 3 = xy$, $0 \le y \le \sqrt{4a^2 - x^2}$, $0 \le x \le 2a$, where a is a positive constant
10. $z = \sqrt{x^2 + y^2}$, $0 \le y \le \sqrt{9a^2 - x^2}$, $-3a \le x \le 3a$, where a is a positive constant

In Exercises 11 through 18, find the area of the surface described.

11. The portion of the plane $\dfrac{x}{a} + \dfrac{y}{b} + \dfrac{z}{c} = 1$ in the first octant, where a, b, and c are positive constants.
12. The lateral surface of a right circular cone of base radius r and altitude h. (*Hint:* Place the coordinate

axes so that the origin is at the base center and a portion of the z-axis coincides with the cone axis. Then show that the surface of the cone has an equation $z = h\left(1 - \dfrac{\sqrt{x^2 + y^2}}{r}\right)$.)

13. Find the area of the part of the cylinder $x^2 + z^2 = 100$ that is inside the cylinder $x^2 + y^2 = 100$.

14. Find the surface area of the upper half of the cone $z^2 = x^2 + y^2$ that is inside the elliptic cylinder $\dfrac{x^2}{16} + \dfrac{y^2}{9} = 1$.

15. Find the surface area of the sphere $x^2 + y^2 + z^2 - a^2 = 0$ that is inside the cylinder $x^2 + y^2 - ax = 0$ in the first octant, where a is a positive constant.

16. Find the surface area of the portion of the cylinder $z^2 + y^2 = b^2$ that is above the triangle in the xy-plane bounded by $x = 0$, $y = 0$, and $x + y = b$, where b is a positive constant.

17. Find the surface area in the first octant cut from the cylindrical surface $z = a^2 - y^2$ by the plane $y = mx$ where a and m are positive constants.

18. Find the area of the part of the cylinder $x^2 + z^2 = a^2$ that is inside the cylinder $x^2 + y^2 = a^2$, where a is a positive constant.

19. **Area of Curved Surfaces in Cylindrical Coordinates.** Suppose that $z = f(x, y)$ is the rectangular equation of a curved surface in three dimensions. If we introduce cylindrical coordinates by the transformations

$$x = r \cos \theta \qquad \text{and} \qquad y = r \sin \theta$$

then $z = f(x, y)$ becomes $z = g(r, \theta)$. Show that

$$z_x^{\,2} + z_y^{\,2} = z_r^{\,2} + \frac{1}{r^2}z_\theta^{\,2}$$

and that in cylindrical coordinates the surface area assumes the form

$$S = \int\!\!\int_R \left(1 + z_r^{\,2} + \frac{1}{r^2}z_\theta^{\,2}\right)^{1/2} r \, dr \, d\theta$$

20. Use the result of Exercise 19 to find the area of the portion of the sphere $x^2 + y^2 + z^2 = a^2$ between the planes $z = b$ and $z = c$, where $0 \le b < c \le a$.

21. Find the area of the paraboloid of revolution $x^2 + y^2 - 4z = 0$ contained within the sphere $x^2 + y^2 + z^2 = 5$. (*Hint:* Apply the result of Exercise 19.)

22. In Section 7.11 we developed and utilized the formula

$$S = 2\pi \int_a^b F(x)[1 + (F'(x))^2]^{1/2} \, dx$$

for the area of the surface of revolution obtained when $y = F(x)$ is rotated about the x-axis. Derive this formula from (5) of this section and thereby establish the consistency of the two formulas. (*Hint:* The equation of the surface of revolution is $y^2 + z^2 = (F(x))^2$.)

18.7 TRIPLE INTEGRALS

We can define integrals for a function $f(x, y, z)$ in a manner analogous to the two-dimensional problem. The simplest kind of region in three space is a rectangular parallelepiped (or box) R, which is bounded by the planes $x = a$, $x = b$, $y = c$, $y = d$, $z = k$, and $z = r$, where $a \le b$, $c \le d$, and $k \le r$. It is assumed that f is continuous on the region R. We form a partition P of this region into subregions $R_1, R_2, R_3, \ldots, R_N$ by means of planes parallel to the coordinate planes so that each subregion also is a box. The norm of this partition $|P|$ is the longest diagonal of all R_i ($i = 1, 2, \ldots, N$). Figure 18.7.1 illustrates R and the subregion R_i with dimensions Δx_i, Δy_i and Δz_i from which the volume ΔV_i of R_i is the product $\Delta x_i \, \Delta y_i \, \Delta z_i$. Let (ξ_i, η_i, ζ_i) be an arbitrary point in R_i and form the sum

$$\sum_{i=1}^{N} f(\xi_i, \eta_i, \zeta_i) \, \Delta V_i \tag{1}$$

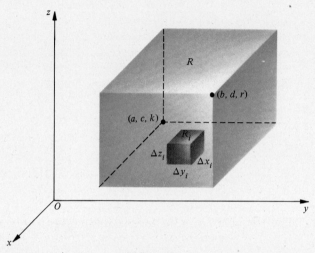

Figure 18.7.1

which is called a **Riemann sum** of f for the partition P. It can be shown that, as the norm $|P| \to 0$, (1) will approach a limit that is independent of the choice of points (ξ_i, η_i, ζ_i). This limit is called the **triple integral** of f over the region R. Hence we write

$$\iiint\limits_{R} f(x, y, z) \, dV = \lim_{|P| \to 0} \sum_{i=1}^{N} f(\xi_i, \eta_i, \zeta_i) \, \Delta V_i \tag{2}$$

Moreover, it can be shown that the triple integral in (2) can be evaluated as an iterated integral. For example,

$$\iiint\limits_{R} f(x, y, z) \, dV = \int_k^r \int_c^d \int_a^b f(x, y, z) \, dx \, dy \, dz \tag{3}$$

where the iterated integral on the right side of (3) is first evaluated with respect to x from $x = a$ to $x = b$ (with y and z fixed), then with respect to y from c to d (with z fixed), and third with respect to z. Also there are five other iterated integrals obtained by reordering the integrations of f over R and the result is independent of the order. Thus

$$\int_k^r \int_a^b \int_c^d f(x, y, z) \, dy \, dx \, dz = \int_a^b \int_c^d \int_k^r f(x, y, z) \, dz \, dy \, dx$$

$$= \cdots = \iiint\limits_{R} f(x, y, z) \, dV \tag{4}$$

EXAMPLE 1   Evaluate $\iiint\limits_{R} 5x^3y^2z\,dV$, where R is bounded by the planes $x = 1$, $x = 3$, $y = 0$, $y = 1$, $z = -1$, and $z = 2$.

Solution   Of the six possible iterated integrals, we shall use the following

$$\int_1^3 \int_{-1}^2 \int_0^1 5x^3y^2z\,dy\,dz\,dx = \int_1^3 \int_{-1}^2 \tfrac{5}{3}x^3y^3z\big|_0^1\,dz\,dx$$

$$= \int_1^3 \int_{-1}^2 \tfrac{5}{3}x^3z\,dz\,dx$$

$$= \int_1^3 \tfrac{5}{6}x^3z^2\big|_{-1}^2\,dx$$

$$= \tfrac{5}{6}[4 - (-1)^2] \int_1^3 x^3\,dx$$

$$= \frac{5}{2}\frac{x^4}{4}\bigg|_1^3 = \frac{5}{8}(81 - 1) = 50 \qquad \bullet$$

The triple integral of a continuous function f of three variables can be defined over a region R in space other than a rectangular parallelepiped. Let R be a region bounded below by $z = F(x, y)$, above by $z = G(x, y)$ and laterally by the cylinders $y = g(x)$ and $y = h(x)$ and the planes $x = a$ and $x = b$ (see Figure 18.7.2). It is further assumed that the four functions F, G, g, h are smooth; that is, F and G have continuous first partial derivatives, and g and h have continuous first derivatives. Let the region R be partitioned by means of planes parallel to the coordinate planes, then the resulting rectangular parallelepipeds $R_1, R_2, \ldots, R_N$, which lie completely within R, form an inner partition P of R. If (ξ_i, η_i, ζ_i) denotes an arbitrary point in R_i and ΔV_i is its volume, then once again the triple integral may be defined by (2), where the limit on the right side of (2) may be proved to

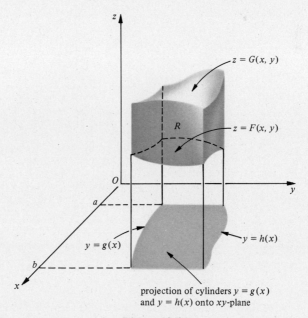

Figure 18.7.2

exist. Furthermore, for a continuous function on R, we have

$$\iiint\limits_R f(x, y, z)\, dV = \int_a^b \int_{g(x)}^{h(x)} \int_{F(x,y)}^{G(x,y)} f(x, y, z)\, dz\, dy\, dx \qquad (5)$$

If the projected region of the cylinders onto the xy-plane is defined instead by $g(y) \le x \le h(y)$ and $c \le y \le d$, then

$$\iiint\limits_R f(x, y, z)\, dV = \int_c^d \int_{g(y)}^{h(y)} \int_{F(x,y)}^{G(x,y)} f(x, y, z)\, dz\, dx\, dy \qquad (6)$$

If $f = 1$ throughout R then the triple integral of f over R reduces to

$$\iiint\limits_R dV = \lim_{|P|\to 0} \sum_{i=1}^N \Delta_i V = V = \text{volume of the region } R \qquad (7)$$

EXAMPLE 2 Find the volume of the solid in the first octant bounded by $z = 20 - x - y$, $z = 0$ and the cylinder $x^2 + y^2 = 144$.

Solution Figure 18.7.3 shows the solid whose volume is required. In our case $G(x, y) = 20 - x - y$, while $F(x, y) = 0$. The limits on y and x are obtained from the equation of the cylinder $x^2 + y^2 = 144$. If we solve for y, then $y = \sqrt{144 - x^2}$, since $y \ge 0$, and the limits of x are 0 and 12. Hence

$$V = \int_0^{12} \int_0^{\sqrt{144-x^2}} \int_0^{20-x-y} dz\, dy\, dx$$

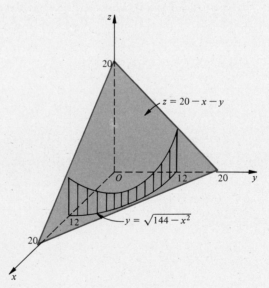

Figure 18.7.3

The iterated integrations may be interpreted as follows. In the first integration, the box $dz\, dy\, dx$ is summed from $z = 0$ to $z = 20 - x - y$ to yield the volume of the column. In the second integration with respect to y, the columns are summed to yield the volume of the slab. In the last integration, the slabs are summed with respect to x to yield the required volume.

$$V = \int_0^{12} \int_0^{\sqrt{144-x^2}} (20 - x - y)\, dy\, dx$$

$$= \int_0^{12} \left(20y - xy - \frac{y^2}{2}\right) \Bigg|_0^{\sqrt{144-x^2}} dx$$

$$= \int_0^{12} [20\sqrt{144 - x^2} - x\sqrt{144 - x^2} - \tfrac{1}{2}(144 - x^2)]\, dx$$

and performance of the integration of the three terms, using tables of integrals, results in

$$V = \left[\frac{20}{2}\left(x\sqrt{144 - x^2} + 144 \sin^{-1}\frac{x}{12}\right) + \frac{1}{3}(144 - x^2)^{3/2} - \frac{1}{2}\left(144x - \frac{x^3}{3}\right)\right] \Bigg|_0^{12}$$

$$= [10(0) + 720\pi - 576 - 576]$$

$$\doteq 1110 \text{ cu units} \qquad \bullet$$

If $f(x, y, z) = \rho(x, y, z)$, the mass density of a body (that is, the mass per unit of volume), and ρ is a continuous function of x, y, and z on R, then the expression for the mass M is easily determined. The volume V is partitioned, as usual, into N pieces with volume ΔV_i, $i = 1, 2, \ldots, N$. The incremental mass ΔM_i is given approximately by

$$\Delta M_i = \rho(\xi_i, \eta_i, \zeta_i)\, \Delta V_i$$

where (ξ_i, η_i, ζ_i) is an arbitrary point in ΔV_i and the approximate mass is given by $\sum_{i=1}^N \Delta M_i$. The mass M is obtained from the limit

$$\lim_{|P| \to 0} \sum_{i=1}^N \rho(\xi_i, \eta_i, \zeta_i)\, \Delta V_i$$

or, equivalently, by

$$M = \iiint_R \rho(x, y, z)\, dV \qquad (8)$$

In particular, if the density is constant, $\rho = \rho_0 = $ constant, that is, the body is homogeneous, then

$$M = \rho_0 V \qquad (9)$$

where V is the volume of the region R.

EXAMPLE 3   Find the mass of the solid in the first octant bounded by the paraboloid of revolution $az = a^2 - x^2 - y^2$, the plane $x + y = a$, and the coordinate planes, where the density ρ is given by $\rho = \rho_0 xyz/a^3$. The quantities ρ_0 and a are positive constants.

Solution   Refer to Figure 18.7.4

$$M = a^{-3}\rho_0 \iiint\limits_R xyz\, dV$$

$$= a^{-3}\rho_0 \int_0^a \int_0^{a-x} \int_0^{a^{-1}(a^2-x^2-y^2)} xyz\, dz\, dy\, dx$$

$$= \frac{a^{-5}\rho_0}{2} \int_0^a \int_0^{a-x} xy(a^2 - x^2 - y^2)^2\, dy\, dx$$

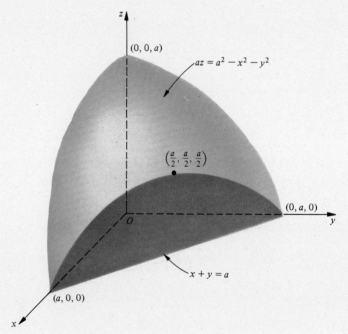

(0, 0, a)

$az = a^2 - x^2 - y^2$

$\left(\frac{a}{2}, \frac{a}{2}, \frac{a}{2}\right)$

O

(0, a, 0)

y

$x + y = a$

(a, 0, 0)

x

Figure 18.7.4

and since $d(a^2 - x^2 - y^2) = -2y\, dy$, the second integral yields

$$M = -\tfrac{1}{12}a^{-5}\rho_0 \int_0^a (a^2 - x^2 - y^2)^3\big|_0^{a-x}\, x\, dx$$

$$= -\tfrac{1}{12}a^{-5}\rho_0 \int_0^a \{[a^2 - x^2 - (a - x)^2]^3 - (a^2 - x^2)^3\}x\, dx$$

$$= -\tfrac{1}{12}a^{-5}\rho_0 \int_0^a [(2ax - 2x^2)^3 - (a^2 - x^2)^3]x\, dx$$

Expansion of the first term and integration of the second results in

$$M = -\tfrac{1}{12}a^{-5}\rho_0 \left[\int_0^a 8(a^3x^3 - 3a^2x^4 + 3ax^5 - x^6)x\, dx + \tfrac{1}{8}(a^2 - x^2)^4\big|_0^a \right]$$

$$= -\tfrac{1}{12}a^{-5}\rho_0 \left[8\left(\frac{a^3x^5}{5} - \frac{3a^2x^6}{6} + \frac{3ax^7}{7} - \frac{x^8}{8} \right) \bigg|_0^a - \frac{a^8}{8} \right]$$

$$= -\tfrac{1}{12}a^3\rho_0[8(\tfrac{1}{5} - \tfrac{1}{2} + \tfrac{3}{7} - \tfrac{1}{8}) - \tfrac{1}{8}] = \tfrac{9}{1120}\rho_0 a^3$$

Exercises 18.7

In Exercises 1 through 10, evaluate the given iterated integral

1. $\displaystyle\int_0^3 \int_0^1 \int_0^2 y \, dz \, dy \, dx$

2. $\displaystyle\int_0^6 \int_{-3}^2 \int_{-1}^1 z \, dz \, dx \, dy$

3. $\displaystyle\int_{-1}^1 \int_0^2 \int_0^x x^2 \, dy \, dx \, dz$

4. $\displaystyle\int_{-2}^0 \int_{-1}^x \int_0^y xy \, dz \, dy \, dx$

5. $\displaystyle\int_0^1 \int_{-y}^y \int_0^{x+y} z \, dz \, dx \, dy$

6. $\displaystyle\int_0^{1/2} \int_z^{z^2} \int_0^{zx} x^2 \, dy \, dx \, dz$

7. $\displaystyle\int_0^\pi \int_0^{\sqrt{\theta}} \int_0^{\sqrt{r}} r^2 z \, dz \, dr \, d\theta$

8. $\displaystyle\int_0^{\pi/2} \int_0^{\sin\theta} \int_0^{r\cos\theta} (r + z) \, dz \, dr \, d\theta$

9. $\displaystyle\int_1^2 \int_0^z \int_0^y e^x \, dx \, dy \, dz$

10. $\displaystyle\int_0^{2\pi} \int_0^a \int_0^{\sqrt{a^2-y^2}} e^{-x^2} xy \, dx \, dy \, dz$

In Exercises 11 through 20, use triple integration to find the volume of the region bounded by the given surfaces.

11. $x = 0, \quad y = 0, \quad z = 0, \quad 3x + 6y + z = 6$
12. $z = 0, \quad z = mx, \quad y = \sqrt{a^2 - x^2}$, where a and m are positive constants (first octant).
13. $y = 2x, \quad y = x^2, \quad z = 0, \quad z = 2x + y$
14. $x^2 + y^2 + z^2 = a^2, \quad a > 0$
15. $z = 8 - (x^2 + y^2)$ and $z = x^2 + y^2$
16. $z = 64 - 3x^3 - 3y^2$ and $z = x^2 + y^2$
17. $z = a, \quad z = b, \quad x = 0, \quad y = c, \quad y = 4x^2$, where a, b and c are positive constants and $a < b$
18. $x = y^2 + z^2$ and $x = 4$
19. $z = 5 + x - x^2, \quad 4x + y = 4$, coordinate planes and in the first octant
20. $x + y = 3, \quad x + 2y = 8, \quad z = 25 - y^2$, coordinate planes and in the first octant
21. Find the mass of a rectangular parallelepiped with dimensions a, b, and c if the mass per unit volume, or density, is proportional to the square of the distance from one vertex.
22. Find the mass of a cube of side a if its density is proportional to the square of its distance from one edge.
23. Find the mass of a tetrahedron bounded by $x + y + z = a$ and the coordinate planes if the density $\rho = k(x^2 + y^2)$, where k and a are positive constants.
24. Find the mass of a tetrahedron bounded by the coordinate planes and $x + y + z = a$ if the density $\rho = k(x + y + z)$, where k and a are positive constants.
25. Find the mass of a solid bounded by the planes $z = 0, y = 0, x - y = 0, x = b$, a positive constant, and the surface $z = \sqrt{x^2 + y^2}$ if the density is proportional to the distance from the z-axis.
26. Find the mass of a solid bounded by the surfaces $z = 3\sqrt{xy}, z = 0, ax = y$, and $x = b$, where a and b are positive constants, if the density is proportional to the distance above the xy-plane.
27. Find the mass of a solid bounded by the surfaces $z = \dfrac{16}{x^2 + 4}$, $z = 0, x - 2y = 0$, and $x = 4$ if the density $\rho = \rho_0 = $ constant.
28. Find the mass of a solid in the first octant bounded by $x = \sqrt{a^2 - y^2 - z^2}$ and the coordinate planes if the density $\rho = kxyz$, where a and k are positive constants.

18.8

APPLICATIONS OF THE TRIPLE INTEGRAL

Consider a solid S which has the shape of a three dimensional region R (as in 18.7) and the density (or mass/unit volume) at x, y, and z is given by $\delta(x, y, z)$, where δ

is assumed to be a continuous function† on R. If a particle of mass m is located at a point (x, y, z) then the moment of m with respect to the xy-, yz- and zx-planes are defined as zm, xm and ym, respectively. We then proceed to partition R into N sub-regions and form the Riemann sum, where the density is taken at an arbitrary point (ξ_i, η_i, ζ_i) in R_i $(i = 1, 2, \dots, N)$. We are led to define the **first moments of mass** or simply **moments** of the solid S with respect to the co-ordinate planes as the limit of such sums

$$M_{xy} = \iiint_R z\, \delta(x, y, z)\, dV, \quad M_{yz} = \iiint_R x\, \delta(x, y, z)\, dV$$

$$M_{zx} = \iiint_R y\, \delta(x, y, z)\, dV$$

(1)

The **center of mass** $(\bar{x}, \bar{y}, \bar{z})$ is defined by

$$\bar{x} = \frac{M_{yz}}{M}, \quad \bar{y} = \frac{M_{zx}}{M}, \quad \text{and} \quad \bar{z} = \frac{M_{xy}}{M}$$

(2)

where

$$M = \text{total mass of the solid } S \text{ given by}$$

$$M = \iiint_R \delta(x, y, z)\, dV$$

(3)

If R is homogeneous then δ is constant and cancels out in (2) and we obtain the **centroid** for such geometric solids. Its value depends only on the shape of the solid.

EXAMPLE 1   Determine the center of mass of the homogeneous solid that is bounded by the plane $x + y + z = a$ (where a is a positive constant) and the coordinate planes.

Solution   The solid is the tetrahedron shown in Figure 18.8.1. Its mass is (for constant δ)

$$M = \int_0^a \int_0^{a-x} \int_0^{a-x-y} \delta\, dz\, dy\, dx$$

$$= \delta \int_0^a \int_0^{a-x} (a - x - y)\, dy\, dx$$

$$= \delta \int_0^a \left[(a-x)y - \frac{y^2}{2} \right] \Big|_0^{a-x} dx$$

$$= \frac{\delta}{2} \int_0^a (a - x)^2\, dx = -\frac{\delta}{6}(a-x)^3 \Big|_0^a = \frac{a^3\delta}{6}$$

By symmetry, $\bar{x} = \bar{y} = \bar{z}$ and it is therefore sufficient to find $\bar{z}$

$$M_{xy} = \int_0^a \int_0^{a-x} \int_0^{a-x-y} z\delta\, dz\, dy\, dx$$

---

† In this section and the next, we use δ to denote the mass density and reserve ρ for a spherical co-ordinate.

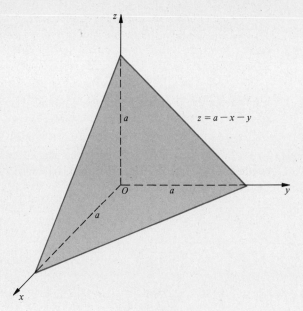

Figure 18.8.1

$$M_{xy} = \delta \int_0^a \int_0^{a-x} \frac{z^2}{2} \bigg|_0^{a-x-y} dy\, dx$$

$$= \frac{\delta}{2} \int_0^a \int_0^{a-x} (a - x - y)^2\, dy\, dx$$

$$= -\frac{\delta}{6} \int_0^a (a - x - y)^3 \bigg|_0^{a-x} dx$$

$$= \frac{\delta}{6} \int_0^a (a - x)^3\, dx$$

$$M_{xy} = \frac{a^4 \delta}{24}$$

Hence the center of mass or centroid is at $\left(\dfrac{a}{4}, \dfrac{a}{4}, \dfrac{a}{4}\right)$. ●

EXAMPLE 2 Find the center of mass of a solid $0 \leq x \leq a$, $0 \leq y \leq b$, and $0 \leq z \leq c$ if its density is proportional to the product of the distances from the coordinate planes.

Solution The solid is a rectangular parallelepiped with density $\delta = kxyz$, where k is a positive constant. Its mass is given by

$$M = \int_0^a \int_0^b \int_0^c kxyz\, dz\, dy\, dx = \frac{ka^2b^2c^2}{8}$$

Now, its moment with respect to the yz-plane is

$$M_{yz} = \int_0^a \int_0^b \int_0^c kx^2yz\, dz\, dy\, dx = \frac{ka^3b^2c^2}{12}$$

so that $\bar{x} = \dfrac{M_{yz}}{M} = \dfrac{2}{3}a$

In a similar manner, $\bar{y} = \dfrac{2}{3}b$ and $\bar{z} = \dfrac{2}{3}c$. Thus $\left(\dfrac{2}{3}a, \dfrac{2}{3}b, \dfrac{2}{3}c\right)$ is the center of mass.

●

EXAMPLE 3   Find the centroid of the hemisphere bounded by $z = \sqrt{a^2 - x^2 - y^2}$ and the xy-plane, where a is a positive constant.

Solution   By symmetry, $\bar{x} = \bar{y} = 0$, that is, the centroid must be on the z-axis and only $\bar{z}$ need be calculated. Furthermore, from symmetry, it is only necessary to consider the region in the first octant.

$$
\begin{aligned}
M_{xy} &= \iiint z\,dV = \int_0^a \int_0^{\sqrt{a^2-x^2}} \int_0^{\sqrt{a^2-x^2-y^2}} z\,dz\,dy\,dx \\
&= \int_0^a \int_0^{\sqrt{a^2-x^2}} \frac{z^2}{2} \Big|_0^{\sqrt{a^2-x^2-y^2}} dy\,dx \\
&= \frac{1}{2} \int_0^a \int_0^{\sqrt{a^2-x^2}} (a^2 - x^2 - y^2)\,dy\,dx \\
&= \frac{1}{2} \int_0^a \left[(a^2 - x^2)y - \frac{y^3}{3} \right] \Big|_0^{\sqrt{a^2-x^2}} dx \\
M_{xy} &= \frac{1}{3} \int_0^a (a^2 - x^2)^{3/2}\,dx
\end{aligned}
$$

Let $x = a\sin\theta$, $dx = a\cos\theta\,d\theta$ so that

$$
M_{xy} = \frac{1}{3}a^4 \int_0^{\pi/2} \cos^4\theta\,d\theta = \frac{1}{3}a^4 \left(\frac{3\pi}{16}\right) = \frac{\pi a^4}{16}
$$

The volume V of the portion of the hemisphere in the first octant is $\dfrac{\pi a^3}{6}$ and

$$
\bar{z} = \frac{M_{xy}}{V} = \frac{\pi a^4}{16} \cdot \frac{6}{\pi a^3} = \frac{3}{8}a
$$

Hence the coordinates of the centroid are

$$
(\bar{x}, \bar{y}, \bar{z}) = \left(0, 0, \frac{3}{8}a\right)
$$

●

Suppose that we have a particle of mass m at the point (x, y, z). Then its moment of inertia (or second moment) with respect to the z-axis is, by definition, $I_z = m(x^2 + y^2)$ (since the square of the distance from the point to the z-axis is $x^2 + y^2$). Next, consider a solid S which occupies a region R in space. Furthermore, we assume that its density $\delta(x, y, z)$ once again is a continuous function on R. As above, we partition R into N regions R_i with volume ΔV_i, $(i = 1, 2, 3, \ldots, N)$ and an element of moment of inertia about the z-axis is given approximately by

$$
\Delta I_z = (\xi_i^2 + \eta_i^2)\,\delta(\xi_i, \eta_i, \zeta_i)\,\Delta V_i
$$

where (ξ_i, η_i, ζ_i) is an arbitrary point in R_i. Taking a limit of such sums of terms as the norm of the partition tends to zero yields the following definition of the *moment of inertia* of the solid S with respect to the z-axis,

$$I_z = \iiint_R (x^2 + y^2)\, \delta(x, y, z)\, dV \tag{4}$$

Similarly, the **moments of inertia** of S about the x- and y-axes are, respectively, given by

$$I_x = \iiint_R (y^2 + z^2)\, \delta(x, y, z)\, dV \quad \text{and} \quad I_y = \iiint_R (z^2 + x^2)\, \delta(x, y, z)\, dV \tag{5}$$

If I is the moment of inertia of the solid S with respect to a particular axis then the **radius of gyration** R with respect to that axis is obtained by solving

$$I = MR^2 \tag{6}$$

where M is the mass of the body (refer to Section 18.4 for more details about this concept).

EXAMPLE 4   Find the moment of inertia and radius of gyration with respect to the z-axis of the solid described in Example 1.

Solution $I_z = \delta \int_0^a \int_0^{a-x} \int_0^{a-x-y} (x^2 + y^2)\, dz\, dy\, dx$ from (4) and the hypothesis that δ is a constant.

$$I_z = \delta \int_0^a \int_0^{a-x} (x^2 + y^2)z \Big|_0^{a-x-y} dy\, dx$$

$$= \delta \int_0^a \int_0^{a-x} (x^2 + y^2)(a - x - y)\, dy\, dx$$

$$= \delta \int_0^a \int_0^{a-x} [(a - x)x^2 + (a - x)y^2 - x^2 y - y^3]\, dy\, dx$$

$$= \delta \int_0^a \left[(a - x)x^2 y + (a - x)\frac{y^3}{3} - \frac{x^2 y^2}{2} - \frac{y^4}{4} \right] \Big|_0^{a-x} dx$$

$$= \delta \int_0^a \left[(a - x)^2 x^2 + \frac{(a - x)^4}{3} - \frac{x^2(a - x)^2}{2} - \frac{(a - x)^4}{4} \right] dx$$

$$= \delta \int_0^a \left[\frac{1}{2}(a - x)^2 x^2 + \frac{1}{12}(a - x)^4 \right] dx$$

$$= \delta \int_0^a \left[\frac{1}{2}(a^2 x^2 - 2ax^3 + x^4) + \frac{1}{12}(a - x)^4 \right] dx$$

$$= \delta \left[\frac{1}{2}\left(\frac{1}{3}a^2 x^3 - \frac{1}{2}ax^4 + \frac{1}{5}x^5 \right) - \frac{1}{60}(a - x)^5 \right] \Big|_0^a$$

$$= \frac{\delta a^5}{30}$$

and from (6) and the expression for the mass in Example 1,

$$R_z = \left[\frac{\delta a^5}{30} \Big/ \frac{\delta a^3}{6} \right]^{1/2} = \frac{a}{\sqrt{5}}$$

Exercises 18.8

In each of Exercises 1 through 12, find the center of mass of the solid having the given density δ and bounded by the surfaces described. The quantities k, a, b and c are positive constants.

1. The planes $x = 2$, $y = 3$, $z = 4$ and the coordinate planes; $\delta = k$.
2. The planes $x = 2$, $y = 3$, $z = 4$ and the coordinate planes; $\delta = kx$.
3. The planes $x = a$, $y = b$, $z = c$ and the coordinate planes; $\delta = ky$.
4. The planes $x = a$, $y = a$, $z = a$ and the coordinate planes if the density δ is proportional to the distance from the yz-plane.
5. The planes $x = a$, $y = a$, $z = a$ and the coordinate planes if the density δ is proportional to the square of the distance from the origin.
6. The plane $\dfrac{x}{a} + \dfrac{y}{b} + \dfrac{z}{c} = 1$ and the coordinate planes; $\delta = k$.
7. The cylinder $y = x^2$ and the planes $z = y$, $y = b$, $x = 0$ and $z = 0$; $\delta = k$.
8. The paraboloid $z = a^2 - x^2 - y^2$ and the plane $z = 0$; $\delta = k$. (*Hint:* Plane polar coordinates are convenient for the evaluation of the multiple integral)
9. The cylinder $x^2 + y^2 = a^2$ and the planes $z = 0$ and $z = a^2$; $\delta = kx^2y^2$.
10. The cylinder $x^2 + y^2 = a^2$ and the planes $z = 0$ and $z = a^2$; $\delta = kx^2y^2z^2$.
11. Upper half of the cone $z^2 = x^2 + y^2$ and the plane $z = a$, $\delta = k$. (The formula for the volume of a right circular cone may be utilized.)
12. Upper half of the ellipsoid $\dfrac{x^2}{a^2} + \dfrac{y^2}{b^2} + \dfrac{z^2}{c^2} = 1$; $\delta = k$. (The formula for the volume of an ellipsoid may be utilized.)

In each of Exercises 13 through 16, find the moment of inertia and radius of gyration of the solid with given boundaries, about the prescribed axis, with specified density δ. The quantities k, a, and b are positive numbers.

13. A cube with side b; about an edge; $\delta = k$.
14. A cube in the first octant with side b and with three faces in the coordinate planes; about the y-axis; $\delta = kxz$.
15. $x + y + z = a$, coordinate planes, about the x-axis; $\delta = k$
16. $x + y + z = a$, coordinate planes, about the x-axis; $\delta = kx$

18.9   TRIPLE INTEGRAL IN CYLINDRICAL AND SPHERICAL COORDINATES

In many applications the triple integral can be evaluated most simply in cylindrical coordinates or in spherical coordinates. In Section 5 of this chapter we found that the differential of area dA in cylindrical coordinates is given by $dA = r\,dr\,d\theta$. Consequently, it follows that in cylindrical coordinates the differential element of volume of a cylindrical wedge is given by $dV = r\,dr\,d\theta\,dz$ (Figure 18.9.1). Therefore

$$\iiint_R f(x, y, z)\, dV = \iiint_R f(r \cos \theta, r \sin \theta, z)\, r\, dr\, d\theta\, dz \tag{1}$$

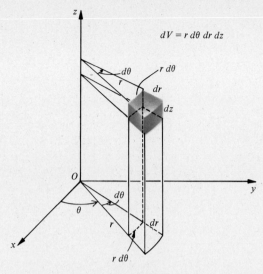

$$dV = r\,d\theta\,dr\,dz$$

Figure 18.9.1

If $f = 1$, we obtain the formula for the volume of a body S occupying a space R. If $f = \delta(x, y, z)$, the mass density, then the mass M in cylindrical coordinates is given by

$$M = \iiint_R \delta(x, y, z)\,dV = \iiint_R \delta(r\cos\theta, r\sin\theta, z)\,r\,dr\,d\theta\,dz \qquad (2)$$

Similarly, the moment of inertia I_z, with respect to the z-axis, is given by (since $x^2 + y^2 = r^2$)

$$I_z = \iiint_R \delta(x, y, z)(x^2 + y^2)\,dV$$

$$= \iiint_R \delta(r\cos\theta, r\sin\theta, z)\,r^3\,dr\,d\theta\,dz \qquad (3)$$

EXAMPLE 1   Find the moment of inertia of a homogeneous solid right circular cylinder ($\delta = k = $ constant) of base radius a and altitude h with respect to its axis (see Figure 18.9.2).

Solution   The equation of the lateral surface of the cylinder with axis coincident with the z-axis is

$$x^2 + y^2 = r^2 = a^2$$

so that

$$I_z = k \int_0^h \int_0^{2\pi} \int_0^a r^3\,dr\,d\theta\,dz$$

and consequently,

$$I_z = \frac{k\pi a^4 h}{2} \qquad (4)$$

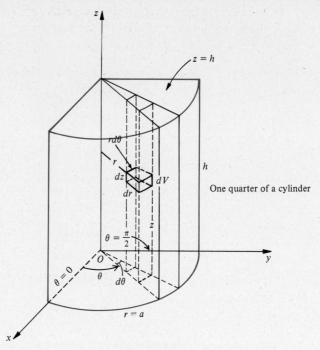

One quarter of a cylinder

Figure 18.9.2

The radius of gyration R_z is found from

$$I_z = MR_z{}^2$$

But $M = k\pi a^2 h$ so that

$$R_z{}^2 = \frac{a^2}{2}$$

from which

$$R_z = \frac{a}{\sqrt{2}} \qquad \bullet \qquad (5)$$

EXAMPLE 2   Find the mass of a solid right circular cylinder of base radius a and altitude h, if the density at each point is proportional to the square of its distance from the axis of the cylinder.

Solution   The density $\delta = kr^2$ where k is a constant and substitution into (2) yields

$$M = k \int_0^h \int_0^{2\pi} \int_0^a r^3 \, dr \, d\theta \, dz$$

$$M = \frac{1}{2} k\pi a^4 h \qquad \bullet$$

Triple integrals and cylindrical coordinates are useful in the determination of the moment of inertia with respect to the z-axis because the square of the distance from the z-axis is given in terms of the cylindrical coordinate by r^2. They are also very helpful in the resolution of other problems involving cylinders (as in Example 2) since the limits of integration are constants in this coordinate system.

EXAMPLE 3   Use cylindrical coordinates to calculate the mass of a homogeneous right circular cone of base radius a and altitude h (see Figure 18.9.3).

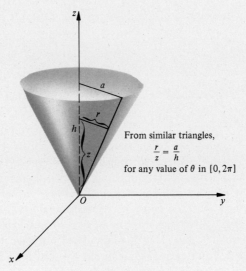

From similar triangles,
$$\frac{r}{z} = \frac{a}{h}$$
for any value of θ in $[0, 2\pi]$

Figure 18.9.3

Solution   The density $\delta = k = $ constant and the equation of the cone is given by $az = hr$. Therefore the mass M is given by

$$M = k \int_0^{2\pi} \int_0^a \int_{hr/a}^h r \, dz \, dr \, d\theta$$

$$= k \int_0^{2\pi} \int_0^a \left[rh - \frac{h}{a} r^2 \right] dr \, d\theta$$

$$= k \int_0^{2\pi} \left(\tfrac{1}{2} hr^2 - \frac{hr^3}{3a} \right) \Big|_0^a \, d\theta$$

$$= \tfrac{1}{3} k\pi ha^2 \qquad \bullet$$

EXAMPLE 4   Find the coordinates of the center of mass of the right circular cone in Example 3.

Solution   By symmetry, $\bar{x} = \bar{y} = 0$ and only $\bar{z}$ need be computed. With the notation of Example 3, the mass moment of the cone with respect to the xy-plane is

$$M_{xy} = k \int_0^{2\pi} \int_0^a \int_{hr/a}^h rz \, dz \, dr \, d\theta$$

$$= k \int_0^{2\pi} \int_0^a \tfrac{1}{2} rz^2 \Big|_{hr/a}^h \, dr \, d\theta$$

$$= \frac{k}{2} \int_0^{2\pi} \int_0^a r(h^2 - a^{-2}h^2r^2) \, dr \, d\theta$$

$$= \frac{kh^2}{2a^2} \int_0^{2\pi} \int_0^a (a^2 r - r^3) \, dr \, d\theta$$

$$= \frac{kh^2}{2a^2} \int_0^{2\pi} \left(\frac{a^2 r^2}{2} - \frac{r^4}{4} \right) \Big|_0^a \, d\theta$$

$$= \frac{kh^2 a^2}{8} \int_0^{2\pi} d\theta = \tfrac{1}{4}\pi kh^2 a^2$$

The mass M of the cone is (from Example 3)

$$M = \tfrac{1}{3}\pi kha^2$$

so that

$$\bar{z} = \frac{M_{xy}}{M} = \tfrac{3}{4}h \qquad \bullet$$

Following the development in cylindrical coordinates we seek an element of volume in spherical coordinates. In Figure 18.9.4 we have displayed an element that is bounded by portions of two spheres with centers at the origin ($\rho =$ constant), two half right circular cones ($\phi =$ constant) and two half planes through the z-axis ($\theta =$ constant). If the increments $\Delta\rho$, $\Delta\phi$, and $\Delta\theta$ are sufficiently small, it is suggested that the volume of the element should be approximately equal to the product of the lengths of the three mutually perpendicular "curvilinear sides" $\Delta\rho$, $\rho\,\Delta\phi$, and $\rho\sin\phi\,\Delta\theta$. Therefore we would have approximately,

$$\Delta V = \rho^2 \sin\phi \, \Delta\rho \, \Delta\phi \, \Delta\theta$$

(More precisely, it can be shown by two applications of the basic law of the mean that

$$\Delta V = \bar{\rho}^2 \sin\bar{\phi} \, \Delta\rho \, \Delta\phi \, \Delta\theta$$

where $\rho < \bar{\rho} < \rho + \Delta\rho$ and $\phi < \bar{\phi} < \phi + \Delta\phi$)

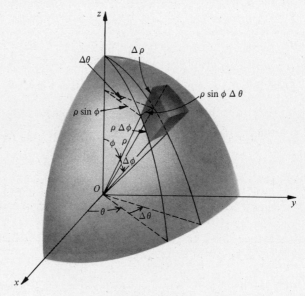

Figure 18.9.4

In either case, by forming the approximating sums and then taking limits in the usual way, we obtain

$$\iiint\limits_{R} f(x, y, z)\, dx\, dy\, dz = \iiint\limits_{R} F(\rho, \phi, \theta)\rho^2 \sin\phi\, d\rho\, d\phi\, d\theta \qquad (6)$$

where (from $x = \rho \sin\phi \cos\theta$, $y = \rho \sin\phi \sin\theta$, $z = \rho \cos\phi$) we have $F(\rho, \phi, \theta) = f(\rho \sin\phi \cos\theta, \rho \sin\phi \sin\theta, \rho \cos\phi)$.

EXAMPLE 5   Find the center of mass of a homogeneous hemisphere (density $\delta = k = $ constant) of radius a and bounded below by the xy-plane.

Solution   With the z-axis as the axis of symmetry, $\bar{x} = \bar{y} = 0$. To find $\bar{z}$, we have $\bar{z} = M_{xy}/M$. The mass M of the hemisphere is obtained from

$$M = k \int_0^{2\pi} \int_0^{\pi/2} \int_0^a \rho^2 \sin\phi\, d\rho\, d\phi\, d\theta$$

$$= k \int_0^{2\pi} \int_0^{\pi/2} \tfrac{1}{3}\rho^3 \Big|_0^a \sin\phi\, d\phi\, d\theta$$

$$= \tfrac{1}{3}ka^3 \int_0^{2\pi} \int_0^{\pi/2} \sin\phi\, d\phi\, d\theta$$

$$= \tfrac{1}{3}ka^3 \int_0^{2\pi} (-\cos\phi) \Big|_0^{\pi/2} d\theta$$

$$= \tfrac{2}{3}\pi a^3 k$$

$$M_{xy} = k \int_0^{2\pi} \int_0^{\pi/2} \int_0^a \rho^2 z \sin\phi\, d\rho\, d\phi\, d\theta$$

and with $z = \rho \cos\phi$

$$M_{xy} = k \int_0^{2\pi} \int_0^{\pi/2} \int_0^a \rho^3 \sin\phi \cos\phi\, d\rho\, d\phi\, d\theta$$

$$= k \int_0^{2\pi} \int_0^{\pi/2} \tfrac{1}{4}\rho^4 \Big|_0^a \sin\phi \cos\phi\, d\phi\, d\theta$$

$$= \tfrac{1}{4}ka^4 \int_0^{2\pi} \tfrac{1}{2}\sin^2\phi \Big|_0^{\pi/2} d\theta$$

$$= \tfrac{1}{8}ka^4 \cdot 2\pi = \tfrac{1}{4}\pi a^4 k$$

Hence $\bar{z} = \tfrac{3}{8}a$ and $(0, 0, \tfrac{3}{8}a)$ is the coordinate of the center of gravity. ●

EXAMPLE 6   Find the moment of inertia and radius of gyration of a homogeneous sphere (density $\delta = k = $ constant) of radius a with respect to a diameter.

Solution   Assume the sphere is centered at the origin and let a part of the z-axis coincide with the diameter. Hence

$$I_z = k \int_0^{2\pi} \int_0^{\pi} \int_0^a \rho^2 (\sin\phi)(x^2 + y^2)\, d\rho\, d\phi\, d\theta$$

$$= k \int_0^{2\pi} \int_0^{\pi} \int_0^a \rho^2 \sin \phi [(\rho \sin \phi \cos \theta)^2 + (\rho \sin \phi \sin \theta)^2] \, d\rho \, d\phi \, d\theta$$

$$= k \int_0^{2\pi} \int_0^{\pi} \int_0^a \rho^4 \sin^3 \phi \, d\rho \, d\phi \, d\theta$$

$$= k \int_0^{2\pi} \int_0^{\pi} \tfrac{1}{5}\rho^5 \Big|_0^a \sin^3 \phi \, d\phi \, d\theta$$

$$= \tfrac{1}{5}ka^5 \int_0^{2\pi} \int_0^{\pi} (1 - \cos^2 \phi) \sin \phi \, d\phi \, d\theta$$

$$= \tfrac{1}{5}a^5 \int_0^{2\pi} [-\cos \phi + \tfrac{1}{3}\cos^3 \phi] \Big|_0^{\pi} d\theta$$

$$= \tfrac{4}{15}ka^5 \int_0^{2\pi} d\theta = \tfrac{8}{15}\pi a^5 k$$

Now $M = \tfrac{4}{3}\pi a^3 k$, so that

$$I_z = \tfrac{2}{5}a^2 M$$

and the radius of gyration R_z is given by

$$R_z = \sqrt{I_z/M} = \sqrt{\tfrac{2}{5}}a \qquad \bullet$$

Exercises 18.9

1. Find the mass of a right circular cylinder with base radius a, height h, if the density is proportional to its distance from the axis of the cylinder.
2. Find the center of mass of the cylinder of Exercise 1.
3. Find the moment of inertia of a homogeneous solid right circular cylinder of base radius a and altitude h with respect to a diameter of the base. Express your answer in terms of the mass of the solid.
4. Find the moment of inertia of the mass of Exercise 3 about a generator. Express your answer in terms of the mass of the solid.
5. A homogeneous solid right circular cone of base radius a and height h has its vertex at the origin and its axis coincident with the z-axis. If α is the semivertical angle (that is, $\tan \alpha = \dfrac{a}{h}$), find an expression for its mass M.
6. Find the mass M of the homogeneous solid bounded by the paraboloid of revolution $z = a^2 - x^2 - y^2$ and the xy-plane.
7. Find the moment of inertia with respect to its axis of symmetry of the homogeneous solid bounded by the paraboloid of revolution $z = a^2 - x^2 - y^2$ and the xy-plane.
8. Find the center of mass of the solid defined in Exercises 6 and 7.
9. Find the moment of inertia of a homogeneous circular cylindrical shell of height h, inner radius a and outer radius b with respect to its axis of symmetry. Express your answer in terms of the mass M of the shell.
10. Find the center of mass of a solid hemisphere of radius a which has a density δ proportional to the nth power of its distance from the center, where n is a nonnegative real number. In particular, find the location of the mass center if $n = 1$ and $n = 2$.
11. A homogeneous solid is bounded above by the sphere $x^2 + y^2 + z^2 - 2az = 0$ and below by the right circular cone $x^2 + y^2 = 3z^2$. Find the mass of the solid.
12. Find the center of mass of the solid in Exercise 11.
13. For the solid in Exercise 11, find the moment of inertia about the z-axis.
14. Find the mass of a homogeneous sphere of radius a using *cylindrical* coordinates.
15. Find the moment of inertia of a homogeneous sphere of radius a with respect to a diameter using *cylindrical* coordinates.

16. Find the moment of inertia of a homogeneous right circular cone of radius a and height h about its axis. Express your answer in terms of the mass M.

17. For the solid in Exercise 16, find the moment of inertia about any line perpendicular to its axis at the vertex.

18. Find the mass of a sphere of radius a if the density δ varies inversely as the square of the distance from the center.

19. In the text, we derived the formula that the moment of inertia of a homogeneous sphere of radius a and mass M with respect to a diameter is given by $\frac{2}{5}Ma^2$. Prove this by taking advantage of spherical symmetry. (*Hint:* $I_x = I_y = I_z$ and the polar moment with respect to the origin of coordinates $I_o = \iiint_R \rho^2 \delta \, dV$ is easily computed.)

20. Find the mass of a homogeneous solid above the xy-plane bounded by the sphere $x^2 + y^2 + z^2 = 4a^2$ and the cylinder $x^2 + y^2 = 2ay$, where a is a positive constant. (*Hint:* Use cylindrical coordinates.)

Review and Miscellaneous Exercises

In Exercises 1 through 20, evaluate each of the repeated integrals and sketch the region R.

1. $\displaystyle\int_0^2 \int_{-1}^1 (3x^2 - 4xy + y^2)\, dx\, dy$

2. $\displaystyle\int_0^4 \int_0^4 xye^{x+y}\, dy\, dx$

3. $\displaystyle\int_0^3 \int_{x^2}^{3x} (x+y)\, dy\, dx$

4. $\displaystyle\int_0^1 \int_0^{(1-y^2)^{1/2}} (x^2 + y^2)\, dx\, dy$, by two methods, using (i) rectangular coordinates and (ii) plane polar coordinates.

5. $\displaystyle\int_a^b \int_0^x \frac{1}{x^2 + y^2}\, dy\, dx$, where $0 < a < b$.

6. $\displaystyle\int_0^c \int_{ax}^{bx} e^{-x^2}\, dy\, dx$, where a, b, and c are constants.

7. $\displaystyle\int_0^3 \int_x^3 \sin y^2\, dy\, dx$

8. $\displaystyle\int_0^4 \int_0^{\sqrt{16-y^2}} (x^2 + y^2)^{3/2}\, dx\, dy$

9. $\displaystyle\int_0^{\pi/2} \int_{\sin\theta}^{\cos\theta} r\, dr\, d\theta$

10. $\displaystyle\int_0^a \int_0^y \sqrt{a^2 - x^2}\, dx\, dy$, where a is a positive constant. Do this in two ways (i) by integrating in the given order and then (ii) by reversing the order of integration.

11. $\displaystyle\int_0^1 \int_0^2 \frac{2xy}{1 + x^2 + y^2}\, dy\, dx$, do this in two ways.

12. $\displaystyle\int_0^\pi \int_0^a \frac{r^4}{r^2 + a^2}\, dr\, d\theta$

13. $\displaystyle\int_0^2 \int_0^{x^2} \int_0^y x^2y\, dz\, dy\, dx$

14. $\displaystyle\int_0^{\pi/2} \int_1^x \int_0^{e^{-y}} (\cos x)e^y\, dz\, dy\, dx$

15. $\int_0^a \int_y^a \int_{-\sqrt{a^2-x^2}}^{\sqrt{a^2-x^2}} dz\,dx\,dy$, where a is a positive constant, by two methods. (*Hint:* Integrate in the order shown and then reverse the order of x and y.)

16. $\int_0^1 \int_z^{2z} \int_0^{y+z} \dfrac{e^y e^z}{e^x} dx\,dy\,dz$

17. $\int_0^1 \int_0^y \int_0^{xy} x^a y^b z^c\,dz\,dx\,dy$, where a, b, and c are nonnegative constants.

18. $\int_1^2 \int_0^x \int_0^y (x + y + z)xyz\,dz\,dy\,dx$

19. $\int_0^a \int_0^{\sqrt{a^2-z^2}} \int_0^{\sqrt{y^2+z^2}} (y + yz^2)\,dx\,dy\,dz$

20. $\int_{-1}^1 \int_{y^2}^y \int_0^{\sqrt{x+4y}} yz\,dz\,dx\,dy$

In Exercises 21 through 24, replace the iterated integral by an equivalent one (or a sum of two iterated integrals) where the order of integration is reversed. Sketch the region of integration showing the elements for both orders of integration.

21. $\int_{-4}^4 \int_{y^2}^{16} f(x, y)\,dx\,dy$

22. $\int_0^1 \int_{-x^2}^x f(x, y)\,dy\,dx$

23. $\int_{-3}^1 \int_{|y|}^{(3-y)/2} f(x, y)\,dx\,dy$

24. $\int_{-a}^a \int_{\sqrt{a^2-x^2}}^a f(x, y)\,dy\,dx$

25. Set up the double integral for the volume of the region bounded by the plane $x + y + z = 6$ and the parabolic cylinder $y = x^2$.

26. Find the volume under the surface $z = 3x + 2$ and over the xy-plane, bounded by $y = e^x$, $y = 0$, $x = 0$, and $x = a$, where a is a positive constant.

27. Evaluate $\int_0^a \int_{-\sqrt{ax-x^2}}^{\sqrt{ax-x^2}} (x^2 + y^2)^{3/2}\,dy\,dx$, where a is a positive constant, by transforming to polar coordinates

28. Evaluate $\int_{-a}^a \int_0^{\sqrt{a^2-y^2}} y^4\,dx\,dy$, where a is a positive constant.

29. Find the average value of a temperature $T(x) = 1 + 3x^2$ defined on a triangular region in the xy-plane bounded by the lines $y = 2 - \dfrac{x}{2}$, $y = \frac{1}{4}(x - 4)$, and $x = 0$. (See Exercise 13 of Section 18.1.)

30. Find the total mass of a lamina in the shape of a trapezoid bounded by the lines $x = 0$, $y = 0$, $x = 3$, and $y = \dfrac{x}{3} + 2$ if its surface density $\rho = \rho_0 xy$, where ρ_0 is a positive constant.

31. Find the total mass of a lamina in the shape of a circle bounded by $r = 2a \cos \theta$ if its surface density $\rho = \rho_0 x^2 y^2$, where a and ρ_0 are positive constants.

32. Find the center of mass of a lamina in the shape of a square bounded by $x = 0$, $x = a$, $y = 0$, and $y = a$, if its surface density $\rho = \rho_0 xy(x + y)$ where a and ρ_0 are positive constants.

33. Find the center of mass of a lamina in the shape of a triangle with vertices $(0, 0)$, $(a, 0)$, and (a, a) if its surface density $\rho = \rho_0(x^2 + y^2)$, where a and ρ_0 are positive constants.

34. Find I_x, I_y, and I_o for the lamina of Exercise 32.

35. Find I_x, I_y, and I_o for the lamina of Exercise 33.

36. Find the area bounded by $r = 3 \csc \theta$, $\theta = \dfrac{\pi}{6}$ and $\theta = \dfrac{3\pi}{4}$. Interpret your result.

37. Find the area inside $r = 2a \cos \theta$ and outside $r = a$, where a is a positive constant.

38. Find the volume of the solid bounded above by the plane $z = x + 2y + 3$, below by the xy-plane, and on the sides by the cylinder $x^2 + y^2 = 16$.

39. Find the area of the portion of $z = 3 + x + \frac{1}{2}y^2$, which lies above $0 \le x \le y$ and $0 \le y \le 4$.

40. Find the area of the piece of $z = (x^2 + y^2)^{1/2}$ which lies above $0 \leq y \leq (4a^2 - x^2)^{1/2}$ and $0 \leq x \leq 2a$, where a is a positive constant.

41. Find the area of the portion of the sphere $x^2 + y^2 + z^2 = a^2$ lying over the circle $r = a \sin \theta$, $z = 0$, where a is a positive constant.

In Exercises 42 through 50, triple integration is to be used.

42. Find the volume of the solid formed in the first octant by the plane $2x + 4y + 3z = 12$ and the coordinate planes.

43. Find the volume of the solid formed by the right circular cylinders $x^2 + y^2 = 9$ and $z^2 + x^2 = 9$.

44. Find the volume of the solid formed by $z = 2a^2 - x^2 - y^2$ and $z = x^2 + y^2$, where a is a positive constant.

45. Find the mass of a cube of side a if its density is proportional to the fourth power of its distance from one edge.

46. Find the mass of a tetrahedron bounded by the coordinate planes and $x + y + z = a$ if the density $\delta = kxyz$, where k and a are positive constants.

47. Find the center of mass of the solid bounded by the planes $x = a$, $y = b$, and $z = c$ and the coordinate planes, where the density $\delta = kx$ and k, a, b, and c are positive constants.

48. For the same boundaries as Exercise 47, find the center of mass if $\delta = kxyz$, where k is a positive constant.

49. Find the mass of a spherical shell with inside radius a and outside radius b, if its density is proportional to the square of its distance from the center.

50. Find the mass of a solid bounded by $x^2 + y^2 + z^2 = 2a^2$ and $az = x^2 + y^2$ if its density is proportional to the distance from the xy-plane and a is a positive constant.

51. A solid homogeneous box occupies the space $0 \leq x \leq a$, $0 \leq y \leq b$, and $0 \leq z \leq c$. Find the moment of inertia and radius of gyration with respect to the x-axis.

52. For the solid in Exercise 51, find the moment of inertia with respect to an axis L parallel to the x-axis and passing through the center of gravity (or centroid in this case). Verify that $I_L = I_x - M\left(\dfrac{b^2 + c^2}{4}\right)$, where M is the mass of the box.

53. Consider a homogeneous solid sphere of radius a with center at the origin of rectangular coordinates in three-space. Find I_L, where L is a line parallel to the x-axis passing through $(0, 0, a)$ and verify that $I_L = I_x + Ma^2$ where M is the mass of the sphere. (*Hint:* $I_x = \frac{2}{5}Ma^2$.)

54. Evaluate $\displaystyle\int_{-\infty}^{\infty} \int_{-\infty}^{\infty} \int_{-\infty}^{\infty} e^{-(x^2+y^2+z^2)} \, dx \, dy \, dz$ by two methods (i) using rectangular coordinates and (ii) using spherical coordinates. $\left(Hint: \displaystyle\int_{0}^{\infty} e^{-t^2} \, dt = \dfrac{\sqrt{\pi}}{2}\right)$

Appendix A
Review Formulas

ALGEBRA

$$\frac{a}{d} + \frac{b}{d} = \frac{a+b}{d} \qquad \frac{a}{b} + \frac{c}{d} = \frac{ad+bc}{bd} \qquad \frac{a}{b} \cdot \frac{c}{d} = \frac{ac}{bd}$$

$$\frac{a}{b} \div \frac{c}{d} = \frac{a}{b} \cdot \frac{d}{c} = \frac{ad}{bc} \qquad -\frac{a}{b} = \frac{a}{-b} = -\frac{a}{b}$$

The solutions of the quadratic equation

$$ax^2 + bx + c = 0$$

are

$$x = \frac{1}{2a}[-b \pm \sqrt{b^2 - 4ac}]$$

The roots are real and unequal if $b^2 - 4ac > 0$; real and equal if $b^2 - 4ac = 0$ imaginary when $b^2 - 4ac < 0$;

$$(a + b)^2 = a^2 + 2ab + b^2 \qquad (a - b)^2 = a^2 - 2ab + b^2$$
$$a^2 - b^2 = (a - b)(a + b) \qquad (a^3 - b^3) = (a - b)(a^2 + ab + b^2)$$
$$(a^3 + b^3) = (a + b)(a^2 - ab + b^2)$$
$$(a^n - b^n) = (a - b)(a^{n-1} + a^{n-2}b + a^{n-3}b^2 + \cdots + b^{n-1})$$

$$a^4 + b^4 = a^4 + 2a^2b^2 + b^4 - 2a^2b^2$$
$$= (a^2 + b^2)^2 - 2a^2b^2$$
$$= (a^2 + b^2 - \sqrt{2}ab)(a^2 + b^2 + \sqrt{2}ab)$$

$$(a + b)^n = a^n + na^{n-1}b + \frac{n(n-1)}{1 \cdot 2}a^{n-2}b^2 + \cdots + nab^{n-1} + b^n \text{ where } n \text{ is a}$$

positive integer.   (Binomial Theorem)

n-factorial: $n! = n(n - 1)(n - 2) \ldots (3)(2)(1)$, n = positive integer, 0-factorial: $0! = 1$

PLANE AND SOLID GEOMETRY

Triangles

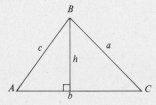

area $\triangle ABC = \dfrac{bh}{2} = \dfrac{1}{2}bc \sin A$

$\qquad\qquad = \sqrt{s(s-a)(s-b)(s-c)}$

where

$\qquad s = \dfrac{1}{2}(a + b + c)$ is the semiperimeter

Trapezoids

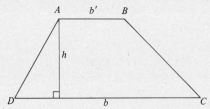

area trapezoid $ABCD = \dfrac{h}{2}(b + b')$

Circles

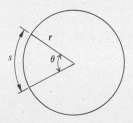

circumference of circle $= 2\pi r$

area of circle $= \pi r^2$

length of circular sector $= s = r\theta$ (θ in radians)

area of circular sector $= \dfrac{1}{2}r^2\theta$

Prisms and Cylinders

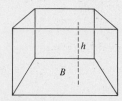

volume of prism $= Bh$ where B is the area of a base and h is the altitude

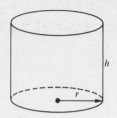

volume of a right circular cylinder $= \pi r^2 h$
where r = radius of base and h = altitude

lateral surface area of a right circular
cylinder $= 2\pi rh$

total surface area of a right circular
cylinder $= 2\pi rh + 2\pi r^2$

Pyramids and Cones

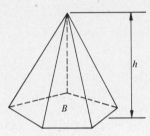

volume of pyramid $= \frac{1}{3}Bh$

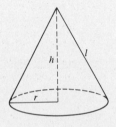

volume of right circular cone $= \frac{1}{3}\pi r^2 h$

lateral surface area of right circular cone
$= \pi rl = \pi r\sqrt{r^2 + h^2}$

total surface area $= \pi rl + \pi r^2$

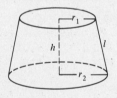

volume of a frustum of a right circular

cone $= \frac{\pi h}{3}(r_1^2 + r_2^2 + r_1 r_2)$

lateral surface area of a frustum of a right
circular cone $= \pi(r_1 + r_2)l$

total surface area of a frustum of a right
circular cone $= \pi[(r_1 + r_2)l + r_1^2 + r_2^2]$

Sphere

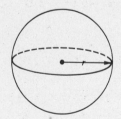

volume of sphere $= \frac{4}{3}\pi r^3$

surface area of sphere $= 4\pi r^2$

TRIGONOMETRY

Definitions

$$\tan \theta = \frac{\sin \theta}{\cos \theta} \qquad \cot \theta = \frac{\cos \theta}{\sin \theta} \qquad \sec \theta = \frac{1}{\cos \theta} \qquad \csc \theta = \frac{1}{\sin \theta}$$

Pythagorean Identities

$$\sin^2 \theta + \cos^2 \theta = 1 \qquad \tan^2 \theta + 1 = \sec^2 \theta \qquad \cot^2 \theta + 1 = \csc^2 \theta$$

Addition Formulas

$$\sin (\theta + \phi) = \sin \theta \cos \phi + \cos \theta \sin \phi$$
$$\sin (\theta - \phi) = \sin \theta \cos \phi - \cos \theta \sin \phi$$
$$\cos (\theta + \phi) = \cos \theta \cos \phi - \sin \theta \sin \phi$$
$$\cos (\theta - \phi) = \cos \theta \cos \phi + \sin \theta \sin \phi$$

$$\tan (\theta + \phi) = \frac{\tan \theta + \tan \phi}{1 - \tan \theta \tan \phi}$$

$$\tan (\theta - \phi) = \frac{\tan \theta - \tan \phi}{1 + \tan \theta \tan \phi}$$

Reduction Formulas

$$\sin (-\theta) = -\sin \theta \qquad \cos (-\theta) = \cos \theta \qquad \tan (-\theta) = -\tan \theta$$

$$\sin \left(\frac{\pi}{2} - \theta\right) = \cos \theta \qquad \sin \left(\frac{\pi}{2} + \theta\right) = \cos \theta$$

$$\cos \left(\frac{\pi}{2} - \theta\right) = \sin \theta \qquad \cos \left(\frac{\pi}{2} + \theta\right) = -\sin \theta$$

$$\tan \left(\frac{\pi}{2} - \theta\right) = \cot \theta \qquad \tan \left(\frac{\pi}{2} + \theta\right) = -\cot \theta$$

$$\sin (\pi - \theta) = \sin \theta \qquad \sin (\pi + \theta) = -\sin \theta$$
$$\cos (\pi - \theta) = -\cos \theta \qquad \cos (\pi + \theta) = -\cos \theta$$
$$\tan (\pi - \theta) = -\tan \theta \qquad \tan (\pi + \theta) = \tan \theta$$

Sum and Difference Formulas

$$\cos \theta \cos \phi = \tfrac{1}{2}[\cos (\theta + \phi) + \cos (\theta - \phi)]$$
$$\sin \theta \sin \phi = \tfrac{1}{2}[\cos (\theta - \phi) - \cos (\theta + \phi)]$$
$$\sin \theta \cos \phi = \tfrac{1}{2}[\sin (\theta + \phi) + \sin (\theta - \phi)]$$

Double and Half-Angle Relations

$$\sin 2\theta = 2 \sin \theta \cos \theta$$
$$\cos 2\theta = \cos^2 \theta - \sin^2 \theta$$
$$= 2 \cos^2 \theta - 1$$
$$= 1 - 2 \sin^2 \theta$$
$$\cos^2 \theta = \frac{1 + \cos 2\theta}{2}$$
$$\sin^2 \theta = \frac{1 - \cos 2\theta}{2}$$

Law of Sines and Cosines

If A, B and C are the interior angles of a triangle and a, b and c, respectively, are the sides opposite to these angles, then

$$\frac{a}{\sin A} = \frac{b}{\sin B} = \frac{c}{\sin C} \qquad \text{(Law of sines)}$$

$$c^2 = a^2 + b^2 - 2ab \cos C \qquad \text{(Law of cosines)}$$

Appendix B
Tables

Table 1   Squares, Cubes, Square Roots, and Cube Roots

n	n^2	n^3	$\sqrt{n}$	$\sqrt{10n}$	$\sqrt[3]{n}$	$\sqrt[3]{10n}$	$\sqrt[3]{100n}$
1	1	1	1.000 000	3.162 278	1.000 000	2.154 435	4.641 589
2	4	8	1.414 214	4.472 136	1.259 921	2.714 418	5.848 035
3	9	27	1.732 051	5.477 226	1.442 250	3.107 233	6.694 330
4	16	64	2.000 000	6.324 555	1.587 401	3.419 952	7.368 063
5	25	125	2.236 068	7.071 068	1.709 976	3.684 031	7.937 005
6	36	216	2.449 490	7.745 967	1.817 121	3.914 868	8.434 327
7	49	343	2.645 751	8.366 600	1.912 931	4.121 285	8.879 040
8	64	512	2.828 427	8.944 272	2.000 000	4.308 869	9.283 178
9	81	729	3.000 000	9.486 833	2.080 084	4.481 405	9.654 894
10	100	1 000	3.162 278	10.000 00	2.154 435	4.641 589	10.000 00
11	121	1 331	3.316 625	10.488 09	2.223 980	4.791 420	10.322 80
12	144	1 728	3.464 102	10.954 45	2.289 428	4.932 424	10.626 59
13	169	2 197	3.605 551	11.401 75	2.351 335	5.065 797	10.913 93
14	196	2 744	3.741 657	11.832 16	2.410 142	5.192 494	11.186 89
15	225	3 375	3.872 983	12.247 45	2.466 212	5.313 293	11.447 14
16	256	4 096	4.000 000	12.649 11	2.519 842	5.428 835	11.696 07
17	289	4 913	4.123 106	13.038 40	2.571 282	5.539 658	11.934 83
18	324	5 832	4.242 641	13.416 41	2.620 741	5.646 216	12.164 40
19	361	6 859	4.358 899	13.784 05	2.668 402	5.748 897	12.385 62
20	400	8 000	4.472 136	14.142 14	2.714 418	5.848 035	12.599 21
21	441	9 261	4.582 576	14.491 38	2.758 924	5.943 922	12.805 79
22	484	10 648	4.690 416	14.832 40	2.802 039	6.036 811	13.005 91
23	529	12 167	4.795 832	15.165 75	2.843 867	6.126 926	13.200 06
24	576	13 824	4.898 979	15.491 93	2.884 499	6.214 465	13.388 66
25	625	15 625	5.000 000	15.811 39	2.924 018	6.299 605	13.572 09
26	676	17 576	5.099 020	16.124 52	2.962 496	6.382 504	13.750 69
27	729	19 683	5.196 152	16.431 68	3.000 000	6.463 304	13.924 77
28	784	21 952	5.291 503	16.733 20	3.036 589	6.542 133	14.094 60
29	841	24 389	5.385 165	17.029 39	3.072 317	6.619 106	14.260 43

Table 1 1109

Table 1   Squares, Cubes, Square Roots, and Cube Roots (*Continued*)

n	n^2	n^3	$\sqrt{n}$	$\sqrt{10n}$	$\sqrt[3]{n}$	$\sqrt[3]{10n}$	$\sqrt[3]{100n}$
30	900	27 000	5.477 226	17.320 51	3.107 233	6.694 330	14.422 50
31	961	29 791	5.567 764	17.606 82	3.141 381	6.767 899	14.581 00
32	1 024	32 768	5.656 854	17.888 54	3.174 802	6.839 904	14.736 13
33	1 089	35 937	5.744 563	18.165 90	3.207 534	6.910 423	14.888 06
34	1 156	39 304	5.830 952	18.439 09	3.239 612	6.979 532	15.036 95
35	1 225	42 875	5.916 080	18.708 29	3.271 066	7.047 299	15.182 94
36	1 296	46 656	6.000 000	18.973 67	3.301 927	7.113 787	15.326 19
37	1 369	50 653	6.082 763	19.235 38	3.332 222	7.179 054	15.466 80
38	1 444	54 872	6.164 414	19.493 59	3.361 975	7.243 156	15.604 91
39	1 521	59 319	6.244 998	19.748 42	3.391 211	7.306 144	15.740 61
40	1 600	64 000	6.324 555	20.000 00	3.419 952	7.368 063	15.874 01
41	1 681	68 921	6.403 124	20.248 46	3.448 217	7.428 959	16.005 21
42	1 764	74 088	6.480 741	20.493 90	3.476 027	7.488 872	16.134 29
43	1 849	79 507	6.557 439	20.736 44	3.503 398	7.547 842	16.261 33
44	1 936	85 184	6.633 250	20.976 18	3.530 348	7.605 905	16.386 43
45	2 025	91 125	6.708 204	21.213 20	3.556 893	7.663 094	16.509 64
46	2 116	97 336	6.782 330	21.447 61	3.583 048	7.719 443	16.631 03
47	2 208	103 823	6.855 655	21.679 48	3.608 826	7.774 980	16.760 69
48	2 304	110 592	6.928 203	21.908 90	3.634 241	7.829 735	16.868 65
49	2 401	117 649	7.000 000	22.135 94	3.659 306	7.883 735	16.984 99
50	2 500	125 000	7.071 068	22.360 68	3.684 031	7.937 005	17.099 76
51	2 601	132 651	7.141 428	22.583 18	3.708 430	7.989 570	17.213 01
52	2 704	140 608	7.211 103	22.803 51	3.732 511	8.041 452	17.324 78
53	2 809	148 877	7.280 110	23.021 73	3.756 286	8.092 672	17.435 13
54	2 916	157 464	7.348 469	23.237 90	3.779 763	8.143 253	17.544 11
55	3.025	166 375	7.416 198	23.452 08	3.802 952	8.193 213	17.651 74
56	3 136	175 616	7.483 315	23.664 32	3.825 862	8.242 571	17.758 08
57	3 249	185 193	7.549 834	23.874 67	3.848 501	8.291 344	17.863 16
58	3 364	195 112	7.615 773	24.083 19	3.870 877	8.339 551	17.967 02
59	3 481	205 379	7.681 146	24.289 92	3.892 996	8.387 207	18.069 69
60	3 600	216 000	7.745 967	24.494 90	3.914 868	8.434 327	18.171 21
61	3 721	226 981	7.810 250	24.698 18	3.936 497	8.480 926	18.271 60
62	3 844	238 328	7.874 008	24.899 80	3.957 892	8.527 019	18.370 91
63	3 969	250 047	7.937 254	25.099 80	3.979 057	8.572 619	18.469 15
64	4 096	262 144	8.000 000	25.298 22	4.000 000	8.617 739	18.566 36
65	4 225	274 625	8.062 258	25.495 10	4.020 726	8.662 391	18.662 56
66	4 356	287 496	8.124 038	25.690 47	4.041 240	8.706 588	18.757 77
67	4 489	300 763	8.185 353	25.884 36	4.061 548	8.750 340	18.852 04
68	4 624	314 432	8.246 211	26.076 81	4.081 655	8.793 659	18.945 36
69	4 761	328 509	8.306 624	26.267 85	4.101 566	8.836 556	19.037 78
70	4 900	343 000	8.366 600	26.457 51	4.121 285	8.879 040	19.129 31
71	5 041	357 911	8.426 150	26.645 83	4.140 818	8.921 121	19.219 97
72	5 184	373 248	8.485 281	26.832 82	4.160 168	8.962 809	19.309 79
73	5 329	389 017	8.544 004	27.018 51	4.179 339	9.004 113	19.398 77
74	5 476	405 224	8.602 325	27.202 94	4.198 336	9.045 042	19.486 95

Table 1   Squares, Cubes, Square Roots, and Cube Roots (*Continued*)

n	n^2	n^3	$\sqrt{n}$	$\sqrt{10n}$	$\sqrt[3]{n}$	$\sqrt[3]{10n}$	$\sqrt[3]{100n}$
75	5 625	421 875	8.660 254	27.386 13	4.217 163	9.085 603	19.574 34
76	5 776	438 976	8.717 798	27.568 10	4.235 824	9.125 805	19.660 95
77	5 929	456 533	8.774 964	27.748 87	4.254 321	9.165 656	19.746 81
78	6 084	474 552	8.831 761	27.928 48	4.272 659	9.205 164	19.831 92
79	6 241	493 039	8.888 194	28.106 94	4.290 840	9.244 335	19.916 32
80	6 400	512 000	8.944 272	28.284 27	4.308 869	9.283 178	20.000 00
81	6 561	531 441	9.000 000	28.460 50	4.326 749	9.321 698	20.082 99
82	6 724	551 368	9.055 385	28.635 64	4.344 481	9.359 902	20.165 30
83	6 889	571 787	9.110 434	28.809 72	4.362 071	9.397 796	20.246 94
84	7 056	592 704	9.165 151	28.982 75	4.379 519	9.435 388	20.327 93
85	7 225	614 125	9.219 544	29.154 76	4.396 830	9.472 682	20.408 28
86	7 396	636 056	9.273 618	29.325 76	4.414 005	9.509 685	20.488 00
87	7 569	658 503	9.327 379	29.495 76	4.431 048	9.546 403	20.567 10
88	7 744	681 472	9.380 832	29.664 79	4.447 960	9.582 840	20.645 60
89	7 921	704 969	9.433 981	29.832 87	4.464 745	9.619 002	20.723 51
90	8 100	729 000	9.486 833	30.000 00	4.481 405	9.654 894	20.800 84
91	8 281	753 571	9.539 392	30.166 21	4.497 941	9.690 521	20.877 59
92	8 464	778 688	9.591 663	30.331 50	4.514 357	9.725 888	20.953 79
93	8 649	804 357	9.643 651	30.495 90	4.530 655	9.761 000	21.029 44
94	8 836	830 584	9.695 360	30.659 42	4.546 836	9.795 861	21.104 54
95	9 025	857 375	9.746 794	30.822 07	4.562 903	9.830 476	21.179 12
96	9 216	884 736	9.797 959	30.983 87	4.578 857	9.864 848	21.253 17
97	9 409	912 673	9.848 858	31.144 82	4.594 701	9.898 983	21.326 71
98	9 604	941 192	9.899 495	31.304 95	4.610 436	9.932 884	21.399 75
99	9 801	970 299	9.949 874	31.464 27	4.626 065	9.966 555	21.472 29

Table 1 1111

Table 2 Natural Logarithms

N	0	1	2	3	4	5	6	7	8	9
1.0	0000	0100	0198	0296	0392	0488	0583	0677	0770	0862
1.1	0953	1044	1133	1222	1310	1398	1484	1570	1655	1740
1.2	1823	1906	1989	2070	2151	2231	2311	2390	2469	2546
1.3	2624	2700	2776	2852	2927	3001	3075	3148	3221	3293
1.4	3365	3436	3507	3577	3646	3716	3784	3853	3920	3988
1.5	4055	4121	4187	4253	4318	4383	4447	4511	4574	4637
1.6	4700	4762	4824	4886	4947	5008	5068	5128	5188	5247
1.7	5306	5365	5423	5481	5539	5596	5653	5710	5766	5822
1.8	5878	5933	5988	6043	6098	6152	6206	6259	6313	6366
1.9	6419	6471	6523	6575	6627	6678	6729	6780	6831	6881
2.0	6931	6981	7031	7080	7129	7178	7227	7275	7324	7372
2.1	7419	7467	7514	7561	7608	7655	7701	7747	7793	7839
2.2	7885	7930	7975	8020	8065	8109	8154	8198	8242	8286
2.3	8329	8372	8416	8459	8502	8544	8587	8629	8671	8713
2.4	8755	8796	8838	8879	8920	8961	9002	9042	9083	9123
2.5	9163	9203	9243	9282	9322	9361	9400	9439	9478	9517
2.6	9555	9594	9632	9670	9708	9746	9783	9821	9858	9895
2.7	9933	9969	°0006	°0043	°0080	°0116	°0152	°0188	°0225	°0260
2.8	1.0296	0332	0367	0403	0438	0473	0508	0543	0578	0613
2.9	0647	0682	0716	0750	0784	0818	0852	0886	0919	0953
3.0	1.0986	1019	1053	1086	1119	1151	1184	1217	1249	1282
3.1	1314	1346	1378	1410	1442	1474	1506	1537	1569	1600
3.2	1632	1663	1694	1725	1756	1787	1817	1848	1878	1909
3.3	1939	1969	2000	2030	2060	2090	2119	2149	2179	2208
3.4	2238	2267	2296	2326	2355	2384	2413	2442	2470	2499
3.5	1.2528	2556	2585	2613	2641	2669	2698	2726	2754	2782
3.6	2809	2837	2865	2892	2920	2947	2975	3002	3029	3056
3.7	3083	3110	3137	3164	3191	3218	3244	3271	3297	3324
3.8	3350	3376	3403	3429	3455	3481	3507	3533	3558	3584
3.9	3610	3635	3661	3686	3712	3737	3762	3788	3813	3838
4.0	1.3863	3888	3913	3938	3962	3987	4012	4036	4061	4085
4.1	4110	4134	4159	4183	4207	4231	4255	4279	4303	4327
4.2	4351	4375	4398	4422	4446	4469	4493	4516	4540	4563
4.3	4586	4609	4633	4656	4679	4702	4725	4748	4770	4793
4.4	4816	4839	4861	4884	4907	4929	4951	4974	4996	5019
4.5	1.5041	5063	5085	5107	5129	5151	5173	5195	5217	5239
4.6	5261	5282	5304	5326	5347	5369	5390	5412	5433	5454
4.7	5476	5497	5518	5539	5560	5581	5602	5623	5644	5665
4.8	5686	5707	5728	5748	5769	5790	5810	5831	5851	5872
4.9	5892	5913	5933	5953	5974	5994	6014	6034	6054	6074
5.0	1.6094	6114	6134	6154	6174	6194	6214	6233	6253	6273
5.1	6292	6312	6332	6351	6371	6390	6409	6429	6448	6467
5.2	6487	6506	6525	6544	6563	6582	6601	6620	6639	6658
5.3	6677	6696	6715	6734	6752	6771	6790	6808	6827	6845
5.4	6864	6882	6901	6919	6938	6956	6974	6993	7011	7029

Table 2 **Natural Logarithms** (*Continued*)

N	0	1	2	3	4	5	6	7	8	9
5.5	1.7047	7066	7084	7102	7120	7138	7156	7174	7192	7210
5.6	7228	7246	7263	7281	7299	7317	7334	7352	7370	7387
5.7	7405	7422	7440	7457	7475	7492	7509	7527	7544	7561
5.8	7579	7596	7613	7630	7647	7664	7681	7699	7716	7733
5.9	7750	7766	7783	7800	7817	7834	7851	7867	7884	7901
6.0	1.7918	7934	7951	7967	7984	8001	8017	8034	8050	8066
6.1	8083	8099	8116	8132	8148	8165	8181	8197	8213	8229
6.2	8245	8262	8278	8294	8310	8326	8342	8358	8374	8390
6.3	8405	8421	8437	8453	8469	8485	8500	8516	8532	8547
6.4	8563	8579	8594	8610	8625	8641	8656	8672	8687	8706
6.5	1.8718	8733	8749	8764	8779	8795	8810	8825	8840	8856
6.6	8871	8886	8901	8916	8931	8946	8961	8976	8991	9006
6.7	9021	9036	9051	9066	9081	9095	9110	9125	9140	9155
6.8	9169	9184	9199	9213	9228	9242	9257	9272	9286	8301
6.9	9315	9330	9344	9359	9373	9387	9402	9416	9430	9445
7.0	1.9459	9473	9488	9502	9516	9530	9544	9559	9573	9587
7.1	9601	9615	9629	9643	9657	9671	9685	9699	9713	9727
7.2	9741	9755	9769	9782	9796	9810	9824	9838	9851	9865
7.3	9879	9892	9906	9920	9933	9947	9961	9974	9988	°0001
7.4	2.0015	0028	0042	0055	0069	0082	0096	0109	0122	0136
7.5	2.0149	0162	0176	0189	0202	0215	0229	0242	0255	0268
7.6	0281	0295	0308	0321	0334	0347	0360	0373	0386	0399
7.7	0412	0425	0438	0451	0464	0477	0490	0503	0516	0528
7.8	0541	0554	0567	0580	0592	0605	0618	0630	0643	0656
7.9	0669	0681	0694	0707	0719	0732	0744	0757	0769	0782
8.0	2.0794	0807	0819	0832	0844	0857	0869	0882	0894	0906
8.1	0919	0931	0943	0956	0968	0980	0992	1005	1017	1029
8.2	1041	1054	1066	1078	1090	1102	1114	1126	1138	1150
8.3	1163	1175	1187	1199	1211	1223	1235	1247	1258	1270
8.4	1282	1294	1306	1318	1330	1342	1353	1365	1377	1389
8.5	2.1401	1412	1424	1436	1448	1459	1471	1483	1494	1506
8.6	1518	1529	1541	1552	1564	1576	1587	1599	1610	1622
8.7	1633	1645	1656	1668	1679	1691	1702	1713	1725	1736
8.8	1748	1759	1770	1782	1793	1804	1815	1827	1838	1849
8.9	1861	1872	1883	1894	1905	1917	1928	1939	1950	1961
9.0	2.1972	1983	1994	2006	2017	2028	2039	2050	2061	2072
9.1	2083	2094	2105	2116	2127	2138	2148	2159	2170	2181
9.2	2192	2203	2214	2225	2235	2246	2257	2268	2279	2289
9.3	2300	2311	2322	2332	2343	2354	2364	2375	2386	2396
9.4	2407	2418	2428	2439	2450	2460	2471	2481	2492	2502
9.5	2.2513	2523	2534	2544	2555	2565	2576	2586	2597	2607
9.6	2618	2628	2638	2649	2659	2670	2680	2690	2701	2711
9.7	2721	2732	2742	2752	2762	2773	2783	2793	2803	2811
9.8	2824	2834	2844	2854	2865	2875	2885	2895	2905	2915
9.9	2925	2935	2946	2956	2966	2976	2986	2996	3006	3016

Table 2 1113

Table 3 **Exponential Functions**

x	e^x	$\log_{10}(e^x)$	e^{-x}	x	e^x	$\log_{10}(e^x)$	e^{-x}
0.00	1.0000	0.00000	1.000000	**0.50**	1.6487	0.21715	0.606531
0.01	1.0101	.00434	0.990050	0.51	1.6653	.22149	.600496
0.02	1.0202	.00869	.980199	0.52	1.6820	.22583	.594521
0.03	1.0305	.01303	.970446	0.53	1.6989	.23018	.588605
0.04	1.0408	.01737	.960789	0.54	1.7160	.23452	.582748
0.05	1.0513	0.02171	0.951229	**0.55**	1.7333	0.23886	0.576950
0.06	1.0618	.02606	.941765	0.56	1.7507	.24320	.571209
0.07	1.0725	.03040	.932394	0.57	1.7683	.24755	.565525
0.08	1.0833	.03474	.923116	0.58	1.7860	.25189	.559898
0.09	1.0942	.03909	.913931	0.59	1.8040	.35623	.554327
0.10	1.1052	0.04343	0.904837	**0.60**	1.8221	0.26058	0.548812
0.11	1.1163	.04777	.895834	0.61	1.8404	.26492	.543351
0.12	1.1275	.05212	.886920	0.62	1.8589	.26926	.537944
0.13	1.1388	.05646	.878095	0.63	1.8776	.27361	.532592
0.14	1.1503	.06080	.869358	0.64	1.8965	.27795	.527292
0.15	1.1618	0.06514	0.860708	**0.65**	1.9155	0.28229	0.522046
0.16	1.1735	.06949	.852144	0.66	1.9348	.28663	.516851
0.17	1.1853	.07383	.843665	0.67	1.9542	.29098	.511709
0.18	1.1972	.07817	.835270	0.68	1.9739	.29532	.506617
0.19	1.2092	.08252	.826959	0.69	1.9937	.29966	.501576
0.20	1.2214	0.08686	0.818731	**0.70**	2.0138	0.30401	0.496585
0.21	1.2337	.09120	.810584	0.71	2.0340	.30835	.491644
0.22	1.2461	.09554	.802519	0.72	2.0544	.31269	.486752
0.23	1.2586	.09989	.794534	0.73	2.0751	.31703	.481909
0.24	1.2712	.10423	.786628	0.74	2.0959	.32138	.477114
0.25	1.2840	0.10857	0.778801	**0.75**	2.1170	0.32572	0.472367
0.26	1.2969	.11292	.771052	0.76	2.1383	.33006	.467666
0.27	1.3100	.11726	.763379	0.77	2.1598	.33441	.463013
0.28	1.3231	.12160	.755784	0.78	2.1815	.33875	.458406
0.29	1.3364	.12595	.748264	0.79	2.2034	.34309	.453845
0.30	1.3499	0.13029	0.740818	**0.80**	2.2255	0.34744	0.449329
0.31	1.3634	.13463	.733447	0.81	2.2479	.35178	.444858
0.32	1.3771	.13897	.726149	0.82	2.2705	.35612	.440432
0.33	1.3910	.14332	.718924	0.83	2.2933	.36046	.436049
0.34	1.4049	.14766	.711770	0.84	2.3164	.36481	.431711
0.35	1.4191	0.15200	0.704688	**0.85**	2.3396	0.36915	0.427415
0.36	1.4333	.15635	.697676	0.86	2.3632	.37349	.423162
0.37	1.4477	.16069	.690734	0.87	2.3869	.37784	.418952
0.38	1.4623	.16503	.683861	0.88	2.4109	.38218	.414783
0.39	1.4770	.16937	.677057	0.89	2.4351	.38652	.410656
0.40	1.4918	0.17372	0.670320	**0.90**	2.4596	0.39087	0.406570
0.41	1.5068	.17806	.663650	0.91	2.4843	.39521	.402524
0.42	1.5220	.18240	.657047	0.92	2.5093	.39955	.398519
0.43	1.5373	.18675	.650509	0.93	2.5345	.40389	.394554
0.44	1.5527	.19109	.644036	0.94	2.5600	.40824	.390628

Table 3   Exponential Functions (*Continued*)

x	e^x	$\log_{10}(e^x)$	e^{-x}	x	e^x	$\log_{10}(e^x)$	e^{-x}
0.45	1.5683	0.19543	0.637628	**0.95**	2.5857	0.41258	0.386741
0.46	1.5841	.19978	.631284	0.96	2.6117	.41692	.382893
0.47	1.6000	.20412	.625002	0.97	2.6379	.42127	.379083
0.48	1.6161	.20846	.618783	0.98	2.6645	.42561	.375311
0.49	1.6323	.21280	.612626	0.99	2.6912	.42995	.371577
0.50	1.6487	0.21715	0.606531	**1.00**	2.7183	0.43429	0.367879
1.00	2.7183	0.43429	0.367879	**1.50**	4.4817	0.65144	0.223130
1.01	2.7456	.43864	.364219	1.51	4.5267	.65578	.220910
1.02	2.7732	.44298	.360595	1.52	4.5722	.66013	.218712
1.03	2.8011	.44732	.357007	1.53	4.6182	.66447	.216536
1.04	2.8292	.45167	.353455	1.54	4.6646	.66881	.214381
1.05	2.8577	0.45601	0.349938	**1.55**	4.7115	0.67316	0.212248
1.06	2.8864	.46035	.346456	1.56	4.7588	.67750	.210136
1.07	2.9154	.46470	.343009	1.57	4.8066	.68184	.208045
1.08	2.9447	.46904	.339596	1.58	4.8550	.68619	.205975
1.09	2.9743	.47338	.336216	1.59	4.9037	.69053	.203926
1.10	3.0042	0.47772	0.332871	**1.60**	4.9530	0.69487	0.201897
1.11	3.0344	.48207	.329559	1.61	5.0028	.69921	.199888
1.12	3.0649	.48641	.326280	1.62	5.0531	.70356	.197899
1.13	3.0957	.49075	.323033	1.63	5.1039	.70790	.195930
1.14	3.1268	.49510	.319819	1.64	5.1552	.71224	.193980
1.15	3.1582	0.49944	0.316637	**1.65**	5.2070	0.71659	0.192050
1.16	3.1899	.50378	.313486	1.66	5.2593	.72093	.190139
1.17	3.2220	.50812	.310367	1.67	5.3122	.72527	.188247
1.18	3.2544	.51247	.307279	1.68	5.3656	.72961	.186374
1.19	3.2871	.51681	.304221	1.69	5.4195	.73396	.184520
1.20	3.3201	0.52115	0.301194	**1.70**	5.4739	0.73830	0.182684
1.21	3.3535	.52550	.298197	1.71	5.5290	.74264	.180866
1.22	3.3872	.52984	.295230	1.72	5.5845	.74699	1.79066
1.23	3.4212	.53418	.292293	1.73	5.6407	.75133	.177284
1.24	3.4556	.53853	.289384	1.74	5.6973	.75567	.175520
1.25	3.4903	0.54287	0.286505	**1.75**	5.7546	0.76002	0.173774
1.26	3.5254	.54721	.283654	1.76	5.8124	.76436	.172045
1.27	3.5609	.55155	.280832	1.77	5.8709	.76870	.170333
1.28	3.5966	.55590	.278037	1.78	5.9299	.77304	.168638
1.29	3.6328	.56024	.275271	1.79	5.9895	.77739	.166960
1.30	3.6693	0.56458	0.272532	**1.80**	6.0496	0.78173	0.165299
1.31	3.7062	.56893	.269820	1.81	6.1104	.78607	.163654
1.32	3.7434	.57327	.267135	1.82	6.1719	.79042	.162026
1.33	3.7810	.57761	.264477	1.83	6.2339	.79476	.160414
1.34	3.8190	.58195	.261846	1.84	6.2965	.79910	.158817
1.35	3.8574	0.58630	0.259240	**1.85**	6.3598	0.80344	0.157237
1.36	3.8962	.59064	.256661	1.86	6.4237	.80779	.155673
1.37	3.9354	.59498	.254107	1.87	6.4483	.81213	.154124
1.38	3.9749	.59933	.251579	1.88	6.5535	.81647	.152590
1.39	4.0149	.60367	.249075	1.89	6.6194	.82082	.151072

Table 3 1115

Table 3   Exponential Functions (*Continued*)

x	e^x	$\log_{10}(e^x)$	e^{-x}	x	e^x	$\log_{10}(e^x)$	e^{-x}
1.40	4.0552	0.60801	0.246597	**1.90**	6.6859	0.82516	0.149569
1.41	4.0960	.61236	.244143	1.91	6.7531	.82950	.148080
1.42	4.1371	.61670	.241714	1.92	6.8210	.83385	.146607
1.43	4.1787	.62104	.239309	1.93	6.8895	.83819	.145148
1.44	4.2207	.62538	.236928	1.94	6.9588	.84253	.143704
1.45	4.2631	0.62973	0.234570	**1.95**	7.0287	0.84687	0.142274
1.46	4.3060	.63407	.232236	1.96	7.0993	.85122	.140858
1.47	4.3492	.63841	.229925	1.97	7.1707	.85556	.139457
1.48	4.3929	.64276	.227638	1.98	7.2427	.85990	.138069
1.49	4.4371	.64710	.225373	1.99	7.3155	.86425	.136695
1.50	4.4817	0.65144	0.223130	**2.00**	7.3891	0.86859	0.135335
2.00	7.3891	0.86859	0.135335	**2.50**	12.182	1.08574	0.082085
2.01	7.4633	.87293	.133989	2.51	12.305	1.09008	.081268
2.02	7.5383	.87727	.132655	2.52	12.429	1.09442	.080460
2.03	7.6141	.88162	.131336	2.53	12.554	1.09877	.079659
2.04	7.6906	.88596	.130029	2.54	12.680	1.10311	.078866
2.05	7.7679	0.89030	0.128735	**2.55**	12.807	1.10745	0.078082
2.06	7.8460	.89465	.127454	2.56	12.936	1.11179	.077305
2.07	7.9248	.89899	.126186	2.57	13.066	1.11614	.076536
2.08	8.0045	.90333	.124930	2.58	13.197	1.12048	.075774
2.09	8.0849	.90756	.123687	2.59	13.330	1.12482	.075020
2.10	8.1662	0.91202	0.122456	**2.60**	13.464	1.12917	0.074274
2.11	8.2482	.91636	.121238	2.61	13.599	1.13351	.073535
2.12	8.3311	.92070	.120032	2.62	13.736	1.13785	.072803
2.13	8.4149	.92505	.118837	2.63	13.874	1.14219	.072078
2.14	8.4994	.92939	.117655	2.64	14.013	1.14654	.071361
2.15	8.5849	0.93373	0.116484	**2.65**	14.154	1.15088	0.070651
2.16	8.6711	.93808	.115325	2.66	14.296	1.15522	.069948
2.17	8.7583	.94242	.114178	2.67	14.440	1.15957	.069252
2.18	8.8463	.94676	.113042	2.68	14.585	1.16391	.068563
2.19	8.9352	.95110	.111917	2.69	14.732	1.16825	.067881
2.20	9.0250	0.95545	0.110803	**2.70**	14.880	1.17260	0.067206
2.21	9.1157	.95979	.109701	2.71	15.029	1.17694	.066537
2.22	9.2073	.96413	.108609	2.72	15.180	1.18128	.065875
2.23	9.2999	.96848	.107528	2.73	15.333	1.18562	.065219
2.24	9.3933	.97282	.106459	2.74	15.487	1.18997	.064570
2.25	9.4877	0.97716	0.105399	**2.75**	15.643	1.19431	0.063928
2.26	9.5831	.98151	.104350	2.76	15.800	1.19865	.063292
2.27	9.6794	.98585	.103312	2.77	15.959	1.20300	.062662
2.28	9.7767	.99019	.102284	2.78	16.119	1.20734	.062039
2.29	9.8749	.99453	.101266	2.79	16.281	1.21168	.061421
2.30	9.9742	0.99888	0.100259	**2.80**	16.445	1.21602	0.060810
2.31	10.074	1.00322	.099261	2.81	16.610	1.22037	.060205
2.32	10.176	1.00756	.098274	2.82	16.777	1.22471	.059606
2.33	10.278	1.01191	.097296	2.83	16.945	1.22905	.059013
2.34	10.381	1.01625	.096328	2.84	17.116	1.23340	.058426

Table 3 Exponential Functions (*Continued*)

x	e^x	$\log_{10}(e^x)$	e^{-x}	x	e^x	$\log_{10}(e^x)$	e^{-x}
2.35	10.486	1.02059	0.095369	**2.85**	17.288	1.23774	0.057844
2.36	10.591	1.02493	.094420	2.86	17.462	1.24208	.057269
2.37	10.697	1.02928	.093481	2.87	17.637	1.24643	.056699
2.38	10.805	1.03362	.092551	2.88	17.814	1.25077	.056135
2.39	10.913	1.03796	.091630	2.89	17.993	1.25511	.055576
2.40	11.023	1.04231	0.090718	**2.90**	18.174	1.25945	0.055023
2.41	11.134	1.04665	.089815	2.91	18.357	1.26380	.054476
2.42	11.246	1.05099	.088922	2.92	18.541	1.26814	.053934
2.43	11.359	1.05534	.088037	2.93	18.728	1.27248	.053397
2.44	11.473	1.05968	.087161	2.94	18.916	1.27683	.052866
2.45	11.588	1.06402	0.086294	**2.95**	19.106	1.28117	0.052340
2.46	11.705	1.06836	.085435	2.96	19.298	1.28551	.051819
2.47	11.822	1.07271	.084585	2.97	19.492	1.28985	.051303
2.48	11.941	1.07705	.083743	2.98	19.688	1.29420	.050793
2.49	12.061	1.08139	.082910	2.99	19.886	1.29854	.050287
2.50	12.182	1.08574	0.082085	**3.00**	20.086	1.30288	0.049787
3.00	20.086	1.30288	0.049787	**3.50**	33.115	1.52003	0.030197
3.01	20.287	1.30723	.049292	3.51	33.448	1.52437	.029897
3.02	20.491	1.31157	.048801	3.52	33.784	1.52872	.029599
3.03	20.697	1.31591	.048316	3.53	34.124	1.53306	.029305
3.04	20.905	1.32026	.047835	3.54	34.467	1.53740	.029013
3.05	21.115	1.32460	0.047359	**3.55**	34.813	1.54175	0.028725
3.06	21.328	1.32894	.046888	3.56	35.163	1.54609	.028439
3.07	21.542	1.33328	.046421	3.57	35.517	1.55043	.028156
3.08	21.758	1.33763	.045959	3.58	35.874	1.55477	.027876
3.09	21.977	1.34197	.045502	3.59	36.234	1.55912	.027598
3.10	22.198	1.34631	0.045049	**3.60**	36.598	1.56346	0.027324
3.11	22.421	1.35066	.044601	3.61	36.966	1.56780	.027052
3.12	22.646	1.35500	.044157	3.62	37.338	57215	.026783
3.13	22.874	1.35934	.043718	3.63	37.713	1.57649	.026516
3.14	23.104	1.36368	.043283	3.64	38.092	1.58083	.026252
3.15	23.336	1.36803	0.042852	**3.65**	38.475	1.58517	0.025991
3.16	23.571	1.37237	.042426	3.66	38.861	1.58952	.025733
3.17	23.807	1.36671	.042004	3.67	39.252	1.59386	.025476
3.18	24.047	1.38106	.041586	3.68	39.646	1.59820	.025223
3.19	24.288	1.38540	.041172	3.69	40.045	1.60255	.024972
3.20	24.533	1.38974	0.040764	**3.70**	40.447	1.60689	0.024724
3.21	24.779	1.39409	.040357	3.71	40.854	1.61123	.024478
3.22	25.028	1.39843	.039955	3.72	41.264	1.61558	.024234
3.23	25.280	1.40277	.039557	3.73	41.679	1.61992	.023993
3.24	25.534	1.40711	.039164	3.74	42.098	1.62426	.023754
3.25	25.790	1.41146	0.038774	**3.75**	42.521	1.62860	0.023518
3.26	26.050	1.41580	.038388	3.76	42.948	1.63295	.023284
3.27	26.311	1.42014	.038006	3.77	43.380	1.63729	.023052
3.28	26.576	1.42449	.037628	3.78	43.816	1.64163	.022823
3.29	26.843	1.42883	.037254	3.79	44.256	1.64598	.022596

Table 3 1117

Table 3 Exponential Functions (*Continued*)

x	e^x	$\log_{10}(e^x)$	e^{-x}	x	e^x	$\log_{10}(e^x)$	e^{-x}
3.30	27.113	1.44317	0.036883	**3.80**	44.701	1.65032	0.022371
3.31	27.385	1.43751	.036516	3.81	45.150	1.65466	.022148
3.32	27.660	1.44186	.036153	3.82	45.604	1.65900	.021928
3.33	27.938	1.44620	.035793	3.83	46.063	1.66335	.021710
3.34	28.219	1.45054	.035437	3.84	46.525	1.66769	.021494
3.35	28.503	1.45489	0.035084	**3.85**	46.993	1.67203	0.021280
3.36	28.789	1.45923	.034735	3.86	47.465	1.67638	.021068
3.37	29.079	1.46357	.034390	3.87	47.942	1.68072	.020858
3.38	29.371	1.46792	.034047	3.88	48.424	1.68506	.020651
3.39	29.666	1.47226	.033709	3.89	48.911	1.68941	.020445
3.40	29.964	1.47660	0.033373	**3.90**	49.402	1.69375	0.020242
3.41	30.265	1.48094	.033041	3.91	49.899	1.69809	.020041
3.42	30.569	1.48529	.032712	3.92	50.400	1.70243	.019840
3.43	30.877	1.48963	.032387	3.93	50.907	1.70678	.019644
3.44	31.187	1.49397	.032065	3.94	51.419	1.71112	.019448
3.45	31.500	1.49832	0.031746	**3.95**	51.935	1.71546	0.019255
3.46	31.817	1.50266	.031430	3.96	52.457	1.71981	.019063
3.47	32.137	1.50700	.031117	3.97	52.985	1.72415	.018873
3.48	32.460	1.51134	.030807	3.98	53.517	1.72849	.018686
3.49	32.786	1.51569	.030501	3.99	54.055	1.73283	.018500
3.50	33.115	1.52003	0.030197	**4.00**	54.598	1.73718	0.018316
4.00	54.598	1.73718	0.018316	**4.50**	90.017	1.95433	0.011109
4.01	55.147	1.74152	.018133	4.51	90.922	1.95867	.010998
4.02	55.701	1.74586	.017953	4.52	91.836	1.96301	.010889
4.03	56.261	1.75021	.017774	4.53	92.759	1.96735	.010781
4.04	56.826	1.75455	.017597	4.54	93.691	1.97170	.010673
4.05	57.397	1.75889	0.017422	**4.55**	94.632	1.97604	0.010567
4.06	57.974	1.76324	.017249	4.56	95.583	1.98038	.010462
4.07	58.577	1.76758	.017077	4.57	96.544	1.98473	.010358
4.08	59.145	1.77192	.016907	4.58	97.514	1.98907	.010255
4.09	59.740	1.77626	.016739	4.59	98.494	1.99341	.010153
4.10	60.340	1.78061	0.016573	**4.60**	99.484	1.99775	0.010052
4.11	60.947	1.78495	.016408	4.61	100.48	2.00210	.009952
4.12	61.559	1.78929	.016245	4.62	101.49	2.00644	.009853
4.13	62.178	1.79364	.016083	4.63	102.51	2.01078	.009755
4.14	62.803	1.79798	.015923	4.64	103.54	2.01513	.009658
4.15	63.434	1.80232	0.015764	**4.65**	104.58	2.01947	0.009562
4.16	64.072	1.80667	.015608	4.66	105.64	2.02381	.009466
4.17	64.715	1.81101	.015452	4.67	106.70	2.02816	.009372
4.18	65.366	1.81535	.015299	4.68	107.77	2.03250	.009279
4.19	66.023	1.81969	.015146	4.69	108.85	2.03684	.009187
4.20	66.686	1.82404	0.014996	**4.70**	109.95	2.04118	0.009095
4.21	67.357	1.82838	.014846	4.71	111.05	2.04553	.009005
4.22	68.033	1.83272	.014699	4.72	112.17	2.04987	.008915
4.23	68.717	1.83707	.014552	4.73	113.30	2.05421	.008826
4.24	69.408	1.84141	.014408	4.74	114.43	2.05856	.008739

Table 3 Exponential Functions (Continued)

x	e^x	$\log_{10}(e^x)$	e^{-x}	x	e^x	$\log_{10}(e^x)$	e^{-x}
4.25	70.105	1.84575	0.014264	**4.75**	115.58	2.06290	0.008652
4.26	70.810	1.85009	.014122	4.76	116.75	2.06724	.008566
4.27	71.522	1.85444	.013982	4.77	117.92	2.07158	.008480
4.28	72.240	1.85878	.013843	4.78	119.10	2.07593	.008396
4.29	72.966	1.86312	.013705	4.79	120.30	2.08027	.008312
4.30	73.700	1.86747	0.013569	**4.80**	121.51	2.08461	0.008230
4.31	74.440	1.87181	.013434	4.81	122.73	2.08896	.008148
4.32	75.189	1.87615	.013300	4.82	123.97	2.09330	.008067
4.33	75.944	1.88050	.013168	4.83	125.21	2.09764	.007987
4.34	76.708	1.88484	.013037	4.84	126.47	2.10199	.007907
4.35	77.478	1.88918	0.012907	**4.85**	127.74	2.10633	0.007828
4.36	78.257	1.89352	.012778	4.86	129.02	2.11067	.007750
4.37	79.044	1.89787	.012651	4.87	130.32	2.11501	.007673
4.38	79.838	1.90221	.012525	4.88	131.63	2.11936	.007597
4.39	80.640	1.90655	.012401	4.89	132.95	2.12370	.007521
4.40	81.451	1.91090	0.012277	**4.90**	134.29	2.12804	0.007477
4.41	82.269	1.91524	.012155	4.91	135.64	2.13239	.007372
4.42	83.096	1.91958	.012034	4.92	137.00	2.13673	.007299
4.43	83.931	1.92392	.011914	4.93	138.38	2.14107	.007227
4.44	84.775	1.92827	.011796	4.94	139.77	2.14541	.007155
4.45	85.627	1.93261	0.011679	**4.95**	141.17	2.14976	0.007083
4.46	86.488	1.93695	.011562	4.96	142.59	2.15410	.007013
4.47	87.357	1.94130	.011447	4.97	144.03	2.15844	.006943
4.48	88.235	1.94564	.011333	4.98	145.47	2.16279	.006874
4.49	89.121	1.94998	.011221	4.99	146.94	2.16713	.006806
4.50	90.017	1.95433	0.011109	**5.00**	148.41	2.17147	0.006738
5.00	148.41	2.17147	0.006738	**5.50**	244.69	2.38862	0.0040868
5.01	149.90	2.17582	.006671	5.55	257.24	2.41033	.0038875
5.02	151.41	2.18016	.006605	5.60	270.43	2.43205	.0036979
5.03	152.93	2.18450	.006539	5.65	284.29	2.45376	.0035175
5.04	154.47	2.18884	.006474	5.70	298.87	2.47548	.0033460
5.05	156.02	2.19319	0.006409	**5.75**	314.19	2.49719	0.0031828
5.06	157.59	2.19753	.006346	5.80	330.30	2.51891	.0030276
5.07	159.17	2.20187	.006282	5.85	347.23	2.54062	.0028799
5.08	160.77	2.20622	.006220	5.90	365.04	2.56234	.0027394
5.09	162.39	2.21056	.006158	5.95	383.75	2.58405	.0026058
5.10	164.02	2.21490	0.006097	**6.00**	403.43	2.60577	0.0024788
5.11	165.67	2.21924	.006036	6.05	424.11	2.62748	.0023579
5.12	167.34	2.22359	.005976	6.10	445.86	2.64920	.0022429
5.13	169.02	2.22793	.005917	6.15	468.72	2.67091	.0021335
5.14	170.72	2.23227	.005858	6.20	492.75	2.69263	.0020294
5.15	172.43	2.23662	0.005799	**6.25**	518.01	2.71434	0.0019305
5.16	174.16	2.24096	.005742	6.30	544.57	2.73606	.0018363
5.17	175.91	2.24530	.005685	6.35	572.49	2.75777	.0017467
5.18	177.68	2.24965	.005628	6.40	601.85	2.77948	.0016616
5.19	179.47	2.25399	.005572	6.45	632.70	2.80120	.0015805

Table 3 1119

Table 3 Exponential Functions (*Continued*)

x	e^x	$\log_{10}(e^x)$	e^{-x}	x	e^x	$\log_{10}(e^x)$	e^{-x}
5.20	181.27	2.25833	0.005517	**6.50**	665.14	2.82291	0.0015034
5.21	183.09	2.26267	.005462	6.55	699.24	2.84463	.0014301
5.22	184.93	2.26702	.005407	6.60	735.10	2.86634	.0013604
5.23	186.79	2.27136	.005354	6.65	772.78	2.88806	.0012940
5.24	188.67	2.27570	.005300	6.70	812.41	2.90977	.0012309
5.25	190.57	2.28005	0.005248	**6.75**	854.06	2.93149	0.0011709
5.26	192.48	2.28439	.005195	6.80	897.85	2.95320	.0011138
5.27	194.42	2.28873	.005144	6.85	943.88	2.97492	.0010595
5.28	196.37	2.29307	.005092	6.90	992.27	2.99663	.0010078
5.29	198.34	2.29742	.005042	6.95	1043.1	3.01835	.0009586
5.30	200.34	2.30176	0.004992	**7.00**	1096.6	3.04006	0.009119
5.31	202.35	2.30610	.004942	7.05	1152.9	3.06178	.0008674
5.32	204.38	2.31045	.004893	7.10	1212.0	3.08349	.0008251
5.33	206.44	2.31479	.004844	7.15	1274.1	3.10521	.0007849
5.34	208.51	2.31913	.004796	7.20	1339.4	3.12692	.0007466
5.35	210.61	2.32348	0.004748	**7.25**	1408.1	3.14863	0.0007102
5.36	212.72	2.32782	.004701	7.30	1480.3	3.17035	.0006755
5.37	214.86	2.33216	.004654	7.35	1556.2	3.19206	.0006426
5.38	217.02	2.33650	.004608	7.40	1636.0	3.21378	.0006113
5.39	219.20	2.34085	.004562	7.45	1719.9	3.23549	.0005814
5.40	221.41	2.34519	0.004517	**7.50**	1808.0	3.25721	0.0005531
5.41	223.63	2.34953	.004472	7.55	1900.7	3.27892	.0005261
5.42	225.88	2.35388	.004427	7.60	1998.2	3.30064	.0005005
5.43	228.15	2.35822	.004383	7.65	2100.6	3.32235	.0004760
5.44	230.44	2.36256	.004339	7.70	2208.3	3.34407	.0004528
5.45	232.76	2.36690	0.004296	**7.75**	2321.6	3.36578	0.0004307
5.46	235.10	2.37125	.004254	7.80	2440.6	3.38750	.0004097
5.47	237.46	2.37559	.004211	7.85	2565.7	3.40921	.0003898
5.48	239.85	2.37993	.004169	7.90	2697.3	3.43093	.0003707
5.49	242.26	2.38428	.004128	7.95	2835.6	3.45264	.0003527
5.50	244.69	2.38862	0.004087	**8.00**	2981.0	3.47436	0.0003355
8.00	2981.0	3.47436	0.0003355	**9.00**	8103.1	3.90865	0.0001234
8.05	3133.8	3.49607	.0003191	9.05	8518.5	3.93037	.0001174
8.10	3294.5	3.51779	.0003035	9.10	8955.3	3.95208	.0001117
8.15	3463.4	3.53950	.0002887	9.15	9414.4	3.97379	.0001062
8.20	3641.0	3.56121	.0002747	9.20	9897.1	3.99551	.0001010
8.25	3827.6	3.58293	0.0002613	**9.25**	10405	4.01722	0.0000961
8.30	4023.9	3.60464	.0002485	9.30	10938	4.03894	.0000914
8.35	4230.2	3.62636	.0002364	9.35	11499	4.06065	.0000870
8.40	4447.1	3.64807	.0002249	9.40	12088	4.08237	.0000827
8.45	4675.1	3.66979	.0002139	9.45	12708	4.10408	.0000787
8.50	4914.8	3.69150	0.0002036	**9.50**	13360	4.12580	0.0000749
8.55	5166.8	3.71322	.0001935	9.55	14045	4.14751	.0000712
8.60	5431.7	3.73493	.0001841	9.60	14765	4.16923	.0000677
8.65	5710.0	3.75665	.0001751	9.65	15522	4.19094	.0000644
8.70	6002.9	3.77836	.0001666	9.70	16318	4.21266	.000613

Table 3 Exponential Functions (*Continued*)

x	e^x	$\log_{10}(e^x)$	e^{-x}	x	e^x	$\log_{10}(e^x)$	e^{-x}
8.75	6310.7	3.80008	0.0001585	**9.75**	17154	4.23437	0.0000583
8.80	6634.2	3.82179	.0001507	9.80	18034	4.25609	.0000555
8.85	6974.4	3.84351	.0001434	9.85	18958	4.27780	.0000527
8.90	7332.0	3.86522	.0001364	9.90	19930	4.29952	.0000502
8.95	7707.9	3.88694	.0001297	9.95	20952	4.32123	0.0000477
9.00	8103.1	3.90865	0.0001234	**10.00**	22026	4.34294	0.0000454

Table 4 Factorials and their
Common Logarithms

n	$n!$	$\log_{10} n!$
1	1	0.00000
2	2	0.30103
3	6	0.77815
4	24	1.38021
5	120	2.07918
6	720	2.85733
7	5040	3.70243
8	40320	4.60552
9	362880	5.55976
10	3628800	6.55976

Table 4 1121

Table 5 Radians to Degrees and Minutes and Vice Versa

Radians	Degrees and Minutes	Degrees and Minutes	Radians
.001	0°3.44′	1′	.00029
.002	0°6.88′	2′	.00058
.003	0°10.31′	3′	.00087
.004	0°13.75′	4′	.00116
.005	0°17.19′	5′	.00145
.006	0°20.63′	6′	.00175
.007	0°24.06′	7′	.00204
.008	0°27.50′	8′	.00233
.009	0°30.96′	9′	.00262
.01	0°34.38′	10′	.00291
.02	1°8.75′	20′	.00582
.03	1°43.13′	30′	.00873
.04	2°17.51′	40′	.01164
.05	2°51.89′	50′	.01454
.06	3°26.26′	1°	.01745
.07	4°0.64′	2°	.03491
.08	4°35.02′	3°	.05236
.09	5°9.40′	4°	.06981
.1	5°43.77′	5°	.08727
.2	11°27.55′	6°	.10472
.3	17°11.32′	7°	.12217
.4	22°55.10′	8°	.13963
.5	28°38.87′	9°	.15708
.6	34°22.65′	10°	.17453
.7	40°6.42′	15°	.26180
.8	45°50.20′	30°	.52360
.9	51°33.97′	60°	1.04720
1.0	57°17.75′	90°	1.57080
2.0	114°35.49′	180°	3.14159
3.0	171°53.24′		
4.0	229°10.99′		
5.0	286°28.73′		
6.0	343°46.48′		
7.0	401°4.25′		
8.0	458°21.97′		
9.0	515°39.72′		

Table 6   Trigonometric Functions for Angles in Degrees

Degrees	Radians	Sin	Cos	Tan	Cot		
0	0.0000	0.0000	1.0000	0.0000		1.5708	90
1	0.0175	0.0175	0.9998	0.0175	57.290	1.5533	89
2	0.0349	0.0349	0.9994	0.0349	28.636	1.5359	88
3	0.0524	0.0523	0.9986	0.0524	19.081	1.5184	87
4	0.0698	0.0698	0.9976	0.0699	14.301	1.5010	86
5	0.0873	0.0872	0.9962	0.0875	11.430	1.4835	85
6	0.1047	0.1045	0.9945	0.1051	9.5144	1.4661	84
7	0.1222	0.1219	0.9925	0.1228	8.1443	1.4486	83
8	0.1396	0.1392	0.9903	0.1405	7.1154	1.4312	82
9	0.1571	0.1564	0.9877	0.1584	6.3138	1.4137	81
10	0.1745	0.1736	0.9848	0.1763	5.6713	1.3963	80
11	0.1920	0.1908	0.9816	0.1944	5.1446	1.3788	79
12	0.2094	0.2079	0.9781	0.2126	4.7046	1.3614	78
13	0.2269	0.2250	0.9744	0.2309	4.3315	1.3439	77
14	0.2443	0.2419	0.9703	0.2493	4.0108	1.3265	76
15	0.2618	0.2588	0.9659	0.2679	3.7321	1.3090	75
16	0.2793	0.2756	0.9613	0.2867	3.4874	1.2915	74
17	0.2967	0.2924	0.9563	0.3057	3.2709	1.2741	73
18	0.3142	0.3090	0.9511	0.3249	3.0777	1.2566	72
19	0.3316	0.3256	0.9455	0.3443	2.9042	1.2392	71
20	0.3491	0.3420	0.9297	0.3640	2.7475	1.2217	70
21	0.3665	0.3584	0.9336	0.3839	2.6051	1.2043	69
22	0.3840	0.3746	0.9272	0.4040	2.4751	1.1868	68
23	0.4014	0.3907	0.9205	0.4245	2.3559	1.1694	67
24	0.4189	0.4067	0.9135	0.4452	2.2460	1.1519	66
25	0.4363	0.4226	0.9063	0.4663	2.1445	1.1345	65
26	0.4538	0.4384	0.8988	0.4877	2.0503	1.1170	64
27	0.4712	0.4540	0.8910	0.5095	1.9626	1.0996	63
28	0.4887	0.4695	0.8829	0.5317	1.8807	1.0821	62
29	0.5061	0.4848	0.8746	0.5543	1.8040	1.0647	61
30	0.5236	0.5000	0.8660	0.5774	1.7321	1.0472	60
31	0.5411	0.5150	0.8572	0.6009	1.6643	1.0297	59
32	0.5585	0.5299	0.8480	0.6249	1.6003	1.0123	58
33	0.5760	0.5446	0.8387	0.6494	1.5399	0.9948	57
34	0.5934	0.5592	0.8290	0.6745	1.4826	0.9774	56
35	0.6109	0.5736	0.8192	0.7002	1.4281	0.9599	55
36	0.6283	0.5878	0.8090	0.7265	1.3764	0.9425	54
37	0.6458	0.6018	0.7986	0.7536	1.3270	0.9250	53
38	0.6632	0.6157	0.7880	0.7813	1.2799	0.9076	52
39	0.6807	0.6293	0.7771	0.8098	1.2349	0.8901	51
40	0.6981	0.6428	0.7660	0.8391	1.1918	0.8727	50
41	0.7156	0.6561	0.7547	0.8693	1.1504	0.8552	49
42	0.7330	0.6691	0.7431	0.9004	1.1106	0.8378	48
43	0.7505	0.6820	0.7314	0.9325	1.0724	0.8203	47
44	0.7679	0.6947	0.7193	0.9657	1.0355	0.8029	46
45	0.7854	0.7071	0.7071	1.000	1.0000	0.7854	45
		Cos	Sin	Cot	Tan	Radians	Degrees

Table 6   1123

Table 7 Trigonometric Functions for Angles in Radians

Radians	Sin	Tan	Cot	Cos	Radians	Sin	Tan	Cot	Cos
.00	.0000	.0000	......	1.0000	.50	.4794	.5463	1.830	.8776
.01	.0100	.0100	99.997	1.0000	.51	.4882	.5594	1.788	.8727
.02	.0200	.0200	49.993	.9998	.52	.4969	.5726	1.747	.8678
.03	.0300	.0300	33.323	.9996	.53	.5055	.5859	1.707	.8628
.04	.0400	.0400	24.987	.9992	.54	.5141	.5994	1.668	.8577
.05	.0500	.0500	19.983	.9988	.55	.5227	.6131	1.631	.8525
.06	.0600	.0601	16.647	.9982	.56	.5312	.6269	1.595	.8473
.07	.0699	.0701	14.262	.9976	.57	.5396	.6410	1.560	.8419
.08	.0799	.0802	12.473	.9968	.58	.5480	.6552	1.526	.8365
.09	.0899	.0902	11.081	.9960	.59	.5564	.6696	1.494	.8309
.10	.0998	.1003	9.967	.9950	.60	.5646	.6841	1.462	.8253
.11	.1098	.1104	9.054	.9940	.61	.5729	.6989	1.431	.8196
.12	.1197	.1206	8.293	.9928	.62	.5810	.7139	1.401	.8139
.13	.1296	.1307	7.649	.9916	.63	.5891	.7291	1.372	.8080
.14	.1395	.1409	7.096	.9902	.64	.5972	.7445	1.343	.8021
.15	.1494	.1511	6.617	.9888	.65	.6052	.7602	1.315	.7961
.16	.1593	.1614	6.197	.9872	.66	.6131	.7761	1.288	.7900
.17	.1692	.1717	5.826	.9856	.67	.6210	.7923	1.262	.7838
.18	.1790	.1820	5.495	.9838	.68	.6288	.8087	1.237	.7776
.19	.1889	.1923	5.200	.9820	.69	.6365	.8253	1.212	.7712
.20	.1987	.2027	4.933	.9801	.70	.6442	.8423	1.187	.7648
.21	.2085	.2131	4.692	.9780	.71	.6518	.8595	1.163	.7584
.22	.2182	.2236	4.472	.9759	.72	.6594	.8771	1.140	.7518
.23	.2280	.2341	4.271	.9737	.73	.6669	.8949	1.117	.7452
.24	.2377	.2447	4.086	.9713	.74	.6743	.9131	1.095	.7385
.25	.2474	.2553	3.916	.9689	.75	.6816	.9316	1.073	.7317
.26	.2571	.2660	3.759	.9664	.76	.6889	.9505	1.052	.7248
.27	.2667	.2768	3.613	.9638	.77	.6961	.9697	1.031	.7179
.28	.2764	.2876	3.478	.9611	.78	.7033	.9893	1.011	.7109
.29	.2860	.2984	3.351	.9582	.79	.7104	1.009	.9908	.7038
.30	.2955	.3093	3.233	.9553	.80	.7174	1.030	.9712	.6967
.31	.3051	.3203	3.122	.9523	.81	.7243	1.050	.9520	.6895
.32	.3146	.3314	3.018	.9492	.82	.7311	1.072	.9331	.6822
.33	.3240	.3425	2.920	.9460	.83	.7379	1.093	.9146	.6749
.34	.3335	.3537	2.827	.9428	.84	.7446	1.116	.8964	.6675
.35	.3429	.3650	2.740	.9394	.85	.7513	1.138	.8785	.6600
.36	.3523	.3764	2.657	.9359	.86	.7578	1.162	.8609	.6524
.37	.3616	.3879	2.578	.9323	.87	.7643	1.185	.8437	.6448
.38	.3709	.3994	2.504	.9287	.88	.7707	1.210	.8267	.6372
.39	.3802	.4111	2.433	.9249	.89	.7771	1.235	.8100	.6294
.40	.3894	.4228	2.365	.9211	.90	.7833	1.260	.7936	.6216
.41	.3986	.4346	2.301	.9171	.91	.7895	1.286	.7774	.6137
.42	.4078	.4466	2.239	.9131	.92	.7956	1.313	.7615	.6058
.43	.4169	.4586	2.180	.9090	.93	.8016	1.341	.7458	.5978
.44	.4259	.4708	2.124	.9048	.94	.8076	1.369	.7303	.5898

Table 7 Trigonometric Functions for Angles in Radians (*Continued*)

Radians	Sin	Tan	Cot	Cos	Radians	Sin	Tan	Cot	Cos
.45	.4350	.4831	2.070	.9004	.95	.8134	1.398	.7151	.5817
.46	.4439	.4954	2.018	.8961	.96	.8192	1.428	.7001	.5735
.47	.4529	.5080	1.969	.8916	.97	.8249	1.459	.6853	.5653
.48	.4618	.5206	1.921	.8870	.98	.8305	1.491	.6707	.5570
.49	.4706	.5334	1.875	.8823	.99	.8360	1.524	.6563	.5487
1.00	.8415	1.557	.6421	.5403	1.30	.9636	3.602	.2776	.2675
1.01	.8468	1.592	.6281	.5319	1.31	.9662	3.747	.2669	.2579
1.02	.8521	1.628	.6142	.5234	1.32	.9687	3.903	.2562	.2482
1.03	.8573	1.665	.6005	.5148	1.33	.9711	4.072	.2456	.2385
1.04	.8624	1.704	.5870	.5062	1.34	.9735	4.256	.2350	.2288
1.05	.8674	1.743	.5736	.4976	1.35	.9757	4.455	.2245	.2190
1.06	.8724	1.784	.5604	.4889	1.36	.9779	4.673	.2140	.2092
1.07	.8772	1.827	.5473	.4801	1.37	.9799	4.913	.2035	.1994
1.08	.8820	1.871	.5344	.4713	1.38	.9819	5.177	.1931	.1896
1.09	.8866	1.917	.5216	.4625	1.39	.9837	5.471	.1828	.1798
1.10	.8912	1.965	.5090	.4536	1.40	.9854	5.798	.1725	.1700
1.11	.8957	2.014	.4964	.4447	1.41	.9871	6.165	.1622	.1601
1.12	.9001	2.066	.4840	.4357	1.42	.9887	6.581	.1519	.1502
1.13	.9044	2.120	.4718	.4267	1.43	.9901	7.055	.1417	.1403
1.14	.9086	2.176	.4596	.4176	1.44	.9915	7.602	.1315	.1304
1.15	.9128	2.234	.4475	.4085	1.45	.9927	8.238	.1214	.1205
1.16	.9168	2.296	.4356	.3993	1.46	.9939	8.989	.1113	.1106
1.17	.9208	2.360	.4237	.3902	1.47	.9949	9.887	.1011	.1006
1.18	.9246	2.427	.4120	.3809	1.48	.9959	10.983	.0910	.0907
1.19	.9284	2.498	.4003	.3717	1.49	.9967	12.350	.0810	.0807
1.20	.9320	2.572	.3888	.3624	1.50	.9975	14.101	.0709	.0707
1.21	.9356	2.650	.3773	.3530	1.51	.9982	16.428	.0609	.0608
1.22	.9391	2.733	.3659	.3436	1.52	.9987	19.670	.0508	.0508
1.23	.9425	2.820	.3546	.3342	1.53	.9992	24.498	.0408	.0408
1.24	.9458	2.912	.3434	.3248	1.54	.9995	32.461	.0308	.0308
1.25	.9490	3.010	.3323	.3153	1.55	.9998	48.078	.0208	.0208
1.26	.9521	3.113	.3212	.3058	1.56	.9999	92.620	.0108	.0108
1.27	.9551	3.224	.3102	.2963	1.57	1.0000	1255.8	.0008	.0008
1.28	.9580	3.341	.2993	.2867					
1.29	.9608	3.467	.2884	.2771					

$\pi = 3.14159265 \ldots \frac{1}{2}\pi = 1.57079632 \ldots \frac{3}{2}\pi = 4.71238898 \ldots 2\pi = 6.28318530 \ldots$

Table 7 1125

Table 8 Hyperbolic Functions

x	Sinh x	Cosh x	Tanh x	x	Sinh x	Cosh x	Tanh x
0.00	0.0000	1.0000	.00000	0.50	0.5211	1.1276	.46212
0.01	0.0100	1.0001	.01000	0.51	0.5324	1.1329	.46995
0.02	0.0200	1.0002	.02000	0.52	0.5438	1.1383	.47770
0.03	0.0300	1.0005	.02999	0.53	0.5552	1.1438	.48538
0.04	0.0400	1.0008	.03998	0.54	0.5666	1.1494	.49299
0.05	0.0500	1.0013	.04996	0.55	0.5782	1.1551	.50052
0.06	0.0600	1.0018	.05993	0.56	0.5897	1.1609	.50798
0.07	0.0701	1.0025	.06989	0.57	0.6014	1.1669	.51536
0.08	0.0801	1.0032	.07983	0.58	0.6131	1.1730	.52267
0.09	0.0901	1.0041	.08976	0.59	0.6248	1.1792	.52990
0.10	0.1002	1.0050	.09967	0.60	0.6367	1.1855	.53705
0.11	0.1102	1.0061	.10956	0.61	0.6485	1.1919	.54413
0.12	0.1203	1.0072	.11943	0.62	0.6605	1.1984	.55113
0.13	0.1304	1.0085	.12927	0.63	0.6725	1.2051	.55805
0.14	0.1405	1.0098	.13909	0.64	0.6846	1.2119	.56490
0.15	0.1506	1.0113	.14889	0.65	0.6967	1.2188	.57167
0.16	0.1607	1.0128	.15865	0.66	0.7090	1.2258	.57836
0.17	0.1708	1.0145	.16838	0.67	0.7213	1.2330	.58498
0.18	0.1810	1.0162	.17808	0.68	0.7336	1.2402	.59152
0.19	0.1911	1.0181	.18775	0.69	0.7461	1.2476	.59798
0.20	0.2013	1.0201	.19738	0.70	0.7586	1.2552	.60437
0.21	0.2115	1.0221	.20697	0.71	0.7712	1.2628	.61068
0.22	0.2218	1.0243	.21652	0.72	0.7838	1.2706	.61691
0.23	0.2320	1.0266	.22603	0.73	0.7966	1.2785	.62307
0.24	0.2423	1.0289	.23550	0.74	0.8094	1.2865	.62915
0.25	0.2526	1.0314	.24492	0.75	0.8223	1.2947	.63515
0.26	0.2629	1.0340	.25430	0.76	0.8353	1.3030	.64108
0.27	0.2733	1.0367	.26362	0.77	0.8484	1.3114	.64693
0.28	0.2837	1.0395	.27291	0.78	0.8615	1.3199	.65271
0.29	0.2941	1.0423	.28213	0.79	0.8748	1.3286	.65841
0.30	0.3045	1.0453	.29131	0.80	0.8881	1.3374	.66404
0.31	0.3150	1.0484	.30044	0.81	0.9015	1.3464	.66959
0.32	0.3255	1.0516	.30951	0.82	0.9150	1.3555	.67507
0.33	0.3360	1.0549	.31852	0.83	0.9286	1.3647	.68048
0.34	0.3466	1.0584	.32748	0.84	0.9423	1.3740	.68581
0.35	0.3572	1.0619	.33638	0.85	0.9561	1.3835	.69107
0.36	0.3678	1.0655	.34521	0.86	0.9700	1.3932	.69626
0.37	0.3785	1.0692	.35399	0.87	0.9840	1.4029	.70137
0.38	0.3892	1.0731	.36271	0.88	0.9981	1.4128	.70642
0.39	0.4000	1.0770	.37136	0.89	1.0122	1.4229	.71139
0.40	0.4108	1.0811	.37995	0.90	1.0265	1.4331	.71630
0.41	0.4216	1.0852	.38847	0.91	1.0409	1.4434	.72113
0.42	0.4325	1.0895	.39693	0.92	1.0554	1.4539	.72590
0.43	0.4434	1.0939	.40532	0.93	1.0700	1.4645	.73059
0.44	0.4543	1.0984	.41364	0.94	1.0847	1.4753	.73522

Table 8 Hyperbolic Functions (*Continued*)

x	Sinh x	Cosh x	Tanh x	x	Sinh x	Cosh x	Tanh x
0.45	0.4653	1.1030	.42190	0.95	1.0995	1.4862	.73978
0.46	0.4764	1.1077	.43008	0.96	1.1144	1.4973	.74428
0.47	0.4875	1.1125	.43820	0.97	1.1294	1.5085	.74870
0.48	0.4986	1.1174	.44624	0.98	1.1446	1.5199	.75307
0.49	0.5098	1.1225	.45422	0.99	1.1598	1.5314	.75736
1.00	1.1752	1.5431	.76159	4.00	27.290	27.308	.99933
1.10	1.3356	1.6685	.80050	4.10	30.162	30.178	.99945
1.20	1.5095	1.8107	.83365	4.20	33.336	33.351	.99955
1.30	1.6984	1.9709	.86172	4.30	36.843	36.857	.99963
1.40	1.9043	2.1509	.88535	4.40	40.719	40.732	.99970
1.50	2.1293	2.3524	.90515	4.50	45.003	45.014	.99975
1.60	2.3756	2.5775	.92167	4.60	49.737	49.747	.99980
1.70	2.6456	2.8283	.93541	4.70	54.969	54.978	.99983
1.80	2.9422	3.1075	.94681	4.80	60.751	60.759	.99986
1.90	3.2682	3.4177	.95624	4.90	67.141	67.149	.99989
2.00	3.6269	3.7622	.96403	5.00	74.203	74.210	.99991
2.10	4.0219	4.1443	.97045	5.10	8.2008	82.014	.99993
2.20	4.4571	4.5679	.97574	5.20	90.633	90.639	.99994
2.30	4.9370	5.0372	.98010	5.30	100.17	100.17	.99995
2.40	5.4662	5.5569	.98367	5.40	110.70	110.71	.99996
2.50	6.0502	6.1323	.98661	5.50	122.34	122.35	.99997
2.60	6.6947	6.7690	.98903	5.60	135.21	135.22	.99997
2.70	7.4063	7.4735	.99101	5.70	149.43	149.44	.99998
2.80	8.1919	8.2527	.99263	5.80	165.15	165.15	.99998
2.90	9.0596	9.1146	.99396	5.90	182.52	182.52	.99998
3.00	10.018	10.068	.99505	6.00	201.71	201.72	.99999
3.10	11.076	11.122	.99595	6.25	259.01	259.01	.99999
3.20	12.246	12.287	.99668	6.50	332.57	332.57	1.0000
3.30	13.538	13.575	.99728	6.75	427.03	427.03	1.0000
3.40	14.965	14.999	.99777	7.00	548.32	548.32	1.0000
3.50	16.543	16.573	.99818	7.50	904.02	904.02	1.0000
3.60	18.286	18.313	.99851	8.00	1490.5	1490.5	1.0000
3.70	20.211	20.236	.99878	8.50	2457.4	2457.4	1.0000
3.80	22.339	22.362	.99900	9.00	4051.5	4051.5	1.0000
3.90	24.691	24.711	.99918	9.50	6679.9	6679.9	1.0000
				10.00	11013	11013	1.0000

Table 8 1127

Table 9 Common Logarithms (base 10)

N	0	1	2	3	4	5	6	7	8	9
10	0000	0043	0086	0128	0170	0212	0253	0294	0334	0374
11	0414	0453	0492	0531	0569	0607	0645	0682	0719	0755
12	0792	0828	0864	0899	0934	0969	1004	1038	1072	1106
13	1139	1173	1206	1239	1271	1303	1335	1367	1399	1430
14	1461	1492	1523	1553	1584	1614	1644	1673	1703	1732
15	1761	1790	1818	1847	1875	1903	1931	1959	1987	2014
16	2041	2068	2095	2122	2148	2175	2201	2227	2253	2279
17	2304	2330	2355	2380	2405	2430	2455	2480	2504	2529
18	2553	2577	2601	2625	2648	2672	2695	2718	2742	2765
19	2788	2810	2833	2856	2878	2900	2923	2945	2967	2989
20	3010	3032	3054	3075	3096	3118	3139	3160	3181	3201
21	3222	3243	3263	3284	3304	3324	3345	3365	3385	3404
22	3424	3444	3464	3483	3502	3522	3541	3560	3579	3598
23	3617	3636	3655	3674	3692	3711	3729	3747	3766	3784
24	3802	3820	3838	3856	3874	3892	3909	3927	3945	3962
25	3979	3997	4014	4031	4048	4065	4082	4099	4116	4133
26	4150	4166	4183	4200	4216	4232	4249	4265	4281	4298
27	4314	4330	4346	4362	4378	4393	4409	4425	4440	4456
28	4472	4487	4502	4518	4533	4548	4564	4579	4594	4609
29	4624	4639	4654	4669	4683	4698	4713	4728	4742	4757
30	4771	4786	4800	4814	4829	4843	4857	4871	4886	4900
31	4914	4928	4942	4955	4969	4983	4997	5011	5024	5038
32	5051	5065	5079	5092	5105	5119	5132	5145	5159	5172
33	5185	5198	5211	5224	5237	5250	5263	5276	5289	5302
34	5315	5328	5340	5353	5366	5378	5391	5403	5416	5428
35	5441	5453	5465	5478	5490	5502	5514	5527	5539	5551
36	5563	5575	5587	5599	5611	5623	5635	5647	5648	5670
37	5682	5694	5705	5717	5729	5740	5752	5763	5775	5786
38	5798	5809	5821	5832	5843	5855	5866	5877	5888	5899
39	5911	5922	5933	5944	5955	5966	5977	5988	5999	6010
40	6021	6031	6042	6053	6064	6075	6085	6096	6107	6117
41	6128	6138	6149	6160	6170	6180	6191	6201	6212	6222
42	6232	6243	6253	6263	6274	6284	6294	6304	6314	6325
43	6335	6345	6355	6365	6375	6385	6395	6405	6415	6425
44	6435	6444	6454	6464	6474	6484	6493	6503	6513	6522
45	6532	6542	6551	6561	6571	6580	6590	6599	6609	6618
46	6628	6637	6646	6656	6665	6675	6684	6693	6702	6712
47	6721	6730	6739	6749	6758	6767	6776	6785	6794	6803
48	6812	6821	6830	6839	6848	6857	6866	6875	6884	6893
49	6902	6911	6920	6928	6937	6946	6955	6964	6972	6981
50	6990	6998	7007	7016	7024	7033	7042	7050	7059	7067
51	7076	7084	7093	7101	7110	7118	7126	7135	7143	7152
52	7160	7168	7177	7185	7193	7202	7210	7218	7226	7235
53	7243	7251	7259	7267	7275	7284	7292	7300	7308	7316
54	7324	7332	7340	7348	7356	7364	7372	7380	7388	7396

Table 9   Common Logarithms (base 10) (*Continued*)

N	0	1	2	3	4	5	6	7	8	9
55	7404	7412	7419	7427	7435	7443	7451	7459	7466	7474
56	7482	7490	7497	7505	7513	7520	7528	7536	7543	7551
57	7559	7566	7574	7582	7589	7597	7604	7612	7619	7627
58	7634	7642	7649	7657	7664	7672	7679	7686	7694	7701
59	7709	7716	7723	7731	7738	7745	7752	7760	7767	7774
60	7782	7789	7796	7803	7810	7818	7825	7832	7839	7846
61	7853	7860	7868	7875	7882	7889	7896	7903	7910	7917
62	7924	7931	7938	7945	7952	7959	7966	7973	7980	7987
63	7993	8000	8007	8014	8021	8028	8035	8041	8048	8055
64	8062	8069	8075	8082	8089	8096	8102	8109	8116	8122
65	8129	8136	8142	8149	8156	8162	8169	8176	8182	8189
66	8195	8202	8209	8215	8222	8228	8234	8241	8248	8254
67	8261	8267	8274	8280	8287	8293	8299	8306	8312	8319
68	8325	8331	8338	8344	8351	8357	8363	8370	8376	8382
69	8388	8395	8401	8407	8414	8420	8426	8432	8439	8445
70	8451	8457	8463	8470	8476	8482	8488	8494	8500	8506
71	8513	8519	8525	8531	8537	8543	8549	8555	8561	8567
72	8573	8579	8585	8591	8597	8603	8609	8615	8621	8627
73	8633	8639	8645	8651	8657	8663	8669	8675	8681	8686
74	8692	8698	8704	8710	8716	8722	8727	8733	8739	8745
75	8751	8756	8762	8768	8774	8779	8785	8791	8797	8802
76	8808	8814	8820	8825	8831	8837	8842	8848	8854	8859
77	8865	8871	8876	8882	8887	8893	8899	8904	8910	8915
78	8921	8927	8932	8938	8943	8949	8954	8960	8965	8971
79	8976	8982	8987	8993	8998	9004	9009	9015	9020	9025
80	9031	9036	9042	9047	9053	9058	9063	9069	9074	9079
81	9085	9090	9096	9101	9106	9112	9117	9122	9128	9133
82	9138	9143	9149	9154	9159	9165	9170	9175	9180	9186
83	9191	9196	9201	9206	9212	9217	9222	9227	9232	9238
84	9243	9248	9253	9258	9263	9269	9274	9279	9284	9289
85	9294	9299	9304	9309	9315	9320	9325	9330	9335	9340
86	9345	9350	9355	9360	9365	9370	9375	9380	9385	9390
87	9395	9400	9405	9410	9415	9420	9425	9430	9435	9440
88	9445	9450	9455	9460	9465	9469	9474	9479	9484	9489
89	9494	9499	9504	9509	9513	9518	9523	9528	9533	9538
90	9542	9547	9552	9557	9562	9566	9571	9576	9581	9586
91	9590	9595	9600	9605	9609	9614	9619	9624	9628	9633
92	9638	9643	9647	9652	9657	9661	9666	9671	9675	9680
93	9685	9689	9694	9699	9703	9708	9713	9717	9722	9727
94	9731	9736	9741	9745	9750	9754	9759	9763	9768	9773
95	9777	9782	9786	9791	9795	9800	9805	9809	9814	9818
96	9823	9827	9832	9836	9841	9845	9850	9854	9859	9863
97	9868	9872	9877	9881	9886	9890	9894	9899	9903	9908
98	9912	9917	9921	9926	9930	9934	9939	9943	9948	9952
99	9956	9961	9965	9969	9974	9978	9983	9987	9991	9996

Table 9   1129

Table 10 Partial Greek Alphabet

α Alpha	λ Lambda	ρ Rho
β Beta	ω Omega	Σ, σ Sigma
δ, Δ Delta	ϕ Phi	θ Theta
ϵ Epsilon	Π, π Pi	ξ Xi
η Eta	ψ Psi	ζ Zeta

Answers to Odd-Numbered Exercises

Exercises 1.1 page 9

1. $x > 3$ **3.** $x < 2$ **5.** $x > -\frac{3}{2}$ **7.** $x > 1$ **9.** all x not in $(1, 6)$
11. no values of x **13.** $x \leq -2$ or $x > 1$ **15.** $-5 \leq x \leq 5$
17. $x \leq -3$ or $-2 \leq x \leq 2$ or $x \geq 3$ **19.** $x > 0$
21. $x < -5$ or $-2 < x < 1$
23. The arithmetic mean of n unequal numbers is greater than the smallest and less than the largest   **27.** all x **29.** no x

Exercises 1.2 page 17

1. $x = \pm\frac{8}{3}$ **3.** $x = 1, -\frac{1}{3}$ **5.** $x = -\frac{8}{3}, -6$ **7.** $x = -4, -\frac{3}{4}$
9. $2 < x < 8$ **11.** $-2 < x < 5$ **13.** $x \leq -12$ or $x \geq -6$ **15.** $x = -\frac{7}{3}$
17. $x < \frac{8}{3}$ **19.** $-\frac{1}{3} \leq x \leq 7$ **21.** $c - a < x < c + a$ **23.** $x \leq -\frac{1}{2}$ or $x \geq \frac{5}{2}$
25. $-4 < x < 3$ **27.** $|x| < |3b|, x = -3b$ **31.** $M = 30$ **33.** $M = \frac{9}{2}$
35. $-\frac{5}{2} \leq x \leq \frac{7}{2}$ **37.** $x < -\frac{2}{3}$ or $x > \frac{4}{3}$

Exercises 1.3 page 24

5. **(a)** line $x = 3$ in the fourth quadrant, $(3, 0)$;   **(b)** points outside the infinite strip $-2 < x < 2$;   **(c)** first and third quadrants   **(d)** y- and x-axes;   **(e)** region to the left and above the lines $x = -2$ and $y = 1$, respectively, boundary lines included; **(f)** region below $y = |x|$;   **(g)** region between the lines $y = \pm x$ and including them   **7.** $(5, -3)$ **9.** $45°$

Exercises 1.4 page 31

1. $(0, 1)$ and $(-\frac{1}{3}, 0)$;   none;   all x and y; none
3. $(0, 0)$; x-axis; $x \leq 0, -\infty < y < \infty$;   none
5. none; y-axis; $x \neq 0, y > 0$; $x = 0, y = 0$
7. $(0, \pm 3)$ and $(\pm 3, 0)$; x-axis, y-axis, origin, $y = x$; $-3 \leq x \leq 3$ and $-3 \leq y \leq 3$;   none
9. $(0, \pm\sqrt{3})$, none;   both axes, origin, $y = x$; $-\infty < x < \infty$ and $|y| \geq \sqrt{3}$;   none

Answers to Odd-Numbered Exercises 1131

11. no y-intercept, $(-3, 0)$; none; $x \neq 0$ and $y \neq 1$, $x = 0$ and $y = 1$
13. no y-intercept, $(5, 0)$; none; $x \neq 0$, $y \leq \frac{1}{20}$; $x = 0$ and $y = 0$
15. $(0, \pm 1), (\pm 1, 0)$; all; $|x| \leq 1$ and $|y| \leq 1$; none
19. one answer: $y = \frac{1}{9}(x^2 - 4)(x^2 - 9)$; one answer: $y = x(x^2 - 9)$

Exercises 1.5 — page 37

1. $d = \sqrt{20} = 2\sqrt{5}$ 3. $d = \sqrt{8} = 2\sqrt{2}$
5. $d = \sqrt{(x_1 + b - x_1)^2 + (0 - 0)^2} = \sqrt{b^2} = |b|$
17. $P_3\left(\dfrac{a}{2}, \dfrac{1}{2}\sqrt{3}a\right)$ or $P_2\left(\dfrac{a}{2}, -\dfrac{1}{2}\sqrt{3}a\right)$

Exercises 1.6 — page 46

1. $y = 3x - 4$ 3. $3x - y = 4$ 5. $x = 5$ 7. $y = -x$ 9. $x + y = -1$
11. $\dfrac{x}{4} + \dfrac{y}{6} = 1$ and $\dfrac{x}{15} - \dfrac{y}{5} = 1$ 13. slope $= -\frac{3}{2}$
15. slope $= -x_1/y_1$ if $y_1 \neq 0$ 21. $v(t) = 36 + 32t$
23. two straight lines, $y = \pm 3x$ 25. $x + y = 2$ or $y - x = 6$ 29. $\tan \phi = 7$

Exercises 1.7 — page 54

1. (a) $x^2 + y^2 = 49$; (b) $(x - 2)^2 + (y - 3)^2 = 25$; (c) $(x + 5)^2 + (y - 1)^2 = 16$;
(d) $(x + 8)^2 + (y + 2)^2 = 11$ 3. $(x + 3)^2 + (y - 2)^2 = 13$
5. $(x + 4)^2 + (y - 5)^2 = 25$ 7. $(x - 1)^2 + (y - 3)^2 = 50$
9. $(x - 3)^2 + (y - 5)^2 = 5$
11. $(x - 1)^2 + (y - 2)^2 = 16$; circle with center at $(1, 2)$ and radius $= 4$
13. $(x - \frac{1}{2})^2 + y^2 = 1$, circle with center at $(\frac{1}{2}, 0)$ and radius $= 1$
15. no locus because $(x + 4)^2 + (y - 3)^2 = -7$
17. $(x - 3)^2 + (y + 4)^2 = 16$ and $(x - 3)^2 + (y - 4)^2 = 16$
19. $2x^2 + 2y^2 + 14x - 11y + 12 = 0$ 21. $x^2 + y^2 - 8x - 4y + 10 = 0$
23. $(3, 2)$, $(-\frac{17}{5}, \frac{26}{5})$, $x + 2y - 7 = 0$ is the equation of the common chord
25. $x^2 + y^2 - 8x + 8y - 36 = 0$
31. if $k = 1$, locus is a straight line—the perpendicular bisector of the line segment
joining the two points

Exercises 1.8 — page 61

1. focus $(2, 0)$, directrix $x = -2$, length of latus rectum $= 8$
3. focus $(0, \frac{1}{4})$, directrix $y = -\frac{1}{4}$, length of latus rectum $= 1$
5. focus $(-\frac{3}{2}, 0)$, directrix $x = \frac{3}{2}$, length of latus rectum $= 6$ 7. $y^2 = 24x$
9. $y^2 = -20x$ 11. $y^2 = 10x$ 13. $x^2 = 7y$ 15. $y^2 = -16x$
17. $(0, 0)$ and $(4, 4)$ 19. $y|_{x=7} = 3$ 21. $y|_{x=-b/2a} = c - \dfrac{b^2}{4a}$

23. $-16(x + 1) = (y - 2)^2$ 25. $y = \dfrac{x^2}{16,000}$
27. $3x - 2y + 7 = 0$ (straight line) or $x^2 + 6y = 0$ (parabola)

Review and Miscellaneous Exercises — page 65

1. ± 100 3. $-9, 3$ 5. $(4, \infty)$ or $(-\infty, -6)$ 7. $(-\infty, \infty)$
9. $(-\infty, -2]$ or $[\frac{1}{3}, \infty)$ 11. $(-\infty, -5]$ or $[1, \infty)$

21. $\frac{2}{3}, \frac{1}{2}(-1 \pm \sqrt{5})$ **27.** $y = 3x - 11$

29. $5x + 3y = 1$; $-2x + 9y = 3$; $-7x + 6y - 2 = 0$;   intersect at $(0, \frac{1}{3})$

31. $(0, \frac{13}{3})$ and $(3, \frac{20}{3})$ **33.** $3x^2 + 3y^2 - 2x + 44y + 91 = 0$, a circle

35. $13x^2 + 13y^2 - 15x - 83y + 42 = 0$

37. $F < 0$ is sufficient but not necessary for the given equation to represent a circle, hence conjecture is false   **39.** $(-\frac{34}{5}, -\frac{12}{5})$ and $(6, 4)$

41. nonintersecting circles   **43.** $x^2 = 16(y + 2)$; $V(0, -2)$; y-axis

45. $(0, 0)$ and (a, a) **47.** $(\frac{3}{5}x_1 + \frac{2}{5}x_2, \frac{3}{5}y_1 + \frac{2}{5}y_2)$ **49.** $\frac{9}{5}\sqrt{5}$

53. $x = 3$ is a vertical asymptote

Exercises 2.1 page 73

1. $D: (-\infty, \infty)$; $R: (-\infty, \infty)$ **3.** $D: (-\infty, \infty)$; $R: [-5, \infty)$

5. $D: [-6, 6]$; $R: [0, 6]$ **7.** $D: x \geq -\frac{3}{2}$; $R: [0, \infty)$

9. D: all real numbers; R: all real numbers except -12

11. D: all real numbers except $x = a$; R: all real numbers except $2a$

13. $D: (-\infty, \infty)$; $R: [0, \infty)$ **15.** D: all $x \neq 0$; R: -1 and 1

17. $D: (-\infty, \infty)$; $R: -3, 0$, and 2 **19.** $D: (-\infty, \infty)$; $R: [0, \infty)$

21. $D: (-\infty, 0]$ or $[3, \infty)$; $R: [0, \infty)$

23. Domain is the set of positive integers, whereas the range is certain positive integers, the first eight of which are shown in the following table.

n	1	2	3	4	5	6	7	8
$n!$	1	2	6	24	120	720	5040	40,320

25. (a) relation is a function. $D: \{1, 2, 3, 4\}$; $R: \{3, 4, 6\}$;   **(b)** relation is not a function. $D: x \in R$; $R: y \in R$;   **(c)** relation is a function. $D: \{-2, -1, 0, 3\}$; $R: \{3, 0, -1, 8\}$ **(d)** $x = 2$ is the relation $\{(x, y) \mid x = 2\}$, it is not a function. For example, $(2, 0)$ and $(2, -3)$ are members of the set domain $\{2\}$ and range $y \in R$.

27. $S = 6x^2$, $x = \sqrt{S/6}$ **29.** $V = (S/6)^{3/2}$; $S = 6V^{2/3}$

Exercises 2.2 page 79

1. (a) -1;   **(b)** 8;   **(c)** 20;   **(d)** $\frac{19}{2}$ **(e)** $3x^2 + 8$;   **(f)** $3x + 17$;   **(g)** $3(x + h) + 8$;

(h) $3(x + h) + 16$;   **(i)** $\dfrac{3(x + h) - 3x}{h} = 3, (h \neq 0)$ **3. (a)** $\sqrt{8} = 2\sqrt{2}$;   **(b)** $\sqrt{3}$;

(c) does not exist;   **(d)** $\sqrt{6x + 17}$;   **(e)** $\dfrac{1}{h}(\sqrt{3(x + h) + 2} - \sqrt{3x + 2}), h \neq 0$

5. (a) 9;   **(b)** $\frac{1}{2}$;   **(c)** $\pi + 1$;   **(d)** $|x|$;   **(e)** $x = \pm\frac{5}{2}$ **7. (a)** $\dfrac{10}{u}$;   **(b)** $\dfrac{10 - 2u}{u}$;

(c) 0   **9. (a)** $\sqrt{x} + x - 1, x \geq 0$;   **(b)** $\sqrt{x} - x + 1, x \geq 0$;

(c) $\sqrt{x}(x - 1), x \geq 0$;   **(d)** $\dfrac{\sqrt{x}}{x - 1}, x \geq 0$ and $x \neq 1$;   **(e)** $\dfrac{x - 1}{\sqrt{x}}, x > 0$;

(f) $\sqrt{x - 1}, x \geq 1$;   **(g)** $\sqrt{x} - 1, x \geq 0$ **11. (a)** $|x + 1| + \dfrac{1}{x}, x \neq 0$;

(b) $|x + 1| - \dfrac{1}{x}, x \neq 0$;   **(c)** $\dfrac{|x + 1|}{x}, x \neq 0$;   **(d)** $x|x + 1|, x \neq 0$;

(e) $\dfrac{1}{x|x + 1|}, x \neq 0, -1$;   **(f)** $\left|\dfrac{1}{x} + 1\right|, x \neq 0$;   **(g)** $\dfrac{1}{|x + 1|}, x \neq -1$

13. (a) odd since $f(-x) = -f(x)$;   **(b)** even;   **(c)** odd;   **(d)** neither;   **(e)** even;

(f) neither   **23.** $V = \dfrac{\pi h}{4}(4a^2 - h^2)$; $S = \pi h(4a^2 - h^2)^{1/2} + \dfrac{\pi}{2}(4a^2 - h^2)$

25. $x > 7$ or $x < \dfrac{5}{3}$ **27.** $g(x) = \dfrac{x + 2}{3 - x}, x \neq 3$; $(g \circ f)(x) = x, x \neq -1$

Exercises 2.3

1. $b \neq \pm 2\sqrt{2}$ 3. $(2m - 3)(m + \frac{1}{2})$
5. quotient $= m^3 - 2m^2 - m + 5$, remainder $= -11$
7. $x^4 + 10x^3 + 33x^2 + 36x$ 9. $\frac{1}{2}, -1$ 11. $1, 1, -3, 4, 4, 4, -4, -4, -4$
13. $y = 2$ 15. $y = 3x^2 - x + 5$ 19. $n =$ odd positive integer
21. $(y - 2)(y + 3)(y^2 + 2y + 4)$ 25. $D: (-\infty, \infty);$ $R: (-\infty, \infty)$
27. $D: (-\infty, \infty);$ $R: \{0, \frac{1}{2}, 1\}$ 29. $D: (-\infty, \infty);$ $R: \{5, 4, 3, \frac{3}{2}, 0\}$

Exercises 2.4

1. $L = 25$ 3. no limit   5. $L = 2$ 7. $L = -5$ 9. $L = 2$
11. $L = -\frac{7}{3}$ 13. $L = -12$ 15. $L = 0$ 17. 6   19. $\frac{3}{11}$ 21. 1

23. no limit   25. 8   27. no limit   29. 2   31. 1   33. $-\dfrac{1}{u^2}$

35. $2t_0 - 4 = 2(t_0 - 2)$ 37. $2a + 7$

Exercises 2.5

1. 0.005   3. 0.001   5. 0.01   7. 0.01   9. $\dfrac{\varepsilon}{2}$ 11. ε 13. $\min\left(\dfrac{4\varepsilon}{5}, \dfrac{1}{4}\right)$

15. $\min(2\varepsilon, 1)$ 17. $\min((\sqrt{15} + 4)\varepsilon, 1)$ 19. $\min\left(\dfrac{\varepsilon}{7}, 1\right)$ 21. no limit

27. $\frac{1}{2}$ 29. $\frac{4}{3}b$

Exercises 2.6

1. (a) 3; (b) 0; (c) does not exist because $\lim\limits_{x \to 5^+} f(x) \neq \lim\limits_{x \to 5^-} f(x)$
3. (a) 0; (b) 0; (c) 0   5. (a) 7; (b) 7; (c) 7   7. (a) 0; (b) 0; (c) 0;
9. (a) 9; (b) 8; (c) does not exist   11. (a) 0; (b) 0; (c) 0
13. (a) 1; (b) 1; (c) 1   15. (a) 0; (b) 0; (c) 0
17. (a) $-\dfrac{1}{x^2}$; (b) $-\dfrac{1}{x^2}$; (c) $-\dfrac{1}{x^2}$

Exercises 2.7

1. $\frac{3}{4}$ 3. 3   5. $\frac{3}{8}$ 7. 3   9. 0   11. $-\frac{1}{2}$ 13. $\frac{8}{15}$ 15. 1

Exercises 2.8

1. ∞ 3. 0   5. ∞ 7. 0   9. $-\frac{5}{3}$ 11. $-\infty$ 13. ∞ 15. ∞
17. $-\infty$ 19. (a) $f(x) = \dfrac{1}{x^2} + L, g(x) = \dfrac{1}{x^2}$; (b) $f(x) = \dfrac{1}{x^4}, g(x) = \dfrac{1}{x^2}$;

(c) $f(x) = \dfrac{1}{x^2} + [\![x]\!], g(x) = \dfrac{1}{x^2}$ are examples

Exercises 2.9

1. $x = 2$ 3. $u = 2$ 5. $x = 3$ 7. none   9. none   11. none
13. $x = 0$ 15. none   17. none   19. $t = 5$ 23. 3   25. n

27. define $f(0) = 1$ **29.** discontinuity at $x = 0$ is not removable
31. $h(x)$ is the only one of the three functions that is continuous at $x = 0$
33. (a) $1, 1, 0, 1$; (b) $-1 \leq x \leq 1$; (c) $f(x) = 1$ if $x \neq 0$ and $f(0) = 0$ for x in $[-1, 1]$, therefore $f(x)$ is continuous for all values of x in $[-1, 1]$ except for $x = 0$ and $x = \pm 1$.

Exercises 2.10 *page 132*

1. 3   **3.** 2   **5.** all x **7.** all $u \neq \pm\sqrt{3}$ **9.** $-3 < y < 3$ **11.** all x
13. all x **15.** all x except $x = -1$ **17.** $f(x) = \begin{cases} (x-2)^{-1}, & \text{if } x \neq 2 \\ 0 & \text{if } x = 2 \end{cases}$

19. $f(x) = \sqrt{100 - x^2}$ **21.** $f(x) = \begin{cases} x & \text{if } x < 0 \\ x + 1 & \text{if } x \geq 0 \end{cases}$ **25.** yes to all four cases

Exercises 2.11 *page 136*

1. discontinuous, continuous, discontinuous, continuous
3. discontinuous, continuous, discontinuous, continuous
5. continuous, discontinuous, discontinuous, discontinuous
7. discontinuous, continuous, continuous, discontinuous
9. continuous, discontinuous, discontinuous, continuous
11. continuous, discontinuous, continuous, discontinuous
13. $(-\infty, \frac{1}{2}), (\frac{1}{2}, 3), (3, \infty)$ **15.** $[a, b]$ **17.** $(-\infty, \infty)$
19. $f(0) = 2$ is the absolute maximum; $f(2) = 0$ is the absolute minimum
21. $f(5) = \frac{25}{3}$ is the absolute maximum; $f(3) = 3$ is the absolute minimum
23. $g(-3) = 20$ is the absolute maximum; $g(1) = 4$ is the absolute minimum
25. $(-2, -1), (2, 3), (5, 6)$

Review and Miscellaneous Exercises *page 138*

1. $x \geq \frac{4}{5}$ **3.** $x \neq 0, 5$ **5.** $(-\infty, \infty)$
7. $f(x) + g(x) = 4 - x, (-\infty, \infty)$; $f(x) \cdot g(x) = 2x(4 - 3x), (-\infty, \infty)$; $\dfrac{f(x)}{g(x)} = \dfrac{2x}{4 - 3x}, x \neq \dfrac{4}{3}$; $\dfrac{g(x)}{f(x)} = \dfrac{4 - 3x}{2x}, x \neq 0$; $f(g(x)) = 2(4 - 3x), (-\infty, \infty)$;
$g(f(x)) = 4 - 6x, (-\infty, \infty)$;   **9.** $f(g(x)) = |x|, (-\infty, \infty)$; $g(f(x)) = x, [0, \infty)$
13. -40 **15.** no limit   **17.** $3u^2$ **19.** $-\frac{1}{4}, -\frac{1}{4}$ **21.** 1   **23.** $1 \pm \sqrt{7}$
25. none   **27.** $\frac{1}{3}$ **29.** all numbers outside $(2, 5)$ **31.** $\frac{10}{3}$ **33.** 0
35. $-\frac{1}{9}$ **37.** $y = \frac{3}{2}, x = \frac{7}{2}$ **39.** $y = 0, x = 1,$ and $x = 6$
41. function monotonically increases and interval is open
43. (a) none; (b) $[0, \infty)$ or $(-\infty, -1)$

Exercises 3.2 *page 149*

1. -4 **3.** 1   **5.** 7   **7.** -1 **9.** $\dfrac{1}{2\sqrt{3}}$
11. tangent at $(-1, 4)$: $5x + y + 1 = 0$; normal: $x - 5y + 21 = 0$
13. tangent at $(-1, -1)$: $y = 3x + 2$; normal: $y = -\dfrac{x}{3} - \dfrac{4}{3}$
15. (a) $(-\frac{2}{3}, \frac{2}{3})$, (b) $(-\frac{1}{3}, 1)$, (c) $(-\frac{7}{12}, \frac{11}{16})$

Exercises 3.3 *page 154*

1. 128 ft/sec   3. 24 ft/sec   5. (a) 5 sec, (b) 3.5 sec, (c) 35/16 sec
7. 8 ft/sec   9. $\frac{3}{8}$ ft/sec   11. $v(t_0) = 20t_0$ ft/sec; $v(2) = 40$ ft/sec
13. $v(t_0) = \dfrac{1}{2\sqrt{t_0 + 5}}$ ft/sec; $v(4) = \frac{1}{6}$ ft/sec
15. (a) $v = -32t$ ft/sec; (b) $v(\sqrt{40}) = -32\sqrt{40} = -64\sqrt{10}$ ft/sec; (c) $\sqrt{20} = 2\sqrt{5}$ sec
17. (a) $v(t) = 20t$ ft/sec; (b) $t = \frac{5}{2}$ sec; (c) $v = 20, 60, 20\sqrt{\dfrac{y}{10}}$ ft/sec
19. (a) 800 ft/sec; (b) $v = 960 - 32t$ ft/sec; $v|_{t=10} = 640$ ft/sec; (c) 30 sec;
(d) 14,400 ft; (e) 60 sec

Exercises 3.4 *page 161*

1. $f'(x) = 2$ 3. $f'(x) = a$ 5. $f'(x) = 4x - 3$ 7. $f'(x) = 2ax + b$
9. $f'(x) = 3x^2$ 11. $f'(x) = -\dfrac{1}{x^2}$ 13. $f'(x) = -\dfrac{3}{2x^2}$ 15. $D_x F = \dfrac{1}{(x + 1)^2}$
17. $D_x F = 2 - \dfrac{9}{x^2}$ 19. $D_x F = -\dfrac{5}{x^2} + \dfrac{16}{x^3}$ 21. $f'(2) = -12$
23. $f'(0) = -3$ 25. not differentiable at $x = 1$
27. g' exists for all $x \neq \frac{3}{2}$; $g' = 2$ if $x > \frac{3}{2}$, $g' = -2$ if $x < \frac{3}{2}$
29. (a) $f'(0)$ does not exist; (b) $g'(0) = 0$; (c) $F'(0) = 0$

Exercises 3.5 *page 168*

1. $f'(x) = 3x^2 + 12x - 8$ 3. $f'(x) = 8x^7 - 30x^4 + 4x^3 - 3$
5. $u'(s) = 9s^2 - 20s - 6$ 7. $g'(y) = y^2 - 2$ 9. $g'(x) = 3x^2 - 10 - \dfrac{2}{x^2}$
11. $g'(x) = 4x^3 - 8x$ 13. $f'(y) = \dfrac{y^2 + 2y}{(y + 1)^2}$ 15. $g'(x) = \dfrac{-23}{(2x + 5)^2}$
17. $u'(x) = -\dfrac{30x^2}{(x^3 - 5)^2}$ 19. $w'(x) = \dfrac{4}{x^2}$ 21. $v'(x) = \dfrac{3x^2 + 4x + 3}{(3x + 2)^2}$
23. $F'(t) = \dfrac{2t(ac + b)}{(c - t^2)^2}$ 25. $g'(t) = \dfrac{3t^4 + 8t^3 + 2t^2 + 8t + 3}{(t + 2)^2}$
27. $f'(x) = 24x^2 - 24x + 6$ 31. $f'(x) = 4x^3 - 3x^2 - 12x$
33. $h'(x) = 10x^4 + 12x^3 - 8$

Exercises 3.6 *page 172*

1. $f'(x) = 21(3x + 2)^6$ 3. $g'(x) = -3(4 + x)^{-4}$
5. $h'(u) = 8(6u - 1)(3u^2 - u + \pi)^7$ 7. $G'(x) = 3(3x - 2)^3(2x + 3)^2(14x + 8)$
9. $F'(x) = 18(x^2 + 2x - 3)^8(3x - 7)^{11}(5x^2 - 13)$
11. $f'(u) = \dfrac{5}{2}\left(u - \dfrac{1}{u^2}\right)^{3/2}\left(1 + \dfrac{2}{u^3}\right)$ 13. $F'(x) = \dfrac{1 + 2\sqrt{x}}{4\sqrt{x}(x + \sqrt{x})}$
15. $y = 3x$ 17. $D_x w = \dfrac{-2(2 - 3x^2)}{x^5}$ 19. $D_x w(-1) = -4$
21. (a) $-3x^{-4}$; (b) $\dfrac{-3(x + 1)^2}{x^4}$; (c) $\dfrac{1}{(x + 1)^2}$; (d) $9x^8$ 23. $D_x \sqrt{x} = \dfrac{1}{2\sqrt{x}}$
25. $g'(x) = \dfrac{1}{x}$ 27. $f'(x) = \begin{cases} \frac{4}{3}(2x)^{-2/3} & \text{if } x > 0 \\ 0 & \text{if } x < 0 \end{cases}$; $f'(0)$ does not exist

1. $f'(x) = \frac{3}{2}x^{1/2} + 7x^{5/2}$ **3.** $F'(x) = \frac{3}{2}x^{1/2} - 3$ **5.** $G'(y) = \frac{2y + 3}{3(y^2 + 3y - 2)^{2/3}}$

7. $f'(u) = \frac{3}{4}(u^2 - u - 2)^{1/2}(2u - 1)$ **9.** $f'(y) = \frac{y(3y^2 - 2a^2)}{\sqrt{y^2 - a^2}}$

11. $G'(x) = \frac{1}{(x^2 + 1)^{3/2}}$ **13.** $h'(t) = \frac{-b^3}{t^2\sqrt{b^2 - t^2}}$ **15.** $F'(x) = \frac{3x^2 + 16x - 10}{2(x + 4)^{3/2}}$

17. $g'(t) = \frac{-4t}{3(t^2 + 1)^{2/3}(t^2 - 1)^{4/3}}$ **19.** $f'(x) = \frac{5}{2}x^{3/2} + 6x + \frac{9}{2}x^{1/2} + 1$

21. $x - 4y + 6 = 0$

23. tangent: $x - 12y + 28 = 0$;   normal: $12x + y - 99 = 0$

25. $g'(x) = \frac{2x(x^2 - 1)}{|x^2 - 1|}$ **27.** $G'(x) = 10|x|$ **29.** $\frac{xh'(x^2 + 5)}{\sqrt{h(x^2 + 5)}}$

1. $-\frac{x}{y}$ **3.** $-\frac{x^2}{y^2}$ **5.** $-\sqrt{\frac{y}{x}}$ **7.** $-\frac{y}{x}$ **9.** $\frac{y - 2x}{3y^2 - x}$ **11.** $\frac{3}{2}\left(\frac{y}{x}\right)^2$

13. $-\left(\frac{y}{x}\right)^{1/3}$ **15.** $\frac{3 - 4y}{4x - 1}$ **17.** $\frac{4x\sqrt{xy} - y}{2\sqrt{xy} + x}$ **19.** $4x - 3y - 50 = 0$

21. tangent: $4x - 7y + 18 = 0$; normal: $7x + 4y - 1 = 0$

1. 20 **3.** $6(v - 1)$ **5.** $3(2y + 1)^{-5/2}$

7. $\frac{2c(bc - ad)}{(cx + d)^3}$; $G''(x) = 0$ if $ad - bc = 0$ **9.** 0 **11.** a **13.** $G(x)$

15. $y' = -\frac{x}{y}$; $y'' = -\frac{a^2}{y^3}$ **17.** $y' = -\frac{y}{x + 2y}$, $y'' = \frac{4}{(x + 2y)^3}$

19. $y' = -\frac{y^{1/2}}{x^{1/2}}$, $y'' = \frac{a^{1/2}}{2x^{3/2}}$ **21.** $y'|_{(a,a)} = -1$; $y''|_{(a,a)} = \frac{2}{a}$; $y'''|_{(a,a)} = -\frac{6}{a^3}$

23. $D^n f = a^n f$ **25.** $f(x) = x^3 - 10x^2 + 31x - 25$ **27.** (a) $f'(x) = \begin{cases} 2x & x > 0 \\ 0 & x \leq 0 \end{cases}$;

(b) yes; $f'(0) = 0$;   **(c)** yes; $f''(x) = \begin{cases} 2 & x > 0 \\ 0 & x < 0 \end{cases}$;   **(d)** no, $f''(0)$ does not exist

29. $v(t) = 3t^2 - 12$; $a(t) = 6t$ force field is to left for $t < 0$ and to the right if $t > 0$. Motion is to right in $(-\infty, -2)$, to left in $(-2, 2)$, and then to right if $t > 2$. $s(-2) = s_{\max}(\text{local})$, $s(2) = s_{\min}(\text{local})$.   **31.** $v^2 D_s^2 v + v(D_s v)^2$

1. $2x + 3x^{1/2} + 1$ **3.** $\frac{-y}{x + 3y^2}$ **5.** $-\frac{y}{x}$ **7.** $\frac{1}{3}\left(1 - \frac{1}{x^2}\right)$

9. $\frac{1}{2}(10x^2 + 7x - 2)^{-1/2}(20x + 7)$ **11.** $-\sqrt{\frac{y}{x}}$ **13.** $\frac{3 + 2\sqrt{x}}{4\sqrt{x}\sqrt{x + 3\sqrt{x}}}$

15. $D_x y = \begin{cases} 0 & \text{if } x > 2 \text{ or if } x < 0 \\ 2 & \text{if } 0 < x < 2 \\ \text{Fails to exist at the corner points } (2, 2) \text{ and } (0, -2) \end{cases}$

17. $(x^3 + x^2 + 5)^2(x^2 - x + 6)^3[17x^4 + x^3 + 44x^2 + 76x - 20]$

19. $-5(2x + 1)(x^2 + x + 1)^{-2}$ **21.** $\dfrac{2x(x^2 - a^2)}{|x^2 - a^2|}$ **23.** $y = \dfrac{x}{3}$

25. $T: 48x - 7y - 75 = 0;\ N: 7x + 48y - 158 = 0$ **27.** $-\dfrac{1}{(x + 1)^2}$

29. $3x^2 - 4$ **31.** $f(x) = x^3 - 10x^2 + 35x - 35$

33. $f(x) = x^3 + 3x^2 - 5x - 10$ **35.** (a) $7(x^2 - 3x + 4)^6(2x - 3)$;

(b) $14x(x^4 + 3x^2 + 4)^6(2x^2 + 3)$;   **(c)** $-7x^{-15}(1 + 3x + 4x^2)^6(3x + 2)$

37. $\dfrac{20(2t + 7)}{(4t^2 + 28t + 54)^2}$ **39.** (a) $2 + 2x^{-3}$;   **(b)** $(10!)x^{-11}$;   **(c)** $(-1)^n(n!)x^{-(n+1)}$

41. $k = 12$ or $k = -\frac{176}{27}$; $(x - 2)^2(x + 3)$ or $(3x + 4)^2(3x - 11)$

43. $\lim\limits_{t \to 0} g(t) = 3f'(a)$

Exercises 4.1 *page 193*

1. $D_r A = 2\pi r;\ D_r A|_{r=5} = 10\pi$ in.^{2}/in.

5. $A = u^2/2;\ D_u A = u;\ D_u A|_{u=8} = 8$ in.^{2}/in.   **7.** $D_h r = -1$ when $r = 2h$

9. $-\frac{1}{8}$ **11.** (a) $4\pi r^2$;   (b) $2\pi r^2$ **13.** \$99.75   **15.** $D_t h = \dfrac{2}{\pi}$ ft/min

17. 40 mi/hr

Exercises 4.2 *page 198*

1. 24π ft^2/sec   **3.** $\dfrac{1}{80\pi}$ ft/min   **5.** $D_t V = \dfrac{k}{4}\sqrt{\dfrac{A}{6}}$ in.^{3}/sec

7. 9.5 ft/sec (approx)   **9.** (a) 3.2 ft/min;   (b) 3.2 ft/min   **11.** 2 ft/sec

13. $D_t y = -\frac{12}{5}\sqrt{5}$ in./sec   **15.** $\dfrac{D_t P/P}{D_t V/V} = -n$ **17.** 10 workers/year

19. $k = 10\sqrt{3} \doteq 17.32$ ft^3/min (approx.)   **21.** 41

Exercises 4.3 *page 207*

1. $f(-3) = -7$ (abs min); $f(5) = 17$ (abs max)

3. $h(0) = 0$ (abs min);   no abs max

5. $G(\frac{3}{2}) = \frac{9}{4}$ (abs max); $G(4) = -4$ (abs min)

7. $f(1) = -8$ (abs min);   no abs max

9. $F(-3) = -\frac{1}{3}$ (abs max); $F(-1) = -1$ (abs min)

11. $H(5) = 0$ (abs min);   no abs max

13. $g(\sqrt{2}) = \dfrac{\sqrt{2}}{4} \doteq 0.35$ (abs max); $g(-1) = -\frac{1}{3}$ (abs min)

15. $F(\frac{2}{3}) = \frac{2}{9}\sqrt{3} \doteq 0.385$ (abs max); $F(0) = F(1) = 0$ (abs min)

17. $g(0) = g(4) = 1 + 3(2^{2/3})$ (abs max); $g(2) = 1$ (abs min)

19. $F(0) = 4$ (abs max); $F(2) = 0$ (abs min)

21. $H(1) = 1$ (abs max); $H(-1) = -\frac{1}{3}$ (abs min)   **23.** no absolute extrema

25. $y = 3x^2 - 6x + 1$ **27.** yes   **29.** not necessarily true   **31.** 19

Exercises 4.4 *page 216*

1. 400 by 200 yd, with the longer side parallel to the river

3. 100 ft by 150 ft   **5.** 40 in. $\times$ 40 in. $\times$ 20 in.

7. $r = \sqrt[3]{\dfrac{V}{2\pi}}$ in.; $h = 2\sqrt[3]{\dfrac{V}{2\pi}}$ in.

9. width $= \dfrac{2R}{\sqrt{3}}$ and depth $= \dfrac{2\sqrt{2}R}{\sqrt{3}}$; $\dfrac{\text{depth}}{\text{width}} = \sqrt{2}$

11. radius $= 6$ in., central angle $= 2$ radians   **13.** $V_{\max} = \dfrac{4\pi R^2 H}{27}$

15. diameter of semicircle $=$ overall height $= \dfrac{2P}{4+\pi}$ **17.** $x = \dfrac{L}{2}$; $\dfrac{qL^4}{384EI}$

19. $20\sqrt{5} \doteq 44.7$ mi/hr   **21.** $x = 5000$; $\$24,000$ **23.** row directly to B

25. $\dfrac{h}{r} = \sqrt{2}$ **27.** $(2,2)$ and $(2,-2)$ **29.** $(3,0)$ and $(-3,0)$ **31.** $V_{\min}{'} = \dfrac{8\pi a^3}{3}$

Exercises 4.5 *page 222*

1. $x = 1$ **3.** $(\frac{1}{4}, \frac{1}{4})$ **5.** $(x^2 y)_{\max} = \dfrac{2\sqrt{3}}{9}$; $(x^2 y)_{\min} = -\dfrac{2\sqrt{3}}{9}$ **9.** $h = 2r$

11. $Q\left(\dfrac{ac + b^2 x_0 - aby_0}{a^2 + b^2}, \dfrac{bc + a^2 y_0 - abx_0}{a^2 + b^2}\right)$ **13.** radius $=$ height of cylinder

15. 1700 and -800

Exercises 4.6 *page 230*

1. $c = 3$ **3.** $c = \sqrt{\frac{7}{3}}$ **5.** $c = \dfrac{a}{\sqrt{3}}$ **7.** $c = 0$

9. (a) No; (b) no; (c) Rolle's theorem remains unblemished

11. $v = 0$ for some value of t in (t_0, t_1) **13.** $c = \frac{11}{3}$ **15.** $c = \frac{64}{27}$

17. $c = \sqrt{6}$ **19.** no; no   **21.** no; yes: $c = 2$

23. yes; $c = \sqrt{ab}, a < \sqrt{ab} < b$

27. (a) $f(x) - g(x) \equiv -9$; (b) $f(x) - g(x) \equiv -2$

29. zero of $f'(x)$ is arithmetic mean of zeros of $f(x)$

35. the velocity of vehicle at some time in (t_0, t_1) is equal to its average velocity in $[t_0, t_1]$

Exercises 4.7 *page 239*

1. $f(3) = -8$ is both a relative and absolute minimum; $f(7) = 8$ is an absolute maximum

3. $h(-1) = -26$ (abs min); $h(5) = 28$ (abs max)

5. $G(1) = 2$ (rel min and abs min); $G(4) = 4\frac{1}{4}$ (abs max)

7. $f(1)$ rel and abs max; $f(3)$ rel and abs min; $f(1) = f(4) = -1$ (abs max); $f(0) = f(3) = -5$ (abs min)

9. $h(-\sqrt{2}) = -2$ (rel and abs min); $h(\sqrt{2}) = 2$ (rel and abs max)

11. $G(-1) = -\frac{1}{2}$ (rel and abs min); $G(1) = \frac{1}{2}$ (rel and abs max)

13. $f(1) = 3$ (rel and abs min); $f(9) = 11$ (abs max)

15. $f(-2) = 1$ (abs min); $f(1) = 4$ (rel and abs max)

17. $h(-1) = -1$ (abs min); $h(2) = 4$ (rel and abs max)

19. $H(0) = 0$ (abs min); $H(1) = 36$ (rel and abs max)   **21.** $f(1) = 5$ is rel min

23. $h(3) = 0$ (rel min)   **25.** $H(\frac{1}{2}) = 0$ (rel min)

27. $f(b) = \frac{1}{2}$ (rel max); $f(-b) = -\frac{1}{2}$ (rel min)   **29.** no extrema

31. $g(1) = g(-1) = 2$ (rel min)   **33.** $F(-\frac{1}{3}) = 0$ (rel min)

35. $H(0) = 4$ (rel max); $H(2) = 0$ (rel min)

Exercises 4.8 *page 250*

1. up $x > 1$;   down $x < 1$;   infl $(1, 0)$
3. up $x > \frac{7}{6}$;   down $x < \frac{7}{6}$;   infl $(\frac{7}{6}, -\frac{379}{54})$ 5. up $x \geq 2$
7. up $x > 2$ or $x < 0$;   down $0 < x < 2$; $(0, 2)$ and $(2, 2)$ infl pts
9. up $x < 3$ $(x \neq 0)$;   down $x > 3$; $(3, \frac{19}{9})$ infl point
11. up $x < 0$;   down $x > 0$; $(0, 0)$ infl point   13. no   15. $a = -\frac{3}{7}$ and $b = \frac{18}{7}$

Exercises 4.9 *page 256*

1. (a) $\dfrac{600}{x} + 40 + \dfrac{90}{x^2}$;   (b) \$39.1;   (c) \$39.18 (approx)
3. (a) \$11.80;   (b) 64
5. (a) $[-\frac{68}{3}, \infty)$;   (b) $\dfrac{x^2}{3} - x - 8 + \dfrac{4}{x}$;   (c) $x^2 - 2x - 8$;   (d) no infl points
7. $x = 4.55$ (approx)
9. (a) $p(x) = 7 - 0.001x$, $R(x) = 7x - 0.001x^2$;   (b) 3500 toys;   (c) \$12,250
11. 625 suits;   \$91,550;   \$250   13. 100;   \$16;   \$800   15. $\dfrac{\alpha - b}{2(\beta + a)}$
17. (a) 6000;   (b) \$5;   (c) \$10,000

Review and Miscellaneous Exercises *page 257*

1. (a) $4\pi r^2$,   (b) $2\pi r^2$ 3. $-\frac{16}{3}$ in/sec   5. \$72,000/day   7. $-\dfrac{9}{32\pi}$ in/sec
9. No absolute extremum
11. $f(8) = 7$ is an abs max and not a rel max; $f(4) = 2$ is both an abs and a
rel min   13. $\frac{3}{2}$ ft/sec and $\frac{11}{2}$ ft/sec   15. (a) $f'(4) = 0$,   (b) $f'(3) = 0$
17. No, however $f'(0) = 0$ 21. $a = -3$ and $b = 2$
23. (a) $x = \sqrt{2a}$,   (b) $x = \sqrt{\dfrac{2a}{3}}$ 25. (a) $a = -\frac{15}{2}$, $b = 12$, c is arbitrary;
(b) $x = 1$ yields relative maximum and $x = 4$ yields a relative minimum;   (c) No
27. $\dfrac{3\sqrt{3}}{4} a^2$ 29. $x = 0$ does not yield a relative extremum for $f(x)$
31. \$700, \$49,000   33. (a) the minimum is 1

Exercises 5.1 *page 266*

1. $dy = dx$ 3. $dy = (4x^3 - 6x^2 + 1)\,dx$ 5. $dy = \dfrac{x}{\sqrt{4 + x^2}}\,dx$
7. $dV = 3e^2\,de$ 9. $dV = 4\pi R^2\,dR$ 11. $dv = \dfrac{3u + 2}{2\sqrt{u + 1}}\,du$ 13. 0.06
15. 0   17. (a) $3\,\Delta x = 3\,dx$;   (b) $dy = 3\,dx$;   (c) $\Delta y - dy = 0$;   (d) $\varepsilon = 0$
19. (a) $\Delta y = (2x - 5)\,\Delta x + (\Delta x)^2$;   (b) $dy = (2x - 5)\,\Delta x$;   (c) $\Delta y - dy = (\Delta x)^2$;
(d) $\varepsilon = \Delta x$ 21. (a) $\Delta y = 3x^2\,\Delta x + 3x(\Delta x)^2 + (\Delta x)^3$;   (b) $dy = 3x^2\,dx = 3x^2\,\Delta x$;
(c) $\Delta y - dy = 3x(\Delta x)^2 + (\Delta x)^3$;   (d) $\varepsilon = 3x\,\Delta x + (\Delta x)^2$ 25. 8.063   27. 9.850
29. 5.892   37. $dy = 2t(3t^4 + 2)\,dt$ 39. $dy = \left(\dfrac{-2ak^2}{t^3} - \dfrac{bk}{t^2}\right)dt$
41. $\dfrac{dy}{dt} = \dfrac{(t^3 - t)(3t^2 - 1)}{\sqrt{(t^3 - t)^2 + 1}}$ 43. $\dfrac{dy}{dt} = \dfrac{4t(3t - y)}{2t^2 + 3y^2}$ 47. $gt\,\Delta t$ ft; $g\,\Delta t$ ft/sec
49. 1.5%   51. $dV \doteq -134$ ft^3

1. $\dfrac{x^5}{5} + C$ **3.** $\frac{4}{5}x^{5/4} + C$ **5.** $\frac{2}{9}x^{9/2} + C$ **7.** $\frac{2}{3}\sqrt{11}\,w^{3/2} + C$

9. $\dfrac{s^5}{5} + 2s^3 + 9s + C$ **11.** $\frac{2}{3}\sqrt{7}x^{3/2} + \dfrac{6}{\sqrt{5}}x^{1/2} + C$ **13.** $\dfrac{x^4}{4} - \dfrac{2x^3}{3} + \dfrac{x^2}{2} + C$

15. $\dfrac{x^4}{4} + 2x^3 + 6x^2 + 8x + C$

17. $f(x) = \dfrac{x^3}{2} - \dfrac{5x^2}{2} + C_1 x + C_2,$ $(C_1, C_2$ consts$)$

19. $\frac{4}{5}x^{5/2} + \frac{16}{3}x^{3/2} + C_1 x + C_2,$ $(C_1, C_2$ consts$)$ **21.** $x^{7/2} + 3x^{1/2}$

23. $f(x) + C$ **27.** $\dfrac{x|x|}{2} + C$

1. $\frac{2}{9}(3x - 5)^{3/2} + C$ **3.** $\frac{2}{9}(x^3 + 4)^{3/2} + C$ **5.** $\frac{1}{5}\sqrt{5t^2 + 4} + C$

7. $\frac{2}{5}(x - 3)^{3/2}(x + 2) + C$ **9.** $\dfrac{2 - x}{(x - 4)^2} + C$ **11.** $-\dfrac{1}{2(x - 6)^2} - \dfrac{2}{(x - 6)^3} + C$

13. $\dfrac{x^5}{5} + 4x^2 - \dfrac{16}{x} + C$ **15.** $-\frac{1}{6}\left(3 + \dfrac{1}{x}\right)^6 + C$

17. $\frac{3}{112}(2y + 1)^{4/3}(8y - 3) + C$ **19.** $\frac{2}{15}(w + a)^{3/2}(3w + 5b - 2a) + C$

21. $3x^3 - 3x^2 + x + C$

1. $f(x) = -2x^2 + 11$ **3.** $s(t) = \sqrt{1 + t} + 3$ **5.** $y(x) = \sqrt{3x^2 + 1}$

7. $g(t) = 2t + \sqrt{t}$ **9.** $F(u) = -\dfrac{1}{u^2} - \dfrac{1}{u}$ **11.** $y = \dfrac{x^3 + 8}{3}$

13. $g(x) = 3x^2 - x + 2$ **15.** $F(x) = \dfrac{1}{x} - \dfrac{2}{x^2} - 2x + 2$

17. 144 ft;   6 sec; -96 ft/sec   **19.** 320 ft/sec   **21.** no—needs 572 ft to stop

25. $\sqrt{30} \doteq 5.48$ sec; $v_0 = 5$ ft/sec   **27.** $\dfrac{2(s_1 t_2 - s_2 t_1)}{t_1 t_2 (t_1 - t_2)}$

Review and Miscellaneous Exercises   *page 286*

1. (a) $\Delta y = dy = 3 \Delta x, \Delta y - dy = 0;$ (b) $\Delta y = dy = 3 \Delta x, \Delta y - dy = 0$

5. $\frac{9}{4}(x^2 + 2)^{4/3} + C$ **7.** $\frac{3}{7}(2 + x)^{7/3} - \frac{3}{2}(2 + x)^{4/3} + C$

9. $\frac{4}{11}t^{11/4} + 4t^{3/4} + C$ **11.** $-\frac{3}{2}(5 + v^2)^{-1/3} + C$ **13.** $\frac{1}{8}(3 - 4x)^{-2} + C$

15. 2.967   **17.** 3%   **19.** 2.001   **21.** $\dfrac{y^2}{2} + \dfrac{4}{x} = 3$

23. $\dfrac{y^2}{2} = \sqrt{2x + 3} + 7$ and in explicit form, from $y = -4$ when $x = -1,$

$y = -\sqrt{2(7 + \sqrt{2x + 3})}$ **25.** $(x + 2)^{3/2} - (3y + 1)^{1/2} = 7$

27. $f - 6g = \dfrac{x^2}{2} + C$ **29.** $\frac{11}{2}$ sec, 242 ft   **31.** 1280 ft/sec; 57,600 ft

1. $[3(1) + 1] + [3(2) + 1] + [3(3) + 1] + [3(4) + 1]$ **3.** $\frac{1}{2} + \frac{2}{3} + \frac{3}{4} + \frac{4}{5} + \frac{5}{6}$

5. $\dfrac{2(1)}{2!} + \dfrac{3(2)}{3!} + \dfrac{4(3)}{4!} + \dfrac{5(4)}{5!} + \dfrac{6(5)}{6!}$

7. $f(1) + f(2) + \cdots + f(n^2 - 1) + f(n^2)$ **9.** $\displaystyle\sum_{k=1}^{n} (2k - 1) = n^2$

11. $\displaystyle\sum_{k=1}^{n} 3^k = \tfrac{3}{2}(3^n - 1)$ **13.** $\displaystyle\sum_{i=1}^{n} y_i \, \Delta x_i$ **15.** 1155   **17.** $\dfrac{n(3n - 1)}{2}$

19. $5^{n+1} - 1$ **21.** $\dfrac{1}{n + 1} - \tfrac{1}{2}$ **23.** $\sqrt{301} - \sqrt{7}$ **25.** $h(m) - h(1)$

27. $\dfrac{(n + 1)(2n + 1)}{6n^2}$ **29.** $\tfrac{15}{2}\dfrac{(n + 1)(2n + 1)}{n^2}$ **39.** $\dfrac{n(n + 1)(n + 5)}{3}$

Exercises 6.2 *page 302*

1. $\tfrac{1}{3}$ sq units   **3.** 8 sq units   **5.** 9 sq units   **7.** 4 sq units
11. $\dfrac{b^2 - a^2}{2}$ sq units; area of trapezoid with bases a and b and altitude $b - a$
13. 20 sq units   **15.** $\tfrac{14}{3}$ sq units

Exercises 6.3 *page 308*

1. (a) 3;   (b) 2;   (c) 2   **3.** -20 **5.** 16
9. 0;   algebraic sum of area and its negative is 0
11. $\displaystyle\int_{-b}^{-a} x^2 \, dx = \int_{a}^{b} x^2 \, dx = \dfrac{b^3}{3} - \dfrac{a^3}{3} = \dfrac{(-a)^3}{3} - \dfrac{(-b)^3}{3}$ **13.** -3.085
15. (a) 14;   (b) 29;   (c) $20\tfrac{3}{4}$ **17.** (a) 2.6;   (b) 3.8;   (c) 3.2
19. (a) 1.897;   (b) 0.924;   (c) 1.400   **21.** $\displaystyle\int_{0}^{1} x^2 \, dx = \tfrac{1}{3}$ **23.** $\displaystyle\int_{0}^{2} \dfrac{1}{1 + x^2} \, dx$
25. $S_n = \tfrac{2}{3} + \dfrac{1}{3n^2} - \dfrac{1}{n^3}$; $\displaystyle\int_{0}^{1} \sqrt{x} \, dx = \tfrac{2}{3}$ **27.** $\tfrac{2}{3}b^{3/2}$

Exercises 6.4 *page 320*

5. $\displaystyle\int_{-10}^{6} f(x) \, dx$ **7.** $\displaystyle\int_{-a}^{b+3} h(x) \, dx$
9. (a) $\tfrac{40}{3}$;   (b) 16;   (c) -8;   (d) 0;   (e) -1 **11.** 0;   45   **13.** 0;   20
15. 20;   36   **17.** 0;   1   **19.** 0;   2   **21.** 2;   6
23. (a) 0;   (b) 1;   (c) $0 + 1 + 2 + \cdots + (n - 1) = \dfrac{(n - 1)(n)}{2}$ **29.** 0
33. 2   **35.** $2(3^{-1/2})$ **37.** 0   **39.** 2.26 (approx)

Exercises 6.5 *page 332*

1. $\dfrac{1}{x}$ **3.** $x^4 - 3x + 7$ **5.** $-(x^2 + 16)^{1/2}$ **7.** $2(x^4 + 9)^{-1/2}$
9. $g(x^{1/2})/2x^{-1/2}$ **11.** $-\tfrac{1}{2}$ **13.** 4   **15.** $\tfrac{43}{6}$ **17.** 21   **19.** $\tfrac{1}{3}(5^{3/2} - 1)$
21. $\dfrac{b^3}{3}$ **23.** $\tfrac{116}{15}$ **25.** 4   **27.** $\tfrac{11}{6}$ **29.** $\tfrac{16}{3}$ sq units   **31.** $\tfrac{158}{3}$ sq units
33. $\tfrac{225}{4}$ sq units   **35.** $\tfrac{123}{2}$ sq units   **37.** $\tfrac{1}{4}(15^{1/2} - 3^{1/2})$ sq units
39. $\tfrac{4}{3}(27 - 5^{3/2})$ sq units

1. $\frac{364}{243} \doteq 1.498$ **3.** $1 - \frac{1}{40} = 0.975$ **5.** $\frac{1}{4}n(n+1)(n^2 + n - 8)$ **9.** $-\frac{5}{2}$
11. 2   **13.** $\frac{2}{15}(7^{3/2} - 2^{3/2})$ **15.** $\sqrt{7} - 2$ **17.** $\frac{4}{3}(5^{3/2} - 8)$ **19.** $2 - 2^{-1/3}$
21. definite integral does not exist   **23.** $-\frac{44}{3}$ **25.** $-\frac{a^3}{48}$ **27.** $\frac{16}{15}a^{5/2}$

29. $2x\sqrt{3 + x^4}$ **31.** $\int_1^3 \frac{du}{u}$

33. (a) $\dfrac{x^9}{3} - \dfrac{5x^3}{3} - 2x$;   **(b)** $3x^8 - 5x^2 - 2$;   **(c)** 0;   **(d)** 0   **35.** $\frac{41}{3}$

37. 200 ft/sec   **39. (a)** the intermediate value theorem;   **(b)** -2 and -1

Exercises 7.1   *page 343*

(All answers, except for Exercise 21, are in square units.)
1. $\frac{32}{3}$ **3.** $\frac{9}{2}$ **5.** 36   **7.** $\frac{32}{3}$ **9.** $\frac{16}{3}$ **11.** $\frac{5}{12}$ **13.** 9   **15.** 9   **17.** $\frac{2}{3}$
19. 16   **21. (a)** $2:1$;   **(b)** $5:1$;   **(c)** $n:1$ if $n \geq 1$; $1:n$ if $0 < n < 1$ **23.** $\frac{16}{3}b^2$

Exercises 7.2   *page 355*

(All answers are in cubic units.)
1. $\dfrac{8\pi}{5}$ **3.** 24π **5.** 18π **7.** $\dfrac{128\pi}{5}$ **9.** $\dfrac{4\pi}{3}$ **11.** 8π **13.** $\dfrac{15\pi}{2}$
15. $\dfrac{2\pi}{3}\left(1 - \dfrac{1}{\sqrt{2}}\right)$ **17.** $\dfrac{128\pi}{5}$ **19.** $\dfrac{8\pi}{3}$ **21.** $V = \frac{4}{3}\pi ab^2$

Exercises 7.3   *page 363*

(All answers are in cubic units.)
1. $\dfrac{81\pi}{2}$ **3.** $\dfrac{128\pi}{3}$ **5.** $\dfrac{4\pi}{15}$ **7.** $\dfrac{3\pi}{10}$ **9.** $\dfrac{16\pi}{3}$ **11.** π **13.** 8π **15.** $\dfrac{40\pi}{3}$
17. $2\pi p^3$ **19.** $\dfrac{\pi h^2}{3}(3r - h)$ **21.** $\dfrac{32\pi}{3}$ **23.** $\dfrac{208\pi}{15}$

Exercises 7.4   *page 368*

(All answers are in cubic units.)
1. 32   **3.** $\frac{2}{3}\pi a^3$ **5.** $\dfrac{2b^3}{3}$ **7.** 80   **9. (a)** $\frac{4}{3}\pi r^3$;   **(b)** $\dfrac{\pi h}{3}(r^2 + R^2 + rR)$
11. $\dfrac{16r^3}{3}$

Exercises 7.5   *page 374*

1. 80 in.-lb   **3.** 8000 in.-lb   **5.** 3300 ft-lb
7. $k\left(\dfrac{1}{b} - \dfrac{1}{a}\right)$ where $k > 0$ is a constant of proportionality   **9.** 10,000 mi-ton

11. $300,000\pi$ ft-lb   **13.** $\frac{2}{3}\pi r^3 k(h + \frac{3}{8}r)$ ft-lb   **15.** 2304 ft-lb

17. $\dfrac{1}{\gamma - 1}(p_2 V_2 - p_1 V_1)$ **19.** $\dfrac{wa^4}{2}\left(1 - \dfrac{\gamma}{w}\right)^2$

Exercises 7.6 page 381

(All answers, except for Exercise 9, are in pounds.)

1. 45,000   **3.** $2666\frac{2}{3}$ **5.** 24,000   **7.** 12,500,000   **9.** 13 ft   **11.** 15,500

13. 61,000

Exercises 7.7 page 389

1. $\frac{68}{15}$ **3.** $\frac{2}{5}$ **5.** 0   **7.** 3.75 ft to right of m_1 **11.** (a) 69; (b) 66

13. 112 slugs; $\frac{24}{7}$ ft from end $x = 0$

15. $\dfrac{(\rho_0 + \rho_1)L}{2}$; $\left(\dfrac{\rho_0 + 2\rho_1}{\rho_0 + \rho_1}\right)\dfrac{L}{3}$ ft from end $x = 0$

Exercises 7.8 page 400

1. $(\frac{16}{7}, \frac{36}{7})$, yes   **3.** $(\frac{11}{8}, \frac{7}{8})$ **9.** $\left(\dfrac{4a}{3}, \dfrac{5b}{6}\right)$ **11.** $(\frac{13}{8}, \frac{17}{8})$ **13.** $(\frac{10}{3}, 2)$

15. $\left(\dfrac{3a}{4}, \dfrac{3a^2}{10}\right)$ **17.** $(\frac{3}{5}, \frac{2}{35})$ **19.** $(\frac{9}{8}, \frac{27}{5})$ **21.** $\left(\dfrac{4a}{3\pi}, \dfrac{4a}{3\pi}\right)$ **23.** $\left(\dfrac{a}{5}, \dfrac{a}{5}\right)$

Exercises 7.9 page 409

1. $(6, 0, 0)$ **3.** $(\frac{3}{4}, 0, 0)$ **5.** $(\frac{27}{16}, 0, 0)$ **7.** $(\frac{9}{8}, 0, 0)$ **9.** $(0, \frac{9}{4}, 0)$

11. $\left(0, \dfrac{5k}{6}, 0\right)$ **13.** $(0, \frac{1}{5}, 0)$ **15.** $(0, \frac{5}{6}, 0)$ **17.** $\left(\dfrac{3a}{8}, 0, 0\right)$, hemisphere

19. On the axis; between bases at a distance $\dfrac{h}{4}\left(\dfrac{3R^2 + 2rR + r^2}{R^2 + rR + r^2}\right)$ from the base of

radius r

Exercises 7.10 page 416

1. $2\sqrt{10}$ **3.** $\sqrt{1 + m^2}\,(d - c)$ **5.** $\frac{8}{27}(10\sqrt{10} - 1) \doteq 9.1$ **7.** $\frac{59}{24}$ **9.** 4

11. $6a$ **13.** $\displaystyle\int_0^2 \sqrt{1 + 4x^2}\,dx$ **15.** $\displaystyle\int_2^5 \sqrt{1 + (7 - 2x)^2}\,dx$

Exercises 7.11 page 423

(All answers are in square units.)

1. $2\pi k(b - a)$ **3.** $84\sqrt{17}\pi$ **5.** $\pi b\sqrt{a^2 + b^2}$ **7.** $\dfrac{32\pi}{3}(2\sqrt{2} - 1)$

9. $\dfrac{\pi b}{6a^2}[(b^2 + 4a^2)^{3/2} - b^3]$ **11.** $S = \pi rL$ **13.** $\dfrac{6\pi a^2}{5}$ **15.** $S = 13\sqrt{2}\pi$

Review and Miscellaneous Exercises   page 423

1. $\frac{16}{3}$ sq units   3. $\frac{2}{15}\pi b^5$ cu units   5. 360 lb-in.   7. 50,000 lb-ft   9. $\frac{\gamma}{6}bh^2$

11. $\frac{45}{22}$ 13. (a) 96 slugs,   (b) $\frac{44}{9}$ ft from end with density 4 slugs/ft

15. $\bar{x} = \bar{y} = \frac{2}{3}$ 17. $(\frac{8}{3}, 0, 0)$ 19. $2\sqrt{1 + a^2}$ 21. $2\pi R(b - a)$ sq units

23. $\frac{9}{2}$ sq units   25. $4a^3$ cu units

27. (a) $\int_0^9 \left(1 + \frac{1}{4x}\right)^{1/2} dx$;   (b) $\int_1^5 \frac{(x^4 + 1)^{1/2}}{x^2} dx$ 29. $\frac{3}{2}$ ft

31. $f(u) = \sqrt{\dfrac{3u - 2}{\pi}}$

Exercises 8.1   page 433

1. (a) 1.3862;   (b) 2.8903;   (c) 2.3025;   (d) 1.3540;   (e) -2.4848 3. $\dfrac{2x}{x^2 + 16}$

5. $\dfrac{2t + 3}{t^2 + 3t}$ 7. $\dfrac{1}{t \ln t}$ 9. $\dfrac{18x}{(x^2 - 8)(x^2 + 1)}$ 11. $\dfrac{10}{2t + 9} \ln^4 (2t + 9)$

13. $\dfrac{2}{t \ln t}$ 15. $\dfrac{(7x^2 + 54)x^2}{2(x^2 + 9)^{3/4}}$ 17. $\dfrac{x^2 - 2x + 5}{(x - 1)^2}$ 19. $\dfrac{(ad - bc)x}{(ax^2 + b)^{1/2}(cx^2 + d)^{3/2}}$

21. 1   23. $\frac{1}{8}(1 - 2 \ln 2) \doteq -0.0483$ 25. $\frac{1}{3} \ln |3t + 1| + C$

27. $\dfrac{x^2}{2} - x + \ln |x + 1| + C$ 29. $\dfrac{t^3}{3} + t^2 + 4t + 8 \ln |t - 2| + C$

31. no value   33. $\ln \left(\dfrac{x}{3}\right)$ 35. $-6 + 8 \ln \frac{5}{3}$

37. $\frac{1}{2} \ln^2 x - 3 \ln x + 14 \ln |3 + \ln x| + C$ 39. $\dfrac{1 - 2x - 4y}{x + 2y - 2}$

41. $y = \dfrac{1 - 2 \ln 2}{4} x - \frac{1}{2}(1 - 3 \ln 2)$ 43. $(-1)^{n-1}(n - 1)! x^{-n}$

45. $x \ln x - x + C$

Exercises 8.2   page 440

1. 1   3. $e^{1/3}$ 5. e^2 7. (b) limit $= e$;   (c) limit $= e$ 11. \$8131.40

Exercises 8.3   page 447

1. $f^{-1}(x) = \dfrac{x}{4}$;   domain of f: $(-\infty, \infty)$; domain of f^{-1}: $(-\infty, \infty)$

3. $f^{-1}(x) = -x + 3$;   domain of f: $(-\infty, \infty)$;   domain of f^{-1}: $(-\infty, \infty)$

5. $f^{-1}(x) = x^2 - 2$;   domain of f: all $x > 0$;   domain of f^{-1}: all $x > \sqrt{2}$

7. no inverse

9. $G^{-1}(x) = \sqrt[5]{x - 1}$;   domain of G: $(-\infty, \infty)$;   domain of G^{-1}: $(-\infty, \infty)$

11. $f^{-1}(x) = \dfrac{5}{2x - 3}$;   domain of f: $x \neq 0$;   domain of f^{-1}: $x \neq \frac{3}{2}$

17. (b) $x \geq 0$;   (c) $F^{-1}(x) = \sqrt[4]{\dfrac{x - 8}{5}}$, $x \geq 8$ 19. $g(x)$ has no inverse

21. (a) domain and range for f: $(-\infty, \infty)$;
(b) f^{-1} exists—domain and range for f^{-1}: $(-\infty, \infty)$ 23. $5^{-1/2}$

25. -1 29. $d = -a$

31. $D_y x = 3^{-1}(x^2 + 1)^{-1}$; $D_y^2 x = -\frac{2}{9}x(x^2 + 1)^{-3}$; $D_y^3 x = -\frac{2}{27}(x^2 + 1)^{-5}(1 - 5x^2)$

1. $3x$ **3.** x^{-2} **5.** 1296 **7.** $2(2x - 5)$ **11.** $3e^{3x-1}$

13. $\frac{1}{2}(x + 5)^{-1/2}\, e^{\sqrt{x+5}}$ **15.** $(1 + 10x)e^{10x}$ **17.** $2bx(e^{bx^2} + e^{-bx^2})$

19. $e^{2x}(x^{-1} + 2\ln 5x)$ **21.** $-4(e^x - e^{-x})^{-2}$ **23.** $\dfrac{e^{3x}}{3} + C$ **25.** $\dfrac{e^{5x}}{5} + C$

27. $-\frac{1}{2}e^{-x^2} + C$ **29.** $\dfrac{x^2}{2} + C$ **31.** $\dfrac{e^{3x}}{3} + \frac{15}{2}e^{2x} + 75e^x + 125x + C$

33. $e^{g(x)+1} + C$ **35.** $2e^4$ **37.** $\frac{1}{6}(e^6 - e^{-6}) - 2$ **39.** $\dfrac{a}{2}(e - e^{-1})$

41. $\ln\left(\dfrac{2}{1 + e^{-1}}\right)$ **43.** $-e^{x-y}$ **45.** $\dfrac{4(5x + 12)}{(2x + 5)(3e^{3y} + 7)}$

47. $F(x) = Ae^{3x}$, (A an arbitrary constant)   **49.** $\dfrac{1}{3\pi}(1 - e^{-6})$ sq units

51. 1000; $100{,}000e^{-1} \doteq \$36{,}787.90$

53. $u(t) = u_0 + Ae^{kt}$, (A an arbitrary constant)

1. $\dfrac{1}{\sqrt{2}}$ **3.** $\log_3 2$ **5.** 3 **7.** $2(3^{2t})\ln 3$ **9.** $-\dfrac{4^{-\sqrt{x}}}{2\sqrt{x}}\ln 4$ **11.** $\dfrac{4}{(4x + 3)\ln 10}$

13. $\dfrac{\log_{10}\left(\dfrac{e}{z}\right)}{z^2}$ **15.** 0 **17.** $\dfrac{\log_b e}{(y + 1)\ln(y + 1)}$

19. $(x^4 + 9)^{\log_{10} x}\left[\log_{10} x\,\dfrac{4x^3}{x^4 + 9} + \dfrac{\log_{10} e}{x}\ln(x^4 + 9)\right]$

21. $z^{(2^z)}\dfrac{2^z}{z}[1 + (\ln 2)z\ln z]$ **23.** $(\log_b x)^{\log_b x}\dfrac{\log_b e}{x}(1 + \ln(\log_b x))$

25. $\dfrac{\pi^x}{\ln \pi} + C$ **27.** $\dfrac{5}{\ln 2}(2^{x/5}) + C$ **29.** $\dfrac{4^{(x^2)}}{2\ln 4} + C$ **31.** $\frac{2}{5}\dfrac{8^{5x/2}}{\ln 8} + C$

33. $\ln b(\ln|\log_b x|) + C$ **35.** $\dfrac{\ln(5 + 2^x)}{\ln 2} + C$

41. $y - 24 = 8(1 + 3\ln 2)(x - 3)$ **43.** (a) $P = 0.5(10^6)e^{t/40}$;   (b) $1{,}414{,}200$

45. 6 years and 306 days

1. $x^2(1 + 3\ln 6x)$ **3.** $\dfrac{e^{-bx}}{x}(1 - bx\ln ax)$ **5.** $\dfrac{1}{(\ln\ln t)(t\ln t)}$

7. $\dfrac{m}{x}\log_a e\,(\log_a x)^{m-1}$

9. $x(x^2 + x + 1)^{x^2-1}[2x^2 + x + 2(x^2 + x + 1)\ln(x^2 + x + 1)]$

11. $-2t\log_{10}(t^8 + 1)$ **13.** $t^{(t^t)}t^t\ln t\left[1 + \ln t + \dfrac{1}{t\ln t}\right]$

15. $(x - a)[4x^2 - (2a + 3b + 3c)x + 2bc + ac + ab]$ **17.** $\frac{1}{20}\ln(10x^2 + 7) + C$

19. $\dfrac{e^{2x}}{2} + 3e^x + \frac{1}{2}e^{-2x} + C$ **21.** $\frac{5}{3}x^3 - 5x^2 + 7\ln|x| + 19x^{-1} + C$

23. $\dfrac{(3 + 5e)^x}{\ln(3 + 5e)} + C$ **25.** $\dfrac{6^{5t+3}}{5\ln 6} + C$ **27.** $\dfrac{6}{\ln 3}$

29. $\dfrac{x^2}{2} - ax + a^2\ln|x + a| + C$ **31.** $\dfrac{3^{(e^x)}}{\ln 3} + C$ **33.** $\dfrac{1 + 2x(x - y)e^{3y}}{1 - 3x^2(x - y)e^{3y}}$

35. $t_1 \doteq 13.5$ years   **37.** $\left(18 - \dfrac{8}{\ln 3}\right)$ sq units   **39.** 0.57

41. $2e^y + e^{-2x} = 3$ or $y = \ln\left[\frac{1}{2}(3 - e^{-2x})\right]$ **43.** $-\ln x, (e^{-3} \le x \le 1)$
45. $f^{-1}(x) = 1 + \sqrt{x - 2}, (3 \le x \le 11)$ **47.** 0.96
49. (a) $(x - 1)e^x + C$, (b) $(x^2 - 2x + 2)e^x + C$

Exercises 9.1   page 477

1. $(0, 1)$ **3.** $(1, 0)$ **5.** $(-1, 0)$

39. $\dfrac{\pi}{4} + n\pi$; $\dfrac{3\pi}{4} + n\pi$, where n is an arbitrary integer

41. $\dfrac{\pi}{3} + 2n\pi$; $\dfrac{5\pi}{3} + 2n\pi$; $\pi + 2n\pi$, where n is an arbitrary integer

43. $4n\pi$ or $(4n + 1)\pi$, where n is an arbitrary integer   **45.** neither   **47.** odd
49. even   **51.** yes   **53.** $\pm\sqrt{a^2 + b^2}$ are absolute extrema

Exercises 9.2   page 486

1. 3   **3.** 4   **5.** no limit   **7.** 1   **9.** 4   **11.** $-\sin a$ **13.** 5   **15.** 0
17. $\cos x \cos 2x - 2 \sin x \sin 2x$ **19.** 0   **21.** $-2 \sin 2\theta$ **23.** $a \sin \theta$

25. $\dfrac{-x \sin (x^2)}{\sqrt{\cos (x^2)}}$ **27.** $e^{3x}(18 \cos 2x + \sin 2x)$ **29.** $b \tan bt$

31. $e^{\sin b\theta}(b \cos b\theta \cos c\theta - c \sin c\theta)$ **33.** $(\sin \theta)^{\sin \theta} \cos \theta (1 + \ln \sin \theta)$
35. $k \sin^{k-1}x \sin (k + 1)x$
37. (a) $3^n(-1)^{\frac{n-1}{2}} \cos 3x$; n odd; $3^n(-1)^{n/2} \sin 3x$, n even;

(b) $2^n(-1)^{\frac{n+1}{2}} \sin 2x$, n odd; $2^n(-1)^{n/2} \cos 2x$, n even   **39.** $\dfrac{-\csc^2 (x + y)}{2 + \csc^2 (x + y)}$

41. $-\left[2 + e^{-(2x+y)} \sin\left(x + \dfrac{\pi}{4}\right)\right]$ **43.** yes   **45.** $3\pi + 1, 1$
47. $-3\sqrt{3}$ is absolute minimum; there is no absolute maximum

49. (b) no; (c) yes   **51.** $\dfrac{\pi}{180} \cos \theta°$

53. (a) $\dfrac{\pi}{2} + k\pi$, k is an arbitrary integer; (b) no   **55.** $f'(0) = 0$

Exercises 9.3   page 493

1. $\frac{1}{3} \sin 3x + C$ **3.** $\frac{1}{5} \sec 5\theta + C$ **5.** $-3 \csc \dfrac{\theta}{3} + C$ **7.** $\tan v - v + C$

9. $t + \frac{1}{6} \cos 6t + C$ **11.** $-\frac{1}{5} \cot 5\theta - \theta + C$ **13.** $\frac{1}{12} \sin^6 2t + C$

15. $-\frac{1}{3} \ln |\csc 3t + \cot 3t| + t + C$ **17.** $\dfrac{1}{2 \cot^2 x} + C$ **19.** $\ln \cos^2 y + C$

21. $-\sin \dfrac{1}{x} + C$ **23.** $2 \tan t - t - 2 \sec t + C$ **25.** $\dfrac{x}{2} - \frac{1}{28} \sin 14x + C$

27. 0   **29.** 2   **31.** $\dfrac{3\pi}{2}$ **33.** $\frac{1}{4}(4 - \pi)$ **35.** 2   **37.** $-\frac{1}{2} \ln (\sqrt{2} - 1)$

39. 0   **41.** $\frac{7}{12}$ **43.** $(\sqrt{2} - 1)$ sq units   **45.** $\left(\dfrac{\sqrt{3}}{2} - \dfrac{\pi}{6}\right)$ sq units

47. $\dfrac{\pi^2}{2}$ cu units   **49.** both

51. The calculation is wrong—the fundamental theorem cannot be applied.   **53.** 3

1. (a) $\frac{\pi}{6}$;   (b) $-\frac{\pi}{4}$;   (c) $\frac{\pi}{4}$ **3.** (a) $\frac{\pi}{3}$;   (b) $\frac{\pi}{3}$;   (c) $\frac{\pi}{6}$

5. (a) $3\sqrt{5}$;   (b) $\frac{240}{289}$ **7.** (a) $\frac{5}{3}$;   (b) $\frac{1}{\sqrt{2}}$ **9.** $\dfrac{2x}{\sqrt{1-4x^2}}$, $-\frac{1}{2} < x < \frac{1}{2}$

11. $\dfrac{1}{\sqrt{x^2-8}}$, $8 < x^2 < 9$ **19.** $0.05\sqrt{5} - 0.5\sqrt{3}$; $0.25\sqrt{3} - 0.1\sqrt{5}$; $\frac{7}{8}$

21. 3

1. $\dfrac{5}{\sqrt{1-25x^2}}$ **3.** $\dfrac{-4}{\sqrt{1-16x^2}}$ **5.** $\dfrac{-1}{1+x^2}$ **7.** $\dfrac{1}{|x|\sqrt{9x^2-1}}$

9. $\dfrac{x}{\sqrt{1-x^2}} + \sin^{-1}x$ **11.** $\dfrac{-2x^2}{\sqrt{1-x^2}}$ **13.** $\dfrac{a}{x^2+a^2}$ **15.** $\dfrac{7}{3|x|\sqrt{x^2-1}}$

17. $-\dfrac{a+b}{|x|\sqrt{x^2-1}}$ **19.** $2\sqrt{9-x^2}$ **21.** $\dfrac{y(x^2+y^2+1)}{x(1-x^2-y^2)}$ **23.** $\dfrac{e^{-(x+y)}}{\sqrt{1-x^2}} - 1$

25. 0.01218 rad/sec   **27.** $\frac{1}{128}$ rad/sec   **29.** $x = \sqrt{b(a+b)}$

31. $3x^2 \tan^{-1}(x^3)$ **35.** $\dfrac{\sin^{-1}(x^2)}{2} + C$

1. $\frac{1}{4}\tan^{-1}\left(\dfrac{x}{4}\right) + C$ **3.** $\sin^{-1}\dfrac{x}{5} + C$ **5.** $x + \frac{1}{2}\tan^{-1}\dfrac{x}{2} + C$

7. $\sec^{-1}|2x| + C$ **9.** $\dfrac{\sqrt{3}}{2}\sec^{-1}\left|\dfrac{x}{2}\right| + C$ **11.** $\frac{1}{5}\sqrt{\frac{5}{7}}\tan^{-1}\sqrt{\frac{5}{7}}t + C$

13. $\sin^{-1}(x-3) + C$ **15.** $\frac{1}{2}\sec^{-1}\left|\dfrac{x-3}{2}\right| + C$ **17.** $\frac{1}{2}(\tan^{-1}x)^2 + C$

19. $\frac{1}{3}\tan^{-1}\left(\dfrac{\tan x}{3}\right) + C$ **21.** $\frac{1}{2}\sin^{-1}\left(\dfrac{2x-1}{3}\right) + C$ **23.** $\dfrac{\pi}{6}$ **25.** $\dfrac{\pi}{8}$

27. Does not exist   **29.** 0   **31.** $\dfrac{\pi}{12}$ **33.** $\frac{1}{12}(\tan^{-1}5 - \tan^{-1}1) \doteq 0.049$

35. $\dfrac{\pi}{12}$ **37.** $\ln 5 + \tan^{-1}2 \doteq 2.82$ **39.** (a) $2\tan^{-1}3$;   (b) $2\tan^{-1}k$;

(c) π, area bounded by the curve $y = \dfrac{1}{1+x^2}$, and the x-axis

41. $\sqrt{3} - \dfrac{\pi}{3} \doteq 0.69$

11. $\cot^2 x + 1 = \csc^2 x$ **13.** $\cos(x+y) = \cos x \cos y - \sin x \sin y$

15. $\tan(x+y) = \dfrac{\tan x + \tan y}{1 - \tan x \tan y}$ **17.** $\tan\dfrac{x}{2} = \dfrac{\sin x}{1 + \cos x}$

19. $\cos x + \cos y = 2\cos\frac{1}{2}(x+y)\cos\frac{1}{2}(x-y)$

21. $\sinh x_1 = \frac{15}{8}$; $\tanh x_1 = \frac{15}{17}$; $\coth x_1 = \frac{17}{15}$; $\operatorname{sech} x_1 = \frac{8}{17}$; $\operatorname{csch} x_1 = \frac{8}{15}$

25. yes

Exercises 9.8

1. $10 \cosh (10x + 9)$ **3.** $\tanh x$ **5.** $\dfrac{x}{\sqrt{x^2 + 1}} \sinh \sqrt{x^2 + 1}$

7. $(5 - 6x^2) \operatorname{csch}^2 (2x^3 - 5x)$ **9.** $\dfrac{-2 \sinh t}{(1 + \cosh t)^2}$ **17.** $\dfrac{4x^3}{\sqrt{x^8 - 1}}$

19. $\dfrac{a \sec^2 ax}{1 - \tan^2 ax}$ **21.** $\sinh^{-1} \dfrac{x}{2} + C$ **23.** $\tfrac{1}{2} \ln \sinh (x^2) + C$

25. $\dfrac{\sinh 6x}{12} - \dfrac{x}{2} + C$ **27.** $\cosh^{-1}\left(\dfrac{x + 4}{2}\right) + C$ **29.** $-\dfrac{\sinh^2 t}{2} + C$

31. $\tanh^{-1} \tfrac{1}{2} \doteq 0.55$ **33.** $\tfrac{1}{2}(\cosh^{-1} 16 - \cosh^{-1} 4) \doteq 0.70$

41. (a) $y = 11.1 \cosh \dfrac{x}{11.1}$, $-15 \leq x \leq 15$;   (b) $H = 33.3$ lb;   (c) $d = 11.7$ ft

Review and Miscellaneous Exercises

1. (a) $\dfrac{2\pi}{3}$,   (b) 8π,   (c) $\dfrac{\pi}{3}$,   (d) none,   (e) 2π **3.** $-\dfrac{\sqrt{5}}{5}$ **5.** $\tfrac{3}{7}$ **7.** 0

9. 64   **11.** $10 \sin 5x \cos 5x$ **13.** $-\dfrac{\sin x}{2\sqrt{\cos x}}$ **15.** $\dfrac{\sin t}{t} + (\cos t)(\ln 3t)$

17. $2 \csc t$ **19.** $8ax \sec^4 (ax^2 + b) \tan (ax^2 + b)$ **21.** $-\dfrac{\sqrt{a^2 - x^2}}{x^2}$

23. Yes, the identity is $\cos 4x = \cos^2 2x - \sin^2 2x$

25. $(\sin a\theta)^{\sin a\theta} a \cos a\theta (1 + \ln \sin a\theta)$ **27.** $\dfrac{1}{2}, \dfrac{\theta}{2}$ **29.** $\dfrac{x}{\sqrt{6x - x^2}}$

31. $3(\cos 2x)^{3x}[\ln \cos 2x - 2x \tan 2x]$ **33.** $\tfrac{1}{3} \sin^{-1} \dfrac{2x}{3} + C$

35. $\dfrac{1}{3\sqrt{5}} \tan^{-1} \dfrac{e^{3t}}{\sqrt{5}} + C$ **37.** $\tfrac{1}{2} \sin^{-1} \dfrac{2x - 3}{4} + C$

39. $x - \tfrac{3}{2} \ln (x^2 + 2x + 3) + \dfrac{1}{\sqrt{2}} \tan^{-1} \dfrac{x + 1}{\sqrt{2}} + C$

41. $-\sqrt{4 - 4x - x^2} - 2 \sin^{-1} \dfrac{x + 2}{2\sqrt{2}} + C$ **43.** $\dfrac{1}{a}\left(\ln \cosh ax - \dfrac{\tanh^2 ax}{2}\right) + C$

45. $\dfrac{\sinh 4t}{32} - \dfrac{t}{8} + C$ **47.** $\ln 2 + \dfrac{3\pi}{8}$ **49.** $\dfrac{\pi}{6\sqrt{3}}$ **51.** $\ln 2 + \dfrac{7\pi}{8}$ **53.** $\dfrac{65}{4}$

55. $\tfrac{1}{8}\left(1 - \dfrac{1}{\cosh^2 4}\right)$ **57.** $(\sinh 3)(1 + \tfrac{2}{3} \sinh^2 3)$ **59.** $\dfrac{1023}{10}$

61. $(a + b) \cosh mx + (a - b) \sinh mx$ **63.** $y = -\dfrac{x}{2}$ **65.** $\sinh a$ **67.** $\dfrac{2}{\pi}$

69. $\theta = \dfrac{\pi}{4} + \dfrac{\alpha}{2}$ **71.** $(3x)^{\sin x}\left(\dfrac{\sin x}{x} + \cos x \ln 3x\right)$

73. $(x^2 + 2)^{\cos x}\left[\dfrac{2x}{x^2 + 2} \cos x - \sin x \ln (x^2 + 2)\right]$

Exercises 10.1

1. $\tfrac{1}{3} \ln |x^3 + 5| + C$ **3.** $-e^{1/x} + C$ **5.** $\tfrac{2}{9}(3x + 7)^{3/2} + C$ **7.** $\dfrac{3^{2x}}{2 \ln 3} + C$

9. $\tfrac{1}{10} \tan^{-1} \dfrac{5x}{2} + C$ **11.** $\dfrac{1}{\sqrt{3}} \ln (x + \sqrt{x^2 - \tfrac{10}{3}}) + C$ **13.** $\dfrac{1}{2a \cos^2 ax} + C$

15. $\tfrac{1}{2} \tan^{-1}\left(\dfrac{e^x}{2}\right) + C$ **17.** $x - \sqrt{10} \tan^{-1}\left(\dfrac{x}{\sqrt{10}}\right) + C$ **19.** $-\dfrac{\sin 2x}{2} + C$

21. $\sin x - \dfrac{\sin^3 x}{3} + C$ **23.** $\frac{1}{2}\tan^{-1}\left(\dfrac{x+3}{2}\right) + C$ **25.** $e^x - 4e^{-x} + C$

27. $\tan\theta - 2\ln|\cos\theta| + C$ **29.** $\frac{1}{18}\tan^{-1}\left(\dfrac{x^3}{6}\right) + C$

31. $\frac{1}{2}\ln(\sin^2\theta + 3) + C$ **33.** $\ln(x - 5 + \sqrt{x^2 - 10x + 61}) + C$

35. $\frac{3}{8}(t^2 + 4)^{4/3} + C$ **37.** $\dfrac{1}{6a}(10 + \sin at)^6 + C$ **39.** $n = 1;\quad -\frac{1}{2}e^{-x^2} + C$

41. $\frac{3}{5}(4 + x)^{5/3} - 6(4 + x)^{2/3} + C$ **43.** $-\frac{1}{8}\ln(3 + 2e^{-4x}) + C$

Exercises 10.2 page 546

1. $\dfrac{\sin^3 x}{3} + C$ **3.** $-\dfrac{\cos^4 2t}{8} + C$ **5.** $-\frac{2}{3}(\cos t)^{3/2} + \frac{2}{7}(\cos t)^{7/2} + C$

7. $\dfrac{\sin 8x}{8} + C$ **9.** $-\dfrac{\cos^{k+1} t}{k+1} + C$, if $k \neq -1$; $-\ln|\cos t| + C$, if $k = -1$

11. $-\cos x + \frac{2}{3}\cos^3 x - \dfrac{\cos^5 x}{5} + C$ **13.** $\dfrac{\theta}{16} + \dfrac{\sin^3 2\theta}{48} - \dfrac{\sin 4\theta}{64} + C$

15. $\dfrac{\cos^{23} t}{23} - \dfrac{\cos^{21} t}{21} + C$ **17.** $-\dfrac{\cos 2x}{4} - \dfrac{\cos 4x}{8} + C$

19. $\dfrac{4\sin^3 x}{3} - \dfrac{4\sin^5 x}{5} + C$ **21.** $\dfrac{\cos^5\theta}{5} - \dfrac{\cos^3\theta}{3} + C$

23. $\dfrac{(a^2 + b^2)}{2}\theta + \dfrac{(a^2 - b^2)}{2}\sin\theta - ab\cos\theta + C$ **25.** $\tan\theta - \sec\theta + C$

27. $\dfrac{\sin^3(t^3)}{9} - \dfrac{\sin^5(t^3)}{15} + C$ **29.** $\dfrac{\pi}{2}$ **33.** $\dfrac{\pi}{32}$

Exercises 10.3 page 552

1. $\dfrac{\tan^2 x}{2} + C$ **3.** $\dfrac{(\tan x + 1)^2}{2} + C$ **5.** $-\dfrac{\csc^3 3x}{9} + C$ **7.** $-\dfrac{\cot^4 5\theta}{20} + C$

9. $-\dfrac{\cot^2\theta}{2} - \ln|\sin\theta| + C$ **11.** $2\left(\tan\dfrac{t}{2} + \frac{1}{3}\tan^3\dfrac{t}{2}\right) + C$

13. $\dfrac{1}{a}\left(\dfrac{\tan^4 a\theta}{4} - \dfrac{\tan^2 a\theta}{2} - \ln|\cos(a\theta)|\right) + C$ **15.** $\dfrac{\sec^5 v}{5} - \dfrac{\sec^3 v}{3} + C$

17. $-\dfrac{\cot^4 x}{4} + C$ **19.** $\dfrac{1}{a}[\frac{1}{2}\tan^2(a\theta + b) + \ln|\cos(a\theta + b)|] + C$

21. $\frac{1}{2}\tan 2x - \frac{1}{2}\cot 2x - 4x + C$ **23.** $\frac{2}{7}\sec^{7/2} w - \frac{2}{3}\sec^{3/2} w + C$

25. $\frac{1}{5}\tan^5\theta + \frac{4}{3}\tan^3\theta + 3\tan\theta + C$ **27.** $\frac{1}{3}(2\sqrt{2} - 1)$ **29.** 0

31. $\dfrac{\sec^{n+4}\theta}{n+4} - \dfrac{2\sec^{n+2}\theta}{n+2} + \dfrac{\sec^n\theta}{n} + C$

Exercises 10.4 page 562

1. $\frac{1}{2}\left(49\sin^{-1}\dfrac{x}{7} + x\sqrt{49 - x^2}\right) + C$ **3.** $(x^2 + 9)^{1/2} + C$

5. $\dfrac{1}{\sqrt{5}}\ln\left(\dfrac{\sqrt{5 + x^2} - \sqrt{5}}{|x|}\right) + C$ **7.** $\sqrt{x^2 - 4} - 2\sec^{-1}\dfrac{x}{2} + C$

9. $\frac{1}{128}\tan^{-1}\dfrac{x}{4} + \dfrac{x}{32(16 + x^2)} + C$ **11.** $\dfrac{-\sqrt{3 - x^2}(2x^2 + 3)}{27x^3} + C$

13. $2\left[\sin^{-1}\dfrac{x}{2} - \dfrac{x(4 - x^2)^{1/2}}{4}\right] + C$ **15.** $\dfrac{x\sqrt{x^2 + 9}}{2} - \frac{9}{2}\sinh^{-1}\dfrac{x}{3} + C$

17. $\frac{1}{10}\tan^{-1}\left(\dfrac{2x + 3}{5}\right) + C$ **19.** $\sin^{-1}\left(\dfrac{x - 2}{2}\right) + C$

Exercises 10.5

page 566

1. $\frac{1}{6}\ln\left|\dfrac{x-3}{x+3}\right| + C$ **3.** $\frac{1}{3}\ln\dfrac{(x-4)^4}{|x-1|} + C$

5. $\frac{3}{4}\ln|x| - \frac{9}{8}\ln|x-2| + \frac{11}{8}\ln|x+2| + C$ **7.** $\ln\left[\dfrac{(x+2)^2}{|x+1|^{1/2}\,|x+3|^{3/2}}\right] + C$

9. $\dfrac{x^4}{4} - \dfrac{7x^3}{3} + 2x + 9\ln|x| + C$ **11.** $\dfrac{x^3}{3} + x + 6\ln|x| - 11\ln|x-4| + C$

13. $3\ln|x| + \dfrac{4}{x} + 8\ln|x+1| + C$ **15.** $\dfrac{-3}{(x-1)^2} + \dfrac{7}{x-1} + 4\ln|x-1| + C$

17. $-\dfrac{1}{2(x+3)^2} + \dfrac{2}{3(x+3)^3} + C$ **19.** $\frac{3}{2}\ln|2x-1| + \ln|x+1| + \dfrac{1}{2x-1} + C$

21. $\frac{1}{27}\ln\left|\dfrac{x-3}{x}\right| + \dfrac{1}{9x} + \dfrac{1}{6x^2} + C$ **23.** $\ln\left|\dfrac{(\sin t + 2)^{1/6}(\sin t - 1)^{1/3}}{(\sin t)^{1/2}}\right| + C$

25. $-\dfrac{\ln 5}{6}$ **27.** integral does not exist   **29.** $\frac{3}{2} - 8\ln 2$

31. $\dfrac{1}{2a}\ln\left|\dfrac{x-a}{x+a}\right| + C$ **33.** $\dfrac{1}{(a-b)(a-c)}\ln|x-a| + \dfrac{1}{(b-a)(b-c)}\ln|x-b|$

$+ \dfrac{1}{(c-a)(c-b)}\ln|x-c| + C$ **35.** $\dfrac{x^2}{2} + 2ax + 3a^2\ln|x-a| - \dfrac{a^3}{x-a} + C$

Exercises 10.6

page 571

1. $\ln\left|\dfrac{(1+x^2)^2}{1+x}\right| + C$ **3.** $\dfrac{x^2}{2} - x - \dfrac{2}{\sqrt{3}}\tan^{-1}\left(\dfrac{2x+1}{\sqrt{3}}\right) + C$

5. $5\ln|x+1| - \frac{1}{2}\ln(x^2+4x+5) + C$ **7.** $-\frac{1}{4}(4x+1)^{-1} + C$

9. $\frac{1}{10}\ln\left(\dfrac{x^2+4}{x^2+9}\right) + C$ **11.** $\ln\dfrac{x^2}{x^2+1} - \frac{1}{2}\left(\tan^{-1}x + \dfrac{x}{x^2+1}\right) + C$

13. $\frac{1}{2}\ln|x-1| - \frac{1}{4}\ln(x^2+1) - \frac{1}{2}\tan^{-1}x + C$

15. $\ln|x-1| - \frac{1}{2}\ln|x+1| - \frac{1}{4}\ln(x^2+1) - \frac{3}{2}\tan^{-1}x + C$

17. $\dfrac{x^3}{3} + x + \frac{3}{2}\ln(x^2+9) - \frac{8}{3}\tan^{-1}\dfrac{x}{3} + C$ **19.** $\dfrac{x^3}{3} - \dfrac{10}{3}\ln|x^3+10| + C$

21. $\ln|x-3| + \frac{1}{2}\ln(x^2+4) + \dfrac{1}{x^2+4} + C$ **23.** $\ln(2\sqrt{\tfrac{2}{5}})$

25. $\frac{1}{4}\ln\frac{9}{5} + \frac{1}{2}\tan^{-1}2$ **27.** $\frac{1}{4}\ln\frac{3}{4}$

29. $n=3:\ -\dfrac{1}{2x^2} + \frac{1}{2}\ln\left(\dfrac{1}{x^2}+1\right) + C;\quad n=4:\ -\dfrac{1}{3x^3} + \dfrac{1}{x} - \tan^{-1}\dfrac{1}{x} + C$

Exercises 10.7

page 576

1. $x\sin x + \cos x + C$ **3.** $-(x+1)e^{-x} + C$ **5.** $x\tan^{-1}x - \frac{1}{2}\ln(1+x^2) + C$

7. $x\tan x + \ln|\cos x| + C$ **9.** $\dfrac{x^3}{9}(3\ln x - 1) + C$ **11.** $\dfrac{x}{2} - \dfrac{\sin 2x}{4} + C$

13. $-\dfrac{\cos x}{3}(2 + \sin^2 x) + C$ **15.** $\dfrac{e^x}{2}(\sin x - \cos x) + C$

17. $\dfrac{e^{ax}(a\sin bx - b\cos bx)}{a^2+b^2} + C$ **19.** $-\dfrac{e^{-x^2}}{2}(1+x^2) + C$

21. $\dfrac{x(x+4)^{21}}{21} - \frac{1}{462}(x+4)^{22} + C$ **23.** $-\dfrac{x^2}{3}(9-x^2)^{3/2} - \frac{2}{15}(9-x^2)^{5/2} + C$

25. $\frac{1}{8}(\sin 3x\cos x - 3\cos 3x\sin x) + C$ **27.** $\dfrac{x^3\tan^{-1}x}{3} - \dfrac{x^2}{6} + \frac{1}{6}\ln(1+x^2) + C$

29. $-\frac{1}{9}$ **31.** $-2e^{-1}$ **33.** $-\frac{1}{90}$

Answers to Odd-Numbered Exercises **1151**

1. $-\cot\frac{x}{2} + C$ **3.** $-2\left(1 + \tan\frac{x}{2}\right)^{-1} + C$ **5.** $\ln\left|\dfrac{1 + \tan\frac{x}{2}}{1 - \tan\frac{x}{2}}\right| + C$

7. $\sqrt{2}\,\tan^{-1}\left(\dfrac{1 + 3\tan\frac{\theta}{2}}{\sqrt{2}}\right) + C$ **9.** $2\tan^{-1}\left(1 + \sqrt{2}\tan\frac{\theta}{2}\right) + C$

11. $\dfrac{1}{\sqrt{5}}\ln\left(\dfrac{\sqrt{5}+1}{\sqrt{5}-1}\right)$ **13.** $\ln(1 + \sqrt{2})$ **15.** $\dfrac{2\pi}{3\sqrt{3}}$

1. $2[\sqrt{x} - \ln(1 + \sqrt{x})] + C$ **3.** $5\ln\left|\dfrac{x^{1/5}}{1 + x^{1/5}}\right| + C$

5. $e^t - 2\ln(e^t + 2) + C$ **7.** $\dfrac{e^{2t}}{2} - e^t + \ln(e^t + 1) + C$

9. $\frac{4}{5}t^{5/4} - t + \frac{4}{3}t^{3/4} - 2t^{1/2} + 4t^{1/4} - 4\ln(1 + t^{1/4}) + C$

11. $\frac{2}{3}[(1 + x^3)^{1/2} + (1 + x^3)^{-1/2}] + C$

15. $\frac{2}{3}(2 + 7x^3)^{1/2} + \dfrac{\sqrt{2}}{3}\ln\left|\dfrac{(2 + 7x^3)^{1/2} - \sqrt{2}}{(2 + 7x^3)^{1/2} + \sqrt{2}}\right| + C$

17. $\frac{4}{3}(\sqrt{2 + \sqrt{3 + x}})^3 - 8\sqrt{2 + \sqrt{3 + x}} + C$

19. $\dfrac{2}{a}\tan^{-1}\left(\dfrac{x + \sqrt{x^2 - a^2}}{a}\right) + C$ **21.** $-\dfrac{1}{a}\ln\left|\dfrac{a + \sqrt{a^2 - x^2}}{x}\right| + C$

23. $\ln\left|\dfrac{\sqrt{x + 2} + \sqrt{x + 1}}{\sqrt{x + 2} - \sqrt{x + 1}}\right| + C$ **25.** $2(2 + \ln 2)$ **27.** $6(e^2 - 1)$

29. integral does not exist   **31.** $\ln 2$

1. $-\dfrac{1}{bx} + \dfrac{a}{b^2}\ln\dfrac{ax + b}{x} + C$ **3.** $\dfrac{\sin 2\theta}{4\cos^2 2\theta} + \frac{1}{4}\ln\left|\tan\left(\theta + \dfrac{\pi}{4}\right)\right| + C$

5. $x\sin^{-1}6x + \frac{1}{6}\sqrt{1 - 36x^2} + C$ **7.** $x(\ln 3x)^3 - 3x(\ln 3x)^2 + 6(x\ln 3x - x) + C$

9. $\dfrac{\sqrt{3x^2 + x - 2}}{3} - \dfrac{1}{6\sqrt{3}}\ln|6x + 1 + 2\sqrt{3}\sqrt{3x^2 + x - 2}| + C$

11. $\dfrac{1}{5\sqrt{2}}\ln\left|\dfrac{\sqrt{x^5 + 2} - \sqrt{2}}{\sqrt{x^5 + 2} + \sqrt{2}}\right| + C$ **13.** $\frac{1}{2}\ln\left|\dfrac{x^2 + 1}{x^2 + 2}\right| + C$

15. $2\sin^{-1}\sqrt{x - 3} + C$ **17.** $-\frac{1}{15}\dfrac{\cos 5x}{\sin^3 5x} - \frac{2}{15}\cot 5x + C$

19. $\dfrac{x^3}{6} + 2x^2 + 2x + \left(\dfrac{x^2}{4} + 2x + \frac{7}{8}\right)\sin 2x + \cos 2x + \dfrac{x\cos 2x}{4} + C$

21. $-x - \frac{5}{3}\cot x - \cot^2 x - \frac{1}{3}\cot x\csc^2 x + C$ **23.** $\frac{1}{5}(4 - x^2)^{-5/2} + C$

25. $(x^2 + 3x + 1)\sinh x - (2x + 3)\cosh x + C$

27. $\sqrt{25 - x^2}(-3x^2 + x + 75) + \dfrac{x^2}{5} + 25\sin^{-1}\dfrac{x}{5} + C$ **29.** $\frac{1}{2}\tan^{-1}(x^2) + C$

31. $\dfrac{x^2}{24(8 - x^3)} + \frac{1}{288}\ln\left(\dfrac{x^2 + 2x + 4}{(2 - x)^2}\right) - \dfrac{1}{48\sqrt{3}}\tan^{-1}\left(\dfrac{x + 1}{\sqrt{3}}\right) + C$

1. approx. $= 8$;   exact $= 8$ **3.** approx. $= 0.253$;   exact $= \frac{1}{4} = 0.250$
5. approx. $\doteq 0.883$;   exact $\doteq 0.901$ **7.** approx. $\doteq 0.784$;   exact $\doteq 0.785$
9. approx. $\doteq 0.890$;   exact $\doteq 0.909$ **11.** approx. $\doteq 0.778$;   exact $\doteq 0.801$
13. approx. $\doteq 1.434$;   exact $\doteq 1.441$ **15.** approx. $\doteq 1.037$;   exact $\doteq 1.034$

17. $E_T = 0$; 8 **19.** $|E_T| \le 0.005$; $0.248 \le \int_0^1 x^3 \, dx \le 0.258$

21. $|E_T| \le 0.0833$; $0.7997 \le \int_0^2 \dfrac{x}{x+1} \, dx \le 0.9663$ **23.** 0.341

1. approx. $= 44$;   exact $= 44$ **3.** approx. $= 1.1$;   exact $\doteq 1.099$
5. approx. $= 56$;   exact $= 56$ **7.** approx. $\doteq 1.386$;   exact $\doteq 1.464$
9. approx. $\doteq 0.803$;   exact $\doteq 0.785$ **11.** approx. $\doteq 0.841$;   exact $\doteq 0.841$
13. approx. $\doteq 0.065$;   exact $\doteq 0.064$ **15.** approx. $\doteq 6.268$;   exact $\doteq 6.229$
17. 0.385 **19.** 1.1107 **21.** 3.155 **23.** approx. $\doteq 0.782a^2$;   exact $\doteq 0.785a^2$

Review and Miscellaneous Exercises   *page 599*

1. $\frac{1}{2}e^{2x} + 6e^x + 9x + C$ **3.** $\dfrac{5^{3x+4}}{3 \ln 5} + C$ **5.** $\dfrac{1}{2\sqrt{3}} \tan^{-1} \dfrac{z^2}{\sqrt{3}} + C$

7. $\frac{2}{3}(w+5)^{3/2} - 10(w+5)^{1/2} + C$
9. $3\sqrt{t^2 + t + 1} + \frac{19}{2} \ln |t + \frac{1}{2} + \sqrt{t^2 + t + 1}| + C$
11. $\frac{1}{2} \ln (x^2 + 8x + 20) - \frac{1}{2} \tan^{-1} \dfrac{x+4}{2} + C$ **13.** $\dfrac{1}{a} \ln (e^{ax} + \sqrt{e^{2ax} + \pi}) + C$

15. $-\dfrac{\cos^7 2\theta}{14} + C$ **17.** $t - \frac{3}{2} \cos \dfrac{2t}{3} + C$ **19.** $\frac{1}{4}\left(\dfrac{3\theta}{2} + \dfrac{\sin 6\theta}{3} + \dfrac{\sin 12\theta}{24} \right) + C$

21. $-\frac{1}{6}(3 + \cosh t)^{-6} + C$ **23.** $\dfrac{\sinh^6 z}{6} + \dfrac{\sinh^8 z}{8} + C$

25. $-\frac{1}{3}\left(\dfrac{\cos^4 3x}{4} - \dfrac{\cos^6 3x}{6} \right) + C$ **27.** $\ln |x - 4| - 4(x-4)^{-1} + C$ **29.** 1

31. $4 \ln 2 - \frac{15}{16}$ **33.** $\dfrac{e^{2x}}{4} - \dfrac{x}{2} + C$ **35.** $\dfrac{e^{ax}}{2}\left[\dfrac{1}{a} + \dfrac{a \cos 2bx + 2b \sin 2bx}{a^2 + 4b^2} \right] + C$

37. $2\sqrt{3}$ **39.** $\ln \left| \dfrac{(x-2)(x+2)}{x} \right| + C$

41. $-\frac{1}{6} \ln |x| - \frac{9}{10} \ln |x+2| + \frac{11}{6} \ln |x+3| + \frac{7}{30} \ln |x-3| + C$

43. $2 \ln |t| + \frac{5}{2} \tan^{-1} t - \dfrac{t}{2(1+t^2)} + C$ **45.** $\frac{2}{3}$ **47.** 0 **49.** $\dfrac{\sqrt{2}}{3}(5 - 2\sqrt{2})$

51. $-\frac{1}{2}(t^2 + 1)e^{-t^2} + C$ **53.** $\dfrac{1}{a} \tan \dfrac{ax}{2} + C$

55. $2\left\{ \dfrac{[\sqrt{x+1} + 2]^3}{3} - 3[\sqrt{x+1} + 2]^2 + 11(\sqrt{x+1} + 2) \right.$
$\left. - 6 \ln [\sqrt{x+1} + 2] \right\} + C$

57. $2\left(\dfrac{x^2}{4} - \dfrac{x^{3/2}}{3} + \dfrac{x}{2} - x^{1/2} + \ln (1 + x^{1/2}) \right) + C$

59. $2 \ln 4 + \frac{3}{2} \ln 10 - \tan^{-1} 3$ **61.** $\dfrac{1}{2c}\left[\frac{1}{2} \ln \left| \dfrac{c+x}{c-x} \right| - \tan^{-1} \dfrac{x}{c} \right] + C$

63. $\dfrac{(2n-1)(2n-3)\ldots 3.1}{(2n)(2n-2)\ldots 4.2} \dfrac{\pi}{2}$

65. $\frac{1}{4}\left[x(a^2-x^2)^{3/2}+\frac{3a^2x}{2}(a^2-x^2)^{1/2}+\frac{3a^4}{2}\sin^{-1}\frac{x}{a}\right]+C$

67. $\frac{1}{4\sqrt{2}}\ln\left|\frac{x^2+\sqrt{2}x+1}{x^2-\sqrt{2}x+1}\right|+\frac{1}{2\sqrt{2}}[\tan^{-1}(\sqrt{2}x+1)+\tan^{-1}(\sqrt{2}x-1)]+C$

69. $\frac{x^4}{4}\left[(\ln x)^2-\frac{\ln x}{2}+\frac{1}{8}\right]+C$ **71.** $3a+7$

75. $\frac{e^x}{2}[x\cos x+(x-1)\sin x]+C$ **77.** (a) 11.643, (b) 11.596   **79.** 3.68

81. $B_2(x)=x^2-x+\frac{1}{6};\quad B_3(x)=x^3-\frac{3x^2}{2}+\frac{x}{2}$ **83.** $\frac{e^{-3}}{3}$ **85.** $\frac{\ln 2}{3}+\frac{\pi}{3\sqrt{3}}$

Exercises 11.2   *page 609*

1. $\langle 6,8\rangle$ **3.** $\langle 5,12\rangle$ **5.** $\langle 28,-16\rangle$ **7.** $\langle -5,2\rangle$ **9.** $\langle -7,-3\rangle$
11. $\langle\frac{5}{2},-5\rangle$ **13.** $\mathbf{a}=\langle 2,3\rangle$, $\mathbf{b}=\langle 4,-2\rangle$ **15.** $\mathbf{a}=\langle 7,14\rangle$, $\mathbf{b}=\langle 5,10\rangle$
17. $k=-3$, $m=-4$ **19.** 9.17 mi., 49° east of south   **21.** $\langle 17,39\rangle$
23. $\sqrt{17}$ **25.** 7

Exercises 11.3   *page 621*

5. -1 **7.** -11 **9.** 0   **11.** $\mathbf{a}\cdot\mathbf{b}=0$; yes   **13.** $\mathbf{a}\cdot\mathbf{b}\neq 0$; no
15. $m=0$ **17.** no value of m **19.** no $\mathbf{a}$ satisfies these equations
21. (a) $\theta\doteq 81°52'$; (b) $\theta=90°$ **23.** $\pm\dfrac{2}{\sqrt{b_1^2+b_2^2}}(b_2\mathbf{i}-b_1\mathbf{j})$

25. both sides are meaningless   **31.** (a) $\frac{8}{5}\mathbf{i}+\frac{4}{5}\mathbf{j}$; (b) $\frac{12}{13}\mathbf{i}-\frac{8}{13}\mathbf{j}$

Exercises 11.4   *page 631*

1. $-3\leq t\leq 3$ **3.** $t\geq 5$ **5.** $-1<t\leq 1$ **7.** $y^2=16x$

9. $x^2+y^2=16, 0\leq y\leq 4$ **11.** $\dfrac{x^2}{9}+\dfrac{y^2}{4}=1$, first quadrant

13. $x=a\cos^{-1}\dfrac{a-y}{a}\pm\sqrt{2ay-y^2}$

15. $x=a_1+t(b_1-a_1);\quad y=a_2+t(b_2-a_2), -\infty<t<\infty$
19. $4(1+2t)\mathbf{i}-6e^{-6t}\mathbf{j}$; $8\mathbf{i}+36e^{-6t}\mathbf{j}$ **21.** $(10\cos 10t)\mathbf{i}+(2t\sinh(t^2))\mathbf{j}$;
$(-100\sin 10t)\mathbf{i}+2(2t^2\cosh(t^2)+\sinh(t^2))\mathbf{j}$ **23.** $12(\mathbf{i}+4\mathbf{j}),\dfrac{\mathbf{i}+4\mathbf{j}}{\sqrt{17}}$

25. $a\left(\frac{1}{2}\mathbf{i}+\frac{\sqrt{3}}{2}\mathbf{j}\right),\frac{1}{2}\mathbf{i}+\frac{\sqrt{3}}{2}\mathbf{j}$ **27.** (a) $5[(\cos 5t)\mathbf{i}-(\sin 5t)\mathbf{j}]$;

(b) $25[(-\sin 5t)\mathbf{i}+(-\cos 5t)\mathbf{j}]$; (c) 25; (d) 0
29. $x=\dfrac{4p}{m^2}$; $y=\dfrac{4p}{m}$, all except $(0,0)$ **31.** the tangent to the circle is

perpendicular to the radius at the point of contact   **33.** $\dfrac{\cos t+\sin t-1}{(\cos t-1)^3}$

37. $x=(a-b)\cos\theta+b\cos\dfrac{a-b}{b}\theta;\quad y=(a-b)\sin\theta-b\sin\dfrac{a-b}{b}\theta$

39. $x=a\cos\theta, y=0$; converts rotational motion into reciprocating motion along a
straight line

Exercises 11.5   *page 640*

1. $(1-3e^{-t})\mathbf{i}+\frac{1}{2}(7-e^{2t})\mathbf{j}$ **3.** $\left(\dfrac{t^4}{12}+2t^2-\frac{10}{3}t+\frac{5}{4}\right)\mathbf{i}+(2(t-1)-\ln|t|)\mathbf{j}$

5. $\langle 30\cos 3t; -30\sin 3t\rangle$; 30; $-9\mathbf{r}(t)$ **7.** $\langle 7,4t\rangle$; $\sqrt{49+16t^2}$; $\langle 0,4\rangle$

9. $\left\langle \dfrac{t^3}{3}, \dfrac{t^4}{4} + 2 \right\rangle$ **11.** $\left\langle 2t^2 - t + 2, \dfrac{3t^2}{2} + 5 \right\rangle$

13. $\mathbf{v} = \sqrt{2.5}(\mathbf{i} + 3\mathbf{j}) \doteq 1.58(\mathbf{i} + 3\mathbf{j}); \quad \mathbf{a} = 0$

15. (a) $\mathbf{v} = \mathbf{r}'(t) = \mathbf{C}, \quad |\mathbf{v}| = |\mathbf{C}|, \quad \mathbf{a} = 0;$ (b) $\mathbf{0};$

(c) straight line passing through $\mathbf{r}(0)$ in the direction of $\mathbf{C}$ **17.** $4033\frac{1}{3}$ lb

19. $\dfrac{v_0^2 \sin 2\alpha}{g}$

Exercises 11.6   page 647

1. $3\sqrt{5}$ **3.** $\frac{1}{27}[(85)^{3/2} - 8]$ **5.** $\dfrac{25\pi}{24}$ **7.** $\sqrt{2}(e^5 - e^{-3})$ **9.** $\dfrac{5\pi^2}{6}$

11. $\frac{1}{27}[(9e^6 + 4)^{3/2} - (13)^{3/2}]$ **13.** $\displaystyle\int_{-1}^{3} (1 + 9t^4)^{1/2}\, dt$

15. $\displaystyle\int_{0}^{\pi} [b^2 + (a^2 - b^2)\sin^2 t]^{1/2}\, dt$ **17.** $(4, -1)$ **19.** $(0, \tfrac{1}{2}(a\,\mathrm{csch}\,a + \cosh a))$

Exercises 11.7   page 655

1. $\dfrac{2t\mathbf{i} + \mathbf{j}}{(4t^2 + 1)^{1/2}}; \quad \dfrac{\mathbf{i} - 2t\mathbf{j}}{(4t^2 + 1)^{1/2}}$ **3.** $\dfrac{e^t\mathbf{i} + \mathbf{j}}{(e^{2t} + 1)^{1/2}}; \quad \dfrac{\mathbf{i} - e^t\mathbf{j}}{(e^{2t} + 1)^{1/2}}$

5. $(\cos t)\mathbf{i} + (\sin t)\mathbf{j}; \quad (-\sin t)\mathbf{i} + (\cos t)\mathbf{j}$ **7.** $\dfrac{4}{(1 + 16x^2)^{3/2}}; \quad (0, 3); \quad \frac{1}{4}$

9. $\dfrac{2x^3}{(x^4 + 1)^{3/2}}; \quad (1, 1); \quad \sqrt{2}$ **11.** $\dfrac{1}{a\cosh^2 \dfrac{x}{a}}; \quad \dfrac{1}{a\cosh^2 1}$

13. $\dfrac{4p^2}{(4p^2 + y^2)^{3/2}}; \quad \dfrac{1}{2|p|}$ **15.** $\dfrac{a^2}{b}$ **17.** $2^{-1/2}\, e^{-t}$ **19.** $\dfrac{6}{[4\sin^2 t + 9\cos^2 t]^{3/2}}$

21. (a) $t, (t^2 + 1)^{1/2};$ (b) $t;$ (c) $1, t$ **23.** 3259 lb (approx.)

25. $a_T = 0; \quad a_N = \dfrac{400}{(1 + 16x^2)^{3/2}}$ **27.** $|a_N| = \sqrt{|\mathbf{a}|^2 - a_T^2}$

29. $Y = \frac{1}{2} + \dfrac{3}{2\sqrt[3]{2}} x^{2/3}$ **31.** $(aX)^{2/3} + (bY)^{2/3} = (a^2 - b^2)^{2/3}$

Review and Miscellaneous Exercises   page 657

1. $\langle -13, -12 \rangle$ **3.** $\langle 2k, -k \rangle$ **5.** $\mathbf{a} = \langle 5, -7 \rangle; \quad \mathbf{b} = \langle -3, 4 \rangle$ **7.** $\langle 4, 12 \rangle$

9. $k = -\frac{9}{7}$ **13.** (a) yes;

(b) two collinear vectors do not serve as a basis for arbitrary vectors in the plane

15. $\mathbf{r} = \mathbf{a} + t\langle 1, m \rangle$, where $\mathbf{a} = \langle a_1, a_2 \rangle$

19. $36°52'$ south of east at speed $= 25$ mph   **23.** a

25. $36°52'$ north of east; $1\frac{1}{2}$ hr

27. $(t^2 + 9)^{1/2}\mathbf{i} + [t - 2\ln|t + 2|]\mathbf{j} + \mathbf{C}$, where $\mathbf{C}$ is an arbitrary constant vector

29. $\mathbf{r}(t) = \left(\dfrac{t^3}{2} - 5t + 6\right)\mathbf{i} + \left(-\dfrac{t^3}{6} + \dfrac{4}{3}\right)\mathbf{j}$ **33.** $y = \dfrac{x^2}{256}$ ft; 3 sec; 192 ft

35. $7\sqrt{26}$

37. (a) $1, 1;$ (b) $\dfrac{\pi}{2};$ (c) $\mathbf{i} = \mathbf{u}(\sin \alpha) + \mathbf{v}(\cos \alpha), \quad \mathbf{j} = \mathbf{u}(\cos \alpha) - \mathbf{v}(\sin \alpha)$

39. $\dfrac{6}{13\sqrt{13}}; \quad \dfrac{13\sqrt{13}}{6}; \quad (-\frac{11}{2}, \frac{16}{3})$ **43.** $\dfrac{3a}{2}$

47. $\mathbf{r}(t) = (t^3 + t - 1)\mathbf{i} + (t^2 + 5)\mathbf{j}$ **49.** $2s$

Exercises 12.1

1. $(2, 0)$ **3.** $(-3, 0)$ **5.** $(\frac{3}{4}\sqrt{2}, \frac{3}{4}\sqrt{2})$ **7.** $(\sqrt{2}, \sqrt{2})$ **9.** $\left(-\dfrac{3\sqrt{2}}{4}, \dfrac{3\sqrt{2}}{4}\right)$

11. $(3, 2n\pi), (-3, (2n+1)\pi)$, n is any integer

13. $\left(2, \dfrac{\pi}{3} + 2n\pi\right), \left(-2, \dfrac{\pi}{3} + (2n+1)\pi\right)$, n is any integer

15. $(5, \theta_o + 2n\pi), (-5, \theta_o + (2n+1)\pi)$, where $\theta_o = \tan^{-1}\frac{3}{4}$ and n is any integer

17. $r\cos\theta = 3$ **19.** $r\sin\theta = b$ **21.** $r^2\cos 2\theta = a^2$

23. $r^2 - 2r(h\cos\theta + k\sin\theta) + h^2 + k^2 - a^2 = 0$ **25.** $x^2 + y^2 = 49$

27. $\sqrt{3}x + y = 0$ **29.** $2(x^2 + y^2) - y = 0$ **31.** $3x - 2y = 1$

33. $(x^2 + y^2)^2 = 12xy$ **35.** $y^2 = 3(3 - 2x)$

Exercises 12.2

23. $(b - a)$ (max.), $(b + a)$ (min.)   **25.** parabola $y^2 = 2x + 1$

27. $r^2 = 2a^2\cos 2\theta$ (lemniscate)

Exercises 12.3

1. $\left(2\sqrt{2}, \dfrac{\pi}{4}\right)$ and the origin   **3.** $(\frac{3}{2}, \cos^{-1}\frac{3}{4}), (\frac{3}{2}, -\cos^{-1}\frac{3}{4})$

5. $(\frac{2}{3}, \sin^{-1}\frac{1}{3}), (\frac{2}{3}, \pi - \sin^{-1}\frac{1}{3})$ and the origin

7. $\left(\dfrac{\sqrt{3}}{2}, \dfrac{\pi}{3}\right), \left(-\dfrac{\sqrt{3}}{2}, -\dfrac{\pi}{3}\right)$ and the origin   **9.** $(3, 0.20)$ (approx.)

11. no points of intersection   **13.** $(2, 0), \left(1, \dfrac{\pi}{2}\right)$

Exercises 12.4

(In 1 through 8 exact answers are in radians, approximate answers in degrees.)

1. $\dfrac{\pi}{3}$ **3.** $\dfrac{\pi}{6}$ **5.** $126°35'$ **7.** $\dfrac{3\pi}{4}$ **11.** $r = ae^{b\theta}$ where $b = \cot\psi$

13. $-\dfrac{2}{\pi}$ **15.** -2 **17.** $\dfrac{\pi}{3}$ **19.** $\dfrac{\pi}{2}$ **21.** $\dfrac{\pi a}{4}$ **23.** $\dfrac{3\sqrt{5}}{2}(e^{2\theta_1} - e^{2\theta_o})$

25. $\dfrac{p}{2}[\sqrt{2} - \ln(\sqrt{2} - 1)]$ **29.** $K = \dfrac{3}{4a}\dfrac{1}{\sin\frac{\theta}{2}}, K_{\min} = \dfrac{3}{4a}$

Exercises 12.5

1. $3\omega, 3\theta\omega$; $3(\dot\omega - \theta\omega^2), 3(\theta\dot\omega + 2\omega^2)$ **3.** $\omega(-b\sin\theta + c\cos\theta)$;

$(b\cos\theta + c\sin\theta)$; $(c\dot\omega - 2b\omega^2)\cos\theta - (b\dot\omega + 2c\omega^2)\sin\theta$;

$(b\dot\omega + 2c\omega^2)\cos\theta + (c\dot\omega - 2b\omega^2)\sin\theta$

5. $ab\sin\theta, ab(1 - \cos\theta)$; $ab^2(2\cos\theta - 1), 2ab^2\sin\theta$

7. $-\frac{5}{2}$ units/sec, $\dfrac{5\sqrt{3}}{2}$ units/sec   **9.** $\dfrac{-2v_o^2}{b}\cos\theta, \dfrac{-2v_o^2}{b}\sin\theta, \dfrac{2v_o^2}{b}$

Exercises 12.6

(All answers are in square units)

1. πb^2 **3.** $\dfrac{3\pi}{2}$ **5.** $\dfrac{\pi b^2}{2}$ **7.** $\dfrac{\pi a^2}{4}$ **9.** a^2 **11.** 11π **13.** $\dfrac{a^2}{4}$

15. $\frac{1}{24}[3(e^{4\pi} - 1) - 4\pi^3]$ **17.** $\frac{9}{8}(\pi - 2)$ **19.** $\frac{1}{2}\left(\dfrac{\pi}{3} + 1 - \sqrt{3}\right)$

21. $a^2(\pi - \frac{3}{2}\sqrt{3})$ **23.** 4π

Review and Miscellaneous Exercises

1. circle with radius 4 and center at the pole

3. straight line parallel to $\theta = \dfrac{\pi}{2}$ axis and 4 units to its left

5. cardioid passing through the pole with polar axis as axis of symmetry

7. lemniscate symmetrical with respect to the pole and passing through the pole

9. parabola with $\theta = \dfrac{\pi}{2}$ as axis of symmetry, vertex at $\left(\dfrac{3}{2}, \dfrac{3\pi}{2}\right)$
and opening upward

11. $r^2 \cos 2\theta = a^2$ **13.** $r^2 = a^2 \cos 2\theta$ **15.** $x^2 + y^2 - 2x + y = 0$

17. $xy = 1$ **19.** $r = \dfrac{4}{1 + \cos \theta}$

23. $\left(a(1 - 2^{-1/2}), \dfrac{\pi}{4}\right), \left(a(1 + 2^{-1/2}), \dfrac{5\pi}{4}\right)$ and the pole

25. $\pi - \frac{3}{2}\sqrt{3}$ square units   **27.** $y = 1$ is an asymptote   **29.** $\dfrac{\pi a^2}{12}$

31. $x\sqrt{3} + y - 2p = 0$ (straight line)   **33.** $-\tan \theta_o$

35. $2a\left[(\sqrt{5} - 2) - \sqrt{3} \ln \left(\dfrac{\sqrt{3} + \sqrt{5}}{\sqrt{2}(\sqrt{3} + 2)}\right)\right]$ **37.** $K = \dfrac{1}{2a} \cos^3 \dfrac{\theta}{2}$

Exercises 13.1

1. $(\pm 3, 0);\ \ (0, \pm 2);\ \ (\pm \sqrt{5}, 0);\ $ and $\dfrac{\sqrt{5}}{3}$

3. $(0, \pm 4);\ \ (\pm 2, 0);\ \ (0, \pm 2\sqrt{3});\ $ and $\dfrac{\sqrt{3}}{2}$

5. $(\pm 5, 0);\ \ (0, \pm 3);\ \ (\pm 4, 0);\ $ and $\frac{4}{5}$

7. $(\pm \sqrt{3}, 0);\ \ (0, \pm \sqrt{2});\ \ (\pm 1, 0);\ $ and $\dfrac{1}{\sqrt{3}}$

9. $(0, \pm \frac{5}{2});\ \ (\pm \frac{3}{4}, 0);\ \ \left(0, \pm \dfrac{\sqrt{91}}{4}\right);\ $ and $\dfrac{\sqrt{91}}{10}$ **11.** $\dfrac{x^2}{25} + \dfrac{y^2}{16} = 1$

13. $\dfrac{x^2}{25} + \dfrac{y^2}{61} = 1$ **15.** $7x^2 + 16y^2 = 256$ **17.** $x + 2y - 8 = 0$ **21.** e

23. $\frac{4}{3}\pi a^2 b$ cu units   **25.** $x + y = 3;\ \ x - 5y - 9 = 0$

27. $y = mx \pm \sqrt{a^2m^2 + b^2}$ **29.** $2ab$ **31.** πab sq units

33. $\dfrac{M - m}{M + m};\ \ 0.0166$

Exercises 13.2

1. $(\pm 3, 0);\ \ (\pm \sqrt{13}, 0);\ \ \dfrac{\sqrt{13}}{3};\ \ y = \pm \frac{2}{3}x$

3. $(0, \pm 4);\ \ (0, \pm 2\sqrt{5});\ \ \dfrac{\sqrt{5}}{2};\ \ y = \pm 2x$

5. $(\pm 2, 0)$; $(\pm \sqrt{5}, 0)$; $\dfrac{\sqrt{5}}{2}$; $y = \pm \dfrac{x}{2}$

7. $(\pm \sqrt{7}, 0)$, $(\pm \sqrt{11}, 0)$; $\dfrac{\sqrt{11}}{\sqrt{7}}$; $y = \pm \dfrac{2}{\sqrt{7}} x$

9. $(\pm 2, \pm 2)$; $(\pm 2\sqrt{2}, \pm 2\sqrt{2})$; $\sqrt{2}$; x- and y-axes   11. $\dfrac{x^2}{64} - \dfrac{y^2}{36} = 1$

13. $\dfrac{y^2}{9} - \dfrac{x^2}{27} = 1$ 15. $\dfrac{x^2}{5} - \dfrac{y^2}{20} = 1$ 17. $y = 3x - 7$

19. $5x - 8y - 80 = 0$ 23. (a, a); $a\sqrt{2}$ 25. b

27. coordinate axes and rectangular hyperbola $xy = \pm 10$

Exercises 13.3 page 728

1. $y' = x'^2$, where $x = x' + 3$ and $y' = y$

3. $y' = x'^2$, where $x = x' + 1$ and $y = y' - 3$

5. $\dfrac{x'^2}{9} + y'^2 = 1$, where $x = x' - 4$ and $y = y' + 7$

7. $x'y' = 10$, where $x = x' + 3$ and $y = y'$

9. $x = x' + 2$, $y = y' + 3$; $x'^2 + y'^2 = 4$, circle with center $(2, 3)$ and radius 2

11. $x = x' - 1$ $y = y' + 5$; $\dfrac{x'^2}{1} + \dfrac{y'^2}{4} = 1$, ellipse with center $(-1, 5)$,

major axis along $x = -1$, $a = 2$, $b = 1$, and $c = \sqrt{3}$; vertices $(-1, 7)$ and $(-1, 3)$; foci $(-1, 5 + \sqrt{3})$ and $(-1, 5 - \sqrt{3})$; eccentricity $= \sqrt{3}/2$

13. $h = 2$, $k = 1$; $x = x' + 2$, $y = y' + 1$; $2x' - 3y' = 0$ and $5x' - 7y' = 0$

15. $x = x' + 3$, $y = y' + 2$; $\dfrac{x'^2}{4} + \dfrac{y'^2}{1} = 1$, ellipse with center $(3, 2)$,

vertices $(5, 2)$ and $(1, 2)$, ends of minor axis at $(3, 3)$ and $(3, 1)$, foci $(3 \pm \sqrt{3}, 2)$

and eccentricity $= \dfrac{\sqrt{3}}{2}$ 17. $x = x' + 2$, $y = y' - 3$, $y' = x'^2$,

parabola with axis $x = 2$, vertex $(2, -3)$, focus $(2, -\tfrac{11}{4})$, and directrix $y = -\tfrac{13}{4}$; curve opens upward

19. $x = x' - 1$, $y = y' + 2$; $x'^2 + 3x'y' + y'^2 - 4 = 0$

21. $x = x' - 1$, $y = y' - 7$; $y' = (x')^4 - 3(x')^3 + 2(x')^2$

23. $x = -\dfrac{b}{2a}$, $y = c - \dfrac{b^2}{4a} - \dfrac{1}{4a}$; $V\left(-\dfrac{b}{2a}, c - \dfrac{b^2}{4a}\right)$, $F\left(-\dfrac{b}{2a}, c - \dfrac{b^2}{4a} + \dfrac{1}{4a}\right)$

Exercises 13.4 page 736

1. $(0, 0)$; $(\pm 9, 0)$; $(\pm \sqrt{65}, 0)$; $\sqrt{65}/9$; $x = \pm \dfrac{81\sqrt{65}}{65}$ 3. $(1, 0)$;

$(5, 0)$ and $(-3, 0)$; $(6, 0)$ and $(-4, 0)$; $\tfrac{5}{4}$; $x = 21/5$; $x = -\tfrac{11}{5}$; $y = \pm\tfrac{3}{4}(x - 1)$

5. $(2, 4)$; $(2, 9)$ and $(2, -1)$; $(2, 7)$ and $(2, 1)$; $\tfrac{3}{5}$; $y = \tfrac{37}{3}$ and $y = -\tfrac{13}{3}$

7. $\dfrac{x^2}{16} + \dfrac{y^2}{7} = 1$ 9. $\dfrac{x^2}{9} - \dfrac{y^2}{27} = 1$ 11. $169x^2 + 144y^2 = 14{,}400$

13. $25x^2 - 144y^2 = 3600$ 15. $9x^2 + 8y^2 - 64y + 56 = 0$

17. $8x^2 - y^2 - 104x + 2y + 319 = 0$

19. $16x^2 + 4xy + 19y^2 + 240x - 120y + 900 = 0$

Exercises 13.5 page 745

1. (a) $x' = 2\sqrt{2}$, $y' = 0$; (b) $x' = 2$, $y' = -2$ 3. $x'y' = 3$

5. $\sin \alpha = \tfrac{3}{5}$ and $\cos \alpha = \tfrac{4}{5}$ 7. $\alpha = 45°$; ellipse with major axis along the y'-axis and minor axis along the x'-axis; in the $x'y'$-system the foci are $(0, \pm 2\sqrt{3})$, ends of major axis are at $(0, \pm\sqrt{14})$, and the ends of the minor axis are at $(\pm\sqrt{2}, 0)$

9. $x' = \pm 4$ **11.** $x'^2 + 2y'^2 + x' = 1$ **13.** $13x''^2 + 3y''^2 = 78$ (ellipse)
15. yes; we may select the x'-axis to be parallel to the straight line;
the equation would then be of the form $y' = $ constant
21. Equilateral hyperbola, center at $(-2, 3)$
23. Ellipse with major and minor axes equal to 4 and $\frac{8}{3}$, respectively

Exercises 13.6 page 750

1. $e = 1$; parabola; $x = -10$ **3.** $e = 3$; hyperbola; $x = \frac{7}{3}$
5. $e = 2$; hyperbola; $y = \frac{9}{2}$ **7.** $e = 5$; hyperbola; $y = -\frac{20}{3}$
9. $e = \frac{c}{b} > 1$; hyperbola; $x = \frac{a}{c}$ **11.** $r = \dfrac{ep}{1 + e\cos\theta}$ **13.** $r = \dfrac{ep}{1 + e\sin\theta}$
15. $r = \dfrac{\frac{4}{3}}{1 + \frac{1}{3}\sin\theta}$ **17.** $r = \dfrac{12}{2 + 3\sin\theta}$
19. $\left(\dfrac{ep}{1 + e}, \pi\right)$; $\left(\dfrac{ep}{1 - e}, 0\right)$; $\dfrac{2ep}{1 - e^2}$ **21.** ellipse with $e = \frac{1}{2}$

Review and Miscellaneous Exercises page 751

1. $\dfrac{x^2}{3^2} + \dfrac{y^2}{6^2} = 1$, origin, $(0, \pm 3\sqrt{3})$, $(0, \pm 6)$, $\dfrac{\sqrt{3}}{2}$, 12 and 6

3. $\dfrac{(x + \frac{3}{2})^2}{1^2} + \dfrac{y^2}{3^2} = 1$, $(-\frac{3}{2}, 0)$, $(-\frac{3}{2}, \pm\sqrt{8})$, $(-\frac{3}{2}, \pm 3)$, $\dfrac{\sqrt{8}}{3}$, 6 and 2

5. no locus   **7.** $e = \dfrac{n}{k}, p = \dfrac{1}{n}$

9. $5x^2 - 4y^2 - 58x + 8y + 157 = 0$, hyperbola

11. $(0, 0)$, $(\pm\sqrt{65}, 0)$, $(\pm 7, 0)$, $\dfrac{\sqrt{65}}{7}$, $y = \pm\frac{4}{7}x$

13. $(5, 0)$, $(5 \pm \sqrt{34}, 0)$, $(8, 0)$ and $(2, 0)$, $\dfrac{\sqrt{34}}{3}$, $y = \pm\frac{5}{3}(x - 5)$

15. $(3, -5)$, $(3 \pm \sqrt{5}, -5)$, $(5, -5)$ and $(1, -5)$, $\dfrac{\sqrt{5}}{2}$, $y + 5 = \pm\frac{1}{2}(x - 3)$

17. $\dfrac{x^2}{25} - \dfrac{y^2}{9} = 1$, hyperbola

19. $x^2 + 2xy + y^2 - 12x + 12y + 36 = 0$, parabola   **21.** Standard position with
center at the origin and an axis of length 6 connecting $(0, \pm 3)$
23. $7x^2 + 8xy + 13y^2 - 40x + 20y = 0$ **25.** $3y^2 + 17y - x + 26 = 0$
27. $y^2 - x'^2 = 2$ **31.** $\frac{1}{2}, r\cos\theta = 9$, 9, ellipse   **35.** 1   **41.** lower half-
plane with non-included boundary $x + y = 0$ **43.** $2a^2$ sq units

Exercises 14.1 page 760

1. 2   **3.** 0   **5.** ∞ **7.** a **9.** $-\dfrac{1}{4\pi^2}$

11. 0; l'Hospital's rule doesn't apply   **13.** $\frac{1}{3}$ **15.** $\frac{1}{6}$ **17.** $\frac{1}{2}$ **19.** 0
21. 1   **23.** $\frac{2}{3}x^{-2/3}$ **25.** na **27.** $a = 1, b = -1, c = 0$ **29.** 0

Exercises 14.2 page 757

1. 0   **3.** 1   **5.** 0   **7.** $\frac{1}{5}$ **9.** 1   **11.** 0   **13.** 1   **15.** e^{-3} **17.** 0
19. 1   **21.** 0   **23.** e^{10} **25.** e^b **27.** $\dfrac{\sqrt{7}}{5}$ **29.** ∞ **31.** 1   **33.** $\frac{1}{8}$
35. ∞ **37.** 0   **39.** 1   **41.** $\frac{1}{2}$

1. $\frac{1}{3}$ **3.** diverges   **5.** diverges   **7.** $\frac{\pi}{2a}$ **9.** 1   **11.** $\frac{1}{4}$ **13.** $\frac{1}{5}$

15. $\frac{1}{2} - \ln \frac{3}{2}$ **17.** $\frac{1}{2e}$ **19.** $\frac{1}{k \ln 2}$ **21.** $\frac{1}{r^2}$ **23.** $\frac{\pi}{4}$ cu units

25. $2\pi a^3 \tan^{-1} \frac{b}{a}$ cu units; $\pi^2 a^3$ cu units

1. 2   **3.** diverges   **5.** diverges   **7.** π **9.** diverges   **11.** diverges

13. $\ln(2 + \sqrt{3})$ **15.** diverges   **17.** $\frac{\pi}{2}$ **19.** diverges   **21.** $3(5^{1/3} + 3^{1/3})$

23. diverges   **25.** converges   **27.** diverges   **29.** converges
31. converges   **33.** converges   **35.** 2   **37.** π

1. $(x - 2)^2 - (x - 2) + 5$ **3.** $(x + 1)^3 - 3(x + 1)^2 + 3(x + 1) - 1$
5. $(x - 3)^3 + 2(x - 3) + 20$

7. $P_2(x) = \frac{1}{2} - \frac{\sqrt{3}}{2}\left(x - \frac{\pi}{3}\right) - \frac{1}{4}\left(x - \frac{\pi}{3}\right)^2$;

$R_2(x) = \frac{\sin c}{3!}\left(x - \frac{\pi}{3}\right)^3$, where c is between $\frac{\pi}{3}$ and x

9. $P_2(x) = 1 - 2x^2$; $R_2(x) = \frac{4}{3}\sin 2c$, where c is between 0 and x

11. $P_5(x) = x - \frac{x^2}{2} + \frac{x^3}{3} - \frac{x^4}{4} + \frac{x^5}{5}$; $R_5(x) = -\frac{(1 + c)^{-6}}{6}x^6$, where c is between 0 and x

13. no expansion about $x = 0$ **15.** $P_4(x) = -\left(x - \frac{\pi}{2}\right) + \frac{1}{6}\left(x - \frac{\pi}{2}\right)^3$;

$R_4(x) = -\frac{1}{120}(\sin c)\left(x - \frac{\pi}{2}\right)^5$, where c is between $\frac{\pi}{2}$ and x

17. $P_2(x) = 2 + 2\sqrt{3}\left(x - \frac{\pi}{3}\right) + 7\left(x - \frac{\pi}{3}\right)^2$;

$R_2(x) = \sec c \tan c(\sec^2 c - \frac{1}{6})\left(x - \frac{\pi}{3}\right)^3$, where c is between $\frac{\pi}{3}$ and x

19. $P_3(x) = (x - 1)^3 - 3(x - 1)^2 - (x - 1) + 3$; $R_3(x) = 0$

21. $P_4(x) = x^2 - \frac{x^4}{3}$; $R_4(x) = \frac{2}{15}(\sin 2c)x^6$, where c is between 0 and x

23. $1 + \frac{x}{3} - \frac{x^2}{9} + R_2(x)$; $R_2(x) = \frac{10}{162}(c + 1)^{-8/3}x^3$, where c is between 0 and x,

$0 \le R_2(x) \le 0.0077$ in $[0, \frac{1}{2}]$ **25.** $-x - \frac{x^2}{2} - \frac{x^3}{3} - \cdots - \frac{x^n}{n} + R_n(x)$;

$R_n(x) = -\frac{x^{n+1}}{(n + 1)(1 - c)^{n+1}}$, where c is between 0 and x and $0 < x < 1$

27. $bx + abx^2 + \frac{b}{6}(3a^2 - b^2)x^3$ **29.** 0.3827

1. ± 3.16 **3.** 2.14   **5.** 2.303   **7.** 3.236   **9.** 0.667   **11.** 1.817

17. $k = \frac{n - 1}{n}$, $g = \frac{1}{n}$ **19.** 0.56   **21.** 0.82   **23.** -0.82

25. $x_1 = r - h$, $x_2 = r + h = x_0$

Review and Miscellaneous Exercises   *page 797*

1. $\frac{3}{2}$ 3. 1   5. e^{10} 7. $\frac{1}{e^2}$ 9. $\frac{1}{4}$ 11. $-\frac{1}{5}$ 13. does not exist

15. $4\sin x$ 17. 0   19. 7   21. 1   23. $\frac{3}{16}$ 25. $\frac{1}{6}$ 27. 0   33. $-\frac{1}{18}$

35. $\frac{\pi}{2\sqrt{3}}$ 37. $-\frac{2}{a^3}$ 39. $\frac{3}{8}\sqrt{\pi}$ 41. 2   43. $\frac{43}{6}$ 45. divergent

47. convergent   49. convergent   51. convergent   53. convergent
55. integrand not defined at 0   57. $\frac{1}{2}$; diverges   59. converges
61. diverges
63. $3, 3-3(x+1)^2, 3-3(x+1)^2+(x+1)^3$; $P_n(x)=P_3(x)$ if $n>3$
65. $1, 1, 1-x^2$ 67. $x, x+\frac{x^3}{3!}, x+\frac{x^3}{3!}+\frac{x^5}{5!}$

69. $\ln a + \frac{x-a}{a} - \frac{(x-a)^2}{2a^2} + \frac{(x-a)^3}{3a^3} - \frac{(x-a)^4}{4a^4} + \frac{(x-a)^5}{5a^5}$

71. $1 - \frac{2}{5}(x-2) + \frac{2}{5^2}(x-2)^2 - \frac{2}{5^3}(x-2)^3 + \frac{2}{5^4}(x-2)^4$

73. not well at all—$R_n(x)=F(x)$ 77. 1.260
79. 1.328,   cubic eq,   the derivative of given cubic is positive for all x 81. 2.027

Exercises 15.1   *page 805*

1. decreasing with 0 as a lower bound and 10 as an upper bound
3. not monotonic with -1 as a lower bound and $\frac{1}{2}$ as an upper bound
5. increasing with -1 as a lower bound and 2 as an upper bound
7. decreasing with 1 as a lower bound and 4 as an upper bound
9. increasing with $\ln\frac{1}{2}$ as a lower bound and 0 as an upper bound
11. not monotonic with $-\frac{1}{\pi}$ as a lower bound and $\frac{1}{\pi^2}$ as an upper bound
13. decreasing with 1 as a lower bound and 21 as an upper bound
15. increasing with e as a lower bound and with no upper bound
17. increasing with $\sinh 1$ as a lower bound and with no upper bound
19. decreasing with 1 as a lower bound and 4 as an upper bound
21. increasing with $\sin 1 \doteq 0.8415$ as a lower bound and 1 as an upper bound
23. decreasing with 0 as a lower bound and $\frac{1}{2}$ as an upper bound
25. decreasing with 0 as a lower bound and 1 as an upper bound

Exercises 15.2   *page 813*

1. converges to 0   3. diverges   5. converges to 0   7. diverges
9. converges to $\frac{1}{2}$ 11. converges to $\frac{1}{2}\sqrt{2}$ 13. converges to $-\frac{4}{7}$
15. converges to 1   17. converges to $\frac{2}{3}$ 19. converges to 0
23. $a_2=ka_1, a_3=k^2a_1, a_4=k^3a_1, \ldots, a_n=k^{n-1}a_1$
25. $s_3=\sqrt{2-\sqrt{2}}, s_4=\sqrt{2-\sqrt{2+\sqrt{2}}}$; 2π

Exercises 15.3   *page 820*

1. $\frac{24}{7}$ 3. $\frac{1}{\sqrt{3}-1}$ 5. $\frac{(xy)^2}{1-xy}$ 7. $\frac{4}{333}$ 9. $\frac{43210}{9999}$ 11. $\sum_{n=1}^{\infty}\frac{1}{3^n}$; 2^{-1}

13. $-\frac{1}{3}+\frac{1}{3}\sum_{n=2}^{\infty}(-1)^{n-1}\frac{1-2n}{n(n-1)}$; 0   15. $\sum_{n=1}^{\infty}\ln\left(1+\frac{1}{n}\right)$; diverges   17. $\frac{7}{6}$

19. divergent 21. $\dfrac{1}{8 - \sqrt{8}}$ 23. $\dfrac{1}{1 + 2x}$ if $|x| < \frac{1}{2}$, divergent otherwise

25. $1 + k^{-2}$ if $k \neq 0$; diverges if $k = 0$ 27. divergent 29. divergent

31. $\frac{1}{4}$ 33. $5a$ ft 35. $\dfrac{n(n + 2)}{3(2n + 1)(2n + 3)}$; $\frac{1}{12}$ 37. $\dfrac{r}{(1 - r)^2}$

Exercises 15.4 page 830

1. convergent   3. convergent   5. divergent   7. convergent
9. divergent   11. convergent   13. convergent   15. convergent
17. divergent   19. convergent   21. convergent   23. divergent
25. convergent   29. divergent

Exercises 15.5 page 835

1. convergent   3. convergent   5. convergent   7. convergent
9. divergent   11. convergent   13. divergent   15. divergent

Exercises 15.6 page 841

1. convergent   3. divergent   5. convergent   7. convergent
9. convergent   11. divergent   13. 0.900 (0.9 is the *exact* result)   15. 0.540
17. conditionally convergent   19. divergent   21. absolutely convergent
23. divergent   25. divergent   27. absolutely convergent
29. absolutely convergent   31. absolutely convergent

Exercises 15.7 page 847

1. $(-2, 2)$ 3. $(-\infty, \infty)$ 5. $(-\frac{1}{2}, \frac{1}{2})$ 7. $(-1, 1]$ 9. $[-b, b)$
11. $[-1, 1]$ 13. $(-\frac{5}{2}, \frac{5}{2})$ 15. $(-\frac{7}{3}, 1)$ 17. $\left(-\dfrac{1}{7e}, \dfrac{1}{7e}\right)$ 19. $[-2, 6)$
21. $\left(1 - \dfrac{1}{\sqrt[4]{2}}, 1 + \dfrac{1}{\sqrt[4]{2}}\right)$ 23. no values of x 25. $\frac{4}{5} < x < 4$

Exercises 15.8 page 856

1. $\displaystyle\sum_{k=0}^{\infty} \dfrac{x^{k+1}}{k + 1} + C$; $\displaystyle\sum_{k=1}^{\infty} kx^{k-1}$; $R = 1$

3. $\displaystyle\sum_{k=1}^{\infty} \dfrac{x^{k+1}}{k^2(k + 1)} + C$; $\displaystyle\sum_{k=1}^{\infty} \dfrac{x^{k-1}}{k}$; $R = 1$

5. $\displaystyle\sum_{k=0}^{\infty} \dfrac{(-1)^k x^{2(k+1)}}{(2k + 2)!} + C$; $\displaystyle\sum_{k=1}^{\infty} \dfrac{(-1)^k x^{2k}}{(2k)!}$; $R = \infty$

7. $\displaystyle\sum_{k=1}^{\infty} \dfrac{2^k(x - 3)^{k+1}}{k(k + 1)} + C$; $\displaystyle\sum_{k=1}^{\infty} 2^k(x - 3)^{k-1}$; $R = \frac{1}{2}$ 9. $\displaystyle\sum_{k=0}^{\infty} x^{2k}$, $-1 < x < 1$

11. $f(x) = \dfrac{1}{10^3} \displaystyle\sum_{k=1}^{\infty} (-1)^{k+1} \left(\dfrac{x}{10}\right)^{3(k-1)}$, $|x| < 10$ 13. $\displaystyle\sum_{k=0}^{\infty} \dfrac{(-1)^k x^{2k+1}}{(2k + 1)k!}$; $R = \infty$

15. 0.4974   **17.** −120   **19.** $f(x) = e^{bx} = \displaystyle\sum_{k=0}^{\infty} \frac{(bx)^k}{k!}$

23. $f(x) = \displaystyle\sum_{k=0}^{\infty} \frac{(-1)^k x^{2k}}{(2k)!}, \quad -\infty < x < \infty$ **25.** 0.797, no

Exercises 15.9   page 863

1. $\displaystyle\sum_{k=0}^{\infty} \frac{(-1)^k x^{2k+1}}{(2k+1)!}$ **3.** $\displaystyle\sum_{k=0}^{\infty} \frac{x^{2k+1}}{(2k+1)!}$ **5.** $\displaystyle\sum_{k=0}^{\infty} (-1)^k x^k$

7. $-1 + 3x - 3x^2 + x^3$ **9.** $\displaystyle\sum_{k=0}^{\infty} \frac{(-1)^k (bx)^{2k}}{(2k)!}$ **11.** $\displaystyle\sum_{k=1}^{\infty} \frac{(-1)^{k+1} 2^{2k-1} x^{2k}}{(2k)!}$

13. $e^a \displaystyle\sum_{k=0}^{\infty} \frac{(x-a)^k}{k!}, \quad R = \infty$ **15.** $\ln 2 + \displaystyle\sum_{k=1}^{\infty} \frac{(-1)^{k+1} (x-2)^k}{2^k k}, \quad R = 2$

17. $\sin a \displaystyle\sum_{k=0}^{\infty} \frac{(-1)^k (x-a)^{2k}}{(2k)!} + \cos a \displaystyle\sum_{k=0}^{\infty} \frac{(-1)^k (x-a)^{2k+1}}{(2k+1)!}, \quad R = \infty$

19. $\displaystyle\sum_{k=0}^{\infty} \frac{(-1)^k x^{2k+1}}{2k+1}, \quad R = 1$ **21.** 1.2214   **23.** 1.54308   **25.** 0.31027

27. $1 + x - \dfrac{3x^2}{2} - \dfrac{11x^3}{6}$ **29.** $e\left(1 - \dfrac{x^2}{2} + \dfrac{x^4}{6}\right)$

Exercises 15.10   page 873

5. $1 + x - \dfrac{x^3}{3} - \dfrac{x^4}{6}, \quad R = \infty$ **7.** $\displaystyle\sum_{k=0}^{\infty} (-1)^k (x-1)^k, \quad R = 1$

9. $1 + \frac{3}{2}(x-1) + \frac{3}{8}(x-1)^2 - \frac{1}{16}(x-1)^3 + \cdots$ **11.** $x - \frac{1}{3}x^3 + \frac{2}{15}x^5 - \frac{17}{315}x^7$

13. $1 + x + \dfrac{x^2}{2} - \dfrac{x^4}{8} + \cdots$ **15.** $1 + x + \frac{3}{2}x^2 + \frac{7}{6}x^3 + \cdots$

17. $x - \dfrac{x^3}{2 \cdot 3} + \dfrac{1 \cdot 3}{2^2 \cdot 5 \cdot 2!}x^5 + \cdots + (-1)^n \dfrac{1 \cdot 3 \cdot 5 \cdots (2n-1)}{2^n (2n+1)n!}x^{2n+1} + \cdots$

19. $(n+1)!$ **21.** $1 + \displaystyle\sum_{k=1}^{\infty} \frac{1 \cdot 3 \cdot 5 \cdots (2k-1)}{2^k k!}x^{3k}, \quad R = 1$

23. $3 + \dfrac{x}{27} + \displaystyle\sum_{k=2}^{\infty} (-1)^{k-1} \frac{2 \cdot 5 \cdot 8 \cdots (3k-4)}{3^{4k} k!}x^k, \quad R = 27$ **25.** 3.072

27. 0.520   **29.** 0.6225

Review and Miscellaneous Exercises   page 875

1. $a_n \to \frac{1}{8}$ **3.** $a_n \to 0$ **5.** diverges to ∞
7. diverges but remains bounded   **9.** $a_n \to -2$ **11.** $a_n \to 0$ **13.** $\frac{1000}{11}$
15. $\frac{1}{2}$ **17.** convergent   **19.** divergent   **21.** convergent   **23.** convergent
25. divergent   **27.** divergent   **29.** convergent if and only if $p > 1$
31. conditionally convergent   **33.** divergent   **35.** divergent
37. absolutely convergent   **39.** absolutely convergent
41. conditionally convergent   **43.** $(-1, 1)$ **45.** $[-4, -2]$ **47.** $(-1, 1)$
49. $(-\infty, \infty)$ **51.** $[-1, 1]$ **53.** $(-\infty, \infty)$ **55.** $|x| \geq 7$

57. $3 \displaystyle\sum_{k=1}^{\infty} (-1)^{k-1} \frac{(x-1)^k}{k}, \quad 0 < x \leq 2$ **59.** no Maclaurin expansion

61. $\frac{1}{c} \sum_{k=0}^{\infty} (-1)^k \frac{(x-c)^k}{c^k}$, $|x - c| < |c|$, $c \neq 0$ **63.** e^{-1} **65.** $\cosh \frac{3}{2}$
67. 2.0408 **69.** 0.95534 **71.** 0.843 **73.** no
81. (a) convergent if and only if $p > 1$; (b) divergent

Exercises 16.1 *page 886*

3. $(1, 0, 0)$, $(1, 2, 0)$, $(0, 2, 0)$, $(0, 2, 4)$, $(0, 0, 4)$, $(1, 0, 4)$
5. $(1, 2, 3)$, $(-2, 2, 3)$, $(-2, 2, 5)$, $(-2, 1, 5)$, $(1, 1, 5)$, $(1, 1, 3)$ **7.** $\sqrt{18} = 3\sqrt{2}$
9. $\frac{1}{2}\sqrt{69}$ **11.** $\sqrt{8b^2 + 36} = 2\sqrt{2b^2 + 9}$ **13.** $\frac{3}{2}\sqrt{29}$ sq units
17. sphere: center $(3, 0, 0)$, radius 5 **19.** no graph
21. sphere: center $(\frac{1}{2}, -1, 1)$, radius 2 **23.** $x^2 + y^2 + z^2 + 6x + 8y - 14z = 0$
25. plane $x = y$
27. (a) $(x_0, y_0, -z_0)$, (b) $(-x_0, y_0, z_0)$, (c) $(x_0, -y_0, z_0)$, (d) $(-x_0, -y_0, -z_0)$
29. (a) $f(x, -y, -z) = f(x, y, z)$, (b) $f(-x, y, -z) = f(x, y, z)$,
(c) $f(-x, -y, z) = f(x, y, z)$

Exercises 16.2 *page 893*

1. $\mathbf{i} - 11\mathbf{j} + 3\mathbf{k}$ **3.** $\frac{1}{2}(5\mathbf{i} + \mathbf{k})$ **5.** 3 **7.** 0 **9.** $\langle 14, 3, -3 \rangle$
11. $3\sqrt{6} - 4\sqrt{38}$ **13.** no meaning **15.** $-\frac{1}{\sqrt{29}}(3\mathbf{i} - 2\mathbf{j} + 4\mathbf{k})$

17. $-10, 6$ **19.** $m = 5$ and $n = 7$ **21.** $\pm\frac{7}{\sqrt{38}}(5\mathbf{i} - 3\mathbf{j} + 2\mathbf{k})$

Exercises 16.3 *page 900*

1. 0 **3.** -9 **5.** 40 **7.** $\frac{2}{3}\sqrt{2}$ **9.** $-\frac{8}{\sqrt{154}}$ **13.** 9 **15.** -2

17. $\frac{3\sqrt{29}}{29}$, $\frac{-2\sqrt{21}}{7}$

19. no, $\mathbf{a} - \mathbf{c}$ must be perpendicular to $\mathbf{b}$ if $\mathbf{a} \neq \mathbf{c}$ and $\mathbf{b} \neq \mathbf{0}$
21. $\frac{1}{\sqrt{3}}, \frac{1}{\sqrt{3}}, \frac{1}{\sqrt{3}}$ **23.** $-\frac{1}{9}, -\frac{8}{9}, -\frac{4}{9}$ **25.** no, sum of squares not equal to 1

29. $\frac{1}{3}\sqrt{6}$ **31.** $\frac{1}{\sqrt{6}}(3\mathbf{i} + \mathbf{j})$

Exercises 16.4 *page 905*

1. $\mathbf{r} = (4 + 3t)\mathbf{i} + (4 + 6t)\mathbf{j} - 3t\mathbf{k}$; $x = 4 + 3t$, $y = 4 + 6t$, $z = -3t$
3. $\mathbf{r} = (-4 + 7t)\mathbf{i} + (1 - 3t)\mathbf{j} + (-2 + 8t)\mathbf{k}$; $x = -4 + 7t$, $y = 1 - 3t$, $z = -2 + 8t$ **5.** $(4, 4, 0)$ **7.** $(-18, 7, -18)$
9. $\mathbf{r} = (3 + u)\mathbf{i} + (-4 + 2u)\mathbf{j} + (6 - u)\mathbf{k}$ **11.** $\mathbf{r} = u(\mathbf{i} + 2\mathbf{j} + 3\mathbf{k})$
13. $\mathbf{r} = (1 + 2t)\mathbf{i} + (1 - t)\mathbf{j} + t\mathbf{k}$ **15.** $\mathbf{r} = (1 - \frac{6}{7}t)\mathbf{i} + (1 - \frac{4}{7}t)\mathbf{j} + 2t\mathbf{k}$
17. $(4, 8, 10)$ **19.** $(11, 10, 9)$ **23.** $\mathbf{r} = (4 + 13t)\mathbf{i} + (-2 + 7t)\mathbf{j} + (3 + t)\mathbf{k}$
25. $(-6, 4, -8)$ **27.** $(-1, 4, 2)$, 1.14 radians **29.** $s_{\min} \doteq 2.16$

Exercises 16.5 *page 913*

1. $2x + 5y + 7z = 0$ **3.** $-3x - y + 7z = 23$ **5.** $7x - y - 2z + 10 = 0$
7. $\left\langle \frac{1}{\sqrt{6}}, \frac{1}{\sqrt{6}}, \frac{2}{\sqrt{6}} \right\rangle$ **9.** $\left\langle \frac{4}{\sqrt{17}}, \frac{-1}{\sqrt{17}}, \frac{0}{\sqrt{17}} \right\rangle$ **11.** $x + y + \frac{1}{2}z = 1$

13. $4x - 3y - z = 22$ **15.** $x - 2y - 3z + 7 = 0$ **17.** $y = -2$

19. $2x + 2y - z = \pm 36$ **21.** $\sqrt{5}$ **23.** $\dfrac{-7}{2\sqrt{31}\sqrt{35}}$ **25.** $4x - 3y + 2z = 13$

27. line determined by $y = \dfrac{11 - 4x}{5}$ and $z = \dfrac{2 - 3x}{5}$

Exercises 16.6 $\qquad$ page 923

1. $-4\mathbf{i} - 5\mathbf{j} + \mathbf{k}$ **3.** $-8\mathbf{i} + 11\mathbf{j} + 9\mathbf{k}$ **5.** $42\mathbf{i} + 42\mathbf{j} + 38\mathbf{k}$ **7.** 28

9. 280   **13.** $\pm \dfrac{5}{\sqrt{3}}(\mathbf{i} + \mathbf{j} - \mathbf{k})$

17. $[(u_2 v_3 - u_3 v_2)^2 + (u_3 v_1 - u_1 v_3)^2 + (u_1 v_2 - u_2 v_1)^2]^{1/2} =$
$[1 - (u_1 v_1 + u_2 v_2 + u_3 v_3)^2]^{1/2}$ **21.** 10 cu units   **23.** 7 cu units

25. $4x - y + 2z = 3$ **27.** $\frac{1}{2}\sqrt{62}$ sq units   **29.** $\dfrac{8}{\sqrt{62}}$ **31.** $\dfrac{28}{\sqrt{26}}$

Exercises 16.7 $\qquad$ page 929

1. right circular cylinder with y-axis as axis and radius $= 2$
3. cubic cylinder with rulings parallel to the y-axis
5. cylinder with sinusoidal function as cross section and rulings parallel to the z-axis
7. right circular cone with axis along the y-axis (one nappe)
9. hyperbolic cylinder with elements parallel to the x-axis
11. paraboloid of revolution with axis along the y-axis   **13.** $z^2 + x^2 = y^2$
15. $x^2 + y^2 = 16$ **17.** $y^2 + z^2 = e^{2x}$ **19.** $y = 2\sqrt{1 - x^2}$, $z = 0$; x-axis
21. $y = 2z$, $x = 0$; z-axis

Exercises 16.8 $\qquad$ page 938

1. 3, 2, 4   **3.** prolate, x-axis   **5.** right circular cone, y-axis
7. elliptic hyperboloid of two sheets   **9.** elliptic paraboloid
11. right circular cylinder, radius $\sqrt{10}$, axis x-axis   **13.** $x = z = 0$ or the y-axis
15. hyperbolic cylinder
17. $y^2 + z^2 = 8x$, elliptic paraboloid of revolution with x-axis as its axis of symmetry
19. $x^2 + y^2 + (z - \frac{1}{2})^2 = (\frac{3}{2})^2$, a sphere with center at $(0, 0, \frac{1}{2})$ and radius $= \frac{3}{2}$
21. $7x^2 + y^2 - z^2 - 7 = 0$, an elliptic hyperboloid of one sheet

Exercises 16.9 $\qquad$ page 945

1. the straight line which is the intersection of the planes $y = 2x$ and $z = 3$
3. half a parabola which is the intersection of the plane $y = \frac{4}{3}x$ and the parabolic
cylinder $z = \frac{1}{3}x^2$, $x \geq 0$ **5.** $\mathbf{i} + 2t\mathbf{j} - 5\mathbf{k}$, $2\mathbf{j}$
7. $t^{-1}\mathbf{i} + 2t^{-3}\mathbf{j} - 3t^{-4}\mathbf{k}$, $-t^{-2}\mathbf{i} - 6t^{-4}\mathbf{j} + 12t^{-5}\mathbf{k}$
9. $(t \cos t + \sin t)\mathbf{i} - 2te^{-t^2}\mathbf{j} + (1 + t^2)^{-1}\mathbf{k}$

$(-t \sin t + 2 \cos t)\mathbf{i} + (4t^2 - 2)e^{-t^2}\mathbf{j} - 2t(1 + t^2)^{-2}\mathbf{k}$ **11.** $\dfrac{1}{\sqrt{89}}(3\mathbf{i} + 4\mathbf{j} + 8\mathbf{k})$

13. $\dfrac{1}{\sqrt{t^2 + 2}}[(t \cos t + \sin t)\mathbf{i} + (\cos t - t \sin t)\mathbf{j} + \mathbf{k}]$ **15.** $3\sqrt{35}$

17. $\sqrt{\frac{13}{2}} + \frac{5}{4}[\ln(1 + \sqrt{\frac{13}{8}}) + \frac{1}{2}\ln \frac{8}{5}]$ **19.** $\dfrac{\pi}{2}$ **21.** $(0, -2, -2)$

23. $|\mathbf{w}(t)| =$ constant, see Example 3

1. $-2e^{-2t}\mathbf{i} - 15(\sin 3t)\mathbf{j} + 15(\cos 3t)\mathbf{k}; \ 4e^{-2t}\mathbf{i} - 45(\cos 3t)\mathbf{j} - 45(\sin 3t)\mathbf{k};$
$\sqrt{4e^{-4t} + 225}, \ \sqrt{16e^{-4t} + 2025}$
3. $b\omega(\cos \omega t)\mathbf{i} - b\omega(\sin \omega t)\mathbf{j} + c\mathbf{k}; \ -b\omega^2(\sin \omega t)\mathbf{i} - b\omega^2(\cos \omega t)\mathbf{j}; \ \sqrt{b^2\omega^2 + c^2}, \ b\omega^2$
5. $\mathbf{i} + 2t\mathbf{j} + 3t^2\mathbf{k}, \ 2\mathbf{j} + 6t\mathbf{k}; \ \sqrt{1 + 4t^2 + 9t^4}, \ 2\sqrt{1 + 9t^2}$
7. $\dfrac{1}{\sqrt{2}}(-(\sin t)\mathbf{i} - (\cos t)\mathbf{j} + \mathbf{k}), \ -(\cos t)\mathbf{i} + (\sin t)\mathbf{j}; \quad x + z = \dfrac{\pi}{2}$
9. $\dfrac{\sqrt{266}}{98}, \ 22(14^{-1/2}), \ \tfrac{1}{7}(266^{1/2})$
11. $(a^2 + b^2)^{-1/2}[-a(\sin t)\mathbf{i} + a(\cos t)\mathbf{j} + b\mathbf{k}], \ -(\cos t)\mathbf{i} - (\sin t)\mathbf{j}$
$(a^2 + b^2)^{-1/2}[b(\sin t)\mathbf{i} - b(\cos t)\mathbf{j} + a\mathbf{k}], \ a(a^2 + b^2)^{-1} \quad$ **15.** $2^{1/2}(2 + t^2)^{-3/2}$
19. $\tau = 2(1 + 2t^2)^{-2} \quad$ **21.** $b(a^2 + b^2)^{-1}$
23. $\tau = 0$ since the curve lies in the plane $x + y - z = 0$

1. $(0, 7, 2), (5, 5\sqrt{3}, -6) \quad$ **3.** $\left(3\sqrt{2}, \dfrac{\pi}{4}, 4\right), \left(2\sqrt{2}, \dfrac{3\pi}{4}, -1\right)$
5. $\left(2\sqrt{2}, \dfrac{\pi}{7}, \dfrac{\pi}{4}\right), \left(10, \dfrac{\pi}{6}, \cos^{-1}\tfrac{3}{5}\right) \quad$ **7.** $r^2 = 3z^2$, circular cone with z-axis as axis
9. $r^2 \cos 2\theta = 1$, hyperbolic cylinder
11. $r = \pm 1$, a circular cylinder of radius 1 with the z-axis as its axis
13. $\rho = a$, a sphere with center at the origin of coordinates and radius a
15. $\phi = \dfrac{\pi}{3}, \dfrac{2\pi}{3}$ a right circular cone with the z-axis as axis and vertex at 0
17. $\rho[3 \sin \phi \cos \theta + 4 \sin \phi \sin \theta + 2 \cos \phi] = 24$, a plane with x-, y- and z-intercepts 8, 6 and 12, resp.
19. $x^2 + y^2 + z^2 = 100$, a sphere with center at the origin and radius 10
21. $x^2 - z = 3$, a parabolic cylinder with rulings parallel to the y-axis
23. $x^2 + y^2 - 3z^2 = 0, z > 0$, one nappe of a right circular cone with vertex angle $\dfrac{\pi}{3}$ and axis the z-axis
25. $x^2 - 3xy - y^2 = 1$, an equation of a hyperbolic cylinder with generators parallel to the z-axis $\quad$ **29.** $2\pi(a^2 + 1)^{1/2} \quad$ **31.** $\tfrac{1}{3}[(4 + t_1^2)^{3/2} - 8]$

Review and Miscellaneous Exercises $\qquad$ *page 960*

1. plane parallel to zx-plane through $(0, 4, 0)$
3. plane through $(0, 0, 0), (1, 0, 1)$ and $(0, 1, 1)$
5. sphere centered at the origin and radius $= 7$
7. straight line through $(3, 2, 0)$ and parallel to the z-axis $\quad$ **9.** x-axis
11. paraboloid of revolution, non-negative half of z-axis is axis of revolution obtained by revolving the parabola $z = x^2, y = 0$ about the z-axis
13. (a) $\sqrt{x^2 + y^2 + z^2}$, (b) $\sqrt{y^2 + z^2}$, (c) $2|z|$, (d) $|x|$
15. $(x + \tfrac{1}{2})^2 + y^2 + z^2 = (\tfrac{3}{2})^2$, a sphere with center at $(-\tfrac{1}{2}, 0, 0)$ and radius $= \tfrac{3}{2}$
17. $-5 \quad$ **19.** $498 \quad$ **21.** always true
23. $s_1 = A_1 + A_3, s_2 = A_2 - A_3, s_3 = -A_3 \quad$ **25.** $\pm\dfrac{1}{\sqrt{3}}(-\mathbf{i} + \mathbf{j} - \mathbf{k})$
27. 54 cu units $\quad$ **29.** $\dfrac{x - 2}{2} = \dfrac{y - 3}{23} = \dfrac{z - 1}{11}$
31. $\dfrac{5}{\sqrt{195}}, \dfrac{7}{\sqrt{195}}, \dfrac{11}{\sqrt{195}}$; sum of squares of direction cosines equals 1

33. $\begin{vmatrix} x - x_0 & y - y_0 & z - z_0 \\ x_1 - x_0 & y_1 - y_0 & z_1 - z_0 \\ a & b & c \end{vmatrix} = 0$ **35.** $67°51'$ **37. (b)** no

39. $5x + y - 4z = 37$ **41.** $(a_1^2 + a_2^2 + a_3^2)^{-1/2}(a_1\mathbf{i} + a_2\mathbf{j} + a_3\mathbf{k}), 0$

43. $(1, 1, 0), \dfrac{\pi}{2}$ **45.** linear momentum $= m\mathbf{v} = $ constant

49. 0, curve lies in the plane $2x + y + z = 7$

51. $\dfrac{x^2}{a^2} + \dfrac{y^2}{b^2} = z$, an elliptic paraboloid

53. (a) cylindrical surface with generators parallel to the x-axis; **(b)** surface of revolution about the z-axis

Exercises 17.1   page 972

1. all x and y; entire xy-plane
3. all x and y for which $y \geq 4 - 3x$; this is a half plane and includes its boundary
5. all x and y for which $4x^2 + y^2 < 16$, that is, the interior of the ellipse
7. all x and y except the parallel lines $y + 2x = n\pi$, $n = $ arbitrary integer
9. (a) 14; **(b)** 16; **(c)** $y^2 - 3xy + 4x^2$
11. (a) $2^{3/2}e^6$; **(b)** not in the domain of the function; **(c)** $e^{x + \frac{1}{x}}, x \neq 0$
13. $\dfrac{3x - 1}{2} > y$—half plane, exclusive of its boundary, $(-\infty, \infty)$
15. $xy \geq 0$, all points in the first and third quadrants including the boundary, $[3, \infty)$
17. the intersection **(i)** of all points on and above the xy-plane; **(ii)** on one side of and on the vertical plane $y = 3 - 4x$; the range is $[0, \infty)$
19. parallel lines of slope -1; $(-\infty, \infty)$
21. hyperbolas if $c \neq 0$; x and y axes if $c = 0$; $(-\infty, \infty)$
23. parabolas $y = 3x^2 - c$; $(-\infty, \infty)$
25. hyperbolas $xy = \frac{1}{3}\ln c, c \neq 1$, x and y axis if $c = 1, (0, \infty)$
27. parallel planes with vector normals $<3, 1, 4>$; a plane passing through the origin
29. paraboloids of revolution about the z axis, particular member passing through $(0, 0, -1)$ **31. (a)** $6^{1/2}$; **(b)** not defined; **(c)** 8; **(d)** u; **(e)** $a^2 - b^2$
33. (a) ellipses $2x^2 + y^2 + 3 = c, (c > 3)$; **(b)** the temperature is largest at the four corners, $T(\pm 10, \pm 8) = 267$ and smallest at $(0, 0)$, where $T(0, 0) = 3$
35. (a) homogeneous, degree 3; **(b)** positively homogeneous, degree 1; **(c)** not homogeneous; **(d)** positively homogeneous, degree 1; **(e)** homogeneous, degree $\frac{1}{3}$

Exercises 17.2   page 979

1. $f_x = y^2 - 3x^2y$; $f_y = 2xy - x^3$ **3.** $h_x = 4x - 5y$; $h_y = -5x - 6y$
5. $z_x = 2x(x^2 + y^2)^{-1}$; $z_y = 2y(x^2 + y^2)^{-1}$
7. $u_x = \dfrac{x}{\sqrt{x^2 + y^2 + 10}}$; $u_y = \dfrac{y}{\sqrt{x^2 + y^2 + 10}}$
9. $f_x = \dfrac{y}{x(x + y)}$; $f_y = -(x + y)^{-1}$
11. $u_x = y(x^2 + y^2)^{-1}$; $u_y = -x(x^2 + y^2)^{-1}$
13. $u_x = \dfrac{(ad - bc)y}{(cx + dy)^2}$; $u_y = \dfrac{(bc - ad)x}{(cx + dy)^2}$; $u_x = 0$ and $u_y = 0$ because $u = $ constant
15. $f_x = yze^{xyz} - \frac{1}{2}(z - x - y)^{-1/2}$; $f_y = xze^{xyz} - \frac{1}{2}(z - x - y)^{-1/2}$;
$f_z = xye^{xyz} + \frac{1}{2}(z - x - y)^{-1/2}$ **17.** 4 **19.** $2(y - x)$ **21.** $3\ln 2$; e
23. $-3^{-1/2}a^{-1}$; $-2^{-1/2}b$

25. $V_r = \dfrac{2\pi}{3} rh$; $V_h = \dfrac{-\pi r^2}{3}$ **27.** 4 **33.** $-e^x$; e^y **35.** $e^{-x^2/2}$; $-e^{-y^2/2}$

37. $\frac{9}{2}\cosh^2\theta \sinh\theta\,(\cosh^3\theta - \sinh^2\phi)^{1/2}$; $-3\sinh\phi\cosh\phi(\cosh^3\theta - \sinh^2\phi)^{1/2}$

39. $2x(y - e^z)^{-1}$; $(3y^2 - z)(y - e^z)^{-1}$ **41.** 1; $(R/x_k)^2$; $k = 1, 2, \ldots, n$

Exercises 17.3 $\hspace{3cm}$ page 983

1. $6x - 2y^2$; $-4xy$; $-2x^2 + 18y$ **3.** $y^2 e^x$; $2(e^y + ye^x)$; $2xe^y$

5. $\sinh 2u \cosh 3v$; $3\sinh 2u \sinh 3v$; $27\sinh^2 u \sinh 3v$

7. $1344(2s + t)^5$; $2688(2s + t)^5$; $336(2s + t)^5$

9. $-x\cos(x + y) - 2\sin(x + y)$; $-x\cos(x + y) - \sin(x + y) + 2y$;
$-x\cos(x + y) + 2x$ **11.** $4x\sec^2(x^2 + y + z)\,[3\sec^2(x^2 + y + z) - 2]$;
$2\sec^2(x^2 + y + z)[3\sec^2(x^2 + y + z) - 2]$ **23.** $b^2 = a^2 + c^2$

29. (a) $-6xe^{-3x^2}$; (b) 0; (c) $6ye^{-3y^2}$

31. (a) ye^{-x^2}; (b) $\displaystyle\int_y^x e^{-t^2}\,dt - ye^{-y^2}$; (c) $-2xye^{-x^2}$; (d) e^{-x^2}; (e) $2(y^2 - 1)e^{-y^2}$

Exercises 17.4 $\hspace{3cm}$ page 991

1. $\frac{4}{3}$ **3.** $\frac{6}{7}$ **13.** continuous **15.** continuous **17.** discontinuous

19. all x and y **21.** all x and y

23. all x and y except for points on the hyperbola $xy = -5$

25. all x and y except $(0, 0)$

27. (a) $x^2 + 6xy^2 + 9y^4 + 2x + 6y^2$; (b) $e^{-(x+3y^2)}$; (c) $u^2 + 2u + 3e^{-2u}$

33. continuous for all (x, y, z) such that $2x - y + z < 4$

35. continuous for all (x, y, z) except when $z = n\pi/2$, $n =$ odd integer or when
$y = 1 \pm \sqrt{3}$

Exercises 17.5 $\hspace{3cm}$ page 997

1. $2x\,dx - 15y^2\,dy$ **3.** $e^{xy}(y\,dx + x\,dy)$ **5.** $\dfrac{1}{y}\,dx - \dfrac{x}{y^2}\,dy$

7. $[3\{2x\cos(x + y) - x^2\sin(x + y)\} - \pi y^3\cos\pi x]\,dx$
$- 3[x^2\sin(x + y) + y^2\sin\pi x]\,dy$ **9.** $\dfrac{x\,dx + y\,dy + z\,dz}{x^2 + y^2 + z^2}$

11. $\tan^{-1}\dfrac{z}{y}\,dx - \dfrac{xz}{y^2 + z^2}\,dy + \dfrac{xy}{y^2 + z^2}\,dz$ **13.** -0.6

15. -0.14 (approximately)

17. (a) $a\,\Delta x + b\,\Delta y$; (b) $a\,\Delta x + b\,\Delta y$; (c) $\varepsilon_1 = \varepsilon_2 = 0$

19. (a) $\dfrac{y\,\Delta x - x\,\Delta y}{y(y + \Delta y)}$; (b) $\dfrac{1}{y}\Delta x - \dfrac{x}{y^2}\Delta y$; (c) $\varepsilon_1 = 0,\ \varepsilon_2 = \dfrac{x\,\Delta y - y\,\Delta x}{y^2(y + \Delta y)}$

21. -4.8π in.^{3} **25.** 0.017 sec **27.** 0.024 **29.** $\frac{47}{60}$ ft^3 **31.** 44.76

Exercises 17.6 $\hspace{3cm}$ page 1003

1. $(-2\sin t + 3\cos t)e^{2\cos t + 3\sin t}$ **3.** $e^t - 2e^{-t} - 4e^{4t}$ **5.** $\dfrac{4t^2 + 3t - 6}{t(t^2 + t - 3)}$

7. $8r^7 e^{4s}$; $4r^8 e^{4s}$ **9.** $2rs^4\cosh r^2 s^4$; $4r^2 s^3\cosh r^2 s^4$

11. 0; $-3\sin(3s + t)$; $-\sin(3s + t)$

13. $2(7r + 3s + 3t)$; $2(3r - t)$; $2(3r - s)$ **15.** $4r + 8rt^2$; $8r^2 t$

23. -2; $4\sqrt{3}$ **25.** 2; -2 **27.** 36; -168 **29.** $\dfrac{8064}{R}$ deg/min

31. $29,433m$ **33.** $-\dfrac{2xy + z^5}{y^3 + 5z^4x}$; $-\dfrac{x^2 + 3y^2z}{y^3 + 5z^4x}$ **35.** $\dfrac{2x(x^2 + y^2)^{-1}}{\pi e^{\pi z} + e^z}$;

$-\dfrac{2y(x^2 + y^2)^{-1}}{\pi e^{\pi z} + e^z}$

Exercises 17.7 page 1012

1. $2x \cos \theta - 4y \sin \theta$; $2 - 6\sqrt{3}$

3. $e^x \cos y \cos \theta - e^x \sin y \sin \theta$; $\frac{1}{4}\sqrt{2}(1 - \sqrt{3})$

5. $\dfrac{2x}{x^2 + y^2} \cos \theta + \dfrac{2y}{x^2 + y^2} \sin \theta$; $\frac{2}{5}\sqrt{2}$

7. $(-y \cos \theta + x \sin \theta)(x^2 + y^2)^{-1}$; $\frac{1}{26}(2 + 3\sqrt{3})$ **9.** $\dfrac{7\pi}{4}$, $4\sqrt{2}$

11. $\theta = \tan^{-1}\frac{1}{4}$, $\sqrt{17}$ **13.** $\tan \theta = -\frac{3}{2}$, $\dfrac{3\pi}{2} < \theta < 2\pi$, $\frac{1}{7}\sqrt{52}$ **15.** $\frac{16}{3}$

17. $\frac{1}{4}(10 - \sqrt{2})$ **19.** (a) $4\mathbf{i} - 5\mathbf{j}$; (b) $(0, 0)$; (c) $\dfrac{1}{\sqrt{41}}(4\mathbf{i} - 5\mathbf{j})$; (d) $\sqrt{41}$

21. (a) $10\mathbf{i} + 7\mathbf{j} + 10\mathbf{k}$; (b) $(\frac{2}{3}, \frac{1}{4}, -\frac{1}{3})$; (c) $\dfrac{1}{\sqrt{249}}(10\mathbf{i} + 7\mathbf{j} + 10\mathbf{k})$; (d) $\sqrt{249}$

23. $f = \text{constant}$, that is, it is independent of x, y and z

Exercises 17.8 page 1017

1. $4\mathbf{i} + 2\mathbf{j} + 4\mathbf{k}$ **3.** $4\mathbf{i} - \mathbf{j} - 2\mathbf{k}$

5. $4x + 3y + 5z = 50$; $\dfrac{x - 4}{8} = \dfrac{y - 3}{6} = \dfrac{z - 5}{10}$

7. $4x + 3y - 2z = 12$; $\dfrac{x - 3}{4} = \dfrac{y - 4}{3} = \dfrac{z - 6}{-2}$

9. $6x + y - 4z = 21$; $\dfrac{x - 2}{6} = y - 1 = \dfrac{z + 2}{-4}$

11. $ex - \frac{1}{2}\sqrt{3} ey - 3z + \frac{1}{6}\sqrt{3} \pi e = 0$;

$(x - \frac{1}{2})/e = \left(y - \dfrac{\pi}{3}\right) \Big/ \left(-\dfrac{\sqrt{3}}{2} e\right) = \left(z - \dfrac{e}{6}\right) \Big/ -3$

13. $3x + 6y - 4z = 13$; $\dfrac{x - 3}{3} = \dfrac{y + 2}{6} = \dfrac{z + 4}{-4}$

15. $6x + 3y + 4z = 108$; $\dfrac{x - 4}{6} = \dfrac{y - 16}{3} = \dfrac{z - 9}{4}$ **19.** a

21. $\dfrac{x - 1}{1} = \dfrac{y - 2}{2} = \dfrac{z - 3}{-.5}$ **23.** $\dfrac{x - 2}{2} = \dfrac{y - 1}{-2} = \dfrac{z - 2}{-1}$

25. $\dfrac{x - x_0}{(f_y g_z - f_z g_y)|_{P_0}} = \dfrac{y - y_0}{(f_z g_x - f_x g_z)|_{P_0}} = \dfrac{z - z_0}{(f_x g_y - f_y g_x)|_{P_0}}$ **27.** $71°$

Exercises 17.9 page 1027

1. $f(1, -2) = -9$, rel min **3.** $F(0, 0) = 10$, saddle point

5. $h(0, 0) = 1$, rel max **7.** $f(0, 0) = 10$, saddle point; $f(\pm\sqrt{3}, 0) = 1$, rel min

9. $g(0, 0) = 0$, saddle point; $g\left(\dfrac{a}{3}, \dfrac{a}{3}\right) = \dfrac{a^3}{27}$, rel max **11.** $\frac{1}{2}\sqrt{6}$ **13.** 4000

15. $V_{\max} = \frac{8}{9}\sqrt{3}abc$ **17.** $y = 3.05x + 4.76$; 35.26

19. $x = \frac{1}{3}\left(\displaystyle\sum_{i=1}^{3} x_i\right)$, $y = \frac{1}{3}\left(\displaystyle\sum_{i=1}^{3} y_i\right)$; $x = \bar{x}$, $y = \bar{y}$, where $(\bar{x}, \bar{y})$ are the coordinates

of the centroid of the three given points
23. (a) $80p_1 - 5p_1^2 - 5p_1p_2 - 60p_2 - 4p_2^2$; (b) $68/11 \doteq \$6.18$, $40/11 \doteq \$3.64$;
(c) $460/11 \doteq 41.82 \doteq 42$ or 4200 units (approx.),
$296/11 \doteq 26.91 \doteq 27$ or 2700 units (approx.); (d) \$35,636

Exercises 17.10 page 1036

1. $f(1, 0) = 3$, (abs max); $f(0, 1) = -1$, (abs min)
3. $h(1, 1) = 3$, (abs max); $h(0, 0) = 0$ (abs min)
5. $G(0, 1) = G(1, 1) = 4$ (abs max); $G(\frac{1}{2}, \frac{1}{4}) = \frac{21}{8}$, (abs min)
7. $f(1, 1) = 1$, (abs max); $f \equiv 0$ on the boundary of the triangular region, (abs min)
9. 8 11. $\frac{16}{9}\sqrt{3}$ 13. 27 15. 625, $\frac{625}{3}$

17. $\sqrt{14} - 4$, $-\sqrt{14} - 4$ 19. $\sum_{i=1}^{n} a_i^2$ 21. diameter = height

Review and Miscellaneous Exercises page 1037

1. all x and y except for the points on the two lines $y = \pm 4x$
3. all x and y in the half plane $x + y - 3 \geq 0$ 5. all x, y, and z such that
$x^2 + y^2 + z^2 \leq a^2$; that is, all points inside and on the sphere of radius a and
center at $(0, 0, 0)$ 7. $3(x^2 - y^3x^{-2})$; $9y^2x^{-1}$; $-9y^2x^{-2}$; $-9y^2x^{-2}$
9. y^2e^{x+y}; $y(y + 2)e^{x+y}$; y^2e^{x+y}; $y(y + 2)e^{x+y}$
11. $y^2z^2 \cosh xyz$; $z^2x^2 \cosh xyz$; $x^2y^2 \cosh xyz$; $xyz^2 \cosh xyz + z \sinh xyz$;
$x^2yz \cosh xyz + x \sinh xyz$; $xy^2z \cosh xyz + y \sinh xyz$
15. $e^t - e^{-t} + (t + \ln 2)(e^t + e^{-t})$
17. $8(r + 4s - 2t)^7$; $32(r + 4s - 2t)^7$; $-16(r + 4s - 2t)^7$
19. $D: -7 \leq xy \leq 7$, $R: 0 \leq u \leq 7$ 21. $D:$ all (x, y), $R: \ln 5 \leq u < \infty$
23. all (x, y) 25. all x and y 37. $\frac{5}{2}\sqrt{2}$, $\sqrt{13}$ 39. $\frac{4}{3}\sqrt{3}$, $\sqrt{6}$

41. $7x + 8y - 3z - 35 = 0$; $\dfrac{x - 2}{7} = \dfrac{y - 3}{8} = \dfrac{z - 1}{-3}$

47. $g(0, 0) = 25$, saddle point 49. $F(3, 3) = -11$, rel min
53. square base $[2C_2(C_1 + C_3)^{-1}V_0]^{1/3}$ on a side and height

$[2C_2(C_1 + C_3)^{-1}]^{-2/3}V_0^{1/3}$ 55. $A = \dfrac{1}{2\pi} \int_{-\pi}^{\pi} f(x)\, dx$; $B = \dfrac{1}{\pi} \int_{-\pi}^{\pi} f(x) \sin x\, dx$;

$C = \dfrac{1}{\pi} \int_{-\pi}^{\pi} f(x) \cos x\, dx$ 57. $f(0, 3) = 9$, (abs max); $f(2, 1) = -3$, (abs min)
59. $f(0, 80) = 560$, (abs max); $f(0, 0) = 0$, (abs min)

Exercises 18.1 page 1049

1. 5 3. $\frac{7}{6}$ 5. $a + \dfrac{a^2}{2}$ 7. 24 cu units 9. $\frac{9}{2}$ cu units 11. $\frac{1}{6}a^5$

13. $8k$ 15. 1500 lb 17. 0.088, 1.763, 0.482

Exercises 18.2 page 1055

1. $\frac{7}{6}$ 3. $\frac{2187}{35}$ 5. $\frac{48}{5}$ 7. $\frac{1}{3}a^3$ 9. $-\frac{208}{15}$ 11. $\frac{38}{3}$ cu units
13. $\frac{368}{7}$ cu units 15. $9(\pi + 1)$ cu units 17. $\frac{1}{6}abc$ cu units
19. $\frac{4}{3}\pi a^3$ cu units 21. $\frac{4}{3}\pi abc$ cu units

Exercises 18.3   page 1061

1. 103.5   3. 156.25   5. $-\frac{11}{60}$ 7. $\frac{1}{4}e^2 - \frac{2}{9}e^3 + \frac{5}{36}$ 9. $e - e^{-1} - 2$

11. $\frac{1}{4}(1 - e^{-4})$ 13. $\frac{32}{3}$ sq units   15. 6 sq units   17. $\int_0^8 \int_{y/4}^{\sqrt[3]{y}} f(x, y)\, dx\, dy$

19. $\int_0^2 \int_0^{\sqrt{y/2}} f(x, y)\, dx\, dy + \int_2^3 \int_0^{3-y} f(x, y)\, dx\, dy$ 21. $\frac{1}{2}\sin 1$ 23. $\frac{13}{18}a^3$

Exercises 18.4   page 1070

1. $\frac{7}{12}a, \frac{7}{12}a$ 3. $\frac{a}{2}, \frac{3}{4}a$ 5. $\frac{6}{7}a, \frac{1}{2}a$ 7. $0, \frac{32a}{15\pi}$ 9. $\frac{5}{12}\rho_0 a^4, \frac{5}{12}\rho_0 a^4, \frac{5}{6}\rho_0 a^4$

11. $\frac{1}{10}\rho_0 a^4, \frac{1}{20}\rho_0 a^4, \frac{3}{20}\rho_0 a^4$ 13. $\frac{2}{21}kb^{7/2}, \frac{2}{5}kb^{5/2}, \sqrt{\dfrac{5b}{21}}$

15. $\frac{2}{3}ka^3, \frac{4}{15}ka^5, \frac{1}{5}\sqrt{10}\,a$ 17. $\frac{1}{4}\pi ab^3\rho_0, \frac{1}{4}\pi a^3 b\rho_0, \frac{b}{2}, \frac{a}{2}$ 21. I_x, I_y

Exercises 18.5   page 1077

1. $\dfrac{a^3}{3}$ 3. $\frac{1}{2}$ 5. $\pi \sin a^2$ 7. $\frac{8}{3}(3\pi - 4)$ sq units   9. $25(8 + \pi)$ sq units

11. $4(2 + \sqrt{3})$ sq units

13. $\left[\pi a^2 + 2(b^2 - a^2)\tan^{-1}\dfrac{a}{b} - 2ab\right]/8$ sq units; $\frac{1}{8}a^2(\pi - 2)$ sq units

15. $\left(\dfrac{4a}{3\pi}, 0\right)$ 17. $\dfrac{3\pi}{64}\rho_0 a^2$ 19. $\dfrac{35\pi}{16}\rho_0 a^4$ 21. $\frac{2}{3}\pi a^3$ cu units

23. $3(3\pi - 4)$ cu units   25. $\frac{1}{12}\pi a^3(8 - 3^{3/2})$ cu units   27. $\dfrac{125\pi}{6}$ cu units

29. $\dfrac{ha^2}{36}(9\pi - 16)$ cu units   31. $\dfrac{4\pi}{3}[a^3 - (a^2 - b^2)^{3/2}]$ cu units

Exercises 18.6   page 1082

1. $8(30^{1/2})$ sq units   3. $\frac{1}{3}(10^{3/2} - 1)$ sq units   5. $\frac{1}{6}(17^{3/2} - 5^{3/2})$ sq units

7. $\frac{1}{4}(66^{3/2} - 2^{3/2})$ sq units   9. $\dfrac{\pi}{6}[(1 + 4a^2)^{3/2} - 1]$ sq units

11. $\frac{1}{2}(a^2b^2 + b^2c^2 + c^2a^2)^{1/2}$ sq units   13. 800 sq units

15. $a^2\left(\dfrac{\pi}{2} - 1\right)$ sq units   17. $\dfrac{1}{12m}[(1 + 4a^2)^{3/2} - 1]$ sq units

21. $\dfrac{8\pi}{3}(2^{3/2} - 1)$ sq units

Exercises 18.7   page 1089

1. 3   3. 8   5. $\frac{1}{3}$ 7. $\frac{1}{24}\pi^3$ 9. $e^2 - e - \frac{5}{2}$ 11. 2 cu units

13. $\frac{24}{5}$ cu units   15. 16π cu units   17. $\frac{1}{3}(b - a)c^{3/2}$ cu units

19. $\frac{31}{3}$ cu units   21. $\dfrac{kabc}{3}(a^2 + b^2 + c^2)$ 23. $\dfrac{ka^5}{30}$ 25. $\frac{1}{3}kb^4$ 27. $4\ln 5$

Exercises 18.8

page 1094

1. $(1, \frac{3}{2}, 2)$ **3.** $\left(\frac{a}{2}, \frac{2b}{3}, \frac{c}{2}\right)$ **5.** $\left(\frac{7a}{12}, \frac{7a}{12}, \frac{7a}{12}\right)$ **7.** $(\frac{5}{12}b^{1/2}, \frac{5}{7}b, \frac{5}{14}b)$

9. $(0, 0, \frac{1}{2}a^2)$ **11.** $(0, 0, \frac{3}{4}a)$ **13.** $\frac{2}{3}kb^5, \sqrt{\frac{2}{3}}b$ **15.** $\frac{1}{30}ka^5; \frac{a}{\sqrt{5}}$

Exercises 18.9

page 1100

1. $\frac{2}{3}\pi a^3 hk$ **3.** $M\left(\frac{a^2}{4} + \frac{h^2}{3}\right)$, where $M = \pi a^2 hk$ is the mass of the cylinder

5. $\frac{1}{6}\pi kh^4\left(\frac{1}{\cos^3\alpha} - 1\right)$ **7.** $\frac{1}{6}\pi a^6 k$ **9.** $\frac{1}{2}M(b^2 + a^2)$ **11.** $\frac{5}{4}\pi a^3 k$

13. $\frac{81}{160}\pi a^5 k$ **15.** $\frac{8}{15}\pi a^5 k$ **17.** $\frac{3M}{5}\left(\frac{a^2}{4} + h^2\right)$

Review and Miscellaneous Exercises

page 1101

1. $\frac{28}{3}$ **3.** $\frac{297}{20}$ **5.** $\frac{\pi}{4}\ln\frac{b}{a}$ **7.** $\frac{1}{2}(1 - \cos 9)$ **9.** 0

11. $3\ln 6 - \ln 2 - \frac{5}{2}\ln 5 + 1$ **13.** $\frac{512}{27}$ **15.** $\frac{2}{3}a^3$

17. $\dfrac{1}{(a + c + 2)(c + 1)(a + b + 2c + 4)}$ **19.** $\frac{1}{4}a^4 + \frac{1}{18}a^6$

21. $\int_0^{16}\int_{-\sqrt{x}}^{\sqrt{x}} f(x, y)\, dy\, dx$ **23.** $\int_0^1\int_{-x}^x f(x, y)\, dy\, dx + \int_1^3\int_{-x}^{3-2x} f(x, y)\, dy\, dx$

25. $\int_{-3}^2\int_{x^2}^{6-x} (6 - x - y)\, dy\, dx$ **27.** $\frac{16}{75}a^5$ **29.** 9 **31.** $\frac{7\pi}{24}a^6\rho_0$

33. $(\frac{4}{5}a, \frac{9}{20}a)$ **35.** $\frac{4}{45}\rho_0 a^6; \frac{2}{9}\rho_0 a^6; \frac{14}{45}\rho_0 a^6$ **37.** $a^2\left(\frac{\pi}{3} + \frac{\sqrt{3}}{2}\right)$ sq units

39. $\frac{1}{3}(18^{3/2} - 2^{3/2})$ sq units **41.** $a^2(\pi - 2)$ sq units **43.** 144 cu units

45. $\frac{28}{45}ka^7$ **47.** $\left(\frac{2}{3}a, \frac{b}{2}, \frac{c}{2}\right)$ **49.** $\frac{4}{5}k\pi(b^5 - a^5)$

51. $I_x = \frac{M}{3}(b^2 + c^2); \quad R_x = \sqrt{\frac{1}{3}(b^2 + c^2)}$ **53.** $I_L = \frac{7}{5}Ma^2$

1172 *Answers to Odd-Numbered Exercises*

Index

Derivative(s) *(cont.)*
  of a constant times a function, 163
  of cos u, 482
  of $\cos^{-1} u$, 504
  of cosh u, 522
  of $\cosh^{-1} u$, 524
  of cot u, 483
  of $\cot^{-1} u$, 506
  of coth u, 523
  of $\coth^{-1} u$, 524
  of csc u, 483
  of $\csc^{-1} u$, 506
  of csch u, 523
  of $\operatorname{csch}^{-1} u$, 525
  definition of, 155–161
  delta form, 155
  directional, 1005–1013
  of an even function, 173
  of exponential functions, 451–452
  of a function, 140–188
  of higher order, 181–188
  of hyperbolic functions, 522–523
  of inverse functions, 446
  of inverse hyperbolic functions, 523–524
  of inverse trigonometric functions, 504–510
  from the left, 159
  linear property, 164
  of logarithmic functions, 427–435
  notation for, 181
  of an odd function, 173
  one-sided, 159
  partial, 974–1041
    higher order, 981–985
  power formula, 162, 167, 174
  of a product, 164–165
  of a quotient, 166
  as a rate of change, 190–193
  from the right, 159
  of sec u, 483
  of $\sec^{-1} u$, 506
  of sech u, 523
  of $\operatorname{sech}^{-1} u$, 525
  of sin u, 482
  of $\sin^{-1} u$, 504
  of sinh u, 522
  of $\sinh^{-1} u$, 524
  of sum of two functions, 163–164
  of tan u, 483
  of $\tan^{-1} u$, 504
  of tanh u, 523
  of $\tanh^{-1} u$, 524
  of trigonometric functions, 479–487
  of vector-valued functions, 938–946
Descartes, René, 18
Determinants, 915
Difference, of functions, 76
Differentiable function, 155
Differentiability
  and continuity, 158–159

functions
  of two variables, 994
  of three variables, 996
  of n variables, 996, 998
Differential(s), 260–267, 994–998
  of dependent or independent variable, 260
Differential calculus, functions of two or more variables, 964–1041
Differential equation, 280–286
  partial, 982
Differential formulas, 262–266
Differential geometry, 948–954
Differential or total differential of functions of two variables, 994
Differentiation, 162
  of algebraic functions, 173–180
  implicit, 177–180
  power series, 848–879
  of vector-valued functions, 938–946
Differentiation formulas, 162–168
Directed distance, 32
Directed distance in space, 882
Direction angles, 899
Direction cosines, 899
Direction numbers, 902
Directional derivative, 1005–1013
Directrix, 55, 730
Discontinuity, essential and removable, 126
Discontinuous function, 124
Discriminant, 81, 742
  invariance, 742, 745
Displacement, 149
Distance
  from a point to a line, 913
  invariance, 729, 743
  on a number line, 11
  between a point and a plane, 912
  between two points, 33, 883
Divergent sequence, 807
Divergent series, 814
Domain, 69, 70
Dominating series, 823
Dot product of vectors, 610, 894
Double integral, 1042–1083
  over a nonrectangular region, volume, 1050–1055
  over a rectangle, volume, 1042–1049
  polar coordinates, 1072–1078
  properties of, 1060–1062

E

e, 435
e^x, 448–457
Eccentricity, 730
  ellipse, 708
  hyperbola, 717
Economic order quantity, 254

Surface of revolution, area, 416–424
  frustrum, 418
  right circular cone, 417
  right circular cylinder, 416
Surface area, 1080
  cylindrical coordinates, 1083
Symmetric equations of a line, 903
Symmetry, axis of, 26
Symmetry of a graph, 26–28

T

Tangent to a curve, 142–149
Tangent function, 474
  derivative of, 483
  integrals involving powers of, 550
  inverse, 499
    derivative of, 505
Tangent line, 142, 144
  ellipse, 710
  slope of curve in polar coordinates, 681
  to graph of a vector function, 630
Tangent plane to a surface, 1015–1018
Tangent vector, unit, 943
Tangent component of acceleration, 654, 950
Taylor, Brooke, 784
Taylor series, 857–879
  remainder, 859
Taylor's formula, 782–789
Taylor's polynomial, 784–789, 800
Taylor's theorem, 782–789
Three-dimensional number space, 882
Total cost function, 251
Total revenue (or income) function, 254
Traces of a plane, 910
Transcendental functions, 84
Transitive law, 2
Translation of axes, 721–729
Transverse axis, hyperbola, 713
Trapezoidal rule
  for approximating an integral, 589–594
  error, 591
  formula, 591
Triangle inequality, 14, 897
Triangle prism, 345
Trigonometric functions, 466–535
  argument, 468
  derivatives of, 479–487
  half-angle formulas, 492
  identities, 470
  inverse trigonometric functions, 494–514
    principal value range, 496
  review, 466–478
Trigonometric integrals
  Part I, 541–547
  Part II, 548–553
Trigonometric series, 1040
Trigonometric substitution, 553–562

Triple integral, 1083–1103
  center of mass, 1090–1103
  centroid of mass, 1090–1103
  cylindrical coordinates, 1094–1103
  first moments of mass, 1090–1103
  mass, 1087–1103
  moment of inertia, 1092–1103
  over a more general region, 1085–1103
  over a rectangular parallelepiped, 1083–1085
  radius of gyration, 1093–1100
  spherical coordinates, 1098–1103
  volume, 1086–1103
Triple secular product, 920
Twisted cubic, 945
Two-dimensional vector, 604
Two-point form, 43

U

Undirected distance, 33
Uniqueness theorem, 102
Uniqueness theorem for limits, 808
Unit vector, 614, 890
Upper half-plane, 19
Upper sum, 299

V

Variables, dependent and independent, 69
Vector(s), 603
  acceleration, 636, 654, 947, 950
  addition of, 605–609, 888–889
  algebraic properties, 889, 894, 915–917
  angle between, 611, 894
  anticommutative property of cross product, 916
  associative law of addition, 604
  basis, 614, 894
  binomial, 951
  bound, 604
  commutative law
    of addition, 604
    of multiplication, 610
  components, 604, 605, 888, 892
  continuous function, 626
  cross product, 914–924
  curvature, 648, 949
  derivative, 627
  difference, 606, 889
  direction angles of, 899
  direction cosines of, 899–901
  displacement, 635, 947
  distributive laws, 917
  dot product, 610, 894
  equal, 605, 888
  equation, 605, 888
  free, 604
  geometric interpretation, 604

FORMS CONTAINING $\sqrt{2au - u^2}$

58. $\int \sqrt{2au - u^2}\, du = \dfrac{u-a}{2}\sqrt{2au - u^2} + \dfrac{a^2}{2}\cos^{-1}\left(1 - \dfrac{u}{a}\right) + C$

59. $\int \dfrac{du}{\sqrt{2au - u^2}} = \cos^{-1}\left(1 - \dfrac{u}{a}\right) + C$

60. $\int u^n \sqrt{2au - u^2}\, du = -\dfrac{u^{n-1}(2au - u^2)^{3/2}}{n+2} + \dfrac{(2n+1)a}{n+2}\int u^{n-1}\sqrt{2au - u^2}\, du$

61. $\int \dfrac{\sqrt{2au - u^2}}{u^n}\, du = \dfrac{(2au - u^2)^{3/2}}{(3-2n)au^n} + \dfrac{n-3}{(2n-3)a}\int \dfrac{\sqrt{2au - u^2}}{u^{n-1}}\, du$

62. $\int \dfrac{u^n\, du}{\sqrt{2au - u^2}} = -\dfrac{u^{n-1}\sqrt{2au - u^2}}{n} + \dfrac{a(2n-1)}{n}\int \dfrac{u^{n-1}}{\sqrt{2au - u^2}}\, du$

63. $\int \dfrac{du}{u^n \sqrt{2au - u^2}} = \dfrac{\sqrt{2au - u^2}}{a(1-2n)u^n} + \dfrac{n-1}{(2n-1)a}\int \dfrac{du}{u^{n-1}\sqrt{2au - u^2}}$

64. $\int \dfrac{du}{(2au - u^2)^{3/2}} = \dfrac{u-a}{a^2 \sqrt{2au - u^2}} + C$

65. $\int \dfrac{u\, du}{(2au - u^2)^{3/2}} = \dfrac{u}{a\sqrt{2au - u^2}} + C$

FORMS CONTAINING TRIGONOMETRIC FUNCTIONS

66. $\int \sin u\, du = -\cos u + C$

67. $\int \cos u\, du = \sin u + C$

68. $\int \tan u\, du = \ln|\sec u| + C = -\ln|\cos u| + C$

69. $\int \cot u\, du = \ln|\sin u| + C$

70. $\int \sec u\, du = \ln|\sec u + \tan u| + C = \ln\left|\tan\left(\dfrac{\pi}{4} + \dfrac{u}{2}\right)\right| + C$

71. $\int \csc u\, du = \ln|\csc u - \cot u| + c = -\ln|\csc u + \cot u| + c$

72. $\int \sin^2 u\, du = \dfrac{u}{2} - \dfrac{\sin 2u}{4} + C$

73. $\int \cos^2 u\, du = \dfrac{u}{2} + \dfrac{\sin 2u}{4} + C$

74. $\int \tan^2 u\, du = \tan u - u + C$

75. $\int \cot^2 u\, du = -\cot u - u + C$

76. $\int \sec^2 u\, du = \tan u + C$

77. $\int \csc^2 u\, du = -\cot u + C$

78. $\int \sec u \tan u\, du = \sec u + C$

79. $\int \csc u \cot u\, du = -\csc u + C$

80. $\int \sin^n u\, du = -\dfrac{1}{n}\sin^{n-1} u \cos u + \dfrac{n-1}{n}\int \sin^{n-2} u\, du$

81. $\int \cos^n u\, du = \dfrac{1}{n}\cos^{n-1} u \sin u + \dfrac{n-1}{n}\int \cos^{n-2} u\, du$

82. $\int \tan^n u\, du = \dfrac{1}{n-1}\tan^{n-1} u - \int \tan^{n-2} u\, du$

83. $\int \cot^n u\, du = -\dfrac{1}{n-1}\cot^{n-1} u - \int \cot^{n-2} u\, du$

84. $\int \sec^n u\, du = \dfrac{1}{n-1}\sec^{n-2} u \tan u + \dfrac{n-2}{n-1}\int \sec^{n-2} u\, du$

85. $\int \csc^n u\, du = -\dfrac{1}{n-1}\csc^{n-2} u \cot u + \dfrac{n-2}{n-1}\int \csc^{n-2} u\, du$

86. $\int \sin mu \sin nu\, du = -\dfrac{\sin(m+n)u}{2(m+n)} + \dfrac{\sin(m-n)u}{2(m-n)} + C,\ (m^2 \neq n^2)$

87. $\int \cos mu \cos nu\, du = \dfrac{\sin(m+n)u}{2(m+n)} + \dfrac{\sin(m-n)u}{2(m-n)} + C,\ (m^2 \neq n^2)$

88. $\int \sin mu \cos nu\, du = -\dfrac{\cos(m+n)u}{2(m+n)} - \dfrac{\cos(m-n)u}{2(m-n)} + C,\ (m^2 \neq n^2)$

89. $\int u \sin u\, du = \sin u - u \cos u + C$

90. $\int u \cos u\, du = \cos u + u \sin u + C$

91. $\int u^2 \sin u\, du = 2u \sin u + (2 - u^2)\cos u + C$

92. $\int u^2 \cos u\, du = 2u \cos u + (u^2 - 2)\sin u + C$

93. $\int u^n \sin u\, du = -u^n \cos u + n\int u^{n-1}\cos u\, du$

94. $\int u^n \cos u\, du = u^n \sin u - n\int u^{n-1}\sin u\, du$

95. $\int \sin^m u \cos^n u\, du = -\dfrac{\sin^{m-1} u \cos^{n+1} u}{m+n} + \dfrac{m-1}{m+n}\int \sin^{m-2} u \cos^n u\, du^{\dagger}$

$\qquad = \dfrac{\sin^{m+1} u \cos^{n-1} u}{m+n} + \dfrac{n-1}{m+n}\int \sin^m u \cos^{n-2} u\, du$

$\dagger$If $m = -n$, use formulas No. 82 or 83.